An Introduction to Quantum Theory

Underpinning the axiomatic formulation of quantum theory presented in this undergraduate textbook is a review of early experiments, a comparison of classical and quantal terminology, a Schrödinger-equation treatment of the one-dimensional quantum box, and a survey of relevant mathematics. Among the many concepts comprehensively discussed are operators; state vectors, wave functions and their interrelation; experimental observables and their quantum counterparts; classical/quantal connections; and symmetry properties. The theory is applied to a wide variety of systems including the non-relativistic H-atom, external electromagnetic fields, and spin. Collisions are described using wave packets. Various time-dependent and time-independent approximations are discussed; applications include electromagnetic transition rates and corrections to the non-relativistic H-atom energies. The final chapter deals with multiparticle systems and identical-particle symmetries with application to the He-atom, the periodic table, and diatomic molecules. There are also brief treatments of advanced subjects such as gauge invariance, observable phase effects, and hidden variables.

The author received his undergraduate degree from Johns Hopkins University and his Ph.D. from the University of Maryland. Following post-doctoral positions at Rice University, Brookhaven National Laboratory, and the United Kingdom Atomic Energy Authority, he accepted an appointment as a faculty member in the Physics Department at Brown University, where he remained, did research and taught for 31 years until his retirement in 1998. Professor Levin's research areas are nuclear-reaction theory, three-body physics, many-body-collision theory, and its applications to nuclear reactions, atomic physics, and molecular structure. He has published widely on these topics, has lectured and done research in many countries, has edited/co-edited several research compilations, is the founder/first chairman of the American Physical Society's Topical Group on Few-Body Systems and Multi-particle Dynamics, is an awardee of the Alexander-von-Humboldt-Stiftung, and is a Fellow of the American Physical Society. Among the large variety of courses he has taught are the year-long graduate and undergraduate courses on quantum mechanics at Brown University. The present text is an outgrowth of these teaching experiences.

An Introduction to Quantum Theory

F. S. Levin

PUBLISHED BY THE PRESS SYNDICATE OF THE UNIVERSITY OF CAMBRIDGE
The Pitt Building, Trumpington Street, Cambridge, United Kingdom

CAMBRIDGE UNIVERSITY PRESS
The Edinburgh Building, Cambridge CB2 2RU, UK
40 West 20th Street, New York, NY 10011-4211, USA
10 Stamford Road, Oakleigh, VIC 3166, Australia
Ruiz de Alarcón 13, 28014 Madrid, Spain
Dock House, The Waterfront, Cape Town 8001, South Africa

http://www.cambridge.org

First published 2002

Printed in the United States of America

Typeface Times 10/12.5pt. *System* 3B2 [KT]

A catalogue record for this book is available from the British Library

Library of Congress Cataloguing in Publication data

Levin, F. S. (Frank S.), 1933–
An introduction to quantum theory / Frank Samuel Levin.
p. cm.
Includes bibliographical references and index.
ISBN 0 521 59161 9 – ISBN 0 521 59841 9 (pb)
1. Quantum theory. I. Title.

QC174.12.L45 2001
530.12–dc21 2001025611

ISBN 0 521 59161 9 hardback
ISBN 0 521 59841 9 paperback

Contents

Preface

Quantum mechanics, the theoretical framework that describes microscopic physical phenomena, is a marvelous intellectual enterprise. Elegant, intriguing, challenging, often counter-intuitive and sprinkled with delightful surprises, it is wholly unlike classical physics. Mastery of quantum theory is an essential ingredient in the education of physicists and chemists. This text, a broad and deep introduction to the theory, is intended not only to help provide that mastery but also to convey to the reader its inherent richness.

The book is designed for use in year-long undergraduate courses, for self-study, and as a supplemental text in graduate courses. Its aims are achieved through a combination of introductory material, including a survey of relevant mathematics; an axiomatic formulation of the theory via six postulates; applications to soluble one-, two-, and three-dimensional systems; theoretical augmentations such as symmetry properties, electromagnetic interactions, spin $\frac{1}{2}$ and generalized angular momentum; and discussion of approximation methods and their application to electromagnetic transitions and systems of identical fermions (atoms and molecules). The reader is assumed to be familiar with material covered in courses on intermediate mechanics, electromagnetism and elementary modern physics, plus mathematics through vector calculus, elements of partial differential equations (their form and separation-of-variables solutions), and some linear algebra (finite dimensions, matrices).

The writing of this book has been strongly influenced by the author's experiences teaching both the graduate and the undergraduate quantum-theory courses at Brown University. Most of the students who take the undergraduate course are junior-level physics majors, many of whom go on to graduate school in physics. Like that undergraduate course, this text is intended to provide a solid preparation for graduate courses on quantum theory. In doing so, it also stresses the fact that quantum mechanics is an evolving subject in which new and sometimes surprising developments have occurred and old controversies have been reinvigorated. Indeed, the second half of the twentieth century has seen developments in quantum mechanics undreamed of in its first *25* years of existence. Among those discussed in this book are various phase effects (Aharonov–Bohm, Berry, etc.); hidden-variable analysis (Bohm, Bell, etc.); identification of a system-dependent, quantal time operator; Feynman's path-integral formulation; and coherent states. Discussions of other sublime topics include exponential time decay, time–energy uncertainty relations, spin regeneration, gauge invariance, minimal coupling, classical limits, scattering of spin-$\frac{1}{2}$ projectiles, fine- and hyperfine-structure effects in hydrogen, and Hartree–Fock theory for atoms.

The book is substantial. This is partly a result of the author's belief that a quantum-mechanics textbook should contain thorough explanations and partly a result of his desire to treat a wide variety of topics in a reasonably full fashion. Since this has led to expanded discussions, one chapter and a number of sections and subsections have been starred: they can be omitted without seriously compromising the pedagogical development. Even with such omissions, however, there is still more material than can be covered in a year-long course. The book contains essentially all the topics the author would include in a course for which the usual academic time constraints could be ignored. Since these constraints cannot be ignored, instructors who adopt this text will be able to select from a wide-ranging menu of possible subjects against which they can structure a course.

Veteran teachers of quantum theory will know how tempting it is to continue a specific analysis beyond what is needed for an adequate treatment of most topics. While some of these additional analyses are presented, the author has tried to avoid undue excess in this regard, choosing instead to cite references that the interested reader may consult. The text is peppered with references to a variety of other material, including many of the original articles on particular topics. This should help to restore to the contemporary scene the names of possibly forgotten scientists whose results were not only the first published but were also inspiring to others.

Analyses involving electromagnetic quantities occur in various places in the text, especially Chapter 12. Although SI units have been adopted universally and are standard in almost all undergraduate books on electromagnetism, cgs units are still found in some widely adopted graduate-level quantum-mechanics texts. Anyone who will read or encounter one of these texts (e.g., those by Baym and by Sakurai) should be familiar with cgs as well as SI units. Consequently, in this book electromagnetic quantities are expressed in terms of constants whose values in both sets of units are explicitly stated. For example, the Coulomb interaction between charges q_1 and q_2 separated by a distance r is written $\kappa_e q_1 q_2 / r$, where $\kappa_e = 1/(4\pi\varepsilon_0)$ in SI units and $\kappa_e = 1$ in cgs units, with the q's defined appropriately.

In addition to this particular piece of pedagogy, the author has indulged in another conceit: the use of a variety of acronyms, examples being w.r.t. for "with respect to"; LHS and RHS for "left-" and "right-hand side"; 1-D, 2-D, and 3-D for "one," "two," and "three dimensions," respectively; the usual abbreviations viz., i.e., and e.g.; etc. For the author their felicity overcomes any lack of grace their use may entail. Correspondingly, the adjective "quantal" is used not only to replace the less-felicitous phrase "quantum-mechanical," but also as a sharp reminder that the subject matter is not "classical." From this viewpoint, "quantal" and "classical" can be considered as "Through the Looking Glass" mirror images.

Although the organization of this book is stated in its table of contents, details and specific items can be discovered only by reading the text. For the newcomer to quantum mechanics who wishes to learn the theory through self-study, the author's advice is to start with Chapter 1 and read sequentially, possibly ignoring all portions set off with an asterisk. All readers should work through the exercises: they are intended as learning tools. Some extend and enrich theoretical concepts, others flesh out various applications; few are of the "plug-in" variety. Some of the exercises contain hints on how to proceed, whereas in others there is a statement of the result to be derived. Instructors should note that not all parts of a long exercise need be assigned.

An instructor who uses this book as a primary text will be melding his/her teaching background and preferences with the material and its presentation. In addition to there being too many topics to fit into a two-semester course, the level of the students obviously needs to be taken into account. Depending on where in the physics curriculum an undergraduate course on quantum theory occurs and what the components of that curriculum are (e.g., a previous modern-physics course or mathematics courses on linear algebra) the first semester could cover topics from the first ten chapters, although it is probably more likely that Chapters 1–7 plus 9 will comprise the first semester.

In structuring a course around this text (or simply when reading it), it may be helpful to know some of the author's intentions concerning the chapters and specific material they contain. For example, in addition to its historical and review aspects, Chapter 1 is designed to start the reader thinking in quantal terms: its exercises dealing with orders of magnitude are especially important here. Chapter 2 continues along the path of thinking quantally. For very-well-prepared students, the topics of these first two chapters could be touched on lightly in lectures, with most of the material assigned for independent reading.

There are many ways of arranging a first meeting with quantum mechanics. The one adopted in Chapter 3, which compares the classical stretched string and the 1-D quantal box, also includes a brief review of waves. Eigenvalue problems are first encountered in this chapter. They are presented in the context of differential equations – which are presumed to be familiar to readers – rather than via matrices. The omittable Section 3.4 introduces the important concepts of Hermiticity and the Dirac delta function, still in the context of differential equations (and Sturm–Liouville systems). These concepts occur again in Chapter 4, where the notions of abstract operators and their matrix realizations are exemplified by the 3-D rotation operator/matrix, which is assumed to be familiar from a course in classical mechanics. Chapter 4 is included to help make the book self-contained, especially if it is used as the text in a junior-year course. Most of this chapter is a not-very-rigorous survey, and much of it will be a review for fourth-year students who have already taken a course on mathematical methods, in which case the instructor can make a selection of topics for lectures and assign the rest as independent reading.

Since this book's approach is axiomatic, the formal theory is presented in Chapter 5 via six postulates. The postulates are illustrated using 1-D (one-degree-of-freedom) systems in Chapters 6 (bound states) and 7 (continuum states). The multiple approaches to the linear harmonic oscillator in Chapter 6 allow an instructor great latitude in dealing with this system: the author's preference is to discuss all of them. Continuum states are the subject of a separate chapter because their Hilbert-space (normalizable, wave-packet) nature is an aspect that, in the author's opinion, needs to be stressed more than the 1-D character of the system. The plane-wave limit is straightforward and the continuum examples then treated are standard. However, by going through the 1-D wave-packet analysis, the stage is set for its 3-D analog in Chapter 15.

How much, if any, of Chapter 8 will be touched on will depend on time constraints and an instructor's taste. While this chapter is omittable, Chapter 9 is NOT: it contains essential theoretical developments. It, along with Chapter 5, is the foundation on which the rest of the text is built. Especially important here are the sections on complete sets of commuting operators and symmetry operations.

With Chapter 10, one enters the 3-D, core portion of undergraduate courses on quantum mechanics. The coordinate-space approach to orbital angular momentum and its eigenvalue problem is deliberate: by postponing the algebraic method to Chapter 14, one focuses in this chapter on the angular dependence, the connection between the $\ell(\ell+1)\hbar^2$ eigenvalue and convergence (discussed in Appendix C), the relationship between ∇^2 and $\hat{\mathrm{L}}^2$, etc. Sections 10.3 and 10.4 may be considered as examples illustrating orbital angular momentum.

In one sense, Chapter 11 contains additional examples of orbital angular momentum. However, these are incidental to its real purpose, viz., to introduce the reader to "realistic" if not real problems, ranging from the treatment of a two-body system as an equivalent one-body system to the discussions of accidental degeneracies. As with the linear oscillator of Chapter 6, the differential equation for the attractive $1/r$ potential is solved and discussed in detail.

Because the quantal treatment of external, classical electromagnetic fields is a subject of great elegance, it deserves a chapter to itself. That elegance, especially made manifest through gauge invariance and the Aharonov–Bohm effect, is discussed in detail in Chapter 12, but, for the time-constrained instructor, its first three sections should suffice. Electromagnetic fields are treated before intrinsic spin. Hence, prior to reading about the Stern–Gerlach experiment, one has learned that, when a spinless particle in a spherically symmetric potential is acted on by a constant (weak) magnetic field, each of its energy levels is split into an odd number of components. The stage is thus set for the spin-$\frac{1}{2}$ interpretation of the Stern–Gerlach experiment, which is discussed in Section 13.1.

The nature of spin $\frac{1}{2}$ makes it one of the most sublime constituents of the quantum treasury, and the analysis in Chapter 13 strives to convey this assessment. That spin $\frac{1}{2}$ is a two-state system helps to substantiate this, as does the solving of the eigenvalue problem for an arbitrary quantization direction. Additional ingredients that the author finds particularly appealing are the direct experimental basis on which two graduate students (!!) constructed their theoretical framework, and the regeneration of extinguished spin components.

Generalized angular momentum need not be included in an undergraduate quantum-mechanics course. However, it and the developments associated with it that are treated in Chapter 14 are crucial elements in this text. Because of this, an instructor who plans to cover the later material in this book dealing with dipole transitions, fine- and/or hyperfine-structure effects in hydrogen, the Pauli principle and two-electron atoms as well as atomic-spectroscopic term values, will need to cover Sections 14.1 and 14.3. Once Section 14.1 has been included, it would be very difficult (at least for the author) to resist including Section 14.2. The final section, on Bell's inequality, is intended for independent study.

It is not unusual to find scattering theory either omitted or treated near the end of undergraduate quantum-mechanics texts. However, since scattering states are the continuum states in 3-D, and Part III is largely concerned with 3-D, for the author it is a logical necessity to include scattering theory in Part III. It is also a logical necessity to begin Chapter 15 with a wave-packet analysis, since only a wave packet can be a proper quantum state in the continuum.

Part IV of this book introduces the reader to some of the approximate procedures for dealing with realistic as well as real systems. The order in which certain material occurs in Part IV is unusual; in particular, time-dependent perturbation theory (Chapter 16) occurs before its time-independent counterpart (Chapter 17). This ordering is based on the author's belief that related topics should be treated as close together as possible. Since electromagnetic fields and the Aharonov–Bohm effect are considered in Chapter 12, time-dependent processes such as electromagnetic transitions and those related to geometric phases are discussed as soon as is feasible, viz., in Chapter 16. Of course, whether either or both of these topics will be included in a course is the instructor's decision. The author's hope is that some of the various geometric-phase material can be touched on, since studies and measurements of quantum phases are an intrinsic part of the physics developed during the latter half of the twentieth century.

Chapter 17 treats most of the standard time-independent approximation methods. Section 17.2 is a beautiful application of perturbation theory to the model H-atom of Chapter 11, and it may well be worth making time for this analysis. On the other hand, the analysis of Section 17.4 cannot be omitted if any of Sections 18.2, 18.4, and 18.5 is to be studied. Section 18.1 is essential: a physicist's undergraduate education is incomplete if it has not included some material on identical-particle symmetries. As for the rest of Chapter 18, an instructor has much to choose from for lectures, and much to assign as possible independent reading. An example is the wide-ranging discussion of approximate calculations on He: the intent here is to show the reader the degrees of accuracy and the quality of the associated physical pictures resulting from simple to highly sophisticated computations.

The selection and organization of the material that forms this book is the author's sole responsibility. Nevertheless he has benefited from, and usually incorporated, the suggestions of various colleagues regarding improvements. In this regard he is indebted to Antal Jevicki, to Janine Shertzer, to the anonymous reviewers, and to Stavros Fallieros, who also made his extensive collection of homework problems available for use in this text. It has taken many years to finish this book, the writing of which has occurred in various locales, on the water as well as on land. The author is grateful to the many marinas and harbors where safe havens and stable environments were provided to CHEERS, on whose chart table a good deal of the writing has taken place. The jocular encouragement of sailing friends, especially Eric and Daisy Broudy and David and Barbara Constance, has been a helpful reminder that even as serious an undertaking as writing a book on quantum mechanics is still a playful activity. The first typing of almost all of the text was done by Winnie Isom, whose role in the preparation of this text has been enormous: she has earned the author's thanks many times over. Additional kudos in this regard go to Nadine Catteral and Elizabeth Peña and to Etta Johnson and Chantée Watts. It is an additional pleasure to acknowledge the splendid support provided by past and present members of the Cambridge University Press community: Simon Capelin, Jo Clegg, Steve Holt, Zoe Naylor, Eoin O'Sullivan, Sue Tuck, and especially Phil Meyler, who convinced his colleagues to take a chance on the author's mammoth text. Finally, and most important of all, this book could not have been written without the unfailing love, patience, and support of the author's wife, Carol Levin, to whom it is dedicated.

Part I

INTRODUCTORY

1

The Need for a Non-Classical Description of Microscopic Phenomena

Quantum mechanics is the theoretical framework that describes physical phenomena on the microscopic scale. In particular, it is used in those situations in which Planck's constant h, approximately equal to 6.62×10^{-34} J s, cannot be assumed to be negligible, despite its apparently tiny magnitude. Macroscopic phenomena, on the other hand, involve magnitudes that are so large that h can be set equal to zero without appreciable error.

As an illustration of this in the macroscopic context, consider the following example: a ladybug (*Hippodamia convergens*) of mass 1 g moves radially at a speed of 0.01 m s^{-1} on a turntable rotating at 10 rpm. In the rotating frame of reference, the ladybug has a kinetic energy of about an erg, and, in the laboratory frame, its angular momentum is roughly 10^{-5} J s when it is at a distance of 0.1 m from the center of the rotating turntable. The latter angular momentum is of the order of $10^{28}h$. Thus, relative to this value, h is well approximated by zero and, indeed, only classical mechanics is needed to describe the ladybug's motion.

In contrast to the preceding example, an electron in the "2p" level of hydrogen has an energy of about 10^{-25} J and an angular momentum whose magnitude is approximately $h/(2\pi) \equiv \hbar$ (pronounced "h-bar"). Here, h cannot be ignored and only a quantum description can accurately account for the properties of this system.

The latter example cites the 2p energy level of the H-atom. One of the goals of this book is to show how quantum mechanics describes the hydrogen atom. This particular "how" involves an understanding both of the postulates of quantum theory and of various technical/mathematical details, topics that are examined beginning in Chapter 5 on the quantal postulates. However, "how" is not the only question that can be raised. Another is "why?" Why is quantum mechanics necessary, i.e., what is the compelling evidence that a non-classical description of microscopic phenomena is needed? This first chapter is concerned with that necessity.

Quantum theory was developed mainly during the first thirty years of the twentieth century. In retrospect, its development can be considered a direct consequence of three key concepts that classical physics has never been able to explain. These are (i), the particle nature of electromagnetic radiation, i.e., the existence of light quanta or photons; (ii), the quantization of energy and of angular momentum; and (iii), the wave nature of particles. Each concept was introduced in order to explain experimental data and each has been verified countless times. All three have major theoretical implications. These concepts are examined in detail in the following subsections via experiments and data that classical physics could not account for. Apart from the discussion of the Bohr theory of the H-atom, the treatment is descriptive, with technical details being left to the problems. This chapter also serves as a partial historical review.

1.1. Photons

There is a sharp distinction in classical physics between particles and waves, so that no confusion is likely to arise when one identifies a macroscopic phenomenon as either particle-like or wave-like. Properties of particles center around the idea of locality: the unique mass, energy and momentum are all localized at a point in space. In contrast, waves are spread out rather than localized, do not possess a definite mass, and propagate in a medium, with each of the frequencies ν which occur in the wave being related to the corresponding wavelengths λ via the relation $\nu\lambda = v_{\text{prop}}$, where v_{prop} is the velocity of propagation in the medium. In addition, rather than having a unique energy and momentum associated with a point in space, waves possess energy and momentum densities. Finally, waves can interfere constructively or destructively, in the latter situation even canceling out completely.

The one type of wave for which the preceding list is not entirely accurate is electromagnetic radiation, which can propagate without benefit of a medium and whose speed is independent of the relative velocity between its source and an observer. These are remarkable properties. More pertinent for quantum mechanics is an even more striking feature of electromagnetic radiation. This is its ability to behave either like a wave or like a particle. Which property is manifested depends on the particular situation; the fact that these two apparently contradictory aspects can exist in the same physical system is accounted for only by quantum theory. Indeed, this explanation is one of the many triumphs of quantum mechanics.

That visible light can be either wave-like or particulate is a relatively recent realization, since, prior to the quantum/photon hypotheses introduced by Planck and exploited by Einstein to account for the photoelectric effect, it had been believed for almost 100 years that electromagnetic radiation was necessarily wave-like. The seminal investigation that established the wave nature of (visible) radiation was the two-slit interference experiment carried out by Thomas Young in 1802. For more than 100 years prior to this experiment, Newton's hypothesis that light was composed of tiny particles – but not of photons in the modern sense of the concept – had been accepted as correct.

There have thus been two revolutions in our understanding of the properties of electromagnetic radiation: the first is the change from a particle to a wave picture, the second is the change from the latter paradigm to a combined particle *and* wave picture.[1] Between these two changes came the crucial steps taken by Planck and Einstein. Before examining the concept of the quantum nature of electromagnetic radiation, we first review Young's experiment and its interpretation as a wave phenomenon. This will not only provide a background against which the classically astounding nature of Planck's work can be contrasted, but will also be an aid in understanding our later conclusion that particles can display wave-like properties.

[1] A third revolution, concerning the absence of the aether and the rise of relativity theory, does not concern us in this book.

The two-slit interference phenomenon

The two-slit experiment is schematically represented in Fig. 1.1. Monochromatic radiation of wavelength $\lambda = c/\nu$, where c is the velocity of light, passes through each of slits 1 and 2 in screen B and falls on screen C, the two screens being separated by a distance D. Point P is an arbitrary observation point on screen C at a distance h away from the centerline; the angle it subtends at the midpoint between the two slits is $\theta = \tan^{-1}(h/D)$. The slits are a distance d apart, and $\mathbf{r}_i$, $i = 1, 2$, is the position of point P from slit i.

As a preliminary, consider what the intensity pattern would be if only slit 1, say, were open. The greatest amount of light would strike the screen at a point opposite the slit and the intensity would then fall off symmetrically from this point. This is schematically represented in Fig. 1.2. Either a particle or a wave picture of light would predict this result, since the intensity pattern is just the distribution of energy that falls on the screen. Hence, such a pattern cannot be used to distinguish between a particle-based and a wave-based description of light (or of electromagnetic radiation in general).

In order to justify this conclusion, we construct a simple particle model of (unpolarized) light. The energy is to be carried by the particles. Since the only parameters are the mass and velocity of each particle, while the light is monochromatic and travels at velocity c, it is reasonable to assume that the particles have identical masses m and speeds c.[2] Each particle has an energy $mc^2/2$. If N of these particles hit a particular point on the screen, the intensity at that point is then $I_{\text{part}} = Nmc^2/2$, distributed over the

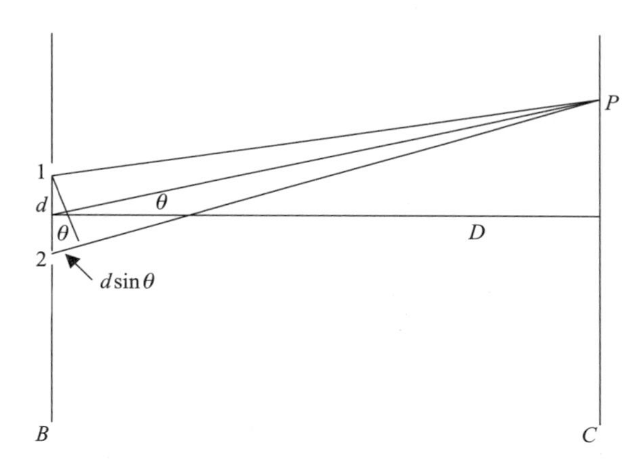

Fig. 1.1 The two-slit experimental arrangement. Monochromatic radiation passes through slits 1 and 2 in screen B and combines on screen C to produce an interference pattern there. Shown is a typical point P on C at distances r_1 and r_2 from slits 1 and 2, respectively, plus the distances d and D, which are the separations between the slits and the screens. $d \sin\theta$ is the extra path length that radiation from 2 must travel to P compared with the path from 1.

[2] Color is not an aspect of this model. For comments on color and Newton's particulate theory of light see Park (1997).

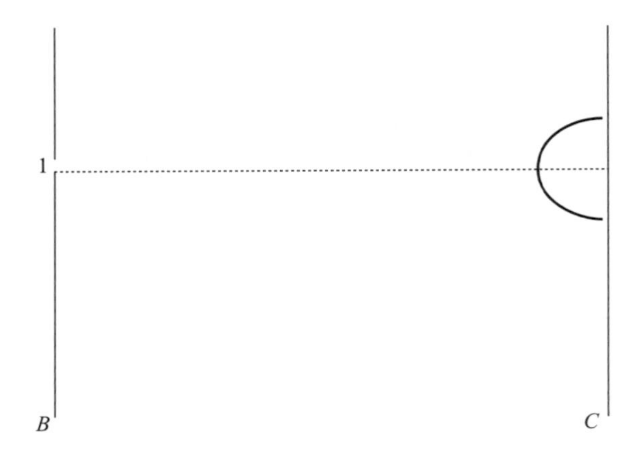

Fig. 1.2 The intensity pattern on screen C that would be seen if slit 2 were closed, so that radiation passes only through slit 1.

screen according to the distribution of the masses as they pass through and are possibly scattered slightly by slit 1; this is in accordance with Fig. 1.2.

In a wave picture, the intensity of light falling on each point of the screen is proportional to the square of the electric field **E** at that point:[3] $I_{\text{wave}} \propto \mathbf{E}^2$. The distribution of E is also as in Fig. 1.2, thus establishing the above claim.

Although a single-slit situation does not distinguish a wave picture from a particle picture, the double-slit arrangement does. Consider the particle description of light first. With slits 1 and 2 both open, the particles will pass through either slit and, assuming that there is a sufficiently low density of particles that they do not strike one another, the resulting pattern on the screen will simply be the sum of the two single-slit intensities, each of whose peaks is opposite the appropriate slit, as in Fig. 1.3. The total intensity will now be

$$I_{\text{total}} = I_{\text{part},1} + I_{\text{part},2}, \tag{1.1}$$

where $I_{\text{part},i}$ is the intensity pattern from particles passing through slit i. There is no interference pattern, just a linear superposition.

This situation changes dramatically when a wave picture is assumed. The intensity is still proportional to the square of the electric field, but now the field is a sum of two contributions, one from each slit:

$$\mathbf{E}_{\text{total}} = \mathbf{E}_1 + \mathbf{E}_2. \tag{1.2}$$

Hence, I_{total} is no longer the sum of two terms, as in Eq. (1.1), but is instead given by

$$I_{\text{total}} \propto (\mathbf{E}_1 + \mathbf{E}_2)^2 = I_1 + I_2 + 2\mathbf{E}_1 \cdot \mathbf{E}_2, \tag{1.3}$$

the factor $\mathbf{E}_1 \cdot \mathbf{E}_2$ in (1.3) being the interference term.

The effect of the preceding interference term is to give a series of intensity maxima and minima, as in Fig. 1.4. To show this, it suffices to consider the scalar part of the field

[3] I is actually proportional to the sum of the squares of the electric and magnetic fields, but in this case the magnetic field $|\mathbf{B}|$ is proportional to the electric field $|\mathbf{E}|$.

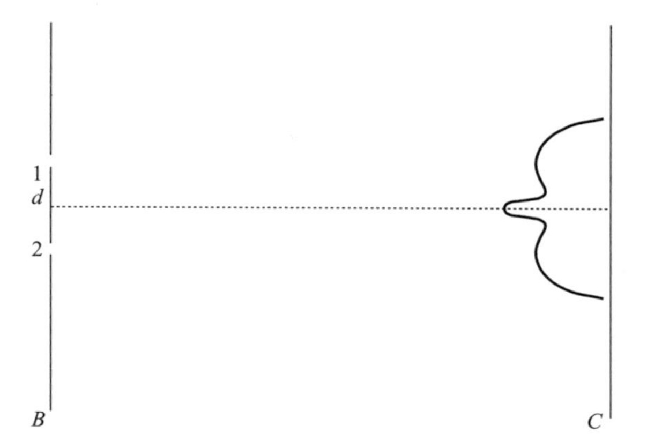

Fig. 1.3 A schematic diagram of the intensity pattern that would arise if the radiation were composed of classical particles. There are no interference effects.

Fig. 1.4 The intensity pattern of light on screen C when both slits are open. The series of maxima and minima are predicted by the wave theory.

arising from each of the two slits. From Huyghen's principle,[4] at large distances the monochromatic field E_j due to slit j behaves like an outgoing wave, whose magnitude is given by the real part of

$$E_j = Ee^{(kr_j - \omega t + \beta)}, \qquad j = 1, 2, \tag{1.4}$$

where E is the electric field strength, k is the wave number ($k = 2\pi/\lambda$), ω is the angular frequency ($\omega = 2\pi\nu = 2\pi c/\lambda$), and β is an arbitrary phase. For the analysis of interference affects in general and especially because of its occurrence in later chapters, it is

[4] Huyghen's principle states that every element in a wave front acts as a source of outgoing spherical waves. See, e.g., Born and Wolf (1980).

advantageous to introduce the complex exponential form and at the end of the analysis either take the real part to get the field or form the absolute square to obtain the intensity.

The intensity I_j resulting from E_j is

$$I_j \propto |E_j|^2 = E^2. \tag{1.5}$$

The total field at P is the sum of the fields from slits 1 and 2, viz.,

$$E_{\text{total}} = E_1 + E_2 = Ee^{-i(\omega t - \beta)}(e^{ikr_1} + e^{ikr_2}), \tag{1.6}$$

and the total intensity I_{total} is thus

$$\begin{aligned} I_{\text{total}} &\propto |E_{\text{total}}|^2 \\ &= 2E^2 + 2E^2 \cos[k(r_1 - r_2)] \\ &= 4E^2 \cos^2[k(r_1 - r_2)/2]. \end{aligned} \tag{1.7}$$

The wave picture, as embodied in Eq. (1.7), therefore leads to the following conclusion: the intensity at screen C is maximum when the argument of the cosine term in Eq. (1.7) is $n\pi$, $n = 0, 1, 2, \ldots$ and is zero when the argument is $(n + \frac{1}{2})\pi$, $n = 0, 1, 2, \ldots$.

Since the distance D is large, the path difference $r_1 - r_2$ is approximately $d \sin\theta$. Hence, the maxima occur at the values ($k = 2\pi/\lambda$)

$$d \sin\theta = n\lambda, \qquad n = 0, 1, 2, \ldots, \tag{1.8}$$

and the minima at

$$d \sin\theta = (n + \tfrac{1}{2})\lambda, \qquad n = 0, 1, 2, \ldots. \tag{1.9}$$

Apart from the decrease in intensity away from the central maximum at $\theta = 0$, this pattern of maxima and minima is the same as that illustrated in Fig. 1.4. Thus, the wave picture predicts an interference pattern like the one that is actually observed, whereas the particle model predicts no interference pattern at all. The two-slit experiment is thus decisive in establishing that light (and electromagnetic radiation at all other wavelengths) is wave-like.

One of the enormous successes of nineteenth-century science was the development and unification of electromagnetic theory (Maxwell's equations). As it turned out, however, this theory could not solve the problem of correctly accounting for the energy density at a given frequency (energy/volume/frequency) of electromagnetic radiation in a blackbody cavity.

Blackbody radiation and the quantum hypothesis

A black body is an object that absorbs and emits electromagnetic radiation perfectly, i.e., without loss of energy. We shall examine the following system: a black body surrounding a cavity containing radiation in equilibrium at absolute temperature T. Such a system can be well approximated in the laboratory by a cavity surrounded by a black absorbing material that is put in a heat bath (e.g., an oven) maintained at the temperature T. A small hole in the material allows the radiation to escape; its energy density $u(\nu)$ at frequency ν can be measured.

Figure 1.5 shows $u(\nu)$ for blackbody radiation as a function of ν for several temperatures. Thermodynamic considerations[5] lead to $\int_0^\infty u(\nu)\,d\nu \propto T^4$, and to Wien's law, viz., $u(\nu) = \nu^3 W(\nu/T)$. The detailed dependence of Wien's function $W(\nu/T)$ on the ratio ν/T cannot be determined from thermodynamic analysis alone, but requires use of radiation theory. From the curves of Fig. 1.5, it is clear that $u(\nu) \to 0$ at both extremes (i.e., for $\nu \to 0$ and $\nu \to \infty$), so that the essential theoretical requirements are that $W(\nu/T) \to 0$ fast enough as $\nu \to \infty$ and $W(\nu/T)$ not diverge so rapidly at small ν as to overwhelm the factor ν^3. A non-terrestrial example of blackbody radiation is the sun: its density of radiant energy is very close to the blackbody curve for a temperature $T = 5{,}800$ K, the latter being approximately equal to the sun's surface temperature.

The classical wave theory of electromagnetic radiation, realized in Maxwell's equations, can be used to determine $u(\nu)$ in the blackbody case. For small ν, the Rayleigh–Jeans approximation is obtained:[6]

$$u(\nu \to 0) \to \frac{8\pi\nu^2}{c^3} k_B T, \tag{1.10}$$

which is in agreement with the observed behavior of the energy density at small values of ν. (Here, k_B is Boltzmann's constant: $k_B = 1.38 \times 10^{-23}$ erg K^{-1}.) However, at large ν, classical electromagnetic theory fails, since it predicts an ultraviolet divergence: $u(\nu \to \infty) \to \infty$ (!). It was not possible, within the classical theory that had previously worked so well, to remove this divergence.

Enter now Planck (1900) and his *ad hoc* quantum hypothesis, contrived as a means by which he was able to eliminate the ultraviolet divergence. (As a nineteenth-century physicist, imbued with classical concepts, the *ad hoc* hypothesis was one Planck never

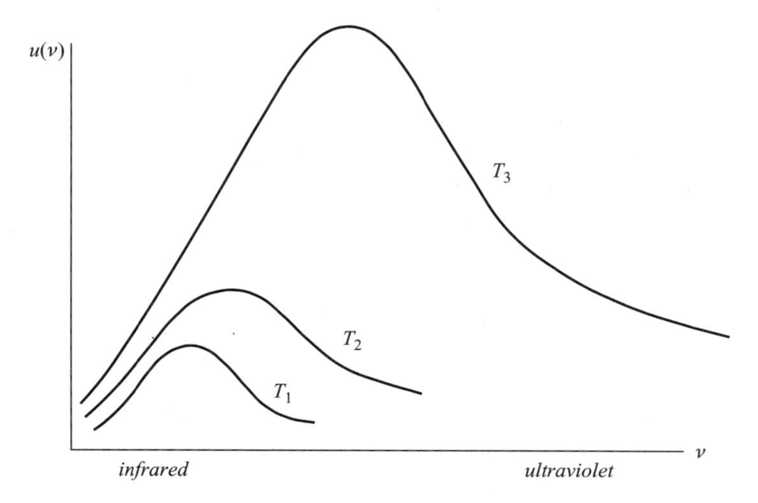

Fig. 1.5 Blackbody radiation curves for several different temperatures $T_1 < T_2 < T_3$.

[5] See, e.g., Born (1957).

[6] See, e.g., Panofsky and Phillips (1955) or Born (1957).

accepted as truly fundamental.) The hypothesis is easily stated: he required that the energy be exchanged between the electromagnetic field at frequency ν and the blackbody walls in discrete amounts, rather than being continuous, as in classical physics. These discrete bundles, or *quanta*, were postulated to occur only as integral multiples of the frequency ν, with a proportionality constant denoted h:

$$E \to E_n = nh\nu, \qquad n = 0, 1, 2, \ldots. \tag{1.11}$$

The hypothesis embodied in Eq. (1.11) led to the following result:[7]

$$u(\nu) = \frac{8\pi h\nu^3}{c^3}\frac{1}{e^{h\nu/(k_B T)} - 1}, \tag{1.12}$$

which not only has the correct behavior at large and small ν but also can be used to account for other (thermodynamic and electromagnetic) aspects of blackbody radiation.

Planck determined the value of h by fitting the formula (1.12) to the blackbody curves, finding the value $h = 6.55 \times 10^{-34}$ J s, in remarkably good agreement with the modern value of 6.62×10^{-34} J s.

That the radiant energy exchanged in the blackbody cavity at frequency ν occurs only in multiples of $h\nu$ is Planck's quantum hypothesis. This energy is thus discretized or *quantized*; $h\nu$ is referred to as a quantum of energy.

The quantum hypothesis was the first step along the path that led to modern quantum mechanics. We now recognize in this hypothesis the signal that classical physics cannot provide the proper framework for describing microscopic phenomena. However, no such recognition was accorded it in the first few years following Planck's publication of his solution to the blackbody problem. This was due in part to the absence of any other application of the quantum concept until Einstein employed it in 1905 to explain the photo-electric effect observed in experiments carried out by R. A. Milliken.

Photons and the photo-electric effect

In the photo-electric effect, electrons are emitted when electromagnetic radiation of intensity I and frequency ν strikes the surface of a metal, as long as ν is greater than a minimum value ν_0 (which varies with the metal). Figure 1.6 schematically represents the process. In addition to the observation that ν must exceed a minimum value ν_0, there are two other results that render this experiment inexplicable from the perspective of the wave theory of radiation. The first is that an increase in the intensity I of the radiation produces an increase in the number N_e of electrons emitted. The second is similar: an increase in the frequency ν of the radiation causes an increase in the energy E_e of the emitted electrons.

From an electromagnetic-theory viewpoint, these results cannot occur. For example, since I measures the energy of the classical electromagnetic wave, E_e should increase with I, not with ν. Furthermore, there should be no connection between ν and either N_e or E_e.

Einstein explained these observations by reverting to a particle picture of electromagnetic radiation that was based on an extrapolation of Planck's quantum hypothesis.

[7] See e.g., Ohanion (1995) for a derivation of Eq. (1.12).

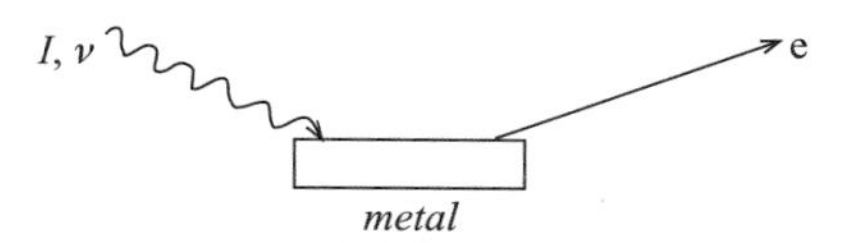

Fig. 1.6 A schematic representation of the photo-electric effect. Radiation of intensity I and frequency ν is absorbed by a metal and electrons are emitted.

Einstein extended that hypothesis by assuming that it also holds for the high-frequency radiation involved in the photo-effect rather than only for the radiation exchanged with the walls of a blackbody cavity. Einstein's assumption can be stated as follows: electromagnetic radiation consists of massless particles denoted light quanta, or, as one now says, of photons; a photon of frequency ν has an energy $E = h\nu$. In this picture, the energy or intensity of radiation can be increased (or decreased) in two ways: by increasing (decreasing) either the number of photons or their frequency (or both). For fixed ν, an increase in I means an increase in the number of photons N_γ (the Greek letter γ often refers to photons).

In Einstein's explanation of the photo-electric effect, the emitted electron is directly knocked out of the metal by a photon, so that an increase in I (or N_γ) means an increase in the number of photo-electrons N_e:

$$N_\text{e} \propto I. \tag{1.13}$$

On the other hand, if ν is increased, then the energy $E_\gamma = h\nu$ of each photon is increased, and hence, in the direct knockout process, the energy E_e of the photon-absorbing electron is increased:

$$E_\text{e} \propto \nu. \tag{1.14}$$

The preceding is a synopsis of Einstein's interpretation of the photo-electric effect. He not only provided an application for Planck's quantum hypothesis, but also re-introduced, via the photon concept, a particle description of radiation. By so doing, he forced his contemporaries to begin to face a distasteful and (to them) unphysical conclusion: electromagnetic radiation, despite its behavior in the two-slit experiment, did not always behave like a wave. The implication is clear: in some situations, light (for example) behaves like a wave, and in other situations it behaves like a collection of particles. This sharing of two disparate properties remained unexplained for many years and was not accepted as a reality by many early twentieth-century physicists until the experiments in 1923 of A. H. Compton on the so-called Compton effect (the scattering of light by atomic electrons), whose successful explanation and predictions were based on the photon concept.[8] How it is possible for electromagnetic radiation to be either particle-like or wave-like was finally explained on the basis of the quantum theory of radiation, a topic beyond the scope of this text.

[8] See, e.g., Born (1957) or French (1968) for a discussion of the Compton effect.

1.2. Quantization of energy and angular momentum

Although we have already encountered the concept of energy quantization, it has occurred only in a limited context, viz., for electromagnetic radiation in special situations. That it can also apply to matter, in particular to the internal energies of atoms, was established primarily by the theoretical work of Bohr (who simultaneously discovered the quantization of angular momentum), and by the experimental work of Franck and Hertz. Their investigations are the topic of this section.

Bohr's theory of the hydrogen atom

The failure of classical physics (and physicists) to provide a framework for understanding various kinds of data in the late nineteenth and early twentieth centuries was the background from which modern quantum theory developed. Among these unexplained data was the vast collection of wavelengths of the radiation emitted by atoms, in particular those emitted by hydrogen. These wavelengths range from the ultraviolet to the far infrared. Despite many efforts to explain or simply to correlate them, only a few empirical relations had resulted. In the case of hydrogen, one of these relations was the 1885 formula of Johann Balmer for the frequencies of the H-atom's spectral lines. Denoting these frequencies by the symbol ν_n and the corresponding wavelengths by λ_n, Balmer's formula is

$$\frac{\nu_n}{c} = \frac{1}{\lambda_n} = \mathfrak{R}\left(\frac{1}{2^2} - \frac{1}{n^2}\right), \qquad n = 3, 4, 5, \ldots, \tag{1.15}$$

where $\mathfrak{R}$, the Rydberg constant, has the value $\mathfrak{R} = 1.096776 \times 10^7\ \text{m}^{-1}$. Not only did (1.15) fit all the frequencies known in Balmer's time (including four in the visible), it also agreed with all the additional ones discovered up to the time of Bohr's model.

There were various attempts to generalize (1.15); the successful one has the form

$$\frac{1}{\lambda_{nm}} = \mathfrak{R}\left(\frac{1}{m^2} - \frac{1}{n^2}\right), \qquad n > m = 1, 2, 3, \ldots, \tag{1.16}$$

and was first verified by Friedrich Paschen in 1908 for the case $m = 3$.

Two features of these formulae require explanation: the value of $\mathfrak{R}$ and the appearance of the integers. Classical physics was unable to account for these points. Niels Bohr was able to do so in much the same way that Planck had accounted for the blackbody-radiation formula, viz., by introducing some *ad hoc* quantum assumptions and then using them in a classical-physics calculation.

Crucial to Bohr's 1913 analysis was the nuclear model of atomic structure deduced by Ernest Rutherford in 1911. Although atoms were known in 1900 to contain both electrons and protons, there was at that time neither a valid nor an accepted model for their overall structure. The first such model was inferred by Rutherford from the results of his experiments on the scattering of α particles[9] by a gold foil. He found that a very small fraction of the α particles was scattered backwards. The only model of the Au atom that

[9] An α particle is the nucleus of a helium atom. It contains two neutrons and two protons.

could explain this was a planetary or nuclear one, that is, one in which a massive, positively charged central core or *nucleus* was somehow surrounded by the electrons at distances well outside the core region.

While this model accounted for the α-particle-scattering data, it made no sense at all from the perspective of classical physics. The main problem arises from the attractive Coulomb force between the electrons and the nucleus. From a classical-physics viewpoint, this attraction means that the electrons would be accelerated towards the nucleus; once they had been accelerated, the electrons would continuously emit electromagnetic radiation, thereby losing energy, and would therefore spiral in closer and closer to the nucleus until they came into contact with it. Now it was known that the size of atoms was of the order of an Ångström unit ($1\ \text{Å} = 10^{-8}$ cm). If the electron in hydrogen were moving in a circle about the proton at a distance of 1 Å then a classical radiation calculation leads to the result that it would fall into the nucleus in about 10^{-11} s! The nuclear model is thus incompatible with classical physics. Nevertheless, Bohr retained it in his theory, adding to it a series of five postulates, which, when they were used in conjunction with a modified classical-physics calculation, explained Eq. (1.16) and the hydrogen spectrum (and thus the structure of H). Despite this major achievement, some of Bohr's conceptual framework and almost all the details of his calculation are incorrect. Furthermore, his theory cannot explain the spectrum of any other atom, not even He, the next atom in the periodic table. Indeed, modern quantum theory is not based on Bohr's theoretical development. Nevertheless, Bohr's contribution is far more profound than the preceding statements suggest; for example, his correspondence principle provides an important link between the classical and quantal pictures.

Bohr's five postulates for the H-atom may be stated as follows.

B1. *Quantized Orbits.* The electron in hydrogen can occupy only discrete or quantized circular orbits; corresponding to the nth orbit of radius r_n are energies E_n, $n = 1, 2, 3, \ldots$.

B2. *Quantum Jumps and Radiation.* Electromagnetic radiation is emitted only when the electron jumps from a larger orbit with a higher energy to one having a lower energy and a smaller radius. The frequency ν of the radiation is related to the orbital energy difference via the energy-conserving Planck–Einstein formula: $h\nu = E_{\text{higher}} - E_{\text{lower}}$. The same relation is also used to determine the photon's frequency in the inverse process of absorption of radiation.

B3. *Stability.* The atom is stable, i.e., it does not radiate, when the electron is in the orbit with the lowest energy (i.e., the one with the smallest radius).

B4. *The Correspondence Principle.* In the limit of large quantum number n, the quantum description must agree with the classical one.

B5. *Dynamics.* Within the constraints of the preceding postulates, the usual laws of classical physics hold, e.g., the electron and proton are bound together by the static Coulomb interaction, energy is conserved, etc.

In Bohr's original analysis the correspondence principle was used to derive the results, among which were specific formulae for $\mathfrak{R}$, r_n, and E_n. In addition, these formulae led to the extremely important conclusion that quantization of orbits necessarily implied that angular momentum L was quantized in units of $\hbar \equiv h/(2\pi)$:

$$r \to r_n \Rightarrow L \to L_n = n\hbar, \; n = 1, 2, 3, \ldots, \tag{1.17}$$

where the symbol $\Rightarrow$ means *implies* or *leads to*.

Although quantization of angular momentum was a by-product of Bohr's work, it is a concept that carries over to modern quantum theory. We shall employ it, in the form of Eq. (1.17), in the following derivation of Bohr's results. Note that B3 makes a virtue of necessity: since neither classical physics nor Bohr himself could explain why the H-atom was stable, he "simply" elevated to the status of a postulate the fact that it did not decay.

The interaction between the proton and the electron (see Fig. 1.7) is the attractive, central, Coulomb potential $V(r) = -\kappa_e e^2/r$, where $\kappa_e = 1$, in cgs units, or $\kappa_e = 1/(4\pi\epsilon_0)$, in SI units. A central force implies conservation (not quantization!) of angular momentum and therefore motion in a plane perpendicular to the direction of the angular momentum. This direction is chosen to be along $\mathbf{z}/z = \mathbf{e}_3$: $\mathbf{L} = L\mathbf{e}_3$, so that the motion is in the x–y plane; hence the co-latitude is given by $\theta = \pi/2$ (see Fig. 1.8). The origin of coordinates is taken to be the center of mass (CM), so that the appropriate mass to use is the reduced mass $\mu = m_e M_p/(m_e + M_p)$ of this two-body system, where m_e (M_p) is the mass of the electron (proton).

Both $\mathbf{r}$ and the corresponding velocity $\mathbf{v}$, which is always perpendicular to $\mathbf{r}$, are in the x–y plane. The magnitude of the angular momentum is $L = \mu v r$; it is quantized via

$$L \to L_n = \mu v_n r_n \equiv n\hbar. \tag{1.18}$$

The second equality in (1.18) is a restatement of Bohr's conclusion, here temporarily

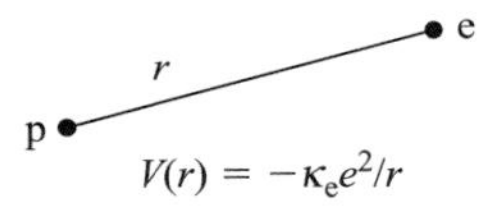

Fig. 1.7 The electron–proton pair forming the H-atom. Their relative separation is r and the potential between them is $V(r) = -\kappa_e e^2/r$.

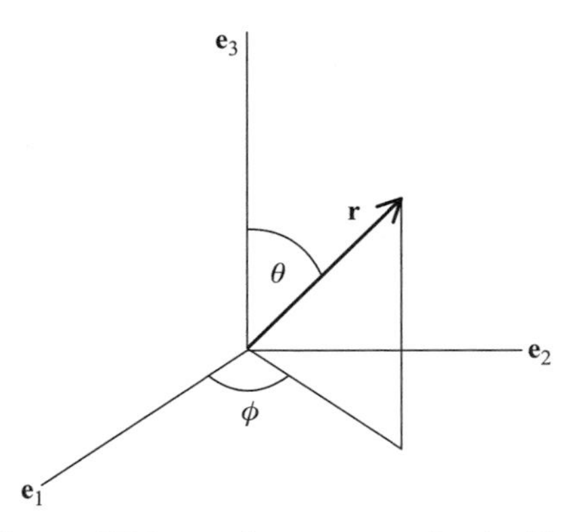

Fig. 1.8 The center of mass (CM) coordinate system for the H-atom. Because of the central force, the motion is confined to a plane, here the x–y ($\mathbf{e}_1$–$\mathbf{e}_2$) plane.

elevated to one of the "laws of motion" in this semi-classical analysis. The quantized values of r_n and v_n are to be determined.

Along with (1.18) two other relations govern the system, viz., the statements of conservation of energy and of Newton's second law:

$$E = K + V = \frac{1}{2}\mu v^2 - \kappa_e \frac{e^2}{r} \longrightarrow E_n = \frac{\mu v_n^2}{2} - \kappa_e \frac{e^2}{r_n}, \tag{1.19}$$

and

$$\mu \frac{v_n^2}{r_n} = -\kappa_e \frac{e^2}{r_n^2}, \tag{1.20}$$

where v_n^2/r_n is the centripetal acceleration.

Equations (1.18) and (1.20) yield the desired result:

$$r_n = n^2 \frac{\hbar^2}{\mu \kappa_e e^2} \equiv n^2 a_H, \tag{1.21}$$

where $a_H = \hbar^2/(\mu\kappa_e e^2)$ is the *reduced-mass* Bohr radius. It is related to the more familiar Bohr radius $a_0 = \hbar^2/(m_e \kappa_e e^2)$ by

$$a_H = a_0\left(1 + \frac{m_e}{M_p}\right), \qquad a_0 = \frac{\hbar^2}{m_e \kappa_e e^2}. \tag{1.22}$$

Since $m_e/M_p \cong 1/1{,}836$, the approximation $a_H \cong a_0$ is a good one, although it is not quite good enough for spectroscopic purposes, since the accuracies in spectral line measurements are one part in 10^6 or better.

The value of a_0 is approximately 0.529×10^{-10} m $\cong \frac{1}{2}$ Å. 1 Å is the typical atomic size, so that the Bohr theory leads to an atomic size of the correct order of magnitude. Indeed, Bohr knew that h must play a role in atomic physics (as in Eq. (1.22)) simply because a quantity of dimension length cannot be formed from the fundamental atomic constants m_e and e without it. That is, m_e and e alone cannot be put together in any way to form a length, and, while $\kappa_e e^2/(m_e c^2)$ is a length (the classical radius of the electron, r_0), c represents effects of relativity, which should not play a role in forming a fundamental atomic length scale.

The smallest Bohr orbit has radius $r_1 = a_H$. All other orbits are larger, scaling by n^2 compared with r_1. An H-atom with an electron in the nth orbit can give rise to radiation when the electron jumps or makes a transition to a smaller orbit, say to $n-1$. To find the frequency of radiation emitted in the transition $r_n \rightarrow r_{n-1}$, we must determine the energy E_n. A simple calculation gives

$$\begin{aligned} E_n &= -\frac{1}{n^2}\left(\frac{1}{2}\frac{\kappa_e e^2}{a_H}\right) = -\frac{1}{n^2}\left(\frac{1}{2}\frac{\mu \kappa_e e^2}{\hbar^2}\right) \\ &\equiv -\frac{1}{n^2}\ \text{Ryd} \\ &= -\frac{1}{2n^2}\ \text{a.u.} \end{aligned} \tag{1.23}$$

Note that E_n is negative, corresponding to a bound system. Equation (1.23) defines two energy units, the Rydberg, denoted Ryd, approximately equal to 13.6 eV ($1\ \text{eV} \cong 1.6 \times 10^{-19}$ J), and the atomic unit, a.u. (also known as a Hartree), equal to 2 Ryd. Since the atom can absorb radiation, let us suppose that it absorbs a photon of energy 1 Ryd when it is in its state of lowest energy, E_1. Adding the value of the absorbed energy to E_1 leads to an energy of zero, i.e., the electron would no longer be bound. Thus, 1 Ryd is both the ionization energy and the binding energy of the H-atom. All other energies E_n, $n \geqslant 2$, are larger than E_1, scaling with n as n^{-2}. An increase in n therefore means a decrease in energy differences, e.g.,

$$E_n - E_{n-1} = \Delta E_n = \frac{2n-1}{n^2(n-1)^2}\,\text{Ryd} \underset{n\to\infty}{\sim} \frac{2}{n^3}\,\text{Ryd}. \tag{1.24}$$

As n increases, the levels cluster together more rapidly than they decrease in magnitude.

In order to visualize the energy-level structure, it is useful to draw what is known as a Grotrian diagram, in which short horizontal lines placed in a column represent the energies, as in Fig. 1.9. These can be ordered as is convenient, but the lowest energy level is conventionally placed at the bottom, with the higher levels sequentially above the lowest or *ground-state* energy level.

These results can be used to calculate the constant $\mathfrak{R}$ of Eqs. (1.15) and (1.16) from the relation $h\nu_{nm} = E_n - E_m$, $n > m$, given in postulate B2. The atom is assumed initially to be in an *excited state* corresponding to the nth orbit ($n > 1$) and subsequently to make a transition to the mth level, $1 \leqslant m < n$, with emission of a photon of energy $h\nu_{nm}$. Expressing E_n in terms of fundamental constants, as in Eq. (1.23), gives for ν_{nm} the form

$$\nu_{nm} = \frac{1}{2}\frac{\mu\kappa_e^2 e^4}{\hbar^3}\left(\frac{1}{m^2} - \frac{1}{n^2}\right). \tag{1.25}$$

The inverse wavelength λ_{nm}^{-1} for the radiation is thus

$$\lambda_{nm}^{-1} = \frac{\nu_{nm}}{c} = \frac{1}{2}\frac{\mu c}{\hbar}\left(\frac{\kappa_e e^2}{\hbar c}\right)^2\left(\frac{1}{m^2} - \frac{1}{n^2}\right), \tag{1.26}$$

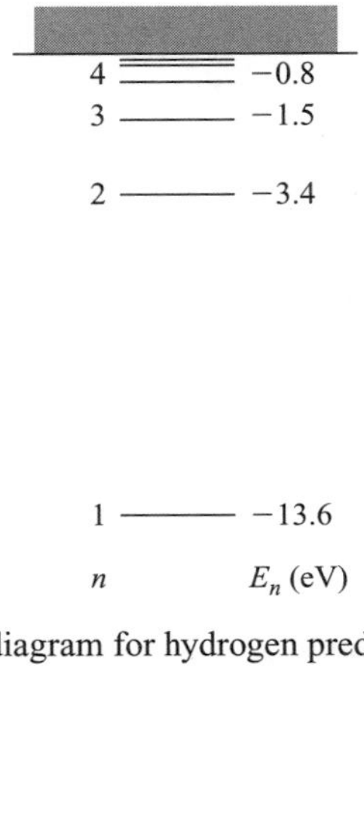

Fig. 1.9 The energy-level diagram for hydrogen predicted by the Bohr theory.

where the ratio $\hbar/(\mu c)$ is denoted the *Compton wavelength* of the H-atom, and $\alpha \equiv \kappa_e e^2/(\hbar c)$ is the fine-structure constant, whose magnitude is approximately 1/137.

Comparison of (1.26) and (1.16) establishes two results of great importance. The first is that Bohr's theory predicts the same *form* for λ_{nm}^{-1} as that encountered in the empirical relations (1.15) and (1.16), especially the difference of inverse squared integers. The second is that the value of the factor $\alpha^2[\mu c/(2\hbar)]$ appearing in Eq. (1.26) is equal to the value deduced for the Rydberg constant from spectroscopic measurements:

$$\frac{1}{2}\frac{\mu c}{\hbar}\alpha^2 = 1.096776 \times 10^7 \text{ m}^{-1} \equiv \mathfrak{R}. \tag{1.27}$$

This collection of results was an enormous triumph, and both the energy-level structure and the orbital picture of Bohr were widely accepted as correct. As we shall see, the set of Bohr energy levels is almost the *same* as is obtained from a modern quantum-mechanical calculation for the hydrogen atom! However, the orbital picture does not follow from such a calculation and, in fact, has *no* quantal basis at all: it is incorrect. Furthermore, the combination of quantized angular momentum and classical physics cannot be used to calculate the energy-level structure of any atoms other than those containing a single electron such as He^+, Li^{2+}, etc. This particular state of affairs was quickly realized soon after Bohr's work when application of his ideas and plausible extensions failed to yield any reasonable results for the next simplest neutral atom, viz., He.[10]

Ultimately, then, Bohr's theory is only a stepping stone on the path to quantum mechanics: postulates B1 and B5 are not valid and thus his atomic theory is generally inapplicable, i.e., it is *wrong*. However, the correspondence principle and the concepts of quantized energy and angular momenta, as well as postulates B3 and B2 (somewhat modified with regard to orbits and jumps), *are* correct.

Many of Bohr's ideas do carry over, as stated, to what we are terming modern quantum theory or simply quantum mechanics. In the period immediately following publication of Bohr's theory and its failure to account for the He spectrum, it was not clear which of Bohr's concepts were valid and which were fallacious. The experiment of James Franck and Gustave Hertz in 1914 helped establish that the concept of energy quantization was valid in microscopic (atomic) physics. We review this experiment in the following.

The Franck–Hertz experiment

The Franck–Hertz experiment established that mercury (chemical symbol Hg), which contains 80 electrons, has a quantized energy-level structure. This was accomplished by demonstrating that, when energetic electrons collide with Hg and energy is transferred to the atom, it occurs not in a continuous fashion but discreetly.

The experimental arrangement is shown schematically in Fig. 1.10. Mercury vapor is placed in an evacuated chamber that has a source of electrons at one end – a hot cathode – and at the other end an anode collector plate just beyond a grid that can be maintained at an adjustable voltage V relative to the cathode. Since the product eV equals the

[10] As noted earlier, this is a conceptual failure, rather than an inability to carry out a sufficiently accurate calculation.

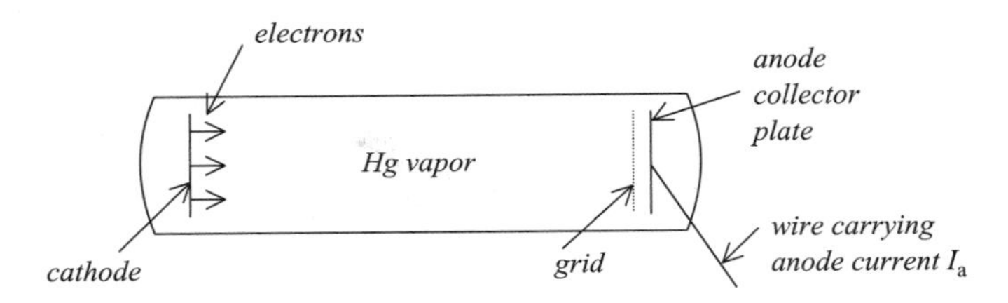

Fig. 1.10 A schematic representation of the Franck–Hertz experimental arrangement.

electron's kinetic energy K_e (in electron volts), adjusting V changes K_e, thus allowing an electron to transfer varying amounts of energy to an atom. Of course, as V increases, the current I_a induced by the electrons collected there will increase.

In the absence of the Hg vapor or of any transfer of electron kinetic energy to the mercury atoms, the anticipated I_a versus V curve would be as indicated in Fig. 1.11. Figure 1.12 represents what was actually measured. The current rises until $K_e = 4.9$ eV, at which point it drops off dramatically, giving rise to a peak followed by a valley as the current again increases with K_e. This peaking/fall-off is seen twice more, at $K_e = 9.8$ and 14.7 eV.

Franck and Hertz correctly interpreted these peaks as evidence of quantized energy levels in Hg, the analogs of the quantized energy-level structure in H. The first peak at 4.9 eV corresponds to an electron transferring that amount of energy to a single Hg atom, thereby putting it into its first excited state (at an energy of 4.9 eV above the ground state). The second (third) peak indicated that a single electron has caused a transition of two (three) Hg atoms to the first excited state.

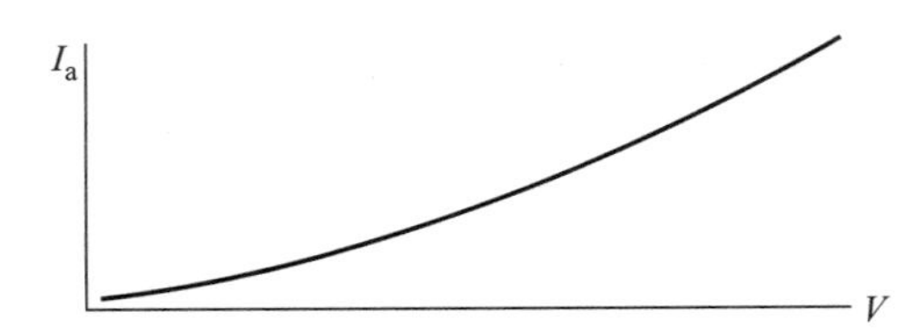

Fig. 1.11 A curve showing the increase in anode current with increasing grid voltage V.

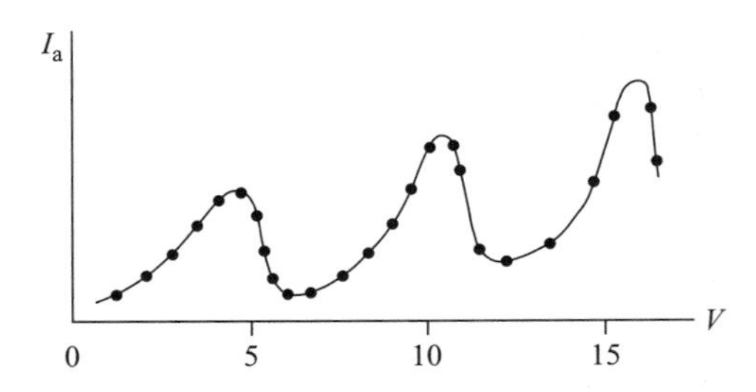

Fig. 1.12 Results of the Franck–Hertz measurement of I_a versus V, showing the maxima and minima in I_a characterizing the transfer of kinetic energy to excitation energy of an atom.

Since 4.9 eV is inferred to be the difference between the ground and first-excited state energies of Hg, then a photon of wavelength $\lambda = 2{,}537$ Å should be emitted when the excited state decays back to the ground state. Ten years after his original experiment with Franck, Hertz detected photons of the correct wavelength $\lambda = 2{,}537$ Å emitted by Hg atoms after they had been excited by electrons transferring 4.9 eV to them. He also detected electrons whose energy loss corresponded to excitation of Hg to higher lying states, as well as the photons corresponding to decay of these states.

The original Franck–Hertz experiment established a quantized energy-level structure for Hg; by implication, this was taken to be true for all other atoms as well. Classical physics could not account for this quantization of energy while the Bohr type of analysis could not explain the details of the energy-level structure of Hg any more than it could those for He. It was evident that the proper theoretical framework remained to be discovered.

1.3. The wave nature of matter

If electromagnetic waves can display particle-like behavior, then one might expect – perhaps from a feeling that the universe of physical phenomena ought to display a kind of symmetry – that particles should somehow possess wave-like properties. The simplest and most fundamental characteristic of a wave, or at least of a monochromatic wave, is its wavelength λ. It was this aspect of waves that Louis de Broglie in 1924 postulated as a fundamental property of any particle (or mass), via the relation

$$\lambda = h/p, \tag{1.28}$$

where p is the magnitude of the particle's momentum. The presence of Planck's constant in (1.28) signifies that the particle's wavelength is a quantum phenomenon.

Equation (1.28) is an extension to particles of the relation between a photon's momentum and its wavelength. Photons are massless, so for them the relativistic equation $E = \sqrt{p^2c^2 + m^2c^4}$ becomes $E = pc$, a result also obtained from classical electrodynamics.[11] Recalling the Planck–Einstein relation $E = h\nu = hc/\lambda$, and equating it to pc, immediately yields (1.28) for the case of photons.

The de Broglie postulate raises an interesting question, viz., what is it in the case of a particle that is waving? The same question arises, of course, in the case of electromagnetic waves, which propagate in a vacuum. Just as in the electromagnetic case, we shall sidestep the question and ask another.[12] Since a particle has a wavelength, and by implication should behave like a wave at least in some circumstances, can its postulated wave-like aspect be verified experimentally?

To attempt to provide a practical answer to this question means seeking phenomena that are exclusive to waves, such as diffraction and interference. Both diffraction and interference have indeed been observed for particles, and we next examine two relevant experiments: one in which electrons have been diffracted by a crystal (the analog of x-ray diffraction) and another in which an interference pattern was observed when electrons were detected after a beam of them had passed through a screen containing two slits.

[11] See, e.g., Jackson (1975) or Panofsky and Phillips (1955).

[12] We return to the original question later in this section.

Electron diffraction

When x-rays are scattered from a crystal, an interference/diffraction pattern can be observed.[13] The pattern results from the interference of coherent waves that have slightly different path lengths, as in the case of two-slit interference. An essential ingredient is that the wavelength of the radiation scattered by the atoms of the crystal be of the order of the atomic spacing in the crystal. Atomic spacings and x-ray wavelengths are typically of the order of Ångström units. If the wave property of an electron is to be made manifest by observing electron diffraction through scattering by a crystal, then it is reasonable to assume that the electron's wavelength λ must also be of the order of Ångström units.

In the experiment of Clinton Davisson and Lester Germer, begun in 1925 and carried out over the next two years (see Fig. 1.13), electrons with wavelengths $\lambda \sim 1$ Å were used. Their first results, obtained before they knew about the de Broglie hypothesis, were a great surprise: they discovered that the intensity I_s of 54.4 eV electrons scattered by a Ni crystal depended in a previously unexpected way on the angle of scattering θ (see Fig. 1.14). The initial fall-off in intensity as θ increases is expected; the problem is that of explaining the peak at $\theta = 50°$. Classical physics could not provide an explanation, but the wavelength hypothesis of de Broglie did so, as follows.

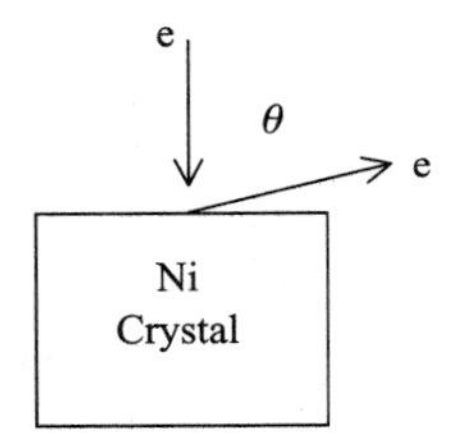

Fig. 1.13 A schematic representation of the Davisson–Germer experimental set-up. Electrons fall perpendicularly on a Ni crystal, are surface scattered through an angle θ, and are then detected.

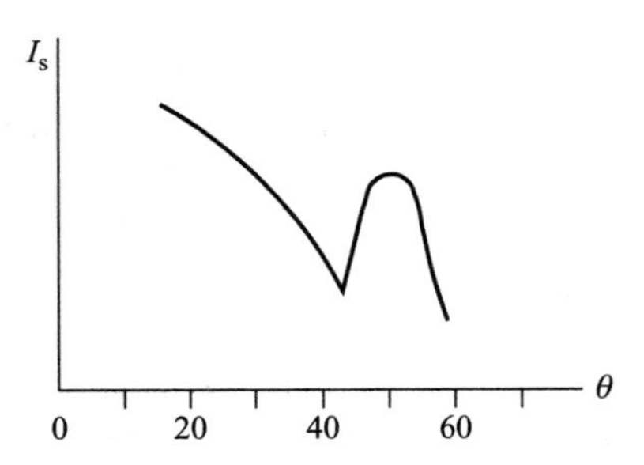

Fig. 1.14 A schematic representation of the Davisson–Germer data on the variation of the intensity I_s of 54 eV electrons scattered by a Ni crystal versus the scattering angle θ, as in Fig. 1.13.

[13] See, e.g., Born (1957) or Ohanian (1995).

As soon as one associates a wavelength with the electrons, the $\theta = 50°$ peak can be understood as a constructive-interference effect, arising in the same way as the maxima from a two-slit experiment with light or in x-ray diffraction patterns. If D is the spacing between atoms in the surface layers of the crystal, then, by using the analysis of the exercises, which is almost identical to that of the two-slit phenomenon in Section 1.1, the following relation is obtained for the angle θ_1 at which the first maximum will occur:

$$\lambda = D \sin \theta_1, \tag{1.29}$$

where λ is the electron wavelength.

For Ni, D was known to be 2.15 Å from x-ray diffraction. Assuming that 50° corresponds to θ_1, a short calculation based on (1.29) gives $\lambda = 1.65$ Å. Is this the wavelength of an electron whose energy is 54.4 eV? To answer this, we simply need to calculate $\lambda = h/p = h/\sqrt{2m_e K_e}$, the value of which is 1.66 Å! The almost perfect agreement is thus compelling evidence for the wave nature of electrons and, by inference, for any other massive object. Corresponding evidence was provided by G. P. Thomson, who in 1928 published results of studies in which a diffraction pattern was obtained by passing an electron beam through a foil. More recently, similar patterns resulting from the scattering of protons, deuterons, α-particles and heavier nuclear projectiles by target nuclei have routinely been obtained in nuclear-physics experiments, although the detailed analysis of these collision phenomena is based on the quantum theory of scattering rather than on the simple de Broglie relation.

Two-slit interference with electrons

In view of the preceding results, one can confidently expect that an interference pattern will be seen on the detection plate after electrons pass through a double-slitted screen. Verification of this expectation was reported by Jönsson (1961), who also obtained three-, four-, and five-slit diffraction patterns.

The two-slit pattern in the electron case is identical to that in the case of light, once again confirming the wave aspect of elementary particles. Given the earlier confirmations, the two-slit results are a bit after the fact, although this does not detract from their impressiveness. The electron experiment is included here, however, not only as additional evidence obtained in a manner strikingly similar to Young's experiment, but also because it implies one of the most important of quantal effects, viz., the ability of a single electron to "interfere" with itself.

To understand the meaning of this claim, let us first recall an essential ingredient of such experiments. This is the need to have a sufficient number of electrons passing through the slits to produce a pattern at the detecting plate or screen. That is, the intensity of electrons must be sufficiently great for a distinct pattern to be detected. This is obvious and corresponds, in the two-slit experiment with light, to there being enough photons for a pattern to be observed: neither one photon nor one electron alone can produce the interference pattern, it takes many.[14] Nevertheless, single photons and single electrons have wave aspects. One property of a wave is that, if it is passed through a two-slit

[14] See, e.g., Feynman, Leighton, and Sands (1965) or French and Taylor (1978) for a discussion of this point.

apparatus, then interference maxima and minima will result. The implication for a single electron is that, if it is incident on a double-slit apparatus, then by virtue of its wave nature some kind of interference phenomenon should occur, i.e., an electron can somehow interfere with itself. Or, putting it more dramatically, a single electron can somehow pass through both slits at once!

What this remarkable statement means experimentally is that, if N electrons are needed to produce an interference pattern, then by allowing the N electrons to fall on the double-slitted screen one at a time, the total pattern at the detector after N of them have been detected will be identical to that which would be obtained if all N were simultaneously incident.

The notion of self-interference brings us back to the question sidestepped earlier in this subsection, viz., what is waving? The answer to this question is expressly a quantal one: while *nothing is waving* in the sense of waves observed macroscopically, "something" indeed ends up producing the *wave-like effect of interference*, including the interference of a particle with itself. This "something" is the particle's state vector, sometimes referred to as its wave function, concepts we shall examine in detail in later chapters. Only a few comments on them are given in this chapter, in order to indicate their connection with wave-like behavior (see brief additional comments in Chapter 2).

State vectors and wave functions play a central role in quantum theory. Among other things, they determine, directly or indirectly, the *probabilities* for the occurrence of various microscopic events, such as a particle being located at a given point in space at a given time, the rate at which photons are emitted by a particular quantum system (i.e., the lifetime of the system against electromagnetic decay), or the scattering of particles into a specified solid angle in a collision. Wave functions, and certain quantities related to them, are sometimes referred to as quantum amplitudes, or simply amplitudes. The absolute square of the amplitude is related to the quantal probability for the occurrence of the event that the amplitude describes.[15] In the case of a particle, there is a separate amplitude for each path it may follow. If more than one path is available in realizing a particular event or outcome, as in the two-slit arrangement, then the quantal amplitudes *must be added before they are squared*, exactly as in calculating the intensity of light on the screen in Young's experiment. This adding of amplitudes and then squaring to obtain the quantal probability is the underlying source of quantal interference.[16] Nothing physically observable is "waving," but the adding of amplitudes followed by the squaring of the sum produces the effect of wave interference. Interference effects are tangible, whereas amplitudes are among the most abstract of physical quantities. Quantum theory connects them.

[15] This is discussed in detail in Chapter 5; see also Appendix A for a short review of probability theory.

[16] We note that, if it is determined which slit the electron passed through, then there is no longer an interference pattern! In terms of the notion of paths and amplitudes, determination of the slit specifies the path and reduces the sum of amplitudes to one only, for which there cannot be interference.

Exercises

1.1 Two narrow slits in a screen are separated by a distance of 6 cm. Electromagnetic radiation of wavelength 3 cm (microwaves) is incident on the screen. At what angles are maxima produced on a detecting screen far away from the slits?

1.2 Light of wavelength 632.8×10^{-9} m (from a He–Ne laser) passes through a two-slit screen and is transmitted to a detecting screen 3.5 m from the slits. If the separation between interference minima on the detection screen is 1.8 cm, what is the distance between the slits?

1.3 The energy density at frequency ν of electromagnetic radiation is given by Eq. (1.12). Show that
(a) $u = \int_0^\infty u(\nu)\mathrm{d}\nu \propto T^4$,
(b) $u(\nu \to 0) \cong (8\pi\nu^3/c^3)k_\mathrm{B}T$,
(c) $u(\nu \to \infty) \to 0$.
(d) If the wavelength λ rather than the frequency ν is used as the independent variable, show that

$$u(\nu)\,d\nu \to u(\lambda)\,d\lambda = g(\lambda)\,d\lambda/(e^{hc/(k_\mathrm{B}\lambda T)} - 1)$$

and determine the quantity $g(\lambda)$.

1.4 (a) Use the result of Exercise 1.3(d) to find an equation obeyed by λ_max, the value for which $u(\lambda_\mathrm{max})$ is a maximum.
(b) Use numerical methods to solve the equation of part (a) so as to determine an approximate expression for λ_max as a function of T.
(c) $T = 2.7$ K for the cosmic blackbody background radiation; what is the corresponding λ_max?

1.5 Show, using conservation of energy and momentum, that a free electron at rest cannot absorb a photon.

1.6 How many photons per second correspond to a power of 2 W for frequencies of
(a) 10^8 Hz (radio waves),
(b) 10^{20} Hz (γ-rays).

1.7 Suppose that a 1-kW He–Ne laser (Exercise 1.2) produces a perfectly collimated beam of photons whose radius is 1.5 cm. To what photon density does this correspond?

1.8 A stationary, 1-g ladybug absorbs a photon and moves involuntarily. What is her velocity if the photon's energy is
(a) 1 eV,
(b) 1 keV,
(c) 1 MeV.

1.9 The atomic unit of length is the Bohr radius a_0, Eq. (1.22). From it atomic units of energy, momentum, etc. are determined.
(a) Show that without $\hbar$, i.e., from the two quantities $\kappa_\mathrm{e}e^2$ and m_e alone, one cannot form a quantity whose dimension is length.
(b) As noted in the text, by adding c to $\kappa_\mathrm{e}e^2$ and m_e, one *can* form a quantity with dimension length. Show that $\kappa_\mathrm{e}e^2/(m_\mathrm{e}c^2)$ is the only such quantity one can

construct. A comparison of this quantity with a_0 should indicate why Bohr argued that $\hbar$ must play a role in the analysis of atomic phenomena.

1.10 The dimensionless quantity $\alpha = \kappa_e e^2/(\hbar c)$ – the fine-structure constant – can be used to establish the scale of many quantities in atomic physics. Let E_1 and v_1 be the lowest energy and orbital velocity in the Bohr model of hydrogen. The proportionality constants C and D in $E_1 = C\mu c^2$ and $v_1 = Dc$ depend on α. Determine that dependence and use it to argue that relativistic effects should be negligible in describing the structure of hydrogen.

1.11 Use the Bohr model to describe an atom consisting of a positive nuclear charge Ze of mass M and a negative charge $-ze$ whose mass is m. In particular, derive expressions for
(a) the analog of a_H, Eq. (1.22),
(b) the energy levels, analogous to Eq. (1.23),
(c) the orbital velocities,
(d) the wavelengths, analogous to Eq. (1.26).

1.12 Using the results of Exercise 1.11, determine the numerical values of the analogs of E_1 and a_H for
(a) a deuterium atom (in which the proton of the H-atom is replaced by a deuteron, $m_d \cong 2M_p$),
(b) a muonium atom (in which the electron of H is replaced by a negative muon, $m_\mu \cong 207m_e$),
(c) a positronium atom (in which the proton of H is replaced by a positron, of mass equal to m_e),
(d) a protonium atom (in which the electron of H is replaced by a negatively charged proton).

1.13 For singly ionized helium (He^+), the nuclear charge and mass are $2e$ and $m_{He} \cong 4M_p$.
(a) Use the Bohr model to show that, for each wavelength λ_{mn} corresponding to a radiative transition in hydrogen, there is a corresponding transition in He^+ of almost identical wavelength.
(b) Determine the ionization energy of He^+ and the difference in energy between its ground state and each of the first two excited states, expressing your answers in electron volts.

1.14 In the Bohr model, the electron moves in circular orbits around the proton, giving rise to a current and a magnetic moment. Assume that $n = 1$ and determine numerically
(a) the current I_1,
(b) the magnetic moment, given in SI units by $I_1 A_1$, where A_1 is the area of the lowest orbit.

Express your answers in terms of amperes (A) and m^2.

1.15 The Coulomb potential is the attractive interaction binding the electron to the proton in hydrogen. The gravitational potential binds the earth to the sun. Assume that the Bohr model applies to the earth–sun system.
(a) Determine an expression for the analog of the Bohr radius for the quantized earth–sun motion, and evaluate it in meters and in Bohr radii a_0.
(b) What is the formula for and the numerical value in electron volts and in joules of the analog of the ionization energy?

The following values may be helpful: $G = 6.67 \times 10^{-11}$, $m_{earth} = 5.97 \times 10^{24}$, $m_{sun} = 1.99 \times 10^{30}$, and earth–sun separation $= 1.5 \times 10^{11}$, all in SI units.

1.16 Apply the Bohr quantization model to the case of a particle of mass m moving in a circular orbit in the potential $V(r) = kr^2/2$.

(a) Express the quantized energy levels in terms of $\hbar\omega$, where $\omega = \sqrt{k/m}$.

(b) Find an expression for r_0, the analog of the Bohr radius, and relate the nth quantized radius r_n to n and r_0.

(c) Determine the constant of proportionality between n and the nth quantized orbital velocity v_n.

1.17 One nuclear analog of the Franck–Hertz experiment would be collisions between a beam of protons and a gas of heavy nuclei. In such a case, one would measure the current I_p due to the protons after they had passed through the gas. Assume that the (identical) nuclei comprising the gas can be excited by proton impact only to levels at 3 and 5 MeV. If the proton kinetic energy K_p can be varied between 0 and 14 MeV, make a sketch of the I_p versus K_p curve one would expect to see, analogous to Fig. 1.12.

1.18 X-rays of wavelength λ fall onto the flat surface of a crystal and are scattered at an angle θ, as in Fig. 1.15. The surface is modeled in this problem as a set of atoms separated by a spacing D, as in the diagrams below.

(a) In Figure 1.15(a), the x-rays are incident vertically. Show that the condition for maximal constructive interference is

$$D \sin\theta = n\lambda.$$

(b) In Figure 1.15(b), the angle of incidence is no longer vertical. Show that the expression for maximal constructive interference of part (a) is changed to

$$D \sin\theta = n\lambda + f(\theta_0),$$

and determine $f(\theta_0)$.

1.19 Determine the de Broglie wavelength for each of the following systems:

(a) electrons, protons and α-particles whose energies are 1 keV,

(b) a Ni-atom at room temperature, $T \cong 300$ K,

(c) a 1-g ladybug moving at a constant speed of 1 cm s^{-1},

(d) an electron with kinetic energy 13.6 eV (compare with a_0),

(e) a proton accelerated to 1 TeV $= 10^{12}$ eV.

1.20 It is stated in the text that the interference pattern produced by a beam of electrons incident on a two-slit screen is a manifestation of its wave nature. The wave-type quantity in this case is the wave function "describing" the electron; in particular, it yields the probability that any electron in the beam will be at a particular point on the

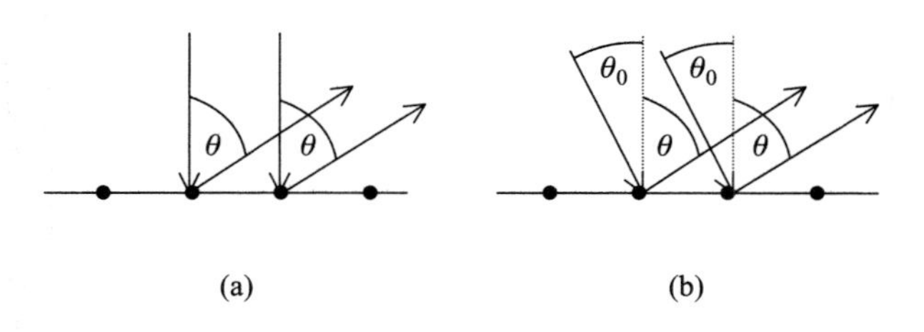

Fig. 1.15

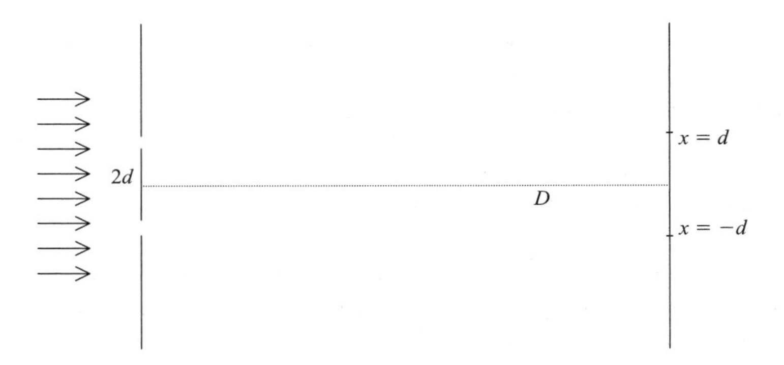

Fig. 1.16

detecting screen. This probability is proportional to the absolute square of the wave function, and, if there is more than one path for an electron to follow, then there is one wave function for each path, with the total or overall wave function being the sum of those for each of the possible paths. The wave function is the analog of the electric field describing light in Section 1.1. We shall use the symbol ψ_j for the wave function corresponding to path j, with the total wave function ψ given by

$$\psi = \sum_j \psi_j,$$

while the probability P is

$$P = |\psi|^2.$$

In this exercise, electrons of momentum $\hbar k$ are incident on the two-slit screen, whose separation is $2d$. The detector screen is a distance D away, as in Fig. 1.16. Denoting the upper (lower) slit as number 1 (2) and the distance along the detector screen by x, the two wave functions at the screen are assumed (for the purposes of this problem) to be

$$\psi_1(x) = Ae^{ik(x-d)^2}/D$$

and

$$\psi_2(x) = Ae^{ik(x+d)^2}/(2D),$$

where A may be considered constant for x not too far from $x = 0$.

(a) What is $P(x)$ when slit 2 is closed?

(b) What is $P(x)$ when both slits are open? If $P(x)$ displays an interference effect, determine the separation Δx between neighboring maxima in the interference pattern.

2

Classical Concepts and Quantal Inequivalences

Chapter 1 was concerned with the need for a quantal description of microscopic phenomena. The thrust of Chapter 3 is an analysis and comparison of a classical and a quantum system; some mathematical generalizations of the analysis are presented as well. Chapter 4 provides an introduction to Hilbert-space concepts that are the mathematical framework of quantum theory, while, in Chapter 5, the axioms of quantum mechanics for pure states are stated and illustrated. The present chapter is intended to be a bridge between those chapters and Chapter 1. Here we review many of the concepts of classical mechanics that have relevance in quantum mechanics. Equally important, if not more so, we stress the lack of equivalence between a number of classical concepts and their quantal counterparts.

This inequivalence is emphasized because of the vocabulary common to classical and quantum mechanics. Classical physics is, of course, the source for much of the vocabulary in quantum physics. However, the ways in which the words are used in the two domains often differ substantially, while the concepts are almost necessarily not identical. Some classical concepts have no direct quantal counterparts. For example, velocity and force are not primary quantal entities. This fact means that the quantal definitions of linear and angular momentum cannot be the same as those in classical mechanics. Similarly, the concept of a "state" in classical physics – the collection of the generalized positions and generalized velocities or momenta – does not carry over to quantum physics. In the quantal case, the state of a quantum system is a highly abstract quantity that is never observable or measurable. Futhermore, quantal states, unlike their classical namesakes, are related to probabilities, as indicated in Section 1.3.

These aspects, and others, stated through the kinematic and dynamical concepts of classical mechanics, are discussed in a series of subsections, starting with some remarks on "particles" and ending with a brief discussion of states, measurements, and Heisenberg's uncertainty principle.

2.1. Particles

Mass is a concept common to classical and quantum physics, as is the point mass or structureless particle. However, the objects that qualify as particles in the quantal domain depend on the energy scale, whereas in classical physics any microscopic mass is regarded a particle. For example, from the macroscopic point of view an atom is a particle with no structure. Microscopically, an atom has structure and therefore cannot be a particle. In non-relativistic quantum mechanics, an atom is composed of three kinds of constituents, viz., electrons, protons and neutrons, and each of them is assumed to be an "elementary" particle. From the perspective of elementary-particle physics, for which the fundamental dynamical framework is relativistic quantum mechanics or field theory – topics with which this book will not be concerned – electrons are taken to be structureless mass points, but protons and neutrons as well as other so-called "elementary" particles, such as mesons, are not. Instead, these entities are currently believed to be composed of point objects denoted quarks. If future theoretical frameworks assign a structure to quarks, they too would no longer be considered particles.

2.2. Coordinate systems: positions, velocities, momenta

Although coordinate systems occur both in classical and in quantum physics, their roles in the two theoretical frameworks differ because the quantities associated with them turn out to differ so much in concept and use.

Positions and velocities

Consider a particle of mass m whose position vector $\mathbf{r}$ has the Cartesian components x, y and z with respect to an arbitrary coordinate system S, as in Fig. 2.1:

$$\begin{aligned}\mathbf{r} &= x\mathbf{e}_x + y\mathbf{e}_y + z\mathbf{e}_z \\ &= x_1\mathbf{e}_1 + x_2\mathbf{e}_2 + x_3\mathbf{e}_3 = \sum_{i=1}^{3} x_i\mathbf{e}_i. \end{aligned} \tag{2.1}$$

In classical mechanics, position is a function of time, $\mathbf{r} = \mathbf{r}(t)$, and in principle it can be determined, i.e., measured, at any t to arbitrary accuracy without disturbing the particle's trajectory. This statement is valid for massive bodies such as planets or spacecraft as well as for single particles. It is, however, not a generally valid statement in the quantal regime, as is explored later in this chapter.

Given $\mathbf{r}(t)$, the particle's velocity $\mathbf{v}(t)$ in Newtonian physics is defined as

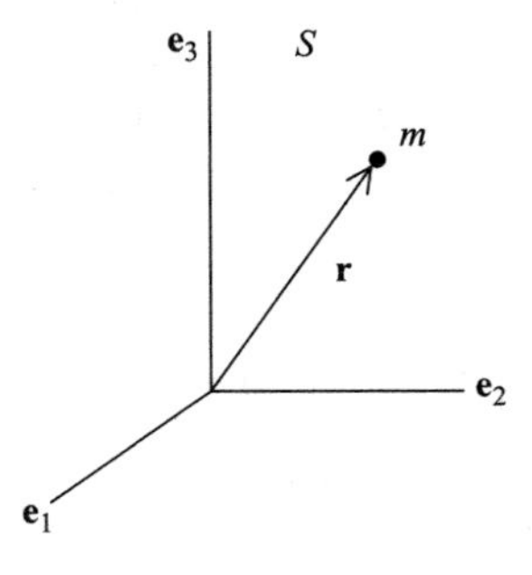

Fig. 2.1 The position $\mathbf{r}$ of a particle of mass m w.r.t. the Cartesian coordinate system S. Unit vectors along the axes x, y, and z or 1, 2, and 3 are denoted $\mathbf{e}_x$, $\mathbf{e}_y$, $\mathbf{e}_z$ or equivalently $\mathbf{e}_1$, $\mathbf{e}_2$, and $\mathbf{e}_3$, respectively.

$$\mathbf{v}(t) = \dot{\mathbf{r}}(t) = d\mathbf{r}(t)/dt$$
$$= \sum_{i=1}^{3} v_i(t)\mathbf{e}_i. \tag{2.2}$$

Just as for $\mathbf{r}(t)$, in the classical domain $\mathbf{v}(t)$ can in principle be determined to arbitrary accuracy by measurement at any time t.

One of the primary goals of classical mechanics is to specify trajectories, in particular the path $\mathbf{r}(t)$ a particle will follow under the influence of the forces acting on it. Apart from chaotic motion, a necessary and sufficient condition for completely predicting the path from the equation of motion, say $\mathbf{F} = m\ddot{\mathbf{r}}$, is that a value of $\mathbf{r}$ and a value of $\mathbf{v}$ at any instant be known, say $\mathbf{r}(t_1)$ and $\mathbf{v}(t_2)$, with t_2 typically set equal to t_1. Thus, two measurements of essentially canonically conjugate quantities suffice to fix $\mathbf{r}(t)$. Although this mode of description is a cornerstone of macroscopic physics – it is, after all, one of the fundamental notions emphasized in courses on Newtonian mechanics – practically *none* of this description carries over to the quantal domain.

First, the concept that a microscopic particle can even have a specific trajectory is foreign to – i.e., is meaningless in – quantum theory. The concept of trajectory exists only as a classical limit. Thus, in quantal bound-state situations, corresponding to orbital motion in classical mechanics, no path is or can be specified, Bohr orbits notwithstanding. In place of a specific $\mathbf{r}(t)$, the quantal equations of motion yield a probability of finding the particle at a particular position $\mathbf{r}$. Second, even if one wanted to introduce a classical-type trajectory $\mathbf{r}(t)$ for the particle, the absence of force as a primary quantal concept militates against this. Third, velocities do not play a fundamental role in quantum mechanics, although *momentum* does. Fourth, and perhaps most important of all, experiments that are designed to determine either the instantaneous position of, or the path taken by, a microscopic particle will result in a change of the position or the path (see, e.g., Section 2.6). In other words, quantum mechanics does not allow one to refer to, much less specify, the actual path followed by a particle.

In view of the foregoing, it will probably be a surprise to the typical reader that one of the major developments of quantum theory makes use of classical trajectories! This development is the path-integral formulation of R. P. Feynman. Since it is discussed in Section 8.2, it is commented on only briefly here. We have seen that no particular path can be singled out quantally as the actual trajectory of a particle. This can be understood as a manifestation of the probabilistic nature of quantum mechanics. Because there is a probability of finding a particle at position $\mathbf{r}$, there will also be a probability for the occurrence of every classical trajectory: in the quantal domain all trajectories are permitted. Feynman brilliantly capitalized on this fact through a formulation of the theory in which all classical paths are summed over. Such a sum is, as we shall discover, a peculiarly quantal feature, which is not known classically.

Linear and angular momentum

For a single particle of mass m, its linear momentum $\mathbf{p}$ is defined classically by

$$\mathbf{p} = m\mathbf{v}. \tag{2.3}$$

The reformulation of Newton's second law $\mathbf{F} = m\ddot{\mathbf{r}}$ to the form

$$\mathbf{F} = \dot{\mathbf{p}}, \tag{2.4}$$

where $\mathbf{F}$ is the total force acting on the particle, serves to emphasize the fundamental role that momentum plays in classical mechanics. The angular momentum $\mathbf{L}$ of the particle, defined by

$$\mathbf{L} = \mathbf{r} \times \mathbf{p} = \mathbf{r} \times m\mathbf{v}, \tag{2.5}$$

is also one of the basic classical concepts. Both $\mathbf{p}$ and $\mathbf{L}$ can be, and often are, conserved quantities.

Linear and angular momenta are also dynamical quantities in quantum mechanics, but since velocity is not, it follows that $\mathbf{p}$ and $\mathbf{L}$ must be redefined when they are used in a quantal context. Their actual definitions and use, and those of other macroscopic concepts that carry over to the microscopic domain, will be seen in Chapter 5 to constitute some of the peculiarities that help make quantum mechanics such an exotic subject and, from a classical point of view, such a bizarre one.

2.3. Dynamical equations, generalized coordinates and conserved quantities

Potentials and dynamical equations

Conservative systems are those for which the force is derivable from a potential energy or potential. Although potentials are normally time-independent, there are situations in which a time dependence can be incorporated into the potential, the electromagnetic case being one example. Under the assumption that the system is governed by one or more potentials – an assumption that will hold throughout this book – there are typically three sets of equations of motion available in classical mechanics. One is Newton's second law, which, for a single particle, is $\mathbf{F} = \dot{\mathbf{p}}$. The other two are Euler's equations, derived from the Langrangian L, and Hamilton's equations, derived from the Hamiltonian H.

Any of the preceding three sets of dynamical equations can be used (at least in principle) to determine the time dependence of the generalized coordinates and momenta in classical mechanics. None carries over to the quantal regime. Replacing them as the fundamental equation determining the time behavior of a quantum system is the time-dependent Schrödinger equation, introduced by Erwin Schrödinger in 1926. This equation contains $\hat{H}$, the quantal analog of the classical Hamiltonian function H; $\hat{H}$ is constructed using the quantal analogs of the generalized momenta and potentials that enter H. The solution to the time-dependent Schrödinger equation is the state vector or the wave function of the (quantal) system, and this latter quantity is often expressed as a function of the generalized coordinates that are used in classical mechanics. However, in a quantal context, the generalized coordinates and the time are all independent variables; hence, as noted before, quantum dynamics does not involve calculation of trajectories (i.e., calculation of the time dependence of the generalized coordinates). Nonetheless, since generalized coordinates do play a role in quantum mechanics, we consider them next.

Generalized coordinates

The notation to be used for a set of generalized coordinates is $\{q_\alpha\}_{\alpha=1}^{N}$. Probably the most common example of generalized coordinates in three dimensions is the set of spherical polar coordinates, denoted as usual[1] by r, θ, and ϕ (i.e., $q_1 = r_1$, $q_2 = \theta$, and $q_3 = \phi$), as in Fig. 2.2, which serves to define them. The angles θ and ϕ are referred to as the co-latitude and the azimuth, respectively. Spherical polar coordinates are useful in any problem for which spherical symmetry is important, e.g., when the potential in the case of a single particle depends only on r, the magnitude of $\mathbf{r}$: $V(\mathbf{r}) = V(r)$.

Corresponding to r, θ, and ϕ are the unit vectors $\mathbf{e}_r$, $\mathbf{e}_\theta$, and $\mathbf{e}_\phi$, indicated in Fig. 2.3. As with the Cartesian unit vectors $\mathbf{e}_1$, $\mathbf{e}_2$, and $\mathbf{e}_3$, the right-handed set $\mathbf{e}_r$, $\mathbf{e}_\theta$, and $\mathbf{e}_\phi$ is an orthonormal one:

$$\mathbf{e}_\alpha \cdot \mathbf{e}_\beta = \delta_{\alpha\beta}, \qquad \alpha, \beta \in \{r, \theta, \phi\}, \tag{2.6}$$

where $\delta_{\alpha\beta}$ is the Kronecker delta, defined by

$$\delta_{\alpha\beta} = \begin{cases} 1, & \alpha = \beta, \\ 0, & \alpha \neq \beta. \end{cases}$$

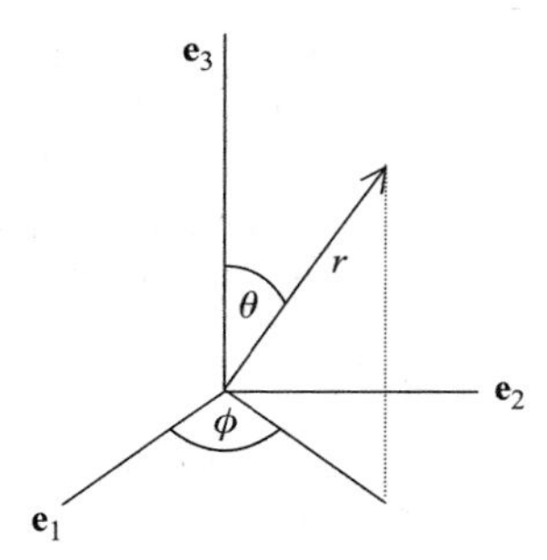

Fig. 2.2 The spherical polar coordinates r, θ, and ϕ.

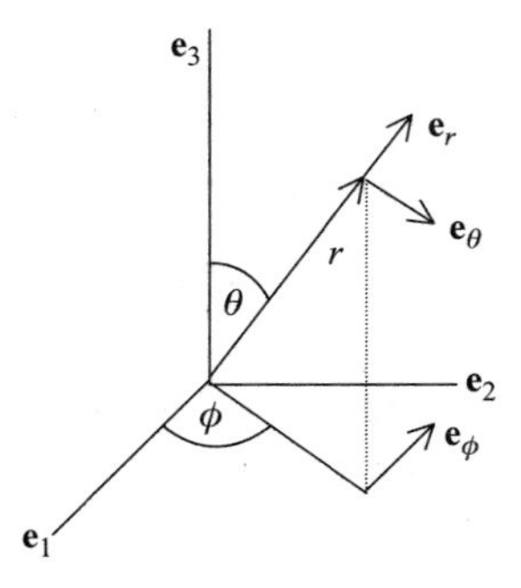

Fig. 2.3 Unit vectors for a spherical polar coordinate system. $\mathbf{e}_r$, $\mathbf{e}_\theta$, and $\mathbf{e}_\phi$ form a right-handed coordinate system.

[1] This symbology is standard in classical mechanics; see, e.g., Fowles and Cassidy (1990), Goldstein (1980), or Marion (1970).

One of the noteworthy aspects of this set of vectors is its position ($\mathbf{r}$) dependence, in contrast to the Cartesian set $\{\mathbf{e}_i\}_{i=1}^3$. Consequently, in calculating the spherical polar components of $\mathbf{v}$, say, it is necessary to evaluate the $\{\dot{\mathbf{e}}_\alpha\}$. The resulting spherical components v_β of $\mathbf{v}$ turn out to be[2]

$$\begin{aligned} v_r &= \dot{r} \\ v_\theta &= r\dot{\theta} \\ v_\phi &= r\sin\theta\,\dot{\phi}. \end{aligned} \tag{2.7}$$

One of the more important properties of this particular set of generalized coordinates is that the products mv_β are *not* equal to the spherical polar components p_β of the momentum vector $\mathbf{p}$: the relation $p_i = mv_i$ holds only in rectangular coordinates. To obtain the p_β, $\beta \in \{r, \theta, \phi\}$, typically requires the introduction of either the Lagrangian L or the Hamiltonian H, although in many cases the kinetic energy K suffices.

Generalized momenta

The Lagrangian L depends on the generalized coordinates $\{q_\alpha\}$ and generalized velocities $\{\dot{q}_\alpha\}$ plus the time, whereas the Hamiltonian, via a canonical transformation, is expressed in terms of the generalized coordinates $\{q_\alpha\}$ and the generalized momenta $\{p_\alpha\}$ plus the time.[2] Here, the subscript α labels members of any set of generalized coordinates including those referring to systems containing more than one particle. As discussed in texts on classical mechanics, L is given by the difference

$$L = K - V, \tag{2.8}$$

with K denoting the sum of all the kinetic energies and V the sum of all the potentials acting on and in the system. It is necessary, of course, that K and V be expressed as functions of the $\{q_\alpha\}$ and $\{\dot{q}_\alpha\}$.

The generalized momentum p_α is defined by

$$p_\alpha \equiv \partial L/\partial \dot{q}_\alpha; \tag{2.9}$$

p_α is the momentum *canonically conjugate* to the generalized coordinate q_α. When the potentials comprising V do not depend on the $\{\dot{q}_\alpha\}$, then from (2.8) it follows that

$$p_\alpha = \frac{\partial K}{\partial \dot{q}_\alpha}, \qquad V \neq V(\{\dot{q}_\alpha\}). \tag{2.10}$$

Apart from the electromagnetic case, the situation defined by (2.10) is the typical one.

The example of spherical polar coordinates is a good illustration of this typical situation. In these coordinates the kinetic energy is given by

$$K = \tfrac{1}{2}m(v_r^2 + v_\theta^2 + v_\phi^2), \tag{2.11}$$

and from Eqs. (2.7) and (2.10) it follows that

[2] These quantities are discussed and/or derived in books on classical mechanics, for example those cited in footnote 1.

$$p_r = m\dot{r} = mv_r$$

$$p_\theta = mr^2\dot{\theta} = mrv_\theta$$

$$p_\phi = mr^2 \sin^2\theta\,\dot{\phi} = mr\sin\theta\, v_\phi. \tag{2.12}$$

Using Eqs. (2.12), K can be re-written as

$$K = \frac{1}{2m}\left(p_r^2 + \frac{p_\theta^2}{r^2} + \frac{p_\phi^2}{r^2\sin^2\theta}\right). \tag{2.13}$$

This re-expressing of the kinetic energy in terms of the momenta serves as a reminder that, by means of a canonical transformation, one can replace the Lagrangian, which depends on the $\{q_\alpha\}$ and $\{\dot{q}_\alpha\}$, by the Hamiltonian, which depends on the $\{q_\alpha\}$ and the $\{p_\alpha\}$:

$$L(\{q_\alpha\}, \{\dot{q}_\alpha\}, t) \to H(\{q_\alpha\}, \{p_\alpha\}, t) = K + V. \tag{2.14}$$

Whereas L has no direct quantal counterpart, the classical Hamiltonian does: it, along with position, momentum, and energy, is converted in quantum theory into fundamental dynamical entities.

As noted, H carries over to the quantal domain whereas the Hamiltonian equations of motion themselves do not (and neither do Euler's equations). In spite of this, a major concept in classical mechanics that arises directly from use of the classical, dynamical equations also occurs in quantum mechanics. This is the notion of the conserved quantity or constant of the motion.

Hamilton's equations of motion provide a particularly elegant means of deriving classical constants of the motion. These equations are

$$\dot{q}_\alpha = \frac{\partial H}{\partial p_\alpha} \tag{2.15}$$

and

$$\dot{p}_\alpha = -\frac{\partial H}{\partial q_\alpha}. \tag{2.16}$$

When $H \neq H(q_\beta)$, then by (2.16) p_β is a constant in time, i.e., it is a conserved quantity. A well-known example is the case of a central potential, $V = V(r)$: since V does not depend on $q_\beta = \phi$, then $p_\beta \equiv L_z$, the z-component of angular momentum, is conserved. In the usual case in which H does not depend explicitly on time, Hamilton's equations (2.15) and (2.16) guarantee that $dH/dt = \dot{H} = 0$, so that H itself is conserved. This is simply an alternate way of stating that energy is conserved, which we may express as

$$H = E, \tag{2.17}$$

where the energy $E = K + V$ is a constant. The quantal analog of Eq. (2.17) is the time-independent Schrödinger equation, the basis for many non-relativistic quantal calculations.

Before turning our attention to the role played by potentials in classical and quantal physics, we recall from Chapter 1 one of the most important differences between the classical and quantal domains. This is the contrast between the continuous ranges of values that are taken by certain classical quantities and the discreteness of the ranges of values of their quantal counterparts. For example, in classical mechanics the magnitude of the z-component of angular momentum may take on any value in the continuum, starting at the value zero. In quantum mechanics, however, the z-component of angular momentum is limited to a denumerable (discrete) set of values, some of which are negative and all of which are multiples of $\hbar = h/(2\pi)$. In other words, just as in the Bohr theory of hydrogen, the z-component of angular momentum becomes *quantized* in the microscopic domain. Furthermore, in quantal bound-state situations (corresponding to classical orbits), the energy is always quantized, as suggested in the preceding chapter.

2.4. Potentials and limits of motion

In general, the potentials occurring in classical physics carry over to non-relativistic quantum mechanics without any change in form, although their specific use (as well as their interpretation) is no longer quite the same. One aspect of potentials in quantum mechanics is especially striking: potentials generally do *not* confine a particle's position (although the probability of finding a particle outside the boundary arising from the potential is often very small).

As an example of this non-confinement, consider a particle of mass m acted on by a one-dimensional harmonic oscillator potential. The potential is $V(q) = \frac{1}{2}kq^2$, where q is the position coordinate and k is the usual force parameter; $\omega = \sqrt{k/m}$ is the angular frequency. In this ideal case of no friction, the equation of motion leads to $q = A\cos(\omega t + \alpha)$, where A is the amplitude of the oscillator and α is a phase angle whose value is determined by the initial condition. The total energy is, of course, $E = p^2/(2m) + kq^2/2 = kA^2/2 = m\omega^2 A^2/2$. (The range of E is continuously infinite, beginning at $E = 0$.) One may regard the maximum (positive or negative) displacement of the oscillator, A, as being determined by the total energy E: $A = \sqrt{2E/k}$.

In this simple example, the particle's trajectory or path is easy to describe: it oscillates between $+A$ and $-A$, with the maximum displacement being larger or smaller as E is larger or smaller. This is represented in Fig. 2.4, wherein the linear trajectory is superposed on a plot of $V(q)$, the maximum extensions being given by the curve $kq^2/2$. That is, the potential forms the boundary on and inside of which the motion occurs. The trajectory cannot extend beyond this boundary because that would imply a negative value

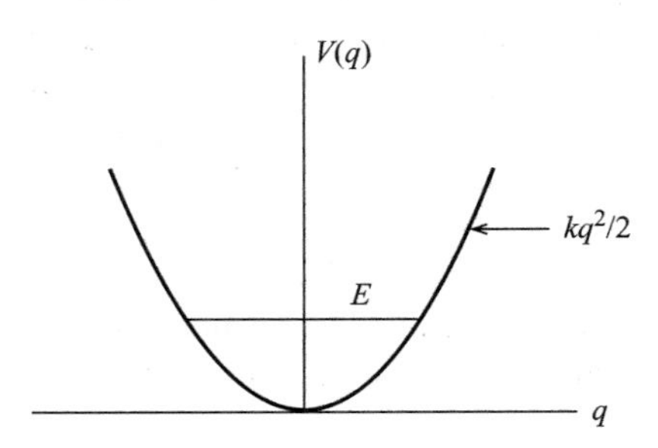

Fig. 2.4 A plot of the potential $V(q) = kq^2/2$ and the linear trajectory (horizontal line) of the oscillator for a given total energy E. The motion occurs on and inside the boundary which the potential creates.

for K, an intrinsically positive quantity, and thus it would imply an imaginary value for the velocity, itself a real measurable quantity.

The linear harmonic oscillator is a system that occurs in quantum physics as well. In what is known as the *coordinate representation*, the quantal version of the potential takes exactly the same form as that in the classical case, viz., $V(q) = kq^2/2$. This is an example of the earlier remark on the carryover of classical potential forms to quantum physics. However, in the present instance of the linear oscillator, the unchanged form of the potential is the only point common to the classical and quantal descriptions. For example, in the quantal case, the energy is quantized, taking on the values $E_n = \hbar\omega(n + \frac{1}{2})$, $n = 0, 1, 2 \ldots$. No other energies are allowed. More pertinent to the theme of this subsection is the "motion" of the quantal oscillator. First, there is no trajectory: it simply is not defined. In its place, one can substitute the *probability* $P(q)$ of finding the particle at the position q. Second, the particle can be at a position q larger than the boundary defined by the potential! That is, if q' is any position larger than the value $\sqrt{2E_n/k}$ (for a given quantal energy E_n), then $P(q') \neq 0$ unless $q' = \infty$. This non-zero probability that a quantal particle can be found beyond a confining boundary of classical physics is the basis for understanding a variety of quantal phenomena such as α-particle decay and tunneling at a Josephson junction.

2.5. States of a system

Both in classical and in quantum physics, measurements are concerned with determining either what the state of a system is at a particular time or how the state develops during an interval of time. Closely related to this, of course, is the aim of developing a theory to predict the outcome of a measurement, which normally means predicting how the system will behave if it is either left undisturbed or influenced by a specified external probe or interaction. A crucial ingredient here is the meaning of the word "state," which we examine in this subsection.

Classical states

In classical mechanics, the state of a system may generally be taken to be the positions and velocities of its constituent particles. The classical state may also be understood as the *trajectory* of the system in its appropriate *phase space*, i.e., as the locus of the sets of points $\{q_\alpha(t), \dot{q}_\alpha(t)\}$ or $\{q_\alpha(t), p_\alpha(t)\}$. For an N-particle system in D dimensions, a point in phase space can be thought of as the tip of a vector in $2ND$ dimensions. In general, therefore, a trajectory in phase space is very non-trivial to visualize unless both N and D are unity, an example of which is the 1-D linear harmonic oscillator, which we now consider. Denoting the oscillator's displacement by q, its momentum by p, and its mass and spring constant by m and k, respectively, then as in Section 2.3 its (conserved) energy E is given by $E = p^2/(2m) + kq^2/2$, a relation that leads to a very simple phase-space trajectory, viz., an ellipse. That is, for fixed E, $E = p^2/(2m) + kq^2/2$ is the equation of an ellipse with semi-axes of length $\sqrt{2mE}$ and $\sqrt{2E/k}$. In Fig. 2.5 are shown several of these ellipses, corresponding to different values of E; each represents states of the 1-D, linear oscillator. Analogous curves can be constructed, e.g., for the relative motion of a two-particle system interacting via an interparticle central force, examples of which are

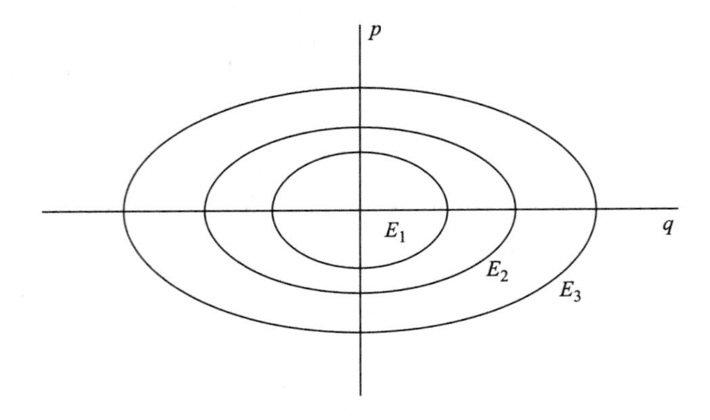

Fig. 2.5 States of a 1-D linear oscillator represented as elliptical trajectories in p–q phase space. The equation of the trajectory is that of energy conservation viz., $E = p^2/(2m) + kq^2/2$, where $p(q)$ is the canonical momentum (position) of the oscillator. Shown are three trajectories corresponding to energies $E_1 < E_2 < E_3$.

discussed in many texts on classical mechanics. In this case, conservation of momentum and of energy together provide constraints that can again lead to trajectories of relatively simple form.

As N or D increases, we are no longer able to represent the state as a trajectory by a simple geometrical curve in a 2-D plot, although we can easily *imagine* that such curves could be constructed if a $2ND$-dimensional space were visualizable. In contrast, this visual representation of a classical state has no quantal counterpart, as is discussed next.

Quantal states

Since the concept of velocity is not a primary one in the quantum realm, and the role played by position is so different there than in classical physics, the definition of "state" in quantum mechanics evidently relies on other concepts. To understand these concepts in detail requires familiarity with material discussed in Chapter 5, but, accepting the risk that the remaining material in this subsection may appear to be an example of the statement "this course is a prerequisite for this course," some further comments on quantal states are in order.

In characterizing quantum states one replaces the classical $\{q_\alpha, \dot{q}_\alpha\}$ or $\{q_\alpha, p_\alpha\}$ by a set of constants known as "good quantum numbers." Depending on the system, these numbers could specify some or all of the following: the energy, the momentum, the (orbital) angular momentum and/or one of its components, as well as purely quantal properties such as spin and parity. Most quantum numbers are related to observables, i.e., to measurable quantities, a few are not, and all correspond to constants of the motion.

To enumerate all of the appropriate quantum numbers signifies that a unique state of a quantal system has been specified. However, there is more to the meaning of "state" in quantum theory than a listing of quantum numbers. The phrase "state of a system" also implies that an abstract vector has been specified in a mathematical vector space denoted a *Hilbert space*. The simpler properties of Hilbert spaces – those relevant to our formulation of quantum theory – are discussed in detail in subsequent chapters.

Much of the formal development of quantum mechanics often occurs via the abstract entities of Hilbert space, but for the purposes of making detailed calculations, the abstract quantities must often be replaced by derivatives, integrals, and specific functions of the kind familiar from calculus and differential equations. Technically, one refers to this replacement as "choosing," "going to," or "working in" a particular representation. In the coordinate or position representation, the abstract state vector is represented by a function of position, say $\Psi(\mathbf{r}, t)$, where $\mathbf{r}$ is the 3-D vector specifying the position of the single particle, which is now assumed to be the system of interest, and t is the time. The function $\Psi(\mathbf{r}, t)$ is one of several means of depicting the particle's quantum state, and as such, is totally different than the $\{q_\alpha(t), p_\alpha(t)\}$ specification of a particle's classical state. Instead, $\Psi(\mathbf{r}, t)$ is analogous to an electric field component $\mathcal{E}_j(\mathbf{r}, t)$ or to the function $y(x, t)$ that designates the displacement y of a stretched string at a position x along its length and at time t: in neither of these two analogs is the position coordinate a function of time. The stretched-string analogy is developed in the next chapter, where we compare and contrast the behavior of a classical stretched string with that of a quantal particle confined to the interior of a 1-D box.

The quantity $\Psi(\mathbf{r}, t)$ is known as the single particle's (coordinate space) *wave function*. It is a solution of the time-dependent Schrödinger (partial differential) equation; examples of this equation and its solutions will be examined in other parts of this book. Wave functions and very simple (1-D) versions of the Schrödinger equation have often been introduced in beginning physics courses in order to acquaint students with some aspects of quantum physics. Thus, some readers may already have learned that, when it is "properly normalized," $|\Psi(\mathbf{r}, t)|^2$ is the probability of finding the particle at position $\mathbf{r}$ at time t. It is tempting to conclude that by knowing $\Psi(\mathbf{r}, t)$ one knows the state of the single-particle system. This is, however, not correct. The wave function is not the state: rather, it is the *coordinate representation* of the state. Furthermore, if instead the *momentum representation* had been chosen, then $\Psi(\mathbf{r}, t)$ would have been replaced by another function, say $\Phi(\mathbf{p}, t)$, which would be both the *momentum-space* wave function and the momentum representative of the state (or the state vector). In general, the two functions have different dependences on their independent variables. Since one of the necessary as well as desirable characteristics of the state is that it should be unique, the preceeding inequality compels us to conclude that wave functions are not states. They are manifestations of that more abstract Hilbert-space quantity, the quantum state vector, to which we shall return in Chapter 5.

These preliminary remarks on states in quantum theory should help to establish how much quantum and classical mechanics differ. Even though only a vague feeling for the concept of quantum states can be conveyed by these comments, they suffice to introduce the concept and allow us to return to the subject of measurement with perhaps a slightly better understanding of the notion of states.

2.6. Measurements and uncertainty

Some aspects of the measurement process for classical and quantal systems are similar, others are notably dissimilar. As an introduction to this topic, we consider some classical measurements. A simple example is the determination of the length of a rod. This normally means fixing the rod in a temporarily unchanging position and then comparing its end points with the marks on a ruler of some type, e.g., a meter or yard stick. Measurement of the (average) velocity of an object means measuring the time it takes a designated point P on it to pass between two positions along its assumed straight-line path, and then dividing the distance traveled by the time elapsed. An assumption here is that the determination of the overlap of P with each of the two designated points does not change the path being followed. One can similarly measure average accelerations or (approximate) instantaneous velocities and acceleration or, in other contexts, frequencies, wavelengths, etc.

The assumption in classical physics is that each member of *any* pair of dynamical variables or pair of observable quantities – which shall henceforth be referred to as "observables" – can be simultaneously measured to within any preassigned level of accuracy. *This assumption does not hold quantally.* The inability to measure simultaneously certain pairs of observables to arbitrary accuracy is the verbalization of that mathematical statement known as Heisenberg's uncertainty relation, which is briefly discussed below and, later, in Chapter 9. It means that one must examine more carefully what is meant by a measurement in micro-physics.

Suppose that we wish to measure both the position and the momentum of an electron by shining light on it; i.e., by the scattering of photons by electrons:

$$\gamma + \mathrm{e} \rightarrow \gamma + \mathrm{e}.$$

The measurement to be considered involves the γ-ray microscope (see, e.g., Heisenberg (1949)), initially described by Heisenberg, whose analysis was subsequently modified by Bohr.

The act of seeing involves, among other phenomena, the scattering of visible light by the object followed by the scattered light passing through the eye's lens. This process can be repeated indefinitely, since the object, being a classical mass, is not noticeably displaced by the photons it scatters. The position and velocity (equivalently, momentum) of the object can be simultaneously determined to arbitrary accuracy, at least in principle.

In order to attempt to carry out the same procedure for the electron, we analyze the following *Gedanken experiment* or thought experiment. A single photon impinges on an isolated, stationary electron and is then scattered through a lens and focused onto a detector, where its presence is to be recorded electronically. The physical arrangement is shown schematically in Fig. 2.6, in which a 2-D, right-handed coordinate system is also displayed.

Since the photon is both probe and information carrier, measurement of its position and momentum determines the corresponding quantities for the electron. It should be clear that the x-coordinate of the photon cannot be known to an arbitrary accuracy: it is limited by the resolving power of the lens, viz., $\lambda/\sin\theta$, where θ is half the angle subtended by the lens at the electron. Denoting the inaccuracy or *uncertainty* in the x-coordinate by δx, we have the estimate

$$\delta x \sim \frac{\lambda}{\sin\theta}, \tag{2.18}$$

where $\sim$ in (2.18) means "of the order of."

The incident photon has momentum p, essentially in the y-direction. In order to be detected it must pass through the lens after scattering from the electron. However, since it

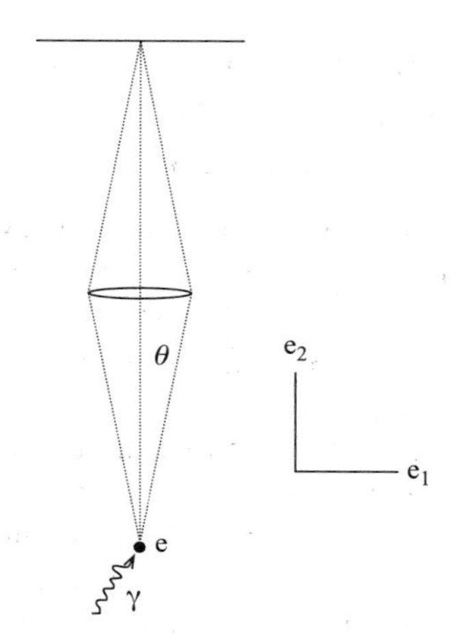

Fig. 2.6 The Heisenberg – γ-ray – microscope.

can pass anywhere within the lens, up to the edges, it can gain an x-component of momentum up to a maximum value $p \sin\theta$. This change or uncertainty in p_x we shall denote δp_x, which is estimated to be of the order of $p \sin\theta$:

$$\delta p_x \sim p \sin\theta. \tag{2.19}$$

Multiplying the two uncertainties, their product is estimated as

$$\delta x\, \delta p_x \sim p\lambda \tag{2.20}$$

The key relation in reducing (2.20) to a form independent of wavelength and initial momentum is that of Planck, viz.,

$$p = h/\lambda, \tag{2.21}$$

whose use transforms (2.20) into

$$\delta x\, \delta p_x \sim h. \tag{2.22}$$

Equation (2.22) is the desired result. It states that, for the γ-ray microscope the product of the inaccuracies or the uncertainties in the measured values of x and its canonically conjugate momentum p_x is of the order of h, a quantity that is negligible in the classical domain but *not in the quantal one*. Because the product of the uncertainties occurs on the left-hand side in Eq. (2.22), it follows that any attempt to reduce either δx or δp_x increases the other. The relation (2.22) thus states that, in the present example, δx and δp_x cannot both be made arbitrarily small, either for the photon or for the electron.

Although (2.22) places restrictions on how well position and momentum can simultaneously be determined in the case of the γ-ray microscope, it turns out that this type of limitation is universal, in that it holds for any pair of quantal observables whose classical counterparts are canonical conjugates, e.g., for y and p_y, for the z-component of angular momentum and the azimuthal angle, and for energy and time.[3]

[3] There are caveats associated with the second and third examples, as discussed later in Sections 10.4 and 16.4, respectively.

Equation (2.22) is a qualitative realization of Heisenberg's uncertainty principle. For the present case of x and p_x, the uncertainty principle takes the form

$$\Delta x \, \Delta p_x \geqslant \hbar/2, \tag{2.23}$$

where Δx (Δp_x) has a specific meaning, which, for the case of the γ-ray microscope, is qualitatively equivalent to δx (δp_x). While the technical definition of Δ will be given later (see Chapter 9), we do not need it here to interpret (2.23): the content of (2.23) is similar to that of (2.22). The essence of (2.23) is that one cannot simultaneously measure both the x-coordinate and the x-component of momentum to arbitrary accuracy. Similar relations hold for other pairs of canonically related observables. Because these inequalities arise from the defining structure of quantum theory, which is the framework in which all microscopic phenomena are to be understood and interpreted, the uncertainty relations place a limit on the accuracy with which certain pairs of quantities can be measured. That limit is *independent* of the experimental method used. In other words, no experiment can be devised that will, for example, "improve" the inequality (2.23) by decreasing its right-hand side. Hence, even though the γ-ray microscope and its analysis may be thought to be crude or even wholly hypothetical, with no connection to actual experiments, the opposite is true: the picture one derives from this *Gedanken experiment* is quite close to reality. The analysis therefore serves the important purpose of fleshing out, i.e., of providing a physical realization of, an inequality derived from fairly abstract considerations.

Before ending this discussion, we point out another aspect wherein classical and quantal measurements are so different. Any visual observation in classical physics can – in principle and in practice – be repeated indefinitely without changing the result: photon energies are too small to disturb the (classical) object from which they scatter. This is no longer true microscopically, since the wavelength of the photon needed to "see" a microscopic object needs to be about the size of that object (or smaller). Suppose the object is an atom, of size ≈ 1 Å. By Planck's relation, $E_\gamma = hc/\lambda$, and for $\lambda \sim 1$ Å, $E_\gamma \sim 10$ keV, enough energy to remove an outer electron from an atom or to kick a stationary, free atom far away. In the first scenario, the target will be changed; in the second, it will no longer be around to be observed!

The latter scenario will also be the case for the electron in the Heisenberg microscope experiment. Hence, in micro-physics, a target might not be – and often is not – available for one to perform a second measurement on it. That quantal measurements are repeatable, just as in classical physics, is a result of the identity of all targets of the same kind in the same state: all electrons are the same, etc.

Exercises

2.1 Figure 2.3 shows both the Cartesian and the spherical polar sets of unit vectors, $\{\mathbf{e}_j\}$ and $\{\mathbf{e}_\beta\}$, respectively. Use the co-latitude θ and the azimuthal angle ϕ to express each of $\mathbf{e}_r$, $\mathbf{e}_\theta$, and $\mathbf{e}_\phi$ in terms of the set $\{\mathbf{e}_x, \mathbf{e}_y, \mathbf{e}_z\}$.

2.2 Derive the relations expressed in Eq. (2.7).

2.3 In analogy to the velocity $\mathbf{v} = d\mathbf{r}/dt$, the acceleration $\mathbf{a} = d\mathbf{v}/dt$ can be expressed in terms of either Cartesiom or spherical polar components. Derive relations for a_r, a_θ, and a_ϕ in terms of a_x, a_y, and a_z.

2.4 Conservative forces are derivable from a potential energy function $V(\mathbf{r})$: $\mathbf{F}(\mathbf{r}) = -\nabla V(\mathbf{r})$, where $\mathbf{F}$ is assumed time-independent. Determine which of the following forces are conservative, and, for those that are, obtain the potential energy function:
(a) $F_x = ax + by^2$, $F_y = az + 2bxy$, $F_z = ay + bz^2$,
(b) $F_x = ay$, $F_y = az$, $F_z = ax$,
(c) $F_r = 2ar \sin\theta \sin\phi$, $F_\theta = ar \cos\theta \sin\phi$, $F_\phi = ar \cos\phi$.
In each of these cases, a and b are constants.

2.5 (a) A particle of mass m moves along the x-axis under the action of a conservative force $F_0(x)$. Starting from Newton's law

$$m\,\frac{d^2x}{dt^2} = F_0(x)$$

derive the energy-conservation relation

$$\tfrac{1}{2}mv^2 + V_0(x) = E = \text{constant}.$$

(b) The same particle is now acted on by the non-conservative force $F(x, v, t) = F_0(x) + G(x, v, t)$. Show that the derivation which yielded conservation of energy in part (a) now leads to

$$\frac{d}{dt}\left[\frac{1}{2}mv^2 + V_0(x)\right] = H(x, v, t)$$

and determine the quantity $H(x, v, t)$, which is not the classical Hamiltonian!

2.6 A particle of mass m moves in 3-D under the influence of a central potential $V(r)$. In this situation, there are two conserved quantities or constants of the motion, viz., the energy E and the angular momentum $\mathbf{L} = \mathbf{r} \times (m\mathbf{v})$. The relation $\mathbf{L}$ = constant means that the particle's motion is confined to a plane (perpendicular to the direction of $\mathbf{L}$), which we choose as the x–z plane. Hence, the spherical coordinates of the particle reduce to r and θ, the circular coordinates.
(a) Starting with either Newton's law or the Lagrangian equations of motion, determine expressions for E and $L = |\mathbf{L}|$ in terms of m, r, $\dot{r}$, θ, and $\dot{\theta}$.
(b) The state of the system is specified by $\mathbf{r}(t)$ and $\mathbf{v}(t)$. It is sometimes simpler to determine the orbit, $r = r(\theta)$. By employing the new variable $u(\theta) = 1/r(\theta)$, show that $u(\theta)$ is the solution of

$$\left(\frac{du}{d\theta}\right)^2 + u^2 = \frac{2m}{L^2}[E - V(r)].$$

(c) Optional. Let $V(r) = \frac{1}{2}kr^2 = \frac{1}{2}m\omega^2 r^2$, with $\omega = \sqrt{k/m}$, and show that the orbit is an ellipse whose center is at the origin.

2.7 In SI units, the force $\mathbf{F}$ acting on a particle of charge q and mass m in a magnetic field $\mathbf{B}$ is

$$\mathbf{F} = q\mathbf{v} \times \mathbf{B}, \qquad \mathbf{v} = d\mathbf{r}/dt.$$

The magnetic field can be obtained from a vector potential $\mathbf{A}$:

$$\mathbf{B} = \nabla \times \mathbf{A}, \qquad \mathbf{A} = \mathbf{A}(\mathbf{r}, t).$$

The Lagrangian in this situation is given by

$$L = \tfrac{1}{2}m\dot{\mathbf{r}}^2 + q\dot{\mathbf{r}} \cdot \mathbf{A}(\mathbf{r}, t).$$

(a) Obtain an expression for the generalized momentum $\mathbf{p}$ from L.
(b) Using cylindrical coordinates ($\mathbf{r} = \rho, \phi, z$) express L in terms of $\rho, \phi, z, \dot{\rho}, \dot{\phi}, \dot{z}$, and the components A_ρ, A_ϕ, and A_z.
(c) Now let $\mathbf{B} = B\mathbf{e}_z$, $B =$ constant. A suitable $\mathbf{A}$ is $A_\rho = 0,\ A_\phi = B\rho/2,\ A_z = 0$ (which can be verified by recourse to the expression for $\nabla \times \mathbf{A}$ in cylindrical coordinates). By substituting this expression for $\mathbf{A}$ into the Lagrangian of part (b), find the three equations of motion for the particle's trajectory, and determine which combination of the variables yields conserved quantities.
(d) Finally, assume that $\rho =$ constant. One of the equations of part (c) should then read $\dot{\phi} = \omega = aB$, where a is a proportionality constant you are to determine. (Your final result for ω should be the cyclotron frequency.)

2.8 The 1-D quantal box is a pedagogic system that confines particles to its interior. The time-dependent wave function $\Psi_0(x, t)$ for a particle of mass m confined to a box located at $0 \leqslant x \leqslant L$ is

$$\Psi_0(x, t) = \begin{cases} \psi_0(x)e^{-iE_0 t/\hbar}, & x \in [0, L], \\ 0, & \text{otherwise,} \end{cases}$$

where $E_0 = \hbar^2\pi^2/(2mL^2)$ and

$$\psi_0(x) = \sqrt{\frac{2}{L}}\sin\left(\frac{\pi x}{L}\right).$$

(a) Determine the functional form of the probability density $\rho_0(x, t)$ that the particle is at position x at time t, using

$$\rho_0(x, t) = |\Psi_0(x, t)|^2.$$

Sketch $\rho_0(x, t)$ as a function of x.
(b) The probability that the particle is in the left-hand half of the box is

$$P_0\left(x \leqslant \frac{L}{2}\right) = \int_0^{L/2} \rho_0(x, t)\, dx.$$

Evaluate $P_0(x \leqslant L/2)$.
(c) Show that Ψ_0 is properly "normalized", i.e., that the probability of finding the particle somewhere in the box, viz.,

$$P_0 = \int_0^L dx\, |\Psi_0(x, t)|^2,$$

is unity.
(d) The most likely position for the particle, i.e., the value one can expect to obtain from a measurement, is

$$\langle x \rangle_0 = \int_0^L dx\, x|\Psi_0(x, t)|^2.$$

Determine $\langle x \rangle_0$ in terms of L. Is this the value you might have guessed from the sketch of part (a)? Explain.

2.9 The momentum of a 1-g ladybug is measured to within an uncertainty of $\Delta p_x = 2 \times 10^{-11}$ kg m s^{-1}. How accurately could its position be measured, at least in principle? Estimate the maximum accuracy one might hope to achieve in practice and comment on its relation to the "in-principle" answer.

2.10 Assume that the position of an electron in an H-atom is uncertain by 1 Å ($\cong 2a_0$). To what uncertainty in momentum does this correspond? The latter value can be transformed into an uncertainty in velocity. Compare this value with the value of the electron's velocity v_0, calculated using Bohr's model. Discuss this comparison.

2.11 Suppose that a radar set generates a pulse of electromagnetic radiation whose duration is 2×10^{-3} s. The uncertainty in position of any of the photons forming this pulse is the length of the pulse.
(a) What is the corresponding uncertainty in the photons' momenta?
(b) Determine the uncertainty in frequency that this implies.

2.12 Suppose that the uncertainty in the speed of an electron moving at 200 m s^{-1} is 10^{-2}%.
(a) What is the uncertainty of its kinetic energy (in electron volts)?
(b) Determine the uncertainty in the electron's position and discuss the consistency of your answer with the concept of the electron being a point particle.

2.13 The density of protons in intergalactic space is about one per cubic meter. Determine such a proton's uncertainty in momentum and in kinetic energy. Compare these numbers with those for a proton confined to a typical nuclear volume, say, a sphere of radius 10^{-14} m.

2.14 β-Decay involves the emission of negative or positive electrons, typically from an atomic nucleus. An isolated neutron has a half life of about 15 min against β-decay into a proton, an electron, and a neutrino. In this process, the maximum kinetic energy of the electron is 0.78 MeV. At one time it was thought that a neutron was a metastable state of a proton and an electron confined to a volume corresponding to a neutron's radius, approximately 10^{-15} m. Use this value in the Heisenberg uncertainty relation and the electron's maximum kinetic energy after the decay to argue for or against a neutron being the metastable system mentioned in the foregoing. That is, argue whether electrons are already present or are created in the β-decay of the neutron (and, by inference, in any nucleus undergoing β-decay).

2.15 The vehicle used in the text to introduce the uncertainty relation $\delta x\, \delta p_x \sim h$ was the γ-ray microscope. Although this type of result was said to be universal, let us investigate whether one can in fact circumvent it by employing a different method of measurement. To determine the position of electrons in a beam, let them fall on a screen with a single slit of width δy. The uncertainty in position is δy, which can be made arbitrarily small. To see whether there is a corresponding increase δp_y in the associated vertical component of momentum, recall that diffraction of electrons has been observed. Therefore, there will be a diffraction pattern on a detector screen, as in Fig. 2.7. If the first minimum occurs at a position y on the screen, at an angle θ to the horizontal, then

$$\sin\theta = \frac{\lambda}{\delta y}.$$

If the electrons in the beam had only a horizontal velocity v_x, then, to get within y, they

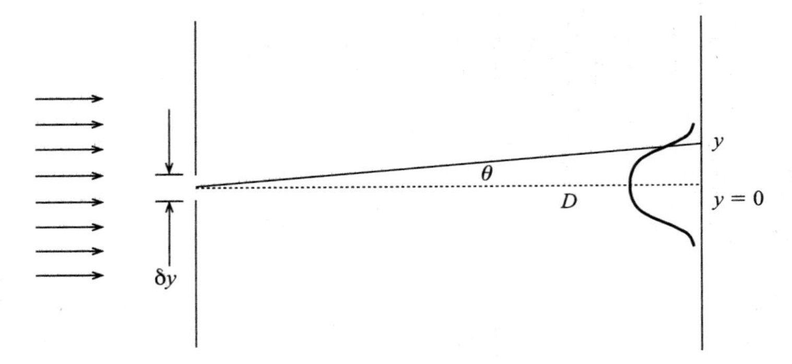

Fig. 2.7

must have been given an uncertain vertical velocity component δv_y. Relate δv_y to $\sin\theta$, use the de Broglie relation, and show that, in this hypothetical example, one again finds $\delta y\,\delta p_y \sim h$.

2.16 The resolving power of a lens and the diffraction effects due to a sharp edge depend on the wavelength. For material objects, λ is determined by h and p. If h were allowed to vary then so would λ. This problem (and the remaining ones) explores the consequences for hypothetical universes in which the laws of physics remain the same but h can take on different values, especially macroscopic ones. For each of the values of h listed below, determine λ for a 1-g ladybug whose speed is 1 cm s^{-1}, for a proton of kinetic energy 1 keV and for an electron of kinetic energy 1 eV:
(a) $h = 6.626 \times 10^{-34}$ J s,
(b) $h = 10^{-12}$ J s,
(c) $h = 10^{-6}$ J s,
(d) $h = 1$ J s.

2.17 In a set of universes with different values of h, Robin Hood is engaged in archery practice. In all cases, he shoots an arrow of mass 0.05 kg at a round target 30 m away. The velocity of the arrow (v_x) is 50 m s^{-1}. If the bow string creates an uncertainty in the arrow's initial position of 0.005 m, what is the minimum diameter the target should have in order that uncertainty will not interfere with the arrow hitting somewhere within the target if
(a) $h = 6.626 \times 10^{-24}$ J s,
(b) $h = 10^{-12}$ J s,
(c) $h = 10^{-3}$ J s,
(d) $h = 1$ J s.

2.18 A ballistic pendulum is a device for measuring the speed of a bullet. The following experiment is carried out in universes of different h: a 5-g bullet is fired 1 m away from the absorbing block. The bullet's speed is 300 m s^{-1}, its position in the rifle's barrel is accurate to 10^{-4} m, and the cross section of the absorbing block, at whose center the rifle is aimed, is a square of side 10 cm. How far to the side of the block must an observer stand in order not to be struck by the bullet due to the action of the uncertainty principle if in the various universes h has the values

(a) $h = 6.626 \times 10^{-24}$ J s,
(b) $h = 10^{-12}$ J s,
(c) $h = 10^{-6}$ J s,
(d) $h = 1$ J s.

3

Introducing Quantum Mechanics: A Comparison of the Classical Stretched String and the Quantal Box

Quantum mechanics was initially formulated as a type of matrix theory by Heisenberg and collaborators. Soon after, it was alternately expressed by Schrödinger as a wave theory through what is now referred to as the Schrödinger wave equation. Schrödinger and Eckart showed that these two approaches were equivalent manifestations of the same underlying formalism, viz., quantum theory.

Quantum theory is formulated in this book (Chapter 5) via six postulates. In order that this postulational approach not be too abstract, it is helpful first to analyze a system that is easily treated by quantum theory. We do so in this chapter, in which the so-called quantal box is studied. The analysis is carried out using the Schrödinger equation, which is introduced in this chapter but without the formal statements of the postulates.

Because the quantal-box problem is similar mathematically to the problem of describing the clamped, vibrating string of classical physics, the string is also analyzed in this chapter. This provides not only a comparison system for the box, but also a brief review of classical waves and wave theory. We also include a section on Hermitian differential operators, Sturm–Liouville systems and the Dirac delta function. All of this material serves as an introduction to Chapters 4 and 5.

We begin with the string. Because many readers will already have seen aspects of the following analysis, we put all of Section 3.1 in small print. This section may be omitted by the knowledgeable reader; the crucial result is Eq. (3.3).

3.1. The uniform stretched string

Consider a uniform string of mass per unit length μ, stretched horizontally under tension T between two fixed points that are separated by a distance L. Figure 3.1(a) shows the equilibrium position of the string, while Fig. 3.1(b) represents a snapshot of the string at some instant after it has been displaced vertically and allowed to vibrate. Coordinates x and $y = y(x, t)$ have also been introduced in Fig. 3.1(b): x is the distance along the horizontal axis, with the clamped left-hand end being at $x = 0$ and the clamped right-hand end being at $x = L$, while $y(x, t)$ is the vertical displacement of the point x on the string at time t. The string can move only in the vertical direction, and gravitational effects are ignored.

The 1-D wave equation

We shall derive the 1-D wave equation, which governs the motion of the string, from Newton's law $F_y = \dot{p}_y$, where F_y is the resultant force and p_y is the momentum in the y-direction. Since the

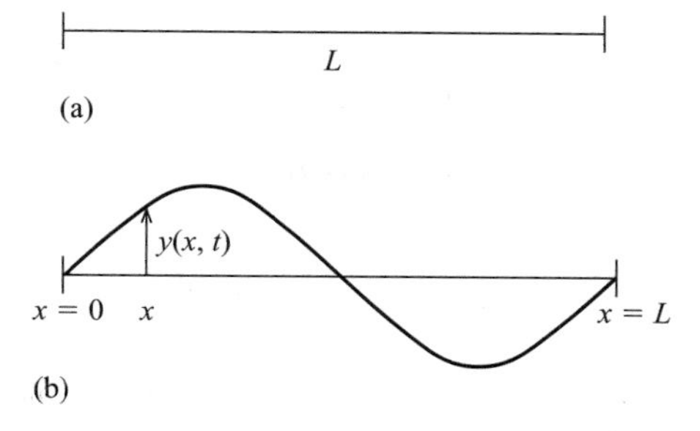

Fig. 3.1 (a) The equilibrium position of a stretched, clamped, uniform string of length L. (b) The instantaneous position of the string after it has been displaced vertically and allowed to oscillate.

string is an extended, continuous body, we consider the behavior of an infinitesimal portion located between x and $x + dx$, as in Fig. 3.2. This segment has a mass $m = \mu\, dx$ and a velocity $\dot{y} = \partial y(x, t)/\partial t$, so that its momentum is

$$p_y = \mu(\partial y(x, t)/\partial t)\, dx. \tag{3.1}$$

The force acting on the segment is most easily obtained using Fig. 3.2, in which the tensions $\mathbf{T}$ at the points $y(x, t)$ and $y(x + dx, t)$ are displayed, along with the angles θ_R and θ_L on the right- and left-hand sides, respectively, of the segment. The horizontal and vertical components of the force $\mathbf{F}^{(L)}$ on the left-hand side of the segment are

$$F_x^{(L)} = -T\cos\theta_L, \qquad F_y^{(L)} = -T\sin\theta_L;$$

a similar pair of relations holds for the right-hand side of the segment. From $\tan\theta_L = \partial y(x, t)/\partial x$, it follows that $\sin\theta_L = [\partial y(x, t)/\partial x]/[1 + (\partial y(x, t)/\partial x)^2]^{1/2}$ and $\cos\theta_L = [1 + (\partial y(x, t)/\partial x)^2]^{1/2}$, with corresponding relations for $\sin\theta_R$ and $\cos\theta_R$ (in which $\partial y(x + dx, t)/\partial x$ occurs).

To proceed further, it is assumed that θ_L and θ_R are small, so that $(\partial y/\partial x)^2$ can be ignored.[1]

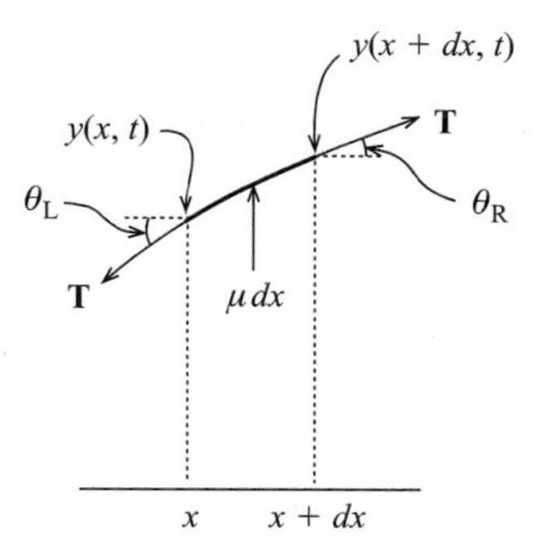

Fig. 3.2 The tension $\mathbf{T}$ acting on the string at its displaced positions $y(x, t)$ and $y(x + dx, t)$ at time t. The mass is $\mu\, dx$. The angles θ_L and θ_R are measured w.r.t. the x-axis.

[1] This is equivalent to the assumption of small oscillations of the string, wherein its maximum amplitude of oscillation is much less than its length.

Then we get

$$F_x^{(L)} = -T = -F_x^{(R)},$$

$$F_y^{(L)} = -T\,\frac{\partial y(x, t)}{\partial x}$$

and

$$F_y^{(R)} = T\,\frac{\partial y(x+dx, t)}{\partial x}.$$

The total horizontal force F_x $(= F_x^{(L)} + F_x^{(R)})$ is zero, as it must be since there is no motion in the x-direction, while the total vertical force is

$$\begin{aligned} F_y = F_y^{(R)} + F_y^{(L)} &= T\left(\frac{\partial y(x+dx, t)}{\partial x} - \frac{\partial y(x, t)}{\partial x}\right) \\ &= T\,\frac{\partial^2 y(x, t)}{\partial x^2}\,dx. \end{aligned} \tag{3.2}$$

From (3.1) and (3.2), Newton's law $F_y = \dot{p}_y$ becomes

$$\frac{\partial^2 y(x, t)}{\partial x^2} = \frac{1}{v^2}\frac{\partial^2 y(x, t)}{\partial t^2}, \tag{3.3}$$

where

$$v = \sqrt{\frac{T}{\mu}} \tag{3.4}$$

is the phase or propagation velocity. That $\sqrt{T/\mu}$ must function as a velocity is easily seen, since its dimension is length/time.

Equation (3.3) is the one-dimensional wave equation; it governs the oscillations of the string. We examine it in the next subsection.

Some general properties of the wave equation and its solutions

The wave equation (3.3) is homogeneous and of second order in its space and time derivatives. Since the unknown function y depends on the two variables x and t, (3.3) is a second-order, *partial* differential equation. It can be thought of as an operator equation:[2]

$$\hat{D}^2(x, t)y = 0, \tag{3.5}$$

where the second-order differential (or derivative) operator $\hat{D}^2(x, t)$, given by

$$\hat{D}^2(x, t) = \frac{\partial^2}{\partial x^2} - \frac{1}{v^2}\frac{\partial^2}{\partial t^2}, \tag{3.6}$$

acts on the unknown function $y(x, t)$.

The structure of $\hat{D}^2(x, t)$ allows one to deduce the general form of the solution to (3.3) or (3.5) in a relatively straightforward fashion. On defining

$$\hat{D}_\pm(x, t) = \frac{\partial}{\partial x} \pm \frac{1}{v}\frac{\partial}{\partial t}, \tag{3.7}$$

(3.6) becomes

$$\hat{D}^2 = \hat{D}_+\hat{D}_- = \hat{D}_-\hat{D}_+. \tag{3.8}$$

Alternately, if we introduce the new variables

$$\xi_\pm = x \pm vt, \tag{3.9}$$

[2] Operators such as (3.6) will always be indicated by a carat ˆ over the symbol denoting the operator.

and the partial derivatives $\partial/\partial\xi_\pm$, Eq. (3.8) is also given by

$$\hat{D}^2 = \frac{\partial^2}{\partial\xi_+\,\partial\xi_-} = \frac{\partial^2}{\partial\xi_-\,\partial\xi_+}. \tag{3.10}$$

Either of (3.8) or (3.10) leads directly to the general solution:

$$y(x,\ t) = f_+(x + vt) + f_-(x - vt), \tag{3.11}$$

where the dependence of f_+ (f_-) is solely on the variable $\xi_+ = x + vt$ ($\xi_- = x - vt$). Other than this particular variable dependence, the functions $f_\pm$ are arbitrary – as long as no special conditions are imposed on $y(x,\ t)$.

Even though the $f_\pm$ are arbitrary, they do have a physical interpretation. Anticipating later terminology, we shall refer to $\xi_+ = x + vt$ as the *phase* of f_+ and to $\xi_- = x - vt$ as the phase of f_-. Although x and t are independent variables insofar as $y(x,\ t)$ is concerned, there is a special pair of relations between them, which has important physical consequences. These are the relations that guarantee that $f_\pm$ is constant, viz.,

$$\xi_\pm = x \pm vt = C_\pm, \tag{3.12}$$

where $C_\pm$ are constants. For those x and t obeying $\xi_+ = C_+$ (or $\xi_- = C_-$), f_+ (f_-) has the same value, even though x and t may have changed.

Points of constant phase obey

$$dC_\pm = 0 = dx \pm v\,dt,$$

from which it follows that such points move or propagate with velocities

$$\frac{dx}{dt} = -v \tag{3.13a}$$

or

$$\frac{dx}{dt} = +v. \tag{3.13b}$$

The value $-v$ goes with $\xi_+ = x + vt$ (f_+), whereas $+v$ is associated with $\xi_- = x - vt$ (f_-).

Putting these results together, we see that f_+ can be interpreted as a disturbance or wave (solution of the wave equation) whose points of constant phase propagate from right to left (R to L) with phase velocity $-v$, whereas f_- is a wave that propagates from left to right (L to R) with phase velocity $+v$. The function $y(x,\ t)$ is thus a linear combination of a left- and a right-moving pair of *running* waves.

The functions $f_\pm(\xi_\pm)$ are waves, or wave forms. Before obtaining the forms appropriate to the clamped, stretched string, we examine a few examples.

EXAMPLE 1

Sinusoidal plane waves. For simplicity we consider a right-moving wave given by a sine function:

$$f_-(x - vt) = A\sin[k(x - vt) + \delta], \tag{3.14}$$

where the amplitude A and phase angle δ are arbitrary, and the wave number k, of dimension $(\text{length})^{-1}$, is introduced in order that the argument of the sine function be dimensionless. Since only a single value of k occurs, Eq. (3.14) defines a *monochromatic* wave.

The wave form of Eq. (3.14) is readily shown not only to be a solution of the 1-D wave equation (1.3), but also to be of infinite extent in space and in time. A portion of $f_-(x - vt)$ for fixed t is shown in Fig. 3.3. In this diagram, the smallest and largest values of f_-, viz., $-A$ and $+A$, occur at x_2 and x_4, while its zeros, or nodes, are at x_1, x_3 and x_5. The shortest spacing between correspondingly identical points, i.e., the repetition distance or *wavelength* λ, is

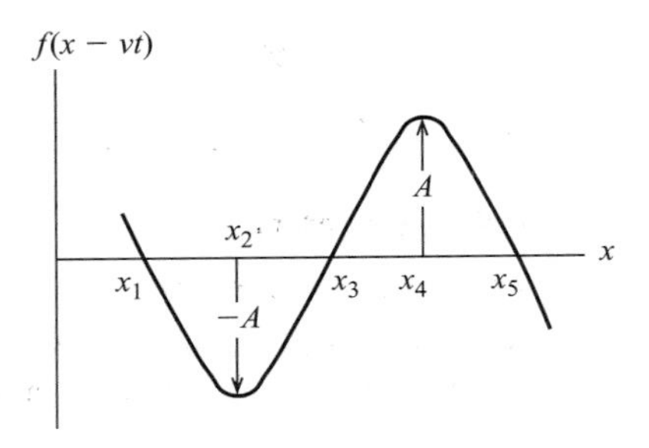

Fig. 3.3 A portion of the sine wave $f(x - vt) = A\sin[k(x - vt) + \delta]$ for fixed t. The zeros or nodes are at x_1, x_3, and x_5; the minimum value is at the trough located at x_2 and the maximum value is at the crest located at x_4.

$\lambda = x_5 - x_1$. From the structure of Eq. (3.14), the first spatial repetition of f_- occurs when $x \rightarrow x + \lambda$ such that $k\lambda = 2\pi$, or

$$\lambda = 2\pi/k. \tag{3.15}$$

It is left as an exercise for the reader to show that, for fixed x, the repetition time or period T is

$$T = \frac{2\pi}{kv}. \tag{3.16}$$

Since the product kv has dimensions $(\text{time})^{-1}$, we define the angular frequency[3] ω by

$$\omega = kv, \tag{3.17}$$

so that (3.16) becomes

$$T = 2\pi/\omega. \tag{3.18}$$

Using these quantities, $f_-(x - vt)$ of (3.14) can be written as

$$f_-(x - vt) = A\sin\left[2\pi\left(\frac{x}{\lambda} - \frac{t}{T}\right) + \delta\right] \tag{3.19a}$$

$$= A\sin(kx - \omega t + \delta). \tag{3.19b}$$

The sine function (3.14) is denoted a *plane wave*. This refers to the fact that the wave propagates along x from L to R, while the oscillation or disturbance is in the y-direction, i.e., in a plane perpendicular to the direction of propagation. (If kx is replaced by $\mathbf{k}\cdot\mathbf{r}$ in the 3-D case, one again refers to a plane wave, since the oscillation is in a plane perpendicular to the direction of propagation $\mathbf{k}$; in this case, the wavelength is $\lambda = 2\pi/k$, where $k = |\mathbf{k}|$.)

The function $\sin(kx - \omega t + \delta)$ encompasses the cosine wave form, obtained by writing δ as $\eta + \pi/2$, where η is also a constant phase angle. Both forms are contained in the complex exponential form

$$\Psi_k(x, t) = Ae^{i(kx - \omega t)}$$
$$= A[\cos(kx - \omega t) + i\sin(kx - \omega t)], \tag{3.20}$$

[3] The circular frequency ν is defined as $\nu = \omega/(2\pi)$, so that $T = 1/\nu$. ν is the number of cycles or oscillations per second.

where the complex constant A includes a phase angle such as δ or η. The wave forms appropriate to the string – and classical physics in general – are obtained by taking the real or the imaginary part of $\Psi_k(x, t)$. However, in quantum mechanics, as we shall see, it is the complex exponential form (3.20) rather than the sine or cosine that is used to describe running waves.

EXAMPLE 2

Standing waves. Because the wave equation is linear, the sum of any two solutions is also a solution. Thus, the sum of the right-propagating wave

$$y_1(x, t) = A\sin(kx - \omega t)$$

and the left-moving wave

$$y_2(x, t) = A\sin(kx + \omega t)$$

(with the same amplitudes, wave numbers, and frequencies), is

$$\begin{aligned} y &= y_1 + y_2 \\ &= A[\sin(kx - \omega t) + \sin(kx + \omega t)], \end{aligned} \tag{3.21}$$

which is also a wave form obeying Eq. (3.3).

Equation (3.21) is the mathematical version of the statement that waves can be added together or superposed, i.e., that they can interfere, as in the examples of the double-slit experiment in Sections. 1.1 and 1.3. Depending on the amplitudes and phases, superposed waves can interfere constructively or destructively. The wave y of (3.21) is an interesting special case, since it represents addition of oppositely running waves. It produces *standing* waves, in which the space and time dependences of $y(x, t)$ are completely separated from one another.

The mathematical transformation to standing waves is based on the formula

$$\sin\theta + \sin\phi = 2\sin\left(\frac{\theta+\phi}{2}\right)\cos\left(\frac{\theta-\phi}{2}\right). \tag{3.22}$$

Applying (3.22) to (3.21) yields

$$y(x, t) = 2A\sin(kx)\cos(\omega t), \tag{3.23}$$

where the separation of the space and time dependences is manifest.

The labeling of (3.23) as a standing wave refers to the spatial portion $\sin(kx)$, which has the same, non-running, shape for all time. The time factor $\cos(\omega t)$ can be absorbed into the amplitude, thereby making it time-dependent:

$$y(x, t) = A(t)\sin(kx), \tag{3.24}$$

where

$$A(t) = 2A\cos(\omega t). \tag{3.25}$$

Hence, as t varies between, say 0 and $T = 2\pi/\omega$, $A(t)$ changes from $2A$ to 0 to $-2A$ and then back to $2A$ through the value 0. Separate plots of $\sin(kx)$ and of $A(t)$, each in the repetition range of the variables x and t, are shown in Fig. 3.4.

We shall soon show that the individual solutions to the wave equation for the clamped string are standing waves of the form $\sin(k_n x)$, where $k_n = n\pi/L$. These waves are confined to the finite region $x \in [0, L]$ and are thus localized in extent. The waves (3.21) or (3.23), being superpositions of two running waves of infinite extent, are also of infinite extent. To obtain general solutions to the wave equation that may propagate over all space but at any instant are spatially localized requires

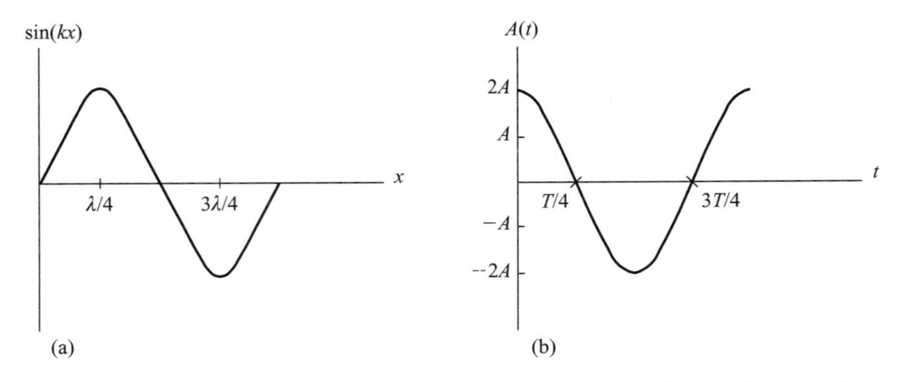

Fig. 3.4 (a) A plot of the standing wave sin(kx) over one length $\lambda = 2\pi/k$. (b) A plot of the amplitude $A(t) = 2A\cos(\omega t)$ over one period $T = 2\pi/\omega$.

that one construct *wave packets*, which are superpositions of an infinite number of individual waves. The global solutions which describe oscillations of the string are examples of wave packets, and we determine such solutions next.

Solution of the wave equation for the clamped, stretched string

The wave packet describing oscillations of the clamped, stretched string is a linear combination of individual solutions whose structure is that of (3.23). They are obtained by applying the method of separation of variables. It is a standard procedure, one which turns the solving of a partial differential equation into the simpler problem of solving ordinary differential equations.

We begin by writing $y(x, t)$ of (3.3) as a product of two functions, one dependent only on x, the other on t:

$$y(x, t) = \chi(x)\tau(t), \tag{3.26}$$

a form analogous to (3.23) or (3.24). Substitution of (3.26) into (3.3) plus the assumption that neither $\chi(x)$ nor $\tau(t)$ vanishes leads to

$$\frac{1}{\chi(x)}\frac{d^2\chi(x)}{dx^2} = \frac{1}{v^2\tau(t)}\frac{d^2\tau(t)}{dt^2}. \tag{3.27}$$

Compared with the wave equation, (3.27) is a remarkable relation, since it says that a function of x equals a function of t. Such a result is possible only if each side is equal to a constant. From the fact that x^2 and v^2t^2 each have dimension (length)^2, the constant must have dimension (length)$^{-2}$. We set it equal to $-k^2$, the negative of a wave number squared, i.e., we write (3.27) as the two equations

$$\frac{d^2\chi(x)}{dx^2} = -k^2\chi(x) \tag{3.28}$$

and

$$\frac{d^2\tau(t)}{dt^2} = -k^2v^2\tau(t) \equiv -\omega^2\tau(t), \tag{3.29}$$

where the angular frequency $\omega = kv$.

Choosing the separation constant ($-k^2$) to be negative is motivated by the fact that the standing wave (3.23) is an example of the type of solution (3.26): here, $\chi(x) = \sin(kx)$, with

$d^2\chi(x)/dx^2 = -k^2\chi(x)$. In Eqs. (3.28) and (3.29), k is an unknown, to be determined as part of the overall solution. If, by solving Eqs. (3.28) and (3.29), we cannot find solutions of the form (3.26) that are consistent with the as-yet-to-be-specified initial and boundary conditions on $y(x, t)$, then we would need to change the separation constant from $-k^2$ to $+k^2$ and try again.

We consider Eq. (3.28) first. Its structure is that of a second-order, homogeneous differential equation with constant coefficients, but it is not a true example of this type of equation because the coefficient of the $\chi(x)$ term, $-k^2$, is an unknown. A more suitable characterization of (3.28) is as an *eigenvalue* equation, i.e., it is in the general form

$$\hat{A}u^{(\lambda)} = \lambda u^{(\lambda)}. \tag{3.30}$$

Here, $\hat{A}$ is an *operator*, in the present case the differential operator $-d^2/dx^2$; $u^{(\lambda)}$ is the desired *eigensolution*; and λ is the *eigenvalue* to be determined, which is equal to k^2 in Eq. (3.28). The reader may have encountered eigenvalue equations as part of a course on linear algebra, most likely in the form of a matrix/vector equation. In that instance, $\hat{A}$ would be an $n \times n$ matrix and u would be a column vector containing n rows. A simple example is

$$\begin{pmatrix} 1 & 0 \\ 0 & -1 \end{pmatrix} \begin{pmatrix} u_1 \\ u_2 \end{pmatrix} = \lambda \begin{pmatrix} u_1 \\ u_2 \end{pmatrix}.$$

In this case of $n = 2$, there are two values of λ, $\lambda = \pm 1$; the two eigenvectors may be written $\binom{1}{0}$ and $\binom{0}{1}$.[4]

Operators, eigenvalues, and eigensolutions play a central role in quantum theory and much of the later chapters of this text is concerned with them. However, they also occur widely in classical physics, the vibrating string being one of numerous examples. We are emphasizing this aspect of the problem of the string/wave equation because it serves as an excellent introduction to this important feature of quantum mechanics.

The eigenvalue relation (3.30) states that, for those special solutions $u^{(\lambda)}$, the action of the operator on the solution is simply to multiply it by a number, λ. In the case of (not-too-large!) matrices, the values of λ are obtained by solving the secular equation; for each λ, a corresponding eigenvalue can be determined (up to a multiplicative constant). For the differential-equation version, the eigenvalues and eigenfunctions are usually determined by means of boundary conditions. If none is specified, then a continuous range of λ and u generally occurs. A simple instance is $\hat{A} = -d/dx$; for each complex constant λ, the corresponding eigenfunction is $u^{(\lambda)}(x) = \exp(-\lambda x)$. In the case of the string to which we now return, the boundary conditions limit k to a denumerably infinite set of values (Eq. (3.36)).

The general solution to Eq. (3.28) is[5]

$$\chi(x) = a\sin(kx) + b\cos(kx). \tag{3.31}$$

Equation (3.31) must describe motion of the string throughout the closed interval $[0, L]$. At $x = 0$ and $x = L$, however, the string, being clamped, cannot move. Therefore the boundary conditions are

$$\chi(x = 0) = 0 \tag{3.32a}$$

and

$$\chi(x = L) = 0. \tag{3.32b}$$

Equations (3.32) state some of the basic physics of this system.

At $x = 0$, (3.31) becomes

$$\chi(x = 0) = b = 0,$$

the second equality arising from (3.32a). With $b = 0$, the general solution is

[4] Proof of these statements is left as an exercise.

[5] An equivalent solution is $\chi(x) = c\exp(ikx) + d\exp(-ikx)$, but its use leads to the same final result, Eq. (3.33).

$$\chi(x) = a\sin(kx); \tag{3.33}$$

it must obey (3.32b):

$$a\sin(kL) = 0. \tag{3.34}$$

One possibility is $a = 0$, which we ignore since it yields the trivial result $\chi(x) = 0$ for all x.

If $a \neq 0$, then we must have

$$\sin(kL) = 0.$$

This is ensured if k is restricted to the following values:

$$k \to k_n = \frac{n\pi}{L}, \qquad n = 1, 2, \ldots, \tag{3.35}$$

with $n = 0$ excluded to again avoid the trivial solution $\chi(x) = 0$, $\forall\ x$. Negative n is excluded, since $\sin(n\pi x/L)$ and $\sin(-n\pi x/L)$ are essentially the same eigensolution.

The allowed wave numbers k_n are given by (3.35); corresponding to each n is the eigenfunction $\chi_n(x)$:

$$\chi_n(x) = A_n \sin\left(\frac{n\pi x}{L}\right), \tag{3.36}$$

where A_n is an as-yet-undetermined constant. Finally, the set of eigenvalues is $\{k_n^2\} = \{n^2\pi^2/L^2\}_{n=1}^{\infty}$.

Since the only values of k are $k_n = n\pi/L$, the quantity ω occurring in (3.29) will be similarly restricted:

$$\omega \to \omega_n = \frac{n\pi v}{L}, \qquad n = 1, 2, \ldots. \tag{3.37}$$

The most general solution to (3.29) for a given ω_n is

$$\tau_n(t) = a_n' \cos(\omega_n t) + b_n' \sin(\omega_n t), \tag{3.38}$$

with a_n' and b_n' being arbitrary constants. The ω_n are known as eigenfrequencies, with $\omega_1 = \pi v/L$ being the fundamental frequency and ω_n, $n > 1$, often denoted the $(n-1)$th harmonic. Sometimes it is helpful to introduce the time $T_L = L/v$, in which case $\omega_n = \pi n/T_L$ so that the arguments in (3.38) become $\pi nt/T_L$, in analogy to the $n\pi x/L$ in Eq. (3.36).

For each $k = k_n$, there is a corresponding solution $y_n(x, t)$ to the wave equation:

$$\begin{aligned} y_n(x, t) &= \chi_n(x)\tau_n(t) \\ &= \sin\left(\frac{n\pi x}{L}\right)[a_n \cos(\omega_n t) + b_n \sin(\omega_n t)], \end{aligned} \tag{3.39}$$

where $a_n = A_n a_n'$ and $b_n = A_n b_n'$. In view of the linearity of the wave equation with its consequent superposition principle, the full solution describing the vibrations of the stretched, uniform, clamped string is the sum of all the y_n of (3.39):

$$\begin{aligned} y(x, t) &= \sum_{n=1}^{\infty} y_n(x, t) \\ &= \sum_{n=1}^{\infty} \sin\left(\frac{n\pi x}{L}\right)[a_n \cos(\omega_n t) + b_n \sin(\omega_n t)]. \end{aligned} \tag{3.40}$$

The spatial portion of this solution is a sum of sine waves, where the wavelength of the nth wave decreases as n increases: $\lambda_n = 1/k_n = 2L/n$. Hence, the number of oscillations increases with n. Plots of the first few spatial solutions are given in Fig. 3.5.

Equation (3.40) contains an infinite number of as-yet-unspecified constants. It might seem therefore that (3.40) is not really a "solution" at all. Such a conclusion is unwarranted, however. *All* the coefficients can be determined once the initial conditions, say at time $t = 0$, are given. That

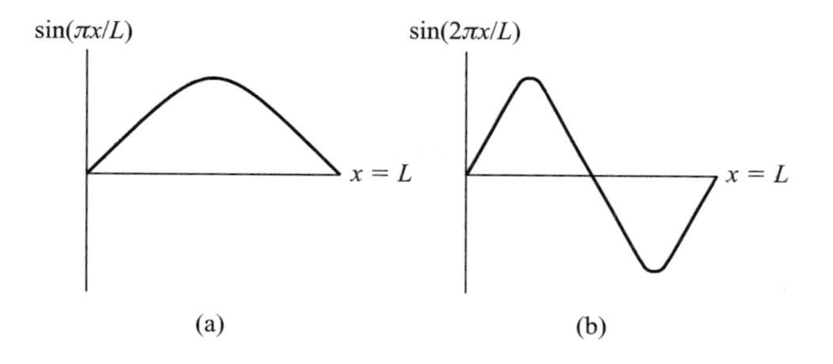

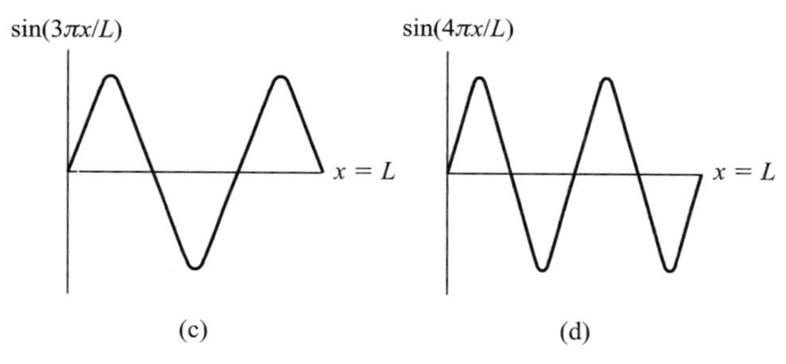

Fig. 3.5 Plots of the first four spatial, standing-wave solutions $\sin(n\pi x/L)$ to the wave equation.

is, knowledge of $y(x, t = 0) = y_0(x)$ and $\partial y(x, t)/\partial t|_{t=0} \equiv v_0(x)$ suffices to specify the $\{a_n\}$ and $\{b_n\}$ via the *orthogonality* properties of the functions $\sin(n\pi x/L)$, as follows.

The initial shape of the string, $y_0(x)$, is related to (3.40) via

$$y_0(x) = \sum_{n=1}^{\infty} a_n \sin\left(\frac{n\pi x}{L}\right), \tag{3.41}$$

while the initial velocity is given by

$$v_0(x) = \frac{\partial y(x, t)}{\partial x}|_{t=0} = \sum_{n=1}^{\infty} b_n \omega_n \sin\left(\frac{n\pi x}{L}\right). \tag{3.42}$$

These two series have the same structure, so only one needs to be examined. We choose the series for $y_0(x)$.

Equation (3.41) is known as a Fourier sine series. It is a special case of the general Fourier series, which we discuss in Appendix B. All we need at present is to assume that $y_0(x)$ is of bounded variation over the interval $[0, L]$, so that (3.41) converges pointwise to $y_0(x)$.

To employ this assumption, we make use of the orthogonality condition obeyed by the $\{\sin(n\pi x/L)\}$ (m and n are integers):[6]

[6] Equation (3.43) is easily obtained using the trigonometric relation $\sin\theta \sin\phi = \frac{1}{2}[\cos(\theta - \phi) - \cos(\theta + \phi)]$. The orthogonality condition (3.43) is one of many examples of such a relation. Orthogonality plays a very important role in quantum mechanics, as will be seen. The condition (3.43) is generalized in Appendix B on Fourier series and then again in Section 3.4, in the context of Sturm–Liouville systems.

$$\int_0^L \sin\left(\frac{n\pi x}{L}\right)\sin\left(\frac{m\pi x}{L}\right)dx = \frac{L}{2}\delta_{mn}, \tag{3.43}$$

where δ_{mn} is the Kronecker delta:

$$\delta_{mn} = \begin{cases} 1, & m = n, \\ 0, & m \neq n. \end{cases} \tag{3.44}$$

Equation (3.43) is the crucial relation, since on multiplying both sides of (3.41) by $(2/L)\sin(m\pi x/L)$, integrating the resulting equality over the interval $[0, L]$, and then interchanging the integral and sum on the RHS – which is allowed by the assumption of pointwise convergence – we get

$$\frac{2}{L}\int_0^L dx \sin\left(\frac{m\pi x}{L}\right) y_0(x) = \sum_{n=1}^{\infty}\frac{2a_n}{L}\int_0^L dx \sin\left(\frac{m\pi x}{L}\right)\sin\left(\frac{n\pi x}{L}\right). \tag{3.45}$$

Substituting (3.43) into the RHS of (3.45) then yields the desired value for a_m:

$$a_m = \frac{2}{L}\int_0^L dx \sin\left(\frac{m\pi x}{L}\right) y_0(x). \tag{3.46}$$

An analogous result holds for b_m of (3.42):

$$b_m = \frac{2}{\omega_m L}\int_0^L dx \sin\left(\frac{m\pi x}{L}\right) v_0(x). \tag{3.47}$$

With Eqs. (3.46) and (3.47), the solution to the clamped-string problem is complete. One might well ask, however, whether the series solution is useful. The answer is YES; often only a few terms in the series are needed to provide a reasonably accurate representation of a particular function. This is especially easy to understand for the case of $y_0(x)$ and $v_0(x)$ each being a slowly varying function of x. Each a_m and b_m is then an integral of a smooth function times $\sin(m\pi x/L)$, for which the number of oscillations increases linearly with m: $\sin(m\pi x/L)$ has $m-1$ nodes or zeros, excluding the ones at $x = 0$ and $x = L$; the first four of these functions are plotted in Fig. 3.5. However, the magnitudes of the integral of a smooth function times ones that vary between positive and negative values become smaller as the number of oscillations increases. Hence, in this often-realized case of smooth $y_0(x)$ and $v_0(x)$, the values of a_m and b_m for large m are usually much smaller than those for small m, and only a relatively small number of terms is needed in the series representation of $y(x, t)$.

Let us consider two examples, in each of which $v_0(x) = 0$; then $b_m = 0, \forall\, m$, so that

$$y(x, t) = \sum_{n=1}^{\infty} a_n \sin\left(\frac{n\pi x}{L}\right)\cos(\omega_n t). \tag{3.48}$$

In the first – trivial – example, we choose $y_0(x) = \sin(m\pi x/L)$. In this case, $a_n = \delta_{nm}$, and we find

$$y(x, t) = \sin\left(\frac{m\pi x}{L}\right)\cos(\omega_m t).$$

If the shape of the string is initially that of one of the harmonics, it retains that shape for all time: no other terms in the series will enter the solution when $t > 0$.

In the second example, the center of the string is initially lifted a distance d ($\ll L$) and is then released. This initial shape is shown in Fig. 3.6 and is given mathematically by

$$y_0(x) = \begin{cases} 2dx/L, & 0 \leqslant x \leqslant L/2, \\ (2d/L)(L-x), & L/2 \leqslant x \leqslant L. \end{cases} \tag{3.49}$$

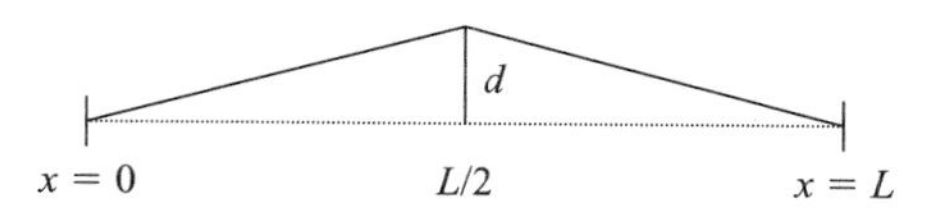

Fig. 3.6 The initial shape of the string lifted a distance d above its center point, $L/2$.

The Fourier coefficients a_m are

$$a_m = \frac{4d}{L^2}\left[\int_0^{L/2} x \sin\left(\frac{m\pi x}{L}\right) dx + \int_{L/2}^{L} (L-x)\sin\left(\frac{m\pi x}{L}\right) dx\right]$$
$$= \frac{8d}{m^2\pi^2}\sin\left(\frac{m\pi}{2}\right), \tag{3.50}$$

where $\int \theta \sin\theta \, d\theta = \sin\theta - \theta\cos\theta$ has been used. Only for odd values of m is $a_m \neq 0$:

$$a_{2n+1} = (-1)^n \frac{8d}{(2n+1)^2\pi^2}, \qquad n = 0, 1, 2, \ldots; \qquad a_{2n} = 0,$$

while the magnitude of a_m rapidly decreases as m increases:

$$\left|\frac{a_{2n+1}}{a_1}\right| = \left(\frac{1}{2n+1}\right)^2.$$

It is left as an exercise to compare the low-order approximations to

$$y_0(x) = \sum_{n=0}^{\infty}(-1)^n \frac{3d}{\pi^2(2n+1)^2}\sin\left(\frac{(2n+1)\pi x}{L}\right)$$

with Eq. (3.49).

3.2. Quantum dynamics

Newton's Law, $\mathbf{F} = \dot{\mathbf{p}}$, is usually considered the primary dynamical equation of classical mechanics. Its analog in quantum mechanics is the time-dependent Schrödinger equation. The Schrödinger equation can be formulated as an abstract operator equation (see Chapter 5). In the so-called *coordinate representation* used in this section, the time-dependent Schrödinger equation is realized as – or takes the form of – a type of wave equation, a quantal wave equation. A wave equation is thus the primary entity in quantum mechanics, as opposed to the case of the classical clamped string, for which the wave equation is a secondary entity, derived from the primary law $\mathbf{F} = \dot{\mathbf{p}}$.

Classical mechanics deals with trajectories, such as the displacement of the string at each point x along its length as a function of time. Thus, in the case of the string, the trajectories become $y = y(x, t)$. In contrast, trajectories do not enter quantum mechanics; instead, the solution of the Schrödinger equation is a *wave function*, which depends on time and on the coordinates of all the particles that comprise the system. Replacing the notion of trajectory is the concept of probability, e.g., the probability that a particle will be found at a particular position, or that its momentum will have a particular value. These probabilities can be obtained from the wave function.

The probabilistic nature of quantum theory means that it is a framework totally unlike Newtonian physics. Nonetheless (and apart from similar vocabularies), there are close connections between classical and quantum mechanics. The one most germane to the treatment in this chapter is the relation of the classical Hamiltonian function $H(\{q_j\}, \{p_j\}, t)$ to quantum dynamics. We consider this next. Bear in mind that much of what is presented informally in the following will be codified into the postulates of quantum theory (Chapter 5).

The time-dependent Schrödinger equation

Many classical systems have exact quantal analogs, for example, the linear harmonic oscillator, the massive spinning object (rotator), and the hydrogen atom. More generally, there is a quantal version of any classical system described by a potential. For such systems, the classical Hamiltonian is "quantized" and turns into the quantal operator $\hat{H}$ that forms part of the Schrödinger equation. We first write out this equation and then indicate how the quantization of the Hamiltonian is induced.

In order to avoid having to deal with many coordinates, the system is assumed to consist of a single particle, whose position is $\mathbf{r}$ and whose mass is m, acted on by some potential $V(\mathbf{r})$. Its quantal behavior is described by the time-dependent Schrödinger equation (Schrödinger 1928)

$$i\hbar \frac{\partial}{\partial t}\Psi(\mathbf{r}, t) = \hat{H}(\mathbf{r}, t)\Psi(\mathbf{r}, t), \tag{3.51}$$

where $\hat{H}$ is the (quantal) Hamiltonian operator and $\Psi(\mathbf{r}, t)$ is the time-dependent wave function for the particle. Since the wave function is the coordinate representation of what is known as the *state vector* of the system (see Chapters 4 and 5), $\Psi(\mathbf{r}, t)$ is sometimes (improperly) referred to as the state of the system. It is also denoted a *probability amplitude*.

INTERPRETATION OF $\Psi(\mathbf{r}, t)$

Let the particle be confined to the volume V, which might be all of space. It is assumed that $\Psi(\mathbf{r}, t)$ is *normalized*, i.e., that[7]

$$\int_V |\Psi(\mathbf{r}, t)|^2\, d^3r = 1, \tag{3.52}$$

where d^3r is an element of volume ($d^3r = dx\,dy\,dz$). The reason for the requirement (3.52) is related to the following interpretational statement: the *probability density* $\rho(\mathbf{r}, t)$ that the particle will be at position $\mathbf{r}$ at time t is

$$\rho(\mathbf{r}, t) = |\Psi(\mathbf{r}, t)|^2, \tag{3.53}$$

while the *probability*[8] $P(\mathbf{r}, t)$ that the particle is in the infinitesimal volume d^3r at position $\mathbf{r}$ is[9]

$$P(\mathbf{r}, t) = \rho(\mathbf{r}, t)\, d^3r. \tag{3.54}$$

[7] If $\int_V |\Psi(\mathbf{r}, t)|^2\, d^3r = N \neq 1$, then $\Psi(\mathbf{r}, t)$ can be made to obey (3.52) by multiplying it by $N^{-1/2}$. We assume that multiplication by $N^{-1/2}$ has already occurred.

[8] An elementary discussion of probability concepts is given in Appendix A.

[9] It would be more appropriate to denote $\rho(\mathbf{r}, t)\, d^3r$ as $dP(\mathbf{r}, t)$, but (3.54) is standard.

Finally, the probability $P_D(t)$ that the particle is in the volume $D \subset V$ is

$$P_D(t) = \int_D \rho(\mathbf{r}, t)\, d^3r. \tag{3.55}$$

Suppose now that $D = V$. Then (3.55) becomes the probability that the particle is in V. However, the particle is *certain* to be somewhere in V, since that is the total volume available to it. By definition, when an outcome is certain, its probability is unity. Equation (3.52) simply states that the probability of finding the particle somewhere in its domain of definition is unity. This is the meaning of the normalization of $\Psi(\mathbf{r}, t)$. Only those Ψ that can be normalized are allowed as solutions. We show by example in Section 3.3 how to normalize allowed solutions.

DETERMINATION OF $\hat{H}(\mathbf{r}, t)$

Let the classical Hamiltonian function be $H(x, y, z, p_x, p_y, p_z)$, assumed time-independent. We write it as $H(\{x_j\}, \{p_j\})$. Then the quantal Hamiltonian operator $\hat{H}(\mathbf{r})$ appearing in (3.51) is obtained from $H(\{x_j\}, \{p_j\})$ by the replacement

$$H(\{x_j\}, \{p_j\}) \rightarrow \hat{H}\left(\{x_j\}, \left\{-i\hbar \frac{\partial}{\partial x_j}\right\}\right). \tag{3.56}$$

Equivalently, one may also *quantize* using the replacements

$$p_{x_j} \rightarrow \hat{P}_{x_j}(\mathbf{r}) \equiv -i\hbar \frac{\partial}{\partial x_j} \tag{3.57}$$

or

$$\mathbf{p} \rightarrow \hat{\mathbf{P}}(\mathbf{r}) \equiv -i\hbar\, \boldsymbol{\nabla}. \tag{3.58}$$

Here, $\hat{\mathbf{P}}(\mathbf{r})$, the coordinate representation of the momentum operator, has been introduced. Note that it does not contain the mass m (!). These replacements make the quantization of H straightforward. Recall that

$$\begin{aligned} H = K + V &= \frac{p^2}{2m} + V(\mathbf{r}) \\ &= \frac{1}{2m}(p_x^2 + p_y^2 + p_z^2) + V(\mathbf{r}) \\ &= \frac{\mathbf{p}\cdot\mathbf{p}}{2m} + V(\mathbf{r}), \end{aligned} \tag{3.59}$$

where the symbol K has been introduced for the kinetic energy $p^2/(2m)$.

Equation (3.58) tells us to replace $\mathbf{p}$ by $-i\hbar\, \boldsymbol{\nabla}$; doing so yields

$$\begin{aligned} H \to \hat{H}(\mathbf{r}) &= -\frac{\hbar^2}{2m}\boldsymbol{\nabla}\cdot\boldsymbol{\nabla} + V(\mathbf{r}) \\ &= -\frac{\hbar^2}{2m}\boldsymbol{\nabla}^2 + V(\mathbf{r}) \\ &= \frac{\hat{P}^2(r)}{2m} + V(\mathbf{r}) \\ &= \hat{K}(\mathbf{r}) + V(\mathbf{r}), \end{aligned} \tag{3.60}$$

where the coordinate representation of the kinetic-energy operator $\hat{K}(r) \equiv \hat{P}^2(\mathbf{r})/(2m) = -\hbar^2\,\boldsymbol{\nabla}^2/(2m)$ has been defined.

With these relations, the time-dependent Schrödinger equation (3.51) becomes

$$i\hbar\,\frac{\partial}{\partial t}\Psi(\mathbf{r},\,t) = \left(-\frac{\hbar^2}{2m}\,\boldsymbol{\nabla}^2 + V(\mathbf{r})\right)\Psi(\mathbf{r},\,t). \tag{3.61}$$

For the following comparison with the 1-D wave equation (3.3), we imagine a particle confined to one spatial dimension, say x, acted on by a 1-D potential $V(x)$. In this case, $\mathbf{r} \to x$, $\boldsymbol{\nabla} \to \partial/\partial x$, and (3.61) reads

$$i\hbar\,\frac{\partial}{\partial t}\Psi(x,\,t) = \left(-\frac{\hbar^2}{2m}\frac{\partial^2}{\partial x^2} + V(x)\right)\Psi(x,\,t). \tag{3.62}$$

Comparison with the 1-D classical wave equation

Before examining general solutions to the time-independent Schrödinger equation, we compare its 1-D version (3.62) with the wave equation (3.3). (Our remarks will also pertain to (3.61) and its multi-particle extensions.)

Equation (3.62) is evidently a quantal equation because it involves Planck's constant $\hbar$. It also contains $i = \sqrt{-1}$, implying that, in contrast to (3.3), the solutions $\Psi(x,\,t)$ are necessarily *complex* functions. This is the reason why the probability density is $|\Psi(x,\,t)|^2$ rather than $\Psi^2(x,\,t)$. Also of significance is the appearance of $\partial/\partial t$ in (3.62), as opposed to the $\partial^2/\partial t^2$ occurring in the classical wave equation. Its most important consequence is that the value of a quantal wave function (or state vector) at only a single earlier time is needed for an in-principle determination of its value at any later time, as shown explicitly in Section 5. This is in sharp contrast to the classical wave equation, in which the $\partial^2/\partial t^2$ operator means that *two* earlier time conditions are needed for an in-principle determination of the solution at a later time, a feature exemplified by Eqs. (3.41) and (3.42). A further consequence will be the absence of either a $\sin(\omega t)$ or a $\cos(\omega t)$ type of time dependence for $\Psi(x,\,t)$ once solution via separation of variables is attempted: only an $\exp(-i\omega t)$-type time dependence will arise. As for the spatial part of Eq. (3.62), we see that $V(x)$ modifies the $\partial^2/\partial x^2$ term. Hence, only for those x such that $V(x) = 0$ (or constant) will the spatial part of $\Psi(x,\,t)$ be of $\sin(kx)$ or $\cos(kx)$ form.

Given this list of differences, it is important to remark on two similarities between (3.3) and (3.62). First, each can be solved using separation of variables; second, the equation for the spatial portion of the unknown solution $y(x,\,t)$ or $\Psi(x,\,t)$ is an eigenvalue equation.

Stationary states; energy eigenvalues

The operator form of (3.62) is

$$i\hbar \frac{\partial}{\partial t}\Psi(x, t) = \hat{H}(x)\Psi(x, t), \tag{3.63}$$

where

$$\hat{H}(x) = -\frac{\hbar^2}{2m}\frac{\partial^2}{\partial x^2} + V(x). \tag{3.64}$$

We shall work with (3.63) and shall seek a solution of the form

$$\Psi(x, t) = \psi(x)T(t). \tag{3.65}$$

Substitution of (3.65) into (3.63) plus the assumption that neither $\psi(x)$ nor $T(t)$ vanishes leads to

$$i\hbar \frac{1}{T(t)}\frac{\partial T(t)}{\partial t} = \frac{1}{\psi(x)}\hat{H}(x)\psi(x). \tag{3.66}$$

As with Eq. (3.27), the equality (3.66) requires that each side of the equation must be a constant. The dimension of the separation constant is energy, as is easily seen from the facts that $\hbar$ has dimension energy $\cdot$ time and $\hat{H}$, obtained from the classical Hamiltonian, has the dimension of energy. We denote the separation constant by E, for energy. Then (3.66) becomes

$$i\hbar \frac{dT(t)}{dt} = ET(t) \tag{3.67}$$

and

$$\hat{H}(x)\psi(x) = E\psi(x). \tag{3.68}$$

The solution to (3.67) is readily found to be

$$T(t) = e^{-iEt/\hbar}; \tag{3.69}$$

because (3.65) is a product, an overall constant that could have multiplied $\exp(-iEt/\hbar)$ is assumed to have been absorbed into $\psi(x)$. Equation (3.61) thus becomes

$$\Psi(x, t) = \psi(x)\exp(-iEt/\hbar). \tag{3.70}$$

Equation (3.69) places no restrictions on the sign or magnitude of E. If any restrictions occur, it will be as a consequence of solving Eq. (3.68), which is an eigenvalue equation, $\hat{H}$ being the operator and E the eigenvalue. Some quantal operators, including $\hat{H}$, enjoy a special property, viz., they are *Hermitian*, a term defined for differential operators in Section 3.4 and more generally in the next chapter. Hermiticity of an operator means that all of its eigenvalues are *real*. Hence, in the present case, E is real: $E^* = E$.

Equation (3.68) has two kinds of solutions: those for which the integral in (3.52) is finite, and those for which it is not. Only the former are acceptable or "physical" solutions; they are referred to as *normalizable* (or "proper").[10] The set of energy

[10] Non-normalizable or improper eigensolutions, corresponding to continuous energies, do play a role in quantum mechanics. Linear combinations of them are used, e.g., to help form normalizable, wave packet solutions to (3.51), as in Section 5.6, Chapter 7, and Appendix B.

eigenvalues associated with normalizable solutions to (3.68) and its equivalent for 3-D or multi-particle systems are discrete. This discrete set may be finite or denumerably infinite. Discreteness is a result of imposing the appropriate boundary conditions on the general solutions to (3.68), just as in the case of the clamped string. In the quantal case, the discreteness of various eigenvalues, especially the energy, is what is meant by the term "quantized". It accounts, e.g., for the quantization of the energy levels of the H-atom, a result obtained by Bohr via his primitive "quantum theory."

The discrete energy eigenvalues and corresponding normalizable eigensolutions are denoted, respectively, *bound-state* energies and wave functions, or simply bound states. They are of paramount importance in quantum mechanics. Among other features, the systems they describe, such as solids, molecules, atoms, and nuclei, are confined – or localized – in space. This means that the constituents of systems in bound states can never be found at infinite spatial separations from one another. We shall see that this holds trivially for the quantum box considered in the next section and shall show it more generally in later chapters. Two other aspects of normalizable eigensolutions are worth pointing out. First, the set of discrete energies constitutes the only ones that are allowed: the system can have no others, i.e., no other energy values can be measured in any (ideal) experiment. Second, each of the eigensolutions represents a quantum state of the system: when the energy is found to be one of the discrete eigenvalues, then the system is in the quantum state whose coordinate representation is just the corresponding bound-state wave function![11]

Let us now see what the existence of bound states implies for the probabilities. We assume that $V(x)$ supports a discrete set of bound states $\{E_n, \psi_n(x)\}$, labeled by the integer quantum number n, $n = 1, 2, \ldots$. Suppose that the system (a single particle in 1-D) is in the nth bound state. Then the solution $\Psi(x, t)$ of Eq. (3.70) becomes $\Psi_n(x, t)$:

$$\Psi_n(x, t) = \psi_n(x)e^{-iE_n t/\hbar},$$

where

$$\hat{H}(x)\psi_n(x) = E_n\psi_n(x), \tag{3.71}$$

and $\Psi_n(x, t)$ is presumed to be normalized. In this case, $\rho(x, t) \to \rho_n(x, t)$:

$$\begin{aligned} \rho_n(x, t) &= |\Psi_n(x, t)|^2 = |\psi_n(x)|^2 \\ &\equiv \rho_n(x), \end{aligned} \tag{3.72}$$

which is time-independent. The probability density has become stationary in time. Eigensolutions to $\hat{H}\psi = E\psi$ are therefore often referred to as *stationary states*. For all other states, i.e., for linear combinations of stationary states, the probability density is time-dependent.

As a final point before turning to the analysis of the quantal box, we note that it can be shown that there is a minimum value of E for bound states. Arranging the set $\{E_n\}$ in ascending order, $E_1 \leqslant E_2 \leqslant E_3 \leqslant \cdots$, then E_1 is this minimum value: it is the *ground-state* energy. The existence of a ground state is the reason why a quantum system cannot

[11] The various properties described in the foregoing will be encountered in Chapter 5 as elements of some of the postulates of quantum theory.

emit photons when it is in that state: there is no state of lower energy for it to decay into. Recall that the absence of radiation from the ground state was an *ad hoc* postulate introduced by Bohr in his model of hydrogen: it made a virtue of necessity, but did not provide an explanation for this aspect of quantum systems.

3.3 The particle in a quantal box

For the quantum systems that occur in nature, the Schrödinger equation is too difficult to solve exactly: the only solutions which can be obtained are approximate ones. Soluble problems, i.e., systems for which the solutions to Eq. (3.51) can be obtained exactly, are obviously important, both as learning tools and also as possible models for more complex systems. The soluble system we examine in this section has been chosen for its ultimate simplicity and close relation mathematically to the clamped string of Section 3.1. It is, however, a contrived example, not one that mirrors a real quantum system.

The potential energy $V(x)$

The system is a quantal particle in a 1-D box. It is the 1-D version of the 3-D quantal box, which is also soluble, but equally unrealistic.[12] These boxes are the quantum counterparts of a classical box whose lid is always fastened, thereby indefinitely trapping the classical particle it contains. By analogy, the quantal box must be such as to confine a particle to its interior. This may seem a trivial requirement, but, because quantal particles can tunnel through any finite-sized "retaining" wall or barrier, the confinement of the particle to the interior of the box means that the walls surrounding the quantum box (or hole in space) must be infinitely high. We shall assume this statement to be true for now; its proof will be given later in this book. For simplicity, we shall also assume these walls to be infinitely thick.

For a particle moving in only 1-D, say along the x-axis, the quantal box must act so as to confine the particle to a finite segment. We take that segment to be the interval $[0, L]$; that this length is the same as that of the clamped string, and positioned between $x = 0$ and $x = L$, is deliberate. The infinitely high and thick walls that confine the particle are described by the potential energy term $V(x)$ in Eq. (3.62). For the 1-D quantal box, $V(x)$ is given by

$$V(x) = \begin{cases} \infty, & x \leqslant 0, \\ 0, & 0 < x < L, \\ \infty, & x \geqslant L. \end{cases} \tag{3.73}$$

A schematic representation is given in Fig. 3.7.

Our claim is that the infinitely high and wide parts of $V(x)$ will confine the particle to the region $[0, L]$. However, $V(x)$ seems to do more than that: its pathological nature outside this interval would seem to make the Schrödinger equation unmanageable. To deal with this seeming problem of infinities in the Schrödinger equation, we make a

[12] This is the fate of some soluble systems. However, the 3-D quantum square well (and its 1-D cousin), which is a finite version of the 3-D quantal box and is studied later in this book, can be considered a valid approximation to a short-range potential.

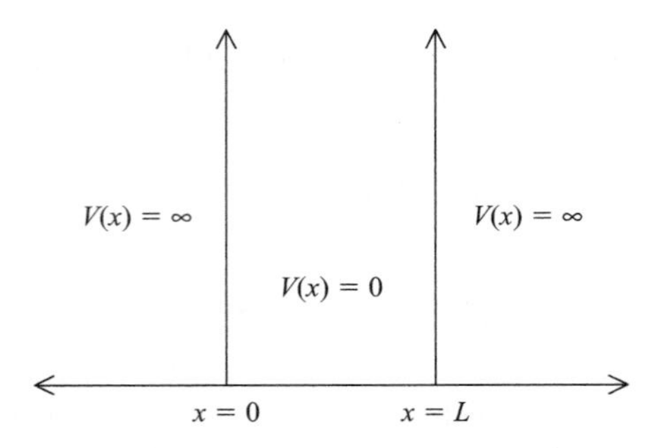

Fig. 3.7 A schematic representation of the 1-D quantum-box potential energy $V(x)$, located between $x = 0$ and $x = L$ (Eq. (3.73)).

standard assumption used in these situations, viz., that, for $V(x) = \infty$, $\psi(x) = 0$ in such a way that

$$V(x)\psi(x) = 0, \qquad x \leqslant 0 \text{ and } x \geqslant L. \tag{3.74}$$

Then Eq. (3.62) is simply the relation $0 = 0$ for $x \notin [0, L]$, and the pathology is no longer a cause for concern.[13]

The time-independent Schrödinger equation

The preceding assumption means that only the interval $[0, L]$ need be considered. Stationary-state wave functions will be of the form $\Psi(x, t) = \psi(x)\exp(-iEt/\hbar)$, Eq. (3.70), with $\psi(x)$ the solution of

$$[\hat{K}(x) + V(x)]\psi(x) = E\psi(x), \qquad x \in [0, L], \tag{3.75}$$

where

$$\hat{K}(x) = -\frac{\hbar^2}{2m}\frac{d^2}{dx^2} \tag{3.76}$$

is the 1-D kinetic-energy operator. However, in the interval $[0, L]$, $V(x) = 0$ – Eq. (3.73) – and therefore (3.75) is simply

$$-\frac{\hbar^2}{2m}\frac{d^2}{dx^2}\psi(x) = E\psi(x), \qquad 0 \leqslant x \leqslant L. \tag{3.77}$$

Multiplying through by $-2m/\hbar^2$, (3.77) becomes

$$\frac{d^2\psi(x)}{dx^2} = -\frac{2mE}{\hbar^2}\psi(x), \qquad 0 \leqslant x \leqslant L. \tag{3.78}$$

With $-2mE/\hbar^2$ in place of $-k^2$, Eq. (3.78) is of the same form as Eq. (3.28), the eigenvalue equation for $\chi(x)$, the spatial part of the vibrating-string displacement. The

[13] The reader may well ask "Why not avoid pathologies by simply dealing with a finite potential?" The answer is that finite potentials are, unfortunately, not so easy to deal with, since the Schrödinger equation (3.68) yields both proper and improper solutions. The box, on the other hand, has only normalizable solutions, which *are* easily determined. Bound states in some finite potentials are discussed in Chapters 6 and 11.

eigenvalue E is to be determined; it can be negative or positive. As with Eq. (3.28), the boundary conditions on $\psi(x)$ will determine the allowed values of E.

What are the appropriate boundary conditions that apply to the time-independent Schrödinger equation (3.78)? The answer uses the facts that $\psi(x)$ and its first derivative must be continuous for all values of x, in order that the term involving $d^2\psi/dx^2$ will be well behaved. However, $\psi(x)$ vanishes outside the interval [0, L] and therefore must obey

$$\psi(x = 0) = 0 \tag{3.79a}$$

and

$$\psi(x = L) = 0. \tag{3.79b}$$

The relations (3.79) are the *same* as those that the string solutions $\chi(x)$ must satisfy, viz., Eqs. (3.32)! The mathematical similarities between the clamped string and the quantum box are beginning to emerge.

We now show that negative values of E are inconsistent with the boundary conditions (3.79). Let $E = -|E|$ and define $\beta^2 = 2m|E|/\hbar^2$, so that (3.78) reads

$$\frac{d^2\psi(x)}{dx^2} = \beta^2\psi(x), \tag{3.80}$$

where we may assume that $\beta > 0$. The linearly independent solutions to (3.80) are $\exp(\pm\beta x)$, from which we construct the general solution:

$$\psi(x) = ae^{\beta x} + be^{-\beta x}, \qquad 0 \leqslant x \leqslant L, \tag{3.81}$$

where a and b are arbitrary constants.

Applying (3.79a) to (3.81) yields $b = -a$, and $\psi(x)$ becomes

$$\begin{aligned}\psi(x) &= a(e^{\beta x} - e^{-\beta x})\\ &= 2a\sinh(\beta x).\end{aligned} \tag{3.82}$$

However, the hyperbolic sine vanishes only when its argument is zero. Hence, if (3.79b) is to be satisfied, i.e., if

$$2a\sinh(\beta L) = 0,$$

then a must be set equal to zero. Thus, only the trivial solution $\psi(x) = 0$ is allowed if E is negative. We therefore conclude that E must be non-negative.

Since E is non-negative, we may introduce a new constant k^2:

$$\frac{2mE}{\hbar^2} = k^2, \qquad E = \frac{\hbar^2 k^2}{2m}, \tag{3.83}$$

whereupon (3.78) becomes identical to (3.28), viz.,

$$\frac{d^2\psi(x)}{dx^2} = -k^2\psi(x), \qquad 0 \leqslant x \leqslant L. \tag{3.84}$$

Because the boundary conditions on $\chi(x)$ and $\psi(x)$ are the same, the eigenvalues and eigensolutions must be the same. That is,

$$\psi(x) \to \psi_n(x) = A_n \sin(k_n x), \qquad x \in [0, L], \tag{3.85}$$

$$k_n = \frac{n\pi}{L}, \qquad n = 1, 2, 3, \ldots \tag{3.86}$$

and

$$E \to E_n = \frac{\hbar^2 k_n^2}{2m} = n^2 \frac{\hbar^2 \pi^2}{2mL^2}. \tag{3.87}$$

Properties of the solutions

Apart from the value of A_n, to be determined below, Eqs. (3.85)–(3.87) are a complete solution of the $\hat{H}\psi = E\psi$ eigenvalue problem for the particle in a quantum box. The eigensolutions and wave numbers are identical to those for the clamped string, thus completing the mathematical connection between the classical clamped string and the quantal particle in the box. In the case of the string, the requirement that $k \to k_n = n\pi/L$ leads to the introduction of the various harmonics and the standing waves. For the particle in the quantum box, however, one uses an entirely different nomenclature, that of *quantization*: via the imposition of boundary conditions, the energy of the particle becomes quantized ($E \to \{E_n\}$), unlike the particle in a classical box, or, for that matter, unlike the oscillation of the clamped string: in either of these cases, E can range continuously upwards from $E = 0$. In contrast, the smallest (ground-state) energy for a particle in the quantal box is $E = E_1 = \pi^2\hbar^2/(2mL^2) > 0$.

Associated with each quantized energy level E_n is a spatial wave function $\psi_n(x)$ and an overall wave function $\Psi_n(x, t) = \psi_n(x)\exp(-iE_n t/\hbar)$. Each is a probability amplitude, which must be normalized:

$$\begin{aligned} 1 &= \int_0^L dx\, |\Psi_n(x, t)|^2 = \int_0^L dx\, |\psi_n(x)|^2 \\ &= |A_n|^2 \int_0^L dx \sin^2\left(\frac{n\pi x}{L}\right) = \frac{L}{2}|A_n|^2, \end{aligned} \tag{3.88}$$

where the relation $\sin^2\theta = [1 - \cos(2\theta)]/2$ has been used. From (3.88) and the assumption (justified in Chapters 5 and 6) that A_n may be chosen real and positive, we get

$$A_n = \sqrt{\frac{2}{L}}, \qquad \forall\ n, \tag{3.89}$$

and thus

$$\psi_n(x) = \sqrt{\frac{2}{L}} \sin\left(\frac{n\pi x}{L}\right). \tag{3.90}$$

Each of the stationary states $\Psi_n(x, t)$, with

$$\Psi_n(x, t) = \sqrt{\frac{2}{L}} \sin\left(\frac{n\pi x}{L}\right) e^{-iE_n t/\hbar}, \tag{3.91}$$

is a possible quantum state of the particle in the box: we need not form linear combinations of the $\Psi_n(x, t)$ to obtain an overall solution. From (3.91), the probability density $\rho_n(x)$ is found to be

$$\rho_n(x) = \begin{cases} |\Psi_n(x, t)|^2 = \dfrac{2}{L}\sin^2\left(\dfrac{n\pi x}{L}\right), & x \in [0, L], \\ 0, & x \notin [0, L]. \end{cases} \tag{3.92}$$

The particle is thus localized, i.e., confined to $[0, L]$. As n increases, $\psi_n(x)$ oscillates more rapidly (recall Fig. 3.5) and the number of nodes in $\rho_n(x)$ increases (linearly with n). On the other hand, $E_n \propto n^2$, so that the energy of the nth state increases rapidly; in particular, the spacing between adjacent levels grows linearly with n: $E_{n+1} - E_n \propto 2n + 1$.

Shown in Fig. 3.8 are a Grotrian diagram representing the first three energy levels plus curves illustrating the associated probability densities. All $\rho_n(x)$ are zero at the walls, viz., at $x = 0$ and $x = L$. Excluding these zeros, the functional form $\sin(n\pi x/L)$ guarantees that $\psi_n(x)$ and $\rho_n(x)$ have a total of $n - 1$ nodes, the mth occurring at $x = mL/n$, $m = 1, 2, \ldots, n - 1$. The nodes are the points in the box where, quantally, the particle can never be found. In the nth state, there are n values of x at which the probability of finding the particle is equally maximal. As $n \to \infty$, one can show that the average of $\rho_n \to L^{-1}$, the classical value (see exercises, Chapter 6).

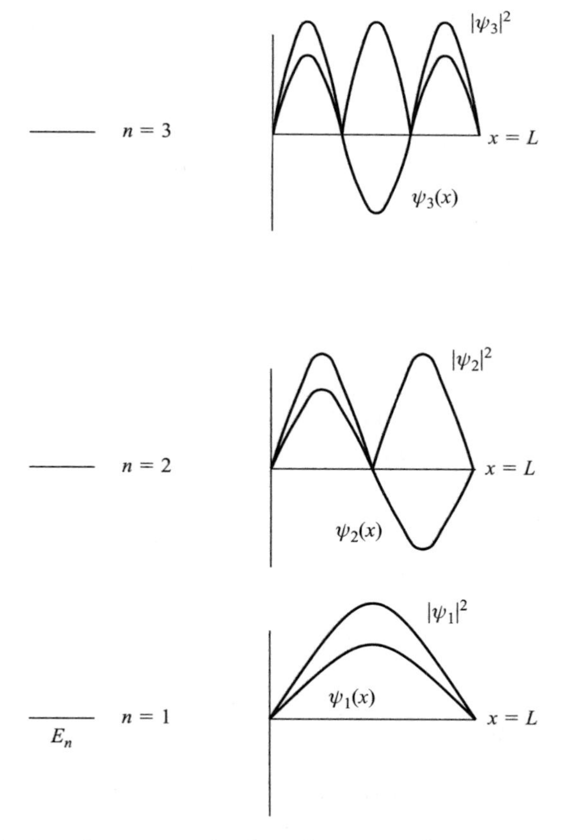

Fig. 3.8 The first three energy levels and corresponding probability amplitudes and probability densities for a particle in a quantum box.

In the language of Section 3.2, the volume V in the present case is the line segment L. By construction $\psi_n(x)$ is normalized so that $P_V(t) = 1$. Let us now consider a specific domain $D \subset V$ and calculate the probability that the particle will be found in D. For D we choose the right-hand half of the box, $[L/2, L]$. Denoting P_D by $P_n(x \geqslant L/2, t)$, we have

$$\begin{aligned} P_n\left(x \geqslant \frac{L}{2}, t\right) &= \int_{L/2}^{L} dx\, |\Psi_n(x, t)|^2 \\ &= \frac{2}{L}\int_{L/2}^{L} dx \sin^2\left(\frac{n\pi x}{L}\right) \\ &= \tfrac{1}{2}. \end{aligned} \tag{3.93}$$

No matter which state the particle is in, the probability that it will be found in the right-hand half of the box is always 0.5. This result should not be a surprise, essentially because the probability densities $\{\rho_n(x)\}$ are all *symmetric* about the midpoint, $x = L/2$. Such symmetry considerations will play a significant role in our later analysis of quantum systems.

Non-stationary states

As a final consideration of the particle in a quantal box, let us look at an instance for which $\rho(x, t)$ is not stationary. Suppose that the wave function $\Psi(x, t)$ for the particle is a linear combination[14] of the wave functions for states 2 and 3:

$$\Psi(x, t) = a_2\Psi_2(x, t) + a_3\Psi_3(x, t), \tag{3.94}$$

where a_2 and a_3 are constants, values for which we shall specify shortly. Before doing so, we remark that

$$|a_2|^2 + |a_3|^2 = 1, \tag{3.95}$$

a relation required by normalization and orthogonality.

The normalization portion of (3.95) simply means that $\Psi(x, t)$, being a quantal probability amplitude, obeys the 1-D version of (3.52):

$$\int |\Psi(x, t)|^2\, dx = 1. \tag{3.96}$$

On substituting (3.94) into (3.96), one finds

$$\begin{aligned} |a_2|^2 \int_0^L \psi_2^2(x)\, dx + |a_3|^2 \int_0^L \psi_3^2(x)\, dx \\ + \frac{4}{L}\,\mathrm{Re}\left(a_2 a_3^* e^{-i(E_2-E_3)t/\hbar}\right)\int_0^L \sin\left(\frac{2\pi x}{L}\right)\sin\left(\frac{3\pi x}{L}\right) dx = 1. \end{aligned} \tag{3.97}$$

The first two integrals in (3.97) are each unity by virtue of the normalization condition

[14] How and why quantal systems may be described by linear combinations of eigenstates is discussed in Chapters 5 and 6.

obeyed by $\psi_n(x, t)$, whereas the third integral vanishes by virtue of the orthogonality condition obeyed by the $\psi_n(x)$, viz., Eq. (3.43). When these relations are used in (3.96), Eq. (3.95) is the result.

Quantum theory assigns a particular meaning to the coefficients a_2 and a_3, and all others like them: each such a_n is a probability amplitude and $|a_n|^2$ is a probability. These are the amplitudes and probabilities that the system will be in the state whose wave function is $\Psi_n(x, t)$ when its overall wave function is a linear combination, e.g., Eq. (3.94). Equivalently, if, as in the present case, the $\psi_n(x, t)$ are energy eigenfunctions, then $|a_n|^2$ is the probability that an energy measurement on the system will yield E_n when the wave function is a linear combination such as (3.94). The various probabilistic interpretations of quantal amplitudes will be formulated in Postulate II in Chapter 5.

The probability density to which (3.94) gives rise is

$$\begin{aligned}\rho(x, t) &= |\Psi(x, t)|^2\\ &= |a_2|^2\rho_2(x) + |a_3|^2\rho_3(x)\\ &\quad + \frac{4}{L}\,\mathrm{Re}(a_2 a_3^* e^{i(E_2-E_3)t/\hbar})\sin\left(\frac{2\pi x}{L}\right)\sin\left(\frac{3\pi x}{L}\right). \end{aligned} \tag{3.98}$$

Comparison of (3.98) with (3.97) shows, of course, that the LHS of (3.97) is just the integral of $\rho(x, t)$ over the interval $[0, L]$. However, whereas the cross term in (3.97) vanishes, that in (3.98) does not do so. Instead, it is present as a time-dependent interference term, whose non-zero contribution to $\rho(x, t)$ makes the probability density move or "slosh" back and forth in the box as t varies. This type of interference effect is an exact analog of that seen in the two-slit experiment.

To illustrate the interference effect, we choose $a_3 = a_2 = \sqrt{\tfrac{1}{2}}$. Then (3.98) becomes

$$\begin{aligned}\rho(x, t) &= \frac{1}{L}\left[\sin^2\left(\frac{2\pi x}{L}\right) + \sin^2\left(\frac{3\pi x}{L}\right)\right]\\ &\quad + \frac{2}{L}\cos[(E_2 - E_3)t/\hbar]\sin\left(\frac{2\pi x}{L}\right)\sin\left(\frac{3\pi x}{L}\right). \end{aligned} \tag{3.99}$$

The cosine term in (3.99) has a period $T = 2\pi\hbar/(E_2 - E_3)$, the result of which is that the time variation in $\rho(x, t)$ repeats itself every period. At $t = 0$, we have

$$\rho(x, 0) = \frac{1}{L}\left[\sin\left(\frac{2\pi x}{L}\right) + \sin\left(\frac{3\pi x}{L}\right)\right]^2$$

whereas at $t = T/2$,

$$\rho\left(x, \frac{T}{2}\right) = \frac{1}{L}\left[\sin\left(\frac{2\pi x}{L}\right) - \sin\left(\frac{3\pi x}{L}\right)\right]^2;$$

$\rho(x, T) = \rho(x, 0)$. Sketched in Fig. 3.9 are plots of $\rho(x, 0)$ and $\rho(x, T/2)$, clearly illustrating that the probability density changes in time.

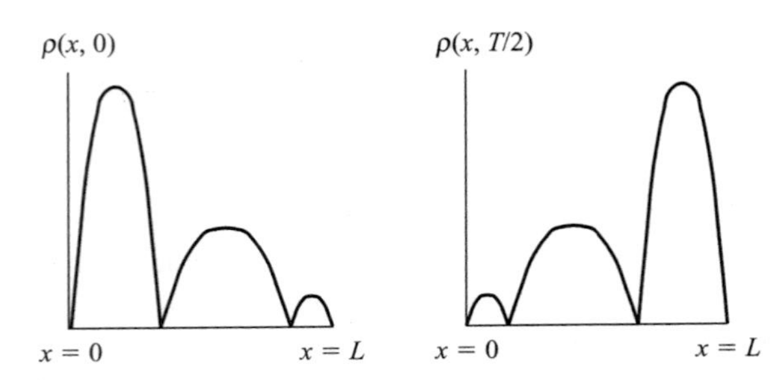

Fig. 3.9 Rough sketches of the probability density $\rho(x, t)$ of Eq. (3.99) at the times $t = 0$ and $t = T/2$. The two densities, which are mirror images of one another, demonstrate the "sloshing" of the probability density with time.

3.4. *The concept of Hermiticity

In the preceding section we mentioned one of the great themes of quantum mechanics: the concept of Hermiticity. It was mentioned but not defined. We do so in this section, which concludes the chapter and our introduction to quantum mechanics.

The concept of Hermiticity, so essential to quantum theory, applies to abstract operators as well as to their possible realizations as matrices or differential operators. Although we shall present a formal definition of Hermiticity for an abstract operator in Chapter 4, it is salutary to introduce a more narrow definition of it using differential operators: after all, $\hat{H}(x)$, which was claimed to be Hermitian, was expressed in Sections 3.2 and 3.3 as a differential operator, and differential operators/equations are relatively easy to understand.

The first step will be to define a Hermitian differential operator. After that, we specialize to a class of second-order differential operators denoted $\hat{L}(x)$ and known as Sturm–Liouville operators, whose Hermiticity we prove under a fairly broad statement of boundary conditions. Almost all of the coordinate-space Hamiltonians of quantum mechanics can be put in Sturm–Liouville form, so this particular specialization is of immense utility. Next, the eigenvalue problem involving Sturm–Liouville operators is addressed: it is shown that, owing to the Hermiticity, the eigenvalues are real and that the eigenfunctions form an orthogonal set, analogous to $\{\sin(n\pi x/L)\}$. The eigenfunctions of $\hat{L}(x)$, of which $\{\sin(n\pi x/L)\}$ is an example, enjoy a property known as completeness, and we use this property to introduce the concept of *closure*, as well as the Dirac delta function, a weird quantity whose pathological nature does not prevent it from being one of the most important mathematical adjuncts in quantum mechanics. The section concludes with a new symbol for the integral that is used in the definition of orthogonality. This symbolism is a precursor to the Dirac bra and ket notation of Section 4.2.

Hermitian differential operators

We consider the concept of Hermiticity for differential operators first. Let $\hat{D}(x)$ be an nth order, linear differential operator in the variable $x \in [a, b]$, where the interval may be finite or infinite in length. $\hat{D}(x)$ is assumed to act on sufficiently well-behaved, complex functions, say $u(x)$, $v(x)$, $w(x)$, etc, all of which (and possibly their derivatives) obey prescribed boundary conditions at the points $x = a$ and $x = b$. An example is $\hat{D}(x) = d^2/dx^2$, $x \in [0, L]$, plus the boundary conditions that the functions all vanish at $x = 0$ and $x = L$, as in the cases of the clamped string and the quantal box.

DEFINITION: HERMITICITY OF $\hat{D}(x)$

Let $u(x)$ and $v(x)$ be two arbitrary functions obeying the prescribed boundary conditions. Then $\hat{D}(x)$ is *Hermitian* if

$$\int_a^b dx\, u^*(x)\hat{D}(x)v(x) = \int_a^b dx\, [\hat{D}(x)u(x)]^* v(x). \tag{3.100}$$

When $\hat{D}(x)$ obeys (3.100), we write $\hat{D}^\dagger(x) = \hat{D}(x)$, where the dagger $\dagger$ signifies the Hermitian conjugate of $\hat{D}(x)$. A formal definition of $\dagger$ is given in Chapter 4. The differential operators $\hat{D}(x)$ that enter classical mechanics are real, but some of those that occur in quantum mechanics involve complex quantities; hence the presence of $\hat{D}^*(x)$ on the RHS of Eq. (3.100).

Expressions like $\hat{D}(x)u(x)$ and $\hat{D}(x)v(x)$ are assumed to be functions that are defined in the interval $[a, b]$ and obey the boundary conditions. As a result, the integrals in (3.100) can be thought of as a certain kind of scalar product (defined below). Because of this, the statement of Hermiticity for $\hat{D}(x)$, Eq. (3.100), is a precise analog of the corresponding definition of Hermiticity for $N \times N$ matrices acting on N-rowed column vectors. This analogy is made explicit in the next chapter.

We now specialize to the case in which $\hat{D}(x) = \hat{L}(x)$, the Sturm–Liouville operator:

$$\hat{L}(x) = \frac{d}{dx}\left(p(x)\frac{d}{dx}\right) - q(x), \qquad x \in [a, b], \tag{3.101}$$

where $p(x)$ and $q(x)$ are sufficiently well-behaved real functions of x.[15] The structure chosen for $\hat{L}(x)$ is deliberate: it is the form taken, e.g., by each of the xyz or $r\theta\phi$ portions of the coordinate-space form of the Hamiltonian operator $\hat{H}(\mathbf{r})$ occurring in the Schrödinger equation for a single particle whose position is $\mathbf{r}$. Equation (3.101) is also the form taken by the differential operator in the equations defining various orthogonal functions such as Legendre polynomials, Hermite functions, Laguerre functions, etc, that occur in applied mathematics. Not coincidentally, these functions play important roles in the solution of some quantal eigenvalue problems.

There is a variety of boundary conditions that can apply to the functions $u(x)$, $v(x)$, $w(x)$, etc., on which $\hat{L}(x)$ acts. A broad class of these is included in the following:

$$[p(x)u^{*\prime}(x)v(x)]_{x=a} = [p(x)u^*(x)v'(x)]_{x=b}, \tag{3.102}$$

where u and v are any members of the set of functions on which $\hat{L}$ acts and $' = d/dx$.

[15] Just how well behaved $p(x)$ and $q(x)$ should be is specified in books dealing with second-order differential equations in the complex plane, e.g., Hassani (1991), Mathews and Walker (1970), Dettman (1988), and Copson (1935), which the interested reader may consult for details. Their specification is unnecessary for our purposes.

Equation (3.102) is the key to proving that $\hat{L}(x)$ is Hermitian, i.e., that $\hat{L}^{\dagger}(x) = \hat{L}(x)$. To show this, form $[\hat{L}(x)u(x)]^*$ and $\hat{L}(x)u(x)$:

$$[\hat{L}(x)u(x)]^* = \frac{d}{dx}\left(p(x)\frac{du^*}{dx}\right) - q(x)u^*(x) \tag{3.103}$$

and

$$\hat{L}(x)v(x) = \frac{d}{dx}\left(p(x)\frac{dv}{dx}\right) - q(x)v(x). \tag{3.104}$$

Next, multiply both sides of (3.103) from the right by $v(x)$ and both sides of (3.104) by $u^*(x)$ from the left, and then subtract the resulting pairs of equalities. This yields

$$u^*(x)\hat{L}(x)v(x) - [\hat{L}(x)u(x)]^* v(x) = u^*(x)\frac{d}{dx}\left(p(x)\frac{dv}{dx}\right) - \left[\frac{d}{dx}\left(p(x)\frac{du^*}{dx}\right)\right]v(x). \tag{3.105}$$

Since the definition of Hermiticity involves an integral (Eq. (3.100)), we integrate both sides of (3.105) on x from a to b:

$$\begin{aligned}\int_a^b dx\{u^*(x)\hat{L}(x)v(x) - [\hat{L}(x)u(x)]^* v(x)\} \\ = \int_a^b dx\left\{u^*(x)\frac{d}{dx}\left(p(x)\frac{dv}{dx}\right) - \left[\frac{d}{dx}\left(p(x)\frac{du^*}{dx}\right)\right]v(x)\right\} \\ = \left\{u^*(x)\left(p(x)\frac{dv}{dx}\right) - \left(p(x)\frac{du^*}{dx}\right)v(x)\right\}_a^b,\end{aligned} \tag{3.106}$$

where the second equality in (3.106) is a result of integrating by parts. Comparison with the boundary-condition statement (3.102) shows that $\{\}_a^b = 0$, thus establishing that (3.106) reduces to a statement of the Hermiticity of $\hat{L}(x)$, i.e.,

$$\int_a^b dx\,[\hat{L}(x)u(x)]^* v(x) = \int_a^b dx\,u^*(x)\hat{L}(x)v(x). \tag{3.107}$$

It is therefore legitimate to write $\hat{L}^{\dagger}(x) = \hat{L}(x)$.

Properties of the Sturm–Liouville eigenvalue equation

We next examine the differential-equation eigenvalue problem involving $\hat{L}(x)$ of Eq. (3.101):

$$\hat{L}(x)u_\lambda(x) = \lambda u_\lambda(x), \qquad x \in [a, b], \tag{3.108}$$

where λ is the eigenvalue and $u_\lambda(x)$ is the eigenfunction corresponding to λ.[16] It is assumed that all the $u_\lambda(x)$ obey the boundary condition (3.102).

[16] There is a more general form for the Sturm–Liouville eigenvalue problem, viz., one in which the RHS of (3.108) is replaced by $\lambda\rho(x)u_\lambda(x)$, with $\rho(x) \geqslant 0$, $x \in [a, b]$. We employ (3.108) because it is simpler: our aim is understanding, not generality.

Because the $u_\lambda(x)$ are allowed to be complex, one might expect the λ to be complex, too. The Hermiticity of $\hat{L}(x)$ prevents this possibility, however: $\hat{L}^\dagger(x) = \hat{L}(x)$ implies that $\lambda^* = \lambda$. Furthermore, if μ and λ are two unequal eigenvalues and $u_\mu(x) \neq u_\lambda(x)$, then $u_\mu(x)$ and $u_\lambda(x)$ ar orthogonal to one another in essentially the same sense as $\sin(n\pi x(L))$ and $\sin(m\pi x/L)$ are orthogonal, when $m \neq n$. Our starting point in the proof of these claims is the pair of relations

$$[\hat{L}(x)u_\mu(x)]^* u_\lambda(x) = \mu^* u_\mu^*(x)u_\lambda(x) \tag{3.109}$$

and

$$u_\mu^*(x)\hat{L}(x)u_\lambda(x) = \lambda u_\mu^*(x)u_\lambda(x), \tag{3.110}$$

equations similar to those obtained in the proof that $\hat{L}^\dagger(x) = \hat{L}(x)$.

Integrating both sides of (3.109) and (3.110) on x over the interval $[a, b]$ and subtracting the results leads, after rearrangement, to

$$(\lambda - \mu^*)\int_a^b dx\, u_\mu^*(x)u_\lambda(x) = \int_a^b dx\, u_\mu^*(x)\hat{L}(x)u_\lambda(x) - \int_a^b dx\, [\hat{L}(x)u_\mu^*(x)]u_\lambda(x). \tag{3.111}$$

However, the integrals on the RHS of (3.111) are equal by virtue of the Hermiticity of $\hat{L}(x)$ and thus

$$(\lambda - \mu^*)\int_a^b dx\, u_\mu^*(x)u_\lambda(x) = 0. \tag{3.112}$$

This relation is the one from which the desired results follow. First, let $\mu = \lambda$, so that $u_\mu(x) = u_\lambda(x)$. Equation (3.112) is then

$$(\lambda - \lambda^*)\int_a^b dx\, |u_\lambda(x)|^2 = 0. \tag{3.113}$$

As long as $u_\lambda(x)$ is non-trivial, the integral in (3.113) cannot vanish, which means that

$$\lambda = \lambda^*. \tag{3.114}$$

All eigenvalues of the Hermitian differential operator $\hat{L}(x)$ are thus real. Recall the claim in Section (3.2) that Hermiticity of the quantal Hamiltonian operator meant that its eigenvalues, the energies E, were real.

Application of Eq. (3.114) to (3.113) gives

$$(\lambda - \mu)\int_a^b dx\, u_\mu^*(x)u_\lambda(x) = 0. \tag{3.115}$$

Assuming that there is no degeneracy, i.e., that there is only one $u_\lambda(x)$ for each λ and that $u_\lambda(x) \neq u_\mu(x)$ when $\lambda \neq \mu$, then (3.115) is equivalent to

$$\int_a^b dx\, u_\mu^*(x)u_\lambda(x) = 0, \qquad \lambda \neq \mu. \tag{3.116}$$

Equation (3.116) states that, for $\lambda \neq \mu$, $u_\mu(x)$ and $u_\lambda(x)$ are *orthogonal* to one another, with orthogonality defined via the integral relation (3.116). As noted, it is a generalization of (3.43) for the case of the Fourier sine functions.

A further result is that the eigenvalues form a denumerably infinite set (Dettman (1988)). They may be indexed by the integer n:

$$\{\lambda\} \rightarrow \{\lambda_n\}_{n=1}^{\infty};$$

the corresponding eigenfunctions $u_\lambda(x)$ will also be labeled by n:

$$\{u_\lambda(x)\} \rightarrow \{u_n(x)\}_{n=1}^{\infty}.$$

It can be shown that the set $\{u_n(x)\}_{n=1}^{\infty}$ is complete (see, e.g., Dettman (1988)), and that any arbitrary function $w(x)$ obeying the same boundary conditions can be expanded in terms of them:

$$w(x) = \sum_{n=1}^{\infty} C_n u_n(x); \tag{3.117}$$

we shall determine the coefficients C_n shortly. The expansion (3.117) is pointwise convergent.

The relation $\lambda = \lambda^*$ leaves the value of the integral in (3.113) unspecified. By multiplication of each u_λ by the proper constant, it is possible to normalize this integral just as we did in the case of the particle in a quantal box. We assume that this has been done, so that orthogonality and normalization of the $u_n(x)$ can be combined:

$$\int_a^b dx\, u_n^*(x) u_m(x) = \delta_{nm}. \tag{3.118}$$

Let us suppose that the $u_n(x)$ in Eq. (3.117) are normalized. Then, the orthonormality relation (3.118) allows us to determine the C_n in analogy to the evaluation of the a_n and b_n in the Fourier sine series expansions, (3.41) and (3.42). On multiplying both sides of (3.117) by $u_m^*(x)$, integrating on x over the interval $[a, b]$, and interchanging integration and summation, we get

$$C_n = \int_a^b dx\, u_m^*(x) w(x). \tag{3.119}$$

The C_n are often referred to as generalized Fourier coefficients.

Closure, the Dirac delta function, and the scalar-product notation

The preceding analysis has been formal: no differential equations were solved. Although it is formal, the analysis is far from empty; important general properties that hold for any Sturm–Liouville system have been established. We now derive a new property obeyed by the complete sets $\{u_n\}$, which in turn introduces a new quantity, the Dirac delta function $\delta(x - x')$. It can be thought of as the continuously indexed analog of the Kronecker delta δ_{nm} that enters the orthonormality relations (3.118).

Let us re-write Eq. (3.119) as

$$C_n = \int_a^b dx'\, u_n^*(x') w(x'). \tag{3.120}$$

The change to x' as the integration variable is needed because the next step is to substitute (3.120) into (3.117):

$$w(x) = \sum_n \int_a^b u_n^*(x')w(x')\,dx'\,u_n(x)$$

$$= \int_a^b dx' \left(\sum_n u_n^*(x')u_n(x) \right) w(x'), \tag{3.121}$$

where the interchange of summation and integration is assumed valid, as is usual in physics.

The summation in large brackets in (3.121) is a function of x and x'. Following the notation of Dirac, who introduced the delta function, this quantity is denoted $\delta(x - x')$:

$$\sum_n u_n^*(x')u_n(x) = \delta(x - x'), \tag{3.122}$$

in terms of which Eq. (3.121) is

$$w(x) = \int_a^b dx'\,\delta(x - x')w(x'). \tag{3.123}$$

A careful examination of Eq. (3.123) leads to a possibly unexpected conclusion: $\delta(x - x')$ cannot be a normal function, since in the integral it must have the property of selecting the one point $x' = x$ in the range $[a, b]$ and assigning it to $w(x')$. This seems to be absurd behavior, given the fact that the contribution of a single point to an integral is zero. In order to achieve this result, Dirac stated that $\delta(x - x')$ must have the following pathological behavior:

$$\delta(x - x') = \begin{cases} \infty, & x' = x, \\ 0, & x' \neq x, \end{cases} \tag{3.124a}$$

such that

$$\int_a^b \delta(x - x')\,dx' = \begin{cases} 1, & x \in [a, b], \\ 0, & x \notin [a, b], \end{cases} \tag{3.124b}$$

and

$$\int_a^b dx'\,\delta(x - x')w(x') = w(x). \tag{3.124c}$$

It follows from (3.124) that $\delta(x) = \delta(-x)$.

These characteristics more or less ensure that $\delta(x - x')$ is too pathological to qualify as a true function. Instead it may be thought of, along with the integral sign and dx', as an operator acting on certain classes of well-behaved "test" functions. This interpretation, and the mathematical theory of distributions which underlies it, was formulated by L. Schwarz (1950). We shall return to this interpretation, and to other properties of the delta function, in particular its formulation as a limit, in the next chapter. With the properties of $\delta(x - x')$ as specified in Eqs. (3.124), (3.122) is known as a *closure* relation. It expresses the completeness of the Sturm–Liouville eigenfunctions as an expansion set. Similar relations hold for many of the eigenfunctions that occur in quantum mechanics. An abstract version of (3.122) will be encountered in Chapter 4, where closure is referred to as a resolution of the identity.

We remarked earlier that $\delta(x - x')$ can be considered a continuum analog of the

Kronecker delta. This can be understood in the context of Fourier transforms. A wide variety of functions $f(x)$ can be expressed as integrals over k of the exponential function $\psi_k(x) = \exp(ikx)/\sqrt{2\pi}$:

$$f(x) = \sqrt{\frac{1}{2\pi}} \int_{-\infty}^{\infty} dk\, e^{ikx} g(k) \tag{3.125a}$$

$$= \int_{-\infty}^{\infty} dk \psi_k(x) g(k). \tag{3.125b}$$

The Fourier transform of $f(x)$, viz. $g(k)$, is given by the inverse transform:

$$g(k) = \sqrt{\frac{1}{2\pi}} \int_{-\infty}^{\infty} dy\, e^{-iky} f(y) \tag{3.126a}$$

$$= \int_{-\infty}^{\infty} dy\, \psi_k^*(y) f(y); \tag{3.126b}$$

note the presence of $\psi_k^*(x)$ in Eq. (3.126b). $f(x)$ and $g(k)$ are referred to as Fourier-transform pairs, and are discussed and derived from Fourier series in Appendix B. The purpose in introducing them here is that they provide a context for considering the functions $\psi_k(x)$.

The continuously indexed set $\{\psi_k(x)\}$ is complete in the same sense as $\{u_n(x)\}$: the elements $\psi_k(x)$ can be used to expand reasonably well-behaved functions $f(x)$. Since the index is continuous, the expansion is given by the Fourier integral of Eqs. (3.125); the $g(k)$ occurring in these integrals is the analog of the expansion coefficient C_n of (3.117) and (3.120). Just as these latter two equations lead to the closure relation (3.122), so do Eqs. (3.125) and (3.126) lead to an equivalent closure relation for the $\psi_k(x)$, one that also behaves somewhat like a delta-function normalization condition on the orthonormal set $\{\psi_k(x)\}$.

These results are obtained by first substituting (3.126a) into (3.125a):

$$f(x) = \int_{-\infty}^{\infty} dy \left(\frac{1}{2\pi} \int_{-\infty}^{\infty} dk\, e^{ik(x-y)} \right) f(y), \tag{3.127}$$

where the interchange of the k and y integrals, which is assumed valid (as usual), has been made. Comparison with (3.122) shows that the integral in large brackets in (3.127) is also a delta function:

$$\frac{1}{2\pi} \int_{-\infty}^{\infty} dk\, e^{ik(x-y)} = \delta(x - y). \tag{3.128a}$$

While Eq. (3.128) is an integral representation for the delta function, one we shall derive via a limiting procedure in Section 4.4, it is more than that, as noted above. Equation (3.128a) is a closure relation, a claim verified by substituting (3.126b) into (3.125b). This leads to

$$\int_{-\infty}^{\infty} dk\, \psi_k^*(y) \psi_k(x) = \delta(x - y). \tag{3.128b}$$

Since $\{\psi_k(x)\}$ is complete, the analogy with the closure relation (3.122) is evident: Eqs. (3.128) are to be considered as closure relations also.

How are these relations to be understood? First, if $x = x'$, then the integral is infinite; this is the value of $\delta(x)$ when its argument is zero. Second, if $x \neq x'$, the integrand on the LHS of (3.128a) oscillates more and more rapidly as k increases, the effect of which is that the positive and negative portions become equal in magnitude and cancel one another out. In the $\psi_k(x)$ case, this behavior is evident from the structure of $\exp(ikx)$; the beauty of the complete-set nature of the Sturm–Liouville solutions $\{u_n(x)\}$ is that the *same* behavior occurs without specification of the functional form of the solutions.

We close this chapter by introducing a short-hand notation for the types of integrals that have appeared in the preceding analyses. It is a notation widely used in mathematics and will serve for us as a precursor to the Dirac bra and ket notation of Chapter 4.

Let $v(x)$ and $w(x)$ be any two sufficiently well-behaved functions, with $x \in [a, b]$. Then $(v, w) = (w, v)^*$ is defined as

$$(v, w) \equiv \int_a^b dx\, v^*(x) w(x). \tag{3.129}$$

Furthermore, if $\hat{D}(x)$ is a differential operator in x, then $(v, \hat{D}w) = (\hat{D}w, v)^*$ is defined by

$$(v, \hat{D}w) \equiv \int_a^b dx\, v^*(x) \hat{D}(x) w(x), \tag{3.130}$$

where $\hat{D}(x)$ acts on the function to its right. This notation leads to a simple definition of the Hermitian conjugate $\hat{D}^\dagger(x)$ of $\hat{D}(x)$. We assume that $v(x)$, $w(x)$, etc, are members of a class of functions (defined, e.g., by continuity and boundary conditions), and that the action of $\hat{D}(x)$ on any member of that class produces a function in the class. Then, for $v(x)$ and $w(x)$ any pair of functions in the class, the Hermitian conjugate $\hat{D}^\dagger(x)$ is defined via

$$(\hat{D}v, w) = (w, \hat{D}v)^* \equiv (v, \hat{D}^\dagger w). \tag{3.131}$$

An operator $\hat{D}(x)$ is Hermitian if $\hat{D}^\dagger(x) = \hat{D}(x)$, as in Eqs. (3.110) and (3.107).

As an example of this symbolism, Eq. (3.118) reads

$$(u_n, u_m) = \delta_{nm}. \tag{3.132}$$

The δ_{nm} in (3.132) functions in the same way as $\delta_{\alpha\beta}$ does in the scalar product of two unit vectors in three-dimensional space, Eq. (2.6). As a consequence, the parenthesis/comma notation (v, w) is often said to define the scalar product of the functions $v(x)$ and $w(x)$, a nomenclature we shall exploit in the next chapter.

Exercises

3.1 Use the secular-determinant method to show that the solutions to the eigenvalue equation

$$\begin{pmatrix} 1 & 0 \\ 0 & -1 \end{pmatrix} \begin{pmatrix} u_1 \\ u_2 \end{pmatrix} = \lambda \begin{pmatrix} u_1 \\ u_2 \end{pmatrix}$$

are $\lambda = \pm 1$ and the vectors $\binom{1}{0}$ and $\binom{0}{1}$. State which of these vectors belongs to $\lambda = -1$.

3.2 Determine the eigenvalues and corresponding eigenvectors that solve

$$\begin{pmatrix} 0 & 1 \\ 1 & 0 \end{pmatrix}\begin{pmatrix} u_1 \\ u_2 \end{pmatrix} = \lambda\begin{pmatrix} u_1 \\ u_2 \end{pmatrix}.$$

(Up to a phase factor whose absolute value is unity, u_1 and u_2 for each value of λ must obey $|u_1|^2 + |u_2|^2 = 1$.)

3.3 The 3×3 matrix that rotates 3-D vectors about the z-axis through an angle θ is

$$R(\mathbf{e}_3, \theta) = \begin{pmatrix} \cos\theta & -\sin\theta & 0 \\ \sin\theta & \cos\theta & 0 \\ 0 & 0 & 1 \end{pmatrix}.$$

Set up the eigenvalue equation for which $R(\mathbf{e}_3, \theta)$ is the matrix, obtain the three distinct eigenvalues, and determine, up to an overall normalization constant in each case, the corresponding eigenvectors. (Note that the two non-real eigenvalues are complex conjugates of each other. Also, if u_1, u_2 and u_3 are the three components of a normalized eigenvector, then they must obey $\sum_1^3 |u_j|^2 = 1$.)

3.4 (a) The separation constant for the 1-D wave equation was chosen to be $-k^2$, with k real. Show that, if k were replaced by $\beta = k + is$, k and s real, then the boundary conditions require $s = 0$.

(b) Prove the claim made in footnote 5 to this chapter.

3.5 Linear combinations of waves with different wave numbers k_j and frequencies ω_j, all related by $\omega_j/k_j = v$, are known as wave packets. Their general form is represented by the functions $f_\pm(x \pm vt)$ of Eq.(3.11) *et seq.* One of the simpler examples of a wave packet is

$$y(x, t) = A\sin(k_1 x - \omega_1 t) + A\sin(k_2 x - \omega_2 t),$$

defined over all of x-space.

(a) Using Eq. (3.22) evaluate the quantities Θ and Φ that occur in

$$y(x, t) = 2A\cos\Theta\sin\Phi.$$

(b) Let $k_1 = k_0 + \Delta k$, $k_2 = k_0 - \Delta k$, $\omega_1 = \omega_0 + \Delta\omega$, and $\omega_2 = \omega_0 - \Delta\omega$ such that $\omega_0/k_0 = v$, while $\Delta k/k_0 \ll 1$ and $\Delta\omega/\omega_0 \ll 1$.

(i) Show that the relations $\omega_1/k_1 = v = \omega_2/k_2$ are valid approximations.

(ii) Set $x = 0$ and sketch on the same graph the t-dependences of $\cos\Theta$ and $\sin\Phi$ for $0 \leqslant t \leqslant T = 2\pi/\omega$.

(iii) Set $t = 0$ and sketch on the same graph the x-dependences of $\cos\Theta$ and $\sin\Phi$ for $0 \leqslant x \leqslant \lambda = 2\pi/k$.

(iv) The preceding graphical analysis should indicate that $y(x, t)$ is a product of a rapidly varying wave form and a slowly varying wave form or envelope. The envelope propagates with the packet's group velocity v_{gr}, which is generally not equal to v (or v_{prop}, the phase velocity). Determine v_{gr} and evaluate it when $\omega = \omega(k)$, $\Delta\omega \to d\omega$, and $\Delta k \to dk$.

3.6 When a stretched string is lifted a distance d ($\ll L$) at its center and then released with no initial velocity, the Fourier expansion of its initial shape is

$$y(x, 0) = \sum_{n=0}^{\infty}(-1)^n \frac{3d}{\pi^2(2n+1)^2}\sin\left(\frac{(2n+1)\pi x}{L}\right).$$

Compare the approximate shapes obtained by truncating the preceding sum at the $n = 0$, the $n = 1$, and the $n = 2$ terms with the exact shape.

3.7 Find the coefficients in the Fourier sine series representation when a stretched string is lifted a distance d ($\ll L$) at the point $x = L/3$ and released with zero initial velocity. Determine how well taking the first three terms of the series approximates the initial shape of the string.

3.8 A stretched string is released with zero velocity from the initial shape

$$y(x, 0) = d\frac{x(L-x)}{L^2}$$

Determine the coefficients a_u in the expansion (3.48) and compare the exact initial shape with the first two and the first three terms of the truncated Fourier series.

3.9 Use the analysis of Appendix B to determine the coefficients a_n and b_n of Eq. (B3) in the Fourier series expansion of

$$f(x) = x, \qquad -L \leqslant x \leqslant L.$$

3.10 When $x \in [-L, L]$ and $f(-L) = f(L)$, with $f(x)$ sufficiently well-behaved, the Fourier coefficients are given by Eq. (B5). Suppose that $x \to \theta$, $[-L, L] \to [-\pi, \pi]$, and $f(\theta = -\pi) = f(\theta = \pi)$. Obtain the general form that Eq. (B3) now takes and determine the new expressions for a_n and b_n.

3.11 Use the result of Exercise 3.10 to find the Fourier expansion of

$$f(\theta) = \pi, \qquad \theta \in [-\pi, \pi].$$

3.12 The following are wave functions for various one-particle quantal systems. Determine the real, positive constant by which each must be multiplied in order that it be normalized over its domain of definition (reference to tables of integrals may be needed).

(a) $\psi(x) = e^{-\lambda|x|}, \quad -\infty \leqslant x \leqslant \infty$,
(b) $\psi(x) = e^{-x^2/(2\lambda^2)}, \quad -\infty \leqslant x \leqslant \infty$,
(c) $\psi(x) = xe^{-x^2/(2\lambda^2)}, \quad 0 \leqslant x \leqslant \infty$,
(d) $\psi(x, t) = [\mathrm{b}^2 + (x - vt)^2]^{-1}, \quad -\infty \leqslant x \leqslant \infty$,
(e) $\psi(\mathbf{r}) = e^{-r/\lambda}$, $0 \leqslant r \leqslant \infty$, $0 \leqslant \theta \leqslant \pi$, $0 \leqslant \phi \leqslant 2\pi$.

3.13 A particle is injected into a 1-D quantal box located between $x = 0$ and $x = L$. For each of the following wave functions, determine the probability $P(t)$ that the particle will be found in the region $L/4 \leqslant x \leqslant 3L/4$:

(a) $\Psi(x, t) = \Psi_n(x, t)$,
(b) $\Psi(x, t) = [\Psi_2(x, t) - \Psi_3(x, t)]/\sqrt{2}$, where $\Psi_n(x, t)$ is given by Eq. (3.91).

If either result depends on time, sketch the behavior of $P(t)$ over the relevant period T of the " motion," i.e., over the range $0 \leqslant t \leqslant T$.

3.14 The quantal box of Section 3.3 is located between $x = 0$ and $x = L$. In this problem the box is located between $x = -L$ and $x = 0$, with all other conditions remaining unchanged. Denoting the eigenfunctions and eigenvalues in this case by $\tilde{\psi}_n(x)$ and $\tilde{E}_n$, determine them both and relate them to $\psi_n(x)$ and E_n, respectively. In view of these relations and that between the two boxes, what symmetry properties of this system, if any, can you deduce, and what is the physical/mathematical basis for your deduction?

3.15 (a) Evaluate the integral I_{nm} for $m = n$ and $m \neq n$:

$$I_{nm} = \int_0^L dx\, \psi_n^*(x)\hat{H}(x)\psi_m(x) = \int_0^L dx\, \psi_n^*(x)\hat{K}(x)\psi_m(x),$$

where $\psi_n(x)$ and $\hat{H}(x)$ are the eigenfunction and Hamiltonian operator for the 1-D box located between $x = 0$ and $x = L$. Your result should be related to an eigenvalue of $\hat{H}$. The integral is related to the "expected" or average value of $\hat{H}$, in particular, when $m = n$, to the average value of $\hat{H}$ when the state of the system is represented by the wave function $\psi_n(x)$.

(b) Carry out the same evaluation when $\hat{H}(x)$ is replaced by $\hat{P}(x) = -i\hbar\, d/dx$, the momentum operator. For which values of n is the momentum-expectation value non-zero?

3.16 The 2-D quantal box is a rectangular hole bounded by infinitely high and wide walls on the four sides of the rectangle. The potential is

$$V(x, y) = \begin{cases} 0, & 0 \leqslant x \leqslant L_1 \text{ and } 0 \leqslant y \leqslant L_2, \\ \infty, & \text{otherwise,} \end{cases}$$

while for a particle of mass m in the box, the kinetic energy operator is

$$\hat{K}(x, y) = -\frac{\hbar^2}{2m}\left(\frac{d^2}{dx^2} + \frac{d^2}{dy^2}\right).$$

(a) The time-independent Schrödinger equation when the particle is in the box is

$$\hat{K}(x, y)\psi(x, y) = E\psi(x, y), \qquad x \text{ and } y \text{ in the box,}$$

and its solutions can be expressed as

$$\left.\begin{aligned} \psi(x, y) \to \psi_{n,m}(x, y) &= \psi_n^{(1)}(x)\psi_m^{(2)}(y) \\ E \to E_{n,m} &= E_n^{(1)} + E_m^{(2)} \end{aligned}\right\} \quad n, m = 1, 2, 3, \ldots.$$

Use the appropriate boundary conditions to determine $\psi_n^{(1)}(x)$, $\psi_m^{(2)}(y)$, $E_n^{(1)}$, and $E_m^{(2)}$, analogous to $\psi_n(x)$ and E_n of Section 3.3.

(b) The total energy $E_{n,m}$ changes every time n or m changes.

(i) Determine the spacings between the levels $E_{n+1,m}$ and $E_{n,m}$, and between $E_{n,m+1}$ and $E_{n,m}$.

(ii) Now let $L_1 = L_2 = L$. Show that in general more than one pair $(\psi_n^{(1)}, \psi_m^{(2)})$ can correspond to the same value of the total energy, a phenomenon known as degeneracy: the level $E_{n,m}$ is degenerate. Define $N = n + m$ and determine the degeneracy in terms of N (e.g., N or N^2 or $\ln N$, etc.).

(iii) Still assuming that $L_1 = L_2 = L$, determine which values of n and m yield an $E_{n,m}$ equal to any one of the E_n for the 1-D box. How are the wave functions for these two cases related, if at all?

(c) The probability $P_{n,m}(x, y)$ that the particle will be found at the "point" x, y is

$$P_{n,m}(x, y) = |\psi_{n,m}(x, y)|^2 dx\, dy.$$

(i) Show that $P_{n,m}(x, y) = P_n(x)P_m(y)$, where P_n and P_m are the individual probabilities, as in Sections 3.2 and 3.3.

(ii) Normalize your solutions so that the probability of finding the particle somewhere in the box is unity. (To do this, you will need to normalize each probability separately. Why?)

(iii) What is the probability that the particle will be at some position $x \in [0, L]$?

3.17 Consider a particle of mass m whose motion is limited to the x-axis and on which no potential acts. Its Hamiltonian is the kinetic energy operator $\hat{K}(x) = \hat{P}^2(x)/(2m)$, with $\hat{P}(x) = -i\hbar\, d/dx$ being the momentum operator.

(a) Show that $\psi(x) = N \exp(ikx)$, $k > 0$, is an eigenfunction of the momentum operator and determine the corresponding momentum eigenvalue.

(b) Show that $\psi(x)$ is also an eigenfunction of the kinetic-energy operator and determine the corresponding energy eigenvalue.

(c) The particle described by $\psi(x)$ is now confined to the region $0 \leqslant x \leqslant L$, subject to the "periodic" boundary condition $\psi(x = 0) = \psi(x = L) \neq 0$. Imposition of this boundary condition has the effect of "quantizing" k: $k \to k_n$. Determine the allowed k_n, expressing your result in terms of L and n, $n = 1, 2, 3 \ldots$. What are the associated quantized energy eigenvalues?

(d) The effect of the periodic boundary condition on $\psi(x)$ is to change it to $\psi_n(x) = N_n \exp(ik_n x)$.

(i) Show that $\int_0^L dx\, \psi_n^*(x)\psi_m(x) = 0$, $m \neq n$.

(ii) Set $n = m$ and find a real, positive value of N_n such that $\int_0^L dx\, |\psi_n(x)|^2 = 1$. (With this choice of N_n, $\{\psi_n(x)\}$ is a complete orthonormal set on the interval [0, L] obeying periodic boundary conditions.)

3.18 A particle of mass μ is constrained to move in a circle of radius a in the $x-y$ plane. No additional potential acts on the particle. Its Hamiltonian is given by

$$\hat{H}(x) = -\frac{\hbar^2}{2\mu a^2}\frac{d^2}{d\phi^2},$$

where ϕ is the azimuthal angle, $\phi \in [0, 2\pi]$.

(a) $\hat{L}_z$, the z-component of the quantal orbital angular-momentum operator, is given by

$$\hat{L}_z = -i\hbar\frac{d}{d\phi}.$$

Find the eigenfunctions $\psi_m(\phi)$ and the eigenvalues of $\hat{L}_z$ (each indexed by the symbol m, whose values you are to obtain) that correspond to the periodic boundary condition $\psi_m(\phi) = \psi_m(\phi + 2\pi)$.

(b) For each $\psi_m(\phi)$ determine a real, positive normalization constant such that $\{\psi_m(\phi)\}$ is orthonormal on the interval $[0, 2\pi]$, i.e., so that $\int_0^L \psi_n^*(\phi)\psi_m(\phi)\, d\phi = \delta_{nm}$ is satisfied.

(c) Show that $\psi_m(\phi)$ is also an eigenfunction of $\hat{H}$ and determine the associated eigenvalue E_m and the degree of degeneracy (number of eigenfunctions) corresponding to the same E_m. (It is helpful in proving that $\psi_m(\phi)$ is an eigenfunction of $\hat{H}$ to express $\hat{H}$ in terms of $\hat{L}_z$.)

3.19 (a) If Eq. (3.100) is satisfied, then $\hat{D}(x)$ is Hermitian. Let $D(x)$ first be equal to $\hat{P}(x) = -i\hbar\, d/dx$ and next be equal to $\hat{K}(x) = \hat{P}^2(x)/(2m)$. Determine, for $u(x) = v(x) = \psi(x)$, whether each of these operators is Hermitian, choosing as $\psi(x)$ the functions (b), then (c), and finally (d) of Exercise 3.12.

(b) Carry out the same analysis for $\psi(x) = \exp(ikx)$, with k real. If, for either choice of

$\hat{D}(x)$, the combination of operator plus wave function is not Hermitian, explain why.

3.20 Replace the RHS of Eq. (3.108) by $\lambda\rho(x)u_\lambda(x)$, as in footnote 16, and
   (a) show that $\lambda = \lambda^*$ still obtains;
   (b) show that the new orthonormality condition, analogous to Eq. (3.118), now involves $\rho(x)$;
   (c) determine whether (and if so, how) the closure relation is changed.

3.21 Fourier transforms play a significant role in quantum theory. The "momentum representation" of a state whose coordinate-space wave function is $\psi_\alpha(x)$ is denoted $\varphi_\alpha(k)$, related to $\psi_\alpha(x)$ by

$$\varphi_\alpha(k) = \sqrt{\frac{1}{2\pi}} \int_{-\infty}^{\infty} dk\, e^{-ikx} \psi_\alpha(x).$$

   (a) Use the properties of the Dirac delta function to show that, if $\psi_\alpha(x)$ is normalized in coordinate space, then $\int_{-\infty}^{\infty} dk\, |\varphi_\alpha(k)|^2 = 1$. (This result validates the interpretation of $|\varphi_\alpha(k)|^2$ as a probability density in k (or momentum) space (see Chapter 5).)
   (b) Determine $\varphi_\alpha(k)$ for the following choices of $\psi_\alpha(x)$:
      (i) $\psi_\alpha(x) = \psi_n(x)$, the 1-D-box eigenfunction,
      (ii) $\psi_\alpha(x) = N_1 x \exp[-x^2/(2\lambda^2)]$, with $-\infty \leqslant x \leqslant \infty$, $\lambda > 0$. (Note: first determine the real, positive constant N_1 that normalizes $\psi_\alpha(x)$.)

3.22 An infinitely high and wide wall occupies the 1-D domain $x \leqslant 0$. A particle moving along the positive x-axis is described by the wave function

$$\psi(x) = \begin{cases} Ne^{-x/\lambda}\sin(k_0 x), & x > 0, \\ 0, & x \leqslant 0. \end{cases}$$

   (a) Evaluate the real, positive normalization constant N and sketch $\psi(x)$ as a function of x.
   (b) Determine the momentum-space wave function $\varphi(k)$, given by

$$\varphi(k) = \sqrt{\frac{1}{2\pi}} \int_{-\infty}^{\infty} dk\, e^{-ikx} \psi(x),$$

   expressing your answer as a difference of two complex ratios, one involving $k - k_0$, the other involving $k + k_0$.
   (c) Now assume that λ is very large ($k_0\lambda \gg 1$), so that, for k in the vicinity of k_0, $\varphi(k)$ is approximately equal to that ratio of part (b) which contains the difference $k - k_0$ in the denominator. Using the latter approximation, sketch $|\varphi(k)|^2$ as a function of k. Your result should be peaked at $k = k_0$ and symmetric about the peak value, and have a width at half its maximum value related to $1/\lambda$. Denoting this width by δk, relate λ to the width (or uncertainty δx) of $\psi(x)$; your result for $\delta k\, \delta x$ should be of order unity. Making use of the fact that $p = \hbar k$ is the quantal momentum, the last result is equivalent to an uncertainty-type relation $\delta p\, \delta x > \hbar$, analogous to the results in Chapters 2 and 9.

3.23 In Section 3.1, it was shown that the wave-equation operator $\hat{D}^2$ factors into a product of operators, Eq. (3.8). A similar factorization arises for some Sturm–Liouville operators that occur in quantum mechanics. An example is the 1-D oscillator, for which $E \to E_n = (n + \frac{1}{2})\hbar\omega$, while the corresponding eigenfunction $\psi_n(x)$ obeys

$$\frac{d^2}{dx^2}\psi_n(x) + (2n + 1 - x^2)\psi_n(x) = 0.$$

(a) This equation can be put in the form $\hat{D}_-(x)\hat{D}_+(x)\psi_n(x) = -2n\psi_n(x)$.
   (i) Determine $\hat{D}_+(x)$ and $\hat{D}_-(x)$.
   (ii) Show that $\hat{D}_-(x)\hat{D}_+(x) \neq \hat{D}_+(x)\hat{D}_-(x)$ and evaluate the *commutator*

$$[\hat{D}_-(x), \hat{D}_+(x)] \equiv \hat{D}_-(x)\hat{D}_+(x) - \hat{D}_+(x)\hat{D}_-(x).$$

(This particular non-vanishing of the commutator is typical for many pairs of quantal operators.)

(b) Define $f_n(x) = \hat{D}_-(x)\psi_n(x)$.
   (i) Use the commutator relation to show that

$$\hat{D}_-(x)\hat{D}_+(x)f_n(x) = -2(n + 1)f_n(x).$$

   (ii) Use the preceding result to determine which one of the oscillator wave functions is proportional to $f_n(x)$.

(c) The preceding proportionality means that only the equation for $\psi_0(x)$ needs to be solved; all other $\psi_n(x)$ can be obtained using $\hat{D}_-(x)$.
   (i) Determine the functional form first of $\psi_0(x)$ from $\hat{D}_-(x)D_+(x)\psi_0(x) = 0$, and then of a second eigenfunction via $\hat{D}_-(x)\psi_0(x)$. (Compare your answers with the functions in (b) and (c) of Exercise 3.12.)

(d) In analogy to $f_n(x)$, the function $g_n(x) = \hat{D}_+(x)\psi_n(x)$ is also proportional to one of the oscillator eigenfunctions. Determine which one it is. (This method of factorization will be used in Chapter 6 to solve the oscillator problem.)

4

Mathematical Background

The quantities in physics that can be measured are conventionally denoted *observables*: entities that can be observed (directly or indirectly) and to which a value in appropriate units can be assigned. Examples include position, momentum, and energy. Unlike in classical physics, the concept of an observable in quantum theory possesses a significance beyond that of the measurable: to each observable there corresponds a unique operator in an abstract, linear vector space. These operators (and others) govern all aspects of quantum systems.

Quantum operators act on abstract vectors in the linear space. Especially important are the eigenvectors and eigenvalues of these operators, in particular because some of the eigenvectors are the *quantum states* mentioned in the preceding chapters. Furthermore, for any operator that is the mathematical image of an observable, its *eigenvalues constitute the entire set of values that measurement of the observable can yield.* Finally, from the quantum states one can obtain the probabilities for the occurrence of events in the microscopic world, for example, the most likely value that measurement of an observable will yield.

The concepts of operators and eigenvalue problems, as well as those of quantal probabilities, of Hermiticity, and of complete sets, were encountered and exemplified in the preceding chapter using the language of differential equations. While much of quantum mechanics could be presented in terms of this language alone, certain fundamental concepts (spin, for instance) cannot be, and it eventually becomes necessary to consider quantum mechanics as a theoretical framework embedded in that abstract linear vector space known as a Hilbert space. The imaging of physical observables by abstract operators in a linear space is an essential ingredient of quantum mechanics. Indeed, this imaging is so basic that it has been incorporated into the group of fundamental postulates of the subject. One cannot, therefore, obtain a proper understanding of quantum mechanics without having some grasp of linear algebra, i.e., of that branch of mathematics dealing with linear operators.[1] The aim of this chapter is to provide enough linear algebra to make quantum theory as it is presented in later chapters comprehensible. The treatment is quantitative but not especially rigorous. References to further material are cited, should the reader wish to pursue the subject beyond what is presented here.

[1] Ordinarily, linear algebra refers to finite-dimensional spaces, but, as we use it here, it includes infinite-dimensional spaces as well.

Our goal is to have the reader feel comfortable with, as well as have some understanding of, Hilbert-space concepts. The very fact that the dimension of Hilbert space is infinite might seem, however, to present an insurmountable barrier of discomfort! We therefore start small – with ordinary 3-D vectors, then move on to their N-dimensional extensions, and finally allow N to become infinite. It might help any hesitant readers to know that the string and quantum-box eigenvalue problems solved in Chapter 3 were actually Hilbert-space eigenvalue problems, and that the complete sets of Sturm–Liouville solutions $\{u_n(x)\}$ can be thought of as an infinite set of unit vectors in Hilbert space. Hence, from this point of view, the expansion (3.118) is of a Hilbert-space vector $w(x)$ in terms of a set of unit vectors $u_n(x)$ and the components C_n, which are given by the *scalar product* (3.119) of $u_n(x)$ and $w(x)$.

Much of what we will present involves the learning of a new point of view in the same way, perhaps, that the character in Molière's play had to learn that it was prose he had always been speaking. Illustrative examples will often involve small (2×2 or 3×3) matrices or differential operators: the goal, as noted, is familiarity and comprehension, not fear and consternation.

4.1. Vectors and matrices in three dimensions

Coordinate systems and orthonormal bases

We start with the 3-D case ($N = 3$) because of its familiarity as well as its relative simplicity. Let $\mathbf{b}$ be an arbitrary 3-D vector, as in Fig. 4.1. Its components with respect to (w.r.t.) the x, y, and z axes of coordinate system S – axes denoted by the unit vectors $\mathbf{e}_1$, $\mathbf{e}_2$, and $\mathbf{e}_3$, respectively (see Fig. 4.2(a)) – are b_1, b_2, and b_3 (Fig. 4.2(b)):

$$\mathbf{b} = \sum_{j=1}^{3} b_j \mathbf{e}_j. \tag{4.1}$$

The set $\{\mathbf{e}_i\}_{i=1}^3$ is orthonormal, that is, the $\mathbf{e}_i$ are both perpendicular to one another and of unit length:

$$\mathbf{e}_j \cdot \mathbf{e}_k = \delta_{jk}, \tag{4.2}$$

where δ_{jk} is the Kronecker delta and the heavy dot signifies the usual scalar product of two 3-D vectors. One obtains the component b_j by forming the scalar product of $\mathbf{b}$ and $\mathbf{e}_j$, i.e., by *projecting* $\mathbf{b}$ onto $\mathbf{e}_j$:

$$b_j = \mathbf{e}_j \cdot \mathbf{b}. \tag{4.3}$$

The set of unit vectors $\{\mathbf{e}_i\}_{i=1}^3$ is a particular *basis* in 3-D space: it *spans* the space, or is *complete*, in that the $\mathbf{e}_i$ are linearly independent and (4.1) holds. Note that, if the $\{\mathbf{e}_j\}$

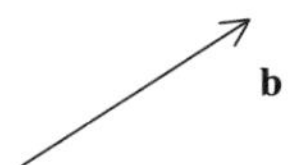

Fig. 4.1 The usual pictorial representation of a 3-D vector $\mathbf{b}$. The arrow defines the direction and the length represents the magnitude of $\mathbf{b}$.

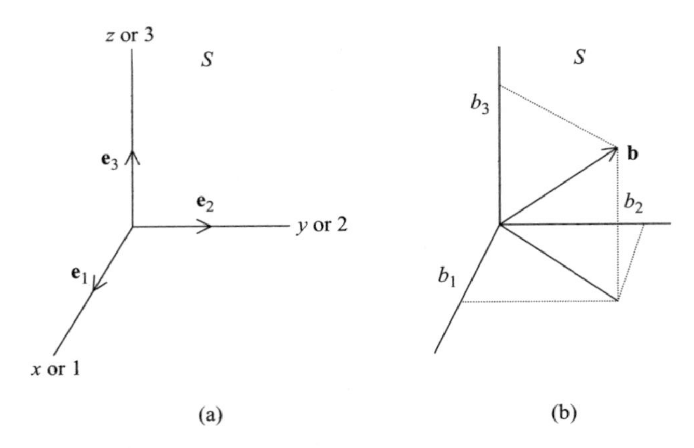

Fig. 4.2 (a) The right-handed coordinate system S, with unit vectors $\mathbf{e}_i$, $i = 1, 2, 3$, along the axes $i = 1, 2, 3$. (b) The components b_i of $\mathbf{b}$ in S.

were not orthogonal, then (4.3) would not hold, and the $\{b_j\}$ would be related by coupled equations.

While $\mathbf{b}$ is independent of the coordinate system S, once a set of vectors $\{\mathbf{e}_k\}$ defining the coordinate axes of S is specified, we say that $\mathbf{b}$ is *equivalent* to the ordered triplet b_1, b_2, b_3. The set $\{\mathbf{e}_k\}$ and the coordinate system S are not unique. There is an infinite number of others, all equivalent, which can be used to express $\mathbf{b}$ via components. Consider the system S$'$, defined by the basis of orthonormal vectors $\{\mathbf{e}'_k\}$ obtained from S by a rotation through the angle θ about $\mathbf{e}_3$ as in Fig. 4.3. The components of $\mathbf{b}$ in S$'$ are denoted $\tilde{b}'_k$:

$$\mathbf{b} = \sum_k \tilde{b}'_k \mathbf{e}'_k = \sum_k b_k \mathbf{e}_k. \tag{4.4}$$

Orthonormality of the bases leads to the standard result

$$\mathbf{b} \cdot \mathbf{b} = b^2 = |\mathbf{b}|^2 = \sum_k \tilde{b}'^2_k = \sum_k b_k^2. \tag{4.5}$$

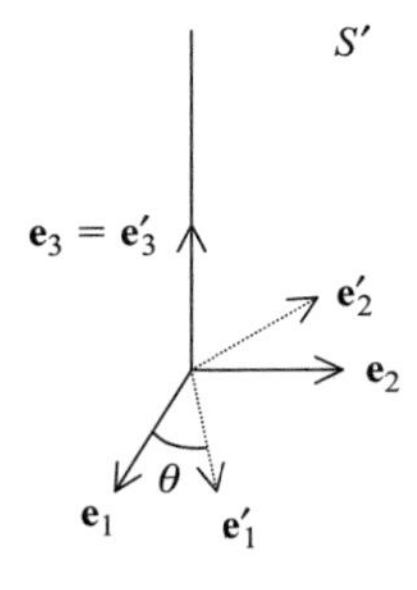

Fig. 4.3 Rotation of the coordinate vectors $\mathbf{e}_1$ and $\mathbf{e}_2$ in S through an angle θ about $\mathbf{e}_3$ to form the new coordinate system S$'$.

Rotation of the coordinate system has resulted in new components and unit vectors; **b** remains unchanged. This is deemed a *passive rotation*. We have employed it to illustrate the non-uniqueness of the coordinate system. The alternate viewpoint is that of the *active rotation*, in which S remains fixed and **b** is rotated in the opposite direction to become the new vector $\mathbf{b}'$, as in Fig. 4.4:

$$\mathbf{b}' = \sum b'_k \mathbf{e}_k, \tag{4.6a}$$

with

$$b'_k = \mathbf{e}_k \cdot \mathbf{b}'. \tag{4.6b}$$

The relation (4.5) also holds for the b'_k:

$$\sum_k b'^2_k = \sum_k b^2_k. \tag{4.7}$$

From now on, rotations will be meant in the active sense: stationary coordinate system, mobile vector.

Matrix representations

Given a basis, any vector **b** can be considered as the ordered triplet b_1, b_2, b_3 of components w.r.t. that basis. Such a representation is easily extended to the case of a vector in a space of N dimensions: in this case one writes **b** as an ordered N-tuplet: $\mathbf{b} = (b_1, b_2, b_3, b_4, \ldots, b_N)$. However, there is an alternate representation that is much more useful, one that we shall utilize extensively. This is the representation in terms of N-rowed, single-column matrices, whose elements are the N components of **b** in a relevant basis. Such matrices are known as column vectors. This representation is introduced here for the case $N = 3$.

From classical mechanics (e.g., Marion (1970), Goldstein (1980), or Fowles and Cassidy (1990)), we know that the components of a rotated vector are related to those of the unrotated vector by the elements of the 3×3 rotation matrix, here denoted R_{kj}:

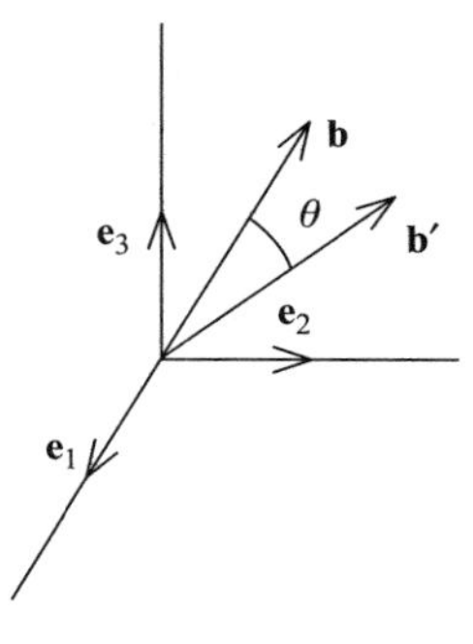

Fig. 4.4 Active rotation of the vector **b** through the angle θ to its new position $\mathbf{b}'$.

$$b'_k = \sum_j R_{kj} b_j, \tag{4.8}$$

or, in detail:

$$\begin{aligned} b'_1 &= R_{11} b_1 + R_{12} b_2 + R_{13} b_3, \\ b'_2 &= R_{21} b_1 + R_{22} b_2 + R_{23} b_3, \\ b'_3 &= R_{31} b_1 + R_{32} b_2 + R_{33} b_3, \end{aligned} \tag{4.9}$$

where the R_{kj} are numbers that depend on the rotation angle θ.

As a simple example, (4.9) takes the following form when the rotation is about the z ($\hat{\mathbf{e}}_3$)-axis through the angle θ:

$$\begin{aligned} b'_1 &= b_1 \cos\theta - b_2 \sin\theta, \\ b'_2 &= b_1 \sin\theta + b_2 \cos\theta, \\ b'_3 &= b_3. \end{aligned} \tag{4.10}$$

Returning to the general case, (4.9) can be written in matrix/single-column-vector form:

$$\begin{pmatrix} b'_1 \\ b'_2 \\ b'_3 \end{pmatrix} = \begin{pmatrix} R_{11} & R_{12} & R_{13} \\ R_{21} & R_{22} & R_{23} \\ R_{31} & R_{32} & R_{33} \end{pmatrix} \begin{pmatrix} b_1 \\ b_2 \\ b_3 \end{pmatrix}. \tag{4.11}$$

The notational replacements to which this corresponds are

$$\mathbf{b} = \sum_j b_j \mathbf{e}_j \rightarrow \begin{pmatrix} b_1 \\ b_2 \\ b_3 \end{pmatrix}, \tag{4.12}$$

with the unit vectors $\mathbf{e}_1$, $\mathbf{e}_2$, and $\mathbf{e}_3$ now being given by

$$\mathbf{e}_1 = \begin{pmatrix} 1 \\ 0 \\ 0 \end{pmatrix}, \qquad \mathbf{e}_2 = \begin{pmatrix} 0 \\ 1 \\ 0 \end{pmatrix}, \qquad \mathbf{e}_3 = \begin{pmatrix} 0 \\ 0 \\ 1 \end{pmatrix}. \tag{4.13}$$

That is, $\mathbf{e}_i$ is represented as a column vector with all row elements equal to zero but for row i, whose value is unity.

Given the matrix representation, we need to address the technical problem of how to express the usual scalar product of two vectors. Suppose that the two vectors are **a** and **b**. In component form, they are given by $\mathbf{a} = \sum_j a_j \mathbf{e}_j$ and $\mathbf{b} = \sum_j b_j \mathbf{e}_j$, and the scalar product is

$$\mathbf{a} \cdot \mathbf{b} = \sum_j a_j b_j. \tag{4.14}$$

In the matrix representation, the vectors are given by

$$\mathbf{a} = \begin{pmatrix} a_1 \\ a_2 \\ a_3 \end{pmatrix}, \qquad \mathbf{b} = \begin{pmatrix} b_1 \\ b_2 \\ b_3 \end{pmatrix}, \tag{4.15}$$

and we must now produce (4.14) from this representation. The procedure is as follows.

One first introduces the transpose $\mathbf{a}^{\mathrm{T}}$ of $\mathbf{a}$, viz.,

$$\mathbf{a}^{\mathrm{T}} = (a_1 \ \ a_2 \ \ a_3), \tag{4.16}$$

which is a single-row, three-column matrix, and then one matrix-multiplies the column vector $\mathbf{b}$ by the row vector $\mathbf{a}^{\mathrm{T}}$ from the left to achieve the desired result:

$$\begin{aligned} \mathbf{a} \cdot \mathbf{b} \to \mathbf{a}^{\mathrm{T}} \cdot \mathbf{b} &= (a_1 \ \ a_2 \ \ a_3) \begin{pmatrix} b_1 \\ b_2 \\ b_3 \end{pmatrix} \\ &= \sum a_j b_j. \end{aligned} \tag{4.17}$$

Note that the order in which the matrices $\mathbf{a}^{\mathrm{T}}$ and $\mathbf{b}$ are multiplied is crucial: had the order been reversed, a 3×3 matrix rather than a number would have been obtained, i.e.,

$$\begin{aligned} \mathbf{b} \cdot \mathbf{a}^{\mathrm{T}} &= \begin{pmatrix} b_1 \\ b_2 \\ b_3 \end{pmatrix} (a_1 \ \ a_2 \ \ a_3) \\ &= \begin{pmatrix} b_1 a_1 & b_1 a_2 & b_1 a_3 \\ b_2 a_1 & b_2 a_2 & b_2 a_3 \\ b_3 a_1 & b_3 a_2 & b_3 a_3 \end{pmatrix}. \end{aligned}$$

The transpose of a column vector is a row vector. In the case of an $n \times m$ matrix, say

$$A = \begin{pmatrix} A_{11} & A_{12} & \cdots & A_{1m} \\ A_{21} & A_{22} & \cdots & A_{2m} \\ \vdots & & & \\ A_{n1} & A_{n2} & \cdots & A_{nm} \end{pmatrix},$$

the transpose of A is obtained by interchanging rows and columns. This is the same as reflecting about the main diagonal. Either way, the result is

$$A^{\mathrm{T}} = \begin{pmatrix} A_{11} & A_{21} & \cdots & A_{n1} \\ A_{12} & A_{22} & \cdots & A_{n2} \\ \vdots & & & \\ A_{1m} & A_{2m} & \cdots & A_{nm} \end{pmatrix}.$$

In a short-hand notation, $A \Leftrightarrow A_{jk}$, $A^{\mathrm{T}} \Leftrightarrow A_{kj}$, i.e., $(A^{\mathrm{T}})_{jk} = A_{kj}$.

The abstract rotation operator

The 3-D Euclidean space of the preceding subsections is an example of a linear vector space.[2] Because of its familiarity, it would appear not to be an "abstract" space at all. Despite this and the specificity of its elements, viz., 3-D vectors, one can still introduce abstract operators into it. Abstract here means that the operators are not yet represented in any way. For example, they are not derivatives, exponentials, or matrices. Yet they can be made to have a specific form or representation. As a precursor to the introduction of abstract operators in N-D linear vector spaces, in Hilbert space, and in quantum theory, we use the 3-D case to consider an abstract operator and its matrix representation.

The operator we choose to examine is the rotation operator $\hat{R}$, which carries the vector $\mathbf{b}$ into the vector $\mathbf{b}'$, as in Fig. 4.4. Note well that the abstract *operator* (indicated by a capital letter overlaid with a carat) is not the rotation *matrix* of Eqs. (4.8) and (4.11), even though the main symbol – R – is the same. All rotations depend ultimately on only two quantities: $\mathbf{n}$, the axis about which the rotation occurs, and θ, the angle of rotation. Therefore, $\hat{R} = \hat{R}(\mathbf{n}, \theta)$ and we may write

$$\mathbf{b}' = \hat{R}(\mathbf{n}, \theta)\mathbf{b}. \tag{4.18}$$

Equation (4.18) is an example of a relationship between abstract quantities. The vectors are abstract, in spite of the possibility of depicting them as directed line segments in figures like Fig. 4.3, and $\hat{R}(\mathbf{n}, \theta)$ is even "more" abstract, since it is not susceptible to depiction. $\hat{R}(\mathbf{n}, \theta)$ is perhaps best thought of as the symbol for an instruction: rotate by an angle θ about the direction $\mathbf{n}$. This transforms $\mathbf{b}$ into $\mathbf{b}'$. $\hat{R}(\mathbf{n}, \theta)$ is thus a transformation, one that is linear; it is this aspect of its operators that causes 3-D Euclidean space to be a linear vector space.

Let us now see how to convert $\hat{R}(\mathbf{n}, \theta)$ from an abstract operator into its realization or representation as a matrix. This is done by introducing the basis/component form for $\mathbf{b}'$ and for $\mathbf{b}$ into (4.18):

$$\mathbf{b}' = \sum_j b'_j \mathbf{e}_j = \sum_j \hat{R}(\mathbf{n}, \theta) b_j \mathbf{e}_j. \tag{4.19}$$

The next step is a crucial one. By definition, $\hat{R}$ acts only on vectors, changing their directions but not their magnitudes. Therefore, $\hat{R}$ can have no effect on the numbers (components) b_j which appear to the right of it in (4.19). This is strictly analogous to the derivative operator acting only on $f(x)$ in the expression $Cf(x)$, where C is a constant. As a consequence of this, Eq. (4.19) can be re-expressed in either of two equivalent forms:

$$\begin{aligned} \sum_j b'_j \mathbf{e}_j &= \sum_j b_j [\hat{R}(\mathbf{n}, \theta)\mathbf{e}_j] \\ &= \sum_j [\hat{R}(\mathbf{n}, \theta)\mathbf{e}_j] b_j. \end{aligned} \tag{4.20}$$

In either form, $\hat{R}(\mathbf{n}, \theta)\mathbf{e}_j$ is a quantity having the character of a vector.

[2] A general definition of linear vector spaces is given, e.g., by Dettman (1988).

The final step in this analysis is to project both sides of (4.20) onto $\mathbf{e}_k$. This yields

$$b'_k = \sum_j [\mathbf{e}_k \cdot \hat{R}(\mathbf{n}, \theta)\mathbf{e}_j] b_j, \tag{4.21}$$

where the vector nature of $\hat{R}(\mathbf{n}, \theta)\mathbf{e}_j$ renders the scalar product in the square brackets meaningful. In other words, $\mathbf{e}_k \cdot \hat{R}(\mathbf{n}, \theta)\mathbf{e}_j$ is a scalar that depends on the direction ($\hat{\mathbf{n}}$) and angle (θ) of rotation as well as on the two subscripts k and j. In a shorthand notation, the term in square brackets becomes

$$\mathbf{e}_k \cdot \hat{R}(\mathbf{n}, \theta)\mathbf{e}_j \equiv R_{kj}(\mathbf{n}, \theta). \tag{4.22}$$

Since $R_{kj}(\mathbf{n}, \theta)$ is a number, the carat appearing over the operator is no longer appropriate and thus does not appear in (4.22). $R_{kj}(\mathbf{n}, \theta)$ is a *matrix representative* of $\hat{R}(\mathbf{n}, \theta)$.

Suppressing the $\mathbf{n}, \theta$ dependence and substituting (4.22) into (4.21) gives

$$b'_k = \sum_j R_{kj} b_j,$$

whose RHS has the standard form of a matrix product and, in fact, is identical to Eq. (4.8). The R_{kj} are elements of a matrix, as in Eq. (4.11). We denote this matrix by R:

$$R = \begin{pmatrix} R_{11} & R_{22} & R_{33} \\ R_{21} & R_{22} & R_{23} \\ R_{31} & R_{32} & R_{33} \end{pmatrix}. \tag{4.23}$$

$R = R(\mathbf{n}, \theta)$ is the matrix representation of the abstract rotation operator $\hat{R}(\mathbf{n}, \theta)$ in the basis $\{\mathbf{e}_j\}$.

We stress that $R \neq \hat{R}(\mathbf{n}, \theta)$: $\hat{R}(\mathbf{n}, \theta)$ acts on vectors in the $N = 3$ space whereas R acts on the components of these vectors expressed in 3×1 column form. R is therefore equivalent but not equal to $\hat{R}(\mathbf{n}, \theta)$. In exactly the same way, the wave functions discussed in Sections 3.2 and 3.3 are equivalent but not equal to the quantum state: the wave functions are representations of the state, a point to which we shall return in later chapters.

4.2. *N* dimensions, Hilbert space, and Dirac notation

It is straightforward in every way (but pictorially!) to extend the 3-D Euclidean space of the preceding section to an N-dimensional one, even when the N-D vectors are allowed to be complex. The further extension of $N \to \infty$ is the one of crucial importance to quantum mechanics. Because there are aspects of this generalization that are not straightforward – ∞, after all, is not an ordinary number – we first examine the finite-N case. Once the ground work of this section has been completed, we will turn in Section 4.3 to a study of operators on these spaces.

N-dimensional vector space

By considering a 3-D vector to be a triplet of components (in some coordinate system), the extension to an N-dimensional vector is almost trivial: one enlarges the number of components from 3 to N. Let $\mathbf{u}$, $\mathbf{v}$, $\mathbf{w}$, etc., be N-dimensional vectors. In column-vector notation, $\mathbf{u}$ is expressed as

$$\mathbf{u} = \begin{pmatrix} u_1 \\ u_2 \\ \vdots \\ u_{N-1} \\ u_N \end{pmatrix}, \tag{4.24}$$

where the u_j are arbitrary numbers. In addition to increasing the number of components, we will also allow them to be complex:

$$\mathbf{u}^* = \begin{pmatrix} u_1^* \\ u_2^* \\ \vdots \\ u_{N-1}^* \\ u_N^* \end{pmatrix}.$$

By assuming that the sum of any two such vectors is also a vector, that multiplication by a complex constant is defined, viz.,

$$a\mathbf{u} = \begin{pmatrix} au_1 \\ au_2 \\ \vdots \\ au_N \end{pmatrix},$$

that the zero vector exists (all components are zero), that the associative and distributive laws hold, etc., these ordered N-tuplets/column vectors form the elements of an N-dimensional vector space, denoted $\mathcal{M}$. $\mathcal{M}$ will later be seen to be a linear space because transformations on it will be assumed to be linear.

In 3-D, the length of a vector $\mathbf{b}$ is a real number obtained from the sum of the squares of its components: $b^2 = \sum_{j=1}^{3} b_j^2$. The length of an N-dimensional complex vector is also a real number, defined in a similar manner:

$$|u|^2 = \sum_{j=1}^{N} |u_j|^2. \tag{4.25}$$

Recalling that b^2 in 3-D is a scalar product, $b^2 = \mathbf{b}^{\mathrm{T}} \cdot \mathbf{b}$, we seek a scalar product form for (4.25). Generalizing from Eqs. (4.16) and (4.17), we see that, on choosing

$$|u|^2 \equiv \mathbf{u}^{\mathrm{T}*} \cdot \mathbf{u}, \tag{4.26}$$

where

$$\mathbf{u}^{\mathrm{T}*} = (u_1^* u_2^* \; \cdots \; u_{N-1}^* u_N^*), \tag{4.27}$$

the definition (4.26) yields the RHS of (4.25).

The quantity $u^{\mathrm{T}*} = u^{*\mathrm{T}}$ is called the *Hermitian conjugate* of $\mathbf{u}$ and is denoted by a dagger, †:

$$\mathbf{u}^\dagger \equiv \mathbf{u}^{*\mathrm{T}} = \mathbf{u}^{\mathrm{T}*}. \tag{4.28}$$

The Hermitian conjugate of a vector is often referred to as the dual vector. It is used to define the scalar product of any two vectors in $\mathcal{M}$, not just that of $\mathbf{u}$ with itself:

$$\mathbf{v}^\dagger \cdot \mathbf{u} = \sum_{j=1}^{N} v_j^* u_j. \tag{4.29}$$

Note that $\mathbf{v}^\dagger \cdot \mathbf{u}$ is complex, so that $\mathbf{u}^\dagger \cdot \mathbf{v} \neq \mathbf{v}^\dagger \cdot \mathbf{u}$.

The definition of $\mathbf{u} \in \mathcal{M}$ as an ordered N-tuplet of components, plus the definition of the scalar product (4.29), suggests introducing a basis of linearly independent, complex unit vectors $\{\hat{\mathbf{e}}_j\}$ that span $\mathcal{M}$ and from which the u_j can be obtained via the scalar product:

$$u_j \equiv \hat{\mathbf{e}}^\dagger \cdot \mathbf{u} \tag{4.30}$$

such that

$$\mathbf{u} = \sum_{j=1}^{N} u_j \hat{\mathbf{e}}_j, \tag{4.31}$$

where the fact that $\{\hat{\mathbf{e}}_j\}$ spans the space implies the uniqueness of the $\{u_j\}$.

Equation (4.31) means that the $\{\hat{\mathbf{e}}_j\}$ are unit vectors in the complex-orthogonality sense:

$$\hat{\mathbf{e}}_j^\dagger \cdot \hat{\mathbf{e}}_k = \delta_{jk}. \tag{4.32}$$

In the column-vector format, the simplest form for each $\hat{\mathbf{e}}_j$ is an extension of Eq. (4.13):

$$\hat{\mathbf{e}}_j = \begin{pmatrix} 0 \\ 0 \\ \vdots \\ 0 \\ e_j \\ 0 \\ \vdots \\ 0 \\ 0 \end{pmatrix}, \tag{4.33}$$

where $|e_j| = 1$. The form (4.33) guarantees that (4.32) holds, but, as in the 3-D case, there are (infinitely) many other bases one could use in place of (4.33). They are related to one another by linear transformations (operators on $\mathcal{M}$) that are unitary, a term we shall define in the next section. In matrix form, they are the analogs of the orthogonal rotational matrix R of (4.23).

Hilbert space

The space $\mathcal{M}$ contains N-dimensional complex vectors for which a scalar product is defined. On going from $\mathcal{M}$ to Hilbert space, denoted $\mathcal{H}$, not only does the dimension become infinite, but also both denumerably and continuously infinite spaces are encompassed by the symbol $\mathcal{H}$: all Hilbert spaces are isomorphic to one another.[3] An example of denumerably indexed elements that form a Hilbert space is the sine and cosine functions of Fourier analysis. An example of continuously indexed elements is the plane waves $\psi_k(x) = \exp(ikx)$, although they are not elements of $\mathcal{H}$, as we shall see later. However, 1-D wave packets can also be labeled by a continuous index; these packets are elements of $\mathcal{H}$.

The difference between these two instances of continuously indexed quantities is that wave packets are normalizable, whereas plane waves are not (recall that they are "normalized" to a delta function). The vectors that constitute a Hilbert space must be normalizable, under whatever definition of a norm is used (e.g., a scalar product), as discussed in various texts, e.g., Richtmyer (1978), Dettman (1988), and Hassani (1991). Let us see how this affects the simplest attempt to construct $\mathcal{H}$, viz., by taking the limit $N \to \infty$ in the vectors of $\mathcal{M}$.

We consider the class of vectors of the form

$$\mathbf{u} = (u_1, u_2, u_3, \ldots, u_N, \ldots), \tag{4.34}$$

where the dots signify an infinite number of terms. Not all such $\mathbf{u}$ can be elements of $\mathcal{H}$: only those with finite length $|u| = (\sum_{j=1}^{\infty} |u_j|^2)^{1/2}$ are allowed. That is, $\mathbf{u} \in \mathcal{H}$ only if

$$|u|^2 = \mathbf{u}^\dagger \cdot \mathbf{u} = \sum_{j=1}^{\infty} |u_j|^2 < \infty. \tag{4.35}$$

The inequality in (4.35) means that the sum converges to a finite quantity. In this case, the norm is $|u|$, the square root of the scalar product. Scalar products are defined for spaces with $N > 3$ by means of a set of criteria they must obey, as noted in Dettman (1988); we shall review these in the next subsection. We simply remark here that the scalar product

$$\mathbf{u}^\dagger \cdot \mathbf{v} = \sum_{j=1}^{\infty} u_j^* v_j \tag{4.36}$$

exists and satisfies these criteria. The space formed by finite-length vectors of the form (4.34) is one example of a Hilbert space (this particular realization of the space is often denoted ℓ^2). We also remark that orthonormal bases that are the infinite-component extensions of the $\hat{\mathbf{e}}_j$ of (4.33) can be introduced, although they are not intrinsic to the present discussion.

A second example of a Hilbert space is the one denoted $L^2(a, b)$; it is a *function space*, whose elements were encountered in Chapter 3. The vectors which constitute $L^2(a, b)$ are complex functions of a real variable, say x, with $x \in [a, b]$; as x varies so do the vectors. The concept of vectors that depend on a variable is familiar from classical mechanics: 3-D position and velocity depend on time, while force, electric and magnetic fields, and angular momentum are all vectors that depend on space and possibly time as well. The dependence on x of the basis vectors of $L^2(a, b)$ is in exact analogy to the position dependence of the unit vectors $\hat{\mathbf{e}}_r$, $\hat{\mathbf{e}}_\theta$, and $\hat{\mathbf{e}}_\phi$ in spherical polar coordinates.

[3] Two spaces with the same properties are isomorphic when the elements of one can be put into a one-to-one correspondence with the elements of the other.

The requirements on the elements $u(x)$, $v(x)$, $w(x)$, etc., of $L^2(a, b)$ are

$$\int_a^b u(x)dx < \infty \tag{4.37a}$$

and

$$\int_a^b |u(x)|^2 \, dx < \infty \tag{4.37b}$$

for all $u(x) \in L^2(a, b)$. Equation (4.37b) is an evident analog of (4.35). Dettman (1988) shows that an $L^2(a, b)$ defined as the set of all $u(x)$ obeying (4.37) is a vector space.

The scalar product of two vectors $u(x)$, $v(x) \in L^2(a, b)$, denoted (u, v), is defined as

$$(u, v) \equiv \int_a^b u^*(x)v(x)dx; \tag{4.38}$$

compare Eq. (3.129) *et seq.*, where the scalar product character of the (,) symbol was introduced. Because $u(x)$ and $v(x)$ are each square integrable (i.e., obey Eq. (4.37b)), it follows that $(u, v) = (v, u)^*$ is finite; proof of this statement is left to the exercises (see, e.g., Dettman (1988)). The finiteness of (4.38) means that, with x regarded as a continuum version of the integer index in (4.34), (4.38) and (4.36) are obvious analogs.

Before going on to consider bases in $L^2(a, b)$, let us examine the definition of the scalar product (4.38) for the vectors in $L^2(a, b)$. For the moment, let (u, v) denote the generic scalar product of any two vectors, including, for example, (4.36). Among the requirements that the symbol (,) must satisfy are

$$(u, v) = (v, u)^*,$$

$$(au, v) = a^*(u, v), \tag{4.39}$$

and

$$(u, u) \geqslant 0,$$

with the equality in the last relation holding only if $u = 0$. It is readily shown that (4.36) and (4.38), as well as $\mathbf{b} \cdot \mathbf{a}$ for two 3-D vectors, all satisfy (4.39). Thus, the definition (4.38) is a consistent one. It is, therefore, reasonable to state that two functions (vectors) $u(x)$, $w(x) \in L^2(a, b)$ are orthogonal if $(u, w) = 0$, and that $v(x)$ is normalized if $(v, v) = 1$.

It can be shown that the two Hilbert spaces ℓ^2 and $L^2(a, b)$ are isomorphic to one another. This suggests the existence of orthonormal bases in $L^2(a, b)$, analogs to the infinite-component set $\{\hat{\mathbf{e}}_j\}$ of ℓ^2. These bases are infinite sets of functions such as the Fourier sine and/or cosine functions or the $\{u_n(x)\}$ of Sturm–Liouville theory. Indeed, $L^2(a, b)$ is a Hilbert space spanned by the $\{u_n(x)\}_{n=1}^{\infty}$.

We shall use the Sturm–Liouville eigenfunctions as an example of an orthonormal basis, since they obey

$$(u_n, u_m) = \delta_{nm}. \tag{4.40}$$

It is reasonable to suppose that the analog of the expansion (4.31) or its $N \to \infty$ extension holds for *any* $v(x) \in L^2(a, b)$, viz.,

$$v(x) = \sum_{n=1}^{\infty} C_n u_n(x) \tag{4.41}$$

with the constants C_n given by

$$C_n = (u_n, v). \tag{4.42}$$

Note that Eq. (4.42) is a precise analog of Eq. (4.3).

Unfortunately, one of the intrinsic features of an infinite-dimensional function space is that the preceding supposition need not be true: in this instance (and others) one must proceed with caution, especially if $a = -\infty$ or $b = \infty$, or both. What might go wrong with (4.41)? One possibility is that it might not hold for every $x \in [a, b]$, i.e., the sum on the RHS might not converge pointwise. Instead, there might be only convergence in the mean, defined (e.g., Dettman (1988)) by

$$\int_a^b dx \left| v(x) - \sum_{n=1}^{\infty} C_n u_n(x) \right|^2 = 0,$$

with C_n given by (4.42). However, Eq. (4.41) will always converge pointwise if the Hilbert space is *separable*, i.e., if $\mathcal{H}$ contains a countable, dense set.[4] Separability is discussed, for example, by Dettman (1988) and by Richtmyer (1978). The separability of $\mathcal{H}$ is the crucial assumption we make, not only in this section, but for quantum theory in general. It guarantees the pointwise convergence of sums like (4.41), from which it follows that the set (or sequence) $\{u_n(x)\}_{n=1}^{\infty}$ is *complete*, i.e., that it is a basis in $\mathcal{H}$. Completeness here means that, for any non-trivial $u(x) \in \mathcal{H}$, at least one $C_n = (u_n, u)$ is non-zero.

Completeness implies the existence of a basis and of pointwise convergence, two features of such importance in quantum mechanics that they are incorporated into the expansion postulate of quantum theory, formulated in Chapter 5 via the statement that the set of eigenstates of every observable is a basis in $\mathcal{H}$. As noted in Chapter 3, completeness of the eigenfunctions of Sturm–Liouville systems has been proved, so that the postulate just noted is a generalization of known results; it also extends the concept from $L^2(a, b)$ to abstract vectors in $\mathcal{H}$.

The goal of this subsection has been to introduce the notion of separability in the context just outlined. We close with the statement of two theorems on orthonormal sequences; the proofs may be found, e.g., in Richtmyer (1978).

THEOREM ON SEPARABILITY

A Hilbert space $\mathcal{H}$ is separable if it contains a complete orthonormal sequence.

Comment. As noted earlier, the Hilbert spaces occurring in quantum theory are assumed to be separable.

[4] A set S of vectors is dense in $\mathcal{H}$ if every neighborhood of any $v \in S$ contains at least one $w \in \mathcal{H}$.

THEOREM ON ORTHONORMAL SEQUENCES

If $\{u_n(x)\}$ is an orthonormal sequence in $\mathcal{H}$, then the following statements are equivalent:

$$\{u_n(x)\}_{n=1}^{\infty} \text{ is complete },\tag{4.43a}$$

$$u(x) = \sum_{n=1}^{\infty} (u_n, u)u(x)\tag{4.43b}$$

$$(u, v) = \sum_{n=1}^{\infty} (u, u_n)(u_n, v), \qquad \forall\, u, v \in \mathcal{H},\tag{4.43c}$$

$$(u, u) = \sum_{n=1}^{\infty} |(u, u_n)|^2, \qquad \forall\, u \in \mathcal{H}.\tag{4.43d}$$

Comments. The relations (4.43) are valid in quantum theory by virtue of the theorem on separability. Note that (4.43d) directly follows from (4.43c) and that the magnitude of each of these sums is assumed to be finite: the vectors/functions in $\mathcal{H}$ are normalizable, by definition. The sum in (4.43c) bears a close relation to the sum of projectors discussed in the next subsection.

Dirac notation

Although quantum mechanics is a Hilbert-space theory involving abstract operators and vectors, the discussion so far has been couched almost entirely in concrete rather than abstract terms: differential operators, vectors that are functions, etc. Little has been said about the abstract quantities, examples of which are the rotation operator $\hat{R}(\mathbf{n}, \theta)$ and the states of a quantum system, whose spatial realizations or coordinate representations are wave functions like $\psi_n(x)$ of the quantal box. The abstract description is initiated in this subsection via the language of bra and ket vectors of Dirac (1947).

Dirac's notation allows us to remove the coordinate dependence from a Hilbert-space vector such as $\psi_n(x)$ or $v(x)$; left behind is an abstract vector. This separation is not without precedence: it was accomplished when (u, v) was written in place of $\int_a^b u^*(x)v(x)\,dx$ in Eq. (4.38). Dirac's symbol for an abstract vector is $|\;\rangle$, denoted a *ket*; its dual or Hermitian conjugate is $\langle\;|$, referred to as a *bra*:

$$\left.\begin{matrix} v(x) \\ \mathbf{u} \end{matrix}\right\} \to |\;\rangle,$$

$$\left.\begin{matrix} v^*(x) \\ \mathbf{u}^\dagger \end{matrix}\right\} \to \langle\;|,$$

$$\langle\;| = |\;\rangle^\dagger.\tag{4.44}$$

Kets and bras are specified by labels placed inside the relevant symbols, e.g., $|\alpha\rangle$ or $\langle\beta| = |\beta\rangle^\dagger$, where α, β, etc., are generic labels. Given two kets $|\alpha\rangle$ and $|\beta\rangle$, their scalar product is written $\langle\beta|\alpha\rangle$, analogous to (u, v):

$$(u, v) \to \langle\beta|\alpha\rangle,\tag{4.45}$$

where, for economy of notation, only one vertical line has been used in forming the scalar product. Like (u, v), $\langle\beta|\alpha\rangle$ is a complex number:

$$\langle\beta|\alpha\rangle = \langle\alpha|\beta\rangle^*. \tag{4.46}$$

It is also a probability amplitude, whose absolute square is a probability or probability density. We shall have much to say in Chapter 5 about the probabilistic interpretation of scalar products like (4.46). As in the case of (u, v), certain of the scalar products $\langle\beta|\alpha\rangle$ can be expressed as integrals; these will be considered later.

Kets will be used not only for vectors in $\mathcal{H}$, but also for improper – non-Hilbert-space – vectors. The former are normalizable, i.e., $|\alpha\rangle \in \mathcal{H} \Rightarrow \langle\alpha|\alpha\rangle < \infty$, and include solutions of the time-dependent and time-independent Schrödinger equations. Among the latter kets are the eigenvectors $|x\rangle$ of the 1-D position operator, denoted $\hat{Q}_x$:

$$\hat{Q}_x|x\rangle = x|x\rangle. \tag{4.47}$$

Equation (4.47) is an example of the labeling of kets by the eigenvalue of the operator of which it is an eigenvector. The non-normalizability arises because the label x is continuous, leading to the scalar product of two such kets being a delta function:

$$\langle x'|x\rangle = \delta(x - x'), \tag{4.48}$$

just as in Eq. (3.128). While improper vectors like $|x\rangle$ are not in $\mathcal{H}$, they are essential ingredients in quantum theory and must be included in the theoretical framework. The bra/ket notation is general enough to encompass both types of vectors. All quantum "states" are proper vectors in $\mathcal{H}$.

Abstract operators have already been introduced, e.g., $\hat{R}(\mathbf{n}, \theta)$, the rotation operator, and the 1-D position operator $\hat{Q}_x$. Let $\hat{A}$ be an arbitrary operator. Its action on a vector $|\alpha\rangle$ is written $\hat{A}|\alpha\rangle$, the result of which is a new vector, say, $|\gamma\rangle$:

$$|\gamma\rangle = \hat{A}|\alpha\rangle. \tag{4.49}$$

The dual of $|\gamma\rangle$ is $\langle\gamma| = |\gamma\rangle^\dagger$, which raises the question of how to treat $(\hat{A}|\alpha\rangle)^\dagger$. The definition is

$$(\hat{A}|\alpha\rangle)^\dagger \equiv \langle\alpha|\hat{A}^\dagger, \tag{4.50}$$

where $\hat{A}^\dagger$ is the Hermitian conjugate of $\hat{A}$. Equation (4.50) mimics the behavior of the transpose of the product of two matrices – the order is reversed – and it carries an implicit statement that, in $\langle\alpha|\hat{A}^\dagger$, $\hat{A}^\dagger$ acts to the left.[5]

A complete definition of $\hat{A}^\dagger$ utilizes the scalar product, as in the case of the Sturm–Liouville operator $\hat{L}$ of (3.107): from Eq. (4.49),

$$\langle\beta|\gamma\rangle = \langle\beta|\hat{A}|\alpha\rangle, \tag{4.51}$$

and then, by definition,

$$\langle\beta|\gamma\rangle^* = \langle\beta|\hat{A}|\alpha\rangle^* \equiv \langle\alpha|\hat{A}^\dagger|\beta\rangle. \tag{4.52}$$

[5] When confusion may exist as to whether an operator acts to the left or to the right, an arrow will be placed under the operator, e.g., $\underset{\leftarrow}{\hat{A}^\dagger}$.

Occasionally, we shall compress the notation and write

$$\langle\beta|\hat{A}|\alpha\rangle = \langle\beta|\hat{A}\alpha\rangle = \langle\hat{A}^\dagger\beta|\alpha\rangle. \tag{4.53}$$

If $\langle\beta|\hat{A}^\dagger|\alpha\rangle = \langle\beta|\hat{A}|\alpha\rangle$ for all $|\alpha\rangle$ and $|\beta\rangle$ in $\mathcal{H}$, then $\hat{A}$ is said to be a *Hermitian* operator and one writes $\hat{A}^\dagger = \hat{A}$. If $\hat{A}$ is Hermitian with respect to vectors in $\mathcal{H}$, it need not be Hermitian w.r.t. to non-normalizable vectors. Care must be exercised and each case examined in detail. Hermitian operators are the fundamental ones in quantum theory, since they are associated with observables. The types and properties of the operators of interest in quantum theory are discussed in the next section.

Both $\langle\beta|\alpha\rangle$ and $\langle\beta|\hat{A}|\alpha\rangle$ are numbers that may also carry a dimension such as momentum or energy. The scalar product $\langle\beta|\alpha\rangle$, sometimes referred to as the *projection* of the ket $|\alpha\rangle$ onto the ket $|\beta\rangle$, is a probability amplitude. The quantity $\langle\beta|\hat{A}|\alpha\rangle$ is a *matrix element* of the operator $\hat{A}$; it is not an operator as long as both $|\alpha\rangle$ and $|\beta\rangle \in \mathcal{H}$. Suppose that $\hat{A}$ is simply a complex number, $\hat{A} = c =$ constant, so that $\hat{A}|\alpha\rangle = c|\alpha\rangle$. In this case $\langle\beta|\hat{A}|\alpha\rangle$ becomes $c\langle\beta|\alpha\rangle$:

$$\langle\beta|c|\alpha\rangle = c\langle\beta|\alpha\rangle; \tag{4.54}$$

only when $\hat{A}$ is a constant – or, as in the case of an eigenvector, when its action produces a constant – can the symbol be removed from between the vertical lines.

In the bra/ket notation, the eigenvalue relation is

$$\hat{A}|a\rangle = a|a\rangle. \tag{4.55}$$

The notation of Eq. (4.55) is typical (but not universal): the same symbol is used for the operator and its eigenvalue, the former in upper case, the latter in lower case. Furthermore, the ket is labeled by the same lower-case symbol as is the eigenvalue. If both sides of (4.55) are projected onto $\langle\beta|$, we get

$$\langle\beta|\hat{A}|a\rangle = \langle\beta|a|a\rangle = a\langle\beta|a\rangle, \tag{4.56}$$

consistent with the comment in the preceding paragraph.

We noted above that forming a scalar product is equivalent to projecting one vector onto another. The Dirac notation allows one to easily construct operators that perform this function; they are denoted, of course, projection operators. We consider them in the context of a discrete basis in $\mathcal{H}$.[6] Let $\{|e_n\rangle\}_{n=1}^{\infty}$ be an orthonormal basis spanning $\mathcal{H}$, where

$$\langle e_n|e_m\rangle = \delta_{nm}, \tag{4.57}$$

and let $|\alpha\rangle \in \mathcal{H}$. Then the expansion of $|\alpha\rangle$ in terms of the $\{|e_n\rangle\}$ is

$$|\alpha\rangle = \sum_n \alpha_n|e_n\rangle, \tag{4.58}$$

with

$$\alpha_n = \langle e_n|\alpha\rangle.$$

[6] The eigenstates of any quantal operator that images a physical observable form a basis (see Postulate VI, Chapter 5), but not all such bases are discrete. A discrete basis is used here for simplicity.

Because α_n is a number, (4.58) can be re-written as

$$|\alpha\rangle = \sum_n |e_n\rangle\alpha_n = \sum_n |e_n\rangle\langle e_n|\alpha\rangle \tag{4.59}$$

and the essential observation is to recall that the scalar product $\langle e_n|\alpha\rangle$ means $\langle e_n|\cdot|\alpha\rangle$. Hence, (4.59) can be re-expressed as

$$|\alpha\rangle = \sum_n \{|e_n\rangle\langle e_n|\}|\alpha\rangle, \tag{4.60}$$

where the *dyadic form* of the first two factors has been isolated via the curly brackets.

The curly brackets define the projection operator $\hat{P}_n$:

$$\hat{P}_n = |e_n\rangle\langle e_n|.$$

Applied to $|\alpha\rangle$, it yields

$$\hat{P}_n|\alpha\rangle = |e_n\rangle\langle e_n|\alpha\rangle = \alpha_n|e_n\rangle, \tag{4.61}$$

which is the *n*th term in the expansion (4.58). $\hat{P}_n$ thus projects the generic vector $|\alpha\rangle$ onto a 1-D subspace spanned by the "unit" vector $|e_n\rangle$. $\hat{P}_n$ also obeys one of the defining relations of a projection operator, viz.,

$$\hat{P}_n^2 = \hat{P}_n\hat{P}_n = |e_n\rangle\langle e_n|e_n\rangle\langle e_n| = |e_n\rangle\langle e_n| = \hat{P}_n:$$

$\hat{P}_n$ is *idempotent*. The foregoing steps have been worked out in detail to help the reader become familiar with manipulations involving bras and kets.

Equation (4.60) does more, however, than lead to $|e_n\rangle\langle e_n|$ being identified as a projection operator. It also leads to a bra/ket version of Eq. (3.122). This is accomplished by interchanging the symbols $\sum_n$ and {, i.e., by again shifting one's point of view:

$$\begin{aligned} |\alpha\rangle &= \sum_n \{|e_n\rangle\langle e_n|\}|\alpha\rangle \\ &= \left\{\sum_n |e_n\rangle\langle e_n|\right\}|\alpha\rangle. \end{aligned} \tag{4.62}$$

The presence of $|\alpha\rangle$ on each side of (4.62) means that the term in curly brackets must be the identity operator $\hat{I}$, since $\hat{I}|\alpha\rangle = |\alpha\rangle$:

$$\sum_n |e_n\rangle\langle e_n| = \sum_n \hat{P}_n = \hat{I}. \tag{4.63}$$

The *closure relation* (4.63) is known as a "resolution of the identity," and it will hold for any complete set, including ones whose index is continuous. It should be evident that Dirac notation is more than a handy symbology!

We end this section with a restatement in Dirac notation of Eq. (4.43).

THEOREM ON ORTHONORMAL SEQUENCES

If $\{|\varphi_n\rangle\}$ is an orthonormal sequence in $\mathcal{H}$, then the following are equivalent:

$$\{|\varphi_n\rangle\} \quad \text{is complete,} \tag{4.64a}$$

$$|\alpha\rangle = \sum_n \langle\varphi_n|\alpha\rangle|\varphi_n\rangle, \qquad \forall\ |\alpha\rangle \in \mathcal{H}, \tag{4.64b}$$

$$\langle\alpha|\chi\rangle = \sum_n \langle\alpha|\varphi_n\rangle\langle\varphi_n|\chi\rangle, \qquad \forall\ |\alpha\rangle,\ |\chi\rangle \in \mathcal{H}, \tag{4.64c}$$

$$\|\ |\alpha\rangle\|^2 = \sum_n |\langle\alpha|\varphi_n\rangle|^2, \qquad \forall\ |\alpha\rangle \in \mathcal{H}. \tag{4.64d}$$

The *norm* of $|\alpha\rangle$, denoted $\|\ |\alpha\rangle\|$, is given by $\|\ |\alpha\rangle\| = \langle\alpha|\alpha\rangle^{1/2}$. Note that the sum on n in (4.64c) is another example of a closure relation.

4.3. Operators

An operator $\hat{A}$ on $\mathcal{H}$ is said to be linear if, for any $|\alpha\rangle$ and $|\beta\rangle \in \mathcal{H}$,

$$\hat{A}[a|\alpha\rangle + b|\beta\rangle] = a\hat{A}|\alpha\rangle + b\hat{A}|\beta\rangle, \tag{4.65}$$

where a and b are arbitrary complex numbers and the relation $\hat{A}a|\alpha\rangle = a\hat{A}|\alpha\rangle$ has been employed. We shall assume linearity to hold if one replaces $|\alpha\rangle$ and/or $|\beta\rangle$ by non-normalizable kets such as $|x\rangle$. With the exception of the time-reversal operator (see, e.g., Sakurai (1994)) all the operators of interest in quantum theory are linear. Examples of linear operators include d^n/dx^n, the integral of a function, and multiplication by a number or a function, whereas $(d/dx)^n$ is non-linear. In addition to (4.65), the following relations hold for linear operators:

$$(\hat{A} + \hat{B})|\alpha\rangle = \hat{A}|\alpha\rangle + \hat{B}|\alpha\rangle$$

and

$$\hat{A}\hat{B}|\alpha\rangle = \hat{A}(\hat{B}|\alpha\rangle).$$

Taxonomy

This subsection reviews some of the types of operators of interest in quantum mechanics. All properties that are listed can be thought of as exemplified by the matrix representation of the operator, using any convenient basis.

(*i*) The unit operator

The simplest and indeed an almost trivial example is the unit operator $\hat{I}$, encountered in the previous subsection. For any ket $|\alpha\rangle$,

$$\hat{I}|\alpha\rangle = |\alpha\rangle. \tag{4.66}$$

Matrix elements of $\hat{I}$ in an orthonormal basis $\{|e_n\rangle\}$ take the form

$$I_{mn} = \langle e_m|\hat{I}|e_n\rangle = \delta_{mn}. \tag{4.67}$$

In an N-dimensional space, I is an $N \times N$ matrix with unity at each main diagonal position and zeros elsewhere:

$$I = \begin{pmatrix} 1 & 0 & 0 & \cdots & 0 \\ 0 & 1 & 0 & & 0 \\ 0 & 0 & 1 & & 0 \\ \vdots & \vdots & \vdots & & \vdots \\ 0 & & & & 0 \\ 0 & & & & 1 \end{pmatrix}.$$

(*ii*) The inverse

The unit operator enters into the definition of the inverse of an operator. Let $\hat{A}$ be an operator on $\mathcal{H}$. Its inverse, denoted $\hat{A}^{-1}$, obeys

$$\hat{A}\hat{A}^{-1} = \hat{A}^{-1}\hat{A} = \hat{I}. \tag{4.68}$$

Equation (4.68) is meaningful *only* if $\hat{A}$ is non-singular; otherwise, $\hat{A}^{-1}$ does not exist.[7] An application of Eq. (4.68) will help us to understand this.

Consider the second equality in (4.68). The product $\hat{A}^{-1}\hat{A}$ can be written $\hat{A}^{-1}\hat{I}\hat{A}$, which becomes, via Eq. (4.63),

$$\hat{A}^{-1}\hat{A} = \sum_k \hat{A}^{-1}|e_k\rangle\langle e_k|\hat{A} = \hat{I}. \tag{4.69}$$

Now forming the m, n matrix element of both sides of (4.69) yields

$$\sum_k \langle e_m|\hat{A}^{-1}|e_k\rangle\langle e_k|\hat{A}|\hat{e}_n\rangle = \delta_{mn}. \tag{4.70}$$

For the present purpose, we will now assume $\mathcal{H}$ to be finite, so that the k-sum in (4.70) is from 1 to N. The matrix elements in the k-sum of (4.70) are those of the operator and its inverse, i.e., the matrices are inverses of one another. Finding the inverse of a matrix involves division by its determinant; when the determinant is zero, A^{-1} is infinite, and A is singular. The singularity of A means that the operator $\hat{A}$ itself is singular. Note that this result is independent of the (orthonormal) basis chosen to obtain the matrix representation of $\hat{A}$, since the determinant is invariant under basis transformations.[8]

(*iii*) Hermitian operators

The *Hermitian conjugate* or *adjoint* of $\hat{A}$, denoted $\hat{A}^\dagger$ as in Section 4.2, is equal to the transpose, complex conjugate of the operator:

$$\hat{A}^\dagger = \hat{A}^{\mathrm{T}*} = \hat{A}^{*\mathrm{T}}. \tag{4.71}$$

In terms of matrix elements in the basis $\{|e_n\rangle\}$, (4.71) means

$$(\hat{A}^\dagger)_{mn} = A^*_{nm}, \tag{4.72}$$

while for arbitrary $|\alpha\rangle$ and $|\beta\rangle$,

$$\langle\alpha|\hat{A}|\beta\rangle = \langle\hat{A}^\dagger\alpha|\beta\rangle = \langle\beta|\hat{A}^\dagger|\alpha\rangle^*.$$

[7] In the example below that illustrates this point, we consider finite dimensions. Singular operators occur in the full $\mathcal{H}$, as in the free-particle Green function encountered in Chapter 15. In addition, there are operators that have a left or a right inverse but not both (e.g., Reed and Simon (1972)). We assume that the latter pathologies do not occur in quantum theory.

[8] Proof of this statement is left as an exercise.

These are important relations and are based on $\langle b|a\rangle = \langle a|b\rangle^*$, which implies that $\langle\beta|\hat{A}\alpha\rangle = \langle A\alpha|\beta\rangle^* = \langle\alpha|\hat{A}^\dagger|\beta\rangle^*$.

In quantum mechanics, the single most important operator class is that of Hermitian operators;[9] for them

$$\hat{A}^\dagger = \hat{A}. \tag{4.73}$$

As noted, all quantal operators that correspond to observables are Hermitian. A few examples of Hermitian operators expressed as matrices are

$$\begin{pmatrix} 1 & 0 & 0 & 0 & \cdots \\ 0 & 1 & 0 & 0 & \cdots \\ 0 & 0 & 1 & 0 & \cdots \\ \vdots & \vdots & \vdots & & \end{pmatrix}, \qquad \begin{pmatrix} 1 & 0 \\ 0 & -1 \end{pmatrix}, \qquad \begin{pmatrix} 0 & -i \\ i & 0 \end{pmatrix}.$$

The two 2-D matrices are realizations of the so-called Pauli spin matrices σ_3 and σ_2, respectively, in the representation in which σ_3 is diagonal, a point discussed in detail in Chapter 13. Notice that neither

$$\begin{pmatrix} 0 & i \\ i & 0 \end{pmatrix} \text{ nor } \begin{pmatrix} 0 & -1 \\ 1 & 0 \end{pmatrix}$$

is Hermitian. On the other hand, any 2-D matrix of the form

$$\begin{pmatrix} a & c \\ c^* & b \end{pmatrix},$$

where a and b are real, *is* Hermitian. It is amusing to note, however, that if, $ab = |c|^2$, this matrix is singular and consequently has no inverse. A non-matrix example of Hermiticity is the Sturm–Liouville operator $\hat{L}(x)$ of Eq. (3.101), as proved in Section 3.4.

(*iv*) Unitary operators

An operator $\hat{U}$ is unitary if its adjoint is equal to its inverse:

$$\hat{U}^\dagger = \hat{U}^{-1}, \tag{4.74}$$

from which it follows that

$$\hat{U}^\dagger\hat{U} = \hat{U}\hat{U}^\dagger = \hat{I}. \tag{4.75}$$

Unitary operators preserve the norm of vectors; hence the name unitary. Let us recall the definition of the norm used herein. The norm of a vector $|\alpha\rangle$ in $\mathcal{H}$ is the square root of the scalar product of the vector with itself:[10]

$$\text{norm of } |\alpha\rangle \equiv \|\,|\alpha\rangle\| = \langle\alpha|\alpha\rangle^{1/2}. \tag{4.76}$$

(This is an extension to higher dimension of the length of a 3-D vector.) Suppose that we transform $|\alpha\rangle$ by applying a unitary operator to it:

[9] Since † is occasionally referred to as the adjoint operation, Hermitian operators are sometimes also termed self-adjoint operators. We shall not use this terminology, since self-adjointness has a slightly different meaning for infinite-dimensional spaces. See Chapter 5 and references cited there for further discussion.

[10] There are many ways of defining norms; see, e.g., Richtmyer (1978). The choice (4.76) is standard for quantum theory.

$$|\alpha\rangle \rightarrow |\beta\rangle = \hat{U}|\alpha\rangle. \tag{4.77}$$

The unitarity of $\hat{U}$ guarantees that the norm of $|\beta\rangle$ is the same as that of $|\alpha\rangle$:

$$\begin{aligned}\langle\beta|\beta\rangle &= \{\langle\alpha|\hat{U}^\dagger\}\{U|\alpha\rangle\} \\ &= \langle\alpha|\hat{U}^\dagger\hat{U}|\alpha\rangle = \langle\alpha|\alpha\rangle,\end{aligned} \tag{4.78}$$

the last equality following from (4.74).

Vectors are said to be normalized (i.e., normalized to unity) when their norm is one. Most vectors are not initially normalized, but can be made so on dividing them by the norm:

$$|\alpha_{\text{norm}}\rangle = |\alpha\rangle/\langle\alpha|\alpha\rangle^{1/2} \tag{4.79}$$

is normalized, as a trivial calculation shows.

Unitary operators are common in quantum mechanics, examples being the time-evolution operator, which evolves a state from one time to another, and the rotation operator, one of whose matrix representations in 3-D was derived in Section 4.1. The general exponential form, of which the latter two operators are examples, will be discussed later in this section.

If in some basis the matrix representation of a unitary operator has only real matrix elements, then that matrix is referred to as *orthogonal*: an orthogonal matrix O obeys

$$O^{\text{T}} = O^{-1}. \tag{4.80}$$

The rotation matrix of Eq. (4.10) is readily shown to be orthogonal.

Eigenvectors and eigenvalues

In the introduction to this chapter, the role of eigenvectors and eigenvalues in quantum theory was outlined. The most commonly occurring quantal eigenvalue equations involve unitary and Hermitian operators. We examine their properties in this subsection, along with a worked example of a 2×2 matrix. As we shall see, if $\hat{A}$ is Hermitian its eigenvalues are real, just as in the case of the Sturm–Liouville operator $\hat{L}(x)$.

(*i*) Unitary operators

Let $\hat{U}$ be a unitary operator; its associated eigenvalue problem is written

$$\hat{U}|\lambda_n\rangle = \lambda_n|\lambda_n\rangle. \tag{4.81}$$

To determine the requisite properties, we first replace the subscript n in (4.81) by m and then apply the operation of Hermitian conjugation to the replacement equation, yielding

$$\langle\lambda_m|\hat{U}^\dagger = \lambda_m^*\langle\lambda_m|. \tag{4.82}$$

The second step is to form the scalar product of the left- and right-hand sides of (4.82) with, respectively, the left- and right-hand sides of (4.81). This gives

$$\langle\lambda_m|\hat{U}^\dagger\hat{U}|\lambda_n\rangle = \lambda_m^*\lambda_n\langle\lambda_m|\lambda_n\rangle. \tag{4.83}$$

A standard assumption in quantum mechanics (and linear algebra) is invoked next, viz., that the effect of $\hat{U}^\dagger$ acting to the left in (4.83) is the same as if it were acting to the right.[11] As a result, the product of operators becomes $\hat{U}^\dagger\hat{U}$, and, by (4.75), this product is the unit operator $\hat{I}$. Hence, (4.83) becomes

$$(\lambda_m^*\lambda_n - 1)\langle\lambda_m|\lambda_n\rangle = 0. \tag{4.84}$$

Equation (4.84) is the key relation. To use it, we set $m = n$ and then assume that $|\lambda_n\rangle$ is not a trivial vector, i.e., that $\langle\lambda_n|\lambda_n\rangle \neq 0$. The latter assumption implies that

$$|\lambda_n|^2 - 1 = 0, \tag{4.85}$$

so that λ_n can be expressed as

$$\lambda_n = \mathrm{e}^{i\mu_n}, \qquad \mu_n \text{ real}. \tag{4.86}$$

In other words, we have proved that, for each non-trivial eigenvector of a unitary operator, the corresponding eigenvalue is of unit magnitude.

The relation (4.86) plus the assumption of non-degeneracy (see below) then implies that the $\{|\lambda_n\rangle\}$ form an *orthonormal* set. To see this, substitute (4.86) into (4.84), which becomes

$$(e^{i(\mu_n-\mu_m)} - 1)\langle\lambda_m|\lambda_n\rangle = 0, \qquad n \neq m. \tag{4.87}$$

Since there are no degeneracies, $\mu_n \neq \mu_m \Rightarrow e^{i(\mu_n-\mu_m)} \neq 1$ and consequently

$$\langle\lambda_m|\lambda_n\rangle = 0, \qquad m \neq n. \tag{4.88}$$

The latter result demonstrates the orthogonality property; normality is trivial to implement via Eq. (4.79). We assume the $|\lambda_n\rangle$ to be normalized, and therefore the orthonormality relation

$$\langle\lambda_m|\lambda_n\rangle = \delta_{mn} \tag{4.89}$$

is established, QED.

This is as much as can be determined from a general analysis; specific values of the μ_n and $|\lambda_n\rangle$ can be obtained only – when it is possible – after $\hat{U}$ has been specified.

(*ii*) Hermitian operators.

The foregoing analyses apply also in the case of Hermitian operators. Only results will be listed here; details of the proofs are left as an exercise. Let $\hat{A}$ be Hermitian: $\hat{A}^\dagger = \hat{A}$. The general eigenvalue problem is

$$\hat{A}|a_n\rangle = a_n|a_n\rangle. \tag{4.90}$$

Assuming that $\langle a_n|a_n\rangle \neq 0$, one can show that

$$a_n^* = a_n = \text{real}, \tag{4.91}$$

and that

$$\langle a_m|a_n\rangle = \delta_{mn}, \tag{4.92}$$

where $|a_n\rangle$ is a normalized eigenvector.

[11] This assumption is discussed in later chapters.

The reality property, Eq. (4.91), is of crucial importance to quantum mechanics. For, in quantum theory, a physical observable (energy, momentum, position) is imaged mathematically by an operator whose eigenvalues constitute the entire set of values which a measurement of the observable can yield. Since the measured values are real, the eigenvalues must be real also. Only Hermitian operators are guaranteed to have real eigenvalues, hence physical observables are imaged by such operators. Note also that, for the unitary and Hermitian cases, the assumption of normalizability means that the norm is *finite:* states are normalizable kets in $\mathcal{H}$. This need not be the case for all *bases,* $\{|x\rangle\}$ being an example.

Reduction to a matrix eigenvalue equation

One of the many computational goals of quantum mechanics is to determine the *spectrum* of a quantal operator, i.e., its set of eigenvalues and the corresponding eigenvectors. We showed how implementation of boundary conditions led to this goal in the case of the 1-D quantum box, for which (4.90) took the form $\hat{H}(x)\psi(x) = E_n\psi_n(x)$ in the coordinate representation. For an operator defined on a low-dimensional subspace of $\mathcal{H}$, the spectrum can often be determined via diagonalization of the operator in any convenient orthonormal basis. We first outline the general method, which converts the eigenvalue equation to an $N \times N$ matrix equation, and then illustrate it with an $N = 2$ example.

The eigenvalue equation will be written

$$\hat{\Lambda}|\lambda\rangle = \lambda|\lambda\rangle. \tag{4.93}$$

Let $\{|e_k\rangle\}_{k=1}^{N}$ be an orthonormal basis in the N-D subspace, so that

$$|\lambda\rangle = \sum_k C_k|e_k\rangle, \tag{4.94}$$

with $C_k = \langle e_k|\lambda\rangle$. Substituting (4.94) into (4.93) and projecting both sides of the resulting equation onto $\langle e_j|$ yields

$$\sum_k \Lambda_{jk}C_k = \lambda C_j, \qquad \forall\ j, \tag{4.95}$$

where

$$\Lambda_{jk} = \langle e_j|\hat{\Lambda}|e_k\rangle \tag{4.96}$$

is the matrix representation of $\hat{\Lambda}$ in the basis $\{|e_j\rangle\}$.

The $N \times N$ set of homogeneous equations (4.95) for the unknown coefficients C_k has a non-trivial solution only if the secular determinant vanishes, i.e., only if

$$\det|\Lambda_{jk} - \lambda\delta_{jk}| = 0. \tag{4.97}$$

Written out in full, (4.97) becomes

$$\begin{vmatrix} \Lambda_{11}-\lambda & \Lambda_{12} & \Lambda_{13} & \cdots & \\ \Lambda_{21} & \Lambda_{22}-\lambda & \Lambda_{23} & \cdots & \\ \Lambda_{31} & \Lambda_{32} & \Lambda_{33}-\lambda & \cdots & \\ \vdots & & & & \\ \Lambda_{N1} & \Lambda_{N2} & \Lambda_{N3} & \cdots & \Lambda_{NN}-\lambda \end{vmatrix} = 0.$$

When the $N \times N$ determinant in (4.97) is expanded,[12] it becomes an Nth-order polynomial equation for λ, of the form

$$a_N \lambda^N + a_{N-1} \lambda^{N-1} + \cdots + a_0 = 0, \tag{4.98}$$

where the a_n are linear combinations of products of the Λ_{jk}. In general, (4.98) has at most N distinct roots $\{\lambda_n\}_{n=1}^{N}$. If M roots are equal, say $\lambda_1 = \lambda_2 = \cdots = \lambda_M$, $M \leqslant N$, then there exists an Mth-order *degeneracy*, a situation that we shall assume (for now) does not occur.

A 2-D example

We close this section with a simple example, that of σ_2, the 2×2 matrix representation of the Pauli spin operator $\hat{\sigma}_2$:

$$\sigma_2 = \begin{pmatrix} 0 & -i \\ i & 0 \end{pmatrix}. \tag{4.99}$$

σ_2 is one of the three matrices introduced below Eq. (4.73).

We write the eigenvalue equation in operator form as

$$\hat{\sigma}_2 |u\rangle = \lambda |u\rangle.$$

Since $\hat{\sigma}_2$ is represented by the 2×2 matrix σ_2, $|u\rangle$ must be represented by a column vector with two rows:

$$|u\rangle \to \begin{pmatrix} u_1 \\ u_2 \end{pmatrix} \equiv \mathbf{u}. \tag{4.100}$$

Equation (4.95) therefore becomes

$$(\sigma_2 - \lambda \hat{I})\mathbf{u} = \begin{pmatrix} -\lambda & -i \\ i & -\lambda \end{pmatrix} \begin{pmatrix} u_1 \\ u_2 \end{pmatrix} = 0. \tag{4.101}$$

Non-trivial solutions to (4.101) exist only when its secular determinant vanishes, i.e., when

$$\begin{vmatrix} -\lambda & -i \\ i & -\lambda \end{vmatrix} = 0. \tag{4.102}$$

In this 2×2 case it is easy to evaluate the determinant, and Eq. (4.102) yields

$$\lambda = \pm 1. \tag{4.103}$$

Corresponding to these two values of λ are two eigenvectors, which we shall denote by $\mathbf{u}^{(+)}$ and $\mathbf{u}^{(-)}$:

$$\mathbf{u}^{(\pm)} = \begin{pmatrix} u_1^{(\pm)} \\ u_2^{(\pm)} \end{pmatrix}.$$

Choosing $\lambda = +1$ and substituting it into (4.101) leads to the pair of equations

[12] Equation (4.98) is obtained by expanding (4.97) via row or column elements and co-factors. See, e.g., Dettman (1988) or Hassani (1991) for details.

$$-u_1^{(+)} = iu_2^{(+)}, \qquad iu_1^{(+)} = u_2^{(+)}, \tag{4.104}$$

which are (necessarily) identical. From (4.104) it follows that

$$\mathbf{u}^{(+)} = \begin{pmatrix} u_1^{(+)} \\ iu_1^{(+)} \end{pmatrix} = u_1^{(+)} \begin{pmatrix} 1 \\ i \end{pmatrix}. \tag{4.105}$$

Normalizing means that

$$\mathbf{u}^{(+)\dagger} \cdot \mathbf{u}^{(+)} = 1, \tag{4.106}$$

which implies that $2|u_1^{(+)}|^2 = 1$, or that

$$u_1^{(+)} = e^{i\alpha^{(+)}} \sqrt{\frac{1}{2}}, \tag{4.107}$$

where $\alpha^{(+)}$ is an undetermined real phase.

The same procedure in the case of $\lambda = -1$ results in $u_2^{(-)} = -iu_1^{(-)}$, so that

$$\mathbf{u}^{(-)} = u_1^{(-)} \begin{pmatrix} 1 \\ -i \end{pmatrix}. \tag{4.108}$$

Imposing normalization leads to

$$u_1^{(-)} = e^{i\alpha^{(-)}} \sqrt{\frac{1}{2}}. \tag{4.109}$$

Hence, the normalized eigenvectors in this particular representation are

$$\mathbf{u}^{(\pm)} = \frac{e^{i\alpha^{(\pm)}}}{\sqrt{2}} \begin{pmatrix} 1 \\ \pm i \end{pmatrix}; \tag{4.110}$$

$\mathbf{u}^{(\pm)\dagger} \cdot \mathbf{u}^{(\mp)} = 0$. Equations (4.103) and (4.110) provide the complete solution to the 2×2 matrix eigenvalue problem (4.101). It is typical of many encountered in quantum mechanics. A few others are given as exercises in the problem portion of this chapter.

4.4. The position operator: delta functions and locality

The eigenvalues of the operators discussed in the last section and of the Hamiltonian for the particle in a box are discrete. In contrast, the eigenvalues of the position and momentum operators are *continuous*. The corresponding eigenkets are thus specified by a continuous label, giving rise to delta-function normalization, as in Eq. (4.48). Because of this, such kets are not vectors in $\mathcal{H}$ and therefore cannot be states of any quantum system. Despite this, the eigenkets of the position and momentum operators play very important roles in quantum theory: they give rise to the coordinate and momentum representations, to coordinate-space and momentum-space wave functions, and to the concept of the *locality* (or non-locality) of operators in the coordinate or momentum representations. In addition, these eigenkets are bases, in terms of which any $|\alpha\rangle \in \mathcal{H}$ can be expanded and continuous resolutions of the identity operator can be introduced. We examine some of these features in this section, using the position operator as the paradigm.

Coordinate-space wave functions and completeness

The position operator for a particle moving in 3-D is denoted $\hat{\mathbf{Q}}$. In 1-D, with the particle constrained to move along the x-axis, the position is x, the operator is $\hat{Q}_x$, and the eigenvalue relation is

$$\hat{Q}_x|x\rangle = x|x\rangle. \tag{4.47}$$

We shall begin our analysis with this 1-D case.

As formalized in Postulate II of Chapter 5, the scalar product of two kets is a quantal probability amplitude. When a position bra is used to form the scalar product, the result is called the coordinate-space wave function or coordinate-space representation of the ket. In this text, coordinate-space wave functions will always be represented by the symbol ψ (or Ψ when time dependence is included). Let $|\alpha\rangle$ be the ket; its coordinate-space wave function is written $\psi_\alpha(x)$):

$$\psi_\alpha(x) \equiv \langle x|\alpha\rangle. \tag{4.111}$$

Then, exactly as in the case of the particle in the box, $|\psi_\alpha(x)|^2$ is the probability density that the particle will be at the point x.

If for $|\alpha\rangle$ we choose $|x'\rangle$, then $\langle x|\alpha\rangle$ becomes $\langle x|x'\rangle$, with

$$\langle x|x'\rangle = \delta(x - x'), \tag{4.48}$$

which is the coordinate-space wave function of the position ket $|x'\rangle$. We shall now show that the delta function in (4.48) is a consequence of using the complete set $\{|x'\rangle\}$ to provide a continuous version of the resolution of the identity.

Since $\{|x'\rangle\}$ is complete, the identity operator $\hat{I}$ has the resolution

$$\hat{I} = \int_{-\infty}^{\infty} dx'\, |x'\rangle\langle x'|, \tag{4.112}$$

which is the integral equivalent of the sum (4.63); it is another form of the closure relation. Applying $\hat{I}$ to $|\alpha\rangle$ and employing (4.112), we get

$$|\alpha\rangle = \int_{-\infty}^{\infty} dx'|x'\rangle\langle x'|\alpha\rangle. \tag{4.113}$$

Next, project both sides of (4.113) onto $\langle x|$:

$$\langle x|\alpha\rangle = \int_{-\infty}^{\infty} dx'\langle x|x'\rangle\langle x'|\alpha\rangle, \tag{4.114}$$

where, as is usual in quantum physics, the scalar product and integral signs have been interchanged.[13] Inserting the definition (4.111) into (4.114) then gives

[13] This interchange is a physicist's assumption, made on the expectation that it is true and on the realization that it is a necessity if the theory is to be practical. Certainly if $|x\rangle$ were a well-behaved function and $\langle x|x'\rangle$ meant an integration involving well-behaved functions, then the interchange could be justified on the usual grounds of interchanging the order of integration in a double integral, although, if the integrals are improper, one might need to be cautious about the infinite limit. As we shall discuss later, the scalar product $\langle x|x'\rangle$ *can* be expressed as an integral, but it is of no help in the present context, as we shall eventually see. We therefore simply accept the interchange as valid.

$$\psi_a(x) = \int_{-\infty}^{\infty} dx' \langle x|x'\rangle \psi_a(x'). \tag{4.115}$$

Finally, the same arguments as those used in connection with Eqs. (3.121), (3.123), and (3.124), but now applied to (4.115), yield Eq. (4.48), thus establishing the delta-function normalization of the ket $|x\rangle$.

This result allows us to derive a generalization of Eq. (3.122). Let $\{|\beta_n\rangle\}$ be any complete, discretely indexed, orthonormal set. It provides a resolution of the identity:

$$\sum_n |\beta_n\rangle\langle\beta_n| = \hat{I}. \tag{4.116}$$

Projecting both sides of (4.116) onto $|x'\rangle$ and then onto $\langle x|$, we find

$$\sum_n \psi_{\beta_n}(x)\psi^*_{\beta_n}(x') = \delta(x - x'), \tag{4.117}$$

where $\psi^*_{\beta_n}(x') \equiv \langle\beta_n|x'\rangle$. Equation (4.117) is the desired generalization. If $|\beta_n\rangle \to |k\rangle$, the 1-D momentum eigenket corresponding to the momentum eigenvalue $p = \hbar k$, then $\langle x|\beta_n\rangle \to \langle x|k\rangle = \psi_k(x) = \exp(ikx)/\sqrt{2\pi}$ and (4.117) goes over to (3.128a).

The Dirac delta function

A rigorous mathematical foundation for quantities like $\langle x|x'\rangle = \delta(x - x')$ that appear under integral signs was first given by Laurent Schwartz (1950) via his theory of distributions. It generalizes the notion of ordinary functions to include integrals whose integrands are products of quantities like $\delta(x - x')$ times functions that are members of certain classes of "test" functions. These integrals are known as distributions. The $\int dx'\, \delta(x - x')$ portion is interpreted as a linear operator or linear functional that, when it acts on a sufficiently well-behaved test function $f(x')$, yields $f(x')$ evaluated at the specific value x of its argument. That is, the linear functional

$$\hat{L}[\bullet] \equiv \int_a^b dx'\, \delta(x - x')\bullet \tag{4.118}$$

means

$$\hat{L}[f] = \int_a^b dx'\, \delta(x - x')f(x') = f(x), \tag{4.119}$$

assuming that $x \in [a, b]$. Discussions of distribution theory can be found in many books, two of which are those by Lighthill (1958) and by Richtmyer (1978).

From the rigorous viewpoint of distribution theory, a quantity like $\delta(x - x')$ has no existence independent of an integral sign (representing the action of the appropriate linear functional on a test function). Sometimes, however, it is useful to *pretend* that quantities like $\delta(x - x')$ can be treated as ordinary functions, bearing in mind in adopting this pretense that there is a rigorous body of mathematics that can be invoked to justify the properties assigned to these improper functions. We adopt this alternate viewpoint here. This posture is attractive to the physicist, since use of $\delta(x - x')$ resembles to some extent a *Gedankenexperiment*, in particular because $\delta(x - x')$ can be regarded as the result of allowing a parameter to go to zero in various well-behaved functions of $x - x'$ that depend on that parameter. We first list some properties of the delta function and then consider it as a limit.

In addition to the properties (3.124), $\delta(x - y)$ also obeys

$$\int_a^b dy\, \delta'(x-y) f(y) = -f'(x), \qquad x \in [a, b], \tag{4.120}$$

with $' = d/dx$, as well as

$$\delta(ax) = a^{-1}\delta(x), \tag{4.121}$$

$$\delta(f(x)) = \sum_j [f'(x_j)]^{-1}\delta(x - x_j) \tag{4.122}$$

and

$$\delta(x-a)\delta(x-b) = \delta(x-a)\delta(a-b), \tag{4.123}$$

where in (4.121) a is a constant and in (4.122) $\{x_j\}$ is the set of points for which $f(x_j) = 0$. The last two relations can be established by inserting each side of the equality into integrals involving a well-behaved function.

We next consider $\delta(x)$ as the limit of well-behaved functions. For example, from the properties (3.124a) and (3.124b), $\delta(x)$ can be thought of as the limit $\eta \to 0$ of a function $R_\eta(x)$ which is zero everywhere except for $x \in [-\eta/2, \eta/2]$, for which $R_\eta(x) = 1/\eta$:

$$R_\eta(x) = \begin{cases} \eta^{-1}, & x \in [-\eta/2, \eta/2], \\ 0, & \text{otherwise}. \end{cases} \tag{4.124}$$

Plots of $R_\eta(x)$ for several η are given in Fig. 4.5. It is evident graphically that

$$\delta(x) = \lim_{\eta \to 0} R_\eta(x). \tag{4.125}$$

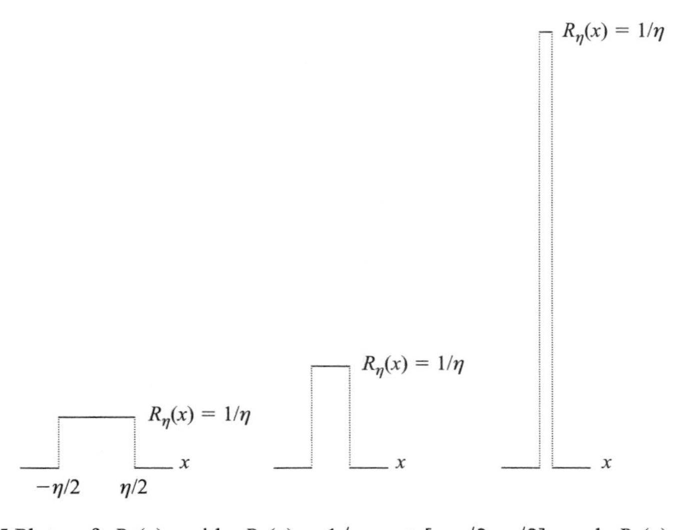

Fig. 4.5 Plots of $R_\eta(x)$, with $R_\eta(x) = 1/\eta$, $x \in [-\eta/2, \eta/2]$, and $R_\eta(x) = 0$, $x \notin [-\eta/2, \eta/2]$, for several values of η. In the limit $\eta \to 0$, $R_\eta(x) \to \delta(x)$.

The preceding representation of $\delta(x)$ is in terms of the limiting behavior of a discontinuous function. There are also representations in terms of continuous functions. One such involves Gaussians, viz.,

$$\delta(x) = \lim_{\eta \to 0} \frac{1}{\eta\sqrt{\pi}} e^{-x^2/\eta^2}. \tag{4.126}$$

Shown in Fig. 4.6 is the function $e^{-x^2/\eta^2}/(\eta\sqrt{\pi})$ for several values of η; the limiting behavior is evident and the integral of this function over all x is easily shown to be unity, as it must be.

Another continuum representation of $\delta(x)$, and one of the most useful, is

$$\delta(x) = \lim_{\eta \to 0} \frac{1}{\pi} \frac{\sin(x/\eta)}{x}. \tag{4.127}$$

Since $\int_{-\infty}^{\infty} dx\,[\sin(x/\eta)]/(\pi x) = 1$ independent of η, the form (4.127) satisfies (3.124b), while the curves of this function shown in Fig. 4.7 for various η indicate that

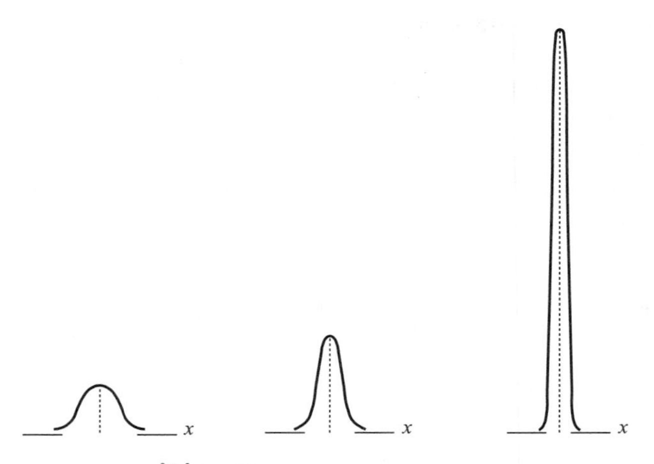

Fig. 4.6 Sketches of $e^{-x^2/\eta^2}/\eta\sqrt{\pi}$ for several (decreasing) values of η.

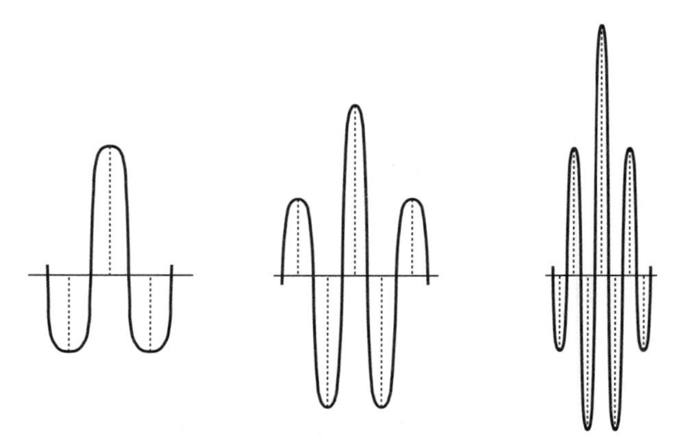

Fig. 4.7 Plots of $(1/\pi)\sin(x/\eta)/x$ for decreasing values of η.

Eq. (3.124a) is obeyed in the limit. From (4.127) it is straightforward to obtain an integral representation for $\delta(x)$:

$$\begin{aligned}\delta(x) &= \lim_{\eta\to 0}\frac{1}{2\pi}\int_{-1/\eta}^{1/\eta} dk\, e^{ikx} \\ &= \frac{1}{2\pi}\int_{-\infty}^{\infty} dk\, e^{ikx}.\end{aligned} \tag{4.128}$$

Thus, $\delta(x)$ is the Fourier transform of unity (recall Eq. (3.128a)). The relation (4.128) will play an important role in the normalization of the eigenfunctions of the momentum operator. Other representations of the delta function are found in the problems.

Delta functions arise in the mathematical descriptions of a variety of physical systems. One should therefore be aware that, in integrals over variables that have a dimension (such as length or velocity), the effect of the delta function is to change the dimensionality of the integral. As an example, consider a 1-D Boltzmann gas, composed of structureless, non-interacting particles of mass m, in equilibrium at absolute temperature T. Its velocity distribution function is

$$f(v_x) = \frac{1}{\sqrt{\pi}v_0}e^{-(v_x/v_0)^2}, \tag{4.129}$$

where $v_0 = \sqrt{2k_B T/m}$. The factor $1/(\sqrt{\pi}v_0)$ on the RHS plays a double role. First, it guarantees that $f(v_x)$ is properly normalized:

$$I = \int_{-\infty}^{\infty} dv_x\, f(v_x) = 1. \tag{4.130}$$

Second, it ensures that, even though the variable v_x has the dimension of velocity, the integral in (4.130) is *dimensionless* – just like its integrand – because v_0 also has the dimension of velocity. Let us now consider the following integral in place of (4.130):

$$I_1 = \int_{-\infty}^{\infty} dv_x\, f(v_x)\delta(v_x) = \frac{1}{v_0\sqrt{\pi}}. \tag{4.131}$$

The effect of the delta function is thus to give I_1 a dimension, viz., an inverse velocity.

Suppose next that $f(x_1, x_2)$ is a sufficiently well-behaved function of x_1 and x_2. Using the properties of the delta function, one has

$$\int dx_1'\, dx_2'\, f(x_1', x_2')\delta(x_1 - x_1') = \int dx_2'\, f(x_1, x_2'), \tag{4.132}$$

with the understanding that the x_1' integration range includes x_1. It therefore follows that

$$\int dx_1'\, dx_2'\, f(x_1', x_2')\delta(x_1 - x_1')\delta(x_2 - x_2') = f(x_1, x_2), \tag{4.133}$$

with a corresponding understanding about the x_2' integration range.

The extension of these results to functions of N variables is straightforward. If the N-D volume element $d^N x$ is defined by

$$d^N x \equiv dx_1\, dx_2 \cdots dx_N \tag{4.134}$$

and the N-variable function $f(x_1, x_2, \ldots, x_N)$ is sufficiently well-behaved, then

$$\int d^N x' f(x'_1, x'_2, \ldots, x'_N) \prod_{j=1}^{N} \delta(x_j - x'_j) = f(x_1, x_2, \ldots, x_N). \tag{4.135}$$

A shorthand notation is often employed to make relations like (4.135) more compact. Let x^N be a single (N-D) variable denoting the N variables $\{x_j\}_1^N$ and *define* the N-D delta function by

$$\delta(\xi^N - x^N) \equiv \prod_{j=1}^{N} \delta(\xi_j - x_j), \tag{4.136}$$

where ξ_j will be an integration variable. Then Eq. (4.135) can be expressed as

$$\int d^N \xi \, f(\xi^N) \delta(\xi^N - x^N) = f(x^N). \tag{4.137}$$

Note that, both in (4.135) and in (4.137), the range of the jth integration variable must include x_j, $\forall\, j$.

The 3-D ($N = 3$) case is often encountered. In place of x^N one writes $\mathbf{r}$, so that (4.137) becomes

$$\int d^3 r' \, f(\mathbf{r}') \delta(\mathbf{r}' - \mathbf{r}) = f(\mathbf{r}), \tag{4.138}$$

where

$$\delta(\mathbf{r}' - \mathbf{r}) = \delta(\mathbf{r} - \mathbf{r}') \equiv \delta(x - x')\delta(y - y')\delta(z - z') \tag{4.139}$$

and x, y, and z must locate a point in the volume containing $\mathbf{r}'$.

It is possible to transform from rectangular coordinates to any other orthogonal set, in which case the RHS of (4.139) changes. Consider spherical polar coordinates. The volume element $d^3 r$ is given by $d^3 r = r^2 \, dr \sin\theta \, d\theta \, d\phi$ and $\delta(\mathbf{r} - \mathbf{r}')$ goes over to

$$\delta(\mathbf{r} - \mathbf{r}') = \frac{1}{rr'} \delta(r - r')\delta(\theta - \theta')\delta(\phi - \phi') \tag{4.140a}$$

$$= \frac{1}{rr'} \delta(r - r')\delta(\Omega - \Omega'), \tag{4.140b}$$

where $\Omega = (\theta, \phi)$ is the solid angle and $d\Omega = \sin\theta \, d\theta \, d\phi$, so that $d^3 r = r^2 \, dr \, d\Omega$.

The analog in 3-D of Eq. (4.128) is

$$\delta(\mathbf{r} - \mathbf{r}') = \left(\frac{1}{2\pi}\right)^3 \int d^3 k \, e^{i\mathbf{k}\cdot(\mathbf{r} - \mathbf{r}')}, \tag{4.141}$$

which is the coordinate-space version of $\hat{I} = \int d^3 k \, |\mathbf{k}\rangle\langle\mathbf{k}|$, where $|\mathbf{k}\rangle$ is a momentum eigenstate (see Chapter 5) and $\langle\mathbf{r}|\mathbf{k}\rangle = (2\pi)^{-3/2} \exp(i\mathbf{k} \cdot \mathbf{r}) \equiv \psi_{\mathbf{k}}(\mathbf{r})$.

Delta functions occur or are useful in a wide variety of physical situations. For example, a point charge q located at $\mathbf{r}_0$ has a charge density $\rho(\mathbf{r})$ proportional to a delta function:

$$\rho(\mathbf{r}) = q\delta(\mathbf{r} - \mathbf{r}_0). \tag{4.142}$$

An analogous expression holds for the current density $\mathbf{j}(\mathbf{r})$ associated with a point charge.

$\delta(\mathbf{r}-\mathbf{r}')$ is also the source term for the Green function $G(\mathbf{r},\mathbf{r}')=G(\mathbf{r}-\mathbf{r}')$ for Poisson's equation, viz.,

$$\nabla^2 G(\mathbf{r}-\mathbf{r}')=\delta(\mathbf{r}-\mathbf{r}'); \tag{4.143}$$

the solution to Eq. (4.143) is

$$G(\mathbf{r}-\mathbf{r}')=-\frac{1}{4\pi}\frac{1}{|\mathbf{r}-\mathbf{r}'|}. \tag{4.144}$$

In the case of the Green function for the Helmholtz equation,

$$(\nabla^2+k^2)G_k(\mathbf{r}-\mathbf{r}')=\delta(\mathbf{r}-\mathbf{r}'), \tag{4.145}$$

the solution is no longer unique; two possibilities are[14]

$$G_k^{(\pm)}(\mathbf{r}-\mathbf{r}')=-\frac{1}{4\pi}\frac{e^{\pm ik|\mathbf{r}-\mathbf{r}'|}}{|\mathbf{r}-\mathbf{r}'|}. \tag{4.146}$$

Locality and the coordinate-space representations of operators

The phrase "in the coordinate representation" was used in Chapter 3 in connection with the momentum operator $\hat{\mathbf{P}}$, the Hamiltonian $\hat{H}$, and the wave function $\psi_n(x)$. A general definition of the last of these quantities was given earlier in this section (Eq. (4.111)): $\psi_\alpha(x)\equiv\langle x|\alpha\rangle$. For a particle moving in 3-D, where $\mathbf{r}$ is the position, one replaces (4.111) by

$$\psi_\alpha(\mathbf{r})\equiv\langle\mathbf{r}|\alpha\rangle. \tag{4.147}$$

In contrast, the coordinate representation of operators has not yet been defined. We do so next, where the operators and vectors will refer to a single particle. In this case

$$\hat{\mathbf{Q}}|\mathbf{r}\rangle=\mathbf{r}|\mathbf{r}\rangle, \tag{4.148}$$

with the "orthonormality" relation

$$\langle\mathbf{r}'|\mathbf{r}\rangle=\delta(\mathbf{r}-\mathbf{r}'), \tag{4.149}$$

while the resolution of the identity is

$$\hat{I}=\int d^3r'\,|\mathbf{r}'\rangle\langle\mathbf{r}'|. \tag{4.150}$$

Let $|\alpha\rangle$ and $|\beta\rangle$ be two vectors in $\mathcal{H}$ related by

$$|\alpha\rangle=\hat{A}|\beta\rangle. \tag{4.151}$$

The coordinate space wave functions $\psi_\alpha(\mathbf{r})$ and $\psi_\beta(\mathbf{r}')$ are given by

$$\psi_\alpha(\mathbf{r})=\langle\mathbf{r}|\alpha\rangle \tag{4.152}$$

and

$$\psi_\beta(\mathbf{r}')=\langle\mathbf{r}'|\beta\rangle. \tag{4.153}$$

[14] These Green functions are discussed in many texts on electrodynamics; for example, Marion (1965) and Jackson (1975). See also the discussion in Chapter 15.

Projecting both sides of (4.151) onto $\langle \mathbf{r}|$ and using (4.152) gives

$$\psi_\alpha(\mathbf{r}) = \langle \mathbf{r}|\hat{A}|\beta\rangle; \tag{4.154}$$

inserting (4.150) to the right of $\hat{A}$ in (4.154) and interchanging the matrix-element symbol and the integration over $\mathbf{r}'$ leads to

$$\psi_\alpha(\mathbf{r}) = \int d^3r' \, \langle \mathbf{r}|\hat{A}|\mathbf{r}'\rangle \psi_\beta(\mathbf{r}'). \tag{4.155}$$

In analogy to $A_{mn} = \langle e_m|\hat{A}|e_n\rangle$, we define $\hat{A}(\mathbf{r}, \mathbf{r}')$ via

$$\hat{A}(\mathbf{r}, \mathbf{r}') \equiv \langle \mathbf{r}|\hat{A}|\mathbf{r}'\rangle, \tag{4.156}$$

so that (4.155) reads

$$\psi_\alpha(\mathbf{r}) = \int d^3r' \, \hat{A}(\mathbf{r}, \mathbf{r}')\psi_\beta(\mathbf{r}'). \tag{4.157}$$

Notice that, despite the matrix-element form of (4.156), the carat $\hat{}$ is still retained: $\hat{A}(\mathbf{r}, \mathbf{r}')$ can be an operator! More on this below.

Equation (4.156) defines the coordinate(-space) representation of the operator. Its dependence is on both spatial positions $\mathbf{r}$ and $\mathbf{r}'$. As a result, the abstract operator relation $|\alpha\rangle = \hat{A}|\beta\rangle$ (Eq. (4.151)) has been transformed into an ordinary 3-D integral or *non-local* relation. The matrix element $\langle \mathbf{r}|\hat{A}|\mathbf{r}'\rangle$ is consequently said to be *non-local*. However, many quantal operators enjoy a special property: they are *local* in the coordinate representation as follows:

$$\text{locality of } \hat{A} \Rightarrow \langle \mathbf{r}|\hat{A}|\mathbf{r}'\rangle = \hat{A}(\mathbf{r}, \mathbf{r}') \to \delta(\mathbf{r} - \mathbf{r}')\hat{A}(\mathbf{r}). \tag{4.158}$$

In the treatment of locality followed in this book, the placing of the $\delta(\mathbf{r} - \mathbf{r}')$ factor to the left of $\hat{A}(\mathbf{r})$ in (4.158) is crucial: the opposite order is not allowed.

Locality of $\hat{A}$ ensures that $\hat{A}$ has only diagonal matrix elements in a coordinate representation. With this property, (4.155) and (4.157) become

$$\psi_\alpha(\mathbf{r}) = \hat{A}(\mathbf{r})\psi_\beta(\mathbf{r}). \tag{4.159}$$

When $\alpha = \beta$, Eq. (4.159) is the typical form taken by quantal eigenvalue equations. For example, the time-independent Schrödinger equation results from $\hat{A}(\mathbf{r}) = \delta_{\alpha\beta}\hat{H}(\mathbf{r})/E_\alpha$. Of course, one must know the functional form of the local realization (the coordinate-space representation) of $\hat{A}$ in order to proceed. For quantal operators, these are given in Chapter 5, Postulate IV. An example is the momentum operator $\hat{\mathbf{P}}$, whose (local) coordinate-space form was stated in Section 3.2: $\hat{\mathbf{P}}(\mathbf{r}) = -i\hbar\nabla$. Other local operators are orbital angular momentum, kinetic energy, and all potentials with classical-physics equivalents. Recall, e.g., that the Hamiltonian operators introduced in Sections 3.2 and 3.3 were all local: $\langle x|\hat{H}|\alpha\rangle = \hat{H}(x)\psi_\alpha(x)$, where $\hat{H}(x)$ is given by Eq. (3.75).

We close this section with some comments concerning Eqs. (4.156) and (4.158). Were the matrix elements in (4.156) formed using normalizable states, then the result would be a number, not an operator. However, the state $|\mathbf{r}\rangle$ (and also the momentum eigenstate $|\mathbf{p}\rangle = |\hbar\mathbf{k}\rangle = |\mathbf{k}\rangle$) is extraordinary: it is not only improper (and thus not in $\mathcal{H}$) but, when it is used to form matrix elements of an operator $\hat{A}$, the result can continue to be an operator. This is consistent with the locality property (4.158), even to the point of writing the coordinate representation of $\langle\alpha|\underset{\leftarrow}{\hat{A}}^\dagger|\mathbf{r}\rangle$, with $\hat{A}$ both Hermitian and local, as

$$\langle\alpha|\underset{\leftarrow}{\hat{A}}^\dagger|\mathbf{r}\rangle = \int d^3r'\, \psi_\alpha^*(\mathbf{r}')\underset{\leftarrow}{\hat{A}}^\dagger(\mathbf{r}')\delta(\mathbf{r}-\mathbf{r}') = \psi_\alpha^*(\mathbf{r})\underset{\leftarrow}{\hat{A}}^\dagger(\mathbf{r}). \tag{4.160}$$

The behavior just described differs from that introduced by Dirac (1947), for whom the definition of locality means the ordering $\hat{A}(\mathbf{r}')\delta(\mathbf{r}'-\mathbf{r})$, with $\hat{A}(\mathbf{r}')$ acting on the delta function. In the case of the 1-D momentum operator $\hat{P}_x(x)$, his coordinate representation is $\langle x|\hat{P}_x|x'\rangle = i\hbar\delta'(x-x')$, where δ' means $d\delta(x-x')/dx'$. Hence, in $\langle x|\hat{P}_x|x'\rangle\psi_\alpha(x')$, the derivative operator does not act on $\psi_\alpha(x')$. Naturally, the end result is the same (as it must be) because of Eq. (4.120). Because of our preference for avoiding where possible derivatives of delta functions, we have introduced the definitions (4.156) and (4.158).

4.5. Functions of operators

Apart from the infinite dimension of $\mathcal{H}$ and the existence of continuously indexed vectors, we have not distinguished between $\mathcal{H}$ and its finite-dimensional counterpart $\mathcal{M}$ in our treatment of complete sets, eigenvalue problems, and operators. This lack of distinction tends, from the physicist's perspective, to be unimportant, and that is largely how we shall regard it. From the mathematician's viewpoint, however, the transition from $\mathcal{M}$ to $\mathcal{H}$ involves identifying and dealing with possible pathologies. These arise because the spectra of operators on $\mathcal{H}$ can be unbounded. Put more technically, an operator $\hat{A}$ is bounded if the following condition involving norms is satisfied: $\|\hat{A}|\alpha\rangle\| \leqslant N\|\,|\alpha\|$, $\forall\ |\alpha\rangle \in \mathcal{H}$ and $N<\infty$. Otherwise, $\hat{A}$ is unbounded. Both bounded and unbounded operators occur in quantum theory, with observables typically being imaged by unbounded operators.

Although we assume in this book that quantal operators on $\mathcal{H}$ are free of the various theoretical pathologies (e.g., we assume that $(\hat{A}^\dagger)^\dagger = \hat{A}$),[15] there is one aspect of unbounded operators that needs to be kept in mind, viz., that, if $\hat{A}$ is unbounded, the power-series expansion of some function $f(\hat{A})$ might not converge, even though the power series for $f(z)$ does converge when z is a complex number. This remark is especially relevant when a unitary operator is expressed as an exponential, viz., $\hat{U}(x) = \exp(i\hat{A}x)$, with $\hat{A}^\dagger = \hat{A}$ and x real. Should the power series for $\exp(i\hat{A}x)$ fail to converge, $\hat{U}(x)$ can still be defined through Stone's theorem, discussed below. Stone's theorem and the lemma preceding it are the only advanced weapons in the Hilbert-space arsenal to which we shall refer, and all the reader will really need to know about them is their existence.

Functions of a single operator

This subsection briefly reviews some operator functions. Probably the simplest of these is the square $\hat{A}^2$ of the operator $\hat{A}$. Its meaning is obvious:

$$\hat{A}^2 = \hat{A}\hat{A}, \tag{4.161}$$

[15] The reader interested in such pathologies can find enlightening discussions in Richtmyer (1978) and Reed and Simon (1972), among many other monographs.

such that

$$\hat{A}^2|u\rangle = \hat{A}(\hat{A}|u\rangle). \tag{4.162}$$

The operator $\hat{A}^n$ is formed by $\hat{A}$ acting n times:

$$\begin{aligned}\hat{A}^n|u\rangle &= \underbrace{\hat{A}\hat{A}\hat{A}\ldots\hat{A}}_{n\text{ factors}}|u\rangle \\ &= \hat{A}^{n-1}(\hat{A}|u\rangle).\end{aligned} \tag{4.163}$$

We next give a definition of an operator function analogous to that of a function of a complex variable z. Just as $f(z)$ is a function of z defined by a functional form, e.g., z^n, e^z, $\sin z$, or by a Taylor's series, etc., so will $f(\hat{A})$ be similarly defined. Suppose that $f(z)$ can be expressed as a Taylor series

$$f(z) = f(z=0) + zf'(z=0) + \frac{z^2}{2!}f''(z=0) + \cdots, \tag{4.164}$$

where $f^{(n)}$ means the nth derivative. Then $f(\hat{A})$ is given by[16]

$$f(\hat{A}) = f(\hat{A}=0) + \hat{A}f'(\hat{A}=0) + \hat{A}^2 f''(\hat{A}=0)/2! + \cdots. \tag{4.165}$$

Equation (4.165), being an infinite power series, has a formidable appearance. A few simple examples are given shortly. Before doing so, it helps to consider the case when $f(\hat{A})$ acts on one of its own eigenfunctions. Let the eigenvalue relation be

$$\hat{A}|a_n\rangle = a_n|a_n\rangle. \tag{4.166}$$

Then it is straightforward to show that

$$f(\hat{A})|a_n\rangle = f(a_n)|a_n\rangle. \tag{4.167}$$

In this case, the operator function is replaced by an ordinary function; this case occurs often in quantum mechanics.

Apart from all coefficients $f^{(n)}$ but one being zero, the next simplest example of (4.165) is an nth-order polynomial, which occurs when $f^{(m)} = 0$, $m \geqslant n+1$:

$$f(\hat{A}) = \sum_{m=0}^{n} a_m \hat{A}^m, \tag{4.168}$$

where $a_m = f^{(m)}(\hat{A}=0)/m!$. The interpretation of $\hat{A}^m$ has already been given. Inverse powers of $\hat{A}$ are allowed as long as $\hat{A}$ is non-singular. When the inverses of operators are needed, then special prescriptions for handling any singularities must be given, as discussed briefly for the case of scattering in Chapter 15.

More common than powers of operators and operator polynomials in quantum mechanics are operators that occur in exponentials, e.g.,

$$\hat{B} = e^{z\hat{A}}, \tag{4.169}$$

where z is a (complex) number. The exponential can be given meaning via its Taylor series, which is *assumed convergent*, as before:

16 The series (4.165) is meaningful only if it converges to $f(\hat{A})$.

$$e^{z\hat{A}} = \sum_{n=0}^{\infty} \frac{z^n}{n!} \hat{A}_n. \tag{4.170}$$

An especially important instance of (4.169) in quantum mechanics is the case in which $\hat{A}$ is Hermitian ($\hat{A}^\dagger = \hat{A}$) and z is purely imaginary ($z = ix$, x real), since $\hat{B}$ then becomes a *unitary* operator $\hat{U}$: $\hat{B} \to \hat{U}$, given by

$$\hat{U} = e^{ix\hat{A}}, \tag{4.171}$$

with

$$\hat{U}^\dagger = e^{-ix\hat{A}} \tag{4.172}$$

such that

$$\begin{aligned} \hat{U}\hat{U}^\dagger &= e^{ix\hat{A}} e^{-ix\hat{A}^\dagger} \\ &= e^{(ix\hat{A} - ix\hat{A}^\dagger)} = e^{\hat{0}} = \hat{I}. \end{aligned} \tag{4.173}$$

Here, $\hat{0}$ is the operator which multiplies all vectors by zero, yielding zero. As will be seen later, (4.171) is the form taken by the quantal operators for evolution in time and for translations and rotations in space.

Stone's theorem

When $\hat{A}$ is unbounded, the power series (4.170) cannot be used to define the exponential in (4.171). It can be given meaning, however, either through the functional calculus described in Chapter VIII of Reed and Simon (1972) or via the polar decomposition discussed in Chapter 9 of Richtmyer (1978). As a consequence, although the power-series expansion cannot be invoked when $\hat{A}$ is unbounded, nevertheless, when $x \to \epsilon$, with ϵ a first-order infinitesimal, (4.171) becomes a meaningful, first-order relation:

$$\hat{U} \cong \hat{I} + i\epsilon\hat{A} + O(\epsilon^2). \tag{4.174}$$

Furthermore, from (4.171) one also gets

$$\frac{d\hat{U}}{dx} = i\hat{A}\hat{U}, \tag{4.175}$$

where we assume that $\hat{A}$ and $\hat{U}$ commute, i.e., that

$$\hat{A}\hat{U} - \hat{U}\hat{A} = 0. \tag{4.176}$$

Regarding (4.175) as an equation defining the operator $\hat{U}$ of (4.171) (subject to $\hat{U} \to \hat{I}$ as $\varepsilon \to 0$), it should be clear from (4.175) that

$$\hat{U}^\dagger = e^{ix\hat{A}}, \tag{4.177}$$

where $\hat{A}^\dagger = \hat{A}$ has been used.

The preceding results suggest that $\hat{U}(x) = \exp(ix\hat{A})$ is unitary: $\hat{U}^\dagger = \hat{U}^{-1}$. One must actually show this; in doing so, one infers (4.173) as a consequence. The desired results are summarized in the following lemma, proved by Reed and Simon (1972).

LEMMA

Let $\hat{A} = \hat{A}^\dagger$, x be a real number and $D(\hat{A})$ be the domain over which $\hat{A}$ is defined. Then $\hat{U}(x) = e^{ix\hat{A}}$ is a unitary operator such that

$$\hat{U}(x+y) = \hat{U}(x)\hat{U}(y), \qquad y = \text{real}, \tag{4.178a}$$

$$\lim_{y\to x} \hat{U}(y)|\alpha\rangle \to \hat{U}(x)|\alpha\rangle, \qquad \forall\ |\alpha\rangle \in \mathcal{H}, \tag{4.178b}$$

$$\lim_{x\to 0} \frac{-i}{x}[\hat{U}(x) - \hat{I}]|\alpha\rangle \to \hat{A}|\alpha\rangle, \qquad \forall\ |\alpha\rangle \in D(\hat{A}), \tag{4.178c}$$

and

$$|\alpha\rangle \in D(\hat{A}) \text{ if } \lim_{x\to 0}\frac{1}{x}[\hat{U}(x) - \hat{I}]|\alpha\rangle \text{ exists.} \tag{4.178d}$$

Equation (4.178c) underlies Eq. (4.174). We shall assume that, for quantal observables, $D(\hat{A}) = \mathcal{H}$.

These results allow us to conclude that $\hat{U}$ of (4.171) is indeed a unitary operator. Equations (4.178a) and (4.178b) are essential ingredients underlying Stone's theorem, which plays an important role in defining the operators for evolution in time and for translations and rotations in space. Before quoting the theorem, we introduce the following definition: *a strongly continuous, one-parameter unitary group is an operator $\hat{U}(x)$ satisfying Eqs. (4.178a) and (4.178b)*. Stone's theorem then postulates that $\hat{U}(x)$ is the exponential of a *unique* operator $\hat{A}$, to whit:

STONE'S THEOREM

Let $\hat{U}(x)$ be a strongly continuous one-parameter group on $\mathcal{H}$. Then there exists an $\hat{A} = \hat{A}^\dagger$ such that $\hat{U}(x) = e^{ix\hat{A}}$, with

$$\hat{A} \equiv \lim_{x\to 0}\left\{\frac{-i}{x}[\hat{U}(x) - \hat{I}]\right\}, \tag{4.179}$$

where (4.179) is understood in terms of the action of both sides on any $|\alpha\rangle \in \mathcal{H}$.

A proof of this theorem can be found, e.g., in Reed and Simon (1972), who show among other things that

$$i\hat{A} = \lim_{x\to 0}\frac{d\hat{U}(x)}{dx}, \tag{4.180}$$

a result that justifies Eq. (4.175).

The operator $\hat{A}$ is referred to as the *infinitesimal generator* of $\hat{U}(x)$. Its importance is in part related to the notion of invariance of a quantal system under symmetry operations, a point discussed in detail in later chapters. We remark here only that, if $\hat{B}$ is an operator whose measured values do not change following some symmetry operation represented by $\hat{U}(x) = \exp(-ix\hat{A})$, then $\hat{B}$ is invariant under $\hat{U}(x)$, and this leads to

$$[\hat{A},\ \hat{B}] = 0, \tag{4.181}$$

as will be shown in Chapter 9.

Commutators and functions of two (or more) operators

The crucial step in going from the first equality to the last in Eqs. (4.173) is the Hermiticity of $\hat{A}$. Were $\hat{A}$ not Hermitian, then the product $\exp(ix\hat{A})\exp(-ix\hat{A}^\dagger)$ could *not* be written as $\exp[ix(\hat{A} - \hat{A}^\dagger)]$, *unless* certain conditions were to hold. We examine this and other features of operator products in the present subsection.

The *commutator* of two operators $\hat{A}$ and $\hat{B}$, denoted by the symbol $[\hat{A},\ \hat{B}]$, is defined as

$$[\hat{A},\ \hat{B}] \equiv \hat{A}\hat{B} - \hat{B}\hat{A} = -[\hat{B},\ \hat{A}]. \tag{4.182}$$

The operators $\hat{A}$ and $\hat{B}$ are said to *commute* when their commutator vanishes:

$$[\hat{A},\ \hat{B}] = 0 \Rightarrow \hat{A} \text{ and } \hat{B} \text{ commute.} \tag{4.183}$$

In this case, $\hat{A}\hat{B} = \hat{B}\hat{A}$. In general, two arbitrary operators do not commute; neither do an operator and either its inverse or its Hermitian conjugate (unless in the latter case the operator is Hermitian). In the case of a product $\hat{A}\hat{B}$ of non-commuting operators $\hat{A}$ and $\hat{B}$, the order in which the two operators occur is obviously important, since $\hat{A}\hat{B}$ means that $\hat{B}$ acts first (recall the $N \times N$ matrix case).

As will be discussed in subsequent chapters, commuting operators play a special role in quantum mechanics. Among their properties is the fact that commuting operators share common eigenvectors and eigenvalues (although care must be exercised in the case of degeneracies). The proof of this statement is straightforward, as indicated below.

Suppose that $[\hat{A},\ \hat{B}] = 0$ and $|a_n\rangle$ is the non-degenerate eigenvector of $\hat{A}$ associated with a_n:

$$\hat{A}|a_n\rangle = a_n|a_n\rangle. \tag{4.184}$$

Acting with $\hat{B}$ (from the left) on both sides of (4.184) gives

$$\hat{B}\hat{A}|a_n\rangle = a_n\hat{B}|a_n\rangle. \tag{4.185}$$

Since $\hat{A}\hat{B} = \hat{B}\hat{A}$, (4.185) can be written

$$\hat{A}(\hat{B}|a_n\rangle) = a_n(\hat{B}|a_n\rangle), \tag{4.186}$$

thus establishing that $\hat{B}|a_n\rangle$ is an eigenvector of $\hat{A}$ corresponding to eigenvalue a_n. However, since $|a_n\rangle$ is non-degenerate, then, from (4.186), $B|a_n\rangle$ must be proportional to $|a_n\rangle$, i.e.,

$$\hat{B}|a_n\rangle = b_m|a_n\rangle, \tag{4.187}$$

where b_m is a number. Equation (4.187) proves that $|a_n\rangle$ is an eigenvector of $\hat{B}$ with eigenvalue b_m. (The case of degenerate eigenvectors will be considered in Chapter 9.) We may therefore relabel $|a_n\rangle$:

$$|a_n\rangle \rightarrow |a_n b_m\rangle, \tag{4.188}$$

thereby indentifying the ket as a simultaneous eigenvector of both $\hat{A}$ and $\hat{B}$.

Let us now return to the product of two exponentiated operators. Consider

$$\hat{A}_1 = e^{\hat{B}_1} \tag{4.189}$$

and

$$\hat{A}_2 = e^{\hat{B}_2}. \tag{4.190}$$

One might assume that, if $[\hat{B}_1, \hat{B}_2] = 0$ then $\hat{A}_1\hat{A}_2 = \exp(\hat{B}_1 + \hat{B}_2)$. This conclusion is indeed correct. However, when $[\hat{B}_1, \hat{B}_2] \neq 0$ one can still write an exponentiated form for the product, but now one in which $[\hat{B}_1, \hat{B}_2]$ appears, viz.,

$$\hat{A}_1\hat{A}_2 = e^{(\hat{B}_1+\hat{B}_2)}e^{[\hat{B}_1,\hat{B}_2]/2}, \tag{4.191}$$

as long as each of $\hat{B}_1$ *and* $\hat{B}_2$ *commutes with the commutator* $[\hat{B}_1, \hat{B}_2]$. The proofs of (4.191) and similar relations are left as exercises.

Exercises

4.1 (a) Determine whether the members of the set of column vectors

$$\left\{ \begin{pmatrix} 1 \\ 0 \\ 1 \\ 0 \end{pmatrix}, \begin{pmatrix} 1 \\ -1 \\ 0 \\ 1 \end{pmatrix}, \begin{pmatrix} 0 \\ 1 \\ -1 \\ 1 \end{pmatrix}, \begin{pmatrix} 1 \\ -1 \\ 1 \\ -1 \end{pmatrix} \right\}$$

are linearly independent. If they are not, use the appropriate ones from the set plus enough others to construct a set of four 4×1 linearly independent, real vectors.

(b) From the preceding result, construct a real orthonormal basis of four 4×1 vectors (Dettmann (1988)).

4.2 (a) Find the eigenvalues and (possibly complex) 2×1 eigenvectors of the two 2×2 matrices

$$M_1 = \begin{pmatrix} 0 & -1 \\ 1 & 0 \end{pmatrix}, \qquad M_2 = \begin{pmatrix} 0 & i \\ i & 0 \end{pmatrix}.$$

(b) Determine whether the members of the sets of eigenvectors of M_1 and of M_2 are each linearly independent and also whether the members of each pair are orthogonal to one another.

(c) If linear independence and orthogonality hold for either pair (or both), create an orthonormal basis and then determine the expansion coefficients of one (or both) vector(s) expanded in a basis created from the other(s).

(d) If it is possible to use each pair to construct orthonormal bases, show that the matrix relating them is unitary.

4.3 (a) Find the eigenvalues and (possibly complex) eigenvectors of

$$M = \begin{pmatrix} 0 & 1 & 0 \\ 1 & 0 & 1 \\ 0 & 1 & 0 \end{pmatrix}.$$

(b) You need not orthonormalize the eigenvectors but show that they are linearly independent.

4.4 Each of the pairs $\mathbf{u}_1 = \binom{1}{1}/\sqrt{2}$, $\mathbf{u}_2 = \binom{1}{-1}/\sqrt{2}$ and $\mathbf{v}_1 = \binom{1}{i}/\sqrt{2}$, $\mathbf{v}_2 = \binom{1}{-i}/\sqrt{2}$ is an orthonormal basis for arbitrary 2×1 vectors $\binom{a}{b}$.

(a) Determine the unitary 2×2 matrix relating these two bases.

(b) (i) Construct the 2×2 projection-operator matrices which project onto $\mathbf{u}_1$, $\mathbf{u}_2$, $\mathbf{v}_1$, and $\mathbf{v}_2$, denoting them P_{u_1}, P_{u_2}, P_{v_1} and P_{v_2}, respectively.

(ii) Show that $P_{u_2}P_{u_1} = 0 = P_{v_1}P_{v_2}$ and that $P_{u_1} + P_{u_2} = P_{v_1} + P_{v_2} = I$, the 2×2 unit matrix.

(c) The operator $\hat{M}$ is represented in the $\mathbf{u}_1$, $\mathbf{u}_2$ basis by the matrix

$$M_u = \begin{pmatrix} 0 & 1 \\ 1 & 0 \end{pmatrix}.$$

What is the matrix M_v representing $\hat{M}$ in the basis $\mathbf{v}_1$, $\mathbf{v}_2$?

4.5 In the basis

$$\mathbf{e}_1 = \begin{pmatrix} 1 \\ 0 \\ 0 \end{pmatrix}, \qquad \mathbf{e}_2 = \begin{pmatrix} 0 \\ 1 \\ 0 \end{pmatrix}, \qquad \mathbf{e}_3 = \begin{pmatrix} 0 \\ 0 \\ 1 \end{pmatrix},$$

the operator $\hat{M}$ is represented by the matrix

$$M_e = \begin{pmatrix} \frac{1}{2} & -\sqrt{\frac{1}{2}} & \frac{1}{2} \\ \frac{1}{\sqrt{2}} & 0 & -\sqrt{\frac{1}{2}} \\ \frac{1}{2} & \sqrt{\frac{1}{2}} & \frac{1}{2} \end{pmatrix}.$$

Find the matrix $M_{e'}$ that represents $\hat{M}$ in the basis

$$\mathbf{e}'_1 = \begin{pmatrix} 1 \\ i \\ 0 \end{pmatrix}/\sqrt{2}, \qquad \mathbf{e}'_2 = \begin{pmatrix} 1 \\ -i \\ 0 \end{pmatrix}/\sqrt{2}, \qquad \mathbf{e}'_3 = \mathbf{e}_3.$$

4.6 In a 2×1 column-vector representation, all the vectors are real, that is, for

$$\mathbf{u} = \begin{pmatrix} u_1 \\ u_2 \end{pmatrix},$$

both u_1 and u_2 are real, $\forall\ \mathbf{u}$.

(a) The action of the matrix M_1 on an arbitrary $\mathbf{u}$ yields

$$M_1 \begin{pmatrix} u_1 \\ u_2 \end{pmatrix} = \begin{pmatrix} u_1 + u_2 \\ u_1 - u_2 \end{pmatrix}.$$

Using the definition $(\mathbf{v}, M\mathbf{u})^* = (\mathbf{u}, M^\dagger \mathbf{v})$, where

$$(\mathbf{v}, \mathbf{u}) = (v_1 v_2)\begin{pmatrix} u_1 \\ u_2 \end{pmatrix} = v_1 u_1 + v_2 u_2,$$

show that

$$M_1^\dagger \begin{pmatrix} u_1 \\ u_2 \end{pmatrix} = \begin{pmatrix} u_1 + u_2 \\ u_1 - u_2 \end{pmatrix},$$

i.e., prove that $M_1^\dagger = M_1$.

(b) In analogy to part (a), the action of M_2 on an arbitrary $\mathbf{u}$ is

$$M_2\begin{pmatrix} u_1 \\ u_2 \end{pmatrix} = \begin{pmatrix} 2u_1 - u_2 \\ u_1 + 2u_2 \end{pmatrix}.$$

Find $M_2^\dagger \mathbf{u}$ and determine whether $M_2^\dagger$ equals M_2.

4.7 Let A and B be square matrices. Determine whether the following statements are correct.
(a) If $AB = 0$, then either $A = 0$ or $B = 0$ (or both).
(b) If $A^\dagger A = 0$, then $A = 0$.

4.8 Show that the number of independent real parameters in an $n \times n$ matrix is
(a) n^2, $A^\dagger = A$, $A_{jk} =$ complex,
(b) n^2, $A^\dagger = A^{-1}$, $A_{jk} =$ complex,
(c) $n(n-1)/2$, $A^{\mathrm{T}} = A$, $A_{jk} =$ real.

4.9 Determine which of the following operators are linear:
(a) $\hat{R}(\mathbf{n}, \theta)$, $\hat{R}(\mathbf{n}_1, \theta_1)\hat{R}(\mathbf{n}_2, \theta_2)$, $\hat{R}$ is a rotation operator;
(b) each $\hat{A}$, as defined below:
(i) $\hat{A}\psi(x) = \psi^2(x)$;
(ii) $\hat{A}\psi(x) = C_1 + C_2 x\psi(x)$, C_1 and C_2 are complex numbers;
(iii) $\hat{A}\psi(x) = x^n\psi(x)$, $n \neq 0$;
(iv) $\hat{A}\psi(x) = C_1 x + C_2\psi(x)$, C_1 and C_2 are complex;
(v) $\hat{A}\psi(x) = [\psi(x)]^\alpha$, $\alpha \neq 0$.

4.10 Derive Eqs.(4.91) and (4.92) when $\hat{A}^\dagger = \hat{A}$.

4.11 It was noted in the text that the expansion (4.41) might converge in the mean but not pointwise. This exercise considers the opposite situation (Byron and Fuller (1970)).
(a) Let $\psi_n(x) = [2(2n)^2/\sqrt{\pi}]^{1/2} n x^2 e^{-nx^2}$, $n = 1, 2, 3, \ldots$. Evaluate

$$\psi(x) = \lim_{n\to\infty} \psi_n(x)$$

and show thereby that $\{\psi_n(x)\}$ converges pointwise.
(b) Next examine

$$\lim_{n\to\infty} \int_{-\infty}^{\infty} dx\, |\psi(x) - \psi_n(x)|^2$$

and determine whether $\{\psi_n(x)\}$ converges in the mean.

4.12 Let $\|\psi\| = (\psi, \psi)^{1/2}$ and assume $\|\ \|$ to be finite for all elements $\psi(x)$, $\varphi(x)$.
(a) Show that Schwarz's inequality holds, viz.,

$$\|\psi\| \cdot \|\varphi\| \geqslant |(\psi, \varphi)|^2.$$

Hint: consider $\psi + \lambda\varphi$, with λ a suitably chosen constant.
(b) Prove the triangle inequality:

$$\|\psi + \varphi\| \leqslant \|\psi\| + \|\varphi\|.$$

Hint: consider manipulations involving $\|\psi + \varphi\|^2$.

4.13 Show that
(a) $\lim_{\eta\to\infty} [1 - \cos(\eta x)]/(\pi\eta x^2) \to \delta(x)$,
(b) $\lim_{\eta\to\infty} \sqrt{\frac{2}{\pi}}\eta \cos(\eta^2 x^2) \to \delta(x)$.

4.14 (a) Demonstrate the validity of Eqs. (4.122) and (4.123).
(b) Show that $\delta(x^2 - a^2) = (a/2)[\delta(x - a) + \delta(x + a)]$.

4.15 Use the results of Exercise 4.14 to evaluate

$$\int_0^{\pi} dx \int_1^2 dy\, \delta(\sin^2 x)\delta(x^2 - y^2).$$

(Jones (1979)).

4.16 Let $\hat{A}$ and $\hat{B}$ be Hermitian, and $\hat{C}$ and $\hat{D}$ be unitary. Show that
(a) $(\hat{A}\hat{B})^\dagger = \hat{B}\hat{A}$,
(b) $i(\hat{A} - \hat{A}^\dagger)$ is Hermitian,
(c) $i(\hat{A}\hat{B} - \hat{B}\hat{A})$ is Hermitian,
(d) $\hat{C}^{-1}\hat{A}\hat{C}$ is Hermitian,
(e) $\hat{D}^{-1}\hat{C}\hat{D}$ is unitary,
(f) if $\hat{A} \to$ a square Hermitian matrix A, then $A_{nn}^2 \geqslant 0$.

4.17 Let $\hat{A}^\dagger = \hat{A}$ and $\hat{A}^4 = \hat{I}$. Find the eigenvalues of $\hat{A}$ (Gasiorowicz (1974)).

4.18 The trace of an operator $\hat{A}$, denoted $\mathrm{Tr}\,\hat{A}$, is the sum of its diagonal elements evaluated in any convenient basis. For an $N \times N$ matrix A, $\mathrm{Tr}\,A = \sum_1^N A_{nn}$. Show that
(a) $\mathrm{Tr}(\hat{A}\hat{B}) = \mathrm{Tr}(\hat{B}\hat{A})$,
(b) $\mathrm{Tr}(\hat{A}\hat{B}\hat{C}) = \mathrm{Tr}(\hat{C}\hat{A}\hat{B}) = \mathrm{Tr}(\hat{B}\hat{C}\hat{A}) \neq \mathrm{Tr}(\hat{B}\hat{A}\hat{C})$,
(c) $\mathrm{Tr}(\hat{U}^{-1}\hat{A}\hat{U}) = \mathrm{Tr}(\hat{A})$, where $\hat{U}$ is unitary.

4.19 Show that
(a) $[\hat{A}\hat{B}, \hat{C}] = \hat{A}[\hat{B}, \hat{C}] - [\hat{C}, \hat{A}]\hat{B}$;
(b) $[\hat{A}, [\hat{B}, C]] + [\hat{B}, [\hat{C}, \hat{A}]] + [\hat{C}, [\hat{A}, \hat{B}]] = 0$;
(c) if $[\hat{A}, \hat{B}] = \hat{C}$ and $[\hat{A}, \hat{C}] = 0 = [\hat{B}, \hat{C}]$, then
  (i) $[\hat{A}, \hat{B}^n] = n\hat{C}\hat{B}^{n-1}$,
  (ii) $[\hat{A}, \exp(\alpha\hat{B})] = \alpha\hat{C}\exp(\alpha\hat{B})$, $\alpha =$ constant.

4.20 Let $\hat{A}$ and $\hat{B}$ be operators defined by $\hat{A}\psi(x) = x\, d\psi(x)/dx$ and $\hat{B}\psi(x) = \int_{-\infty}^{x} dy\, y\psi(y)$, where $\psi(x)$ is normalizable.
(a) Evaluate $[\hat{A}, \hat{B}]$.
(b) Determine the eigenvalues and eigenfunctions of $\hat{B}$. (Hint: convert the integral equation into a differential one.) (Kok and Visser (1987)).

4.21 Assuming that the relevant Taylor series is well defined, show that
(a) $e^{i\hat{A}}\hat{B}e^{-i\hat{A}} = \hat{B} + i[\hat{A}, \hat{B}] + \{(i)^2/2!\}[\hat{A}, [\hat{A}, \hat{B}]] + \{(i)^3/3!\}[\hat{A}, [\hat{A}, [\hat{A}, \hat{B}]]] + \cdots$.
Hint: consider $\hat{C}(x) = e^{ix\hat{A}}\hat{B}e^{-ix\hat{A}}$.
(b) $e^{\hat{A}}\hat{B}e^{-\hat{A}} = \hat{B} + [\hat{A}, \hat{B}] + \{1/2!\}[\hat{A}, [\hat{A}, \hat{B}]] + \cdots$.

4.22 Let $[\hat{A}, \hat{B}] = \hat{C}$, with $[\hat{A}, \hat{C}] = 0 = [\hat{B}, \hat{C}]$.
(a) Use Exercise 4.21 to prove that

$$e^{x\hat{A}}\hat{B} = \{\hat{B} + x\hat{C}\}e^{x\hat{A}}.$$

(b) Show, using part (a), that

$$e^{\hat{A}}e^{\hat{B}} = e^{(\hat{A}+\hat{B})}e^{\hat{C}/2}.$$

Hint: let $\hat{D}(x) = e^{x\hat{A}}e^{x\hat{B}}e^{-x(\hat{A}+\hat{B})}$ and show that $d\hat{D}(x)/dx = x\hat{G}(x)\hat{D}(x)$, where $\hat{G}(x)$ is an operator you are to determine in solving this problem.

Part II

THE CENTRAL CONCEPTS

5

The Postulates of Quantum Mechanics

In this chapter the basic concepts of quantum theory are formulated via a self-consistent set of six postulates. These six postulates deal with the following: operator images of physical observables; properties of state vectors and wave functions; the connection with experimental measurements; coordinate-space forms of those operators having classical analogs; the primary dynamical equation of quantum mechanics; and the completeness of the eigenvectors of observables. The postulates are the framework on which the edifice of quantum mechanics is constructed. A separate section is devoted to the statement of each postulate and to a few examples chosen to illustrate or apply it. A comprehensive set of applications is presented in Chapters 6 and 7, where various 1-D systems are used as the vehicles for illustrating the postulates. The basic theory of this chapter is extended and developed in Chapters 8 and 9. Since probability is a key interpretational concept, a brief introduction to this topic is given in an appendix.

We emphasize that our formulation of quantum theory (in this chapter and throughout this book) is restricted to the case of pure states. This is not a major restriction. It means that, when we consider ensembles of individual quantum systems (e.g., single particles, atoms, nuclei, etc.), all of the systems in the ensemble are assumed to be in the same quantum state, not necessarily an eigenstate of a quantal operator. Ensembles for which this assumption does not hold are described by a density operator or density matrix rather than by a pure state; treatments of such cases can be found in references cited later.

5.1. Observables

The first postulate deals with the mathematical imaging of observables by Hermitian operators.

POSTULATE I.

To every physical observable, e.g., position, energy, linear and angular momentum, intrinsic spin, there corresponds a Hermitian linear operator (also denoted an observable) acting on vectors in a separable Hilbert space $\mathcal{H}$. Let $\hat{A}$ be such an observable for some physical system. Its eigenvalues a_n and eigenvectors $|a_n\rangle$ play a special role: the only values of $\hat{A}$ which can be obtained in an ideal measurement are its eigenvalues $\{a_n\}$, while $|a_n\rangle$ is the quantum state of the system when the value of the observable has been measured to be a_n.

Table 5.1 *Physical observables and the mathematical symbols corresponding to their Hilbert-space operators*

Observable	Hilbert-space symbol
Energy	$\hat{H}$
Position	$\hat{Q}_x$ (in 1-D) or $\hat{\mathbf{Q}}$ (in 3-D)
Linear Momentum	$\hat{P}_x$ (in 1-D) or $\hat{\mathbf{P}}$ (in 3-D)
(Orbital) Angular Momentum	$\hat{\mathbf{L}} = \hat{\mathbf{Q}} \times \hat{\mathbf{P}}$
Intrinsic Spin	$\hat{\mathbf{S}}$
Kinetic Energy[a]	$\hat{K} = \hat{\mathbf{P}} \cdot \hat{\mathbf{P}}/(2m)$

[a] Kinetic energy is not an observable classically, but, because it is an operator in quantum mechanics, it is included in this list in order to display its connection to $\hat{\mathbf{P}}$.

This first postulate states that the possible values one can measure for an observable $\hat{A}$ are the eigenvalues of the equation

$$\hat{A}|a_n\rangle = a_n|a_n\rangle. \tag{5.1}$$

Furthermore, if a_n is the measured value of the observable, then $|a_n\rangle$ is the state of that system following the measurement which has yielded a_n. This statement is made under the assumption that only the quantum number a_n is needed to specify the state. We shall see in Chapter 9 how this part of Postulate I is modified when more than one quantum number is needed to specify the state.

Table 5.1 lists both the observables of classical mechanics that have quantal counterparts and the symbols of the Hilbert-space operators that are their mathematical images. Operators such as spin, for which there is no classical equivalent, will be introduced in later chapters. As noted previously, operators are represented by capital letters overlaid with a carat: ˆ; their eigenvalues are generally indicated by the same symbols in lower case, one exception being the energy, the operator for which is $\hat{H}$ whereas the eigenvalue is E, possibly enhanced by sub- or superscripts.

Examples of the operators that image observables and their corresponding eigenvalues and eigenvectors have been encountered in the preceding chapters. The 1-D position operator $\hat{Q}_x$ and its eigenvalue equation (4.47) were introduced in Chapter 4; this led directly to the Dirac delta function, the coordinate representation, and the concept of locality. In Chapter 3 the Hamiltonian operator $\hat{H}$ and the coordinate-space form of the time-independent Schrödinger equation $\hat{H}(x)\psi_n(x) = E_n\psi_n(x)$ for the particle in a 1-D box were explored in detail. The box energy $E_n = n^2\hbar^2\pi^2/(2mL^2)$ and normalized eigenfunction $\psi_n(x) = (2/L)^{1/2}\sin(n\pi x/L^2)$ correspond, respectively, to a_n and to the coordinate-space form of $|a_n\rangle$ in Eq. (5.1).

Measurement (detection) implies that the system undergoes an interaction of some kind. Until the system is altered by a subsequent interaction (which may occur as part of a subsequent measurement), it remains in the state $|a_n\rangle$ when a_n is the measured value of the observable $\hat{A}$.

This forcing of the system into state $|a_n\rangle$ via measurement is, however, not always possible. It will occur only if the detection (or production) apparatus is capable of sufficient discrimination that it selects just the eigenvalue a_n. For example, a_n will be selected if the *resolution* of the apparatus is finer (smaller) than the separation between a_n and $a_{n\pm1}$. Suppose that we have N systems for which $\hat{A}$ is to be measured. If the resolution is fine enough, then the value of $\hat{A}$ for every one of the N systems can be a_n, and in this case each system will be in state $|a_n\rangle$.

Let us next assume that the resolution of the apparatus remains fine enough to isolate each of the eigenvalues, but that the experiment measuring $\hat{A}$ allows each of the N systems to end up with any one of the three values a_{n-1}, a_n, and a_{n+1} in such a way that for each system the state $|\alpha\rangle$ is

$$|\alpha\rangle = \sum_{j=n-1}^{n+1} w_j|a_j\rangle. \tag{5.2}$$

In order for $|\alpha\rangle$ of (5.2) to be a pure state, *both the magnitude and the phase (or relative phase) of each w_j must be known.* An example of (5.2) is Eq. (3.94) with the values $a_2 = a_3 = \sqrt{\frac{1}{2}}$ that led to Eq. (3.99). We emphasize that, whenever a superposition like (5.2) is used to represent a pure state, each w_j is assumed to be known.

Finally, suppose that the apparatus cannot or is not set to discriminate between, say, a_n and $a_{n\pm1}$ (but can distinguish the other a_j from them). In this case, the relative populations of the three eigenvalues among the N systems might be known, but not the relative phase relations between the states $|a_n\rangle$ and $|a_{n\pm}\rangle$. Without knowledge of these phase relations, i.e., if $|w_j|$ but not arg w_j is known, then (5.2) cannot be used to describe each system. The ensemble is then in a *mixed* state and the appropriate tool for describing this situation is the density operator or density matrix. Discussions of mixed states can be found, e.g., in Baym (1976), Cohen-Tannoudji, Diu, and Laloë (1977), Gottfried (1966), and Sakurai (1994). In Section 6.1, as an application of the 1-D quantum-box eigenvalue problem, we discuss a specific instance wherein a *Gedanken* measurement leads to non-fully-determined w_j's, and we also indicate how full knowledge of the w_j's can be attained.

Many observables, of course, can be measured in such a way as to yield a unique a_n and $|a_n\rangle$. However, in the case of a *continuous* spectrum, a linear superposition analogous to (5.2) *always* occurs. The reason is simple. Let $\hat{B}$ be the observable and $\{b\}$ its continuous set of eigenvalues; the eigenvalue equation is

$$\hat{B}|b\rangle = b|b\rangle. \tag{5.3}$$

In order to measure a single – or "sharp" – value b, itself a point in a continuum, the apparatus must be capable of isolating a point, i.e., of distinguishing a value having no extent or "width." However, all detectors are of finite size and will therefore admit not a single point-value b but a range of them. Corresponding to this measured range of b values will be a linear superposition of the states $|b\rangle$. Suppose that the values of b lie in the closed interval $[b_1, b_2]$. Then in this case the pure quantum state $|\beta\rangle$ will be given by

$$|\beta\rangle = \int_{b_1}^{b_2} db\, w(b)|b\rangle, \tag{5.4}$$

where $w(b)$ is a known weight function.

Equation (5.4) may appear to be a trivial extension of (5.2). Its consequences, rather than being trivial, are of the utmost importance. The reason is that (5.4) converts *non-normalizable*, *continuum* states $|b\rangle$ into normalizable ones. That is, the relation $\langle b'|b\rangle = \delta(b - b')$ – illustrated by the example of the states $|x\rangle$ of the 1-D position operator $\hat{Q}_x$ discussed in Section 4.4, where delta-function normalization was introduced – means that $|b\rangle$ is not a Hilbert space vector: $|b\rangle \notin \mathcal{H}$. The effect of the linear superposition (5.4), on the other hand, will be to produce a *normalizable* state $|\beta\rangle$ as long as the following condition holds:

$$\int_{b_1}^{b_2} |w(b)|^2 \, db = \text{finite}. \tag{5.5}$$

Even when $b_2 = -b_1 = \infty$, the weight functions occurring in physics obey (5.5). The end result is that, via (5.4), $|\beta\rangle \in \mathcal{H}$, in contrast to $|b\rangle$. Thus, whereas a non-normalizable $|b\rangle$ cannot be a quantum state (even though b is a value of the observable $\hat{B}$), a linear superposition of the $|b\rangle$ can be such a state. Despite this, one often works directly with the non-normalizable or *improper* states $|b\rangle$, particularly in the case of collisions, as described in Chapters 7 and 15.

5.2. States, wave functions, and probabilities

The state of a quantum system is a vector in $\mathcal{H}$. As such, it is a highly abstract quantity. Its realization as a wave function makes it easier to work with, and its interpretation in terms of probabilities gives it meaning. These aspects are explored in the present section via application of the second postulate to a system consisting of a single particle. The extension to multi-particle systems is straightforward and is discussed later in this book.

Quantal probabilities are the absolute squares of various scalar products, themselves often referred to as *amplitudes*. Two of these scalar products are so important that they are given the special name of *wave functions*. Wave functions are the scalar product of a Hilbert-space state with the eigenstates of the position operator $\hat{\mathbf{Q}}$ or the momentum operator $\hat{\mathbf{P}}$ and, as such, define the position or momentum representation of the state, or, equivalently, the position (or coordinate) and momentum wave functions (recall Section 4.4). The relevant eigenstates and eigenvalues obey

$$\hat{\mathbf{Q}}|\mathbf{r}\rangle = \mathbf{r}|\mathbf{r}\rangle \tag{5.6}$$

and

$$\hat{\mathbf{P}}|\mathbf{p}\rangle = \mathbf{p}|\mathbf{p}\rangle, \tag{5.7}$$

where $\mathbf{r}(\mathbf{p})$ is the position (momentum) of the particle. Since $\mathbf{r}$ and $\mathbf{p}$ are continuous, $|\mathbf{r}\rangle$ and $|\mathbf{p}\rangle$ are each normalized to a delta function:

$$\langle \mathbf{r}'|\mathbf{r}\rangle = \delta(\mathbf{r}' - \mathbf{r}) \tag{5.8}$$

and

$$\langle \mathbf{p}'|\mathbf{p}\rangle = \delta(\mathbf{p}' - \mathbf{p}). \tag{5.9}$$

To obtain the wave functions, and to formulate Postulate II, we introduce the generic single-particle state, denoted $|\alpha\rangle$. It is a Hilbert-space state, which is *assumed to be*

normalized. The coordinate-space representative of $|\alpha\rangle$ is the scalar product $\langle \mathbf{r}|\alpha\rangle$, which we shall denote by the symbol $\psi_\alpha(\mathbf{r})$:

$$\psi_\alpha(\mathbf{r}) \equiv \langle \mathbf{r}|\alpha\rangle \tag{5.10}$$

is the coordinate-space wave function. Similarly, the momentum-space representative of $|\alpha\rangle$ is the scalar product $\langle \mathbf{p}|\alpha\rangle$, for which we use the special symbol $\varphi_\alpha(\mathbf{p})$:

$$\varphi_\alpha(\mathbf{p}) \equiv \langle \mathbf{p}|\alpha\rangle \tag{5.11}$$

is the momentum-space wave function.

These definitions deliberately avoid the usage $\alpha(\mathbf{r})$ and $\alpha(\mathbf{p})$ because the functions labeled by ψ_α and φ_α are different, while calculus tells us that $\alpha(\mathbf{r})$ and $\alpha(\mathbf{p})$ are the same functions of two different variables. That is, if $\mathbf{s}$ is an arbitrary vector variable,

$$\psi_\alpha(\mathbf{s}) \neq \varphi_\alpha(\mathbf{s}),$$

a result that should be no surprise since $\langle \mathbf{r}|$ and $\langle \mathbf{p}|$ are eigenstates of two different operators. The distinction is essential.

POSTULATE II.

The state of a quantum system is a (normalizable) vector in $\mathcal{H}$. It is labeled by the eigenvalues of all operators for which it is a simultaneous eigenstate, and it may be a linear combination of eigenvectors of quantal operators. Scalar products of a state with other vectors (suitably normalized), in $\mathcal{H}$ or not, yield quantal probability amplitudes, the absolute square of each being a probability or a probability density, as follows.

(a) For a quantum system consisting of a single particle, the position- or coordinate-space wave function $\psi_\alpha(\mathbf{r}) = \langle \mathbf{r}|\alpha\rangle$ is the *probability amplitude* that the particle will be at position $\mathbf{r}$. The *probability* $P_\alpha(\mathbf{r})$ for finding the particle between $\mathbf{r}$ and $\mathbf{r} + d\mathbf{r}$ is

$$P_\alpha(\mathbf{r}) = |\langle \mathbf{r}|\alpha\rangle|^2 d^3r \equiv |\psi_\alpha(\mathbf{r})|^2 d^3r; \tag{5.12}$$

hence, $\rho_\alpha(\mathbf{r}) = |\psi_\alpha(\mathbf{r})|^2$ is a standard *probability density*.[1] Corresponding remarks hold for the momentum-space wave function $\varphi_\alpha(\mathbf{p})$. In the case of a particle in 1-D, say x, the analogous quantities are $\psi_\alpha(x) = \langle x|\alpha\rangle$, $P_\alpha(x) = |\psi_\alpha(x)|^2\, dx$, etc.

(b) If the system is in state $|\alpha\rangle$, then the probability amplitude $C_{\beta\alpha}$ for finding it in any other (normalized) quantal state $|\beta\rangle$ is

$$C_{\beta\alpha} = \langle \beta|\alpha\rangle; \tag{5.13}$$

the corresponding (discrete) probability $P_{\beta\alpha}$ is

$$P_{\beta\alpha} = |\langle \beta|\alpha\rangle|^2. \tag{5.14}$$

Neither $|\beta\rangle$ nor $|\alpha\rangle$ need be an eigenstate of a quantal operator.

[1] See Appendix A for a brief treatment of probability theory. The assumption that $|\alpha\rangle$ is normalized to unity, i.e., that $\langle\alpha|\alpha\rangle = 1$, means that the total probability is unity: $\int |\psi_\alpha|^2\, d^3r = 1$.

(c) In the event that a quantal system can evolve to a state $|\alpha, t\rangle$ at time t by more than one path, as, for example, in the case of the two-slit experiment with electrons, then, as long as one employs no device that can determine which path was actually followed, $|\alpha, t\rangle$ is a linear combination of states $|\alpha_j, t\rangle$, where j specifies the particular path:[2]

$$|\alpha, t\rangle = \sum_j w_j |\alpha_j, t\rangle; \tag{5.15}$$

the constant w_j is a known weight. The amplitude $C_{\beta\alpha}(t) = \langle\beta|\alpha, t\rangle$ now becomes a sum of amplitudes $C_{\beta\alpha_j}(t)$ and the probability $P_{\beta\alpha}(t)$ contains interference terms:

$$\begin{aligned} P_{\beta\alpha}(t) = |\langle\beta|\alpha, t\rangle|^2 &= \left|\sum_j w_j \langle\beta|\alpha_j, t\rangle\right|^2 \\ &= \left|\sum_j w_j C_{\beta\alpha_j}(t)\right|^2. \end{aligned} \tag{5.16}$$

On the other hand, the sum in (5.15) collapses to a single term[3] if a determination of the path k that is actually followed is made, in which case $w_j \to \delta_{jk}$.

This postulate is a statement of the probabilistic interpretation of quantum mechanics and each of its three portions has already been exemplified in Section 3.3 via the quantal-box eigenvalue problem. Postion-space probability densities $\rho_n(x)$ (Eq. (3.92)) have been plotted in Fig. 3.8. Parts (b) and (c) of Postulate II are illustrated for the box problem by the superposition wave function (3.94). For instance, corresponding to $C_{\beta\alpha}$ of (5.13) are either of the coefficients a_2 or a_3 in Eq. (3.94): the amplitudes are $a_j = (\Psi_j|\Psi)$, $j = 2$ or 3, and the probability that Ψ_j is present in Ψ is $|a_j|^2$. Furthermore, as noted in the paragraph below Eq. (3.97), $|a_j|^2$ is also the probability that an energy measurement will yield E_j. Finally, since Ψ of (3.94) is a superposition – a coordinate-space version of Eq. (5.15) – the interference noted in connection with (5.16) shows up in the probability density $\rho(x, t)$ of (3.98) and (3.99).

Postulate II is a cornerstone of the theory, from which follows both the uncertainty principle of Heisenberg and a limitation on the kinds of questions it is meaningful to ask in a quantum context. It does *not* mean, however, that quantum theory is imprecise – far from it! Consider, e.g., the deuteron, a stable nucleus formed from a neutron (n) and a proton (p). Although one cannot specify the distance between n and p in the ground state of the deuteron – quantum mechanics yields only the probabilities of various separations – it is possible using quantum theory to calculate the average separation, the binding energy, the total angular-momentum eigenvalue, and the values of the magnetic moment

[2] "Path" as used in this context need not refer to a physical path in coordinate space, although this is the typical interpretation. Thus, some "path" which produced the linear superposition (3.94) of box eigenfunctions was followed.

[3] Equation (5.15) represents $|\alpha, t\rangle$ as a "wave packet," and this collapse or "reduction" of the wave packet has long been a source of controversy and debate. The interested reader can find discussions in Wheeler and Zurek (1983), Bell (1988), Cini and Levy-Leblond (1990), and Peierls (1991). See also Cushing (1994).

and of the quadrupole moment, all to high accuracy. Thus, the inability to know the values of some quantities that in classical physics are known to arbitrary accuracy does not prevent us from using quantum theory to determine many observables with precision. Furthermore, the agreement between quantum theory and experiment is usually so good that discrepancies are attributed not to a failure of quantum theory itself but typically to neglect of dynamical effects that were deemed for some reason to be negligible – and occasionally to errors in measurement.

The ability to calculate certain observables with great accuracy coupled with the inability to know precisely the values of other observables, e.g., exactly where the electron in an H-atom is located relative to the proton, means that one can draw no detailed picture to represent the spatial configuration of an atom or a nucleus. There is no reliable answer to the question "What does a particular microscopic system really look like?"

This inability to provide a picture or likeness of microscopic "reality" is often an unsettling feature when quantum theory is first encountered. It can be discomforting to realize that one must give up the pictures and/or models that one's eyes and classical physics supply for macroscopic systems. This inability to state in detail (to "describe") what is "out there" when one is discussing a microscopic system, for example the H-atom, naturally leads a newcomer to quantum theory to ask what the phrase "a quantum system is *described* by its state vector" is supposed to mean.

The answer is typically quantal, in that the ordinary meanings of "describe" and of "description" are altered: a quantum description of a system is composed of quantum numbers, probability amplitudes, expectation values of relevant operators, rates of transition to various states, and possibly other quantum features, all derivable from the state vector of the system. A pictorial representation is often used as well, although it is never to be interpreted as "what is actually out there." It is based on the probability of finding the system in some spatial volume, or, in the case of a single particle, finding it at a particular point. In the case of bound states, these probability distributions are usually localized in such a way that the vast majority of the distribution is confined to a relatively small spatial volume. (Recall the distributions for a particle in a box, Figs. 3.8 and 3.9.) For an atom, this results in the electrons being localized as a "charge cloud," as we indicate shortly in the case of the ground state of hydrogen. In general, these charge-cloud distributions for the H-atom provide a basis for understanding the structures of atoms and many molecules; included in the latter case are angles and lengths of chemical bonds (which are measurable quantities). Such visualizations are helpful in gaining a "feel" for the results of often complex computations, i.e., an understanding of the system, and we shall thus "pictorialize" the calculations arising in later chapters of this book. It is in these pictorializations and the construction of quantal models for more complex systems, as opposed to solving the equations, that the physics is often apprehended.

To illustrate the foregoing comments, we consider a realistic example (as opposed to the quantal box): the ground state of the H-atom. The proton, here assumed infinitely massive, is taken as the origin of the coordinate system, so that H becomes a one-particle (i.e., a one-electron) system. It is further assumed that effects due to the intrinsic spin of the electron (discussed in Chapter 13) and to special relativity may be ignored. On ignoring the latter two effects, the quantal energy spectrum of H becomes identical to that of the simple Bohr theory, viz., $E_n = -\kappa_e e^2/(2a_0 n^2)$. The associated eigenstates are labeled by three quantum numbers, one of which is n. The other two quantum numbers are denoted ℓ and m_ℓ (see Chapter 11 for details); for fixed n, the allowed values of ℓ and m_ℓ are $\ell = 0, 1, 2, \ldots, n-1$ and, for each ℓ, $m_\ell = -\ell, -\ell+1, -\ell+2, \ldots, \ell-1, \ell$. Hence, for this model of hydrogen, the label α in the generic state $|\alpha\rangle$ is replaced by the three integers $n\ell m_\ell$:

$$|\alpha\rangle \rightarrow |n\ell m_\ell\rangle, \text{ non-relativistic hydrogen atom.} \tag{5.17}$$

Correspondingly, the coordinate-space wave function becomes

$$\psi_{n\ell m_\ell}(\mathbf{r}) = \langle \mathbf{r}|n\ell m_\ell\rangle, \tag{5.18}$$

where $\mathbf{r}$ is the position of the electron relative to the proton.

By assumption, $|n\ell m_\ell\rangle$ is normalized, i.e.,

$$\langle n\ell m_\ell|n\ell m_\ell\rangle = 1. \tag{5.19}$$

This implies that

$$\int d^3r\, |\psi_{nlm_\ell}(\mathbf{r})|^2 = 1, \tag{5.20}$$

a relation that follows from insertion into (5.19) of

$$\hat{I} = \int d^3r\, |\mathbf{r}\rangle\langle\mathbf{r}|, \tag{5.21}$$

one of several possible resolutions of the identity. Equation (5.20) states that

$$\rho_{n\ell m_\ell}(\mathbf{r}) = |\psi_{n\ell m_\ell}(\mathbf{r})|^2 \tag{5.22}$$

is a normalized probability distribution.

Before specializing to the ground state, we note from Appendix A that normalization of $\psi_{n\ell m_\ell}(\mathbf{r})$ means that the probability $P_{n\ell m_\ell}(V)$ of finding the electron somewhere in a volume V centered on the proton when its state is $|n\ell m_\ell\rangle$ is

$$P_{n\ell m_\ell}(V) = \int_V d^3r\, |\psi_{n\ell m_\ell}(\mathbf{r})|^2. \tag{5.23}$$

Furthermore, if V is a sphere of radius a, then the probability $P_{n\ell m_\ell}(r \leqslant a)$ that the electron's radial coordinate is less than or equal to a is

$$P_{n\ell m_\ell}(r \leqslant a) = \int_0^a r^2\, dr \int_{\text{sphere}} d\Omega\, |\psi_{n\ell m_\ell}(r, \theta, \phi)|^2, \tag{5.24}$$

where r, θ, and ϕ are the polar coordinates of $\mathbf{r}$ and $d\Omega = \sin\theta\, d\theta\, d\phi$ is the differential of solid angle. This expression for $P_{n\ell m_\ell}(r \leqslant a)$ is based on an important principle: if the probability density depends on several variables and one (or more) of these is not specified (or measured), then all allowed values of each such variable must be summed on or integrated over in order to obtain the relevant probability. This rule means that, if an observable is not measured, then the probability associated with it will be unity since one of its values must occur. In the present case, any angles θ and ϕ may occur but none is specified, so we sum (integrate) over all of them to get the angle-independent result $P_{n\ell m_\ell}(r \leqslant a)$.

Let us now consider the ground state, for which E_n takes on its smallest value. This occurs for $n = 1$, which means that $\ell = 0 = m_\ell$. Thus the ground state is $|100\rangle$. It is shown in Chapter 11 that the coordinate-space wave function $\langle r\theta\phi|100\rangle$ is

$$\psi_{100}(r\theta\phi) = \left(\frac{1}{\pi a_0^3}\right)^{1/2} e^{-r/a_0}, \tag{5.25}$$

a result independent of the polar angles. The Bohr radius a_0 is equal to $\hbar^2/(m_e\kappa_e e^2)$ (Eq. (1.22)), and the factor $[1/(\pi a_0^3)]^{1/2}$ ensures normalization.

Thus, the ground-state wave function is normalized and spherically symmetric (no θ or ϕ dependence). It gives rise to a probability density proportional to $\exp(-2r/a_0)$, whose maximum is at $r = 0$ and that falls to $1/e$ of its $r = 0$ value at $r = a_0/2 \cong 0.25$ Å, as shown in Fig. 5.1. Since the factor r^2 occurs in the volume integral, one often introduces the *radial probability amplitude* $r\psi_{n\ell m_\ell}(\mathbf{r})$, from which one obtains the *radial probability distribution* $\rho^{\mathrm{rad}}_{n\ell m_\ell}(\mathbf{r})$, which in the case of the ground state is

$$\rho^{\mathrm{rad}}_{100}(\mathbf{r}) = |r\psi_{100}(\mathbf{r})|^2. \tag{5.26}$$

Figure 5.2 shows the radial probability density $|r\psi_{100}(r)|^2$ plotted as a function of r. In contrast to $|\psi_{100}(r)|^2$, it is zero at $r = 0$ and its maximum $r_{\max}$ is larger than zero. An

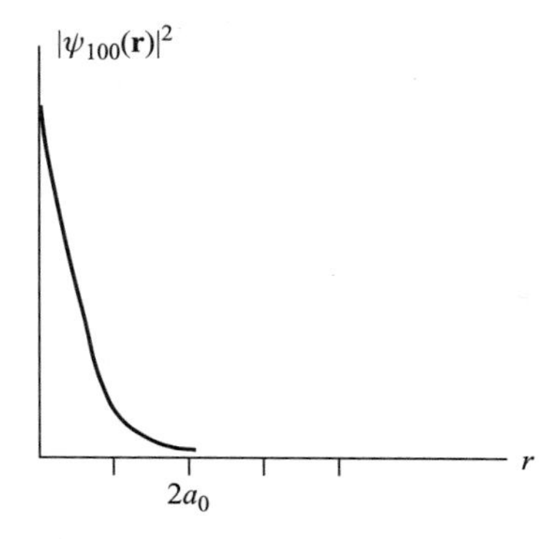

Fig. 5.1 The probability density $|\psi_{100}(\mathbf{r})|^2$ for the ground state of the H-atom.

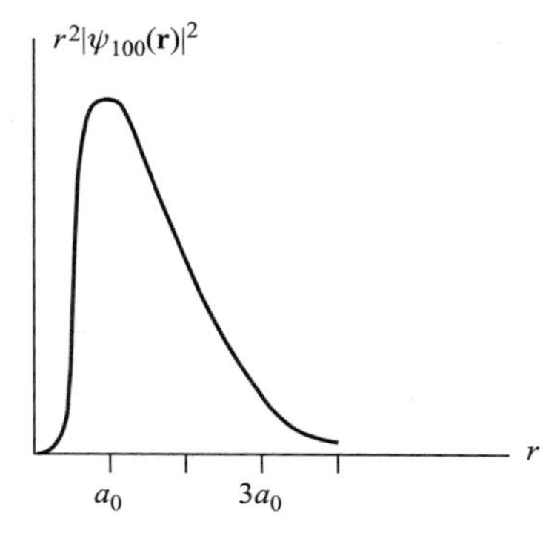

Fig. 5.2 The radial probability distribution $\rho_{\mathrm{rad}}(\mathbf{r}) = r^2|\psi_{100}(\mathbf{r})|^2$ for the ground state of the H-atom.

easy calculation leads to $r_{max} = a_0$, the radius of the lowest orbit in Bohr's model of hydrogen. Of course, this is not the distance of the electron from the proton, but only the radius at which $\rho_{100}^{rad}(\mathbf{r})$ is maximal. As shown in Fig. 5.2, the radial probability density $r^2|\psi_{100}(r)|^2$ is concentrated in a spherical shell lying between $r \cong 0.5a_0$ and $r \cong 3a_0$. This shell is the electronic charge cloud noted earlier; it underlies the picture in which the electron in the H-atom is *localized* in the vicinity of a_0 even though it is not *at* the radius a_0. Indeed, a_0 is not even equal to the average or the expected value of r: that quantity, which we denote $\bar{r}$, is larger than a_0; its value will later be shown to be $\bar{r} = 3a_0/2$. This is a further argument against accepting a planetary-orbit picture for the structure of hydrogen (as noted before, the Bohr model is simply not valid). Nevertheless, the values r_n of the nth Bohr orbits play much the same role for $n > 1$ as that in the case $n = 1$: when $n > 1$, then for those states $|n, \ell = n - 1, m_\ell = 0\rangle$, the radial probability distribution $|r\langle r\theta\phi|n, \ell = n - 1, m_\ell = 0\rangle|^2$ has its maximum at $r_{max} = r_n = n^2 a_0$. Furthermore, for such states the average value $\bar{r}$ will also be greater than r_n, and $\rho_{n\ell m_\ell}^{rad}(\mathbf{r})$ will continue to be concentrated in a small shell, though not necessarily a spherical one. These points will be established in Chapter 11.

Other probability amplitudes and densities will be evaluated in subsequent sections of this book. In the case of bound states, the probability distributions are usually localized in such a way that the vast majority of the distribution is confined to a relatively small spatial volume. For an atom, this results in the electrons being characterized as a charge cloud. In general, the charge-cloud distributions for the H-atom provide the basis for understanding the structure of heavier atoms and many molecules. Aspects of this are discussed in Chapter 18.

5.3. Measurements/connection with experimental data

Postulates I and II have dealt with some theoretical aspects of observables: their abstract operator images, states of quantum systems as the eigenvectors of quantal operators, and the probabilistic interpretation of states. Postulate III is concerned with experimental aspects of observables: their measurement and the relation of measured values to theoretical predictions. Ideal measurements of eigenvalues play but a limited role here, since measurements are generally not ideal, while the states of quantum systems are not always eigenvectors of the operator whose eigenvalues one might wish to measure. Furthermore, we may wish to compare a measurement with a theoretical prediction, often arising from an approximate calculation. How to connect theory and experiment in these more general situations is the subject of the present section.

Measurements yield numbers. The implication of this seemingly trivial statement is that the final results of a theoretical analysis on some quantum system should include predictions of the numbers that have been or will be obtained in an experiment. These numbers can include energies, lifetimes of decaying states, electric and magnetic moments, etc. On the theoretical side, every such number involves the matrix element of an operator. The simplest example of this is the eigenvalue a_n of an observable $\hat{A}$. Starting with Eq. (5.1),

$$\hat{A}|a_n\rangle = a_n|a_n\rangle,$$

it follows from projecting both sides of this equation onto $\langle a_n|$ that

$$a_n = \langle a_n|\hat{A}|a_n\rangle.$$

Thus, a_n is a diagonal matrix element of the operator $\hat{A}$.

More generally, one can always associate with every experimental measurement the matrix element – not necessarily a diagonal one – of a quantal operator. That is, quantal predictions will involve theoretical quantities of the form $B_{\beta\alpha} = \langle\beta|\hat{B}|\alpha\rangle$, *where* $|\alpha\rangle$ *and* $|\beta\rangle$ *are quantal states and* $\hat{B}$ *is a quantal operator, which may but need not be an observable. Some predictions will be proportional to* $B_{\beta\alpha}$*; others may involve* $|B_{\beta\alpha}|^2$.

In order to relate theoretical predictions to experimental measurements in a relatively simple way, our formulation of Postulate III will be in terms of measurements of an observable $\hat{A}$ when the system is in quantal state $|\alpha\rangle$. This assumption is not too restrictive, since it allows us to deal with a variety of theoretical and experimental possibilities. Furthermore, the basic concepts are readily extended to cases in which measurements involve off-diagonal matrix elements and/or operators that are not observables, as in the cases of 3-D scattering and electromagnetic radiation. We consider experimental aspects first.

Two ingredients (among several) are essential to achieving accurate measurements. First, the apparatus must have sufficient resolution to discriminate between adjacent values of the observable.[4] This ingredient, essential to an ideal measurement, is discussed in Section 5.1. Second, enough counts (observations) must be made that fluctuations away from the desired value are small. That is, one must have good statistics. Our concern here is only with the latter aspect. To achieve it means that N, the number of times the observable is measured, must be large. In the ideal situation $N \to \infty$, just as in the case of the *a priori* determination of the probability of achieving a particular outcome for some event (see Appendix A). This analogy to probability holds in practice as well, since neither in experimental measurements nor in *a posteriori* determinations of the probability of the occurrence of a particular outcome is an infinite number of observations made.

We now assume good statistics, i.e., N large enough that $\sqrt{N} \ll N$. If the N measurements of the observable $\hat{A}$ yield $\{a^{(j)}\}_{j=1}^{N}$, then the "value" of the observable extracted from these measurements is the average $\langle\hat{A}\rangle_\alpha$, defined as

$$\langle\hat{A}\rangle_\alpha = \sum_{j=1}^{N} P_j a^{(j)}, \tag{5.27a}$$

where the subscript α on the LHS of (5.27a) signifies that all N of the systems measured were each in the state $|\alpha\rangle$ when the experiment was performed, while P_j is the probability that the value $a^{(j)}$ will occur. It is often the case that these probabilities are *a priori* equal, i.e., $P_j = N^{-1}$, $\forall\ j$, in which case

$$\langle\hat{A}\rangle_\alpha = \frac{1}{N}\sum_{j=1}^{N} a^{(j)}. \tag{5.27b}$$

[4] By "apparatus" is meant both the devices which generate the probe and those which detect the final systems, including photons. Note also that control of resolution is meant to encompass items such as systematic errors, finite-size effects, various biases, etc.

The relation of the measured average value of (5.27) to that predicted theoretically is given by Postulate III.

POSTULATE III.

Assume that the physical system for which the observable $\hat{A}$ is to be measured is initially prepared in quantum state $|\alpha\rangle$. Then *the theoretical prediction for the average or expected value* $\langle\hat{A}\rangle_\alpha$ *of the observable* $\hat{A}$ *when the system is in state* $|\alpha\rangle$ *is given by the matrix element* $\langle\alpha|\hat{A}|\alpha\rangle$, i.e.,

$$\langle\hat{A}\rangle_\alpha = \langle\alpha|\hat{A}|\alpha\rangle. \tag{5.28}$$

The relation of this quantity to the values $a^{(j)}$ observed in experiments is

$$\langle\alpha|\hat{A}|\alpha\rangle = \langle\hat{A}\rangle_\alpha = \sum_{j=1}^{N} P_j a^{(j)}, \tag{5.29}$$

where an "in-principle" limit of $N \to \infty$ is implied but not observed in practice.

Note the difference between Postulates I and III: the former states that $\{a_n\}$ comprises the entire set of values that an observable $\hat{A}$ can have (for simplicity here, all a_n are assumed discrete), whereas Postulate III, via Eq. (5.29), relates the *average* value obtained in a measurement of $\hat{A}$ to that predicted by theory. Aspects of this are illustrated in the following examples.

Eigenstates

Assume that at least a portion of the spectrum of $\hat{A}$ is discrete, i.e.,

$$\hat{A}|a_m\rangle = a_m|a_m\rangle, \qquad m = 1, 2, \ldots.$$

Let the system on which $\hat{A}$ is to be measured be prepared in state $|a_n\rangle$, i.e., $|\alpha\rangle = |a_n\rangle$. Then $\langle\alpha|\hat{A}|\alpha\rangle = a_n$ and (5.29) becomes

$$a_n = \sum_{j=1}^{N} P_j a^{(j)}. \tag{5.30}$$

One expects not only that $P_j = 1/N$, but also that the results of very accurate measurements will yield $a^{(j)} = a_n \pm \delta a^{(j)}$, where $\delta a^{(j)}$ contains the effects of experimental error, with $\delta a^{(j)}/a_n \ll 1$.

As a specific example, let us consider a particle in the 1-D quantal box: it is not a realistic physical system but displays many of the characteristics of one. Let $|n, t\rangle$ be the Hilbert-space vector whose coordinate-space representation is $\Psi_n(x, t)$ of Eq. (3.91). If the particle's state is $|n, t\rangle$, then the theoretical value of the energy for that state is $E_n = \langle n, t|\hat{H}|n, t\rangle$, where $\hat{H}$ is the relevant Hamiltonian. Were it possible to have N particles in the box, all in state $|n, t\rangle$, and also to measure the average energy of such a system, then a_n on the LHS of (5.30) would be replaced by E_n, while the $a^{(i)}$ on the RHS would be replaced by the measured values of the energy, $E^{(j)}$.

Next, suppose that the state in which the systems are prepared is a linear combination of eigenstates of $\hat{A}$, say

$$|\alpha\rangle = \sum_{n=1}^{n_0} C_n |a_n\rangle, \tag{5.31}$$

with $C_n = \langle a_n|\alpha\rangle$. If there are N systems for which (5.31) is the state, it follows from Postulate II that, of the N systems, $N_n = |C_n|^2 N$ are in eigenstate $|a_n\rangle$, while Postulate III leads to the following theoretical prediction for $\langle\hat{A}\rangle_\alpha$:

$$\begin{aligned}\langle\hat{A}\rangle_\alpha &= \sum_{n=1}^{n_0} |C_n|^2 a_n \\ &= \frac{1}{N}\sum_{n=1}^{n_0} N_n a_n.\end{aligned} \tag{5.32}$$

Equation (5.32) is itself a probabilistically weighted sum. By extending this type of analysis to "preparation" of final states, we shall show below how $|C_n|$ can be extracted – to within experimental error – from certain measurements.

The particle in a quantal box again provides an illustration of the foregoing. Corresponding to $|\alpha\rangle$ of (5.31) is the superposition wave function (3.94), so that the C_n of (5.31) take on the values $C_1 = C_4 = C_j = \cdots = 0$, $C_2 = a_2$, and $C_3 = a_3$. Let us assume that an energy measurement were to be made, with all N particles being in the state $|\alpha\rangle$ of (5.31). Corresponding to $\langle\hat{A}\rangle_\alpha$ is $\langle\hat{H}\rangle_\alpha = \langle\alpha|\hat{H}|\alpha\rangle$; a straightforward calculation yields $\langle\hat{H}\rangle_\alpha \equiv \bar{E} = |a_2|^2 E_2 + |a_3|^2 E_3$. Choosing $a_2 = a_3 = \sqrt{\frac{1}{2}}$, we get $\bar{E} = (E_2 + E_3)/2 = 13\pi^2\hbar^2/(4mL^2)$.

Finally, let us suppose that the prepared state $|\alpha\rangle$ of the system is $|b_m\rangle$, an eigenstate of a second observable $\hat{B}$ that does not commute with $\hat{A}$, whose spectrum is assumed discrete. Since $[\hat{A}, \hat{B}] \neq 0$, $|b_m\rangle$ is not an eigenstate of $\hat{A}$ and therefore the value $\langle b_m|\hat{A}|b_m\rangle$ now predicted for $\langle\hat{A}\rangle_\alpha$ can no longer be expressed as a finite sum of a_n's, as in Eq. (5.32). Indeed, using

$$\hat{I} = \sum_{n=1}^{\infty} |a_n\rangle\langle a_n| \tag{5.33}$$

as a resolution of the identity (see Postulate VI) and employing (5.33) in $\langle b_m|\hat{A}|b_m\rangle$, we find that

$$\langle\hat{A}\rangle_\alpha = \sum_{n=1}^{\infty} |\langle b_m|a_n\rangle|^2 a_n, \tag{5.34}$$

which is an infinite sum on the a_n's. In general, none of the coefficients $\langle b_m|a_n\rangle$ can be expected to be zero, leaving an awkward infinite sum to evaluate. This suggests attempting to determine $\langle b_m|\hat{A}|b_m\rangle$ directly. A possible means for doing so is to convert $\langle b_m|\hat{A}|b_m\rangle$ into an integral by introducing a representation. Let us assume that $\hat{A}$ is local in the coordinate representation. Its use leads to

$$\langle\hat{A}\rangle_\alpha = \langle b_m|\hat{A}|b_m\rangle = \int d^3r\, \psi^*_{b_m}(\mathbf{r})\hat{A}(\mathbf{r})\psi_{b_m}(\mathbf{r}),$$

where $\hat{A}$ and $|b_m\rangle$ are presupposed to refer to a single-particle system. The preceding integral may well be easier to evaluate than the infinite sum in (5.34): for example, the integral might accurately be approximated numerically.

Further examples of the foregoing, for the quantal box and other systems, are discussed in the next chapter.

With these remarks concerning the connection with theoretical predictions, we turn to experimental aspects of Postulate III, in particular to measurements involving binding energies, decaying states, and collision phenomena.

Example: binding energy

Some experiments lead to the determination of more than one observable, others are unique in this regard. As an example of the latter, we consider the determination of the energy of a photon just sufficient to dissociate a deuteron – the stable nucleus consisting of a neutron and a proton bound together by the strong nuclear interaction – into a free (unbound) neutron and a free proton, each having zero kinetic energy in their center-of-mass coordinate system. To achieve this, the energy E_γ of the photon must be equal to the *binding energy* $B_{\rm d}$ of the deuteron. The goal of the measurement is to determine $B_{\rm d}$.

The experiment may be thought of as the shining of photons of variable energy E_γ onto deuterium gas at room temperature; owing to the Boltzmann factor, the average kinetic energy of the deuterons is about 0.025 eV, a quantity small enough for our purposes to ignore.

When $E_\gamma < B_{\rm d}$, no protons and neutrons are produced, but when the threshold value $E_\gamma = B_{\rm d}$ is reached, the photon can be absorbed, dissociate the deuteron, and detectors that distinguish protons can then record their presence. The jth measurement of the threshold energy produces a value $B_{\rm d}^{(j)}$; their totality is the set $\{B_{\rm d}^{(j)}\}_{j=1}^N$.

The value for $B_{\rm d}$ that is extracted from the measured ones is, from (5.27a), the *average* or expected value, viz.,

$$B_{\rm d} = \sum_{j=1}^{N} P_{\rm d}^{(j)} B_{\rm d}^{(j)}, \tag{5.35}$$

where $P_{\rm d}^{(j)}$ is the probability that $B_{\rm d}^{(j)}$ is the outcome. In such experiments, it is assumed that each $B_{\rm d}^{(j)}$ is equally likely to occur, so that $P_j = 1/N$, with the result that[5]

$$B_{\rm d} = \frac{1}{N} \sum_{j=1}^{N} B_{\rm d}^{(j)}. \tag{5.36}$$

The experimental value of $B_{\rm d}$ is 2.2245 MeV; (5.36) might yield $B_{\rm d} = 2.23$ MeV.

The same kind of analysis would apply to the determination of, say, the minimum energy $B_{\rm e}$ needed to ionize an atom initially in its ground state, i.e., to remove one of its electrons from a state having negative energy to a state with zero energy.

Among the common ingredients in the preceding processes, viz.,

$$\gamma + {\rm d} \rightarrow {\rm n} + {\rm p}$$

[5] We need not address the question of assigning errors, determining a dispersion, etc.

and

$$\gamma + \mathrm{A} \rightarrow \mathrm{A}^+ + \mathrm{e},$$

where A^+ represents the singly ionized atom, is the fact that in all measurements for both cases, the target is always in its ground state. Hence, the measurement will always yield a value of B_d or B_e. That is, in order for the experiment to be repeatable, the systems on which the measurements are being carried out must each be in the same initial state, so that the measurement determines an energy relevant to the desired state. This condition is essential, and we henceforth assume it to hold even if no statement to that effect is made.

As a final point, we call the reader's attention to the fact that the LHS of (5.36) is a "call to arms": B_d, and eigenvalues in general, are not known in advance and need to be *calculated*. Very few physical systems are simple enough that the states $|a_n\rangle$ and/or their associated wave functions can be obtained exactly, textbook examples notwithstanding. Indeed, the soluble systems treated in textbooks tend to be the *only* ones for which exact solutions are available. Models and/or approximation methods, both of which will be considered in this book, are thus essential ingredients in analyses of real systems. Furthermore, the inability to solve dynamical equations exactly is often mirrored by an ignorance of some of the operators needed to describe one or another physical system, so that models and approximations occur here as well.

Example: decaying states/branching ratios

When the first excited state of an atom or particle-stable nucleus decays, only one final state in the atom or nucleus can be reached, viz., the ground state, in which case a photon of unique energy is emitted. However, for higher excited states, there is often a variety of final states that can be reached by photon emission. Figure 5.3 illustrates this situation using a Grotrian diagram and wavy lines for the photons. It shows a state of energy E_e that can decay via photon emission to one of three final states E_j, $j = 1, 2, 3$. The normalized probabilities P_j for the decay to occur via the "branch" $E_\mathrm{e} \rightarrow E_j$, $j = 1, 2, 3$, obey $\sum_j P_j = 1$. The ratio P_j/P_k is denoted the "branching" ratio for the decays $E_\mathrm{e} \rightarrow E_j$ and $E_\mathrm{e} \rightarrow E_k$. If N measurements are made on the decays from level E_e, then the number of decays N_j that proceed via branch j is $N_j = NP_j$. Hence, measurement of the relative numbers of photons corresponding to levels E_j and E_k, viz., N_j and N_k, yields the relative probabilities:

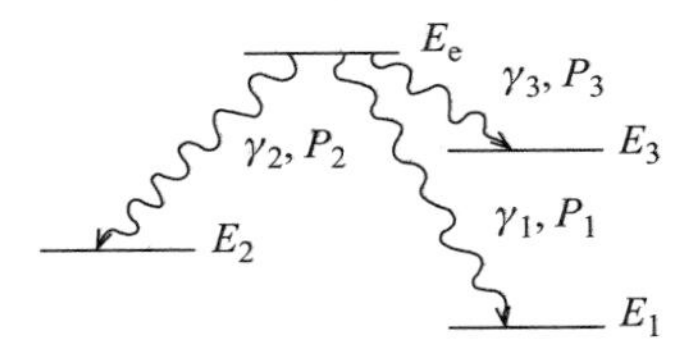

Fig. 5.3 An illustration of the different "branches" by which a hypothetical excited state of energy E_e can decay via photon emission to any one of three final states with energies $E_1 < E_2 < E_3$. The wavy lines represent the photons; the symbol γ_j denotes the photon in the transition $E_\mathrm{e} \rightarrow E_j$, while P_j is the probability that the decay will occur via branch j, $j = 1, 2, 3$.

$$\frac{N_j}{N_k} = \frac{P_j}{P_k}. \tag{5.37}$$

Each P_j, $j = 1, 2, 3$, is the absolute square of a quantal amplitude C_j and the task of theory is to determine these amplitudes, usually via models and approximation, and then to use the calculated values of P_j/P_k to compare with the LHS of (5.37) in an attempt to determine whether the model/approximation can be considered valid.

Branching ratios are not limited to photon emission: they occur when any unstable state can decay in more than one way. Examples are found in almost all areas of microphysics. One such concerns the short-lived, neutral K meson, denoted K_S^0. Its lifetime is 0.89×10^{-10} s and its primary decay modes are

$$K_S^0 \to \pi^+ + \pi^- \tag{5.38}$$

and

$$K_S^0 \to \pi^0 + \pi^0, \tag{5.39}$$

where π^0 and $\pi^\pm$ are the neutral, and the positively and negatively charged pi mesons or "pions." The branching ratio for the decays (5.38) and (5.39) is

$$\frac{K_S^0 \to \pi^+ + \pi^-}{K_S^0 \to \pi^0 + \pi^0} = \frac{69}{31}, \tag{5.40}$$

where the RHS of (5.40), the relative number of decays into the two branches, is experimentally determined. Note that the sum of the numerator plus the denominator in (5.40) adds up to 100, suggesting that only the two branches (5.38) and (5.39) can occur. In fact, other decay modes have been observed, but their occurrence is down by a factor of 10^3, so that, to an accuracy of better than within 1%, the ratio in (5.40) is valid. Correct prediction of the lifetime and the branching ratio (5.40) is a requirement of any theory of elementary particles claiming to be valid.

There is a long-lived partner to K_S^0, the symbol for which is K_L^0; its lifetime is 5.2×10^{-8} s. The K_S^0 and K_L^0 can be thought of as two manifestations of a single particle that can exist in one of only two states, denoted $|K_S^0\rangle$ and $|K_L^0\rangle$. We shall examine this situation in detail in the exercises to Chapter 13 (two-state systems).

Example: collision processes

A further instance in which a variety of final states can arise from a single initial state is afforded by collisions between a single-particle projectile and a structured target. Included here are electron–atom and electron–molecule scattering, and proton–nucleus collisions involving both scattering and reactions. If the energy E of the projectile is high enough for inelastic scattering to occur, assuming that the target has particle-stable excited states (with energies E_j^{ex}, $j = 1, 2, \ldots$), then inelastically scattered projectiles with final energies $E - E_j^{\text{ex}}$ can be observed.

An inelastically scattered projectile with final energy $E - E_j^{\text{ex}}$ corresponds to the target having been excited to a level with energy E_j^{ex}.[6] An analogous statement holds for a

[6] We assume for simplicity that negligible recoil energy is imparted to the target during the collision and that non-relativistic kinematics may be used. In addition, conservation of energy is taken for granted.

reaction, the only change being that $E - E_j^{\text{ex}}$ is replaced by $E - E_j^{\text{f}}$, the energy of the ejectile, where E_j^{f} is the energy of the jth excited state of the residual system f (e.g., a molecule or nucleus) reached via the reaction. Note that, in each case, the collision has "prepared" the final system in an excited state, the probability for which can be extracted from experiment, as described in the following.

After the collision – inelastic scattering or a reaction – the ejectile can be observed at an angle θ relative to the initial direction of the projectile and also with some final energy E_{f} (equal to $E - E_j^{\text{ex}}$ in the case of inelastic scattering or to $E - E_j^{\text{f}}$ in the case of a reaction). Either the final energy is fixed and the number $N(\theta)$ of ejectiles emitted at angle θ is measured, or θ is fixed and the number $N(E_{\text{f}})$ of ejectiles emitted that leave the target or residual system in the jth excited level is measured.

An example of the latter situation is given by the proton-in, neutron-out nuclear reaction

$$\text{p} + {}^{14}\text{C} \rightarrow {}^{14}\text{N} + \text{n}, \tag{5.41}$$

where C (N) is the chemical symbol for carbon (nitrogen) and the left-hand superscript 14 refers to the total number of neutrons plus protons in the nucleus. Once again, p (n) refers to the incident proton (ejected neutron). Shown in Fig. 5.4 is the number of ejected neutrons $N(E_{\text{f}})$ observed at a scattering angle of 40° having energies E_{f} in the range of 26 MeV $\leqslant E_{\text{f}} \leqslant$ 34 MeV. The neutrons are produced by protons of incident energy $E = 35$ MeV bombarding the ^{14}C target. Ten pronounced peaks are seen in this energy

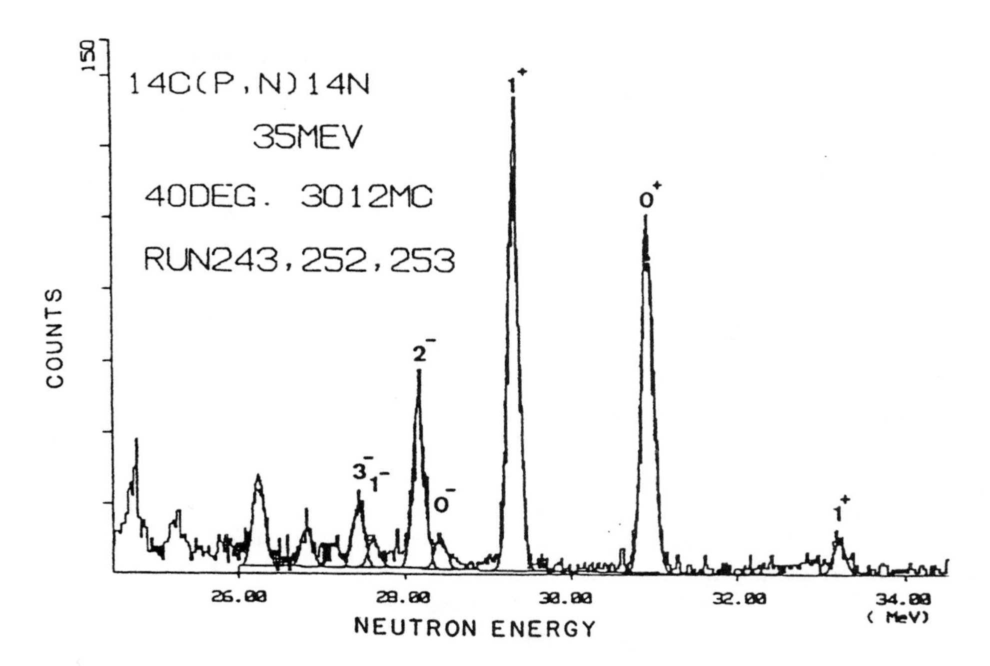

Fig. 5.4 A portion of the energy spectrum or "excitation function" of neutrons emerging from the reaction $\text{p} + {}^{14}\text{C} \rightarrow {}^{14}\text{N} + \text{n}$. The energy of the incident protons was 35 MeV and the neutrons were detected at a scattering angle of 40° relative to the direction of incidence of the proton beam. (From Orihara *et al.* (1983).)

range. They correspond to the states in ^{14}N which the outgoing neutrons leave behind. That is, for each peak, ^{14}N is left in a discrete excited state whose energy is determined by conservation of energy. Since the neutron energy E_f increases to the right on the abscissa of Fig. 5.4, the excited-state energies increase to the left. Seven peaks are labeled by the value of the angular momentum and the value of a new quantum number denoted parity, which can be $+$ or $-$, and is discussed in Chapter 9.

The height h_j of the jth peak in Fig. 5.4 is proportional to the probability that this state is populated in the reaction when $\theta = 40°$. Thus, the 1^+ level corresponding to $E_f \cong 29.5$ MeV has the greatest probability of being excited, while the 1^-, the 0^-, and the other 1^+ levels have the least probability of being populated. Again, the aim of theory is to predict these results.

5.4. The coordinate representation for observables

Operators, along with states and their associated wave functions, are the means by which one can calculate the quantities that are measured in experiments. The determination of states/wave functions is thus a primary goal of quantum theory. To do this requires solving the relevant dynamical equations, viz., eigenvalue equations of the form (5.1) or the time-dependent Schrödinger equation, which is discussed in the next section.

These dynamical equations involve the operators that image physical observables. Since the operators are formulated as abstract entities in Hilbert space, some means for working directly with them is required. One procedure is to employ the commutation relations obeyed by the operators. This is relatively straightforward to do for simple systems like the harmonic oscillator and for the simplest model of the H-atom. It leads to the eigenstate representation, in which the operators are represented as infinite and, often, discrete matrices. We shall later treat the 1-D harmonic oscillator in this way.

While the commutation relation approach is useful, it tends to be limited to the simpler systems. Furthermore, if one needs the probability amplitudes that the system is at particular spatial points, then a coordinate representation must be introduced as well, in order to convert state vectors into wave functions. Coordinate-space wave functions are often utilized to help provide a physical feeling for the behavior and interpretation of quantal systems, an aspect of special importance when many degrees of freedom are involved.

Coordinate-space wave functions are usually obtained as the solutions of the coordinate-space forms of dynamical equations. Such equations must obviously contain quantal operators expressed in the coordinate representation. Examples are the momentum and energy (Hamiltonian) operators, which were informally introduced in Chapter 3 in connection with the quantal-box problem. Postulate IV formalizes and generalizes these definitions by stating the coordinate form of a number of (single-particle) quantal operators. From these one can then derive both the commutation relations obeyed by the operators and the momentum representation for the operators.

The single-particle operators whose coordinate representations are stated in Postulate IV are all local. Locality is discussed in Section 4.4, in particular via Eqs. (4.158)–(4.160). As an example of a relation involving local operators, consider the eigenvalue equation (5.1),

Table 5.2 *Some quantal operators and their (local) coordinate representations for a particle of mass m*

Operator	Coordinate representation
Position, $\hat{\mathbf{Q}}$	$\mathbf{r}$
Classical potential energy, $\hat{V}_{\text{cl}}(\hat{\mathbf{Q}})$	$V_{\text{cl}}(\mathbf{r})$
Linear momentum, $\hat{\mathbf{P}}$	$\hat{\mathbf{P}}(\mathbf{r}) = -i\hbar\,\nabla$
(Orbital) angular momentum, $\hat{\mathbf{L}}$	$\hat{\mathbf{L}}(\mathbf{r}) = -i\hbar\mathbf{r}\times\nabla$
Kinetic energy, $\hat{K}$	$\hat{K}(\mathbf{r}) = \hat{P}^2/(2m) = -[\hbar^2/(2m)]\,\nabla^2$
Energy, $\hat{H} = \hat{K} + \hat{V}$, $\hat{V} = \hat{V}_{\text{cl}}(\hat{\mathbf{Q}})$	$\hat{H}(r) = -[\hbar^2/(2m)]\,\nabla^2 + V_{\text{cl}}(\mathbf{r})$

$$\hat{A}|a_n\rangle = a_n|a_n\rangle,$$

where $\hat{A}$ and $|a_n\rangle$ now refer to a single particle in three dimensions. Locality of $\hat{A}$ means that the coordinate-space form of (5.1) becomes, in analogy to (4.159),

$$\hat{A}(\mathbf{r})\psi_n(\mathbf{r}) = a_n\psi_n(\mathbf{r}), \tag{5.42}$$

where $\mathbf{r}$ is the position of the particle and $\psi_n(\mathbf{r}) \equiv \langle\mathbf{r}|a_n\rangle$. In Eq. (5.42), $\hat{A}(\mathbf{r})$ stands for any of the operators in Tables 5.1 and 5.2, the latter of which is introduced in connection with Postulate IV.

In addition to their being local, operators discussed in this section are the quantum versions of classical quantities. The quantum operators such as spin and parity do not have coordinate representations, and are not included with those of Table 5.2.

POSTULATE IV.

The operators which are the quantal analogs of classical quantities are all local in the coordinate representation.

(a) The coordinate representation of the position operator $\hat{\mathbf{Q}}$ and of the operator $\hat{V}(\hat{\mathbf{Q}})$ corresponding to any classical potential energy or electromagnetic potential $V_{\text{cl}}(\mathbf{r})$ are

$$\hat{\mathbf{Q}}(\mathbf{r}) = \mathbf{r} \tag{5.43}$$

and

$$\hat{V}(\hat{\mathbf{Q}}) = V_{\text{cl}}(\mathbf{r}), \tag{5.44}$$

i.e., $\hat{\mathbf{Q}}$ becomes $\mathbf{r}$, and, for a classical potential, $\hat{V}(\hat{\mathbf{Q}})$ becomes $V_{\text{cl}}(\mathbf{r})$. Examples are given below.

(b) The coordinate representation of the linear momentum operator $\hat{\mathbf{P}}$ is

$$\hat{\mathbf{P}}(\mathbf{r}) = -i\hbar\,\nabla, \tag{5.45}$$

where the derivatives in the gradient operator are taken with respect to the components of $\mathbf{r}$.

(c) The coordinate representations of the (orbital) angular-momentum operator $\hat{\mathbf{L}} = \hat{\mathbf{Q}} \times \hat{\mathbf{P}}$ and kinetic-energy operator $\hat{K} = \hat{\mathbf{P}}\cdot\hat{\mathbf{P}}/(2m)$ are

$$\hat{\mathbf{L}}(\mathbf{r}) = -i\hbar\mathbf{r} \times \nabla \tag{5.46}$$

and

$$\hat{K}(\mathbf{r}) = -\frac{\hbar^2}{2m}\nabla^2. \tag{5.47}$$

(d) The Hamiltonian or energy operator, $\hat{H}$, is defined non-relativistically[7] as the sum of the kinetic- and potential-energy operators, $\hat{K}$ and $\hat{V}$, respectively:

$$\hat{H} = \hat{K} + \hat{V}. \tag{5.48}$$

For the case of a particle of mass m acted on by a classical potential $V_{\mathrm{cl}}(\mathbf{r})$, the coordinate representation of $\hat{H}$ is

$$\hat{H}(\mathbf{r}) = \hat{K}(\mathbf{r}) + V_{\mathrm{cl}}(\mathbf{r}) \tag{5.49a}$$

$$= -\frac{\hbar^2}{2m}\nabla^2 + V_{cl}(\mathbf{r}). \tag{5.49b}$$

These statements are summarized in Table 5.2. Probably the most remarkable aspect of Postulate IV is given by Eq. (5.45), since it defines the coordinate representation of the momentum operator as being mass-independent. It is an unexpected contrast to the classical-physics situation, a contrast that would be maintained if one were to introduce $\hat{\mathbf{P}}/m$ as a velocity operator, since the coordinate representation of this operator, unlike the classical-physics analog, *is* mass-dependent.

The preceding statements concern a single particle in 3-D. If a 1-D situation is envisaged, then, e.g., $\mathbf{r} \to x$, $\hat{\mathbf{Q}}(\mathbf{r}) \to \hat{Q}_x(x) = x$, $\hat{\mathbf{P}}(\mathbf{r}) \to \hat{P}_x(x) = -i\hbar\, d/dx$, $\hat{\mathbf{L}}(\mathbf{r}) \to 0$, and $\hat{K}(\mathbf{r}) \to \hat{K}_x(x) = -[\hbar^2/(2m)]\, d^2/dx^2$, with corresponding changes in $V(\mathbf{r})$ and $\hat{H}(\mathbf{r})$. Furthermore, in a system containing N particles labeled 1, 2, ..., N, with coordinates $\mathbf{r}_1, \mathbf{r}_2 \ldots, \mathbf{r}_N$ and masses $m_1, m_2, \ldots, m_N$, one can introduce operators for position, potential, linear and angular momentum, kinetic energy, and total energy for each particle j, viz.,

$$\hat{\mathbf{Q}}_j,\ \hat{V}_j,\ \hat{\mathbf{P}}_j,\ \hat{\mathbf{L}}_j,\ \hat{K}_j,\ \hat{H}_j,$$

whose coordinate representations are as given in Postulate IV, the only change being the addition of the subscript j wherever an unsubscripted $\mathbf{r}$, ∇, or m appears. We shall consider these operators in the later sections of this book dealing with N-particle systems, $N \geqslant 2$.

Each of the coordinate representations of parts (*b*), (*c*), and (*d*) of Postulate IV allow one (at least in principle) to solve the eigenvalue problem typified by Eq. (5.1), thus yielding both eigenvalues and eigenfunctions. Much of this text is concerned with solution of the stationary-state equation

$$\hat{H}\psi = E\psi, \tag{5.50}$$

where the coordinate-dependence is suppressed to allow for the possibility of this

[7] Only non-relativistic quantum mechanics is considered in this book.

equation referring to multi-particle systems. Equation (5.50) is known as the time-independent Schrödinger equation; we shall derive it in the next section.

The orbital-angular-momentum eigenvalue problem plays a key role in many 3-D situations; it is examined in Chapter 10. In the remainder of this section we introduce some examples of potentials, solve the linear momentum eigenvalue problem and then derive the momentum representations of the operators $\hat{\mathbf{Q}}$, $\hat{K}$, and $\hat{V}$.

Examples of potentials

(a) Potentials from Classical Physics

Probably the three most frequently encountered forces in classical physics are the constant force (the electric field between capacitor plates), the linear restoring force (the harmonic oscillator) and the inverse square force (Coulomb attraction or repulsion and Newtonian gravitation). The classical potentials corresponding to these forces are identical to the coordinate representations of their quantal analogs. These analogs are $\hat{V}_{\rm const}(\hat{\mathbf{Q}}) = -\mathbf{F}\cdot\hat{\mathbf{Q}}$, $\hat{V}_{\rm h.o.}(\hat{\mathbf{Q}}) = \frac{1}{2}m\omega_0^2\hat{Q}^2$, and $\hat{V}_{\rm coul}(\hat{\mathbf{Q}}) = \kappa_{\rm e}q_1q_2/|\hat{\mathbf{Q}}|$, where $\mathbf{F}$ is the constant force, m and ω_0 are the particle's mass and oscillator frequency, and q_1 and q_2 are the two charges, respectively, with $\hat{\mathbf{Q}}$ being the relative position operator (the separation between the two charges). Notice that, as long as $\hat{Q}^{-1}$ in $\hat{V}_{\rm coul}$ acts on a (relative) position state, as it will when the coordinate representation is invoked, it can simply be replaced by the magnitude of the relative separation, viz., r. Non-linearities are easily included in the harmonic-oscillator case, e.g., in 1-D, $\hat{V}_{\text{non-lin h.o.}}(\hat{\mathbf{Q}}) = \frac{1}{2}m\omega_0^2\hat{Q}^2 + f(|\hat{\mathbf{Q}}|)$, where $f(|\hat{\mathbf{Q}}|)$ could be of the form $f(\hat{Q}) = b\hat{Q}^3 + g\hat{Q}^4$.

Another instance in which classical potentials occur is the case of the electromagnetic field. Just as in classical mechanics, scalar and vector potentials, rather than the electric and magnetic fields, occur in the quantal Hamiltonian. Furthermore, in analogy to the classical case, the vector potential enters via alteration of the momentum operator. Aspects of the electromagnetic field are examined in Chapters 12 and 16.

(b) Some Quantal Potentials

Potentials – or interactions – in quantum mechanics are often in a form not encountered in classical physics. An example is given by Eq. (3.73), the potential acting on a particle confined to the interior of the 1-D quantal box. A more radical example is the so-called one-term separable potential, proportional to a projection operator projecting onto an arbitrary Hilbert-space state, say $|g\rangle$:

$$\hat{V}_{\rm sep} = \lambda|g\rangle\langle g|.$$

The ket $|g\rangle$ is often referred to as a *form factor*; it can be, e.g., 1-D or 3-D. Specification of the coordinate or momentum representation of $|g\rangle$ and the value of λ determines $\hat{V}_{\rm sep}$ uniquely. Separable potentials have been used widely in analyses of nucleon–nucleon scattering and of the three-nucleon systems (^{3}H and ^{3}He), and we examine this potential in the exercises. Other potentials involving operators will be considered in later chapters.

Many potentials that occur in quantum mechanics have coordinate representations that could be classical interactions but often seem not to have been used in classical mechanics. That is, the quantal operator becomes a simple multiplicative quantity: $V_{\rm quantal}(\hat{\mathbf{Q}}) \to f(\mathbf{r})$. In the typical situation, the potential is *central*, i.e., $f(\mathbf{r}) = f(r)$. Many different functional forms $f(r)$ have been studied, and we shall examine some of them in Chapters 6 and 11. We mention a few here, for example, the Yukawa (or screened Coulomb) potential, $V_{\text{Yukawa}}(r) = -V_0 e^{-\mu r}/(\mu r)$, where μ^{-1} is the range of the potential; the Gaussian potential $V_{\text{Gauss}}(r) = -V_0 e^{-(\beta r)^2}$, where β^{-1} is the range; and the Woods–Saxon potential $V_{\text{W-S}}(r) = -V_0\{1 + \exp[(r - R)/a]\}^{-1}$. The last one has played an important role in the analysis of the elastic scattering of light nuclei by other nuclei and of electrons by neutrons or protons. None of these three potentials allows analytic solutions to the corresponding eigenvalue problems.

For $V_0 > 0$, the negative sign means that each of V_{Yukawa}, V_{Gauss} and $V_{\text{W-S}}$ is attractive. Shown in Fig. 5.5(a) are curves depicting the general dependence of each potential on r, while Fig. 5.5(b) shows the square-well potential, defined by

$$V_{\text{sq}}(r) = \begin{cases} -V_0, & r \leqslant r_0 \\ 0, & r \geqslant r_0. \end{cases} \tag{5.51}$$

The square-well potential has a discontinuity at $r = r_0$, and represents a certain idealized limit of the Gaussian and Woods–Saxon potentials. Unlike the latter potentials, in the

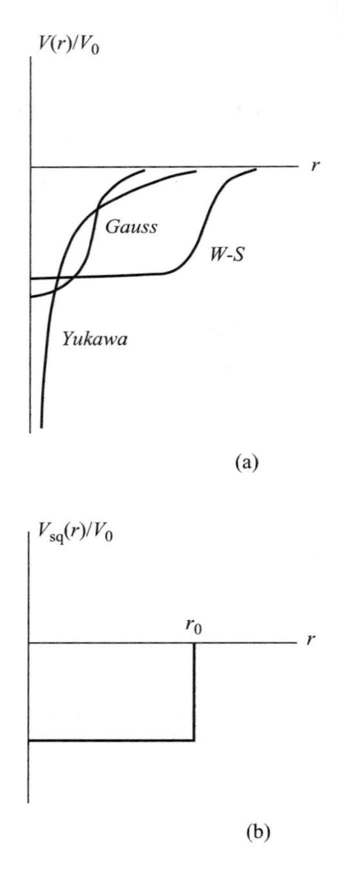

Fig. 5.5 (a) Comparisons of the smoothly varying, attractive Gaussian, Woods-Saxon and Yukawa potentials with the idealized square well potential (b). Each potential has a strength V_0.

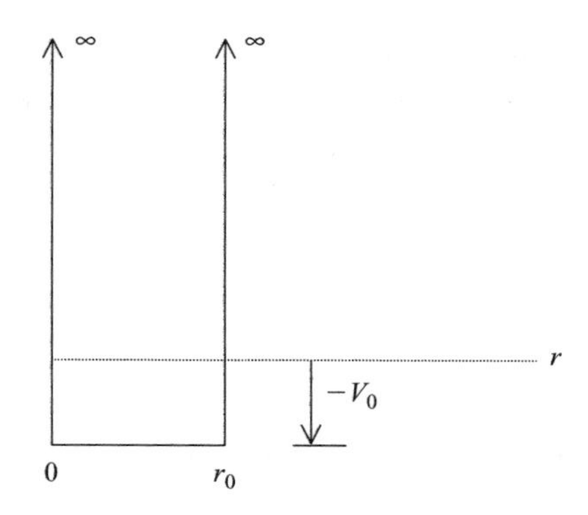

Fig. 5.6 The infinite square well, or quantum box, for a particle in three dimensions. Here, r is the radial coordinate ($r \geqslant 0$; $r = |\mathbf{r}|$).

square-well eigenvalue problem – assuming that V_0 is large enough to support bound states – there is an analytic representation for the eigenvalues, given in terms of the solutions of a transcendental equation, as will be shown in Chapters 6 and 11.

The finite square well of Eq. (5.51) is the quantum analog of a squash court: as long as the ball (particle) does not have enough energy to rise above the top of the wall (barrier at $r = r_0$), the ball cannot leave the court, although it is here that the analogy fails, since, in the quantum case, the particle can be found *outside* the well; however, the probability of its being at $r \gg r_0$ decreases exponentially with r. These points are discussed in Chapter 6.

The square-well potential is, pictorially, the simplest of all the finite-range potentials: it is constant up to a point, and then vanishes. However, as noted in the foregoing paragraph, it does not confine the particle to the region $r \leqslant r_0$. A 1-D confining potential is that of the box, Eq. (3.73). Its 3-D analog is the infinite square well, defined via

$$V_{\infty\mathrm{sq}}(r) = \begin{cases} -V_0, & r \leqslant r_0, \\ \infty, & r > r_0. \end{cases} \tag{5.52}$$

It is shown in Fig. 5.6, and often is referred to as the "3-D quantal box." The infinite square well enjoys the property of leading to a solvable eigenvalue problem, just like the 1-D case.

The linear momentum eigenvalue problem

Linear momentum is an observable that takes on a continuous range of values, hence the delta-function normalization (5.9) of the momentum eigenstates $|\mathbf{p}\rangle$ in

$$\hat{\mathbf{P}}|\mathbf{p}\rangle = \mathbf{p}|\mathbf{p}\rangle. \tag{5.7}$$

This leads to $|\mathbf{p}\rangle \notin \mathcal{H}$.

The coordinate-space version of (5.7) is

$$-i\hbar\,\nabla\psi_{\mathbf{p}}(\mathbf{r}) = \mathbf{p}\psi_{\mathbf{p}}(\mathbf{r}), \tag{5.53}$$

where $\psi_{\mathbf{p}}(\mathbf{r}) = \langle \mathbf{r}|\mathbf{p}\rangle$, as in (5.10), and Eq. (5.45) has been used for $\hat{\mathbf{P}}(\mathbf{r})$. It is convenient (and conventional) at this point to introduce the wave vector $\mathbf{k}$ via

$$\mathbf{k} = \mathbf{p}/\hbar \tag{5.54}$$

(recall the de Broglie relation $p = h/\lambda$ and the wavelength/wave-number relation $\lambda = k/(2\pi)$). Rather than write $\psi_{\mathbf{p}} = \psi_{\hbar\mathbf{k}}$, one simply drops the $\hbar$ in the subscript, so that (5.54) becomes

$$\nabla\psi_{\mathbf{k}}(\mathbf{r}) = i\mathbf{k}\psi_{\mathbf{k}}(\mathbf{r}). \tag{5.55}$$

In accord with this change, we shall henceforth write momentum eigenstates as $|\mathbf{k}\rangle$, such that

$$\hat{\mathbf{P}}|\mathbf{k}\rangle = \hbar\mathbf{k}|\mathbf{k}\rangle. \tag{5.56}$$

The wave-function form $\psi_{\mathbf{k}}(\mathbf{r}) = \langle \mathbf{r}|\mathbf{k}\rangle$ is thus a direct consequence of (5.56).

It is straightforward to show that, as long as $\mathbf{r}$ may range over all space, the solution to (5.56) is a plane wave:

$$\psi_{\mathbf{k}}(\mathbf{r}) = N_{\mathbf{k}}e^{i\mathbf{k}\cdot\mathbf{r}}, \tag{5.57}$$

whose 1-D version is $\psi_k(x) = \exp(ikx)/\sqrt{2\pi}$. Using the normalization condition $\langle \mathbf{k}'|\mathbf{k}\rangle = \delta(\mathbf{k} - \mathbf{k}')$ and Eq. (4.141) with $\mathbf{k}$ and $\mathbf{r}$ interchanged, one finds

$$N^*_{\mathbf{k}'}N_{\mathbf{k}} = (2\pi)^{-3}. \tag{5.58}$$

Since the RHS of (5.58) is independent of $\mathbf{k}$ and $\mathbf{k}'$, so is the LHS. It is standard to choose $N_{\mathbf{k}}$ to be the real number $(2\pi)^{-3/2}$: $N_{\mathbf{k}} = (2\pi)^{-3/2}$, and, combining these results, we have

$$\psi_{\mathbf{k}}(\mathbf{r}) = (2\pi)^{-3/2}e^{i\mathbf{k}\cdot\mathbf{r}}. \tag{5.59}$$

The relation between $\psi_{\mathbf{k}}(\mathbf{r})$ and $\psi_{\mathbf{p}}(\mathbf{r})$ is

$$\psi_{\mathbf{p}}(\mathbf{r}) = \hbar^{-3/2}\psi_{\mathbf{k}}(\mathbf{r}/\hbar).$$

Neither $\mathbf{p}$ nor $\mathbf{k}$ is restricted in any way except that they are real.

The eigenstates of $\hat{\mathbf{P}}$ are also eigenstates of the kinetic-energy operator $\hat{K}$:

$$\hat{K}|\mathbf{k}\rangle = \frac{\hat{\mathbf{P}}\cdot\hat{\mathbf{P}}}{2m}|\mathbf{k}\rangle = E_k|\mathbf{k}\rangle = \frac{\hbar^2k^2}{2m}|\mathbf{k}\rangle, \tag{5.60}$$

where the eigenvalue E_k is $\hbar^2k^2/(2m)$. For a free particle, i.e., one for which $\hat{V} = 0$ in all of space, $\hat{K}$ is often referred to as the free-particle Hamiltonian; its eigensolution (5.60) corresponds to a plane wave propagating in the direction of $\mathbf{k}$. Obviously, $|-\mathbf{k}\rangle$ is also an eigenstate of $\hat{K}$ with the same eigenvalue, so that a degeneracy exists. In the case of the 1-D quantal box, both the $|k\rangle$ and $|-k\rangle$ states are required in order to achieve coordinate-space solutions that vanish at the edges of the box.

The exponential form clearly demonstrates that the momentum wave functions are not Hilbert-space vectors. Hence, $\psi_{\mathbf{k}}(\mathbf{r})$ cannot itself describe a physical situation. This is evident, since $|\psi_{\mathbf{k}}(\mathbf{r})|^2 = (2\pi)^{-3}$, which means that, if a particle's wave function were $\psi_{\mathbf{k}}(\mathbf{r})$, then the probability density for its being at any point $\mathbf{r}$ would be independent of $\mathbf{r}$, while the total probability of its being found somewhere in space, $\int d^3r\,|\psi_{\mathbf{k}}(\mathbf{r})|^2$, would be infinite.

Despite the latter situation, momentum wave functions are very useful in quantum mechanics. For instance, they relate the coordinate and momentum representations of a state via a Fourier integral transform. To see this, we start from the definition $\psi_\alpha(\mathbf{r}) = \langle\mathbf{r}|\alpha\rangle$ and then insert a momentum-state resolution of the identity (i.e., $\hat{I} = \int d^3k\,|\mathbf{k}\rangle\langle\mathbf{k}|$) into $\langle\mathbf{r}|\alpha\rangle$. This leads to

$$\psi_\alpha(\mathbf{r}) = \int d^3k\,\langle\mathbf{r}|\mathbf{k}\rangle\langle\mathbf{k}|\alpha\rangle$$
$$= \left(\frac{1}{2\pi}\right)^{3/2}\int d^3k\,e^{i\mathbf{k}\cdot\mathbf{r}}\varphi_\alpha(\mathbf{k}), \tag{5.61}$$

where

$$\varphi_\alpha(\mathbf{k}) = \langle\mathbf{k}|\alpha\rangle. \tag{5.62}$$

The inverse relation follows from Fourier-transform theory (see Appendix B):

$$\varphi_\alpha(\mathbf{k}) = \left(\frac{1}{2\pi}\right)^{3/2}\int d^3r\,e^{-i\mathbf{k}\cdot\mathbf{r}}\psi_\alpha(\mathbf{r}), \tag{5.63}$$

a result that can also be derived by starting with the definition (5.62) and inserting a position-state resolution of the identity into the scalar product $\langle\mathbf{k}|\alpha\rangle$. Equation (5.63) embodies the relation $\langle\mathbf{k}|\mathbf{r}\rangle = \langle\mathbf{r}|\mathbf{k}\rangle^*$ or, equivalently, $\psi_{\mathbf{k}}^*(\mathbf{r}) = (2\pi)^{-3/2}\exp(-i\mathbf{k}\cdot\mathbf{r})$.

$\langle\mathbf{r}|\mathbf{k}\rangle$ not only connects the coordinate and momentum representations of state vectors, but also relates these two representations for operators. Because of this, Dirac referred to $\langle\mathbf{r}|\mathbf{k}\rangle$ as a transformation coefficient (Dirac 1947). We employ it in the next subsection to obtain the momentum representation of operators, in particular of $\hat{\mathbf{Q}}$.

The momentum representation

Obtaining the momentum representation for operators like $\hat{\mathbf{P}}$ and $\hat{K} = \hat{P}^2/(2m)$ is trivial; for others, like $\hat{\mathbf{Q}}$, it is not so straightforward. We examine both situations in this subsection.

As a first point, let us recall that a quantal operator $\hat{A}$ acts on vectors in Hilbert space and yields other Hilbert-space vectors. For example, for $|\beta\rangle \in \mathcal{H}$, $\hat{A}|\beta\rangle = |\alpha\rangle \in \mathcal{H}$. Thus, the coordinate representation of $|\alpha\rangle$, viz., $\psi_\alpha(\mathbf{r}) = \langle\mathbf{r}|\alpha\rangle$, is related to $\hat{A}|\beta\rangle$ via

$$\psi_\alpha(\mathbf{r}) = \int d^3r'\langle\mathbf{r}|\hat{A}|\mathbf{r}'\rangle\psi_\beta(\mathbf{r}'). \tag{4.155}$$

When $\hat{A}$ is local in coordinate space, (4.155) becomes $\psi_\alpha(\mathbf{r}) = \hat{A}(\mathbf{r})\psi_\beta(\mathbf{r})$.

Next, consider $\hat{A}|\mathbf{k}\rangle$. It might be thought that, if $\hat{A}|\mathbf{k}\rangle$ cannot be immediately evaluated and $\langle\mathbf{r}'|A|\mathbf{r}\rangle$ is known, then one could obtain the momentum representation by an analogous use of (5.21) in $\langle\mathbf{k}'|\hat{A}|\mathbf{k}\rangle$:

$$\langle \mathbf{k}'|\hat{A}|\mathbf{k}\rangle = \int d^3r'\, d^3r \langle \mathbf{k}'|\mathbf{r}'\rangle\langle \mathbf{r}'|\hat{A}|\mathbf{r}\rangle\langle \mathbf{r}|\mathbf{k}\rangle$$

$$= \left(\frac{1}{2\pi}\right)^3 \int d^3r'\, d^3r e^{-i\mathbf{k}'\cdot\mathbf{r}'}\langle \mathbf{r}'|\hat{A}|\mathbf{r}\rangle e^{i\mathbf{k}\cdot\mathbf{r}},$$

which, in the case of a local operator, goes over to

$$\langle \mathbf{k}'|\hat{A}|\mathbf{k}\rangle = \left(\frac{1}{2\pi}\right)^3 \int d^3r\, e^{-i\mathbf{k}'\cdot\mathbf{r}}\hat{A}(\mathbf{r})e^{i\mathbf{k}\cdot\mathbf{r}}. \tag{5.64}$$

Unfortunately, because $\psi_{\mathbf{k}}(\mathbf{r}) = (2\pi)^{-3/2}\exp(i\mathbf{k}\cdot\mathbf{r})$ is not a Hilbert-space wave function, (5.64) can be ambiguous, implying that one may need to find an alternate approach. This turns out to be true when $\hat{A} = \hat{\mathbf{Q}}$, as will be shown below.

There are no such problems when $\hat{A} = \hat{\mathbf{P}}$ or $\hat{K}$, since $\hat{P}|\mathbf{k}\rangle = \hbar\mathbf{k}|\mathbf{k}\rangle$, implying that

$$\langle \mathbf{k}'|\hat{\mathbf{P}}|\mathbf{k}\rangle = \delta(\mathbf{k} - \mathbf{k}')\hbar\mathbf{k} \tag{5.65}$$

and

$$\langle \mathbf{k}'|\hat{K}|\mathbf{k}\rangle = \delta(\mathbf{k} - \mathbf{k}')\frac{\hbar^2 k^2}{2m}. \tag{5.66}$$

An intermediate case concerns local, finite-range potentials that are multiplicative. For such potentials we write

$$\langle \mathbf{r}'|\hat{V}|\mathbf{r}\rangle = \delta(\mathbf{r} - \mathbf{r}')V(\mathbf{r}), \tag{5.67}$$

and substitution of (5.67) into (5.64), with $\hat{A} \to \hat{V}$, produces

$$\langle \mathbf{k}'|\hat{V}|\mathbf{k}\rangle = \left(\frac{1}{2\pi}\right)^3 \int d^3r\, e^{-i\mathbf{k}'\cdot\mathbf{r}}V(\mathbf{r})e^{i\mathbf{k}\cdot\mathbf{r}}$$

$$= \left(\frac{1}{2\pi}\right)^{3/2}\left[\left(\frac{1}{2\pi}\right)^{3/2}\int d^3r\, e^{i(\mathbf{k}-\mathbf{k}')\cdot\mathbf{r}}V(\mathbf{r})\right]. \tag{5.68}$$

The second line of (5.68) follows from the first because $V(\mathbf{r})$ is simply a multiplicative function.

Since $V(\mathbf{r})$ is of finite range, i.e., is zero for large but finite r, the integral in (5.68) exists. In fact, the quantity in square brackets is the Fourier transform $\tilde{V}(\mathbf{k}' - \mathbf{k})$ of $V(\mathbf{r})$. Hence, in the momentum representation, a local potential $V(\mathbf{r})$ goes over to its Fourier transform, with argument $\mathbf{k} - \mathbf{k}'$:

$$\langle \mathbf{k}'|V|\mathbf{k}\rangle = \left(\frac{1}{2\pi}\right)^{3/2}\tilde{V}(\mathbf{k} - \mathbf{k}'), \tag{5.69}$$

where $\tilde{V}$ signifies the Fourier transform of V. Although $\langle \mathbf{k}'|V|\mathbf{k}\rangle$ is not local in momentum space, its dependence on the two wave vectors is only through their difference. An analogous result holds in 1-D.

We illustrate the foregoing analysis with the example of a screened Coulomb (or modified Yukawa) potential, $V^{\mu}_{\text{s.c.}}(\mathbf{r})$:

$$\hat{V}^{(\mu)}_{s.c.}(\mathbf{r}) = V_0 \frac{e^{-\mu r}}{r}.$$

Here, μ is the screening parameter, such that, in the limit $\mu \longrightarrow 0$, $\hat{V}^{(\mu)}_{\rm s.c}(r)$ goes over to an ordinary or unscreened Coulomb potential. It is not difficult to show that

$$\langle \mathbf{k}' | \hat{V}^{(\mu)}_{\rm s.c.} | \mathbf{k} \rangle = \frac{V_0}{2\pi^2} \frac{1}{(\mathbf{k} - \mathbf{k}')^2 + \mu^2}, \tag{5.70}$$

whose $\mu \longrightarrow 0$ limit is well defined, though it is not well behaved when $\mathbf{k}' = \mathbf{k}$.

Finally, let us consider the position operator $\hat{\mathbf{Q}}$. Since it is local in a coordinate representation, replacement of $\hat{A}$ by $\hat{\mathbf{Q}}$ in (5.64) gives

$$\langle \mathbf{k}' | \hat{\mathbf{Q}} | \mathbf{k} \rangle = \left(\frac{1}{2\pi} \right)^3 \int d^3 r \, e^{-i\mathbf{k}'\cdot\mathbf{r}} \mathbf{r} e^{i\mathbf{k}\cdot\mathbf{r}}. \tag{5.71}$$

The RHS of (5.71) is, however, not defined, and thus straightforward use of (5.64) has failed to yield an unambiguous result. A procedure that will eventually prove successful is to replace $\mathbf{r}$ by the gradient w.r.t. $\mathbf{k}$. However, if such a replacement is used directly in (5.71), it, too, fails. In each case, the cause of the problem is the absence from (5.71) of any quantity that leads to convergence, i.e., neither $\mathbf{r}$ nor $\psi_{\mathbf{k}}(\mathbf{r})$ is normalizable.

Instead of blindly substituting into (5.64), we proceed in analogy to (4.155). Starting with $\mathbf{Q}|\alpha\rangle$, one finds

$$\langle \mathbf{k} | \hat{\mathbf{Q}} | \alpha \rangle = \int d^3\mathbf{r} \, \langle \mathbf{k} | \mathbf{r} \rangle \langle \mathbf{r} | \hat{\mathbf{Q}} | \alpha \rangle. \tag{5.72}$$

Now, $\langle \mathbf{r} | \hat{\mathbf{Q}} | \alpha \rangle = \hat{\mathbf{Q}}(\mathbf{r}) \psi_\alpha(\mathbf{r}) = \mathbf{r} \psi_\alpha(\mathbf{r})$ and $\langle \mathbf{k} | \mathbf{r} \rangle = (2\pi)^{-3/2} \exp(-i\mathbf{k} \cdot \mathbf{r})$, so that (5.72) becomes

$$\langle \mathbf{k} | \hat{\mathbf{Q}} | \alpha \rangle = \left(\frac{1}{2\pi} \right)^{3/2} \int d^3 r e^{-i\mathbf{k}\cdot\mathbf{r}} \mathbf{r} \psi_\alpha(\mathbf{r}), \tag{5.73}$$

and, because $\psi_\alpha(\mathbf{r})$ is a Hilbert-space wave function, the integral in (5.73) may be assumed to exist and therefore to be manipulable.

One of the desired manipulations involves

$$e^{-i\mathbf{k}\cdot\mathbf{r}} \mathbf{r} = \mathbf{r} e^{-i\mathbf{k}\cdot\mathbf{r}} = i \nabla_{\mathbf{k}} e^{-i\mathbf{k}\cdot\mathbf{r}}, \tag{5.74}$$

where $\nabla_{\mathbf{k}}$ means the gradient operator taken w.r.t. the components of $\mathbf{k}$. Substitution of (5.74) into (5.73) leads to

$$\langle \mathbf{k} | \hat{\mathbf{Q}} | \alpha \rangle = \int d^3 r \, i \nabla_{\mathbf{k}} \langle \mathbf{k} | \mathbf{r} \rangle \langle \mathbf{r} | \alpha \rangle.$$

Invoking the standard assumption that $\nabla_{\mathbf{k}}$ can be taken outside the integral sign and then employing the relation (5.21) converts (5.73) into

$$\langle \mathbf{k} | \hat{\mathbf{Q}} | \alpha \rangle = i \nabla_{\mathbf{k}} \langle \mathbf{k} | \alpha \rangle. \tag{5.75}$$

Equation (5.75) implies that the momentum-space representation of $\hat{\mathbf{Q}}$ is local, i.e., $\langle \mathbf{k}' | \hat{\mathbf{Q}} | \mathbf{k} \rangle = \delta(\mathbf{k} - \mathbf{k}') \hat{\mathbf{Q}}(\mathbf{k})$, such that

$$\hat{\mathbf{Q}}(\mathbf{k}) = i \nabla_{\mathbf{k}}. \tag{5.76}$$

This is indeed the case; a formal proof is left as an exercise. Notice that $i = \sqrt{-1}$ enters (5.76) preceded by a plus sign, in contrast to the minus sign in Eq. (5.45).

5.5. The fundamental dynamical equation

Among the vectors in Hilbert space will always be one that is the state of a system. Many physical properties of the system can be obtained once its state is known; the question of how to determine the state is evidently a crucial one. The answer is given in Postulate V.

A special notation will be used for that ket which is the state of the system, viz., $|\Gamma, t\rangle$. The symbol Γ will generally be a collective label consisting of all relevant quantum numbers, while t indicates the time dependence: states evolve in time, their temporal development being governed by the time-dependent Schrödinger equation, formulated by Erwin Schrödinger in 1926 and already encountered in Chapter 3. $|\Gamma, t\rangle$ is a specific example of the generic ket $|\alpha, t\rangle$ introduced earlier in this chapter. We shall use capital Greek letters for the time-dependent wave functions, which for a single particle are

$$\Psi_\Gamma(\mathbf{r}, t) = \langle \mathbf{r}|\Gamma, t\rangle \tag{5.77}$$

and

$$\Phi_\Gamma(\mathbf{k}, t) = \langle \mathbf{k}|\Gamma, t\rangle. \tag{5.78}$$

Two operators appear in the time-dependent Schrödinger equation, a first-order time derivative and the Hamiltonian $\hat{H}$. The Hamiltonian is the sum of kinetic- and potential-energy operators,

$$\hat{H} = \hat{K} + \hat{V}.$$

For a single particle of mass m, $\hat{K} = \hat{P}^2/(2m)$, as in Table 5.1, whereas for N particles, with masses $m_1, m_2, \ldots, m_N$ and momentum operators $\hat{\mathbf{P}}_1, \hat{\mathbf{P}}_2 \ldots, \hat{\mathbf{P}}_N$, $\hat{K}$ is given by

$$\hat{K} = \sum_{j=1}^{N} \frac{\hat{P}_j^2}{2m_j}. \tag{5.79}$$

Thus, for any quantal system whose potential-energy operator $\hat{V}(\{\hat{\mathbf{Q}}_j\}_{j=1}^N)$ is obtained from a classical-physics potential energy, $\hat{H}(\{\hat{\mathbf{P}}_j\}, \{\hat{\mathbf{Q}}_j\})$ depends on the quantal operators $\{\hat{\mathbf{P}}_j\}$ and $\{\hat{\mathbf{Q}}_j\}$ in the same way as that in which the Hamiltonian function of classical mechanics, viz., $H_{\text{cl}}(\{p_k\}, \{q_k\})$, depends on the canonical positions and momenta. This prescription is contained in the definition of $\hat{H}$ given in part (d) of Postulate IV.

POSTULATE V.

The state $|\Gamma, t\rangle$ of any non-relativistic quantum system is a solution of the time-dependent Schrödinger equation

$$i\hbar \frac{\partial}{\partial t}|\Gamma, t\rangle = \hat{H}|\Gamma, t\rangle, \tag{5.80}$$

where $\hat{H}$, the Hamiltonian operator for the system, is the sum of the relevant kinetic and potential operators (defined in Postulate IV).

As written, (5.80) is an equation involving abstract Hilbert-space quantities. It becomes an equation in space and time on going into the position representation. Consider a spinless, single-particle system. By projecting both sides of (5.80) onto the position state $\langle \mathbf{r}|$, one obtains

$$i\hbar \frac{\partial}{\partial t}\langle \mathbf{r}|\Gamma,\, t\rangle = \langle \mathbf{r}|\hat{H}|\Gamma,\, t\rangle. \tag{5.81}$$

Next, assume that $\hat{H}$ is local in the coordinate representation, in which case (5.81) is transformed into

$$i\hbar \frac{\partial}{\partial t}\Psi_\Gamma(\mathbf{r},\, t) = \hat{H}(\mathbf{r})\Psi_\Gamma(\mathbf{r},\, t). \tag{5.82}$$

On using (5.48) and (5.47), and assuming $\hat{V}$ to be a classical-type potential, Eq. (5.82) becomes

$$i\hbar \frac{\partial}{\partial t}\Psi_\Gamma(\mathbf{r},\, t) = \left(-\frac{\hbar^2}{2m}\nabla^2 + V(\mathbf{r})\right)\Psi_\Gamma(\mathbf{r},\, t), \tag{5.83}$$

which we first met in Chapter 3 as Eq. (3.61). Equation (5.83) is the wave-function form of the time-dependent Schrödinger equation for a single particle experiencing a local, multiplicative potential. It is sometimes claimed to be the fundamental dynamical equation of quantum mechanics for the preceding single-particle system. As indicated here, it is the coordinate-space realization of the abstract, Hilbert-space relation (5.80) for such a system; the latter equation is the truly fundamental one.

The 1-D version of (5.83) was studied in Chapter 3 via the method of separation of variables. This led to the stationary-state solution $\Psi(x, t) = \psi(x)\exp(-iEt/\hbar)$, where $\psi(x)$ is a solution of the 1-D eigenvalue equation $\hat{H}(x)\psi(x) = E\psi(x)$. Particular solutions $\psi_n(x) = (2/L)^{1/2}\sin(n\pi x/L)$ and $E_n = n^2\pi^2\hbar^2/(2mL)$ were obtained for the case of a particle in a box and the probabilistic nature of the solutions was studied. We shall employ these solutions later in this section and also again in Section 5.6. However, given that we already possess a solution of the time-dependent Schrödinger equation in the special case of stationary states, we focus next on a general means for obtaining the time-dependences of $|\Gamma, t\rangle$ and $\Psi_\Gamma(\mathbf{r}, t)$, one that does not require the assumption of stationary states.

Time evolution

The time-dependent Schrödinger equation governs the temporal development of quantal systems; i.e., it determines the evolution $|\Gamma, t_0\rangle \rightarrow |\Gamma, t\rangle$, $t > t_0$. This propagation in time can be expressed by means of a *unitary time-evolution operator*[8] $\hat{U}(t, t_0)$:

$$|\Gamma,\, t\rangle = \hat{U}(t,\, t_0)|\Gamma,\, t_0\rangle, \tag{5.84}$$

where, necessarily,

$$\lim_{t\to t_0}\hat{U}(t,\, t_0) = \hat{I}. \tag{5.85}$$

[8] The structure of (5.84) confirms the remarks made in the subsection below Eq. (3.62) to the effect that knowledge of states or wave functions at only a single time is needed in order to obtain them at a later time.

That the time-evolution operator *is* unitary will be established in the following, where we determine the dependence of $\hat{U}(t, t_0)$ on $\hat{H}$. Two cases will be considered: $\hat{H} \neq \hat{H}(t)$, the more commonly occurring situation, at least pedagogically, and $\hat{H} = \hat{H}(t)$, such that $[\hat{H}(t), \hat{H}(t')] = 0$, $t \neq t'$. The reason for the latter condition will be stated when the case $\hat{H} = \hat{H}(t)$ is treated. The mathematical underpinning of the analysis is Stone's theorem, Section 4.5.

We consider the situation $\hat{H} \neq \hat{H}(t)$ first. Substituting (5.84) into (5.80) and rearranging yields

$$\frac{\partial}{\partial t}\hat{U}(t, t_0)\,|\Gamma, t_0\rangle = -i\frac{\hat{H}}{\hbar}\hat{U}(t, t_0)|\Gamma, t_0\rangle.$$

Since $|\Gamma, t_0\rangle$ is an arbitrary non-zero state, we may re-write this equation as one for $\hat{U}(t, t_0)$ alone, i.e.,

$$\frac{\partial}{\partial t}\hat{U}(t, t_0) = -i\frac{\hat{H}}{\hbar}\hat{U}(t, t_0). \tag{5.86}$$

The structure of (5.86) is identical to that of Eq. (4.175); hence, the form of the solution to (4.175), viz. (4.171), applies to (5.86). The initial condition (5.85) means that we replace x of (4.175) by $t - t_0$, in which case we have

$$\hat{U}(t, t_0) = \hat{U}(t - t_0) = e^{-i\hat{H}(t-t_0)/\hbar}, \qquad \hat{H} \neq \hat{H}(t). \tag{5.87}$$

Since $\hat{H} \neq \hat{H}(t)$, $\hat{U}(t, t_0)$ depends only on the difference $t - t_0$, while the form of Eq. (5.87) guarantees that it is a unitary operator ($\hat{H}^\dagger = \hat{H}$) and also that Eq. (5.85) is satisfied. Substituting (5.87) into (5.84) gives

$$|\Gamma, t\rangle = e^{-i\hat{H}(t-t_0)/\hbar}|\Gamma, t_0\rangle. \tag{5.88a}$$

The corresponding result for the one-particle, coordinate-space wave function $\Psi_\Gamma(\mathbf{r}, t)$ is

$$\Psi_\Gamma(\mathbf{r}, t) = e^{-i\hat{H}(\mathbf{r})(t-t_0)/\hbar}\Psi_\Gamma(\mathbf{r}, t_0). \tag{5.88b}$$

The concise form of $\hat{U}$ as the exponential of $\hat{H}$ may suggest that Eqs. (5.88a) and (5.88b) are explicit solutions, respectively, of (5.80) and (5.82). They are not! The former equations are simply re-writings of the latter, and are "solutions" only in the sense that an integral is the inverse of a derivative: until the actual effect of $\hat{H}$ on $|\Gamma, t\rangle$ or of $\hat{H}(\mathbf{r})$ on $\Psi_\Gamma(\mathbf{r}, t)$ is known in detail, (5.88a) and (5.88b) are just formulae, i.e., "formal" solutions. They are analogous to an integral before it is evaluated.

Lack of explicitness does not render these formal solutions useless, however. For example, they yield the complete time dependence for the so-called "stationary-state" solutions, discussed in the next subsection. In some instances, it is actually possible to evaluate the time-evolution operator $\hat{U}(t - t_0)$, thus providing direct insight into the behavior of the quantal system. Furthermore, these formal solutions are a means to derive the Feynman path-integral formulation of quantum mechanics (Feynman (1948)), discussed in Section 8.3. Finally, and this is demonstrated below, the formal solutions allow us to conclude that, once a state has been normalized, it remains so for all time: temporal evolution via the Hamiltonian operator does not alter the normalization.

To understand this statement, consider the scalar product $\langle\Gamma, t|\Gamma, t\rangle$:

$$\langle \Gamma, t|\Gamma, t\rangle = \langle \Gamma, t_0|\hat{U}^\dagger(t-t_0)\hat{U}(t-t_0)|\Gamma, t_0\rangle.$$

However, $\hat{U}^\dagger(t-t_0) = e^{i\hat{H}^\dagger(t-t_0)/\hbar} = e^{i\hat{H}(t-t_0)/\hbar}$, from which one easily finds $\langle \Gamma, t|\Gamma, t\rangle = \langle \Gamma, t_0|\Gamma, t_0\rangle$: once normalized, always normalized.

Let us now consider the case $\hat{H} = \hat{H}(t)$, with $[\hat{H}(t), \hat{H}(t')] = 0$, $t \neq t'$. Most of the preceding carries over to this case, although the structure of $\hat{U}(t, t_0)$ is no longer given by Eq. (5.87). To determine the new structure, we need only replace $\hat{H}$ in (5.86) by $\hat{H}(t)$:

$$\frac{\partial}{\partial t}\hat{U}(t, t_0) = -i\frac{\hat{H}(t)}{\hbar}\hat{U}(t, t_0). \tag{5.89}$$

The similarity of (5.89) to the first-order differential equation $df(t)/dt = g(t)f(t)$, whose solution is $f(t) = \exp[\int_{t_0}^t dt'\, g(t')]\, f(t_0)$, suggests that the solution to (5.89) should be

$$\hat{U}(t, t_0) = e^{-(i/\hbar)\int_{t_0}^t dt'\, \hat{H}(t')}, \qquad [\hat{H}(t), \hat{H}(t')] = 0, \tag{5.90}$$

formulated so that it obeys the initial condition (5.85).

Although the initial condition is satisfied, it must still be shown that (5.90) obeys (5.89). This may seem like a silly point to raise, on the ground that

$$\frac{\partial}{\partial t}e^{-(i/\hbar)\int_{t_0}^t dt'\, \hat{H}(t')} = e^{-(i/\hbar)\int_{t_0}^t dt'\, \hat{H}(t')}\left(-\frac{i}{\hbar}\hat{H}(t)\right). \tag{5.91}$$

However, (5.91) is valid only if $[\hat{H}(t), \hat{H}(t')] = 0$, $t \neq t'$, in which case the order of the exponential and $\hat{H}(t)$ on its RHS can be interchanged. That interchange means that Eqs. (5.91) and (5.89) are identical. Hence, Eq. (5.90) *is* the desired solution when $[\hat{H}(t), \hat{H}(t')] = 0$.

Equation (5.91) encompasses (5.87), since it reduces to the latter equation when $\hat{H} \neq \hat{H}(t)$. Notice, however, that the time dependence in (5.91) is no longer $t - t_0$. The result (5.91) is obviously unitary, implying that, even when the Hamiltonian is time-dependent, an initially normalized state remains so for all time unless other interactions come into play. Furthermore, as in the case $\hat{H} \neq \hat{H}(t)$, the expressions

$$|\Gamma, t\rangle = e^{-(i/\hbar)\int_{t_0}^t dt'\, \hat{H}(t')}|\Gamma, t_0\rangle \tag{5.92a}$$

and

$$\Psi_\Gamma(\mathbf{r}, t) = e^{-(i/\hbar)\int_{t_0}^t dt'\, \hat{H}(\mathbf{r},t')}\Psi_\Gamma(\mathbf{r}, t_0) \tag{5.92b}$$

are only "formal" solutions. We shall encounter the need for Eq. (5.92b) in Section 12.6, where the relevant integral $\int_{t_0}^t dt'\, \hat{H}(\mathbf{r}, t')$ will be evaluated in detail.

As a final point, we note that, if $[\hat{H}(t), \hat{H}(t')] \neq 0$, then the exponential form (5.90) is not the solution, neither is the derivative, as given by Eq. (5.91), even well defined. The correct form for $\hat{U}(t, t_0)$ when $[\hat{H}(t), \hat{H}(t')] \neq 0$ is an infinite expansion known as the Dyson series after its originator, F. J. Dyson (1949). The Dyson series is examined in texts more advanced than this one, e.g., Galindo and Pascual (1990), Gottfried (1966), and Sakurai (1994), to which the interested reader is referred.

Stationary states

For any $\Psi_\Gamma(\mathbf{r}, t)$ obeying (5.83), the probability density $\rho_\Gamma(\mathbf{r}, t)$ that the particle will be at position $\mathbf{r}$ at time t is

$$\rho_\Gamma(\mathbf{r}, t) = |\Psi_\Gamma(\mathbf{r}, t)|^2. \tag{5.93}$$

In contrast to the norm, the probability density $\rho_\Gamma(\mathbf{r}, t)$ can vary with time. There is one circumstance, however, in which $\rho_\Gamma(\mathbf{r}, t)$ will remain constant in time. This occurs when, at time $t_0 = 0$, $\Psi_\Gamma(\mathbf{r}, t_0 = 0)$ is an eigenfunction of $\hat{H}(\mathbf{r})$ or, correspondingly, when $|\Gamma, t_0 = 0\rangle$ is an eigenstate of $\hat{H}$, just as in Section 3.2.

Let us assume that the spectrum of $\hat{H}$ includes a discrete set of eigenvalues $\{E_n\}$ labeled, for simplicity, by the quantum number $n = 1, 2, 3 \ldots, N$, where N need not be finite:

$$\hat{H}|n, \gamma_n\rangle = E_n|n, \gamma_n\rangle, \tag{5.94}$$

with n having been separated from the collective index Γ, the remainder being denoted γ_n. The coordinate-space wave function describing the single particle is

$$\psi_{n,\gamma_n}(\mathbf{r}) = \langle \mathbf{r}|n, \gamma_n\rangle. \tag{5.95}$$

The states $|n, \gamma_n\rangle$ are normalized.

At $t_0 = 0$, the assumption that $|\Gamma, t_0 = 0\rangle$ is an eigenstate of $\hat{H}$ means that

$$|\Gamma, t_0 = 0\rangle = |n, \gamma_n\rangle, \tag{5.96}$$

where the nth eigenstate has arbitrarily been selected as the $t_0 = 0$ state. On substituting this relation into (5.88a), $|\Gamma, t\rangle$ becomes

$$\begin{aligned}|\Gamma, t\rangle &= e^{-i\hat{H}t/\hbar}|n, \gamma_n\rangle \\ &= e^{-iE_n t/\hbar}|n, \gamma_n\rangle,\end{aligned} \tag{5.97}$$

where the second equality follows from Eq. (4.167). Equation (5.97) shows that, apart from the phase factor $\exp(-iE_n t/\hbar)$, the state remains $|n, \gamma_n\rangle$ for all time. Projecting both sides of Eq. (5.97) onto $\langle \mathbf{r}|$, the coordinate-space wave function for the single-particle system is

$$\Psi_\Gamma(\mathbf{r}, t) = e^{-iE_n t/\hbar}\psi_{n,\gamma_n}(\mathbf{r}), \tag{5.98}$$

and $\rho_\Gamma(\mathbf{r}, t)$ is now easily seen to be constant in time:

$$\rho_\Gamma(\mathbf{r}, t) = |\Psi_\Gamma(\mathbf{r}, t)|^2 = |\psi_{n,\gamma_n}(\mathbf{r})|^2. \tag{5.99}$$

As in Section 3.2, eigenstates of $\hat{H}$ are referred to as *stationary states*, a nomenclature that refers to the probability rather than to the state. Since the time dependence of $|\Gamma, t\rangle$ is given by $\exp(-iE_n t/\hbar)$ for the case of an eigenstate, $\partial|\Gamma, t\rangle/\partial t = -iE_n|\Gamma, t\rangle/\hbar$ and the time-dependent Schrödinger equation reduces to a time-independent, energy eigenvalue problem:

$$\hat{H}|n, \gamma_n\rangle = E_n|n, \gamma_n\rangle. \tag{5.100}$$

It is from (5.100) that the eigenstates $|n, \gamma_n\rangle$ and energy eigenvalues E_n are to be determined. Note that the use of the "formal" time-evolution equation (5.88a) obviates the need to employ the separation of variables method.

In the most general formulation (5.100) may be written as

$$\hat{H}|E_\gamma, \gamma\rangle = E_\gamma|E_\gamma, \gamma\rangle, \tag{5.101}$$

with the equation for the single-particle coordinate-space wave function taking the form

$$\hat{H}(\mathbf{r})\psi_{E_\gamma,\gamma}(\mathbf{r}) = E_\gamma\psi_{E_\gamma,\gamma}(\mathbf{r}). \tag{5.102}$$

Equations (5.101) and (5.102) are each denoted the *time-independent Schrödinger equation*. Much of Chapters 6–15 is devoted to exploring properties of (5.102) and solving it for a variety of one- and two-particle systems.

5.6. Eigenstate expansions

Eigenstates are essential ingredients in quantal descriptions, but they enjoy a special feature that enhances their usefulness in ways that will be explored later in this book. This feature is the subject of the sixth postulate.

POSTULATE VI.

The eigenstates of every quantal observable $\hat{A}$ are complete and form an orthonormal basis in Hilbert space.

This postulate is a generalization of the corresponding statement concerning solutions of the Sturm–Liouville eigenvalue problem (Chapter 3), exemplified by the Fourier series, which includes the quantal-box eigensolutions. The postulate means that the eigenstates of any observable $\hat{A}$ provide a resolution of the identity operator. For example, if the spectrum of $\hat{A}$ is discrete,

$$\hat{A}|a_n\rangle = a_n|a_n\rangle, \qquad n = 1, 2, \ldots,$$

then Postulate VI is equivalent to

$$\hat{I} = \sum_{n=1}^{\infty} |a_n\rangle\langle a_n|. \tag{5.103}$$

If the spectrum of $\hat{B}$ is continuous, $\hat{B}|b\rangle = b|b\rangle$, as in the case of position and momentum, then

$$\hat{I} = \int db\, |b\rangle\langle b|. \tag{5.104}$$

Finally, if the spectrum of $\hat{C}$ is mixed, i.e., partly discrete and partly continuous, then

$$\hat{C}\left\{\begin{matrix} |c_n\rangle \\ |c\rangle \end{matrix}\right\} = \left\{\begin{matrix} c_n|c_n\rangle \\ c|c\rangle \end{matrix}\right\}, \qquad n = 1, \ldots, N, \tag{5.105}$$

where $c \geqslant 0$, and

$$\hat{I} = \sum_{n=1}^{N} |c_n\rangle\langle c_n| + \int_0^\infty dc\, |c\rangle\langle c|. \tag{5.106a}$$

Sometimes (5.106a) is re-expressed in the short-hand form

$$\hat{I} = \sum\!\!\!\!\!\!\int |c_\mu\rangle\langle c_\mu|, \tag{5.106b}$$

where μ implies both discrete and continuous values for c. All of these expressions for $\hat{I}$ are closure relations; they can be generalized to the case in which more than one quantum number is needed to specify the states (see Chapter 9).

Postulate VI has amusing implications. For example, on assuming that one has a discrete spectrum, insertion of $\hat{I}$ given by Eq. (5.103) to the left and right of an arbitrary quantal operator $\hat{A}$ yields

$$\hat{A} = \sum_{n,m} |a_n\rangle\langle a_n|\hat{A}|a_m\rangle\langle a_m| = \sum_n a_n |a_n\rangle\langle a_n|. \tag{5.107}$$

That is, each quantal operator $\hat{A}$ can be expressed as a sum of products of its eigenvalues times the projectors onto the corresponding eigenstates. We shall employ this result in the exercises for Chapter 13. The generalization of (5.107) to the other two types of spectra is straightforward.

A further consequence of this postulate is that the eigenstates of any observable $\hat{A}$ can be used to expand an arbitrary quantum state $|\alpha\rangle$. Again let the spectrum of $\hat{A}$ be discrete. Then use of (5.103) gives

$$|\alpha\rangle = \sum_n C_n |a_n\rangle, \tag{5.108}$$

where $C_n = \langle a_n|\alpha\rangle$. Furthermore, if $\hat{A}$ refers to a single-particle system and has a coordinate representation, then projecting both sides of (5.108) onto $\langle \mathbf{r}|$ gives

$$\psi_\alpha(\mathbf{r}) = \sum_n C_n \psi_{a_n}(\mathbf{r}). \tag{5.109}$$

In other words, Postulate VI states that the convergence of the sum in (5.109) is pointwise.

The quantal-box eigenfunctions provide a ready illustration of this. Suppose that, at $t_0 = 0$, a particle is injected into the box and that its Hilbert-space state is $|\alpha\rangle$. By Postulate VI, the box eigenfunctions $\psi_n(x) = (2/L)^{1/2}\sin(n\pi x/L)$ are a basis. Employing them, the version of (5.109) relevant to this situation is

$$\psi_\alpha(x) = \sum_{n=1}^{\infty} C_n \psi_n(x), \tag{5.110}$$

where

$$C_n = \sqrt{\frac{2}{L}} \int_0^L dx\, \psi_\alpha(x) \sin(n\pi x/L). \tag{5.111}$$

Although these results bear a striking resemblance to Eqs. (3.41) and (3.46) for the initial shape of the uniform, clamped string, the physics described is very different and so is the behavior in time of the box and string solutions, $\Psi_\alpha(x, t)$ and $y(x, t)$, respectively, as we shall show in the following subsection.

Postulate VI has been used in connection with the eigenstates of the position and momentum operators, but the idea is here generalized to *any* observable. Applications of Postulate VI abound in quantum mechanics, two areas being approximation theory and collision theory. We apply it in the following subsection to help elucidate the time-dependences of non-stationary states.

Non-stationary states

From Postulate V, the state $|\Gamma, t\rangle$ of a non-relativistic quantum system obeys (5.80), whose formal solution for the case in which $\hat{H} \neq \hat{H}(t)$ is given by (5.88a). Let us choose $t_0 = 0$ in Eq. (5.88a), giving

$$|\Gamma, t\rangle = e^{i\hat{H}t/\hbar}|\Gamma\rangle, \tag{5.112}$$

where $|\Gamma\rangle$ is the state at $t = 0$.

By assumption, $|\Gamma\rangle$ is not an eigenstate of $\hat{H}$. However, the eigenstates of $\hat{H}$ can be used to expand $|\Gamma\rangle$. For simplicity, the spectrum of $\hat{H}$ is assumed to be discrete and non-degenerate, i.e., only one integer quantum number n is needed to label the states and eigenvalues:

$$\hat{H}|n\rangle = E_n|n\rangle, \qquad n = 1, 2, \ldots. \tag{5.113}$$

By Postulate VI, $\{|n\rangle\}$ is a basis, so that

$$|\Gamma\rangle = \sum_n C_n|n\rangle, \tag{5.114}$$

where $C_n = \langle n|\Gamma\rangle$ and Eq. (5.114) follows from $\sum_n |n\rangle\langle n| = \hat{I}$.

Substitution of (5.114) into (5.112) leads to

$$|\Gamma, t\rangle = \sum_n C_n|n\rangle e^{-iE_nt/\hbar}, \tag{5.115}$$

where it has been assumed – as usual – that $e^{-i\hat{H}t/\hbar}\sum_n = \sum_n e^{-i\hat{H}t/\hbar}$. The time-dependence of $|\Gamma, t\rangle$ has now been made manifest, albeit via an infinite sum of phase factors $e^{-iE_nt/\hbar}$, each weighted by $C_n|n\rangle$.

Equation (5.115) means that the corresponding coordinate-space probability density is not stationary. For instance, let $|\Gamma, t\rangle$ refer to a spinless single-particle system, whose position is $\mathbf{r}$. Then the position-space form of (5.115) is

$$\Psi_\Gamma(\mathbf{r}, t) = \sum_n C_n\psi_n(\mathbf{r})e^{-iE_nt/\hbar}, \tag{5.116}$$

whose probability density $\rho_\Gamma(\mathbf{r}, t) = |\Psi_\Gamma(\mathbf{r}, t)|^2$ becomes

$$\rho_\Gamma(\mathbf{r}, t) = \left|\sum_n C_n\psi_n(\mathbf{r})e^{-iE_nt/\hbar}\right|^2. \tag{5.117}$$

Equation (5.117) clearly establishes that $\rho_\Gamma(\mathbf{r}, t)$ varies in time.

Although (5.117) demonstrates the non-stationarity property of a linear superposition, it is too formal for illustrative purposes. To flesh out the construction we consider two examples.

EXAMPLE 1

A wave packet. Suppose that an accelerator produces particles whose normalizable state at time $t = 0$ is $|\beta\rangle$, which is assumed known. Once a particle begins to move towards the target, its $t > 0$ behavior is typically governed by the free-particle Hamiltonian $\hat{H}_0 = \hat{K} = \hat{P}^2/(2m)$, so that the state $|\beta, t\rangle$ is given by

$$|\beta, t\rangle = e^{-i\hat{H}_0 t/\hbar}|\beta\rangle. \tag{5.118}$$

Since $|\beta\rangle$ is normalizable, it cannot be an eigenstate of $\hat{H}_0$ and (5.118) is only a formal relation. However, if a momentum-space resolution of the identity operator is inserted into (5.118), we get

$$\begin{aligned}|\beta, t\rangle &= \int d^3k\, e^{-i\hat{H}_0 t/\hbar}|\mathbf{k}\rangle\langle\mathbf{k}|\beta\rangle \\ &= \int d^3k\, e^{-i\hbar k^2 t/2m}|\mathbf{k}\rangle\varphi_\beta(\mathbf{k}),\end{aligned} \tag{5.119}$$

where Eqs. (5.60) and (5.62) have been used. Going into a coordinate representation, (5.119) becomes ($\langle\mathbf{r}|\beta, t\rangle = \Psi_\beta(\mathbf{r}, t)$)

$$\Psi_\beta(\mathbf{r}, t) = \left(\frac{1}{2\pi}\right)^{3/2}\int d^3k\, e^{i(\mathbf{k}\cdot\mathbf{r} - E_k t/\hbar)}\varphi_\beta(\mathbf{k}), \tag{5.120}$$

where $E_k = \hbar^2 k^2/(2m)$. The momentum-space wave function $\varphi_\beta(\mathbf{k})$ is assumed known, and therefore it may be possible to evaluate the integral in (5.120), thereby explicitly expressing $\Psi_\beta(\mathbf{r}, t)$ as a function of $\mathbf{r}$ and t. This would not be possible without the complete set expansion (5.119), thus demonstrating its utility. We shall have more to say about quantal wave packets in Chapters 7 and 15, where we deal with collisions.

EXAMPLE 2

The box eigenstates. We turn to the case of a particle injected into a quantal box at time $t_0 = 0$, in state $|\alpha\rangle$, which is assumed known. The coordinate-space wave function is $\psi_\alpha(x)$, given by Eq. (5.110). Once it is in the box, the behavior of the particle in time is governed by the box Hamiltonian $\hat{H}(x)$:

$$\begin{aligned}\Psi_\alpha(x, t) &= e^{-i\hat{H}(x)t/\hbar}\psi_\alpha(x) \\ &= \sum_{n=1}^{\infty} C_n e^{-iE_n t/\hbar}\psi_n(x),\end{aligned} \tag{5.121}$$

where $E_n = n^2\pi^2\hbar^2/(2mL^2)$ and C_n is given by Eq. (5.111).

Let us compare the box solution (5.121) with the $v_0(x) = 0$ string solution $y(x, t)$ of Eq. (3.48), slightly re-written as

$$y(x, t) = \sum_{n=1}^{\infty} a_n \psi_n(x) \cos(\tilde{\omega}_n t), \tag{5.122}$$

where $\tilde{\omega}_n = n\pi v/L$. On writing $E_n = \hbar\omega_n$, (5.121) becomes

$$\Psi_\alpha(x, t) = \sum_{n=1}^{\infty} C_n e^{-i\omega_n t} \psi_n(x), \tag{5.123}$$

where $\omega_n = n^2\pi^2\hbar/(2mL^2)$. In each case, the time dependence involves angular frequencies, but there is no reason to expect $\tilde{\omega}_n = \omega_n$, and we do not. What is strikingly dissimilar in these two expansions is the actual time dependence. The classical physics solution *must be real*; the $\cos(\tilde{\omega}_n t)$ factors mirror this. The $\exp(-i\omega_n t)$ time dependence in $\Psi_\alpha(x, t)$ is a consequence of the factor $i\hbar$ in the time-dependent Schrödinger equation: $\Psi_\alpha(x, t)$ is necessarily complex. Furthermore, a_n in Eq. (5.122) is simply a Fourier coefficent that indicates the magnitude with which $\psi_n(x) = (2/L)^{1/2} \sin(n\pi x/L)$ enters $y(x, t = 0)$, but C_n is more than this. It is a probability amplitude: $|C_n|^2$ is the probability that E_n would be the result of a measurement of the energy of the particle in the box (assuming for the moment that such a measurement could be performed). Finally, because $\Psi_\alpha(x, t)$ is a wave function (or quantal amplitude), $|\Psi_\alpha(x, t)|^2 = \rho_\alpha(x, t)$ is the probability density for finding the particle at $x \in [0, L]$, and it is time dependent. In the simple case in which

$$C_n = \begin{cases} 0, & n = 1, 4, 5, \ldots, \\ \sqrt{\tfrac{1}{2}}, & n = 2, 3, \end{cases} \tag{5.124}$$

we end up with Eq. (3.99) and a manifest demonstration of the sloshing back and forth of the probability density. Recall also that the total probability is unity and thus time-independent (Eq. (3.97)). This result is guaranteed by the unitary nature of the time-evolution operator, which expresses the conservation of probability.

Exercises

5.1 In a particular representation, the operator image of the spin-$\frac{1}{2}$ observable A is given by the matrix

$$A = \frac{\hbar}{2}\begin{pmatrix} \dfrac{1}{\sqrt{2}} & \dfrac{-i}{\sqrt{2}} \\ \dfrac{i}{\sqrt{2}} & \dfrac{-1}{\sqrt{2}} \end{pmatrix}.$$

(a) What are the values of A that can be measured in an ideal experiment?

(b) If the largest possible value of A is measured, express the (normalized) state of the system as a column vector.

5.2 In a particular representation, the operator image of the spin-1 observable A is given by the matrix

$$A = \frac{\hbar}{2}\begin{pmatrix} 0 & -\sqrt{2}i & 0 \\ \sqrt{2}i & 0 & -\sqrt{2}i \\ 0 & \sqrt{2}i & 0 \end{pmatrix}.$$

(a) Determine the values of A that can be obtained in an ideal measurement of A.

(b) If a measurement yields the minimum value of A, express the normalized state of the system as a column vector.

5.3 Let $\{|1\rangle, |2\rangle\}$ be an orthonormal basis.

(a) The observable A is represented by the operator

$$\hat{A} = A_{11}|1\rangle\langle 1| + A_{22}|2\rangle\langle 2|,$$

where A_{11} and A_{22} are real. Determine the possible measured values of A (expressed in terms of A_{11} and A_{22}) and the normalized states (expressed in terms of $|1\rangle$ and $|2\rangle$) corresponding to them.

(b) The operator image $\hat{B}$ of a second observable B has a more complicated representation in the same basis:

$$\hat{B} = B_{11}|1\rangle\langle 1| + B_{22}|2\rangle\langle 2| + B_{12}[|1\rangle\langle 2| + |2\rangle\langle 1|]$$

with B_{11}, B_{22} and B_{12} real. Analogously to part (a), determine the possible measured values of B and the states corresponding to them. Also show that these results reduce to those of part (a) when $B_{12} = 0$, $B_{11} = A_{11}$, and $B_{22} = A_{22}$.

5.4 In this problem, the $n = 1$ state of the 1-D box is used in modeling five excited states of a hypothetical nucleus. The fundamental $n = 1$ box state can be occupied by a neutron (n) or a proton (p). The two energies are $E_1^{(n)} = \hbar^2\pi^2/(2m_n L^2)$ and $E_1^{(p)} = \hbar^2\pi^2/(2m_p L^2)$, with $L = 15$ fm (1 fm $= 10^{-15}$ m) and $m_n c^2 = 939.565$ MeV and $m_p c^2 = 938.272$ MeV ($\hbar c = 197.33$ MeV fm). The five nuclear excited states correspond to excitation of a single nucleon or of a pair of nucleons, and the energies of these five states relative to the ground state are $E_1 = E_1^{(p)}$, $E_2 = E_1^{(n)}$, $E_3 = 2E_1^{(p)}$, $E_4 = E_1^{(p)} + E_1^{(n)}$, and $E_5 = 2E_1^{(n)}$, each of which is assumed for simplicity to be populated by absorption of a photon of energy $\hbar\omega_j = E_j$, j = 1, 2, . . ., 5.

(a) Determine the numerical values of the energies E_j, $j = 1, 2, \ldots, 5$.

(b) Photons (γ-rays) of energy 1.819 MeV $\pm\, \Delta E$ are absorbed by the nucleus. What is the maximum value of ΔE that will allow population of individual excited states?

(c) Suppose that the greatest percentage accuracy which could be achieved for the photon energies is 99.8%. Under this circumstance, and assuming that there are no other excited nuclear levels below 3 MeV, how many energy levels would be "seen" in photon absorption by the hypothetical nucleus?

5.5 A 1-D particle is described by a wave packet of momentum states:

$$|\beta\rangle = \int_{-\infty}^{\infty} dk\, A(k)|k\rangle,$$

such that the coordinate-space wave function $\psi_\beta(x)$ is given by

$$\psi_\beta(x) = \langle x|\beta\rangle = \int_{-\infty}^{\infty} dk\, A(k)\langle x|k\rangle = \sqrt{\frac{1}{2\pi}}\int_{-\infty}^{\infty} dk\, A(k)e^{ikx} = N\frac{e^{ik_0x}}{\sqrt{b^2+x^2}},$$

k_0 and b real.

(a) Determine N, N real and positive.

(b) Determine the probability amplitude $\langle k'|\beta\rangle$ that the momentum $\hbar k'$ is present in $|\beta\rangle$.

(c) Determine the probability $P(|x| \leqslant b/\sqrt{3})$ that the particle will be found in the region $x \in [-b/\sqrt{3},\, b/\sqrt{3}]$.

(d) Evaluate $\langle \hat{P}_x\rangle_\beta$, the expectation value of $\hat{P}_x$, given by

$$\langle \hat{P}_x\rangle_\beta = \langle\beta|\hat{P}_x|\beta\rangle = \int_{-\infty}^{\infty} dx\, \psi_\beta^*(x)\left(-i\hbar\frac{d}{dx}\right)\psi_\beta(x).$$

5.6 Let $\hat{A}$ and $\hat{B}$ be two observables of a quantal system, with

$$\left.\begin{aligned}\hat{A}|a_n\rangle &= a_n|a_n\rangle\\ \hat{B}|b_n\rangle &= b_n|b_n\rangle\end{aligned}\right\} \qquad n = 1, 2, \ldots,$$

and $\langle a_n|a_m\rangle = \delta_{nm} = \langle b_n|b_m\rangle$. None of the a_n or the b_n is degenerate, none of the eigenstates $|a_n\rangle(|b_n\rangle)$ is also an eigenstate of $\hat{B}(\hat{A})$, and $\langle a_n|b_m\rangle \neq 0,\ \forall\ n, m$.

(a) Prior to the measurement of B, the system is in state $|a_n\rangle$. What is the probability $P_{a_n}(b_m)$ that measurement of B will yield b_m?

(b) If the preceding measurement yielded b_j, what is the state of the system immediately following the measurement?

5.7 A system consists of two non-interacting particles confined to a 1-D box located between $x = 0$ and $x = L$. The particles are labeled 1 and 2, have masses m_1 and m_2, and are governed by the Hamiltonian $\hat{H} = -\{[\hbar^2/(2m_1)]\, d^2/dx_1^2 + [(\hbar^2/2m_2)]\, d^2/dx_2^2\}$, where x_j is the position coordinate of particle j, $j = 1, 2$.

(a) The first wave function for the system is

$$\psi_\alpha(x_1, x_2) = \psi_{n_1}(x_1)\psi_{n_2}(x_2),$$

where $\psi_n(x)$ is a 1-D box eigenfunction.

(i) If the energy of the system is measured, what value will be found?

(ii) What is the probability for finding particle 1 in the interval $[0,\ L/3]$?

(b) The system is next put into a second state, whose wave function is

$$\psi_\beta(x_1, x_2) = \frac{2\psi_1(x)\psi_2(x_2) - 6\psi_3(x_1)\psi_4(x_2)}{2\sqrt{10}},$$

with $\psi_n(x_j)$ again being a box eigenfunction.

(i) If the energy of the system is measured, what values will be found, and with what probabilities will they occur?

(ii) Suppose that the energy measurement yields $E_{1,2} = E_1 + E_2 = [\hbar^2\pi^2/(2L^2)](1/m_1 + 4/m_2)$. What is the wave function subsequent to the measurement?

(iii) Using the preceeding wave function, what is the probability of finding particle 1 in the interval $[0,\ L/3]$?

5.8 The wave function for a particle in 3-D is

$$\psi(x, y, z) = Ne^{-|x|/(2a)}e^{-|y|/(2b)}e^{-|z|/(2c)}.$$

(a) Calculate N, choosing it to be real and positive.
(b) Determine the probability that the particle will be found in the range $x \in [0, a]$.
(c) A simultaneous measurement of the y- and z-positions is made. What is the (joint) probability that y will lie in the interval $[-b, b]$ and z will be in the interval $[-c, c]$?

5.9 The probability $P(r, \theta, \phi)$ that a particle, acted on by a 3-D isotropic harmonic oscillator potential, will be found in the vicinity of the point defined by the spherical polar coordinates r, θ, and ϕ is

$$P(r, \theta, \phi) = Ar^2 e^{-\beta^2 r^2} \cos^2\theta \, d^3 r,$$

where $d^3 r = r^2 \, dr \sin\theta d\theta d\phi$.
(a) Evaluate the real, positive normalization constant A.
(b) Find r_m, the most probable value of the radius.
(c) Determine the probability density $\rho(r)$ that the particle will be at the radius r.
(d) The radial probability density is $\rho^{(\text{rad})} = r^2\rho(r)$. Sketch on the same graph both $\rho(r)$ and $\rho^{(\text{rad})}(r)$.
(e) Calculate $\langle r \rangle$, the expected or average value of r, defined by

$$\langle r \rangle = \int_0^\infty dr r \rho^{(\text{rad})}(r),$$

and indicate on the graph of part (d) both r_m and $\langle r \rangle$.

5.10 This problem, a continuation of the preceding one, uses the same $P(r, \theta, \phi)$.
(a) Determine the probability density $\rho(\theta)$ that the particle will be found at an angle θ.
(b) Define, in analogy to $\rho^{(\text{rad})}(r)$, the angular probability density $\rho^{(\text{ang})}(\theta) = \sin\theta \, \rho(\theta)$, and sketch on the same graph $\rho(\theta)$ and $\rho^{(\text{ang})}(\theta)$, $\theta \in [0, \pi]$.
(c) Evaluate $\langle \cos\theta \rangle$, the expected or average value of $\cos\theta$, defined by

$$\langle \cos\theta \rangle = \int_0^\pi d\theta \cos\theta \, \rho^{(\text{ang})}(\theta).$$

(d) Obtain $\langle \cos^2\theta \rangle$, the variance of $\cos\theta$, defined analogously to $\langle \cos\theta \rangle$.
(e) The quantal uncertainty in $\cos\theta$ (see Chapter 9) is $\Delta\cos\theta$, defined by

$$\Delta\cos\theta = [\langle \cos^2\theta \rangle - \langle \cos\theta \rangle^2]^{1/2}.$$

Use the results of parts (c) and (d) to evaluate $\Delta\cos\theta$.

5.11 Protons (and neutrons) are now believed to be composite objects consisting of three fundamental particles denoted quarks, which give rise to all of a proton's observed properties. Among these are its size and shape, which in this problem is assumed to be a sphere of radius $R = 10^{-15}$ m. Since isolated quarks are believed to be unobservable, one might think to verify their presence in a proton via their influence on the behavior of the electron in the ground state of the H-atom. In order for this influence to be non-negligible, the electron would need to have a non-trivial probability of being found inside the proton. Use Eq. (5.25) to determine that probability, i.e., the probability $P(r \leqslant R)$ that the electron will be found inside the proton. (Hint: a "quick" way to determine $P(r \leqslant R)$ is to employ an approximation related to the small size of R. If you use this approximation, state clearly what is meant by "small," i.e., why the approximation is valid in this problem. Note: if you do not use the approximation, the numerical value of $P(r \leqslant R)$ is more difficult to determine, yet the gain in accuracy is negligible.)

5.12 The wave function for a particle of mass m in a 1-D box located between $x = 0$ and $x = L$ is $\psi_1(x) = \sqrt{2/L}\sin(\pi x L)$.

(a) The right-hand wall is suddenly shifted from $x = L$ to $x = 2L$.

(i) What is the probability that the particle will be in the ground state of the new box?

(ii) Ditto for the first excited state of the new box.

(iii) In which state of the new box is the particle most likely to be found?

(b) Both walls of the box are suddenly and simultaneously removed. What is the probability density $\rho(k)$ that the particle will have momentum $\hbar k$? Sketch $\rho(k)$ and discuss it qualitatively.

5.13 At time $t = 0$, the state vector $|\Gamma\rangle$ describing a particle of mass m in a 1-D box is a linear superposition of box eigenstates:

$$|\Gamma\rangle = \sqrt{0.2}|1\rangle - \sqrt{0.4}|2\rangle + \sqrt{0.4}|3\rangle.$$

(a) Use the above superposition to determine the state $|\Gamma, t\rangle$ at any time $t > 0$.

(b) If the energy of the particle were measured, what values could be observed and with what probabilities would they occur?

(c) What is the expected value of the energy if a large number of energy measurements were made, the particle always being in the state of part (a). Is your result time-dependent? Explain why.

(d) If the particle's position is measured, what value is expected to occur? Is this result time-dependent? Explain why, and compare with the answer to part (c).

5.14 Let $|E_n\rangle$ and E_n be the non-degenerate eigenstates and eigenvalues of the Hamiltonian operator $\hat{H}$:

$$\hat{H}|E_n\rangle = E_n|E_n\rangle, \qquad n = 1, 2, 3, \ldots.$$

Let $\hat{B}$ be an observable for the system such that

$$\hat{B}|E_1\rangle = |E_2\rangle; \qquad \hat{B}|E_2\rangle = |E_1\rangle; \qquad \hat{B}|E_n\rangle = 0, \qquad n > 2.$$

(a) Determine the eigenvalues and eigenstates of $\hat{B}$, using $\{|E_n\rangle\}$ as a basis to express the latter eigenstates.

(b) Suppose that B is measured at time $t = 0$ and is found to have the value 1. After a time t, with the system remaining undisturbed, B is measured again: what is the probability that the value 1 will be found?

5.15 In the orthonormal basis $\{|1\rangle, |2\rangle\}$, the Hamiltonian for a system has the 2×2 form

$$H = \begin{pmatrix} a & b \\ b & a \end{pmatrix},$$

i.e., $\langle j|\hat{H}|j\rangle = a$, $j = 1, 2$; $\langle 1|\hat{H}|2\rangle = \langle 2|\hat{H}|1\rangle = b$; a and b real and positive.

(a) Find the eigenvalues E_1 and $E_2 > E_1$ of $\hat{H}$, and their corresponding (normalized) eigenvectors $|E_1\rangle$ and $|E_2\rangle$, expressing the $|E_j\rangle$ as linear combinations of $|1\rangle$ and $|2\rangle$ with real coefficients you are to determine.

(b) Let the solution of the relevant time-dependent Schrödinger equation be $|\Gamma, t\rangle$. If at time $t = 0$, $|\Gamma, t = 0\rangle = |1\rangle$, determine $|\Gamma, t\rangle$, expressing it as a linear combination of $|1\rangle$ and $|2\rangle$ with time-dependent coefficients.

(c) Determine the numerical value of the probability $P_2(t = \pi\hbar/(2b))$ that the system will be found in state $|2\rangle$ at time $t = \pi\hbar/(2b)$.

5.16 Three of the six quarks used in describing "elementary" particles are the "up," the "down," and the "strange" quarks, whose states, respectively, are denoted $|u\rangle$, $|d\rangle$, and $|s\rangle$. A system composed of these objects is governed by the Hamiltonian

$$\hat{H} = \hbar\omega_0[|u\rangle\langle d| + |d\rangle\langle u|] + \hbar\omega_1|s\rangle\langle s|.$$

(a) Determine the eigenvalues and the orthonormalized eigenvectors of $\hat{H}$, expressing the latter in terms of $|u\rangle$, $|d\rangle$, and $|s\rangle$.

(b) At time $t = 0$, the normalized eigenstate of the system $|\Gamma\rangle$ is

$$|\Gamma\rangle = a|d\rangle + b|s\rangle, \qquad a \text{ and } b \text{ real}.$$

At a later time, the state is $|\Gamma, t\rangle$. Determine the probabilities of finding the system in each of the states $|u\rangle$, $|d\rangle$, and $|s\rangle$ at time $t > 0$.

(c) What is the $t > 0$ expectation value of the energy when the state is $|\Gamma, t\rangle$ of part (b)?

(d) State whether and why each of the following two truncations of $\hat{H}$ is an acceptable Hamiltonian (for all or part of the original system):

$$\hat{H} \to \hat{H}_1 = \hbar\omega_0[|u\rangle\langle d| + |d\rangle\langle u|], \qquad \hat{H} \to \hat{H}_2 = \hbar\omega_0|u\rangle\langle d| + \hbar\omega_1|s\rangle\langle s|.$$

5.17 A non-degenerate quantal system is governed by a Hamiltonian $\hat{H}$:

$$\hat{H}|E_n\rangle = E_n|E_n\rangle, \qquad n = 1, 2, \ldots,$$

with $\langle E_n|E_m\rangle = \delta_{nm}$. In addition, $\hat{A}$ is an observable for the system such that

$$\hat{A}|a_\alpha\rangle = a_\alpha|a_\alpha\rangle, \qquad \alpha = 1, 2, \ldots,$$

with $\langle a_\alpha|a_\beta\rangle = \delta_{\alpha\beta}$ and the $\{a_\alpha\}$ non-degenerate. Furthermore $[\hat{A}, \hat{H}] \neq 0$, so that none of the $|E_n\rangle$ ($|a_\alpha\rangle$) is an eigenstate of $\hat{A}$ ($\hat{H}$).

(a) Prior to measurement of $\hat{A}$, the system is in state $|E_n\rangle$. What is the probability $P_n(a_\alpha)$ that measurement of $\hat{A}$ will yield a_α?

(b) If the preceding measurement were made at time $t = 0$ and yielded the result a_α, what would the state of the system immediately following the measurement be, i.e., what is $|\Gamma, t = 0\rangle$?

(c) Following the $t = 0$ measurement of $\hat{A}$ that yielded a_α, the system was isolated and then at time $t > 0$, $\hat{A}$ was again measured. What is the probability $P(a_\alpha, t)$ that the value a_α will again be found? Express your answer in terms of the $P_n(a_\alpha)$.

5.18 The non-normalizable wave function

$$\psi(x, y, z) = \sin(k_1 x)\cos(k_2 y)\, e^{ik_3 z}$$

is a superposition of momentum wave functions.

(a) Determine which values of the x-, y-, and z-components of momentum are present in $\psi(x, y, z)$ and the (relative) probabilities with which they occur.

(b) Determine whether $\psi(x, y, z)$ is an eigenfunction of the kinetic-energy operator $\hat{K}_x(x)$. If it is, to what energy does it belong? If it is not, which energies occur in $\psi(x, y, z)$ and with what probabilities?

5.19 The momentum-space wave function $\varphi_a(k)$ of a 1-D free particle is a real, positive constant for $k \in [k_1, k_2]$ and is zero otherwise.

(a) Determine the normalized form of $\varphi_a(k)$.

(b) The coordinate-space form of the wave function is $\psi_a(x)$. Define $K = k_1 + k_2$ and $\Delta k = k_2 - k_1$ and determine the unique functions $f(Kx)$ and $g(\Delta k\, x)$ in the expression

$$\psi_a(x) = \sqrt{\frac{\Delta k}{2\pi}} f(Kx) g(\Delta k\, x).$$

(c) Prove that your $\psi_a(x)$ is normalized.

(d) Sketch the probability density $\rho_a(x) = |\psi_a(x)|^2$, measuring x in units of $(\Delta k)^{-1}$.

(e) Use symmetry arguments to determine the value of $\langle x \rangle_a = \int_{-\infty}^{\infty} dx\, x \rho_a(x)$, the average or expected value of x.

5.20 (a) Starting with the normalized coordinate-space wave functions, determine their momentum-space counterparts for

(i) the 1-D box eigenfunctions $\psi_n(x)$,

(ii) the H-atom ground-state wave function $\psi_{100}(\mathbf{r})$ of Eq. (5.25),

(iii) the (un-normalized) harmonic oscillator function

$$\psi(r, \theta, \phi) \propto r e^{-\beta^2 r^2/2} \cos\theta.$$

(b) Sketch the behaviors of the momentum-space probability densities for each of the three functions of part (a), choosing whatever reasonable means you wish to present the result for the oscillator function, part (a) (iii).

5.21 (a) Using the expressions given in Table 5.2 evaluate the commutators

(i) $[\hat{Q}_x(x), \hat{P}_x(x)]$,

(ii) $[\hat{\mathbf{Q}}(\mathbf{r}), \hat{\mathbf{P}}(\mathbf{r})]$.

(Hint: apply (i) to $\psi(x) \in \mathcal{H}$ and (ii) to $\psi(\mathbf{r}) \in \mathcal{H}$.)

(b) $\hat{P}_x$ and $\hat{\mathbf{P}}(\mathbf{r})$ are postulated to be local operators. Show that locality of $\hat{Q}_x$ and the value of $[\hat{Q}_x, \hat{P}_x]$ imply that $\hat{P}_x$ is local.

5.22 A particle of mass m moves in 1-D in the absence of a potential. The momentum eigenstates $|k\rangle$, $\langle x|k\rangle = e^{ikx}/\sqrt{2\pi}$, form a complete set:

$$\hat{I} = \int_{-\infty}^{\infty} dk\, |k\rangle\langle k|,$$

where $\langle k'|k\rangle = \delta(k - k')$. Introduce two new sets of states, denoted $|k, \pi\rangle$:

$$|k, \pi\rangle = \frac{1}{\sqrt{2}}[|k\rangle + \pi| - k\rangle], \qquad k \geqslant 0, \qquad \pi = \pm 1.$$

(a) Prove that the set $\{|k, \pi\rangle\}$ is orthonormal, i.e., that

$$\langle k', \pi'|k, \pi\rangle = \delta_{\pi'\pi}\delta(k - k').$$

(b) Show that $\{|k, \pi\rangle\}$ is complete, i.e., that

$$\hat{I} = \sum_{\pi} \int_0^{\infty} dk\, |k, \pi\rangle\langle k, \pi|.$$

(c) Evaluate $\psi_{\pi,k}(x) = \langle x|k, \pi\rangle$ and determine whether it is an eigenfunction of $\hat{K}_x(x)$.

5.23 (a) Starting with the result $\hat{P}_x(x) = -i\hbar\, d/dx$ and the assumptions that $\psi_\alpha(x) \in \mathcal{H}$ and $\psi_\beta(x) \in \mathcal{H}$, prove that $\hat{P}_x$ is Hermitian.

(b) Discuss whether $\hat{P}_x$ remains Hermitian if either or both of $\psi_\alpha(x)$ and $\psi_\beta(x)$ is or are replaced by non-normalizable plane-wave states.

5.24 Using the result of part (a) (i) of Exercise 5.21 evaluate the following commutators:

(a) $[\hat{Q}_x, \hat{P}_x^2]$,
(b) $[\hat{Q}_x^2, \hat{P}_x]$,
(c) $[\hat{Q}_x^n, \hat{P}_x]$,
(d) $[Q_x^2, \hat{P}_x^2]$,
(e) $[\hat{K}_x, \hat{P}_x]$,
(f) $[V(r), \hat{\mathbf{P}}(\mathbf{r})]$.

(Note: it will be shown in Chapter 9 that, if two operators commute, then they share a common set of eigenstates.)

5.25 (a) In rectangular coordinates, $\hat{L}_z(x, y) = -i\hbar(x\,\partial/\partial y - y\,\partial/\partial x)$. Use the transformation from rectangular to spherical polar coordinates to re-express $\hat{L}_z(x, y)$ in terms of r, θ, and ϕ.

(b) Show that $[\hat{L}_z, \hat{K}(\mathbf{r}) + V(r)] = 0$, thereby establishing that eigenfunctions of a one-particle, spherically-symmetric Hamiltonian are also eigenfunctions of $\hat{L}_z$.

(c) Evaluate $[\hat{L}_z, \hat{\mathbf{P}}(\mathbf{r})]$.

5.26 Evaluate $\tilde{V}(\mathbf{k} - \mathbf{k}')$ of Eq. (5.69) for

(a) the screened Coulomb or Yukawa potential, i.e., prove Eq. (5.70);
(b) the exponential potential $V_{\text{exp}} = -Ve^{-r/r_0}$;
(c) the Gaussian potential $V_{\text{Gauss}} = -V_0 e^{-(\beta r)^2}$;
(d) the square well, Eq. (5.51).
(e) Let $\mathbf{q} = \mathbf{k} - \mathbf{k}'$, and sketch the preceding $\tilde{V}(\mathbf{q}) = \tilde{V}(q)$ for $q \in [0, \infty]$.

5.27 The one-term separable potential, $V_{\text{sep}} = \lambda|g\rangle\langle g|$, is non-local both in the coordinate representation and in the momentum representation. Its 1-D eigenvalue problem is considered in the exercises to Chapter 6.

(a) In 1-D, an often-used form factor is $\langle k|g\rangle = (k^2 + \beta^2)^{-1}$, with β real and positive. Determine the coordinate-space form $\langle x|g\rangle$ for this form factor.

(b) If $|g\rangle$ refers to 3-D, with $\langle \mathbf{k}|g\rangle = (k^2 + \beta^2)^{-1}$, evaluate $\langle \mathbf{r}|g\rangle$.

5.28 Prove that Eq. (5.75) does imply Eq. (5.76).

5.29 Let $\hat{C}$ be the complex conjugation operator, i.e., $\hat{C}\psi = \psi^*$, where ψ may be a function in 1-D, 2-D, or 3-D.

(a) Determine whether $\hat{C}^\dagger = \hat{C}$.
(b) Find the eigenvalues and eigenfunctions of $\hat{C}$.
(c) Determine whether the eigenfunctions corresponding to different eigenvalues of $\hat{C}$ are orthogonal.
(d) Determine whether the eigenfunctions of $\hat{C}$ form a complete set.

5.30 Let $\hat{H}$ be the Hamiltonian for a 1-D particle and $\Psi_\Gamma(x, t)$ a solution of the time-dependent Schrödinger equation which is real for some time t_0:

$$\Psi_\Gamma^*(x, t_0) = \Psi_\Gamma(x, t_0).$$

This solution is also real for a second time $t_1 = t_0 + \Delta t$:

$$\Psi^*_\Gamma(x,\ t_0 + \Delta t) = \Psi_\Gamma(x,\ t_0 + \Delta t).$$

Show that the system is periodic and find the period T. That is, show that a time interval such that $\Psi_\Gamma(x,\ t + T) = \Psi_\Gamma(x,\ t)$ exists for any T. Is this possible when $V(x) = 0$? (Hint: use the set $\{\psi_{E_n}(x)\}$ of complete, orthonormal, *real* eigenfunctions of $\hat{H}$ as an expansion basis. Your solution will include a condition obeyed by the eigenvalues E_n.)

5.31 (a) The single-particle, time-dependent Schrödinger equation for the coordinate-space wave function $\Psi_\Gamma(\mathbf{r},\ t)$ is given by Eq. (5.83).
   (i) Show that its analog for the momentum-space wave function $\Phi_\Gamma(\mathbf{k},\ t)$ is an integro-differential equation.
   (ii) Assuming that $\Phi(\mathbf{k},\ t) = \phi_\Gamma(\mathbf{k})\exp(-iEt/\hbar)$, determine the time-independent, momentum-space equation obeyed by $\phi_\Gamma(\mathbf{k})$.
(b) Let $\Psi_\Gamma(\mathbf{r},\ t) = Ne^{-\beta r}e^{-i\omega t}$ be the wave function of a 3-D particle of mass m acted on by a potential $V(r)$, with β and ω real.
   (i) From the fact that $\Psi_\Gamma(\mathbf{r},\ t)$ obeys Eq. (5.83) determine $V(r)$ as a function of r, β, and ω.
   (ii) Evaluate N, choosing it to be real and positive, and then determine $\Phi_\Gamma(\mathbf{k},\ t)$ and the time-dependent equation it obeys.

5.32 In the non-relativistic approximation and suppressing the spin-$\frac{1}{2}$ degree of freedom, the ground-state wave function of the electron in the H-atom is $\psi_{100}(\mathbf{r})$ of Eq. (5.25).
(a) Use this wave function to determine the following expectation values:
   (i) $\bar{r} = \langle\hat{Q}_r\rangle = (\psi_{100}(\mathbf{r}),\ r\psi_{100}(\mathbf{r}))$,
   (ii) $r_{\text{rms}} = \langle\hat{Q}_r^2\rangle^{1/2}$,
   (iii) $\langle\hat{Q}_x\rangle$ (compare with $\bar{r}$ and explain any significant discrepancies),
   (iv) $\langle\hat{K}\rangle$ and $\langle V\rangle = -\langle\kappa_e e^2/\hat{Q}_r\rangle$ (compare their sum with E_1 of the Bohr model). Hint: express $\hat{K}$ in spherical polar coordinates.
(b) Evaluate the probability $P(r > a_0)$ that the electron will be found at radii beyond the Bohr radius a_0. Explain whether and why the value of $P(r > a_0)$ is consistent with the values of $\bar{r}$ and r_{rms}.

6

Applications of the Postulates: Bound States in One Dimension

In this chapter, the postulates are applied to a variety of soluble problems in 1-D, i.e., to systems having one degree of freedom. Each system consists of a particle acted on by either a *confining* or a *non-confining* potential. Confining potentials are those which, both for positive and for negative values of their arguments, become infinite. An example is the 1-D quantal box, Eq. (3.73). The wave functions describing a particle in such a potential are zero everywhere the potential is infinite. The effect is to trap or confine the particle in the potential: it can never behave as a free particle asymptotically, i.e., like one whose stationary-state wave function is a plane wave. All eigenstates in confining potentials are thus bound (normalizable), in contrast to the case of non-confining potentials, in which particles may be either bound or asymptotically free. Because only bound states occur in the case of confining potentials, we examine these cases first. We first revisit the quantum box and then consider the linear harmonic oscillator, which we solve using two different methods. Bound states in non-confining potentials are studied in Sections 6.3–6.6.

The major emphases in the present chapter are on solving the eigenvalue problem and using the results to illustrate aspects of the postulates. Section 6.1 concentrates especially on this feature. In general, the mathematical image of the potential energy is a Hilbert-space operator $\hat{V}$, but, in this chapter and the next, $\hat{V}$ is assumed to be local and multiplicative in the coordinate representation. The 1-D coordinate will be denoted x, assumed to lie in the interval $[a, b]$: $x \in [a, b]$. Either $a = -\infty$, or $b = +\infty$, or both, will be allowed.

The operator form of the time-independent Schrödinger equation is

$$\hat{H}\,|\Gamma\rangle = (\hat{K} + \hat{V})|\Gamma\rangle = E_\Gamma|\Gamma\rangle, \tag{6.1}$$

whose coordinate respresentation becomes

$$\left(-\frac{\hbar^2}{2m}\frac{d^2}{dx^2} + V(x)\right)\psi_\Gamma(x) = E_\Gamma\psi_\Gamma(x), \tag{6.2}$$

where $V(x)$ is assumed to be real.

Before considering the specific examples, we list a few general properties of the solutions $\psi_\Gamma(x)$ to (6.2).

(i) Equation (6.2) has two linearly independent solutions, linear combinations of which are also solutions.

(ii) Both $\psi_\Gamma(x)$ and $\psi'_\Gamma(x) = d\psi_\Gamma(x)/dx$ must be continuous in order that $\psi''_\Gamma(x) = d^2\psi_\Gamma(x)/dx^2$ exists. If $V(x)$ is discontinuous at some points, then $\psi''_\Gamma(x)$ will be also; if $V(x)$ is infinite at a single point, x_1, e.g., $V(x) = V_0\delta(x - x_1)$, then $\psi'_\Gamma(x_1)$ is discontinuous and $\psi''_\Gamma(x_1)$ is undefined. Finally, if $V(x)$ is infinite for a range of x, then $\psi_\Gamma(x)$ is zero in that range, a condition that follows from (6.2).

(iii) Because $|\psi_\Gamma(x)|^2$ is a probability density and particles can be at only one point at a time, $\psi_\Gamma(x)$ must be single-valued.

(iv) In general, Eq. (6.2) can have at most two types of solutions: those that go to zero when $x \to \pm\infty$ and are thus in $\mathcal{H}$, and those that do not go to zero when $x \to \pm\infty$, and are consequently non-normalizable, improper eigensolutions ($\notin \mathcal{H}$). The former solutions are denoted "bound states" and are obtained by solving an eigenvalue problem; the associated eigenvalues are the bound state energies. In the case of non-normalizable solutions, the continuous values of E satisfy $E \geqslant 0$, and the solutions $\psi_\Gamma(x)$ are specified by means of asymptotic boundary conditions, as discussed in Chapter 7.

(v) For the case of bound states, the eigensolutions are non-degenerate and are specified by means of an integer quantum number n, $n \geqslant 0$. The eigenvalues E_n can always be arranged so that they increase with increasing n: $E_n < E_{n+1}$, for all allowed n. Furthermore, if $V_{\min}$ is the minimum value of $V(x)$, then $E_n > V_{\min}$, for all n.

(vi) The number of nodes (zeros) in $\psi_n(x)$ depends on the minimum value of n and whether the potential is confining. For confining potentials, ψ_n has $n + 2$ zeros when $n \geqslant 0$ and $n + 1$ zeros when $n \geqslant 1$. In other words, the wave function corresponding to the lowest-lying energy has two nodes, the next three, etc.

These various properties will be useful in our subsequent analysis. We turn first to the quantum box.

6.1. The quantum box, revisited

Although the eigenvalue problem for the quantal box has been solved, it remains a felicitous system for illuminating features both of the postulates and of the preceding general comments on the properties of eigensolutions. We examine some of these in this section.

The quantum box is also known as the 1-D infinite square well: square, because the edges of the potential well are sharp, rather than rounded; and infinite, because the sides of the well are infinitely high and wide. In the "regional" terminology used with non-confining potentials, the functional form of $V(x)$ for the well is

$$V(x) = \begin{cases} \infty, & x \leqslant 0 \qquad \text{(Region I)}, \\ 0, & 0 < x < L \qquad \text{(Region II)}, \\ \infty, & x \leqslant L \qquad \text{(Region III)}; \end{cases} \tag{6.3}$$

it is schematically indicated in Fig. 6.1. Having infinitely high and wide walls means that the particle cannot leave the box: it is bound forever, since walls that are infinitely wide and high are impenetrable.[1]

[1] Impenetrability poses the non-practical problem of getting the particle into the box in the first place, a problem we ignore with impunity.

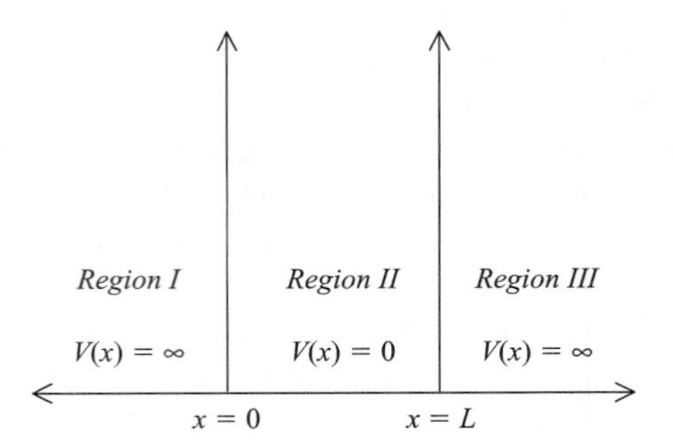

Fig. 6.1 The 1-D quantum box, or infinite square-well potential.

To understand the behavior of this system, it is necessary to solve (6.2). Since $V(x) = \infty$ in regions I and III, property (ii) tells us that $\psi_n(x)$ must be zero in these regions (where the integer n is replacing the general quantum number Γ). This behavior is easy to verify, since, if $\psi_n(x) \neq 0$ in these regions, then the term $V(x)\psi_n(x)$ will be infinite in Eq. (6.2), while the other terms, in particular $E_n\,\psi_n(x)$, will be finite, thus contradicting the equality embodied in the relation. An inference here is that the zero value of $\psi_n(x)$ in regions I and III overwhelms the corresponding ∞ in $V(x)$, so that the product $V(x)\psi_n(x)$ is zero. We assume this necessary condition.

The foregoing comments were used in Section 3.3 along with the relevant boundary conditions to obtain the desired solutions of Schrödinger's equations, viz.,

$$\psi_n(x) = \begin{cases} \sqrt{\dfrac{2}{L}}\sin\left(\dfrac{n\pi x}{L}\right), & x \in [0,\, L], \\ 0, & x \notin [0,\, L], \end{cases} \tag{6.4}$$

$$E_n = n^2 E_1 = n^2 \frac{\pi^2\hbar^2}{2mL^2}, \tag{6.5}$$

$$\Psi_n(x,\, t) = e^{-iE_n t/\hbar}\psi_n(x), \tag{6.6}$$

and

$$k_n = n\pi/L, \qquad E_n = \hbar^2 k_n^2/(2m), \tag{6.7}$$

where, in all these formulae, $n = 1, 2, \ldots$.

The Hilbert-space solutions are the kets $|n\rangle$, such that $\psi_n(x) = \langle x|n\rangle$, where $\langle n|m\rangle = \int dx\, \psi_n^*(x)\psi_m(x) = \delta_{nm}$. In principle, if $\psi_n(x) = A_n \sin(k_n x)$, normalization yields

$$A_n = e^{i\phi_n}\sqrt{\frac{2}{L}}, \tag{6.8}$$

with ϕ_n an arbitrary phase. Values of $k_n = -n\pi/L$, $n = 1, 2, 3, \ldots$, are not ruled out *ab initio*, but, because $\sin(-n\pi x/L) = -\sin(n\pi x/L)$, and, for $n = -m$, $-\exp(i\phi_{-m})$ can be set equal to $\exp(i\phi_m)$, owing to the arbitrariness of the phases, and only positive n need be taken into account. Furthermore, since only quantal phase *differences* can be determined experimentally, and then only in special circumstances normally involving interference effects, it is customary to set the arbitrary, overall phase of any quantum state equal to zero. Doing so produces the wave function (6.4).

Repeated in Fig. 6.2 are the Grotrian diagram representing the first three energy levels plus curves illustrating the associated probability amplitudes $\psi_n(x)$ and densities $|\psi_n(x)|^2$ originally plotted in Fig. 3.8. All wave functions are zero at the walls, viz., at $x = 0$ and $x = L$. Excluding these zeros, the functional form $\sin(n\pi x/L)$ guarantees that $\psi_n(x)$ has a total of n nodes, the mth occurring at $x = mL/n$, $m = 1, 2, \ldots, n-1$ (recall property (vi)). Thus $\psi_1(x)$, the lowest or ground-state wave function, is referred to as "nodeless," while $\psi_2(x)$ is said to have one node, etc. The larger the number of nodes the more rapidly $\psi_n(x)$ oscillates. The nodes are the points in the box where, quantally, the particle can never be found. In the nth state, there are n values of x at which the probabilities for finding the particle are equally maximal. It is relatively straightforward to show that, as $n \to \infty$, the average of the probability density tends to L^{-1}, the classical value. The proof of this is left to the problems.

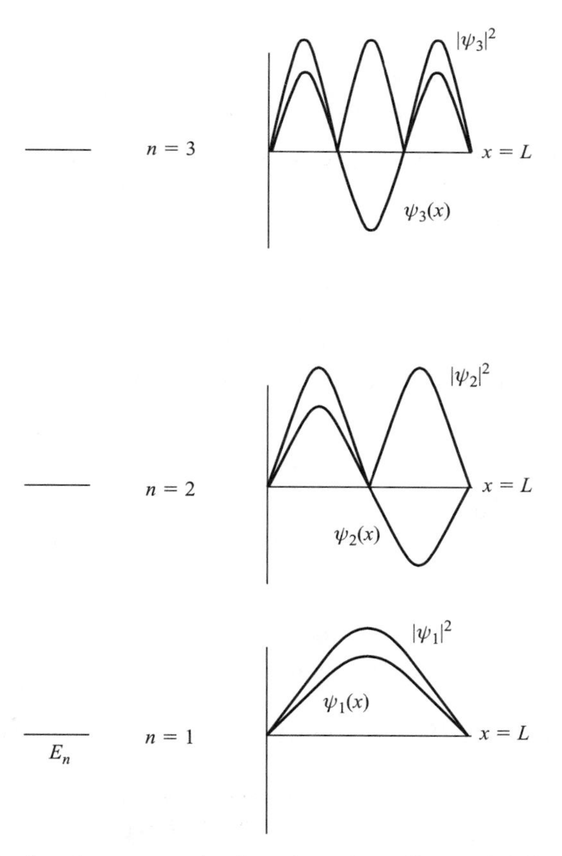

Fig. 6.2 The first three energy levels and corresponding probability amplitudes (ψ_n) and densities ($|\psi_n|^2$) for a particle in a quantum box.

From the relation

$$E_n = \langle n|\hat{H}|n\rangle = \int_0^L dx\, \psi_n^*(x)\left(-\frac{\hbar^2}{2m}\frac{d^2}{dx^2}\right)\psi_n(x),$$

it is evident that the expectation value of the kinetic energy increases (quadratically) with n. Since $\psi_n(x)$ oscillates more rapidly as a function of x than does $\psi_m(x)$ if $n > m$, then, as n increases, so does the curvature of $\psi_n(x)$. Correlating to E_n, we conclude that the greater the curvature in $\psi_n(x)$, the larger the expectation value of the kinetic energy will be, a result that holds in general, not just in the present case of the 1-D quantum box.

This result also follows from a semi-classical argument: the de Broglie relation $p = h/\lambda$ suggests the following qualitative behavior for the kinetic energy (E_k): $E_k \sim 4\pi^2\hbar^2/(2m\lambda^2)$. The "wavelength" λ_n of $\psi_n(x)$ clearly depends on the number of nodes, viz., $\lambda_n = 2L/n$, and substituting this into the expression for E_k yields exactly E_n of Eq. (6.5)! In general, the greater the number of oscillations in the wave function the larger the expectation value of the kinetic energy.

In the remainder of this section, we treat two additional topics: some properties of the system when it is in a single eigenstate, and the generalization to the case in which the system is described by a linear combination of eigenstates.

A single eigenstate

Since the quantal box is a pedagogic system, we may assume that measurements on a particle in the box are feasible. Suppose, then, that, at time $t = 0$, the particle's state is determined to be $|n\rangle$, perhaps by use of an energy filter (recall Postulate I). At any later time t, its coordinate-space wave function is $\Psi_n(x, t)$ of Eq. (6.6). Because of the stationary-state character of $\Psi_n(x, t)$, it gives rise to the same probability density as does $\psi_n(x)$, viz., $|\psi_n(x)|^2$. Since $\psi_n(x)$ is distributed over the range [0, L], we can ask for the probability that the particle is in any portion of [0, L]. In Section 3.3, we found that P_n ($x \geqslant L/2$, t), the probability that at time t the particle will be found in the right-hand half of the box, was 0.5:

$$P_n(x \geqslant L/2,\, t) = \tfrac{1}{2}, \tag{6.9}$$

independent of time (and quantum number n) and consistent with the symmetry of $|\Psi_n(x, t)|^2$ about the point $x = L/2$. Indeed, as noted in Section 3.3, this symmetry property suffices to *predict* the result (6.9), since $|\Psi_n(x, t)|^2$ has the same shape for $x \leqslant L/2$ as it does for $x \geqslant L/2$.

Two other quantities that are readily determined from $\Psi_n(x, t)$ are the average or expected values of the position and of the momentum, $\langle\hat{Q}_x(t)\rangle_n$ and $\langle\hat{P}_x(t)\rangle_n$, respectively, as in Postulate III.

We remarked that the value P_n ($x \leqslant L/2$, t) $= \frac{1}{2}$ can be predicted from the symmetry of $|\Psi_n(x, t)|^2$ about the point $x = L/2$. There are many situations in which invoking symmetry considerations can simplify analyses, even leading to answers without the need for a detailed calculation. Clearly the attempt to use a symmetry property is worth undertaking, assuming that the system enjoys such a property; one should always seek to obtain an answer – or an order of magnitude estimate – in advance or in place of a numerical calculation. In the present case, the symmetry of $|\Psi_n(x, t)|^2$ about the point $x = L/2$ suggests that the average or the expected value of position should be $L/2$:

$$\langle \hat{Q}_x(t) \rangle_n \stackrel{?}{=} L/2.$$

It is straightforward to verify that the equality holds by performing the relevant integral:

$$\begin{aligned} \langle \hat{Q}_x(t) \rangle_n &= \langle n, t | \hat{Q}_x | n, t \rangle \\ &= \frac{2}{L} \int_0^L dx \, x \sin^2\left(\frac{n\pi x}{L}\right). \end{aligned} \tag{6.10}$$

Although $\psi_n^2(x)$ is symmetric, the integrand in (6.10) is not, so the integration must be carried out. (Note that, if the lower limit were $-L$ instead of 0, the integral would vanish, since the integrand is an odd function and the interval would be even.) The same trigonometric identity as that used in connection with $P_n(x \geqslant L/2, t)$ yields

$$\langle \hat{Q}_x(t) \rangle_n = \frac{1}{L} \int_0^L dx \left[x - x \cos\left(\frac{2n\pi x}{L}\right) \right]. \tag{6.11}$$

The first term in (6.11) gives the result $L/2$; the second can be evaluated by employing the relation $\int \theta \cos\theta \, d\theta = \cos\theta + \theta \sin\theta$, whose use leads to the value zero. Hence, we find, as predicted, that

$$\langle \hat{Q}_x(t) \rangle_n = L/2, \tag{6.12}$$

which is another t- and n-independent result. The average position is always the center of the box when $\Psi(x, t) = \Psi_n(x, t)$.

The average momentum is obtained from

$$\begin{aligned} \langle \hat{P}_x(t) \rangle_n &= \langle n, t | \hat{P}_x | n, t \rangle \\ &= \frac{2}{L} \int_0^L \sin\left(\frac{n\pi x}{L}\right) \left(-i\hbar \frac{d}{dx}\right) \sin\left(\frac{n\pi x}{L}\right) dx, \end{aligned} \tag{6.13}$$

where the coordinate representation of $\hat{P}_x$ has been used. Once again, the expectation value is independent of t.

Carrying out the differentiation in (6.13), $\langle \hat{P}_x(t) \rangle_n$ becomes

$$\langle \hat{P}_x(t) \rangle_n = -\frac{2n\pi i\hbar}{L^2} \int_0^L \sin\left(\frac{n\pi x}{L}\right) \cos\left(\frac{n\pi x}{L}\right) dx. \tag{6.14}$$

The integral in (6.14) is zero – a result following from Fourier-series analysis, or from an integral table, or from the fact that, over the interval $[0, L]$, the sine and cosine of the same argument are 90° out of phase, so that their product has equal positive and negative parts, which exactly cancel out when they are combined as in the integral. Hence, we get

$$\langle \hat{P}_x(t) \rangle_n = 0, \tag{6.15}$$

i.e., the average momentum is zero. Eq. (6.15) should not be a surprising result, since a moment's reflection indicates a firm classical basis for it, viz., a particle bouncing back and forth with constant kinetic energy between two rigid walls will suffer a change of direction of its momentum at each (elastic) encounter with a wall; the average momentum of an assembly of such particles will be zero, since half will go to the left and the other half to the right. We note in passing that Eq. (6.15) holds for bound states in any 1-D potential, as proved in the exercises.

These, and other quantities, have simple forms when a particle in the quantum box is described by an eigenstate $|n\rangle$. The situation is more complicated, and thus is more interesting, in the case of a superposition, examined next.

A superposition of eigenstates

If the impenetrability is removed from a quantum box just long enough for a particle to be injected into it and is then reimposed at time $t = 0$, the state $|\Gamma\rangle$ of the particle will be a linear superposition of the box eigenstates $|n\rangle$:

$$|\Gamma\rangle = \sum_{n=1}^{\infty} C_n |n\rangle, \tag{6.16}$$

a result that follows from Postulate VI.

In (6.16), C_n is the probability amplitude that $|n\rangle$ is present in $|\Gamma\rangle$; $|C_n|^2$ is the probability for this, as prescribed by Postulate II. Normalization of $|\Gamma\rangle$ means that $\sum_n |C_n|^2 = 1$. At any time $t > 0$, the state becomes $|\Gamma,\ t\rangle$:

$$|\Gamma,\ t\rangle = \sum_n C_n e^{-iE_n t/\hbar} |n\rangle, \tag{6.17}$$

whose coordinate-space wave function is

$$\Psi_\Gamma(x,\ t) = \sum_n C_n \psi_n(x) e^{iE_n t/\hbar}. \tag{6.18}$$

In examining this example of a superposition state, we shall assume that N particles (rather than one) are injected into the box, each being described by (6.17) or (6.18). Furthermore, we also assume that measurement yields the result that N_2 of the N particles have energy E_2 while the remaining $N_3 = N - N_2$ particles have energy E_3. This restriction to two energies greatly simplifies the analysis without unduly sacrificing the consequences of the expansions (6.17) and (6.18); recall Eq. (3.94).

The purpose in this section is to demonstrate the effect that a superposition of eigenstates has on the values of, for example, the average energy $\langle \hat{H} \rangle_\Gamma$, the average position $\langle \hat{Q}_x \rangle_\Gamma$, and the average momentum $\langle \hat{P}_x \rangle_\Gamma$, at any time t. To do this requires knowledge of $|\Gamma,\ t\rangle$ or $\Psi_\Gamma(x,\ t)$. These quantities are completely determined once the values of C_n in either of the expansions (6.17) and (6.18) are specified. Instead of simply stating these values, let us first investigate whether the assumption (via energy measurements) that N_2 particles have energy E_2 and $N_3 = N_2 - N$ particles have energy E_3 suffices to determine the C_n's. Such an assumption mimics information yielded by actual experiments. We are therefore attempting to use this information to construct $|\Gamma,\ t\rangle$ or $\Psi_\Gamma(x,\ t)$. As we shall see, the information given is insufficient to carry out the construction in full, thereby rendering use of (6.17) or (6.18) inappropriate and forcing a density-operator description, which we shall not pursue. However, were we to use the incorrect eigenstate expansion (with incompletely determined C_n), one result would be an inability to predict the average values noted above, as we shall show.

To begin the construction, we remark that, since only E_2 and E_3 are obtained in an energy measurement, only $|n_2\rangle$ and $|n_3\rangle$ can occur in (6.17). For, if any other state $|n\rangle$ were present, then there would be some non-zero probability of measuring its energy E_n, $n \neq 2$ or 3. We therefore conclude that

$$C_n = 0, \qquad n \neq 2 \text{ or } 3. \tag{6.19}$$

A further conclusion can now be drawn. Since $|C_n|^2$ is the probability that $|n\rangle$ is present in the state, then

$$N_n = |C_n|^2 N \tag{6.20a}$$

is the number of particles that will be measured to have energy E_n, by Postulate III. However, since N_2 and N_3 are assumed known, (6.20a) gives

$$|C_n|^2 = N_n/N, \qquad n = 2,\, 3, \tag{6.20b}$$

exactly as in the discussion following Postulate II.

Unfortunately, no further information is available: (6.20b) yields $|C_n|$ but not its phase. The most we can do is write

$$C_n = e^{i\delta_n}\sqrt{N_n/N}, \qquad n = 2,\, 3, \tag{6.21}$$

where δ_n is a real but unknown number. Lack of knowledge of δ_n may seem unimportant: after all, we have set the phase of each normalization coefficient A_n equal to zero with impunity (or so we have claimed; see Eq. (6.8) and the subsequent discussion). The difference between the C_n of (6.21) and the A_n of (6.8) is that $\psi_n(x)$ is a *single* state whose overall phase, not being measurable and not affecting matrix elements, can be set to zero, whereas δ_n, not being an overall phase in the state $|\Gamma, t\rangle$, *will* affect both probabilities and matrix elements, as demonstrated next.

From Eqs. (6.19) and (6.21), and assuming a pure-state, wave-function description to be valid, $\Psi_\Gamma(x, t)$ becomes

$$\Psi_\Gamma(x, t) = \sqrt{\frac{N_2}{N}}e^{i\delta_2}\psi_2(x)e^{-iE_2 t/\hbar} + \sqrt{\frac{N_3}{N}}e^{i\delta_3}\psi_3(x)e^{-iE_3 t/\hbar}. \tag{6.22}$$

δ_2 and δ_3 are each unknown, but it is only their difference that is important, as can be seen by re-writing (6.22) as

$$\Psi_\Gamma(x, t) = e^{i\delta_3}\left(\sqrt{\frac{N_2}{N}}e^{i(\delta_2-\delta_3)}\psi_2(x)e^{-iE_2 t/\hbar} + \sqrt{\frac{N_3}{N}}\psi_3(x)e^{-iE_3 t/\hbar}\right), \tag{6.23}$$

and then recalling that, in matrix elements of operators and in probabilities involving $\Psi_\Gamma(x, t)$, the general form of integrand that occurs is $\Psi_\Gamma^*(x, t)\hat{A}(x)\Psi_\Gamma(x, t)$, where $\hat{A}(x) = 1$ in the case of probabilities. The factors $\exp(-i\delta_3)$ and $\exp(i\delta_3)$ in such expressions will cancel out, leaving only $\exp[\pm i(\delta_2 - \delta_3)]$, itself still an unknown.

To repeat our earlier comment, although the phase factors that arise when individual eigenstates are normalized do not produce any significant consequences and therefore can be set equal to unity, the fact that $\delta_2 - \delta_3$ is indeterminate has immediate consequences. As an illustration of this, we re-calculate the probability, at time $t = 0$, that any one of the N particles is in the right-hand-half portion of the box, using the invalid $\Psi_\Gamma(x,\ t)$ of (6.23). The probability is now given by

$$\begin{aligned} P_\Gamma(x \geqslant L/2,\ t=0) &= \int_{L/2}^{L} dx\,|\Psi_\Gamma(x,\ t=0)|^2 \\ &= \int_{L/2}^{L} dx \left(\frac{N_2}{N}\psi_2^2(x) + \frac{N_3}{N}\psi_3^2(x) \right. \\ &\quad \left. + 2\sqrt{\frac{N_2 N_3}{N^2}}\psi_2(x)\psi_3(x)\cos(\delta_2-\delta_3) \right), \end{aligned} \tag{6.24}$$

where the reality of $\psi_n(x)$ has been employed.

The first two integrals in (6.24) have already been evaluated, while the integrand of the third contains the factor $\sin(2\pi x/L)\sin(3\pi x/L)$, a product most easily dealt with by using the trigonometric identity

$$\sin\theta\sin\phi = \tfrac{1}{2}[\cos(\theta-\phi) - \cos(\theta+\phi)]. \tag{6.25}$$

On carrying out the integrations, the probability is found to be

$$P_\Gamma\left(x \geqslant \frac{L}{2},\ t=0\right) = \frac{1}{2} - \frac{8}{5\pi}\sqrt{\frac{N_2 N_3}{N^2}}\cos(\delta_2-\delta_3). \tag{6.26}$$

Equation (6.26) clearly displays an "interference" structure due to the factor $\cos(\delta_2-\delta_3)$. What it means in concrete terms is that $P_\Gamma(x \geqslant L/2,\ t=0)$ is indeterminate, in contrast to the single-eigenstate case. Such an indeterminancy is not permitted for pure states; it is one of the reasons why a wave-function description is invalid in the present case. Ignoring this, and continuing to use (6.23), the best that one can do is place bounds on $P_\Gamma(x \geqslant L/2,\ t=0)$. For example, if $N_2 = N_3 = N/2$, then P_Γ lies somewhere in the following range:

$$0.245 \leqslant P_\Gamma\left(x \geqslant \frac{L}{2},\ t=0\right) \leqslant 0.755;$$

choosing $N = 10^4$, the number of particles N_{RH} in the right-hand half of the box lies between 2450 and 7550 but is otherwise unknown. The uncertainty here is huge.

This uncertainty in $P_\Gamma(x \geqslant L/2,\ t=0)$ will carry over to any quantity whose value is non-zero, thereby, in effect, rendering $\Psi_\Gamma(x,\ t)$ of (6.22) or (6.23) *useless* w.r.t. its predictive powers. On the other hand, *were* N_{RH} *also to be measured*, then $\delta_2 - \delta_3$ could be inferred (modulo 2π), the predictive power would be restored, and $\Psi_\Gamma(x,\ t)$ of Eq. (6.23) would be a valid wave function. A conclusion of greater generality than in this immediate context is that *measurements that involve interference effects can often determine relative phases.*

To proceed further, we now assume that $\delta_2 = \delta_3 = 0$. Hence, $\Psi_\Gamma(x,\ t)$ becomes

$$\Psi_\Gamma(x,\ t) = \sqrt{\frac{N_2}{N}}\psi_2(x)e^{-iE_2 t/\hbar} + \sqrt{\frac{N_3}{N}}\psi_3(x)e^{-iE_3 t/\hbar}. \tag{6.27}$$

Equation (6.27) gives rise to the probability density

$$\rho_\Gamma(x,\,t) = \frac{N_2}{N}\psi_2^2(x) + \frac{N_3}{N}\psi_3^2(x) + 2\sqrt{\frac{N_2N_3}{N^2}}\psi_2(x)\psi_3(x)\cos\left(\frac{(E_3-E_2)t}{\hbar}\right), \tag{6.28}$$

which is closely related to Eqs. (3.98) and (3.99). Defining the angular frequency ω_{23} via $\hbar\omega_{23} = E_3 - E_2$, $\rho_\Gamma(x,\,t)$ is seen to have the same period $T = 2\pi/\omega_{23}$ as does ρ of Eq. (3.99), and the same properties. See the last equations in Section 3.3.

Two examples of the time-dependent interference property – the "sloshing" in time – are $P_\Gamma(x \geqslant L/2,\,t)$ and $\langle\hat{Q}_x(t)\rangle_\Gamma$. Following the procedures encountered previously, one finds

$$P_\Gamma\left(x \geqslant \frac{L}{2},\,t\right) = \frac{1}{2} - \frac{8}{5\pi}\sqrt{\frac{N_2N_3}{N^2}}\cos(\omega_{23}t) \tag{6.29}$$

and

$$\langle\hat{Q}_x(t)\rangle_\Gamma = \frac{L}{2}\left(1 - \frac{8}{\pi^2}\frac{24}{25}\sqrt{\frac{N_2N_3}{N^2}}\cos(\omega_{23}t)\right). \tag{6.30}$$

The right-hand sides of (6.29) and (6.30) are similar in structure to that of Eq. (6.26), but, unlike the latter result, the oscillating interference term $\cos(\omega_{23}t)$ of the first two equations becomes a *known* function of time once L is specified. Thus, the probability $P_\Gamma(x \geqslant L/2,\,t)$ and the average position $\langle\hat{Q}_x(t)\rangle_\Gamma$ can be determined in practice for all t after the values of N_2, N_3, and L have been stated.

Figure 6.3 shows the x-dependence of $L\rho_\Gamma(x,\,t)$ from Eq. (6.28) for various values of t

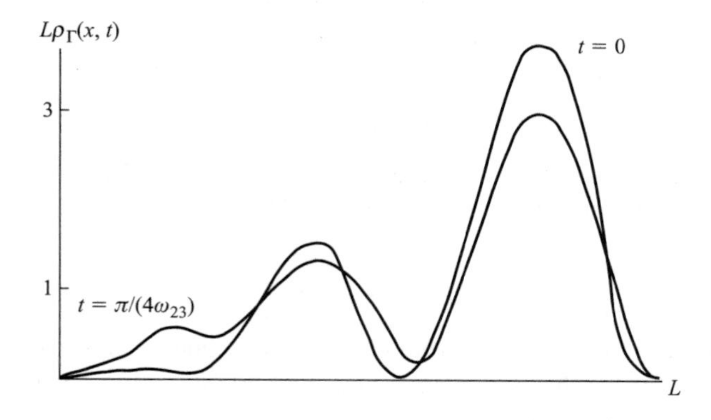

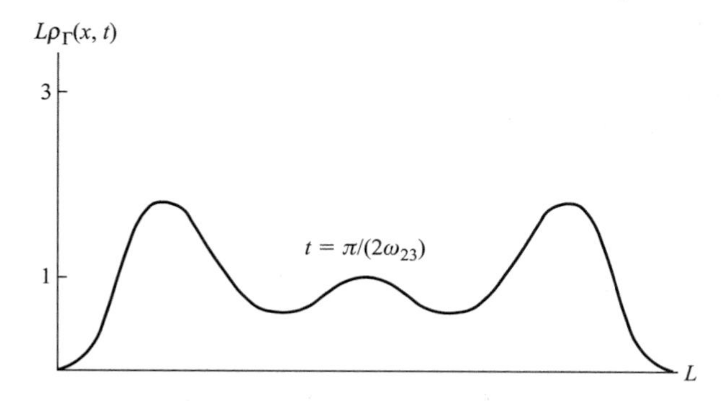

Fig. 6.3 Sketches of $L\rho_\Gamma(x,\,t)$ for three values of t, Eq. (6.28).

(in units of π/ω_{23}). All results are based on the choice $N_2 = N_3 = N/2$. The sloshing of the maxima of $L\rho_\Gamma(x, t)$, and of $\rho_\Gamma(x \geqslant L/2, t)$ and $L^{-1}\langle \hat{Q}_x(t)\rangle_\Gamma$ from the RHS to the LHS of the box as t goes from 0 to π/ω_{23} is evident from Fig. 6.3. Note that, because $\rho_\Gamma(x, t)$ is a probability density in a finite region, $L\rho_\Gamma(x, t)$ need not be, and in Fig. 6.3 is not, less than unity for all x: the only requirement is that $\rho_\Gamma(x, t)$ be normalized to unity, which it is. What these results demonstrate so conclusively, of course, is that a superposition leads to quantities that depend on time, in contrast to the single-eigenstate case.

As a final point, we note that the expectation value of the momentum $\langle \hat{P}(t)\rangle_\Gamma$ remains zero in the superposition case, while $\langle \hat{H}\rangle_\Gamma$ displays neither interference nor a time dependence. Proofs of these statements are left as exercises.

6.2. The linear harmonic oscillator

In classical mechanics, linear oscillations are often introduced by means of a mass attached to a spring that is then stretched within its elastic limit. In a more general analysis, one employs a potential-energy function $V(x)$ assumed to have a minimum at a point x_0. By keeping $x - x_0$ small, $V(x)$ can be expanded in a Taylor series about x_0 and then approximated by retaining only the first two non-vanishing terms:

$$V(x) = V(x_0) + \tfrac{1}{2}(x - x_0)^2 V''(x_0), \tag{6.31}$$

where the prime means differentiation w.r.t. x. In place of the force constant k used in Section 2.4, it is useful to introduce the oscillator frequency ω_0 defined by $\omega_0 = [V''(x_0)/m]^{1/2}$, with m being the mass of the particle, so that (6.31) takes the form

$$V(x) = V(x_0) + \tfrac{1}{2}m\omega_0^2(x - x_0)^2. \tag{6.32}$$

Equation (6.32) can be simplified by assuming that the zero of potential energy is at $x = x_0$ and that x_0 coincides with the origin of coordinates. Under these assumptions, $V(x)$ becomes

$$V(x) = \tfrac{1}{2}m\omega_0^2 x^2, \tag{6.33}$$

which is the standard form for the potential energy for the linear harmonic oscillator. An arbitrary potential $V(x)$ that has a minimum equal to zero at $x = 0$ is shown in Fig. 6.4, along with its parabolic approximation (6.33).

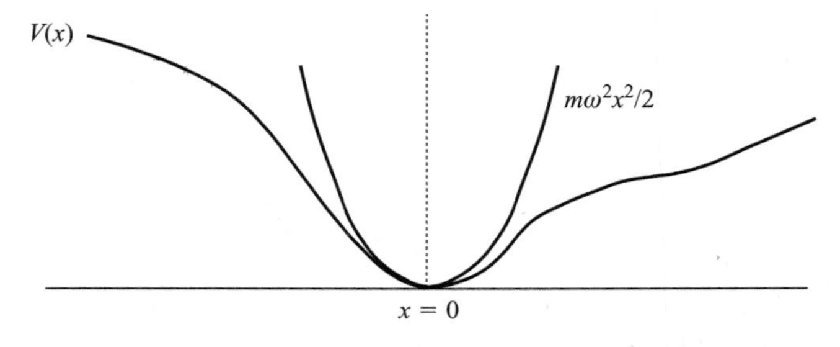

Fig. 6.4 A schematic representation of the harmonic-oscillator approximation $m\omega^2x^2/2$ to an arbitrary potential having a minimum equal to zero at $x = 0$.

The transition from the classical to the quantal oscillator problem involves the potential-energy operator

$$\hat{V}(\hat{Q}) = \tfrac{1}{2}m\omega_0^2\hat{Q}^2, \tag{6.34}$$

whose coordinate-space representation is Eq. (6.33). Then, from Eqs. (6.1) and (6.2), the operator and coordinate-space forms of the time-independent Schrödinger equation are

$$\hat{H}|\Gamma\rangle = \left(\frac{\hat{P}_x^2}{2m} + \frac{1}{2}m\omega_0^2\hat{Q}_x^2\right)|\Gamma\rangle = E_\Gamma|\Gamma\rangle \tag{6.35}$$

and

$$\hat{H}(x)\psi_\Gamma(x) = \left(-\frac{\hbar^2}{2m}\frac{d^2}{dx^2} + \frac{1}{2}m\omega_0^2x^2\right)\psi_\Gamma(x) = E_\Gamma\psi_\Gamma(x). \tag{6.36}$$

The 1-D linear harmonic oscillator is possibly the most widely used solvable system in quantum mechanics, and we devote almost half of this chapter to its treatment. It is a system well worth understanding in detail. The oscillator is typical of many physical systems in that it can be solved in more than one way, and we shall use both (6.35) and (6.36) to analyze the 1-D quantum-oscillator eigenvalue problem. As a preliminary, we introduce the "natural" energy and length units for the quantal linear harmonic oscillator and then show that there are no eigenstates that correspond to asymptotically free particles: only bound states occur.

There are only three fundamental constants occurring in the oscillator Hamiltonian from which constants having the dimensions energy and length can be constructed. These are $\hbar$, ω_0, and m. The product of the first two, $\hbar\omega_0$, provides the natural unit of energy. A natural length unit is less obvious, but, if we recall that the dimension of $\hbar$ is energy-time, i.e., $m\ell^2/t$, then $\hbar$ can be converted into a quantity with the dimension of length by dividing it by a mass, multiplying by a time, and then taking the square root of the result. The oscillator provides both a mass – m – and a time – ω_0^{-1}. Hence, the natural unit of length is $[\hbar/(m\omega_0)]^{1/2}$, which we denote by the symbol λ_0:

$$\lambda_0 = \sqrt{\frac{\hbar}{m\omega_0}}. \tag{6.37}$$

Because these quantities occur naturally in the system, it is reasonable to expect that E and x can be expressed in units of $\hbar\omega_0$ and λ_0, respectively. We therefore write $E_\Gamma = \epsilon_\Gamma\hbar\omega_0$, with ϵ_Γ dimensionless, and then extract the factor $\hbar\omega_0$ from the Hamiltonians in (6.35) and (6.36). These equations then become

$$\frac{1}{\hbar\omega_0}\hat{H}|\Gamma\rangle = \left(\frac{\hat{P}_x^2}{2\hbar^2/\lambda_0^2} + \frac{\hat{Q}_x^2}{2\lambda_0^2}\right)|\Gamma\rangle = \epsilon_\Gamma|\Gamma\rangle \tag{6.38}$$

and

$$\frac{1}{2}\left[-\frac{d^2}{d(x/\lambda_0)^2} + \left(\frac{x}{\lambda_0}\right)^2\right]\psi_\Gamma(x) = \epsilon_\Gamma\psi_\Gamma(x). \tag{6.39}$$

Note that, on expressing E in natural units, λ_0 automatically becomes the unit for x.

We turn next to the nature of the eigensolutions for the 1-D linear harmonic oscillator.

From the analysis in Section 5.4 on the eigenvalue problem for linear momentum, one may speculate that unbound (continuum or non-normalizable) states will behave, at least asymptotically, like exp(ikx). This speculation will be verified in Chapter 7. In view of this later verification we now show that (6.39) has no continuum solutions and that normalizability requires that all solutions behave asymptotically like $\exp(-x^2/\lambda_0^2)$.

The latter statement is easily demonstrated by performing an asymptotic analysis on (6.39). That is, we examine (6.39) in the limit x (or $x/\lambda_0) \to \infty$. In this limit $(x/\lambda_0)^2 \gg \epsilon_\Gamma$, so that (6.39) is, to an excellent approximation, just

$$\frac{d^2\psi_\Gamma(\eta)}{d\eta^2} = \eta^2\psi_\Gamma(\eta), \qquad \eta \gg 1, \tag{6.40}$$

where

$$\eta \equiv x/\lambda_0.$$

Because $\eta \gg 1$, solutions to (6.40) that hold asymptotically are

$$\psi_\Gamma(\eta) \underset{\eta \to \infty}{\sim} c_\pm e^{\pm\eta^2/2} \underset{x\to\infty}{=} c_\pm e^{\pm x^2/(2\lambda_0^2)}, \tag{6.41}$$

as is readily verified. The term $\eta^2\psi_\Gamma(\eta)$ prevents a factor like $\exp(ikx)$ from appearing in $\psi_\Gamma(x)$.

Although both signs in the exponentials of (6.41) are allowed mathematically, the probalistic interpretation of $|\psi_\Gamma(x)|^2$ requires that $c_+ = 0$, so that the asymptotic behavior of all $\psi_\Gamma(x)$ is uniquely

$$\psi_\Gamma(x) \underset{x\to\infty}{\to} c_- e^{-x^2/(2\lambda_0^2)}. \tag{6.42}$$

As a final step in this preliminary discussion, we prove that the asymptotic form (6.42) is actually one of the eigenfunctions of (6.39): the corresponding eigenvalue is $\epsilon_\Gamma = \frac{1}{2}$. Differentiation of $\exp[-x^2/(2\lambda_0^2)]$ yields the proof, since (6.39) becomes

$$\tfrac{1}{2}c_- e^{-x^2/(2\lambda_0^2)} = \epsilon_\Gamma c_- e^{-x^2/(2\lambda_0^2)}, \tag{6.43}$$

which establishes that $\epsilon_\Gamma = \frac{1}{2}$. This result shows that $(\lambda_0\sqrt{\pi})^{-1/2}\exp[-x^2/(2\lambda_0^2)]$ is a normalized eigenfunction of the quantal oscillator Hamiltonian corresponding to the energy eigenvalue $\hbar\omega_0/2$. In anticipation of future results, we denote this eigenfunction $\psi_0(x)$ and the energy eigenvalue E_0:

$$\psi_0(x) = \frac{1}{\sqrt{\lambda_0\sqrt{\pi}}} e^{-x^2/(2\lambda_0^2)} \tag{6.44}$$

and

$$E_0 = \tfrac{1}{2}\hbar\omega_0. \tag{6.45}$$

Energies and states of the oscillator will eventually be labeled by the single integer n, whose minimum value is 0; hence E_0 and ψ_0 are the lowest-lying energy and its corresponding wave function, although, since this claim has yet to be proved, the index 0 on E_0 and ψ_0 is at this point only an arbitrary choice. Although one needs to prove that ψ_0 is the ground-state wave function, comparison with the ground state $\psi_1(x) = \sqrt{2/L}\sin(\pi x/L)$ of the quantum box suggests it: both functions have the fewest possible nodes that are allowed, viz., the ones that occur at the extreme values of their arguments.

Since $\langle \hat{K} \rangle$ will increase with an increase in the number of nodes in the wave function, states with the minimum value of $\langle \hat{K} \rangle$ should lie lowest; ψ_0 of (6.44) is one such.

Although asymptotic analysis has yielded one solution, which is conjectured to be the ground state, a full-scale investigation of the eigenvalue problem is required in order to go further. We shall do this employing first Eq. (6.38) and then Eq. (6.39). The procedures in the two approaches differ; each involves powerful methods that are applicable to other systems.

Solution via creation and annihilation operators

The fact that the reduced Hamiltonian in (6.38) is a sum of the squares of $\hat{P}_x$ and $\hat{Q}_x$ leads to an elegant solution of the eigenvalue problem. The crucial step is the re-writing of the reduced Hamiltonian in terms of a product of operators. If $[\hat{P}_x, \hat{Q}_x]$ were zero, then $\hat{H}/(\hbar\omega_0)$ could be written as

$$\frac{1}{\hbar\omega_0}\hat{H} \stackrel{?}{=} \left(\frac{\hat{Q}_x}{\lambda_0\sqrt{2}} + i\frac{\lambda_0}{\hbar\sqrt{2}}\hat{P}_x\right)\left(\frac{\hat{Q}_x}{\lambda_0\sqrt{2}} - i\frac{\lambda_0}{\hbar\sqrt{2}}\hat{P}_x\right), \tag{6.46}$$

a result that would be an equality were $\hat{P}_x$ and $\hat{Q}_x$ numbers rather than operators. However, $[\hat{P}_x, \hat{Q}_x] \neq 0$, and "$\stackrel{?}{=}$" rather than "$=$" has been used in (6.46).

Before expanding the product in (6.46), we introduce some useful notation. Following standard nomenclature, we define the operator $\hat{a}$:

$$\hat{a} \equiv \frac{1}{\lambda_0\sqrt{2}}\hat{Q}_x + i\frac{\lambda_0}{\hbar\sqrt{2}}\hat{P}_x; \tag{6.47}$$

$\hat{a}$ is the first factor on the RHS of (6.46). Because $\hat{P}_x$ and $\hat{Q}_x$ are Hermitian, it follows that the second factor on the RHS of (6.46) is $\hat{a}^\dagger$:

$$\hat{a}^\dagger = \frac{1}{\lambda_0\sqrt{2}}\hat{Q}_x - i\frac{\lambda_0}{\hbar\sqrt{2}}\hat{P}_x. \tag{6.48}$$

The RHS of (6.46) is the product $\hat{a}\hat{a}^\dagger$, which equals $\hat{H}/(\hbar\omega_0)$ plus a commutator:

$$\hat{a}\hat{a}^\dagger = \frac{1}{\hbar\omega_0}\hat{H} + \frac{i}{2\hbar}(\hat{P}_x\hat{Q}_x - \hat{Q}_x\hat{P}_x), \tag{6.49}$$

where the parenthesis in (6.49) is the commutator $[\hat{P}_x, \hat{Q}_x]$ written out in full.

To evaluate the commutator, we let it act on an arbitrary Hilbert-space vector $|\alpha\rangle$, setting the result equal to $|\beta\rangle$:

$$|\beta\rangle = [\hat{P}_x, \hat{Q}_x]|\alpha\rangle. \tag{6.50}$$

Choosing the momentum representation for the evaluation, (6.50) becomes

$$\begin{aligned} \langle k|\beta\rangle \equiv \varphi_\beta(k) &= \int dk' \langle k|\hat{P}_x\hat{Q}_x - \hat{Q}_x\hat{P}_x|k'\rangle\varphi_\alpha(k') \\ &= [\hat{P}_x(k)\hat{Q}_x(k) - \hat{Q}_x(k)\hat{P}_x(k)]\varphi_\alpha(k), \end{aligned} \tag{6.51}$$

where the second line is a consequence of the locality of $\hat{P}_x$ and $\hat{Q}_x$ in the momentum representation. Since $\hat{P}_x(k) = \hbar k$ and $\hat{Q}_x(k) = i\, d/dk$ (from Eq. (5.76)), use of the latter two relations in (6.51) gives

$$\begin{aligned} \varphi_\beta(k) &= i\hbar\left(\frac{d}{dk}\,k - k\,\frac{d}{dk}\right)\varphi_\alpha(k) \\ &= -i\hbar\varphi_\alpha(k), \end{aligned} \tag{6.52}$$

where the equality $[(d/dk)k\varphi_\alpha(k)] = \varphi_\alpha(k) + k\, d\varphi_\alpha(k)/dk$ has been employed. It is readily shown that the identical result obtains in the coordinate representation, viz.,

$$\langle x|[\hat{P}_x, \hat{Q}_x]|\alpha\rangle = -i\hbar\psi_\alpha(x).$$

In each representation, the commutator is equal to $-i\hbar$, from which we conclude that

$$[\hat{P}_x, \hat{Q}_x] = -i\hbar\,\hat{I}. \tag{6.53}$$

We note in passing that, if m and n refer to rectangular components, then (6.53) is easily generalized to

$$[\hat{P}_m, \hat{Q}_n] = -i\hbar\hat{I}\delta_{mn}\,, \tag{6.54}$$

a result that will be employed numerous times.

Substituting (6.53) into Eq. (6.49) gives

$$aa^\dagger = \frac{1}{\hbar\omega_0}\,\hat{H} + \frac{1}{2}\hat{I}, \tag{6.55}$$

clearly demonstrating that (6.46) is not an equality: the $\stackrel{?}{=}$ is indeed justified.

Suppose that the order of factors were reversed, i.e., let us assume that $\hat{a}^\dagger\hat{a}$ is written in place of $\hat{a}\hat{a}^\dagger$. Then, steps analogous to the preceding ones yield

$$\hat{a}^\dagger\hat{a} = \frac{1}{\hbar\omega_0}\,\hat{H} - \frac{1}{2}\hat{I}; \tag{6.56}$$

subtraction of (6.56) from (6.55) leads to a new commutation relation, viz.,

$$[\hat{a}, \hat{a}^\dagger] = \hat{I}. \tag{6.57}$$

We shall soon make use of (6.57).

Either (6.55) or (6.56) could be used to express $\hat{H}$, but, since it is advantageous to work with a sum rather than a difference of operators, we choose (6.56). Defining the Hermitian operator $\hat{N}$ by

$$\hat{N} \equiv \hat{a}^\dagger\hat{a}, \tag{6.58}$$

the Hamiltonian takes the form

$$\hat{H} = \left(\hat{N} + \frac{\hat{I}}{2}\right)\hbar\omega_0. \tag{6.59}$$

The solution of the quantal oscillator problem has been transformed into finding the eigenvalues and eigenvectors of $\hat{N} = \hat{a}^\dagger\hat{a}$. This is a relatively straightforward task, as we now demonstrate.

The eigenvalue problem for $\hat{N}$ is

$$\hat{N}|\gamma\rangle = \hat{a}^\dagger\hat{a}|\gamma\rangle = \gamma|\gamma\rangle, \tag{6.60}$$

where the restriction $\gamma = n =$ integer or zero will soon be demonstrated. We infer the existence of at least one non-trivial $|\gamma\rangle$, since we already know that a non-trivial eigenstate exists: its coordinate representation is given by (6.44).

Acting with (6.59) on $|\gamma\rangle$ gives

$$\hat{H}|\gamma\rangle = (\gamma + \tfrac{1}{2})\hbar\omega_0|\gamma\rangle, \tag{6.61}$$

from which follow the identifications

$$E_\Gamma \equiv E_\gamma = (\gamma + \tfrac{1}{2})\hbar\omega_0 \tag{6.62}$$

and

$$|\Gamma\rangle = |\gamma\rangle. \tag{6.63}$$

From (6.62), the spectrum of $\hat{H}$ is determined once the allowed values of γ are obtained. To accomplish this, we start by assuming that $\langle\gamma|\gamma\rangle = 1$, and then project both sides of Eq. (6.60) onto $\langle\gamma|$:

$$\begin{aligned}\langle\gamma|\hat{a}^\dagger\hat{a}|\gamma\rangle &= \langle\hat{a}\gamma|\hat{a}\gamma\rangle \\ &= \gamma\langle\gamma|\gamma\rangle = \gamma.\end{aligned} \tag{6.64}$$

The scalar product $\langle\hat{a}\gamma|\hat{a}\gamma\rangle$ is semi-positive-definite,

$$\langle\hat{a}\gamma|\hat{a}\gamma\rangle \geqslant 0, \tag{6.65}$$

so that (6.64) plus (6.65) yield the important result that

$$\gamma \geqslant 0. \tag{6.66}$$

Because $\langle\hat{a}\gamma|\hat{a}\gamma\rangle = \gamma$, it follows that $\hat{a}|0\rangle = 0$ only if $\gamma = 0$, while $\hat{a}|\gamma\rangle > 0$, $\gamma \neq 0$. However, $\gamma = 0$ must occur, since the relation

$$E_\gamma = (\gamma + \tfrac{1}{2})\hbar\omega_0 \geqslant \tfrac{1}{2}\hbar\omega_0 \tag{6.67}$$

is compatible with the result (6.45) only if $\gamma = 0$. Furthermore, the possibility that $|0\rangle$ is the null vector, a possibility consistent with $\hat{a}\,|\gamma\rangle = 0$, is ruled out because $|0\rangle$, the eigenstate corresponding to the energy $\hbar\omega_0/2$, is known to have the non-zero, coordinate-space representative (6.44).[2]

To obtain the non-zero values of γ, we examine $\hat{a}|\gamma\rangle$ and $\hat{a}^\dagger|\gamma\rangle$, each of which is, in fact, an eigenvector of $\hat{N}$. Verification of this claim will make use of

$$[\hat{N}, \hat{a}] = -\hat{a} \tag{6.68}$$

and

$$[\hat{N}, \hat{a}^\dagger] = \hat{a}^\dagger, \tag{6.69}$$

results that follow from the commutation relation (6.57).

Let us consider $\hat{N}[\hat{a}|\gamma\rangle]$ first:

$$\begin{aligned}\hat{N}\hat{a}|\gamma\rangle &= (\hat{a}\hat{N} - \hat{a})|\gamma\rangle \\ &= \hat{a}(\hat{N} - 1)|\gamma\rangle \\ &= (\gamma - 1)\hat{a}|\gamma\rangle,\end{aligned} \tag{6.70}$$

[2] These remarks provide an *a posteriori* justification for using "0" as the subscript to label $\psi_0(x)$ and E_0 of Eqs. (6.44) and (6.45).

where (6.68) has been used so as to allow $\hat{N}$ to act directly on $|\gamma\rangle$. The structure of (6.70) shows that, as stated, $\hat{a}|\gamma\rangle$ is an eigenvector of $\hat{N}$; its eigenvalue is $\gamma - 1$. We thus conclude that

$$\hat{a}|\gamma\rangle = |\gamma - 1\rangle, \tag{6.71}$$

where a constant of proportionality has been suppressed. It will be restored later.

Exactly the same analysis leads to $\hat{a}|\gamma - 1\rangle \equiv \hat{a}^2|\gamma\rangle$ being an eigenstate of $\hat{N}$ whose eigenvalue is $\gamma - 2$: $\hat{N}\hat{a}^2|\gamma\rangle = (\gamma - 2)\hat{a}^2|\gamma\rangle$, which means that $\hat{a}^2|\gamma\rangle = |\gamma - 2\rangle$. More generally, it follows that

$$\hat{a}^n|\gamma\rangle = |\gamma - n\rangle, \tag{6.72}$$

with

$$\hat{N}\hat{a}^n|\gamma\rangle = (\gamma - n)\hat{a}^n|\gamma\rangle. \tag{6.73}$$

In analogy to $\hat{a}|\gamma\rangle = |\gamma - 1\rangle$, one can show that

$$\hat{a}^\dagger|\gamma\rangle = |\gamma + 1\rangle, \tag{6.74}$$

since

$$\hat{N}[\hat{a}^\dagger|\gamma\rangle] = (\gamma + 1)[\hat{a}|\gamma\rangle]. \tag{6.75}$$

The analogs of (6.72) and (6.73) are then readily shown to be

$$(a^\dagger)^n|\gamma\rangle = |\gamma + n\rangle, \tag{6.76}$$

with

$$\hat{N}(\hat{a}^\dagger)^n|\gamma\rangle = (\gamma + n)(\hat{a}^\dagger)^n|\gamma\rangle. \tag{6.77}$$

In Eqs. (6.72) and (6.76) constants of proportionality have again been suppressed.

The state with the smallest eigenvalue is $|0\rangle$, corresponding to $\gamma = 0$. Note that, if $\hat{a}|0\rangle$ were not zero, then, according to (6.70), the state $\hat{a}|0\rangle$ would be an eigenstate of $\hat{N}$ with eigenvalue $\gamma = -1$. Since the latter result would violate $\gamma = 0$ being the smallest value, it follows that $\hat{a}|0\rangle$ must be zero. Equation (6.76), on the other hand, states that

$$(\hat{a}^\dagger)^n|0\rangle = |n\rangle, \tag{6.78}$$

which, by virtue of the original definition (6.60), is an eigenstate of $\hat{N}$ corresponding to eigenvalue n:

$$\hat{N}|n\rangle = n|n\rangle, \qquad n = 0, 1, 2, \ldots. \tag{6.79}$$

Repeated application of $\hat{a}^\dagger$ to $|0\rangle$ generates an infinite sequence of eigenstates $\{|n\rangle\}_0^\infty$, each obeying $\hat{N}|n\rangle = n|n\rangle$. Since the set is infinite, one may (and should!) ask whether it includes *all* the eigenstates of $\hat{N}$, i.e., (by Postulate VI) is it complete? The answer is YES, a statement we prove by showing that states $|\gamma\rangle$, with $\gamma \neq$ integer, do not occur. Thus, let $\gamma = m + s$, $m =$ integer, $0 \leqslant s < 1$. Then

$$|s\rangle = \hat{a}^m|m + s\rangle$$

obeys

$$\hat{N}|s\rangle = s|s\rangle$$

such that (as in (6.64))

$$s = \frac{\langle \hat{a}s|\hat{a}s\rangle}{\langle s|s\rangle}.$$

However, $\hat{a}|s\rangle$ must be zero, for if it were not, then $\hat{a}|s\rangle$ would be an eigenstate of $\hat{N}$ with eigenvalue $s - 1 < 0$. Negative eigenvalues are not allowed (by (6.66)), and, to avoid them, we must have $\hat{a}|s\rangle = 0$, from which it follows that

$$s = 0.$$

Hence, only integer γ occur: $\gamma = n$, $n = 0, 1, 2, \ldots$; and therefore $\{|n\rangle\}_0^\infty$ contains all the eigenstates of $\hat{N}$.

The operator $\hat{N} = \hat{a}^\dagger \hat{a}$ is denoted the *number operator*, nomenclature suggested by Eq. (6.79). Since $\hat{a}^\dagger|n\rangle$ produces a state with eigenvalue $n + 1$, and $\hat{a}|n\rangle$ yields a state with eigenvalue $n - 1$, the latter two operators are also referred to by graphical names: $\hat{a}^\dagger$ is denoted a *creation operator* and $\hat{a}$ an *annihilation operator.*

We next determine the suppressed constants of proportionality between $\hat{a}^\dagger|n\rangle$ and $|n + 1\rangle$ and between $\hat{a}|n\rangle$ and $|n - 1\rangle$. This is most easily accomplished by specifically assuming that $\langle 0|0\rangle = 1 = \langle n|n\rangle$ and then evaluating the normalization constant C_n occurring in

$$(\hat{a}^\dagger)^n|0\rangle = C_n|n\rangle, \tag{6.80}$$

with C_0 chosen to be unity: $C_0 = 1$. From $\langle n|C_n^* C_n|n\rangle$ and (6.80) the equalities

$$\begin{aligned} |C_n|^2 &= \langle 0|\hat{a}^n(\hat{a}^\dagger)^n|0\rangle \\ &= \{\langle 0|\hat{a}^{n-1}\}(\hat{a}\hat{a}^\dagger)\{(\hat{a}^\dagger)^{n-1}|0\rangle\} \\ &= |C_{n-1}|^2\langle n-1|\hat{a}\hat{a}^\dagger|n-1\rangle \\ &= |C_{n-1}|^2\langle n-1|\hat{N} + \hat{I}|(n-1)\rangle \\ &= n|C_{n-1}|^2 \end{aligned} \tag{6.81}$$

are deduced with the aid of (6.79) and the commutation relation (6.57).

An immediate consequence of (6.81) is that $|C_{n-1}|^2 = (n-1)|C_{n-2}|^2$, etc.; hence, one ends up with

$$|C_n|^2 = n!. \tag{6.82}$$

We choose the phase factor so that C_n is real and positive, viz.,

$$C_n = \sqrt{n!}, \tag{6.83}$$

consistent with $C_0 = 1$. The normalized state $|n\rangle$ is thus related to $|0\rangle$ by

$$|n\rangle = \frac{1}{(n!)^{1/2}}(\hat{a}^\dagger)^n|0\rangle. \tag{6.84}$$

Application of the preceding analysis to $\hat{a}^\dagger|n\rangle$ produces the result

$$\hat{a}^\dagger|n\rangle = \sqrt{n+1}|n+1\rangle. \tag{6.85}$$

Finally, by acting on (6.85) with $\hat{a}$ and employing (6.57) and (6.79), we deduce another significant relation,

$$\hat{a}|n\rangle = \sqrt{n}|n-1\rangle. \tag{6.86}$$

With the preceding sets of results in hand, we have the capability of determining not only the spectrum of the oscillator Hamiltonian, but also the matrix elements of various operators, e.g., $\hat{P}_x$, $\hat{P}_x^2$, $\hat{Q}_x$, etc. We begin with the quantal eigenvalue problem. The $\{|n\rangle\}_0^\infty$ are the eigenstates of $\hat{H}$, each $|n\rangle$ corresponding to the eigenvalue ($\hat{H} = (\hat{N} + \hat{1}/2)\hbar\omega_0$)

$$E_n = (n + \tfrac{1}{2})\hbar\omega_0, \qquad n = 0, 1, 2, \ldots. \tag{6.87}$$

As with the quantum box, the lowest oscillator state has the *non-zero* energy $E_0 = \hbar\omega_0/2$. Since $|1\rangle$, $|2\rangle$, ... are obtained from $|0\rangle$ via application of $\hat{a}^\dagger$, $(\hat{a}^\dagger)^2$, ..., then, starting with the first excited state $|1\rangle$, we may say that each state created by the action of $\hat{a}^\dagger$ has one more quantum of excitation energy $\hbar\omega_0$ than the previous one. For example, state $|n\rangle$ has n quanta (with energy $n\hbar\omega_0$) more than the ground state and one quantum of energy $\hbar\omega_0$ more than $|n-1\rangle$. In other words, $\hat{a}^\dagger$ *creates* a quantum of energy; correspondingly, $\hat{a}$ *annihilates* it.

The spacing between adjacent 1-D harmonic-oscillator energy levels is uniform, equal to $\hbar\omega_0$. Shown in Fig. 6.5 is a Grotrian diagram representing a portion of the infinite set of 1-D oscillator levels. Apart from its being a completely solvable quantum system and thus important in illuminating quantal concepts, the oscillator also serves as a model approximately describing some of the states of a few real systems, e.g., molecules and nuclei in certain portions of their spectra. To use the oscillator as a model in such cases, the minimal requirement is the existence of a portion of the spectrum in which the levels are equally spaced. Concurrent with this must be a physical picture in which the system behaves – at least in part – like an oscillator. Examples include "vibrations" of a diatomic molecule along the axis joining the two nuclei, and "harmonic oscillations" of the surface of a nucleus about its equilibrium shape.

Displayed in Fig. 6.6 are two sets of spectra in Grotrian-diagram form, one from the diatomic molecule HCl and the other from the nucleus ^{76}Se. Each can be reasonably well

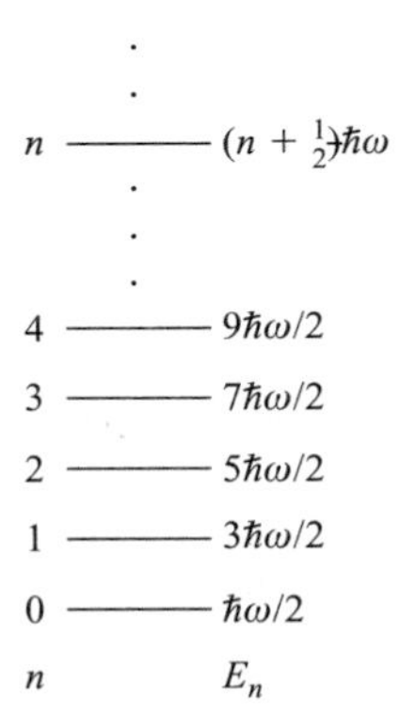

Fig. 6.5 The Grotrian diagram for the 1-D harmonic oscillator.

(a) Oscillator Approximations for Diatomics

$\overline{\nu} = \nu/c$, $h\nu = hc\overline{\nu}$, $hc = 12400$ eV Å $= 1.24$ eV $\times 10^{-4}$ cm

HCl	$\hat{\nu}_{obs}$ (cm^{-1})	$\tilde{\nu}_{calc}$ (cm^{-1})
0→1	2885.9	2885.9
0→2	5668.0	5771.8
0→3	8347.0	8657.7
0→4	10923.1	11543.6
0→5	13396.5	14429.5

HCl	$h\nu_{obs}$ (eV)	$h\nu_{calc}$ (eV)	$\Delta E/E_{obs}$
0→1	0.358	0.358	0
0→2	0.703	0.715	0.017
0→3	1.035	1.073	0.037
0→4	1.354	1.431	0.057
0→5	1.661	1.789	0.077

Fig. 6.6 (a) From Barrow (1962), p. 45. Note that an anharmonic oscillator, with an x^3 term and thus one additional parameter, yields an almost perfect fit.

approximated, for some range of states, as an oscillator. Since these are multi-particle systems, the oscillator-model description eventually breaks down as other dynamical effects become important, thus altering the behavior away from the simplistic model. Note, in particular, that the second "state" of ^{76}Se is actually a triplet of states, the separations of which are small relative to the oscillator spacing, $\hbar\omega_0 \simeq 0.56$ MeV; already the simple oscillator model is breaking down. This is clear evidence that a more sophisticated model (or models) is (are) needed; such exist.[3]

In addition to serving as a model for portions of the spectra for some molecules and nuclei, the oscillator plays a fundamental role in our understanding of the physics both of electromagnetic and of certain condensed-matter phenomena. In each case the states (of the electromagnetic field and of crystal lattices) are those of linear harmonic oscillators. For the electromagnetic field, the quanta of excitation are photons, almost the same as enter the early Planck and Einstein descriptions of blackbody radiation and the photoelectric effect, the only difference being the energy of the photons, which is given by $(n+\frac{1}{2})h\nu$ (or $(n+\frac{1}{2})\hbar\omega$) rather than by $nh\nu$. In the case of lattice vibrations, the low-lying, linear-harmonic quanta of excitation are denoted phonons; they play an important role in behavior ranging from conduction of heat to superconductivity. The oscillator is far more than just a soluble quantum-mechanics problem.

[3] The literature on nuclear models is vast. The models employ quantal procedures this book will not delve into. A few references that readers of this book may find adequate to their background are Chapter 5 of Elton (1966), Chapters 16–18 of Frauenfelder and Henley (1991), and Chapters 5–7 of Cottingham and Greenwood (1986).

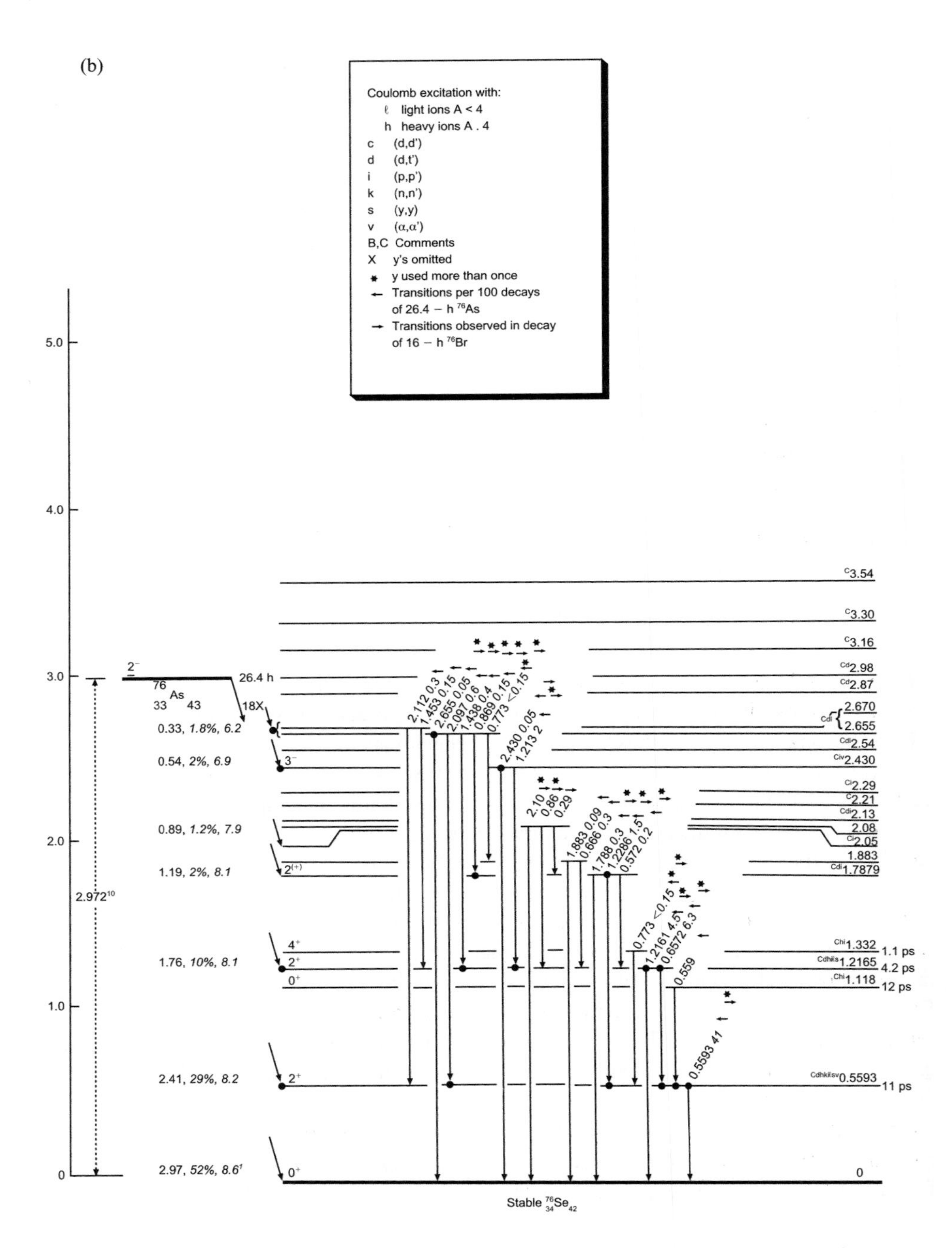

Fig. 6.6 (b) The oscillator-like energy-level structure for the five lowest states in $^{76}_{34}$Se, from Ikegami, Ewbank, and Way (1966).

The eigenstate representation

Equation (6.36) defines the coordinate representation $\hat{H}(x)$ of the oscillator Hamiltonian. $\hat{H}(x)$ can be thought of as the diagonal element, at position x, of an infinite matrix, indexed by two continuous numbers, say x and x'. Were a transformation to the momentum representation to be made, then the Hamiltonian would again be represented as an infinite matrix, also continuously indexed. The solution of the oscillator eigenvalue problem in terms of the eigenstates $|n\rangle$ of the number operator $\hat{N}$ provides a third representation for the oscillator Hamiltonian, one referred to as the eigenstate representation. We examine this representation before going on to the coordinate-space form of the solutions.

In the eigenstate representation, $\hat{H}$ is again an infinite diagonal matrix, indexed not continuously but by integers. Correspondingly, $\hat{P}_x$ and $\hat{Q}_x$ are also represented by denumerably infinite matrices, although these matrices are not diagonal. Let us see how this comes about.

The oscillator eigenstate representation for an operator $\hat{A}$ is obtained by forming the matrix element $\langle m|\hat{A}|n\rangle$ with $|m\rangle$ and $|n\rangle$ being oscillator eigenstates. If $\hat{A} \to \hat{H}$, the oscillator Hamiltonian, then the time-independent Schrödinger equation immediately leads to

$$\langle m|\hat{H}|n\rangle = \delta_{mn}(n + \tfrac{1}{2})\hbar\omega_0. \tag{6.88}$$

Equation (6.88) defines a diagonal matrix H whose structure is as follows:

$$H = \begin{pmatrix} \hbar\omega_0/2 & & & & \\ & 3\hbar\omega_0/2 & & 0 & \\ & & 5\hbar\omega_0/2 & & \\ & 0 & & 7\hbar\omega_0/2 & \\ & & & & \ddots \end{pmatrix}, \tag{6.89}$$

where the rows and columns are labeled by m and n, respectively, with each of m and $n = 0, 1, 2, \ldots$. In this representation, the state $|n\rangle$ becomes an infinite-rowed column vector, all of whose elements are zero except for the nth-row element, which is unity; this is the infinite analog of the finite vectors discussed in Section 3.2:

$$|n\rangle \to \begin{pmatrix} 0 \\ 0 \\ 0 \\ \vdots \\ 0 \\ 1 \\ 0 \\ 0 \\ 0 \\ \vdots \\ \vdots \end{pmatrix} \leftarrow n\text{th row}. \tag{6.90}$$

Note that the most general form of the nth-row element is e_n, where $|e_n|^2 = 1$, as in Eq. (4.33). We have elected to work with real eigenvectors, and have chosen $e_n = 1$.

With $\hat{H}$ and $|n\rangle$ represented by infinite matrices with elements indexed by the integers (and zero), one may expect similar results for $\hat{P}_x$ and $\hat{Q}_x$. Evaluation of $\langle m|\hat{P}_x|n\rangle$ and $\langle m|\hat{Q}_x|n\rangle$ is most simply carried out by expressing these operators in terms of $\hat{a}$ and $\hat{a}^\dagger$. From (6.47) and (6.48) one finds

$$\hat{Q}_x = \frac{\lambda_0}{\sqrt{2}}(\hat{a}^\dagger + \hat{a}) \tag{6.91}$$

and

$$\hat{P}_x = \frac{i\hbar}{\lambda_0\sqrt{2}}(\hat{a}^\dagger - \hat{a}). \tag{6.92}$$

Matrix elements of $\hat{a}^\dagger$ and $\hat{a}$ are easily obtained using Eqs. (6.85) and (6.86):

$$\langle m|\hat{a}^\dagger|n\rangle = \sqrt{n+1}\langle m|n+1\rangle = \sqrt{n+1}\delta_{m,n+1} \tag{6.93}$$

and

$$\langle m|\hat{a}|n\rangle = \sqrt{n}\langle m|n-1\rangle = \sqrt{n}\delta_{m,n-1}. \tag{6.94}$$

These results, when they are employed in the expressions for $\hat{Q}_x$ and $\hat{P}_x$, lead to

$$\langle m|\hat{Q}_x|n\rangle = \frac{\lambda_0}{\sqrt{2}}(\sqrt{n+1}\delta_{m,n+1} + \sqrt{n}\delta_{m,n-1}) \tag{6.95}$$

and

$$\langle m|\hat{P}_x|n\rangle = \frac{i\hbar}{\lambda_0\sqrt{2}}(\sqrt{n+1}\delta_{m,n+1} - \sqrt{n}\delta_{m,n-1}). \tag{6.96}$$

Each of the preceding four matrices has only off-diagonal elements, the elements in each case lying just above and/or just below the main diagonal, as assigned by the Kronecker deltas $\delta_{m,n\pm1}$. They are readily shown to have the following structures:

$$a^\dagger = \begin{pmatrix} 0 & 0 & 0 & 0 & 0 & & \\ \sqrt{1} & 0 & 0 & 0 & 0 & & \\ 0 & \sqrt{2} & 0 & 0 & 0 & & \\ 0 & 0 & \sqrt{3} & 0 & 0 & & \\ 0 & 0 & 0 & \sqrt{4} & 0 & & \\ 0 & 0 & 0 & 0 & \sqrt{5} & 0 & \\ & & & & & \ddots & \\ & & & & & & \ddots \end{pmatrix}, \tag{6.97}$$

$$a = \begin{pmatrix} 0 & \sqrt{1} & 0 & 0 & 0 & 0 & & \\ 0 & 0 & \sqrt{2} & 0 & 0 & 0 & & \\ 0 & 0 & 0 & \sqrt{3} & 0 & 0 & & \\ 0 & 0 & 0 & 0 & \sqrt{4} & 0 & & \\ 0 & 0 & 0 & 0 & 0 & \sqrt{5} & 0 & \\ & & & & & & \ddots & \\ & & & & & & & \ddots \end{pmatrix}, \tag{6.98}$$

$$Q_x = \frac{\lambda_0}{\sqrt{2}} \begin{pmatrix} 0 & \sqrt{1} & 0 & 0 & 0 & 0 & & \\ \sqrt{1} & 0 & \sqrt{2} & 0 & 0 & 0 & & \\ 0 & \sqrt{2} & 0 & \sqrt{3} & 0 & 0 & & \\ 0 & 0 & \sqrt{3} & 0 & \sqrt{4} & 0 & & \\ 0 & 0 & 0 & \sqrt{4} & 0 & \sqrt{5} & & \\ 0 & 0 & 0 & 0 & \sqrt{5} & 0 & & \\ & & & & & & \ddots & \\ & & & & & & & \ddots \end{pmatrix}, \tag{6.99}$$

and

$$P_x = \frac{i\hbar}{\lambda_0\sqrt{2}} \begin{pmatrix} 0 & -\sqrt{1} & 0 & 0 & 0 & \\ \sqrt{1} & 0 & -\sqrt{2} & 0 & 0 & \\ 0 & \sqrt{2} & 0 & -\sqrt{3} & 0 & \\ 0 & 0 & \sqrt{3} & 0 & -\sqrt{4} & \\ 0 & 0 & 0 & \sqrt{4} & 0 & \\ & & & & & \ddots \end{pmatrix}. \tag{6.100}$$

Matrix multiplications of Q_x and P_x and of P_x and Q_x using Eqs. (6.99) and (6.100) each yield a diagonal matrix. Their difference is an infinite-matrix form of the commutator $[\hat{Q}_x, \hat{P}_x]$:

$$Q_x P_x - P_x Q_x = i\hbar \begin{pmatrix} 1 & & & & & \\ & 1 & & & & \\ & & 1 & & & \\ & & & 1 & & \\ & & & & 1 & \\ & & & & & \ddots \end{pmatrix}. \tag{6.101}$$

Matrix representations such as those of Eqs. (6.89), (6.99), and (6.100) were the foundation of the "matrix mechanics" approach to quantum theory introduced by Heisenberg (1925) and elaborated by him, Born, and Jordan (Born *et al.* (1925)). That the matrix-mechanics and the Schrödinger-equation approaches are each aspects of the same, abstract dynamical system was demonstrated by Schrödinger (1928) and Eckart (1926). The preceding analysis indicates this for the case of the 1-D harmonic oscillator, and also shows that no one of the three representations – coordinate, momentum, or eigenstate – is more "fundamental" than the other two. Each is a particular realization of an abstract, Hilbert-space operator, whose utility will depend upon the application at hand. For example, the eigenstate representation is crucial in dealing with the angular momentum $\hat{\mathbf{J}}$, a generalization of the orbital angular-momentum operator $\hat{\mathbf{L}} = \hat{\mathbf{Q}} \times \hat{\mathbf{P}}$. What the eigenstate representation does not provide, however, is the probability that the particle in the oscillator potential will be found at a particular position x. For this, the coordinate representation is needed, and we turn to this next.

The coordinate representation

Through use of $\hat{a}|0\rangle = 0$, (6.86), and the definitions (6.47) and (6.48), one can generate all the wave functions for the 1-D quantal oscillator. The procedure is straightforward, as follows. In a coordinate representation, $\hat{a}|0\rangle = 0$ can be put into the form

$$\left(x + \lambda_0^2 \frac{d}{dx}\right) \psi_0(x) = 0, \tag{6.102}$$

where $\hat{P}_x(x) = -i\hbar d/dx$ and the notation $\langle x|0\rangle = \psi_0(x)$ have been used.

Equation (6.102) is first order in d/dx; its solution is readily found to be

$$\psi_0(x) \propto e^{-x^2/(2\lambda_0^2)}, \tag{6.103}$$

the normalized form of which is

$$\psi_0(x) = \frac{1}{\sqrt{\lambda_0 \pi^{1/2}}} e^{-x^2/(2\lambda_0^2)}, \tag{6.104}$$

where the real normalization constant $\sqrt{\lambda_0 \sqrt{\pi}}$ has been used. Our analysis has reproduced (6.44), as it must.[4]

To obtain $\psi_1(x)$, we apply the coordinate form of $\hat{a}^\dagger$ to $\psi_0(x)$ and use (6.84):

$$\begin{aligned} \psi_1(x) &= \frac{1}{\lambda_0 \sqrt{2}} \left(x - \lambda_0^2 \frac{d}{dx}\right) \frac{1}{\sqrt{\lambda_0 \pi^{1/2}}} e^{-x^2/(2\lambda_0^2)} \\ &= \sqrt{\frac{2}{\lambda_0 \sqrt{\pi}}} \frac{x}{\lambda_0} e^{-x^2/(2\lambda_0^2)}. \end{aligned} \tag{6.105}$$

This is the first-excited-state wave function, belonging to the eigenvalue $E_1 = 3\hbar\omega_0/2$. $\psi_1(x)$ has one node at finite x and is an odd function of its argument, as it must be in order to satisfy $\langle 1|0\rangle = 0$. It also contains a first-order polynomial in x.

The second-excited-state wave function $\psi_2(x)$ is obtained from $\psi_1(x)$ by means of (6.85), viz.,

[4] An alternate way of introducing the coordinate representation is through the "reduced" or dimensionless coordinate $\eta = x/\lambda_0$. Use of η changes the coordinate-space forms of $\hat{P}_x$, $\hat{Q}_x$, $\hat{a}$, $\hat{a}^\dagger$ and the wave function, now given by $\langle \eta|0\rangle = \tilde{\psi}_0(\eta)$, the tilde serving as a reminder that a scale transformation has been employed. Note that $\tilde{\psi}_0(\eta) = \pi^{-1/4} e^{-\eta^2/2} \neq \psi_0(\eta)$, i.e., more than a simple change of (or re-naming of the) independent variable has occurred. Our final results will always be expressed in terms of x.

$$\psi_2(x) = \frac{1}{\sqrt{2}} \frac{1}{\lambda_0\sqrt{2}} \left(x - \lambda_0^2 \frac{d}{dx} \right) \sqrt{\frac{2}{\lambda_0\sqrt{\pi}}} \frac{x}{\lambda_0} e^{-x^2/(2\lambda_0^2)}$$

$$= \frac{1}{\sqrt{2\lambda_0\sqrt{\pi}}} \left[2\left(\frac{x}{\lambda_0}\right)^2 - 1 \right] e^{-x^2/(2\lambda_0^2)}. \tag{6.106}$$

In contrast to $\psi_1(x)$, $\psi_2(x)$ is an even function of x, a second-order polynomial involving only even powers of the argument. There seems to be a pattern here, viz., ψ_0 and ψ_2 are even functions, whereas ψ_1 is odd in x. It is not difficult to show that, for all n, $\psi_{2n}(x)$ contains an even polynomial of order $2n$, whereas $\psi_{2n+1}(x)$ is an odd polynomial of order $2n+1$ (each times $\exp[-x^2/(2\lambda_0^2)]$). The argument is based on $\hat{a}^\dagger(x) = (\lambda_0^{-1}/\sqrt{2})(x - \lambda_0^2 d/dx)$ being an odd function of x, i.e.,

$$\hat{a}^\dagger(-x) = -\hat{a}^\dagger(x), \tag{6.107}$$

from which one deduces that $[\hat{a}^\dagger(x)]^n$ is even or odd as n is even or odd. Since $\psi_n(x) \propto [\hat{a}^\dagger(x)]^n \psi_0(x)$ and $\psi_0(x)$ is an even function, $\psi_n(x)$ is also even or odd as n is even or odd. The polynomial character follows from the structure of $\hat{a}^\dagger(x)$.

All the $\psi_n(x)$ can be generated via application of $(\hat{a}^\dagger)^n$ to $\psi_0(x)$. Because $\{\psi_n(x)\}$ is complete and each $\psi_n(x)$ is a product of a polynomial and $\exp[-x^2/(2\lambda_0^2)]$, the set of polynomials should also form a complete set. In fact, they are an *orthonormal* set on $[-\infty, \infty]$, with weight function $\rho(x) = \exp(-x^2/\lambda_0^2)$. The functions of this set are the Hermite polynomials, whose properties are well known and documented in the mathematical literature. We derive them now by solving the time-independent Schrödinger equation. In doing so, we shall see how the imposition of appropriate boundary conditions on the solutions of the Schrödinger equation compels the eigenvalues to have the structure of Eq. (6.87). We will employ an infinite-series expansion to solve the oscillator eigenvalue problem in differential-equation form. This is a technique well worth mastering, for it is a standard procedure and will be encountered several times in this book.

As in the asymptotic analysis leading to (6.40), the independent variable is again chosen to be $\eta = x/\lambda_0$, in terms of which the Schrödinger equation (6.39) becomes

$$\psi''(\eta) - \eta^2\psi'(\eta) + 2\epsilon\psi(\eta) = 0, \tag{6.108}$$

where $' \equiv d/d\eta$, $'' \equiv d^2/d\eta^2$ and the subscript Γ has been suppressed. Since the asymptotic behavior is already known to be $\exp(-\eta^2/2)$, which appears in each solution via $(\hat{a}^\dagger)^n|0\rangle$, we factor it out of $\psi(\eta)$:

$$\psi(\eta) = H(\eta)e^{-\eta^2/2}, \tag{6.109}$$

where $H(\eta)$ is a function to be determined (and use of "H" anticipates the symbol for the Hermite polynomials).

On substituting (6.109) into (6.108), $H(\eta)$ is found to obey

$$H''(\eta) - 2\eta H'(\eta) + (2\epsilon - 1)H(\eta) = 0. \tag{6.110}$$

Our previous analysis has led to $\epsilon \to \epsilon_n = n + \frac{1}{2}$, but we now pretend ignorance of this. Hence, (6.110) is here regarded as a true eigenvalue equation: both $H(\eta)$ and ϵ are to be determined. We proceed by expressing $H(\eta)$ as a power series in η:

$$H(\eta) = \eta^s \sum_{m=0}^{\infty} a_m \eta^m. \tag{6.111}$$

Since the oscillator potential is centered at $x = 0$, the expansion (6.111) is about the point $\eta = 0$ ($x = 0$). The coefficients a_m and the (indicial) exponent s are obtained by substituting (6.111) into (6.110) and analyzing the result.

A power series such as (6.111) is used to represent the general form of the solution. Finiteness of $\psi(\eta)$ and thus of $H(\eta)$ for all η is the reason why the sum includes no negative powers of η and also why $s \geqslant 0$ is to be expected.

Using (6.111) and carrying out the differentiations in (6.110) yields the following:

$$\eta^s \sum_{m=0}^{\infty} (m+s)(m+s-1) a_m \eta^{m-2} - \eta^s \sum_{m=0}^{\infty} [2(m+s)+1-2\epsilon] a_m \eta^m = 0. \tag{6.112}$$

Although the infinity of terms renders (6.111) and (6.112) complicated, a simplification of (6.112) results from use of the following lemma: if an infinite power series in x is zero, then the coefficient of each power of x separately vanishes[5] (e.g., $b_n = 0, \forall\ n$, in $\sum_n b_n x^n = 0$).

Among the terms obtained on applying the preceding lemma to (6.112) is the pair

$$s(s-1)a_0 = 0 \tag{6.113}$$

and

$$s(s+1)a_1 = 0, \tag{6.114}$$

the LHSs of which are the coefficients of η^{s-2} and η^{s-1}. This pair are the *indicial equations* for the solution (6.111).

The presence of the two powers η^{m-2} and η^m means that Eq. (6.112) leads to a two-term recursion relation, involving a_{m+2} and a_m. Since a_0 and a_1 are, respectively, the coefficients on which all higher even and odd coefficients will depend, we choose to solve the indicial pair by keeping a_0 and a_1 each non-zero. Only the value $s = 0$ satisfies the preceding pair of equations, and, with this choice, (6.112) reads

$$\sum_{m=0}^{\infty} m(m-1) a_m \eta^{m-2} - \sum_{m=0}^{\infty} (2m+1-2\epsilon) a_m \eta^m = 0. \tag{6.115}$$

Because neither the $m = 0$ nor the $m = 1$ term in the first sum on the LHS of (6.115) contributes, we re-index this sum, replacing m by $m - 2$ and starting again with $m = 0$. Equation (6.115) then becomes

$$\sum_{m=0}^{\infty} [(m+1)(m+2) a_{m+2} - (2m+1-2\epsilon) a_m] \eta^m = 0, \tag{6.116}$$

from which we conclude that

$$a_{m+2} = \frac{2m+1-2\epsilon}{(m+1)(m+2)} a_m. \tag{6.117}$$

[5] See, e.g., Mathews and Walker (1970) or Hassani (1991).

This relation, which contains the as-yet-unknown value of ϵ, verifies the previous assertion that all even (odd) a_m are determined once values of a_0 (a_1) are given. Hence, $H(\eta)$ will be an infinite series depending on a_0, a_1, and ϵ. For this series to be an acceptable representation of $H(\eta)$, it must converge to a function that grows less quickly asymptotically than does $\exp[-x^2/(2\lambda_0^2)]$, in order to ensure that $\psi(x \to \pm\infty) \to 0$.

The asymptotic behavior can be estimated by examining the large-m limit of the ratio a_{m+2}/a_m:

$$\lim_{m\to\infty} \frac{a_{m+2}}{a_m} = \lim_{m\to\infty} \frac{2m+1-2\epsilon}{(m+1)(m+2)} \simeq \frac{2}{m}. \tag{6.118}$$

Comparison with the ratio of successive terms in the Taylor-series expansion of

$$e^{+2\eta^2} = \sum_{m=0}^{\infty} \frac{2^m}{m!} \eta^{2m} \tag{6.119}$$

shows that the latter ratio is the same as the large-m limit (6.118). Hence, apart from a finite sum of terms, the infinite series for $H(\eta)$ behaves like $\exp(2\eta^2)$, therefore rendering $H(\eta)\exp(-\eta^2/2)$ non-normalizable.

Such unacceptable behavior is avoided only if the infinite series *terminates* after a finite number of terms, thus becoming a polynomial. Polynomial status is achieved in two steps: first, one sets either a_1 or a_0 equal to zero, in which case the series contains only even or odd powers of η, respectively; and second, one chooses $2\epsilon - 1 = 2n$, where n is even or odd depending, respectively, on whether $a_1 = 0$ or $a_0 = 0$. Because of the factor $2m + 1 - 2\epsilon$ in (6.117), a_n will now be non-zero, but all the higher-order coefficients will vanish, i.e., $a_{n+2} = a_{n+4} = a_{n+6} = \cdots = 0$. The process of turning the series for $H(\eta)$ into an nth–order polynomial $H_n(\eta)$ via

$$\epsilon \to \epsilon_n = n + \tfrac{1}{2}, \qquad n = 0, 1, 2, \ldots \tag{6.120}$$

ensures that $\psi(x)$ is normalizable and determines the eigenvalues to be $E_n = \epsilon_n \hbar\omega_0 = (n + \frac{1}{2})\hbar\omega_0$; they are, of course, the same as those found previously.

The polynomials $H_n(\eta)$, denoted the Hermite polynomials, are even or odd as n is even or odd. The constants a_0 and a_1 are chosen so that the $H_n(\eta)$ are identical to the polynomials $H_n(\eta)$ that occur in the Taylor series for the *generating function* $\exp[\eta^2 - (\xi - \eta)^2] = \exp(-\xi^2 + 2\eta\xi)$, viz.,

$$e^{-\xi^2+2\xi\eta} = \sum_{n=0}^{\infty} \frac{H_n(\eta)}{n!} \xi^n. \tag{6.121}$$

It is straightforward to show from (6.121) that

$$H_n(\eta) = (-1)^n e^{\eta^2} \frac{d^n}{d\eta^n} e^{-\eta^2}, \tag{6.122}$$

from which one finds

$$H_0(\eta) = 1, \tag{6.123a}$$

$$H_1(\eta) = 2\eta, \tag{6.123b}$$

$$H_2(\eta) = 4\eta^2 - 2, \tag{6.123c}$$

Table 6.1 *Some of the low-lying eigenvalues and eigenfunctions of the quantum oscillator*

n	$E_n = (n + \frac{1}{2})\hbar\omega_0$	$\psi_n(x) = A_n H_n\left(\frac{x}{\lambda_0}\right) e^{-x^2/(2\lambda_0^2)}$
0	$\frac{1}{2}\hbar\omega_0$	$\frac{1}{(\lambda_0\sqrt{\pi})^{1/2}} e^{-x^2/(2\lambda_0^2)}$
1	$\frac{3}{2}\hbar\omega_0$	$\frac{1}{(2\lambda_0\sqrt{\pi})^{1/2}} 2\left(\frac{x}{\lambda_0}\right) e^{-x^2/(2\lambda_0^2)}$
2	$\frac{5}{2}\hbar\omega_0$	$\frac{1}{(8\lambda_0\sqrt{\pi})^{1/2}} \left[4\left(\frac{x}{\lambda_0}\right)^2 - 2\right] e^{-x^2/(2\lambda_0^2)}$
3	$\frac{7}{2}\hbar\omega_0$	$\frac{1}{(48\lambda_0\sqrt{\pi})^{1/2}} \left[8\left(\frac{x}{\lambda_0}\right)^3 - 12\left(\frac{x}{\lambda_0}\right)\right] e^{-x^2/(2\lambda_0^2)}$

$$\lambda_0 = \left(\frac{\hbar}{m\omega_0}\right)^{1/2}$$

$$A_n = (n!2^n\lambda_0\sqrt{\pi})^{-1/2}$$

$$H_n(\zeta) = \sum_{m=0}^{[n/2]} \frac{(-1)^m n!(2\zeta)^{n-2m}}{m!(n-2m)!}, \qquad [n/2] = \text{largest integer in } n/2$$

$$H_3(\eta) = 8\eta^3 - 12\eta, \tag{6.123d}$$

etc.

The full eigenfunction solutions for the quantal oscillator are thus given by

$$\psi_n(x) = A_n H_n(x/\lambda_0) e^{-x^2/(2\lambda_0^2)}, \tag{6.124}$$

where A_n is a normalization constant to be determined. Comparison of (6.124) and the set (6.123) with (6.44), (6.105), and (6.106) shows that, for each of $n = 0$, 1, and 2, $A_n = [n!2^n\lambda_0\sqrt{\pi}]^{-1/2}$. This result holds in general:

$$A_n = (n!2^n\lambda_0\sqrt{\pi})^{-1/2}, \qquad \forall\, n. \tag{6.125}$$

Proof of Eq. (6.125) is left to the exercises, although we note here that the factor $(n!)^{-1/2}$ can be predicted from (6.84).

In contrast to the eigenfunctions for the quantal box, which are contained entirely inside the box, the eigenfunctions for the oscillator extend beyond the edges of the oscillator potential, a possibility anticipated in Section 2.4. Thus, there is a non-zero probability that the particle will be found in the classically forbidden regions, a typical quantum-theoretical result. This is illustrated in Fig. 6.7, using the first three eigenfunctions of Table 6.1. That $\psi_0(x)$ and $\psi_2(x)$ are even functions and $\psi_1(x)$ is an odd function of x is evident from Fig. 6.7.

Because $\psi_0(x)$ is even and the interval $[-\infty, \infty]$ is symmetric about $x = 0$, the average or expected value of the position $\langle\hat{Q}_x\rangle_0$ in the state ψ_0 is zero:

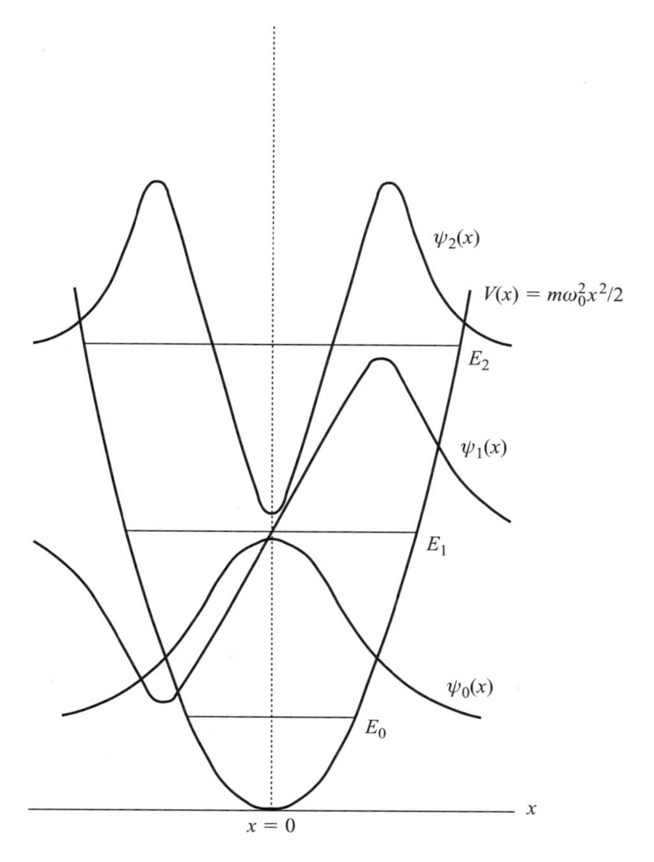

Fig. 6.7 A schematic representation of the oscillator potential, the first three energy eigenvalues, and the first three eigenfunctions, with the ordinate for each ψ_n coinciding with the horizontal line representing E_n. Each $\psi_n(x)$ extends beyond the edge of the potential.

$$\begin{aligned}\langle \hat{Q}_x \rangle_0 &= \langle 0|\hat{Q}_x|0\rangle \\ &= \int_{-\infty}^{\infty} dx\, x|\psi_0(x)|^2 = 0.\end{aligned} \tag{6.126}$$

The expectation value of the momentum in the oscillator ground state, $\langle \hat{P}_x \rangle_0$, is also zero:

$$\begin{aligned}\langle \hat{P}_x \rangle_0 &= \langle 0|\hat{P}_x|0\rangle \\ &= \int_{-\infty}^{\infty} dx\, \psi_0^*(x)\left(-i\hbar\frac{d}{dx}\right)\psi_0(x) \\ &\propto \int_{-\infty}^{\infty} dx\, x|\psi_0(x)|^2 = \langle \hat{Q}_x \rangle_0 = 0.\end{aligned} \tag{6.127}$$

The latter result can also be proved by using Fourier transforms to convert from a matrix element in x-space to one in k-space. However, in this special case of the linear oscillator, the formulation of the eigenvalue problem in terms of the operators $\hat{a}$ and $\hat{a}^\dagger$ allows an alternate procedure for determining the momentum-space eigenfunction $\varphi_0(k)$. Instead of obtaining it via a Fourier transform, we use $\hat{a}|0\rangle = 0$ and invoke the momentum representation:

$$\hat{a}(k)\varphi_0(k) = 0. \tag{6.128}$$

However,

$$\hat{a}(k) = \frac{i\lambda_0}{\sqrt{2}}\left(k + \lambda_0^2 \frac{d}{dk}\right),$$

where the 1-D version of (5.76) has been employed, so that (6.128) becomes

$$\left(k + \lambda_0^{-2}\frac{d}{dk}\right)\varphi_0(k) = 0. \tag{6.129}$$

This equation is the precise analog of (6.102). Its normalized solution resembles (6.104):

$$\varphi_0(k) = \sqrt{\frac{\lambda_0}{\sqrt{\pi}}}e^{-\lambda_0^2 k^2/2}, \tag{6.130}$$

which tells us that the Fourier transform of a Gaussian is again a Gaussian. In terms of $\varphi_0(k)$, the expectation value (6.127) becomes $\langle \hat{P}_x\rangle_0 = \hbar\int_{-\infty}^{\infty} dk\, k|\varphi_0(k)|^2 = 0$, as expected.

Since the average values of position and momentum vanish in the state $|0\rangle$, we next consider the expectation values of their squares, $\langle \hat{Q}_x^2\rangle_0$ and $\langle \hat{P}_x^2\rangle_0$, quantities that are proportional to the average values of the potential and kinetic energies, respectively. Using the preceding analysis, these quantities can be expressed as

$$\langle \hat{Q}_x^2\rangle_0 = \frac{1}{\lambda_0\sqrt{\pi}}\int_{-\infty}^{\infty} dx\, x^2 e^{-x^2/\lambda_0^2} \tag{6.131}$$

and

$$\langle \hat{P}_x^2\rangle_0 = \frac{\hbar^2\lambda_0}{\sqrt{\pi}}\int_{-\infty}^{\infty} dk\, k^2 e^{-\lambda_0^2 k^2}. \tag{6.132}$$

Each integral has the same structure, viz.,

$$I_2 = \int_{-\infty}^{\infty} d\xi\, \xi^2 e^{-a\xi^2},$$

whose value, from Table 6.2, is

$$I_2 = \frac{1}{2a}\sqrt{\frac{\pi}{a}}.$$

The two expectation values are now trivially obtained:

$$\langle \hat{Q}_x^2\rangle_0 = \lambda_0^2/2 = \frac{\hbar}{2m\omega_0} \tag{6.133}$$

and

Table 6.2 *Some Gaussian integrals*

$$I_{2n} = \int_0^\infty x^{2n} e^{-ax^2}\,dx = \frac{1}{2}\int_{-\infty}^\infty x^{2n} e^{-ax^2}\,dx = \frac{(2n-1)!!}{2^{n+1}a^n}\sqrt{\frac{\pi}{a}}$$

$$I_{2n+1} = \int_0^\infty x^{2n+1} e^{-ax^2}\,dx = \frac{n!}{2a^{n+1}}$$

$$\int_{-\infty}^\infty x^{2n+1} e^{-ax^2}\,dx = 0$$

Note: $(2n-1)!! = (2n-1)(2n-3)(2n-5)\cdots 1$

$(-1)!! = 1$

$0! = 1$

$$\langle \hat{P}_x^2\rangle_0 = \frac{\hbar^2}{2\lambda_0^2} = \frac{\hbar m\omega_0}{2}. \tag{6.134}$$

Recalling that $\hat{V} = m\omega_0^2\hat{Q}_x^2/2$ and $\hat{K} = \hat{P}_x^2/(2m)$, we readily find

$$\langle \hat{K}_0\rangle = \langle \hat{V}_0\rangle = \tfrac{1}{4}\hbar\omega_0. \tag{6.135}$$

Hence, the total ground-state energy $E_0 = \langle \hat{H}\rangle_0 = \hbar\omega_0/2$ is shared equally between the averages of its kinetic and potential forms. This equality is a realization, in the oscillator case, of the quantal virial theorem, which is discussed as an exercise in Chapter 8.

The expectation values $\langle \hat{Q}_x^2\rangle_0$ and $\langle \hat{P}_x^2\rangle_0$ can be used to evaluate the Heisenberg uncertainty relation for the case of the oscillator ground state (a general analysis is given in Chapter 9). The uncertainty principle states that, if $\hat{A}$ and $\hat{B}$ are two non-commuting operators, then the product $\Delta A\,\Delta B$ of their uncertainties ΔA and ΔB obeys the inequality

$$\Delta A\,\Delta B \geqslant \tfrac{1}{2}|\langle[\hat{A},\hat{B}]\rangle| \tag{6.136}$$

where

$$\Delta A = \sqrt{\langle \hat{A}^2\rangle - \langle \hat{A}\rangle^2}, \tag{6.137}$$

and $\langle\,\rangle$ implies expectation w.r.t. a particular state. Choosing $\hat{A} = \hat{Q}_x$ and $\hat{B} = \hat{P}_x$, and writing ΔQ_x as Δx, use of the preceding expectation values leads to

$$\Delta x\,\Delta P_x = \hbar/2 \tag{6.138}$$

for the ground state of the oscillator. The Gaussian wave function $\exp[-x^2/(2\lambda_0^2)]$ is thus seen to produce the minimum uncertainty, a result that is rederived in Chapter 9.

Coherent (quasi-classical) states

In the preceding subsections, the quantal harmonic oscillator has been studied from a variety of perspectives. One topic so far omitted is the determination of the eigenstates of the operator $\hat{a}$. These eigenstates, known as coherent or quasi-classical states (Glauber (1963)), are important in their own right, since they play a role in many areas of physics, especially quantum optics. We close this section with a brief introduction to these states; various of their properties are left to the exercises in this and later chapters.

Denoting the coherent state by $|\zeta\rangle$, the eigenvalue problem to be solved is

$$\hat{a}|\zeta\rangle = \zeta|\zeta\rangle. \tag{6.139}$$

Since $\hat{a}^\dagger \neq \hat{a}$, we cannot expect that only real ζ will occur. Expanding $|\zeta\rangle$ in terms of the oscillator states $|n\rangle$, viz.,

$$|\zeta\rangle = \sum_{n=0}^{\infty} C_n|n\rangle, \tag{6.140}$$

(6.139) becomes

$$\sum_{n=0}^{\infty} C_{n+1}\sqrt{n+1}|n\rangle = \sum_{n=0}^{\infty} \zeta C_n|n\rangle,$$

from which we get

$$C_{n+1} = \frac{\zeta C_n}{\sqrt{n+1}}.$$

This relation is reminiscent of Eq. (6.81) and, analogous to that case, one finds

$$C_n = \frac{\zeta^n}{\sqrt{n!}} C_0,$$

so that

$$|\zeta\rangle = C_0 \sum_{n=0}^{\infty} \frac{\zeta^n}{\sqrt{n!}}|n\rangle. \tag{6.141}$$

The constant C_0 is obtained from normalization:

$$\begin{aligned} 1 &= \langle\zeta|\zeta\rangle \\ &= |C_0|^2 \sum_n \frac{|\zeta|^{2n}}{n!} \\ &= |C_0|^2 e^{|\zeta|^2}. \end{aligned}$$

Choosing C_0 to be real and positive, this relation yields

$$C_0 = e^{-|\zeta|^2/2},$$

and thus (6.141) becomes

$$|\zeta\rangle = e^{-|\zeta|^2/2} \sum_{n=0}^{\infty} \frac{\zeta^n}{\sqrt{n!}}|n\rangle. \tag{6.142}$$

Equation (6.142) is a state with unusual properties, some of which we explore next.

First, the set $\{|\zeta\rangle\}$ is continuous, since ζ can be any complex number. However, for finite ζ and $\zeta' \neq \zeta$, $|\zeta\rangle$ and $|\zeta'\rangle$ are not orthogonal. A straightforward calculation yields

$$\langle\zeta|\zeta'\rangle = e^{-(|\zeta|^2/2-\zeta^*\zeta'-|\zeta'|^2/2)},$$

which can be made arbitrarily small by choosing ζ or ζ' very large, but is not zero except in the limit, in which case $\lim|\zeta\rangle \to \infty$. While the set $\{|\zeta\rangle\}$ is thus overcomplete, it nevertheless can be used as an expansion set. That is, one can show that

$$\pi\int d^2\zeta\,|\zeta\rangle\langle\zeta| = \hat{I},$$

where $d^2\zeta = d(\mathrm{Re}\,\zeta)\,d(\mathrm{Im}\,\zeta) = |\zeta|\,d|\zeta|\,d\theta$, with θ being the polar angle in the complex-ζ plane.

A second point is that the average number of quanta $\bar{n}$ in $|\zeta\rangle$, obtained from $\langle\zeta|\hat{N}|\zeta\rangle = \langle\zeta|\hat{a}^\dagger\hat{a}|\zeta\rangle$, is found via Eq. (6.139) to be

$$\overline{n} = \langle\zeta|\hat{N}|\zeta\rangle = |\zeta|^2.$$

The probability P_n that $|n\rangle$ occurs in $|\zeta\rangle$, given by $P_n = |\langle n|\zeta\rangle|^2$, is therefore

$$P_n = e^{-\overline{n}}\overline{n}/n!,$$

i.e., a Poisson distribution. Thus, $|\zeta\rangle$ does not contain a fixed number of quanta. Incidentally, since $\overline{n} = |\zeta|^2$, the average energy in $|\zeta\rangle$ is just $(|\zeta|^2 + \frac{1}{2})\hbar\omega_0$.

A third aspect of $|\zeta\rangle$ concerns the value it yields for the product of the uncertainties $\Delta Q_x\,\Delta P_x$. Even though $|\zeta\rangle$ is an infinite sum over the $|n\rangle$, and $\Delta Q_x\,\Delta P_x = (n+\frac{1}{2})\hbar$ for the nth oscillator state, for $|\zeta\rangle$ one finds $\Delta Q_x\,\Delta P_x = \hbar/2$, the minimum value. This result, which is the subject of an exercise in Chapter 8, is one of the reasons why $|\zeta\rangle$ is known as a coherent or quasi-classical state.

A final characteristic of $|\zeta\rangle$ is that its expansion (6.142) can be summed and replaced by a "simple" operator form. It is obtained by recalling that $|n\rangle = (\hat{a}^\dagger)^n|0\rangle/\sqrt{n!}$, Eq. (6.84). Hence (6.142) becomes

$$\begin{aligned}|\zeta\rangle &= e^{-|\zeta|^2/2}\sum_n \frac{(\zeta\hat{a}^\dagger)^n}{n!}|0\rangle \\ &= e^{-|\zeta|^2/2}e^{\zeta\hat{a}^\dagger}|0\rangle,\end{aligned} \tag{6.143}$$

a compact form that is useful in various applications.

We continue the analysis of coherent states in the exercises and in Chapter 8.

6.3. General remarks on non-confining potentials

For confining potentials, such as the box and the oscillator, all of the eigensolutions of the time-independent Schrödinger equation describe bound states: the coordinate-space eigenfunctions are zero asymptotically, behavior that renders them normalizable. This asymptotic behavior is a consequence of the intrinsic nature of a confining potential, which is or becomes infinite for asymptotic values of its spatial coordinate(s). Potentials that do not enjoy this property are deemed "non-confining."

Non-confining potentials are, or go to, a finite constant (often zero) when one or more of their coordinates become asymptotic. Excluding systems of infinite extent, such potentials are the type encountered in the real quantum systems on which experiments are performed. A simple case to analyze, and the one explored in the next three sections, is the class of 1-D potentials $V(x)$, which are zero for $x \leqslant x_1$ and/or $x \geqslant x_2 > x_1$. For such potentials, there is at least one region where the time-independent Schrödinger equation has the form $\hat{K}(x)\psi(x) = E\psi(x)$. As we shall show in Chapter 7, this allows a new type of potential-well solution when $E > 0$, viz., one that is not normalizable. Portions of these improper vectors are momentum eigenstates confined to the regions $x \leqslant x_1$ and/or $x \geqslant x_2$, wherein the particle is free. Here, "free" means that the particle is not confined to a finite portion of the interval $[-\infty, \infty]$. The occurrence of these positive-energy, non-normalizable, asymptotically free solutions is in sharp contrast to the positive-energy eigenstates of confining potentials, since the latter states are normalizable. Because these asymptotically free or *continuum* solutions are both qualitatively and quantitatively different than bound states, we postpone further discussion of them to Chapter 7.

The solutions of interest in the remainder of this chapter therefore correspond to bound states in non-confining potentials. These solutions are normalizable and their energies are negative, just as in the Bohr model of hydrogen. However, bound states can occur *only* if the minimum value of $V(x)$ does not occur asymptotically. When $V(x) = 0$ for $x \leqslant x_1$ and/or $x \geqslant x_2$, this means that $V(x)$ must be negative somewhere in configuration space. Furthermore, as we shall prove in the following, if the minimum value of $V(x)$ is $-V_0$ and bound states can occur, then V_0 is a lower bound to the set of eigenenergies $\{E_n\}$: $E_{\min} > V_0$, where $E_{\min}$ is the smallest eigenvalue. When a non-confining potential can support bound states, the set of (negative-energy) eigenvalues defines the bound-state part of the spectrum. Shown in Fig. 6.8 is a Grotrian-diagram representation of the spectrum of a one-particle Hamiltonian $\hat{H}$ that contains both bound and continuum portions.

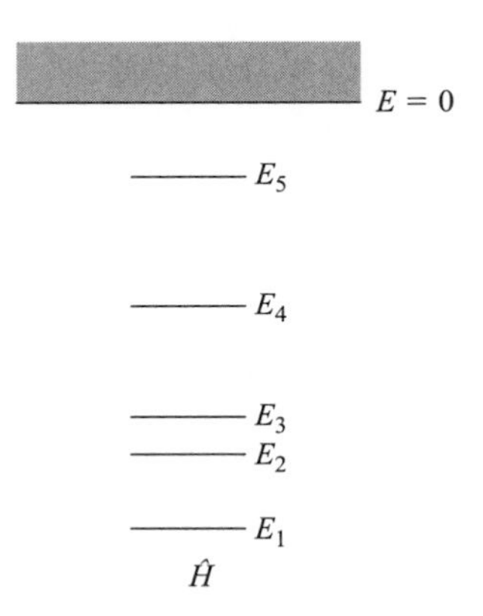

Fig. 6.8 The spectrum of a one-particle Hamiltonian $\hat{H}$ that can support a finite number of bound states. All bound-state energies E_n are negative; $n = 1$ is the ground state, and $|E_1|$ is the dissociation energy. The continuum begins at $E = 0$.

Boundary conditions and eigenvalues

Although there are many kinds of potentials that support bound states, two of the three examples considered in the rest of this chapter are instances of a generic type, viz., an attractive local potential $V(x)$ that is non-zero only in the finite interval $[x_1, x_2]$:

$$V(x) = \begin{cases} 0, & x \leqslant x_1 \quad \text{(Region I)}, \\ <0, & x_1 \leqslant x \leqslant x_2 \quad \text{(Region II)}, \\ 0, & x \geqslant x_2 \quad \text{(Region III)}. \end{cases} \tag{6.144}$$

For the present purpose of discussing the general properties of solutions, no specific form for $V(x)$ in region II is required. A schematic representation of one such $V(x)$ is shown in Fig. 6.9.

Eqation (6.144) is not overly restrictive, inasmuch as all non-confining potentials will be zero either for $x \leqslant x_1$ or for $x \geqslant x_2$ (or for both, as in (6.144)), so that at least part of the following analysis will always apply. We concentrate on regions I and III, where $V(x) = 0$, because the corresponding stationary-state solutions are almost trivally easy to obtain. In either region the time-independent Schrödinger equation is

$$-\frac{\hbar^2}{2m}\psi''(x) = E\psi(x), \qquad x \leqslant x_1 \text{ or } x \geqslant x_2. \tag{6.145}$$

Since $E < 0$ (the possibility of a zero-energy bound state is excluded from this discussion), we write $E = -\hbar^2\alpha^2/(2m)$, where α is a wave number whose (quantized) values determine the bound energies. Using this expression for E, (6.145) becomes

$$\psi''(x) - \alpha^2\psi(x) = 0, \qquad x \leqslant x_1 \text{ or } x \geqslant x_2. \tag{6.146}$$

In either region, the most general solution is a linear combination of $\exp(\pm\alpha x)$, but in neither region is such a linear combination satisfactory. A simple calculation establishes this. Consider region I, $x \leqslant x_1$. The general solution $\psi_{\text{I}}(x)$ is given by

$$\psi_{\text{I}}(x) = Ae^{\alpha x} + Be^{-\alpha x}, \qquad x \leqslant x_1, \tag{6.147}$$

where A and B are constants. We want the complete wave function $\psi(x)$ – which is the sum $\psi_{\text{I}}(x) + \psi_{\text{II}}(x) + \psi_{\text{III}}(x)$ of the solutions in regions I, II, and III – to be normalized. The contribution to the normalization integral from ψ_1 is

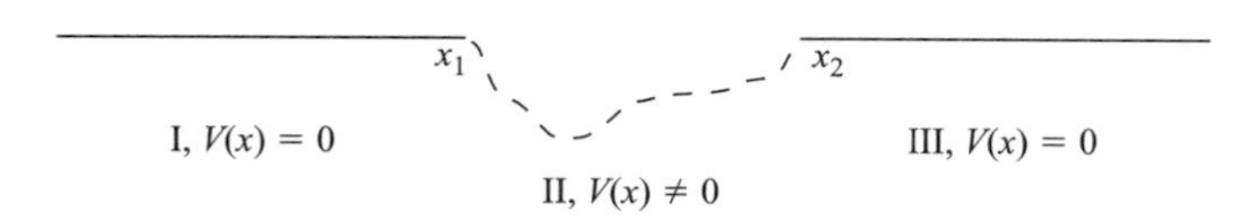

Fig. 6.9 A schematic representation of an arbitrary potential $V(x)$ that is zero for $x \leqslant x_1$ (region I) and for $x \geqslant x_2$ (region III), and is negative somewhere in region II ($x_1 \leqslant x \leqslant x_2$). The values x_1 and x_2 and the dotted line are each completely arbitrary; the latter can be replaced by any other locus that is negative somewhere in region II.

$$I_{\rm I} = \int_{-\infty}^{x_1} dx |\psi_{\rm I}(x)|^2$$
$$= \int_{-\infty}^{x_1} dx |Ae^{\alpha x} + Be^{-\alpha x}|^2;$$

the presence of the term $Be^{-\alpha x}$ in $\psi_{\rm I}(x)$ renders $I_{\rm I} = \infty$, an unacceptable result.

A finite result, and thus a provisionally acceptable one, requires $B = 0$. Hence, normalizability of $\psi(x)$ may be achieved only if (6.147) is replaced by

$$\psi_{\rm I}(x) = Ae^{\alpha x}. \tag{6.148}$$

A similar conclusion holds for $x \geqslant x_2$. In this region, the most general solution $\psi_{\rm III}$ is

$$\psi_{\rm III}(x) = Ce^{\alpha x} + De^{-\alpha x}, \qquad x \geqslant x_2, \tag{6.149}$$

but the contribution $I_{\rm III}$ to the normalization integral from $\psi_{\rm III}(x)$, viz.,

$$I_{\rm III} = \int_{x_2}^{\infty} dx |\psi_{\rm III}(x)|^2,$$

is also infinite, unless $C = 0$. Thus, only if (6.149) is superseded by

$$\psi_{\rm III}(x) = De^{-\alpha x} \tag{6.150}$$

will the full wave function be normalizable.

With the choices (6.148) and (6.150), the statement of normalization

$$\int_{-\infty}^{x_1} dx |\psi_{\rm I}(x)|^2 + \int_{x_1}^{x_2} dx |\psi_{\rm II}(x)|^2 + \int_{x_2}^{\infty} dx |\psi_{\rm III}(x)|^2 = 1$$

reduces to

$$\frac{|A|^2}{2\alpha} e^{2\alpha x_1} + I_{\rm II} + \frac{|D|^2}{2\alpha} e^{-2\alpha x_2} = 1, \tag{6.151}$$

where, by the mean-value theorem of calculus,

$$I_{\rm II} = \int_{x_1}^{x_2} dx |\psi_{\rm II}(x)|^2$$

is finite.

Equation (6.151) is one of several relations involving the coefficients A and D as well as analogous ones that will occur in $\psi_{\rm II}(x)$. The other relations arise from imposing boundary conditions: $\psi(x)$ and $\psi'(x)$ must be continuous at x_1 and x_2. These relations are

$$\psi_{\rm I}(x_1) = \psi_{\rm II}(x_1), \tag{6.152a}$$
$$\psi_{\rm I}'(x_1) = \psi_{\rm II}'(x_1) \tag{6.152b}$$

and

$$\psi_{\rm II}(x_2) = \psi_{\rm III}(x_2), \tag{6.152c}$$
$$\psi_{\rm II}'(x_2) = \psi_{\rm III}'(x_2). \tag{6.152d}$$

The set of equations (6.152) defines the boundary conditions on $\psi(x)$ for the finite values x_1 and x_2. Equations (6.148) and (6.150) can also be thought of as boundary conditions on $\psi(x)$ in the asymptotic regions $x \to \pm\infty$:

$$\psi(x \to \pm\infty) \to A_\pm e^{\mp\alpha x}, \tag{6.153}$$

where A_+ (A_-) has been used in place of D (A).

The specific values of α will depend on the behavior of $V(x)$ in region II, i.e., on $\psi_{\mathrm{II}}(x)$. These values are determined by solving the boundary-matching equations (6.152) or their analogs for non-confining potentials differing in form from (6.144). However, the *existence* of bound states is due not to the matching of ψ and ψ' at x_1 and x_2, but to the asymptotic condition (6.153). An analysis of how a computer would generate a solution $\psi(x)$ to the Schrödinger equation illustrates this remark fairly simply. The typical procedure involves a number of steps, among which are the following: choose an algorithm for numerically solving a second-order differential equation,[6] select an E, guess the (starting) values of $\psi(x)$ at two values of $x < x_1$, use the algorithm and the starting values to generate and propagate the numerical solution to a point $x > x_2$, and then compare it with the asymptotic form of ψ_{III}.

Since $\psi_{\mathrm{I}}(x) \propto \exp(\alpha x)$, $\alpha = \sqrt{2m|E|/\hbar^2}$, the starting values can be determined once an E has been selected. Doing so, and then employing the algorithm, will eventually produce a numerical solution $\psi_{\mathrm{III}}^{(\mathrm{num})}(x)$ in region III. The form of that solution will be

$$\psi_{\mathrm{III}}^{(\mathrm{num})}(x) = C(E)\mathrm{e}^{\alpha x} + D(E)\mathrm{e}^{-\alpha x}, \tag{6.154}$$

in other words, a linear combination of the independent solutions $\exp(\pm\alpha x)$, each multiplied by a numerical coefficient that depends on the chosen value of E.

Equation (6.154) is not an acceptable form, as has already been noted. In order for it to be one, the coefficient $C(E)$ must vanish. In general, this can happen only for a finite set of values, say $\{E_n\}$:

$$C(E_n) = 0, \qquad \forall\ E_n. \tag{6.155}$$

Thus, as long as $E = E_n$, $\psi_{\mathrm{II}}(x)$ will join smoothly onto $\psi_{\mathrm{III}} = D(E_n)e^{-\alpha_n x}$ at $x = x_2$. For any other value of E, $\psi_{\mathrm{II}}(x)$ will join smoothly onto $\psi_{\mathrm{III}}^{(\mathrm{num})}(x)$ of (6.154), with the result that $\psi(x \to \infty) \to \infty$. Figure 6.10 shows, in the neighborhood of $x = x_2$, two solutions generated from $\psi(x) = \exp(\alpha x)$ at some $x < x_1$: one for $\alpha = \alpha_n =$

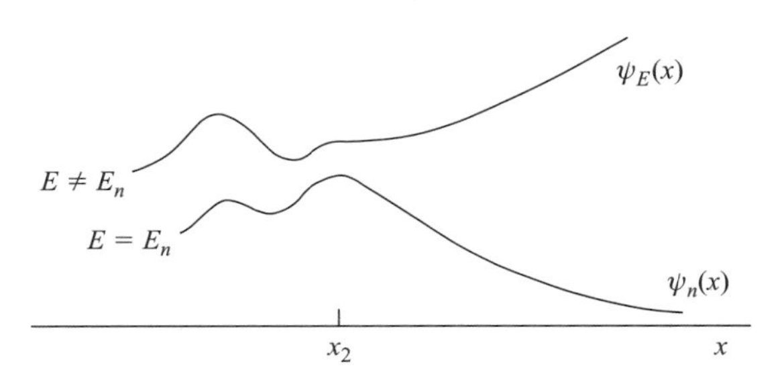

Fig. 6.10 Behavior near $x = x_2$ of two solutions to the Schrödinger equation generated from the starting solutions $\exp(\alpha x)$ and $\exp(\alpha_n x)$ at some $x < x_1$. Here $\alpha = \sqrt{2m|E|/\hbar^2}$ does not correspond to an eigenvalue, while $\alpha_n = \sqrt{2m|E_n|/\hbar^2}$ is the bound-state wave number associated with the bound-state eigenvalue E_n.

[6] Numerical procedures, such as those of Numerov or Runge–Kutta, are discussed in many books, e.g. Press *et al.* (1986) and Rice (1983).

$\sqrt{2m|E_n|/\hbar^2}$, the other for $\alpha = \sqrt{2m|E|/\hbar^2}$, $E \neq E_n$. The divergence of the non-eigenvalue solution is evident.

We note in passing that arguments and conclusions analogous to those of the preceding can be generated for other non-confining potentials. Specific examples are discussed in the next sections.

Lower bounds on eigenvalues

In this subsection, we prove the following.

THEOREM ON LOWER BOUNDS

If $-V_0$ is the minimum value of $V(x)$, then $-V_0$ is a lower bound to the set of eigenvalues $\{E_n\}$ of the corresponding Schrödinger equation.

Proof:

The time-independent Schrödinger equation is

$$\frac{\hat{P}_x^2(x)}{2m}\psi_n(x) + V(x)\psi_n(x) = E_n\psi_n(x). \tag{6.156}$$

If both sides of (6.156) are multiplied by $\psi_n^*(x)$ and the resulting expression is integrated over the range of x where $V(x)$ and/or $\psi(x)$ are non-vanishing, one has[7]

$$\int \psi_n^*(x)\frac{\hat{P}_x^2(x)}{2m}\psi_n(x)dx + \int \psi_n^*(x)V(x)\psi_n(x)dx = E_n\int dx\,|\psi_n(x)|^2. \tag{6.157}$$

We consider the three terms in (6.157) one at a time, noting first that the kinetic-energy-operator term is positive definite. This is most easily seen by reverting to Dirac notation:

$$\begin{aligned}\int \psi_n^*(x)\frac{\hat{P}_x^2(x)}{2m}\psi_n(x)dx &= \left\langle \psi_n \middle| \frac{\hat{P}_x^2}{2m} \middle| \psi_n \right\rangle \equiv \langle \hat{K}\rangle_n \\ &= \frac{1}{2m}\langle \hat{P}_x\psi_n|\hat{P}_x\psi_n\rangle \\ &= \frac{1}{2m}\|\hat{P}_x|\psi_n\rangle\|^2 > 0. \end{aligned} \tag{6.158}$$

Next, since $-V_0 = \min V(x)$, one has that

$$\int dx\,\psi_n^*(x)V(x)\psi_n(x) \geqslant -V_0\int dx|\psi_n(x)|^2.$$

Using this result in (6.157) leads to

$$E_n\langle\psi_n|\psi_n\rangle = \langle\hat{K}\rangle_n + \langle\psi_n|V|\psi_n\rangle \geqslant \langle\hat{K}\rangle_n - V_0\langle\psi_n|\psi_n\rangle. \tag{6.159}$$

If the positive quantity $\langle K_n\rangle$ is subtracted from the RHS of the inequality (6.159), this inequality is strengthened:

[7] Recall that, if in some region $V(x) = \infty$, then $\psi(x) = 0$ such that $V(x)\psi(x) = 0$ in that region.

$$E_n\langle\psi_n|\psi_n\rangle > -V_0\langle\psi_n|\psi_n\rangle.$$

By assumption, $\psi_n(x)$ is a non-trivial solution, so that $\langle\psi_n|\psi_n\rangle \neq 0$, from which we conclude that

$$E_n > -V_0, \tag{6.160}$$

QED.

This inequality is certainly satisfied for the eigenvalues of the quantum box and the linear harmonic oscillator, since $V_0 = 0$ and $E_n > 0$ in each case. For a potential whose general form is given by (6.144), all the E_n are negative, a result that, combined with (6.160), yields

$$-V_0 < E_n \leqslant 0 \tag{6.161}$$

($V(x)$ of (6.144)).

We next turn to the technical problem of determining eigenvalues and wave functions using the method of boundary matching for three non-confining potentials. These three potentials will also be used to generate specific continuum (stationary-state) solutions in Chapter 7, where we shall show how one can use these solutions to obtain the corresponding bound-state eigenvalues. Only one of the three potentials examined here yields an explicit expression for the bound-state wave numbers α_n: in general, only a few non-confining potentials allow this, so that one usually ends up with implicit rather than direct expressions for the eigenvalues.

6.4. The half-space square well

Rather than start with a potential well having the general form of Eq. (6.144), we choose instead to analyze an eigenvalue problem related to the quantum box. This is the half-space square well, obtained by eliminating the right-hand half of the box and replacing it by a potential well $V(x)$ that is of constant depth $-V_0$ out to the point $x = L$ and is zero for $x \geqslant L$:

$$V(x) = \begin{cases} \infty, & x \leqslant 0 \quad \text{(Region I)}, \\ -V_0, & 0 \leqslant x \leqslant L \quad \text{(Region II)}, \\ 0, & x \geqslant L \quad \text{(Region III)}. \end{cases} \tag{6.162}$$

The half-space square well is shown in Fig. 6.11, in which regions I, II, and III are also specified. This potential is the analog of the 3-D, radially symmetric square-well potential, viz., $V(r) = -V_0$, $r \leqslant a$; $V(r) = 0$, $r \geqslant a$. The eigenvalue problem for $\psi(x)$ in the 1-D case is identical to that of the so-called "S-wave radial wave function" for the 3-D square well, which will be examined in Chapter 11. Square wells are idealizations of short-range, attractive potentials, as noted in Section 5.4; they are useful and informative "soluble" approximations.

Although the wave functions in each region are easily obtained, the eigenvalues are determined only by solving (numerically) a transcendental equation. This complication is due to the apparent "simplification" of eliminating the right-hand infinite wall of the quantum box. In region I, the solution is the trivial one, viz.,

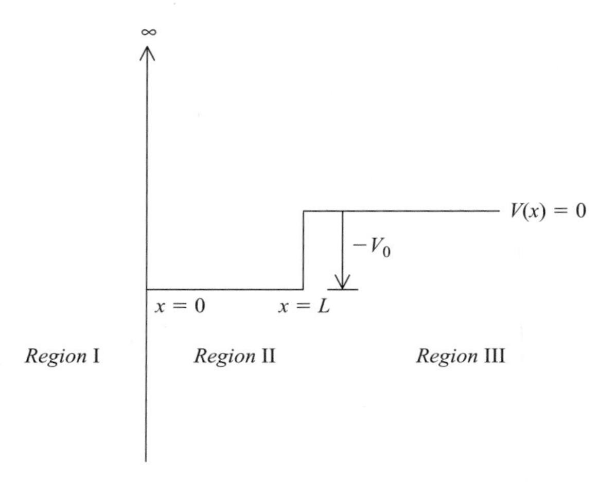

Fig. 6.11 The half-space square well.

$$\psi_{\rm I}(x) = 0, \qquad x \leqslant 0,$$

as required by $V(x) = \infty$, $x \leqslant 0$ (see Section 6.3).

In region II, the time-independent Schrödinger equation reads

$$-\frac{\hbar^2}{2m}\psi''(x) - (V_0 - |E|)\psi(x) = 0, \qquad 0 \leqslant x \leqslant L, \tag{6.163}$$

where $E = -|E|$ has been used. It is not hard to show that Eq. (6.161) holds also for the half-space square well, so that the quantity $V_0 - |E| > 0$. Introducing the wave number γ by the relation

$$V_0 - |E| = \frac{\hbar^2\gamma^2}{2m}, \tag{6.164}$$

Eq. (6.163) becomes

$$\psi_\gamma''(x) + \gamma^2\psi_\gamma(x) = 0, \qquad 0 \leqslant x \leqslant L. \tag{6.165}$$

The latter equation is identical in structure to (3.84) and, since $\psi_\gamma(x = 0) = 0$, the solution to (6.165) must be identical to (3.85):

$$\psi_\gamma(x) = A_\gamma \sin(\gamma x), \qquad 0 \leqslant x \leqslant L. \tag{6.166}$$

The constants A_γ and γ have yet to be determined. Note that

$$\gamma = \sqrt{\frac{2mV_0}{\hbar^2} - \alpha^2}, \tag{6.167}$$

where $|E| = \hbar^2\alpha^2/(2m)$.

Equation (6.166) is the general form of solution in region II. We already know the region-III solution:

$$\psi_{\rm III}(x) = De^{-\alpha x}, \tag{6.168}$$

where D and the bound-state wave number α are still unknown (D will depend on α (or γ)).

Our earlier claim, that the form of wave function in each region is easily obtained, is now verified. To obtain the eigenvalues, or equivalently, the bound-state wave numbers, we turn to the boundary matching relations, Eqs. (6.152c) and (6.152d). Substituting (6.166) and (6.168) into these equations and replacing x_2 by L yields

$$A_\gamma \sin(\gamma L) = De^{-\alpha L} \tag{6.169a}$$

and

$$\gamma A_\gamma \cos(\gamma L) = -\alpha De^{-\alpha L}. \tag{6.169b}$$

Equations (6.169a) and (6.169b) are homogeneous in A_γ and D. This suggests dividing one by the other, thus eliminating the normalization constants and obtaining a relation involving only α, γ, and L. On doing so, we find

$$\alpha L = -\gamma L \cot(\gamma L). \tag{6.170}$$

This is one of two relations involving α and γ, the other being (6.167), i.e.,

$$\alpha L = \sqrt{\frac{2mL^2}{\hbar^2} V_0 - (\gamma L)^2}. \tag{6.171}$$

Comparing (6.170) and (6.171) yields

$$\frac{\sqrt{2mL^2 V_0/\hbar^2 - \rho^2}}{\rho} = -\cot\rho, \tag{6.172}$$

which is the transcendental "eigenvalue" equation predicted above. Here, of course, $\rho = \gamma L$ is a dimensionless parameter.

Once V_0, L, and m are specified, Eq. (6.172) can be solved numerically for the allowed values of γ, which, via (6.170) or (6.171), then give the bound-state wave numbers α and thus the eigenvalues $E = -\hbar^2\alpha^2/(2m)$. Such a procedure, however, does not permit us to draw any general conclusions, for example, on whether there are conditions whose satisfaction ensures the existence of at least one bound state. There is, however, a simple *graphical* analysis that does allow this, as follows.

Let us define the functions $f(\rho)$ and $g(\rho)$ by

$$f(\rho) = \rho^{-1}\sqrt{2mL^2 V_0/\hbar^2 - \rho^2}$$

and

$$g(\rho) = -\cot\rho.$$

Then Eq. (6.172) is the same as $f(\rho) = g(\rho)$. We therefore plot on one graph both $f(\rho)$ and $g(\rho)$ as functions of ρ; any points ρ_n where they intersect determine the eigenvalues. Shown in Fig. 6.12 are separate plots of $f(\rho)$, $g(\rho)$, and the two together.

Because both αL and $\rho = \gamma L$ are positive, the minimum value of $f(\rho)$ is zero, which occurs at $\rho_{\max} = (2mL^2 V_0/\hbar^2)^{1/2}$. In order for $f(\rho)$ to intersect $g(\rho)$, ρ must be at least equal to $\pi/2$, as demonstrated by Fig. 6.12(c). Hence, if $\rho_{\max} < (2n+1)\pi/2$, or if

$$V_0 < (2n+1)^2 \frac{\pi^2}{4} \frac{\hbar^2}{2mL^2}, \tag{6.173}$$

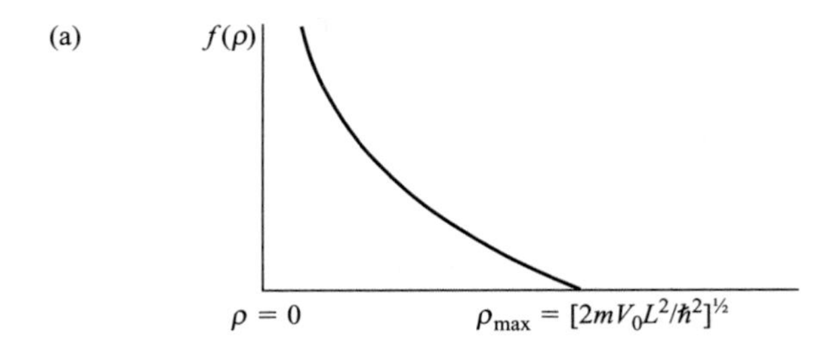

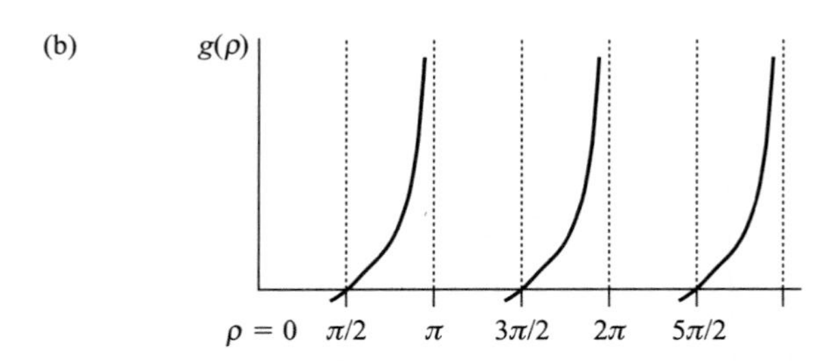

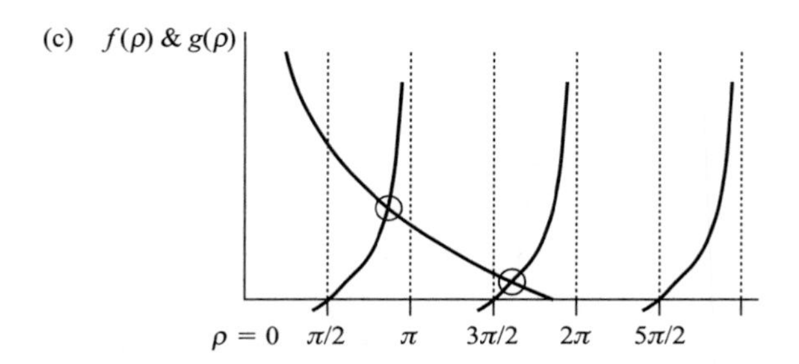

Fig. 6.12 (a) A plot of $f(\rho)$ versus ρ, (b) a plot of $g(\rho)$ versus ρ, and (c) a plot of $f(\rho)$ and $g(\rho)$ together versus ρ. Since the allowed values obey $\alpha L \geqslant 0$ and $\rho = \gamma L \geqslant 0$, only the positive portion of $g(\rho)$ is shown. The circles indicate the points where the two sets of curves cross; in the present case these are the only values of ρ for which this occurs; these determine the eigenvalues, as noted in the text.

then there are n crossing points ρ_j in Fig. 6.12(c) and thus there are n bound-state eigenvalues E_j given by

$$E_j = -V_0 + \frac{\hbar^2}{2mL^2}\rho_j^2, \qquad j = 0, 1, 2, \ldots, n,$$

where each ρ_j is a solution of Eq. (6.172). In particular, for $n = 0$, there are no bound states! In other words, V_0 or L can be too small for the half-space square-well potential to support a bound state. Even when there *are* bound states, particular values of ρ_i can generally be obtained only numerically. An example of this is to be found in the exercises.

Let us assume that there is at least one bound state. For each E_i there is a wave function (now denoted) $\psi_i(x)$ given by the sum of the corresponding region-II and region-III portions; $\psi_i(x)$ contains the two as-yet-undetermined constants A_i and D ($= D_i$). Each may be chosen real; choosing A_i to be positive, D_i will then have the sign of $\sin(\gamma_i L)$, since (6.169a) yields

$$D_i = A_i e^{\alpha_i L} \sin(\gamma_i L). \tag{6.174}$$

The constant A_i is obtained from (6.174) and the normalization integral

$$\begin{aligned} 1 &= \int_0^\infty dx\, |\psi_i(x)|^2 \\ &= A_i^2 \int_0^L dx \sin^2(\gamma_i x) + D_i^2 e^{-2\alpha_i L}/(2\alpha_i), \end{aligned}$$

the second line of which is the form (6.151) takes in the present instance. Evaluating the integral and using (6.174), we find as one of several possible forms the result

$$A_i = \left[\left(\frac{L}{2} + \frac{1}{4\alpha_i} \right) - \frac{1}{4} \left(\frac{\sin(2\gamma_i L)}{\gamma_i} + \frac{\cos(2\gamma_i L)}{\alpha_i} \right) \right]^{-1/2}.$$

Insertion of this result into (6.174) yields D_i.

In region II, $\psi_i(x)$ is a sine function whose detailed shape depends on the value of γ_i. Whatever its shape, it should be clear from the form of (6.168) that the probability of finding the particle well beyond the point $x = L$ is very small. Thus, while the half-space square well does not wholly confine the particle to the region $x \leqslant L$, the particle cannot be too far from L, just as in the case of the harmonic oscillator. Since $\alpha_{i+1} > \alpha_i$, it is also evident that, the more highly excited the state, the further to the right of the edge of the well can one expect to find the particle. This behavior for an arbitrary ψ_i is illustrated in Fig. 6.13.

The foregoing provides all the information needed to determine both the eigenvalues and the eigenfunctions once m, V_0, and L are given. It is thus a complete theoretical solution to the bound-state portion of the half-space square-well problem. We close with a final comment, viz., if the left-hand edge of the well were at $x = x_1$ rather than at $x = 0$, the above analysis would go through in the same fashion, the only changes being those that result from replacing $\sin(\gamma_i x)$ by $\sin[\gamma(x - x_1)]$ as the region-II solution.

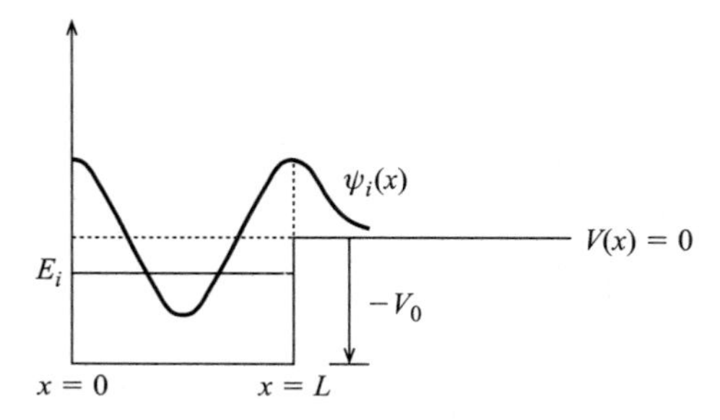

Fig. 6.13 A schematic illustration of a possible eigenvalue (E_i) in the half-space square well and the corresponding wave function $\psi_i(x)$, showing the smooth join of $\psi_{\rm II}(y)$ to $\psi_{\rm III}(x)$ at $x = L$. The zero values of $V(x)$ and $\psi(x)$ are chosen to coincide.

6.5. The square well

As the next eigenvalue problem, we briefly consider the square well, which is related to the previous potential by eliminating the left-hand infinite wall and placing the left and right edges of the well at L_1 and L_2, respectively. That is, the 1-D square-well potential is defined by

$$V(x) = \begin{cases} 0, & x \leqslant L_1 \quad \text{(Region I)}, \\ -V_0, & L_1 \leqslant x \leqslant L_2 \quad \text{(Region II)}, \\ 0, & x \geqslant L_2 \quad \text{(Region III)}. \end{cases} \tag{6.175}$$

It is pictured in Fig. 6.14. Although the square-well eigenvalue problem is a bit tedious to solve, it is employed as an example here because it contains some new features relative to the quantum box and the half-space square well.

The wave functions in regions I and III are already known (see Eqs. (6.148) and (6.150)), so we need consider only region II. Using the notation of Eq. (6.164), the wave function $\psi_\gamma(x)$ in region II obeys (6.165), which is now valid in the interval $[L_1, L_2]$. However, the solution to (6.165) that applies in the present case is the analog of (3.31), viz.,

$$\psi_\gamma(x) = B_\gamma \sin(\gamma x) + C_\gamma \cos(\gamma x), \qquad L_1 \leqslant x \leqslant L_2. \tag{6.176}$$

One (slightly) new feature here is the presence of the cosine term, which must occur because the boundary condition $\psi_\gamma(x = 0) = 0$ is inapplicable. The only conditions to be imposed on the wave function are that its three portions and their first derivatives match smoothly at L_1 and L_2.

The latter conditions are expressed by Eqs. (6.152), which become in the present case

$$Ae^{\alpha L_1} = B_\gamma \sin(\gamma L_1) + C_\gamma \cos(\gamma L_1) \tag{6.177a}$$

$$\alpha Ae^{\alpha L_1} = \gamma[B_\gamma \cos(\gamma L_1) - C_\gamma \sin(\gamma L_1)] \tag{6.177b}$$

$$De^{-\alpha L_2} = B_\gamma \sin(\gamma L_2) + C_\gamma \cos(\gamma L_2) \tag{6.177c}$$

$$-\alpha De^{-\alpha L_2} = \gamma[B_\gamma \cos(\gamma L_2) - C_\gamma \sin(\gamma L_2)]. \tag{6.177d}$$

This is a group of four homogeneous equations for the four unknowns A, B, C, and D. Because they are homogeneous, the set (6.177) has a non-trivial solution only if its secular determinant vanishes. Imposing this requirement means that

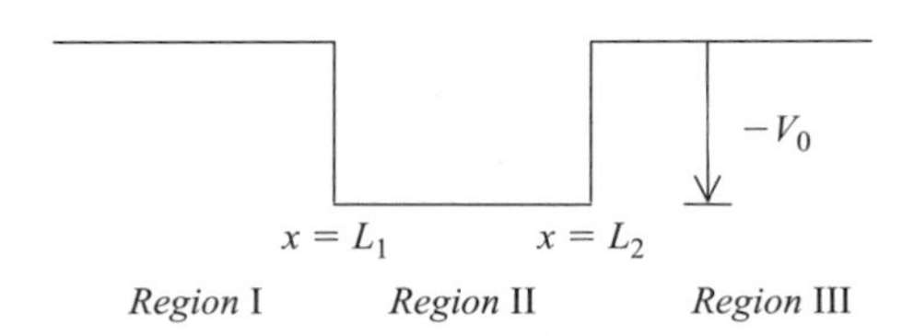

Fig. 6.14 The 1-D square-well potential of Eq. (7.185).

$$\begin{vmatrix} e^{\alpha L_1} & -\sin(\gamma L_1) & -\cos(\gamma L_1) & 0 \\ \alpha e^{\alpha L_1} & -\gamma\cos(\gamma L_1) & \gamma\sin(\gamma L_1) & 0 \\ 0 & \sin(\gamma L_2) & \cos(\gamma L_2) & -e^{-\alpha L_2} \\ 0 & \gamma\cos(\gamma L_2) & -\gamma\sin(\gamma L_2) & \alpha e^{-\alpha L_2} \end{vmatrix} = 0. \qquad (6.178)$$

4×4 determinants are not too difficult to evaluate, and this one is made relatively simple by the presence of the zeros. By expanding the determinant by means of the elements in column one, collecting terms, and then employing the trigonometric identities

$$\cos(\theta_2 - \theta_1) = \cos\theta_2 \cos\theta_1 + \sin\theta_2 \sin\theta_1$$

and

$$\sin(\theta_2 - \theta_1) = \sin\theta_2 \cos\theta_1 - \sin\theta_1 \cos\theta_2,$$

Eq. (6.178) can be put into the form

$$(\gamma^2 - \alpha^2)\sin[\gamma(L_2 - L_1)] - 2\gamma\alpha\cos[\gamma(L_2 - L_1)] = 0, \qquad (6.179)$$

which is an analog of the transcendental relation (6.170) whose solution yields the eigenenergy wave numbers for the half-space square well.

To determine the allowed values of γ (and thus α), one must solve (6.179). As with (6.170), a general understanding of these quantities can be obtained using graphical methods, a procedure whose implementation is the substance of one of the exercises. However, there is another aspect to (6.179) – it is also a new feature — which we comment on here and whose theoretical significance is examined in Chapter 9.

Let us define the distance between the two edges of the square well as L:

$$L = L_2 - L_1,$$

so that (6.179) reads

$$(\gamma^2 - \alpha^2)\sin(\gamma L) - 2\gamma\alpha\cos(\gamma L) = 0. \qquad (6.180)$$

We now claim that Eq. (6.180) is the transcendental eigenvalue equation for a square well of depth V_0 whose left- and right-hand edges are at $x = 0$ and $x = L$, a claim whose verification is also left to the exercises.

That the eigenvalue problem of a square well located between L_1 and L_2 is equivalent to that of a square well located between 0 and $L_2 - L_1 = L$ should be no surprise: since the domain is the infinite interval $[-\infty, \infty]$, it cannot make a difference where in that interval one chooses the origin. Furthermore, as can be seen from solving Exercise 6.27, one has an easier eigenvalue problem to solve when one edge of the well is at $x = 0$, so it is certainly helpful to locate the square well between $x = 0$ and $x = L$. That one can shift the well to a more desirable location (desirable at least after the fact) is an example of what is known as *translational invariance*. One can introduce a *translation operator* $\hat{U}_T$, whose action on a wave function (or on an eigenket of the position operator) is to *translate* the coordinate by a given amount, as discussed in Chapter 9. If the Hamiltonian is invariant under this operation, its eigenvalues are unchanged by the act of translation, as in the case of the square well. The formalism for, and the meaning of, this and similar *symmetry operators* is an especially important aspect of quantum theory and we devote a section of Chapter 9 to this topic.

Determination of the eigenvalues is, of course, only part of the problem: the wave

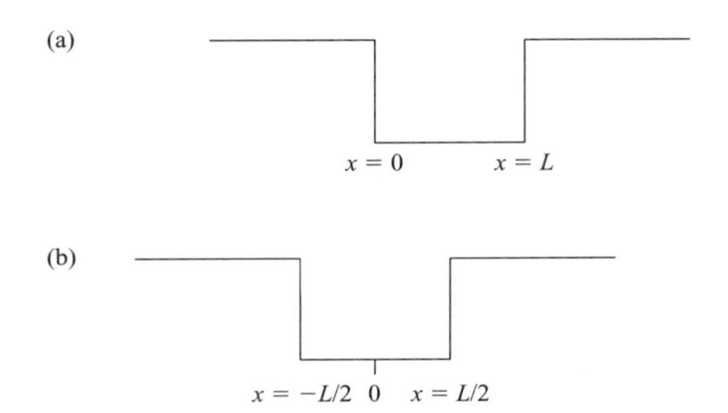

Fig. 6.15 Translation of the origin of coordinates from the left-hand edge of the square well (a) to its center (b).

functions remain to be obtained. This is algebraically more complex than deriving the eigenvalue relations and we leave it also to the exercises for the case in which the left-hand edge of the well is at $x = 0$.

Given the translational invariance of the square well, it might be thought that the origin of coordinates could be put anywhere without changing the set of eigensolutions and/or eigenvalues. Although this is true for the eigenvalues, there is one point in space where the form of the eigensolutions changes dramatically. This is the point midway between the two well edges. For example, if we start with the well in the interval $[0,\ L]$, translating the origin to the midpoint means that the well is now in the interval $[-L/2,\ L/2]$, as in Fig. 6.15(b). The latter well occupies a unique position: it is the only square well which is *symmetrically placed*, i.e., for which $V(-x) = V(x)$. The operation of replacing x by $-x$ involves a reflection about $x = 0$; the operator which changes coordinates into their negatives is called the parity operator, denoted $\hat{U}_{\mathrm{P}}$. When a Hamiltonian is invariant under reflection (or w.r.t. parity), then its eigenfunctions are also eigenfunctions of $\hat{U}_{\mathrm{P}}$. Since the symmetrically placed square well is invariant under reflection, its eigenfunctions will also be eigenfunctions of $\hat{U}_{\mathrm{P}}$. The effect of this is to divide the symmetric square-well solutions into two distinct classes (as shown in Chapter 9): even solutions ($\cos(\gamma x)$) and odd solutions ($\sin(\gamma x)$), whereas for the asymmetric well the solution is a linear combination of the even and odd ones, as in Eq. (6.176). The role of symmetry is of sufficient importance that detailed investigations into its properties and effects are postponed to Chapter 9.

6.6. The delta-function potential

A mild exception to the postponement noted above concerns the "pathological" delta-function potential

$$V(x) = g\delta(x), \tag{6.181}$$

where g is a strength parameter having the dimension energy–length, the "length" part resulting from the delta function's effect of reducing the dimensionality (see Section 4.4).

Recalling Eq. (4.125), we conclude that a negative delta-function potential is the limiting case of a symmetric square well in which the depth becomes infinite while the width becomes zero. One reason for including (6.181) among our examples is that it leads to one of the few analytically soluble eigenvalue problems. Another reason, however, is that $\delta(x)$ is the 1-D version of the ultimate, short-ranged, 3-D potential. Suppose that two particles, whose positions are $\mathbf{r}_1$ and $\mathbf{r}_2$, interact via a very-short-range interaction $V(\mathbf{r}_{12})$, $\mathbf{r}_{12} = \mathbf{r}_1 - \mathbf{r}_2$. A reasonable approximation for such a short-ranged potential is $V(\mathbf{r}_{12}) \to V_0\delta(\mathbf{r}_{12})$, the effect of which, in an integral over $\mathbf{r}_1$ and $\mathbf{r}_2$, is to set $\mathbf{r}_2 = \mathbf{r}_1$. One instance of this is a highly simplified form of the interaction between two nucleons. Delta-function potentials are useful in establishing some general properties of certain nuclear states.

If m is the mass of the particle acted on by (6.181), the Schrödinger equation for the wave function $\psi(x)$ is

$$-\frac{\hbar^2}{2m}\psi''(x) + g\delta(x)\psi(x) = E\psi(x) \equiv -\frac{\hbar^2\alpha^2}{2m}\psi(x). \tag{6.182}$$

As in the preceding two examples, there are three regions in this problem, with region II being the point $x = 0$. Regions I and III are the infinite intervals $x < 0$ and $x > 0$, respectively, and the solutions to (6.182) in these two regions are given by Eqs. (6.148) and (6.150):

$$\psi(x) = \begin{cases} Ae^{\alpha x}, & x < 0, \\ De^{-\alpha x}, & x > 0. \end{cases} \tag{6.183}$$

To find $\psi(0)$, the wave function in region II, we recall that $\psi(x)$ must be both continuous and finite. Extrapolating $\psi_{\text{I}}(x) = Ae^{\alpha x}$ and $\psi_{\text{III}}(x) = De^{-\alpha x}$ each to $x = 0$ yields

$$\lim_{x\to 0}\psi_{\text{I}}(x) = A$$

and

$$\lim_{x\to 0}\psi_{\text{III}}(x) = D.$$

These two expressions will yield the same value and thus make $\psi(x)$ continuous at $x = 0$ only if $D = A$. Thus, setting $D = A$ and combining the two functional forms of (6.183), we have

$$\psi(x) = Ae^{-\alpha|x|}. \tag{6.184}$$

Choosing A to be real and positive, normalization yields $A = \sqrt{\alpha}$.

Apart from not yet knowing the value(s) of α, the wave function $\psi(x)$ is now completely determined. As is required on physical grounds, it is continuous. For a finite potential, we have demanded also that $\psi'(x)$ be continuous. In the present case we easily find that $\psi'(x)$ does not obey this condition: $\psi'(\mp 0) = \pm\alpha A$, depending on whether we approach $x = 0$ from negative $(+\alpha A)$ or positive values $(-\alpha A)$. Such behavior should be no surprise: since $\psi(0)$ is finite, Eq. (6.182) tells us that $\psi''(0)$ must be proportional to a delta function, which implies that $\psi'(x)$ must be discontinuous at $x = 0$. This fact, plus the knowledge that the delta function is actually a linear operator whose definition requires it to appear under an integral sign, suggests integrating the Schrödinger equation over an interval containing the point $x = 0$, say $[-a, b]$. The integral of ψ'' will yield $\psi'(b) - \psi'(-a)$, and, if we let b and a each approach zero, this difference will yield $-2\alpha A$, a quantity involving the unknown wave number α. The other integrals will allow us to relate α to g, as we now show.

On integrating both sides of (6.182) between $-a$ and b, we get

$$-\frac{\hbar^2}{2m}\int_{-a}^{b}\frac{d}{dx}\psi'(x)dx + g\psi(0) = E\int_{-a}^{b}\psi(x)\,dx. \tag{6.185}$$

On the LHS of (6.185), the integral equals $[-\hbar^2/(2m)][\psi'(b) - \psi'(-a)]$, which, from (6.184), is $[\hbar^2/(2m)]A\alpha(e^{-\alpha a} + e^{-\alpha b})$. Inserting this expression into (6.185) and taking the limits $a \to 0$ and $b \to 0$ gives

$$\frac{\hbar^2\alpha}{m}A + gA = 0,$$

or

$$\alpha = -mg/\hbar^2. \tag{6.186}$$

Since a bound state requires $\alpha > 0$, the eigenvalue equation (6.186) refers to a bound state only if $g < 0$. If we write

$$g = -\frac{\hbar^2}{mx_0}, \tag{6.187}$$

where x_0 is a length, then it follows that

$$\alpha = 1/x_0. \tag{6.188}$$

Hence, as long as g is negative and non-zero, no matter how small, the potential $g\delta(x)$ will support one (and only one) bound state. This result is an example of the theorem proved in Chapter 17, that, in 1-D, any wholly attractive potential, no matter how weak, will support a bound state.

The potential (6.181) is symmetric (invariant) under the transformation $x \to -x$. Examination of $\psi(x)$ shows that it, too, is symmetric under this transformation. It is thus an eigenfunction of the parity operator (with eigenvalue unity). One can also solve the asymmetric eigenvalue problem involving $\delta(x \pm a)$ analytically; an exercise is devoted to this.

In ending this chapter, we remind the reader that the five potentials examined in it were chosen because they led to tractable eigenvalue problems. The exercises contain additional soluble eigenvalue problems. There are, of course, plenty of potentials for which only non-graphical numerical solutions can be obtained. We have not included any such in the foregoing because their solution does not provide any special insight, and they are thus pedagogically much less useful.

Exercises

6.1 The edges of a 1-D quantal box are at $x = 0$ and $x = L$. The probabilities that a particle of mass m will be observed in the ground, in the first-excited, and in the second-excited state are, respectively, 0.6, 0.3, and 0.1.

(a) If each term contributing to the particle's wave function has phase factor unity at $t = 0$, what is the wave function for $t > 0$?

(b) What is the probability that the particle will be found at the positions $x = L/3$ and $x = L/2$ at $t = 0$, using the wave function of part (a)?

(c) What is the expected value of the energy that a measurement would yield at $t = mL^2/(\hbar\pi)$?

6.2 One means of determining the factor $\cos(\delta_3 - \delta_2)$ in Eq. (6.24) is through measurement of $N_{\rm RH}$, the number of particles in the right-hand half of the box, as noted in the text. An alternate means of obtaining the value of $\cos(\delta_3 - \delta_2)$ is by determining the average position of the particles in the box. (It is assumed that the relevant measurements are feasible.) Very different Ψ_Γ's could result from these two methods. This problem compares them for the case $N_2 = N_3 = N/2 = 5000$.

(a) Suppose first that measurement yields $N_{\rm RH} = 6000$ at $t = 0$.

(i) Determine $\cos(\delta_3 - \delta_2)$ and a value for $\delta_3 - \delta_2$.

(ii) Calculate the $t = 0$ value of $\langle\hat{Q}_x\rangle_\Gamma$.

(b) For the same system, we now suppose that, at $t = 0$, measurement yields $\langle\hat{Q}_x\rangle_\Gamma = 0.5L$.

(i) Determine $\cos(\delta_3 - \delta_2)$ and a value for $\delta_3 - \delta_2$.

(ii) Calculate the $t = 0$ value of $N_{\rm RH}$.

(c) On the same graph, sketch the two different $\rho_\Gamma(x, t)$ obtained from parts (a) and (b) at $t = 0$ and $t = \pi/\omega_{23}$.

6.3 Use Eq. (6.23) to show that $\langle\hat{P}_x(t)\rangle_\Gamma$ is zero for all t and that $\langle H\rangle_\Gamma$ exhibits neither an interference pattern nor a time dependence.

6.4 The classical analog of a particle confined to a quantal box of width L is a particle of mass m moving with constant speed v along a straight line perpendicular to two parallel walls separated by a distance L, in an environment with no friction and zero gravity. The aim of this problem is to compare analogous classical and quantal probabilities, showing that, as $n \to \infty$, they become equal, i.e., that Bohr's correspondence principle (B4, Chapter 1) is fulfilled.

(a) Show that $P_{\rm cl}(\Delta x)$, the probability that the classical particle will be found in the interval Δx centered at x_0, is

$$P_{\rm cl}(\Delta x) = \frac{\Delta x}{L}.$$

(Hint: $P_{\rm cl}(\Delta x)$ is the ratio of the time Δt spent in the interval Δx to T, the time it takes the particle to travel the distance L.)

(b) When the quantal particle is in state ψ_n the probability $P_n(\Delta x)$ that it will be found in the interval Δx centered at x_0 is obtained from $\rho_n(x) = |\psi_n(x)|^2$.

(i) On the same graph sketch $\rho_3(x)$ and $\rho_{15}(x)$.

(ii) Extrapolate from the preceding result to infer an average value of $\rho_n(x)$ as n becomes very large, and use the mean-value theorem of calculus to estimate $P_n(\Delta x)$.

(c) Calculate $P_n(\Delta x)$ and show that

$$\lim_{n\to\infty} P_n(\Delta x) \to P_{\rm cl}(\Delta x),$$

thus establishing the classical limit for this quantum system and validating Bohr's correspondence principle for it. Your estimate for part (b)(ii) should agree with this limit, one we shall refer back to in Chapter 8.

6.5 Prove that, if $|\Gamma, t\rangle$ is any 1-D bound state, then

$$\langle \Gamma, t|\hat{P}_x|P, t\rangle = 0.$$

6.6 Let $\hat{a}$ and $\hat{a}^\dagger$ be the annihilation and creation operators for the 1-D harmonic oscillator and let ν and ω be constants. Determine the eigenvalue spectrum of the Hamiltonian

$$\hat{H} = \hbar\omega\hat{a}^\dagger\hat{a} + \nu(\hat{a}^\dagger + \hat{a}).$$

6.7 Let $\hat{H}$ be the 1-D oscillator Hamiltonian, with

$$\hat{H}|n\rangle = \hbar\omega(n + \tfrac{1}{2})|n\rangle,$$

as in Section 6.2.

(a) Evaluate $e^{\beta\hat{a}}|\alpha\rangle$, i.e., determine the coefficients C_n in

$$\rho^{\beta\hat{a}}|\alpha\rangle = \sum_n C_n|n\rangle,$$

where β is constant, possibly complex.

(b) Evaluate the matrix element $\langle n|e^{\beta\hat{a}^\dagger}|\alpha\rangle$.

(c) Define $A(\beta) = \exp(\beta x)$, where x is the 1-D coordinate, and evaluate

$$\sum_{n=0}^{\infty}\langle n|A(ik)|\alpha\rangle^2, \qquad k = \text{real}.$$

6.8 In contrast to the creation and annihilation operators of the 1-D oscillator, the "fermion" creation and annihilation operators, $\hat{b}^\dagger$ and $\hat{b}$, respectively, obey the "anti-commutation" relations

$$\hat{b}^\dagger\hat{b} + \hat{b}\hat{b}^\dagger \equiv \{\hat{b}^\dagger, \hat{b}\} = \hat{I}$$

and

$$\{\hat{b}, \hat{b}\} = \{\hat{b}^\dagger, \hat{b}^\dagger\} = 0,$$

where $\hat{I}$ is the unit operator. Consider a fermion system governed by the Hamiltonian

$$\hat{H} = E\hat{b}^\dagger\hat{b} \equiv E\hat{N},$$

where E is an energy unit. Since $\hat{H}$ and $\hat{N}$ share common eigenstates, we need only consider $\hat{N}$.

(a) Show that $\hat{N}^2 = \hat{N}$.

(b) Determine the allowed values of n in $\hat{N}|n\rangle = n|n\rangle$.

(c) The $\{|n\rangle\}$ of part (b) are the energy eigenstates. Determine the matrix element $N_{mn} = \langle m|\hat{N}|n\rangle$, and write out the full matrix N.

(d) From the results of part (c), determine the column vector form for each $|n\rangle$ in the eigenstate representation; assume that real and positive normalization constants apply.

(e) Next, determine the eigenstate representations of $\hat{b}^\dagger$ and $\hat{b}$.

(f) Finally, using the relevant portions of the preceding results, determine $\hat{b}^\dagger|n\rangle$ and $\hat{b}|n\rangle$ in the eigenstate representation.

6.9 The relation (6.33) follows in part from (6.32) on assuming that the origin of coordinates occurs at x_0, i.e., that $x_0 = 0$. In this exercise, it is assumed that $x_0 \neq 0$.

(a) Determine the form of the eigenfunctions and the energy spectrum for the 1-D oscillator when $x_0 \neq 0$.

(b) Calculate the probability that the $x_0 = 0$ eigenfunction $\psi_0(x)$ will be present in the ground state of part (a).

(c) Ditto for the case in which $\psi_0(x)$ is replaced by $\psi_1(x)$. (Note: your results for parts (b) and (c) should depend on x_0.)

6.10 Let the oscillator potential of Eq. (6.33) be replaced by

$$V(x) = \begin{cases} \frac{1}{2}m\omega_0^2 x^2, & x > 0, \\ \infty, & x \leqslant 0. \end{cases}$$

Determine the eigenfunctions, the energy spectrum, and the spacing of adjacent energy levels for this "half-space" oscillator.

6.11 The generating function for the Hermite polynomials $H_n(\eta)$ is given by Eq. (6.121).

(a) Show that Eq. (6.122) follows from (6.121).

(b) Use Eq. (6.122) to derive the "recursion relations" obeyed by the $H_n(\eta)$, viz.,

$$H_{n+1}(\eta) - 2\eta H_n(\eta) + 2nH_{n-1}(\eta) = 0$$

and

$$H_n'(\eta) - 2nH_{n-1}(\eta) = 0,$$

where $' \equiv d/d\eta$.

(c) Prove that, with proper choice of phase, Eq. (6.125) is valid. (Hint: use the first recursion relation of part (a), first for $n \to n-1$ and then as it stands; multiply each of these relations by the appropriate Hermite polynomial; next subtract one of the resulting pair of equations from the other, multiply the result by $\exp(-\lambda_0 x^2)$, and integrate it over the range $x \in [-\infty, \infty]$. Use of the orthogonality relation plus a straightforward analysis leads to Eq. (6.125).)

6.12 Use the momentum representation to convert (6.35) into a differential equation for $\varphi_\Gamma(k)$ involving d^2/dk^2. Compare this equation with (6.108) and from it derive the functional form of the solutions $\varphi_n(k)$, where Γ has been replaced by n, the oscillator quantum number.

6.13 Defining $\langle \hat{A} \rangle_n \equiv \langle n|\hat{A}|n\rangle$, where $|n\rangle$ is the nth oscillator state, show that Eq. (6.135) generalizes to $\langle \hat{K} \rangle_n = \langle \hat{V} \rangle_n$, where $\hat{V} = m\omega_0^2 \hat{Q}_x^2/2$. (Hint: employ the results of Exercise 6.12.)

6.14 The set of oscillator eigenfunctions $\{\psi_n(x)\}$ is a basis and can be used to expand a function $f(x)$, $x \in [-\infty, \infty]$.

(a) Determine the functional form of the coefficients C_n in the expansion

$$f(x) = \sum_{n=0}^{\infty} C_n A_n H_n(x/\lambda_0) e^{-x^2/(2\lambda_0^2)},$$

with A_n given by (6.125).

(b) Now let $f(x) \to \delta(x - x_0)$, in which case the preceding $C_n \to C_n(x_0)$. Evaluate the functional form of these $C_n(x_0)$.

(c) Set $x_0 = 0$ and $\lambda_0 = 1$ (in appropriate units), and then plot the resulting approximations to $\delta(x)$ obtained by keeping the $n = 0$, then the $n = 0$ and $n = 1$, and finally the $n = 0$, 1, and 2 terms in the oscillator expansion of the delta function.

6.15 The physical picture underlying the concept of ionic bonding in a diatomic molecule AB is that an electron from the neutral atom A is transferred to the neutral atom B, forming a pair of ions, i.e., A^+B^-. Thus, HCl may be thought of as H^+Cl^- and NaCl as Na^+Cl^-. As noted in the text, diatomics can undergo vibrational motion along the line forming them, and in this problem such vibrations will be approximated by 1-D linear harmonic oscillations. Let x be the relative separation between A^+ and B^-. The vibrational potential energy $V_{\mathrm{vib}}(x)$ will be modeled as a sum of a short-range repulsion γ/x^n and a long-range Coulomb attraction $-\kappa_e e^2/x$:

$$V_{\mathrm{vib}}(x) = \frac{\gamma}{x^n} - \frac{\kappa_e e^2}{x};$$

one goal of this exercise is to determine values of the strength γ and the integer n from experimental data when V_{vib} is approximated harmonically.

(a) Determine the value x_0 which minimizes V_{vib}, expressing your answer in terms of γ, n, κ_e, and e^2.

(b) Expanding V_{vib} as in Eqs. (6.31) and (6.32), and using for m the reduced mass of AB, viz., $m = m_A m_B/(m_A + m_B)$, derive an expression for ω_0 in terms of the parameters of the potential and m.

(c) Two pieces of information are needed in order to fix n and γ. In this exercise they are taken to be ν, the frequency of radiation emitted or absorbed in transitions between adjacent levels, and x_0, the equilibrium separation. Using the following information, determine n and γ for HCl and for NaCl: $\nu(\mathrm{HCl}) = 8.65 \times 10^{13}$ Hz, $x_0(\mathrm{HCl}) = 1.28$ Å; $\nu(\mathrm{NaCl}) = 1.14 \times 10^{13}$ Hz, $x_0(\mathrm{NaCl}) = 2.51$ Å.

6.16 At $t = 0$, the wave function for a particle is $\Psi_0(x, t = 0) = N_0 \exp(-x^2/\lambda_0^2)$. For $t > 0$, the wave function becomes $\Psi_1(x, t) = N_1 \exp[-(x - x_0)^2/(2\lambda_0^2)]$, where $N_0^2 = |N_1|^2$ and x_0 and λ_0 are real. Let $\hat{H}$ be the energy operator for the system. Is the preceding information consistent with

(a) $\hat{H}^\dagger = \hat{H}$;

(b) $[\hat{P}_x, \hat{H}] = 0$;

(c) $V(x) = -Ax$, where A is a real constant;

(d) $d\langle \hat{K}_x \rangle_t/dt = 0$ for $t = 0$ and $t > 0$?

6.17 Calculate the probability that a particle in the ground state of a 1-D harmonic oscillator will be found in the classically forbidden region, i.e., that it will be in the regions outside the classical turning points. (The numerical answer is to be obtained by evaluating the so-called "error function" or probability integral.)

6.18 Like that of Exercise 6.4, the goal of this problem is to show that the large-n limit of the 1-D harmonic oscillator probability goes over to the corresponding classical result. The potential energy is given by Eq. (6.33), λ_0 is defined by Eq. (6.37), and the energy unit is taken to be $\hbar\omega_0/2$, so that $E = \gamma\hbar\omega_0/2$, with $\gamma \cong 2n$ in the quantal case for n large.

(a) Let Δx be a small interval about the point x_0 on the classical oscillator's path.

(i) Show that the probability $P_{\mathrm{cl}}(x_0, \Delta x)$ that the classical particle will be found in the interval Δx centered on x_0 is

$$P_{\mathrm{cl}}(x_0, \Delta x) = \frac{2\,\Delta x}{v(x_0)T},$$

where T is the period of the motion and $v(x_0)$ is the particle's speed at the point x_0 (compare this with part (a) of Exercise 6.4).

(ii) Show that conservation of energy leads to

$$P_{\text{cl}}(x_0, \Delta x) = \frac{\Delta x}{\lambda_0 \pi} \sqrt{\frac{1}{\gamma - x_0^2/\lambda_0^2}} \equiv \rho_{\text{cl}}(x_0)\,\Delta x,$$

where $\rho_{\text{cl}}(x_0)$ is the classical probability density.

(b) The quantal probability density, $\rho_n(x_0) = |\Psi_n(x_0)|^2$, will be compared with $\rho_{\text{cl}}(x_0)$ in the large-n limit using a large-n approximation to $H_n(x_0/\lambda_0)$.

(i) The relevant large-n approximation to $H_n(y)$ is (Abramowitz and Stegun (1965))

$$H_n(y) \cong \frac{2^{n+1}(n/2e)^{n/2}}{\sqrt{2\cos\phi}} e^{n\phi^2} \cos\left[\left(2n + \frac{1}{2}\right)\phi - \frac{n\pi}{2}\right],$$

where $\phi = \sin^{-1}(y/\sqrt{2n})$. Use this result plus the large-n approximation $n! \cong n^n e^{-n}$ (Sterling's formula) to show that, for large n, the 1-D oscillator probability density becomes

$$\rho_n(x_0) \cong \frac{2}{\lambda_0 \pi} \sqrt{\frac{1}{2n - x_0^2/\lambda_0^2}} \cos^2\left(\frac{\sqrt{2n}x_0}{\lambda_0} - \frac{n\pi}{2}\right).$$

(ii) On the same graph, compare $\rho_{\text{cl}}(x_0)$ with $\rho_n(x_0)$ for the case $n = 20$, paying special attention to the behavior of $\rho_{\text{cl}}(x_0)$ at the values $x_0 = \pm A$, where A is the amplitude of oscillation. Except in the vicinity of the points $\pm A$, your results should show that $\rho_{\text{cl}}(x_0)$ behaves like the average of $\rho_{20}(x_0)$. (Recall that $\gamma = 2n$.)

(iii) The quantal probability $P_n(x_0, \Delta x)$ is

$$P_n(x_0, \Delta x) = \int_{x_0 - \Delta x/2}^{x_0 + \Delta x/2} \rho_n(x)\,dx.$$

Show that, in the limit $n \to \infty$, $P_n(x_0, \Delta x) \to P_{\text{cl}}(x_0, \Delta x)$, as required by Bohr's correspondence principle of Chapter 1.

6.19 Let $\{\psi_n(x)\}$ be the normalized eigenfunctions of the 1-D oscillator.

(a) If $f(x)$ is a continuous function of x, evaluate $[\hat{a}(x), f(x)]$ and $[\hat{a}^\dagger(x), f(x)]$.

(b) Choose $f(\hat{Q}_x) = \exp(ik\hat{Q}_x)$ and show that

(i) $\langle n|f(\hat{Q}_x)|0\rangle \propto \langle n-1|f(\hat{Q}_x)|0\rangle$,

(ii) $\langle n|f(\hat{Q}_x)|0\rangle \propto \langle 0|f(\hat{Q}_x)|0\rangle$;

(iii) calculate the proportionality constants in each case and evaluate $\langle 0|f(\hat{Q}_x)|0\rangle$.

6.20 Let $\langle x|\alpha\rangle = [\exp(ikx)]\psi_0(x) = \Psi(x, t = 0)$ be the state of a harmonic oscillator at time $t = 0$, where $\psi_0(x)$ is the ground-state wave function.

(a) Is $|\alpha\rangle$ normalized? If not, normalize it.

(b) Evaluate $\langle \alpha|\hat{Q}_x|\alpha\rangle$ and $\langle \alpha|\hat{P}_x|\alpha\rangle$.

(c) Determine the probability that, at time $t = 0$, the energy will be measured to be $E_n = (n + \frac{1}{2})\hbar\omega_0$.

(d) Determine both the amplitude and the probability that an energy measurement at time $t > 0$ will yield $E_n = (n + \frac{1}{2})\hbar\omega_0$.

6.21 At time $t = 0$ the state of a harmonic oscillator is the coherent state $|\zeta\rangle$. Show that $|\zeta, t\rangle$, the state at time $t > 0$, has the structure

$$|\zeta, t\rangle = f(t)|\zeta f^2(t)\rangle$$

and determine the factor $f(t)$.

6.22 Let $\langle\hat{A}\rangle_\zeta = \langle\zeta|\hat{A}|\zeta\rangle$, where $|\zeta\rangle$ is a coherent state.
(a) Evaluate $\langle\hat{P}_x\rangle_\zeta$ and $\langle\hat{Q}_x\rangle_\zeta$.
(b) Express $\langle\hat{N}^2\rangle_\zeta$ in terms of $\langle\hat{N}\rangle_\zeta^2$, and $\langle\hat{N}^3\rangle_\zeta$ in terms of $\langle\hat{N}\rangle_\zeta^3$.

6.23 Assume that the 1-D Hamiltonian $\hat{H}(x)$ has a bound state $\psi_n(x)$ with n nodes, where $n \geqslant 1$ (excluding the nodes at $x = \pm\infty$). Prove that this system must support at least one more bound state $\psi_{n'}(x)$, having n' nodes in $x \in (-\infty, \infty)$, where $n' < n$. Use this result to prove that the ground state $\psi_0(x)$ of $\hat{H}(x)$ must be nodeless for $x \in (-\infty, \infty)$.

6.24 Suppose that the potential energy $V(x)$ of a 1-D Hamiltonian has no infinite discontinuities. From this assumption, show that the bound states (if any exist) are not degenerate, i.e., that, if $\tilde{\psi}_E(x)$ and $\psi_E(x)$ are assumed to be different eigenstates of $\hat{H}(x)$ corresponding to the same energy E, then $\tilde{\psi}_E(x) \propto \psi_E(x)$. (Hint: show that the Wronskian $W(\psi_E, \tilde{\psi}'_E) \equiv \psi_E(x)\tilde{\psi}'_E(x) - \psi'_E(x)\tilde{\psi}_E(x) = 0$, where $' = d/dx$.)

6.25 Consider a 1-D half-space square well located to the left of the origin:

$$V(x) = \begin{cases} \infty, & x \geqslant 0, \\ -V_0, & 0 \geqslant x \geqslant -L, \\ 0, & x \geqslant -L. \end{cases}$$

Determine the eigenvalue equation and the forms of the solution in the preceding three regions, stating in particular the analog of the equation for A_i of Section 6.4. Compare your results with those of Section 6.4

6.26 The "holely" box. Let

$$V(x) = \begin{cases} \infty, & x \leqslant 0, \\ -V_0, & 0 \leqslant x \leqslant L_1, \\ 0, & L_1 \leqslant x \leqslant L_2, \\ \infty, & x \geqslant L_2. \end{cases}$$

Derive a transcendental eigenvalue relation for the case $E > 0$. Show that your result yields the eigenvalues of the infinite square well when $V_0 \to 0$ and $L_2 \to L_1 = L$.

6.27 Carry through the analysis of Section 6.5 when the edges of the square well are at $L_1 = 0$ and $L_2 = L$, and use the results to verify the claim made below Eq. (6.180).

6.28 Determine the complete solution to the 1-D delta-function-potential problem when $V(x) = g\delta(x - a)$.

6.29 Solve the eigenvalue problem for the half-space delta-function potential

$$V(x) = \begin{cases} \infty, & x \leqslant 0, \\ -g\delta(x - a), & x > 0,\ g > 0. \end{cases}$$

6.30 Let a 1-D potential be an asymmetric sum of delta functions:

$$V(x) = g_1\delta(x) + g_2\delta(x - a).$$

(a) Derive a transcendental relation from which a possible bound-state wave number can be determined.

(b) Show that, if either g_1 or $g_2 = 0$, the results of part (a) reproduce the results of the text or Exercise 6.28.

(c) Set $g_2 = -g_1$, and discuss the existence of a bound state.

6.31 Let $V(x)$ be an attractive delta-function potential centered at the origin: $V(x) = -g\delta(x)$. The object of this exercise is to solve the eigenvalue problem in momentum space. Let $\varphi(k) = [1/(2\pi)]^{1/2} \int dx \exp(-ikx)\, \psi(x)$.

(a) Show that

$$\varphi(k) = \psi(x = 0) f(k^2, \alpha^2),$$

where $f(k^2, \alpha^2)$ is a function you are to determine and $\psi(x)$ is the coordinate-space wave function.

(b) Integrate both sides of the preceding equation on k, recall the expression for $\psi(x = 0)$ as a Fourier transform, and derive an equation in which $\psi(x = 0)$ appears both on the LHS and on the RHS.

(c) Assume that $\psi(x = 0) \neq 0$ in the preceding relation, carry out the relevant integration, and show that Eq. (6.186) is the result.

6.32 The separable potential $\hat{V}_{\text{sep}} = \lambda|g\rangle\langle g|$ was introduced in Section 5.4. In this exercise, the form factor $|g\rangle$ refers to the 1-D case: $\langle k|g\rangle = g(k)$, $\langle x|g\rangle = g(x)$. Let $|\alpha\rangle$ be the eigenstate and $E_\alpha = -\hbar^2\alpha^2/(2m)$ its energy, where m is the particle's mass.

(a) Project both sides of the time-independent Schrödinger equation for $|\alpha\rangle$ onto $\langle k|$ and solve the resulting equation for $\langle k|\alpha\rangle$, the momentum-space wave function. (Note: $\langle g|\alpha\rangle$ is a non-zero number.)

(b) From the result of part (a), determine the coordinate-space wave function $\langle x|\alpha\rangle = \psi_\alpha(x)$.

(c) Multiply both sides of the equation derived in part (a) by $\langle g|k\rangle$, integrate all members of the resulting equation on k, cancel out the factor $\langle g|\alpha\rangle$ and obtain an implicit relation (involving an integral) from which α may be obtained.

(d) From the result of part (c), show that

(i) consistency requires $\lambda = -|\lambda|$,

(ii) $\hat{V}_{\text{sep}}$ can support at most one bound state,

(iii) not all $|\lambda|$ yield a bound state.

(e) Obtain an explicit formula for α, if

(i) $g(k) = 1,\ |k| \leqslant k_0;\ g(k) = 0,\ |k| > k_0;$

(ii) $g(k) = \sqrt{k},\ |k| \leqslant k_0;\ g(k) = 0,\ |k| > k_0.$

7

Applications of the Postulates: Continuum States in One Dimension

Non-confining potentials can support two types of stationary states: normalizable, negative-energy bound states, discussed in the preceding chapter, and asymptotically free, positive-energy *continuum* states, which are the subject of the present chapter. Because continuum states are asymptotically free, particles in such states can be observed an arbitrary distance from the region in which the non-confining potential acts. There is a wide variety of phenomena that continuum states describe, including radio-active decay (e.g., α, β, and γ radiation), photon-induced ejection of particles (e.g., a photon plus a neutral atom leading to an electron plus an ionized atom), creation of particles such as pions and heavier mesons and baryons, and all collision phenomena. We examine the physics of 1-D collision/scattering phenomena in this chapter. Because stationary-state scattering is described by time-independent wave functions that are asymptotic to $\exp(ikx)$ and/or $\exp(-ikx)$, such wave functions are not normalizable and are thus inadmissible: like the eigenstates of $\hat{\mathbf{P}}$ or $\hat{P}_x$, they are "improper" states. Hilbert-space – or proper – scattering states are wave packets, formed from linear combinations of improper eigenstates. They are normalizable solutions of the time-dependent Schrödinger equation. We consider them in Section 7.1.

Wave-packet states provide the only true description of scattering and reactions. Nevertheless, collisions are typically described using improper stationary states, since such a formulation is, in almost all ways, simpler. How the physics is properly accounted for by plane-wave states (as these stationary states are usually called) is the subject of Section 7.2. The link between the wave-packet and plane-wave descriptions is the quantal equation of continuity, which relates probability density and probability current density in the same way as electric charge and current densities are related.

With the use of a plane-wave description justified in Section 7.2, the final two sections of the chapter analyze scattering by a variety of attractive (Section 7.3) and repulsive (Section 7.4) potentials, the latter concentrating especially on that peculiarly quantum phenomenon of tunneling through walls.

7.1. Wave-packet description of scattering

The physical system considered in this chapter is a single particle of mass m and positive energy acted on by a non-confining 1-D potential. The generic potential $V(x)$ is a generalization of the one described by Eq. (6.144), which is reformulated and re-illustrated here as Eq. (7.1) and Fig. 7.1:

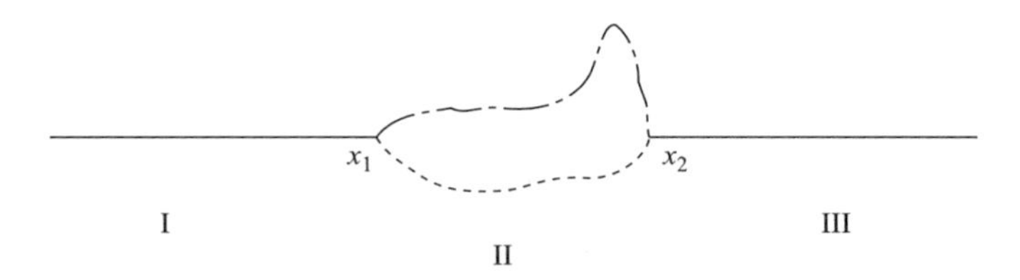

Fig. 7.1 The generic non-confining potential $V(x)$ of Eq. (7.1). The dotted portion in region II indicates an attractive case, while the dash–dot line represents a schematic repulsive potential.

$$V(x) = \begin{cases} 0, & x \leqslant x_1 \quad \text{(Region I)}, \\ \text{non-zero}, & x_1 < x < x_2 \quad \text{(Region II)}, \\ 0, & x \geqslant x_2 \quad \text{(Region III)}, \end{cases} \tag{7.1}$$

where the "non-zero" value of $V(x)$ in region II allows for both attractive and repulsive possibilities.

The (normalizable) wave packet describing the scattering event is a solution of the time-dependent Schrödinger equation

$$i\hbar \frac{\partial \Psi_\Gamma(x, t)}{\partial t} = \hat{H}(x)\Psi_\Gamma(x, t), \tag{7.2}$$

with $\hat{H}(x) = \hat{K}(x) + V(x)$, where $V(x)$ is of the type given by Eq. (7.1). The packet will be formed as a linear combination of plane-wave, stationary states. We consider them first.

Stationary-state solutions

A stationary-state wave function $\Psi(x, t)$ is given by

$$\Psi(x, t) = \psi(x)e^{-iEt/\hbar}, \tag{7.3}$$

where the subscript Γ has been suppressed. $\psi(x)$ obeys

$$\hat{H}(x)\psi(x) = E\psi(x), \tag{7.4}$$

and it is assumed that $E > 0$. Since $E > 0$, we write $E = \hbar^2 k^2/(2m)$, $k > 0$. In regions I and III, the domains of interest, (7.4) becomes

$$\psi''(x) + k^2\psi(x) = 0, \qquad x \leqslant x_1 \text{ or } x \geqslant x_2. \tag{7.5}$$

The linearly independent solutions to (7.5) are the pairs $\exp(\pm ikx)$ and $\sin(kx)$ plus $\cos(kx)$. We work with the former pair because they are eigenstates of the momentum operator $\hat{P}_x$. This is the logical choice, since, in regions I and III, $\hat{H}(x) = \hat{K}(x)$ and $[\hat{P}_x(x), \hat{K}(x)] = 0$, thereby allowing $\psi(x)$ to be simultaneous eigenfunctions of $\hat{H}(x)$ and $\hat{P}_x(x)$ in these regions.

The solutions to (7.5) are the momentum eigenfunctions $\psi_k(x)$, labeled by the wave number k:

$$\psi_{\pm k}(x) = A_{\pm k}e^{\pm ikx}, \tag{7.6}$$

where $A_{\pm k}$ are constants. The time-dependent wave function (7.3) becomes

$$\Psi_{\pm k}(x,\ t) = A_{\pm k} e^{\pm i(kx \mp \omega_k t)}, \qquad x \leqslant x_1,\ x \geqslant x_2, \tag{7.7}$$

with

$$\omega_k = E/\hbar = \hbar k^2/(2m). \tag{7.8}$$

The solutions (7.7) for regions I and III guarantee that, for $E > 0$, the stationary-state wave function is not normalizable, independent of the form taken by the solution $\psi_{\mathrm{II}}(x)$. For example, substituting (7.7) into I_{III}, the portion of the normalization integral arising from region III, we have

$$I_{\mathrm{III}} = \int_{x_2}^{\infty} |\Psi_{\pm k}(x,\ t)|^2\, dx = \infty. \tag{7.9}$$

The same result is obtained if $\Psi_{\pm k}$ is replaced by a linear combination of Ψ_k and Ψ_{-k}. Hence, $\Psi_{\pm k} \notin \mathcal{H}$ and thus cannot be states of a quantum system. Despite this, as has already been noted, the continuum eigenstates of $\hat{H}(x)$ are used *as if* they were proper eigenstates of the system. We adopt this pretense in Sections 7.3 and 7.4. To understand why this procedure is meaningful and how its results may be interpreted, we must first understand the nature of the proper Hilbert-space solutions in the unbound case, which means analyzing the wave-packet solution.

Wave packets in regions I and III are constructed as linear combinations of $\Psi_k(x,\ t)$ or of $\Psi_{-k}(x,\ t)$. Since these wave forms are the elements of the wave packet, we examine them in a little more detail. The phase $\Phi_k^{(\pm)}$ of $\Psi_{\pm k}(x,\ t)$ is

$$\Phi_k^{(\pm)} = kx \mp \omega_k t. \tag{7.10}$$

Paired points x and t at which the phase remains constant move with one of the two phase velocities $v_{\mathrm{ph}}^{(\pm)}$, given by

$$v_{\mathrm{ph}}^{(\pm)} = \frac{dx}{dt} = \pm v_k = \pm\frac{\omega_k}{k} = \pm\frac{\hbar k}{2m}, \tag{7.11}$$

a result that follows from $\Phi_k^{(\pm)} = \text{constant} \Rightarrow d\Phi_k^{(\pm)} = 0 = k\, dx \mp \omega_k\, dt$, as in Section 3.1.

Equation (7.11) implies that $\Psi_{+k}(x,\ t)$ is a plane wave propagating from left to right (L to R), while $\Psi_{-k}(x,\ t)$ is a plane wave propagating from right to left (R to L). Although the propagation characteristics are evident, the wave functions $\Psi_{\pm k}(x,\ t)$ cannot truly correspond to a moving particle because the probability of finding a particle described by such a wave is everywhere constant: the momentum (k) and energy ($E = \hbar^2 k^2/(2m)$) are both known, but neither of $\Psi_{\pm k}$ is localized – even though the domain of localization at any instant need not (and would not) correspond to the particle's size. In the language of the uncertainty relation, sharp momentum ($\hbar k$) means totally uncertain position ($\Delta k = 0 \Rightarrow \Delta x = \infty$), which is why a plane wave is spread over all of space. It is *localization*, i.e., $\Delta x =$ finite, that guarantees normalizability.

The localization requirement in classical physics is satisfied by forming linear combinations of waves with differing wave numbers and frequencies (i.e., $\Delta k \neq 0$). It is these linear combinations that are denoted wave packets, and they are solutions of the classical wave equation because that equation is linear. An example is analyzed in Appendix B. As has already been noted, a similar resolution of the localization/normalization problem is employed in quantum theory: one works with wave packets, as in Section 5.5. The asymptotic behavior will determine the form of the quantum wave packet. We show how this occurs by first examining the physics of a 3-D situation and then transferring the relevant aspects of the analysis to the 1-D case.

A physical description of scattering

In the 3-D arena of experimental physics, particles can achieve positive energies in a wide variety of ways. Of relevance to the 1-D cases we shall consider is the use of an accelerator that prepares particles to be projectiles in a collision experiment. The accelerator is assumed to produce a beam of particles of mass m at an early time $-T$ with average momentum $\hbar\mathbf{k}_0$ (directed towards the target) and average energy $E_0 = \hbar^2 k_0^2/(2m)$. Each particle is described by a wave packet whose central, or average, momentum is $\hbar\mathbf{k}_0$.

As an example, consider elastic scattering of a neutron by a nucleus, for which the interaction potential is short-ranged, i.e., is concentrated in a very small region of space. Its 1-D analog is the potential (7.1). After the collision, the neutron is described by a scattered wave packet whose average momentum is $\hbar\mathbf{k}'_0$. Since energy is conserved, we require $k'_0 = k_0 = (2mE_0/\hbar^2)^{1/2}$; however, the direction of the scattered neutron can range from forward to backward, as shown schematically via the momentum diagrams of Fig. 7.2.

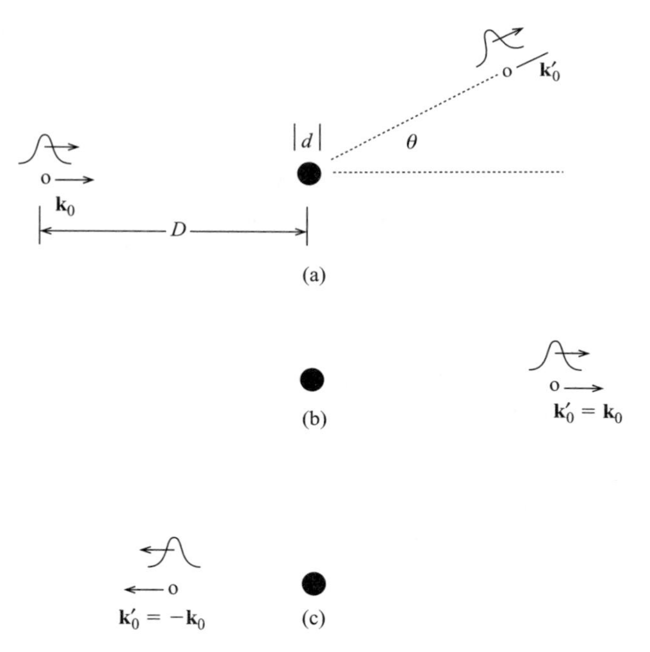

Fig. 7.2 A representation of neutron–nucleus elastic scattering via momentum vectors: from initial average momentum $\mathbf{k}_0$ to final average momentum $\mathbf{k}'_0$. (a) Scattering through an angle θ, with incident and scattered particles shown simultaneously. (b) Forward scattering, $\theta = 0$, $\mathbf{k}'_0 = \mathbf{k}_0$. (c) Backward scattering, $\theta = \pi$, $\mathbf{k}'_0 = -\mathbf{k}_0$. Above the small circle representing the neutron is, in each illustration, a schematic wave packet.

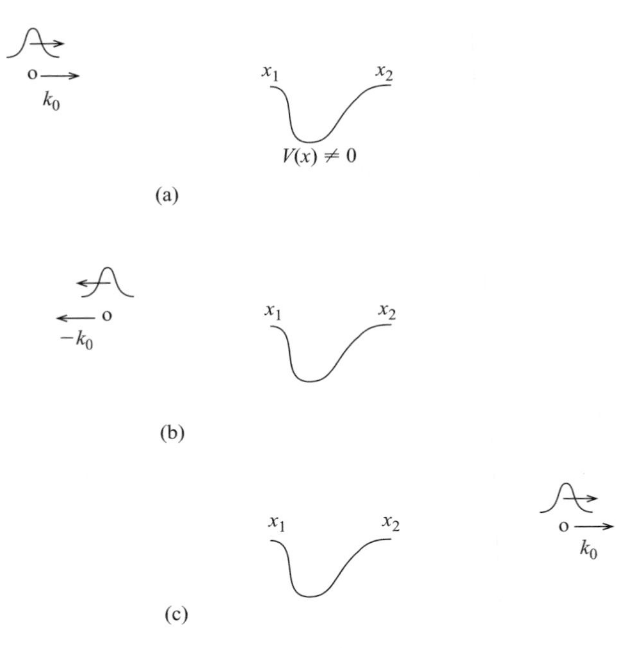

Fig. 7.3 1-D scattering in the presence of an attractive potential that is non-zero only for $x \in [x_1, x_2]$: (a) the incident particle, moving from L to R; (b) "reflection," corresponding to scattering through $\theta = \pi$ in the 3-D case; and (c) "transmission," corresponding to forward scattering ($\theta = 0$) in the 3-D case. All symbols are the 1-D counterparts of the 3-D symbols of Fig. 7.2.

Figure 7.2 has the neutrons emerging from an accelerator located well to the left of the target, moving from left to right towards the nuclei with average momentum $\mathbf{k}_0$, and elastically scattering through an angle θ, $0 \leqslant \theta \leqslant \pi$. If the number of incident neutrons crossing a unit area normal to $\mathbf{k}_0$ is N, then the number $N(\theta)$ observed at the angle θ is related to the probability that a neutron will be scattered through that angle.

The foregoing description easily carries over to the 1-D situation. The accelerator (source of particles) is to the left of the potential well, i.e., in region I, at a position $-x_a \ll x_1$. The incident particle moves from L to R and, after experiencing the potential, can be scattered either forward ($\theta = 0$) or backward ($\theta = \pi$). Forward scattering means the particle moves from L to R in region III, while backward scattering means the particle is moving from R to L in region I. The former process is denoted *transmission*, the latter, *reflection*. This description is illustrated in Fig. 7.3, the 1-D analog of Fig. 7.2.

Some general properties of wave packets

To describe the behavior of the particle in regions I and III, the plane waves $\exp[\pm i(kx \mp \omega_k t)]$ must be replaced by wave packets. Although we shall eventually construct the incident, reflected, and transmitted packets corresponding to the physics just outlined, we first examine some general aspects of a wave packet obeying the free-particle (zero-potential) Schrödinger equation in all space. We choose the case of a particle moving from L to R in the region $[-\infty, \infty]$.

Some illustrations of wave packets at time $t = 0$ are shown in Fig. 7.4. Each happens to be symmetric, is localized, has a well-defined maximum, and can be represented, for all t, as a linear combination of plane waves moving from L to R. This generic form for the packet $\Psi(x, t)$ is[1]

$$\Psi(x, t) = \sqrt{\frac{1}{2\pi}} \int_{-\infty}^{\infty} dk\, A(k) e^{i(kx - \omega_k t)}, \tag{7.12}$$

where, as before, $\omega_k = \hbar k^2/(2m)$ and the combination $kx - \omega_k t$ ensures the L to R nature of the motion.

The profile function $A(k)$ determines the coordinate-space form of the packet. If, as in

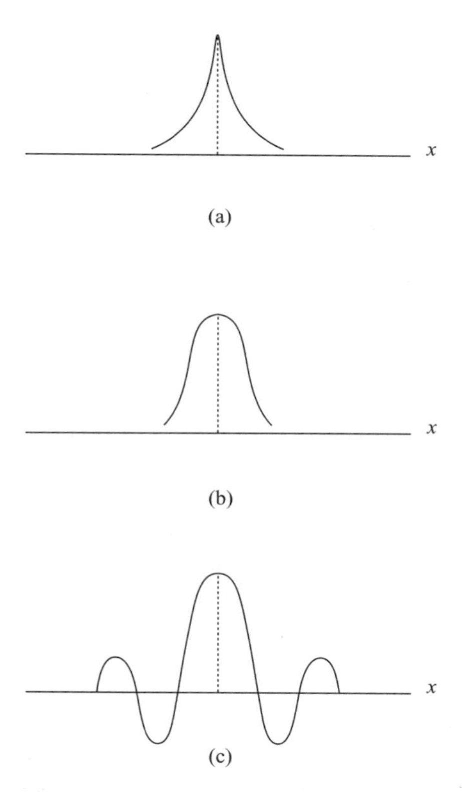

Fig. 7.4 Three examples of symmetric wave packets at time $t = 0$: (a) $\exp(-\alpha|x|)$, (b) $\exp(-\alpha^2 x^2)$, and (c) $\sin(\alpha x)/(\alpha x)$.

[1] The integration in (7.12) is from $-\infty$ to $+\infty$ because the integral represents a complete-set expansion. In terms of kets, and at $t = 0$, (7.12) is equivalent to $|\Gamma, t = 0\rangle = |A\rangle = \int_{-\infty}^{\infty} dk\, \langle k|A\rangle |k\rangle$, such that $\langle x|\Gamma, t = 0\rangle \equiv \Psi(x, t = 0) = \sqrt{1/(2\pi)} \int_{-\infty}^{\infty} dk\, A(k) \exp(ikx)$. The left-moving wave packet introduced below is actually a dual or bra vector, developed in the same way, viz., $\langle \Gamma, t = 0| = \langle \tilde{A}| = \int_{-\infty}^{\infty} dk\, \langle \tilde{A}|k\rangle \langle k|$ such that $\langle \Gamma, t = 0|x\rangle \equiv \tilde{\Psi}(x, t = 0) = \sqrt{1/(2\pi)} \int_{-\infty}^{\infty} dk\, \tilde{A}^*(k) \exp(-ikx)$; choosing $\tilde{A}(x) = A^*(x)$ makes $\tilde{\Psi}(x, t)$ a left-moving wave packet with the same profile function as that of $\Psi(x, t)$.

the examples of Fig. 7.4, $\Psi(x, t)$ is spatially symmetric, viz., $\Psi(x, t) = \Psi(-x, t)$, than it is straightforward to show that $A(k)$ is also symmetric. Furthermore, $A(k) = \sqrt{1/(2\pi)} \int dx\, \Psi(x, 0) \exp(-ikx)$, and, since $\Psi(x, t)$ is normalized for all t, it follows that

$$\int_{-\infty}^{\infty} dk\, |A(k)|^2 = 1. \tag{7.13}$$

Proof of these two results is left to the problems.

Apart from satisfying (7.13), the only requirements placed on $A(k)$ are that it be concentrated in the vicinity of and be symmetric about its average or central value k_0, e.g.,

$$A(k) \cong 0, \qquad k < k_0 - \delta k \text{ or } k > k_0 + \delta k. \tag{7.14}$$

The behavior (7.14) allows us to derive two approximate results, one of which is

$$\Psi(x, t) \cong e^{i\omega_{k_0} t} \Psi(x - v_0 t, 0), \qquad t \ll \frac{2m}{\hbar (\delta k)^2}, \tag{7.15}$$

where the semi-classical speed v_0 is given by

$$v_0 = \hbar k_0 / m. \tag{7.16}$$

Thus, apart from a trivial phase factor, if $t \ll 2m/[(\delta k)^2 \hbar]$, then the wave packet propagates so that, at time t, its maximum is at position $v_0 t$ (we assume that we have a symmetric packet whose maximum is at $x = 0$ when $t = 0$): the center of the packet moves classically; only its phase changes as t becomes greater than zero.

Equation (7.15) is obtained by noting that $\omega_k t = \hbar k^2 t/(2m)$ can be re-expressed as

$$\begin{aligned} \omega_k t &= \hbar (k - k_0)^2 t/(2m) - \hbar k_0^2 t/(2m) + \hbar k k_0 t/m \\ &\cong -\omega_{k_0} t + k v_0 t, \qquad t \ll 2m/(\hbar\, \delta k^2), \end{aligned} \tag{7.17}$$

where the validity of the second line in (7.17) relies on (7.14). Substitution of (7.17) into (7.12) yields (7.15). We observe in passing that, from the expression $[d\omega/dk]_{k_0}$, v_0 is the group velocity of the packet (7.12).

In order for the initial ($t = 0$) probability distribution to describe the projectile as a particle traveling the macroscopic distance D from accelerator to target with the semi-classical speed v_0, i.e., for the approximation (7.15) to be valid, the time $t_{\delta k} = 2m/(\hbar \delta k^2)$ should not be larger than the semi-classical traveling time $t_D = D/v_0$. To get a feeling for the effect on the wave packet of this conclusion, we set $t_D = t_{\delta k}$, which yields the following estimate for δk: $\delta k = \sqrt{2k_0/D}$. This implies that $\delta k \ll k_0$. and thus that the spatial extent of the packet, $\delta x \sim 1/\delta k$, will be much larger than the typical unit of length for the system. Two examples suffice to demonstrate this. We consider an electron of energy 1 keV and a proton of energy 10 MeV; in each case, $(v_0/c)^2 \ll 1$, thus justifying the preceding use of non-relativistic kinematics. For the electron, $k_0 = 16$ Å^{-1}; for the proton, $k_0 = 0.7$ fm^{-1}, where 1 fm $= 10^{-15}$ m is the nuclear unit of length (1 fm is a "fermi"). Choosing $D = 3$ m, we find $\delta k \cong 10^{-5}$ Å$^{-1} \cong 2 \times 10^{-6} k_0$ for the electron and $\delta k = 0.7 \times 10^{-8}$ fm$^{-1} \cong 10^{-8} k_0$ for the proton, corresponding to $\delta x \cong 3 \times 10^4$ Å and $\delta x \cong 1.5 \times 10^8$ fm for the electron and proton, respectively. The wave packets are evidently sharply peaked in momentum space and much larger in coordinate space than typical atomic or nuclear dimensions. They thus approximate plane-wave behavior for $t \leqslant t_{\delta k} = t_D$, but remain normalizable.

The second approximate result, similar to (7.15), can be derived by replacing k in the defining integral (7.12) by $k - k_0$. This leads to

$$\Psi(x,\, t) \cong \frac{1}{\sqrt{2\pi}} e^{i(k_0 x - \omega_{k_0} t)} \int_{-\infty}^{\infty} dq\, A(q + k_0) e^{iq(x - v_0 t)}, \qquad t \ll \frac{2m}{\hbar(\delta k)^2}, \tag{7.18}$$

which shows that, for $t \ll 2m/[(\delta k)^2 \hbar]$, the packet is approximately equal to a plane wave of momentum $\hbar k_0$ and energy $\hbar\omega_{k_0}$ multiplied by a modulating function of x and t (the integral in (7.18)), which ensures the normalizability of $\Psi(x,\, t)$.

Two other useful results can be obtained. First, again assuming that $A(k)$ is symmetric and sharply peaked around k_0, the wave packet

$$\Psi(x,\, t) = \sqrt{\frac{1}{2\pi}} \int_{-\infty}^{\infty} dk\, A(k) e^{i[k(x - x_0) - \omega_k t]} \tag{7.19}$$

will have its maximum at x_0 when $t = 0$ and will propagate in time via

$$\Psi(x,\, t) \cong e^{i\omega_{k_0} t} \Psi(x - x_0 - v_0 t,\, 0), \qquad t \ll \frac{2m}{\hbar\, \delta k^2}. \tag{7.20}$$

Second, the wave packet

$$\tilde{\Psi}(x,\, t) = \sqrt{\frac{1}{2\pi}} \int_{-\infty}^{\infty} dk\, A(k) e^{-i[k(x - x_0) + \omega_k t]} \tag{7.21}$$

propagates from R to L and, for $A(k)$ the same as above, is also maximal at $x = x_0$ when $t = 0$. In particular, the approximations leading to (7.15) now give

$$\tilde{\Psi}(x,\, t) \cong e^{i\omega_{k_0} t} \Psi(x - x_0 + v_0 t,\, 0), \qquad t \ll 2m/(\hbar\, \delta k^2). \tag{7.22}$$

We next examine the temporal evolution of the simplest and one of the most important wave packets, viz., the Gaussian.

Example of the Gaussian wave packet

The Gaussian wave packet is generated from a Gaussian profile function, e.g.,

$$A(k) = \left(\frac{d}{\sqrt{\pi}}\right)^{1/2} e^{-(k - k_0)^2 d^2/2}, \tag{7.23}$$

where the normalization factor $(d/\sqrt{\pi})^{1/2}$ guarantees that (7.23) obeys (7.13). The constant d characterizes the width of the packet in coordinate space, while d^{-1} does this in momentum space. $A(k)$ is symmetric about the central or average "momentum" k_0.

From (7.12) the $t = 0$ (Gaussian) wave packet $\Psi_G(x,\, 0)$ is

$$\begin{aligned}\Psi_G(x,\, 0) &= \sqrt{\frac{d}{2\pi^{3/2}}} \int_{-\infty}^{\infty} dk\, e^{-(k - k_0)^2 d^2/2} e^{ikx} \\ &= \sqrt{\frac{d}{2\pi^{3/2}}} e^{ik_0 x} \int_{-\infty}^{\infty} dq\, e^{-q^2 d^2/2} \cos(qx),\end{aligned} \tag{7.24}$$

where the new variable $q = k - k_0$ has been introduced and the $\sin(qx)$ portion of

$\exp(iqx)$ has been eliminated due to the symmetry of $\exp(-q^2d^2)$ about $q = 0$. The dq integral appears in many tables of integrals (e.g., Gradshteyn and Ryzhik (1965)):

$$\int_{-\infty}^{\infty} dq\, e^{-q^2d^2/2}\cos(qx) = \frac{\sqrt{2\pi}}{d}e^{-x^2/(2d^2)}, \tag{7.25}$$

so that $\Psi_G(x, 0)$ is finally given by

$$\Psi_G(x, 0) = \sqrt{\frac{1}{d\sqrt{\pi}}}e^{ik_0x}e^{-x^2/(2d^2)}. \tag{7.26}$$

The $t = 0$ spatial wave packet is the plane wave $\exp(ik_0x)$ modified by the Gaussian factor $\exp[-x^2/(2d^2)]$, thus exemplifying the comment below Eq. (6.130) that the Fourier transform of a Gaussian is also a Gaussian. Note the resemblance to the form (7.18) when $t = 0$. A straightforward calculation shows that, as expected, the average value of the momentum for (7.25) is

$$\langle \hat{P}_x \rangle = \hbar k_0.$$

The probability density $\rho(x, 0) = |\Psi_G(x, 0)|^2$ is identical to that of the ground state of the linear harmonic oscillator:

$$\rho(x, 0) = \frac{1}{d\sqrt{\pi}}e^{-x^2/d^2}, \tag{7.27}$$

which is obviously symmetric about $x = 0$. The expected values of x and x^2 are $\langle x \rangle = 0$ and $\langle x^2 \rangle = d^2/2$, so that $d/\sqrt{2}$ is the standard deviation. $\rho(x, 0)$ is plotted as Fig. 7.5(a); d is the width of the packet.

It is easy to see that the profile function $A(k)$ is just the $t = 0$ momentum-space wave function $\varphi(k)$:

$$\varphi(k) = \sqrt{\frac{1}{2\pi}}\int_{-\infty}^{\infty} dx\, \Psi(x, 0)e^{-ikx} \equiv A(k).$$

We have refrained from using the symbol $\varphi(k)$ in order to emphasize our assumption that $A(k)$ is assigned an *a priori* form that is supposed to describe the momentum distribution produced by the accelerator. (Actually it is $|A(k)|^2$ that mimics the beam characteristic, but we shall continue to refer to $A(k)$ as describing the beam of incident particles.) $|A(k)|^2 \propto \exp[-(k - k_0)^2d^2]$ is thus the probability density that the particle's momentum will be $\hbar k$. This density is symmetric about k_0 and has a width of $\sqrt{2}/d$.

The time-developed wave function $\Psi_G(x, t)$ is obtained from $\Psi_G(x, 0)$ using the time-evolution operator $\hat{U}(t) = \exp(-i\hat{K}_x t/\hbar)$, with the kinetic-energy operator $\hat{K}_x$ being the full Hamiltonian for this case of a free particle:

$$\Psi_G(x, t) = e^{-i\hat{K}_x t/\hbar}\Psi_G(x, 0). \tag{7.28}$$

Since $\Psi_G(x, 0)$ is a linear combination of momentum/kinetic-energy eigenstates, $\Psi_G(x, t)$ is most easily evaluated by substituting (7.24) into (7.28) and allowing $\hat{K}_x$ to act directly on $\exp(ikx)$. This yields

$$\Psi_G(x, t) = \sqrt{\frac{d}{2\pi^{3/2}}} \int_{-\infty}^{\infty} dk\, e^{-(k-k_0)^2 d^2/2} e^{i[kx-\hbar k^2 t/(2m)]}. \tag{7.29}$$

Owing to the presence of the factor $\exp[-\hbar k^2 t/(2m)]$, the integral in (7.29), no longer being a simple Fourier transform such as occurs in $\Psi_G(x, 0)$, is not straightforward to evaluate. On the other hand, were the factor $i = \sqrt{-1}$ absent from the integral (thus making it real), then application of the procedure often referred to as "completing the square" would lead to an integral that has already been evaluated. We shall now develop this procedure; it will turn out to be applicable to (7.29).

The integral we shall consider is

$$I(\alpha, \beta) = \int_{-\infty}^{\infty} dx\, e^{-\alpha^2 x^2} e^{\beta x}, \qquad \alpha \text{ and } \beta \text{ real}. \tag{7.30}$$

By writing $\alpha^2 x^2 - \beta x$ as $\alpha^2 x^2 - \beta x + [\beta/(2\alpha)]^2 - [\beta/(2\alpha)]^2 = [\alpha x - \beta/(2\alpha)]^2 - [\beta/(2\alpha)]^2$, $I(\alpha, \beta)$ is equal to

$$I(\alpha, \beta) = e^{-[\beta/(2\alpha)]^2} \int_{-\infty}^{\infty} dx\, e^{-\alpha^2[x-\beta/(2\alpha^2)]^2} \tag{7.31a}$$

$$= \frac{1}{\alpha} e^{-[\beta/(2\alpha)]^2} \int_{-\infty}^{\infty} dy\, e^{-y^2}$$

$$= \frac{\sqrt{\pi}}{\alpha} e^{-[\beta/(2\alpha)]^2}, \tag{7.31b}$$

where Table 6.2 has been employed.

Ignoring for the moment the presence of the factor i, the integrand in (7.29) can be put in the $\alpha^2 x^2 - \beta x$ form of (7.31a) by using $q = k - k_0$ as a new integration variable. A few lines of algebra lead to

$$\Psi_G(x, t) = \sqrt{\frac{d}{2\pi^{3/2}}} e^{i(k_0 x - \omega_{k_0} t)} I, \tag{7.32}$$

where

$$I = \int_{-\infty}^{\infty} dq\, e^{-q^2[d^2/2 - i\hbar t/(2m)] + iq(x - v_0 t)}, \tag{7.33}$$

with $v_0 = \hbar k_0/m$, as before.

Comparison of Eqs. (7.31a) and (7.33) shows that they have the same general form, with $\alpha = \sqrt{d^2/2 + i\hbar t/(2m)}$ and $\beta = i(x - v_0 t)$. Completing the square in (7.33) yields

$$I = e^{-(x-v_0 t)^2/\{2d^2[1+i\hbar t/(md^2)]\}} \int_{-\infty}^{\infty} dq \exp\left[-q^2\left(\frac{d^2}{2} + \frac{i\hbar t}{2m}\right)\left(q - i\frac{x - v_0 t}{2\left(\frac{d^2}{2} + \frac{i\hbar t}{2m}\right)}\right)^2\right]. \tag{7.34}$$

The transformation from (7.33) to (7.34) is the same as that from (7.31a) to (7.31b). If the same step is also valid for complex α and β, as in (7.34), then I is easily evaluated. However, *is* such a step valid when β and α are complex? The answer to this question is

YES, although it is non-trivial to show it. We shall simply accept this "yes" answer and employ the result.[2] It gives

$$I = \sqrt{\frac{2\pi}{d^2[1 + i\hbar t/(md^2)]}} e^{-(x-v_0 t)^2/\{4d^2[1+i\hbar t/(md^2)]\}}. \tag{7.35}$$

The factor $1 + i\hbar t/(md^2)$ contains the ratio $t/(md^2/\hbar)$, whose magnitude governs the change of shape or the "spreading" of the wave packet with time. The time $t_{\mathrm{d}} = md^2/\hbar$ can be thought of as a spreading or decay time; it is intrinsic to the packet and the particle through its dependence on d and m and is an example for the Gaussian packet of the time $t_{\delta k}$ of the preceding subsection. Replacing $md^2/\hbar$ in (7.35) by t_{d} and then substituting into (7.32), $\Psi_{\mathrm{G}}(x, t)$ becomes

$$\Psi_{\mathrm{G}}(x, t) = \sqrt{\frac{1}{d\sqrt{\pi}(1 + it/t_{\mathrm{d}})}} e^{i(k_0 x - \omega_{k_0} t)} \exp\left(-\frac{(x - v_0 t)^2}{2d^2(1 + it/t_{\mathrm{d}})}\right). \tag{7.36}$$

Equation (7.36) is the fully time-developed, free, Gaussian wave packet. It bears a close resemblance to the approximate form (7.18), yet (7.36) is an exact result. Furthermore, as in the approximation (7.15), at any time t the maximum of $|\Psi_{\mathrm{G}}(x, t)|$ is located at the position $x = v_0 t$. The value of $|\Psi_{\mathrm{G}}(x, t)|$ at any x and t depends, among other things, on the ratio t/t_{d}. Let us estimate t_{d} for the case of an electron ($m = m_{\mathrm{e}}$) incident on a hydrogen atom. Using for d the value 10^5 Å estimated for δx in the preceding subsection ($E = 1$ keV), we find $t_{\mathrm{d}} \cong 10^{-6}$ s. Thus, for values of t less than this impact time, the packet suffers little spreading or diminution in magnitude as it propagates.

To get a better feeling for the general behavior in time, we consider the probability density

$$\begin{aligned} \rho(x, t) &= |\Psi_{\mathrm{G}}(x, t)|^2 \\ &= \frac{1}{d\sqrt{\pi(1 + t^2/t_{\mathrm{d}}^2)}} \exp\left(-\frac{(x - v_0 t)^2}{d^2(1 + t^2/t_{\mathrm{d}}^2)}\right). \end{aligned} \tag{7.37}$$

For any value of t, the maximum of $\rho(x, t)$ is at $x = v_0 t$. However, as t increases so does $1 + t^2/t_{\mathrm{d}}^2$, the effect of which is to reduce the height of $\rho(v_0 t, t)$ and at the same time increase the width of the packet, via $d\sqrt{1 + t^2/t_{\mathrm{d}}^2}$. Shown in Fig. 7.5 are plots of $\rho(x, t)$ of Eq. (7.37) for several values of t (in increments of t_{d}). Depending on the values of d, k_0, and m, a coordinate-space Gaussian wave packet can exhibit appreciable spreading as it propagates. In contrast, in momentum space, there is no spreading of the packet. For this reason, it may be advantageous to work in momentum space when one is using wave packets to describe some collision processes (Kuruoglu and Levin (1990), (1992)).

The Gaussian shape is only one example of a wave packet: many others can be and have been employed. We examine a few of them in the exercises rather than pursue them

[2] The integral in question can be evaluated by using the method of contour integration and the theory of residues of complex-variable analysis. Hassani (1991), e.g., in Example 7.2.11, applies this procedure in a way that can be extended to (7.34).

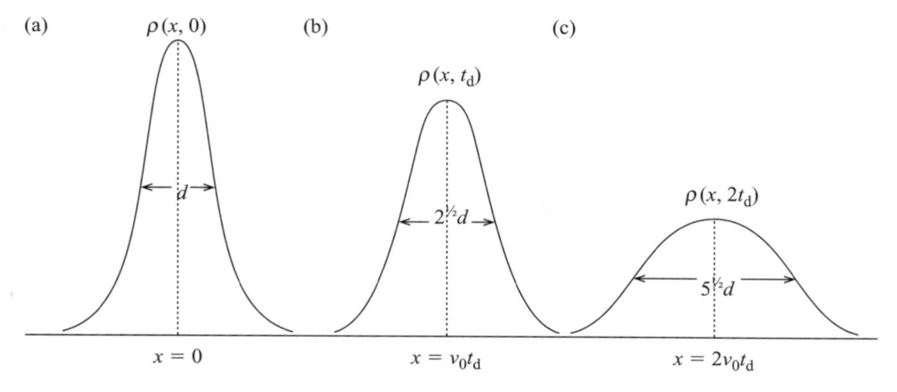

Fig. 7.5 Spreading and diminishing of the probability density for a Gaussian wave packet as time increases.

here. Instead we turn next to the construction of the incident, reflected, and transmitted packets corresponding to scattering by the generic non-confining potential of Eq. (7.1). We use as an interpretative guide the results of the analysis leading to Eqs. (7.15) and (7.18)–(7.20).

Wave-packet analysis of scattering

The packets describing the incident, reflected, and transmitted waves are denoted, respectively, $\Psi_{\mathrm{inc}}(x,\ t)$, $\Psi_{\mathrm{refl}}(x,\ t)$ and $\Psi_{\mathrm{tr}}(x,\ t)$. The first two are in region I, the third in region III. Each obeys the time-dependent Schrödinger equation

$$i\hbar\frac{\partial}{\partial t}\Psi(x,\ t)=\hat{K}_x(x)\Psi(x,\ t), \tag{7.38}$$

which is valid in regions I and III. The profile function of each packet is assumed, as before, to be symmetric about the central value k_0.

The incident particle leaves the accelerator at point $x=-x_{\mathrm{a}}$ at time $t=-T$. From the foregoing results, its wave packet $\Psi_{\mathrm{inc}}(x,\ t)$ will be

$$\Psi_{\mathrm{inc}}(x,\ t)=\int_{-\infty}^{\infty}dk\,A(k)e^{i[k(x+x_{\mathrm{a}})-\omega_k(t+T)]}, \qquad -x_{\mathrm{a}}\leqslant x\leqslant x_1, \tag{7.39}$$

where $A(k)$ is presumed to be a known function. The symmetry of $A(k)$ means that the packet's center is at $x=-x_{\mathrm{a}}$ when $t=-T$, consistent with the physics described. Furthermore, the center of the packet moves with the average velocity v_0, reaching the point x_1 in time $t_1=(x_1+x_{\mathrm{a}})/v_0-T$.

Once the packet reaches the edge of the potential well at x_1, part of it will be reflected and part of it will be transmitted. This same behavior characterizes the propagation of the initially transmitted packet throughout the well. In region III, only a transmitted wave, moving from L to R, is present.

Since the reflected wave is confined to region I, $x\leqslant x_1$, it represents the total effect of all reflections in region II. It should also be negligible until $t\simeq t_1-T$ in order to avoid violating causality: an effect should not precede its cause. The wave packet for a localized wave moving from R to L is

$$\Psi_{\text{refl}}(x,\ t)=\int_{-\infty}^{\infty}dk\,B(k)e^{-i[k(x-x_{\text{a}})+\omega_k(t+T)]},\qquad x\leqslant x_1, \tag{7.40}$$

in which the factor $\exp(ikx_{\text{a}})$ takes account of the fact that the initial position is $-x_{\text{a}}$, as in Eq. (7.39). The quantity $B(k)$, which is proportional to $A(k)$ and depends on the details of the potential, is determined via boundary matching, which is discussed below.

Using the average-behavior analysis, the center of the packet – assuming $B(k)$ also to be sharply peaked at k_0 – is located at $x = x_{\text{a}} - v_0(t + T)$. This implies that, at $t = -T$, the packet's center would be at the (forbidden) position $x = x_{\text{a}} > x_1$, at which time the incident packet's center is at $x = x_{\text{a}}$. Because $x_{\text{a}} > x_1$, there is no reflected packet present in region I at time $t = -T$. However, this analysis needs modification, since the k-dependent phase $\delta_{\text{R}}(k)$ (the R stands for reflected) of $B(k)$ has not been accounted for. To do so, we write $B(k) = |B(k)|\exp(i\delta_{\text{R}}(k))$. $\Psi_{\text{refl}}(x,\ t)$ now becomes

$$\Psi_{\text{refl}}(x,\ t)=\int_{-\infty}^{\infty}dk\,|B(k)|e^{-i[k(x-x_{\text{a}})-\delta_{\text{R}}(k)+\omega_k(t+T)]},\qquad x\leqslant x_1. \tag{7.41}$$

The approximation that leads to the average behavior also implies that $\delta_{\text{R}}(k)$ is to be expanded around k_0, with terms involving $(k - k_0)^n$, $n \geqslant 2$, being ignored due to the sharpness of $A(k)$:

$$\delta_{\text{R}}(k)\cong\delta_{\text{R}}(k_0)+(k-k_0)\delta'_{\text{R}}(k_0). \tag{7.42}$$

Substituting (7.42) into (7.41) leads to

$$\Psi_{\text{refl}}(x,\ t)\cong e^{i[\omega_{k_0}(t+T)+\delta_{\text{R}}(k_0)-k_0\delta'_{\text{R}}(k_0)]}\Psi_{\text{refl}}(x-x_{\text{a}}-\delta'_{\text{R}}(k_0)+v_0(t+T),\ 0),\qquad x\leqslant x_1, \tag{7.43}$$

whose center (maximum) is at $x = x_{\text{a}} + \delta'_{\text{R}}(k_0)$ when $t = -T$. We make the reasonable assumption that the phase length $\delta'_{\text{R}}(k_0)$, whose value depends on the details of the microscopically sized well, is also microscopic in size. Thus, at $t = -T$, there is still no reflected wave packet in region I.

At $t + T = t_1$, the center of the reflected packet is at $\delta'_{\text{R}}(k_0) - x_1$, a quantity that could be to the left or to the right of x_1 (by a microscopic distance), depending on the sign of $\delta'_{\text{R}}(k_0)$, that is, it depends on the detailed property of the potential well. For some $T + t > t_1$, the center of the reflected packet will certainly be in region I (and propagating to the left, as it should).[3]

For $t \geqslant t_1 + T$ the leading edge of the initial packet encounters the potential well, where it must smoothly join onto the solution $\Psi_{\text{II}}(x,\ t)$ of the time-dependent Schrödinger equation involving the non-zero potential $V(x)$. We shall consider $\Psi_{\text{II}}(x,\ t)$ soon.

The wave function in region III is the transmitted packet $\Psi_{\text{tr}}(x,\ t)$, which moves to the right. In analogy to Eqs. (7.40) and (7.41), it has the form

[3] The term $\delta'_{\text{R}}(k_0)/v_0$ is often denoted the *time delay* associated with the reflected packet. It is the extra time it takes the packet to begin reflecting, in comparison with the propagation time were there no well at all. A similar term can be introduced in the case of the transmitted packet (see below).

$$\Psi_{\rm tr}(x,\ t) = \int_{-\infty}^{\infty} dk\, C(k) e^{i[k(x+x_{\rm a})-\omega_k(t+T)]} \tag{7.44a}$$

$$= \int_{-\infty}^{\infty} dk\, |C(k)| e^{i[k(x+x_{\rm a})+\delta_{\rm T}(k)-\omega_k(t+T)]}, \qquad x \geqslant x_2, \tag{7.44b}$$

where the phase $\delta_{\rm T}(k)$ (the T stands for transmitted) of $C(k)$ has been made explicit in Eq. (7.44b). $C(k)$ is to be determined via boundary matching, which will be discussed shortly.

Treating $\delta_{\rm T}(k)$ in the same way as $\delta_{\rm R}(k)$ – see Eq. (7.42) – gives

$$\delta_{\rm T}(k) \cong \delta_{\rm T}(k_0) + (k - k_0)\delta'_{\rm T}(k_0); \tag{7.45}$$

use of the same analysis as that in the case of the reflected wave leads, again assuming that $C(k)$ is strongly peaked around k_0, to the center of $\Psi_{\rm tr}(x,\ t)$ being approximately at

$$x = -x_{\rm a} - \delta'_{\rm T}(k_0) + v_0(t + T). \tag{7.46}$$

Since $\delta'_{\rm T}(k_0)$ can also be expected to be microscopic while $x_{\rm a}$ is macroscopic, it follows that the requirement of $x > x_2$ in (7.46) means there is no possibility of a transmitted packet until $t \approx t_1$, a result again consistent with causality.

The forms of the incident, reflected, and transmitted packets have now been introduced and their general behaviors elucidated. The full region-I (normalizable) solution is

$$\Psi_{\rm I}(x,\ t) = \Psi_{\rm inc}(x,\ t) + \Psi_{\rm refl}(x,\ t),$$

$$= \int_{-\infty}^{\infty} dk\, e^{-ikx_{\rm a}}[A(k)e^{ikx} + B(k)e^{-ikx}]e^{-i\omega_k(t+T)}, \qquad x \leqslant x_1, \tag{7.47}$$

while that in region III is

$$\Psi_{\rm III}(x,\ t) = \Psi_{\rm tr}(x,\ t) = \int_{-\infty}^{\infty} dk\, e^{-ikx_{\rm a}} C(k)e^{ikx}e^{-i\omega_k(t+T)}, \qquad x \geqslant x_2. \tag{7.48}$$

These describe the physics of the problem, i.e., they answer the questions of why the two independent solutions $\exp(\pm ikx)$ are linearly combined in $\Psi_{\rm I}(x,\ t)$ but only $\exp(ikx)$ is present in $\Psi_{\rm III}(x,\ t)$. In the case of region II, the form analogous to Eqs. (7.47) and (7.48) must contain the two linearly independent solutions of the time-independent Schrödinger equation. For energy $E = \hbar^2 k^2/(2m)$, these solutions obey

$$[\hat{K}_x(x) + V(x)]\psi_k^{(j)}(x) = E\psi_k^{(j)}(x), \qquad x_1 \leqslant x \leqslant x_2, \tag{7.49}$$

and $j = 1$ or 2. For our present purposes, $\psi_k^{(1)}(x)$ and $\psi_k^{(2)}(x)$ are assumed to be known. Because E is given, (7.49) is not an eigenvalue equation.

The general stationary-state solution in region II is the linear combination

$$\zeta_k(x) = D(k)\psi_k^{(1)}(x) + F(k)\psi_k^{(2)}(x), \tag{7.50}$$

where $D(k)$ and $F(k)$ are eventually to be determined by boundary matching. The time-dependent solution corresponding to (7.50) is

$$Z_k(x,\ t) = \zeta_k(x)e^{-i\omega_k(t+T)},$$

and, inserting the phase factor $\exp(ikx_{\rm a})$, the full region-II solution is

$$\Psi_{\mathrm{II}}(x,\, t) = \int_{-\infty}^{\infty} dk\, e^{ikx_{\mathrm{a}}}[D(k)\psi_k^{(1)}(x) + F(k)\psi_k^{(2)}(x)]e^{-i\omega_k(t+T)}, \qquad x_1 \leqslant x \leqslant x_2. \tag{7.51}$$

Among the five coefficients $A(k)$, $B(k)$, $C(k)$, $D(k)$ and $F(k)$, only $A(k)$ is known: the other four are determined by requiring the relevant solutions and their derivatives to match smoothly at the points x_1 and x_2, exactly as in the bound-state case. In particular, one has

$$\Psi_{\mathrm{I}}(x_1,\, t) = \Psi_{\mathrm{II}}(x_1,\, t), \tag{7.52a}$$

$$\Psi_{\mathrm{I}}'(x_1,\, t) = \Psi_{\mathrm{II}}'(x_1,\, t), \tag{7.52b}$$

$$\Psi_{\mathrm{II}}(x_2,\, t) = \Psi_{\mathrm{III}}(x_2,\, t), \tag{7.52c}$$

and

$$\Psi_{\mathrm{II}}'(x_2,\, t) = \Psi_{\mathrm{III}}'(x_2,\, t), \tag{7.52d}$$

where the prime means the derivative w.r.t. the spatial variable, evaluated at x_1 or x_2.

The four equations (7.52) all have the same form: an equality between two integrals over the wave number k. For simplicity, let us consider Eq. (7.52c); the conclusions we draw from our analysis will apply to the other three equations. Substituting the forms (7.48) and (7.51) into (7.52c) and rearranging, one gets

$$\int_{-\infty}^{\infty} dk\, e^{-i\omega_k(t+T)+ikx_{\mathrm{a}}}[D(k)\psi_k^{(1)}(x_2) + F(k)\psi_k^{(2)}(x_2) - C(k)e^{ikx_2}] = 0. \tag{7.53}$$

The preceding integral, which is a function of x_2, x_{a}, and t, is zero. It is reasonable to expect that, in order for it to be zero, the term in square brackets must vanish for each k. We expect this on physical grounds: since the Schrödinger equation is linear, the matching conditions should hold for each value of the wave number k. This would certainly be the case if $A(k)$ were proportional to a delta function: $A(k) \propto \delta(k - k_0)$. We may reach the conclusion $[\ ] = 0$ by noting that, since x_{a} is an arbitrary point, we may multiply both sides of (7.53) by $\exp(ik'x_{\mathrm{a}})$ and integrate on x_{a} from $-\infty$ to $+\infty$. On the LHS, the integration on x_{a} yields $2\pi\delta(k - k')$, while the RHS remains zero; the conclusion is $[\ ] = 0$. Hence, we find that (7.52c) is equivalent to

$$D(k)\psi_k^{(1)}(x_2) + F(k)\psi_k^{(2)}(x_2) = C(k)e^{ikx_2}. \tag{7.54c}$$

Similar equivalences hold[4] for the other equations forming (7.52):

$$A(k)e^{ikx_1} + B(k)e^{-ikx_1} = D(k)\psi_k^{(1)}(x_1) + F(k)\psi_k^{(2)}(x_1), \tag{7.54a}$$

$$ik[A(k)e^{ikx_1} - B(k)e^{-ikx_1}] = D(k)\,\frac{d\psi_k^{(1)}(x_1)}{dx_1} + F(k)\,\frac{d\psi_k^{(2)}(x_1)}{dx_1}, \tag{7.54b}$$

and

$$D(k)\,\frac{d\psi_k^{(1)}(x_2)}{dx_2} + F(k)\,\frac{d\psi_k^{(2)}(x_2)}{dx_2} = ikC(k)e^{ikx_2}. \tag{7.54d}$$

[4] In the derivation of (7.54b) and (7.54d) an interchange of differentiation and integration has occurred. We assume that this interchange is always allowed for the cases arising in physics.

The set (7.54) can be solved for the unknown coefficients B, C, D, and F in terms of the remaining (known) quantities in these equations, i.e., $A(k)$, $\psi_k^{(i)}(x_j)$, $\psi_k^{(i)\prime}(x_j)$, and $\exp(\pm ikx_i)$, i and j each equal to 1 and 2. The resulting expressions are algebraically complex, and the only feature we will need is that, since Eqs. (7.54) are linear, each of $B(k)$, $C(k)$, $D(k)$, and $F(k)$ is proportional to $A(k)$. Particular values of these coefficients will be determined for several simple potentials in Sections (7.3) and (7.4). We close this section by noting that, once $B(k)$ and $C(k)$ have been obtained, it is still necessary to assign a physical interpretation to them, beyond their appearance in the region-I and -II wave functions. This is crucial when improper eigenstates $\exp(\pm ikx)$ are used, and consideration of this topic via the equation of continuity is the subject of the next section. No additional aspects of wave packets are essential for the remaining development and we leave further exploration of them to the exercises.

7.2. The equation of continuity and the plane-wave limit for scattering

When a quantal system is described by a wave function that is a superposition of stationary states, the probability density becomes time-dependent, and, for bound states, the position where the probability density is maximized moves back and forth within a (more or less) well-defined interval. In the case of a superposition forming a continuum wave packet, the maximum in the probability density is not confined to a particular interval but propagates in time. For the situation described in the preceding section, the incident wave packet, propagating from L to R in region I, eventually disappears, becoming transformed into the reflected and transmitted packets. Mathematically, this behavior is expressed as

$$\rho_{\text{inc}}(x,\, t) = |\Psi_{\text{inc}}(x,\, t)|^2 \to 0, \; t \text{ sufficiently large,} \tag{7.55a}$$

$$\left.\begin{array}{c} \rho_{\text{refl}}(x,\, t) = |\Psi_{\text{refl}}(x,\, t)|^2 \\ \text{and} \\ \rho_{\text{tr}}(x,\, t) = |\Psi_{\text{tr}}(x,\, t)|^2 \end{array}\right\} \text{become non-zero, } t \text{ sufficiently large.} \tag{7.55b}$$

The relations (7.55) represent a new physical development: the transformation of probability from one form to others, in particular, the decrease and eventual extinction in time of the initial (incident) probability density and the increase from zero to full strength of the reflected and transmitted probability densities.

These changes in probability density are analogous to the increase or decrease of electric charge in a region, a change always accompanied by a flow of electric current. This electric phenomenon is described by the equation of continuity, relating changes in the charge and current densities. Given the foregoing analogy between probability density and electric charge, we might expect there to be a probability current density, related to the change in probability density via a quantal equation of continuity. There is such an equation, and the concept of current that occurs in it is an important ingredient in the interpretation of scattering phenomena, in particular that arising from a plane-wave description.

The equation of continuity

In electrodynamics, the equation of continuity relates the time derivative of the charge density to the gradient of the current density. The analogous quantal equation involves the time derivative of the probability density. This quantity can be obtained from manipulations involving the time-dependent Schrödinger equation; one consequence of these manipulations is the direct generation of a probability current density.

We consider a particle of mass m (in 3-D) governed by the Hamiltonian $\hat{H} = \hat{K} + \hat{V}$. The particle's state vector $|\Gamma, t\rangle$ obeys

$$i\hbar \frac{\partial|\Gamma, t\rangle}{\partial t} = \hat{H}|\Gamma, t\rangle. \tag{7.56}$$

Taking the Hermitian conjugate of both sides of (7.56) yields the equation for the dual vector $\langle\Gamma, t|$:

$$-i\hbar\langle\Gamma, t| \underset{\leftarrow}{\frac{\partial}{\partial t}} = \langle\Gamma, t|\, \underset{\leftarrow}{\hat{H}}, \tag{7.57}$$

where the action of the operators to the left is made explicit.

On multiplying both sides of (7.56) from the right by $\langle\Gamma, t|$ and both sides of (7.57) from the left by $|\Gamma, t\rangle$, and then subtracting one of the resulting equations from the other, one finds

$$i\hbar \frac{\partial}{\partial t}\{|\Gamma, t\rangle\langle\Gamma, t|\} = \{\hat{H}|\Gamma, t\rangle\}\langle\Gamma, t| - |\Gamma, t\rangle\{\langle\Gamma, t|\, \underset{\leftarrow}{\hat{H}}\}. \tag{7.58}$$

Defining the density operator $\hat{\rho}(t)$ by[5]

$$\hat{\rho}(t) = |\Gamma, t\rangle\langle\Gamma, t|, \tag{7.59}$$

Eq. (7.58) can be re-written as

$$\frac{\partial\hat{\rho}(t)}{\partial t} = \frac{1}{i\hbar}[\hat{H}, \hat{\rho}(t)]. \tag{7.60}$$

Equation (7.60) is a particular example of the role played by the Hamiltonian and the commutation relation in determining the rates of change with time of operators and their expectation values, a feature explored in detail in Chapter 8.

We proceed towards the standard form of the quantal equation of continuity by first going into a coordinate representation. Let $|\mathbf{r}\rangle$ be an eigenstate of the 3-D position operator $\hat{\mathbf{Q}}$. Projecting both sides of (7.60) onto the bra $\langle\mathbf{r}|$ from the left and onto its dual $|\mathbf{r}\rangle$ (with the same $\mathbf{r}$ in each) from the right, Eq. (7.60) becomes

$$\frac{\partial}{\partial t}|\Psi_\Gamma(\mathbf{r}, t)|^2 = \frac{1}{i\hbar}\,[\Psi_\Gamma^*(\mathbf{r}, t)\{\hat{H}(\mathbf{r})\Psi_\Gamma(\mathbf{r}, t)\} - \{\hat{H}(\mathbf{r})\Psi_\Gamma^*(\mathbf{r}, t)\}\Psi_\Gamma(\mathbf{r}, t)], \tag{7.61}$$

where $\hat{H}(\mathbf{r})$ acts only on the wave function inside the curly brackets.

The LHS of (7.61) is the rate of change with time of the probability density. In order to

[5] Equation (7.59) defines the density operator for the case of pure states, the only ones being considered in this book. Its generalization to the case of mixed states is discussed in a variety of references, e.g., Cohen-Tannoudji, Diu, and Laloë (1977), Sakurai (1994), and Schiff (1968).

convert the RHS into the divergence of a vector, we next make the important assumption that $\hat{V}(\mathbf{r})$ is a multiplicative function of $\mathbf{r}$:

$$\hat{V}(\mathbf{r}) \to V(\mathbf{r}), \tag{7.62}$$

a reduction that holds for many potential-energy operators. Under this assumption, the terms on the RHS of (7.61) containing $V(\mathbf{r})$ cancel out, and we find

$$\frac{\partial}{\partial t}|\Psi_\Gamma(\mathbf{r},\, t)|^2 = \frac{1}{i\hbar}[\Psi_\Gamma^*(\mathbf{r},\, t)\{\hat{K}(\mathbf{r})\Psi_\Gamma(\mathbf{r},\, t)\} - \{\hat{K}(\mathbf{r})\Psi_\Gamma^*(\mathbf{r},\, t)\}\Psi_\Gamma(\mathbf{r},\, t)]. \tag{7.63}$$

Using

$$\hat{K}(\mathbf{r}) = \frac{\hat{\mathbf{P}}^2(\mathbf{r})}{2m} = -\frac{\hbar^2}{2m}\nabla^2 = -\frac{\hbar^2}{2m}\nabla\cdot\nabla, \tag{7.64}$$

it is readily shown that (7.63) is equal to

$$\frac{\partial}{\partial t}|\Psi_\Gamma(\mathbf{r},\, t)|^2 = -\frac{\hbar}{2mi}\nabla\cdot[\Psi_\Gamma^*(\mathbf{r},\, t)\{\nabla\Psi_\Gamma(\mathbf{r},\, t)\} - \{\nabla\Psi_\Gamma^*(\mathbf{r},\, t)\}\Psi_\Gamma(r,\, t)]. \tag{7.65}$$

We rewrite Eq. (7.65) in the final form

$$\frac{\partial}{\partial t}\rho_\Gamma(\mathbf{r},\, t) + \nabla\cdot\mathbf{J}_\Gamma(\mathbf{r},\, t) = 0, \tag{7.66}$$

which is the desired equation of continuity. Equation (7.66) contains the usual probability density

$$\rho_\Gamma(\mathbf{r},\, t) = |\Psi_\Gamma(\mathbf{r},\, t)|^2, \tag{7.67}$$

plus the new quantity $\mathbf{J}_\Gamma(\mathbf{r},\, t)$, which we interpret as the *probability current density*:

$$\mathbf{J}_\Gamma(\mathbf{r},\, t) = \frac{\hbar}{2mi}[\Psi_\Gamma^*(\mathbf{r},\, t)\{\nabla\Psi_\Gamma(\mathbf{r},\, t)\} - \{\nabla\Psi_\Gamma^*(\mathbf{r},\, t)\}\Psi_\Gamma(\mathbf{r},\, t)] \tag{7.68a}$$

$$\equiv \operatorname{Im}\left(\Psi_\Gamma^*(\mathbf{r},\, t)\frac{\hbar\nabla}{m}\Psi_\Gamma(\mathbf{r},\, t)\right). \tag{7.68b}$$

Note that $\mathbf{J}_\Gamma(\mathbf{r},\, t)$ is a function, not an operator.

The interpretation of $\mathbf{J}_\Gamma$ as a probability current density is based on the analogy with the equation of continuity in electrodynamics. Thus, on integrating both sides of Eq. (7.66) over a fixed volume V bounded by a surface S, one gets

$$\frac{\partial}{\partial t}\int_V d^3r\rho_\Gamma(\mathbf{r},\, t) = -\int_S d\mathbf{S}\cdot\mathbf{J}_\Gamma(\mathbf{r},\, t), \tag{7.69}$$

where the divergence theorem has been used to convert the integral containing $\mathbf{J}_\Gamma$. Equation (7.69) states that the rate of change with time of the probability of finding the particle in the volume V (usually a negative quantity) is equal to the integral of $\mathbf{J}_\Gamma$ over the surface. Since the dimension $[\rho_\Gamma]$ is length^{-3}, it follows that $[\mathbf{J}_\Gamma]$ equals length$^{-2}\cdot$time^{-1}, i.e., $\mathbf{J}_\Gamma$ has the dimension of number per area per time, which is the same dimension as that of current density per unit charge. The interpretation of $\mathbf{J}_\Gamma$ as a probability current density is no surprise.

If one is considering a 1-D situation, then $\mathbf{r} \to x$, $\nabla\cdot\mathbf{J}_\Gamma \to \partial J_\Gamma/\partial x$, and Eq. (7.66) reduces to

$$\frac{\partial}{\partial t}\rho_\Gamma(x,\, t) + \frac{\partial}{\partial x}J_\Gamma(x,\, t) = 0, \tag{7.70}$$

which is the form that will be used in this section.

Any solution of the 1-D time dependent Schrödinger equation must obey the continuity equation. When Ψ_Γ is a stationary state, (7.70) is trivially satisfied, since ρ_Γ is constant in time and J_Γ is a constant in space (or zero). Two examples demonstrate this.

Consider first a bound eigenstate:

$$\Psi_\Gamma(x,\, t) \to \psi_n(x)e^{-iE_n t/\hbar},$$

for which $\partial\rho_n(x,\, t)/\partial t = \partial|\psi_n(x)|^2/\partial t = 0$; on the other hand, $J_n = 0$ due to the fact that $\psi_n(x)$ can always be chosen real, so that (7.70) reduces to $0 - 0 = 0$. As a second example, let $\Psi_\Gamma(x,\, t) \to \Psi_{k_0}(x,\, t)$, a renormalized plane-wave state:

$$\Psi_{k_0}(x,\, t) = \frac{A_0}{\sqrt{2\pi}}e^{i(k_0 x - \omega_{k_0} t)}, \tag{7.71}$$

with A_0 a constant. As usual, the probability density is a constant,

$$\rho_{k_0}(x,\, t) = \frac{|A_0|^2}{2\pi}, \tag{7.72}$$

and so is J_{k_0}:

$$J_{k_0}(x,\, t) = \frac{|A_0|^2}{2\pi}\frac{\hbar k_0}{m} \equiv \rho_{k_0}(x,\, t)v_{k_0}, \tag{7.73}$$

where the semi-classical velocity $v_{k_0} = \hbar k_0/m$ has been re-introduced.

While (7.70) is again trivially satisfied for the choice (7.71), the relation $J_{k_0} = \rho_{k_0}v_{k_0}$ is in precise analogy to the relation between the charge density $\rho_{\text{ch}}(\mathbf{r})$ and the current density $\mathbf{J}(\mathbf{r})$ for a point charge, viz., $\mathbf{J}(\mathbf{r}) = \rho_{\text{ch}}(\mathbf{r})\mathbf{v}_{k_0}$. Also, although the concept of a normalizable probability density fails for plane waves, the interpretation of J_{k_0} being the product of the (constant) number of particles and their velocity is an acceptable concept. This feature carries over to the case of a wave packet with a sharply peaked profile function and is the key to understanding why one can use plane waves in place of wave packets to describe scattering phenomena. We therefore apply these ideas to the free-particle wave packet of Eq. (7.12), from which we obtain the following probability and current densities:

$$\rho(x,\, t) = \frac{1}{2\pi}\int_{-\infty}^{\infty} dk'\, dk\, A^*(k')A(k)e^{i[(k-k')x - (\omega_k - \omega_{k'})t]} \tag{7.74}$$

and

$$J(x,\, t) = \frac{1}{2\pi}\int_{-\infty}^{\infty} dk'\, dk\, A^*(k')A(k)\left(\frac{\hbar}{2m}(k+k')\right)e^{i[(k-k')x - (\omega_k - \omega_{k'})t]}, \tag{7.75}$$

which are easily shown to obey Eq. (7.70), as they must.

We now prove that, to a good approximation, J and ρ of Eqs. (7.74) and (7.75) are related in the same way as their plane-wave counterparts are, viz., by Eq. (7.73). To establish this relation, we re-write k and k' in (7.75) via

$$k = k_0 + (k - k_0)$$

and

$$k' = k_0 + (k' - k_0)$$

and then note that, for values of k or k' well away from k_0, the factors $(k - k_0)A(k)$ and $(k' - k_0)A^*(k)$ are each small relative to $k_0 A(k)$ and $k_0 A^*(k')$ due to the fact that the profile functions are sharply peaked about k_0, which here is assumed not equal to zero. Hence, to a good approximation, $J(x, t)$ becomes

$$\begin{aligned} J(x, t) &\cong \frac{\hbar k_0}{2\pi m}\int_{-\infty}^{\infty} dk'\, dk\, A^*(k')A(k)e^{i[(k-k')x-(\omega_k-\omega_{k'})t]} \\ &= v_{k_0}\rho(x, t), \end{aligned} \tag{7.76}$$

where the narrower the profile function, the smaller are the corrections to (7.76).

Equation (7.76) states that the probability current density is approximately equal to the product of the probability density and the average or central velocity of the wave packet. Since $\rho(x, t)$ is localized, so is $J(x, t)$. If the initial beam contains N particles, each described by the packet (7.12), then $N(x, t) = N\rho(x, t)$ is the density of particles at position x contained in the range x to $x + dx$, and $NJ(x, t)$, the number of particles passing the point x in unit time, is just $v_{k_0}N(x, t)$. Current densities will be used to interpret the scattering parameters $B(k)$ and $C(k)$.

The transition to the plane-wave description

The application of the preceding analyses to the wave packets describing scattering employs a modification of the generic potential (7.1). The modification leads to a 1-D analog of inelastic scattering, which is a collision wherein some of the incident particle's kinetic energy is used to excite the target, resulting in an ejectile with a smaller kinetic energy and a smaller wave number. The reduction to the potential of Eq. (7.1) from this modification is trivially accomplished. The overall goal of the analysis is to show that the same final results are obtained when plane waves are used in place of wave packets.

In the "inelastic-scattering" scenario, particles are still incident from L to R starting at $x = -x_a$ (at $t = -T$) in region I, but the potential acting on each particle is modified away from (7.1) by adding a constant potential $V_1 > 0$ in region III. The modified potential $V^M(x)$ thus has the form

$$V^M(x) = \begin{cases} 0, & x \leqslant x_1 \quad \text{(Region I)}, \\ < 0, & x_1 < x < x_2 \quad \text{(Region II)}, \\ V_1 > 0, & x \geqslant x_2 \quad \text{(Region III)}, \end{cases} \tag{7.77}$$

and is shown schematically in Fig. 7.6. Included in Fig. 7.6 is a straight line at height $E_0 = \hbar^2 k_0^2/(2m)$, the average energy in the incident and reflected wave packets. It is explicitly assumed that E_0 is bigger than V_1 by at least a factor of two.

The time-independent Schrödinger equation in region III now changes from (7.5) to

$$\psi_{\rm III}(x) + K^2\psi_{\rm III}(x) = 0, \qquad x \geqslant x_2, \tag{7.78}$$

where

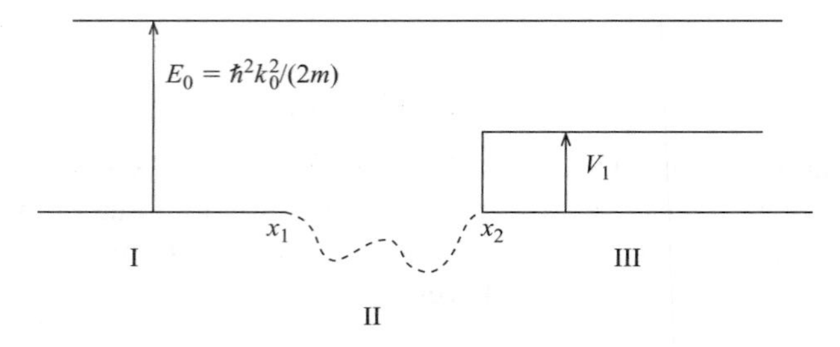

Fig. 7.6 The modified generic potential

$$V^{\mu}(x) = \begin{cases} 0, & x \leqslant x_1 & \text{(Region I)}, \\ <0, & x_1 < x < x_2 & \text{(Region II)}, \\ V_1, & x \geqslant x_2 & \text{(Region III)}, \end{cases}$$

with $V_1 = \text{constant} > 0$.

$$K^2 = \frac{2m}{\hbar^2}(E - V_1) = k^2 - \frac{2m}{\hbar^2}V_1,$$

with $E = \hbar^2 k^2/(2m)$, as before. Solutions to (7.78) are of the form

$$\psi_{\text{III}}(x) = C(k)e^{\pm iKx},$$

from which we find the stationary-state solution to the time-dependent Schrödinger equation in region III which moves from L to R to be

$$\Psi(x,\, t) = C(k)e^{i(Kx-\omega_k t)}, \tag{7.79}$$

where $\omega_k = \hbar k^2/(2m)$. Notice that only in the x-dependent part of the solution is the wave number changed: inclusion of the potential V_1 in region III does not alter the energy $E = \hbar^2 k^2/(2m)$.

For the modified potential, the region-III wave packet $\Psi_{\text{tr}}(x,\, t)$ is

$$\Psi_{\text{tr}}(x,\, t) = \sqrt{\frac{1}{2\pi}}\int_{-\infty}^{\infty} dk\, C(k)e^{i[K(x-x_a)-\omega_k(t+T)]}, \qquad x \geqslant x_2. \tag{7.80}$$

As a result, the boundary matching equations (7.54c) and (7.54d) become

$$D(k)\psi_k^{(1)}(x_2) + F(k)\psi_k^{(2)}(x_2) = C(k)e^{iKx_2} \tag{7.54c$'$}$$

and

$$D(k)\psi_k^{(1)\prime}(x_2) + F(k)\psi_k^{(2)\prime}(x_2) = iKC(k)e^{iKx_2}. \tag{7.54d$'$}$$

Note that, although $C(k)$ could have been re-written as $C(K)$, $C(k)$ is retained as a reminder that the scattering parameters are linked to the total energy.

Let us next assume that Eqs. (7.54a), (7.54b), (7.54c$'$), and (7.54d$'$) have been solved for $D(k)$, $F(k)$, and the scattering parameters $B(k)$ and $C(k)$. From the latter two quantities we can construct the reflected and transmitted wave packets, which in turn will produce the reflected and transmitted current densities $J_{\text{refl}}(x,\, t)$ and $J_{\text{tr}}(x,\, t)$, respectively. Correspondingly, $\Psi_{\text{inc}}(x,\, t)$ will give rise to $J_{\text{inc}}(x,\, t)$. Since $NJ_{\text{inc}}(x,\, t_{\text{i}})$ is the number of incident-beam particles crossing the point x ($> -x_{\text{a}}$) at an early time t_{i}, then, at a time t_f well after the time t_1 it takes the center of the incident packet to reach x_1, there should only be reflected and transmitted packets, with $NJ_{\text{refl}}(-x_a, t_f)$ and $NJ_{\text{tr}}(x_a, t_f)$ being the numbers of reflected and transmitted particles crossing $-x_a$ and $+x_a$, respectively. Furthermore, in the present circumstance that particles neither decay nor are absorbed, we expect $NJ_{\text{inc}}(x_a, t_i) = NJ_{\text{refl}}(-x_a, t_f) + NJ_{\text{tr}}(x_a, t_f)$. The values of $B(k)$ and $C(k)$ must reflect this statement of particle conservation. We now use the equation of continuity to demonstrate this, wherein we shall see that particle conservation is characterized by two new quantities, the reflection coefficient $R(k)$ and the transmission coefficient $T(k)$, whose sum is unity: $T(k) + R(k) = 1$. These quantities are 1-D analogs of the scattering cross section, a concept introduced in Chapter 15, where we discuss scattering in 3-D.

The wave function $\Psi(x, t)$ describing the 1-D scattering process is the sum of the contributions from regions I, II, and III:

$$\Psi(x, t) = \Psi_{\text{I}}(x, t) + \Psi_{\text{II}}(x, t) + \Psi_{\text{III}}(x, t), \tag{7.81}$$

where Ψ_{I} is the sum of the incident and reflected packets (see Eq. (7.47)), Ψ_{II} is given by Eq. (7.51), and Ψ_{III} is the transmitted packet, Eq. (7.80). $\Psi(x, t)$ obeys the equation of continuity, viz.,

$$\frac{\partial \rho(x, t)}{\partial t} + \frac{\partial}{\partial x} J(x, t) = 0, \tag{7.82}$$

where $\rho(x, t)$ and $J(x, t)$ are obtained from $\Psi(x, t)$ of (7.81).

We integrate both sides of (7.82) over a *macroscopic* portion of the x-axis, between, say, $-x_a$ and $+x_a$. The ρ-dependent term is

$$\int_{-x_a}^{x_a} dx \frac{\partial}{\partial t} \rho(x, t) = \frac{\partial}{\partial t} \int_{-x_a}^{x_a} dx\, \rho(x, t) = 0, \tag{7.83}$$

since $\int_{-x_a}^{x_a} dx\, \rho(x, t)$ is essentially the normalization integral – which it will be as long as x_a is very large and t is smaller than the time it takes either Ψ_{refl} or Ψ_{tr} to reach the boundaries. We maintain this restriction on t.

The other term in the integral of (7.82) thus obeys

$$\int_{-x_a}^{x_a} dx \frac{\partial}{\partial x} J(x, t) = 0$$

or

$$J(-x_a, t) = J(x_a, t). \tag{7.84}$$

This is the relation from which we shall derive the equation $R(k) + T(k) = 1$. (One could also integrate both sides of (7.84) over all times and obtain the same relations. However, since infinite times correspond to infinite spatial distances, this would necessitate replacing $\pm x_a$ by $\pm\infty$, a replacement that would complicate the resulting expressions unnecessarily; so we retain (7.84).)

The point $-x_a$ is in region I, so that $J(-x_a, t)$ is to be computed from $\Psi_{\text{I}}(-x_a, t)$:

$$J(-x_{\mathrm{a}},\,t) = \frac{\hbar}{2mi}\Bigg([\Psi^*_{\mathrm{inc}}(-x_{\mathrm{a}},\,t) + \Psi^*_{\mathrm{refl}}(-x_{\mathrm{a}},\,t)] \times \frac{\partial}{\partial x_{\mathrm{a}}}[\Psi_{\mathrm{inc}}(-x_{\mathrm{a}},\,t) + \Psi_{\mathrm{refl}}(-x_{\mathrm{a}},\,t)] - \text{c.c.}\Bigg), \tag{7.85}$$

where c.c. means "complex conjugate". Equation (7.85) can be written as the sum of the incident current, the reflected current, and an interference term:

$$J(-x_{\mathrm{a}},\,t) = J_{\mathrm{inc}}(-x_{\mathrm{a}},\,t) + J_{\mathrm{refl}}(-x_{\mathrm{a}},\,t) + J_{\mathrm{interf}}(-x_{\mathrm{a}},\,t), \tag{7.86}$$

where

$$J_{\mathrm{interf}}(-x_{\mathrm{a}},\,t) = \frac{\hbar}{2mi}\left[\left(\Psi^*_{\mathrm{inc}}(-x_{\mathrm{a}},\,t)\,\frac{\partial}{\partial x_{\mathrm{a}}}\Psi_{\mathrm{refl}}(-x_{\mathrm{a}},\,t) + \Psi^*_{\mathrm{refl}}(-x_{\mathrm{a}},\,t)\,\frac{\partial}{\partial x_{\mathrm{a}}}\Psi_{\mathrm{inc}}(-x_{\mathrm{a}},\,t)\right) - \text{c.c.}\right]. \tag{7.87}$$

Owing to Ψ_{inc} and Ψ_{refl} being localized wave packets, J_{interf} is essentially zero, since there is no overlap between Ψ^*_{inc} and $\partial\Psi_{\mathrm{refl}}/\partial x_{\mathrm{a}}$, etc. The lack of overlap between Ψ_{refl} and Ψ_{inc} – at least for times $t \neq t_1$ – was discussed in Section 7.1. Although the derivatives of the wave packets will not have the same shape as the wave packets themselves, they are both non-zero in the same localized region (at the same time), and each moves with the same average, semi-classical speed v_{k_0}. Hence, as long as we do not choose $t \cong t_1$, the transit time from $-x_{\mathrm{a}}$ to x_1, none of the products in (7.87) will contain overlapping terms and therefore

$$J_{\mathrm{interf}}(-x_{\mathrm{a}},\,t) = 0. \tag{7.88}$$

In view of (7.88), only the first two terms on the RHS of (7.86) contribute, and (7.84) becomes

$$J_{\mathrm{inc}}(-x_{\mathrm{a}},\,t) + J_{\mathrm{refl}}(-x_{\mathrm{a}},\,t) = J_{\mathrm{tr}}(x_{\mathrm{a}},\,t). \tag{7.89}$$

Because Ψ_{refl} is a packet that moves from R to L, the form of (7.76) taken by the reflected current is

$$J_{\mathrm{refl}}(x,\,t) \cong -v_{k_0}\rho_{\mathrm{refl}}(x,\,t). \tag{7.90}$$

In other words, $-J_{\mathrm{refl}}(x,\,t)$ is positive, from which we obtain the expected conservation-law form of (7.89):

$$J_{\mathrm{inc}}(-x_{\mathrm{a}},\,t) = -J_{\mathrm{refl}}(-x_{\mathrm{a}},\,t) + J_{\mathrm{tr}}(x_{\mathrm{a}},\,t). \tag{7.91}$$

Note that all effects of the interaction region (II) are contained in the RHS of (7.91).

Equation (7.91) is a relation involving wave packets. Our ultimate interest is in describing scattering by means of plane waves. We now show how this may be accomplished, employing approximations based on properties of very narrow profile functions. Substituting (7.39), (7.40), and (7.80) into the relevant expressions for the current densities, Eq. (7.91) becomes

$$\int dk\, dk'\, A^*(k')A(k)\frac{\hbar}{2m}(k+k')e^{-i(\omega_k-\omega_{k'})(t+T)}$$

$$= \int dk\, dk' \left(B^*(k')B(k)\frac{\hbar}{2m}(k+k')e^{2i(k-k')x_a} + C^*(k')C(k)\frac{\hbar}{2m}(K+K')e^{2i(K-K')x_a} \right) e^{-i(\omega_k-\omega_{k'})(t+T)}, \tag{7.92}$$

where the common factor $(2\pi)^{-1}$ has been canceled out.

To reduce (7.92) to the desired form, we assume $A(k)$ to be a very narrow function of k, symmetric and sharply peaked about the value k_0 (the corresponding value of K is $K_0 = \sqrt{k_0^2 - 2mV_1/\hbar^2}$). Because of this assumption, we approximate any function of k that $A(k)$ multiplies by its value at k_0. For example, we have noted that $B(k)$ and $C(k)$ are proportional to $A(k)$, relations we write as

$$B(k) = r(k)A(k) \tag{7.93a}$$

and

$$C(k) = t(k)A(k); \tag{7.93b}$$

these are approximated by

$$B(k) \cong r(k_0)A(k) \tag{7.94a}$$

and

$$C(k) \cong t(k_0)A(k). \tag{7.94b}$$

These relations include a specific assumption, viz., the exclusion of potentials that cause $B(k)$ or $C(k)$ to vary rapidly with k. In a similar fashion, we write

$$kA(k) \cong k_0 A(k) \tag{7.95a}$$

and

$$KC(k) \cong K_0 C(k). \tag{7.95b}$$

Finally, $\exp[2i(k-k')x_a]$ and $\exp[2i(K-K')x_a]$ are each approximated by unity.[6]

With these approximations, Eq. (7.92) takes the following form:

$$v_{k_0}\int dk\, dk'\, A^*(k')A(k)e^{i(\omega_k-\omega_{k'})(t+T)}$$

$$= [v_{k_0}|r(k_0)|^2 + v_{K_0}|t(k_0)|^2]\int dk\, dk'\, A^*(k')A(k)e^{i(\omega_k-\omega_{k'})(t+T)}. \tag{7.96}$$

The same product of integrals appears on each side of Eq. (7.96). Because $A(k)$ is localized, the k and the k' integrals are each finite and non-zero (the real and imaginary parts cannot simultaneously be zero). Therefore dividing the RHS of (7.96) by its LHS yields

$$R(k_0) + T(k_0) = 1, \tag{7.97}$$

where we have introduced the reflection coefficient

[6] This result may *not* be achieved by setting $x_a = 0$, since x_a is necessarily a large distance, never to be replaced by zero.

$$R(k_0) \equiv |r(k_0)|^2 \tag{7.98}$$

and the transmission coefficient

$$T(k_0) \equiv \frac{v_{K_0}}{v_{k_0}} |t(k_0)|^2 = \frac{K_0}{k_0} |t(k_0)|^2. \tag{7.99}$$

Equation (7.97) is a statement of particle (and current) conservation. Although it has been obtained by assuming a wave packet with a very narrow profile function, so that corrections to it will be very small, it no longer refers to the wave-packet analysis. In fact, (7.97)–(7.99) could just as well have been derived by working with plane waves *ab initio*, as long as the relevant expressions are substituted into (7.89) or (7.91), *NOT* into Eq. (7.84): the use of wave packets is essential in setting the term J_{interf} equal to zero, Eq. (7.88). If plane waves are used, $J_{\text{interf}} \neq 0$.

Although the preceding results follow if one re-calculates everything starting with plane waves in place of wave packets, it is simpler to note that a plane wave is obtained as the limit of an infinitely narrow profile function, that is, via the replacements

$$A(k) \to 2\pi A\delta(k - k_0) \tag{7.100}$$

$$\Rightarrow$$

$$\Psi_{\text{inc}}(x, t) \to Ae^{i(k_0 x - \omega_{k_0} t)}, \tag{7.101}$$

$$\Psi_{\text{refl}}(x, t) \to B(k_0)e^{-i(k_0 x + \omega_k t)} = r(k_0)Ae^{-i(k_0 x + \omega_{k_0} t)}, \tag{7.102}$$

and

$$\Psi_{\text{tr}}(x, t) \to C(k_0)e^{i(K_0 x - \omega_{k_0} t)} = t(k_0)Ae^{i(K_0 x - \omega_{k_0} t)}, \tag{7.103}$$

so that

$$J_{\text{inc}} \to |A|^2 v_{k_0}, \tag{7.104}$$

$$J_{\text{refl}} \to -|r(k_0)|^2|A|^2 v_{k_0} = -R(k_0)|A|^2 v_{k_0}, \tag{7.105}$$

and

$$J_{\text{tr}} \to |t(k_0)|^2|A|^2 v_{K_0} = T(k_0)|A|^2 v_{k_0}. \tag{7.106}$$

Since plane waves spread over all space, references to the initial position $-x_{\text{a}}$ and initial time $-T$ have been eliminated from Eqs. (7.101)–(7.103). A similar reduction occurs for the solution in region II:

$$\Psi_{\text{II}}(x, t) \to [D(k_0)\psi_{k_0}^{(1)}(x) + F(k_0)\psi_{k_0}^{(2)}(x)]e^{-i\omega_{k_0} t}. \tag{7.107}$$

Let us recapitulate these last sets of results. Assumption of a very narrow profile function leads to the approximate Eqs. (7.96)–(7.99); these results are exact, however, if an infinitely sharp (delta-function) profile function is used. Correspondingly, for a broad profile function, one should work with conservation of current in the form of (7.91). The use of plane waves, as in Eqs. (7.101)–(7.103), means that

$$\Psi_{\text{I}}(x, t) = Ae^{i(k_0 x - \omega_{k_0} t)} + B(k_0)e^{-i(k_0 x - \omega_{k_0} t)} \tag{7.108a}$$

and

$$\Psi_{\rm III}(x, t) = C(k_0)e^{i(k_0x-\omega_{k_0}t)}, \tag{7.108b}$$

while $\Psi_{\rm II}(x, t)$ is given by (7.107). The coefficients B, C, D, and F are still obtained from the matching conditions (7.54), with the replacement of k everywhere in these equations by k_0. Once expressions for these quantities have been obtained and the ratios $r(k_0) = B/A$ and $t(k_0) = C/A$ determined, the reflection and transmission coefficients $R(k_0)$ and $T(k_0)$ of Eqs. (7.98) and (7.99) can be calculated.

The interpretation of these coefficients is straightforward: $NR(k_0)$ and $NT(k_0)$ are the numbers of particles reflected from and transmitted by the potential out of the total number N of incident particles. These are, in principle, measurable quantities. Note that the distance $x_{\rm a}$, which occurs in the wave-packet description, is macroscopic, corresponding to the accelerators and detectors in real collision experiments being macroscopic distances from the target (potential well). At such (asymptotic) distances, the wave functions are those of free particles, which are $\exp(\pm ik_0x)$ in 1-D. Scattering experiments are carried out in order to probe the target and its quantal behavior, features that determine the coefficients $D(k_0)$ and $F(k_0)$. These quantities influence $r(k_0)$ and $t(k_0)$ but are not observable, in contrast to $R(k_0)$ and $T(k_0)$, whose behaviors as a function of k_0 would be the only exact information (in a 1-D situation) from which one could attempt to infer details of the potential. To this might be added physical intuition, etc. In the applications discussed in the next two sections, $V(x)$ is given, and we solve for $R(k_0)$ and $T(k_0)$, so that their behavior can be discussed in detail. Note that, if the added potential V_1 is zero, then $K_0 \to k_0$ and $T(k_0) \to |t(k_0)|^2$.

7.3. Plane-wave scattering: attractive potentials

In this section, we study plane-wave scattering by the three purely attractive, local potentials whose bound states were analyzed in Section 6.3. We consider first the square well (Eq. (6.175)) and then the delta-function potential (Eq. (6.181)). The former is more than a soluble scattering problem: it also provides a 1-D model of the Ramsauer–Townsend effect, a phenomenon observed both in atomic and in nuclear collisions. Scattering by the half-space square well (Eq. (6.162)) is the final example considered. Since in this case the region $x < 0$ is forbidden to the particle, the incident beam must originate from the far right rather than from the far left, as has been our previous assumption. As we shall see, this aspect creates no difficulty. The last topic in this subsection will be the establishing of a connection between the bound and scattering states in the same potential. "Inelastic" scattering is examined in the next section.

The attractive square well

This potential is given by Eq. (6.175), which we repeat here:

$$V(x) = \begin{cases} 0, & x \leqslant L_1 \quad \text{(Region I)}, \\ -V_0, & L_1 \leqslant x \leqslant L_2 \quad \text{(Region II)}, \\ 0, & x \geqslant L_2 \quad \text{(Region III)}. \end{cases} \tag{7.109}$$

Shown in Fig. 7.7 is the square-well potential, a horizontal line representing the value of

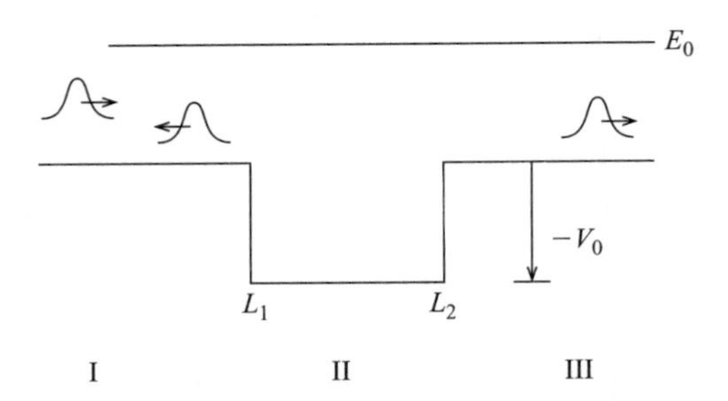

Fig. 7.7 The square well of Eq. (7.109), the three regions, the incident energy E_0, and the schematic wave packets from which the incident, reflected, and transmitted plane waves are abstracted.

the incident energy $E_0 = \hbar^2 k_0^2/(2m)$, and schematic wave packets for which the plane waves are the abstractions.

In this example, the initial beam of particles moves from L to R, so that the incident, reflected, and transmitted plane waves are given by Eqs. (7.101)–(7.103), while Eqs. (7.108a) and (7.108b) are the region-I and region-III wave functions, with K_0 replaced by k_0 in Eqs. (7.103) and (7.108b).

The region-II wave function is the solution of the Schrödinger equation

$$-\frac{\hbar^2}{2m}\psi''(x) - V_0\psi(x) = \frac{\hbar^2 k_0^2}{2m}\psi(x), \qquad L_1 \leqslant x \leqslant L_2, \tag{7.110a}$$

which can be re-written as

$$\psi''(x) + k_1^2\psi(x) = 0, \qquad L_1 \leqslant x \leqslant L_2, \tag{7.110b}$$

where

$$k_1^2 = k_0^2 + \frac{2m}{\hbar^2}V_0.$$

Here, k_1 is the wave number for the energy measured from the bottom of the potential well.

The two solutions to (7.110b) are the pairs $\exp(\pm ik_1x)$ or their real and imaginary parts, $\sin(k_1x)$ and $\cos(k_1x)$. Since the Hamiltonian in (7.110a) commutes with the momentum operator $\hat{P}_x(x)$, we again choose the exponentials as the linearly independent solutions:

$$\psi_{k_0}^{(1)}(x) = e^{ik_1x}$$

and

$$\psi_{k_0}^{(2)}(x) = e^{-ik_1x},$$

so that $\psi_{\rm II}(x)$ is

$$\psi_{\rm II}(x) = De^{ik_1x} + Fe^{-ik_1x}, \tag{7.111}$$

where the k_0 dependence of D and F is suppressed.

Matching functions and first derivatives at $x = L_1$ and $x = L_2$, as in Eqs. (7.54), yields

$$Ae^{ik_0L_1} + Be^{-ik_0L_1} = De^{ik_1L_1} + Fe^{-ik_1L_1}, \tag{7.112a}$$

$$k_0(Ae^{ik_0L_1} - Be^{-ik_0L_1}) = k_1(De^{ik_1L_1} - Fe^{-ik_1L_1}), \tag{7.112b}$$

$$Ce^{ik_0L_2} = De^{ik_1L_2} + Fe^{-ik_1L_2}, \tag{7.112c}$$

and

$$k_0Ce^{ik_0L_2} = k_1(De^{ik_1L_2} - Fe^{-ik_1L_2}). \tag{7.112d}$$

Although the set (7.112) can be solved for B, C, D, and F using determinants, this is cumbersome, given the structure of these equations. A simpler procedure is to solve Eqs. (7.112c) and (7.112d) for D and F in terms of C and then use the resulting expressions to eliminate D and F in (7.112a) and (7.112b) in terms of C. Doing so leaves two somewhat complicated algebraic equations, which can be solved to give the scattering parameters B and C in terms of A, k_0, k_1, L_1, and L_2. On carrying out these steps, one finds

$$r(k_0) = \frac{B}{A} = \frac{i[(k_1^2 - k_0^2)/(2k_0k_1)]\sin(k_1L)\,e^{2ik_0L_1}}{d} \tag{7.113}$$

and

$$t(k_0) = \frac{C}{A} = \frac{e^{-ik_0L}}{d}, \tag{7.114}$$

where

$$d = \cos(k_1L) - i[(k_1^2 + k_0^2)/(2k_0k_1)]\sin(k_1L), \tag{7.115}$$

and $L = L_2 - L_1$.

The quantities in the square brackets in (7.113) and (7.115) are easily converted into ratios involving E_0 and V_0:

$$\frac{k_1^2 - k_0^2}{2k_0k_1} = \frac{V_0}{2\sqrt{E_0(E_0 + V_0)}}$$

and

$$\frac{k_1^2 + k_0^2}{2k_0k_1} = \frac{2E_0 + V_0}{2\sqrt{E_0(E_0 + V_0)}}.$$

In terms of them, the reflection and transmission coefficients are

$$R(k_0) = |r(k_0)|^2 = \left(\frac{V_0^2}{4E_0(E_0 + V_0)}\sin^2(k_1L)\right)T(k_0) \tag{7.116}$$

and

$$T(k_0) = \frac{1}{1 + \{V_0^2/[4E_0(E_0 + V_0)]\}\sin^2(k_1L)}, \tag{7.117}$$

from which it is readily seen that conservation of current is satisfied, i.e., that $R(k_0) + T(k_0) = 1$.

The dependences of $R(k_0)$ and $T(k_0)$ on the interval length L are only through

$\sin^2(k_1L)$, the quantity $V_0^2/[4E_0(E_0 + V_0)]$ being a factor that can be treated independently of the $\sin^2(k_1L)$ term. Keeping E_0 and V_0 constant, it is evident that, as L varies, $R(k_0) \to 0$ as $T(k_0) \to 1$, but $R(k_0)$ always remains less than unity for any $E_0 \neq 0$. That is, transmission always occurs as long as $k_0 \neq 0$. However, for $k_1L = n\pi$, $n = 1, 2, \ldots$, there will be no reflection at all, and the entire "beam" of particles is transmitted through the square well.

The behaviors of $T(k_0)$ and $R(k_0)$ as functions of L are shown in Fig. 7.8 for a fixed value of $f_0 = V_0^2/[4E_0(E_0 + V_0)]$ and L between $n\pi/k_1$ and $(n+1)\pi/k_1$. When k_1L/π is an integer, only transmission occurs, and the square well becomes transparent to the incident beam. That is, the square well becomes transparent to the incident beam when L equals an integer number of half wavelengths, i.e., when $L = n(\lambda_1/2)$, where $\lambda_1 = 2\pi/k_1$. Transparency is a constructive interference effect among the waves inside the well that contribute to transmission; correspondingly, there is destructive interference among the waves that contribute to reflection, i.e., the waves reflecting back from L_2 and transmitted at L_1 are π radians out of phase with the waves reflecting directly from L_1; this gives rise to perfect cancelation and $R(k_0) = 0$.

This interference phenomenon in the present case of the square well serves as a 1-D model for and analog of the Ramsauer–Townsend effect, viz., a minimum in the scattering cross section (a quantity proportional to the number of particles scattered, Chapter 15) seen at very low energies in some electron–atom and neutron–nucleus collisions. This minimum corresponds to $T(k_0) = 1$ in the 1-D case, and may be thought of as a transparency of the target, which acts like a sphere with a well-defined radius (see, e.g., Massey and Burhop (1952) for the atomic case and Hodgson (1971) for the nuclear case).

In addition to $T(k_0)$ periodically achieving the value unity, two other limiting aspects of $T(k_0)$ and $R(k_0)$ are of interest. These are their behaviors for low energy and high energy, defined here by the following relations:

$$\text{Low energy:} \quad E_0/V_0 \ll 1,$$
$$\text{High energy:} \quad E_0/V_0 \gg 1.$$

We can invoke semi-classical arguments to predict the behaviors of $T(k_0)$ and $R(k_0)$ in each of these limiting cases. Consider high energies first. When E_0 is large, the semi-

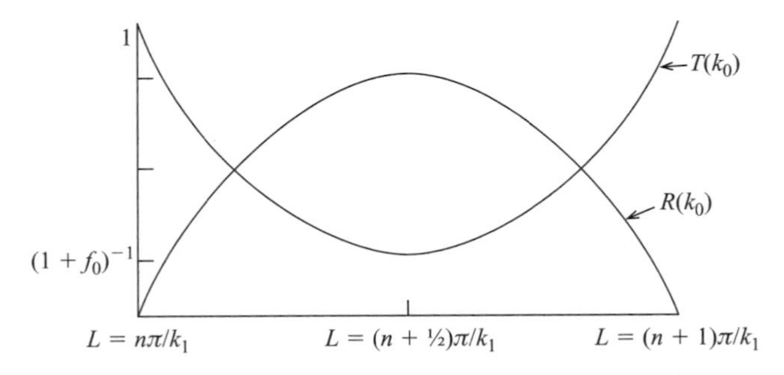

Fig. 7.8 Plots of $T(k_0)$ and $R(k_0)$ as functions of the width L of the square well, for L in the range $[n\pi/k_1, (n+1)\pi/k_2]$. The quantity f_0 is the ratio $V_0^2/[4E_0(E_0 + V_0)]$, assumed fixed (and approximately equal to 4).

classical velocity $v_{k_0} = \hbar k_0/m$ is large, implying that the incident particle will pass through the potential well in such a short time that it will scarcely feel the interaction. In other words, it should "sail through," implying that $T(k_0) \cong 1$ and consequently $R(k_0) \cong 0$ at high energies. For very low energies, on the other hand, the particle takes such a long time to traverse the well that transmission should be inhibited in comparison with the reflection that occurs when the particle encounters the sharp edge of the square well at L_1. In this low-energy case, therefore, we expect $R(k_0) \gg T(k_0)$, which suggests that $T(k_0)$ should be zero for $E_0 = 0$.

The foregoing conclusions are supported by analysis. In the case of high energies, one has

$$\frac{V_0^2}{4E_0(E_0 + V_0)} = \frac{V_0^2}{4E_0^2}\frac{1}{1 + V_0/E_0} \cong \frac{V_0^2}{4E_0^2},$$

so that

$$T(k_0) \cong 1 - \frac{V_0^2 \sin^2(k_1 L)}{4E_0^2} \underset{E_0/V_0 \to \infty}{\longrightarrow} 1$$

and

$$R(k_0) \cong \frac{V_0^2 \sin^2(k_1 L)}{4E_0^2} \underset{E_0/V_0 \to \infty}{\longrightarrow} 0.$$

For low energies, we find

$$\frac{V_0^2}{4E_0^2(1 + V_0/E_0)} \gg 1,$$

in which case

$$T(k_0) \cong \frac{4E_0}{V_0 \sin^2(k_1 L)}$$

and thus

$$R(k_0) \cong 1 - \frac{4E_0}{V_0 \sin^2(k_1 L)},$$

where it is explicitly assumed that $k_1 L \neq n\pi$, where n is an integer.

An amusing point is that, at low energies, $T(k_0)$ goes to zero linearly in the small parameter E_0/V_0, whereas at high energies $R(k_0)$ goes to zero quadratically in the now-small parameter V_0/E_0. One cannot eliminate transmission except at exceedingly low energies.

The delta-function potential

From the definition (4.125) of the delta function as the limit of a rectangular distribution, one should be able to derive expressions for the reflection and transmission coefficients for an attractive delta-function potential by taking the appropriate limits in Eqs. (7.116) and (7.117). Although this is possible, such a procedure circumvents the problem of determining the relevant (plane-wave) solutions. We therefore proceed *ab initio*.

The delta-function potential is given by Eq. (6.181). In keeping with the expression (6.187) deduced for the attractive, bound-state case, we write (6.181) in the form

$$V(x) = -\frac{\hbar^2}{mx_0}\delta(x), \tag{7.118}$$

where the point $x = 0$ has been chosen as the position of the potential.

Under our standard assumption that the incident energy is $E_0 = \hbar^2 k_0^2/(2m)$, and using (7.118), the time-independent Schrödinger equation for the eigenfunction $\psi(x)$ takes the form

$$\psi''(x) + \frac{2}{x_0}\delta(x)\psi(x) = -k_0^2\psi(x). \tag{7.119}$$

As in the bound-state problem, region II is the point $x = 0$, so that the analysis in the delta-function-potential portion of Section 6.6 carries over to the scattering regime.

The spatial parts of the incident, reflected, and transmitted plane waves are

$$\psi_{\text{inc}}(x) = Ae^{ik_0x},$$

$$\psi_{\text{refl}}(x) = Be^{-ik_0x},$$

and

$$\psi_{\text{tr}}(x) = Ce^{ik_0x},$$

where the incident beam is assumed to move from L to R. These lead to the following region-I and region-III eigenfunctions:

$$\psi_{\text{I}}(x) = Ae^{ik_0x} + Be^{-ik_0x}, \qquad x<0 \tag{7.120}$$

and

$$\psi_{\text{III}}(x) = Ce^{ik_0x}, \qquad x>0, \tag{7.121}$$

which are schematically illustrated in Fig. 7.9.

As in the bound-state example, $\psi(x)$ must be continuous, which means that

$$\psi_{\text{I}}(x=0) = \psi_{\text{III}}(x=0)$$

or

$$A + B = C. \tag{7.122}$$

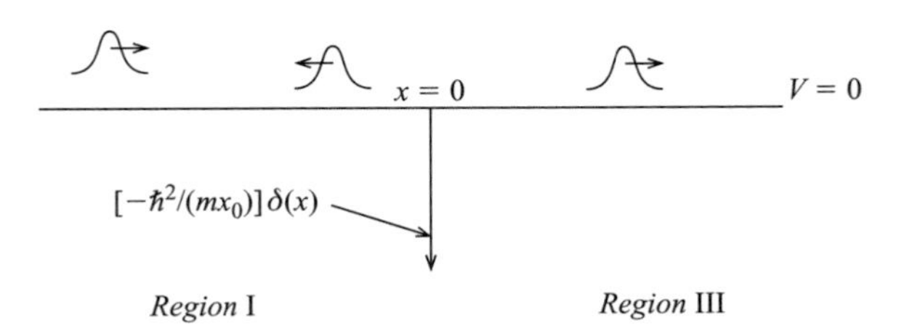

Fig. 7.9 The attractive delta-function potential, with a schematic representation of the wave packets which the incident, reflected, and transmitted plane waves represent.

This single relation replaces the pair (7.54a) and (7.54c).

On the other hand, the presence of the delta function in Eq. (7.119) implies that $\psi'(x)$ has a discontinuity across the point $x = 0$. This means that we may not take the derivatives of ψ_{I} and of ψ_{III}, and then equate them at $x = 0$. Instead, we follow the same procedure as in the bound-state case and integrate each term in (7.119) on x between the points $-\epsilon$ and $+\epsilon$, and then let $\epsilon \to 0$. The result is

$$ik_0 C - ik_0(A - B) + \frac{2}{x_0}(A + B) = 0. \tag{7.123}$$

Since $C = A + B$ by Eq. (7.122), this relation yields

$$B = \frac{-A}{1 + ik_0 x_0}, \tag{7.124}$$

from which one gets

$$C = \frac{ik_0 x_0}{1 + ik_0 x_0} A, \tag{7.125}$$

as well as

$$R(k_0) = \frac{1}{1 + (k_0 x_0)^2} \tag{7.126}$$

and

$$T(k_0) = \frac{(k_0 x_0)^2}{1 + (k_0 x_0)^2}. \tag{7.127}$$

It is obvious that current is conserved ($R + T = 1$), although it requires a short calculation to establish that (7.124) and (7.125) do lead to a discontinuity in $\psi'(x)$ across the point $x = 0$. The high- and low-energy limits of R and T are the same as in the square-well case, viz.,

$$\lim_{k_0 \to \infty} \begin{cases} R(k_0) \to 0 \\ T(k_0) \to 1 \end{cases}$$

and

$$\lim_{k_0 \to 0} \begin{cases} R(k_0) \to 1 \\ T(k_0) \to 0. \end{cases}$$

The half-space square well

Bound states in the half-space square well were studied in Section 6.4. The functional form of the potential is

$$V(x) = \begin{cases} \infty, & x \leqslant 0 \quad \text{(Region I)}, \\ -V_0, & 0 \leqslant x \leqslant L \quad \text{(Region II)}, \\ 0, & x \geqslant L \quad \text{(Region III)}. \end{cases} \tag{7.128}$$

In this case of scattering, both the incident and the reflected particles must be in region

III, since the infinite value of $V(x)$ in region I requires the wave function to be zero there. Hence, our solution must produce $|R(k_0)| = 1$.

Since the incident beam is in region III, $\psi_{\text{inc}}(x)$ must refer to a wave moving from R to L:

$$\psi_{\text{inc}}(x) = Ae^{-ik_0x}, \tag{7.129}$$

where, as usual, A is assumed known. The negative sign in (7.129) is a statement that the particle moves from R to L: (7.129) is for the incident, not for the reflected, wave. Therefore, it is $\psi_{\text{refl}}(x)$ that must represent motion from L to R:

$$\psi_{\text{refl}}(x) = Be^{ik_0x}, \qquad x > L. \tag{7.130}$$

The total wave function in region III is the sum of ψ_{inc} and ψ_{refl}:

$$\psi_{\text{III}}(x) = Ae^{-ik_0x} + Be^{ik_0x}. \tag{7.131}$$

We have already remarked that $\psi_{\text{I}}(x)$ must be zero. Since $\psi(x)$ must be continuous at $x = 0$, $\psi_{\text{II}}(x)$ must vanish at $x = 0$:

$$\psi_{\text{II}}(x = 0) = 0. \tag{7.132}$$

The analysis of Section 6.4 indicates that $\psi_{\text{II}}(x)$ should be a sine function. Its argument must be k_1x, where $k_1^2 = 2m(E_0 + V_0)/\hbar^2$, as in the square-well case, a claim easily verified by examining the Schrödinger equation in region II. We have

$$-\frac{\hbar^2}{2m}\psi_{\text{II}}''(x) - V_0\psi_{\text{II}}(x) = E_0\psi_{\text{II}}(x) = \frac{\hbar^2 k_0^2}{2m}\psi_{\text{II}}(x), \tag{7.133}$$

or

$$\psi_{\text{II}}''(x) + k_1^2\psi_{\text{II}}(x) = 0, \tag{7.134}$$

with

$$k_1^2 = k_0^2 + \frac{2m}{\hbar^2}V_0,$$

as noted in the foregoing. The only solution to (7.134) that satisfies (7.132) is

$$\psi_{\text{II}}(x) = D\sin(k_1x), \qquad 0 \leqslant x \leqslant L; \tag{7.135}$$

the analog with the bound-state case, Eq. (6.166), is exact.

Both B and D can be expressed in terms of A using the matching conditions (7.54c) and (7.54d), with $x_2 = L$. From (7.134) and (7.135) one gets

$$D\sin(k_1L) = Ae^{-ik_0L} + Be^{ik_0L}$$

and

$$k_1D\cos(k_1L) = -ik_0(Ae^{-ik_0L} - Be^{ik_0L}).$$

This pair yields the result

$$r(k_0) = \frac{B}{A} = -e^{-2ik_0L}\frac{k_1L\cot(k_1L) + ik_0L}{k_1L\cot(k_1L) - ik_0L}. \tag{7.136}$$

It was remarked above that $R(k_0) = |r(k_0)|^2 = 1$ must occur; because of this, the general form of (7.136) is a *necessity*, i.e., the numerator and denominator of the ratio occurring in $r(k_0)$ must be complex conjugates of one another, for only this relation leads to $|r(k_0)| = 1$. To see this, we write

$$k_1 L \cot(k_1 L) + ik_0 L \equiv K e^{i\theta}, \tag{7.137}$$

where

$$K = \{[k_1 L \cot(k_1 L)]^2 + k_0^2 L^2\}^{1/2}$$

and

$$\theta = \tan^{-1}\{k_0 L/[k_1 L \cot(k_1 L)]\}.$$

Using (7.137), Eq. (7.136) becomes

$$B = -Ae^{2i\delta(k_0)} = r(k_0)A, \tag{7.138}$$

where the phase (shift) $\delta(k_0)$ is given by

$$\delta(k_0) = \theta - k_0 L. \tag{7.139}$$

We see that B differs from A only by a shift or change of phase: $|B| = |A|$, i.e., $|r(k_0)| = 1$, as it must be in order to ensure conservation of current. This, however, has a profound effect on $\psi_{\mathrm{III}}(x)$, as substitution of (7.139) into (7.131) shows:

$$\begin{aligned}\psi_{\mathrm{III}}(x) &= A(e^{-ik_0x} - e^{2i\delta(k_0)}e^{ik_0x}) \\ &= 2iAe^{i\delta(k_0)}\sin(k_0x + \delta(k_0)).\end{aligned} \tag{7.140}$$

It is amusing to compare this form for $\psi_{\mathrm{III}}(x)$ with the corresponding result for the case of there being no attractive well, just an infinitely wide and infinitely high wall extending from $x = 0$ to $x = -\infty$. A calculation like the preceding one leads to

$$\psi_{\mathrm{III}}(x) = 2iA\sin(k_0x), \qquad V_0 = 0, \tag{7.141}$$

a sine wave for all $x > 0$, and in particular, for $x \geqslant L$. We see that the only change in the (sine) functional form of ψ_{II} on going from the $V_0 = 0$ case to the $V_0 \neq 0$ case is the appearance of the phase shift $\delta(k_0)$. Apart from the factor $\exp(i\delta(k_0))$, this shift in the phase of the sine function is the only manifestation of the existence of a non-zero potential. As we show in Chapter 15 on 3-D scattering, this phenomenon occurs in that case as well.

It is straightforward to obtain the coefficient D in Eq. (7.135):

$$D = 2iAe^{i\delta(k_0)}\frac{\sin(k_0L + \delta(k_0))}{\sin(k_1L)},$$

a result also forced by continuity of $\psi(x)$ at $x = L$. These results are illustrated in Fig. 7.10.

The high-energy limit, viz., $E_0 \gg V_0$, is readily seen to yield $\delta(k_0) \to 0$, i.e., for sufficiently large incident momentum, the particle travels too fast to be disturbed by the potential. That is, for $E_0 \gg V_0$, the physics is the same as if $V_0 = 0$. The behaviors of $\delta_0(k_0)$ and $\psi_{\mathrm{III}}(x)$ in the low-energy limit are left as an exercise.

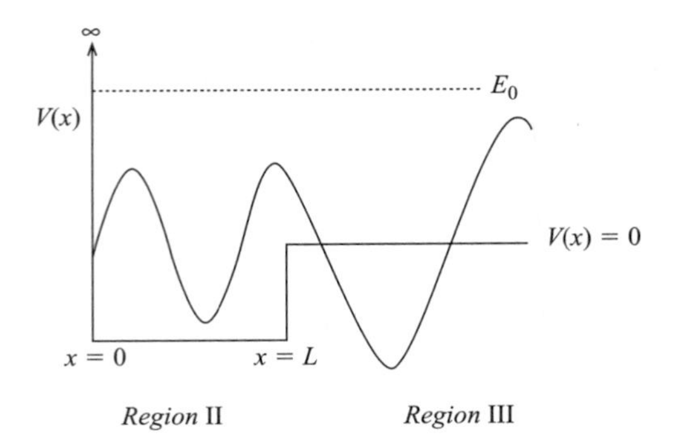

Fig. 7.10 The half-space square well, a straight line representing the incident energy E_0, and the wave functions $\psi_{\rm II}(x)/[i\exp(i\delta)]$ and $\psi_{\rm III}(x)/[i\exp(i\delta)]$ for arbitrary choices of k_0, k_1, and L. Note that the periods and amplitudes of these two wave functions differ.

The bound-state/scattering-state connection

In regions I and III, where the attractive confining potential is zero, the wave function $\psi(x)$ obeys

$$\psi''(x) + \frac{2m}{\hbar^2}E\psi(x) = 0. \tag{7.142}$$

For bound states, for which $E = -\hbar^2\alpha^2/(2m) < 0$, (7.142) becomes

$$\psi''(x) - \alpha^2\psi(x) = 0, \tag{7.143}$$

and the solutions are the appropriate choice of

$$\psi_\alpha(x) = e^{+\alpha x} \text{ and } e^{-\alpha x}.$$

In the case of scattering, $E = \hbar^2 k_0^2/(2m) > 0$, (7.142) goes to

$$\psi''(x) + k_0^2\psi(x) = 0, \tag{7.144}$$

and the solutions are appropriate linear combinations of

$$\psi_{k_0}(x) = e^{ik_0x} \text{ and } e^{-ik_0x}.$$

It is clear that, if we replace k^2 by $-\alpha^2$, then (7.144) is identical to (7.143). There is thus a simple relation between the Schrödinger equation for scattering and the Schrödinger equation for bound states (in regions I and III). While the structure of the differential equation is determined by the sign of the energy, the wave functions depend on the square root of E, i.e., on α and k_0. If we replace positive E by negative E, it is not clear which of $i\alpha$ and $-i\alpha$ should replace k_0. This ambiguity is apparently magnified when we consider the scattering solution in the region containing the incident and reflected waves. For, no matter which of the two substitutions $k_0 \to \pm i\alpha$ is made, the scattering solution appears not to go over to the bound-state solution: either the incident or the reflected wave will be transformed into the growing exponential $\exp(\alpha|x|)$, which is not allowed. This apparent failure to produce only a decaying exponential might suggest that one cannot connect the scattering and bound-state *solutions*, in contrast to the $E \to -E$ connection between the Schrödinger *equation* for scattering and bound states.

Such a conclusion is, in fact, false: not only can one go from the scattering to the correct bound-state solutions, but, almost as a bonus, one also obtains the bound-state energies (wave numbers) as well. However, it is essential in doing this to take account of the coefficient $r(k_0)$. Let us see how this happens.

We consider the case in which the particle is incident from L to R, as in the square-well and delta-function examples. The region-I wave function is

$$\psi_{\rm I}(x) = A[e^{ik_0x} + r(k_0)e^{-ik_0x}], \qquad x < x_1, \tag{7.145}$$

where $r(k_0) = B/A$. Since $x < x_1$, (7.145) can be written

$$\psi_{\rm I}(x) = A[e^{-ik_0|x|} + r(k_0)e^{ik_0|x|}], \qquad x < 0, \tag{7.146}$$

Furthermore, in the $x > 0$ portion of region III ($x = |x|$), the wave function is

$$\psi_{\rm III}(x) = A[t(k_0)e^{ik_0|x|}]. \tag{7.147}$$

The achievement of a proper bound-state wave function in region III is guaranteed only by the prescription $k_0 \to i\alpha$, since then

$$\psi_{\rm III}(x) \underset{k_0 \to i\alpha}{\longrightarrow} A[t(i\alpha)e^{-\alpha|x|}],$$

as required on physical grounds.

The same prescription must be used in $\psi_{\rm I}(x)$, which becomes

$$\psi_{\rm I}(x) \underset{k_0 \to i\alpha}{\longrightarrow} A[e^{\alpha|x|} + r(i\alpha)e^{-\alpha|x|}]. \tag{7.148}$$

Equation (7.148) now has the (unacceptable) growing exponential term. This term will dominate (7.148) unless the factor $r(i\alpha)$ becomes *infinite* at those $\alpha = \alpha_n$ which correspond to bound-state wave numbers, since, for any finite $|x|$, $r(i\alpha_n)\exp(-\alpha_n|x|)$ would then overwhelm $\exp(\alpha_n|x|)$. Precisely this situation occurs! In the language of complex variable theory,[7] if k_0 is allowed to become a complex variable, then $\psi_{\rm I}(x)$ becomes a complex function (of k_0) that has a singularity at $k_0 = i\alpha_n$; the residue of the singularity (which generally is a simple pole) is the bound-state wave function $\exp(-\alpha_n|x|)$.

We demonstrate this for the three examples of scattering analyzed in the preceding subsections. As will be seen, in each case – and in general – $r(k_0)$ (or the coefficient B) can be expressed as a ratio:

$$r(k_0) = \frac{N(k_0)}{D(k_0)}, \tag{7.149}$$

such that

$$D(i\alpha_n) = 0 \tag{7.150}$$

while $N(i\alpha_n) \neq 0$.

[7] Readers who are not familiar with the elements of the theory of complex variables can find brief but adequate introductions in, e.g., Dettman (1988), Hassani (1991), and Wyld (1976). However, complex-variable theory is not essential: it suffices to simply use the prescription $k_0 \to i\alpha$.

Referring to Eqs. (7.113), (7.124), and (7.136), we find the following expressions for $D(k_0)$:

$$D(k_0) = [2k_0k_1\cos(k_1L) - i(k_1^2 + k_0^2)\sin(k_1L)]/(2k_0k_1) \qquad \text{(square well)}, \tag{7.151}$$

$$D(k_0) = -(1 + ik_0x_0), \qquad \text{(delta function)}, \tag{7.152}$$

$$D(k_0) = k_1L\cot(k_1L) - ik_0L \qquad \text{(half-space square well)}. \tag{7.153}$$

In Eqs. (7.151) and (7.153), the quantity k_1 is given by

$$k_1 = \sqrt{\frac{2m}{\hbar^2}V_0 + k_0^2}.$$

Furthermore, for these two square-well cases, the region-II bound-state wave number γ is given by

$$\gamma = \sqrt{\frac{2m}{\hbar^2}V_0 - \alpha^2}.$$

Hence, when $k_0 \to i\alpha$, k_1 is also transformed:

$$k_1 \underset{k_0 \to i\alpha}{\longrightarrow} \gamma. \tag{7.154}$$

With the replacement (7.154), the limit $k_0 \to i\alpha$ of each of the three expressions for $D(k_0)$ yields

$$D(i\alpha) = [2\alpha\gamma\cos(\gamma L) - (\gamma^2 - \alpha^2)\sin(\gamma L)]/(2\alpha\gamma) \qquad \text{(square well)}, \tag{7.155}$$

$$D(i\alpha) = -(1 - \alpha x_0) \qquad \text{(delta function)}, \tag{7.156}$$

$$D(i\alpha) = \gamma L\cot(\gamma L) + \alpha L \qquad \text{(half-space square well)}. \tag{7.157}$$

Comparison of (7.155), (7.156), and (7.157) with (6.180), (6.188), and (6.170), respectively, shows that, in each case, $D(i\alpha) = 0$ at exactly these values of α at which bound states occur. Therefore, as claimed, the zeros of $D(k_0)$ – or equivalently, the poles of $r(k_0)$ – occur at the bound-state wave numbers. On the basis of these examples, we further conclude that, if one can determine $r(k_0)$ for any attractive well, then one obtains the bound-state wave numbers at the same time via $r(i\alpha) = \infty$ or $D(i\alpha) = N(i\alpha)/r(i\alpha) = 0$.

The present analysis began by examining the case of a particle incident from L to R, as exemplified by the square-well and delta-function potentials. However, the prescription $k_0 \to +i\alpha$ is valid also when the projectile moves from R to L, as indicated by the result for $D(k_0)$ in the case of the half-space square well. That the corresponding scattering wave function also has the proper behavior in the limit $k_0 \to i\alpha$ is readily shown. This scattering-state/bound-state relationship holds for any potential that can support at least one bound state, and in 3-D as well.

7.4. Plane-wave scattering: potential barriers and tunneling phenomena

The converse of an attractive potential is termed a repulsive potential or a potential barrier. Thus, the opposite of a square well is a square barrier, a repulsive delta function is the opposite of an attractive one. These two contrasting pairs are shown in Fig. 7.11.

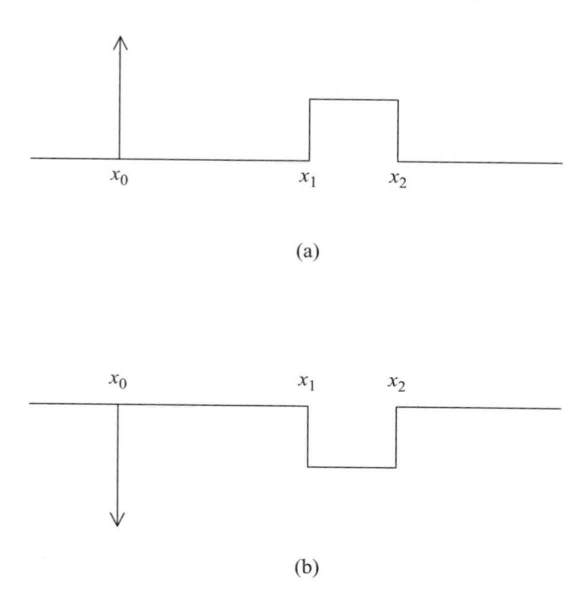

Fig. 7.11 The repulsive delta function and square barriers of (a) are converse, respectively, to their attractive counterparts of (b).

Attractive (1-D) potentials defined throughout $[-\infty, \infty]$ have the special feature of supporting bound states. While repulsive potentials cannot sustain bound states, they enjoy the unique feature of *tunneling*: particles can penetrate not only into but through any potential barrier of finite width. (This behavior is the positive-energy counterpart of the non-zero probability that a particle bound in a finite potential can be in the classically forbidden region, i.e., beyond the edge(s) of the well.) Put more dramatically, from a quantal perspective, tunneling means that there is a non-zero probability that, instead of rebounding from a wall, a ball thrown at it will go through it!

Tunneling through repulsive barriers is commonplace in micro-physics. An example is the penetration of the Coulomb barrier between two positively charged nuclei (with charges q_1 and q_2), a necessary ingredient in the occurrence of nuclear reactions. Another example is the penetration of the angular-momentum barrier that exists in 3-D for particles whose wave functions are not spherically symmetric (i.e. that depend on $\mathbf{r}$ rather than solely on r).

The potential step

To set the stage for our study of barrier-penetration phenomena, we consider a particularly simple, idealized situation, viz., an infinitely wide step potential, whose functional form is

$$V(x) = \begin{cases} 0, & x \leqslant 0 \quad \text{(Region I)}, \\ V_0, & x \geqslant 0 \quad \text{(Region II)}, \end{cases} \tag{7.158}$$

where $V_0 > 0$ is the height of the step. This potential is sketched in Fig. 7.12, along with

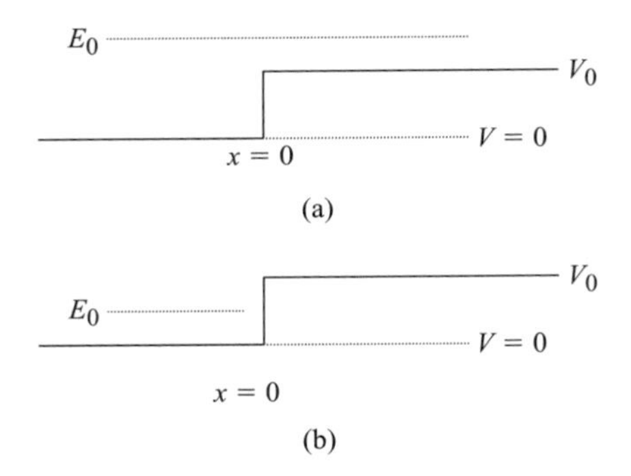

Fig. 7.12 Scattering by a potential step of height V_0: (a) incident $E_0 > V_0$ and (b) incident energy E_0 less than the barrier height.

two values for the incident energy E_0: $E_0 > V_0$ and $E_0 < V_0$. The former will give rise to reflection and transmission, the latter to reflection and penetration but not to transmission, as we now show.

Since $V(x) = 0$ for $x \leqslant 0$, the region-I wave function $\psi_{\text{I}}(x)$ is the usual sum of an incident (L to R) and a reflected (R to L) wave:

$$\psi_{\text{I}}(x) = Ae^{ik_0x} + Be^{-ik_0x}. \tag{7.159}$$

The form of the region-II wave function $\psi_{\text{II}}(x)$ will depend on the relative size of E_0 and V_0. $\psi_{\text{II}}(x)$ obeys

$$\psi''_{\text{II}}(x) + \frac{2m}{\hbar^2}(E_0 - V_0)\psi_{\text{II}}(x) = 0. \tag{7.160}$$

When $E_0 > V_0$, the solution is a plane wave, but when $E_0 < V_0$, the solution is a decaying exponential. We consider these two cases separately.

(a) $E_0 > V_0$. When $E_0 > V_0$, then $K_0^2 = 2m(E_0 - V_0)/\hbar^2 > 0$ and $\psi_{\text{II}}(x)$ is a transmitted wave with wave number $K_0 < k_0$:

$$\psi_{\text{II}}(x) = Ce^{iK_0x}, \qquad x \geqslant 0, \tag{7.161}$$

as in Eq. (7.108b). Both $\psi(x)$ and $\psi'(x)$ must be continuous at the matching point $x_1 = 0$. This leads to the following analogs of Eqs. (7.54c′) and (7.54d′):

$$A + B = C \tag{7.162a}$$

and

$$k_0(A - B) = K_0C, \tag{7.162b}$$

which yield

$$r(k_0) = \frac{B}{A} = \frac{k_0 - K_0}{k_0 + K_0} \tag{7.163}$$

and

$$t(k_0) = \frac{C}{A} = \frac{2k_0}{k_0 + K_0}.$$

Finally, the reflection and transmission coefficients are

$$R(k_0) = |r(k_0)|^2 = \left(\frac{k_0 - K_0}{k_0 + K_0}\right)^2 \tag{7.164}$$

and

$$T(k_0) = \frac{K_0}{k_0}|t(k_0)|^2 = \frac{4k_0 K_0}{(k_0 + K_0)^2}, \tag{7.165}$$

which obey the conservation law $R(k_0) + T(k_0) = 1$. Here, transmission corresponds to inelastic scattering, reflection to elastic scattering. Note that, *classically*, there would be no reflection, but conservation of flux requires it quantally. These are standard results, the only unusual feature perhaps being the numerator in $R(k_0)$: it goes to zero as $E_0 - V_0 \to 0$. However, one must not extrapolate to the latter limit by merely examining $R(k_0)$ and $T(k_0)$, since, when $E_0 = V_0$, the form of the solution $\psi_{\rm II}(x)$ changes dramatically; exploration of which is left to the exercises. These comments suggest that, for $E_0 < V_0$, the physics will alter drastically, as noted above. We consider this case next.

(b) $E_0 < V_0$. When $V_0 > E_0$, then $2m(E_0 - V_0)/\hbar^2 \equiv -\beta^2 < 0$, and (7.160) becomes

$$\psi''_{\rm II}(x) - \beta^2 \psi_{\rm II}(x) = 0, \qquad x \geqslant 0, \tag{7.166}$$

which resembles the equation for a bound state. The general solution to (7.166) is a linear combination of $\exp(\pm\beta x)$, but only the decaying exponential is allowed:

$$\psi_{\rm II}(x) = Ce^{-\beta x}, \tag{7.167}$$

a solution that diminishes to $1/e$ of its $x = 0$ value when $x = \beta^{-1}$, a length that depends on the difference of V_0 and E_0.

The $x = 0$ boundary-matching equations, corresponding to (7.162), are

$$A + B = C \tag{7.168a}$$

and

$$k_0(A - B) = i\beta C, \tag{7.168b}$$

whose solutions are

$$B = \frac{k_0 - i\beta}{k_0 + i\beta} A \tag{7.169}$$

and

$$C = \frac{2k_0}{k_0 + i\beta} A. \tag{7.170}$$

From (7.169), $|r(k_0)|$ is seen to be unity, and therefore

$$R(k_0) = 1. \tag{7.171}$$

This means that the transmission coefficient $T(k_0)$ vanishes, even though $C \neq 0$:

$$T(k_0) = 1 - R(k_0) = 0. \tag{7.172}$$

That the transmission coefficient must be zero also follows from the fact that the current density corresponding to a decaying exponential like (7.167) vanishes, quite independent of the value of the factor C.

The potential step is the second example of a potential for which there is only reflection, the other being the half-space square well analyzed in the preceding section. Just as in that case, we can re-express $\psi_{\mathrm{I}}(x)$ of (7.159) as a standing wave.

We first write $k_0 + i\beta$ in polar form, viz.,

$$k_0 + i\beta = \sqrt{k_0^2 + \beta^2}\, e^{i\delta}, \tag{7.173}$$

where

$$k_0^2 + \beta^2 = \frac{2m}{\hbar^2} V_0 \tag{7.174}$$

and

$$\tan\delta = \frac{\beta}{k_0} = \sqrt{\frac{V_0^2 - E_0^2}{E_0^2}}. \tag{7.175}$$

Using this notation, we have

$$B = e^{-2i\delta} A \tag{7.176}$$

and therefore $\psi_{\mathrm{I}}(x)$ becomes

$$\psi_{\mathrm{I}}(x) = 2Ae^{-i\delta}\cos(k_0 x + \delta). \tag{7.177}$$

The penetrating wave $\psi_{\mathrm{II}}(x)$ is readily found to be

$$\psi_{\mathrm{II}}(x) = 2Ae^{-i\delta}\sqrt{\frac{E_0}{V_0}}\, e^{-\beta x}; \tag{7.178}$$

matching these two solutions at $x = 0$ yields

$$\cos\delta = \sqrt{E_0/V_0}, \tag{7.179}$$

a result that also follows from (7.175).

Either (7.175) or (7.179) means that

$$0 < \delta < \pi/2,$$

with $\pi/2$ corresponding to $E_0 = V_0$. The maximum value of $\psi_{\mathrm{I}}(x)/[2A\exp(-i\delta)]$ is unity, which is achieved at $x = -\delta/k_0$. $\psi_{\mathrm{II}}(x)/[2A\exp(-i\delta)]$ has its maximum at $x = 0$, and it is smaller than the corresponding maximum in $\psi_{\mathrm{I}}(x)/[2A\exp(-i\delta)]$ by the factor $\sqrt{E_0/V_0}$. In addition, $\psi_{\mathrm{II}}(x)$ decays as x increases. A comparison of the region-I and region-II solutions is sketched in Fig. 7.13. The sketch is an illustration of the claim made earlier in this section, and ensured by the solution (7.167), that the incident wave penetrates into the classically forbidden region. The penetration depth β^{-1} is the quantal analog of the skin depth to which an electromagnetic wave penetrates a conductor. The sketch in Fig. 7.13 suggests a possibility that the mathematical solution may not: if the width of the step were made finite, the decaying solution inside the finite step might be

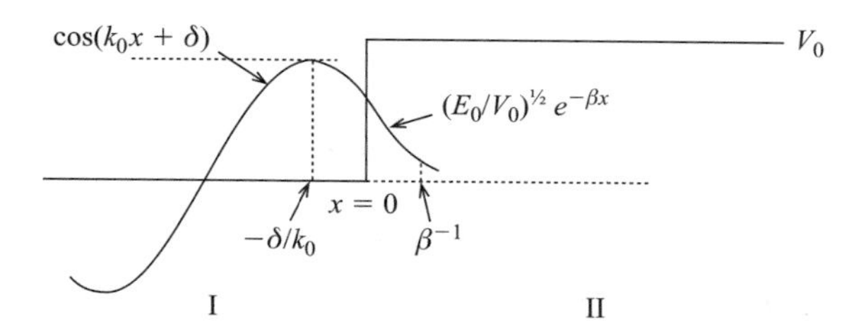

Fig. 7.13 A sketch of the renormalized region-I wave function $\cos(k_0x + \delta)$ and the region-II wave function $\sqrt{E_0/V_0}\exp(-\beta x)$ superimposed on the potential step for the case $E_0 < V_0$. The horizontal dotted line in region I is both the value of E_0 for the energy part of the diagram and the maximum value (unity) for the region-I wave function $\cos(k_0x + \delta)$.

able to leak through the right-hand edge of the step and turn into a transmitted wave. This would be "tunneling" through the barrier. We next show that this indeed happens.

The square barrier

The square barrier is illustrated in Fig. 7.14. If the barrier height is V_0, then, as in the case of the potential step, the form of solution depends on whether the incident energy E_0 is larger or smaller than V_0. We consider only the latter case here. For simplicity we have put the left-hand edge of the barrier at $x = 0$ and the right-hand edge at $x = L$. This potential has the following functional form:

$$V(x) = \begin{cases} 0, & x \leqslant 0 \quad \text{(Region I)}, \\ V_0 > 0, & 0 \leqslant x \leqslant L \quad \text{(Region II)}, \\ 0, & x \geqslant L \quad \text{(Region III)}. \end{cases} \tag{7.180}$$

The incident beam is assumed to come from the left, thus fixing the forms of solution in regions I and III. Again using the notation $\beta^2 = 2m(V_0 - E_0)/\hbar^2$, the solution in region II obeys (7.166), but, since the right-hand edge of the barrier is at the finite value $x = L$, the growing exponential $\exp(\beta x)$ must be included in $\psi_{\mathrm{II}}(x)$. We therefore have

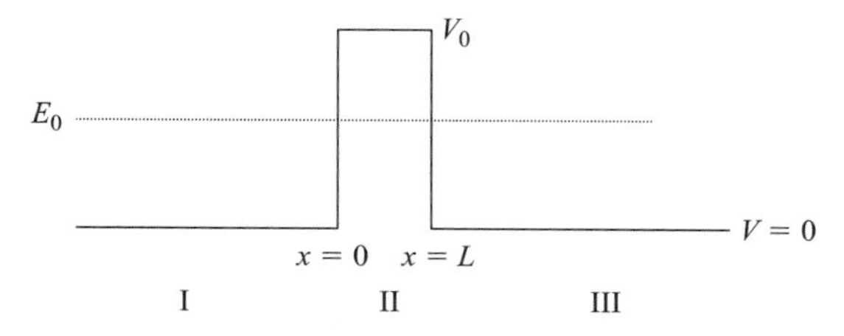

Fig. 7.14 A square barrier situated between $x = 0$ and $x = L$. The incident energy, shown by the dotted line, obeys $E_0 < V_0$. The three regions are those of Eq. (7.180).

$$\psi_{\mathrm{I}}(x) = Ae^{ik_0 x} + Be^{-ik_0 x}, \tag{7.181a}$$

$$\psi_{\mathrm{II}}(x) = De^{-\beta x} + Fe^{\beta x}, \tag{7.181b}$$

and

$$\psi_{\mathrm{III}}(x) = Ce^{ik_0 x}. \tag{7.181c}$$

The four coefficients B, C, D, and F are obtained by using the matching conditions (7.54) at the points $x = 0$ and $x = L$. This leads to

$$A + B = D + F, \tag{7.182a}$$

$$ik_0(A - B) = -\beta(D - F), \tag{7.182b}$$

$$De^{-\beta L} + Fe^{\beta L} = Ce^{ik_0 L}, \tag{7.182c}$$

and

$$-\beta(De^{-\beta L} - Fe^{\beta L}) = ik_0 Ce^{ik_0 L}. \tag{7.182d}$$

A straightforward calculation then yields the following expressions for the coefficients:

$$\frac{C}{A} = t(k_0) = \frac{e^{-ik_0 L}}{\cosh(\beta L) + i\left(\dfrac{\beta^2 - k_0^2}{2\beta k_0}\right)\sinh(\beta L)}, \tag{7.183a}$$

$$\frac{B}{A} = r(k_0) = -i\left(\frac{\beta^2 + k_0^2}{2\beta k_0}\right)\sinh(\beta L)\, e^{ik_0 L}\, t(k_0), \tag{7.183b}$$

$$\frac{D}{A} = \frac{\left(\dfrac{\beta - ik_0}{2\beta}\right)[\cosh(\beta L) + \sinh(\beta L)]}{\cosh(\beta L) + i\left(\dfrac{\beta^2 - k_0^2}{2\beta k_0}\right)\sinh(\beta L)}, \tag{7.183c}$$

and

$$\frac{F}{A} = \frac{\left(\dfrac{\beta + ik_0}{2\beta}\right)[\cosh(\beta L) - \sinh(\beta L)]}{\cosh(\beta L) + i\left(\dfrac{\beta^2 - k_0^2}{2\beta k_0}\right)\sinh(\beta L)}. \tag{7.183d}$$

These relations are not especially easy to interpret and we shall soon look at some limiting cases. First, however, we evaluate the transmission and reflection coefficients:

$$T(k_0) = |t(k_0)|^2 = \frac{1}{1 + \left(\dfrac{\beta^2 + k_0^2}{2\beta k_0}\right)^2 \sinh^2(\beta L)},$$

where the relation $\cosh^2\theta - \sinh^2\theta = 1$ has been used. If we recall that $\beta^2 + k_0^2 = 2mV_0/\hbar^2$ and that $\beta k_0 = (2m/\hbar^2)[E_0(V_0 - E_0)]^{1/2}$, the last relation becomes

$$T(k_0) = \frac{1}{1 + \dfrac{V_0^2}{4E_0(V_0 - E_0)} \sinh^2(\beta L)}. \tag{7.184}$$

Equation (7.183b) then gives

$$R(k_0) = \frac{V_0^2 \sinh^2(\beta L)}{4E_0(V_0 - E_0)} T(k_0). \tag{7.185}$$

This pair of equations (necessarily) satisfies $T(k_0) + R(k_0) = 1$. In addition to these results, the following is obtained for $\psi_{\rm II}(x)$:

$$\psi_{\rm II}(x) = \frac{A\{\cosh[\beta(L - x)] - i(k_0/\beta)\sinh[\beta(L - x)]\}}{\left[\cosh(\beta L) + i\left(\dfrac{\beta^2 - k_0^2}{2\beta k_0}\right)\sinh(\beta L)\right]}. \tag{7.186}$$

In order to gain some insight into these results, we examine three limiting cases: (a) the low-energy limit, $E_0 \ll V_0$; (b) a very narrow barrier, for which $\beta L \ll 1$; and (c) a wide barrier, for which $\beta L \gtrsim 2$. Note that "width" is somewhat of a misnomer, since $\beta L \ll 1$ might mean $V_0 - E_0$ very small: the size of βL depends on V_0 and E_0.

(a) $E_0 \ll V_0$. In this case, $V_0^2/[4E_0(V_0 - E_0)] \cong V_0/(4E_0)$ and the leading approximations to $T(k_0)$ and $R(k_0)$ are

$$T(k_0) \cong \frac{4E_0}{V_0 \sinh^2(\beta L)} \tag{7.187}$$

and

$$R(k_0) \cong 1 - \frac{4E_0}{V_0 \sinh^2(\beta L)}. \tag{7.188}$$

In other words, for very low energies, reflection predominates, just as in the attractive-square-well case.

(b) $\beta L \ll 1$. We assume that neither E_0 nor $V_0 - E_0$ is very small, and then use the approximations $\cosh(\beta L) \cong 1$ and $\sinh(\beta L) \cong \beta L$ to get

$$T(k_0) \cong 1 - \frac{V_0^2}{4E_0(V_0 - E_0)} \beta L \tag{7.189}$$

and

$$R(k_0) \cong \frac{V_0^2}{4E_0(V_0 - E_0)} \beta L. \tag{7.190}$$

In contrast to the low-energy limit and also to the wide-barrier case discussed next, in the limit of a narrow barrier, transmission is favored over reflection.

(c) $\beta L \gtrsim 2$. For these values of βL, one finds $\cosh(\beta L) \cong \exp(\beta L)/2 \cong \sinh(\beta L)$ and thus

$$T(k_0) \cong \frac{1}{1 + \dfrac{V_0^2 e^{2\beta L}}{16E_0(V_0 - E_0)}} \cong \frac{16E_0(V_0 - E_0)}{V_0^2} e^{-2\beta L} \tag{7.191}$$

and

$$R(k_0) \cong 1 - \frac{16E_0(V_0 - E_0)}{V_0^2} e^{-2\beta L}. \tag{7.192}$$

The latter two approximations, which are good even for βL somewhat less than 2, show that, for a wide barrier, transmission is inhibited by the factor $\exp(-2\beta L)$. This factor evidently represents the diminution of the wave function as it traverses the barrier, a conclusion we can check using Eq. (7.186). A measure of the diminution is $|\psi_{\mathrm{II}}(L)/\psi_{\mathrm{II}}(0)|$, for which we get

$$\left| \frac{\psi_{\mathrm{II}}(L)}{\psi_{\mathrm{III}}(0)} \right| \cong \frac{2e^{-\beta L}}{\sqrt{1 + k_0^2/\beta^2}};$$

the factor $\exp(-\beta L)$ appears, as expected. Furthermore, $T(k_0)|A|^2$ should be equal to $|\psi_{\mathrm{II}}(L)|^2$. The latter quantity is easily seen to be

$$|\psi_{\mathrm{II}}(L)|^2 \cong \frac{4E_0(V_0 - E_0)e^{-2\beta L}}{V_0^2} |A|^2,$$

which is the same as $T(k_0)$ of (7.191), multiplied by $|A|^2$. The general behavior of this case is sketched in Fig. 7.15.

The essential point is not, however, that transmission is inhibited, but rather that it occurs at all. Classically, it is forbidden, but quantally it is allowed. Furthermore, as long as $\exp(-2\beta L)$ is not so tiny that only a few particles in a beam will tunnel through the barrier, tunneling should be an observable process. In fact, it is, as we have noted at the beginning of this section. While the 1-D square barrier establishes the phenomenon of tunneling, this potential remains a pedagogic example rather than a realistic one. Tunneling in realistic situations takes place through barriers that are not square; it may also depend on factors other than βL, V_0, E_0, etc. A general approach to tunneling through non-square barriers can be formulated via the WKB approximation, which is discussed briefly in Chapter 17. However, one can approach tunneling through non-square barriers in the context of this section by introducing a small-βL approximation that leads to a radically different expression for $T(K_0)$ than that of Eq. (7.189). We explore the latter approach in the last of the exercises to this chapter.

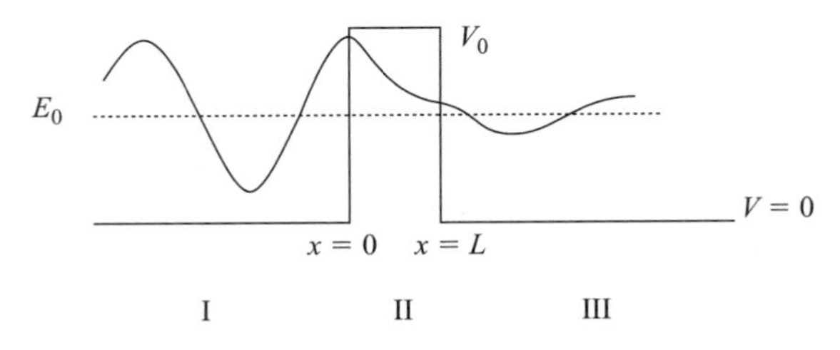

Fig. 7.15 Transmission through a repulsive square well. Shown are sketches of the real parts of $\psi_{\mathrm{inc}}(x)$, $\psi_{\mathrm{II}}(x)$, and $\psi_{\mathrm{III}}(x)$ for the case $A = 1$. Nodes of the wave function occur along the dotted line representing the incident energy E_0.

Exercises

7.1 Let $A(k)$ be the profile function of Eq. (7.12).

(a) Show that $\Psi(x, t) = \Psi(-x, t)$ implies that $A(k) = A(-k)$.

(b) Show that normalization of $\Psi(x, t)$ leads to Eq. (7.13).

7.2 One of the simplest free-particle profile functions is

$$A(k) = \begin{cases} N, & k_1 \leqslant k \leqslant k_2, \\ 0, & \text{otherwise.} \end{cases}$$

(a) Evaluate N.

(b) Determine the functional form of $\Psi(x, t = 0)$ and its "size."

(c) Plot $A(k)$ and $\Psi(x, t = 0)$, and compare their characteristics.

(d) Determine the expected values of the momentum and the kinetic energy associated with $A(k)$.

7.3 A profile function that yields a somewhat more complicated $\Psi(x, t = 0)$ than that in Exercise 7.2 is

$$A(k) = \frac{N}{k^2 + k_0^2}, \qquad k_0 > 0.$$

Apply items (a)–(d) of Exercise 7.2 to this case. Also, what is the probability that the momentum lies in the range $[-\pi/6, \pi/6]$?

7.4 A smoother cut-off profile function than the square wave of Exercise 7.2 is

$$A(k) = \begin{cases} 1 - \cos\left[2\pi\left(\dfrac{k_2 - k}{k_2 - k_1}\right)\right], & k_1 \leqslant k \leqslant k_2 \\ 0, & \text{otherwise.} \end{cases}$$

(a) If $A(k)$ is not normalized, find its real, positive normalization constant.

(b) Plot $A(k)$ as a function of k, stating the values of the position k_0 of its maximum and of $2\,\Delta k$, its full width at half the maximum value.

(c) Determine the average momentum and kinetic energy that $A(k)$ yields.

(d) The coordinate-space packet $\Psi(x, t = 0)$ associated with $A(k)$ is straightforward but a bit tedious to calculate. The optional part of this problem is the determination of $\Psi(x, t = 0)$.

7.5 (a) Let $N_2 = N_3 = N/2$ in Eq. (6.27) and determine the $J_\Gamma(x, t)$ it produces. As a check on your answer, show that it (and the corresponding $P_\Gamma(x, t)$) obeys the quantal equation of continuity.

(b) Now carry out the same analysis for the case $N_2 = 3N/4$, $N_3 = N/4$.

7.6 Let $\Psi_n(x, t) = \psi_n(x)\exp(-iE_n t/\hbar)$, where $\psi_n(x)$ and E_n are the 1-D oscillator eigenfunctions and energies. If $\Psi_\Gamma(x, t) = [\Psi_0(x, t) + \Psi_1(x, t)]/\sqrt{2}$, determine

(a) $\rho_\Gamma(x, t)$,

(b) $J_\Gamma(x, t)$.

7.7 The time delay for the reflected packet is $\tau_R = v_0^{-1}\, d\delta_R(k_0)/dk_0$ (footnote 3, Chapter 7).

(a) Re-express this result using as the derivative $d\delta_R(E_0)/dE_0$, $E_0 = \hbar^2 k_0^2/(2m)$.

(b) Determine the form of τ_T, the time delay for the transmitted packet, using Eq. (7.45).

(c) Determine τ_R and τ_T for the delta-function potential and discuss the cases $x_0 = 0$ and $x_0 \neq 0$.

7.8 (a) Determine the time delay $\tau_R = v_0^{-1}\, d\delta_R(k_0)/dk_0$ of a reflected packet for the half-space square well.

(b) The half-space square well can be used as a crude model for low-energy neutron–nucleus scattering. Evaluate τ_R numerically for this potential, assuming that $V_0 = 40$ MeV, $L = 4 \times 10^{-15}$ m, and $E_0 = 1$ keV.

7.9 (a) Determine the appropriate limits under which Eq. (7.116) $\rightarrow$ Eq. (7.126) and Eq. (7.117) $\rightarrow$ Eq. (7.127).

(b) Show that Eqs. (7.124) and (7.125) imply a discontinuity at $x = 0$ in $\psi'(x)$ for the delta-function potential $V(x) = -\hbar^2\delta(x)/(mx_0)$.

7.10 A 1-D particle is incident from the right on the half-space delta-function potential of Exercise (6.29).

(a) Separately determine $R(k_0)$ and $T(k_0)$, and show that their sum is unity.

(b) Demonstrate that, for $x > a$, $\psi(x) \propto \sin(k_0 x + \delta(k_0))$ and determine an expression for $\delta(k_0)$ analogous to Eq. (7.139).

(c) Show that the pole in $r(k_0)$ yields the bound-state wave number for this potential.

7.11 A 1-D particle is incident from the left on the double-delta-function potential

$$V(x) = g[\delta(x) - \delta(x - a)], \qquad g > 0.$$

(a) Determine $R(k_0)$ and $T(k_0)$ separately, showing that their sum is unity.

(b) Prove that $r(k_0)$ has a pole that yields the bound-state equation obtained in Exercise 6.30(c).

(c) Analyze the high- and low-energy limits of $R(k_0)$ and $T(k_0)$.

7.12 A 1-D particle is incident from the right on the following potential:

$$V(x) = \begin{cases} \infty, & x \leqslant 0 \\ -V_0, & 0 \leqslant x \leqslant a \\ g\delta(x - b), & x > a,\ b > a. \end{cases}$$

(a) Determine $R(k_0)$ and $T(k_0)$ separately, distinguishing between the cases $g > 0$ and $g < 0$.

(b) Show that, for $x > b$, $\psi(x) \propto \sin(k_0 x + \delta(k_0))$ and determine an expression for $\delta(k_0)$ analogous to Eq. (7.139), distinguishing between the cases $g > 0$ and $g < 0$.

(c) Analyze the high- and low-energy limits of $R(k_0)$ and $T(k_0)$.

7.13 The potential in this problem is the infinitely wide step defined by Eq. (7.158).

(a) A 1-D particle is incident on the step from the left; its energy E_0 is equal to the step's height: $E_0 = V_0$. Determine (i) $\psi(x)$ everywhere and (ii) $R(k_0)$.

(b) Now let the particle be incident from $x = +\infty$. Determine (i) $R(k_0)$ and (ii) $T(k_0)$.

7.14 A 1-D particle of energy $E_0 > V_0$ is incident from the left on the square barrier of Eq. (7.180).

(a) Separately determine $R(k_0)$ and $T(k_0)$, comparing your answers with the results for the attractive-square-well case, Eqs. (7.116) and (7.117).

(b) Determine the high-energy behavior of your results for $R(k_0)$ and $T(k_0)$.

(c) Evaluate your results in the limit $E_0 \to V_0$ and compare them with the answers obtained by a direct solution of the Schrödinger equation for this case.

7.15 For the step potential of Eq. (7.158) let the particle's energy obey $E_0 > V_0$, and define K_0 by $K_0^2 = 2m(E_0 - V_0)/\hbar^2$. The wave function in this exercise has the form

$$\psi(x) = \begin{cases} Ae^{ik_0x} + Be^{-ik_0x}, & x \leqslant 0, \\ Ce^{iK_0x} + De^{-iK_0x}, & x \geqslant 0, \end{cases}$$

where A and D are assumed known.

(a) Determine the four elements of the matrix S defined by

$$\begin{pmatrix} \sqrt{K_0}C \\ \sqrt{k_0}B \end{pmatrix} = S \begin{pmatrix} \sqrt{k_0}A \\ \sqrt{K_0}D \end{pmatrix}.$$

(b) Show that S is unitary and that its unitarity is related to conservation of current. If the square-root factors in the definition of S were all set equal to unity, would the resulting S still be unitary?

(c) Express S in terms of the quantal index of refraction $n_0 = K_0/k_0$.

(d) Set $D = 0$ and relate $R(k_0)$ and $T(k_0)$ to the elements of S, i.e., to k_0 and K_0, or to n_0. (Compare your results with Eqs. (7.164) and (7.165).)

(e) Now set $A = 0$ and redo part (d), comparing also with the results of exercise 7.13(b).

(f) Finally, show that the nature of the results of parts (d) and (e) is a consequence of $S^\dagger S = I$. (Note: the matrix S of this exercise (and the next) is a 1-D example of the scattering or "S" matrix of 3-D collision theory, which is discussed in texts on scattering theory and in graduate-level texts on quantum mechanics, e.g., Newton (1982), Taylor (1972), and Sakurai (1994).)

7.16 The analysis of Exercise 7.15 also applies to the potential (7.1), as follows. Assume that

$$\psi(x) = \begin{cases} Ae^{ik_0x} + Be^{-ik_0x}, & x < x_1, \\ Ce^{ik_0x} + De^{-ik_0x}, & x > x_2 \end{cases}$$

for this potential, with the unkown coefficients B and C also related to the "known" coefficients A and D by a 2×2 "S" matrix:

$$\begin{pmatrix} B \\ C \end{pmatrix} = S \begin{pmatrix} A \\ D \end{pmatrix}.$$

(a) Assume that S is unitary and determine $R(k_0)$ and $T(k_0)$ first for the case $D = 0$ and then for the case $A = 0$.

(b) Comment on the relation between the two $R(k_0)$'s and the two $T(k_0)$'s determined in part (a).

(c) Now reverse the analysis by calculating $R(k_0)$ and $T(k_0)$ for each of the two cases of part (a) and showing that conservation of current implies that $S^+S = I$.

7.17 This exercise introduces an alternate to Eq. (7.189) when $\beta L \ll 1$, which allows one to treat the case of a smoothly varying, non-square barrier.

(a) The exact result (7.183a) can be expressed as $t(k_0) = F(\beta, k_0, L)\exp(-\beta L)$. Determine the functional form of $F(\beta, k_0, L)$.

(b) Show that, for $\beta L \ll 1$, $|F(\beta, k_0, L)|^2 \cong 1$, and hence that $T(k_0) \cong e^{-2\beta L}$, $\beta L \ll 1$.

(c) The preceding result can be applied to a smoothly varying barrier $V(x)$ by approximating the portion of the barrier for which $V(x) > E_0$ by a collection of narrow rectangles of width Δx, as in Fig. 7.16. (In the limit $\Delta x \to 0$, the rectangular approximants tend to $V(x)$.) The endpoints a and b, which are the classical turning points, depend on E.

(i) By analyzing the wave functions (7.181a) and (7.181c) for the single square barrier, show that $|T(k_0)|^2$ is the relative – or transmission, or barrier-penetration – probability that the particle will be found in the region $x > L$.

(ii) Assuming that the probability for penetration of each of the rectangular barriers of Fig. 7.16 is an independent entity, fill in the details of the analysis leading to the result

$$P(E_0, a, b) = e^{-2\int_a^b [2m(V(x)-E_0)/\hbar^2]^{1/2} dx},$$

where $P(E_0, a, b)$ is the probability of a particle of energy E_0 passing or tunneling through the barrier $V(x)$.

7.18 The preceding expression can be applied to the phenomenon known as field-induced emission of electrons from a metal. For the purpose of providing a simplified treatment, the metal is modeled as a 1-D quantum square well of height V_0 and the highest energy level of any of the (non-interacting) electrons is denoted E_0. The work function W of the metal, the least energy needed to remove an electron, is thus $W = V_0 - E_0$. Tunneling can (and does) occur when the metal is immersed in a uniform electric field $\mathcal{E}$. An energy diagram is shown in Fig. 7.17; it is a sketch of the following potential:

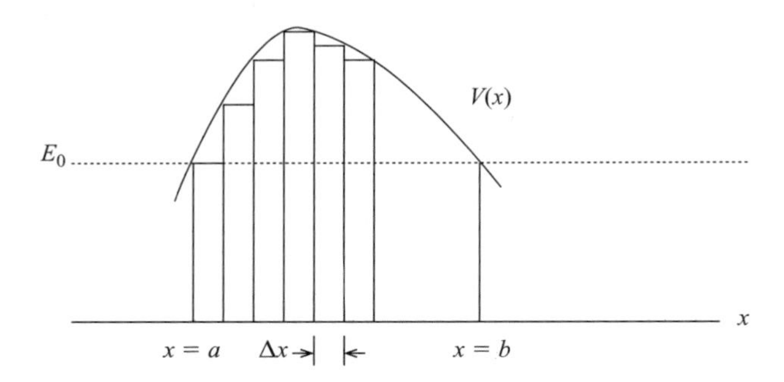

Fig. 7.16

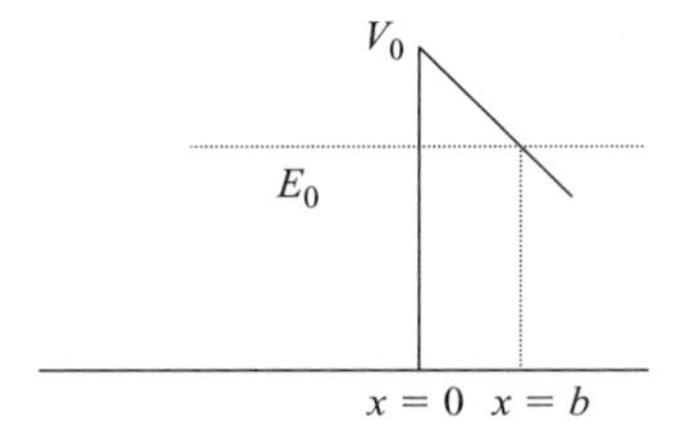

Fig. 7.17

$$V(x) = \begin{cases} 0, & x < 0, \\ V_0 - e\mathcal{E}x, & x > 0, \end{cases}$$

for which tunneling means passage through the triangular barrier located between $x = 0$ and $x = b$.

(a) Using the result of the preceding exercise, show that the probability of tunneling $P(W, \mathcal{E})$ in this case is given by

$$P(W, \mathcal{E}) \cong e^{-[4\sqrt{2m_e}W^{3/2}/(3\hbar e\mathcal{E})]}.$$

(b) Tunneling via field-induced emission is highly improbable for distances $b > 10$ Å. Determine the $\mathcal{E}$ required to achieve tunneling if $W = 4$ eV.

7.19 Radio-active decay by emission of α-particles is a process that can be modeled as a 1-D barrier-penetration event using the result of Exercise 7.17. The 1-D potential seen by the α-particle – which is taken to be structureless – is approximated as a square-well nuclear attraction out to a radius R, after which it becomes a Coulomb interaction between the α-particle and the charge Ze of the residual or daughter nucleus:

$$V(x) = \begin{cases} -V_0, & x \leqslant R, \\ +2Ze^2\kappa_e/x, & x > R, \end{cases}$$

as in Fig. 7.18.

(a) Typical observed α-particle energies E_0 are in the range 5–10 MeV. Show that, for heavy nuclei, the height V_M of the barrier at $x = R$ is much larger than E_0.

(b) Using the result of Exercise 7.17, show that the probability of tunneling $P(Z, E_0)$ is

$$P(Z, E_0) \cong e^{-[2\pi Ze^2\kappa_e\sqrt{2M_\alpha}/(\hbar E_0^{1/2})]},$$

where a small correction term has been ignored in the exponential.

(c) For a single species undergoing decay by emission of a photon or an α-particle, the number $N(t)$ present at any time t is related to the number present at time $t = 0$ by $N(t) = N(0)\exp(-t/\tau)$, where τ is the lifetime of a single member of the species

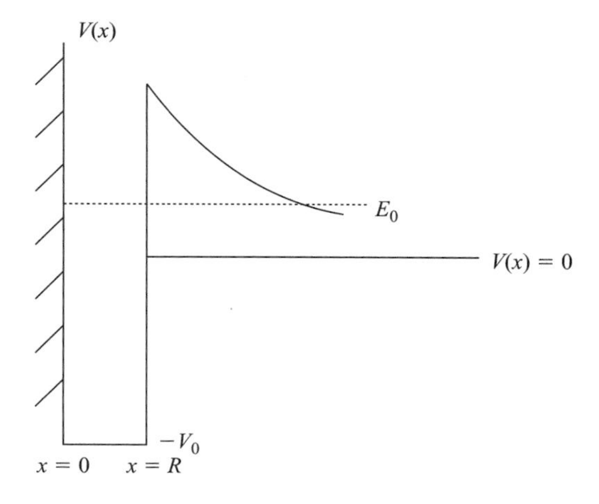

Fig. 7.18

against the decay. (This result follows from assuming that the rate at which $N(t)$ changes in time is proportional to $N(t)$.)

(i) If v_0 is the velocity associated with E_0 ($E_0 = m_\alpha v_0^2/2$), use qualitative arguments to infer that, in this case of α-decay, the lifetime is approximately $\tau_\alpha \cong 2RT(Z, E_0)/v_0$.

(ii) The connection between τ_α and experimental information is that, when $t = t_{1/2}$, the measured half-life of the α-decaying nucleus (i.e., $N(t_{1/2})/N(0) = \frac{1}{2}$), then $\tau_\alpha = \ln 2/t_{1/2}$. Compare the values of τ_α obtained from the empirical data listed below with the values from part (c) (i) above: ^{226}Ra, for which $E_0 = 4.8$ MeV and $t_{1/2} = 1.6 \times 10^3$ years; ^{218}Po, for which $E_0 = 6.1$ MeV and $t_{1/2} = 3$ min; and ^{212}Po, for which $E_0 = 11.7$ MeV and $t_{1/2} = 3 \times 10^{-7}$ s.

8

*Quantal/Classical Connections

The axiomatic approach to quantum theory, with the Schrödinger equation being included as one of the postulates, is not the only way to introduce the subject. Among others is one that has found application to various areas of physics: the path-integral formulation of R. P. Feynman, which was developed from an insight of Dirac (1947), who observed that a particular quantal probability amplitude corresponds to an exponential of the classical action. Feynman not only established what that correspondence is but also used it to derive the time-dependent Schrödinger equation. One of the topics dealt with in this chapter is Feynman's formulation of quantum theory. Because it establishes a connection between classical and quantum mechanics, we also examine, in Section 8.3, aspects of the classical limit of quantum mechanics, an analysis that functions a little like the inverse of the Dirac/Feynman formulation. Since this approach involves time-dependent amplitudes, we begin by studying some temporal aspects of quantum theory. Neither the Feynman formulation nor the discussion of the classical limit is a necessary ingredient for learning how to deal with quantal systems involving either a particle moving in more than 1-D or more than two particles. Nevertheless, the topics dealt with in Sections 8.2 and 8.3 are fundamental to one's eventual understanding of quantum theory, which is more than a sufficient ground for treating them herein.

8.1. Some temporal aspects of the theory

The time dependence of (generalized) expectation values

We consider an arbitrary quantum system, whose behavior is governed by some Hamiltonian $\hat{H}$. Let $\{|\Gamma_j, t\rangle\}$ be a set of possible states of the system, where $|\Gamma_j\rangle = |\Gamma_j, t=0\rangle$ need not be an eigenstate of $\hat{H}$. $|\Gamma_j, t\rangle$ obeys the time-dependent Schrödinger equation

$$i\hbar \frac{\partial}{\partial t}|\Gamma_j, t\rangle = \hat{H}|\Gamma_j, t\rangle. \tag{8.1}$$

If the system is in an eigenstate of the observable $\hat{A}$, then measurement of $\hat{A}$ in an ideal experiment will yield the eigenvalue associated with the eigenstate, as per Postulate I. For states such as $|\Gamma_j, t\rangle$ of (8.1), the measured value of observable $\hat{A}$ is proportional either to a matrix element of $\hat{A}$ or to the absolute square of such a matrix element. Generalizing Postulate III, these matrix elements will typically be of the form

$$\langle \hat{A} \rangle_{jk}(t) \equiv \langle \Gamma_j,\, t|\hat{A}|\Gamma_k,\, t\rangle, \tag{8.2}$$

where j need not be equal to k. When $j = k$, (8.2) becomes an expectation value, and is the time-independent analog of Eq. (5.128).

We shall refer to $\langle \hat{A} \rangle_{jk}(t)$ of (8.2) as a *generalized expectation value.*[1] The goal of our analysis is to determine how $\langle \hat{A} \rangle_{jk}(t)$ changes with time. To do so, we evaluate $d\langle \hat{A} \rangle_{jk}(t)/dt$. Since $\langle \hat{A} \rangle_{jk}(t)$ depends only on t, the total derivative can be replaced by a partial derivative:

$$\begin{aligned} \frac{d}{dt}\langle \hat{A} \rangle_{jk}(t) &= \frac{\partial}{\partial t}\langle \hat{A} \rangle_{jk}(t) \\ &= \left\{\frac{\partial}{\partial t}\langle \Gamma_j,\, t|\right\}\hat{A}|\Gamma_k,\, t\rangle + \langle \Gamma_j,\, t|\frac{\partial \hat{A}}{\partial t}|\Gamma_k,\, t\rangle \\ &\quad + \langle \Gamma_j,\, t|\hat{A}\left\{\frac{\partial}{\partial t}|\Gamma_k,\, t\rangle\right\}. \end{aligned} \tag{8.3}$$

Using the Schrödinger equation (8.1) to transform the time derivatives in the first and third terms, (8.3) becomes

$$\frac{d}{dt}\langle \hat{A} \rangle_{jk}(t) = \langle \Gamma_j,\, t|\left\{\frac{1}{i\hbar}[\hat{A},\, \hat{H}] + \frac{\partial \hat{A}}{\partial t}\right\}|\Gamma_k,\, t\rangle, \tag{8.4}$$

where $[\hat{A},\, \hat{H}] = \hat{A}\hat{H} - \hat{H}\hat{A}$ is the commutator of $\hat{A}$ and $\hat{H}$.

Equation (8.4) is the desired result. It is the analog for generalized expectation values of the time-dependent Schrödinger equation (8.1) for states. Equation (8.4) requires only that $|\Gamma_j,\, t\rangle$ and $|\Gamma_k,\, t\rangle$ obey (8.1). Thus, $\hat{H}$ itself could be time-dependent, in contrast to Eq. (5.88a), which requires $\hat{H} \neq \hat{H}(t)$.

Schrödinger and Heisenberg pictures

Suppose that the states $|\Gamma_j,\, t\rangle$ and $|\Gamma_k,\, t\rangle$ in Eq. (8.4) are stationary, so that both are of the form $[\exp(-iEt/\hbar)]|\Gamma\rangle$. Then each side of (8.4) has the structure $\langle \Gamma_j|\hat{O}\,(t)|\Gamma_k\rangle$, where $\hat{O}\,(t)$ is a time-dependent operator. Since in this special case of stationary states, $|\Gamma_j\rangle$ and $|\Gamma_k\rangle$ are time-independent, Eq. (8.4) appears to hold as an equation between certain time-dependent operators as well as between matrix elements of these operators. That a relation analogous to (8.4) may hold for operators motivates our introduction of the Heisenberg "picture," which we analyze next. The results obtained, plus Eq. (8.4), will play an important role in understanding the classical limit, which is discussed in Section 8.3.

Most operators in quantum mechanics, considered as abstract Hilbert-space entities, are time-independent. For these cases, the time dependence of matrix elements occurs because the state vectors or wave functions are functions of time. When the time dependence arises from the state vectors, one is said to be working in or using the *Schrödinger picture*. Because of the relation $|\Gamma,\, t\rangle = \hat{U}(t)|\Gamma\rangle$, (Eq. (5.84) with $t_0 = 0$), where $i\hbar\, \partial\hat{U}(t)/\partial t = \hat{H}\hat{U}(t)$, it is possible to shift the time-dependence in $\langle \hat{A} \rangle_{jk}(t)$ from

[1] When $k \neq j$, the generalized expectation value is an off-diagonal matrix element, often denoted a transition-matrix element, the reason for which will become clear in Chapter 16.

the states to the operators.[2] Doing so defines the *Heisenberg picture*, wherein all operators depend on time.

One easily goes from one picture to another, as follows. We write

$$|\Gamma_i,\, t\rangle = \hat{U}(t)|\Gamma_i\rangle, \qquad i = j \text{ or } k, \tag{8.5}$$

so that $\langle \hat{A}\rangle_{jk}(t)$ becomes

$$\langle \hat{A}\rangle_{jk}(t) = \langle \Gamma_j|\hat{U}^\dagger(t)\hat{A}\hat{U}(t)|\Gamma_k\rangle. \tag{8.6}$$

How one regards the RHS of (8.6) determines which picture is operative:

Schrödinger picture

$$|\Gamma\rangle \to U(t)|\Gamma\rangle = |\Gamma,\, t\rangle \Rightarrow \langle \hat{A}\rangle_{\rm jk}(t) = \langle \Gamma_{\rm j},\, t|\hat{A}|\Gamma_{\rm k},\, t\rangle \tag{8.7a}$$

Heisenberg picture

$$\hat{A} \to \hat{A}^{\rm (H)}(t) \equiv \hat{U}^\dagger(t)\hat{A}\hat{U}(t) \Rightarrow \langle \hat{A}\rangle_{jk}(t) = \langle \Gamma_j|\hat{A}^{\rm (H)}(t)|\Gamma_k\rangle. \tag{8.7b}$$

The result $\langle \hat{A}\rangle_{jk}(t)$ is unchanged, of course, but in the Heisenberg picture the operators are transformed via (8.7b) so that *they* carry the full time dependence.

Some authors prefer to write $\hat{A}$ and $|\Gamma,\, t\rangle$ as $\hat{A}^{\rm (S)}$ and $|\Gamma,\, t\rangle^{\rm (S)}$, respectively, and to further label $|\Gamma\rangle$ as $|\Gamma\rangle^{\rm (H)}$ in order to emphasize which picture is being employed. It suffices for us to use only the notation $\hat{A}^{\rm (H)}(t)$ and then to remember that, in the Heisenberg picture, the state vectors (and corresponding wave functions) are time independent.

Let us compare $d\hat{A}^{\rm (H)}(t)/dt$ with $d\langle \hat{A}\rangle_{jk}(t)/dt$ of Eq. (8.4). Differentiating each term of (8.7b) w.r.t time and employing the relation $i\hbar\, \partial\hat{U}(t)/\partial t = \hat{H}\hat{U}(t)$, we find

$$\frac{d\hat{A}^{\rm (H)}(t)}{dt} = \frac{1}{i\hbar}\left[\hat{A}^{\rm (H)},\, \hat{H}^{\rm (H)}(t)\right] + \frac{\partial \hat{A}^{\rm (H)}}{\partial t}, \tag{8.8a}$$

where

$$\hat{H}^{\rm (H)}(t) = \hat{U}^\dagger(t)\hat{H}\hat{U}(t) \tag{8.9}$$

and

$$\frac{\partial \hat{A}^{\rm (H)}}{\partial t} \equiv \hat{U}^\dagger(t)\,\frac{\partial \hat{A}}{\partial t}\,\hat{U}(t). \tag{8.10}$$

The term $\partial\hat{A}^{\rm (H)}/\partial t$ in (8.8a) is no surprise, but the presence in the commutator of $\hat{H}^{\rm (H)}(t)$, the Heisenberg form of the Hamiltonian, might be. It occurs because of the

[2] We use the general form $\hat{U}(t)$ rather than $\exp(-i\hat{H}t/\hbar)$ in order to allow for the possibility of a time dependence in $\hat{H}$, in which case (5.88a) does not hold.

possibility that $\hat{H} = \hat{H}(t)$, in which case $[\hat{H}, \hat{U}(t)]$ need not be zero. We note in passing that, if (8.4) had been the starting point of this analysis, $\hat{H}^{(\mathrm{H})}$ would again have been obtained. However, although time-dependent Hamiltonians do occur in quantum mechanics, e.g., as time-varying, external electromagnetic fields, we shall concentrate here on the simpler (and more widely occurring) case for which $\hat{H} \neq \hat{H}(t)$. Then (8.8a) becomes

$$\frac{d\hat{A}^{(\mathrm{H})}(t)}{dt} = \frac{1}{i\hbar}\left[\hat{A}^{(\mathrm{H})}(t), \hat{H}\right] + \frac{\partial \hat{A}^{(\mathrm{H})}}{\partial t}, \qquad \hat{H} \neq \hat{H}(t). \tag{8.8b}$$

Like Eq. (8.4), Eq. (8.8b) will be used in analyzing the classical limit of quantum mechanics. In addition, there are numerous examples for which (8.8a) allows a relatively direct means of solving certain quantal problems. We illustrate this via application to the 1-D harmonic oscillator.

EXAMPLE

The 1-D Harmonic Oscillator.

Our goal is to determine $\hat{Q}_x^{(\mathrm{H})}(t)$ and $\hat{P}_x^{(\mathrm{H})}(t)$. This will be done using the $\hat{a}$, $\hat{a}^\dagger$ forms for $\hat{P}_x$ and $\hat{Q}_x$. We have for $\hat{a}^{(\mathrm{H})}(t)$ from (8.8b) the equation

$$\frac{d\hat{a}^{(\mathrm{H})}(t)}{dt} = \frac{1}{i\hbar}\left[\hat{a}^{(\mathrm{H})}(t), \hat{H}\right] = \frac{1}{i\hbar}\left[\hat{a}, \hat{H}\right]^{(\mathrm{H})}, \tag{8.11}$$

where the second equality follows from $[\hat{H}, \hat{U}] = 0$. Recalling that $\hat{H} = (\hat{N} + \frac{1}{2})\hbar\omega_0$, with $\hat{N} = \hat{a}^\dagger \hat{a}$, the commutator in (8.11) yields $\hbar\omega_0 \hat{a}^{(\mathrm{H})}(t)$, from which one gets

$$\frac{d\hat{a}^{(\mathrm{H})}(t)}{dt} = -i\omega_0 \hat{a}^{(\mathrm{H})}(t), \tag{8.12}$$

whose solution is

$$\hat{a}^{(\mathrm{H})}(t) = \hat{a}^{(\mathrm{H})}(0)e^{-i\omega_0 t} = \hat{a}e^{-i\omega_0 t}. \tag{8.13}$$

Although $\hat{a}^{\dagger(\mathrm{H})}(t)$ can be obtained in the same way, it is simpler to note that $[\hat{a}^{(\mathrm{H})}(t)]^\dagger = \hat{a}^{\dagger(\mathrm{H})}$; applying the † operation to both sides of (8.13) produces the desired result:

$$\hat{a}^{\dagger(\mathrm{H})}(t) = \hat{a}^\dagger e^{i\omega_0 t}. \tag{8.14}$$

While Eqs. (8.13) and (8.14) are of interest in their own right, we regard them here simply as intermediaries in obtaining $\hat{Q}_x^{(\mathrm{H})}(t)$ and $\hat{P}_x^{(\mathrm{H})}(t)$, which are related to $\hat{a}^{(\mathrm{H})}(t)$ and $\hat{a}^{\dagger(\mathrm{H})}(t)$ by Eqs. (6.47) and (6.48). These plus (8.13) and (8.14) lead to

$$\frac{1}{\lambda_0}\hat{Q}_x^{(\mathrm{H})}(t) + \frac{i\lambda_0}{\hbar}\hat{P}_x^{(\mathrm{H})}(t) = \left(\frac{1}{\lambda_0}\hat{Q}_x + \frac{i\lambda_0}{\hbar}\hat{P}_x\right)e^{-i\omega_0 t} \tag{8.15}$$

and

$$\frac{1}{\lambda_0}\hat{Q}_x^{(\mathrm{H})}(t) - \frac{i\lambda_0}{\hbar}\hat{P}_x^{(\mathrm{H})}(t) = \left(\frac{1}{\lambda_0}\hat{Q}_x - \frac{i\lambda_0}{\hbar}\hat{P}_x\right)e^{i\omega_0 t}. \tag{8.16}$$

$\hat{Q}_x^{(\mathrm{H})}(t)$ and $\hat{P}_x^{(\mathrm{H})}(t)$ are obtained by adding and subtracting the last two equations:

$$\hat{Q}_x^{(\mathrm{H})}(t) = \hat{Q}_x \cos(\omega_0 t) + \frac{1}{m\omega_0}\hat{P}_x \sin(\omega_0 t) \tag{8.17a}$$

and

$$\hat{P}_x^{(\mathrm{H})}(t) = -m\omega_0 \hat{Q}_x \sin(\omega_0 t) + \hat{P}_x \cos(\omega_0 t). \tag{8.17b}$$

Use of the equations for $\hat{a}^{(\mathrm{H})}(t)$ and $\hat{a}^{\dagger(\mathrm{H})}(t)$ have led to straightforward evaluation of $\hat{Q}_x^{(\mathrm{H})}(t)$ and $\hat{P}_x^{(\mathrm{H})}(t)$.

These expressions for the operators $\hat{Q}_x^{(\mathrm{H})}(t)$ and $\hat{P}_x^{(\mathrm{H})}(t)$ are strikingly similar to their classical analogs for $x(t)$ and $p_x(t) = m\dot{x}(t)$, the displacement and momentum of a particle of mass m acted on by the restoring force $-m\omega_0^2 x$, viz.

$$x(t) = x_0 \cos(\omega_0 t) + \frac{1}{m\omega_0} p_0 \sin(\omega_0 t) \tag{8.18a}$$

and

$$p_x(t) = -m\omega_0 x_0 \sin(\omega_0 t) + p_0 \cos(\omega_0 t), \tag{8.18b}$$

where $x_0 = x(t = 0)$ and $p_0 = p_x(t = 0)$. In contrast to all other quantal results determined so far, $\hat{Q}_x^{(\mathrm{H})}(t)$ and $\hat{P}_x^{(\mathrm{H})}(t)$ and their classical analogs behave exactly alike. Does this mean that, via the Heisenberg picture, at least for the harmonic oscillator, we have achieved the classical limit without reference to large quantum numbers and Bohr's correspondence principle? The answer is NO, essentially because one must compare *like* quantities, not *unlike* ones. That is, instead of examining $x(t)$ and $\hat{Q}_x^{(\mathrm{H})}(t)$, we should compare $x(t)$ with $\langle\hat{Q}_x^{(\mathrm{H})}\rangle(t)$, etc.

To evaluate the diagonal matrix elements $\langle\hat{Q}_x^{(\mathrm{H})}\rangle(t)$ and $\langle\hat{P}_x^{(\mathrm{H})}\rangle(t)$ needed for the comparison with $x(t)$ and $p_x(t)$, we choose the oscillator eigenstate $|n\rangle$:

$$\langle\hat{Q}_x^{(\mathrm{H})}\rangle_n(t) = \langle n|\hat{Q}_x^{(\mathrm{H})}(t)|n\rangle$$

and

$$\langle\hat{P}_x^{(\mathrm{H})}\rangle_n(t) = \langle n|\hat{P}_x^{(\mathrm{H})}(t)|n\rangle.$$

Each of the preceding matrix elements, from Eqs. (8.17), involves $\langle\hat{Q}_x\rangle_n$ and $\langle\hat{P}_x\rangle_n$. A glance at Eqs. (6.95) and (6.96) tells us, however, that the diagonal matrix elements $\langle\hat{Q}_x\rangle_n$ and $\langle\hat{P}_x\rangle_n$ vanish. Hence, no matter how large an n we choose, both $\langle\hat{Q}_x^{(\mathrm{H})}\rangle_n(t)$ and $\langle\hat{P}_x^{(\mathrm{H})}\rangle_n(t)$ are zero, in contrast to $x(t)$ and $p_x(t)$: single eigenstate matrix elements do not yield the classical limit.

To obtain non-vanishing values of $\langle\hat{Q}_x^{(\mathrm{H})}\rangle(t)$ and $\langle\hat{P}_x^{(\mathrm{H})}\rangle(t)$, one must use superpositions of oscillator states whose labels include both n and $n+1$: the combination $|\Gamma_1\rangle = C_n|n\rangle + C_{n+1}|n+1\rangle$ works in this regard, while $|\Gamma_2\rangle = C_n|n\rangle + C_{n+2}|n+2\rangle$ does not. However, although a superposition such as $|\Gamma_1\rangle$ will yield $\langle\hat{Q}_x^{(\mathrm{H})}\rangle(t)$ and $\langle\hat{P}_x^{(\mathrm{H})}\rangle(t)$ that mimic the classical behaviors of $x(t)$ and $p_x(t)$, we shall show in Section 8.3 that it does not lead to the classical limit. One must be cautious in leaping to conclusions.

Time evolution via propagators

In this subsection we re-examine the evolution of wave functions from time t' to time $t > t'$. It will be shown that the temporal evolution is given as a spatial integral involving a quantity known as a propagator, which is the coordinate-space matrix element of the time-evolution operator. This seemingly simple concept has far-reaching consequences, providing in particular the underpinning for Feynman's path-integral approach to quantum mechanics, which is discussed in Section 8.2. The system we consider as the vehicle for these developments is a single particle of mass m whose quantal behavior is governed by a time-independent Hamiltonian $\hat{H}$.

As in Chapter 5, the state vectors for the system at times t' and $t > t'$, viz., $|\Gamma, t'\rangle$ and $|\Gamma, t\rangle$, are related by the time-evolution equation

$$|\Gamma, t\rangle = e^{-i\hat{H}(t-t')/\hbar}|\Gamma, t'\rangle. \tag{5.88a}$$

Projecting both sides of Eq. (5.88a) onto the position bra $\langle \mathbf{r}|$ and inserting $\hat{I} = \int d^3r' |\mathbf{r}'\rangle\langle\mathbf{r}'|$ to the left of $|\Gamma, t'\rangle$, one gets the wave function form of (5.79):

$$\Psi_\Gamma(\mathbf{r}, t) = \int d^3r' \, K(\mathbf{r}, t; \mathbf{r}', t')\Psi_\Gamma(\mathbf{r}', t'), \tag{8.19}$$

where the *propagator* $K(\mathbf{r}, t; \mathbf{r}', t')$ is given by

$$K(\mathbf{r}, t; \mathbf{r}', t') = \langle \mathbf{r}|e^{-i\hat{H}(t-t')/\hbar}|\mathbf{r}'\rangle. \tag{8.20}$$

Equation (8.19) expresses $\Psi_\Gamma(\mathbf{r}, t)$ as an integral over all space of a kernel function – here the propagator – times the wave function at the earlier time t'. We shall later reinterpret the propagator as a probability amplitude, so that this integral has the form of a weighted sum of probability amplitudes, an interpretation crucial to the path-integral formulation.

For the present, however, the propagator is just the coordinate-space matrix element of the time-evolution operator. It takes on a familiar form if the eigenstates of $\hat{H}$ are used as a resolution of the identity. For instance, if $\hat{H}$ has only bound states, say $|\gamma_n, E_n\rangle$, where $(\hat{H} - E_n)|\gamma_n, E_n\rangle = 0$ and $\hat{I} = \sum_n |\gamma_n, E_n\rangle\langle\gamma_n, E_n|$, then (8.20) is readily transformed into

$$K(\mathbf{r}, t; \mathbf{r}', t') = \sum_n \psi_n(\mathbf{r})\psi_n^*(\mathbf{r}')e^{-iE_n(t-t')}/\hbar. \tag{8.21}$$

Substitution of (8.21) into (8.19) gives $\Psi_\Gamma(\mathbf{r}, t)$ as an infinite sum of the $\psi_n(\mathbf{r})$ times time-dependent coefficients, exactly as in Eq. (5.116).

The discrete-spectrum case, (8.21), yields, via closure, a simple demonstration that

$$\lim_{t\to t'} K(\mathbf{r}, t; \mathbf{r}', t') = \delta(\mathbf{r} - \mathbf{r}'). \tag{8.22}$$

Inspection of (8.19) confirms (8.22), since the RHS of (8.21) must yield $\Psi_\Gamma(\mathbf{r}, t)$ in the limit $t \to t'$.

In the discussion below (5.88b), it was noted that that equation was not an explicit solution of the time-dependent Schrödinger equation, although in a few cases the time-evolution operator can be explicitly evaluated. The latter remark evidently applies to the propagator as well, and we examine one of the systems for which an evaluation is possible. The system is that of a free particle moving in 1-D over the infinite interval $-\infty \leqslant x \leqslant \infty$.

In this case, $\hat{H} \to \hat{K}_x = \hat{P}_x^2/(2m)$ and $K(\mathbf{r}, t; \mathbf{r}', t') \to K^{(\text{free})}(x, t; x', t')$, where

$$K^{\text{free}}(x, t; x', t') = \langle x|e^{-i\hat{K}_x(t-t')/\hbar}|x'\rangle. \tag{8.23}$$

Since the normalized eigenstates of the kinetic energy operator $\hat{K}_x$ are the (1-D) plane waves $|k\rangle$, the analog of (8.21) becomes

$$K^{\text{free}}(x,\, t;\, x',\, t') = \frac{1}{2\pi}\int_{-\infty}^{\infty} dk\, e^{ik(x-x')} e^{-i\hbar k^2(t-t')}/(2m). \tag{8.24}$$

The integral in (8.24) is reminiscent of the k-integral of Eq. (7.29), in the subsection on the Gaussian wave packet. Not surprisingly, the same methods as those that led to (7.35) via Eqs. (7.30) and (7.31b) apply in the case of Eq. (8.24); it can be shown[3] that

$$K^{\text{free}}(x,\, t;\, x',\, t') = \sqrt{\frac{m}{2\pi i\hbar(t-t')}} \exp\left(i\frac{m(x-x')^2}{2\hbar(t-t')}\right). \tag{8.25}$$

Comparison of the $t \to t'$ limit of (8.25) with Eq. (4.126) establishes that the 1-D version of (8.22) is (necessarily) satisfied by (8.25).

In one sense, Eq. (8.25) is almost a bit of physics esoterica: while it is a simple expression, actually using it to obtain a time-evolved, 1-D wave function via Eq. (8.19) seems like an inordinately complicated effort. For instance, if the wave function at time t' is a plane wave, $\Psi_k(x',\, t') = \psi_k(x')\exp[-i\hbar k^2 t'/(2m)]$, then it is evidently far easier to apply $\hat{U}(t-t') = \exp[-i\hat{H}(t-t')/\hbar]$ to $\Psi_k(x',\, t')$, thereby directly obtaining $\Psi_k(x,\, t)$, than it is to carry out the integration in (8.19), though it can be done. Also, while the integrations involved are similar, the time-evolved Gaussian wave packet of Section (7.1) has already been obtained without first going through the analysis establishing that it, too, can be calculated using the propagator form for time evolution.

In another sense, however, these are only crotchety remarks related to performing integrals: they do not touch on some of the essential qualities of propagators. We mention three such here, all connected with work of Richard Feynman. First, they were a vital ingredient in Feynman's interpretation of positrons (the positively charged antiparticle to the electron) as electrons moving backward in time, and they also led to his introducing Feynman diagrams, a major tool for carrying out analyses in quantum electrodynamics (Feynman (1962)). Second, the square-root factor multiplying the exponential in (8.25) is a key to determining the correct normalization factor in the Feynman path integral. Third, as noted in the foregoing, the propagator can be re-interpreted as a probability amplitude, an aspect we consider next.

The propagator as a probability amplitude

To say that the propagator can be re-interpreted as a probability amplitude means that it can be re-expressed as the scalar product of a bra and a ket. This is easy to demonstrate; its consequences are profound.

The structure of Eq. (8.20) indicates that the kets involved in the scalar product noted above must be

$$|\mathbf{r},\, t\rangle \equiv e^{i\hat{H}t/\hbar}|\mathbf{r}\rangle \tag{8.26a}$$

and

[3] The details of the proof are left to the exercises for this chapter.

$$|\mathbf{r}', t'\rangle \equiv e^{i\hat{H}t'/\hbar}|\mathbf{r}'\rangle, \tag{8.26b}$$

from which it follows (trivially) that

$$\hat{K}(\mathbf{r}, t; \mathbf{r}', t') = \langle \mathbf{r}, t|\mathbf{r}'t'\rangle. \tag{8.27}$$

In order for these relations to be more than empty mathematical equalities, some meaning must be ascribed to the kets defined by (8.26). The significance of $|\mathbf{r}, t\rangle$ is that it is an eigenket of the position-operator in the Heisenberg picture. To show this, we act on both sides of the position-operator eigenvalue equation

$$\hat{\mathbf{Q}}|\mathbf{r}\rangle = \mathbf{r}|\mathbf{r}\rangle$$

with $\exp(i\hat{H}t/\hbar)$ and then insert $\hat{I} = \exp(-i\hat{H}t/\hbar)\exp(i\hat{H}t/\hbar)$ to the left of $|\mathbf{r}\rangle$. The transformed eigenvalue equation is

$$\hat{\mathbf{Q}}^{(\mathrm{H})}(t)|\mathbf{r}, t\rangle = \mathbf{r}(t)|\mathbf{r}, t\rangle, \tag{8.28a}$$

where the eigenvalue $\mathbf{r}(t)$ is given by

$$\mathbf{r}(t) = e^{i\hat{H}t/\hbar}\mathbf{r}e^{-i\hat{H}t/\hbar}. \tag{8.28b}$$

Equation (8.28a) demonstrates that $|\mathbf{r}, t\rangle$ is indeed a Heisenberg-picture eigenket. This may at first sight seem inconsistent if not absurd, since, in constructing the Heisenberg picture, we *removed* the time dependence from the kets. However, that construction was based on matrix elements formed from the solutions $|\Gamma, t\rangle$ of the time-dependent Schrödinger equation. When $\hat{H} \neq \hat{H}(t)$, as in the present case, then

$$|\Gamma, t\rangle = e^{-i\hat{H}t/\hbar}|\Gamma\rangle,$$

and application of $\exp(i\hat{H}t/\hbar)$ produces $|\Gamma\rangle$ by exact cancelation:

$$|\Gamma\rangle = e^{i\hat{H}t/\hbar}|\Gamma, t\rangle.$$

However, for any other ket such as $|\mathbf{r}\rangle$, action of $\exp(i\hat{H}t(\hbar))$ *introduces* a time dependence: it is a necessary though perhaps unexpected consequence. The apparent absurdity is seen to be a result of not previously having considered general eigenvalue equations in the Heisenberg picture. It suffices to be able to interpret the ket $|\mathbf{r}, t\rangle$: we do not need its detailed form. Indeed, solving for it and its eigenvalue $\mathbf{r}(t)$ is typically non-trivial,[4] as is suggested by inspection of the 1-D version of $\hat{\mathbf{Q}}^{(\mathrm{H})}(t)$, viz., Eq. (8.17a).

Since $|\mathbf{r}, t\rangle$ is the position ket in the Heisenberg picture, we shall interpret it as a ket that specifies the position $\mathbf{r}$ at time t, with an analogous statement for $|\mathbf{r}', t'\rangle$. The propagator of (8.27) is thus the probability amplitude that, in the Heisenberg picture, a particle at position $\mathbf{r}'$ at time t' will be at position $\mathbf{r}$ at time t. This interpretation suggests that $|\mathbf{r}\rangle$ is the position eigenket at time $t = 0$, a nomenclature consistent with Eq. (8.28a).

If $|\mathbf{r}'', t''\rangle$ is the position eigenket at time t'', we may ask whether the closure relation is still valid. The answer is YES:

[4] Not all Heisenberg-operator eigenvalue problems are non-trivial; for example, those concerning coherent states are not.

$$\int d^3r''|\mathbf{r}'', t''\rangle\langle\mathbf{r}'', t''| = e^{i\hat{H}t''/\hbar}\left(\int d^3r''|\mathbf{r}''\rangle\langle\mathbf{r}''|\right)e^{-i\hat{H}t''/\hbar}$$
$$= \hat{I}. \tag{8.29}$$

Hence, the set of kets $\{|\mathbf{r}'', t''\rangle\}$ is complete.

Because of (8.29), the propagator obeys what is sometimes referred to as a "composition" property and at other times as a "decomposition" property.[5] Let t'' be intermediate between t and t': $t' < t'' < t$. Then insertion of (8.29) into $\hat{K}(\mathbf{r}, t; \mathbf{r}', t') = \langle\mathbf{r}, t|\mathbf{r}', t'\rangle$ yields

$$\hat{K}(\mathbf{r}, t; \mathbf{r}', t') = \int d^3r''\,\langle\mathbf{r}, t|\mathbf{r}'', t''\rangle\langle\mathbf{r}'', t''|\mathbf{r}', t'\rangle$$
$$= \int d^3r''\,\hat{K}(\mathbf{r}, t; \mathbf{r}'', t'')\hat{K}(\mathbf{r}'', t''; \mathbf{r}', t'). \tag{8.30}$$

Equation (8.30) expresses the desired property. It is easily generalized. Let the interval $t - t'$ be divided into N segments:

$$t - t' = \sum_{n=1}^{N}(t_n - t_{n-1}),$$

where $t_0 = t'$ and $t_N = t$. Insertion of $N - 1$ resolutions of the identity using the analog of (8.29) then leads to

$$K(\mathbf{r}, t; \mathbf{r}', t') = \langle\mathbf{r}, t|\mathbf{r}', t'\rangle = \int d^3r_1 \ldots d^3r_{N-1}\prod_{n=1}^{N}\langle\mathbf{r}_n, t_n|r_{n-1}, t_{n-1}\rangle, \tag{8.31}$$

with $\mathbf{r}_N = \mathbf{r}$ and $\mathbf{r}_0 = \mathbf{r}'$. Notice that, in the decomposition expressions (8.30) and (8.31), there is no integration over time. In quantum theory, time is a parameter, not a dynamical variable: t enters $|\mathbf{r}, t\rangle$ via $\exp(i\hat{H}t/\hbar)$, not as the eigenvalue of an operator.

8.2. Path-integral formulation of quantum mechanics

The original formulations of quantum mechanics were as a matrix theory by Heisenberg and collaborators and as a wave theory by Schrödinger. The equivalence of the two theories was quickly established, and the probabilistic interpretation of the wave function was articulated soon after by Born (1926). Over many years, the subject was developed and codified, leading to the postulational approach of this book.

In 1948 Feynman presented his space–time approach to non-relativistic quantum mechanics (Feynman (1948)), which provides a third formulation of quantum theory. This formulation, now commonly referred to as the path-integral approach, links certain quantal probability amplitudes to the classical action. Via this link, classical and quantum physics are seen to be profoundly interconnected. Feynman's method has been applied in many areas, including quantum-field theory, statistical mechanics, and semi-classical quantization analyses.

[5] "Composition" and "decomposition" bear the same relation to one another in this context as do "flammable" and "inflammable" in ordinary English.

Feynman was led to this formulation by an observation of Dirac (1947), namely that – in the language of Section 8.1 – $\langle \mathbf{r}, t|\mathbf{r}', t'\rangle$ is related to the classical action $S(\mathbf{r}, t; \mathbf{r}', t')$. In particular, $\langle \mathbf{r}, t|\mathbf{r}', t'\rangle$ corresponds to $\exp(iS(\mathbf{r}, t; \mathbf{r}', t')/\hbar)$. Feynman's analysis gave a precise meaning to the phrase "corresponds to" and also led to a new derivation of the time-dependent Schrödinger equation, the topics of this section. With this as the foundation for his work on quantum electrodynamics, Feynman was awarded the Nobel prize in 1965, sharing it with J. Schwinger and S. Tomonoga. Feynman's inspiration was Dirac's book, which has also been the inspiration for L. Schwartz's development of the theory of distributions (Schwartz (1950–51)). One can only speculate that many persons, subsequent to learning of Schwartz's and Feynman's achievements, may have spent countless hours poring over Dirac's treatise in the (so far) vain hope of finding a third lode of glory to mine.

Path integrals and the classical action

As a preliminary to specifying the "corresponds-to" relation noted in the preceding, let us recall the definition of the classical action. Throughout this section we shall consider only a single particle moving in 1-D, here chosen to be x. In 1-D, the action is $S(x, t; x', t')$, defined by

$$S(x, t; x', t') = \int_{t'}^{t} dt'' L(x''(t''), \dot{x}''(t''), t''), \tag{8.32}$$

where $L(x'', \dot{x}'', t'')$ is the Lagrangian, given by

$$L(x'', \dot{x}'', t'') = K(x'', \dot{x}'', t'') - V(x'', \dot{x}'', t''), \tag{2.8}$$

with $K(V)$ being the kinetic (potential) energy.

The principle of least action states that the integral in (8.32) is to be carried out over that path (trajectory) which minimizes the action; from this, one derives the Euler equation of motion $d(\partial L/\partial \dot{x})/dt - \partial L/\partial x = 0$. In Eq. (8.32), x is the value of the position at time t, and x' is its value at time t'. The path or trajectory is, of course, $x(t)$, so that $S(x, t; x', t')$ could just as well have been written as $S(x; t, t')$. Note that the dimension of S is energy–time, the same as $\hbar$.

The "corresponds-to" relation is

$$\langle x, t|x' t'\rangle \sim e^{iS(x,t;x',t')/\hbar}, \tag{8.33}$$

where "$\sim$" in (8.33) means "corresponds to" or "behaves like." It cannot mean equality, however. For suppose that "$\sim$" in (8.33) were replaced by "=." Then, since

$$\begin{aligned} S(x, t; x', t') &= \int_{t'}^{t} dt''\, L(t'') = \sum_{n=1}^{N} \int_{t_{n-1}}^{t_n} dt''\, L(t'') \\ &= \sum_{n=1}^{N} S(x_n, t_n; x_{n-1}, t_{n-1}), \end{aligned} \tag{8.34}$$

where $t_0 = t'$, $t_N = t$, and the x'' and $\dot{x}''$ dependences are suppressed in $L(t'')$, it follows that (8.33) would become

$$\langle x,\, t|x',\, t'\rangle = \prod_{n=1}^{N} e^{iS(x_n,t_n;x_{n-1},t_{n-1})/\hbar} = \prod_{n=1}^{N} \langle x_n,\, t_n|x_{n-1},\, t_{n-1}\rangle. \tag{8.35}$$

Comparison with the 1-D version of (8.33) shows why "∼"≠"=": Eq. (8.35) does not contain any integrals and is thus inconsistent with (8.33).

Dirac, realizing this, left his statement of the relation in the "∼" form. Feynman's achievement was in finding a way to equate the two sides of Eq. (8.33). Although we know that this has been accomplished (how it is done will be explained below)), it is worth pausing a moment to ask whether there is a simple case for which the "∼" symbol can somehow be turned into an equals sign. A YES answer provides a motive beyond Dirac's intuition for anyone (Feynman and/or us) seeking a more general solution to the equality problem. The system for which a YES answer holds is the free particle, whose propagator is given by

$$\langle x,\, t|x',\, t'\rangle^{(\text{free})} = \sqrt{\frac{m}{2\pi i\hbar(t-t')}}\exp\left(i\frac{m(x-x')^2}{2\hbar(t-t')}\right). \tag{8.25}$$

As we now show, the exponential in (8.25) is just $i/\hbar$ times the classical action.

For a free particle, $V = 0$ and $T = mv^2/2$, with v a constant, in which case

$$S^{(\text{free})}(x,\, t;\, x',\, t') = \tfrac{1}{2}mv^2(t-t').$$

However, $v = (x-x')/(t-t')$, so that $S^{(\text{free})}$ becomes

$$S^{(\text{free})}(x,\, t;\, x',\, t') = \frac{m(x-x')^2}{2(t-t')}. \tag{8.36}$$

Comparison with (8.25) yields

$$\langle x,\, t|x',\, t'\rangle^{(\text{free})} = \sqrt{\frac{m}{2\pi i\hbar(t-t')}}e^{iS^{(\text{free})}(x,t;x',t')/\hbar}, \tag{8.37}$$

as claimed.

Hence, for this particular case and with the normalization provided by $\sqrt{m/[2\pi i\hbar(t-t')]}$, the "corresponds-to" relation is an equality. The reason for success has been the constancy of L for the case of a free particle. However, it is a special case, which is not obviously generalizable to the case $V \neq 0$. We understand this by noting that the classical action always involves an integral of L over that path which minimizes S, whereas Eq. (8.30), for example, involves an integral over *all* possible paths $\mathbf{r}''$. The latter statement is purely quantal: since we cannot observe which path was actually followed microscopically, we must sum (integrate) over all of them.[6] Only for the (finite) straight-line path of a free particle do the quantal and classical factors agree.

What Feynman realized is that, if the time intervals in (8.34) are infinitesimal, the path is a straight line even when $V \neq 0$. That is, a straight line path occurs if one sets $t_n = t_0 + n\epsilon$ and then takes the limit $\epsilon \to 0$ (in which case $N \propto \epsilon^{-1} \to \infty$ (but $t_N = t$ is finite)). For finite ϵ, a straight-line-segment approximation to an arbitrary path $x(t)$ is shown in Fig. 8.1. It is evident that, as $\epsilon \to 0$, each straight portion $\Delta x_n = x_n - x_{n-1} \to 0$

[6] Equation (8.30) represents a kind of superposition principle.

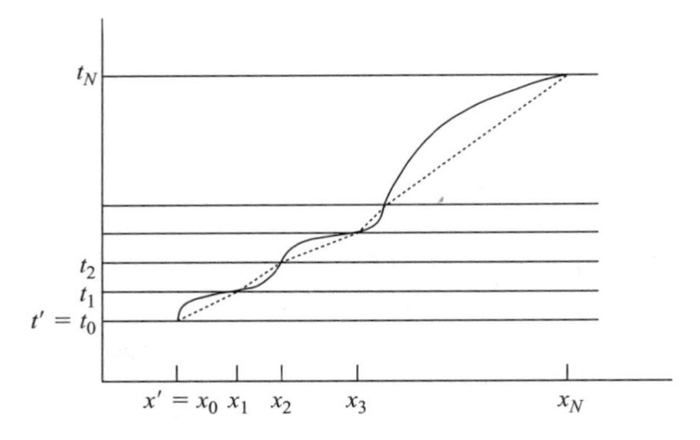

Fig. 8.1 Solid line: a typical path $x(t)$. Dotted lines: straight-line-segment approximations to $x(t)$.

and we recover $x(t)$. The infinitesimal time interval was incorporated by Feynman by equating $\langle x_n, t_n|x_{n-1}, t_{n-1}\rangle$ to $\exp(iS(x_n, t_n; x_{n-1}, t_{n-1})/\hbar)$ *only when* $t_n - t_{n-1} = \epsilon \to 0$. Finally, in place of the normalization $\sqrt{m/[2\pi i\hbar(t-t')]}$, a new factor $1/A$ is introduced; an additional A^{-1} multiplies the overall integral. The value of A depends on L.

Putting these comments into mathematical form, we have

$$\langle x, t|x', t'\rangle = \lim_{\substack{N\to\infty \\ \epsilon\to 0}} \int dx_1 \int dx_2 \ldots \int dx_N \prod_{n=1}^{N} \langle x_n t_n|x_{n-1}, t_{n-1}\rangle, \tag{8.38a}$$

with

$$\langle x_n, t_n|x_{n-1}, t_{n-1}\rangle \underset{\epsilon\to 0}{=} \frac{1}{A} e^{iS(x_n,t_n;x_{n-1},t_{n-1})/\hbar}. \tag{8.38b}$$

Substituting (8.38b) into (8.38a) and multiplying the product of integrals by A^{-1}, one obtains Feynman's relation between $\langle x, t; x', t'\rangle$ and the classical action:

$$\langle x, t|x', t'\rangle = \lim_{\substack{N\to\infty \\ \epsilon\to 0}} \frac{1}{A} \int \frac{dx_1}{A} \int \frac{dx_2}{A} \ldots \int \frac{dx_N}{A} \prod_{n=1}^{N} e^{iS(x_n,t_n;x_{n-1},t_{n-1})/\hbar}. \tag{8.38c}$$

Shorthand notations are often employed for the RHS of (8.38c). Two of them are

$$\langle x, t|x', t'\rangle = \int_{x'}^{x} \mathcal{D}x(t)\, e^{iS(x,t;x',t')/\hbar} \tag{8.39a}$$

and

$$\langle x, t|x', t'\rangle = \int_{x'}^{x} [dx]\, e^{iS(x,t;x',t')/\hbar}. \tag{8.39b}$$

The relations (8.39) serve to define the symbols $\mathcal{D}x(t)$ and $[dx]$.

In his initial publication on the path-integral approach, Feynman (1948) introduced a postulate, which is quoted below.

"The paths contribute equally in magnitude, but the phase of their contribution is the classical action (in units of $\hbar$), i.e., the time integral of the Lagrangian taken along the path."

Equation (8.38c) is the mathematical equivalent of this statement. Despite its being a non-trivial expression to evaluate (analytically or numerically), Eq. (8.38c) has been widely applied, as noted above. For a fascinating account of the development of his space–time approach and the thought processes and analysis that went into it, the interested reader may consult Feynman's Nobel-prize-acceptance speech (reprinted in *Physics Today*, August 1966 issue, page 31). Good sources for the path-integral method are the original article by Feynman (1948) and the book by Feynman and Hibbs (1965).

Derivation of the time-dependent Schrödinger equation

The application of Eq. (8.38c) that we make here is to the derivation of the time-dependent Schrödinger equation. In doing so, we show how to work with the infinitesimal time interval.

The starting point is Eq. (8.19) with t replacing t', $t+\epsilon$ replacing t, and $\mathbf{r} \to x$:

$$\Psi_\Gamma(x,\, t+\epsilon) = \int dx'\, \langle x,\, t+\epsilon | x',\, t\rangle \Psi_\Gamma(x',\, t). \tag{8.40}$$

From (8.38b), the propagator is

$$\langle x,\, t+\epsilon | x',\, t\rangle = \frac{1}{A} e^{iS(x,t+\epsilon;x',t)/\hbar}, \tag{8.41}$$

and ϵ is explicitly assumed to be a first-order infinitesimal, so the limit $\epsilon \to 0$ is omitted.

Application of Eq. (8.32) to (8.41) gives

$$S(x,\, t+\epsilon;\, x',\, t) = \int_t^{t+\epsilon} dt''\, L(x'',\, \dot{x}'',\, t''), \tag{8.42}$$

which, by virtue of the mean-value theorem of calculus, becomes

$$S(x,\, t+\epsilon;\, x',\, t) = \epsilon L_{\text{av}},$$

where L_{av} is an average value for the integrand in (8.42).

L_{av} can be approximated in various ways, each involving an approximation for t'', $x''(t'')$, and $\dot{x}''(t'')$. Some are

$$x'' \cong \begin{cases} (x+x')/2, \\ x', \\ x, \end{cases}$$

$$t'' \cong \begin{cases} t+\epsilon, \\ t, \\ t+\epsilon/2 = (t+\epsilon+t)/2, \end{cases}$$

and

$$\dot{x}'' \cong \frac{x-x'}{\epsilon}.$$

We choose $x'' \cong (x + x')/2$, $t'' \cong t$, and $\dot{x}'' = (x - x')/\epsilon$, which lead to

$$L_{\text{av}} \cong \frac{m}{2}\frac{(x - x')^2}{\epsilon^2} - V\left(\frac{x + x'}{2}, t\right),$$

$$S(x, t+\epsilon; x', t) \cong \frac{m\epsilon}{2}\left(\frac{x - x'}{\epsilon}\right)^2 - \epsilon V\left(\frac{x + x'}{2}, t\right),$$

and

$$\Psi(x, t+\epsilon) \cong \frac{1}{A}\int_{-\infty}^{\infty} dx' \exp^{\{i\epsilon/\hbar[(m/2)(x-x'/\epsilon)^2 - V((x+x')/2, t)]\}}\Psi(x', t). \tag{8.43}$$

The first term in the exponential in (8.43) goes like $\exp[im(x - x')^2/(2\hbar\epsilon)]$, which oscillates very rapidly due to the $1/\epsilon$ term, unless x' is very near to x. We therefore set $x' = x + \eta$, with η^2 infinitesimal, whence (8.43) becomes

$$\Psi(x, t+\epsilon) \cong \frac{1}{A}\int_{-\infty}^{\infty} d\eta \exp^{\{i[m\eta^2/2\hbar\epsilon - (\epsilon/\hbar)V(x+(\eta/2), t)]\}}\Psi(x + \eta, t). \tag{8.44}$$

Although the integral in (8.44) is from $-\infty$ to ∞, the main contribution comes from $\eta \sim (\hbar\epsilon/m)^{1/2}$. Introducing a separate Taylor-series expansion for each term in Eq. (8.44) that depends on η or ϵ, and keeping no powers higher than ϵ or η^2, we have

$$\Psi(x, t+\epsilon) \cong \Psi(x, t) + \epsilon\frac{\partial\Psi(x, t)}{\partial t},$$

$$\epsilon V(x + \eta/2, t) \cong \epsilon V(x, t),$$

and

$$\Psi(x + \eta, t) \cong \Psi(x, t) + \eta\frac{\partial\Psi(x, t)}{\partial x} + \frac{\eta^2}{2}\frac{\partial^2\Psi(x, t)}{\partial x^2}.$$

With these results, (8.44) becomes (valid to $O(\epsilon^{3/2})$)

$$\Psi(x, t) + \epsilon\frac{\partial\Psi(x, t)}{\partial t} = \frac{1}{A}e^{-i\epsilon V(x,t)/\hbar}\int_{-\infty}^{\infty} d\eta\, e^{im\eta^2/(2\hbar\epsilon)}\left(\Psi(x, t) + \eta\frac{\partial\Psi(x, t)}{\partial x} + \frac{\eta^2}{2}\frac{\partial^2\Psi(x, t)}{\partial x^2}\right). \tag{8.45}$$

Next, making a Taylor-series expansion of $\exp(-i\epsilon V(x, t)/\hbar)$ and noting that the integral of the second term in large parentheses on the RHS of (8.45) vanishes, (8.45) is transformed into

$$\Psi(x, t) + \epsilon\frac{\partial\Psi(x, t)}{\partial t} = \frac{1}{A}\left(1 - \frac{i\epsilon}{\hbar}V(x, t)\right)\left[\Psi(x, t)\int_{-\infty}^{\infty} d\eta \exp\left(i\frac{m\eta^2}{2\hbar\epsilon}\right) + \frac{1}{2}\frac{\partial^2\Psi(x, t)}{\partial x^2}\int_{-\infty}^{\infty} d\eta\, \eta^2 \exp\left(i\frac{m\eta^2}{2\hbar\epsilon}\right)\right]. \tag{8.46}$$

Finally, since the integrals in (8.46) can be shown to be

$$\int_{-\infty}^{\infty} d\eta \exp\left(i\frac{m\eta^2}{2\hbar\epsilon}\right) = \sqrt{\frac{2\pi i\hbar\epsilon}{m}}$$

and

$$\int_{-\infty}^{\infty} d\eta\, \eta^2 \exp\left(-i\frac{m\eta^2}{2\hbar\epsilon}\right) = \frac{\sqrt{\pi}}{2}\left(\frac{2i\hbar\epsilon}{m}\right)^{3/2},$$

we get

$$\Psi(x, t) + \epsilon\frac{\partial\Psi(x, t)}{\partial t} = \frac{1}{A}\left(1 - \frac{i\epsilon}{\hbar}V(x, t)\right)\sqrt{\frac{2\pi i\hbar\epsilon}{m}}\left(\Psi(x, t) + \frac{i\hbar\epsilon}{2m}\frac{\partial^2\Psi}{\partial x^2}\right). \tag{8.47}$$

Dropping terms of order $\epsilon^{3/2}$ and higher, and then equating coefficients of ϵ^0 and ϵ^1 on each side of (8.47), we obtain $A = \sqrt{2\pi i\hbar\epsilon/m}$ and

$$i\hbar\frac{\partial\Psi(x, t)}{\partial t} = -\frac{\hbar^2}{2m}\frac{\partial^2\Psi(x, t)}{\partial x^2} + V(x, t)\Psi(x, t) \equiv \hat{H}(x, t)\Psi(x, t), \tag{8.48}$$

the time-dependent Schrödinger equation (!).

Feynman's relation between the propagator and the classical action has led to the fundamental equation of quantum mechanics, thereby justifying the claim that the path-integral method is an independent way of formulating quantum theory. Further details can be found in the references cited in the foregoing.

8.3. The classical limit

Bohr's correspondence principle proclaims both a fundamental aspect of, and a requirement on, quantum theory. Generalizing from Postulate B4 of Section 1.2 on the Bohr atom, the principle may be stated as follows.

> By taking an appropriate limit, the quantal description of a macroscopic system must go over to the classical one.

This principle simply recognizes that classical mechanics/electrodynamics, being a valid framework for understanding macroscopic phenomena, must be obtained from quantum theory when it is applied to classical systems. In particular, one must be able to recover Newton's second law or an equivalent dynamical relation.

The correspondence principle implies that a suitable limit will always exist, though none is specified. Three instances of one such limit, viz., $n \to \infty$, have already been encountered. One is the equality of the quantal and classical radiation frequencies in Bohr's model of the H-atom, the other two are the equality of the average quantal and the classical probabilities for the cases of a particle in the 1-D quantal box and a 1-D harmonic oscillator potential (Exercises 6.4 and 6.18).

The limit $n \to \infty$, however, is not universally appropriate: many systems have only a finite number of bound states, thus precluding an $n \to \infty$ limit. Furthermore, classical behavior is not restricted to bound systems – it obviously occurs for unbound systems via collision processes, and an $n \to \infty$ limit is again not relevant.

Universal limits do exist, however. One, the limit $\hbar \to 0$, is suggested by the first paragraph in Chapter 1: $\hbar$ must be negligible for macroscopic systems. A second limit is related to the first, and is suggested by the absence of uncertainty relations for any pair of simultaneous or sequential measurements in classical physics. In contrast, canonically conjugate observables in quantum mechanics lead to non-vanishing commutators and the uncertainty relations, e.g., $\Delta Q \Delta P \geqslant \hbar$, where Q and P are a generalized coordinate and its conjugate momentum, respectively. Since ΔQ is a measure of the fluctuation away from the average (quantal) value Q_{av}, we take the unimportance or absence of fluctuations to be the second limit, i.e., the limit in which both $\Delta Q/Q_{\text{av}} \ll 1$ and $\Delta P/P_{\text{av}} \ll 1$.

The first of these universal limits ($h \to 0$) will give rise to the semi-classical or WKB approximation and the recovery of the Hamilton–Jacobi equation of classical mechanics. The second will be embodied in the classical limit of Ehrenfest's equations and also in the semi-classical behavior of positive-energy wave packets. We begin our analysis with a derivation of Ehrenfest's theorem. As in the preceding section, only a single particle is considered.

Ehrenfest's theorem

The two relations known as Ehrenfest's theorem are specific examples of Eq. (8.4). For $\hat{A}$, we select first $\hat{\mathbf{Q}}$ and then $\hat{\mathbf{P}}$, the (3-D) position and momentum operators for the single particle. A local Hamiltonian $\hat{H}(\mathbf{r})$ of the form

$$\hat{H}(\mathbf{r}) = \frac{\hat{P}^2(\mathbf{r})}{2m} + V(\mathbf{r}) \tag{8.49}$$

is assumed to govern the motion, and only diagonal matrix elements

$$\langle \hat{A} \rangle_\Gamma(t) = \langle \Gamma,\, t | \hat{A} | \Gamma,\, t \rangle$$

are considered. $|\Gamma,\, t\rangle$ is typically a linear superposition, or wave packet.

For each choice of $\hat{A}$ stated above, $\partial \hat{A}/\partial t = 0$, and the commutators of (8.4) are easily evaluated, so that we get

$$\frac{d\langle \hat{\mathbf{Q}} \rangle_\Gamma(t)}{dt} \equiv \frac{d\langle \mathbf{r} \rangle_\Gamma(t)}{dt} = \frac{1}{i\hbar} \langle [\hat{\mathbf{r}},\, \hat{H}] \rangle_\Gamma = \frac{1}{\mathrm{m}} \langle \hat{\mathbf{P}} \rangle_\Gamma(\mathrm{t}) \tag{8.50a}$$

and

$$\frac{d\langle \hat{\mathbf{P}} \rangle_\Gamma(t)}{dt} = \frac{1}{i\hbar} \langle [\hat{\mathbf{P}},\, \hat{H}] \rangle_\Gamma = -\langle \nabla_{\mathbf{r}} V \rangle_\Gamma, \tag{8.50b}$$

where $\nabla_{\mathbf{r}}$ means gradient w.r.t. $\mathbf{r}$, and the $\mathbf{r}$-dependence has been made explicit in the matrix elements, thereby contravening Dirac notation (and leading to a difficulty later – but read on!).

Equations (8.50) constitute Ehrenfest's theorem. These relations bear a striking resemblance to the classical-mechanics relations $\dot{\mathbf{r}} = \mathbf{p}/\mathrm{m}$ and $\dot{\mathbf{p}} = \mathbf{F} = -\nabla V$. Differentiating both sides of (8.50a) w.r.t. time and then substituting into (8.50b) yields

$$m \frac{d^2}{dt^2} \langle \mathbf{r} \rangle_\Gamma(t) = \langle -\nabla_{\mathbf{r}} V \rangle_\Gamma. \tag{8.51}$$

Equation (8.51) is suggestive of classical physics. To further the analogy, we define $\mathbf{F}(\mathbf{r})$ by

$$\mathbf{F}(\mathbf{r}) \equiv -\nabla_{\mathbf{r}} V(\mathbf{r}). \tag{8.52}$$

For any $V(\mathbf{r})$ that is a classical-physics potential energy, $\mathbf{F}(\mathbf{r})$ is the corresponding classical force. Substituting (8.52) into (8.51) gives

$$m\,\frac{d^2}{dt^2}\langle\mathbf{r}\rangle_\Gamma(t) = \langle\mathbf{F}(\mathbf{r})\rangle_\Gamma, \tag{8.53}$$

which at first glance would appear to be the quantal version of Newton's second law, $m\mathbf{a} = m\ddot{\mathbf{r}} = \mathbf{F}(\mathbf{r})$.

Unfortunately, appearances can be deceiving, and in Eq. (8.53) we have such a situation: (8.53) fails to be a quantal version of Newton's second law. The reason is simple: in order to become the second law, the RHS of Eq. (8.53) would need to be $\mathbf{F}(\langle\mathbf{r}\rangle_\Gamma(t))$; it is not. Although $\langle\mathbf{F}\rangle(\mathbf{r})_\Gamma \neq \mathbf{F}(\langle\mathbf{r}\rangle_\Gamma(t))$, there *are* circumstances in which these two quantities are approximately equal; these circumstances *define* the classical limit.

As a means of introducing the approximation of interest, let us recall that $\langle\mathbf{F}(\mathbf{r})\rangle_\Gamma$ is not actually a function of $\mathbf{r}$: our notation deviated from the usual in order to emphasize certain aspects of the expectation values. In terms of an integral – i.e., going into the coordinate representation – we have

$$\begin{aligned}\langle\mathbf{F}(\mathbf{r})\rangle_\Gamma &= -\int d^3r\,\Psi_\Gamma{}^*(\mathbf{r},\,t)[\nabla_{\mathbf{r}}V(\mathbf{r})]\Psi(\mathbf{r},\,t).\\ &= \int d^3r\,\Psi_\Gamma{}^*(\mathbf{r},\,t)\mathbf{F}(\mathbf{r})\Psi(r,\,t).\end{aligned} \tag{8.54}$$

To get $\mathbf{F}(\langle\mathbf{r}\rangle_\Gamma(t))$ into our equation, we expand $F(\mathbf{r})$ in (8.54) around the value $\langle\mathbf{r}\rangle_\Gamma(t) \equiv \mathbf{r}_\Gamma$:

$$\begin{aligned}\mathbf{F}(\mathbf{r}) = \mathbf{F}(\mathbf{r}_\Gamma) &+ (\mathbf{r}-\mathbf{r}_\Gamma)\cdot(\nabla F)_{\mathbf{r}_\Gamma}\\ &+\frac{1}{2}\sum_{j,k}(r_j - r_{j,\Gamma})(r_k - r_{k,\Gamma})\left(\frac{\partial^2 F}{\partial r_j\,\partial_r k}\right)_{r_{j,\Gamma}r_{k,\Gamma}} + \cdots.\end{aligned} \tag{8.55}$$

Now, the terms $\mathbf{F}(\mathbf{r}_\Gamma)$, $(\nabla F)_{\mathbf{r}_\Gamma}$, etc., are constants w.r.t to the integration variable $\mathbf{r}$, so that, on substituting (8.55) into (8.54), we find

$$\langle\mathbf{F}(\mathbf{r})\rangle_\Gamma \cong \mathbf{F}(\mathbf{r}_\Gamma) + \frac{1}{2}\sum_{j,k}[\langle r_j r_k\rangle_\Gamma - \langle r_j\rangle_\Gamma\langle r_k\rangle_\Gamma]\left(\frac{\partial^2 F}{\partial r_j\,\partial r_k}\right)_{r_{j,\Gamma}r_{k,\Gamma}}, \tag{8.56}$$

where the time dependence of the expectation values of r_j and/or r_k has been suppressed and the higher-order terms in the Taylor series have been dropped on the assumption that third and higher powers of $(r_j - r_{j,\Gamma})$ can be neglected in the matrix element (8.54). If, in addition to the latter assumption, the correction term to $\mathbf{F}(\mathbf{r}_\Gamma)$ in (8.56) is also neglected, then these approximations transform Eq. (8.53) into

$$m\,\frac{d^2\mathbf{r}_\Gamma}{dt^2} \cong \mathbf{F}(\mathbf{r}_\Gamma), \qquad \mathbf{r}_\Gamma \equiv \langle\mathbf{r}\rangle_\Gamma(\mathrm{t}), \tag{8.57}$$

which *is* Newton's second law.

The key ingredient in the approximations leading to (8.57) is the neglect of the expectation value of the terms in the Taylor series beyond the leading term $\mathbf{F}(\mathbf{r}_\Gamma)$, in particular neglect of the second term on the RHS of (8.56). The term in square brackets is roughly a measure of the width of the quantal wave packet, so that the validity of the classical limit (8.57) requires the product of the width and the second derivative of the force (evaluated at $\mathbf{r}_\Gamma$) to be much smaller than unity. This condition is expected to hold for packets of very narrow width (as well as for potentials whose spatial variation is slow over small but macroscopic distances). Assuming that one has narrow widths, and recalling that a measure of the width is the uncertainty, our criterion is that the classical limit is reached for those wave packets whose uncertainties are very small relative to the typical values $\mathbf{r}_\Gamma(t)$ of the classical trajectory, e.g. $\Delta x/x_\Gamma \ll 1$, as noted above. This criterion clearly suggests that the particle described by such a packet can be observed to follow a classical trajectory/orbit.

The preceding analysis is sound as far as it goes, but does not go far enough. Omitted is any mention of the fact that quantities like Δx are time-dependent and generally increase in value as the wave packet spreads. We must therefore modify the argument by requiring quantities like $\Delta x = \Delta x(t)$ to remain small over macroscopic periods of time, which in turn means that the wave packet $|\Gamma,\, t\rangle$ must be well localized over such times. This is equivalent to stating that macroscopic objects will always be locatable/localizable. As an example, suppose that a 1-g mass is to be described by a (non-spreading) wave packet with $d \sim \delta x \sim 10^{-3}$ cm. Using the 1-D wave-packet analyses of Section 7.1 as an estimate, the time $T = md^2/\hbar$ or $T = 2m/(\hbar\delta k^2) \sim 2m(\delta x)^2/\hbar$ will be of the order of 10^{23} s. This is a comfortably long, ultra-macroscopic time! It may be contrasted to the value $T{\sim}10^{-15}$ s for description of a proton by a non-spreading wave packet with $\delta x \sim d \sim 10^{-10}$ cm $= 10^3$ fm.

The Gaussian wave packet

We concluded the preceding subsection with remarks on the need to work with narrow wave packets in order to obtain Newton's second law. Our earlier comments noted that the limit $h \to 0$ also implies classical behavior. Let us take this limit in a specific wave packet, viz. the Gaussian packet, as an example of how quantum mechanics goes over to classical mechanics when $h \to 0$.

Rather than consider the wave function $\Psi_G(x,\, t)$ of Eq. (7.36), we examine the corresponding probability density $\rho_G(x,\, t)$:

$$\rho_G(x,\, t) = \left(\frac{1}{\pi\left(d^2 + \dfrac{h^2 t^2}{m^2 d^2}\right)}\right)^{1/2} \exp\!\left(-\frac{(x - v_0 t)^2}{[d^2 + \hbar^2 t^2/(m^2 d^2)]}\right). \tag{8.58}$$

On taking the limit $h \to 0$, $\rho_G(x,\, t)$ becomes

$$\lim_{\hbar\to 0} \rho_G(x,\, t) \to \frac{1}{d\sqrt{\pi}} e^{(x-v_0 t)^2/d^2}, \tag{8.59}$$

which describes a free particle whose initial velocity is v_0 and whose position is distributed around the origin via a Gaussian of width d. In this $\hbar \to 0$ limit, the momentum is fixed, and the position can vary. The uncertainty relation $\Delta P_x \Delta x \geqslant \hbar/2$ is no longer of consequence (it is trivially satisfied), since ΔP_x and $\hbar$ are both zero, while $\Delta x \sim d$.

To achieve the exact behavior of a classical free particle, whose trajectory is $x = v_0 t$, we must take the limit $d \to 0$. Doing so in (8.59) yields

$$\lim_{d\to 0}\left[\lim_{h\to 0} \rho_G(x,\, t)\right] \to \delta(x - v_0 t), \tag{8.60}$$

where Eq. (4.126) has been employed. Equation (8.60) is the means by which quantum mechanics specifies that the trajectory is that of a classical, free, point particle. The double limit, $h \to 0$ and packet width $\to 0$, leads precisely to classical motion.

The semi-classical or WKB approximation

The Dirac/Feynman prescription establishes a relation between the propagator and the classical action; the latter, via the principle of least action, gives rise to Euler's equations in classical mechanics. The wave function is related to the propagator through an integral over all paths. By introducing a specific form for the wave function itself, substituting into the time-dependent Schrödinger equation, and then taking the limit $h \to 0$, we can obtain another fundamental relation of classical mechanics, the Hamilton–Jacobi equation (see, e.g. Fetter and Walecka (1980) or Goldstein (1980)).

We start by writing the one-particle wave function $\Psi_\Gamma(\mathbf{r},\, t)$ in the form

$$\Psi_\Gamma(\mathbf{r},\, t) = N e^{iS(\mathbf{r},t)/\hbar}, \tag{8.61}$$

where N is a normalization constant and the phase function $S(\mathbf{r},\, t)$ is not to be confused with the action $S(\mathbf{r},\, t;\, \mathbf{r}',\, t')$, although the resemblance to (8.33) is evident. Equation (8.61) puts $\Psi_\Gamma(\mathbf{r},\, t)$ in the "eikonal" form, which is used in electrodynamics to go from wave to geometrical optics. Approximations involving (8.61) were introduced by Jeffries, Wentzel, Kramers, and Brillouin; these are referred to in physics as the WKB approximation, which is examined in Chapter 17. Our interest here is only in the equation obeyed by $S(\mathbf{r},\, t)$ in the limit $h \to 0$.

Assuming that the particle's behavior is governed by a local, time-independent Hamiltonian $\hat{H}(\mathbf{r}) = \hat{K}(\mathbf{r}) + V(\mathbf{r})$, we find, on substituting (8.61) into the time-dependent Schrödinger equation, that $S(\mathbf{r},\, t)$ obeys

$$\frac{\partial S(\mathbf{r},\, t)}{\partial t} = -\frac{1}{2m}[(\nabla S)^2 - i\hbar\, \nabla^2 S] - V. \tag{8.62}$$

The equations of the WKB approximation are obtained by expanding $S(\mathbf{r},\, t)$ as a power series in $\hbar$; see Chapter 17. As has just been noted, however, we consider here only the $h \to 0$ (classical) limit of (8.62). Introducing the notation

$$S_{\rm cl}(\mathbf{r},\, t) = \lim_{h\to 0} S(\mathbf{r},\, t), \tag{8.63}$$

$S_{\rm cl}(\mathbf{r},\, t)$ is readily seen to obey

$$\frac{\partial S_{\rm cl}(\mathbf{r},\, t)}{\partial t}+\frac{1}{2m}(\nabla S_{\rm cl})^2+V=0, \tag{8.64}$$

which is the Hamilton–Jacobi equation. Thus, in the limit $h \to 0$, the phase $S(\mathbf{r},\, t)$ (of the wave function expressed in the form (8.61)) becomes Hamilton's principal function of classical mechanics. If we knew $S_{\rm cl}(\mathbf{r},\, t)$ – from which one obtains the momentum $\mathbf{p}(t)$ by $\mathbf{p}(t) = \nabla S_{\rm cl}(\mathbf{r},\, t)$ – then using it in (8.61) would give the zeroth-order term in a semi-classical approximation to the wave function $\Psi_\Gamma(\mathbf{r},\, t)$. Nevertheless, as we show in Chapter 17, the approximation $S(\mathbf{r},\, t) \cong S_{\rm cl}(\mathbf{r},\, t)$ is not sufficient to yield a wave function with desirable characteristics. This fact notwithstanding, we have demonstrated that the Hamilton–Jacobi equation of classical mechanics plays a role in helping to determine the wave function in the (semi-)classical limit: the inverse of the Dirac/Feynman methodology has been achieved.

The harmonic oscillator, revisited

The (1-D, linear) harmonic oscillator was the example used in Section 8.1 to illustrate equations of motion in the Heisenberg picture. $\hat{Q}_x^{(\rm H)}(t)$ and $\hat{P}_x^{(\rm H)}(t)$ were seen to have the same form as the classical quantities $x(t)$ and $p_x(t)$. However, the only non-vanishing expectation values of $\hat{Q}_x^{(\rm H)}(t)$ and $\hat{P}_x^{(\rm H)}(t)$ were those taken between certain linear superpositions of harmonic oscillator states, and we cautioned against associating these non-vanishing expectation values with classical behavior. There are several reasons for being cautious. For one thing, there is an infinite number of such linear superpositions: do all of them correspond to classical behavior? If they do, then quantum mechanics would not lead to uniqueness classically, which would be an unacceptable result. If not, then do *any* of them correspond to classical behavior, and, if the answer here is YES, which, how, and why (a question to which we shall return later)? Furthermore, forming linear superpositions that give non-vanishing expectation values does not involve any limiting process, so that Bohr's correspondence principle is not even relevant.

There is, of course, a limit in which the quantal oscillator does yield classical behavior. As shown in Exercise 6.18, one forms the average of $\lim_{n\to\infty}\rho_n(x)$, where $\rho_n(x) = |\psi_n(x)|^2$; the resulting quantity then turns out to be the classical probability distribution for a particle oscillating harmonically between two fixed points.

Our previous discussion has focused on the linear superpositions but has not involved the above-mentioned limit. If no limit is taken, then we should re-examine what is meant by classical behavior of such superpositions. One way of phrasing this is as follows: are there other ways (not involving the limit $n \to \infty$) in which *any* linear superposition of oscillator states can or will display behavior characteristic of a classical oscillator? An affirmative answer would indicate a classical *analog*, rather than a classical *limit*. Not surprisingly, there is a linear superposition for which the answer is YES: it is the linear superposition denoted $|\zeta\rangle$, the coherent or "quasi-classical" state – Eq. (6.142) – whose probability density will be shown to correspond at all times to oscillation between two fixed points. This phenomenon is closely connected with the fact that the time-evolved state does not spread in time. It is for these reasons that $|\zeta\rangle$ is known as a quasi-classical state. We now prove this assertion.

To demonstrate the desired behavior, we make use of two intermediate results, stated as Lemmas 8.1 and 8.2.

LEMMA 8.1

If $|\zeta\rangle$ of Eq. (6.142) is the coherent state at time $t = 0$, then the state at time t, denoted $|\zeta,\, t\rangle$, is given by

$$|\zeta,\, t\rangle = e^{-i\omega_0 t/2}|\zeta(t)\rangle, \tag{8.65}$$

where

$$\zeta(t) \equiv \zeta e^{-i\omega_0 t} \tag{8.66}$$

is the label of the ket on the RHS of Eq. (8.65). The latter equation results from applying the time-evolution operator $\exp(-i\hat{H}t/\hbar)$, with $\hat{H} = (\hat{a}^\dagger\hat{a} + \frac{1}{2})\hbar\omega_0$, to (6.142):

$$\begin{aligned}|\zeta,\, t\rangle &= e^{-i\hat{H}t/\hbar}|\zeta\rangle \\ &= e^{-|\zeta|^2/2}\sum_n \frac{\zeta^n e^{-i(n+1/2)\omega_0 t}|n\rangle}{\sqrt{n!}} \\ &= e^{-i\omega_0 t/2}|\zeta e^{-i\omega_0 t}\rangle, \quad \textbf{QED}.\end{aligned}$$

LEMMA 8.2

An alternate to Eq. (6.143) is the equivalent compact form

$$|\zeta\rangle = e^{-(|\zeta|^2-\zeta^2)/2}e^{-i\lambda_0\zeta\sqrt{2}\hat{P}_x/\hbar}|0\rangle, \tag{8.67}$$

where $\lambda_0 = \sqrt{\hbar/(m\omega_0)}$ is the length parameter for the quantal oscillator. The first step in proving (8.67) is to recall the equality $\hat{P}_x = [i\hbar/(\lambda_0\sqrt{2})](\hat{a}^\dagger - \hat{a})$, from which it follows that

$$e^{-i\lambda_0\zeta\sqrt{2}\hat{P}_x/\hbar}|0\rangle = e^{\zeta(\hat{a}^\dagger-\hat{a})}|0\rangle. \tag{8.68}$$

Next, we wish to introduce the product $\exp(\zeta\hat{a}^\dagger)\exp(-\zeta\hat{a})$ into the RHS of (8.68). This can be accomplished by employing Eq. (4.191), with $\hat{B}_1 = \zeta\hat{a}^\dagger$ and $\hat{B}_2 = -\zeta\hat{a}$; its use leads to

$$e^{\zeta(\hat{a}^\dagger-\hat{a})}|0\rangle = e^{-\zeta^2/2}e^{\zeta\hat{a}^\dagger}e^{-\zeta\hat{a}}|0\rangle. \tag{8.69}$$

However, $\hat{a}|0\rangle = 0$ means that $\exp(-\zeta\hat{a})\,|0\rangle = |0\rangle$, and hence (8.69) becomes

$$\begin{aligned}e^{\zeta(\hat{a}^\dagger-\hat{a})}|0\rangle &= e^{-\zeta^2/2}e^{\zeta\hat{a}^\dagger}|0\rangle \\ &= e^{(|\zeta|^2-\zeta^2)/2}|\zeta\rangle.\end{aligned} \tag{8.70}$$

The second equality in (8.70) results from inserting (6.143) into the first equality. Finally, equating the RHS of (8.70) to the LHS of (8.68) yields (8.67), **QED**.

COROLLARY

A consequence of (8.67) is that, if d is a parameter of dimension length, then $\exp(-id\hat{P}_x/\hbar)\,|0\rangle$ produces a coherent state with ζ given by $d/(\lambda_0\sqrt{2})$:

$$e^{-id\hat{P}_x/\hbar}|0\rangle \propto \left|\frac{d}{\lambda_0\sqrt{2}}\right\rangle, \tag{8.71}$$

i.e.,

$$\left[\hat{a} - \left(\frac{d}{\lambda_0\sqrt{2}}\right)\right]e^{-id\hat{P}_x/\hbar}|0\rangle = 0. \tag{8.72}$$

Proofs of these two relations are left to the exercises.

The probability density we wish to examine is

$$\rho_\zeta(x,\, t) = |\langle x|\zeta(t)\rangle|^2. \tag{8.73}$$

An explicit form for $\psi_\zeta(x,\, t) \equiv \langle x|\zeta(t)\rangle$ is obtained by combining Eqs. (8.65) and (8.67):

$$\psi_\zeta(x,\, t) = e^{-i\omega_0 t/2}e^{(|\zeta|^2-\zeta^2 e^{-2i\omega_0 t})/2}\langle x|e^{-ix_0\hat{P}_x/\hbar}|0\rangle, \tag{8.74}$$

where

$$x_0 = x_0(t) \equiv \lambda_0\zeta e^{-i\omega_0 t}\sqrt{2} \tag{8.75}$$

is a time-dependent complex length parameter.

Determination of $\psi_\zeta(x,\, t)$ requires evaluation of the matrix element $\langle x|\exp(-ix_0\hat{P}_x/\hbar)|0\rangle$. Rather than work out the meaning of the operator $\exp(-ix_0\hat{P}_x/\hbar)$ here, we make use of a result that will be derived in Section 9.4, viz.,

$$\langle x|e^{-ix_0\hat{P}_x/\hbar} = \langle x - x_0|. \tag{8.76}$$

That is, $\exp(-ix_0\hat{P}_x/\hbar)$, when it is acting on 1-D position kets $|x\rangle$ and position bras $\langle x|$, changes them to $|x - x_0\rangle$ and $\langle x - x_0|$, respectively. Using (8.76) in (8.74), we find

$$\psi_\zeta(x,\, t) = e^{-i\omega_0 t/2}e^{-(|\zeta|^2-\zeta^2 e^{-2i\omega_0 t})/2}\sqrt{\frac{1}{\lambda_0\sqrt{\pi}}}e^{-(x-x_0)^2/2\lambda_0^2}, \tag{8.77}$$

where $\langle x - x_0|0\rangle \equiv \psi_0(x - x_0)$ is the coordinate-space function for the ground state of the oscillator, as in Eq. (6.44).

Although the complex length x_0 has been introduced as a notational shorthand, it actually plays a much more important role: through its real and imaginary parts, which occur in the second and third exponentials of Eq. (8.77), we can introduce two quantities of special significance for the coherent state. These quantities are the expectation values of $\hat{Q}_x$ and $\hat{P}_x$. Straightforward evaluation leads to

$$\langle\zeta,\, t|\hat{Q}_x|\zeta,\, t\rangle \equiv \bar{x} = \lambda_0\sqrt{2}[(\mathrm{Re}\,\zeta)\cos(\omega_0 t) + (\mathrm{Im}\,\zeta)\sin(\omega_0 t)] \tag{8.78a}$$

and

$$\langle\zeta,\, t|\hat{P}_x|\zeta,\, t\rangle \equiv \hbar\bar{k} = m\omega_0\lambda_0\sqrt{2}[(\mathrm{Im}\,\zeta)\cos(\omega_0 t) - (\mathrm{Re}\,\zeta)\sin(\omega_0 t)]x, \tag{8.78b}$$

where the ζ and t dependences are suppressed in $\bar{x}$ and $\bar{k}$.

On the other hand, the real and imaginary parts of x_0 are readily found to be

$$\mathrm{Re}\, x_0 = \lambda_0\sqrt{2}[(\mathrm{Re}\,\zeta)\cos(\omega_0 t) + (\mathrm{Im}\,\zeta)\sin(\omega_0 t)] \tag{8.79a}$$

and

$$\mathrm{Im}\, x_0 = \lambda_0\sqrt{2}[(\mathrm{Im}\,\zeta)\cos(\omega_0 t) - (\mathrm{Re}\,\zeta)\sin(\omega_0 t)]. \tag{8.79b}$$

Comparison of the last two pairs of equations yields the important relations

$$\bar{x} = \mathrm{Re}\, x_0 \tag{8.80a}$$

and

$$\bar{k} = (\mathrm{Im}\, x_0)/\lambda_0^2. \tag{8.80b}$$

Using Eqs. (8.80) for the real and imaginary parts of x_0, $\psi_\zeta(x,\, t)$ takes on a familiar (and suggestive) form:

$$\psi_\zeta(x,\, t) = e^{i[\bar{k}(x-\bar{x})-\omega_0 t]/2}\sqrt{\frac{1}{\lambda_0\sqrt{\pi}}}e^{-(x-\bar{x})^2/(2\lambda_0^2)}, \tag{8.81}$$

which is a Gaussian wave packet of central momentum $\bar{k}/2$ and energy $\hbar\omega_0/2$, but with constant (time-independent) width. $\psi_\zeta(x,\, t)$ therefore does not spread in time. This property becomes evident on forming the probability density:

$$\rho_\zeta(x,\, t) = \frac{1}{\lambda_0\sqrt{\pi}}e^{-(x-\langle\zeta,\, t|\hat{Q}_x|\zeta,\, t\rangle)^2/\lambda_0^2}. \tag{8.82}$$

It is a Gaussian distribution whose peak position is $\langle\zeta,\, t|\hat{Q}_x|\zeta,\, t\rangle = \bar{x}$. Equations (8.78a) and (8.18a) show that both $\bar{x}$ and the position $x(t)$ of a classical oscillator have the same time dependence, viz., an oscillation between two fixed points. This behavior therefore characterizes $\rho_\zeta(x,\, t)$: its peak position oscillates between two fixed points. Coherent states do indeed display classical behavior, though such states are not a classical limit.

Exercises

8.1 Suppose that $\hat{A}$ in Eq. (8.4) is replaced by the product of two operators: $\hat{A} \to \hat{B}\hat{C}$.
   (a) Show that, for this case, the analog of the { } term in (8.4) involves two time derivatives and two commutators, one featuring $[\hat{H},\, \hat{B}]$, the other featuring $[\hat{H},\, \hat{C}]$.
   (b) Let $\hat{B} = \hat{Q}_x$, $\hat{C} = \hat{P}_x$, $\hat{H}$ be the 1-D linear harmonic oscillator Hamiltonian, and $|\Gamma_j,\, t\rangle = |\Gamma_k,\, t\rangle = |n,\, t\rangle$, where $|n\rangle$ is the nth oscillator eigenstate. Show that substitution of these quantities into the result of part (a) leads to $\langle\hat{K}_x\rangle_n = \langle\hat{V}\rangle_n$, thus providing an alternate means of obtaining the result of Exercise 6.13.

8.2 The quantal virial theorem, which is a generalization of the result of Exercise 8.1(b), is the subject of this exercise. Let $\hat{H}(\mathbf{r}) = \hat{K}(\mathbf{r}) + V(\mathbf{r})$ be the Hamiltonian for a particle in 3-D mass m, and let $|\Gamma,\, t\rangle = |\Gamma\rangle\exp(-iE_\Gamma t/\hbar)$ be a time-dependent stationary state of the system.

(a) Work through the expression of $d\langle\hat{\mathbf{Q}}\cdot\hat{\mathbf{P}}\rangle_\Gamma(t)/dt$ to obtain the result $2\langle\hat{K}\rangle_\Gamma = \langle\hat{\mathbf{Q}}\cdot\nabla_{\hat{\mathbf{Q}}}V(\hat{\mathbf{Q}})\rangle\Gamma \equiv \langle\mathbf{r}\cdot\nabla_\mathbf{r}V(\mathbf{r})\rangle_\Gamma$, which is a statement of the quantal virial theorem.

(b) Show that, for 1-D bound states, with $V(\mathbf{r}) \to V(x)$, the preceding result becomes $2\langle\hat{K}_x\rangle_\Gamma = \langle x\, dV(x)/dx\rangle_\Gamma$.

(c) In the 1-D case assume that $V(x) \propto x^n$, $n = 1, 2, 3, \ldots$, and relate $\langle\hat{K}\rangle_\Gamma$ to $\langle V\rangle_\Gamma$ and each of them to E_Γ. Verify $\langle\hat{K}_x\rangle_n = \langle V\rangle_n$ for the linear harmonic oscillator.

(d) Carry out the same analysis as in part (c) for the attractive, central, 3-D potential $V(r) \propto r^n$, $n = -2, -1, 1, 2, \ldots$. Note especially the results for $n = -2$ and $n = -1$ (Coulomb/gravitational attraction), and discuss the $n = -2$ case (see comments on this case in Landau and Lifshitz (1958)).

8.3 Correlation/coherence functions.

(a) Determine the dependences on t, t', ω_0 and λ_0 of $C(t, t') = \langle 0|\hat{Q}_x^{(\mathrm{H})}(t')|0\rangle$, where $|0\rangle$ is the ground state for a 1-D oscillator whose frequency is ω_0, and $\lambda_0 = [\hbar/(m\omega_0)^{1/2}]$.

(b) Let $|\zeta\rangle$ be the coherent state of Eq. (6.142). Evaluate the coherence function γ_ζ defined by $\gamma_\zeta = \langle\zeta|\hat{a}'|\zeta\rangle\langle\zeta|\hat{a}|\zeta\rangle/\langle\zeta|\hat{a}'\hat{a}|\zeta\rangle$.

8.4 Let $\hat{H}(x)$ be the 1-D harmonic oscillator Hamiltonian and $|\zeta\rangle$ be a coherent state. Show that $\langle\zeta|\hat{Q}_x|\zeta\rangle$ displays a sinusoidal behavior.

8.5 In general, the Heisenberg-picture operator $\hat{A}^{(\mathrm{H})}$ has an – assumed convergent – infinite series expansion in t. Show that, for a free particle, $\hat{\mathbf{Q}}^{(\mathrm{H})}(t)$ contains terms involving only the powers t^0 and t, and evaluate the operator coefficient of the power t.

8.6 Let $\hat{A}^{(\mathrm{H})}(t)$ be the Heisenberg-picture operator of Eq. (8.7b), with $\hat{H} \neq \hat{H}(t)$.

(a) State the condition(s) under which Hermiticity of $\hat{A}$ implies the Hermiticity of $\hat{A}^{(\mathrm{H})}(t)$.

(b) Show that, if $[\hat{A}, \hat{H}] = 0$, then $\hat{A}^{(\mathrm{H})}(t) = \hat{A}$.

(c) Prove that, if $\hat{A}$ is Hermitian, then $2i\,\mathrm{Im}[\langle\alpha|\hat{A}\hat{A}^{(\mathrm{H})}(t)|\alpha\rangle] = \langle\alpha|[\hat{A}, \hat{A}^{(\mathrm{H})}(t)]|\alpha\rangle$.

8.7 (a) Show that the Heisenberg equation of motion (8.8b) for $\hat{\mathbf{Q}}$ of the 3-D isotropic harmonic oscillator ($V(\mathbf{r}) = m\omega^2r^2/2$) can be written in the classical form: viz., $\mathbf{F} = m\mathbf{a}$.

(b) Use this result to show that matrix elements of $\hat{\mathbf{Q}}^{(\mathrm{H})}(t)$ between stationary states of this system vanish unless the energies of the two states differ by $\hbar\omega_0$.

8.8 Let $\hat{H}(x)$ be the Hamiltonian for a 1-D constant force: $\hat{H}(x) = \hat{K}_x - xF$, $F > 0$.

(a) Determine $\hat{Q}_x^{(\mathrm{H})}(t)$ and $\hat{P}_x^{(\mathrm{H})}(t)$

(b) Redo part (a) for $F \to F(t)$.

8.9 A 1-D linear oscillator is acted on by a spatially uniform, time-dependent force $F(t)$ – the forced, linear harmonic oscillator, for which $\hat{H}(t) = (\hat{N} + \frac{1}{2})\hbar\omega_0 - \hat{Q}_xF(t)$.

(a) Determine $\hat{a}^{(\mathrm{H})}(t)$ by solving the differential equation it obeys.

(b) Evaluate, for $t \neq t'$, $[\hat{a}^{(\mathrm{H})}(t), \hat{a}^{(\mathrm{H})}(t')]$ and $[\hat{Q}_x^{(\mathrm{H})}(t), \hat{Q}_x^{(\mathrm{H})}(t')]$, and interpret your results.

8.10 This exercise introduces the "interaction picture," which is intermediate between the Schrödinger and Heisenberg pictures. Let a time-dependent Hamiltonian be a sum of a time-independent part $\hat{H}_0$ and a time-dependent portion $\hat{H}_1(t)$: $\hat{H}_0(t) = \hat{H}_0 + \hat{H}_1(t)$. Let $|\Gamma, t\rangle$ be a solution of the time-dependent Schrödinger equation and $\hat{A}(t)$ be a Schrödinger-picture operator . Their *interaction-picture* analogs are defined, respectively, as

$$|\Gamma,\ t\rangle^{(\mathrm{I})} = e^{-i\hat{H}_0(t-t_0)/\hbar}|\Gamma,\ t_0\rangle \text{ and } \hat{A}^{(\mathrm{I})}(t) = e^{i\hat{H}_0(t-t_0/\hbar}\hat{A}(t_0)e^{-i\hat{H}_0(t-t_0)/\hbar}.$$

(a) Determine the differential equations obeyed by $|\Gamma,\ t\rangle^{(\mathrm{I})}$ and $\hat{A}^{(\mathrm{I})}(t)$.
(b) Show that the solution to the first of the equations obtained in part (a) is

$$|\Gamma,\ t\rangle^{(\mathrm{I})} = |\Gamma,\ t_0\rangle^{(\mathrm{I})} + \frac{1}{i\hbar}\int_{t_0}^{t} \mathrm{d}t'\ \hat{H}_1^{(\mathrm{I})}(t')|\Gamma,\ t'\rangle^{(\mathrm{I})}.$$

8.11 The forced linear oscillator was introduced in Exercise 8.9. An explicit expression for its state vector $|\Gamma,\ t\rangle$ can be determined using the interaction-picture of Exercise 8.10 and the relation $\hat{H}_0 = (\hat{N} + 1/2)\hbar\omega$.
(a) Show, using Eqs. (8.13) and (8.14), that, up to a factor that you are to determine, the operator relating $d|\Gamma,\ t\rangle^{(\mathrm{I})}/dt$ to $|\Gamma,\ t\rangle^{(\mathrm{I})}$ is the sum $[\hat{a}^{(\mathrm{H})}(t) + \hat{a}^{(\mathrm{H})\dagger}(t)]$.
(b) Define $G(t) = (2m\hbar\omega)^{-1/2}\int dt'\ F(t')e^{i\omega t}$, and show that $d|\Gamma,\ t\rangle^{(\mathrm{I})}/dt = [i\hat{a}\, dG^*/dt + i\hat{a}^\dagger\, dG/dt]|\Gamma,\ t\rangle^{(\mathrm{I})}$. A direct solution to this equation is not possible because $[\hat{a},\ \hat{a}^\dagger] \neq 0$. This problem can be avoided by introducing a second interaction-picture state via $|\Gamma,\ t\rangle^{(\mathrm{II})} = e^{-iG(t)\hat{a}^\dagger}|\Gamma,\ t\rangle^{(\mathrm{I})}$. Obtain the solution to the equation obeyed by $|\Gamma,\ t\rangle^{(\mathrm{II})}$, expressing it as an operator acting on $|\Gamma,\ 0\rangle^{(\mathrm{II})} = |\Gamma,\ 0\rangle^{(\mathrm{I})} = |\Gamma,\ 0\rangle$.
(c) Use the definitions of $|\Gamma,\ t\rangle^{(\mathrm{II})}$ and $|\Gamma,\ t\rangle^{(\mathrm{I})}$ to show that

$$|\Gamma,\ t\rangle = \hat{O}(t)e^{-\int_0^t dt'\ G(t')[dG(t')/dt']}|\Gamma,\ 0\rangle,$$

where $\hat{O}(t)$, a product of three operators, is to be determined as part of your solution.
(d) Finally, assume that the oscillator ground state is $|\Gamma,\ 0\rangle = |0\rangle$, and prove that the probability of finding the oscillator in state $|n\rangle$ at time $t > 0$ is $P_{n0}(t) = (n!)^{-1}|G(t)|^{2n}e^{-|G(t)|^2}$.

8.12 Carry out the integration leading from Eq. (8.24) to Eq. (8.25).

8.13 Prove the following properties of $K^{\text{free}}(x,\ t;\ x',\ t')$.
(a) In the limit $t \to t'$, $K^{\text{free}}(x,\ t;\ x',\ t') \to \delta(x - x')$.
(b) It obeys the 1-D version of Eq. (8.30).
(c) For $t' = 0$, it is a solution of $[i\hbar\partial/\partial t - \hat{K}_x(x)]K^{\text{free}}(x,\ t;\ x',\ 0) = i\hbar\delta(x - x')\delta(t)$.

8.14 Show that

$$\psi(x,\ t) = \int dx'\ K^{\text{free}}(x,\ t;\ x',\ t')\psi(x',\ t_0) + \frac{1}{i\hbar}\int_{t_0}^{t} dt' \int dx'\ K^{\text{free}}(x,\ t;\ x',\ t')V(x')\psi(x',\ t')$$

satisfies the time-dependent Schrödinger equation with $\hat{H}(x) = \hat{K}_x(x) + V(x)$.

8.15 The momentum-space analog of Eq. (8.19) is $\Phi_\Gamma(\mathbf{k},\ t) = \int d^3k\ K(\mathbf{k},\ t;\ \mathbf{k}',\ t')\Phi_\Gamma(\mathbf{k}',\ t')$.
(a) Determine whether $K(\mathbf{k},\ t;\ \mathbf{k}',\ t')$ obeys the analog of the composition property (8.30).
(b) Evaluate $K^{\text{free}}(k,\ t;\ k',\ t')$, the 1-D, momentum-space, free propagator.
(c) Let a particle of mass m in 1-D be acted on by a constant force F: $V(x) = -xF$. (i) Write out the time-independent, momentum-space Schrödinger equation for this case and determine its eigenvalues $E_\Gamma(k)$, eigenfunctions $\varphi_\Gamma(k)$, and stationary-state

wave functions $\Phi_\Gamma(K, t)$. (ii) Evaluate the dependences of $K(k, t; k', t')$ on k, k', t, t', and F for this system.

8.16 The propagator for the 1-D linear harmonic oscillator is (Sakurai 1994)

$$K^{\mathrm{ho}}(x, t; x', t') =$$
$$\frac{m\omega_0}{2\pi i\hbar \sin[\omega_0(t-t')]^{1/2}} \exp\left(\frac{i\{m\omega_0\{(x^2+x'^2)\cos[\omega_0(t-t')]\} - 2xx'\}}{2\hbar \sin[\omega_0(t-t')]}\right).$$

(a) Show that $K^{\mathrm{ho}}(x, t; x', t') = \{m\omega_0/2\pi i\hbar \sin(\omega_0 T]\}^{1/2} e^{(i/\hbar)S^{\mathrm{ho}}(x,x',T)}$, where $S^{\mathrm{ho}}(x, x', T) = m \int dt[v^2 - \omega_0^2 x^2]/2$, with $x(t)$ being the position coordinate of the 1-D classical harmonic oscillator, $v(t) = dx(t)/dt$, and $T = t - t'$.

(b) Set $t' = 0$ in the definition of $K^{\mathrm{ho}}(x, t; x', t')$ and relate $K^{\mathrm{ho}}(x, t+T; x', 0)$ to $K^{\mathrm{ho}}(x, t; x', 0)$ for the cases $T = 2n\pi/\omega_0$ and $T = (2n+1)\pi/\omega_0$. Relate this behaviour to that obtained from the expansion (8.21).

(c) Find the $\omega_0 \to 0$ limit of $K^{\mathrm{ho}}(x, t; x', t')$ and explain why the result should or should not be equal to $K^{\mathrm{free}}(x, t; x', t')$.

(d) In analogy to the result stated in exercise 8.13(e), determine the equation obeyed by $K^{\mathrm{ho}}(x, t; x', t')$

8.17 In analogy to the procedure involved in going from Eq. (7.34) to Eq. (7.35), show explicitly that the integrals in (8.46) have the values listed above (8.47).

8.18 The time-dependent Schrödinger equation (8.48) was derived from the path-integral formulation by using in L_{av} the values $x'' = (x + x')/2$, $t'' = t$ and $x'' = (x - x')/2$. Show that the other choices for x'', t'', and x'' also yield Eq. (8.48).

8.19 (a) Determine the probability current density $\mathbf{J}(\mathbf{r}, t)$ that results from using Eq. (8.61).

(b) Replace the RHS of (8.61) by the expression $(\rho^{1/2}(\mathbf{r}, t)e^{iS(\mathbf{r},t)}$, where $\rho(\mathbf{r}, t) = |\Psi(\mathbf{r}, t)|^2$, and determine the proportionality function relating $\mathbf{J}(\mathbf{r}, t)$ to $\nabla S(\mathbf{r}, t)$.

(c) Set $N_2 = N_3 = N/2$ in Eq. (6.27) and evaluate $J(x, t)$. Then determine both $J(0, t)$ and $J(\varepsilon, t)$, where ε is a positive infinitesimal; the latter result will indicate how the value of $J(0, t)$ is reached.

8.20 Prove the relations (8.71) and (8.72).

8.21 (a) Evaluate the probability current density $J_\xi(x, t)$ associated with the coherent state $\psi_\xi(x, t)$ and relate it to the probability density $\rho_\xi(x, t)$.

(b) Discuss the relevance of this result to the statement that $\psi_\xi(x, t)$ does not spread in time.

9

Commuting Operators, Quantum Numbers, Symmetry Properties

Chapters 5–7 have shown us why and how states play such an essential role in quantum theory. States are distinguished through their labeling, which consists of a set of quantum numbers specific to the state. In general, each degree of freedom gives rise to a quantum number, where, in the present context, the phrase "degrees of freedom" refers to the number of spatial coordinates characterizing a given quantum system. The 1-D systems employed in the two chapters illustrating the postulates were thus single-degree-of-freedom systems: all the eigenstates encountered there were labeled by a single quantum number (n for bound states and k for continuum states) related to the energy of the system.

Having studied systems with a single degree of freedom, it might seem that our next step would be to consider systems with more than one degree of freedom, for example, two 1-D particles or a particle in 3-D subject to a potential, etc. Among the goals of such studies would be the assigning of labels to the eigenstates. If the assignment of quantum numbers were based solely on spatial degrees of freedom, then the preceding endeavor would indeed be a way to proceed with our development of quantum theory. However, spatial degrees of freedom are not the only source of the labels which specify eigenstates. An additional source is the sets of eigenvalues of the operators that commute with $\hat{H}$ and with each other. Such operators are the main subject of this chapter.

The consequences for quantum theory of operators that commute with $\hat{H}$ and/or with each other are profound and are not limited to the supplying of labels for quantum states, essential as that may be. Some of the topics we consider in this chapter make important use of commuting operators, especially symmetry operators. These topics serve in part as bridging material to later chapters as well as to flesh out the theory beyond what has been done so far.

Although commuting operators are our main interest, there are, of course, many pairs of quantal operators that have non-vanishing commutators. We consider this topic first, showing in Section 8.1 that the non-commutativity of two operators also leads to a profound consequence, viz., the Heisenberg uncertainty principle and its important theoretical and experimental implications. After that, we turn to the concept of "constants of the motion" in quantum mechanics; then to the fundamental concept of "complete sets of commuting operators" – on which the labeling of states is based; and finally, to an introductory but sometimes lengthy discussion of symmetries and symmetry operators, with illustrative examples. The study of symmetries in quantum mechanics is of great importance, not least because many quantal systems are invariant under one or more symmetry operations.

9.1. The Heisenberg uncertainty principle

If physics, among its other attributes, is the science of measurement, then quantum theory is the science of measurements on the microscopic scale. That measurements in macro- and micro-physics can lead to very different consequences was noted in Chapters 1 and 2, especially in the summary in Section 2.6 of Heisenberg's analysis of the γ-ray microscope. The qualitative analysis performed therein led to the estimate

$$\delta x\, \delta p_x \sim h,$$

where $\delta x\, (\delta p_x)$ is the uncertainty in the measured value of the position (momentum) of the photon. It was also noted there that this relation is a version of the more technical expression $\Delta x\, \Delta p_x \geqslant \hbar/2$. Our purpose in the present section is to derive and discuss the theory underlying this more technical relation – viz., the Heisenberg uncertainty relation – and to use it as the foundation for the statement of Heisenberg's uncertainty principle. We begin with a statement of the latter concept.

A statement of the uncertainty principle

Postulate I tells us that, when a measurement of the observable $\hat{A}$ yields the eigenvalue a_n, the resulting state of the physical system is $|a_n\rangle$. Hence, if $\hat{A}$ is remeasured, the value a_n is certain to be obtained. Let us now assume that the second measurement is of $\hat{B}$ (rather than of $\hat{A}$) and that the eigenvalue b_m is found; the state of the system thus changes from $|a_n\rangle$ to $|b_m\rangle$. With the system in state $|b_m\rangle$, a new measurement of $\hat{A}$ is next carried out. The crucial questions for our analysis are the following. (1) Under what circumstances will a_n again be the certain outcome? (2) If a_n is not the result of re-measuring $\hat{A}$, what range of values can one expect to find? Or, put another way, how accurately can one remeasure $\hat{A}$ when the system is in an eigenstate of $\hat{B}$? A related question is: (3) How accurately can two observables be simultaneously measured (as in the case of the γ-ray microscope)?

The answer to the questions (2) and (3) regarding accuracy are embodied in Heisenberg's uncertainty principle (Heisenberg (1927)). Before stating it, we define the terms *compatible* and *incompatible* measurements.

DEFINITION

> Measurements of any two observables $\hat{A}$ and $\hat{B}$ are deemed compatible if $[\hat{A}, \hat{B}] = 0$. When $[\hat{A}, \hat{B}] \neq 0$, the measurements are said to be incompatible.

As we shall show, compatibility means that measuring $\hat{B}$ subsequent to $\hat{A}$ does not alter the value originally measured for $\hat{A}$. In other words, when $[\hat{A}, \hat{B}] = 0$, the second measurement of $\hat{A}$ in the process "measure $\hat{A}$, measure $\hat{B}$, remeasure $\hat{A}$" yields with certainty the value first obtained for $\hat{A}$ (which is therefore the answer to question (1)).

Our statement of Heisenberg's uncertainty principle is as follows:

The uncertainty principle: If the measurements of two observables are compatible, then, in principle, the accuracy that can be obtained for each is perfect, i.e., the uncertainties in the measured values can be zero (whether they *will be* depends on the experiment). However, if the measurements are incompatible, the product of the uncertainties resulting from a simultaneous measurement can never be less than a minimum value, independent of the experimental method used.

There is a cautionary note to bear in mind regarding the preceding statement: the term "uncertainty" used in connection with incompatible measurements is a technical one that remains to be defined. Until we do so (below), the quantitative meaning of the uncertainty principle is uncertain; qualitatively, however, its significance is clear.

Compatible measurements

Commuting operators lead to compatible measurements. The underlying reason is simple: when $[\hat{A}, \hat{B}] = 0$, $\hat{A}$ and $\hat{B}$ share a common set of eigenstates, and the measurement $\hat{A} \to a_n$ followed by $\hat{B} \to b_m$ does not alter the initial state, thus ensuring that a remeasurement of $\hat{A}$ will yield a_n with certainty. That $\hat{A}$ and $\hat{B}$ share a common set of eigenstates when $[\hat{A}, \hat{B}] = 0$ was established for the case of no degeneracies in Section 4.5, Eqs. (3.145)–(3.188). This analysis will now be extended, by explicit construction, to the degenerate case.[1] We shall show that, by using the eigenvalues of $\hat{B}$ as a second label for the eigenstates (originally only of $\hat{A}$), the degeneracy is removed. Our only additional assumption is that there are no other operators that commute with $\hat{A}$ or $\hat{B}$, a restriction that will be removed in Section 9.3.

So, we assume that $[\hat{A}, \hat{B}] = 0$; that the nth eigenvalue is N-fold degenerate, i.e.,

$$\hat{A}|a_n^{(i)}\rangle = a_n|a_n^{(i)}\rangle, \qquad i = 1, \ldots, N; \tag{9.1}$$

and that the eigenstates of $\hat{B}$ remain to be determined. The $\{|a_n^i\rangle\}_{i=1}^N$ are linearly independent and span an N-dimensional subspace $\mathcal{M}_N \subset \mathcal{H}$ and, though they need not be orthogonal or normalized, we presume that they have been made so via the Gram–Schmidt process; i.e., we assume that $\langle a_n^{(j)}|a_n^{(k)}\rangle = \delta_{jk}$.

Although a measurement of $\hat{A}$ that yields a_n means that the subsequent state of the system is an arbitrary linear combination of the N $|a_n^{(i)}\rangle$, we now show that subsequent measurement of $\hat{B}$ and then of $\hat{A}$ will necessarily yield both the value a_n and a unique state, rather than an *arbitrary* linear combination of the N $|a_n^{(i)}\rangle$. This is in contrast to the result of measuring $\hat{A}$ only. The unique state noted above is the eigenstate $|b_m\rangle$ of $\hat{B}$ corresponding to the eigenvalue b_m obtained in the second measurement; it is a specific linear combination of the $|a_n^{(i)}\rangle$'s.

To establish this, we first remark that, as in Section 4.5, $[\hat{A}, \hat{B}] = 0$ implies

$$\hat{A}[\hat{B}|a_n^{(i)}\rangle] = a_n[\hat{B}|a_n^{(i)}\rangle], \qquad \forall\, i, \tag{9.2}$$

i.e., $\hat{B}|a_n^{(i)}\rangle \in \mathcal{M}_N$. This result already ensures that the second measurement of $\hat{A}$, subsequent to the $\hat{A} \to a_n$ and $\hat{B}$ measurements, will yield a_n with certainty. However,

[1] Far from being mere mathematical curiosities, degeneracies occur widely in quantum mechanics and must be taken into account if there exist no situations (additional interactions or operators) that remove them.

because $\hat{B}|a_n^{(i)}\rangle \in \mathcal{M}_N$, it follows that N linear combinations of the $|a_n^{(i)}\rangle$'s which are each eigenstates of $\hat{B}$ can be found. That is, the eigenvalue equation

$$\hat{B}|b\rangle = b|b\rangle \tag{9.3}$$

can be solved for the N $|b_m\rangle \in \mathcal{M}_N$ via

$$|b_m\rangle = \sum_{i=1} C_{mi}|a_n^{(i)}\rangle, \qquad m = 1, 2, \ldots, N, \tag{9.4}$$

with b_m and C_{mi} determined in the usual matrix/determinant fashion. Since $\hat{B}^\dagger = \hat{B}$, then $b_m^* = b_m$ and it follows that the $|b_m\rangle$ can be made orthonormal, $\langle b_m|b_\ell\rangle = \delta_{m\ell}$. Because of the assumption that there are no other operators that commute with $\hat{A}$ and with $\hat{B}$, the spectrum $\{b_m\}$ is taken to be non-degenerate. This situation need not hold if there is a $\hat{C}$ such that $[\hat{C}, \hat{A}] = [\hat{C}, \hat{B}] = 0$, a situation addressed in Section 9.3.

From the construction (9.4), it is clear that every $|b_m\rangle$ is an eigenstate of $\hat{A}$ with eigenvalue a_n. Hence, the states of the system characterized by the operators $\hat{A}$ and $\hat{B}$ may be labeled by the two quantum numbers a_n and b_m, viz., $|a_n, b_m\rangle$, such that

$$\hat{A}|a_n, b_m\rangle = a_n|a_n, b_m\rangle \tag{9.5}$$

and

$$\hat{B}|a_n, b_m\rangle = b_m|a_n, b_m\rangle. \tag{9.6}$$

There is thus a total of N orthonormal states $|a_n, b_m\rangle$, each distinguished by a different value of b_m. Hence, diagonalization of $\hat{B}$ via linear combinations of the $|a_n^{(i)}\rangle$ has resulted in a lifting (removing) of the N-fold degeneracy, as predicted. Measurement of $\hat{A}$ followed by $\hat{B}$ not only yields a_n and b_m, but also leaves the system in the unique state $|a_n, b_m\rangle$. Subsequent measurements of $\hat{A}$ or $\hat{B}$ will yield a_n or b_m with certainty; the resulting state will always be $|a_n, b_m\rangle$.

Incompatible measurements and the uncertainty relation

When $[\hat{A}, \hat{B}] = 0$, not only are sequential measurements of $\hat{A}$ and $\hat{B}$ compatible, but so also are simultaneous measurements of $\hat{A}$ and $\hat{B}$, a statement that is easily proved via the analysis of the preceding subsection. In contrast, however, neither sequential nor simultaneous measurements are compatible when $[\hat{A}, \hat{B}] \neq 0$. As a result of the non-commutativity of $\hat{A}$ and $\hat{B}$, the result $\hat{A} \to a_n$ is no longer guaranteed if the measurements $\hat{A} \to a_n$ followed by $\hat{B} \to b_m$ have been made first. In this context it is useful to devise criteria for estimating the deviation from a particular outcome when $[\hat{A}, \hat{B}] \neq 0$. The estimate that is universally used is the *product of the uncertainties* of $\hat{A}$ and of $\hat{B}$, denoted ΔA and ΔB, respectively. It is this choice that leads to

$$\Delta P_x \, \Delta Q_x \geqslant \hbar/2, \tag{9.7}$$

the relation that is the underpinning for the inequality (2.23), $\Delta x \, \Delta p_x \geqslant \hbar/2$. The presence of the product means that the underlying measurements are of the simultaneous rather than the sequential type.

The uncertainty ΔA is actually a state-dependent quantity, defined as

$$\Delta A \equiv \sqrt{\langle (\hat{A} - \langle \hat{A}_\gamma \rangle)^2 \rangle_\gamma}$$
$$= \sqrt{\langle \hat{A}^2 \rangle_\gamma - \langle \hat{A} \rangle_\gamma^2}, \tag{9.8}$$

where the dependence on the state $|\gamma\rangle$ is suppressed on the LHS of (9.8).

The quantity $\langle (\hat{A} - \langle \hat{A} \rangle_\gamma)^2 \rangle_\gamma$ is known as the dispersion or variance of $\hat{A}$ for the state $|\gamma\rangle$. In statistical analyses and in probability theory, the square root of the variance is the standard deviation (see Appendix A); in quantum mechanics, the square root (9.8) is denoted the uncertainty. ΔA is clearly a measure of the deviation of $\hat{A}$ from its expected value. An alternate way of characterizing ΔA is that it is an estimate of the fluctuation of $\hat{A}$ about its average value for the state $|\gamma\rangle$. Because $\hat{A}^\dagger = \hat{A}$, $\langle (\hat{A} - \langle \hat{A} \rangle_\gamma)^2 \rangle_\gamma$ is real and non-negative, as is its square root. Furthermore, even if $\langle \hat{A} \rangle_\gamma = 0$, $\langle \hat{A}^2 \rangle_\gamma$ is generally non-zero, so that ΔA remains a valid measure of the deviation from the average value. Finally, and most important, the choice of the product $\Delta A \, \Delta B$ as the estimate allows one to derive an extremely useful relation, referred to as the Heisenberg uncertainty relation.

The derivation is based on Schwartz's inequality, viz.,

$$\| |\alpha\rangle \| \cdot \| |\beta\rangle \| \geqslant |\langle \beta | \alpha \rangle|^2, \tag{9.9}$$

whose proof is Exercise 4.12(a). (The norm, $\| \cdot \|$, is defined in Chapter 4.) Before applying (9.9), we note that the equality in (9.9) holds only when $|\alpha\rangle$ and $|\beta\rangle$ are proportional.

In the application of (9.9) to the present situation we take $|\alpha\rangle$ and $|\beta\rangle$ to be

$$|\alpha\rangle = (\hat{A} - \langle \hat{A} \rangle_\gamma)|\gamma\rangle \tag{9.10a}$$

and

$$|\beta\rangle = (\hat{B} - \langle \hat{B} \rangle_\gamma)|\gamma\rangle, \tag{9.10b}$$

so that (9.9) becomes

$$\Delta A^2 \, \Delta B^2 \geqslant |\langle (\hat{A} - \langle \hat{A} \rangle_\gamma)(\hat{B} - \langle \hat{B} \rangle_\gamma) \rangle_\gamma|^2. \tag{9.11}$$

To treat the product of operators on the RHS of (9.11) we proceed as follows. Let $\hat{F}$ and $\hat{G}$ be two arbitrary operators. Defining their anticommutator $\{\hat{F}, \hat{G}\}$ by

$$\{\hat{F}, \hat{G}\} \equiv \hat{F}\hat{G} + \hat{F}\hat{G}, \tag{9.12}$$

the product $\hat{F}\hat{G}$ can be written as

$$\hat{F}\hat{G} = \tfrac{1}{2}[\hat{F}, \hat{G}] + \tfrac{1}{2}\{\hat{F}, \hat{G}\}. \tag{9.13}$$

Although the anticommutator is Hermitian, $\{\hat{F}, \hat{G}\}^\dagger = \{\hat{F}, \hat{G}\}$, the commutator is anti-Hermitian, viz., $[\hat{A}, \hat{B}]^\dagger = -[\hat{A}, \hat{B}]$. The commutator becomes Hermitian when it is multiplied by *i*, so that (9.13) can be re-written as

$$\hat{F}\hat{G} = -\frac{i}{2}(i[\hat{F}, \hat{G}]) + \frac{1}{2}(\{\hat{F}, \hat{G}\}), \tag{9.14}$$

where each term in parentheses in (9.14) is now a Hermitian operator.

Equation (9.14) can be used to transform (9.11) by setting $\hat{F} = \hat{A} - \langle \hat{A} \rangle_\gamma$ and

$\hat{G} = \hat{B} - \langle \hat{B} \rangle_\gamma$, and noting that the expectation values make no contribution to the commutator:

$$\Delta A^2 \, \Delta B^2 \geqslant \tfrac{1}{4} | - i \langle i[\hat{A}, \hat{B}] \rangle_\gamma + \langle \{\hat{F}, \hat{G}\} \rangle_\gamma |^2. \tag{9.15}$$

Each expectation value in (9.15), being the diagonal element of a Hermitian operator, is a real number. Hence, the absolute square of the sum becomes a sum of absolute squares:

$$\Delta A^2 \, \Delta B^2 \geqslant \tfrac{1}{4} | \langle [\hat{A}, \hat{B}] \rangle_\gamma |^2 + \tfrac{1}{4} \langle \{F, G\} \rangle_\gamma^2. \tag{9.16}$$

Finally, since $\langle \{\hat{F}, \hat{G}\} \rangle_\gamma$ is semi-positive definite, the inequality in (9.16) is only strengthened by dropping this term from the RHS of the relation, giving

$$\Delta A^2 \, \Delta B^2 \geqslant \tfrac{1}{4} | \langle [\hat{A}, \hat{B}] \rangle_\gamma |^2$$

or

$$\Delta A \, \Delta B \geqslant \tfrac{1}{2} | \langle [\hat{A}, \hat{B}] \rangle_\gamma |, \tag{9.17}$$

an inequality involving only $\hat{A}$, $\hat{B}$, their commutator, and the chosen state $|\gamma\rangle$.

Equation (9.17), the technical version of the uncertainty principle, is the standard form of the *Heisenberg uncertainty relation*. It provides an estimate of the deviations of two observables from their expected values when the two are measured simultaneously, the state of the system being $|\gamma\rangle$. More importantly, it states that, if $[\hat{A}, \hat{B}] \neq 0$, one can reduce the uncertainty in the measured value of one observable (in a simultaneous measurement) only by increasing the uncertainty in the measured value of the other:

$$\Delta A \geqslant \tfrac{1}{2} | \langle [\hat{A}, \hat{B}] \rangle_\gamma | / \Delta B. \tag{9.18}$$

No experiment that circumvents this result can be devised: it is an inescapable consequence of quantum theory, the framework used to describe microscopic phenomena.

The position–momentum form of the uncertainty relation

The most famous example of the uncertainty relation involves a position coordinate $\hat{Q}_j$ and its conjugate momentum coordinate $\hat{P}_j$, for which case we have $[\hat{Q}_j, \hat{P}_j] = i\hbar$, so that (9.17) takes the form

$$\Delta P_j \, \Delta Q_j \geqslant \hbar/2; \tag{9.19}$$

compare this with Eq. (9.7). This example is of special interest because it can be used in two important ways. First, it allows one to determine the wave packet $\langle x|\gamma\rangle = \psi_\gamma(x)$ for which the equality in (9.19) is satisfied. This is the minimum-uncertainty packet. Second, on replacing the inequality $\geqslant \hbar/2$ by the order of magnitude $\sim \hbar$ (or $\hbar/2$), it permits one to make various quantal estimates.

For the equality

$$\Delta P_x \, \Delta Q_x = \hbar/2 \tag{9.20}$$

to be valid, two other equalities must hold. First, the term $\langle \{(\hat{P}_x - \langle \hat{P}_x \rangle_\gamma), (\hat{Q}_x - \langle \hat{Q}_x \rangle_\gamma)\} \rangle_\gamma$, which had previously been discarded, must be set equal to zero, and second, the relation $|\alpha\rangle \propto |\beta\rangle$, which ensures an equality in (9.9), must be satisfied.

Going into a coordinate representation, recalling Eqs. (9.10), and choosing $\hat{A} = \hat{P}_x$ and $\hat{B} = \hat{Q}_x$, the preceding proportionality becomes

$$\eta\left(-i\hbar\frac{d}{dx} - \hbar k_0\right)\psi_\gamma(x) = (x - x_0)\psi_\gamma(x_0), \tag{9.21}$$

where η is the proportionality constant relating $|\alpha\rangle$ to $|\beta\rangle$, $\hbar k_0 = \langle\hat{P}_x\rangle_\gamma$, and $x_0 = \langle\hat{Q}_x\rangle_\gamma$.

The solution to (9.21) is a plane wave modified by a pure imaginary Gaussian:

$$\psi_\gamma(x) = Ne^{ik_0x}e^{i(x-x_0)^2/(2\eta\hbar)}, \tag{9.22}$$

where N is a normalization constant. However, the relation $\langle\{\hat{P}_x - \hbar k_0, \hat{Q}_x - x_0\}\rangle_\gamma = 0$ can be used to show that $\eta = -i|\eta|$ (normalization eliminates $+i|\eta|$), while $|\eta|$ can be related to Δx. Proof of these statements is left to the exercises. Using $\eta = -i|\eta|$, the analysis shows that, apart from phase factors, the minimum-uncertainty wave function is equal both to the ground-state of the harmonic oscillator and to the coordinate-space form of the coherent or quasi-classical state, $\psi_\xi(x, t)$ of Eq. (8.81). The former equality is consistent with the result of Exercise 6.20, while the latter suggests that, for $|\xi, t\rangle$ – Eq. (8.65) – the minimum-uncertainty relation holds. Proof of the latter statement is also left to the exercises; its validity is consistent with the fact that $\rho_\xi(x, t)$, Eq. (8.82), does not spread in time.

The second application of the position–momentum uncertainty relation is to the fine art of making order-of-magnitude estimates in quantum mechanics. Changing $\Delta Q_x\,\Delta P_x \geqslant \hbar/2$ into

$$\Delta Q_x\,\Delta P_x \sim \hbar \tag{9.23}$$

provides one important means of doing so. In (9.23), the symbol "$\sim$" means "is of the order of magnitude of." We illustrate this relation by estimating the energies of three quantum systems. In each case, the classical-mechanics expression for E is converted into a semi-classical one via use of (9.23) and suitable expressions for ΔQ_x.

EXAMPLE 1.

The quantum box. We seek an order-of-magnitude estimate of the energy of a particle in a quantum box of width L. It is obtained by using (9.23) in the classical energy relation $E = K + V$, where K and V are the classical kinetic and potential energies, respectively. In the box, $K = p_x^2/(2m)$ and $V = 0$, so that

$$E_{\rm box} = p_x^2/(2m). \tag{9.24}$$

There are three steps involved in making the estimate. First, p_x is replaced by the quantal uncertainty ΔP_x. Second, (9.23) is used to estimate ΔP_x:

$$\Delta P_x \sim \hbar/\Delta Q_x. \tag{9.25}$$

Finally, ΔQ_x is estimated by setting it equal to L, the range of values for which (6.4) holds. Putting these steps together yields

$$E_{\rm box} \sim \frac{\hbar^2}{2mL^2}. \tag{9.26}$$

Use of the uncertainty relation in the term (9.23) plus $\Delta Q_x \sim L$ has changed (9.24) to the semi-classical relation (9.26). Note that it is an *estimate* of the energy of a particle in a box: it gives the correct order of magnitude, but not an exact result. With hindsight, we could replace the RHS of (9.24) by h; then E_{box} would be equal to $E_1 = \pi^2\hbar^2/(2mL^2)$, but that presumes a solution to the problem. The relation is valuable because it does not presume a solution yet nevertheless yields a reliable order-of-magnitude estimate.

EXAMPLE 2.

The 1-D oscillator. We again start with the classical expression for the energy, which for the oscillator is

$$E_{\text{osc}} = \frac{p_x^2}{2m} + \frac{m\omega_0^2 x^2}{2}. \tag{9.27}$$

On making the replacements $p_x \to \Delta P_x$ and $\Delta P_x \sim \hbar/\Delta Q_x$, (9.27) becomes

$$E_{\text{osc}} \sim \frac{\hbar^2}{2m\,\Delta Q_x^2} + \frac{m\omega_0^2 x^2}{2}. \tag{9.28}$$

An estimate of ΔQ_x is now required. Since the range of x is $[-\infty, \infty]$, we choose $\Delta Q_x = x$, a choice that allows (9.28) to be *minimized* w.r.t. x, thus leading to an estimate of the ground-state energy of the oscillator:

$$E_{\text{osc}} \sim E(x) = \frac{\hbar^2}{2mx^2} + \frac{m\omega_0^2 x^2}{2}, \tag{9.29}$$

the minimum of which occurs at $x = \lambda_0 = \sqrt{\hbar/(m\omega_0)}$, the "natural" length unit for the quantal oscillator. Hence, for the oscillator, the uncertainty-relation estimate yields

$$E_{\min} = E(\lambda_0) = \hbar\omega_0, \tag{9.30}$$

which differs from the exact result by a factor of two, and is certainly a reliable estimate.

EXAMPLE 3.

The hydrogen atom. Following the procedure used in Example 2, the classical energy for a particle of mass m_e and charge $q = -e$ in the Coulomb field of an infinitely massive positive charge is

$$E_{\text{Coul}} = \frac{\mathbf{p}^2}{2m_e} - \frac{\kappa_e e^2}{r}, \tag{9.31}$$

where $\mathbf{p}$ and $\mathbf{r}$ are the particle's 3-D momentum and position, respectively. In place of $\Delta P_x \sim \hbar/\Delta Q_x$, we use instead $\mathbf{p}^2 \to \Delta P^2 \sim (\hbar/\Delta Q)^2$, and then estimate ΔQ by setting it equal to r, the separation of the two charges. Eq. (9.31) then becomes

$$E_{\text{Coul}} \sim E(r) = \frac{\hbar^2}{2m_e r^2} - \frac{\kappa_e e^2}{r}. \tag{9.32}$$

$E(r)$ can be minimized w.r.t. r, leading to an estimate for the ground-state energy of the H atom. The minimum occurs at

$$r_{\min} = \frac{\hbar^2}{m_e \kappa_e e^2} \equiv a_0, \tag{9.33}$$

where a_0 is the Bohr radius: once again the "natural" unit of length for the system is the value which minimizes the energy. The minimum energy is

$$E_{\min} = -\frac{(\kappa_e e^2)^2}{2(\hbar^2 m_e)} = -\frac{\kappa_e e^2}{2a_0}(!), \tag{9.34}$$

i.e., just the ground-state energy of the Bohr model of the H-atom. Once more the method has produced a reliable estimate.

It should be clear from these examples that estimation via the uncertainty principle is a useful method. Other examples as well as explicit evaluations of $\Delta P_x \, \Delta Q_x$ are the subject of some of the exercises. Uncertainty relations involving other operators are considered in Section 10.4 and especially Section 16.4.

9.2. Quantal constants of the motion

Our discussion of constants of the motion in quantum mechanics is based on Eq. (8.4), re-written below:

$$\frac{d}{dt}\langle \hat{A} \rangle_{jk}(t) = \langle \Gamma_j, t | \left\{ \frac{1}{i\hbar}[\hat{A}, \hat{H}] + \frac{\partial \hat{A}}{\partial t} \right\} | \Gamma_k, t \rangle, \tag{9.35}$$

where $[\hat{A}, \hat{H}] = \hat{A}\hat{H} - \hat{H}\hat{A}$ is the commutator of $\hat{A}$ and $\hat{H}$. The matrix element $\langle \hat{A} \rangle_{jk}(t)$ is the generalized expectation value of Chapter 8.[2]

Equation (9.35) has widespread consequences, one of which will be explored in this chapter. Among other applications, we used it in examining the classical limit of quantum mechanics in Section 8.3. That it might be an appropriate tool for doing so is suggested by its close structural resemblance to the expression for the rate of change with time of a classical quantity $A_{\text{cl}}(\{q_i\}, \{p_i\}, t)$ that depends on the generalized coordinates $\{q_i\}$ and momenta $\{p_i\}$ as well as on t. As is shown in texts on classical mechanics (e.g., Fowles and Cassidy (1990), Fetter and Walecka (1980) and Goldstein (1980)), this expression is

$$\frac{dA_{\text{cl}}}{dt} = [A_{\text{cl}}, H]_{\text{cl}} + \frac{\partial A_{\text{cl}}}{\partial t}, \tag{9.36}$$

where $H = H(\{q_i\}, \{p_i\})$ is the classical Hamiltonian and $[A, H]_{\text{cl}}$ is the classical Poisson bracket, given by

$$[A, H]_{\text{cl}} = \sum_i \left(\frac{\partial A}{\partial p_i}\frac{\partial H}{\partial q_i} - \frac{\partial A}{\partial q_i}\frac{\partial H}{\partial p_i} \right). \tag{9.37}$$

[2] When $k \neq j$, the generalized expectation value is an off-diagonal matrix element, often denoted a transition-matrix element, the reason for which is clarified in Chapter 16.

The obvious resemblance between (9.35) and (9.36) indicates that there may be a limit in which quantal dynamics – at least in terms of expectation values – goes over to classical dynamics. Furthermore, it is also strongly suggestive that one may go from classical dynamics to quantum mechanics by replacing the Poisson bracket by the commutator. A nice discussion of the latter point can be found in Dirac's monograph on quantum mechanics (Dirac 1947).

Some of the eventual consequences of (9.35) that we shall examine arise from making specific choices for $\hat{A}$. In this subsection, however, we restrict ourselves to results of a general nature concerning constants of the motion or conserved quantities. We assume both that

$$[\hat{A}, \hat{H}] = 0 \tag{9.38a}$$

and that

$$\frac{\partial A}{\partial t} = 0. \tag{9.38b}$$

Then the RHS of (9.35) vanishes, and therefore

$$\langle \hat{A} \rangle_{jk}(t) = C_{jk}, \tag{9.39}$$

where the constant C_{jk} will generally depend on the state labels Γ_j and Γ_k.

When the conditions (9.38) hold $\hat{A}$ is said to be a "constant of the motion"; all of its matrix elements (9.39) thus define conserved quantities. In particular, if $\hat{A}$ and $\hat{H}$ are the only observables associated with the system, then (9.38a) implies that $\hat{A}$ and $\hat{H}$ have joint eigenstates, labeled by the eigenvalues of each operator. It follows that, if $|\Gamma_j, t\rangle$ and $|\Gamma_k, t\rangle$ are common eigenstates of $\hat{A}$ and $\hat{H}$, then $\langle \hat{A} \rangle_{jk}(t) = \delta_{jk} a_j$, where a_j is the jth eigenvalue of $\hat{A}$. The conserved quantities are just the $\{a_j\}$, which is the reason why they can be used to label the states: among the labels included in the set represented by Γ_k is a_k.

We may carry this argument one step further and assume that there exist two operators $\hat{A}$ and $\hat{B}$, which commute with $\hat{H}$ and are each time-independent. Then each is a constant of the motion, but, unless the relation $[\hat{A}, \hat{B}] = 0$ is valid, the eigenvalues of only one of them can be selected to label eigenstates of $\hat{H}$. This situation is the subject of the theorem on complete sets of commuting operators, which is treated in the next section.

The preceding discussion of quantal constants of the motion is mirrored in the classical case, since the quantity $A_{\rm cl}$ is also denoted a constant of the motion when the classical analogs of (9.38) hold, viz., when $[A_{\rm cl}, H]_{\rm cl} = 0 = \partial A_{\rm cl}/\partial t$. A trivial example of this concept occurs when $\hat{A} = \hat{H} \neq \hat{H}(t)$. Then, for $j \neq k$ and neither state an eigenstate of $\hat{H}$, we find that $\langle \hat{H} \rangle_{jk}$ is constant in time, a result that also follows by expanding $|\Gamma_j, t\rangle$ and $|\Gamma_k, t\rangle$ in the eigenstates of $\hat{H}$. On the other hand, if both states are eigenstates, say, $|\Gamma_j, t\rangle = |\gamma_j, E_j, t\rangle$, then we find $\langle \hat{H} \rangle_{jk} = \delta_{jk} E_j =$ constant, as must happen. Hence, when $\hat{H} \neq \hat{H}(t)$, quantum theory leads to conservation of energy.

If (9.38b) but not (9.38a) holds, and the states are eigenstates of $\hat{H}$, viz., $|\Gamma_j, t\rangle = |\gamma_j, E_j, t\rangle$, then (9.35) is uninteresting since its solution is the same as the detailed form of the matrix element, viz., $\langle \hat{A} \rangle_{jk} = e^{-i(E_k - E_j)t/\hbar} \langle \gamma_j, E_j | \hat{A} | \gamma_k, E_k \rangle$. Neither is anything new obtained when $|\Gamma_j, t\rangle$ is a linear combination of eigenstates. To go further requires choosing specific instances for $\hat{A}$ and then evaluating $[\hat{A}, \hat{H}]$, examples of which were considered in Section 8.3.

9.3. Complete sets of commuting operators

In Section 9.1 it was shown that $[\hat{A}, \hat{B}] = 0$ plus the restriction that no other operators had vanishing commutators with $\hat{A}$ and $\hat{B}$ meant that $\hat{A}$ and $\hat{B}$ shared a common set of eigenstates labeled by two quantum numbers, one an eigenvalue of $\hat{A}$, the other an eigenvalue of $\hat{B}$. The eigenstates of $\hat{A}$ need not, of course, be eigenstates of $\hat{B}$ also. If they are not, suitable linear combination of the degenerate eigenstates of $\hat{A}$ can be formed, as in Section 9.1, which become eigenstates of $\hat{B}$; these are then the common eigenstates of $\hat{A}$ and $\hat{B}$, labeled as in the foregoing.

Elimination of degeneracies

We now drop the above restriction and assume that, for the system under consideration, there is a third operator, $\hat{C}$, which commutes both with $\hat{A}$ and with $\hat{B}$:

$$[\hat{A}, \hat{C}] = 0, \tag{9.40a}$$

$$[\hat{B}, \hat{C}] = 0, \tag{9.40b}$$

$$[\hat{A}, \hat{B}] = 0. \tag{9.40c}$$

Applying the analysis involved in going from Eqs. (9.1) to (9.6) successively to (9.40a), to (9.40b), and to (9.40c), it is evident that each of the operator pairs $(\hat{A}, \hat{B})$, $(\hat{B}, \hat{C})$, and $(\hat{A}, \hat{C})$ share common eigenstates, with members of each of the three sets labeled by two eigenvalues, one for each of the two operators of which they are eigenstates. Methods for determining these various eigenstates (and their degeneracies) will be given as specific examples in Parts III and IV.

Because of the pairwise nature of the labeling of the eigenstates, it should also be evident that we can simply combine the three sets into a single set, labeled by the three quantum numbers a_n, b_m, and c_p:

$$\hat{A}|a_n, b_m, c_p\rangle = a_n|a_n, b_m, c_p\rangle, \qquad n = 1, \ldots, N_n, \tag{9.41a}$$

$$\hat{B}|a_n, b_m, c_p\rangle = b_m|a_n, b_m, c_p\rangle, \qquad m = 1, \ldots, M_m, \tag{9.41b}$$

and

$$\hat{C}|a_n, b_m, c_p\rangle = c_p|a_n, b_m, c_p\rangle, \qquad p = 1, \ldots, P_p, \tag{9.41c}$$

where N_n, M_m, and P_p are the degeneracies of the nth, mth, and pth eigenvalues of $\hat{A}$, $\hat{B}$, and $\hat{C}$, respectively.

Once the individual sets of quantum numbers $\{a_n\}$, $\{b_m\}$, and $\{c_p\}$ have been determined, the state $|a_n, b_m, c_p\rangle$ can equally well be expressed either as

$$|a_n, b_m, c_p\rangle = |a_n\rangle|b_m\rangle|c_p\rangle \tag{9.42a}$$

or as

$$|a_n, b_m, c_p\rangle = |a_n\rangle \otimes |b_m\rangle \otimes |c_p\rangle. \tag{9.42b}$$

In Eq. (9.42b), the symbol $\otimes$ denotes an outer product. This form corresponds to the overall Hilbert space $\mathcal{H}$ of the system being the direct product of the Hilbert spaces for each of the three operators $\hat{A}$, $\hat{B}$, and $\hat{C}$:

$$\mathcal{H} = \mathcal{H}_A \otimes \mathcal{H}_B \otimes \mathcal{H}_C, \tag{9.43}$$

in a self-evident notation.

Before proceeding, we remark that all three of Eqs. (9.40) must hold for $\hat{A}$, $\hat{B}$, and $\hat{C}$ to have common eigenstates. For example, $[\hat{P}_x, \hat{Q}_x] \neq 0 \neq [\hat{K}_x, \hat{Q}_x]$, and we cannot construct common eigenstates of $\hat{P}_x$ and $\hat{Q}_x$ or of $\hat{K}_x$ and $\hat{Q}_x$, although this can be done for $\hat{K}_x$ and $\hat{P}_x$ since $[\hat{K}_x, \hat{P}_x] = 0$.

We next suppose that there are three operators $\hat{A}$, $\hat{B}$, and $\hat{C}$ obeying Eqs. (9.40) and that a set of eigenstates $\{|a_n, b_n, c_n\rangle\}$ whose elements obey Eqs. (9.41) has been found. In this situation there are two possibilities: (i), at least one of the set $\{|a_n, b_m, c_p\rangle\}$ is degenerate; and (ii), every $|a_n, b_m, c_p\rangle$ is non-degenerate. In the former case, the existence of a degeneracy implies that one or more operators $\hat{D}$, $\hat{F}$, etc., each of which commutes with $\hat{A}$, $\hat{B}$, and $\hat{C}$, can be found such that the totality $\hat{A}$, $\hat{B}$, $\hat{C}$, $\hat{D}$, $\hat{F}$, ... generates a set of non-degenerate eigenstates $\{|a_n, b_m, \ldots, f_r, \ldots\rangle\}$. This turns case (i) above effectively into case (ii), albeit now with more than three mutually commuting operators. Both situations are encompassed by writing the set of non-degenerate eigenstates as $\{|a_n, b_m, \ldots\rangle\}$, with

$$\hat{A}|a_n, b_m, \ldots\rangle = a_n|a_n, b_m, \ldots\rangle, \tag{9.44a}$$

$$\hat{B}|a_n, b_m, \ldots\rangle = b_m|a_n, b_m, \ldots\rangle, \tag{9.44b}$$

and

$$|a_n, b_m, \ldots\rangle = |a_n\rangle \otimes |b_m\rangle \otimes \cdots. \tag{9.45}$$

Equations (9.44) and (9.45) constitute a statement of one of the two major results of this section. Because of Postulate VI, each of the individual eigenstates, e.g., $|b_m\rangle$, forms a complete set spanning the associated Hilbert space. Equation (9.45) then tells us that $\{|a_n, b_m, \ldots\rangle\}$ is a complete set of states spanning the full Hilbert space $\mathcal{H} = \mathcal{H}_A \otimes \mathcal{H}_B \otimes \cdots$. Hence, $\{|a_n, b_m, \ldots\rangle\}$ is a complete set of states for the particular quantum system under consideration. These states are generated by the operators $\{\hat{A}, \hat{B}, \ldots\}$. A special name is given to the latter set: they are called a "complete set of commuting operators," often written as the acronym CSCO.

By use of the construction outlined in the foregoing, we have reached one of the tenets of quantum theory, viz., for every quantum system there exists a CSCO whose common eigenstates form a non-degenerate basis for the relevant Hilbert space. We close this subsection with three remarks. First, although in many cases an initial degeneracy can be lifted by including as one of the CSCOs a more or less obvious symmetry operator, in other cases the use of such "obvious" symmetry operators (e.g., the generators of the rotation operator when $V(\mathbf{r}) = V(r)$ – see Section 9.7 and Chapter 10) does *not* lift the degeneracy. One then refers to the degeneracy as "accidental," and a more involved investigation into symmetries is normally required. Second, apart from any possible accidental degeneracies, it is often non-trivial to find the entire set of CSCOs. This is especially true for an interacting N-particle system, wherein at least $3N$ operators are needed in the 3-D case, one for each spatial degree of freedom for each particle. On the other hand, if the N particles do not interact, the task of finding the CSCO can be much simplified. This is one reason among many that (non-interacting) independent-particle approximations are used to model systems with many degrees of freedom, a point to which we return in Part IV. Third, because all members of the CSCO commute with one another, it is possible to measure all the observables for which they are the mathematical image without altering the state $|a_n, b_m, \ldots\rangle$ of the system. Hence, repeated measurements of one or more of them will always yield the same value(s).

"Good" or conserved quantum numbers

By Postulate V, the fundamental dynamical equation of quantum theory is the time-dependent Schrödinger equation. It contains the Hamiltonian $\hat{H}$, and therefore $\hat{H}$ is necessarily one of the CSCOs for every quantal system. With this fact in mind, we can draw an important conclusion. Let all the operators in the CSCO be time-independent. In this case the conditions under which Eq. (9.39) is satisfied hold. On setting $|\Gamma_k\rangle = |\Gamma_j\rangle = |a_n, b_m, \ldots\rangle$ and choosing $\hat{A}$ in (9.39) to be any of the CSCOs, it follows that C_{jk} in (9.42) is one of the eigenvalues/quantum numbers appearing in $|a_n, b_m, \ldots\rangle$. Hence, when the CSCO consists of time-independent operators, all of the eigenvalues/quantum numbers are constants of the motion: they are thus conserved or "good" quantum numbers. This is the underlying reason why the eigenvalues of the CSCO are used to label the eigenstates of the system: because they are constant, they uniquely define each state.

If $\hat{H} = \hat{H}(t)$, then energy is not conserved, i.e., energy can be put into or removed from the system. Even so, there are numerous situations in which one can still refer to conservation of energy in portions of the system, e.g., if an atom, molecule, or nucleus interacts with the electromagnetic field, the initial and final states of the atom, molecule, or nucleus are typically ones of fixed energy (see Chapter 16 on time-dependent approximation methods). Furthermore, there are occasions when $\hat{A} = \hat{A}(t)$ and the *form* of the eigenvalue remains the same even though it becomes time-dependent. We consider one such example in Chapter 12.

9.4. General remarks on symmetries and symmetry operators

Symmetric objects are familiar sights in the everyday, macroscopic world. One says that an object is symmetric if something can be done to it that does not alter its shape or appearance. Many geometric shapes are symmetric, two examples being a sphere and an equilateral triangle. The sphere is the perfectly symmetric object in 3-D, since it can be either rotated through *any* angle or reflected across *any* plane perpendicular to *any* axis through its center, without altering its appearance. There is, therefore, an infinite number of symmetry operations under the action of which the sphere is invariant. The equilateral triangle has a far lower degree of symmetry, since there are only three rotations and three reflections in its plane of definition that leave the appearance of the triangle invariant (see Fig. 9.1).

Some macroscopic objects are *approximately* symmetric, the human body being an example. It is almost invariant under reflection through a plane bisecting its left and right halves. That the human body is almost symmetric in this manner is the reason why it is

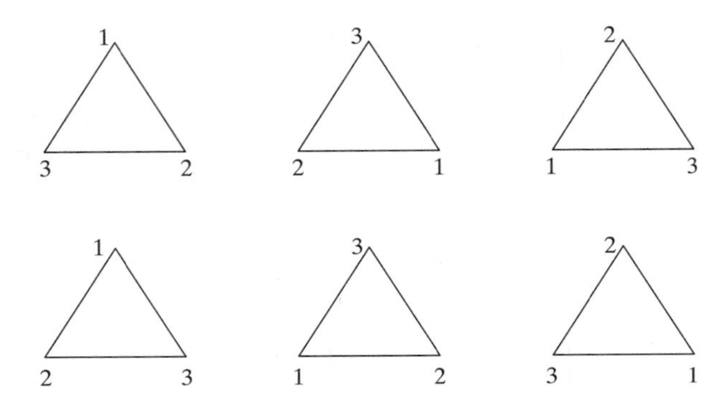

Fig. 9.1 Results of the six symmetry operations performed in the plane of the paper that leave the shape of an equilateral triangle invariant. The vertices are labeled in order to make identification of the angles and axes of reflection/rotation obvious.

often so difficult to tell which is the photograph of a person and which is the photograph of his or her image in a plane mirror. This lack of distinguishability means that our notion of left and right is a convention, not an intrinsic property, and it is the basis for using the words "port" and "starboard" on a boat or ship.

Rotations and reflections are two of the symmetry operations that also occur in the study of symmetries in micro-physics. The former is a continuous operation, the latter a discrete one. Other continuous symmetry operations are translations in space and in time, while additional discrete operations are time reversal ($t \to -t$, a type of inversion or reflection) and interchange of identical particles. When a quantal system is invariant under these symmetry operations, two things are implied: on the one hand, if the symmetry operation can be applied directly to a physical system, then measurements carried out before and after must yield identical results, while on the other, the Hamiltonian of the system must be invariant when the symmetry transformation is applied to it. In the latter case, the relevant symmetry operator commutes with $\hat{H}$ and symmetry-based eigenvalues occur as eigenstate labels, a claim we now prove.

Let $|\Gamma\rangle = |E_\gamma, \gamma\rangle$ be an eigenstate of $\hat{H}$:

$$\hat{H}|\Gamma\rangle = E_\gamma|\Gamma\rangle, \tag{9.46}$$

and let $\hat{A}$ be an arbitrary operator that can act on $|\Gamma\rangle$, producing a state in $\mathcal{H}$. Since $|\Gamma\rangle$ obeys (9.46), we ask for the Schrödinger-type equation obeyed by the transformed state $\hat{A}|\Gamma\rangle$. It is found by acting on both sides of (9.46) with $\hat{A}$ from the left:

$$\hat{A}\hat{H}|\Gamma\rangle = E_\gamma\hat{A}|\Gamma\rangle. \tag{9.47}$$

The state $\hat{A}|\Gamma\rangle$ appears on the RHS but not the LHS of Eq. (9.47). It will occur on both sides if $\hat{A}^{-1}$ exists, and we therefore assume that $\hat{A}$ is non-singular, in which case (9.47) becomes

$$(\hat{A}\hat{H}\hat{A}^{-1})[\hat{A}|\Gamma\rangle] = E_\gamma[\hat{A}|\Gamma\rangle]. \tag{9.48}$$

Hence, $\hat{A}|\Gamma\rangle$ is an eigenstate with energy E_γ of the *similarity-transformed* Hamiltonian $\hat{H}' = \hat{A}\hat{H}\hat{A}^{-1}$:

$$\hat{H} \underset{\hat{A}}{\rightarrow} \hat{H}' = \hat{A}\hat{H}\hat{A}^{-1}. \tag{9.49}$$

We now suppose that $\hat{A}$ is a symmetry operator and that the physical system is invariant under the symmetry operation (such as spatial reflection). Invariance means that the Hamiltonian is unchanged, so that $\hat{H}' = \hat{H}$, i.e.,

$$\hat{H} = \hat{A}\hat{H}\hat{A}^{-1} \qquad \text{(invariance statement)}. \tag{9.50}$$

Acting on both sides of (9.50) from the right with $\hat{A}$ leads to

$$\hat{H}\hat{A} = \hat{A}\hat{H}, \tag{9.51a}$$

which we restate as

$$[\hat{A}, \hat{H}] = 0. \tag{9.51b}$$

If, in addition to (9.51), $\hat{A} \neq \hat{A}(t)$ – which will be true in general – then $\hat{A}$ is a quantal constant of the motion and thus its eigenvalues are candidates for the set of good quantum numbers with which states are labeled. They *will* be included among the labels if $\hat{A}$ is one of CSCOs for the system.

We also remark that the preceding analysis generalizes from the Schrödinger equation to *any* eigenvalue equation. That is, if

$$\hat{B}|b\rangle = b|b\rangle, \tag{9.52}$$

then, under the transformation

$$|b\rangle \rightarrow \hat{A}|b\rangle, \tag{9.53a}$$

one easily finds that

$$\hat{B} \rightarrow \hat{B}' = \hat{A}\hat{B}\hat{A}^{-1} \tag{9.53b}$$

such that

$$(\hat{B}' - b)[\hat{A}|b\rangle] = 0. \tag{9.53c}$$

The invariance of $\hat{H}$ under the action of an operator $\hat{A}$, leading to Eq. (9.51b), and the structure of the operator $\hat{B}'$ of Eq. (9.53b) are essential ingredients in the following analysis of symmetries and symmetry operators. The quantal systems chosen to illustrate the role played by the symmetries are ones for which Eq. (9.51b) holds exactly. In Part IV we shall encounter situations in which approximate symmetries occur. It should be clear that the study of symmetry is of fundamental importance in quantum mechanics, the somewhat introductory level of our treatment notwithstanding. We begin with the discrete symmetry operation of spatial reflections, and then consider the continuous cases of translations and rotations in space. Only one-particle systems will be examined, with illustrative examples generally confined to motion involving a single degree of freedom.

9.5. Reflections

Starting with the behavior of a 3-D vector under a spatial reflection (an inversion of coordinates), we consider some aspects of reflections in classical physics and then go on to the quantal case, ending up with an analysis of the 1-D symmetric quantum box.

Reflections: geometry and classical physics

Consider an arbitrary 3-D vector $\mathbf{b}$ having components $b_j = \mathbf{e}_j \cdot \mathbf{b}$ w.r.t. to a coordinate system S, as in Fig. 9.2(a). A spatial reflection or inversion can be thought of either as the passive transformation $\{\mathbf{e}_j\} \to \{-\mathbf{e}_j\}$, which turns a right-handed into a left-handed coordinate system, or as the active transformation $\{b_j\} \to \{-b_j\}$. In the latter case, the $\mathbf{e}_j$ remain unchanged. Each transformation yields $\mathbf{b} \to -\mathbf{b}$, but, as with rotations in 3-D, we shall consider only the latter transformation. The inverted vector is shown in Fig. 9.2(b).

There is further correspondence with 3-D rotations. The transformation $\mathbf{b} \to -\mathbf{b}$ can be thought of as being effected by an operator, called the *parity* operator, denoted $\hat{\mathcal{P}}$:

$$\hat{\mathcal{P}}\mathbf{b} = -\mathbf{b}. \tag{9.54}$$

In a 3-D matrix/column-vector notation, $\hat{\mathcal{P}}$ is the 3×3 matrix

$$\hat{\mathcal{P}} \to \mathcal{P} = \begin{pmatrix} -1 & 0 & 0 \\ 0 & -1 & 0 \\ 0 & 0 & -1 \end{pmatrix} \equiv -I, \tag{9.55}$$

where I is the 3-D unit matrix.

Considered from the viewpoint of an eigenvalue problem, $\mathbf{b}$ is a 3-D eigenvector of $\hat{\mathcal{P}}$ with parity eigenvalue or "parity" $\pi = -1$. Many quantities are parity eigenvectors; the parity is either $+1$ or -1, a claim that follows from $\hat{\mathcal{P}}^2\mathbf{b} = \pi^2\mathbf{b} = \mathbf{b}$. However, not all 3-D vectors have negative parity (i.e., $\pi = -1$). Those that do are referred to as "polar" vectors. Those that do not are labeled "axial" or "pseudo-" vectors, an example being the cross product of two polar vectors. Let $\mathbf{a}$ and $\mathbf{b}$ be two polar vectors. Their cross product is the vector $\mathbf{c}$,

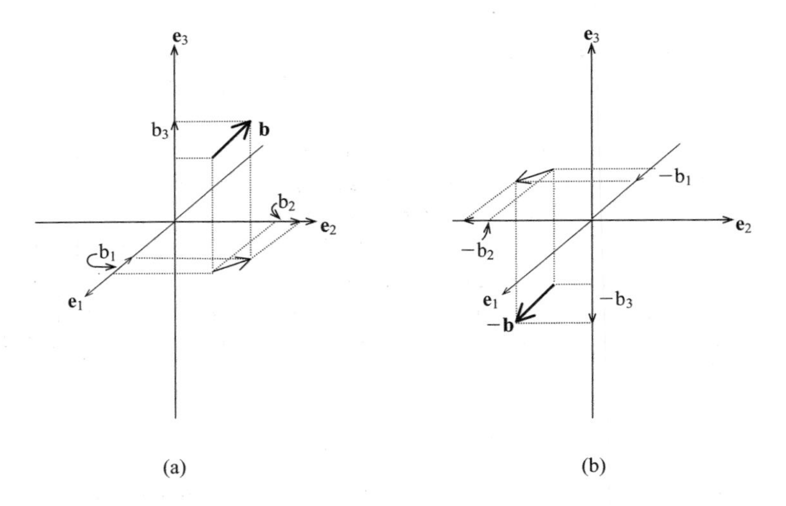

Fig. 9.2 (a) A vector $\mathbf{b}$ with components $\{b_j\}$ in the coordinate system S and (b) its reflection $-\mathbf{b}$, with components $\{-b_j\}$ in the same coordinate system.

$$\mathbf{c} = \mathbf{a} \times \mathbf{b}, \tag{9.56}$$

and from

$$\hat{\mathcal{P}}(\mathbf{a} \times \mathbf{b}) = (\hat{\mathcal{P}}\mathbf{a}) \times (\hat{\mathcal{P}}\mathbf{b}) \tag{9.57}$$

it follows that

$$\hat{\mathcal{P}}\mathbf{c} = \mathbf{c}. \tag{9.58}$$

In other words, $\pi_\mathbf{c} = +1$, as noted above.

Most of the kinematic vectors of classical mechanics are polar. For example, if $\mathbf{r}$ is the position vector of a particle, then

$$\hat{\mathcal{P}}\mathbf{r} = -\mathbf{r}, \tag{9.59}$$

from which it follows that velocity $\mathbf{v}$, momentum $\mathbf{p} = m\mathbf{v}$, and acceleration are also polar. So, too, is electric current density $\mathbf{j}$, a conclusion based on $\mathbf{j}(\mathbf{r}) = q\mathbf{v}\delta(\mathbf{r} - \mathbf{r}_0)$ for a point charge q located at position $\mathbf{r}_0$ and moving with velocity $\mathbf{v}$. On the other hand the (orbital) angular momentum $\mathbf{L} = \mathbf{r} \times \mathbf{p}$ is an axial vector.

Non-Cartesian coordinates can be used to describe the effects of inversion, for instance, spherical polar coordinates $(r\theta\phi)$, $0 \leqslant \theta \leqslant \pi$, $0 \leqslant \phi \leqslant 2\pi$. The spherical polar equivalent to $x \rightarrow -x$, $y \rightarrow -y$, and $z \rightarrow -z$ is

$$\left.\begin{aligned} r &\rightarrow r \\ \theta &\rightarrow \pi - \theta \\ \phi &\rightarrow \pi + \phi \end{aligned}\right\} \text{behavior under inversion.} \tag{9.60}$$

One need not invert all coordinates, in which case a partial reflection occurs. If only x is inverted, $x \rightarrow -x$, $y \rightarrow y$, and $z \rightarrow z$, then one has a reflection through the y–z plane, and a *mirror image* is obtained. This behavior is illustrated in Figs. 9.3(a)–(c). The first of these shows the reflection of a point particle located at position x, $y = z = 0$; the second (Fig. 9.3(b)) displaying the effect of the mirror reflection on a y–z planar figure located at x. The third of the three (Fig. 9.3(c)) is the most illuminating, showing in detail not only how the x–y portion of a 3-D object is reflected in a mirror but also how and why – even under a mirror reflection – the current density changes sign.

If a quantal system is invariant under reflections, then not only is parity a good quantum number, but also the invariance can have important dynamical consequences, as we shall show in the next subsection. In classical physics, however, reflection symmetry plays an almost insignificant role, its main function being to supply classical analogs and pictures that can help illustrate quantal effects of reflection invariance or non-invariance. Nonetheless, it is worthwhile to study some classical-physics examples. We consider two situations, one in which the dynamics is invariant under a mirror reflection, the other in which it is not. In the first instance there is no essential dynamical change, whereas in the second there is.

The first example consists of a block sliding down a frictionless mechanical plane as in Fig. 9.4 on which the mirror reflection $x \rightarrow -x$, $y \rightarrow y$, $z \rightarrow z$ is imposed. Since the force is in the z-direction, both the magnitude and the direction of the acceleration remain unaffected by the reflection, even though the block slides in the opposite direction. In the second example of a thin, infinitely long wire oriented along $\mathbf{e}_1$ and carrying a current in the positive-x-direction, the mirror reflection *does* produce a change in a measurable

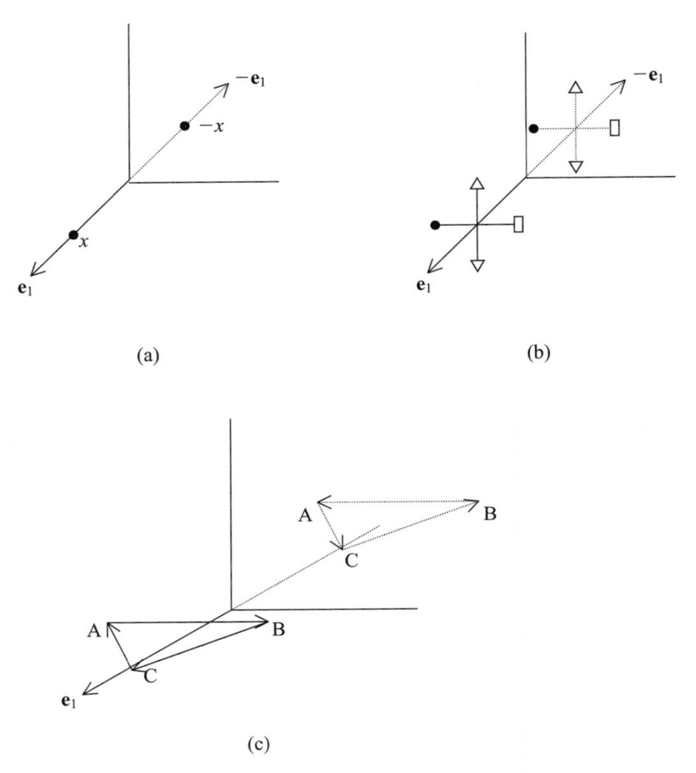

Fig. 9.3 (a) The 1-D inversion $x \to -x$. (b) Reflection through the y–z plane (i.e., $x \to -x$) of a figure located in the y–z plane. (c) Reflection of a current-carrying wire (solid-line triangle) located in the y–z plane into the mirror image (dotted-line) triangle. The current remains in the x–y plane after the reflection, but its direction in the mirror image is opposite to that of the original, i.e., the current density $\mathbf{j}(\mathbf{r}) \to -\mathbf{j}(\mathbf{r})$ under reflection, as noted below Eq. (9.59).

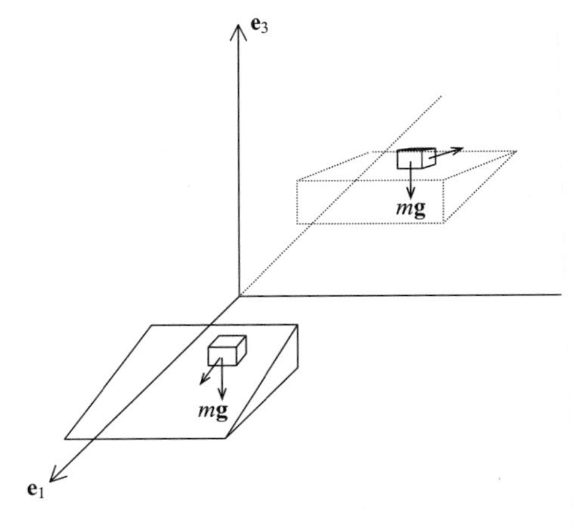

Fig. 9.4 A block sliding down an inclined plane and its reflected version under the inversion $x \to -x$. The inversion has no effect on the force of gravity; hence the acceleration of the block is unaffected by the inversion.

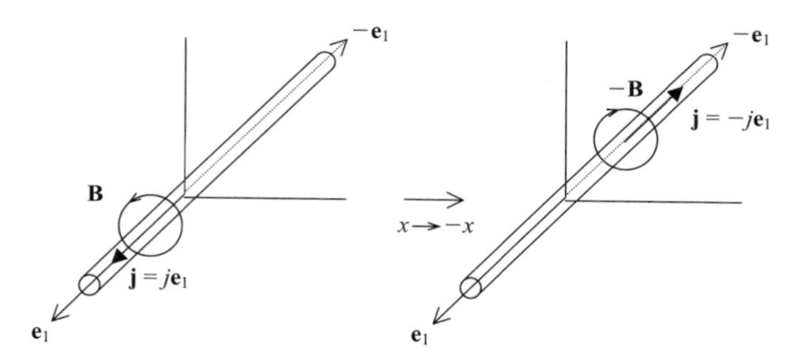

Fig. 9.5 A change in direction of the magnetic field generated by an infinitely long, thin cylinder oriented along the x-axis due to the inversion $x \rightarrow -x$.

quantity, viz., the induced magnetic field. For, under the inversion $x \rightarrow -x$, $y \rightarrow y$, $z \rightarrow z$, the current density changes sign and the magnetic field $\mathbf{B}$, although it remains in the y–z plane, nevertheless changes its direction (Fig. 9.5), too.

Reflections: quantum physics

In classical physics, i.e., for ordinary space, the parity operator is $\hat{\mathcal{P}}$: $\hat{\mathcal{P}}\mathbf{r} = -\mathbf{r}$. Since $\hat{\mathcal{P}}$ acts on 3-D (or 1-D) vectors, it cannot be the quantal parity operator, because that operator acts in Hilbert space on operators and states, but has no effect on eigenvalues such as $\mathbf{r}$. We shall denote the quantal parity operator by $\hat{U}_{\mathrm{P}}$; it is not only unitary, but also its own inverse. We are ultimately interested in the behavior of quantal operators, states, and quantal systems under spatial reflections.

The relation between $\hat{U}_{\mathrm{P}}$ and $\hat{\mathcal{P}}$ is

$$\hat{U}_{\mathrm{P}}|\mathbf{r}\rangle = |\hat{\mathcal{P}}\mathbf{r}\rangle = |-\mathbf{r}\rangle, \tag{9.61}$$

where an arbitrary phase factor has been set equal to unity. Acting with $\hat{U}_{\mathrm{P}}$ from the left on both sides of (9.61) yields

$$\hat{U}_{\mathrm{P}}^2 = \hat{I}, \tag{9.62}$$

from which we infer that

$$\hat{U}_{\mathrm{P}}^{-1} = \hat{U}_{\mathrm{P}}, \tag{9.63}$$

as claimed. Furthermore, if $|\alpha\rangle$ is an eigenstate of $\hat{U}_{\mathrm{P}}$, viz.,

$$\hat{U}_{\mathrm{P}}|\alpha\rangle = \pi_\alpha|\alpha\rangle, \tag{9.64}$$

then acting on both sides of (9.64) with $\hat{U}_{\mathrm{P}}$ and then employing (9.63) leads to

$$\pi_\alpha^2 = 1, \tag{9.65a}$$

i.e., to

$$\pi_\alpha = \pm 1. \tag{9.65b}$$

The eigenvalues of $\hat{U}_{\mathrm{P}}$ are thus real, and the eigenstates of $\hat{U}_{\mathrm{P}}$ have either positive parity

($\pi_a = +1$) or negative parity ($\pi_a = -1$). They are sometimes said to be even ($\pi_a = +1$) or odd ($\pi_a = -1$) under reflection.

In addition to the preceding relations we require, in analogy to $\hat{\mathcal{P}}\mathbf{r} = -\mathbf{r}$, that

$$\hat{U}_{\mathrm{P}}\hat{\mathbf{Q}}\hat{U}_{\mathrm{P}}^{-1} = -\hat{\mathbf{Q}}, \tag{9.66}$$

a relation that holds for each component of $\hat{\mathbf{Q}}$. Substitution of (9.63) into (9.66) tells us that $\hat{U}_{\mathrm{P}}$ and $\hat{\mathbf{Q}}$ *anticommute*:

$$\hat{U}_{\mathrm{P}}\hat{\mathbf{Q}} + \hat{\mathbf{Q}}\hat{U}_{\mathrm{P}} \equiv \{\hat{U}_{\mathrm{P}}, \hat{\mathbf{Q}}\} = 0, \tag{9.67}$$

where { , } is the symbol for the anticommutator. Equation (9.61) may be taken as a fundamental relation. Equation (9.66), which is the (quantal) operator analog of the classical relation $\hat{\mathcal{P}}\mathbf{r} = -\mathbf{r}$, is, in fact, consistent with (9.61), since we can use it in the form of (9.67) to derive (9.61):

$$\begin{aligned} \hat{\mathbf{Q}}\hat{U}_{\mathrm{P}}|\mathbf{r}\rangle &= -\hat{U}_{\mathrm{P}}\hat{\mathbf{Q}}|\mathbf{r}\rangle \\ &= -\mathbf{r}\hat{U}_{\mathrm{P}}|\mathbf{r}\rangle. \end{aligned} \tag{9.68}$$

Hence, $\hat{U}_{\mathrm{P}}|\mathbf{r}\rangle$ is, up to an arbitrary phase factor that was set equal to unity in Eq. (9.61), the eigenstate of $\hat{\mathbf{Q}}$ corresponding to the eigenvalue $-\mathbf{r}$.

It is left to the exercises to show that $\hat{\mathbf{P}}$, like $\hat{\mathbf{Q}}$, is an *odd* operator:

$$\hat{U}_{\mathrm{P}}\hat{\mathbf{P}}U_{\mathrm{P}}^{-1} = -\hat{\mathbf{P}}. \tag{9.69}$$

From Eqs. (9.66) and (9.67) it follows that the orbital angular-momentum operator $\mathbf{L} = \hat{\mathbf{Q}} \times \hat{\mathbf{P}}$ is an even operator under parity:

$$\hat{U}_{\mathrm{P}}\hat{\mathbf{L}}\hat{U}_{\mathrm{P}}^{-1} = \hat{\mathbf{L}}. \tag{9.70a}$$

That is, $\hat{U}_{\mathrm{P}}$ and $\hat{\mathbf{L}}$ commute:

$$[\hat{U}_{\mathrm{P}}, \hat{\mathbf{L}}] = 0. \tag{9.70b}$$

$\hat{\mathbf{Q}}$ and $\hat{\mathbf{P}}$ are polar-vector operators, whereas $\hat{\mathbf{L}}$ is a pseudo-vector operator. Note that $\hat{\mathbf{L}}$ and $\hat{U}_{\mathrm{P}}$ can have common eigenstates, but that $\hat{\mathbf{P}}$ and $\hat{U}_{\mathrm{P}}$ cannot.

It is readily shown that all powers and products of operators even under inversion are themselves even. Moreover, if $\hat{A}$ and $\hat{B}$ are odd operators, i.e., if

$$\{\hat{A}, \hat{U}_{\mathrm{P}}\} = \{\hat{B}, \hat{U}_{\mathrm{P}}\} = 0, \tag{9.71}$$

then

$$[\hat{A}\hat{B}, \hat{U}_{\mathrm{P}}] = 0 \tag{9.72a}$$

and

$$[\hat{A}^{2n}, \hat{U}_{\mathrm{P}}] = 0, \tag{9.72b}$$

while

$$\{\hat{A}^{2n+1}, \hat{U}_{\mathrm{P}}\} = 0, \tag{9.72c}$$

where $n = 1, 2, 3, \ldots$ – as is easily verified.

An immediate consequence is that the kinetic energy operator is even, since it is proportional to $\hat{\mathbf{P}} \cdot \hat{\mathbf{P}}$ (or to a sum of the squares of individual momentum operators for a system containing more than one particle). The behavior of the Hamiltonian under inversion therefore depends on the corresponding behavior of the potential-energy operator $\hat{V}$. It suffices to consider single-particle systems, for which $\hat{V} = \hat{V}(\hat{\mathbf{Q}})$. Only if $\hat{V}$ is even under reflections will eigenstates of $\hat{H}$ also be eigenstates of $\hat{U}_{\mathrm{P}}$, that is,

$$(\hat{H} - E_\gamma)|\gamma, E_\gamma\rangle = 0 \tag{9.73a}$$

and

$$[\hat{V}, \hat{U}_{\mathrm{P}}] = 0 \tag{9.73b}$$

imply that π_γ is included among the set of quantum numbers $\{\gamma\}$.

It may also happen that specification of π_γ is unnecessary due to π_γ being related to one of the existing quantum numbers. This is the case for the linear harmonic oscillator, $\hat{V}(\hat{\mathbf{Q}}) = m\omega_0^2\hat{Q}_x^2/2$, where the parity π_n of $\langle x|n\rangle = \psi_n(x) = A_n H_n(x/\lambda_0)\exp[-x^2/(2\lambda_0^2)]$, Eq. (6.124), is $\pi_n = (-1)^n$. This relation follows from a basic property of the Hermite polynomials, viz., $H_n(-x) = (-1)^n H_n(x)$: see Eq. (6.107) and the ensuing discussion. In contrast, if a cubic term is added to the quadratic one, say $\hat{V}(\hat{\mathbf{Q}}) = \alpha\hat{Q}_x^2 + \beta\hat{Q}_x^3$, then $\hat{V}(\hat{\mathbf{Q}})$ is of mixed symmetry character under inversion, being neither even nor odd. The corresponding eigenstates do not have well-defined parity. The asymmetric quantum box, infinite everywhere but for $0 < x < L$, also has solutions $\psi_n(x)$ that are not parity eigenstates, since $V(x) = \infty$ for $x < 0 \Rightarrow \psi_n(x) = 0$, $x < 0$; one cannot conclude that $\psi_n(-x) = -\psi_n(x)$, as suggested by $\psi_n(x) = \sqrt{2/L}\sin(n\pi x/L)$.

The preceding remarks have been based in part on the coordinate representation. We consider this further. Let $|\alpha\rangle$ be a generic single-particle state, assumed to be an eigenstate of $\hat{U}_{\mathrm{P}}$, Eq. (9.64). Its coordinate representation is

$$\psi_\alpha(\mathbf{r}) = \langle \mathbf{r}|\alpha\rangle, \tag{9.74}$$

with

$$\psi_\alpha(-\mathbf{r}) = \langle -\mathbf{r}|\alpha\rangle. \tag{9.75}$$

However, from application of the Hermitian conjugate to (9.61),

$$\langle -\mathbf{r}| = \langle \mathbf{r}|\hat{U}_{\mathrm{P}}^\dagger, \tag{9.76}$$

and the substitution of (9.76) into (9.75) plus the use of Eq. (9.64) leads to

$$\psi_\alpha(-\mathbf{r}) = \pi_\alpha\psi_\alpha(\mathbf{r}). \tag{9.77a}$$

Equation (9.77a) is equivalent to

$$\hat{U}_{\mathrm{P}}\psi_\alpha(\mathbf{r}) = \psi_\alpha(-\mathbf{r}) = \pi_\alpha\psi_\alpha(\mathbf{r}), \tag{9.77b}$$

where we stress that $|\alpha\rangle$ must be a parity eigenstate. From the (known) functional form of $\psi_\alpha(\mathbf{r})$ one can generally deduce the value of π_α. Two points are worth noting. First a (local) coordinate-space form $\hat{U}_{\mathrm{P}}(\mathbf{r})$ need not be (and has not been) stated. Second, even though the 3-D coordinate vector has been introduced, the parity operator is not changed from $\hat{U}_{\mathrm{P}}$ to $\hat{\mathcal{P}}$: it remains $\hat{U}_{\mathrm{P}}$ because $\psi_\alpha(\mathbf{r})$ is still a Hilbert-space vector.

Next, let $\hat{A}'$ be the space-reflected version of the single-particle operator $\hat{A}$:

$$\hat{A}' = \hat{U}_{\mathrm{P}}\hat{A}\hat{U}_{\mathrm{P}}^{-1} = \hat{U}_{\mathrm{P}}\hat{A}\hat{U}_{\mathrm{P}}; \tag{9.78}$$

see Eq. (9.53b). It is left as an exercise to show that

$$\hat{A}'(\mathbf{r}, \mathbf{r}') \equiv \langle \mathbf{r}|\hat{A}'|\mathbf{r}'\rangle = \hat{A}(-\mathbf{r}, -\mathbf{r}'), \tag{9.79}$$

which is equivalent to

$$\hat{A}'(\mathbf{r}, \mathbf{r}') = \hat{U}_{\mathrm{P}}\hat{A}(\mathbf{r}, \mathbf{r}')\hat{U}_{\mathrm{P}}^{-1}. \tag{9.80}$$

For a local operator $\hat{A}$, one has

$$\hat{A}'(\mathbf{r}) = \hat{A}(-\mathbf{r}) = \hat{U}_{\mathrm{P}}\hat{A}(\mathbf{r})\hat{U}_{\mathrm{P}}^{-1}. \tag{9.81}$$

In particular, if

$$[\hat{V}(\hat{\mathbf{Q}}), \hat{U}_{\mathrm{P}}] = 0,$$

where

$$\hat{H} = \hat{K} + \hat{V}(\mathbf{Q}),$$

then

$$\hat{H}(\mathbf{r}) = \hat{H}(-\mathbf{r}), \tag{9.82}$$

where (9.82) follows on inspection of

$$\hat{H}(\mathbf{r}) = -\hbar^2\nabla^2/(2m) + V(\mathbf{r}), \qquad V(-\mathbf{r}) = V(\mathbf{r}).$$

Our final consideration prior to examining a specific Schrödinger-equation eigenvalue problem concerns the *parity-selection rule* for matrix elements. Define $A_{\beta\alpha}$ as

$$A_{\beta\alpha} = \langle\beta|\hat{A}|\alpha\rangle \tag{9.83}$$

and assume that

$$\hat{U}_{\mathrm{P}}|\alpha\rangle = \pi_\alpha|\alpha\rangle, \tag{9.84a}$$

$$\hat{U}_{\mathrm{P}}|\beta\rangle = \pi_\beta|\beta\rangle, \tag{9.84b}$$

and

$$\hat{A}' = \hat{U}_{\mathrm{P}}\hat{A}\hat{U}_{\mathrm{P}}^{-1} = \pi_A\hat{A}. \tag{9.84c}$$

Equation (9.84c) is crucial; there *are* operators for which it holds, e.g., $\hat{\mathbf{P}}$, $\hat{\mathbf{Q}}$, $\hat{\mathbf{L}}$, and $\hat{K}$. Since $\hat{U}_{\mathrm{P}}^{-1} = \hat{U}_{\mathrm{P}}^\dagger$ and $\hat{U}_{\mathrm{P}}^\dagger\hat{U}_{\mathrm{P}} = \hat{I}$, Eq. (9.83) can be re-expressed as

$$\begin{aligned} A_{\beta\alpha} &= \langle\beta|\hat{U}_{\mathrm{P}}^\dagger(\hat{U}_{\mathrm{P}}\hat{A}\hat{U}_{\mathrm{P}}^{-1})\hat{U}_P|\alpha\rangle \\ &= \pi_\beta\pi_A\pi_\alpha A_{\beta\alpha}, \end{aligned} \tag{9.85a}$$

where Eqs. (9.84) have been used to produce the second equality.

Equation (9.85a) requires that

$$\pi_\beta\pi_A\pi_\alpha = 1. \tag{9.85b}$$

Should the product be negative, then $A_{\beta\alpha} = 0$:

$$A_{\beta\alpha} = \langle\beta|\hat{A}|\alpha\rangle = 0 \qquad \text{if } \pi_\beta\pi_A\pi_\alpha = -1. \tag{9.85c}$$

Equations (9.85b) and (9.85c) express the parity-selection rule. An immediate applica-

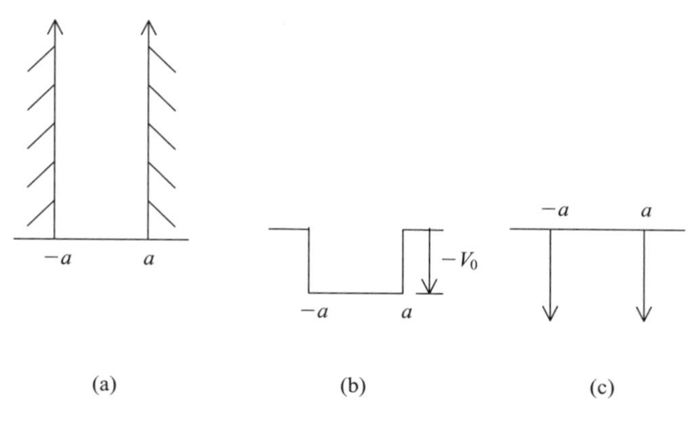

Fig. 9.6 Examples of 1-D, spatially symmetric potential wells: (a) the symmetric quantum box, (b) the symmetric attractive square well, and (c) symmetrically placed delta functions of equal strength.

tion is to scalar products (probability amplitudes), i.e., to the case $\hat{A} = \hat{I}$, for which $\pi_I = 1$. If $|\alpha\rangle$ and $|\beta\rangle$ are each eigenstates of $\hat{U}_P$, then the parity-selection rule tells us that $\langle\beta|\alpha\rangle = 0$ if $\pi_\beta \neq \pi_\alpha$. That is, $|\alpha\rangle$ and $|\beta\rangle$ must behave in the same way under parity in order for $\langle\beta|\alpha\rangle$ to have a chance of being non-zero.[3] When parity is a good quantum number and $\langle\beta|\alpha\rangle \neq 0$, we may further conclude that, w.r.t. scalar products, it makes no difference whether one computes in the framework of a coordinate system or its inverse: the scalar product cannot distinguish between them.

An important example of (9.85c) involves the position operator (or any of its components). We know that $\hat{\mathbf{Q}}' = -\hat{\mathbf{Q}}$, so that $\pi_Q = -1$. Hence, in this case, $A_{\beta\alpha} = 0$ unless $\pi_\beta\pi_\alpha = -1$. One application of this rule concerns the existence of a static (or permanent) dipole moment in inversion-symmetric systems whose structures are governed by parity-conserving (or even-parity) interactions. The dipole operator is proportional to $\hat{\mathbf{Q}}$, while "static" here means $|\beta\rangle = |\alpha\rangle$, so that, both for the neutron and for homopolar molecules, there should be no permanent dipole moment because $\pi_{\hat{\mathbf{Q}}} = -1$ and $\pi_\alpha^2 = +1$. If a non-vanishing static dipole moment is measured, it means that there is a parity-non-conserving interaction present in the system. The weak interaction, which is responsible (among other things) for β-decay, does not conserve parity, as was suggested by Lee and Yang (1956) and verified experimentally by Wu *et al.* (1957). This was a major surprise – one of the rare epiphanies in physics – because each of the four fundamental interactions (strong, electromagnetic, weak, and gravitational) had been thought to be parity conserving. A less spectacular but no less important application of the parity-selection rule is to electromagnetic transitions in quantum systems, which we consider briefly in Chapter 16.

The 1-D, symmetric quantum box

Since we were not then prepared to discuss the notion of invariance under reflections, only the delta function, among the non-confining, 1-D potentials considered in Chapter 6,

[3] Of course, if either $|\alpha\rangle$ or $|\beta\rangle$ (or both) is not an eigenstate of $\hat{U}_P$, then the parity-selection rule is inapplicable and no parity-based conclusions about $\langle\beta|\alpha\rangle$ can be drawn.

was of even character under inversion, and parity did not enter into our considerations. This was a deliberate omission, since there are many single-particle potentials that are even under reflection. Any 3-D potential of the form $V(r)$ is an example, and we consider some such in Chapter 11. A few 1-D examples are shown in Fig. 9.6; we work out below the case of the symmetric quantum box, Fig. 9.6(a).

The potential for the 1-D symmetric box is

$$V(x) = \begin{cases} \infty, & x \leqslant -a, \\ 0, & -a < x < a, \\ \infty, & x \geqslant a, \end{cases} \tag{9.86}$$

and the eigensolutions of the Schrödinger equation are non-zero only for $x \in (-a, a)$. From the analysis of Chapter 3, the possible solutions to the Schrödinger equation

$$\left(\frac{d^2}{dx^2} + k^2\right)\psi(x) = 0, \qquad -a < x < a, \tag{9.87}$$

are $\sin(kx)$ and $\cos(kx)$, subject to the boundary condition that

$$\psi(x = \pm a) = 0. \tag{9.88}$$

However, because $V(-x) = V(x)$, the CSCO for the symmetric box is $\{\hat{H}, \hat{U}_{\mathrm{P}}\}$, and parity is therefore a good quantum number. Since $\cos(kx)$ and $\sin(kx)$ have opposite parities, $\cos(kx)$ being even and $\sin(kx)$ being odd, the solution to (9.88) cannot be a linear combination of $\sin(kx)$ and $\cos(kx)$, as was the case for the asymmetric box. Instead they must be considered separately. The even solutions, $\cos(kx)$, are denoted $\psi^{(\mathrm{e})}(x)$, while the odd solutions, $\sin(kx)$, are denoted $\psi^{(\mathrm{o})}(x)$.

As with the asymmetric box, the boundary conditions (9.88) determine the eigenvalues. For the even solutions, we have

$$\cos(ka) = 0,$$

from which we conclude that

$$k \to k_n^{(\mathrm{e})} = (n + \tfrac{1}{2})\pi/a, \qquad n = 0, 1, 2, \ldots. \tag{9.89}$$

The normalized even solutions are

$$\psi_n^{(\mathrm{e})}(x) = \sqrt{1/a}\cos\left(\frac{(2n+1)\pi x}{2a}\right), \qquad n = 0, 1, 2, \ldots, \tag{9.90a}$$

where the normalization constant has been chosen real and positive; the corresponding energy eigenvalue is

$$E_n^{(\mathrm{e})} = (n + \tfrac{1}{2})^2 \frac{\hbar^2\pi^2}{2ma^2}, \qquad n = 0, 1, 2, \ldots. \tag{9.90b}$$

For the odd states, generated by

$$\sin(ka) = 0,$$

one finds

$$k \to k_n^{(\mathrm{o})} = n\pi/a, \qquad n = 1, 2, 3, \ldots, \tag{9.91}$$

$$\psi_n^{(o)}(x) = \sqrt{1/a}\sin\left(\frac{n\pi x}{a}\right) \tag{9.92a}$$

and

$$E_n^{(o)} = n^2 \frac{\hbar^2\pi^2}{2ma^2}, \tag{9.92b}$$

where the normalization constant in $\psi_n^{(o)}(x)$ has again been set equal to $\sqrt{1/a}$.

Since $E = \langle\hat{K}\rangle$, the state with the least curvature in coordinate space lies lowest. That state is $\psi_1^{(e)}$; its energy is $E_0/4$, where $E_0 = \hbar^2/(2ma^2)$. The first excited state is $\psi_1^{(o)}(x)$, with energy E_0, etc. The energies alternate between even and odd values, with the spacing Δ_n between adjacent odd (higher) and even (lower) energies being

$$\begin{aligned}\Delta_n &= [(n+1)^2 - (n+\tfrac{1}{2})^2]E_0 \\ &= (n+\tfrac{3}{4})E_0,\end{aligned}$$

while the spacings $\Delta_n^{(e)}$ and $\Delta_n^{(o)}$ between neighboring even and odd energies, respectively, are

$$\Delta_n^{(e)} = (2n+2)E_0$$

and

$$\Delta_n^{(o)} = (2n+1)E_0.$$

In Fig. 9.7 are plotted the four lowest-lying energy levels and the corresponding wave functions for the symmetric box. Excluding the nodes at $x = \pm a$, we find that the nth excited state has n nodes; the ground-state (zeroth excited state) is nodeless. The pattern seen here, viz., that the ground-state is nodeless while the first excited state has one node (at least in the vicinity of the well), tends to be a universal feature of symmetric potential wells.

As a final point, we compare the solution of the symmetric box with that of the

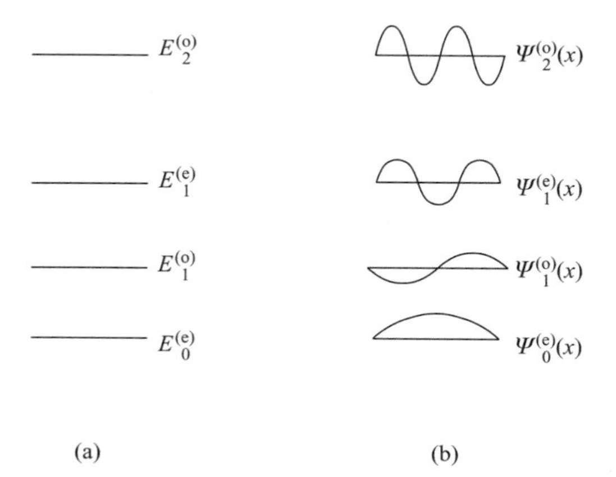

Fig. 9.7 The low-lying spectrum of the symmetric quantum box, located between $x = -L/2$ and $x = L/2$: (a) the first four energy levels (Grotrian diagram) and (b) sketches of the first four wave functions.

asymmetric box. The comparison is facilitated by setting $a = L/2$, so that each well has a width L. With this redefinition, we find

$$\psi_n^{(e)}(x) \to \sqrt{2/L}\cos\left(\frac{(2n+1)\pi x}{L}\right), \qquad x \in \left[-\frac{L}{2}, \frac{L}{2}\right], \tag{9.93a}$$

$$E_n^{(e)} \to (2n+1)^2 E_1, \tag{9.93b}$$

$$\psi_n^{(o)}(x) \to \sqrt{2/L}\sin\left(\frac{2n\pi x}{L}\right), \qquad x \in \left[-\frac{L}{2}, \frac{L}{2}\right], \tag{9.93c}$$

and

$$E_n^{(o)} \to (2n)^2 E_1, \tag{9.93d}$$

with $E_1 = \hbar^2\pi^2/(2mL^2)$, the ground-state energy in the asymmetric box, Eq. (6.5). Comparison of Eqs. (9.93b) and (9.93d) with Eq. (6.5) shows that the energy-level spectra for symmetric and asymmetric boxes of width L are the same: the location of the box cannot affect its energy spectrum. On the other hand, the wave functions in the two wells *are* different, and this is a consequence as well as a manifestation of the presence versus absence of the symmetry property. The solutions to other symmetric potential problems, in particular the symmetric square well, which was briefly discussed in Section 6.6, below Eq. (6.179), are left for the exercises.

9.6. Translations

By "translation" one means a displacement in coordinate space, e.g.,

$$\mathbf{r} \to \mathbf{r}' = \mathbf{r} + \mathbf{a}, \tag{9.94}$$

where $\mathbf{a}$ is an arbitrary constant vector. The result (9.94) can be achieved *actively* by physically moving the vector, or *passively* by displacing the origin of the coordinate system through $-\mathbf{a}$, as in Fig. 9.8. Probably the simplest application of invariance under

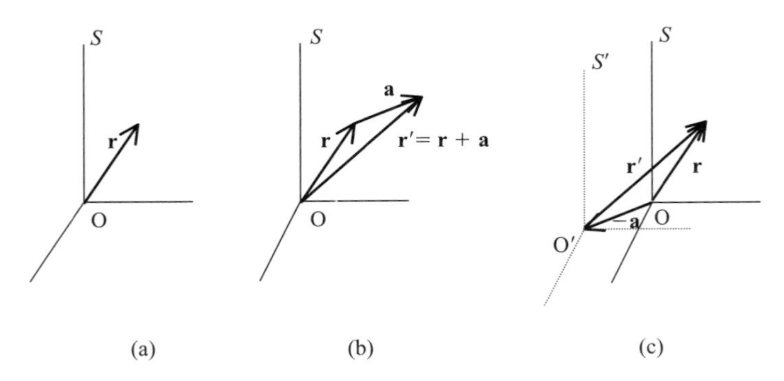

Fig. 9.8 Translation of a 3-D vector. (a) The original vector to the point $\mathbf{r}$ in the coordinate system S, whose origin is O. (b) The translated vector to the point $\mathbf{r}' = \mathbf{r} + \mathbf{a}$ in the coordinate system S. (c) The vector to the point $\mathbf{r}$ as seen in the coordinate system S′, whose origin O′ is at the position $-\mathbf{a}$ in S.

translation is to empty space: empty space is both isotropic and homogeneous, that is, it does not change under rotations and translations, respectively.

In the case of empty (infinite) space, homogeneity means that an origin of coordinates can be chosen arbitrarily. If a physical system is to be invariant under translations, then it must be either looplike (its edges must joined) or infinite in extent, and it must also be periodic in structure. A trivial example is a 1-D plane wave, e.g., $\cos(kx - \omega t)$, which extends over all space and is periodic with wave length $\lambda = 2\pi/k$. Much more physically interesting is an infinite lattice, 1-D and 2-D examples of which are illustrated in Fig. 9.9.

In concordance with parity and rotation, one can introduce a classical-physics translation operator $\hat{\mathcal{T}}(\mathbf{a})$, which, when it acts on an arbitrary 3-D vector $\mathbf{b}$, translates it through $\mathbf{a}$, as in Eq. (9.94):

$$\mathbf{b} \to \mathbf{b}' = \hat{\mathcal{T}}(\mathbf{a})\mathbf{b} = \mathbf{b} + \mathbf{a}. \tag{9.95}$$

The quantal analog of $\hat{\mathcal{T}}(\mathbf{a})$ is the (unitary) translation operator $\hat{U}_{\mathrm{T}}(\mathbf{a})$, which transforms states, wave functions, and operators. In the remainder of this section, we derive the form of $\hat{U}_{\mathrm{T}}(\mathbf{a})$, and then use it to examine translational invariance for free particles and for particles acted on by a periodic, 1-D potential. In contrast to the rotation operator $\hat{U}_{\mathrm{R}}(\mathbf{n}, \theta)$ and its generator $\hat{\mathbf{L}}$, whose eigenvectors and eigenvalues are sufficiently complicated to analyze that this task is postponed to Chapter 10, the translation operator

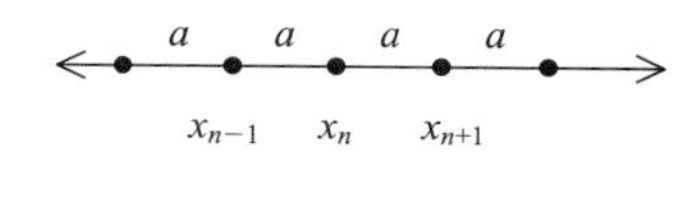

(a)

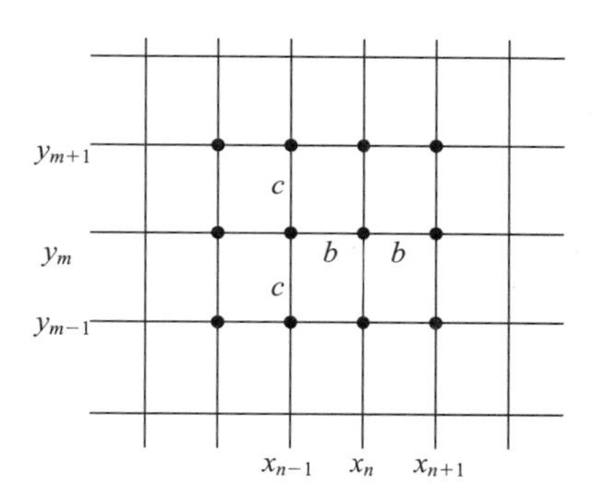

(b)

Fig. 9.9 Finite portions of infinite lattice structures. (a) A 1-D lattice having period a in the x-direction: $x_{n+1} = x_n + a$. (b) A 2-D lattice with period b in the x-direction and period c in the y-direction: $x_{n+1} = x_n + b$, $y_{m+1} = y_m + c$.

and its properties can be determined straightforwardly; its eigenvectors often can be, as well. This is true for the particular example of the 1-D periodic potential that we solve in detail. The solution of this problem provides a 1-D model for the behavior of electrons in 3-D periodic structures (crystal lattices), i.e., a highly simplified model of conductors and insulators.

The quantal translation operator

In order to derive the form of the quantal operator $\hat{U}_{\mathrm{T}}(\mathbf{a})$ for translation by a displacement $\mathbf{a}$, as in Eq. (9.95), we consider a single-particle system described by a so-called "scalar function," i.e., a system whose coordinate-space wave function can be written $\psi(\mathbf{r})$. That is, no quantum numbers are needed to label the state (the same system will be used in the next section to derive the quantal rotation operator). Although a quantum system described by an unlabeled state is an invention – a creature of our imagination – $\psi(\mathbf{r})$ itself is a mathematically well-defined, Hilbert-space vector; it functions like a *Gedanken experiment*, but in the theoretical arena. Elimination of the labels allows us to concentrate solely on the effect of changing the coordinate. However, in the present case of translations, the result for $\hat{U}_{\mathrm{T}}(\mathbf{a})$ will be valid for any particle, including those with spin, not just for scalar particles.

Under a translation through $\mathbf{a}$,

$$\mathbf{r} \to \mathbf{r}' = \hat{\mathcal{T}}(\mathbf{a})\mathbf{r} \tag{9.96a}$$

and

$$\psi(\mathbf{r}) \to \psi'(\mathbf{r}') = \hat{U}_{\mathrm{T}}(\mathbf{a})\psi(\mathbf{r}), \tag{9.96b}$$

where $\psi'(\mathbf{r}')$ is the translated function evaluated at $\mathbf{r}' = \mathbf{r} + \mathbf{a}$. The crucial point to recognize now is that ψ is a *scalar* and therefore the translated scalar function evaluated at the displaced point must be the *same* as the untranslated scalar function evaluated at the original point:

$$\psi'(\mathbf{r}') = \psi(\mathbf{r}). \tag{9.97}$$

Using Eq. (9.94) to express $\mathbf{r}$ in terms of $\mathbf{r}'$ and then dropping primes yields

$$\psi'(\mathbf{r}) = \psi(\mathbf{r} - \mathbf{a}). \tag{9.98}$$

The derivation of $\hat{U}_{\mathrm{T}}(\mathbf{a})$ follows on combining Eqs. (9.96b), (9.97), and (9.98) and then allowing the displacement $\mathbf{a}$ to become an infinitesimal length ϵ, leading to

$$\begin{aligned} \hat{U}_{\mathrm{T}}(\epsilon)\psi(\mathbf{r}) &= \psi(\mathbf{r} - \epsilon) \\ &\cong \psi(\mathbf{r}) - \epsilon \cdot \nabla\psi(\mathbf{r}) \\ &= (\hat{I} - \epsilon \cdot \nabla)\psi(\mathbf{r}). \end{aligned} \tag{9.99}$$

We shall use Stone's theorem to identify the infinitesimal generator of translations, but, before doing so, we relate the quantity $-\epsilon \cdot \nabla$ in (9.99) to a by-now well-known quantal operator, the momentum $\hat{\mathbf{P}}$:

$$-\epsilon \cdot \nabla = -i\epsilon \cdot \hat{\mathbf{P}}/\hbar. \tag{9.100}$$

Equation (9.100) leads to the important result that $\hat{\mathbf{P}}$ is the infinitesimal generator of spatial translations; hence, by Stone's theorem, $\hat{U}_{\mathrm{T}}(\mathbf{a})$ is

$$\hat{U}_{\mathrm{T}}(\mathbf{a}) = e^{-i\mathbf{a}\cdot\hat{\mathbf{P}}/\hbar}, \tag{9.101}$$

an expression guaranteeing unitarity: $\hat{U}_{\mathrm{T}}^{\dagger} = \hat{U}_{\mathrm{T}}^{-1}$. We stress that **a** is arbitrary.

Translations are continuously generated from the identity operator, with $\hat{U}_{\mathrm{T}}(\mathbf{a})$ obeying

$$\hat{U}_{\mathrm{T}}(\mathbf{a} = 0) = \hat{I}, \tag{9.102a}$$

$$[\hat{U}_{\mathrm{T}}(\mathbf{a}),\ \hat{U}_{\mathrm{T}}(\mathbf{b})] = 0, \tag{9.102b}$$

and

$$\hat{U}_{\mathrm{T}}(\mathbf{a})\hat{U}_{\mathrm{T}}(\mathbf{b}) = \hat{U}_{\mathrm{T}}(\mathbf{a} + \mathbf{b}), \tag{9.102c}$$

each of which follows directly from Eq. (9.101).

The preceding analysis used the coordinate representation and the scalar property of $\psi(\mathbf{r})$. It may be helpful to any readers who might have misgivings about this procedure to rederive the first equality in Eq. (9.99) by working directly with the state vectors $|\mathbf{r}\rangle$ and $|\psi\rangle$, with $\psi(\mathbf{r}) = \langle\mathbf{r}|\psi\rangle$. The fundamental relation, analogous to (9.61), is

$$\hat{U}_{\mathrm{T}}(\mathbf{a})|\mathbf{r}\rangle = |\hat{T}(\mathbf{a})\mathbf{r}\rangle = |\mathbf{r} + \mathbf{a}\rangle, \tag{9.103}$$

where an arbitrary phase factor has been set equal to unity. Since $\hat{U}_{\mathrm{T}}^{\dagger} = \exp(i\mathbf{a}\cdot\hat{\mathbf{P}}/\hbar)$, the Hermitian conjugate of (9.103) must be

$$\langle\mathbf{r}|\hat{U}_{\mathrm{T}}^{\dagger}(\mathbf{a}) = \langle\hat{U}_{\mathrm{T}}^{\dagger}(\mathbf{a})\mathbf{r}| = \langle\mathbf{r} - \mathbf{a}|, \tag{9.104}$$

and therefore

$$\langle\mathbf{r}|\hat{U}_{\mathrm{T}}(\mathbf{a})|\psi\rangle = \hat{U}_{\mathrm{T}}(\mathbf{a})\psi(\mathbf{r}) = \psi(\mathbf{r} - \mathbf{a}). \tag{9.105}$$

Setting $\mathbf{a} = \epsilon$, Eq. (9.105) leads to (9.99) and ultimately to (9.101).

Translational symmetry is of particular importance in condensed-matter physics when the solid-state structures are crystal lattices having specific periodic arrangements of atoms and/or molecules. One often assumes a macroscopic-sized solid to be infinite in extent, so as to eliminate the need to deal with edge effects, and in this case translational symmetry applies. We consider two infinite cases in the next two subsections and then conclude with a detailed investigation of the eigenvalue problem for one of these two cases.

Plane waves

The simplest example of a translationally invariant system is a free particle, wherein $V(\mathbf{r}) = 0$ and

$$\hat{H} = \hat{K} = \frac{\hat{P}^2}{2m}. \tag{9.106}$$

Since $[\hat{\mathbf{P}},\ \hat{H}] = 0$, the eigenvalue problem is

$$\hat{H}|\mathbf{k}\rangle = E_k|\mathbf{k}\rangle, \tag{9.107}$$

where $E_k = \hbar^2 k^2/(2m)$ and $(\hat{\mathbf{P}} - \hbar\mathbf{k})|\mathbf{k}\rangle = 0$.

Given the relation $[\hat{\mathbf{P}}, \hat{H}] = 0$, then $[\hat{U}_T(\mathbf{a}), \hat{H}] = 0$ and hence

$$\hat{U}_T(\mathbf{a})|\mathbf{k}\rangle = \mu|\mathbf{k}\rangle, \tag{9.108}$$

where the dependence of the eigenvalue μ on $\mathbf{a}$ and $\mathbf{k}$ is suppressed. Equation (9.108) is a special case of the more general relation

$$\hat{U}_T(\mathbf{a})|\alpha\rangle = |\alpha\rangle' = |\alpha\rangle_{\mathbf{a}}.$$

The structure of $\hat{U}_T(\mathbf{a})$ leads to immediate determination of μ:

$$\begin{aligned}\hat{U}_T(\mathbf{a})|\mathbf{k}\rangle &= e^{-i\mathbf{a}\cdot\hat{\mathbf{P}}/\hbar}|\mathbf{k}\rangle \\ &= e^{-i\mathbf{a}\cdot\mathbf{k}}|\mathbf{k}\rangle,\end{aligned} \tag{9.109}$$

i.e.,

$$\mu = e^{-i\mathbf{k}\cdot\mathbf{a}}, \tag{9.110}$$

which, for given $\mathbf{k}$ and $\mathbf{a}$, is a constant of absolute magnitude unity. The exponential form for μ and its dependence on $\mathbf{k}\cdot\mathbf{a}$ are almost completely dictated by the form of $\hat{U}_T(\mathbf{a})$. The quantum number $\mathbf{k}$, which labels the energy eigenket $|\mathbf{k}\rangle$, also refers to the fact that $|\mathbf{k}\rangle$ is an eigenket of $\hat{U}_T(\mathbf{a})$. The label $\mathbf{a}$ appears in μ but not in $|\mathbf{k}\rangle$, essentially because $\mathbf{a}$, being arbitrary, characterizes $\hat{U}_T(\mathbf{a})$ but not $\hat{H}$: $\mathbf{k}$ is the common link.

Although the abstract Hilbert-space approach produced Eq. (9.110), it is instructive to derive this result using coordinate-space methods. Projecting both sides of (9.108) onto $\langle\mathbf{r}|$ gives

$$\langle\mathbf{r}|\hat{U}_T(\mathbf{a})|\mathbf{k}\rangle = \mu\langle\mathbf{r}|\mathbf{k}\rangle = \left(\frac{1}{2\pi}\right)^{3/2}\mu e^{i\mathbf{k}\cdot\mathbf{r}}. \tag{9.111}$$

From (9.104), the LHS of (9.111) is $\langle\mathbf{r}-\mathbf{a}|\mathbf{k}\rangle$, and we may write

$$e^{i\mathbf{k}\cdot(\mathbf{r}-\mathbf{a})} = \mu e^{i\mathbf{k}\cdot\mathbf{r}},$$

which again gives $\mu = \exp(-i\mathbf{k}\cdot\mathbf{a})$.

The same analysis applies to the free particle in 1-D, say x, and yields

$$\hat{U}_T(a_x)|k_x\rangle = e^{-ik_x a_x}|k_x\rangle. \tag{9.112}$$

We shall see that a similar result occurs in the case of a 1-D, periodic potential, which is studied next.

The 1-D periodic potential, Bloch waves

Metals conduct, insulators do not; i.e., the electrons in a metal (insulator) can (cannot) create a current when an external electric field is applied. Both metals and insulators are typically crystal-lattice structures, with periodicities characteristic of the element or molecule which forms the solid. Whether an external electric field produces a measurable flow of electrons is partly a consequence of the material being a periodic lattice, the latter feature giving rise to the so-called "band structure" of the quantal energy spectrum. A simple model in which band structure is expected is a particle confined to 1-D and experiencing an array of equally spaced, identical potentials. The potentials, which form a 1-D lattice structure, imitate the effect of the atoms or molecules located at the lattice sites in a real solid.

In this subsection we show (i) that the energy eigenstates are of the form known as Bloch functions or Bloch waves, and (ii) that it is possible for the energy spectrum of this system to display a band structure. In the subsequent subsection, a specific model for the periodic array of potentials is chosen, and the corresponding wave functions/energies are determined in detail.

The 1-D Hamiltonian $\hat{H}(x)$ is

$$\hat{H}(x) = \hat{K}_x(x) + V(x), \tag{9.113}$$

where the potential is periodic with period a:

$$V(x \pm a) = V(x). \tag{9.114}$$

For simplicity, it is assumed that each one of the infinite array of potentials is both localized and symmetric about its own central point; these central points are at $0, \pm a, \pm 2a, \pm 3a, \ldots$. The array of potentials is schematically indicated in Fig. 9.10.

In view of (9.114), $\hat{H}(x)$ is invariant w.r.t. a limited set of translations, viz., those through the lattice-spacing distance a or any integer multiple of a:

$$\hat{U}_{\mathrm{T}}(a)\hat{H}(x)\hat{U}^{\dagger}_{\mathrm{T}}(a) = \hat{H}(x). \tag{9.115}$$

The eigenfunctions of $\hat{H}$ are therefore eigenfunctions of $\hat{U}_{\mathrm{T}}(a)$ as well:

$$(\hat{H}(x) - E)\psi_E(x) = 0 \tag{9.116a}$$

and

$$\hat{U}_{\mathrm{T}}(a)\psi_E(x) = \mu\psi_E(x) = \psi_E(x - a). \tag{9.116b}$$

Since $|\mu| = 1$, we may write $\mu = \exp(i\lambda(a))$, where $\lambda(a)$ is real. The phase λ is taken to be linear in a, consistent with the limit $a \to \epsilon =$ infinitesimal applied to (9.116b). Following Eq. (9.110) or (9.112), we write $\lambda = -ka$:

$$\mu = e^{-ika}, \tag{9.117}$$

with k arbitrary but not at this point related to the energy.

Inserting $\mu = \exp(-ika)$ into Eq. (9.116b) gives

$$\psi_E(x - a) = e^{-ika}\psi_E(x), \tag{9.118}$$

which is the same relation as that obeyed by the free-particle solution $\exp(ikx)$. Non-zero $V(x)$ means that $\psi_E(x)$ cannot be a pure plane wave, but the appearance of $\exp(-ika)$ in

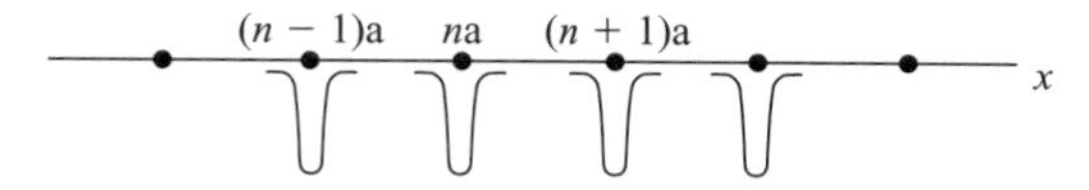

Fig. 9.10 A portion of an infinite array of identical (attractive) potentials whose centers are at the lattice sites $0, \pm a, \pm 2a, \ldots$. The lattice spacing or periodicity is a.

(9.118) suggests including the plane wave as part of $\psi_E(x)$, thereby turning it into a *modified* plane wave:

$$\psi_E(x) = e^{ikx} u_{E(k)}(x). \tag{9.119a}$$

The possible k-dependence of the auxiliary function u is made explicit by writing $E(k)$ (see comments in the paragraph below Eq. (9.119b)).

If (9.119a) is to be valid, then it must satisfy (9.118), and this can occur only if

$$u_{E(k)}(x - a) = u_{E(k)}(x), \tag{9.119b}$$

as substitution of (9.119a) into (9.118) shows. That is, $u_{E(k)}(x)$ must be periodic, with period a.

The modified plane wave described by (9.119a) and (9.11b) is known as a Bloch wave or Bloch function, named after the physicist Felix Bloch, who first derived it (Bloch (1928)). Mathematicians know the solution as a Floquet function; see, e.g., Whittaker and Watson (1965). The wave number k is now a parameter that enters the Schrödinger equation for $u_{E(k)}(x)$; since it influences E, we have incorporated the dependence of E on k via $E(k)$. Note that the "solution" (9.119a) is formal: $u_{E(k)}(x)$ remains to be determined.

Although $u_{E(k)}(x)$ is still unknown, the mere structure of (9.119a) and (9.119b) allows us to draw a far-reaching conclusion. The analysis we use is general; it will be applied in the next subsection as well. It is based on two conditions obeyed by solutions of the Schrödinger equation: (i) any solution can be expressed as the sum of two linearly independent solutions; (ii) all solutions and their first derivatives must be continuous (except for delta-function potentials; recall Section 6.6).

Condition (i) means that

$$\psi_E(x) = A_1 \psi_E^{(1)}(x) + A_2 \psi_E^{(2)}(x), \tag{9.120a}$$

where $\psi_E^{(1)}(x)$ and $\psi_E^{(2)}(x)$ are linearly independent and A_1 and A_2 are constants to be determined. Condition (ii) may be expressed as

$$\lim_{\eta \to 0} \psi_E(x - \eta) = \lim_{\eta \to 0} \psi_E(x + \eta) \tag{9.120b}$$

and

$$\lim_{\eta \to 0} \left. \frac{d\psi_E}{dx} \right|_{x-\eta} = \lim_{\eta \to 0} \left. \frac{d\psi_E}{dx} \right|_{x+\eta}. \tag{9.120c}$$

Our derivation proceeds as follows. Set $x = a$ in Eqs. (9.120b) and (9.120c). On the LHS of the latter two equations use Eq. (9.120a), but on the RHS introduce the solution evaluated at $x = 0$ via Eq. (9.118). Doing so yields

$$A_1 \psi_E^{(1)}(a) + A_2 \psi_E^{(2)}(a) = e^{ika} \left[A_1 \psi_E^{(1)}(0) + A_2 \psi_E^{(2)}(0) \right] \tag{9.121a}$$

and

$$A_1 \psi_E^{(1)\prime}(a) + A_2 \psi_E^{(2)\prime}(a) = e^{ika} \left[A_1 \psi_E^{(1)\prime}(0) + A_2 \psi_E^{(2)\prime}(0) \right]. \tag{9.121b}$$

These two relations are a pair of equations that the unknown coefficients A_1 and A_2

must satisfy.[4] Non-trivial A_1 and A_2 occur only if the determinant of coefficients in Eqs. (9.121) is zero, i.e., if

$$\begin{vmatrix} \psi_E^{(1)}(a) - e^{ika}\psi_E^{(1)}(0) & \psi_E^{(2)}(a) - e^{ika}\psi_E^{(2)}(0) \\ \psi_E^{(1)\prime}(a) - e^{ika}\psi_E^{(1)\prime}(0) & \psi_E^{(2)\prime}(a) - e^{ika}\psi_E^{(2)\prime}(0) \end{vmatrix} = 0. \tag{9.122}$$

Expanding the determinant in (9.122), rearranging terms, and using the fact that the Wronskian W_E of the two solutions,

$$W_E = \psi_E^{(1)}(x)\psi_E^{(2)\prime}(x) - \psi_E^{(1)\prime}(x)\psi_E^{(2)}(x),$$

is a constant (i.e., is independent of x) eventually leads to a relation we write as

$$\cos(ka) = \frac{N_E}{2W_E}, \tag{9.123}$$

where

$$N_E = [\psi_E^{(1)}(a)\psi_E^{(2)\prime}(0) + \psi_E^{(1)}(0)\psi_E^{(2)\prime}(a)] - [\psi_E^{(1)\prime}(a)\psi_E^{(2)}(0) + \psi_E^{(1)\prime}(0)\psi_E^{(2)}(a)]. \tag{9.124}$$

The far-reaching conclusion is based on Eq. (9.123). At first glance, this relation may appear innocuous rather than momentous. The reader, however, should not be fooled by appearances: because of the factor $\cos(ka)$, Eq. (9.123) serves as a filter for the eigenvalues E. That is, only those E for which the RHS of (9.123) lies in the range $[-1, 1]$ are allowed. We therefore conclude that any E for which $|N_E/(2W_E)| > 1$ is excluded from the energy spectrum of a particle in an infinite array of periodic potentials. The latter statement is not an empty one, since such exclusions actually exist: they form gaps in an otherwise continuous energy spectrum (as shown in the next subsection via use of a simple model). The allowed energies fall into distinct, continuous groups known as energy bands.

The existence of bands and (band) gaps is indicated by Eq. (9.123).[5] We next use a simple, soluble potential model to show how they may occur.

An infinite array of 1-D delta-function potentials

To show that energy bands and band gaps occur, we need to solve the relevant Schrödinger equation, with $V(x)$ of Eq. (9.113) obeying (9.114), and this requires a soluble model, of which few exist, e.g., delta functions, square wells, etc.

While any soluble model will suffice, we choose the potential to be an array of repulsive delta functions, since this leads to a relatively simple solution. The potential is taken to be

$$V(x) = \sum_{n=-\infty}^{\infty} \frac{\hbar^2}{mx_0}\delta(x - na), \tag{9.125}$$

where m is the particle's mass, x_0 is the length we have associated with our previous use

[4] Recall that $\psi_E^{(1)}(x)$ and $\psi_E^{(2)}(x)$ are assumed to be known.

[5] An interesting proof of their existence, using mathematical methods more sophisticated than we wish to consider here, can be found in Galindo and Pascual (1990).

of delta-function potentials (Chapters 6 and 7) and a is the spacing. The Schrödinger equation is

$$\left(-\frac{\hbar^2}{2m}\frac{d^2}{dx^2} + \sum_{n=-\infty}^{\infty} \frac{\hbar^2}{mx_0}\delta(x - na) - E\right)\psi(x) = 0, \tag{9.126}$$

which becomes, with $E = \hbar^2 q^2/(2m)$,

$$\psi''(x) - \frac{2}{x_0}\sum_n \delta(x - na)\psi(x) + q^2\psi(x) = 0. \tag{9.127}$$

For $x \neq na$, $n = 0, \pm1, \pm2, \ldots$, the solution is

$$\psi(x) = Ae^{iqx} + Be^{-iqx}, \qquad x \neq na,\ n = 0, \pm1, \ldots, \tag{9.128}$$

so that the $\psi_E^{(1)}$ and $\psi_E^{(2)}$ of the prior subsection are just the plane waves $\exp(\pm iqx)$. The periodic function, here denoted $u(x)$, is

$$u(x) = e^{-ikx}\psi(x) = Ae^{i(q-k)x} + Be^{-i(q+k)x}. \tag{9.129}$$

The constants A and B are determined from imposing the requirements that $u(x)$ (i.e., $\psi(x)$) be continuous at both $x = 0$ and $x = a$ and that $\psi(x)$ satisfy any integral of both sides of (9.127).

Continuity of $u(x)$ implies that

$$\lim_{\eta\to 0} u(-\eta) = \lim_{\eta\to 0} u(\eta),$$

while the requirement of periodicity means that

$$u(\eta) = u(\eta + a).$$

These two relations plus Eq. (9.129) lead to

$$A + B = Ae^{i(q-k)a} + Be^{-i(q+k)a}. \tag{9.130a}$$

The integral requirement is the typical one:

$$\int_{-\eta}^{\eta} dx\left(\frac{d^2\psi(x)}{dx^2} - \frac{2}{x_0}\sum \delta(x - na)\psi(x) + q^2\psi(x)\right) = 0.$$

Since η is an infinitesimal, the second term in the integral gives a non-zero contribution only at $x = 0$, while the last term is identically zero. Evaluating the first term in the limit $\eta \to 0$, the integral goes over to

$$\left.\frac{d\psi(\eta)}{d\eta}\right|_{\eta=0} - \left.\frac{d\psi(-\eta)}{d\eta}\right|_{\eta=0} = \frac{2}{x_0}\psi(0). \tag{9.131}$$

A consequence of periodicity is that

$$\frac{d\psi(-\eta)}{d\eta} = e^{-ika}\frac{d\psi(-\eta + a)}{d\eta},$$

and, on inserting this result plus Eq. (9.128) into (9.131), we find

$$iq[A - B - Ae^{i(q-k)a} + Be^{-i(q+k)a}] = \frac{2}{x_0}(A + B). \tag{9.130b}$$

Equations (9.130a) and (9.130b) are precise analogs to Eqs. (9.121a) and (9.121b), respectively. As with the latter relations, Eqs. (9.130) have non-trivial solutions for A and B only if the determinant of the coefficients vanishes, i.e., only if

$$\begin{vmatrix} 1 - e^{i(q-k)a} & 1 - e^{-i(q+k)a} \\ 1 - e^{i(q-k)a} + 2i/(qx_0) & -1 + e^{-i(q+k)a} + 2i/(qx_0) \end{vmatrix} = 0. \tag{9.132}$$

Solving Eq. (9.132) for ka leads to

$$\cos(ka) = \cos(qa) + \frac{a}{x_0}\frac{\sin(qa)}{qa}, \tag{9.133}$$

an explicit version of Eq. (9.123).

As a preliminary to analyzing this result, we point out that the no-potential, free-particle case corresponds to $x_0 = \infty$, for which (9.133) yields $q = k$ and $E = \hbar^2 k^2/(2m)$, the free-particle energy. To be a non-trivial model, however, x_0 must be finite, in which case (9.133) is an implicit equation defining $q = q(k)$. That there may be values of q that are physically unacceptable, leading to energy gaps, is easily understood. It suffices to consider non-negative values of k and q. No restrictions have been placed on k, so that, as k varies, $\cos(ka)$ will attain its maximum and minimum values:

$$-1 \leqslant \cos(ka) \leqslant 1. \tag{9.134}$$

Denoting the RHS of (9.133) by $F(q)$,

$$F(q) = \cos(qa) + \frac{a}{x_0}\frac{\sin(qa)}{qa}, \tag{9.135}$$

it is obvious by inspection that $|F(q)|$ can exceed unity. Therefore, all q for which $|F(q)| > 1$ are forbidden, as are the corresponding energies $E = \hbar^2 q^2/(2m)$.

This behavior is readily demonstrated graphically. Since the allowed $F(q)$ obey $|F(q)| \leqslant 1$, we plot in Fig. 9.11 the two straight lines ± 1 on a graph of $F(q)$ versus q; physically forbidden values of q are those for which $|F(q)| > 1$. The forbidden q, viz., those outside the intervals $[q_1, q_2]$, $[q_3, q_4]$, etc., give rise to the forbidden energy bands. Figure 9.12 shows schematic plots of $E(q(k))$ versus k, with the band gap j designated ΔE_j. Since $ka = 0$ corresponds to $q_1 > 0$, $E(q_1(k)) > 0$. The endpoint of each energy band in Fig. 9.12(a) is at one of the points $k = n\pi/a$ $(n = 1, 2, 3, \ldots)$ of the curve $\hbar^2 k^2/(2m)$, which is the energy in the zero-potential case. Because of the gaps, all endpoints but that for $k = 0$ are double-valued, with the slope of each energy band at its endpoints being zero, i.e., $dE/dk|_{k=n\pi/a} = 0$, a result whose proof is left to the exercises. For $k < 0$, the corresponding energy bands are the mirror image of those in Fig. 9.12(a), so only the positive values are shown. Furthermore, the whole energy spectrum can be plotted on the $0 \leqslant k \leqslant \pi/a$ portion of the $E(q)$ versus k curve if all the results for $k > \pi/a$ are reflected and/or slid back to begin at $k = 0$. This procedure gives rise to Fig. 9.12(b).

An important feature of the infinite array of periodically spaced potentials (attractive or repulsive) is that there are no bound states: only a continuous energy spectrum exists. For proofs of this statement, the reader may consult Galindo and Pascual (1990) or Volume IV of Reed and Simon (1972). In other words, this system yields no states that are localized in space – all are de-localized. Let us suppose for the moment that the

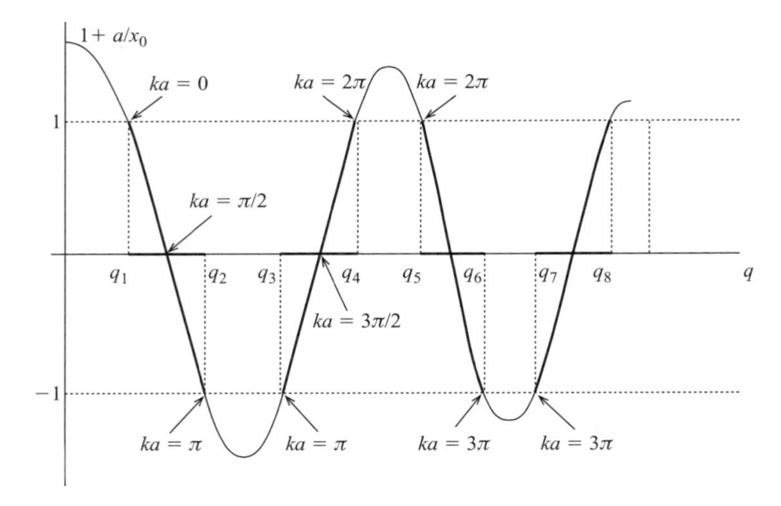

Fig. 9.11 A sketch of $F(q) = \cos(qa) + (a/x_0)\sin(qa)/(qa)$ versus q. The dark lines along the q-axis indicate the allowed values of q, since only for q in these ranges is $|F(q)| \leqslant 1$. The associated $F(q)$ are the dark portions of the curve. $|F(q)| > 1$ for those q such that $q < q_1$, $q_2 < q < q_3$, $q_4 < q < q_5$, etc.; these are the forbidden q which gave rise to the energy gaps, while the allowed q gave rise to the energy bands. The associated *ka* values are also shown.

particles in such a potential are electrons and that the quantal effects of the identity of all electrons (the Pauli principle – see Chapter 18) can be ignored.[6] Then, does delocalization of the electron wave function correspond to a flow of current? The answer is NO, essentially because the electron motion is random, which in this 1-D model means that half move to the right and half to the left, giving rise to net zero current.

The foregoing negative answer does not invalidate the claim that the 1-D, infinite array of periodic potentials can be a model of actual 3-D systems. It is a valid model because the existence of the energy-band spectrum predicts the correct (3-D) behavior in an applied electric field, again assuming the particles to be elctrons. The effect of the electric field will be to add energy to the electrons in the lattice, moving an electron to a higher energy state, *assuming that there is one to which it can go*. In this respect, the set of energy bands behaves like the individual energy levels in an isolated atom, and as is also true for atoms (as will be shown in Chaper 18), only a fixed number of electrons can be in any particular band.[7] The electrons can absorb energy from the applied electric field only if there is a higher energy state for them to go into, either in the same band or in a higher-lying band. In the former case, it takes only a tiny field to cause excitation and a resulting flow of current as electrons move from one site to another without further change of energy level. Unfilled bands, those that do not contain all of the fixed number of electrons which they can accomodate, are denoted "conduction bands"; every

[6] This is a non-essential point for the purposes of this discussion.

[7] The finite size of real solids is incorporated into models such as the present one by making it of finite, not infinite, length and then imposing so-called periodic boundary conditions on $\psi_E(x)$. The effect is to replace a continuum of energies in the band by a discrete set; each member of which is an "energy level." See Exercise 18.18.

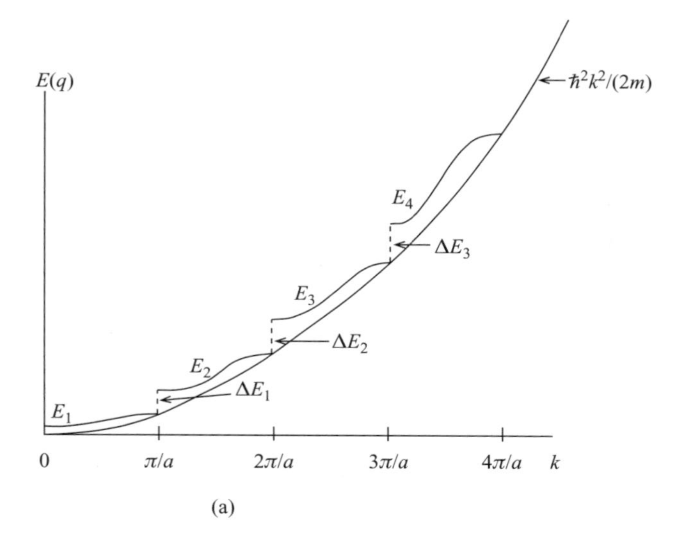

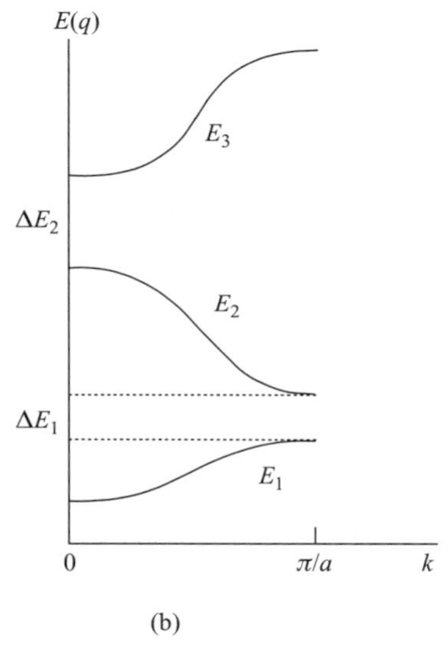

Fig. 9.12 The energy-band/energy-gap structure of the energy spectrum of the infinite array of delta functions, $V(x) \propto \sum_{n=-\infty}^{\infty} \delta(x + na)$. (a) $E(q)$ versus k, $q = q(k)$, with k measured in units of $n\pi/a$, $n = 0, 1, 2, \ldots$. Superposed on the plot of the energy bands E_j is the parabolic curve $\hbar^2 k^2/(2m)$, the value of E when $V = 0$. (b) An enlarged portion of the $E(q)$ of (a), with the second curve reflected about $k = \pi/a$, and the third translated to the left by a distance $2\pi/a$. In this manner, it suffices to plot the curves only out to $k = \pi/a$.

conductor obviously has an unfilled conduction band. On the other hand, insulators are those materials which have completely filled energy bands plus an energy gap between the highest-lying filled band and the lowest-lying but empty condution band that ordinary electric fields cannot bridge because ΔE of the gap is too large. As a result, no current flows. These various band structures are sketched in Fig. 9.13.

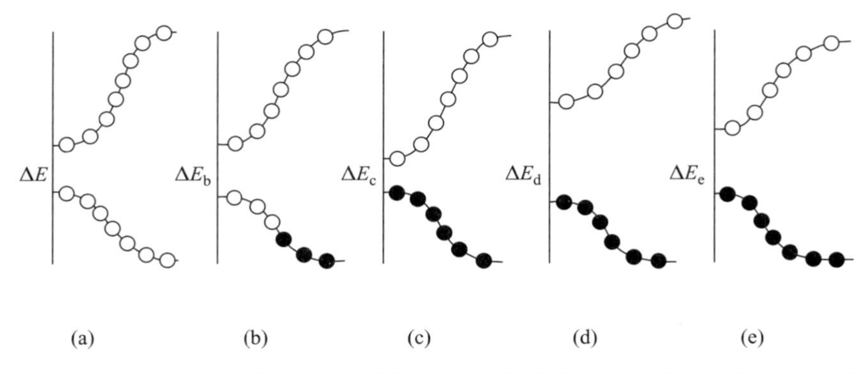

Fig. 9.13 A graphical illustration of how the 1-D, finite array of periodic potentials accounts for the division of crystal structures into conductors, insulators, and intrinsic semiconductors. (a) Adjacent energy bands separated by a gap ΔE. Since the array is finite, the infinite-array continuous band is replaced by a finite set of energy levels, indicated by the open circles. (b) A conductor, in which the lower-lying levels are partly occupied (solid circles) and the higher-lying ones are empty. A weak electric field can move electrons from a filled to an empty level, both in the lower band, thereby causing a current to flow. (c) A conductor in which the lower band is filled and the higher band is empty, with a gap ΔE_c small enough that a weak field can excite electrons to the upper (conduction) band; a current results in this case also. (d) An insulator, with filled and empty bands as in (c), but with a gap ΔE_d so large that typical electric fields cannot supply enough energy to excite electrons into the upper band. (e) An intrinsic semiconductor, with filled and empty bands as in (c) and (d), but with an energy gap ΔE_e small enough that strong electric fields can excite electrons into the unfilled band, thereby causing a current to flow.

The preceding picture is strictly valid only at absolute temperature $T = 0$. For $0 < T < 300\text{K}$, some of the electrons in an insulator can be thermally excited into the conduction band (out of the filled "valence band") but in general there are too few to provide a measurable current when an electric field is applied. Conduction *can* occur at $T \sim 300\text{K}$ (room temperature) for that class of materials known as (intrinsic) *semiconductors*, for which the energy gap is small enough and the number of thermally excited electrons large enough (Fig. 9.13(e)).

We stress again that the basis for this behavior is the energy-band/energy-gap spectrum characteristic of periodic potentials. The 1-D delta-function array provides a soluble, qualitative model that displays the relevant behavior (conductor, insulator, semiconductor), seen now to be a purely quantal phenomenon. A quantitative description requires a much more detailed and complicated analysis in which electron identity, three-dimensionality, density-of-states factors, etc., are taken into account. These matters are discussed in condensed-matter texts.

9.7. Rotations

The classical-physics rotation operator $\hat{R}(\mathbf{n}, \theta)$ – analogous to $\hat{\mathcal{P}}$ and $\hat{\mathcal{T}}$ – was introduced in Chapter 4 in the context of 3-D vectors. There it was shown that, while a 3-D vector $\mathbf{b}$

is transformed into a new vector $\mathbf{b}'$ via rotation, its length – the scalar product of $\mathbf{b}$ with itself – remains invariant, Eq. (4.7). Indeed, the scalar product of any two 3-D vectors is invariant under rotations, a point elaborated on shortly. The possibility of invariance under rotations arises in microscopic phenomena as well: there are many quantal systems that respect this symmetry. As with reflections and translations, our treatment of rotations begins with consideration of the classical case, then moves on to quantum mechanics, in particular the unitary quantal rotation operator $\hat{U}_{\mathrm{R}}(\mathbf{n}, \theta)$, and concludes with a quantal application.

Rotations: geometry and classical physics

The discussion in this subsection will set the stage for the subsequent treatment of the quantal case, providing both important input information and results whose general structure are common to both cases. We begin by using the elementary definition of the scalar product to establish that the scalar product of any two 3-D vectors is invariant under rotations. If $\mathbf{a}$ and $\mathbf{b}$ are the vectors and $\theta_{\mathbf{ab}}$ is the angle between them, then their scalar product is

$$\mathbf{a} \cdot \mathbf{b} = ab \cos \theta_{\mathbf{ab}}. \tag{9.136}$$

However, under a rotation each vector is rotated through the same angle, thus preserving the value of $\theta_{\mathbf{ab}}$ (Fig. 9.14) and therefore of $\mathbf{a} \cdot \mathbf{b}$.

The invariance of the scalar product places restrictions on $\hat{R}(\mathbf{n}, \theta)$. In the column-vector/matrix formulation, the scalar product is written $\mathbf{a}^{\mathrm{T}} \cdot \mathbf{b}$, which becomes $\mathbf{a}'^{\mathrm{T}} \cdot \mathbf{b}'$ after rotation, with

$$\mathbf{a}'^{\mathrm{T}} \cdot \mathbf{b}' = \mathbf{a}^{\mathrm{T}} \cdot \mathbf{b}. \tag{9.137}$$

For a rotation about $\mathbf{n}$ through the angle θ, the new and old vectors are related via

$$\mathbf{b}' = \hat{R}(\mathbf{n}, \theta)\mathbf{b} \tag{9.138a}$$

and

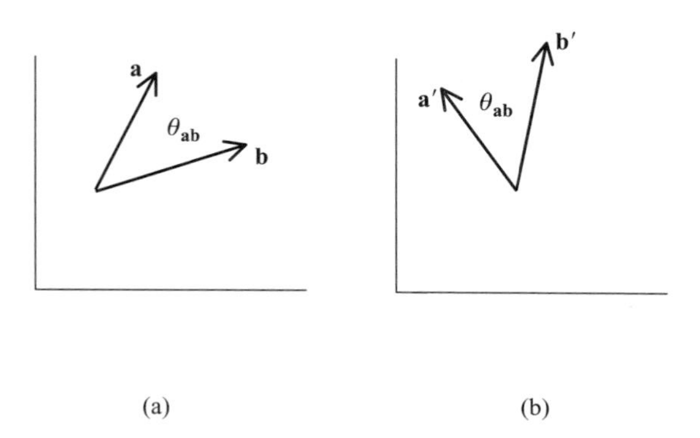

Fig. 9.14 Under a rotation about an axis perpendicular to the plane of the paper, $\mathbf{a}$ and $\mathbf{b}$ of (a) go over to $\mathbf{a}'$ and $\mathbf{b}'$ of (b), but the angle $\theta_{\mathbf{ab}}$ between them is unchanged.

$$\mathbf{a}'^{\mathrm{T}} = \mathbf{a}^{\mathrm{T}}\hat{R}^{\mathrm{T}}(\mathbf{n},\,\theta), \tag{9.138b}$$

where the transposed version of the relation $\mathbf{a}' = \hat{R}\mathbf{a}$ necessarily inverts the order of the operator and the vector, as in (9.138b). The latter point is crucial, because it transforms Eq. (9.137) into

$$\mathbf{a}^{\mathrm{T}}\hat{R}^{\mathrm{T}}(\mathbf{n},\,\theta)\cdot\hat{R}(\mathbf{n},\,\theta)\mathbf{b} = \mathbf{a}^{\mathrm{T}}\cdot\mathbf{b}, \tag{9.139}$$

a relation that can be satisfied only if

$$\hat{R}^{\mathrm{T}}(\mathbf{n},\,\theta)\hat{R}(\mathbf{n},\,\theta) = \hat{I}. \tag{9.140}$$

Equation (9.140) is the operator version of a standard result of classical mechanics: the inverse of the rotation matrix is its transpose, thereby establishing that $R(\mathbf{n},\,\theta)$ is orthogonal. However, because $\hat{R}(\mathbf{n},\,\theta)$ acts on and produces real 3-D vectors,

$$\hat{R}^{*}(\mathbf{n},\,\theta) = R(\mathbf{n},\,\theta) \tag{9.141}$$

and (9.140) can consequently be re-written as

$$\hat{R}^{\dagger}(\mathbf{n},\,\theta)\hat{R}(\mathbf{n},\,\theta) = \hat{I}. \tag{9.142}$$

This is the classical analog of the corresponding quantal result, viz., the rotation operator is unitary:

$$\hat{R}^{\dagger}(\mathbf{n},\,\theta) = \hat{R}^{-1}(\mathbf{n},\,\theta). \tag{9.143}$$

The inverse of a rotation is an operation that restores the original vector. This is uniquely achieved by rotating through the negative of the original angle; hence,

$$\hat{R}^{\dagger}(\mathbf{n},\,\theta) = \hat{R}^{-1}(\mathbf{n},\,\theta) \equiv \hat{R}(\mathbf{n},\,-\theta). \tag{9.144}$$

This is another result that carries over to the quantal case. Equation (9.144) is readily verified using the matrix version of $\hat{R}(\mathbf{n},\,\theta)$ and choosing $\mathbf{n}$ to be one of the unit coordinate vectors, for instance, $\mathbf{n} = \mathbf{e}_3$, in which case

$$R(\mathbf{e}_3,\,\theta) = \begin{pmatrix} \cos\theta & -\sin\theta & 0 \\ \sin\theta & \cos\theta & 0 \\ 0 & 0 & 1 \end{pmatrix}, \tag{9.145a}$$

$$R(\mathbf{e}_3,\,-\theta) = \begin{pmatrix} \cos\theta & \sin\theta & 0 \\ -\sin\theta & \cos\theta & 0 \\ 0 & 0 & 1 \end{pmatrix}, \tag{9.145b}$$

from which one readily finds that

$$R(\mathbf{e}_3,\,\pm\theta)R(\mathbf{e}_3,\,\mp\theta) = I.$$

Another feature of the classical case that carries over to the quantal one is that, if two different directions of rotation are chosen, then the result of carrying out the two rotations depends on their order: rotations generally do not commute with one another. This is illustrated in Fig. 9.15 and expressed mathematically by

$$\hat{R}(\mathbf{n}_1,\,\theta_1)\hat{R}(\mathbf{n}_2,\,\theta_2) - \hat{R}(\mathbf{n}_2,\,\theta_2)\hat{R}(\mathbf{n}_1,\,\theta_1) \neq 0. \tag{9.146}$$

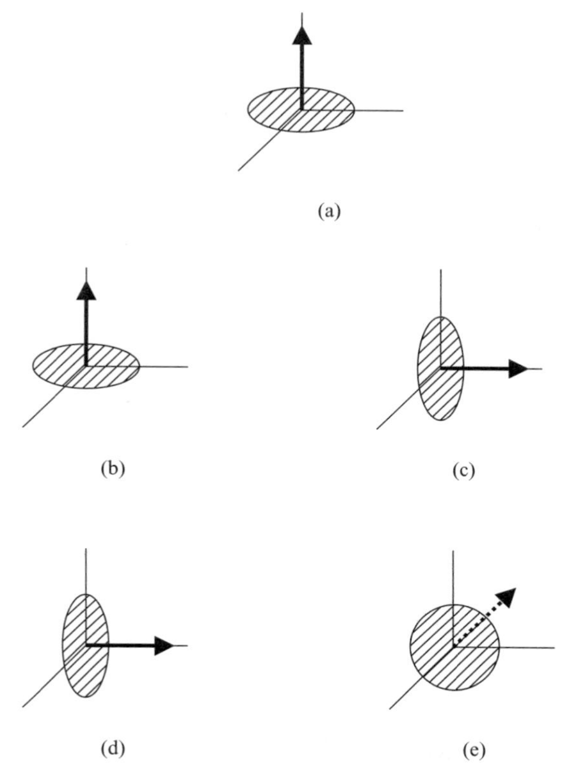

Fig. 9.15 The non-commutativity of rotations. (a) A disk in the x–y plane spinning about $\mathbf{e}_3$. (b)The effect of a rotation of 90° about $\mathbf{e}_3$ on the disk of (a). (c) The effect of a rotation of 90° about $\mathbf{e}_1$ on the disk of (b). (d) The effect of a rotation of 90° about $\mathbf{e}_1$ on the disk of (a). (e) The effect of a rotation of 90° about $\mathbf{e}_3$ on the disk of (d). The order in which rotations are carried out affects the result: (a) → (b) → (c) ≠ (a) → (d) → (e).

There are two exceptions to (9.146). The first concerns infinitesimal angles: if $\theta_1 = \theta_2 = \varepsilon$, then, to first order in ε, the RHS of (9.146) is zero.[8] The second exception occurs when $\mathbf{n}_1 = \mathbf{n}_2 = \mathbf{n}$, in which case

$$\hat{R}(\mathbf{n}, \theta_1)\hat{R}(\mathbf{n}, \theta_2) = \hat{R}(\mathbf{n}, \theta_2)\hat{R}(\mathbf{n}, \theta_1) = \hat{R}(\mathbf{n}, \theta_1 + \theta_2). \tag{9.147}$$

In other words, two successive rotations about the same axis are the same as a single rotation through one angle equal to the sum of the two, and the order is unimportant. A consequence of (9.147) is that, by choosing $\theta_2 = -\theta_1$ and invoking (9.146), we get

$$\hat{R}(\mathbf{n}, \theta = 0) = \hat{I}, \tag{9.148}$$

another result that carries over to the quantal case, i.e., to the Hilbert-space rotation operator $\hat{U}_{\mathrm{R}}(\mathbf{n}, \theta)$.

Before examining rotations in quantum mechanics, we briefly consider a manifestation

[8] By the retention of ε^2 terms in the quantum case one can derive the commutation relations obeyed by the quantal generators of rotations. We shall follow such an approach in Chapter 14.

of rotational invariance in classical dynamics. Let $V(\mathbf{r})$ be the potential (energy) of a single particle. Rotational invariance means that $V(\mathbf{r}) = V(r)$. A major consequence of this spherical symmetry is conservation of angular momentum:

$$\frac{d}{dt}\mathbf{L} = \frac{d}{dt}(\mathbf{r} \times \mathbf{p}) = \mathbf{r} \times \dot{\mathbf{p}} = -\mathbf{r} \times \nabla V = 0,$$

since $\mathbf{r} = r\mathbf{e}_r$ and $\nabla V(r) = (\partial V/\partial r)\mathbf{e}_r$. A further consequence, due directly to $\mathbf{L} =$ constant, is that the trajectory is in a plane perpendicular to $\mathbf{L}$. These features are hallmarks of rotationally invariant potentials. For such cases, rotation of the system about $\mathbf{L}$ will change the initial point of the orbit but not its form.

Rotations: quantum physics

The subject of rotations and rotational invariance in quantum mechanics is far more complicated than the corresponding studies of parity and translations, and in this chapter we deal only with the simplest case, that of a spinless particle. Our main goal is to determine the form of the rotation operator as well as some of its ramifications. The form we derive is that of a unitary operator expressed as an exponential (recall Stone's theorem of Chapter 4); the infinitesimal generator obtained here for spinless particles will be generalized in Chapter 14.

Because the quantal rotation operator $\hat{U}_{\mathrm{R}}(\mathbf{n}, \theta)$ may affect quantum-number labels (see below), we again consider a single-particle system described in coordinate space by a "scalar" function, i.e., a system whose coordinate-space wave function can be written $\psi(\mathbf{r})$. Recall that no quantum numbers are needed to label the state. The results we derive will be applicable to any spinless particle (or system thereof), in particular those described by a labeled Hilbert-space state. As with the translation operator, the analysis will be carried out for $\psi(\mathbf{r}) = \psi(x, y, z)$ and then for its abstract ket, denoted $|\psi\rangle$.

Under a rotation through θ about $\mathbf{n}$

$$\mathbf{r} \to \mathbf{r}' = \hat{R}(\mathbf{n}, \theta)\mathbf{r} \tag{9.149a}$$

and

$$\psi(\mathbf{r}) \to \psi'(\mathbf{r}') = \psi'(x', y', z'), \tag{9.149b}$$

where ψ' is the rotated function. Because ψ is a *scalar*, the rotated scalar function evaluated at the rotated point must be the *same* as the unrotated scalar function evaluated at the original point:

$$\psi'(\mathbf{r}') = \psi(\mathbf{r}). \tag{9.150}$$

This may be more easily grasped in the framework of a passive rotation: the *value* of $\psi(\mathbf{r})$ cannot change under rotation of the coordinate system, even though our point of view(ing) is changed.

To proceed further, we re-express $\mathbf{r}$, appearing in $\psi(\mathbf{r})$, in terms of $\mathbf{r}'$, so that the same coordinate vector appears in ψ' and ψ. The inverse of (9.149a) is

$$\mathbf{r} = \hat{R}(\mathbf{n}, -\theta)\mathbf{r}', \tag{9.151}$$

by which Eq. (9.150) becomes

$$\psi'(\mathbf{r}') = \psi(\hat{R}(\mathbf{n}, -\theta)\mathbf{r}').$$

Dropping the prime on $\mathbf{r}'$, the latter relation reads

$$\psi'(\mathbf{r}) = \psi(\hat{R}(\mathbf{n}, -\theta)\mathbf{r}); \tag{9.152}$$

i.e., the rotated function evaluated at $\mathbf{r}$ is the same as the unrotated function evaluated at the point $\hat{R}(\mathbf{n}, -\theta)\mathbf{r}$.

The rotated function evaluated at $\mathbf{r}$ must also be the result of applying $\hat{U}_{\mathrm{R}}(\mathbf{n}, \theta)$ to $\psi(\mathbf{r})$:

$$\begin{aligned} \hat{U}_{\mathrm{R}}(\mathbf{n}, \theta)\psi(\mathbf{r}) &= \psi'(\mathbf{r}) \\ &= \psi(\hat{R}(\mathbf{n}, -\theta)\mathbf{r}), \end{aligned} \tag{9.153}$$

where the second line provides an operational definition of $\hat{U}_{\mathrm{R}}(\mathbf{n}, \theta)$, one that depends on $\hat{R}(\mathbf{n}, -\theta)$.

An expression for $\hat{U}_{\mathrm{R}}(\mathbf{n}, \theta)$ is most easily obtained by choosing $\mathbf{n}$ along one of the unit vectors $\mathbf{e}_j$, $j = 1, 2,$ or 3, and then letting θ be a first-order infinitesimal (recall Eqs. (4.174) *et seq.*). Choosing $\mathbf{n} = \mathbf{e}_3$, and $\theta \to \epsilon$, we find from Eq. (9.145b) that

$$\hat{R}(\mathbf{e}_3, -\epsilon)\mathbf{r} \to \begin{pmatrix} x + \epsilon y \\ y - \epsilon x \\ z \end{pmatrix}, \tag{9.154}$$

and hence that

$$\hat{U}_{\mathrm{R}}(\mathbf{e}_3, \epsilon)\psi(\mathbf{r}) \cong \psi(x + \epsilon y, y - \epsilon x, z). \tag{9.155}$$

The arrow in (9.154) and the "$\cong$" sign in (9.155) indicate that only first-order terms in ϵ have been kept.

Equation (9.155) is simplified by introducing a Taylor-series expansion for its RHS and keeping only first-order terms:

$$\begin{aligned} \psi(x + \epsilon y, y - \epsilon x, z) &\cong \psi(\mathbf{r}) + \epsilon\left(y\frac{\partial}{\partial x} - x\frac{\partial}{\partial y}\right)\psi(\mathbf{r}) \\ &= \left[1 + \epsilon\left(y\frac{\partial}{\partial x} - x\frac{\partial}{\partial y}\right)\right]\psi(\mathbf{r}). \end{aligned} \tag{9.156}$$

Comparing Eqs. (9.155) and (9.156), we deduce that

$$\hat{U}_{\mathrm{R}}(\mathbf{e}_3, \epsilon) \cong \hat{I} + \epsilon\left(y\frac{\partial}{\partial x} - x\frac{\partial}{\partial y}\right). \tag{9.157}$$

Equation (9.157) is almost the desired result for $\hat{U}_{\mathrm{R}}(\mathbf{e}_3, \epsilon)$. It will become so once we identify the operator in the large parentheses as being proportional to $\hat{L}_z$. By direct evaluation of Eq. (5.46), the z-component of the orbital angular-momentum operator $\hat{\mathbf{L}}(\mathbf{r})$ is

$$\hat{L}_z(\mathbf{r}) = -i\hbar\left(x\frac{\partial}{\partial y} - y\frac{\partial}{\partial x}\right). \tag{9.158}$$

$\hat{U}_{\mathrm{R}}(\mathbf{e}_3, \epsilon)$ is therefore related to $\hat{L}_z$:

$$\hat{U}_{\rm R}(\mathbf{e}_3, \epsilon) \cong \hat{I} - i\epsilon\hat{L}_z/\hbar$$
$$= \hat{I} - i\epsilon\mathbf{e}_3 \cdot \hat{\mathbf{L}}/\hbar. \tag{9.159}$$

For a spinless particle, the infinitesimal generator of rotations about the z-axis is the z-component of orbital angular momentum. It is left as an exercise to show that the choices $\mathbf{n} = \mathbf{e}_1$ and $\mathbf{n} = \mathbf{e}_2$ lead to

$$\hat{U}_{\rm R}(\mathbf{e}_j, \epsilon) \cong \hat{I} - i\epsilon\mathbf{e}_j \cdot \hat{\mathbf{L}}/\hbar, \qquad j = 1, 2. \tag{9.160}$$

The generalization from $\mathbf{n}$ being one of the coordinate axes to its being an arbitrary direction is

$$\hat{U}_{\rm R}(\mathbf{n}, \epsilon) \cong \hat{I} - i\epsilon\mathbf{n} \cdot \hat{\mathbf{L}}/\hbar. \tag{9.161}$$

In other words, $\mathbf{n} \cdot \hat{\mathbf{L}}$ is the infinitesimal generator in this case. Finally, employing Stone's theorem for the case of finite rotations, we get

$$\hat{U}_{\rm R}(\mathbf{n}, \theta) = e^{-i\theta\mathbf{n}\cdot\hat{\mathbf{L}}/\hbar}, \tag{9.162}$$

the exponential, unitary expression for the quantal rotation operator in the case of spinless particles.

In the preceding derivation of this important result, Hilbert-space kets have played no role. We now rederive Eq. (9.153), from which (9.162) follows, starting with the kets $|\mathbf{r}\rangle$ and $|\psi\rangle(\psi(\mathbf{r}) = \langle\mathbf{r}|\psi\rangle)$. Like its sibling in Section 9.6, this analysis should lay to rest any lingering concerns some readers may have concerning Eq. (9.153).

As with the parity and translation operators and Eq. (9.61), a fundamental equation relates $\hat{U}_{\rm R}(\mathbf{n}, \theta)$ and $\hat{R}(\mathbf{n}, \theta)$:

$$\hat{U}_{\rm R}(\mathbf{n}, \theta)|\mathbf{r}\rangle = |\hat{R}(\mathbf{n}, \theta)\mathbf{r}\rangle, \tag{9.163a}$$

where an arbitrary phase has again been set equal to zero. Note that (9.163a) also applies to the momentum ket $|\mathbf{k}\rangle$:

$$\hat{U}_{\rm R}(\mathbf{n}, \theta)|\mathbf{k}\rangle = |\hat{R}(\mathbf{n}, \theta)\mathbf{k}\rangle. \tag{9.163b}$$

The rotated ket $|\psi'\rangle$ is defined by

$$|\psi'\rangle = \hat{U}_R(\mathbf{n}, \theta)|\psi\rangle, \tag{9.164}$$

with

$$\psi'(\mathbf{r}) = \langle\mathbf{r}|\psi\rangle'$$
$$= \langle\mathbf{r}|\hat{U}_{\rm R}(\mathbf{n}, \theta)|\psi\rangle; \tag{9.165}$$

Eq. (9.165) is the other fundamental relation.

The derivation proceeds by evaluating the RHS of (9.165) in two different ways. In the first, $\hat{U}_{\rm R}(\mathbf{n}, \theta)$ is taken to act to the right and to be local (an assumption justified *a posteriori* by consistency), in which case (9.165) reads

$$\psi'(\mathbf{r}) = \hat{U}_{\rm R}(\mathbf{n}, \theta)\psi(\mathbf{r}), \tag{9.166}$$

where the $\mathbf{r}$-dependence of $\hat{U}_R$ is suppressed. Equation (9.166) is the same as the LHS of

(9.153), thereby justifying the latter equation, should a need for justification have been experienced.

In the second evaluation of $\langle \mathbf{r}|\hat{U}_{\mathrm{R}}|\psi\rangle$, we want $\hat{U}_{\mathrm{R}}(\mathbf{n},\,\theta)$ to act to the left, on $\langle \mathbf{r}|$:

$$\langle \mathbf{r}|\overset{\hat{U}}{\leftarrow}_{\mathrm{R}}(\mathbf{n},\,\theta)|\psi\rangle = \langle \hat{U}^{\dagger}_{\mathrm{R}}(n,\,\theta)\mathbf{r}|\psi\rangle, \tag{9.167}$$

(using the not wholly satisfactory condensed notation). The bra $\langle \hat{U}^{\dagger}_{\mathrm{R}}(\mathbf{n},\,\theta)\mathbf{r}|$ is the same as $\langle \mathbf{r}\overset{\leftarrow}{\hat{U}}{}^{\dagger}_{\mathrm{R}}(\mathbf{n},\,\theta)|$, which is obtained from (9.167). Since $\hat{U}_{\mathrm{R}}(\mathbf{n},\,\theta)$ is the quantal analog of $\hat{R}(\mathbf{n},\,\theta)$, and $\hat{R}^{\dagger}(\mathbf{n},\,\theta) = \hat{R}(\mathbf{n},\,-\theta)$, we require that, under Hermitian conjugation, Eq. (9.163a) becomes

$$\begin{aligned}\langle \mathbf{r}\hat{U}^{\dagger}_{\mathrm{R}}(\mathbf{n},\,\theta)| &= \langle \hat{R}^{\dagger}(\mathbf{n},\,\theta)\mathbf{r}| \\ &= \langle \hat{R}(n,\,-\theta)\mathbf{r}|.\end{aligned} \tag{9.168}$$

Equation (9.168) is a key relation. It transforms (9.167) into

$$\langle \mathbf{r}|\hat{U}_{\mathrm{R}}(\mathbf{n},\,\theta)|\psi\rangle = \psi(\hat{R}(\mathbf{n},\,-\theta)\mathbf{r}), \tag{9.169}$$

which, together with (9.166), means that

$$\begin{aligned}\hat{U}_{\mathrm{R}}(\mathbf{n},\,\theta)\psi(\mathbf{r}) &= \psi'(\mathbf{r}) \\ &= \psi(\hat{R}(\mathbf{n},\,\theta)\mathbf{r}).\end{aligned} \tag{9.153}$$

We have therefore justified our claim of being able to derive Eq. (9.153) using a Hilbert-space analysis involving kets.

In the preceding subsection, it was claimed that $\hat{U}_{\mathrm{R}}(\mathbf{n},\,\theta)$ satisfied the same properties as did $\hat{R}(\mathbf{n},\,\theta)$. The structure of Eq. (9.162) allows easy verification of most of this claim:

$$\hat{U}_{\mathrm{R}}(\mathbf{n},\,\theta = 0) = \hat{I}, \tag{9.170a}$$

$$\hat{U}^{\dagger}_{\mathrm{R}}(\mathbf{n},\,\theta) = \hat{U}_{\mathrm{R}}(\mathbf{n},\,-\theta) = \hat{U}^{-1}_{\mathrm{R}}(\mathbf{n},\,\theta), \tag{9.170b}$$

and

$$\hat{U}_{\mathrm{R}}(\mathbf{n},\,\theta_1)\hat{U}_{\mathrm{R}}(\mathbf{n},\,\theta_2) = \hat{U}_{\mathrm{R}}(\mathbf{n},\,\theta_1+\theta_2). \tag{9.170c}$$

Given these relations, the analog of Eq. (9.146) can confidently be expected to hold also. The proof, however, is not quite so straightforward, since it involves the commutation relations between $\mathbf{n_1}\cdot\hat{\mathbf{L}}$ and $\mathbf{n_2}\cdot\hat{\mathbf{L}}$. In general, any two different components of $\hat{\mathbf{L}}$ do not commute; this fact will yield the expected result, viz.,

$$\left[\hat{U}_{\mathrm{R}}(\mathbf{n}_1,\,\theta_1),\,\hat{U}_{\mathrm{R}}(\mathbf{n}_2,\,\theta_2)\right] \neq 0. \tag{9.170d}$$

To show that $[\mathbf{n_1}\cdot\hat{\mathbf{L}},\,\mathbf{n_2}\cdot\hat{\mathbf{L}}] \neq 0$, it suffices to consider the x, y, and z components. The jth component of $\hat{\mathbf{L}} = \hat{\mathbf{Q}}\times\hat{\mathbf{P}}$ is

$$\hat{L}_j = -i\hbar(\hat{Q}_k\hat{P}_\ell - \hat{Q}_\ell\hat{P}_k), \tag{9.171}$$

where j, k, and ℓ take on the values 1, 2, and 3 in cyclic order (for $j = 3 \Rightarrow z$, (9.171) is the same as (9.158)). From (9.171), and retaining the cyclic order of j, k, and ℓ, one finds

$$[\hat{L}_j,\,\hat{L}_k] = i\hbar\hat{L}_\ell, \qquad \text{cyclic ordering of } j,\,k,\,\ell, \tag{9.172}$$

which is the commutation relation for orbital angular momentum.

Equation (9.172) is one of the reasons why rotations and rotational invariance are so much more complicated to deal with than are parity and invariance under reflection. In place of the single parity operator $\hat{U}_\mathrm{P}$ or the three commuting generators of $\hat{U}_\mathrm{T}(\mathbf{a})$, we have three non-commuting operators ($\hat{L}_x$, $\hat{L}_y$, and $\hat{L}_z$) that are used to construct $\hat{U}_\mathrm{R}$:

$$\hat{U}_\mathrm{R}(\mathbf{n},\,\theta) = e^{-i\theta(n_1\hat{L}_1 + n_2\hat{L}_2 + n_3\hat{L}_3)/\hbar}, \tag{9.173}$$

where $n_j = \mathbf{e}_j \cdot \mathbf{n}$. Hence, even if $\hat{H}$ is invariant under rotations, implying that

$$[\hat{L}_j,\,\hat{H}] = 0, \qquad j = 1,\,2,\,3, \tag{9.174}$$

eigenstates of $\hat{H}$ cannot also be simultaneous eigenstates of the three operators $\hat{L}_1$, $\hat{L}_2$, and $\hat{L}_3$ due to the fact that $[\hat{L}_j,\,\hat{L}_k] \neq 0$. Instead, eigenstates of $\hat{H}$ can be eigenstates of at most one of these three components. However, as will be shown in the next chapter, each component of $\hat{L}$ *does* commute with $\hat{L}^2 = \hat{L}_x^2 + \hat{L}_y^2 + \hat{L}_z^2$, and therefore the eigenstates of a rotationally invariant $\hat{H}$ can be eigenstates both of $\hat{L}^2$ and of one component of $\hat{\mathbf{L}}$, typically chosen to be $\hat{L}_z$. However, since an eigenstate of $\hat{L}_z$ is not also an eigenstate of $\hat{L}_x$ or $\hat{L}_y$, the application of $\hat{U}_\mathrm{R}$ in the form of (9.173) to an eigenstate of $\hat{H}$ will not preserve its $\hat{L}_z$ character. This possible failure to preserve the $\hat{L}_z$ character of a rotationally invariant eigenstate of $\hat{H}$ is what was meant by the first sentence of the paragraph above Eq. (9.149a), and was thus the reason why the preceding analysis was formulated in terms of a scalar (quantum-numberless) wave function and its Hilbert-space ket.

Orbital angular-momentum eigenstates are denoted $|\ell m_\ell\rangle$, where the quantum numbers ℓ and m_ℓ are related to the eigenvalues of $\hat{L}^2$ and $\hat{L}_z$, respectively. For rotationally invariant single-particle systems, $\hat{H}$, $\hat{L}^2$, and $\hat{L}_z$ are the CSCOs and the (bound) eigenstates of $\hat{H}$ are standardly labeled $|n\ell m_\ell\rangle$, with n being the so-called principal quantum number, while the energy eigenvalues typically depend on n and ℓ but not on m_ℓ: $E \to E_{n\ell}$.[9] Determination of the $|\ell m_\ell\rangle$, of $|n\ell m_\ell\rangle$ for particular $V(r)$'s, of their coordinate-space forms, of their behavior under rotations, etc, are topics we will address in Chapters 10 and 11.

Detailed knowledge of the $\hat{L}^2$ and $\hat{L}_z$ eigenstates is unnecessary for considering some of the effects of $\hat{U}_\mathrm{R}(\mathbf{n},\,\theta)$ in general. For example, the unitarity of $\hat{U}_\mathrm{R}$ means that, just like the classical case, quantal scalar products are invariant under rotations. Although this is true for any system, we consider only spinless particles. Following the previous notation, let

$$|\alpha\rangle' = \hat{U}_\mathrm{R}(\mathbf{n},\,\theta)|\alpha\rangle \tag{9.175a}$$

and

$$'\langle\beta| = \langle\beta|\hat{U}_\mathrm{R}^\dagger(\mathbf{n},\,\theta), \tag{9.175b}$$

from which we immediately obtain the aforementioned invariance:

$$'\langle\beta|\alpha\rangle' = \langle\beta|\alpha\rangle. \tag{9.176}$$

[9] One exception to E depending on two of the three quantum numbers occurs when $V(r) \propto 1/r$, e.g., the H-atom, for which $E = E_n = -\kappa_\mathrm{e} e^2/(2a_\mathrm{H} n^2)$, the same as in the Bohr model. $E \neq E_{n\ell}$ in this case means a high degree of degeneracy – see Chapter 11.

Under a rotation, a quantal operator $\hat{A}$ becomes

$$\hat{A}' = \hat{U}_{\mathrm{R}}(\mathbf{n}, \theta)\hat{A}\hat{U}^{\dagger}_{\mathrm{R}}(\mathbf{n}, \theta). \tag{9.177}$$

Some operators will have well-defined properties under rotations, others will not. A special case is the so-called "scalar" operator – the operator analog of the scalar function $\psi(\mathbf{r})$ – for which $\hat{A}' = \hat{A}$:

$$\hat{A}' = \hat{U}_{\mathrm{R}}(\mathbf{n}, \theta)\hat{A}\hat{U}^{\dagger}_{\mathrm{R}}(\mathbf{n}, \theta) = \hat{A}, \qquad \hat{A} = \text{ a scalar operator.} \tag{9.178}$$

Rotationally invariant Hamiltonians are examples of scalar operators.

The effects of rotations on matrix elements can be dealt with in one of two ways. We may either rotate the operator and keep the states the same, or vice versa. These two ways can lead to different answers. Suppose that the matrix element is

$$A_{\beta\alpha} = \langle\beta|\hat{A}|\alpha\rangle. \tag{9.179}$$

The matrix element of the rotated operator evaluated between unrotated states is

$$A'_{\beta\alpha} = \langle\beta|\hat{U}_{\mathrm{R}}(\mathbf{n}, \theta)\hat{A}\hat{U}^{\dagger}_{\mathrm{R}}(\mathbf{n}, \theta)|\alpha\rangle, \tag{9.180a}$$

from Eq. (9.178). Keeping $\hat{A}$ the same but rotating the states, as in Eqs. (9.175), gives $A_{\beta'\alpha'}$:

$$\begin{aligned} A_{\beta'\alpha'} &= {}^{\backprime}\langle\beta|\hat{A}|\alpha\rangle' \\ &= \langle\beta|\hat{U}^{\dagger}_{\mathrm{R}}(\mathbf{n}, \theta)\hat{A}\hat{U}_{\mathrm{R}}(\mathbf{n}, \theta)|\alpha\rangle. \end{aligned} \tag{9.180b}$$

The matrix elements $A'_{\beta\alpha}$ and $A_{\beta'\alpha'}$ obviously need not be equal; whether one uses $A'_{\beta\alpha}$ or $A_{\beta'\alpha'}$ to define the effect of rotation on the matrix element $A_{\beta\alpha}$ depends on the circumstance. Note that simultaneous rotation of the states and the operator does not alter $A_{\beta\alpha}$. Exploration of these ideas is left to the exercises.

The two-dimensional rotor

While solution of the full $\hat{L}^2$, $\hat{L}_z$ eigenvalue problem is postponed to the next chapter, we examine here a simple case involving only $\hat{L}_z$: the quantal analog of a mass m_o moving in a circle of constant radius a.

For a rotating system in classical mechanics, the rotational contribution to the kinetic energy, K_{Rot}, is

$$K_{\mathrm{Rot}} = \frac{L^2}{2\mathcal{I}},$$

where $\mathbf{L}$ is the angular momentum and $\mathcal{I}$ is the moment of inertia. We assume the motion to be in the x–y plane, so that $\mathbf{L} \to L_z$; for this system, $\mathcal{I} = m_0 a^2$ and K_{Rot} becomes

$$K_{\mathrm{Rot}} = \frac{L_z^2}{2m_0 a^2}.$$

Since there is no potential energy in this problem, K_{Rot} is the classical Hamiltonian H:

$$H = K_{\text{Rot}} = \frac{L_z^2}{2m_0a^2}. \tag{9.181}$$

Although the motion takes place on a 2-D plane surface, there is only one degree of freedom: the azimuthal angle ϕ, with the position ρ on the circle being given by $\rho = a\phi$, as in Fig. 9.16. Indeed, $L_z = \mathcal{I}\dot{\phi}$ and K_{Rot} becomes $ma^2\dot{\phi}^2/2$. Quantally, the 2-D rotor will also be a system with only one degree of freedom, as we show next.

Quantization of the system means replacing L_z in (9.181) by $\hat{L}_z$:

$$\hat{H} = \frac{1}{2m_0a^2}\hat{L}_z^2. \tag{9.182}$$

In rectangular coordinates, $\hat{L}_z$ is given by Eq. (9.158), whose x, y-dependence implies two degrees of freedom rather than a single one. However, the azimuthal (rotational) symmetry of the system plus the fact that classically only $\dot{\phi}$ enters H suggests re-expressing $\hat{L}_z$ in spherical polar rather than rectangular coordinates, in the hope that $\hat{L}_z$ will then depend only on ϕ. Either from the results of Exercise 5.25 or by a straightforward calculation of $\hat{L}_z$ in terms of spherical polar coordinates via $x = r\sin\theta\cos\phi$, $y = r\sin\theta\sin\phi$, and $z = r\cos\theta$, Eq. (9.158) becomes

$$\hat{L}_z(\mathbf{r}) \to \hat{L}_z(\phi) = -i\hbar\frac{\partial}{\partial\phi} \text{ (!)}. \tag{9.183}$$

In spherical polar coordinates, $\hat{L}_z$ contains only an angular derivative; similar remarks pertain to $\hat{L}_x$ and $\hat{L}_y$, as shown in the next chapter. This result should be no surprise, since the dimension of $\hbar$ is energy–time, the same as that for angular momentum, while $x\,\partial/\partial y$ and $y\,\partial/\partial x$ are each dimensionless.

Equation (9.183) leads to

$$\hat{H} \to \hat{H}(\phi) = -\frac{\hbar^2}{2m_0a^2}\frac{\partial^2}{\partial\phi^2}. \tag{9.184}$$

Use of (9.184) in

$$\hat{H}(\phi)\psi(\phi) = E\psi(\phi) \tag{9.185}$$

leads to

$$\psi(\phi) = Ne^{i\mu\phi}, \tag{9.186}$$

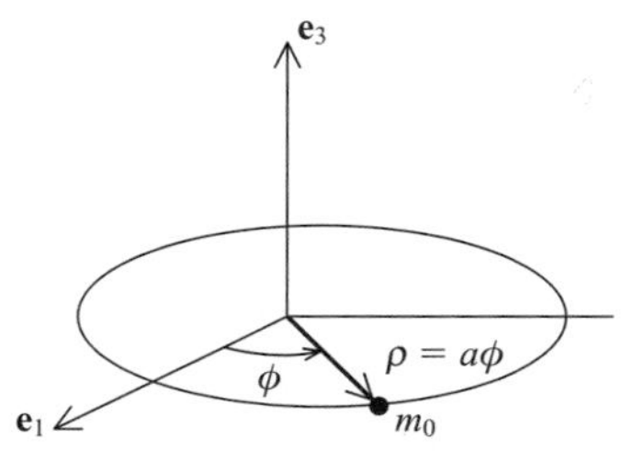

Fig. 9.16 A mass m_0 moving in a circle of radius $\rho = a =$ constant located in the x–y plane. Only the z-component of angular momentum is non-zero.

with

$$E = \frac{\hbar^2 \mu^2}{2m_0 a^2}. \tag{9.187}$$

The quantum number μ is obtained from the boundary condition

$$\psi(\phi = 0) = \psi(\phi = 2\pi), \tag{9.188}$$

i.e., $\psi(\phi)$ must be single-valued (periodic in ϕ with period 2π). The only way (9.186) can satisfy (9.188) is if μ is an integer or zero:

$$\mu \to m = 0, \pm 1, \pm 2, \pm 3, \ldots.$$

Choosing N real, the solution to the eigenvalue problem (9.185) is therefore

$$\psi(\phi) \to \psi_m(\phi) = \sqrt{\frac{1}{2\pi}} e^{im\phi}, \qquad m = 0, \pm 1, \ldots \tag{9.189a}$$

and

$$E \to E_m = m^2 \frac{\hbar^2}{2m_0 a^2}, \qquad m = 0, \pm 1, \pm 2, \ldots. \tag{9.189b}$$

Except for $m = 0$, every energy level is doubly degenerate. A Grotrian diagram for the 2-D rotor is shown in Fig. 9.17. The spacing ΔE_m between levels is

$$\Delta E_m = E_{m+1} - E_m = (2m + 1) \frac{\hbar^2}{2m_0 a^2},$$

whose increase is proportional to m.

These results are a full solution to the 2-D rotor problem, one that displays a limited version of rotational symmetry. Two aspects of this problem are worth noting. First, since $[\hat{L}_z, \hat{H}] = 0$, the $\psi_m(\phi)$'s of (9.189a) are also eigenfunctions of $\hat{L}_z$ with eigenvalue $m\hbar$,

$$\hat{L}_z \psi_m(\phi) = m\hbar \psi_m(\phi). \tag{9.190}$$

This is another example of a symmetry-operator eigenvalue being used to label the energy eigenfunction. Equation (9.190) is valid in general, not just for the 2-D rotator. Second, if the motion is not confined to the plane, i.e., if the rotor is 3-D, then the

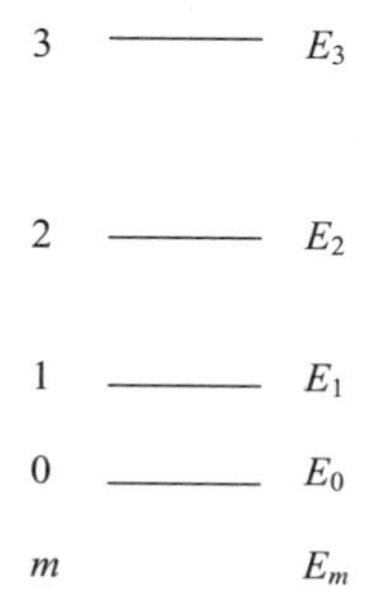

Fig. 9.17 The first four levels in a Grotrian diagram for the 2-D rotor, $E_m = m^2\hbar^2/(2m_0a^2)$.

eigenvalue problem itself changes, and neither (9.189a) nor (9.189b) is valid. We shall examine the symmetric 3-D rotor in the exercises to Chapter 11.

Exercises

9.1 A quantal system consists of a single particle of mass m in 3-D. By evaluating the relevant commutators, determine whether measurements of the following pairs of observables are compatible or incompatible:

(a) orbital angular momentum and linear momentum,

(b) orbital angular momentum and position,

(c) kinetic energy and orbital angular momentum.

9.2 Evaluate $\Delta P_x \Delta Q_x$ for a particle of mass m in the ground state of a 1-D box located between $x = 0$ and $x = L$.

9.3 Determine $\Delta P_x \Delta Q_x$ for a particle of mass m in the ground state of a 1-D linear harmonic oscillator of frequency ω_0.

9.4 The z- and x-components of the intrinsic spin of a neutron are to be measured simultaneously. These components are here represented, respectively, by the matrices

$$S_z = \frac{\hbar}{2}\begin{pmatrix} 1 & 0 \\ 0 & -1 \end{pmatrix}, \qquad S_x = \frac{\hbar}{2}\begin{pmatrix} 0 & 1 \\ 1 & 0 \end{pmatrix}.$$

If the spin state of the neutron is the column vector

$$\frac{1}{2}\begin{pmatrix} 1 \\ \sqrt{3} \end{pmatrix},$$

what is the minimum uncertainty involved in this measurement?

9.5 Show, by explicit calculation, that $\Delta P_x \Delta Q_x$ for the 1-D coherent state $\psi_\zeta(x, t)$ of Eq. (8.81) has the value $\hbar/2$, independent of ζ and t.

9.6 Prove that the quantity η occurring in Eq. (9.22) is pure imaginary, and relate $|\eta|$ to Δx.

9.7 Compare $\Delta P_x \Delta Q_x$ for the 1-D wave packets of Exercises 7.2 and 7.3.

9.8 Determine the time-dependence, if any, of $\Delta P_x \Delta Q_x$ for the Gaussian wave packet (7.36). If your result *is* time-dependent, evaluate it for the same values of t as those specified in Fig. 7.5 and comment on any connections between the spreading of the packet and the uncertainty product.

9.9 Use the position–momentum uncertainty product to estimate the ground-state energy for a particle of mass m

(a) at rest on a horizontal, massless table top of adjustable height, located a short distance above the earth's surface;

(b) acted on by the 1-D potential $V(x) = \gamma|x|$.

9.10 Double commutators.

(a) Let $\hat{A}$ be Hermitian, and $|\alpha\rangle$ and $|\beta\rangle$ be eigenstates of a single-particle Hamiltonian $\hat{H}$ with energies E_α and E_β, respectively. Prove that $\langle\alpha|[\hat{A}, [\hat{H}, \hat{A}]]/2|\alpha\rangle = \sum_\beta (E_\beta - E_\alpha)|\langle\alpha|\hat{A}|\beta\rangle|^2$, where $\sum_\beta$ implies an integral on any continuum state quantum numbers.

(b) Show that the LHS of the preceding indentity equals $\frac{1}{2}[\partial^2\langle\alpha|e^{i\lambda\hat{A}}\hat{H}e^{-i\lambda\hat{A}}|\alpha\rangle/\partial\lambda^2]_{\lambda=0}$, $\lambda=$ real.

(c) Now let $\hat{H}=\hat{K}+V(\mathbf{r})$ and set $\hat{A}=\mathbf{e}\cdot\hat{\mathbf{Q}}$, where $\mathbf{e}$ is an arbitrary unit vector. (i) Calculate $[\hat{A},[\hat{H},\hat{A}]]$. (ii) Show that $[e^{i\lambda\hat{A}}\hat{H}e^{-i\lambda\hat{A}}]=(\hat{\mathbf{P}}-\lambda\hbar\mathbf{e})^2/(2m)$, where m is the mass of the particle and λ is again real, and then use this result to re-calculate $[\hat{A},[\hat{H},\hat{A}]]$. (iii) Finally, show that the foregoing leads to $\sum_\beta(E_\beta-E_\alpha)|\langle\beta|\mathbf{e}\cdot\hat{\mathbf{Q}}|\alpha\rangle|^2=\hbar^2/(2m)$, a relation known as the Thomas–Reiche–Kuhn (TRK) sum rule. (Hint: use the fact that $\hat{\mathbf{P}}=m[\hat{H},\hat{\mathbf{Q}}]/\hbar$.)

9.11 In analogy to the Heisenberg picture, define an operator $\hat{A}(\lambda)=e^{i\lambda\hat{B}}\hat{A}e^{-i\lambda\hat{B}}$, where λ is real and $\hat{B}$ is Hermitian such that $[\hat{B},[\hat{B},\hat{A}]]=\omega^2\hat{A}$, $\omega=$ real.

(a) Show that $\hat{A}(\lambda)=\hat{A}\cos(\omega\lambda)+(i/\omega)[\hat{B},\hat{A}]\sin(\omega\lambda)$.

(b) Let $\hat{B}|b\rangle=b|b\rangle$. Prove that $\langle b'|\hat{A}|b\rangle=0$ unless $b'=b\pm\omega$.

(c) Set $\lambda=t$ (the time) and $\hat{B}=\hat{H}/\hbar$, where $\hat{H}=\hat{K}(\mathbf{r})+m\omega^2r^2/2$, the 3-D harmonic oscillator Hamiltonian for a particle of mass m. Apply the preceding results to the cases $\hat{A}=\hat{\mathbf{Q}}$ and $\hat{A}=\hat{\mathbf{P}}$ to show that, exactly as in 1-D, $\hat{\mathbf{Q}}^{(\mathrm{H})}(t)$ and $\hat{\mathbf{P}}^{(\mathrm{H})}(t)$ are periodic functions of time.

(d) Finally, let $\hat{A}=\hat{Q}^2$ and then $\hat{A}=\hat{P}^2$, and show that, with minor modifications, the preceding results carry over to this case as well, as does the condition for the vanishing of $\langle b'|\hat{A}|b\rangle$.

9.12 Assume that the set $\{|a_n\rangle\otimes|b_m\rangle\}$ (orthonormal eigenkets of $\hat{A}$ and $\hat{B}$) is complete. Is it true that this always implies that $[\hat{A},\hat{B}]=0$? Either prove that this is the case or give a counter-example.

9.13 In the orthonormal basis $\{|1\rangle,|2\rangle\}$ the operators $\hat{A}$ and $\hat{B}$ are represented by the matrices

$$A=\begin{pmatrix}a & 0\\ 0 & a\end{pmatrix},\qquad B=\begin{pmatrix}0 & b\\ b & 0\end{pmatrix},$$

where a and b are real.

(a) Determine the eigenvalues of $\hat{A}$ and $\hat{B}$, and identify any degeneracies.

(b) Find an orthonormal basis of eigenvectors common to A and B, and specify the eigenvalues for each eigenvector.

(c) Which of the six sets $\{\hat{A}\}$, $\{\hat{B}\}$, $\{\hat{A},\hat{B}\}$, $\{\hat{A}^2,\hat{B}\}$, $\{\hat{A},\hat{B}^2\}$, and $\{\hat{A}^2,\hat{B}^2\}$ form CSCO for the system described by $\hat{A}$ and $\hat{B}$, and which do not? Prove your assertions.

(d) Prove that $[\hat{A},\hat{B}]=0$.

9.14 In the orthonormal basis $\{|n\rangle\}_{n=1}^3$, the operators $\hat{A}$ and $\hat{B}$ are represented by the matrices

$$A=\begin{pmatrix}a & 0 & 0\\ 0 & -a & 0\\ 0 & 0 & -a\end{pmatrix},\ B=\begin{pmatrix}b & 0 & 0\\ 0 & 0 & -ib\\ 0 & ib & 0\end{pmatrix},$$

a and b real. Perform the same analysis as in parts (a)–(d) of Exercise 9.13 (Sakurai 1994).

9.15 Assume that the measured values of the observable $\hat{B}$ are unchanged following a symmetry operation (such as rotation or translation) whose Hilbert-space operator is $\hat{U}(\lambda)=e^{-i\lambda\hat{A}}$, $\lambda=$ real and $\hat{A}$ Hermitian. Prove that $[\hat{A},\hat{B}]=0$.

9.16 Prove the validity of each of Eq. (9.69), the triad of Eqs. (9.72), and Eq. (9.79).

9.17 A particle in 3-D moves under the influence of an even potential, viz., $V(-\mathbf{r}) = V(\mathbf{r})$. At time t_0 its wave function $\Psi_\alpha(\mathbf{r}, t_0)$ is an eigenfunction of $\hat{U}_\text{P}$ with eigenvalue π_α.
   (a) Prove that, for all $t \geqslant t_0$, $\Psi_\alpha(\mathbf{r}, t)$ remains an eigenfunction of $\hat{U}_\text{P}$ with the same parity π_α.
   (b) Show that $\Psi_\alpha(\mathbf{r}, t)$ need not be an eigenfunction of $\hat{U}_\text{p}$ in order for $\langle \hat{\mathbf{P}} \rangle_\alpha$ to be time-independent; and that only $V(\mathbf{r}) = V(-\mathbf{r})$ is required.

9.18 Let $|\Gamma, t\rangle$ be a solution of the time-dependent Schrödinger equation for a single particle in 3-D acted on by the even potential $V(\mathbf{r}) = V(-\mathbf{r})$.
   (a) Prove (i) that $\langle \Gamma, t|\hat{U}_\text{P}|\Gamma, t\rangle$ need not vanish, (ii) that, if $\mathbf{J}_\Gamma(\mathbf{r}, t)$ is the current density associated with $|\Gamma, t\rangle$ then the current density associated with $\hat{U}_\text{P}|\Gamma, t\rangle$ is $-\mathbf{J}_\Gamma(-\mathbf{r}, t)$.
   (b) Assume that $|\Gamma, t\rangle$ is an eigenstate of $\hat{U}_\text{P}$ with parity π_Γ. Prove that $\Phi_\Gamma(-\mathbf{k}, t) = \pi_\Gamma \Phi_\Gamma(\mathbf{k}, t)$, and hence that $|\Phi_\Gamma(-\mathbf{k}, t)|^2 = |\Phi_\Gamma(\mathbf{k}, t)|^2$.

9.19 A particle of mass m is in a 1-D quantal box whose impenetrable walls are at $x = \pm a$ and that contains a delta function of strength $g > 0$ at $x = 0$. Determine the eigenvalues and eigenfunctions (to within a normalization constant) for this potential. If you find eigenvalues given as a solution of a transcendental equation, derive upper and lower bounds on the nth such eigenvalue by comparing it with those obtained when $g = 0$ and $g = \infty$. Contrast your solutions to those of the $g = 0$ case.

9.20 The 1-D potential acting on a particle of mass m is $V(x) = -g[\delta(x-a) + \delta(x+a)]$, $g > 0$. This potential, which provides a 1-D model for a diatomic molecule, supports at most two bound states. Obtain the transcendental equation from which the bound-state energies can be obtained, determine the condition under which two bound states will occur, and, if this does occur, specify which of the two states lies lower in energy.

9.21 Discuss the possible bound states in the symmetric 1-D square well ($V_0 > 0$)

$$V(x) = \begin{cases} -V_0, & |x| \leqslant a, \\ 0, & |x| > a, \end{cases}$$

and obtain criteria, analogous to Eq. (6.173), for the numbers of even and odd bound states.

9.22 A single particle of mass m is acted on by a 3-D potential $V(\mathbf{r})$. Its propagator is $K(\mathbf{r}, t; \mathbf{r}', t')$ – Eqs. (8.19) and (8.20).
   (a) Show that, if $V(\mathbf{r} + \mathbf{a}) = V(\mathbf{r})$, then $K(\mathbf{r} + \mathbf{a}, t; \mathbf{r}' + \mathbf{a}, t') = K(\mathbf{r}, t; \mathbf{r}', t')$.
   (b) Prove that, if a propagator satisfies the preceding relation for all $\mathbf{r}$, t, $\mathbf{r}'$, and t', then $V(\mathbf{r} + \mathbf{a}) = V(\mathbf{r})$.
   (c) If $V(\mathbf{r})$ has even parity, what is the corresponding symmetry property of the propagator?

9.23 On the same figure, plot $F(q)$ of Eq. (9.135) for the values $a/x_0 = 4$, 1, and $\frac{1}{4}$, for $qa \in [0, 9\pi/2]$. Comment on how the potential strength influences these results.

9.24 It was noted below Eq. (9.135) that $q(k_n)$ is double-valued at each band endpoint $k_n = n\pi/a$, where $n \neq 0$.
   (a) Show that one of these two values of $q(k_n)$ is k_n itself.
   (b) The other value may be straightforwardly determined when $n\pi$ is large, in which case one may write $q(k_n) = n\pi/a + \delta_n$, $|\delta_n| \ll 1$. Prove this assertion and evaluate δ_n for $n\pi$ large.

(c) From the results of parts (a) and (b), evaluate the difference of the upper and lower values of the band energies at $k = n\pi/a$, i.e., determine the band gap (at the band edge) and show that it is independent of n.

9.25 A particle of mass m is subject to the 1-D potential of Eq. (9.125).

(a) Verify the claim made in the paragraph below Eq. (9.135) that $[dE/dk]_{k=n\pi/a} = 0$.

(b) The wave function $\psi(x)$ for this potential is a linear combination of left- and right-moving plane waves, Eq. (9.128). Show that, at the band edge ($k = n\pi/a$), $\psi(x)$ becomes a standing wave proportional to $\sin(qx + \theta_n)$, and evaluate θ_n as an arctangent of a function of qa.

(c) At the band edge, one of the two values of q is $q = k_n + \delta_n$. Using the relation $\tan(\pi/2 + x) = -\cot x$ and the approximate value of δ_n obtained in part (b) of Exercise 9.24, show that $\theta_n \cong \alpha + \beta/(n\pi)$, and evaluate the constants α and β.

9.26 In addition to Eq. (9.133), a second realization of the dispersion relation (9.123) can be obtained when the infinite array consists of square-well potentials, viz.,

$$V(x) = \sum_{n=\infty}^{\infty} V_n(x),$$

with

$$V_n(x) = \begin{cases} V_0 > 0, & na \leqslant x \leqslant na + b, \\ 0, & na + b \leqslant x \leqslant (n+1)a, \end{cases}$$

as in Fig. 9.18.

(a) Assume that $V_0 < E = \hbar^2 q^2/(2m)$ and define $q_0 = [2m(E - V_0)]^{1/2}/\hbar$. Show that the dispersion relation for this case takes the form $\cos(ka) = \cos(q_0 b)\cos[q(a-b)] - f(q, q_0)\sin(q_0 b)\sin[q(a-b)]$, and determine the function $f(q, q_0)$. (Hints: choose $n = 0$; apply Eqs. (9.120b) and (9.120c) just as in the delta-function case but only at $x = 0$; match solutions and derivatives at $x = b$.)

(b) By taking appropriate limits, show that the dispersion relation of part (a) goes over to that of the delta-function array.

(c) Now assume that $E < V_0$, let $q_0 \to i|q_0|$, and obtain the analog of the result of part (a).

9.27 A particle of mass m is acted on by the 1-D infinite, periodic potential

$$V = \sum_{n=-\infty}^{\infty} V_n,$$

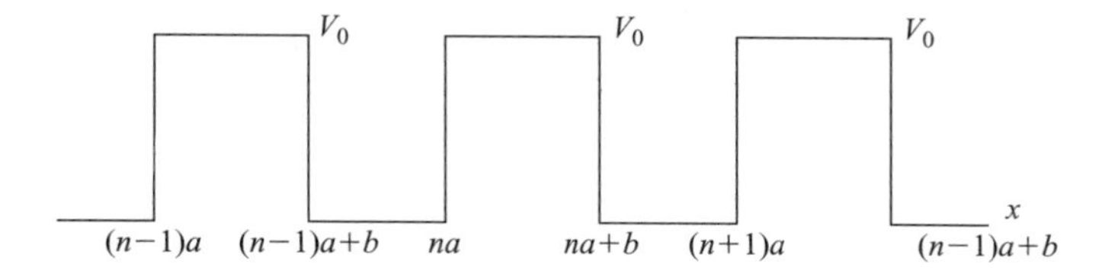

Fig. 9.18

where V_n is the potential at site n, whose position is $x = na$, such that $\hat{U}_\mathrm{T}(\pm a)V_n\hat{U}_T(\pm a) = V_{n\pm 1}$. Each V_n is assumed to be very short-ranged and attractive. The Hamiltonian for V_n in isolation is $\hat{H}_n = \hat{K}_x + V_n$, whose ground state $|n, E_0\rangle$ obeys $(\hat{H}_n - E_0)|n, E_0\rangle = 0$.

(a) Show that $\hat{U}_\mathrm{T}(\pm a)|n, E_0\rangle = |n \pm 1, E_0\rangle$.

(b) Let $\hat{H} = \hat{K}_x + V$ be the exact Hamiltonian for the system. Prove that its eigenstates $|\Gamma\rangle$ are also eigenstates of $\hat{U}_\mathrm{T}(\pm a)$.

(c) In general the eigentstates $|\Gamma\rangle$ cannot be determined exactly; this exercise is concerned with the approximation

$$|\Gamma_\theta\rangle = \sum_{-\infty}^{\infty} e^{in\theta}|n, E_0\rangle.$$

(i) Prove that $|\Gamma_\theta\rangle$ is an eigenstate of $\hat{U}_\mathrm{T}(a)$ and determine the associated eigenvalue. (ii) From part (i), $\langle x|\Gamma_\theta\rangle$ can be expressed in the Bloch-state form $u_k(x)e^{ikx}$. Identify the portions of $\langle x|\Gamma_\theta\rangle$ corresponding to k and to $u_k(x)$.

(d) To convert $|\Gamma_\theta\rangle$ into an *approximate eigenstate* of $\hat{H}$, the following "tight-binding" approximation is employed:

$$\langle m, E_0|\hat{H}|n, E_0\rangle = \begin{cases} E_0, & m = n, \\ E_1, & m = n + 1, \\ 0, & \text{otherwise.} \end{cases}$$

Show that these results are independent of the site label n.

(e) The preceding relation means that $\hat{H}|n, E_0\rangle = E_0|n, E_0\rangle + E_1[|n + 1, E_0\rangle + |n - 1, E_0\rangle]$ in the tight-binding approximation. Show that the last equation implies that $|\Gamma_\theta\rangle$ is an approximate eigenstate of $\hat{H}$ and determine the corresponding approximate eigenvalue $E_{\Gamma_\theta} = E_{\Gamma_\theta}(k)$.

(f) From the result of part (b), etc., and restricting k to lie in the range $[-\pi/a, \pi/a]$, show that $E_{\Gamma_\theta}(k)$ gives rise to a continuous energy band, determine the maximum and minimum values of $E_{\Gamma_\theta}(k)$, and sketch $E_{\Gamma_\theta}(k)$ versus k for $k \in [0, \pi/a]$. (Hint: since n in $\sum_n V_n$ is a dummy summation index, one can also use $\sum_m V_m$, $m = n \pm p$, $p =$ any finite integer.)

9.28 A particle of mass m is subjected to the potential $V(x)$ of Exercise 9.27. Its energy is much larger that the maximum of $V(x)$, and the zero of energy is adjusted so that $\int_0^a dx\, V(x) = 0$. This exercise introduces an approximate solution to the time-independent Schrödinger equation different than that of Exercise 9.27. Here, the Bloch function is expanded as a Fourier series: $u_k(x) = \sum_{-\infty}^{\infty} A_n e^{-2n\pi ix/a}$, with $A_n = A_n(k)$; $V_n \equiv a^{-1}\int_0^a dx\, V(x)e^{2n\pi ix/a}$ and $k_n \equiv k - 2n\pi/a$ are defined for later use.

(a) For $E \gg V(x)$, $u_k(x)$ varies slowly with x, so that $A_0 \gg A_n$ and $A_nV(x) \cong 0$, $n \neq 0$. Show that, under these assumptions, $E \cong E_0$ and $A_n \cong F_nA_0$, where E is the energy and E_0 and F_n are to be expressed in terms of k and any other relevant quantities.

(b) When $k \cong n\pi/a$, it can be shown that A_n is no longer negligible, in which case $u_k(x) \cong A_0 + A_n e^{-2n\pi ix/a}$. (i) Using the value of E_0 from part (a), show that the approximate energy $E = E(k)$ can be expressed as $E = (E_0 + E_n)/ \pm W_n$, and evaluate W_n in terms of E_0, $E_n = \hbar^2k_n^2/(2m)$, V_n, and V_n^*. (ii) There are two branches to $E(k)$: $E_< = (E_0 + E_n)/2 - W_n$, $k < n\pi/a$, and $E_> = (E_0 + E_n)/2 + W_n$, $k > n\pi/a$. Evaluate $E(k = 0)$, $E(k < n\pi/a)$, $E(k > n\pi/a)$, and then

$E(k = n\pi/a)$. You should find that $E(k = n\pi/a)$ is double-valued, say $E^{\pm}(n\pi/a)$. Determine both the band gap $\Delta_n = E^+(n\pi/a) - E^-(n\pi/a)$ and $E(k = 2n\pi/a)$. (iii) Evaluate $[dE(k)/dk]_{k=n\pi/a}$ and make a rough sketch of $E(k)$ versus k for $k \in [0, 2n\pi/a]$, labeling the points $E(k = 0)$, $E_<(k = n\pi/a)$, $E_>(k = n\pi/a)$, and $E_>(k = 2n\pi/a)$. Compare your curve with that for the delta-function array.

9.29 Starting with Eq. (9.153) and expressions for $R(\mathbf{e}_1, \theta)$ and $R(\mathbf{e}_2, \theta)$ analogous to Eqs. (9.145), derive Eq. (9.160).

9.30 A single particle in 3-D is acted on by a central potential $V(r)$. As noted below Eq. (9.174), the CSCO for this system is $\{\hat{H}, \hat{L}^2, \hat{L}_z\}$ and its eigenstates $|n\ell m_\ell\rangle$ obey $(\hat{L}_z - \hbar m_\ell)|n\ell m_\ell\rangle = 0$.
(a) Prove that the $|n\ell m_\ell\rangle$ are degenerate.
(b) Prove that $\langle n'\ell' m_\ell|\hat{H}|n\ell m_\ell\rangle$ is diagonal in m_ℓ.

9.31 (a) Determine whether it is possible for a single particle in 3-D to be simultaneously in eigenstates of (i) $\hat{U}_{\text{P}}$ and $\hat{U}_{\text{T}}(a)$, (ii) $\hat{U}_{\text{P}}$ and $\hat{U}_{\text{R}}(\mathbf{n}, \theta)$, (iii) $\hat{U}_{\text{T}}(a)$ and $\hat{U}_{\text{R}}(\mathbf{n}, \theta)$, (iv) $\hat{H}(\mathbf{r}) = \hat{K}(\mathbf{r}) + V(r)$ and either $\hat{U}_{\text{P}}$ or $\hat{U}_{\text{T}}(a)$.
(b) Let $\mathbf{r} = \mathbf{r}_1 - \mathbf{r}_2$, where $\mathbf{r}_1$ and $\mathbf{r}_2$ are the 3-D coordinates of particles 1 and 2, and define $\hat{H}_{12}(\mathbf{r}) = -\hbar^2\nabla_{\text{r}}^2/(2\mu) + V(\mathbf{r})$, where μ is the reduced mass. Determine the circumstances under which eigenstates of $\hat{H}_{12}(\mathbf{r})$ can also be eigenstates of $\hat{U}_{\text{P}}$, $\hat{U}_{\text{R}}(\mathbf{n}, \theta)$, or $\hat{U}_{\text{T}}(\mathbf{a})$.

9.32 The plane wave $\psi_{\mathbf{k}}(\mathbf{r}) = (2\pi)^{-3/2}e^{i\mathbf{k}\cdot\mathbf{r}}$ is an eigenstate of $\hat{\mathbf{P}}$ with eigenvalue $\hbar\mathbf{k}$. Let $\mathbf{k} = k\mathbf{e}_1$. Show that $\psi'(\mathbf{r}) = \exp[(i\pi\hat{L}_z/(2\hbar)]\psi_{\mathbf{k}}(\mathbf{r})$ is also an eigenfunction of $\hat{\mathbf{P}}$ by determining the corresponding eigenvalue. (Hint: first consider the operator $\exp(i\theta\hat{L}_z/\hbar)$ and then let $\theta = \pi/2$.)

9.33 A particle of mass m and charge q in 2-D is bound by the symmetric linear oscillator potential $m\omega^2r^2/2$, where $\mathbf{r} = r(\mathbf{e}_1\cos\phi + \mathbf{e}_2\sin\phi)$. The eigenfunctions are products of a radial function and the eigenfunctions of $\hat{L}_z$ that entered the analysis of the 2-D rotor. In particular, the ground-state wave function is $\psi_{0,0}(r, \phi) = \exp[(-r^2/(2\lambda^2)]/(\lambda\pi^{1/2})$, while the two-fold-degenerate, first-excited-state eigenfunctions are $\psi_{0,\pm1}(r, \phi) = r\{\exp[(-r^2/(2\lambda^2)]\}[\exp(\pm i\phi)]/\lambda^2\pi^{1/2})$, where $\lambda = [\hbar/(m\omega)]^{1/2}$ is the usual oscillator length parameter (see Section 11.7 for quantum numbers, etc.). In principle, electromagnetic transitions can occur between these two levels, and, in one of the simplest approximations, the excited state lifetime against decay via photon emission to the ground state involves the absolute square of the matrix element $M_{0,\pm1} = \langle 0, 0|\mathbf{n}\cdot\hat{\mathbf{Q}}|0, \pm1\rangle$, where the unit polarization vector is $\mathbf{n} = n_1\mathbf{e}_1 + n_2\mathbf{e}_2$, and $\hat{\mathbf{Q}} = \hat{Q}_x\mathbf{e}_1 + \hat{Q}_y\mathbf{e}_2$.
(a) Evaluate $M_{0,\pm1}$, expressing it in terms of n_1, n_2, and λ.
(b) Now assume that a rotation about $\mathbf{e}_3$ through the angle θ is performed *without altering* $\mathbf{n}$. (i) Evaluate the matrix element $M_{0',\pm1'}$ corresponding to $A'_{\beta\alpha}$ of Eq. (9.180a). (ii) Next, determine the matrix element $M_{0',\pm1'}$ corresponding to $A_{\beta'\alpha'}$ of Eq. (9.180b).
Compare the two results of this part and, if they differ, comment on whether and why they would give rise to different lifetimes.

9.34 Suppose that the charged 2-D oscillator of Exercise 9.33 could be prepared in the state $\psi_\alpha = a\psi_{0,+1} + b\psi_{0,-1}$, with $a^2 + b^2 = 1$.
(a) Evaluate the transition-matrix element $M_{0,\alpha} = \langle 0, 0|\mathbf{n}\cdot\hat{\mathbf{Q}}|\alpha\rangle$ in terms of a, b, n_1, n_2, and λ (in the notation of Exercise 9.33).

(b) The same rotation as in part (b) of the preceding exercise is now carried out. Evaluate the two "rotated" matrix elements $M'_{0,\alpha}$ and $M_{0',\alpha'}$ and indicate whether, and, if so, why, a rotation in this case could alter the "observed" lifetime.

9.35 Redo Exercises 9.33 and 9.34 assuming that the polarization vector also rotates.

Part III

SYSTEMS WITH FEW DEGREES OF FREEDOM

10

Orbital Angular Momentum

Our goal in this chapter is to solve the orbital angular-momentum eigenvalue problem postponed from Section 9.7. This is accomplished in Sections 10.1 and 10.2. The orbital angular-momentum eigenstates form a complete set, and we use them in Section 10.3 as a means of expanding the plane wave ($\exp(i\mathbf{k}\cdot\mathbf{r})$). Contained in such an expansion are particular r-dependent coefficients – the spherical Bessel functions, which will be determined as part of the analysis. These functions play an important role in solutions of the 3-D square-well problem encountered in Chapter 11, and also in the analysis of 3-D scattering, Chapter 15. Section 10.3 may be thought of as a first application of the orbital angular-momentum eigenstates. In the final section of the chapter we examine aspects of the uncertainty relations involving the angular-momentum operators, in particular, the so-called ladder operators and the angle operator conjugate to $\hat{L}_z$.

10.1. The orbital angular momentum operator

In this section we discuss a variety of topics connected with the orbital angular momentum operator $\hat{\mathbf{L}}$, such as its dependence on the spherical polar coordinates r, θ, and ϕ; the appearance of $\hat{L}^2$ in the kinetic-energy operator when $\hat{K}(\mathbf{r})$ is expressed in terms of r, θ, and ϕ; and the so-called radial-momentum operator $\hat{P}_{\text{rad}}(r)$.

Some properties of $\hat{\mathbf{L}}$

We first derive some properties of $\hat{\mathbf{L}}$, defined as the quantal analog of the classical quantity $\mathbf{r}\times\mathbf{p}$: $\hat{\mathbf{L}} = \hat{\mathbf{Q}}\times\hat{\mathbf{P}}$. $\hat{\mathbf{L}}$ has rectangular components $\hat{L}_j$, $j = 1, 2, 3$, given by

$$\hat{L}_j = \hat{Q}_k\hat{P}_\ell - \hat{Q}_\ell\hat{P}_k, \qquad jk\ell \text{ taken in cyclic order,} \tag{10.1a}$$

or, in terms of the rank-3 antisymmetric tensor $\epsilon_{jk\ell}$,

$$\hat{L}_j = \sum_{k,\ell}\epsilon_{jk\ell}\hat{Q}_k\hat{P}_\ell, \tag{10.1b}$$

where

$$\epsilon_{jk\ell} = \begin{cases} +1, & jk\ell = \text{a cyclic permutation of 123,} \\ -1, & jk\ell = \text{a non-cyclic permutation of 123,} \\ 0, & \text{otherwise.} \end{cases}$$

As noted in Section 9.7, the commutator of different components of $\hat{\mathbf{L}}$ is non-zero:

$$[\hat{L}_j,\ \hat{L}_k] = i\hbar\epsilon_{jk\ell}\hat{L}_\ell, \tag{10.2}$$

a result that is readily obtained using (10.1b). Because of (10.2), only one of the three components of $\hat{\mathbf{L}}$ can be included in the CSCO for a quantal system described by a spherically symmetric Hamiltonian. That component is conventionally chosen to be $\hat{L}_z$. Since spherical symmetry means that $[\hat{L}_j,\ \hat{H}] = 0,\ \forall\ j$, it follows that $[\hat{L}^2,\ \hat{H}] = 0$, where $\hat{L}^2 = \sum_{j=1}^{3}\hat{L}_j^2$. Furthermore, as was noted in Section 9.7, it can be shown – and we show it next – that $[\hat{L}^2,\ \hat{L}_z] = 0$. Hence, $\hat{H}$, $\hat{L}^2$, and $\hat{L}_z$ form the CSCO for a spherically symmetric, spin-independent Hamiltonian.

The proof of $[\hat{L}^2,\ L_j] = 0$ is straightforward. Given that $[\hat{L}_j^2,\ \hat{L}_j] = 0$, we need only consider the commutator

$$\hat{D} = \left[\hat{L}_k^2 + \hat{L}_\ell^2,\ \hat{L}_j\right]. \tag{10.3}$$

Using $[\hat{A}\hat{B},\ \hat{C}] = \hat{A}[\hat{B},\ \hat{C}] + [\hat{A},\ \hat{C}]\hat{B}$ and the commutation relation (10.2), $\hat{D}$ becomes

$$\hat{D} = i\hbar\varepsilon_{kj\ell}\big(\hat{L}_k\hat{L}_\ell + \hat{L}_\ell\hat{L}_k\big) + i\hbar\varepsilon_{\ell jk}\big(\hat{L}_\ell\hat{L}_k + \hat{L}_k\hat{L}_\ell\big). \tag{10.4}$$

However, $\varepsilon_{\ell jk} = \varepsilon_{k\ell j} = -\varepsilon_{kj\ell}$, the two terms on the RHS of (10.4) therefore cancel out, $\hat{D} = 0$, and Eq. (10.3) is established, QED.

Although the eigenvalues of $\hat{L}^2$ and $\hat{L}_z$ have not yet been determined, it is possible to derive a very useful inequality relating them. The relation between these eigenvalues follows from the more general result

$$\langle\hat{L}^2\rangle_\gamma \geqslant \langle\hat{L}_z^2\rangle_\gamma, \tag{10.5}$$

where, for any operator $\hat{A}$,

$$\langle\hat{A}\rangle_\gamma \equiv \langle\gamma|\hat{A}|\gamma\rangle \tag{10.6}$$

and $|\gamma\rangle$ is *any* Hilbert-space state. On choosing $|\gamma\rangle$ to be the common eigenstate of $\hat{L}^2$ and $\hat{L}_z$, Eq. (10.5) states that the square of the eigenvalue of $\hat{L}_z$ is never greater than the eigenvalue of $\hat{L}^2$, a result we shall employ to advantage in the next section.

To prove (10.5), set $\hat{A} = \hat{B}^2$ in (10.6) and let $\hat{B}^\dagger = \hat{B}$, so that $\hat{A} = \hat{B}^\dagger\hat{B}$. Then (10.6) is just the norm of $\hat{B}$ w.r.t. $|\gamma\rangle$:

$$\langle\hat{B}^2\rangle_\gamma = \|\hat{B}|\gamma\rangle\| \geqslant 0; \tag{10.7}$$

hence, any $\hat{B} = \hat{B}^\dagger$ is a semi-positive-definite operator. From this it follows that $\langle\hat{L}_j^2\rangle \geqslant 0$ and therefore that

$$\langle\hat{L}^2\rangle_\gamma = \langle\hat{L}_x^2\rangle_\gamma + \langle\hat{L}_y^2\rangle_\gamma + \langle\hat{L}_z^2\rangle_\gamma \geqslant \langle\hat{L}_z^2\rangle_\gamma,$$

which is Eq. (10.5), QED.

Spherical-polar-coordinate representation

We saw in Section 9.7, Eq. (9.183), that, if spherical polar coordinates were used to express the coordinate representation, then $\hat{L}_z(\mathbf{r}) \to \hat{L}_z(\phi)$, with

$$\hat{L}_z(\phi) = -i\hbar \frac{\partial}{\partial \phi}, \tag{9.183}$$

where ϕ is the usual azimuthal angle. The dependence of $\hat{L}_z$ on the dimensionless quantity ϕ is a natural consequence of $\hbar$ having the dimension of angular momentum. Analogous results hold for $\hat{L}_x$ and $\hat{L}_y$.

One way of determining the θ, ϕ dependence of $\hat{L}_x$ and $\hat{L}_y$ is through use of the chain rule for partial derivatives. An alternate procedure is to express ∇ in terms of its components along $\mathbf{e}_r$, $\mathbf{e}_\theta$, and $\mathbf{e}_\phi$, then evaluate the vector product $r\mathbf{e}_r \times \nabla$, and finally project the resulting expression for $\hat{\mathbf{L}}$ onto the Cartesian unit vectors $\mathbf{e}_j$. From the expression

$$\nabla = \mathbf{e}_r \frac{\partial}{\partial r} + \mathbf{e}_\theta \frac{1}{r}\frac{\partial}{\partial \theta} + \mathbf{e}_\phi \frac{1}{r \sin\theta}\frac{\partial}{\partial \phi}, \tag{10.8}$$

$\hat{\mathbf{L}}(\mathbf{r}) = -i\hbar \mathbf{r} \times \nabla$ becomes

$$\begin{aligned}\hat{\mathbf{L}}(\mathbf{r}) &= -i\hbar\left(\mathbf{e}_r \times \mathbf{e}_\theta \frac{\partial}{\partial \theta} + \mathbf{e}_r \times \mathbf{e}_\phi \frac{1}{\sin\theta}\frac{\partial}{\partial \phi}\right) \\ &= -i\hbar\left(-\mathbf{e}_\theta \frac{1}{\sin\theta}\frac{\partial}{\partial \phi} + \mathbf{e}_\phi \frac{\partial}{\partial \theta}\right).\end{aligned} \tag{10.9}$$

Since $\hat{L}_j = \mathbf{e}_j \cdot \hat{\mathbf{L}}$, we need $\mathbf{e}_j \cdot \mathbf{e}_\theta$ and $\mathbf{e}_j \cdot \mathbf{e}_\phi$, $j = x, y, z$. The requisite scalar products are (see Fig. 10.1)

$$\begin{aligned}\mathbf{e}_x \cdot \mathbf{e}_\theta &= \cos\theta \cos\phi, & \mathbf{e}_x \cdot \mathbf{e}_\phi &= -\sin\phi, \\ \mathbf{e}_y \cdot \mathbf{e}_\theta &= \cos\theta \sin\phi, & \mathbf{e}_y \cdot \mathbf{e}_\phi &= \cos\phi, \\ \mathbf{e}_z \cdot \mathbf{e}_\theta &= -\sin\theta, & \mathbf{e}_z \cdot \mathbf{e}_\phi &= 0,\end{aligned} \tag{10.10}$$

which lead to

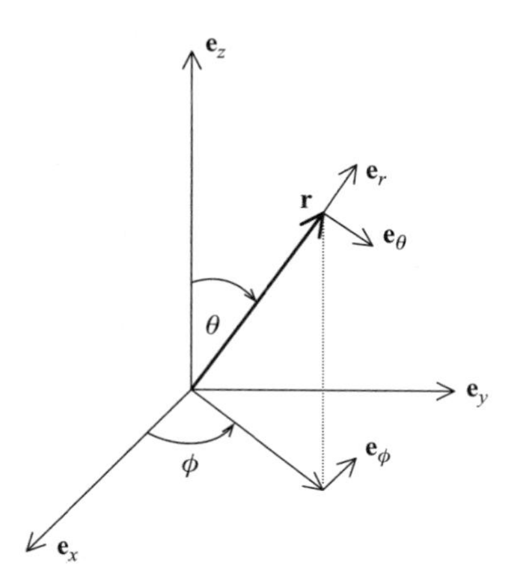

Fig. 10.1 Unit vectors in Cartesian ($\mathbf{e}_x$, $\mathbf{e}_y$, $\mathbf{e}_z$) and spherical-polar ($\mathbf{e}_r$, $\mathbf{e}_\theta$, $\mathbf{e}_\phi$) coordinates.

$$\hat{L}_x(\theta, \phi) = i\hbar\left(\sin\phi\,\frac{\partial}{\partial\theta} + \cot\theta\cos\phi\,\frac{\partial}{\partial\phi}\right), \tag{10.11a}$$

$$\hat{L}_y(\theta, \phi) = i\hbar\left(-\cos\phi\,\frac{\partial}{\partial\theta} + \cot\theta\sin\phi\,\frac{\partial}{\partial\phi}\right), \tag{10.11b}$$

and

$$\hat{L}_z(\theta, \phi) = \hat{L}_z(\phi) = -i\hbar\,\frac{\partial}{\partial\phi}. \tag{10.11c}$$

These are the Cartesian *components* of $\hat{\mathbf{L}}$ expressed in spherical polar coordinates, just as $\hat{L}_j = -i\hbar(x_k\,\partial/\partial x_\ell - x_\ell\,\partial/\partial x_k)$ is a Cartesian component expressed in Cartesian (or rectangular) coordinates. One could also calculate the spherical polar components of $\hat{\mathbf{L}}$ in either coordinate system, but these are not advantageous when one is evaluating $\hat{L}^2 = \hat{\mathbf{L}}\cdot\hat{\mathbf{L}}$. Using the Cartesian components (10.11), a straightforward calculation yields

$$\hat{L}^2(\theta, \phi) = -\hbar^2\left(\frac{1}{\sin\theta}\frac{\partial}{\partial\theta}\sin\theta\frac{\partial}{\partial\theta} + \frac{1}{\sin^2\theta}\frac{\partial^2}{\partial\phi^2}\right). \tag{10.12}$$

The expression in large parentheses in Eq. (10.12) occurs in the spherical-polar-coordinate version of the Laplacian, which we examine next.

The Laplacian in spherical polar coordinates

The two most widely used coordinate systems for expressing $\hat{P}^2(\mathbf{r})/\hbar^2 = -\nabla^2$ are Cartesian and spherical polar. The Cartesian form for the Laplacian,

$$\nabla^2 = \frac{\partial^2}{\partial x^2} + \frac{\partial^2}{\partial y^2} + \frac{\partial^2}{\partial x^2},$$

is most appropriate when the potential and/or the boundary conditions are stated in terms of the x, y, z behavior of wave functions. For the spherically symmetric situation, $V(\mathbf{r}) = V(r)$, spherical polar coordinates provide a more useful form for ∇^2:

$$\nabla^2 = \frac{1}{r^2}\frac{\partial}{\partial r}r^2\frac{\partial}{\partial r} + \frac{1}{r^2}\left(\frac{1}{\sin\theta}\frac{\partial}{\partial\theta}\sin\theta\,\frac{\partial}{\partial\theta} + \frac{1}{\sin^2\theta}\frac{\partial^2}{\partial\phi^2}\right). \tag{10.13}$$

However, the term in large parentheses in (10.13) is proportional to $\hat{L}^2$; hence, by use of Eq. (10.12), ∇^2 can be re-written as

$$\nabla^2 = \frac{1}{r^2}\frac{\partial}{\partial r}r^2\,\frac{\partial}{\partial r} - \frac{1}{r^2}\,\frac{\hat{L}^2(\theta, \phi)}{\hbar^2}. \tag{10.14a}$$

$\hat{L}^2(\theta, \phi)$ is thus the "angular" part of $\hat{P}^2(\mathbf{r})$ in a spherical-polar-coordinate representation:

$$r^2\left(\hat{P}^2(\mathbf{r}) + \hbar^2\frac{1}{r^2}\frac{\partial}{\partial r}r^2\,\frac{\partial}{\partial r}\right) = \hat{L}^2(\theta, \phi). \tag{10.14b}$$

Correspondingly, the kinetic-energy operator $\hat{K}(\mathbf{r}) = \hat{K}(r, \theta, \phi)$ takes the form

$$\hat{K}(\mathbf{r}) = -\frac{\hbar^2}{2m}\frac{1}{r^2}\frac{\partial}{\partial r}r^2\frac{\partial}{\partial r} + \frac{1}{2mr^2}\hat{L}^2(\theta, \phi). \tag{10.15}$$

Equation (10.15) can be used to re-write the (single-particle) Hamiltonian for the case of a spherically symmetric potential:

$$\begin{aligned}\hat{H} &= \hat{K}(\mathbf{r}) + V(r) \\ &= \hat{H}_{\text{rad}}(r) + \frac{1}{2mr^2}\hat{L}^2(\theta, \phi),\end{aligned} \tag{10.16a}$$

where

$$\hat{H}_{\text{rad}}(r) = -\frac{\hbar^2}{2m}\frac{1}{r^2}\frac{\partial}{\partial r}r^2\frac{\partial}{\partial r} + V(r). \tag{10.16b}$$

One of the advantages of the separation of $\hat{H}(\mathbf{r})$ into radial ($\hat{H}_{\text{rad}}(r)$) and angular ($\hat{L}^2(\theta, \phi)$) parts is that the invariance of $\hat{H}(\mathbf{r})$ under rotations is made manifest: rotations are generated by $\hat{\mathbf{L}}$, and $[\hat{\mathbf{L}}(\theta, \phi), r] = 0 = [\hat{\mathbf{L}}(\theta, \phi), \hat{L}^2(\theta, \phi)]$. The structure of (10.16a) also means that, in

$$(\hat{H}(\mathbf{r}) - E_\Gamma)\psi_\Gamma(\mathbf{r}) = 0, \tag{10.17}$$

the search for the eigenfunction $\psi_\Gamma(\mathbf{r})$ can proceed from an assumed product form:

$$\psi_\Gamma(\mathbf{r}) = R_\Gamma(r)Y_\Gamma(\theta, \phi), \tag{10.18}$$

where $Y_\Gamma(\theta, \phi)$ is the simultaneous eigenfunction of $\hat{L}^2(\theta, \phi)$ and $\hat{L}_z(\phi)$ that we will determine in Section 10.2, and $R_\Gamma(r)$ is the radial solution. The dependence of $R_\Gamma(r)$ on r is mainly determined by $V(r)$, and, as we shall show in Section 10.3, $R_\Gamma(r)$ remains to be determined even when $V(\mathbf{r}) = 0$. This is the free-particle case, in which the $R_\Gamma(r)$ become the spherical Bessel functions referred to in the introductory part of this chapter. Several non-zero $V(r)$ will be examined in Chapter 11.

Radial momentum

A consequence of $\hat{\mathbf{L}}(\mathbf{r}) = r\mathbf{e}_r \times \hat{\mathbf{P}}(\mathbf{r})$ is that only the $\mathbf{e}_\theta$ and $\mathbf{e}_\phi$ portions of $\hat{\mathbf{P}}(\mathbf{r})$ contribute to $\hat{\mathbf{L}}$, Eqs. (10.8) and (10.9). Also, the square of $\hat{\mathbf{P}}$, Eq. (10.14b), contains the $\mathbf{e}_\theta$ and $\mathbf{e}_\phi$ portions of $\hat{\mathbf{P}}(\mathbf{r})$ in the form $\hat{L}^2(\theta, \phi)/r^2$. These facts suggest the possibility that the term $-\hbar^2 r^{-2}\,\partial(r^2\,\partial/\partial r)/\partial r$ in $\hat{P}^2(\mathbf{r})$ might be the square of an as-yet unconsidered quantity, viz., $\hat{P}_{\text{rad}}(\mathbf{r}) \equiv \hat{P}_{\text{rad}}(r)$, the radial component of momentum (in the coordinate representation). If such a quantity were to exist, for example

$$\hat{P}_{\text{rad}}(r) \stackrel{?}{=} \mathbf{e}_r \cdot \hat{\mathbf{P}}(\mathbf{r}), \tag{10.19}$$

then $\hat{P}^2_{\text{rad}}(r)$ should not only be equal to $-\hbar^2 r^{-2}\,\partial(r^2\,\partial/\partial r)/\partial r$, but must also be Hermitian: $\hat{P}^\dagger_{\text{rad}}(r) = \hat{P}_{\text{rad}}(r)$. However, $\mathbf{e}_r \cdot \hat{\mathbf{P}}(\mathbf{r})$ is not Hermitian, as the replacement $\mathbf{e}_r = \mathbf{r}/r$ shows, and the question mark in (10.19) is germane.

Although (10.19) fails as a possible definition of $\hat{P}_{\text{rad}}(r)$, appropriate symmetrization of the RHS of Eq. (10.19) leads to a Hermitian operator, and indeed the correct one, as is shown next. We consider

$$\hat{P}_{\text{rad}}(\mathbf{r}) = \frac{1}{2}\left[\left(\frac{\mathbf{r}}{r}\right)\cdot\hat{\mathbf{P}}(\mathbf{r}) \;+\; \hat{\mathbf{P}}(\mathbf{r})\cdot\left(\frac{\mathbf{r}}{r}\right)\right]. \tag{10.20}$$

The crucial term in (10.20) is

$$\begin{aligned}\hat{\mathbf{P}}(\mathbf{r})\cdot\left(\frac{\mathbf{r}}{r}\right) &= -i\hbar\left(\frac{3}{r}-\frac{r}{r^2}\right)\\ &= -2i\hbar/r,\end{aligned}$$

by which (10.20) is transformed into

$$\hat{P}_{\text{rad}}(r) = -i\hbar\left(\frac{\partial}{\partial r}+\frac{1}{r}\right). \tag{10.21}$$

If (10.21) is correct, then its square must be equal to the first term on the RHS of Eq. (10.14a). A short calculation gives

$$\begin{aligned}\hat{P}^2_{\text{rad}}(r) &= -\hbar^2\left(\frac{\partial^2}{\partial r^2}+\frac{2}{r}\frac{\partial}{\partial r}\right)\\ &= -\hbar^2\frac{1}{r^2}\frac{\partial}{\partial r}r^2\frac{\partial}{\partial r},\end{aligned}$$

thereby establishing that

$$\hat{P}^2(\mathbf{r}) = \hat{P}^2_{\text{rad}}(r) + \frac{1}{r^2}\hat{L}^2(\theta,\phi), \tag{10.22}$$

with $\hat{P}_{\text{rad}}(r)$ defined by Eq. (10.21).

We consider the eigenvalue problem for $\hat{P}_{\text{rad}}(r)$ next. In contrast to $\hat{\mathbf{P}}(\mathbf{r})$, the $1/r$ term makes $\hat{P}_{\text{rad}}(r)$ of (10.21) a singular operator. It is therefore strictly defined only for $r \in (0, \infty]$, a feature discussed in more detail by, e.g., Galindo and Pascual (1990). This feature will not be a problem for us, other than its yielding eigenfunctions that are themselves singular at $r = 0$, a point to which we shall return shortly. Since the eigenvalue/eigenfunction relation for $\hat{\mathbf{P}}(\mathbf{r})$ is

$$\hat{\mathbf{P}}(\mathbf{r})[e^{i\mathbf{k}\cdot\mathbf{r}}] = \hbar\mathbf{k}[e^{i\mathbf{k}\cdot\mathbf{r}}], \tag{10.23}$$

and $\hat{P}_{\text{rad}}(r)$ is the radial part of $\hat{\mathbf{P}}(\mathbf{r})$, we might expect the analog of (10.23) for $P_{\text{rad}}(r)$ to be

$$\hat{P}_{\text{rad}}(r)e^{ikr} \overset{?}{=} \hbar k e^{ikr}. \tag{10.24a}$$

Were $\hat{P}_{\text{rad}}(r)$ equal to $-i\hbar\partial/\partial r$, then (10.24a) would be correct, but the presence of the $-i\hbar/r$ term in (10.21) renders (10.24a) invalid. The proper relation is

$$\hat{P}_{\text{rad}}(r)\frac{e^{\pm ikr}}{r} = \pm\hbar k\frac{e^{\pm ikr}}{r}, \tag{10.24b}$$

i.e., the same eigenvalue as in (10.24a) but now with the correct eigenfunctions ($\exp(\pm ikr)/r$), known as outgoing (+) and ingoing (−) spherical waves.

"Outgoing" here means that, as t increases in $\exp[i(kr-\omega t)]$, the points of constant phase on the spherical wave fronts propagate to larger values of r with a phase velocity $v_r = \partial r/\partial t = \omega/k$. A 2-D sketch of this behavior is shown in Fig. 10.2(a). These

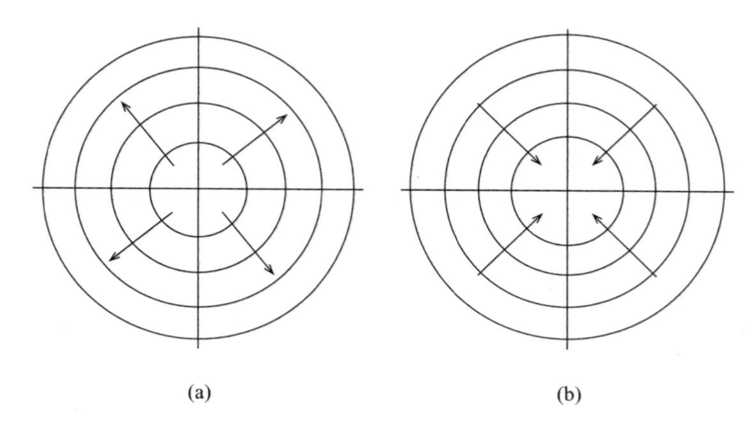

Fig. 10.2 2-D representation of spherical waves. (a) Outgoing spherical waves, which propagate from smaller to larger values of r, as in a spatially uniform explosion. (b) Ingoing spherical waves, which propagate from larger to smaller values of r, corresponding to an implosion.

outgoing spherical waves correspond to an explosion, whereas an *ingoing* spherical wave $\exp[-i(kr+\omega t)]$ corresponds to an implosion, as in Fig. 10.2(b). We shall encounter outgoing and ingoing spherical waves in Section 10.3 and again in Chapter 15 on 3-D scattering theory. Equation (10.24b) can be converted into an abstract, Hilbert-space expression, details of which are left to the exercises. We note in passing that the solutions $\exp(\pm ikr)/r$ are simultaneous eigenfunctions of $\hat{P}_{\text{rad}}(r)$ and $\hat{P}^2_{\text{rad}}(r)$, while their real and imaginary parts, $\cos(kr)/r$ and $\sin(kr)/r$, are eigenfunctions of $\hat{P}^2_{\text{rad}}$ only.

Since $\hat{P}_{\text{rad}}(r)$ is the radial component of momentum and r is the radial component of position, it is not surprising that their commutator gives the same result as $[\hat{Q}_j, \hat{P}_j]$, $j = 1, 2, 3$:

$$\left[r, \hat{P}_{\text{rad}}(r)\right] = -i\hbar. \tag{10.25}$$

Defining $\hat{Q}_r$ to be the radial component of the position operator, viz.,

$$\hat{Q}_r = \mathbf{e}_r \cdot \hat{\mathbf{Q}}, \tag{10.26}$$

the Hilbert-space version of (10.25) is

$$\left[\hat{Q}_r, \hat{P}_{\text{rad}}\right] = i\hbar, \tag{10.27}$$

from which we conclude that

$$\Delta Q_r \, \Delta P_{\text{rad}} \geqslant \hbar/2 \tag{10.28}$$

is the relevant uncertainty relation. Furthermore, if (10.28) were replaced by

$$\Delta Q_r \, \Delta P_{\text{rad}} \sim \hbar, \tag{10.29}$$

then it is this relation that would have been the more appropriate one to have used in Example 3 of Section 9.1. We were obviously unable to have done so then.

10.2. The orbital angular momentum eigenvalue problem

General features

In this section we discuss the $\hat{L}^2/\hat{L}_z$ eigenvalue problem and its solution. The reader may have wondered why $\hat{L}^2$ has been chosen as the second operator (in addition to $\hat{L}_z$), or, for that matter, why only two such operators are considered. The choice of $\hat{L}^2$ is essentially dictated by Eqs. (10.16): $\hat{L}^2$ contains the entire θ, ϕ dependence of a spherically symmetric Hamiltonian. Furthermore, as there is a quantum number for each spatial degree of freedom, then the two angles θ and ϕ imply two quantum numbers, and two quantum numbers imply two commuting operators, which must be $\hat{L}^2$ and one of its components. $\hat{L}_z$ is the conventionally chosen component.

The common eigenket of $\hat{L}^2$ and $\hat{L}_z$ was denoted $|\ell m_\ell\rangle$ in Section 9.7, with ℓ related but not equal to the eigenvalue of $\hat{L}^2$. Since $|\ell m_\ell\rangle$ is the end product, we begin by writing the common eigenket as $|\lambda\mu\rangle$,

$$\hat{L}^2|\lambda\mu\rangle = \hbar^2\lambda|\lambda\mu\rangle, \tag{10.30a}$$

$$\hat{L}_z|\lambda\mu\rangle = \hbar\mu|\lambda\mu\rangle; \tag{10.30b}$$

we shall derive the relations between λ and ℓ, and between μ and m_ℓ. When $|\gamma\rangle = |\lambda\mu\rangle$ in (10.5), it reads

$$\lambda \geqslant \mu^2 \geqslant 0, \tag{10.31}$$

thereby suggesting that μ may depend on λ: we shall show that it does.

There are two standard procedures for solving Eqs. (10.30). In one, the abstract form of these equations is retained and commutation relations are used to achieve the desired results. This is the method utilized to solve the eigenvalue problem for the case of generalized angular momentum (of which $\hat{\mathbf{L}}$ is a particular example), and we employ it in Chapter 14. The other procedure invokes the coordinate representation. Knowledge of $\hat{\mathbf{L}}^2(\theta, \phi)$ and $\hat{L}_z(\phi)$ readily turns Eqs. (10.30) into differential equations whose solutions are the coordinate-space angular wave functions. We follow the coordinate-space method here because these functions are needed later in this chapter as well as in the next.

Introducing the temporary notation[1]

$$Y_\lambda^\mu(\theta, \phi) = \langle\theta\phi|\lambda\mu\rangle, \tag{10.32}$$

the coordinate-space form of Eqs. (10.30) is

$$\hat{L}^2(\theta, \phi)Y_\lambda^\mu(\theta, \phi) = \lambda\hbar^2 Y_\lambda^\mu(\theta, \phi) \tag{10.33a}$$

and

$$\hat{L}_z(\phi)Y_\lambda^\mu(\theta, \phi) = \mu\hbar Y_\lambda^\mu(\theta, \phi), \tag{10.33b}$$

with $\hat{L}^2(\theta, \phi)$ given by (10.12) and $\hat{L}_z(\phi) = -i\hbar\,\partial/\partial\phi$.

[1] Coordinate-space wave functions have generally been denoted ψ, but Y is the standard symbol, and we use it *ab initio*. The notation Y_λ^μ is temporary because, after identifying $\lambda \to \ell(\ell+1)$ and $\mu \to m_\ell$, we shall finally write $Y_\ell^{m_\ell}(\theta, \phi)$.

We consider Eq. (10.33b) first:

$$-i\hbar \frac{\partial}{\partial\phi} Y_\lambda^\mu(\theta, \phi) = \mu\hbar Y_\lambda^\mu(\theta, \phi),$$

a solution of which is a product:

$$Y_\lambda^\mu(\theta, \phi) = N_\mu P_\lambda^\mu(\theta) e^{i\mu\phi}. \tag{10.34}$$

However, as in Section 9.7, $\exp(i\mu\phi)$ can be a single-valued eigenfunction of $\hat{L}_z$ only if the real number $\mu \to m = 0, \pm 1, \pm 2, \pm 3, \ldots$, in which case $N_\mu \to N_m = \sqrt{1/(2\pi)}$, where an overall phase factor has, as usual, been set equal to unity. These conclusions lead to

$$Y_\lambda^\mu(\theta, \phi) \to Y_\lambda^m(\theta, \phi) = P_\lambda^m(\theta) e^{im\phi}/\sqrt{2\pi}, \qquad m = 0, \pm 1, \pm 2, \ldots. \tag{10.35}$$

Two points are worth stressing. First, the ϕ dependence of $Y_\lambda^m(\theta, \phi)$ occurs through the eigenstate $\Phi_m(\phi) = \langle\phi|m\rangle$ of $\hat{L}_z(\phi)$:

$$[\hat{L}_z(\phi) - m\hbar]\Phi_m(\phi) = 0,$$

with

$$\Phi_m(\phi) = \sqrt{\frac{1}{2\pi}} e^{im\phi}, \qquad m = 0, \pm 1, \pm 2, \ldots, \tag{10.36}$$

exactly as in Section 9.7. Second, the product form for $Y_\lambda^m(\theta, \phi)$, Eqs. (10.34) or (10.35), does not mean that the μ or m dependence of P can be dropped just because it also occurs in $\exp(i\mu\phi)$ or $\exp(im\phi)$: until shown otherwise, this dependence on the $\hat{L}_z$ eigenvalue must be retained in P.

In light of Eq. (10.35), (10.33a) becomes an equation for $P_\lambda^m(\theta)$ alone:

$$\left(\frac{1}{\sin\theta}\frac{d}{d\theta}\sin\theta\frac{d}{d\theta} - \frac{m^2}{\sin^2\theta} + \lambda\right) P_\lambda^m(\theta) = 0, \qquad 0 \leqslant \theta \leqslant \pi. \tag{10.37}$$

This equation is easier to treat by changing to the new variable $x = \cos\theta$, $x \in [-1, 1]$, whence $P_\lambda^m(\theta) \to P_\lambda^m(x)$ obeying

$$(1 - x^2)P_\lambda^{m\prime\prime}(x) - 2xP_\lambda^{m\prime}(x) + \left(\lambda - \frac{m^2}{1 - x^2}\right) P_\lambda^m(x) = 0, \tag{10.38}$$

with $' \equiv d/dx$. We shall refer to (10.38) as Legendre's equation.

It is helpful in solving (10.38) to recall that $\hat{\mathbf{L}}$ (and therefore $\hat{L}^2$) is invariant under inversion. Therefore, $P_\lambda^m(\theta)$ and $\Phi_m(\phi)$ must each be eigenstates of the parity operator. In the coordinate representation we readily find

$$\hat{U}_{\mathrm{P}}\Phi_m(\phi) = e^{im\pi}\Phi_m(\phi) = (-1)^m\Phi_m(\phi), \tag{10.39}$$

where (9.60) had been used. Hence, the parity of $\Phi_m(\phi)$ (or of $|m\rangle$) is

$$\pi_m = (-1)^m. \tag{10.40}$$

Although the parity of $P_\lambda^m(\theta)$ has yet to be determined, the fact that $[\hat{U}_{\mathrm{P}}, \hat{L}^2] = 0$ means that the solutions to (10.38) as well as to (10.37) can be classified as even or odd. This is obvious by inspection of (10.38) under the replacement $x \to -x$ ($x = \cos\theta \to -x$ when $\cos\theta \to \cos(\pi - \theta)$).

Equation (10.38) is singular at $x = \pm 1$, corresponding to $\theta = 0$ and $\theta = \pi$, the endpoints of the θ domain. From the theory of linear second-order differential equations in the complex plane (Copson 1935, Wyld 1976, Mathews and Walker 1970, Hassani 1991) one can show that there is one class of solutions to Eq. (10.38) that are regular at $x = \pm 1$ and another in which the solutions are singular at $x = \pm 1$. Only the former class is of interest in the present context. The solutions in this class are the ℓth-degree Legendre polynomials denoted $P_\ell(x = \cos\theta)$ when $m = 0$, whereas they are the associated Legendre functions denoted $P_\ell^m(x = \cos\theta)$ when $m \neq 0$. In each case, one finds $\lambda = \ell(\ell+1)$, $\ell = 0, 1, 2, 3, \ldots$. The reader may have encountered the $P_\ell(x)$ in electrostatics; they occur in a Taylor-series expansion of $(1 - 2x\rho + \rho^2)^{-1/2}$:

$$(1 - 2x\rho + \rho^2)^{-1/2} = \sum_{\ell=0}^{\infty} \rho^\ell P_\ell(x), \qquad \rho < 1. \tag{10.41}$$

The solving of Eq. (10.38) is carried out in Appendix C. We use a power-series expansion when $m = 0$, deriving both the eigenvalue relation $\lambda = \ell(\ell+1)$ and the functional form of $P_\ell(x)$. The $P_\ell^m(x)$, $m \neq 0$, are obtained via differentiation. Relevant properties of the solutions are discussed in the following.

Legendre functions

When $m = 0$, the solutions to (10.38) are the Legendre polynomials $P_\ell(x)$, where $\lambda = \ell(\ell+1)$, with one form of $P_\ell(x)$ given by Rodrigues' formula (Eq. (C9)):

$$P_\ell(x) = \frac{1}{2^\ell \ell!}\frac{d^\ell}{dx^\ell}(x^2 - 1)^\ell. \tag{10.42}$$

Much of the information one might desire concerning the Legendre polynomials can be obtained from this formula. For example, since the constant $(2^\ell \ell!)^{-1}$ in (10.42) ensures that $P_\ell(1) = 1$, it follows that

$$P_\ell(-1) = (-1)^\ell; \tag{10.43a}$$

furthermore

$$\hat{U}_\mathrm{P} P_\ell(x) = (-1)^\ell P_\ell(x), \tag{10.43b}$$

$$P_\ell(x) \text{ has } \ell \text{ zeros in } -1 \leqslant x \leqslant 1, \tag{10.43c}$$

and

$$\int_{-1}^{1} P_\ell^2(x)\, dx = \frac{2}{2\ell+1}, \tag{10.43d}$$

so that the "orthonormality" condition becomes

$$\int_{-1}^{1} dx\, P_\ell(x) P_n(x) = \delta_{\ell n}\frac{2}{2\ell+1}. \tag{10.43e}$$

The proof of Eq. (10.43e), which uses integration by parts, is left to the exercises.

Equal in importance to the preceding results are the functional forms of $P_\ell(x)$, which are also obtainable from Eq. (10.42); the first few are

$$P_0(x) = 1, \tag{10.44a}$$

$$P_1(x) = x, \tag{10.44b}$$

and

$$P_2(x) = \tfrac{1}{2}(3x^2 - 1). \tag{10.44c}$$

Other values of $P_\ell(x)$ are given in Table 10.1 later.

For $m = 0$, we have seen that $\lambda \to \ell(\ell+1)$; the corresponding change in $|\lambda m\rangle$ is to $|\ell 0\rangle$, with

$$P_\ell(\cos\theta) = \langle\theta\phi|\ell 0\rangle, \tag{10.45}$$

while

$$\hat{L}^2(\theta, \phi)P_\ell(\cos\theta) = \hbar^2\ell(\ell+1)P_\ell(\cos\theta). \tag{10.46}$$

The Legendre polynomials $\{P_\ell(x)\}$ are a complete set on the interval $[-1, 1]$:

$$f(x) = \sum_\ell f_\ell P_\ell(x), \qquad x \in [-1, 1], \tag{10.47a}$$

where

$$f_\ell = \frac{2\ell+1}{2}\int_{-1}^{1} dx\, P_\ell(x) f(x), \tag{10.47b}$$

with $f(x)$ being at least piece-wise continuous. In other words, the set $\{|\ell 0\rangle\}$ is complete:

$$\sum_\ell \frac{2\ell+1}{2}|\ell 0\rangle\langle\ell 0| = \hat{I}. \tag{10.47c}$$

When $m \neq 0$, the eigenvalue is still given by $\lambda = \ell(\ell+1)$, while the $m \neq 0$ solutions are now re-written as $P_\ell^m(x) = P_\ell^m(\cos\theta)$, the associated Legendre functions. They also form a complete set. As shown in Appendix C, $P_\ell^m(x)$ can be expressed as

$$P_\ell^m(x) = (1-x^2)^{|m|/2}\frac{d^{|m|}P_\ell(x)}{dx^{|m|}}; \tag{10.48}$$

both the derivative and power of $(1 - x^2)$ involve non-negative quantities. In texts written by mathematicians, a factor $(-1)^{|m|}$ is usually inserted on the RHS of (10.48), but physicists typically omit this phase factor, reserving it for the definition of the spherical harmonics $Y_\ell^m(\theta, \phi)$ introduced shortly. When looking up the $P_\ell^m(x)$ not inferrable from Table 10.1, the reader should be aware of possible phase-factor differences between (10.48) and other definitions.

Note that, because the eigenvalue in (10.38) is $\lambda = \ell(\ell+1)$ rather than m, the relevant orthogonality relation is

$$\int_{-1}^{1} dx\, P_\ell^m(x)P_{\ell'}^m(x) = 0, \qquad \ell' \neq \ell. \tag{10.49}$$

That is, the same value of m must occur in each associated Legendre function, otherwise the orthogonality relation (10.49) fails. The normalization integral for $P_\ell^m(x)$ turns out to be $2[(\ell+|m|)!]/(2\ell+1)[(\ell-|m|)!]$, so that the orthonormality relation is

$$\int_{-1}^{1} dx\, P_\ell^m(x) P_{\ell'}^m(x) = \frac{2}{2\ell+1}\frac{(\ell+|m|)!}{(\ell-|m|)!}\delta_{\ell\ell'}. \tag{10.50}$$

For $|m|$ an odd integer, $(1-x^2)^{|m|/2}$ is the $|m|$th power of $\sqrt{1-x^2}$. However, the variable x is shorthand for $\cos\theta$, and in terms of $\sin\theta$ and $\cos\theta$, we find

$$P_\ell^m(\theta) = (\sin\theta)^{|m|}\frac{d^{|m|}P_\ell(\cos\theta)}{d(\cos\theta)^{|m|}}. \tag{10.51}$$

As an illustration of this relation, let $\ell = 1$, $|m| = 1$. Then $dP_1(x)/dx$ is unity and $P_1^1(\theta) = \sin\theta$. The values of other P_ℓ^m's can be inferred from the spherical harmonics of the next subsection. We also note that, when $m = 0$, $P_\ell^0(x) = P_\ell(x)$.

It is shown in Appendix C that, for a given ℓ, m can take on only the values $m = -\ell, -\ell+1, \ldots, -1, 0, 1, 2, \ldots, \ell$. Therefore, consonant with the remark below (10.31), the allowed values of m in the ket $|\ell m\rangle$ depend on ℓ. A useful reminder of this is achieved by relabeling: $m \to m_\ell$, with

$$m_\ell = -\ell, -\ell+1, -\ell+2, \ldots, -2, -1, 0, 1, \ldots, \ell. \tag{10.52}$$

With this notational change, we have $P_\ell^m \to P_\ell^{m_\ell}(x)$, with

$$P_\ell^{m_\ell}(x) = (1-x^2)^{|m_\ell|/2}\frac{d^{|m_\ell|}}{dx^{|m_\ell|}}P_\ell(x).$$

From here on, $P_\ell^{m_\ell}(\cos\theta)$ will be the standard symbology for what we had initially denoted as $P_\ell^m(\theta)$:

$$\hat{L}^2(\theta,\phi)P_\ell^{m_\ell}(\cos\theta) = \hbar^2\ell(\ell+1)P_\ell^{m_\ell}(\cos\theta). \tag{10.53}$$

Spherical harmonics

In terms of the notational changes for the eigenvalues – $\lambda \to \ell(\ell+1)$, $\mu \to m_\ell$ – the Hilbert-space form of the orbital angular momentum eigenvalue equations becomes

$$\hat{L}^2|\ell, m_\ell\rangle = \hbar^2\ell(\ell+1)|\ell, m_\ell\rangle, \qquad \ell = 0, 1, 2, \ldots \tag{10.54a}$$

and

$$\hat{L}_z|\ell, m_\ell\rangle = \hbar m_\ell|\ell, m_\ell\rangle, \qquad |m_\ell| \leqslant \ell. \tag{10.54b}$$

In place of the temporary notation $Y_\lambda^\mu(\theta,\phi) = \langle\theta,\phi|\mu\lambda\rangle$, we now revert to the standard symbol[2] $Y_\ell^{m_\ell}$ for the coordinate-space eigenfunction:

$$Y_\ell^{m_\ell}(\theta,\phi) = \langle\theta,\phi|\ell, m_\ell\rangle, \tag{10.55}$$

where

$$Y_\ell^{m_\ell}(\theta,\phi) = N_\ell^{m_\ell}P_\ell^{m_\ell}(\cos\theta)e^{e^{im_\ell\phi}}. \tag{10.56}$$

[2] Although the use of Y is standard, the placement of the quantum numbers ℓ and m_ℓ as in (10.55) is not. Other notations include $Y_{\ell m_\ell}(\theta,\phi)$, $Y_{m_\ell}^{(\ell)}(\theta,\phi)$, and $Y_\ell^{(m_\ell)}(\theta,\phi)$. Many others also use m (or μ(!)) in place of our m_ℓ. *Caveat lector.*

The $Y_\ell^{m_\ell}(\theta, \phi)$ are known in physics as *spherical harmonics*; they are the coordinate-space eigenfunctions of $\hat{L}^2$ and $\hat{L}_z$. Since $\{|\ell, m_\ell\rangle\}$ is complete, i.e.,

$$\sum_{\ell, m_\ell} |\ell, m_\ell\rangle\langle\ell, m_\ell| = \hat{I}, \tag{10.57}$$

and $\langle\theta', \phi'|\theta, \phi\rangle \equiv \langle\Omega'|\Omega\rangle = \delta(\Omega' - \Omega)$, with $d\Omega = \sin\theta\, d\theta\, d\phi$ and $\delta(\Omega - \Omega') = \delta(\theta - \theta')\delta(\phi - \phi')$, then $N_\ell^{m_\ell}$ in (10.56) must be chosen so as to normalize $Y_\ell^{m_\ell}(\theta, \phi)$ to unity:

$$\int d\Omega\, |Y_\ell^{m_\ell}(\theta, \phi)|^2 = 1. \tag{10.58}$$

From Eq. (10.50) the magnitude of $N_\ell^{m_\ell}$ is easy to obtain, while the phase is chosen so that

$$Y_\ell^{m_\ell *}(\theta, \phi) = (-1)^{m_\ell} Y_\ell^{-m_\ell}(\theta, \phi). \tag{10.59}$$

The most widely used phase convention is that of Condon and Shortley:[3]

$$Y_\ell^{m_\ell} = \begin{cases} (-1)^{m_\ell}\sqrt{\dfrac{2\ell+1}{4\pi}\dfrac{(\ell-m_\ell)!}{(\ell+m_\ell)!}}P_\ell^{m_\ell}(\cos\theta)e^{im_\ell\phi}, & m_\ell \geqslant 0, \\ \sqrt{\dfrac{2\ell+1}{4\pi}\dfrac{(\ell-|m_\ell|)!}{(\ell+|m_\ell|)!}}P_\ell^{|m_\ell|}(\cos\theta)e^{im_\ell\phi}, & m_\ell < 0, \end{cases} \tag{10.60}$$

and it is the one adopted in this text. The normalization constants of (10.60) ensure that

$$\int d\Omega\, Y_{\ell'}^{m'_\ell *}(\Omega) Y_\ell^{m_\ell}(\Omega) = \delta_{\ell\ell'}\delta_{m_\ell m'_\ell}, \tag{10.61}$$

where the $\delta_{m_\ell m'_\ell}$ factor comes from the $\exp(im_\ell\phi)$ portion of $Y_\ell^{m_\ell}(\theta, \phi)$.

If we recall Eq. (9.70b), viz. $[\hat{U}_{\rm P}, \hat{\mathbf{L}}] = 0$, then parity must be a good quantum number for $|\ell, m_\ell\rangle$ and $Y_\ell^{m_\ell}(\theta, \phi)$. On applying Eq. (9.60) to Eqs. (10.60) via Eq. (10.51), we find that

$$\hat{U}_{\rm P} Y_\ell^{m_\ell}(\theta, \phi) = (-1)^\ell Y_\ell^{m_\ell}(\theta, \phi). \tag{10.62}$$

The parity of $Y_\ell^{m_\ell}$ (and thus of $|\ell m_\ell\rangle$) is $\pi_Y = (-1)^\ell$; hence, there is no need for an additional parity quantum number: the parity is specified by ℓ through $(-1)^\ell$.

When the eigenstate of a single particle is $|\ell, m_\ell\rangle$, the particle is said to have orbital angular momentum ℓ and z-component (or magnetic quantum number) m_ℓ, the parenthetical notation being clarified later. Thus, the probability density for finding a particle with orbital angular momentum ℓ and z-component m_ℓ at (θ, ϕ) is

$$\rho_{\ell m_\ell}(\theta, \phi) = |Y_\ell^{m_\ell}(\theta, \phi)|^2 = \rho_{\ell m_\ell}(\theta). \tag{10.63}$$

Evaluation of $\rho_{\ell m_\ell}(\theta)$ requires knowledge of $Y_\ell^{m_\ell}(\theta, \phi)$; the $\ell = 0$ and $\ell = 1$ spherical harmonics are given below, while Table 10.1 lists all of them for $\ell = 0$, 1, 2, and 3:

[3] Named after E. U. Condon and G. H. Shortley, whose monograph *Theory of Atomic Spectra* (Condon and Shortley (1963)) was the "bible" of atomic spectroscopy for several decades, and still is a useful reference.

Table 10.1 *Selected values of $Y_\ell^{m_\ell}(\theta, \phi)$ and $P_\ell(\cos\theta)$*

	$Y_\ell^{m_\ell}(\theta, \phi)$	$P_\ell(\cos\theta) = \sqrt{\frac{4\pi}{2\ell+1}} Y_\ell^0(\theta, \phi)$
$\ell = 0,\ m_\ell = 0$	$\sqrt{\frac{1}{4\pi}}$	1
$\ell = 1,\ m_\ell = 0$	$\sqrt{\frac{3}{4\pi}} \cos\theta$	$\cos\theta$
$\ell = 1,\ m_\ell = \pm 1$	$\mp\sqrt{\frac{3}{8\pi}} \sin\theta\, e^{\pm i\phi}$	
$\ell = 2,\ m_\ell = 0$	$\frac{1}{2}\sqrt{\frac{5}{4\pi}}(3\cos^2\theta - 1)$	$\frac{1}{2}(3\cos^2\theta - 1)$
$\ell = 2,\ m_\ell = \pm 1$	$\mp\sqrt{\frac{15}{8\pi}} \sin\theta\cos\theta\, e^{\pm i\phi}$	
$\ell = 2,\ m_\ell = \pm 2$	$\frac{1}{2}\sqrt{\frac{15}{8\pi}} \sin^2\theta\, e^{\pm 2i\phi}$	
$\ell = 3,\ m_\ell = 0$	$\frac{1}{2}\sqrt{\frac{7}{4\pi}} \cos\theta\,(5\cos^2\theta - 3)$	$\frac{1}{2}\cos\theta\,(5\cos^2\theta - 3)$
$\ell = 3,\ m_\ell = \pm 1$	$\mp\frac{1}{2}\sqrt{\frac{21}{4\pi}} \sin\theta\,(5\cos^2\theta - 1) e^{\pm i\phi}$	
$\ell = 3,\ m_\ell = \pm 2$	$\frac{1}{2}\sqrt{\frac{105}{8\pi}} \sin^2\theta\cos\theta\, e^{\pm 2i\phi}$	
$\ell = 3,\ m_\ell = \pm 3$	$\mp\frac{1}{2}\sqrt{\frac{35}{4\pi}} \sin^3\theta\, e^{\pm 3i\phi}$	
$\ell = 4,\ m_\ell = 0$	$\frac{1}{8}\sqrt{\frac{9}{4\pi}}\,(35\cos^4\theta - 30\cos^2\theta + 3)$	$\frac{1}{8}\,(35\cos^4\theta - 30\cos^2\theta + 3)$

$$Y_0^0(\theta, \phi) = \sqrt{\frac{1}{4\pi}}, \tag{10.64a}$$

$$Y_1^0(\theta, \phi) = \sqrt{\frac{3}{4\pi}} \cos\theta, \tag{10.64b}$$

$$Y_1^{\pm 1}(\theta, \phi) = \mp\sqrt{\frac{3}{8\pi}} \sin\theta\, e^{\pm i\phi}. \tag{10.64c}$$

From the definition (10.60) it follows that

$$Y_\ell^0(\theta, \phi) = \sqrt{\frac{2\ell+1}{4\pi}} P_\ell(\cos\theta), \tag{10.65}$$

so that the value of $P_\ell(\cos\theta)$ can be obtained from Y_ℓ^0, as in Table 10.1.

We shall show in Section 10.4 how $|\ell, m_\ell \pm 1\rangle$ is related to $|\ell m_\ell\rangle$, leading to operator relations that generate $Y_\ell^{m_\ell \pm 1}(\theta, \phi)$ from $Y_\ell^{m_\ell}(\theta, \phi)$. This will turn out to be the analog of the creation- and annihilation-operator relations for the eigenfunctions of the 1-D linear harmonic oscillator. It is clear from Table 10.1, however, that, for a given ℓ, the simplest functional forms of $Y_\ell^{m_\ell}$ are those with the lowest values of m_ℓ. To illustrate graphically the probability densities $\rho_{\ell m_\ell}(\theta)$, we choose the cases $\ell = 1$, $m_\ell = 0, \pm 1$ and $\ell = 2$, $m_\ell = 0$, displayed in Figs. 10.3 and 10.4. The probability distributions are all renormalized so that their maximum values are unity. They may be thought of as angular values distributed over a sphere of radius unity; since they are uniform in ϕ, only the $\phi = \pi/2$ and $3\pi/2$ values are shown in the figures. The directional character of these distributions is evident, those for $\ell = 1$ being characteristic of a dipole whereas that for $\ell = 2$, $m_\ell = 0$ is a typical quadrupole distribution. This directionality is an important feature of some chemical bonds, a point discussed in Chapter 18.

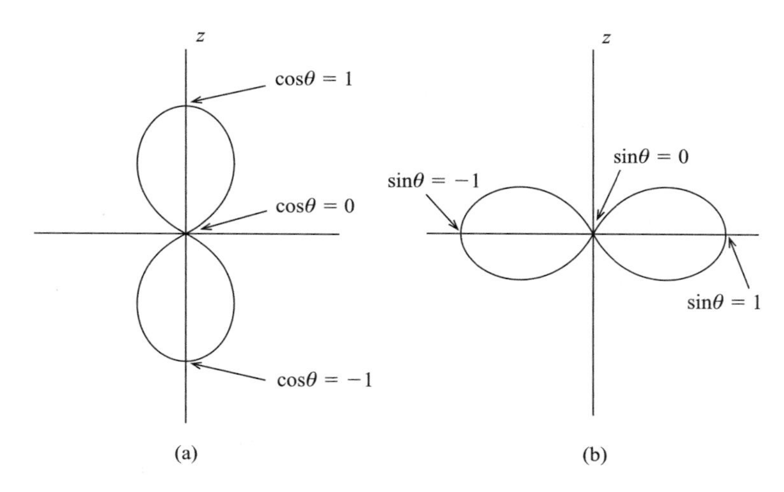

Fig. 10.3 Plots of the renormalized angular probability densities $\rho_{\ell m_\ell}(\theta)$: (a) $\rho_{10}(\theta) = \cos^2\theta$ and (b) $\rho_{1\pm1}(\theta) = \sin^2\theta$. Since $\rho_{\ell m_\ell}$ is uniform in ϕ, only the $\phi = \pi/2$ and $\phi = 3\pi/2$ values ($y = +1$ and $y = -1$, respectively) are displayed.

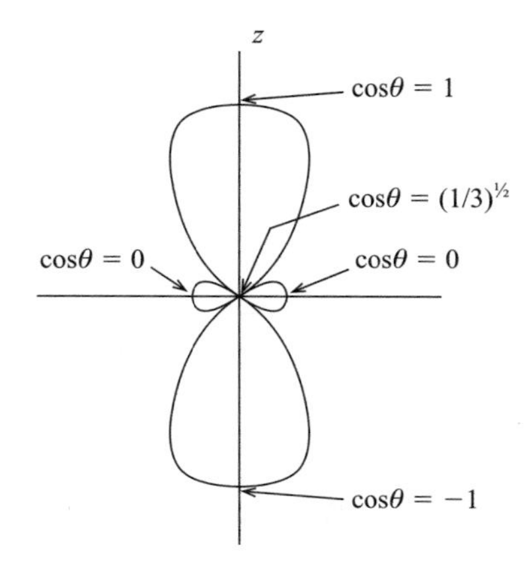

Fig. 10.4 The renormalized angular probability density $\tilde{\rho}_{20}(\theta) = \frac{1}{4}(3\cos^2\theta - 1)^2$, shown for $\phi = \pi/2$ and $\phi = 3\pi/2$.

Some properties of spherical harmonics

We begin by returning to Eqs. (10.17) and (10.18). Since $\hat{H}(\mathbf{r})$ is assumed to be rotationally invariant, i.e., $V(\mathbf{r}) = V(r)$, then the solution to (10.17), initially written in the form of (10.18) can now be re-expressed as

$$\psi_\Gamma(\mathbf{r}) = R_\Gamma(r) Y_\ell^{m_\ell}(\theta, \phi). \tag{10.18a}$$

The equation obeyed by $R_\Gamma(r)$ is obtained by substituting Eqs. (10.16) and (10.18a) into (10.17) and taking the scalar product of each term of the resulting expression with $Y_\ell^{m_\ell}$, which yields

$$\left[-\frac{\hbar^2}{2m}\left(\frac{1}{r^2}\frac{d}{dr}r^2\frac{d}{dr} - \frac{\ell(\ell+1)}{r^2}\right) + V(r)\right] R_\Gamma(r) = E_\Gamma R_\Gamma(r). \tag{10.66a}$$

In Eq. (10.66a) we see that a spherically symmetric potential not only implies a wave function of the form (10.18a), but also leads to a radial (eigenvalue) equation independent of m_ℓ. Let us consider bound states. For each ℓ, there may be sets of bound-state energies, say $\{E_{n\ell}\}$, and bound-state wave functions correspondingly labeled $\{R_{n\ell}(r)\}$. Assuming such to exist, the form of Eq. (10.66a) for bound states is

$$\left[-\frac{\hbar^2}{2m}\left(\frac{1}{r^2}\frac{d}{dr}r^2\frac{d}{dr} - \frac{\ell(\ell+1)}{r^2}\right) + V(r)\right] R_{n\ell}(r) = E_{n\ell} R_{n\ell}(r), \tag{10.66b}$$

with $E_{n\ell} \neq E_{n\ell m_\ell}$ and $R_{n\ell}(r) \neq R_{n\ell m_\ell}(r)$. Thus, for each ℓ, there is a $(2\ell + 1)$-fold degeneracy arising from the m_ℓ dependence of the full wave function. It will be shown in Section 12.2 that immersion of the system in a magnetic field lifts this degeneracy.

The label n is traditionally referred to as the principal quantum number. The preceding results imply that, for bound states in a spherically symmetric potential,

$$\psi_\Gamma(\mathbf{r}) \to \psi_{n\ell m_\ell}(r, \theta, \phi) = R_{n\ell}(r) Y_\ell^{m_\ell}(\theta, \phi). \tag{10.18b}$$

We shall consider a positive-energy case in Section 10.3.

Properties of the $Y_\ell^{m_\ell}(\Omega)$ are clearly important in quantum mechanics. The $m_\ell = 0$ behavior is given by Eq. (10.65). A further simple relation occurs when $\theta = 0$, i.e., when $(\theta, \phi) = \mathbf{e}_3$, the z-axis:

$$Y_\ell^{m_\ell}(\mathbf{e}_3) = \delta_{m_\ell 0}\sqrt{\frac{2\ell+1}{4\pi}}. \tag{10.67}$$

Simple relations between $Y_1^{m_\ell}(\theta, \phi)$ and the x, y, and z components of $\mathbf{r}$ also exist:

$$(x \pm iy)/r = \mp\sqrt{\frac{8\pi}{3}} Y_1^{\pm 1}(\theta, \phi) \tag{10.68a}$$

and

$$z/r = \sqrt{\frac{4\pi}{3}} Y_1^0(\theta, \phi). \tag{10.68b}$$

The combinations $(x + iy)/\sqrt{2}$, z, and $(x - iy)/\sqrt{2}$ are often referred to as the "spherical components of the vector $\mathbf{r}$," written as

$$r_1^{\pm 1} = \mp(x \pm iy)/\sqrt{2} = \sqrt{\frac{4\pi}{3}} r Y_1^{\pm 1}(\theta, \phi) \tag{10.69a}$$

and

$$r_1^0 = z = \sqrt{\frac{4\pi}{3}} r Y_1^0(\theta, \phi). \tag{10.69b}$$

Equations (10.69) not only relate the spherical and Cartesian components of **r**, but also show that **r** behaves like a spherical harmonic of rank 1. Similar relations hold for an arbitrary 3-D vector.

The orthonormality/completeness property of the $\{Y_\ell^{m_\ell}(\theta, \phi)\}$ also provides another useful relation. Let $\psi_\alpha(\mathbf{r})$ be the wave function for a 3-D particle whose Hamiltonian need not be spherically symmetric. Expansion of $\psi_\alpha(\mathbf{r})$ in terms of the spherical harmonics results in

$$\psi_\alpha(\mathbf{r}) = \sum_{\ell, m_\ell} R_{\alpha,\ell,m_\ell}(r) Y_\ell^{m_\ell}(\theta, \phi), \tag{10.70a}$$

where the radial function $R_{\alpha,\ell,m_\ell}(r)$ is given by

$$R_{\alpha,\ell,m_\ell}(r) = \int d\Omega\, Y_\ell^{m_\ell *}(\theta, \phi) \psi_\alpha(\mathbf{r}). \tag{10.70b}$$

A variety of probability densities can now be constructed from this expansion, wherein it is assumed that $\psi_\alpha(\mathbf{r})$ is normalizable. For example, $|R_{\alpha,\ell,m_\ell}(r)|^2 = |\int d\Omega\, Y_\ell^{m_\ell *}(\Omega)\psi_\alpha(\mathbf{r})|^2$ is the probability density that the particle will be at r and in angular momentum state $|\ell, m_\ell\rangle$. The probability density that the particle in state $|\alpha\rangle$ will have orbital angular momentum ℓ and z-component m_ℓ, and be found at position (r, θ, ϕ) is $|R_{\alpha,\ell,m_\ell}(r) Y_\ell^{m_\ell}(\theta, \phi)|^2$. Or, the probability density that the particle in state $|\alpha\rangle$ has z-component of angular momentum equal to m_ℓ and is at position (r, θ, ϕ) is $|\sum_{\ell \geqslant m_\ell} R_{\alpha,\ell,m_\ell}(r) Y_\ell^{m_\ell}(\theta, \phi)|^2$. Other probability distributions (and also probabilities) can be formulated as well.

The last property we discuss concerns a portion of the full completeness relation (10.57) expressed in coordinate space form:

$$\sum_{\ell, m_\ell} Y_\ell^{m_\ell *}(\Omega_1) Y_\ell^{m_\ell}(\Omega_2) = \delta(\Omega_1 - \Omega_2), \tag{10.71}$$

where Ω_1 and Ω_2 are arbitrary solid angles. Although Ω_1 and Ω_2 are arbitrary, we may associate them with radius vectors $\mathbf{r}_1$ and $\mathbf{r}_2$, as in Fig. 10.5. The key aspect of this figure is the angle

$$\theta = \cos^{-1}\left(\frac{\mathbf{r}_1 \cdot \mathbf{r}_2}{r_1 r_2}\right);$$

it is invariant under rotations.

The portion of Eq. (10.71) that we focus on is the sum on m_ℓ for a fixed value of ℓ. Our claim is that this latter sum – written out in Eq. (10.72) below – is, like θ, an invariant under rotations.

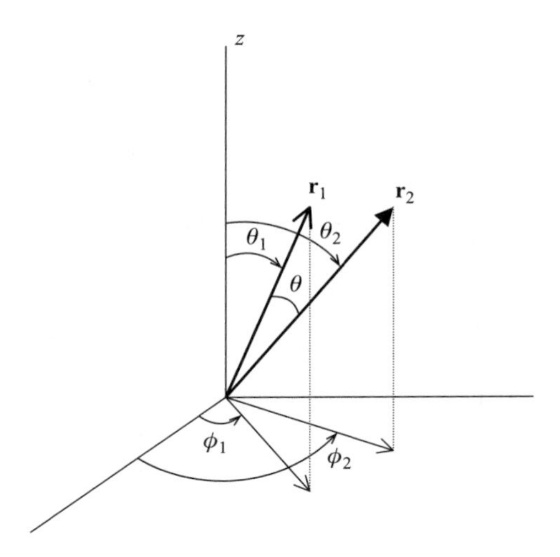

Fig. 10.5 Radius vectors $\mathbf{r}_1$ and $\mathbf{r}_2$, with their associated solid (or spherical) angles $\Omega_1 = (\theta_1, \phi_1)$ and $\Omega_2 = (\theta_2, \phi_2)$. Also shown is the angle θ between the two vectors: $\cos\theta = \mathbf{r}_1 \cdot \mathbf{r}_2/(r_1 r_2)$.

Because the sum is an invariant, it must depend only on θ, the angle "between" Ω_1 and Ω_2. This dependence is essentially given by $P_\ell(\cos\theta)$:

$$\sum_{m_\ell=-\ell} Y_\ell^{m_\ell *}(\Omega_1) Y_\ell^{m_\ell}(\Omega_2), = \frac{2\ell+1}{4\pi} P_\ell(\cos\theta), \tag{10.72a}$$

or equivalently,

$$Y_\ell^0(\theta) = \sqrt{\frac{4\pi}{2\ell+1}} \sum_{m_\ell=-\ell}^{\ell} Y_\ell^{m_\ell}(\Omega_1) Y_\ell^{m_\ell *}(\Omega_2). \tag{10.72b}$$

Notice that the complex conjugate is applied differently in the two forms of (10.72), a feature introduced to reinforce the fact that the sum on m_ℓ is real. Equation (10.72) is known as the spherical-harmonic-addition theorem; it is a generalization of the relation $\cos\theta = \cos\theta_1 \cos\theta_2 + \sin\theta_1 \sin\theta_2 \cos(\phi_1 - \phi_2)$ of spherical trigonometry. Indeed, the latter relation is just Eq. (10.72a) for the case $\ell = 1$, as is readily verified.

The essential and probably surprising aspect of the m_ℓ sum in Eq. (10.72) is that it is a rotational invariant; obviously the entire sum on ℓ and m_ℓ is as well. We shall make use of Eq. (10.72) in the next section. The proof of Eq. (10.72) is not difficult, but it is most easily accomplished via matrix elements of the rotation operator $\hat{U}_\mathrm{R}(\mathbf{n}, \chi) = \exp(-i\mathbf{n} \cdot \hat{\mathbf{L}}\chi/\hbar)$, where χ is the angle of rotation about the axis $\mathbf{n}$. As we do not make other use of these matrix elements in this book, the reader is referred to textbooks on angular momentum for the proof, e.g., Rose (1957).

10.3. Spherical-harmonics expansion of plane waves

In this section we use the spherical harmonics to expand the wave function for a free particle ($V(\mathbf{r}) = 0$). In working through the details involving the angular-momentum aspects, we will confront those radial functions known as spherical Bessel functions, the symbol for which is j_ℓ. The j_ℓ are encountered in 3-D bound-state and scattering problems, so our treatment of them here helps lay the foundation for their use later.

The Hamiltonian for a free particle is the kinetic-energy operator, the eigenvalue problem for which is

$$\hat{K}|\mathbf{k}\rangle = E_k|\mathbf{k}\rangle, \qquad E_k = \frac{\hbar^2 k^2}{2m}, \tag{10.73}$$

where $|\mathbf{k}\rangle$ is an eigenstate of the momentum operator $\hat{\mathbf{P}}$. If Eq. (10.73) were the only relevant relation, then $[\hat{\mathbf{L}}, \hat{K}] = 0$ would suggest that $|\mathbf{k}\rangle$ is also an orbital angular-momentum eigenstate. This is impossible, of course, because the other operator generating $|\mathbf{k}\rangle$ is $\hat{\mathbf{P}}$, and $[\hat{\mathbf{L}}, \hat{\mathbf{P}}] \neq 0$. While a plane wave is not an eigenstate of $\hat{\mathbf{L}}$, it can be expanded in angular-momentum eigenstates. Our immediate goal is to determine the "radial coefficients" in a spherical harmonic expansion of $\langle\mathbf{r}|\mathbf{k}\rangle$.

Coordinate representation, spherical Bessel functions

The essential part of $\langle\mathbf{r}|\mathbf{k}\rangle = \psi_{\mathbf{k}}(\mathbf{r})$ is the function $\exp(i\mathbf{k}\cdot\mathbf{r}) = \exp(ikr\cos\theta)$, where θ is the angle between $\mathbf{k}$ and $\mathbf{r}$. Since θ is the only angle that occurs in $\psi_{\mathbf{k}}(\mathbf{r})$, we may choose the wave vector $\mathbf{k}$ to lie along the z-axis of the coordinate system specifying $\mathbf{r}$. Furthermore, since the x- and y-axes of this coordinate system are arbitrary, we may also choose $\mathbf{r}$ to lie in the x–z plane, in which case the azimuthal angle ϕ of $\mathbf{r}$ is zero. Figure 10.6 illustrates these remarks; θ is seen to be the co-latitude. In creating such a coordinate system, we are "quantizing along the direction of the wave vector $\mathbf{k}$." Quantizing along a particular direction can be done only once in any particular analysis: should one or more other directions occur in a problem, e.g., that of a constant magnetic field $\mathbf{B}$, all such other directions must be referred to the chosen quantization axis, here $\mathbf{k}$.

With the only angle being the co-latitude θ, and thus no need for an azimuthal angle ϕ, the Legendre polynomials are the natural complete set of functions in which to expand $\exp(i\mathbf{k}\cdot\mathbf{r})$:

$$e^{i\mathbf{k}\cdot\mathbf{r}} = \sum_{\ell=0}^{\infty} R_\ell^{\mathrm{F}}(kr) P_\ell(\cos\theta), \tag{10.74}$$

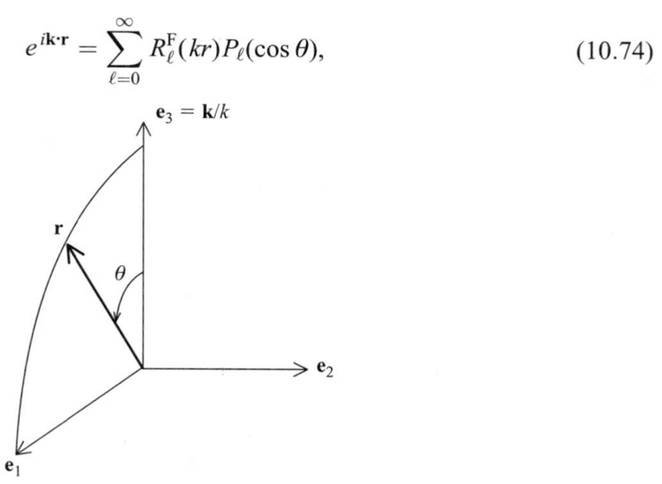

Fig. 10.6 The coordinate system in which $\mathbf{k}$ defines the z-axis and $\mathbf{r}$ lies in the x–z plane, making an angle θ (its co-latitude) with $\mathbf{k}$.

where the free-particle radial coefficient $R_\ell^{\rm F}(kr)$ is given by

$$R_\ell^{\rm F}(kr) = \frac{2\ell+1}{2}\int_0^\pi d\theta \sin\theta \, P_\ell(\cos\theta)e^{ikr\cos\theta}$$
$$= \frac{2\ell+1}{2}\int_{-1}^{1} dx\, P_\ell(x)e^{ikrx}. \tag{10.75}$$

Equation (10.75) is an integral expression for $R_\ell^{\rm F}(kr)$ that will be evaluated shortly. Before doing so, we remark that, since $\exp(ikr\cos\theta)$ is a solution of the time-independent Schrödinger equation for a free particle ($V = 0$) having energy $E = \hbar^2k^2/(2m)$, then $R_\ell^{\rm F}(kr)$ also obeys the following version of Eq. (10.66a):

$$\left(\frac{1}{r^2}\frac{d}{dr}r^2\frac{d}{dr} - \frac{\ell(\ell+1)}{r^2} + k^2\right)R_\ell^{\rm F}(kr) = 0,$$

or, dividing by k^2,

$$\left(\frac{1}{(kr)^2}\frac{d}{d(kr)}(kr)^2\frac{d}{d(kr)} - \frac{\ell(\ell+1)}{k^2r^2} + 1\right)R_\ell^{\rm F}(kr) = 0. \tag{10.66c}$$

In contrast to the case of bound states, the principal quantum number for the plane wave (and other continuum states) is the wave number k; rather than being appended to the radial coefficient as $R_{k\ell}^{\rm F}(r)$ it occurs as part of the argument of the function, $R_\ell^{\rm F}(kr)$. That the combination must be kr is evident both from (10.75) and from (10.66c).

Rather than attempting to solve Eq. (10.66c), a task taken up in the next chapter, we deal instead with Eq. (10.75). This expression can be evaluated by substituting Rodrigues' formula, Eq. (10.42), into (10.75). Doing so, and integrating by parts ℓ times, $R_\ell^{\rm F}(kr)$ becomes

$$R_\ell^{\rm F}(\rho) = \frac{2\ell+1}{\ell!2^{\ell+1}}(-i\rho)^\ell\int_{-1}^{1} dx\, e^{i\rho x}(x^2-1)^\ell, \tag{10.76}$$

where $\rho = kr$. Further integration by parts finally yields

$$R_\ell^{\rm F}(\rho) = (2\ell+1)i^\ell\left[(-\rho)^\ell\left(\frac{1}{\rho}\frac{d}{d\rho}\right)^\ell\frac{\sin\rho}{\rho}\right]. \tag{10.77}$$

The quantity in the square brackets of (10.77) defines $j_\ell(\rho)$, the spherical Bessel function of order ℓ:

$$j_\ell(\rho) \equiv (-\rho)^\ell\left(\frac{1}{\rho}\frac{d}{d\rho}\right)^\ell\frac{\sin\rho}{\rho}. \tag{10.78a}$$

Comparison of (10.77) and (10.75) shows that Eq. (10.78a) is equivalent to

$$j_\ell(\rho) = \frac{i^{-\ell}}{2}\int_{-1}^{1} dx\, P_\ell(x)e^{i\rho x}. \tag{10.78b}$$

From Eq. (10.78a), we see that $j_\ell(\rho)$ is a polynomial in $\sin\rho$ and $\cos\rho$. The three lowest-order $j_\ell(\rho)$ are

$$j_0(\rho) = \frac{\sin\rho}{\rho} \tag{10.79a}$$

$$j_1(\rho) = \frac{1}{\rho}\left(\frac{\sin\rho}{\rho} - \cos\rho\right) \tag{10.79b}$$

$$\begin{aligned} j_2(\rho) &= \frac{3}{\rho^2}\left(\frac{\sin\rho}{\rho} - \cos\rho\right) - \frac{\sin\rho}{\rho} \\ &= \frac{3}{\rho} j_1(\rho) - j_0(\rho). \end{aligned} \tag{10.79c}$$

Equation (10.79c) is an example of a general recursion-relation formula linking j_ℓ's of different order; it is discussed in the exercises. Each of the three preceding j_ℓ's is finite at $\rho = 0$; this is a general feature of the spherical Bessel functions. In particular, the power-series expansion of $j_\ell(\rho)$, which is discussed in the next chapter, yields the following small-ρ behavior:

$$j_\ell(\rho) \underset{\rho\to 0}{\to} \frac{2^\ell \ell!}{(2\ell+1)!}\rho^\ell. \tag{10.80}$$

This non-singular behavior at $\rho = 0$ is to be expected, since $\exp(i\mathbf{k}\cdot\mathbf{r})$ is finite at k or $r = 0$.

Using Eqs. (10.77) and (10.78), the spherical-harmonics expansion of the plane wave becomes

$$\begin{aligned} e^{i\mathbf{k}\cdot\mathbf{r}} &= \sum_{\ell=0}^{\infty} (2\ell+1) i^\ell P_\ell(\cos\theta) j_\ell(kr) \\ &= \sum_{\ell=0}^{\infty} \sqrt{4\pi(2\ell+1)}\, i^\ell Y_\ell^0(\Omega) j_\ell(kr), \end{aligned} \tag{10.81}$$

where $\Omega \to \theta(\phi = 0)$, and $\theta = \cos^{-1}[\mathbf{k}\cdot\mathbf{r}/(kr)]$. The (orbital) angular-momentum expansion (10.81) of the plane wave is also referred to as a "partial-wave" expansion, the ℓth term of the sum being the ℓth partial wave. This nomenclature arises in scattering theory (see Chapter 15), one of the areas in which (10.81) is very useful.

It was shown in Chapter 7 that careful use of (1-D) plane waves led to essentially the same results as those obtained via use of narrow wave packets. As will be discussed in Chapter 15, a similar situation holds for scattering in 3-D, wherein $\exp(i\mathbf{k}\cdot\mathbf{r})$ is the wave function describing a particle incident on a short-ranged potential. Of interest in this context is the behavior of $\exp(i\mathbf{k}\cdot\mathbf{r})$ for asymptotic values of r. The expansion (10.81) plus the asymptotic form of $j_\ell(\rho)$ is the standard way of analyzing this large-r behavior.

From the same power series representation as that which leads to (10.80), it can also be shown (Chapter 11) that

$$\begin{aligned} \lim_{\rho\to\infty} j_\ell(\rho) &\sim \frac{\sin(\rho - \ell\pi/2)}{\rho} \\ &= \frac{i}{2\rho}\left[e^{-i(\rho-\ell\pi/2)} - e^{i(\rho-\ell\pi/2)}\right], \end{aligned} \tag{10.82}$$

where "$\sim$" indicates asymptotic values of the argument (here ρ). The asymptotic form of

$j_\ell(\rho)$ is thus a difference of ingoing and outgoing spherical waves; apart from the minus sign, these waves are equally weighted.

The full plane wave behaves in a like manner:

$$e^{i\mathbf{k}\cdot\mathbf{r}} \underset{r\to\infty}{\sim} \frac{i}{kr}\sum_\ell \sqrt{\pi(2\ell+1)}i^\ell Y_\ell^0(\Omega)(e^{-i(kr-\ell\pi/2)} - e^{i(kr-\ell\pi/2)}). \tag{10.83}$$

Although this result is not easy to anticipate, it plays a crucial role in 3-D scattering theory: the amplitude for scattering is obtained from the asymptotic behavior of the solution of the time-independent Schrödinger equation, which is given by a sum of a plane wave and a scattered wave. We explore this in detail in Chapter 15.

The appearance in (10.81) of only $m_\ell = 0$ eigenfunctions is a signal that $\mathbf{k}$ has been chosen as the z (quantization)-axis. This restriction can be dropped, i.e., an arbitrary coordinate system can be introduced by using the spherical-harmonic-addition theorem, Eq. (10.72), in (10.81). Let the solid angles of $\mathbf{k}$ and $\mathbf{r}$ in such a coordinate system be $\Omega_k = (\theta_k, \phi_k)$ and $\Omega_r = (\theta_r, \phi_r)$, respectively, as per Fig. 10.7. In terms of these angles, we have[4]

$$P_\ell(\cos\theta) = \frac{4\pi}{2\ell+1}\sum_{m_\ell} Y_\ell^{m_\ell}(\Omega_r)Y_\ell^{m_\ell *}(\Omega_k). \tag{10.84}$$

Hence, Eq. (10.81) takes the alternate form

$$e^{i\mathbf{k}\cdot\mathbf{r}} = 4\pi\sum_{\ell,m_\ell} i^\ell Y_\ell^{m_\ell}(\Omega_r)Y_\ell^{m_\ell *}(\Omega_k)j_\ell(kr). \tag{10.85}$$

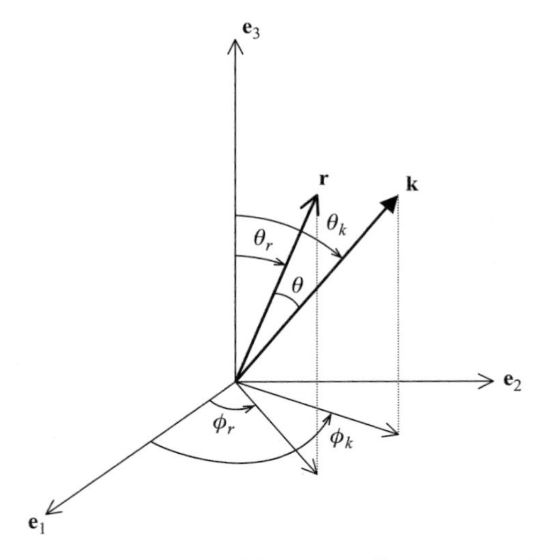

Fig. 10.7 The vectors $\mathbf{k}$ and $\mathbf{r}$ in an arbitrary coordinate system defined by the unit vectors $\mathbf{e}_1$, $\mathbf{e}_2$, and $\mathbf{e}_3$. The co-latitudes (θ_k and θ_r) and azimuths (ϕ_k and ϕ_r) are displayed, along with the angle θ between $\mathbf{k}$ and $\mathbf{r}$.

[4] In view of the reality of $P_\ell(\cos\theta)$, the assigning of the complex conjugate symbol "$*$" in (10.84) is not of major significance. We place it on $Y_\ell^{m_\ell}(\Omega_k)$ because $\exp(i\mathbf{k}\cdot\mathbf{r}) \propto \langle\mathbf{r}|\mathbf{k}\rangle$, while for $|\alpha\rangle \in \mathcal{H}$, $\langle\mathbf{r}|\alpha\rangle = \psi_\alpha(\mathbf{r}) = \sum_{\ell,m_\ell} R_{\alpha\ell m_\ell}(r)Y_\ell^{m_\ell}(\Omega_r)$, not $\sum_{\ell,m_\ell} R_{\alpha\ell m_\ell}(\mathbf{r})Y_\ell^{m_\ell *}(\Omega_r)$.

10.4. Orbital angular momentum uncertainty relations

The basic uncertainty relation is given by Eq. (9.17):

$$\Delta A\,\Delta B \geqslant \tfrac{1}{2}|\langle[\hat{A},\,\hat{B}]\rangle_\gamma|, \tag{9.17}$$

with

$$\Delta A = \sqrt{\langle\hat{A}^2\rangle_\gamma - \langle\hat{A}\rangle_\gamma^2}. \tag{9.8}$$

In the two subsections that follow, Eq. (9.17) is applied to components of the orbital angular-momentum operator. The system will be a spinless particle bound to a spherically symmetric potential $V(r)$, for which the state is $|\gamma\rangle = |n,\,\ell,\,m\rangle$, with n being the principal quantum number – recall Eq. (10.18b).

The two cases considered are $\hat{A} = \hat{L}_x$, $\hat{B} = \hat{L}_y$ and $\hat{A} = \hat{L}_z$, with $\hat{B}$ its canonically conjugate operator, which in a coordinate representation is expected to be the angle ϕ. As the following subsections indicate, these cases are non-trivial.

Ladder operators and the uncertainty relation $\Delta L_x\,\Delta L_y$

We consider the case $\hat{A} = \hat{L}_x$, $\hat{B} = \hat{L}_y$ first. From $[\hat{L}_x,\,\hat{L}_y] = i\hbar\hat{L}_z$, Eq. (9.17) becomes

$$\Delta L_x\,\Delta L_y \geqslant \tfrac{1}{2}\hbar\langle n,\,\ell,\,m_\ell|\hat{L}_z|n,\,\ell,\,m_\ell\rangle = m_\ell\hbar^2/2. \tag{10.86}$$

The LHS of the inequality (10.86) involves matrix elements of $\hat{L}_x$ and $\hat{L}_y$ and their squares, taken between angular-momentum eigenstates. These are previously undetermined matrix elements, and it is their evaluation that makes the $\Delta L_x\,\Delta L_y$ uncertainty relation an especially important one.

The two quantities on which we shall concentrate are $\hat{L}_x|n,\,\ell,\,m_\ell\rangle$ and $\hat{L}_y|n,\,\ell,\,m_\ell\rangle$. It is not straightforward to evaluate them directly because $|n,\,\ell,\,m_\ell\rangle$ is an eigenstate of neither $\hat{L}_x$ nor $\hat{L}_y$. An indirect evaluation is possible, however. A clue as to how one might proceed is the fact that each of the two preceeding quantities is an eigenstate of $\hat{L}^2$:

$$\hat{L}^2\left\{\begin{array}{l}\hat{L}_x|n,\,\ell,\,m_\ell\rangle\\ \hat{L}_y|n,\,\ell,\,m_\ell\rangle\end{array}\right\} = \ell(\ell+1)\hbar^2\left\{\begin{array}{l}\hat{L}_x|n,\,\ell,\,m_\ell\rangle\\ \hat{L}_y|n,\,\ell,\,m_\ell\rangle\end{array}\right\}, \tag{10.87}$$

a result that follows from $[\hat{L}^2,\,\hat{L}_x] = 0 = [\hat{L}^2,\,\hat{L}_y]$. Since $\hat{L}_x$ and $\hat{L}_y$ do not commute with $\hat{L}_z$, neither $\hat{L}_x|n,\,\ell,\,m_\ell\rangle$ nor $\hat{L}_y|n,\,\ell,\,m_\ell\rangle$ is an eigenstate of $\hat{L}_z$, but we shall now introduce an additional commutation relation that suggests applying $\hat{L}_z$ to linear combinations of these two quantities.

The additional commutation relation is derived from

$$[\hat{L}_z,\,\hat{L}_x] = i\hbar\hat{L}_y \tag{10.88a}$$

and

$$[\hat{L}_z,\,\hat{L}_y] = i\hbar\hat{L}_x \tag{10.88b}$$

by multiplying both sides of (10.88b) by $\pm i$ and then adding the result to Eq. (10.88a). This yields

$$[\hat{L}_z,\,\hat{L}_\pm] = \pm\hbar\hat{L}_\pm, \tag{10.89}$$

where

$$\hat{L}_\pm = \hat{L}_x \pm i\hat{L}_y. \tag{10.90}$$

For reasons that will become evident shortly, the (non-Hermitian) operators $\hat{L}_\pm$ are known as "ladder" or "raising/lowering" operators.

In view of Eq. (10.89), we consider in place of $\hat{L}_x|n, \ell, m_\ell\rangle$ and $\hat{L}_y|n, \ell, m_\ell\rangle$ the states

$$|\pm\rangle \equiv \hat{L}_\pm|n, \ell, m_\ell\rangle. \tag{10.91}$$

Acting on $|\pm\rangle$ with $\hat{L}_z$, and employing Eq. (10.91), we readily find that

$$\hat{L}_z|\pm\rangle = (m_\ell \pm 1)\hbar|\pm\rangle. \tag{10.92}$$

In other words, $|\pm\rangle$ is an (un-normalized) eigenstate of $\hat{L}_z$ with eigenvalue $m_\ell \pm 1$:

$$\hat{L}_\pm|n, \ell, m_\ell\rangle = C_\pm\hbar|n, \ell, m_\ell \pm 1\rangle, \tag{10.93}$$

where $C_\pm$ is a constant to be determined. Equations (10.92) and (10.93) underlie the appellations "ladder" and "raising/lowering" operators for the $\hat{L}_\pm$, since the effect of $\hat{L}_+/\hat{L}_-$ acting on $|n, \ell, m_\ell\rangle$ is to increase (raise)/decrease (lower) the value of m_ℓ by one unit. In other words, repeated action of $\hat{L}_\pm$ on $|n, \ell, m_\ell\rangle$ takes one up and down the "ladder" of m_ℓ values.

Use of the ladder operators in conjunction with $\hat{L}^2$ and $\hat{L}_z$ provides a purely algebraic (non-differential-equation) alternate for solving the $\hat{L}^2$, $\hat{L}_z$ eigenvalue problem, thereby yielding Eqs. (10.54). Aspects of this are addressed in the exercises and a full-scale use of this type of method is employed in Chapter 14 in connection with the eigenvalue problem for generalized angular momentum. Such an analysis is unnecessary at this stage, and we turn next to the determination of the constants $C_\pm$ and the ultimate evaluation of $\Delta L_x \Delta L_y$.

By construction,

$$\hat{L}_\pm^\dagger = \hat{L}_\mp; \tag{10.94}$$

using this in Eq. (10.93) we get

$$\begin{aligned} |C_\pm|^2 &= \langle n, \ell, m_\ell|\hat{L}_\mp\hat{L}_\pm|n, \ell, m_\ell\rangle \\ &= \langle n, \ell, m_\ell|\hat{L}_x^2 + \hat{L}_y^2 \pm i[\hat{L}_x, \hat{L}_y]|n, \ell, m_\ell\rangle \\ &= \langle n, \ell, m_\ell|\hat{L}^2 - \hat{L}_z^2 \mp \hbar\hat{L}_z|n, \ell, m_\ell\rangle \\ &= [\ell(\ell+1) - m_\ell(m_\ell \pm 1)], \end{aligned} \tag{10.95}$$

where $\langle n, \ell, m_\ell|n, \ell, m_\ell\rangle = 1$ has been used. It is customary to choose $C_\pm$ real and positive, so that, from (10.95),

$$\begin{aligned} C_\pm &= \sqrt{\ell(\ell+1) - m_\ell(m_\ell \pm 1)} \\ &= \sqrt{(\ell \mp m_\ell)(\ell \pm m_\ell + 1)}. \end{aligned} \tag{10.96}$$

Notice that $C_\pm = 0$ when $m_\ell = \pm\ell$, so that $\hat{L}_\pm$ cannot create a state in which $|m_\ell| > \ell$. That is, $\hat{L}_\pm$ and $C_\pm$ respect the limit $|m_\ell| \leqslant \ell$.

Equations (10.90), (10.93), and (10.96) are the keys to evaluating ΔL_x and ΔL_y. Solving (10.90) for $\hat{L}_x$ and $\hat{L}_y$ leads to

$$\hat{L}_x = \tfrac{1}{2}(\hat{L}_+ + \hat{L}_-) \tag{10.97}$$

and

$$\hat{L}_y = \frac{1}{2i}(\hat{L}_+ - \hat{L}_-), \tag{10.98}$$

from which we learn – via Eq. (10.93) – that diagonal matrix elements of $\hat{L}_x$ and $\hat{L}_y$ vanish:

$$\langle n,\, \ell,\, m_\ell|\hat{L}_x|n,\, \ell,\, m_\ell\rangle = \langle n,\, \ell,\, m_\ell|\hat{L}_y|n,\, \ell,\, m_\ell\rangle = 0. \tag{10.99}$$

Since only those matrix elements of $\hat{L}_\pm$ in which the magnetic quantum numbers differ by unity are non-zero, i.e.,

$$\langle n',\, \ell',\, m'_\ell|\hat{L}_\pm|n,\, \ell,\, m_\ell\rangle = \delta_{n'n}\delta_{\ell'\ell}\delta_{m'_\ell, m_{\ell\pm1}}\sqrt{(\ell \mp m_\ell)(\ell \pm m_\ell + 1)}, \tag{10.100}$$

one readily finds

$$\begin{aligned}\langle n,\, \ell,\, m'_\ell|\hat{L}_x|n,\, \ell,\, m_\ell\rangle = \tfrac{1}{2}\Big[&\delta_{m'_\ell, m_{\ell+1}}\sqrt{(\ell - m_\ell)(\ell + m_\ell + 1)} \\ &+ \delta_{m_\ell, m_{\ell-1}}\sqrt{(\ell + m_\ell)(\ell - m_\ell + 1)}\Big]\end{aligned} \tag{10.101}$$

and

$$\begin{aligned}\langle n,\, \ell,\, m'_\ell|\hat{L}_y|n,\, \ell,\, m_\ell\rangle = \frac{1}{2i}\Big[&\delta_{m'_\ell, m_{\ell+1}}\sqrt{(\ell - m_\ell)(\ell + m_\ell + 1)} \\ &-\delta_{m'_\ell, m_{\ell-1}}\sqrt{(\ell + m_\ell)(\ell - m_\ell + 1)}\Big].\end{aligned} \tag{10.102}$$

Evaluation of ΔL_x and ΔL_y is now straightforward. In view of (10.99) we have

$$\Delta L_j = \sqrt{\langle \hat{L}_j^2\rangle_{n,\ell,m_\ell}}, \qquad j = x \text{ or } y.$$

Making use of Eqs. (10.97) and (10.98) and the commutation relation

$$[\hat{L}_+,\, \hat{L}_-] = 2\hbar\hat{L}_z, \tag{10.103}$$

it is straightforward to show that

$$\Delta L_x = \Delta L_y = \hbar\sqrt{[\ell(\ell+1) - m_\ell^2]/2}. \tag{10.104}$$

Substituting Eq. (10.104) into Eq. (10.86) leads to

$$\ell(\ell+1) \geqslant m_\ell(m_\ell + 1), \tag{10.105}$$

a relation valid for all m_ℓ, with the equality holding only when $m_\ell = \ell$. It should be no surprise that the uncertainty relation is verified.

As an aside, we remark that Eqs. (10.54) and (10.100)–(10.102) allow one to use the eigenstate representation to convert the various angular-momentum operators into a set of matrices, in analogy to what was done in Section 6.2 for the case of the linear harmonic oscillator. We leave this construction to the exercises.

$\hat{L}_z$ and the angle operator

The commutation relation

$$[\hat{P}_j(\mathbf{r}), x_j] = -i\hbar \tag{10.106a}$$

is the coordinate-space realization of

$$[\hat{P}_j, \hat{Q}_j] = -i\hbar, \tag{10.106b}$$

given that $\hat{P}_j(\mathbf{r}) = -i\hbar\, d/dx_j$. We say that $\hat{P}_j$ and $\hat{Q}_j$ are canonically conjugate Hilbert-space operators, and that $\hat{P}_j(\mathbf{r})$ and x_j are canonically conjugate quantities in a Cartesian-coordinate representation.

It would seem as if this characterization also applies to the azimuthal angle ϕ and the z-component of $\hat{\mathbf{L}}$ in the spherical-polar-coordinate representation, wherein $\hat{L}_z(\phi) = -ih\, d/d\phi$, since for these quantities the analog of Eq. (10.106a) is

$$[\hat{L}_z(\phi), \phi] = -i\hbar. \tag{10.107}$$

Despite appearances, Eq. (10.107) is an *anomaly*, not a coordinate-space realization of a Hilbert-space commutation relation. As we shall demonstrate, Eq. (10.107) leads to inconsistencies, a sure sign that a seemingly valid mathematical relationship is, in fact, a pathological one.

The flaw invalidating Eq. (10.107) is that ϕ cannot be assigned a unique value – i.e., it cannot be measured – and therefore there is no Hilbert-space operator to which it corresponds. Non-uniqueness is a consequence of the relations

$$x = r\sin\theta\cos\phi \tag{10.108a}$$

and

$$y = r\sin\theta\sin\phi, \tag{10.108b}$$

since any of the infinite number of angles

$$\phi_n = \phi + 2n\pi, \qquad n = 0, \pm 1, \pm 2, \ldots,$$

can be used in Eqs. (10.108) to yield the unique values of x and y. It might be thought that this problem could be overcome by restricting ϕ to values in the interval $[0, 2\pi]$, but, as Peierls (1979) has pointed out, such a restriction would make ϕ discontinuous, with adverse consequences for the relation $\hat{L}_z(\phi) = -i\hbar\, d/d\phi$.

Are these problems conceptual rather than practical, i.e., could we not employ (10.107) anyway, say as the basis for deriving a meaningful uncertainty relation? The answer is NO. To show this, let us assume, in analogy to $\hat{\mathbf{Q}}$ or $\hat{Q}_j$, that there is an angle operator $\hat{\mathbf{\Phi}}$ whose coordinate representation is ϕ:

$$\hat{\mathbf{\Phi}}|\phi\rangle \stackrel{?}{=} \phi|\phi\rangle, \tag{10.109a}$$

such that

$$\langle\phi'|\hat{\mathbf{\Phi}}|\phi\rangle \stackrel{?}{=} \phi\delta(\phi' - \phi). \tag{10.109b}$$

Because of the $\phi \to \phi + 2n\pi$ anomaly, the $\stackrel{?}{=}$ symbol indicates our doubts concerning the existence of an unambiguous $\hat{\mathbf{\Phi}}$.

If $\hat{\Phi}$ is meaningful, then the operator analog of Eq. (10.107) is

$$[\hat{L}_z, \hat{\Phi}] = -i\hbar. \tag{10.110}$$

However, Eq. (10.110) cannot be correct, since sandwiching both sides of it between $\langle n, \ell, m_\ell'|$ and $|n, \ell, m_\ell\rangle$ leads to an invalid result, viz.,

$$\begin{aligned} -i\hbar\delta_{m_\ell' m_\ell} &= \langle n, \ell, m_\ell'|[\hat{L}_z, \hat{\Phi}]|n, \ell, m_\ell\rangle \\ &= (m_\ell' - m_\ell)\hbar\langle n, \ell, m_\ell'|\hat{\Phi}|n, \ell, m_\ell\rangle. \end{aligned} \tag{10.111}$$

Equation (10.111) is invalid because its LHS is non-zero only for those values of m_ℓ' and m_ℓ which cause its RHS to vanish, thereby establishing an inconsistency, as claimed below Eq. (10.107). The ambiguous definition (10.109) of the angle operator also implies an inconsistency when one uses (10.110) in an attempt to set up an uncertainty relation, viz.,

$$\Delta L_z\, \Delta\Phi \overset{?}{\geqslant} \hbar/2; \tag{10.112}$$

a demonstration of the inconsistency associated with (10.112) is left to the exercises.

This analysis serves to caution the reader against too readily generalizing results of a particular representation – here, spherical polar coordinates. However, the failure of Eqs. (10.110) and (10.112) does not preclude there being an operator more appropriate than $\hat{\Phi}$. All that is needed is an operator whose coordinate (ϕ-)representation is unambiguous. In fact, there are many operators that one might meaningfully use in place of the $\hat{\Phi}$ defined by Eq. (10.109), the variety of choice mirroring the variety expressed by the set $\phi_n = \phi + 2n\pi$. Possibilities include $\exp(i\hat{\Phi})$, $\cos\hat{\Phi}$, and $\sin\hat{\Phi}$. Suppose that we choose $\exp(i\hat{\Phi})$ as the angle operator, now denoted $\hat{\Upsilon}$:

$$\hat{\Upsilon} = e^{i\hat{\Phi}}. \tag{10.113}$$

The fact that $\hat{\Upsilon}$ is unitary rather than Hermitian is not a problem, since the angle is not an observable.

To see that (10.113) yields consistent results, we evaluate its commutator with $\hat{L}_z$, using a spherical polar coordinate representation:

$$[\hat{L}_z(\phi), \hat{\Upsilon}(\phi)] = \hbar\hat{\Upsilon}(\phi) = \hbar e^{i\phi}. \tag{10.114}$$

Formation of the matrix element that led to Eq. (10.111) now leads from Eq. (10.114) to

$$(m_\ell' - m_\ell)\hbar\langle n, \ell, m_\ell'|\hat{\Upsilon}|n, \ell, m_\ell\rangle = \hbar\langle n, \ell, m_\ell'|\hat{\Upsilon}|n, \ell, m_\ell\rangle, \tag{10.115}$$

i.e., to

$$m_\ell' = m_\ell + 1. \tag{10.116}$$

In other words, $\hat{\Upsilon}$ is a raising operator:

$$\hat{\Upsilon} = C_+^{-1}\hat{L}_+,$$

where $C_+ = \sqrt{(\ell - m_\ell)(\ell + m_\ell + 1)}$. Correspondingly, $\hat{\Upsilon}^\dagger$ is a lowering operator:

$$\hat{\Upsilon}^\dagger = C_-^{-1}\hat{L}_-.$$

Further properties of $\hat{\Upsilon}$ and its consequent uncertainty relations are developed via the exercises.

Exercises

10.1 Starting with the formulae for $\hat{L}_x(x, y, z)$ and $\hat{L}_y(x, y, z)$, express x, y, z in terms of r, θ and ϕ and then use the chain rule for partial derivatives to derive Eqs. (10.11a) and (10.11b).

10.2 In 2-D circular coordinates, the gradient and Laplacian operators are $\nabla = \mathbf{e}_r\,\partial/\partial r + \mathbf{e}_\phi\,\partial/r\,\partial\phi$ and $\nabla^2 = (1/r)\,\partial(r\,\partial/\partial r)/\partial r - [1/(\hbar^2 r^2)]\hat{L}_z^2(\phi)$.

(a) Determine $\hat{P}_{\text{rad}}^{(2)}(r)$, the 2-D analog of the radial-momentum operator of Eq. (10.21), and then evaluate $[r, \hat{P}_{\text{rad}}^{(2)}(r)]$.

(b) What are the 2-D analogs $\psi_{\pm k}(r)$ of the 3-D radial-momentum eigenfunctions $[\exp(\pm ikr)]/r$?

(c) Express the 2-D kinetic-energy operator $\hat{K}(r, \phi)$ in terms of $\hat{P}_{\text{rad}}^{(2)}(r)$ and $\hat{L}_z(\phi)$, and then establish whether the products $\psi_{\pm k}(r)\Phi_m(\phi)$ are eigenfunctions of $\hat{K}(r, \phi)$. If so, find the eigenvalues.

10.3 Use the values $x = 1$, 0, and -1 in the LHS of Eq. (10.42) to verify (i) Eq. (10.43a), (ii) the results $P_\ell(1) = 1$ and $P_{2n+1}(0) = 0$, and (iii) the relation $P_{2n}(0) = (-1)^n[(2n-1)!!]/f(n)$, where $f(n)$ is to be determined, with $n = 0, 1, 2, \ldots$.

10.4 Use integration by parts to establish the RHS of Eq. (10.43d). Also show that $\int_{-1}^{1} dx\, x^n P_m(x) = 0$, $n < m$.

10.5 Let $F(x, \rho) = (1 - 2x\rho + \rho^2)^{-1/2}$, the generating function for the P_ℓ's, Eq. (10.41).

(a) Derive an equation relating F and $\partial F/\partial\rho$.

(b) Substitute the RHS of (10.41) into the result of part (a) and show that it leads to the recursion relation $(\ell+1)P_{\ell+1}(x) - (2\ell+1)xP_\ell(x) + \ell P_{\ell-1}(x) = 0$.

10.6 Use $P_0(x) = 1$, $P_1(x) = x$ and the recursion relation of the preceding exercise to verify the $P_\ell(x)$ entries in Table 10.1 ($x = \cos\theta$) for the cases $\ell = 2$, 3, and 4, and from them determine $P_5(x)$.

10.7 (a) Evaluate the coefficients $C_\ell(x_0)$ in the expansion $\delta(x - x_0) = \sum_\ell C_\ell(x_0)P_\ell(x)$.

(b) By setting $x_0 = 0$, show that the properties of $C_\ell(0)$ ensure that only even values of ℓ enter the preceding expansion of $\delta(x)$.

(c) Sketch on the same graph the approximations to $\delta(x)$ that occur when the first two and then the first three terms in the Legendre polynomial expansion are kept, where $x \in [-1, 1]$.

10.8 Set $\rho = 1$ in Eq. (10.41) and sketch together the approximations to $(2 - 2x)^{-1/2}$ obtained by keeping the first two, the first three, and then the first four terms on the RHS of (10.41).

10.9 Construct plots of $|Y_2^m(\theta, \phi)|^2$ for $m = 1$ and 2, analogous to those of Figs. 10.3 and 10.4, and discuss the three $\ell = 2$ curves ($m = 0$, 1, and 2).

10.10 (a) Equations (10.68) and (10.69) are relations between x, y, and z and $Y_1^m(\theta, \phi)$, $m = 0, \pm 1$. Find analogous relations for the five $Y_2^m(\theta, \phi)$ in terms of x, y, and z.

(b) Using the expressions from Table 10.1 and Eq. (10.12), verify by explicit calculation that $Y_2^0(\theta, \phi)$ and $Y_2^{\pm 1}(\theta, \phi)$ are eigenfunctions of $L^2(\theta, \phi)$ with eigenvalues $6\hbar^2$.

10.11 Let $\hat{H}(\mathbf{r}) = \hat{K}(\mathbf{r}) + V(r)$, such that $(\hat{H} - E_n)|\Gamma_n\rangle = 0$, $n = 1, 2$, where the two bound-state energies are different. Show that $\langle\Gamma_1|\hat{\mathbf{L}}|\Gamma_2\rangle = 0$.

10.12 Assume that $\Psi_\Gamma(\mathbf{r}, t)$ is a solution of the time-dependent Schrödinger equation for a single particle in a spherically symmetric potential such that, at $t = 0$, $\Psi_\Gamma(\mathbf{r}, 0)$ is an eigenstate of $\hat{L}^2$ and $\hat{L}_z$ but not of the Hamiltonian $\hat{H}(\mathbf{r}) = \hat{K}(\mathbf{r}) + V(r)$.
(a) Let $m\hbar$ be the eigenvalue of $\hat{L}_z$ and $\langle H\rangle_t$ be the expectation value of $\hat{H}$ for the single particle at $t > 0$. Is $\langle\hat{H}\rangle_t$ (i) independent of m? (ii) linear in m? or (iii) neither of the foregoing? Justify your conclusions.
(b) Calculate the expectation values of $\cos\theta$ and $\cos\phi$ for all $t \geqslant 0$.

10.13 Suppose that $\hat{A}$ is an operator such that $[\hat{A}, \hat{L}_x] = 0 = [\hat{A}, \hat{L}_y]$. Determine whether $[\hat{A}, \hat{L}^2] = 0$ also.

10.14 Let $|\alpha, m\rangle$ and $|\beta, m'\rangle$ be eigenstates of $\hat{L}_z$ with eigenvalues $m\hbar$ and $m'\hbar$, respectively.
(a) For an arbitrary operator $\hat{A}$, define $\hat{B} = [\hat{L}_z, \hat{A}]$, and then show that $\langle\beta, m'|\hat{B}|\alpha, m\rangle = \hbar(m' - m)\langle\beta, m'|\hat{A}|\alpha, m\rangle$.
(b) Prove that $\langle\beta, m|\mathbf{e}_1|\alpha, m\rangle = 0 = \langle\beta, m|\mathbf{e}_2|\alpha, m\rangle$.
(c) Correspondingly, prove that, if $\hat{\mathbf{O}} = \hat{\mathbf{Q}}, \hat{\mathbf{P}}$, or $\hat{\mathbf{L}}$, then $\langle\beta, m|\hat{\mathbf{O}}|\alpha, m\rangle = \langle\beta, m|\hat{O}_z|\alpha, m\rangle\mathbf{e}_3$.

10.15 A system consisting of a 3-D particle of mass m and charge q, acted on by a central potential $V(r)$, is placed in a weak, uniform magnetic field $\mathbf{B} = B\mathbf{e}_3$. As will be shown in Chapter 12, a good approximation to the Hamiltonian is often $\hat{H}(\mathbf{r}) = \hat{K}(\mathbf{r}) + V(r) + \omega_L\hat{L}_z$, where $\omega_L \propto B$ is the Larmor frequency.
(a) Assume that, at $t = t_0$, $\Psi_\Gamma(r, \theta, \phi, t_0)$ is an eigenstate of $\hat{H}_0 = \hat{H} - \omega_L\hat{L}_z$, but not of $\hat{L}_z$ itself. Show that the probability density $\rho_\Gamma(r, \theta, \phi, t) = |\Psi_\Gamma(r, \theta, \phi, t)|^2$ for all $t \geqslant t_0$ can be written as $\rho_\Gamma(r, \theta, \phi(t), t_0)$, where $\phi(t) = t - \omega_L(t - t_0)$, corresponding to the system rotating about $\mathbf{e}_3$.
(b) Solve the Heisenberg equations of motion for each of the components $\hat{L}_j^{(H)}(t - t_0) = \{\exp[i\hat{H}(t - t_0)/\hbar]\}\hat{L}_j\{\exp[-i\hat{H}(t - t_0)/\hbar]\}$, $j = 1, 2, 3$. Establish which of these operators, if any, also corresponds to a rotation about $\mathbf{e}_3$.
(c) Redo part (b) with $\hat{Q}_j$ and then $\hat{P}_j$ replacing $\hat{L}_j$.

10.16 For a particle of mass m acted on by the 3-D harmonic oscillator potential $V(r) = m\omega^2r^2/2$, the eigenfunctions are the (normalized) products $R_{n_r\ell}(r)Y_\ell^{m_\ell}(\Omega)$; they belong to the eigenvalues $E_{n_r\ell} = (2n_r + \ell + \frac{3}{2})\hbar\omega$, where $n_r = 0, 1, 2, \ldots$ and $\ell = 0, 1, 2, \ldots$. At time $t = 0$, the particle is prepared in the state $\Psi_\Gamma(\mathbf{r}, 0) = [(6)^{-1/2}R_{01}(r)Y_1^1(\Omega) - (3)^{-1/2}R_{10}(r)Y_0^0(\Omega) + (3)^{-1/2}R_{12}(r)Y_2^2(\Omega) - (6)^{-1/2}R_{13}(r)Y_3^2(\Omega)]$.
(a) What is $\Psi_\Gamma(\mathbf{r}, t)$, the wave function at times $t > 0$?
(b) Determine the probabilities that, at time t, the particle will be observed in an eigenstate (i) of $\hat{L}^2$ with eigenvalues $2n\hbar^2$, $n = 0, 1, 2, 3$; and (ii) of $\hat{L}_z$ with eigenvalues $\hbar, 2\hbar, 3\hbar$.
(c) (i) What is the probability density that, at time, t, the particle will be found at position r, θ, ϕ when its m_ℓ value is 2? (ii) What is the total probability that the particle will be observed at time t with the same m_ℓ value?
(d) Determine the probabilites that, at time t, an energy measurement on the particle would yield the values $5\hbar\omega/2$, $7\hbar\omega/2$, $9\hbar\omega/2$, and $11\hbar\omega/2$.
(e) If $\hat{L}^2$ or $\hat{L}_z$ were measured at time t, what values would be expected?

(f) If the particle's energy were measured at time t, what value would one expect to observe?

(g) What would $\Psi_\Gamma(\mathbf{r}, t)$ be if, at $t = 0$, measurement of $\hat{L}_z$ yielded (i) $\hbar$ and (ii) $2\hbar$?

10.17 The probability density that a particle in 3-D will be found at position r, θ, ϕ is $\rho_a(r, \theta, \phi) = Nr^2[\exp(-r^2/b^2)]\cos^2\theta$.

(a) Determine r_m, the most probable value of the radius, expressing the result in terms of b.

(b) Evaluate (i) $\langle\cos\theta\rangle_a$, the expected value of $\cos\theta$; (ii) $\langle\cos^2\theta\rangle_a$, the variance of $\cos\theta$.

10.18 A particle in 3-D is described by the wave function $\psi(\mathbf{r}) = N(ax^2 + by^2 + czx)\exp(-r^2/\beta^2)$, where a, b, c, and β are real constants.

(a) Let $a = b = c = 1$. What are the probabilities that (i) $\ell = 0, 1, 2, \ldots$ will occur if $\hat{L}^2$ is measured, (ii) $m_\ell = 0, \pm1, \pm2, \ldots$ will occur if $\hat{L}_z$ is measured, (iii) $\pm\hbar$ will occur if $\hat{L}_y$ is measured?

(b) Redo part (a) for the case $c = 0$.

10.19 (a) Show that $\psi_\pm(\mathbf{r}) = N(x \pm iy)f(r)$ is an eigenfunction both of $\hat{L}^2$ and of $\hat{L}_z$, and state the corresponding eigenvalues.

(b) Construct a wave function $\psi_0(\mathbf{r})$ that is an eigenfunction of $\hat{L}^2$ whose eigenvalue is the same as in part (a), but whose $\hat{L}_z$ eigenvalue differs by one unit from those found in (a).

(c) Find an eigenfunction $\hat{L}^2$ and $\hat{L}_x$, analogous to those of parts (a) and (b), that has the same $\hat{L}^2$ eigenvalue but whose $\hat{L}_x$ eigenvalue is maximal.

(d) If the wave function for a particle is that of part (c), what are the probabilities that it will be found in each of the states described by the wave functions ψ_0, ψ_+, and ψ_- of parts (b) and (a)?

10.20 The simplest quantal rigid rotator in 3-D is a particle of mass m constrained to move on a sphere, whose radius is here taken to be a. The Hamiltonian for such a system is the angular part of the kinetic energy operator, viz., $\hat{H}(\theta, \phi) = \hat{L}^2(\theta, \phi)/(2ma^2)$, where ma^2 is the moment of inertia.

(a) Find the eigenvalues and normalized eigenfunctions of $\hat{H}(\theta, \phi)$, state the degeneracy of each eigenvalue, and specify the spacing between adjacent energy levels.

(b) At $t = 0$, the wave function for a rigid rotator is $\Psi_\Gamma(\theta, \phi, t = 0) = C_0 + iC_1\cos\theta$, with C_0 and C_1 real. What are the $t \geqslant 0$ expectation values of (i) $\mathbf{r} = a\mathbf{e}_r$, (ii) $\hat{\mathbf{L}}$, and (iii) $\hat{L}^2$?

10.21 A system of N, non-interacting rigid rotators, each having moment of inertia I (see Exercise 10.20 for details), is put through two successive measuring devices. The effect of the first is to limit the range of ℓ values to the set $\{0, 1, \ldots, 5\}$. The effect of the second, for each rotator, is to fix the value of m_ℓ to unity, the parity to positive, and the average energy to $6\hbar^2/I$. How many rotators N_ℓ are there with each orbital angular-momentum quantum number ℓ?

10.22 The $t = 0$ wave function for a rigid rotator (Exercise 10.20) is $\Psi_1(\theta, \phi) = [3/(4\pi)]^{1/2}\sin\theta\cos\phi$.

(a) What would the $t \geqslant 0$ outcomes be if (i) $\hat{L}_x$, (ii) $\hat{L}_y$, and (iii) $\hat{L}_z$ were to be measured?

(b) What are the expectation values of (i) $\hat{L}_y$, (ii) $\hat{L}^2$, and (iii) $\mathbf{e}_r$ for all $t \geqslant 0$?

(c) If Ψ_1 were replaced at $t = 0$ by $\Psi_2(\theta, \phi, 0) = (8\pi)^{-1/2}[1 + 3^{1/2} \sin\theta \cos\phi]$, what would the $t \geqslant 0$ expectation value of $\mathbf{e}_r$ now be?

10.23 Using integration by parts, prove that Eq. (10.77) follows from Eq. (10.75).

10.24 (a) Show, starting with Eq. (10.78), that (i) $d[\rho^{\ell+1} j_\ell(\rho)]/d\rho = f_1(\rho) j_{\ell-1}(\rho)$, and (ii) $d[\rho^{-1} j_\ell(\rho)]/d\rho = f_2(\rho) j_{\ell+1}(\rho)$, where f_1 and f_2 are functions you are to determine.

(b) Using the preceding results, derive the recursion relation $j_{\ell-1}(\rho) + j_{\ell+1}(\rho) = [(2\ell + 1)/\rho] j_\ell(\rho)$.

10.25 (a) Use the foregoing recursion relation and Eqs (10.79b) and (10.79c) to express $j_3(\rho)$ in terms of $\sin\rho$ and $\cos\rho$.

(b) From the answer to part (a) obtain the $\rho \to 0$ and $\rho \to \infty$ limits of $j_3(\rho)$ and compare them with the results of Eqs. (10.80) and (10.82).

10.26 (a) The ladder operators $\hat{L}_\pm$ are defined by Eq. (10.90). Determine their coordinate representations $\hat{L}_\pm(\theta, \phi)$.

(b) Equation (10.93) plus $m_\ell \leqslant \ell$ means that $\hat{L}_+|\ell, \ell\rangle = 0$. Use the latter relation to determine the θ dependence of $P_\ell^\ell(\theta)$.

(c) From the result of part (b) and Eq. (10.93), find the functional dependences of $Y_\ell^{\ell-1}(\theta, \phi)$ on θ and ϕ. (This procedure can be used to obtain the functional dependences of all the $Y_\ell^{m_\ell}(\theta, \phi)$'s.)

(d) Sketch the behavior of $|Y_\ell^\ell(\theta, \phi)|^2$ for $\phi = \pi/2$ and $\phi = 3\pi/2$, analogous to Fig. 10.3.

10.27 (a) Set $\ell = 1$ and construct the 3×3 matrices that represent $\hat{L}^2$, $\hat{L}_x$, $\hat{L}_y$, and $\hat{L}_z$ in the eigenstate representation.

(b) Determine the (real-element) 3×1 column vectors corresponding to the $\hat{L}_z$ eigenvalues $0, \pm\hbar$.

(c) Solve the 3×3 matrix eigenvalue problem for $\hat{L}_x$, normalizing the eigenvectors.

(d) For each of these $\hat{L}_x$ eigenvectors, what is the probability that a measurement of $\hat{L}_z$ yields 0?

10.28 Let the angular part of the Hamiltonian be $\hat{H} = (E_1/\hbar^2)\hat{L}_z^2 + (E_2/\hbar^2)(\hat{L}_x^2 - \hat{L}_y^2)$ for a particle with orbital angular momentum unity ($\ell = 1$); the radial part can be ignored. Without using a coordinate representation, determine the eigenvectors and normalized eigenvectors for this $\hat{H}$. (Hint: expand the general eigenket via the complete set $\{|1, m_\ell\rangle\}$ and express $\hat{L}_x$ and $\hat{L}_y$ in terms of the ladder operators $\hat{L}_\pm$.)

10.29 Prove that Eq. (10.112), like (10.111), is inconsistent.

10.30 As noted in the text, $\cos\hat{\Phi}$ or $\sin\hat{\Phi}$ could be used in place of Eq. (10.113).

(a) Evaluate $[\hat{L}_z, \cos\hat{\Phi}]$ and $[\hat{L}_z, \sin\hat{\Phi}]$.

(b) Use the results of part (a) to evaluate the matrix elements of $\cos\hat{\Phi}$ and $\sin\hat{\Phi}$ analogous to (10.115), and then demonstrate their consistency.

(c) Evaluate the three uncertainty products $\Delta L_z\, \Delta\Upsilon$, $\Delta L_z\, \Delta\cos\Phi$, and $\Delta L_z\, \Delta\sin\Phi$, and establish their consistency.

11

Two-Particle Systems, Potential-Well Bound-State Problems

This chapter is concerned with solutions to the bound-state Schrödinger equation for a variety of central potentials in 2-D and 3-D. The obvious candidate for analysis is a system consisting of a single particle, in which, for the 3-D case, the angular part of the wave function is $Y_\ell^{m_\ell}$, as discussed in the preceding chapter. One-particle systems are excellent pedagogical devices, but they seldom occur in nature. In contrast, two-particle systems are common, for example, the hydrogen atom and its one-electron relatives (He^+, Li^{2+}, etc.), and the deuteron, which is the neutron–proton bound state. It is therefore of profound importance that two-particle systems can often be simplified in ways that make them equivalent to one-particle systems. Precisely because of this, we are able to treat one-electron atoms and the deuteron exactly, thereby determining their bound-state wave functions and energies, at least for solvable interparticle potentials. In general, no such simplifications can be made for systems of three or more particles. Hence the emphasis on two-particle systems.

Reduction of the two-particle problem to an equivalent one-particle problem, plus properties of the 3-D radial solutions are the subjects of Sections 11.1–11.3, following which we specify the relevant potentials and, for each of them, determine the appropriate radial wave functions and energy eigenvalues.

11.1. Some features of the two-particle system

The most general two-particle Hamiltonian $\hat{H}(1, 2)$ is

$$\hat{H}(1, 2) = \hat{H}_1 + \hat{H}_2 + V_{12}, \tag{11.1}$$

where 1 and 2 are the particle labels, and

$$\hat{H}_j = \hat{K}_j + V_j, \qquad j = 1, 2, \tag{11.2}$$

with $\hat{K}_j$ being the jth particle's kinetic-energy operator and V_j the external potential acting on particle j, while V_{12} is the interparticle interaction, which is dependent on their relative separation in a coordinate representation. That is, the potentials are assumed local, in particular

$$\langle \mathbf{r}_j | V_j | \mathbf{r}'_j \rangle = V_j(\mathbf{r}_j)\delta(\mathbf{r}_j - \mathbf{r}'_j) \tag{11.3}$$

and

$$\langle \mathbf{r}_1, \mathbf{r}_2 | V_{12} | \mathbf{r}_1', \mathbf{r}_2' \rangle = V_{12}(\mathbf{r}_1 - \mathbf{r}_2)\delta(\mathbf{r}_1 - \mathbf{r}_1')\delta(\mathbf{r}_2 - \mathbf{r}_2'), \tag{11.4}$$

with $\mathbf{r}_j$ being the position vector of particle j, $j = 1, 2$, as illustrated in Fig. 11.1.

With few exceptions, the eigenvalue problem for the Hamiltonian (11.1) is impossible to solve exactly even if the $\hat{H}_j$ eigenvalue problem itself has been (or can be) solved. An example is a two-electron atom, with $\hat{H}_j$ a spin-independent, hydrogenic Hamiltonian (whose eigenfunctions/eigenvalues are known) and V_{12} the Coulomb repulsion between the two electrons: no exact solution, for instance, of the He-atom eigenvalue problem exists, although approximate solutions of extraordinary accuracy have been determined. We shall examine approximation methods and the He-atom in Part IV.

While $\hat{H}(1, 2)$ generally leads to intractable eigenvalue problems, there are two modifications of $\hat{H}(1, 2)$ for which the eigenvalue problems *are* tractable (at least for solvable potentials). These modifications are obtained by setting one or two of the interactions in (11.1) to zero, and they are the modifications of interest in this chapter. The first of these yields $\hat{H}_{\text{IP}}(1, 2)$, which is $\hat{H}(1, 2)$ with V_{12} set to zero:

$$\hat{H}(1, 2) \underset{V_{12}\to 0}{\longrightarrow} \hat{H}_{\text{IP}}(1, 2) = \hat{H}_1 + \hat{H}_2. \tag{11.5}$$

The other modification, denoted $\hat{H}_{12}$, is the portion of $\hat{H}(1, 2)$ that remains when both V_1 and V_2 are eliminated:

$$\hat{H}(1, 2) \underset{V_1, V_2\to 0}{\longrightarrow} \hat{H}_{12} = \hat{K}_1 + \hat{K}_2 + V_{12}. \tag{11.6}$$

We note that a tractable Hamiltonian is not usually obtained if just V_1 or V_2 is eliminated from $\hat{H}(1, 2)$; furthermore, setting all three interactions in (11.1) equal to zero leads to a trivial (and thus uninteresting) case we need not consider further.

Each of $\hat{H}_{\text{IP}}(1, 2)$ and $\hat{H}_{12}$ allows exact solutions. We consider $\hat{H}_{\text{IP}}$ in this section, $\hat{H}_{12}$ in the next. The subscript IP stands for "independent particle," and refers to the fact that, lacking the mutual interaction V_{12}, the system described by $\hat{H}_{\text{IP}}(1, 2)$ is of two particles acting independently of one another.[1] This independence is a consequence of

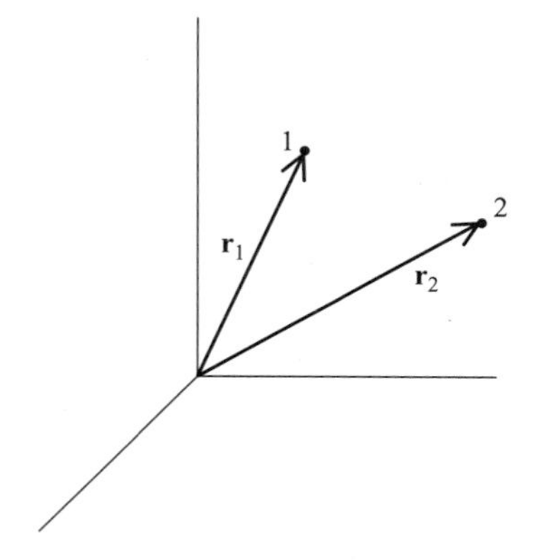

Fig. 11.1 Position vectors for particles 1 and 2 measured in an arbitrary coordinate system.

[1] This independent behavior is valid as long as the particles are not identical "fermions," i.e., are not spin-$\frac{1}{2}$ particles such as pairs of electrons, protons, neutrons, ^{3}He nuclei, etc. Identical particles are treated in Part IV, and spin $\frac{1}{2}$ is introduced in Chapter 13. In this chapter, 1 and 2 are assumed to be distinguishable.

$$[\hat{H}_1,\ \hat{H}_2] = 0, \tag{11.7}$$

a result that necessarily holds because the labels refer to different particles. From the commuting-operator theorem of Chapter 9 and the assumption that there already exists a CSCO for each of particles 1 and 2, it follows that eigenstates of $\hat{H}_{\rm IP}(1, 2)$ are labeled by the quantum numbers appropriate to $\hat{H}_1$ and $\hat{H}_2$.

Let the $\hat{H}_j$ eigenvalue problem be

$$\hat{H}_j|\gamma_j\rangle_j = E_{\gamma_j}|\gamma_j\rangle_j, \qquad j = 1, 2, \tag{11.8}$$

where $\gamma_j = n_j\ell_j m_{\ell_j}$ for bound states in a central potential and the subscript j on $|\ \rangle_j$ identifies the particle which the ket describes. By assumption, $\hat{V}_j$ permits (11.8) to be solved exactly, so the $|\gamma_j\rangle$ (and thus $\psi_{\gamma_j}(\mathbf{r}_j) = \langle\mathbf{r}_j|\gamma_j\rangle$) are known. All the potentials discussed in the later sections of this chapter fall into this category. A solution to the $\hat{H}_{\rm IP}(1, 2)$ eigenvalue problem

$$(\hat{H}_{\rm IP}(1, 2) - E_\Gamma)|\Gamma\rangle_{1,2} = 0 \tag{11.9}$$

is therefore

$$|\Gamma\rangle_{1,2} \equiv |\gamma_1\rangle_1|\gamma_2\rangle_2 \tag{11.10a}$$

and

$$E_\Gamma = E_{\gamma_1} + E_{\gamma_2}. \tag{11.10b}$$

The order of subscripts on $|\Gamma\rangle_{1,2}$ is significant:

$$|\Gamma\rangle_{2,1} \equiv |\gamma_1\rangle_2|\gamma_2\rangle_1 \tag{11.11}$$

is a second solution to (11.9) corresponding to the same energy (11.10b), as is easily shown by substituting both (11.5) and (11.11) into (11.9). Hence, there is a *particle-label degeneracy* for $\hat{H}_{\rm IP}(1, 2)$ above and beyond any degeneracies that may characterize $\hat{H}_1$ and/or $\hat{H}_2$ individually. Clearly, the state

$$|\Gamma\rangle = a|\Gamma\rangle_{1,2} + b|\Gamma\rangle_{2,1}, \tag{11.12}$$

with a and b arbitrary, is also a solution to (11.9). Particle-label degeneracy is an extremely important aspect of independent-particle Hamiltonians that we will encounter and explore in detail later in the book but not any further in this chapter, whose major concern is in solving few-degree-of-freedom eigenvalue problems.

Let us now return to the "IP" designation of (11.5). Since 1 and 2 are assumed distinguishable, there are none of the particle-interchange symmetry constraints on $|\Gamma\rangle_{1,2}$ or $|\Gamma\rangle_{2,1}$ that would arise were 1 and 2 a pair of identical particles. Therefore, the sets of quantum numbers γ_1 and γ_2 can be chosen arbitrarily and independently of one another. The quantal motion of particles 1 and 2 is thus completely decoupled: they are truly independent of each other. This means that measuring devices that would select, for example, $|\gamma_1\rangle$ for particle 1 and $|\gamma_2\rangle$ for particle 2 can be constructed. No restrictions apply, at least in principle.

Before turning to $\hat{H}_{12}$, there are several additional points worth stressing. First, solution of the $\hat{H}_{\rm IP}(1, 2)$ two-particle eigenvalue problem involves solving two unrelated one-particle eigenvalue problems. The analyses of Sections 11.3 *ff*, as appropriate, apply to $\hat{H}_1$ and $\hat{H}_2$ individually. Second, while the labels 1 and 2 refer to two particles, and an assumption of 3-D behavior has been made, neither feature is intrinsic to the foregoing treatment. The subscripts 1 and 2 could also refer, for example, to the x and the y spatial components of a single particle. Or, 1 and 2 might continue to denote the particles, but the dimensionality of the system could be 1-D or 2-D. In each case, the formal results hold; only the specific details of the solution will change. Third, the general structure of the solutions for any $\hat{H}_{\mathrm{IP}}(1, 2)$ that is a sum of two commuting Hamiltonians, viz., a product of two individual eigenstates for $|\Gamma\rangle$ and a sum of two individual energies for E_Γ, is maintained for a system of N distinguishable particles described by an analogous N-particle Hamiltonian, say $\hat{H}_{\mathrm{IP}}(1, 2, \ldots, N)$, which is a sum of N individual mutually commuting Hamiltonians, one for each particle.

That is, if $\hat{H}_{\mathrm{IP}}(1, 2, \ldots, N)$ is given by

$$\begin{aligned} \hat{H}_{\mathrm{IP}}(1, 2, \ldots, N) &= \hat{H}_1 + \hat{H}_2 + \hat{H}_3 + \cdots + \hat{H}_N \\ &= \sum_{j=1}^{N} \hat{H}_j, \end{aligned} \tag{11.13}$$

with

$$[\hat{H}_j, \hat{H}_k] = 0, \tag{11.14}$$

then in

$$(\hat{H}_{\mathrm{IP}}(1, 2, \ldots, N) - E_\Gamma^{(N)})|\Gamma\rangle = 0, \tag{11.15}$$

one of the $N!$ solutions corresponding to

$$\begin{aligned} E_\Gamma^{(N)} &= E_{\gamma_1} + E_{\gamma_2} + \cdots + E_{\gamma_N} \\ &= \sum_{j=1}^{N} E_j \end{aligned} \tag{11.16}$$

is

$$\begin{aligned} |\Gamma\rangle = |\Gamma\rangle_{1,2,3,\ldots,N} &= |\gamma_1\rangle_1 |\gamma_2\rangle_2 \cdots |\gamma_N\rangle_N \\ &= \prod_{j=1}^{N} |\gamma_j\rangle_j, \end{aligned} \tag{11.17}$$

where

$$(\hat{H}_j - E_{\gamma_j})|\gamma_j\rangle_j = 0, \qquad j = 1, 2, \ldots, N. \tag{11.18}$$

The proof of this claim is left to the exercises. Notice that the solution of this special N-particle problem reduces to the solution of N single-particle eigenvalue problems, to which the later analyses of this chapter could be applicable. The general structure of Eqs. (11.13)–(11.17) will be encountered in our treatment of the atomic-shell model, Part IV.

EXAMPLE

A 2-D oscillator. We close this section with a simple example illustrating the features of $\hat{H}_{\mathrm{IP}}(1, 2)$ described in the foregoing. The example is that of a single particle of mass m moving in 2-D under the influence of an asymmetric linear harmonic oscillator potential. Its Hamiltonian is

$$\hat{H} = \left(\frac{1}{2m}\hat{P}_x^2 + \frac{1}{2}m\omega_x^2\hat{Q}_x^2\right) + \left(\frac{1}{2m}\hat{P}_y^2 + \frac{1}{2}m\omega_y^2\hat{Q}_y^2\right)$$
$$\equiv \hat{H}_x + \hat{H}_y, \tag{11.19}$$

$\omega_x \neq \omega_y$. Given that $\hat{H}_x$ and $\hat{H}_y$ refer to different coordinates, then $[\hat{H}_x, \hat{H}_y] = 0$, and solutions to the eigenvalue problem for $\hat{H}$ of (11.19), symbolized by Eqs. (11.10a), are products:

$$|\Gamma\rangle = |n_x\rangle|n_y\rangle,$$

where $|n\rangle$ is a 1-D oscillator state, and

$$E_\Gamma = (n_x + \tfrac{1}{2})\hbar\omega_x + (n_y + \tfrac{1}{2})\hbar\omega_y,$$

with n_x and n_y taking on the values 0, 1, 2, ... Unlike the case $\omega_x = \omega_y$, which is studied later, this system is non-degenerate.

Although this example is simple, it serves quite well for emphasizing the independence described above. The x and the y motions are indeed unconnected – decoupled – and $|n_x\rangle$ and $|n_y\rangle$ can be chosen arbitrarily. The same results hold if 1 and 2 were to refer to two 1-D particles, each acted on by an oscillator potential $m_j\omega_j^2x_j^2/2$, $j = 1$ and 2.

11.2. Reduction of the $\hat{H}_{12}$ two-body problem to an effective one-body problem

The subscript 12 on $\hat{H}_{12}$ refers to the fact that V_{12} is the only potential present in the Hamiltonian. Because of this, $\hat{H}_{12}$ describes a two-particle system that is isolated from all external influences. This suggests that one should be able to treat the relative-motion interparticle behavior in a way that yields the spectrum of the isolated system. It is clear from

$$[\hat{K}_j(\mathbf{r}_j), V_{12}(\mathbf{r}_1 - \mathbf{r}_2)] \neq 0 \tag{11.20}$$

that such a procedure fails if $\hat{H}_{12}$ is in the form of (11.6). That is, Eq. (11.20) prevents Eq. (11.6) from being decomposed into two commuting portions, one of which contains V_{12}. This, however, is an artifact of using the coordinates and operators of each individual particle. By using relative and center-of-mass (CM) coordinates and the corresponding operators, the desired decomposition can be effected, and the V_{12} eigenvalue problem solved, at least in principle. We discuss this procedure in the following, first for a classical and then for the quantal system.

Classical mechanics

If there are no external forces acting on an isolated pair of classical particles, then their classical Hamiltonian is

$$H_{\text{cl}}(1, 2) = \frac{p_1^2}{2m_1} + \frac{p_2^2}{2m_2} + V_{12}, \tag{11.21}$$

and Newtonian mechanics tells us that both the momentum and the angular momentum associated with the CM of this system are conserved. Furthermore, if the interparticle potential is central, then the relative-motion angular momentum is also conserved. Because of the foregoing, analysis of a two-particle system is often simpler if one makes a canonical transformation from the individual particle coordinates $\mathbf{r}_1$ and $\mathbf{r}_2$ to those describing relative ($\mathbf{r}$) and CM ($\mathbf{R}$) motion (Fig. 11.2):

$$\mathbf{r} \equiv \mathbf{r}_{12} = \mathbf{r}_1 - \mathbf{r}_2 \tag{11.22a}$$

and

$$\mathbf{R} = \frac{1}{M}(m_1\mathbf{r}_1 + m_2\mathbf{r}_2), \tag{11.22b}$$

where m_j is the mass of particle j and $M = m_1 + m_2$ is the total mass.

Since the coordinate transformation (11.22) is canonical, the kinetic energy $p_1^2/(2m_1) + p_2^2/(2m_2)$ can be expressed using relative ($\mathbf{p}$) and CM ($\mathbf{P}$) momenta:

$$\frac{p_1^2}{2m_1} + \frac{p_2^2}{2m_2} = \frac{p^2}{2\mu} + \frac{P^2}{2M}, \tag{11.23}$$

where

$$\mathbf{p} = \frac{1}{M}(m_2 p_1 - m_1 p_2) = \mu\dot{\mathbf{r}} \tag{11.24a}$$

and

$$\mathbf{P} = \mathbf{p}_1 + \mathbf{p}_2 = M\dot{\mathbf{R}}; \tag{11.24b}$$

the reduced mass μ is

$$\mu = m_1 m_2/M. \tag{11.25}$$

Finally, writing the interparticle potential as $V_{12}(\mathbf{r})$, the classical Hamiltonian $H_{\rm cl}(1, 2)$ can be separated into two independent parts,

$$H_{\rm cl}(1, 2) = H_{\rm cl}(\mathbf{r}, \mathbf{p}) + H_{\rm cl}(\mathbf{P}), \tag{11.26a}$$

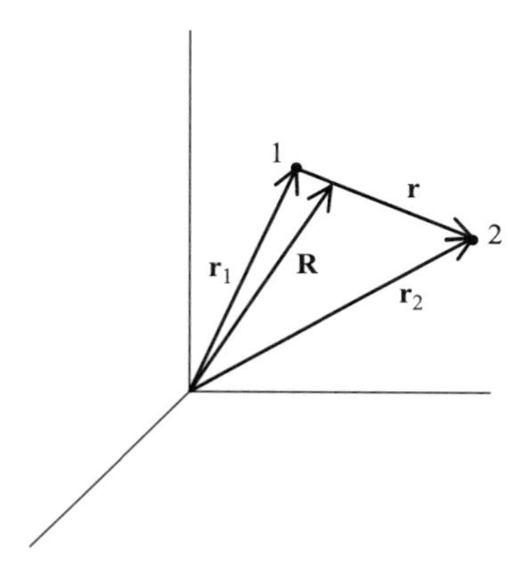

Fig. 11.2 Relative ($\mathbf{r} \equiv \mathbf{r}_1 - \mathbf{r}_2$) and center-of-mass ($\mathbf{R} = (m_1\mathbf{r}_1 + m_2\mathbf{r}_2)/(m_1 + m_2)$) position vectors for an arbitrary two-particle system of masses m_1 and m_2.

where

$$H_{\text{cl}}(\mathbf{r},\,\mathbf{p}) = \frac{p^2}{2\mu} + V_{12}(\mathbf{r}) \tag{11.26b}$$

and

$$H_{\text{cl}}(\mathbf{P}) = \frac{1}{2M}P^2. \tag{11.26c}$$

This is the Hamiltonian version of the usual Lagrangian analysis leading to the classical treatment of the internal motion (trajectory) of a pair of bodies experiencing only a mutual interaction, for instance, interactions of the form $r_{12}^{\pm n}$, Coulomb or gravitational potentials, being one example.

The quantum case: kinetic energy and Hamiltonians

Many quantal Hamiltonians can be obtained from their classical counterparts by quantizing the momenta of the individual particles (Postulate IV, Chapter 5). This cannot be done in a straightforward fashion for the Hamiltonian $H_{\text{cl}}(1, 2)$ of Eq. (11.26a), because the momentum $\mathbf{p} = \mu\dot{\mathbf{r}}$ corresponding to the relative-motion coordinate $\mathbf{r}$ is not a direct observable: one can measure $\mathbf{p}_1$ and $\mathbf{p}_2$ and then form the linear combination (11.24a), but there is no particle whose mass is μ. This is in contrast to $\mathbf{P}$ and $\mathbf{R}$, which are both observables (for a bound-state system), implying the existence of their quantal counterparts $\hat{\mathbf{P}}_{\text{CM}}$ and $\hat{\mathbf{Q}}_{\text{CM}}$. Despite this aspect of $\mathbf{p}$ and $\mathbf{r}$, we shall show that it is still possible to introduce Hermitian operators $\hat{\mathbf{P}}_{\text{rel}}$ and $\hat{\mathbf{Q}}_{\text{rel}}$ that are both the quantal counterparts of $\mathbf{p}$ and $\mathbf{r}$ and the appropriate operators to supplement $\hat{\mathbf{P}}_{\text{CM}}$ and $\hat{\mathbf{Q}}_{\text{CM}}$. One need only remember that not all quantal operators are the mathematical images of observables, even if they seem to be so.

To set the stage for the development, we first show that the quantal kinetic-energy operator can be transformed exactly as the classical kinetic energy, Eq. (11.23). To do this, we employ the coordinate representation to transform the kinetic-energy operator

$$\hat{K}(\mathbf{r}_1,\,\mathbf{r}_2) = \hat{K}_1(\mathbf{r}_1) + \hat{K}_2(\mathbf{r}_2) = -\frac{\hbar^2}{2m_1}\nabla_1^2 - \frac{\hbar^2}{2m_2}\nabla_2^2 \tag{11.27}$$

from an expression involving derivatives taken w.r.t. components of $\mathbf{r}_1$ and $\mathbf{r}_2$ to one in which the derivatives are taken w.r.t. the components (separately) of $\mathbf{r}$ and $\mathbf{R}$. It suffices to consider only one component, say x_j in $\mathbf{r}_j = x_j\mathbf{e}_x + y_j\mathbf{e}_y + z_j\mathbf{e}_z$, $j = 1$ or 2. Using the chain rule (i.e., the linear independence of $\mathbf{r}$ and $\mathbf{R}$) and Eqs. (11.22), the relations among the first derivatives are

$$\frac{\partial}{\partial x_1} = \frac{\partial}{\partial x} + \frac{m_1}{M}\frac{\partial}{\partial X}, \qquad \frac{\partial}{\partial x_2} = -\frac{\partial}{\partial x} + \frac{m_2}{M}\frac{\partial}{\partial X}. \tag{11.28}$$

An easy calculation then yields

$$-\frac{\hbar^2}{2m_1}\frac{\partial^2}{\partial x_1^2} - \frac{\hbar^2}{2m_2}\frac{\partial^2}{\partial x_2^2} = -\frac{\hbar^2}{2\mu}\frac{\partial^2}{\partial x^2} - \frac{\hbar^2}{2M}\frac{\partial^2}{\partial X^2}; \tag{11.29a}$$

in other words,

$$\hat{K}_{1,x}(\mathbf{r}_1) + \hat{K}_{2,x}(\mathbf{r}_2) = \hat{K}_x(\mathbf{r}) + \hat{K}_x(\mathbf{R}). \tag{11.29b}$$

Two points about Eqs. (11.29) are noteworthy. First, in transforming the second derivatives, no cross-terms of the form $\partial^2/\partial x\,\partial X$ appear, even though they might have been expected from Eq. (11.28). The absence of such cross-terms is a consequence of (11.22) being a canonical transformation (unit Jacobian), a point explored in the exercises. Second, the mass parameters associated with $\partial^2/\partial x^2$ and $\partial^2/\partial X^2$ are μ and M, respectively – i.e., just those which go along with $\mathbf{p}$ and $\mathbf{P}$ in the transformation (11.23) of the classical kinetic energy.

Similar results hold for the y and z portions of $\hat{K}(\mathbf{r}_1, \mathbf{r}_2)$, so that the complete transformation is

$$\begin{aligned}\hat{K}(\mathbf{r}_1, \mathbf{r}_2) &= -\frac{\hbar^2}{2m_1}\nabla_1^2 - \frac{\hbar^2}{2m_2}\nabla_2^2 \\ &= -\frac{\hbar^2}{2\mu}\nabla_{\mathbf{r}}^2 - \frac{\hbar^2}{2M}\nabla_{\mathbf{R}}^2 \\ &= \hat{K}_{\text{rel}}(\mathbf{r}) + \hat{K}_{\text{CM}}(\mathbf{R}) \equiv \hat{K}(\mathbf{r}, \mathbf{R}).\end{aligned} \tag{11.30}$$

The quantal version of the two-particle classical Hamiltonian (11.21) is given by (11.6):

$$\hat{H}_{12}(\mathbf{r}_1, \mathbf{r}_2) = \hat{K}_1(\mathbf{r}_1) + \hat{K}_2(\mathbf{r}_2) + V_{12}(\mathbf{r}).$$

Using (11.30), $\hat{H}_{12}(\mathbf{r}_1, \mathbf{r}_2)$ can now be expressed in the desired form:

$$\hat{H}_{12}(\mathbf{r}_1, \mathbf{r}_2) \to \hat{H}_{12}(\mathbf{r}, \mathbf{R}) = \hat{K}_{\text{CM}}(\mathbf{R}) + \hat{H}_{\text{rel}}(\mathbf{r}), \tag{11.31a}$$

where

$$\hat{H}_{\text{rel}}(\mathbf{r}) = \hat{K}_{\text{rel}}(\mathbf{r}) + V_{12}(\mathbf{r}), \tag{11.31b}$$

with

$$[\hat{K}_{\text{CM}}(\mathbf{R}), \hat{H}_{\text{rel}}(\mathbf{r})] = 0. \tag{11.31c}$$

Unlike $\hat{H}_{12}(\mathbf{r}_1, \mathbf{r}_2)$, whose simplest eigenfunctions cannot be written as a one-term product of the form $\xi_\alpha(\mathbf{r}_1)\xi_\beta(\mathbf{r}_2)$, Eq. (11.31c) guarantees that the simplest eigenfunctions of $\hat{H}_{12}(\mathbf{r}, \mathbf{R})$ *do* have a product form. That is, there are solutions to

$$(\hat{H}_{12}(\mathbf{r}, \mathbf{R}) - E_\Gamma)\psi_\Gamma(\mathbf{r}, \mathbf{R}) = 0 \tag{11.32a}$$

of the form

$$\psi_\Gamma(\mathbf{r}, \mathbf{R}) = \psi_\alpha(\mathbf{r})\psi_\beta(\mathbf{R}), \tag{11.32b}$$

where

$$\hat{H}_{\text{rel}}(\mathbf{r})\psi_\alpha(\mathbf{r}) = E_\alpha\psi_\alpha(\mathbf{r}), \tag{11.32c}$$

$$\hat{K}_{\text{CM}}(\mathbf{R})\psi_\beta(\mathbf{R}) = E_\beta\psi_\beta(\mathbf{R}), \tag{11.32d}$$

and

$$E_\Gamma = E_\alpha + E_\beta.$$

These results are obvious analogs of Eqs. (11.9) and (11.10).

Equation (11.32c) defines an eigenvalue problem for a single "particle" of mass μ whose position coordinate is $\mathbf{r}$. It is a generic type, examples of which we shall solve later in this chapter. Unlike the true one-particle problem for which the operators $\hat{\mathbf{P}}$, $\hat{\mathbf{Q}}$, and (in 3-D) $\hat{\mathbf{L}} = \hat{\mathbf{Q}} \times \hat{\mathbf{P}}$ exist and help to determine relevant constants of the motion and quantum numbers, the corresponding operators that would be associated with Eq. (11.32c) are so far undefined; likewise for Eq. (11.32d): $\hat{\mathbf{P}}_{\mathrm{CM}}$, etc., have not been introduced. In other words, we do not yet have a CSCO for the relative/CM motion problem. We address this concern in the next two subsections.

The quantum case: relative/CM position and momentum operators

Although one can simply postulate the existence of operators $\hat{\mathbf{Q}}_{\mathrm{rel}}$, $\hat{\mathbf{Q}}_{\mathrm{CM}}$, $\hat{\mathbf{P}}_{\mathrm{rel}}$ and $\hat{\mathbf{P}}_{\mathrm{CM}}$ and their associated eigenvalue relations, it is rather more transparant to argue from the structure of the kinetic-energy-operator relation (11.30). Consider $\hat{K}(\mathbf{r}_1, \mathbf{r}_2)$ first:

$$\hat{K}(\mathbf{r}_1, \mathbf{r}_2) = \hat{K}_1(\mathbf{r}_1) + \hat{K}_2(\mathbf{r}_2),$$

where

$$\hat{K}_j(\mathbf{r}_j) = \frac{1}{2m_j} \hat{P}_j^2(\mathbf{r}_j), \qquad j = 1, 2,$$

and the CSCO is $\{\hat{K}_1, \hat{K}_2, \hat{\mathbf{P}}_1, \hat{\mathbf{P}}_2\}$. The eigenfunctions of $\hat{K}(\mathbf{r}_1, \mathbf{r}_2)$ are plane waves,

$$(\hat{K}(\mathbf{r}_1, \mathbf{r}_2) - E_{k_1 k_2}) e^{i(\mathbf{k}_1 \cdot \mathbf{r}_1 + \mathbf{k}_2 \cdot \mathbf{r}_2)} = 0, \tag{11.33}$$

where $E_{k_1 k_2} = E_{k_1} + E_{k_2}$, $E_{k_j} = \hbar^2 k_j^2/(2m_j)$. That is, the eigenstates are $|\mathbf{k}_1, \mathbf{k}_2\rangle$, whose eigenfunctions are $\langle \mathbf{r}_1, \mathbf{r}_2 | \mathbf{k}_1, \mathbf{k}_2\rangle = \langle \mathbf{r}_1 | \mathbf{k}_1\rangle\langle \mathbf{r}_2 | \mathbf{k}_2\rangle$, with

$$\hat{I} = \int d^3k_1\, d^3k_2\, |\mathbf{k}_1, \mathbf{k}_2\rangle\langle \mathbf{k}_1, \mathbf{k}_2| \tag{11.34a}$$

$$= \int d^3r_1\, d^3r_2\, |\mathbf{r}_1, \mathbf{r}_2\rangle\langle \mathbf{r}_1, \mathbf{r}_2|. \tag{11.34b}$$

In view of Eq. (11.30), the product of plane waves $\langle \mathbf{r}_1, \mathbf{r}_2 | \mathbf{k}_1, \mathbf{k}_2\rangle$ must also be an eigenfunction of $\hat{K}(\mathbf{r}, \mathbf{R})$, but, since the differential operators are now $\nabla_{\mathbf{r}}^2$ and $\nabla_{\mathbf{R}}^2$, $\exp[i(\mathbf{k}_1 \cdot \mathbf{r}_1 + \mathbf{k}_2 \cdot \mathbf{r}_2)]$ must be transformable to an expression involving $\mathbf{r}$ and $\mathbf{R}$ and wave numbers $\mathbf{k}_{\mathrm{rel}}$ and $\mathbf{k}_{\mathrm{CM}}$. Equations (11.24) suggest what the relations between $\mathbf{k}_1$ and $\mathbf{k}_2$ and $\mathbf{k}_{\mathrm{rel}}$ and $\mathbf{k}_{\mathrm{CM}}$ are expected to be. Using Eqs. (11.22) to express $\mathbf{r}_1$ and $\mathbf{r}_2$ in terms of $\mathbf{r}$ and $\mathbf{R}$ completely determines the desired transformation of the plane wave $\langle \mathbf{r}_1, \mathbf{r}_2 | \mathbf{k}_1, \mathbf{k}_2\rangle$:

$$e^{i(\mathbf{k}_1 \cdot \mathbf{r}_1 + \mathbf{k}_2 \cdot \mathbf{r}_2)} = e^{i(\mathbf{k}_{\mathrm{rel}} \cdot \mathbf{r} + \mathbf{k}_{\mathrm{CM}} \cdot \mathbf{R})}, \tag{11.35}$$

where

$$\mathbf{k}_{\mathrm{rel}} = \frac{1}{M}(m_2 \mathbf{k}_1 - m_1 \mathbf{k}_2) \tag{11.35a}$$

and

$$\mathbf{k}_{\mathrm{CM}} = \mathbf{k}_1 + \mathbf{k}_2, \tag{11.35b}$$

relations consonant with the expectation based on Eqs. (11.24). Also, by acting on the RHS of (11.35a) with $\hat{K}(\mathbf{r}, \mathbf{R})$, we find that

$$E_{k_1 k_2} = \frac{\hbar^2 k_{\text{rel}}^2}{2\mu} + \frac{\hbar^2 k_{\text{CM}}^2}{2M}, \tag{11.35c}$$

whose RHS, from (11.35a) and (11.35b), is indeed equal to $\hbar^2 k_1^2/(2m_1) + \hbar^2 k_2^2/(2m_2)$.

The foregoing coordinate-representation results provide the desired background. By remaining in this representation it is clear that local operators

$$\hat{\mathbf{P}}_{\text{rel}}(\mathbf{r}) = -i\hbar\,\nabla_{\mathbf{r}} \tag{11.36a}$$

and

$$\hat{\mathbf{P}}_{\text{CM}}(\mathbf{R}) = -i\hbar\,\nabla_{\mathbf{R}} \tag{11.36b}$$

can be introduced such that the following relations hold:

$$\hat{K}(\mathbf{r}, \mathbf{R}) = \frac{1}{2\mu}\hat{\mathbf{P}}_{\text{rel}}^2(\mathbf{r}) + \frac{1}{2M}\hat{P}_{\text{CM}}^2(\mathbf{R}), \tag{11.37}$$

$$(\hat{\mathbf{P}}_{\text{rel}}(\mathbf{r}) - \hbar\mathbf{k}_{\text{rel}})e^{i(\mathbf{k}_{\text{rel}}\cdot\mathbf{r})} = 0, \tag{11.38a}$$

and

$$(\hat{\mathbf{P}}_{\text{CM}}(\mathbf{R}) - \hbar\mathbf{k}_{\text{CM}})e^{i(\mathbf{k}_{\text{CM}}\cdot\mathbf{R})} = 0. \tag{11.38b}$$

Furthermore, from the structure of $\nabla_{\mathbf{r}}^2$ and $\nabla_{\mathbf{R}}^2$ in spherical polar coordinates ($\mathbf{r} \Leftrightarrow r,\ \theta_{\text{rel}},\ \phi_{\text{rel}}$, and $\mathbf{R} \Leftrightarrow R,\ \theta_{\text{CM}},\ \phi_{\text{CM}}$), it is evident that the squares of relative and CM orbital angular-momentum operators $\hat{L}_{\text{rel}}^2(\theta_{\text{rel}}, \phi_{\text{rel}})$ and $\hat{L}_{\text{CM}}^2(\theta_{\text{CM}}, \phi_{\text{CM}})$ can also be meaningfully introduced.

The remaining analysis follows straightforwardly from the preceding results. For example, Eq. (11.35) implies the relations

$$\hat{\mathbf{P}}_{\text{rel}} = \frac{1}{M}(m_2\hat{\mathbf{P}}_1 - m_1\hat{\mathbf{P}}_2) \tag{11.39a}$$

and

$$\hat{\mathbf{P}}_{\text{CM}} = \hat{\mathbf{P}}_1 + \hat{\mathbf{P}}_2, \tag{11.39b}$$

from which one readily finds

$$[\hat{\mathbf{P}}_{\text{rel}}, \hat{\mathbf{P}}_{\text{CM}}] = 0. \tag{11.39c}$$

Furthermore, the functional form of the one-particle transformation bracket (scalar product)

$$\langle \mathbf{r}|\mathbf{k}\rangle = (2\pi)^{-3/2}e^{i\mathbf{k}\cdot\mathbf{r}} \tag{11.40}$$

implies that $\exp[i(\mathbf{k}_{\text{rel}} \cdot \mathbf{r} + \mathbf{k}_{\text{CM}} \cdot \mathbf{R})]$ is itself a product of such scalar products:

$$\begin{aligned}(2\pi)^{-3/2}e^{i(\mathbf{k}_{\text{rel}}\cdot\mathbf{r}+\mathbf{k}_{\text{CM}}\cdot\mathbf{R})} &= \langle \mathbf{r}|\mathbf{k}_{\text{rel}}\rangle\langle \mathbf{R}|\mathbf{k}_{\text{CM}}\rangle \\ &= \langle \mathbf{r}, \mathbf{R}|\mathbf{k}_{\text{rel}}, \mathbf{k}_{\text{CM}}\rangle.\end{aligned} \tag{11.41}$$

The kets on the RHS of (11.41) are clearly eigenvectors of $\hat{\mathbf{P}}_{\text{rel}}$ and $\hat{\mathbf{P}}_{\text{CM}}$, while the bras are similarly eigenvectors belonging to the relative and CM position operators $\hat{\mathbf{Q}}_{\text{rel}}$ and $\hat{\mathbf{Q}}_{\text{CM}}$, defined, in analogy to (11.22) by

$$\hat{\mathbf{Q}}_{\text{rel}} \equiv \hat{\mathbf{Q}}_1 - \hat{\mathbf{Q}}_2 \tag{11.42a}$$

and

$$\hat{\mathbf{Q}}_{\text{CM}} \equiv \frac{1}{\text{M}}(m_1\hat{\mathbf{Q}}_1 + m_2\hat{\mathbf{Q}}_2), \tag{11.42b}$$

for which

$$[\hat{\mathbf{Q}}_{\text{rel}}, \hat{\mathbf{Q}}_{\text{CM}}] = 0. \tag{11.42c}$$

These various operators give rise to the abstract eigenvalue relations

$$(\hat{\mathbf{P}}_{\text{rel}} - \hbar\mathbf{k}_{\text{rel}})|\mathbf{k}_{\text{rel}}\rangle = 0, \tag{11.43a}$$

$$(\hat{\mathbf{P}}_{\text{CM}} - \hbar\mathbf{k}_{\text{CM}})|\mathbf{k}_{\text{CM}}\rangle = 0, \tag{11.43b}$$

$$(\hat{\mathbf{Q}}_{\text{rel}} - \mathbf{r})|\mathbf{r}\rangle = 0, \tag{11.43c}$$

and

$$(\hat{\mathbf{Q}}_{\text{CM}} - \mathbf{R})|\mathbf{R}\rangle = 0. \tag{11.43d}$$

The commutation relations between components of these operators are especially interesting. Owing to the canonical nature of the transformation from individual-particle to relative and CM operators, one finds

$$[\hat{Q}_{\text{rel},j}, \hat{P}_{\text{rel},k}] = i\hbar\delta_{jk}, \tag{11.44a}$$

$$[\hat{Q}_{\text{CM},j}, \hat{P}_{\text{CM},k}] = i\hbar\delta_{jk} \tag{11.44b}$$

$$[\hat{Q}_{\text{rel},j}, \hat{P}_{\text{CM},k}] = 0, \qquad \forall\, j \text{ and } k, \tag{11.44c}$$

and

$$[\hat{Q}_{\text{CM},j}, \hat{P}_{\text{rel},k}] = 0, \qquad \forall\, j \text{ and } k. \tag{11.44d}$$

Thus, their structure, expressed in terms of the individual-particle operators, ensures that the relative and the CM operators are truly independent of one another, a statement that follows, of course, from the commutation relations (11.44c) and (11.44d). Evidently, the canonical type of transformation, whose origin is classical mechanics, is equally important in quantum mechanics.

The relative and CM operators allow alternate resolutions of the identity. Using the relations (11.34) and (11.35), the transformations

$$|\mathbf{r}_1, \mathbf{r}_2\rangle = |\mathbf{r}, \mathbf{R}\rangle \tag{11.45a}$$

and

$$|\mathbf{k}_1, \mathbf{k}_2\rangle = |\mathbf{k}_{\text{rel}}, \mathbf{k}_{\text{CM}}\rangle, \tag{11.45b}$$

with, e.g., $\mathbf{r} = \mathbf{r}(\mathbf{r}_1, \mathbf{r}_2)$ and $\mathbf{R} = \mathbf{R}(\mathbf{r}_1, \mathbf{r}_2)$, and the fact that a canonical transformation implies a unit Jacobian, we obtain

$$\hat{I} = \int d^3r\, d^3R\, |\mathbf{r}, \mathbf{R}\rangle\langle\mathbf{r}, \mathbf{R}| \tag{11.46a}$$

and

$$\hat{I} = \int d^3k_{\text{rel}}\, d^3k_{\text{CM}}\, |\mathbf{k}_{\text{rel}}, \mathbf{k}_{\text{CM}}\rangle\langle\mathbf{k}_{\text{rel}}, \mathbf{k}_{\text{CM}}|. \tag{11.46b}$$

Since $|\mathbf{r}, \mathbf{R}\rangle = |\mathbf{r}\rangle|\mathbf{R}\rangle$ and $|\mathbf{k}_{\text{rel}}, \mathbf{k}_{\text{CM}}\rangle = |\mathbf{k}_{\text{rel}}\rangle|\mathbf{k}_{\text{CM}}\rangle$, the expansions (11.46) are products of integrals: in precise analogy to $\{|\mathbf{r}_1\rangle\}$ and $\{|\mathbf{r}_2\rangle\}$, the kets $\{|\mathbf{r}\rangle\}$ and $\{|\mathbf{R}\rangle\}$ span separate Hilbert spaces, and similarly for the momentum eigenkets. That is, we may write either

$$\mathcal{H} = \mathcal{H}_1 \otimes \mathcal{H}_2$$

or

$$\mathcal{H} = \mathcal{H}_{\text{rel}} \otimes \mathcal{H}_{\text{CM}},$$

as in Eq. (9.43).

Only orbital angular-momentum operators remain to be considered. Their existence has already been inferred below (11.38b). The formal definitions are straightforward:

$$\hat{\mathbf{L}}_{\text{rel}} = \hat{\mathbf{Q}}_{\text{rel}} \times \hat{\mathbf{P}}_{\text{rel}} \tag{11.47a}$$

and

$$\hat{\mathbf{L}}_{\text{CM}} = \hat{\mathbf{Q}}_{\text{CM}} \times \hat{\mathbf{P}}_{\text{CM}} \tag{11.47b}$$

are operators in $\mathcal{H}_{\text{rel}}$ and $\mathcal{H}_{\text{CM}}$, respectively, just as $\hat{\mathbf{L}}_j = \hat{\mathbf{Q}}_j \times \hat{\mathbf{P}}_j$, $j = 1$ or 2, is an operator in $\mathcal{H}_j$. One of the beauties of the canonical relations (11.39) and (11.42) is that

$$\hat{\mathbf{L}}_{\text{rel}} + \hat{\mathbf{L}}_{\text{CM}} = \hat{\mathbf{L}}_1 + \hat{\mathbf{L}}_2, \tag{11.47c}$$

thereby establishing that (product) eigenstates of $\hat{L}_{1,z} + \hat{L}_{2,z}$ are also (product) eigenstates of $\hat{L}_{\text{rel},z} + \hat{L}_{\text{CM},z}$. However, this is a curiosity, since we do not need to consider this type of product eigenstate: only kets such as $|\ell, m_\ell\rangle_{\text{rel}}$ and $|\ell, m_\ell\rangle_{\text{CM}}$ are generally of interest.

Central potentials, CSCO, and the reduced radial wave function

Let us now apply these results to the generic 3-D eigenvalue problem we wish to solve, viz., Eqs. (11.32), and let us assume that the interparticle potential is central:

$$V_{12}(\mathbf{r}) = V_{12}(r).$$

Then the CSCO for the CM motion is $\{\hat{K}_{\text{CM}}, \hat{\mathbf{P}}_{\text{CM}}\}$, while that for the relative motion is $\{\hat{H}_{\text{rel}}, \hat{L}^2_{\text{rel}}, \hat{L}_{\text{rel},z}\}$, exactly as in the single-particle case. The solution for the CM motion is a plane wave,

$$\psi_\beta(\mathbf{R}) \to \psi_{\mathbf{k}_{\text{CM}}}(\mathbf{R}) = \left(\frac{1}{2\pi}\right)^{3/2} e^{i\mathbf{k}_{\text{CM}}\cdot\mathbf{R}}, \tag{11.48}$$

while that for bound-state relative motion is

$$\psi_\alpha(\mathbf{r}) \to \psi_{n\ell m_\ell}(\mathbf{r}) = R_{n\ell}(r) Y_\ell^{m_\ell}(\theta_{\text{rel}}, \phi_{\text{rel}}). \tag{11.49}$$

The relative-motion radial function in this case of a 3-D spherically symmetric potential obeys ($V_{12}(r) \to V(r)$)

$$\left[-\frac{\hbar^2}{2\mu}\left(\frac{1}{r^2}\frac{d}{dr}r^2\frac{d}{dr} - \frac{\ell(\ell+1)}{r^2}\right) + V(r)\right] R_{n\ell}(r) = E_{n\ell} R_{n\ell}(r). \tag{11.50}$$

On comparing this with Eq. (10.66b) we see that the only difference is the mass parameter.

The Schrödinger equation for the isolated two-particle system has thus been reduced to an effective one-body radial equation. Equation (11.50) differs from its 1-D counterpart (6.2) in the more complicated nature of the derivatives and the presence of the angular-momentum term $\ell(\ell+1)/r^2$. By introducing the "reduced" radial wave function, we shall show how to convert the derivative term in (11.50) from $r^{-2}\, d(r^2\, d/dr)$ to d^2/dr^2. This will then allow us to concentrate in the next subsection on the term $\ell(\ell+1)/r^2$ and how it affects the structure of 3-D radial wave functions.

The reduced radial wave function $u_{n\ell}(r)$ is defined by

$$u_{n\ell}(r) \equiv r R_{n\ell}(r)$$

or (11.51)

$$R_{n\ell}(r) = \frac{u_{n\ell}(r)}{r}.$$

On substituting (11.51) into (11.50) one readily finds that $u_{n\ell}(r)$ obeys

$$\left[-\frac{\hbar^2}{2\mu}\left(\frac{d^2}{dr^2} - \frac{\ell(\ell+1)}{r^2}\right) + V(r)\right] u_{n\ell}(r) = E_{n\ell} u_{n\ell}(r), \tag{11.52}$$

an equation involving only a second derivative, as noted earlier.

Equation (11.52) is converted into standard Sturm–Liouville form by introducing the "reduced" effective (ℓ-dependent) potential

$$v_\ell(r) \equiv \frac{2\mu}{\hbar^2} V(r) + \frac{\ell(\ell+1)}{r^2} \tag{11.53a}$$

and the eigenvalue

$$\lambda_{n\ell} \equiv -\frac{2\mu}{\hbar^2} E_{n,\ell}: \tag{11.53b}$$

$$\left(\frac{d^2}{dr^2} - v_\ell(r)\right) u_{n\ell}(r) = \lambda_{n\ell} u_{n\ell}(r). \tag{11.54}$$

Comparing (11.54) with the Sturm–Liouville eigenvalue equation (3.108), the analog of the operator $\hat{L}(x)$ appearing in the latter equation is

$$\hat{L}(r) = \frac{d^2}{dr^2} - v_\ell(r), \tag{11.55}$$

i.e., in (11.54) the Sturm–Liouville functions $p(r)$ and $q(r)$ are $p(r) = 1$ and $q(r) = v_\ell(r)$. Thus, all the results derived for the (1-D) Sturm–Liouville system apply directly to (11.54) and indirectly to (11.50). If the radial interval is $[0, a]$, then for each ℓ the orthogonality relation (3.116) becomes ($u_{n\ell}$ can always be chosen real)

$$\int_0^a dr\, u_{n\ell}(r)u_{m\ell}(r) = \int_0^a dr\,[rR_{n\ell}(r)][rR_{m\ell}(r)] = 0, \qquad n \neq m. \tag{11.56}$$

Furthermore, $\{u_{n\ell}^{(0)}\}_{n=1}^{\infty}$ forms a complete set for each ℓ on the interval $[0, a]$. The $\lambda_{n\ell}$ are obviously real. Notice that $\int_0^a dr\, u_{n\ell}(r)u_{n\ell'}(r) \neq 0$, $\ell \neq \ell'$, but that this is no problem overall because of the orthogonality property of differing $Y_\ell^{m_\ell}$.

While the orthogonality properties of the 3-D radial wave functions are best expressed in terms of the $u_{n\ell}(r)$, it is often easiest to solve directly for the $R_{n\ell}(r)$, as in Sections 11.4 *ff.* Either Eq. (11.50) or Eq. (11.54) suffices to discuss the properties of the 3-D radial solutions, an undertaking we turn to next.

11.3. Some properties of the solutions in three dimensions

The quantity $v_\ell(r)$, defined by Eq. (11.53a), is an ℓ-dependent effective potential. It contains the term $\ell(\ell+1)/r^2$, whose presence makes 3-D so different than 1-D. In analyzing the effects of this so-called *centrifugal barrier* term, we shall make the simplifying assumption that $V(r)$ is nowhere repulsive ($V(r) \leqslant 0$, $\forall\ r$), that for small r, $V(r)$ is not more singular than r^{-1} (a Coulomb or gravitational potential), and that, for some $r_1 > 0$, $2\mu|V(r)|/\hbar^2 > \ell(\ell+1)/r^2$, $r > r_1$. Under these conditions, which are applicable to the potentials of Sections 11.4–11.6, the effective potential $v_\ell(r)$ has the general appearance sketched in Fig. 11.3.

Several aspects of Fig. 11.3 are of interest. First, as long as $\ell \neq 0$, $v_\ell(r \to 0) \to +\infty$. An infinite potential at some point means that the wave function must vanish there, i.e., $u_{n\ell}(r \to 0) \to 0$, $\ell \neq 0$. Second, the minimum value of $v_\ell(r)$, occurring at the point $r_{\rm min}$, might not be deep enough for $v_\ell(r)$ to support *any* bound states, even if $\ell = 0$. (This

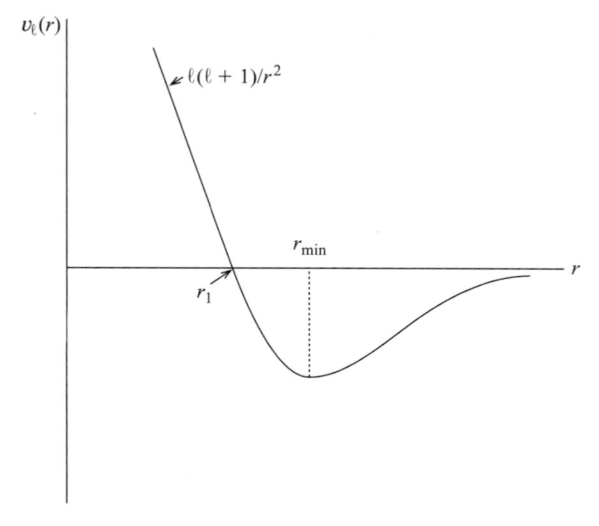

Fig. 11.3 A sketch of $v_\ell(r)$ under the conditions described in the text. The dip in $v_\ell(r)$, i.e., the region for $r > r_1$ but near to r_1, provides an attractive potential well that may be able to support bound states. Notice that the repulsive centrifugal barrier term $l(l+1)/r^2$ vanishes for $l = 0$.

feature of 3-D is explored in Section 11.5.) Third, since $v_\ell(r \to \infty) \to 0$, the asymptotic absence of a potential means that any bound state $u_{n\ell}(r)$ must behave as $\exp(-\alpha_{n\ell} r)$, not as $\exp(ik_{n\ell} r)$, $\alpha_{n\ell}$ and $k_{n\ell}$ each real. Fourth, the centrifugal-barrier term $\ell(\ell+1)/r^2$ grows as ℓ^2 with increasing ℓ, and it will therefore eventually overwhelm the (attractive) $2\mu V(r)/\hbar^2$ term, generally making the occurrence of bound states impossible for sufficiently large ℓ.

Let us now determine the effect of the centrifugal barrier on the radial wave functions. It is simplest to work with Eq. (11.54) for $u_{n\ell}(r)$ and then use (11.51) to transfer the result to $R_{n\ell}(r)$.

For asymptotic values of r, $v_\ell(r) \to 0$, and (11.54) reads

$$u''_{n\ell}(r) \cong \lambda_{n\ell} u_{n\ell}(r), \qquad r \text{ very large.} \tag{11.57a}$$

Since we are interested in bound states (i.e., normalizable $u_{n\ell}$),

$$u_{n\ell}(r) \sim e^{-\alpha_{n\ell} r}, \qquad r \text{ very large;} \tag{11.57b}$$

that is, the $\exp(+\alpha_{n\ell} r)$ term is not an allowed solution. Substitution of (11.57b) into (11.57a) yields

$$\alpha_{n\ell}^2 = \lambda_{n\ell} = -\frac{2\mu}{\hbar^2} E_{n\ell}, \tag{11.58a}$$

or

$$E_{n\ell} = -\frac{\hbar^2 \alpha_{n\ell}^2}{2\mu} < 0. \tag{11.58b}$$

The asymptotic form is consistent with negative energy eigenvalues.

For very small r, both $2\mu|V(r)|/\hbar^2$ and $\lambda_{n\ell}$ are much smaller than the centrifugal-barrier term $\ell(\ell+1)/r^2$, and (11.54) becomes

$$u''_{n\ell}(r) - \frac{\ell(\ell+1)}{r^2} u_{n\ell}(r) \cong 0, \qquad r \text{ very small.} \tag{11.59a}$$

The two linearly independent solutions $u_{n\ell}^{(1)}(r)$ and $u_{n\ell}^{(2)}(r)$ of (11.59a) therefore behave in the small-r limit as

$$u_{n\ell}^{(1)}(r) \underset{r\to 0}{\longrightarrow} r^{\ell+1} \tag{11.59b}$$

and

$$u_{n\ell}^{(2)}(r) \underset{r\to 0}{\longrightarrow} r^{-\ell}. \tag{11.59c}$$

The first of these, $u_{n\ell}^{(1)}$, is the *regular* solution, which is well-behaved at $r = 0$, whereas the second, $u_{n\ell}^{(2)}$, which is singular at the origin, is the *irregular* solution. As with the asymptotic form $\exp(+\alpha_{n\ell} r)$, so too for the irregular solution $u_{n\ell}^{(2)}(r)$: normalizability of bound-state wave functions eliminates each of them. We retain only $u_{n\ell}^{(1)}$, and henceforth drop the superscript "1."

Combining the foregoing results, one may always write the regular solution as

$$u_{n\ell}(r) = A_{n\ell} r^{\ell+1} e^{-\alpha_{n\ell} r} F_{n\ell}(r) \tag{11.60a}$$

and

$$R_{n\ell}(r) = A_{n\ell} r^{\ell} e^{-\alpha_{n\ell} r} F_{n\ell}(r), \tag{11.60b}$$

where $A_{n\ell}$ is a normalization constant, $F_{n\ell}(r)$ is a (well-behaved) function to be determined, and $\alpha_{n\ell}$ is the bound-state wave number, which is related to the energy eigenvalue $E_{n\ell}$ by Eq. (11.58b).

Other properties of the solutions can be inferred, and we state a few more, arguing heuristically. First, assuming that $V(r)$ is sufficiently attractive to support at least several bound states, it follows that $v_0(r)$ is more attractive than $v_\ell(r)$, $\ell > 0$. Hence, we must have

$$E_{1,\ell=0} < E_{1,\ell>0}.$$

That is, the energy of the lowest-lying $\ell = 0$ (S) state is lower than the energies of all of the $\ell \neq 0$ states.

We next derive a lower bound on the energy, the analysis essentially being that of Section 6.3; details are left to the exercises. Let V_0 be the minimum value of the potential in the range $r \in [0, \infty]$ and recall that $\hat{K}$ is essentially a positive-definite operator ($\langle \hat{K} \rangle > 0$).[2] Then, using $\langle V \rangle > V_0$ and $\langle \hat{K} \rangle > 0$, we have

$$E = \langle \hat{K} \rangle + \langle V \rangle > \langle V \rangle > V_0, \tag{11.61}$$

i.e., E is bounded from below, exactly as in the 1-D case.

Third, we fix ℓ and note from (11.61) that only in the case of a $1/r$ potential or an infinite potential (e.g., a quantal box or an oscillator) can there be an infinite number of bound states. Hence the number of bound states $N_{n\ell}$ is generally a finite number. The simplest $u_{n\ell}$ will have zeros only at $r = 0$ ($\ell \neq 0$) and at $r = \infty$. The next simplest $u_{n\ell}$ will have one node between $r = 0$ and $r = \infty$, and so on. Since $\langle \hat{K} \rangle = [1/(2\mu)]\langle \hat{P} \cdot \hat{P} \rangle$ behaves like an average of the square of the slope of $u_{n\ell}$ over the range of the integral ($[0, \infty]$ for $V(r)$ non-infinite over this range), then the more nodes in $u_{n\ell}$, the greater the square of the slope and therefore the larger $\langle \hat{K} \rangle$ will be. On the other hand, as the number of nodes increases, the number of positive and negative values of $R_{n\ell}$ also increases, with a corresponding decrease in $\langle V \rangle$. Hence, as the number of nodes increases, $E = \langle \hat{K} \rangle + \langle V \rangle$ also increases. Following standard practice, we now let n specify the number of nodes for a general ℓ, excluding the one at the origin (if there is one). Then the lowest-lying state has one node $\Rightarrow R_{1\ell}(r)$; the next higher-lying state has two nodes $\Rightarrow R_{2\ell}(r)$; etc. This behavior will be verified in subsequent sections.

So far, we have assumed that $V(r)$ is the type of potential that will support bound states (not all attractive 3-D potentials do so: see Section 11.5 for an example). Suppose that $V(r)$ does yield bound states. It is clearly of interest either to know how many bound states will occur or else to determine an upper limit on that number. This topic has been the subject of a variety of investigations and different upper limits have been obtained. A comprehensive review is that of B. Simon in the collection of essays edited by Lieb, Simon, and Wightman (1976); see also Galindo and Pascual (1990). We state only one of them below and refer the reader to the references cited for others, for proofs, and for further citations to the literature.

[2] $\langle \hat{K} \rangle = 0$ only when the state itself is zero everywhere, a trivial case we exclude (the state cannot be a non-zero constant for $r \in [0, \infty]$ since it is then not normalizable).

Following Simon, we let $n_\ell(V)$ be the number of bound states corresponding to orbital angular momentum ℓ, supported by the potential $V(r)$. The first bound ever derived is that of Bargmann (1952):

$$n_\ell(V) \leqslant \frac{2\mu}{\hbar^2(2\ell+1)} \int_0^\infty r|V(r)|\,dr, \tag{11.B}$$

where μ is the reduced mass. For this bound to hold, $V(r \to 0)$ must be less singular than r^{-2}, while, for $r \to \infty$, $V(r)$ must go to zero more quickly than r^{-3}. Furthermore, the following integral must be finite:

$$\int_0^\infty r^2|V(r)|\,dr < \infty.$$

Several aspects of (11.B) are evident on inspection. First, the presence of the factor $2\ell+1$ in the denominator is a quantitative expression of our previous qualitative argument that, for a finite-ranged potential, the number of bound states decreases as ℓ increases: the repulsive angular-momentum factor eventually overwhelms $V(r)$. Second, if $(2\mu/\hbar^2)\int_0^\infty r|V(r)|\,dr < 1$, then the (3-D) potential $V(r)$ supports no bound states. Third, Coulomb potentials, behaving as r^{-1}, can support – at least in principle – an infinite number of bound states for every ℓ. We prove that this is the case in Section 11.6. This result might suggest that any (3-D) central potential of the form r^{-n}, $n \geqslant 2$, would also support an infinite number of bound states. Such is not the case: for $n \geqslant 3$ there are no bound states (these potentials are too pathological), while for $n = 2$ the bound-state character depends critically on the strength of the potential, as discussed in detail by Landau and Lifshitz (1958).

11.4. Two quantal boxes

As in the 1-D case, a quantal box in 2-D or 3-D is a potential that, because of its infinitely thick and high walls, absolutely confines a particle of mass m to a finite region of space. The boundaries of the box can be of any shape, though the ones examined in this section have radial symmetry, rectangular boxes being considered only in the exercises. We start with a spherical (3-D) box and then consider its 2-D analog. In the latter case, the radial eigenfunctions are Bessel functions, whereas in the former, they are the spherical Bessel function j_ℓ encountered in Chapter 10.

The spherical box

In 3-D, the box is essentially a spherical hole at the center of an infinitely repulsive domain. Taking the radius of the box to be a, the spherically symmetric central potential $V(r)$ is

$$V(r) = \begin{cases} 0, & r < a \\ \infty, & r \geqslant a, \end{cases} \tag{11.62}$$

and is sketched in Fig. 11.4. The Schrödinger wave function $\psi_{n\ell m_\ell}(r, \theta, \phi)$ is the product $R_{n\ell}(r)Y_\ell^{m_\ell}(\Omega)$, where $\Omega = (\theta, \phi)$ is the pair of spherical polar angles for the relative-

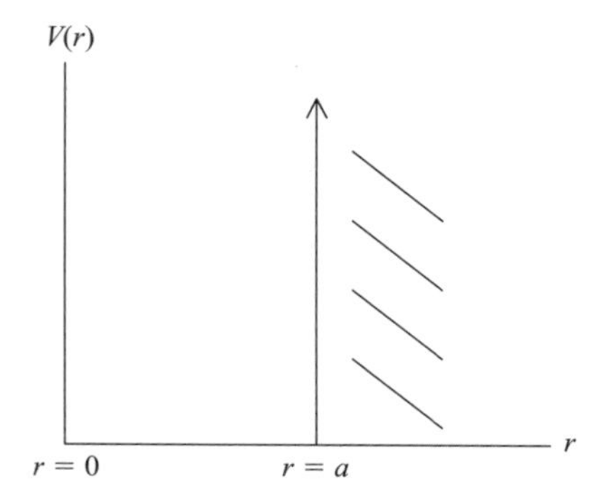

Fig. 11.4 A sketch of a spherical box of radius a: $V(r)$ is zero for $r < a$ and is ∞ for $r \geqslant a$, the latter behavior being represented by the hatched area to the right of the vertical line at $r = a$.

separation vector $\mathbf{r} = \mathbf{r}_{12}$, and $R_{n\ell}(r)$ obeys (11.50) with $V(r)$ given by (11.62) and $\mu = m$.

Outside the box, $R_{n\ell}(r)$ must vanish,

$$R_{n\ell}(r) = 0, \qquad r \geqslant a, \tag{11.63}$$

while inside the box it is that solution of

$$-\frac{\hbar^2}{2m}\left(\frac{1}{r^2}\frac{d}{dr}r^2\frac{d}{dr} - \frac{\ell(\ell+1)}{r^2}\right)R_{n\ell}(r) = E_{n\ell}(r)R_{n\ell}(r), \tag{11.64}$$

which is finite at the origin.[3]

Since the minimum value of V is zero, then all the $E_{n\ell}$ are positive (min $E_{n\ell} >$ min V), and we may write

$$E_{n\ell} = \frac{\hbar^2}{2m}k_{n\ell}^2, \tag{11.65}$$

with $k_{n\ell} > 0$. Using (11.65) for $E_{n\ell}$ and defining $\rho_{n\ell} = k_{n\ell}r$, Eq. (11.54) becomes

$$\left(\frac{1}{\rho_{n\ell}^2}\frac{d}{d\rho_{n\ell}}\rho_{n\ell}^2\frac{d}{d\rho_{n\ell}} - \frac{\ell(\ell+1)}{\rho_{n\ell}^2} + 1\right)R_{n\ell}(\rho_{n\ell}) = 0. \tag{11.66}$$

Given that Eq. (11.66) has the same structure as Eq. (10.66c) on p. 384, that the spherical Bessel function is a regular solution to the latter equation (finiteness of $j_\ell(\rho)$ at $\rho = 0$ was discussed in Section 10.3), and that only one of the two linearly independent solutions of (11.50) *is* regular, we immediately conclude that

$$R_{n\ell}(\rho_{n\ell}) = j_\ell(\rho_{n\ell}) = j_\ell(k_{n\ell}r), \ \forall\ \ell. \tag{11.67a}$$

[3] Just as with Eq. (11.52), it is typical that only one of the two linearly independent solutions to second-order ordinary differential equations is regular at the origin, the other being irregular (infinite) there. Since wave functions are probability amplitudes, they must be finite everywhere in their domain of definition. Therefore only regular solutions are allowed. For domains that exclude the origin, the irregular solution is also allowed, as shown in Section 11.5. See also the comments below Eq. (10.40).

Table 11.1 *Values of $z_{n\ell}$ that yield $j_\ell(z_{n\ell}) = 0$*

	ℓ			
n	0	1	2	3
1	π	1.430π	1.835π	2.224π
2	2π	2.549π	2.895π	3.316π
3	3π	3.471π	3.923π	4.360π
4	4π	4.478π	4.939π	5.387π

The role played by the index n is determined by imposition of the boundary condition (11.63), i.e., we require that

$$j_\ell(k_{n\ell}a) = 0. \tag{11.67b}$$

Equation (11.67b) states that $k_{n\ell}a$ is a *zero* of the ℓth-order spherical Bessel function. Since $j_\ell(\rho)$ is a linear combination of sines and cosines (with ρ-dependent coefficients), and each of sine and cosine has an infinite number of zeros along the real axis, we infer that each $j_\ell(\rho)$ also has an infinite number of zeros for $\rho \in [0, \infty]$. That is, the zeros $z_{n\ell}$ of $j_\ell(\rho)$ are obtained from

$$j_\ell(z_{n,\ell}) = 0, \qquad n = 1, 2, \ldots, \infty, \forall\, \ell. \tag{11.67c}$$

The first four zeros for $\ell = 0$, 1, 2, and 3 are displayed in Table 11.1, while the entire set of zeros for $\ell = 0$ is given below, in Eq. (11.69b).

Comparison of Eqs. (11.67b) and (11.67c) yields

$$k_{n\ell} = \frac{z_{n\ell}}{a}, \tag{11.68a}$$

from which the $E_{n\ell}$ of a particle of mass m in a spherical box are seen to be

$$E_{n\ell} = \frac{\hbar^2 z_{n\ell}^2}{2ma^2}, \qquad \begin{cases} \ell = 0, 1, 2, \ldots, \\ n = 1, 2, \ldots \text{ for each } \ell. \end{cases} \tag{11.68b}$$

The proportionality of $E_{n\ell}$ to $\hbar^2/(ma^2)$ is reminiscent of the result for the 1-D box, which was studied in Chapters 3 and 6. This is no accident, for in each of these cases $E_{n\ell}$ is just the expectation value of the kinetic energy operator, and, since $\hbar$ and m occur as fixed quantities in $\hat{K}$, the only quantity left to form an expression whose dimension is energy is the size of the box (a or L). Although this is a generic result based on dimensional analysis, the structural similarity of the 1-D and 3-D energies is actually an equality when $\ell = 0$, i.e., for S-states if we choose $a = L$. The validity of this remark follows from Eq. (10.79a):

$$j_0(\rho) = \frac{\sin\rho}{\rho},$$

the zeros of which are given by

$$z_{n0} = n\pi, \qquad n = 1, 2, 3, \ldots. \tag{11.69a}$$

Substituting (11.69a) into (11.68b) yields

$$E_{n0} = n^2 \frac{\hbar^2 \pi^2}{2ma^2}, \tag{11.69b}$$

which is identical to Eq. (3.87) when $a = L$, the width of the 1-D box.

In addition to equality of the energies when $a = L$, the solutions $\psi_n(x)$ to the 1-D box and the S-state reduced-radial functions $u_{n0}(\rho)$ for the 3-D box enjoy the same sine-wave functional form, a result that should be unsurprising. The reader may wish to refer to Section 3.3 and Fig. 3.8 for details that apply to the 3-D, S-wave case.

The simple n^2 dependence of the energies E_{n0} of Eq. (11.69b) for $\ell = 0$ does not carry over to higher ℓ, as illustrated by the spacing shown in Fig. 11.5, in which the four lowest-lying energies $E_{n\ell}$ for $\ell = 0$, 1, 2, and 3 are displayed in units of $E_0 = \hbar^2\pi^2/(2ma^2)$. Correspondingly, the eigenfunctions $j_\ell(k_{n\ell}r)$ for $\ell > 0$ are also more complex; discussion of the $\ell = 1$ case, which is analogous to the foregoing, will be found in the exercises.

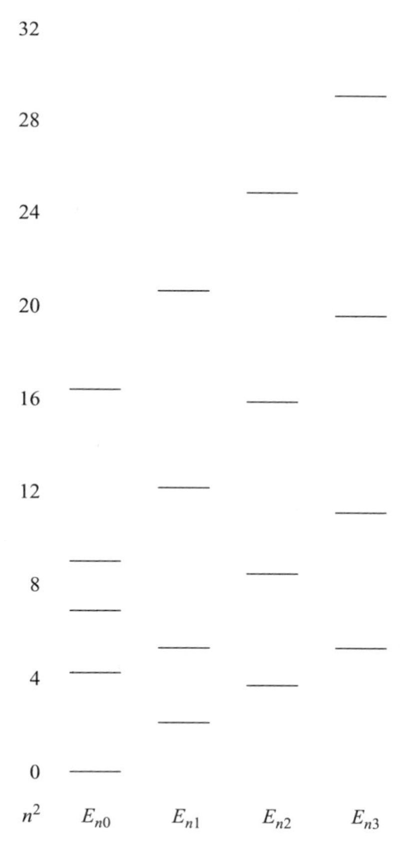

Fig. 11.5 Energy levels (Grotrian diagram) for a particle of mass m in a spherical box of radius a. $E_{n\ell}$ is expressed in units of $E_0 = \hbar^2\pi^2/(2ma^2)$; shown are the lowest four levels ($n = 1$, 2, 3, and 4) for $\ell = 0$, 1, 2, and 3.

We close with a brief discussion of the $j_\ell(\rho)$. The Sturm–Liouville property (11.56) means that

$$\int_0^a dr\, r^2 j_\ell(k_{n\ell}r) j_\ell(k_{m\ell}r) = 0, \qquad n \neq m. \tag{11.70}$$

Of greater interest is the normalization integral. It can be shown (see, e.g., Hassani, (1991), Vol. I of Morse and Feshbach (1953), and especially Wyld (1976)) that

$$\int_0^a dr\, r^2 j_\ell^2(k_{n\ell}r) = \frac{a^3}{2}\left(\left.\frac{dj_\ell(\rho)}{d\rho}\right|_{\rho=k_{n\ell}a}\right)^2, \tag{11.71}$$

a result that is based on the relation between the spherical and the ordinary Bessel functions; the latter, denoted $J_\nu(\rho)$, are considered in the next subsection.

The power-series expansion for $j_\ell(\rho)$ is

$$j_\ell(\rho) = \sqrt{\frac{\pi}{2\rho}}\left(\frac{\rho}{2}\right)^{\ell+1/2}\sum_{n=0}^{\infty}\frac{(-1)^n}{n!\Gamma(n+\ell+\frac{3}{2})}\left(\frac{\rho}{2}\right)^{2n}, \tag{11.72}$$

a result that can be proved by a mildly tedious effort after substituting (11.72) into (11.66), but is more easily verified by referring to one or more of the mathematical physics books cited below Eq. (11.70). Note that $\Gamma(x)$ is the so-called gamma function[4] (e.g., Wyld (1976)), with $\Gamma(n+\frac{3}{2}) = (n+\frac{1}{2})\Gamma(n+\frac{1}{2})$, n is an integer or zero, and $\Gamma(\frac{1}{2}) = \sqrt{\pi}$. The small-$\rho$ behavior of $j_\ell(\rho)$ given in Eq. (10.80) follows from (11.72); see Wyld (1976) for a (non-trivial) derivation of the asymptotic behavior, Eq. (10.82). Further properties of the $j_\ell(\rho)$ are discussed in Section 11.5 along with details of the spherical Neumann function $n_\ell(\rho)$, the irregular solution to Eq. (11.66).

The confining circle

The confining circle is the 2-D analog of the spherical box and is also the zero-height, 2-D version of the cylindrical box considered in the exercises. 2-D systems are more than pedagogical devices illustrating aspects of quantum mechanics (e.g., symmetry properties, in Section 11.7; and new radial functions, in this subsection): to an excellent approximation, 2-D systems occur in nature. One instance is the set of high-T_c organic superconductors. These materials can be thought of as unlinked 2-D arrays of atoms, with the conduction electrons of one array being confined to that array: there is no crossing from one array to another. Our use of 2-D examples is thus an introduction to an intriguing area of physics.

We let x and y designate the 2-D rectangular coordinates, but, because the potential has rotational symmetry in the x–y plane, it is more appropriate to use the polar coordinates r and ϕ:

$$x = r\cos\phi,$$

$$y = r\sin\phi,$$

where ϕ is the usual azimuthal angle.

[4] The gamma function is defined as an integral in Eq. (11.84).

These two sets are shown in Fig. 11.6(a). In terms of r and ϕ, the potential for the confining circle is

$$V(r) = \begin{cases} 0, & r < b, \\ \infty, & r \geqslant b, \end{cases} \tag{11.73}$$

as sketched in Fig. 11.6(b). It is, as claimed, the 2-D analog of the spherical box.

Since $V(\mathbf{r}) = V(r)$, we need an expression for the Laplacian in terms of 2-D polar coordinates. The result,

$$\begin{aligned} \nabla^2 &= \frac{\partial^2}{\partial r^2} + \frac{1}{r}\frac{\partial}{\partial r} + \frac{1}{r^2}\frac{\partial^2}{\partial \phi^2} \\ &= \frac{1}{r}\frac{\partial}{\partial r} r \frac{\partial}{\partial r} + \frac{1}{r^2}\frac{\partial^2}{\partial \phi^2}, \end{aligned} \tag{11.74}$$

leads to the 2-D, polar-coordinate Hamiltonian

$$\hat{H}(r, \phi) = -\frac{\hbar^2}{2m_0}\left(\frac{1}{r}\frac{\partial}{\partial r} r \frac{\partial}{\partial r} + \frac{1}{r^2}\frac{\partial}{\partial \phi^2}\right) + V(r), \tag{11.75}$$

where we assume that a single particle of mass m_0 is confined by $V(r)$.

The time-independent Schrödinger equation is

$$\hat{H}(r, \phi)\psi(r, \phi) = E\psi(r, \phi), \tag{11.76}$$

and our first task is to ascertain the CSCO. Only two operators are needed ($D = 2$), one of which is $\hat{H}(r, \phi)$, while the other must be connected with the fact that $\hat{H}(r, \phi)$ is invariant w.r.t. rotations in the x–y plane. However, from the 3-D analysis of Chapter 10, the infinitesimal generator of such rotations is $\hat{L}_z(\phi) = -i\hbar\, \partial/\partial\phi$. That is,

$$[e^{-i\phi \hat{L}_z/\hbar}, \hat{H}(r, \phi)] = 0, \tag{11.77}$$

and therefore the CSCO is $\{\hat{L}_z, \hat{H}(r, \phi)\}$.

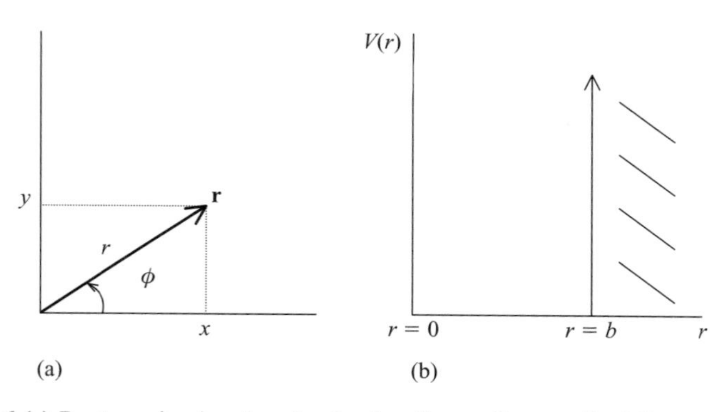

Fig. 11.6 (a) Rectangular (x, y) and polar (r, ϕ) coordinates of a 2-D vector $\mathbf{r}$. (b) A sketch of the 2-D confining-circle potential of radius b: $V(r)$ is zero for $r < b$ and ∞ for $r \geqslant b$. Except for the edge being at b rather than a, this sketch is identical to that of Fig. 11.5. It is essential to keep in mind that, in this figure, r is the magnitude of a 2-D, not a 3-D, vector.

From the commuting operator results of Chapter 9, it next follows that the eigenfunctions $\exp(im\phi)/\sqrt{2\pi}$, $m = 0, \pm 1, \pm 2, \ldots$, of $\hat{L}_z$ are also eigenfunctions of $\hat{H}(r, \phi)$, i.e.,

$$\psi(r, \phi) \to \psi_m(r, \phi) = R_m^{(2)}(r)\frac{e^{im\phi}}{\sqrt{2\pi}}, \qquad m = 0, \pm 1, \pm 2, \ldots \tag{11.78a}$$

and

$$E \to E_m. \tag{11.78b}$$

The equation obeyed by the 2-D radial wave function $R_m^{(2)}(r)$, obtained by substituting Eqs. (11.75) and (11.78) into (11.76), is

$$\left(\frac{d^2}{dr^2} + \frac{1}{r}\frac{d}{dr} - \frac{m^2}{r^2} - V(r) + \frac{2m_0}{\hbar^2}E_m\right)R_m^{(2)} = 0. \tag{11.79}$$

In view of the fact that $V(r) = \infty$ for $r \geqslant b$, $R_m^{(2)}(r)$ vanishes in the latter region,

$$R_m^{(2)}(r) = 0, \qquad r \geqslant b, \tag{11.80}$$

and we need to solve Eq. (11.79) only for $r < b$, for which $V(r) = 0$. As in the preceding subsection, $\min V = 0 \Rightarrow E_m > 0$, so that we may introduce the wave number k_m (> 0):

$$k_m^2 = \frac{2m_0 E_m}{\hbar^2}. \tag{11.81}$$

Thus, in the interval $r < b$, $R_m^{(2)}(r)$ obeys

$$\left(\frac{d^2}{dr^2} + \frac{1}{r}\frac{d}{dr} - \frac{m^2}{r^2} + k_m^2\right)R_m^{(2)}(r) = 0, \tag{11.82a}$$

or, re-expressed in terms of the redefined dimensionless variable $\rho_m = k_m r$,

$$\left(\frac{d^2}{d\rho_m^2} + \frac{1}{\rho_m}\frac{d}{d\rho_m} + 1 - \frac{m^2}{\rho_m^2}\right)R_m^{(2)}(\rho_m) = 0. \tag{11.82b}$$

Equation (11.82b) is known as Bessel's equation, and among its various solutions are the Bessel functions $(J_m(\rho_m))$, the Neumann functions $(N_m(\rho_m))$, and the Hankel functions $(H_m^{(1)}(\rho_m)$ and $H_m^{(2)}(\rho_m))$. The only one of interest as a radial wave function in this subsection is $J_m(\rho_m)$, but, because both it and the Neumann function N_m are related to the solutions of the 3-D spherically symmetric square well of Section 11.5 (spherical Bessel and spherical Neumann functions) and both are also needed to solve Exercise 11.12, we examine both J_m and N_m next. The discussion will be brief.

The generic form of (11.82b) is

$$\left(\frac{d^2}{dx^2} + \frac{1}{x}\frac{d}{dx} + 1 - \frac{m^2}{x^2}\right)v_m(x) = 0, \tag{11.82c}$$

where $x \in [0, \infty]$ is assumed real, as is the index m, which we shall restrict to the values appropriate to (11.82b), viz., $m = 0, \pm 1, \pm 2, \ldots$, although both non-integer and complex-number values are possible.

Bessel functions $J_m(x)$ are the regular solutions to (11.82c); their standard power-series expansion is (Hassani (1991), Mathews and Walker (1970), Wyld (1976)).

$$J_m(x) = \left(\frac{x}{2}\right)^m \sum_{n=0}^{\infty} \frac{(-1)^n}{n!\Gamma(n+m+1)} \left(\frac{x}{2}\right)^{2n}, \qquad m \geqslant 0. \tag{11.83}$$

As in Eq. (11.72), $\Gamma(n)$ denotes the gamma function, one of whose definitions is (see, e.g., Wyld (1976))

$$\Gamma(y) = \int_0^\infty dt\, e^{-t} t^{y-1} = (y-1)\Gamma(y-1), \tag{11.84}$$

so that, when the real variable y is equal to an integer, say $y = n + 1$, $n = 0, 1, \ldots$, then $\Gamma(n+1) = n!$ That the series (11.83) is a solution to (11.82c) is easily established by direct substitution.[5]

By the ratio test, the series for $J_m(x)$ converges for all $x \in [0, \infty)$. For x small, one finds

$$J_m(x) \underset{x\to 0}{\longrightarrow} \frac{1}{m!}\left(\frac{x}{2}\right)^m, \tag{11.85a}$$

while for x asymptotic, the behavior is

$$J_m(x) \underset{x\to\infty}{\sim} \sqrt{\frac{2}{\pi x}} \cos\left(x - \frac{m\pi}{2} - \frac{\pi}{4}\right). \tag{11.85b}$$

Were m not an integer, then a second, linearly independent, irregular solution to (11.82c) would be $J_{-m}(x)$, which is obtained from (11.83) by changing the sign of m. Because m *is* an integer, however, it can be shown (Wyld 1976) that

$$J_{-m}(x) = (-1)^m J_m(x), \qquad m = 0, 1, 2, \ldots. \tag{11.85c}$$

Hence, in the present circumstances, J_{-m} is not the irregular solution to (11.82c). The standard function employed in place of J_{-m} for the integer case is the Neumann function $N_m(x)$, defined by a limiting process (Wyld 1976):

$$\begin{aligned} N_m(x) &= \lim_{\nu\to \text{integer}\, m} \left(\frac{J_\nu(x)\cos(\nu\pi) - J_{-\nu}(x)}{\sin(\nu\pi)}\right) \\ &\to \frac{1}{\pi}\left(\frac{dJ_m(x)}{dx} - (-1)^m \frac{dJ_{-m}(x)}{dx}\right). \end{aligned} \tag{11.86}$$

This function is strongly singular at $x = 0$: Wyld (1976), for example, shows that $N_m(x)$ is a sum of functions, one being $(2/\pi)J_m(x)\ln(x/2)$, which makes manifest the pathological behavior of $N_m(x)$ at the origin. We will not need the general form of $N_m(x)$, although its asymptotic behavior, of use later, is analogous to that of $J_m(x)$:

[5] Unlike functions such as sine or cosine, which have both a geometric interpretation and a power-series expansion, $J_m(x)$ has no "simpler," geometric character. This is in contrast to the spherical Bessel function $j_\ell(x)$, whose power series – Eq. (11.72) – can be re-expressed as a linear combination of sines and cosines, as is clear from the discussion of Section 10.3. The reader, who may be encountering for the first time a function whose power series is not reducible to simpler functions, should not be put off by this: their properties have been investigated and tabulated, and plots of them are available (see especially Abramowitz and Stegun (1965)). Indeed, use of the power series, integral expressions, and recursion relations, not to mention direct numerical integration of the relevant differential equation, allows anyone with programming skills and access to a good enough computer to generate such plots themselves.

$$N_m(x) \underset{x\to\infty}{\sim} \sqrt{\frac{2}{\pi x}} \sin\left(x - \frac{m\pi}{2} - \frac{\pi}{4}\right). \tag{11.87a}$$

Let us now return to the confining-circle potential. Inside the potential, the radial wave function $R_m^{(2)}(\rho_m)$ must be the regular solution J_m:

$$R_m^{(2)}(\rho_m) = J_m(\rho_m) = J_m(k_m r), \qquad r < b. \tag{11.87b}$$

As with the spherical box, the eigenvalues E_{mn} $(= \hbar^2 k_m^2/(2m_0))$ are obtained by imposing the boundary condition that

$$J_m(k_m b) = 0. \tag{11.88}$$

Ordinary Bessel functions, like their spherical counterparts, possess an infinite number of zeros. For the mth order, we denote the nth zero by Z_{mn}:

$$J_m(Z_{mn}) = 0, \qquad n = 1, 2, \ldots, \tag{11.89}$$

from which we find

$$k_m \to k_{mn} = Z_{mn}/b, \tag{11.90a}$$

and therefore that

$$E_m \to E_{mn} = \frac{\hbar^2 Z_{mn}^2}{2m_0 b^2}. \tag{11.90b}$$

The proportionality of E_{mn} to the energy unit $E_0^{(2)} = \hbar^2/(2m_0 b^2)$ occurs for the reason stated in the preceding subsection. Displayed in Table 11.2 are the values of the first four Z_{mn}'s for each of $m = 0$, 1, 2, and 3, while Fig. 11.7 is a Grotrian diagram of E_{mn} expressed in units of $E_0^{(2)}$, for the same values of m and n as those listed in Table 11.2. Although the specific values in Figs. 11.5 and 11.7 are different, the trends are the same. Modifications to Fig. 11.7 that occur when the confining circle becomes a 3-D cylindrical box are left to the exercises.

Although we shall not plot the radial functions $J_m(r)$ in this subsection, it is convenient to gather together a few of their integral properties. The orthonormality relations for the region $r \in [0, b]$ and the boundary condition (11.88) are

$$\int_0^b dr\, r J_m\left(Z_{mn}\frac{r}{b}\right) J_m\left(Z_{mn'}\frac{r}{b}\right) = 0,\ n \neq n'$$

and

$$\int_0^b dr\, r\left[J_m\left(Z_{mn}\frac{r}{b}\right)\right]^2 = \frac{b^2}{2}\left(\frac{dJ_m(\rho)}{d\rho}\right)^2_{\rho = Z_{mn}}.$$

These are derived and discussed in many texts, e.g., Morse and Feshbach (1953) and Wyld (1976). We also mention that $\{J_m(Z_{mn}r/b)\}_{n=1}^{\infty}$ is a complete set on the interval $r \in [0, b]$ so that any well-behaved function $f(r)$ defined on this interval and obeying the condition $f(r = b) = 0$ can be expanded in terms of the $J_m(Z_{mr}r/b)$. The resulting series converges uniformly in the closed interval. However, for a well-behaved function $g(r)$ defined on this interval but for which $g(r = b) \neq 0$, the convergence is not uniform

Table 11.2 *Values of Z_{nm} that yield $J_m(Z_{nm}) = 0$*

	m			
n	0	1	2	3
1	0.766π	1.220π	1.635π	2.031π
2	1.757π	2.233π	2.679π	3.107π
3	2.755π	3.238π	3.699π	4.143π
4	3.754π	4.241π	4.710π	5.164π

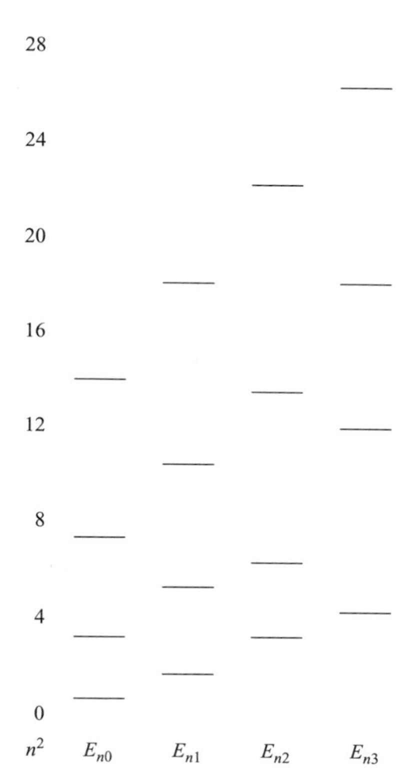

Fig. 11.7 The energy-level scheme for a particle of mass m_0 in a 2-D confining-circle potential of radius b. E_{nm} is expressed in units of $E_0^{(2)} = \hbar^2\pi^2/(2m_0b^2)$; displayed are the lowest four levels ($n = 1, 2, 3$, and 4) for $m = 0, 1, 2$, and 3.

at $r = b$. One evidently must be careful: blind use of complete expansion sets could be a trap for the unwary.

11.5. The symmetric three-dimensional square well

In contrast to the preceding section and the 1-D examples of Chapter 6, the 3-D potentials of this section (on the 3-D square well) and the next (on Coulomb interaction) arise in the description of actual two-body systems. This, of course, is one of the reasons for our study of the eigenvalue problems they generate. The symmetric square-well potential considered in the following provides a model for the nuclear interaction between two protons, two neutrons, or a neutron and a proton. Despite decades of investigation, the functional form of this interaction is still not fully known, although several of its most salient features, especially its short-range character, are. Soluble models of this interaction are therefore of great help in helping us to understand properties of the deuteron (the neutron–proton bound state) and of the collisions between two nucleons (a "nucleon" is a neutron or a proton). Furthermore, since it is one of the few soluble 3-D potentials, the square-well potential would be of interest even if it were not a model for a short-range interaction.

The symmetric square well is illustrated in Fig. 11.8; its functional form is

$$V(r) = \begin{cases} -V_0, & r \leqslant r_0, \\ 0, & r \geqslant r_0. \end{cases} \tag{11.91}$$

If it supports any bound states, their energies $E_{n\ell}$ must be larger than the depth $-V_0$, i.e., $V_0 > |E_{n\ell}|$.

Only the radial part of the Schrödinger equation – (11.50) – concerns us, and we write it twice, once for $r < r_0$ (the interior region), and again for $r > r_0$ (the exterior region):

$$\left[-\frac{\hbar^2}{2\mu}\left(\frac{1}{r^2}\frac{d}{dr}r^2\frac{d}{dr} - \frac{\ell(\ell+1)}{r^2}\right) - V_0\right]R_{n\ell}(r) = E_{n\ell}R_{n\ell}(r), \qquad r \leqslant r_0 \tag{11.92a}$$

and

$$\left[-\frac{\hbar^2}{2\mu}\left(\frac{1}{r^2}\frac{d}{dr}r^2\frac{d}{dr} - \frac{\ell(\ell+1)}{r^2}\right)\right]R_{n\ell}(r) = E_{n\ell}R_{n\ell}(r), \qquad r \geqslant r_0. \tag{11.92b}$$

The asymptotic analysis of Section 11.3 tells us that $E_{n\ell}$ in Eq. (11.92b) must be negative, i.e.,

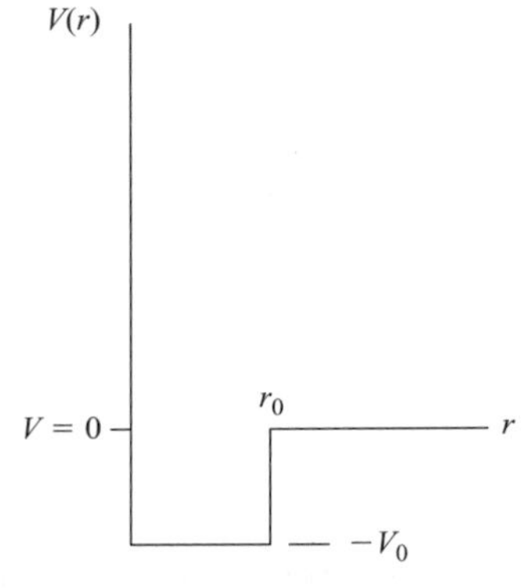

Fig. 11.8 A sketch of the 3-D symmetric square well, with a depth $-V_0$ and a range r_0. The regions $r < r_0$ and $r > r_0$ are the *interior* and the *exterior* intervals, respectively.

$$E_{n\ell} = -\frac{\hbar^2}{2\mu}\alpha_{n\ell}^2, \qquad \alpha_{n\ell} > 0, \tag{11.93a}$$

in order that

$$R_{n\ell}(r) \underset{r\to\infty}{\sim} e^{-\alpha_{n\ell} r}. \tag{11.93b}$$

Apart from the change of sign in $E_{n\ell}$, Eqs. (11.92b) and (11.64) are identical, raising the possibility that j_ℓ could be the solution to (11.92b). However, because of the change of sign, the analog in (11.92b) of the quantity $k_{n\ell}$ appearing in (11.64) is the factor $i\alpha_{n\ell}$: $E_{n\ell} = \hbar^2(i\alpha_{n\ell})^2/(2\mu)$. Therefore, were j_ℓ to be the solution to (11.92b), its argument would be $i\alpha_{n\ell}r$, and, from Eq. (10.82), we would find

$$j_\ell(i\alpha_{n\ell}r) \underset{r\to\infty}{\sim} \frac{1}{2\alpha_{n\ell}r}(e^{\alpha_{n\ell}r}e^{i\ell\pi/2} - e^{-\alpha_{n\ell}r}e^{-i\ell\pi/2}). \tag{11.94}$$

The growing exponential $e^{\alpha_{n\ell}r}$ in (11.94) renders $j_\ell(i\alpha_{n\ell}r)$ unacceptable.

Since j_ℓ has the wrong asymptotic behavior, we must consider the other solution to (11.64), viz., the irregular one (whose pathology at $r = 0$ is of no concern). This second solution is known as a spherical Neumann function, written as n_ℓ. The spherical Bessel and Neumann functions are related to their "big" siblings the ordinary Bessel and Neumann functions by (see, e.g., Wyld (1976) or Morse and Feshbach (1953))

$$j_\ell(\rho) = \sqrt{\frac{\pi}{2\rho}}J_{\ell+1/2}(\rho) \tag{11.95a}$$

and

$$n_\ell(\rho) = \sqrt{\frac{\pi}{2\rho}}N_{\ell+1/2}(\rho). \tag{11.95b}$$

It can be shown (see the exercises) that, in analogy to Eq. (10.78a), $n_\ell(\rho)$ can be expressed as derivatives of $\cos\rho/\rho$:

$$n_\ell(\rho) = -(-\rho)^\ell\left(\frac{1}{\rho}\frac{d}{d\rho}\right)^\ell\frac{\cos\rho}{\rho}. \tag{11.96}$$

The first three spherical Neumann functions are

$$n_0(\rho) = -\frac{\cos\rho}{\rho}, \tag{11.97a}$$

$$n_1(\rho) = -\frac{\sin\rho}{\rho} - \frac{\cos\rho}{\rho^2}, \tag{11.97b}$$

$$n_2(\rho) = -\left(\frac{3}{\rho^3} - \frac{1}{\rho}\right)\cos\rho - \frac{3}{\rho^2}\sin\rho$$

$$= \frac{3}{\rho^2}n_1(\rho) - n_0(\rho); \tag{11.97c}$$

compare with Eqs. (10.79). For ρ small, one finds

$$n_\ell(\rho) \underset{\rho\to 0}{\longrightarrow} -\frac{(2\ell)!}{2^\ell\ell!}\frac{1}{\rho^{\ell+1}}, \tag{11.98a}$$

consistent with Eq. (11.59c). The asymptotic behavior follows on substituting (11.87) into (11.95b):

$$n_\ell(\rho) \underset{\rho\to\infty}{\sim} -\cos(\rho - \ell\pi/2)/\rho. \tag{11.98b}$$

Like $j_\ell(k_{n\ell}r)$, $n_\ell(k_{n\ell}r)$ satisfies (11.64). Hence, if n_ℓ is to be considered a solution of (11.92b), its argument must also be $i\alpha_{n\ell}r$, just as in our attempt to have $j_\ell(i\alpha_{n\ell}r)$ as the solution. However, the same pathology as that which we saw in Eq. (11.94) besets $n_\ell(i\alpha_{n\ell}r)$. That is, setting $\rho = i\alpha_{n\ell}r$ in (11.98b) leads to

$$n_\ell(i\alpha_{n\ell}r) \underset{r\to\infty}{\sim} \frac{i}{2\alpha_{n\ell}r}(e^{\alpha_{n\ell}r}e^{i\ell\pi/2} + e^{-\alpha_{n\ell}r}e^{-i\ell\pi/2}), \tag{11.99}$$

and the growing exponential eliminates $n_\ell(i\alpha_{n\ell}r)$ as a potential exterior solution ($r > r_0$) to the symmetric square well, just as it did $j_\ell(i\alpha_{n\ell}r)$.

While neither j_ℓ nor n_ℓ individually can be the exterior solution, comparison of the asymptotic forms (11.94) and (11.99) shows that the linear combination $j_\ell + in_\ell$ does display the requisite behavior. This combination defines the spherical Hankel function $h_\ell^{(1)}$:[6]

$$h_\ell^{(1)}(\rho) \equiv j_\ell(\rho) + in_\ell(\rho), \tag{11.100a}$$

such that

$$h_\ell^{(1)}(i\alpha_{n\ell}r) \underset{r\to\infty}{\sim} -(-i)^\ell \frac{e^{-\alpha_{n\ell}r}}{\alpha_{n\ell}r}. \tag{11.100b}$$

Therefore, to within a normalization constant, $h_\ell^{(1)}(i\alpha_{n\ell}r)$ is the exterior solution to the 3-D symmetric square well. The first three of these functions are

$$h_0(i\alpha_{n\ell}r) = -\frac{e^{-\alpha_{n\ell}r}}{r}, \tag{11.101a}$$

$$h_1(i\alpha_{n\ell}r) = i\left(\frac{1}{\alpha_{n\ell}r} + \frac{1}{\alpha_{n\ell}^2r^2}\right)e^{-\alpha_{n\ell}r}, \tag{11.101b}$$

$$h_2(i\alpha_{n\ell}r) = \left(\frac{1}{\alpha_{n\ell}r} + \frac{3}{\alpha_{n\ell}^2r^2} + \frac{3}{\alpha_{n\ell}^3r^3}\right)e^{-\alpha_{n\ell}r}. \tag{11.101c}$$

So far, we have determined only the functions that form the exterior solution; the energies/bound-state wave numbers are undetermined. To obtain these, we must solve the interior equation, (11.92a), and then match the interior and exterior solutions and the derivatives, in precise analogy to the 1-D square wells, Sections (6.4) and (6.5), especially the half-space, 1-D square well, whose notation we follow. For simplicity, we drop the subscript n and write $E_{n\ell} \to E_\ell = -|E_\ell|$, whereupon (11.92a) goes to

$$-\frac{\hbar^2}{2\mu}\left(\frac{1}{r^2}\frac{d}{dr}r^2\frac{d}{dr} - \frac{\ell(\ell+1)}{r^2}\right)R_\ell(r) = (V_0 - |E_\ell|)R_\ell(r), \qquad r \leqslant r_0. \tag{11.102}$$

[6] The spherical Hankel functions $h_\ell^{(1)}(\rho)$ and $h_\ell^{(1)^*}(\rho)$ are a second pair of linearly independent solutions of the free-particle equation (11.92b). We will encounter this pair in Section 15.4.

Since $V_0 - |E_\ell|$ is always non-negative, we introduce the interior wave number γ_ℓ ($\gamma_\ell > 0$) by writing

$$V_0 - |E_\ell| \equiv \frac{\hbar^2 \gamma_\ell^2}{2\mu} \tag{11.103a}$$

and then define the dimensionless variable ρ_ℓ:

$$\rho_\ell \equiv \gamma_\ell r. \tag{11.103b}$$

The same manipulations as those that led to Eq. (11.66) now produce

$$\left(\frac{1}{\rho_\ell^2} \frac{d}{d\rho_\ell} \rho_\ell^2 \frac{d}{d\rho_\ell} - \frac{\ell(\ell+1)}{\rho_\ell^2} + 1 \right) R_\ell(\rho_\ell) = 0, \qquad \rho_\ell \leqslant \gamma_\ell r_0, \tag{11.104}$$

whose structure is again that of Eq. (10.66c) on p. 384 and whose regular solution is therefore $j_\ell(\rho_\ell)$. In other words, the interior solution is

$$R_\ell(r) = A_\ell j_\ell(\gamma_\ell r), \qquad r \leqslant r_0, \tag{11.105}$$

where A_ℓ is a normalization constant.

Given that the exterior solution is

$$R_\ell(r) = B_\ell h_\ell^{(1)}(i\alpha_\ell r), \tag{11.106}$$

where the principal quantum number n is again suppressed, continuity requires that

$$\lim_{\epsilon \to 0} \begin{cases} R_\ell(r_0 - \epsilon) = R_\ell(r_0 + \epsilon), \\ \left. \dfrac{dR_\ell}{dr} \right|_{r_0 - \epsilon} = \left. \dfrac{dR_\ell}{dr} \right|_{r_0 + \epsilon}, \end{cases} \tag{11.107}$$

the same type of matching condition as that used in the 1-D cases. On dividing the derivative equality by the function equality, the normalization constants drop out and we get

$$\frac{1}{j_\ell(\gamma_\ell r_0)} \frac{dj_\ell(\gamma_\ell r_0)}{dr_0} = \frac{1}{h_\ell^{(1)}(\alpha_\ell r_0)} \frac{dh_\ell^{(1)}(\alpha_\ell r_0)}{dr_0}. \tag{11.108a}$$

This transcendental relation is one of two needed to determine α_ℓ, the other being

$$\alpha_\ell r_0 = \sqrt{\frac{2\mu r_0^2}{\hbar^2} V_0 - (\gamma_\ell r_0)^2}, \tag{11.108b}$$

which follows from (11.103a). The reader may wish to compare with the 1-D half-space square-well relations, Eqs. (6.170) and (6.171).

Let us assume that there is at least one bound state for several of the ℓ values. Equations (11.108), being transcendental, are generally solved numerically once V_0 and r_0 have been specified. The $\{\alpha_{n\ell}\}$ and $\{E_{n\ell}\}$ can be determined to whatever accuracy is desired, the limitations being those of computer-memory round-off error. The latter can be significant when ℓ is large, especially if one plans to use recursion relations – see the exercises – rather than calculating j_ℓ and h_ℓ directly.

Once the $\alpha_{n\ell}$ are obtained, we may need to calulate the normalization constants, in order, say, to obtain average values of $\mathbf{r}$ or $\hat{\mathbf{P}}$. The desired relations are

$$A_\ell j_\ell(\gamma_{n\ell} r_0) = B_\ell h_\ell^{(1)}(i\alpha_{n\ell} r_0) \tag{11.109a}$$

and

$$|A_\ell|^2 \int_0^{r_0} dr\, r^2 j_\ell^2(\gamma_{n\ell} r) + |B_\ell|^2 \int_{r_0}^{\infty} dr\, r^2 |h_\ell^{(1)}(i\alpha_{n\ell} r)|^2 = 1. \tag{11.109b}$$

A_ℓ can always be chosen real and positive, so that B_ℓ is either real or pure imaginary, depending on the value of ℓ.

EXAMPLE

The S-state case, $\ell = 0$. Not only is $\ell = 0$ the simplest to analyze, but also it is the case which provides a simple model for the deuteron, whose ground state is largely[7] $\ell = 0$. For $\ell = 0$ it is advantageous to work directly with the reduced wave function $u \equiv u_0$.

In the interior region we have ($\alpha_0 \to \alpha$, $\gamma_0 \to \gamma$)

$$u(r) = A \sin(\alpha r), \qquad r \leqslant r_0, \tag{11.110a}$$

while in the exterior interval we get

$$u(r) = B\, e^{-\alpha r}, \qquad r \geqslant r_0. \tag{11.110b}$$

The matching conditions (11.108) now become

$$\alpha r_0 = -\gamma r_0 \cot(\gamma r_0) \tag{11.111a}$$

and

$$\alpha r_0 = \sqrt{\frac{2\mu r_0^2}{\hbar^2} V_0 - \gamma^2 r_0^2}. \tag{11.111b}$$

However, apart from the different ranges (L versus r_0), Eqs. (11.111) are identical to those of the 1-D half-space square well, Eqs. (6.170) and (6.171). The claim made in the first paragraph of Section 6.4 is hereby verified. The graphical analysis of that section applies directly to the present case, and we refer the reader not only to Fig. 6.12, but also to Eq. (6.173).

In the present context the latter inequality states that, if

$$V_0 < (2n+1)^2 \frac{\hbar^2 \pi^2}{8\mu r_0^2} \tag{11.112}$$

then there are n $\ell = 0$ bound states. Hence, for $V_0 < \hbar^2\pi^2/(8\mu r_0^2)$, the 3-D square well supports no bound states at all!

Let us compare this with the Bargmann bound, Eq. (11.B) on p. 412:

$$n_\ell < \frac{2\mu}{\hbar^2(2\ell+1)} \int_0^\infty r V(r)\, dr.$$

[7] Because neutrons and protons are spin-$\frac{1}{2}$ particles, the deuteron state vector contains spin components in addition to the spatial terms. Through the former, a small amount of an $\ell = 2$ state is present in the deuteron. We ignore it in our discussion here of the deuteron. See Chapter 14 for further comments.

For the square well this becomes

$$n_\ell < \frac{2\mu}{\hbar^2(2\ell+1)} V_0 \int_0^{r_0} r\,dr,$$

or

$$V_0 > (2\ell+1)\frac{\hbar^2}{\mu r_0^2} n_\ell. \qquad (11.113)$$

When $\ell = 0$ the Bargmann inequality predicts the existence of a bound state only if the well depth V_0 obeys ($n_0 = 1$)

$$V_0 > \frac{\hbar^2}{\mu r_0^2},$$

whereas (11.112) states that there are no bound states if

$$V_0 < \frac{\pi^2}{8}\frac{\hbar^2}{\mu r_0^2}.$$

Since $\pi^2/8 \cong 1.234$, the Bargmann bound is consistent with (11.112).

Application of this general $\ell = 0$ square-well analysis to a spinless $\ell = 0$ model of the deuteron is explored in the exercises.

11.6. The $1/r$ potential

In this section we solve the Schrödinger-equation eigenvalue problem for an attractive $1/r$ potential. The classic example is the H-atom, but any one-electron atom or a two-particle system bound by gravitational forces will have a set of states and energy spectrum whose structure is the analog of those for hydrogen. A single electron bound to a nucleus of charge Z is the system we will consider. It is illustrated in Fig. 11.9, where the H-atom corresponds to $Z = 1$. The Coulomb potential for this system is

$$V(r) = -\kappa_e \frac{Ze^2}{r}, \qquad \kappa_e = \begin{cases} 1, & \text{cgs units,} \\ 1/(4\pi\epsilon_0), & \text{SI units.} \end{cases} \qquad (11.114)$$

Sketched in Fig. 11.10 are $2\mu V(r)/\hbar^2$, the centrifugal barrier $\ell(\ell+1)/r^2$, and the reduced effective potential

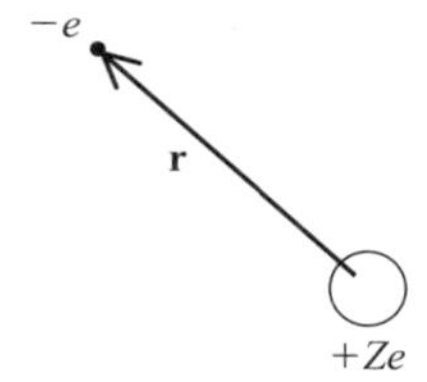

Fig. 11.9 The one-electron atom, whose nucleus (shown as the large circle, but considered as a point particle) has charge Ze. $\mathbf{r}$ is the vector from the nucleus to the electron.

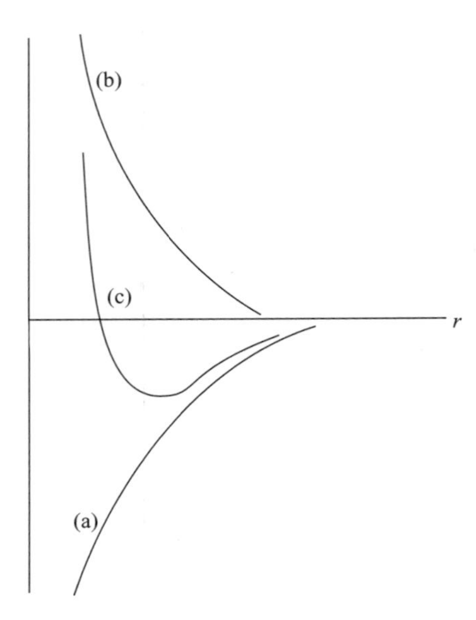

Fig. 11.10 Reduced potentials for the one-electron atom: (a) the Coulomb attraction, $-2\mu\kappa_e Ze^2/(\hbar^2 r)$, (b) the centrifugal barrier, $\ell(\ell+1)/r^2$, and (c) the ℓ-dependent effective potential $\upsilon_\ell(r) = \ell(\ell+1)/r^2 - 2\mu\kappa_e Ze^2/(\hbar^2 r)$.

$$\upsilon_\ell(r) = \frac{\ell(\ell+1)}{r^2} - \frac{2\mu\kappa_e Ze^2}{\hbar^2 r}. \tag{11.115}$$

$\upsilon_\ell(r)$ appears sufficiently attractive to support bound states and, in fact, supports an infinite number of them, as allowed by the Bargmann bound. All the bound-state energies are negative: $E_{n\ell} < 0$.

Determination of the eigenvalues

It is slightly simpler to work with the reduced radial function $u_{n\ell}(r)$; dropping the subscript n, Eq. (11.52) becomes

$$\left[-\frac{\hbar^2}{2\mu}\left(\frac{d^2}{dr^2} - \frac{\ell(\ell+1)}{r^2}\right) - \frac{\kappa_e Ze^2}{r}\right] u_\ell(r) = E_\ell(r) u_\ell(r). \tag{11.116}$$

As in previous bound-state problems, we introduce the bound-state wave number α_ℓ via[8]

$$E_\ell = -\frac{\hbar^2\alpha_\ell^2}{2\mu}, \tag{11.117}$$

in terms of which Eq. (11.116) becomes

$$\left(\frac{d^2}{dr^2} - \frac{\ell(\ell+1)}{r^2} + \frac{2\mu\kappa_e Ze^2}{\hbar^2 r} - \alpha_\ell^2\right) u_\ell(r) = 0. \tag{11.118}$$

[8] We are deliberately *not* expressing E_ℓ in terms of the Rydberg unit $\kappa_e e^2/a_0$. We will determine α_ℓ^2 by solving the eigenvalue equation (11.118).

Like the other terms in the preceding square bracket, $2\mu\kappa_e Ze^2/(\hbar^2 r)$ must have dimension (length)$^{-2}$. One factor of (length)$^{-1}$ arises from r^{-1}. The analysis of the Bohr atom in Chapter 1 provides the other factor of (length)$^{-1}$: it is the quantity $\hbar^2/(\mu\kappa_e e^2)$, which is the Bohr radius a_0 when $\mu = m_e$. We denote this radius as a:

$$a \equiv \frac{\hbar^2}{\mu\kappa_e e^2}, \tag{11.119}$$

whereupon (11.118) reads

$$\frac{d^2 u_\ell(r)}{dr^2} + \left(\frac{2Z}{ar} - \frac{\ell(\ell+1)}{r^2} - \alpha_\ell^2\right) u_\ell(r) = 0. \tag{11.120}$$

Division of all terms in (11.120) by α_ℓ^2 yields a differential equation in the dimensionless variable $\alpha_\ell r$. Rather than use this quantity, we take account of prior analysis of (11.120), on the basis of which we choose $2\alpha_\ell r$ as the dimensionless variable:

$$\rho \equiv 2\alpha_\ell r; \tag{11.121a}$$

we also define

$$\lambda \equiv \frac{Z}{\alpha_\ell a}. \tag{11.121b}$$

The final form taken by (11.120) is thus

$$u_\ell''(\rho) - \left[\frac{\ell(\ell+1)}{\rho^2} - \left(\frac{\lambda}{\rho} - \frac{1}{4}\right)\right] u_\ell(\rho) = 0, \tag{11.122}$$

where $' = d/d\rho$.

Equation (11.122) is the equation we will solve. The wave number α_ℓ appears in ρ and also in λ, which plays the role of the eigenvalue for (11.122). Applying the results of Section 11.3 (in connection with Eq. (11.52)), $u_\ell(\rho)$ can be expressed as

$$u_\ell(\rho) = \rho^{\ell+1} e^{-\rho/2} F(\rho), \tag{11.123}$$

with $F(\rho)$ as the function to be determined. Substitution of (11.123) into (11.122) leads, after a short exercise in differentiation, to

$$\rho F''(\rho) + (2\ell + 2 - \rho)F'(\rho) + [\lambda - (\ell+1)]F(\rho) = 0. \tag{11.124}$$

An infinite power-series expansion will be used to express $F(\rho)$ and to find the eigenvalue λ. As with the equations for the Hermite and Legendre polynomials, the solution to (11.124) will also turn out to be a polynomial, obtained by equating $\lambda - (\ell+1)$ to an integer.

We proceed by recalling that since the small-ρ behavior of $u_\ell(\rho)$ has already been extracted in Eq. (11.123), $F(\rho \to 0) \to$ constant. Hence the power-series expansion of $F(\rho)$ is

$$F(\rho) = \sum_{m=0}^{\infty} C_m \rho^m. \tag{11.125}$$

Substitution of this expansion into (11.124) leads to

$$\sum_{m=0}^{\infty} m(m-1)C_m\rho^{m-1} + \sum_{m=0}^{\infty} m(2\ell+2)C_m\rho^{m-1}$$

$$- \sum_{m=0}^{\infty} mC_m\rho^m + \sum_{m=0}^{\infty}(\lambda - \ell - 1)C_m\rho^m = 0. \tag{11.126a}$$

The first sum in Eq. (11.126a) starts at $m = 2$, while the next two begin with $m = 1$. Relabeling $m \to m + 2$ in the first and $m \to m + 1$ in the second pair, and rearranging, we get

$$\sum_{m=0}^{\infty}\{[(m+1)(m+2)C_{m+2} - (m+1)C_{m+1}]\rho^{m+1}.$$

$$+ [(m+1)(2\ell+2)C_{m+1} + (\lambda - \ell - 1)C_m]\rho^m\} = 0. \tag{11.126b}$$

This equation can be satisfied only if the coefficient of each power of ρ is separately set equal to zero, as in the cases of the Hermite and Legendre polynomials. Thus, for $m = 0$, one has

$$C_1 = \frac{\ell + 1 - \lambda}{2\ell + 2} C_0; \tag{11.127a}$$

in the general case the recursion relation between coefficients C_{m+1} and C_m is

$$C_{m+1} = \frac{m + (\ell + 1) - \lambda}{(m+1)(m + 2\ell + 2)} C_m. \tag{11.127b}$$

To represent a quantal wave function, the series (11.125) must converge. The large-m limit of (11.127b) leads to problems (which should come as no surprise, recalling our previous infinite-series expansions!):

$$\lim_{m\to\infty} C_{m+1} = \frac{1}{m} C_m,$$

implying that the series (11.125) behaves asymptotically like $\exp \rho$ – a claim left to the reader to verify. Hence, the full series (11.125) would yield a $u_\ell(\rho)$ behaving like $\exp(\rho/2)$ for large ρ, which is unacceptable: $u_\ell(\rho)$ must go like $\exp(-\rho/2)$, Eq. (11.123).

To avoid the pathology of a non-normalizable bound-state wave function, the infinite series for $F(\rho)$ must truncate to a polynomial, which will happen if $\lambda - (\ell + 1)$ is set equal to an integer (or zero). Let the integer be n_r:

$$\lambda - \ell - 1 = n_r, \tag{11.128a}$$

so that

$$\sum_{m=0}^{\infty} \to \sum_{m=0}^{n_r}, \tag{11.128b}$$

i.e., $n_r = m_{\max}$. Since ℓ is an integer (or zero), λ must be an integer as well:

$$\lambda = n_r + \ell + 1 \equiv n, \qquad n = 1, 2, 3, \ldots. \tag{11.128c}$$

However, $\lambda = Z/(\alpha_\ell a)$ and we therefore find

$$\frac{Z}{\alpha_\ell a} = n, \qquad n = 1, 2, 3, \ldots,$$

or

$$\alpha_\ell \to \alpha_n \equiv \frac{Z}{na}, \qquad n = 1, 2, 3, \ldots. \tag{11.129}$$

This is a remarkable result: the ℓ dependence of α_ℓ is replaced by a dependence on $n = n_r + \ell + 1$, and, since $n_r \geqslant 0$, we conclude that

$$\ell + 1 \leqslant n. \tag{11.128d}$$

Hence, for a fixed value of the principal quantum number n ($n = 1, 2, 3 \ldots$), no orbital angular-momentum states with $\ell > n - 1$ can occur. There is a strong suggestion of degeneracy, whose existence is discussed later.

Using the bound-state wave number relation (11.129), the bound-state energies are

$$E_\ell \to E_n = -\frac{\hbar^2}{2\mu}\frac{Z^2}{n^2 a^2}$$
$$= -\frac{1}{n^2}\frac{Z^2}{2}\frac{\kappa_e e^2}{a}, \tag{11.130}$$

which is a generalization of Bohr's expression for the H-atom energy levels. In fact, if we set $Z = 1$ and $\mu = m_e M_p/(m_e + M_p) \cong m_e$, then (11.130) becomes

$$E_n = -\frac{1}{n^2}\frac{\kappa_e e^2}{2a_0} \qquad Z = 1,\ M_p = \infty,$$

which is just Eq. (1.23) for the Bohr energies! That the Bohr energy levels are also the non-relativistic quantal ones for the spinless H-atom is truly remarkable. A Grotrian diagram of the levels is postponed until the wave functions are obtained. Although the wave functions are probability amplitudes rather than orbits, Bohr or otherwise, some of them yield results related to the Bohr orbits, as mentioned in Section 5.2 and discussed in more detail in the next subsection.

Equation (11.130) is, as noted, a non-relativistic result. Let us express it in terms of the reduced-mass rest energy μc^2. Inserting the expression (11.119) for a into (11.130) and assuming that $n = 1$, we get

$$E_1 = -Z^2\frac{1}{2}\mu c^2\left(\frac{\kappa_e e^2}{\hbar c}\right)^2, \tag{11.130$'$}$$

where the fine-structure constant $\kappa_e e^2/(\hbar c)$ is written out to avoid any possible confusion with the wave number α_n. Since $\kappa_e e^2/(\hbar c) \approx 1/137$ and $\mu c^2 \cong m_e c^2 \cong 0.5$ MeV, then, for small Z, E_1 is of the order of tens of electron-volts. That is, for small Z, $E_1/(\mu c^2) \ll 1$, implying that relativistic effects are indeed negligible. However, for larger Z, say $Z \gtrsim 40$, $E_1/(\mu c^2)$ is 0.1 or larger, implying that relativistic effects are no longer ignorable. This implication is borne out in calculations of the ground-state energies of non-light atoms, a point to which we return in Chapter 18.

The wave functions

Restricting the eigenvalue λ to integers, $\lambda = n$, $n = 1, 2, 3, \ldots$, leads to normalizable polynomial solutions to (11.124). Restoring the subscripts n and ℓ, Eq. (11.124) becomes

$$\rho F''_{n\ell}(\rho) + (2\ell + 2 - \rho)F'_{n\ell}(\rho) + [n - (\ell + 1)]F_{n\ell}(\rho) = 0. \qquad (11.131)$$

The (real) polynomials which satisfy this equation are known as the associated Laguerre polynomials. Unfortunately for readers who may wish to consult other treatments of the quantal $1/r$ potential, there is neither a uniformly accepted notation for indicating the n and ℓ dependences of these polynomials nor do all authors use the same sign or normalization when stating their definitions of the polynomials. We shall introduce and discuss aspects of these polynomials before relating $F_{n\ell}(\rho)$ to them. Many physics and mathematics books deal with the Laguerre polynomials. Two that readers may wish to consult are those of Lebedev (1972) and Galindo and Pascual (1990), who use the same definitions as those presented in the following.

The associated Laguerre polynomial $L_p^k(x)$ is defined[9] by the Rodrigues' formula

$$L_p^k(x) \equiv \frac{e^x}{p!x^k}\frac{d^p}{dx^p}(e^{-x}x^{p+k}), \qquad p = 0, 1, 2, \ldots. \qquad (11.132a)$$

Only integer k (> 0) occur in the $1/r$-potential problem, in which case Eq. (11.132a) can be written as a sum:

$$L_p^k(x) = \sum_{m=0}^{p} \frac{(-1)^m (p + k)!x^m}{(p - m)!(k + m)!m!}. \qquad (11.132b)$$

(Expressions for low-order L_p^k will be given shortly.)

With the normalization of (11.132a), the orthonormality relation for these functions, which are regular for all $x \in [0, \infty]$, is

$$\int_0^\infty dx\, e^{-x}x^k L_p^k(x)L_{p'}^k(x) = \delta_{pp'}\frac{(p + k)!}{p!}. \qquad (11.133)$$

Note that the same superscript – k – appears in each of the two polynomials. This is necessary, since k is a parameter rather than the eigenvalue p in the equation that $L_p^k(x)$ satisfies, viz.

$$x\frac{d^2}{dx^2}L_p^k(x) + (k + 1 - x)\frac{d}{dx}L_p^k(x) + pL_p^k(x) = 0. \qquad (11.134)$$

The functions $F_{n\ell}(\rho)$ that enter $R_{n\ell}(r)$ are obtained by comparing Eqs. (11.134) and (11.131); we see that

$$F_{n\ell}(\rho) = L_{n-\ell-1}^{2\ell+1}(\rho), \qquad (11.135)$$

where the normalization leading to (11.133) is to be retained. (The reader will no doubt have noticed that the identification (11.135) has been made without our having gone back to Eqs. (11.125), (11.127), and (11.128). That is, we have not made use of the original

[9] Some authors omit the factor $(p!)^{-1}$ in (11.132); this is not a crucial omission for $R_{n\ell}$ since the latter quantity, being a quantal amplitude, is itself normalized to unity.

series expansion of $F_{n\ell}$ or the recursion relation for the coefficients C_m. It is left as an exercise to show that, apart from a normalization constant, i.e., apart from the value of C_0, the truncated series for $F_{n\ell}$ is the same as that given by Eq. (11.132b) for $k = 2\ell + 1$ and $p = n - \ell - 1$.)

Which L_p^k actually enter the quantal wave function? To answer this question, recall that, for a given n, the allowed ℓ values are $\ell = 0, 1, 2, \ldots, n - 1$, Eq. (11.128d). Hence it is clear that only odd superscripts appear, the minimum value being 1, while the smallest subscript is 0. However, from (11.132a), $L_0^k(x) = 1$, so that functional forms are needed only for $n - \ell > 2$. A few representative examples are listed below:

$$
\begin{aligned}
n, \ell = n - 1&: \quad L_0^{2\ell+1} = 1,\\
n = 2, \ell = 0&: \quad L_1^1(x) = 2 - x,\\
n = 3, \ell = 0&: \quad L_2^1(x) = 3 - 3x - x^2/2,\\
n = 4, \ell = 0&: \quad L_3^1(x) = 4 - 6x + 2x^2 - x^3/6,\\
n = 3, \ell = 1&: \quad L_1^3(x) = 4 - x,\\
n = 4, \ell = 1&: \quad L_2^3(x) = 10 - 5x + x^2/2.
\end{aligned}
$$

These and others (without further specification) will be incorporated into the functional forms for the $R_{n\ell}(\rho)$, whose general form is

$$R_{n\ell}(\rho) = N_{n\ell}\, \rho^\ell e^{-\rho/2} L_{n-\ell-1}^{2\ell+1}(\rho), \qquad \rho = \frac{2Zr}{na}, \tag{11.136}$$

while the normalization constant $N_{n\ell}$ is obtained from

$$
\begin{aligned}
1 &= |N_{n\ell}|^2 \int_0^\infty dr\, r^2 e^{-\rho} \rho^{2\ell} (L_{n-\ell-1}^{2\ell+1}(\rho))^2\\
&= |N_{n\ell}|^2 \left(\frac{na}{2Z}\right)^3 \int_0^\infty d\rho\, e^{-\rho} \rho^{2\ell+2} (L_{n-\ell-1}^{2\ell+1}(\rho))^2.
\end{aligned} \tag{11.137}
$$

The fact that $\rho^{2\ell+2}$ enters (11.137) means that the normalization integral (11.133) cannot be used to determine the constant $N_{n\ell}$. An evaluation of the integral in (11.137) and ones like it was first carried out by Schrödinger (1928) when he postulated the quantal wave equation that now bears his name and applied it to the hydrogen atom. Tabulation of these types of integrals can be found in various references, e.g., Schrödinger, *op cit.*, and Galindo and Pascual (1990). The result needed to determine $N_{n\ell}$ is

$$\int_0^\infty d\rho\, e^{-\rho} \rho^{2\ell+2} (L_{n-\ell-1}^{2\ell+1}(\rho))^2 = \frac{2n(n+\ell)!}{(n-\ell-1)!}, \tag{11.138}$$

from which one obtains ($N_{n\ell}$ being chosen real and positive)

$$N_{n\ell} = \frac{2}{n^2}\left(\frac{Z}{a}\right)^{3/2} \sqrt{\frac{(n-\ell-1)!}{(n+\ell)!}}. \tag{11.139}$$

Hence, the normalized, real, radial wave function is

$$R_{n\ell}(r) = \frac{2}{n^2}\left(\frac{Z}{a}\right)^{3/2}\sqrt{\frac{(n-\ell-1)!}{(n+\ell)!}}e^{-Zr/(na)}\left(\frac{2Zr}{na}\right)^{\ell}L_{n-\ell-1}^{2\ell+1}\left(\frac{2Zr}{na}\right), \tag{11.140}$$

and the complete 3-D wave function, written with subscripts n, ℓ, and m_ℓ as $\psi_{n\ell m_\ell}(\mathbf{r})$, is

$$\psi_{n\ell m_\ell}(\mathbf{r}) = R_{n\ell}(r)Y_\ell^{m_\ell}(\Omega), \qquad \begin{cases} n = 1, 2, 3, \ldots, \\ \ell = 0, 1, \ldots, n-1. \end{cases} \tag{11.141}$$

The spherical harmonics are known quantities and their absolute squares for low values of ℓ were sketched in Chapter 10. A few of the low-n normalized radial wave functions are stated below:

$$R_{10}(r) = 2\left(\frac{Z}{a}\right)^{3/2} e^{-Zr/a}, \tag{11.142a}$$

$$R_{20}(r) = 2\left(\frac{Z}{2a}\right)^{3/2} e^{-Zr/(2a)}\left(1 - \frac{Zr}{2a}\right), \tag{11.142b}$$

$$R_{21}(r) = \frac{2}{\sqrt{3}}\left(\frac{Z}{2a}\right)^{3/2} e^{-Zr/(2a)}\left(\frac{Zr}{2a}\right), \tag{11.142c}$$

$$R_{30}(r) = 2\left(\frac{Z}{3a}\right)^{3/2} e^{-Zr/(3a)}\left[1 - 2\left(\frac{Zr}{3a}\right) + \frac{2}{3}\left(\frac{Zr}{3a}\right)^2\right], \tag{11.142d}$$

$$R_{31}(r) = \frac{4\sqrt{2}}{3}\left(\frac{Z}{3a}\right)^{3/2} e^{-Zr/(3a)}\left(\frac{Zr}{3a}\right)\left[1 - \frac{1}{2}\left(\frac{Zr}{3a}\right)\right], \tag{11.142e}$$

$$R_{32}(r) = \frac{4}{3\sqrt{10}}\left(\frac{Z}{3a}\right)^{3/2} e^{-Zr/(3a)}\left(\frac{Zr}{3a}\right)^2. \tag{11.142f}$$

These suffice to give the reader a feeling for how the $R_{n\ell}$ behave as functions of n and ℓ. Equations (11.142) have been written as functions of the variable $Zr/(na)$ that appears in the exponential term, with the factor $[Z/(na)]^{3/2}$ extracted from the overall normalization constant in order to help convert the normalization integral into one involving dimensionless variables. The polynomial character of $R_{n\ell}(r)$ is given by $L_{n-\ell-1}^{2\ell+1}$, whose order, and therefore whose total number of nodes (zeros), is $n - \ell - 1$. Between $r = 0$ and $r = \infty$, R_{20} and R_{31} each have two nodes while R_{30} has three. This nodal character is essential for guaranteeing the orthogonality of differing $R_{n\ell}$'s: $R_{n\ell}$ has different signs on either side of a node.

The radial wave functions $R_{n\ell}$, the full wave functions $\psi_{n\ell m_\ell}(\mathbf{r})$, and the corresponding state vectors $|n\ell m_\ell\rangle$ are sometimes referred to by a special spectroscopic notation. A letter is used to indicate the value of ℓ, while the numerical value of n is stated. For the special case of one-electron atoms, in particular hydrogen, either capital or lower-case letters can be used, although, for individual electron states in any N-electron atom ($N \geqslant 2$), only lower-case letters are employed. We will use the lower-case notation. Table 11.3 lists the correspondence. States with $\ell = 0$ are denoted ns, e.g., 1s, 2s, and 3s; those with $\ell = 1$ are referred to as np, e.g., 2p, 3p, 4p; etc. Thus, the state vector $|n = 5, \ell = 4, m_\ell\rangle$ is denoted the 5g state, and so forth. This notation will be used to label energy levels in the Grotrian diagram of the spectrum, Fig. 11.13.

Table 11.3 *Spectroscopic notation for orbital angular momentum ℓ, $\ell = 0, \ldots, 6$*

	ℓ						
	0	1	2	3	4	5	6
Letter designations	s	p	d	f	g	h	i

The $R_{n\ell}(r)$ are the probability amplitudes, and the $|R_{n\ell}(r)|^2 = R^2_{n\ell}(r)$ are the probability densities, that the electron will be found at radius r. Radial probabilities involve the volume factor $r^2\,dr$ and, as in Section 5.2 in connection with the ground state $|n = 1, \ell = 0, m_\ell = 0\rangle$, it is most convenient to extract from $r^2\,dr$ its r^2 factor and then work directly with $u^2_{n\ell}(r)$, the radial probability density $\rho^{\text{rad}}_{n\ell}(r)$:

$$\rho^{\text{rad}}_{n\ell}(r) = u^2_{n\ell}(r) = r^2 R^2_{n\ell}(r). \tag{11.143}$$

It follows that the full radial probability $P_{n\ell}(r)$ is

$$P^{\text{rad}}_{n\ell}(r) = \rho^{\text{rad}}_{n\ell}(r)\,dr = u^2_{n\ell}\,dr. \tag{11.144}$$

Plotted in Fig. 11.11 as functions of r measured in units of a/Z are the three $\rho^{\text{rad}}_{n\ell}(r)$ corresponding to $n = 1$ and $n = 2$. Since $\rho^{\text{rad}}_{2\ell}$ is non-zero for a much larger range of r than is ρ^{rad}_{10}, the maximum magnitude of each $\rho^{\text{rad}}_{2\ell}$ is less than that of ρ^{rad}_{10}. In each case, and this is true for all the n, ℓ values, the probability density that the electron will be found at $r = 0$, i.e., that it will sit on the attractive charge center, is zero. On the other hand, the positions of the maxima in the probability densities decrease as Z increases, due to the factor Z^{-1}. This mirrors the increase of binding energy with Z ($E_n \propto Z^2$); each of these circumstances is a consequence of larger Z leading, classically, to more strongly attractive forces. Note also that, since the reduced mass enters a as μ^{-1}, a heavier particle such as a negative muon or an anti-proton, will be located much closer to the charge center than will an electron, due to the decrease in the scale parameter a/Z.

None of the foregoing withstanding, the three curves of Fig. 11.11 display the general characteristic of $\rho^{\text{rad}}_{n\ell}$ for one-electron systems: the probability density – and thus the likeliest position of the electron – is in a shell centered about the maximum (or the largest maximum, if there is more than one). The lower the value of n the narrower the range of r characterizing the shell; for large n, the electron is likely to be, relatively speaking, far from the charge center. These remarks can be understood by comparing the average $\bar{r}_{n\ell}$ and root-mean-square $r^{\text{rms}}_{n\ell}$ expectation values of the radial position operator $\hat{Q}_r$ for different n's and ℓ's:

$$\bar{r}_{n\ell} = \langle \hat{Q}_r \rangle_{n\ell} \tag{11.145a}$$

and

$$r^{\text{rms}}_{n\ell} = \langle \hat{Q}^2_r \rangle^{1/2}_{n\ell}, \tag{11.145b}$$

where $\langle\,\rangle$ implies $\int dr\, r^2$.

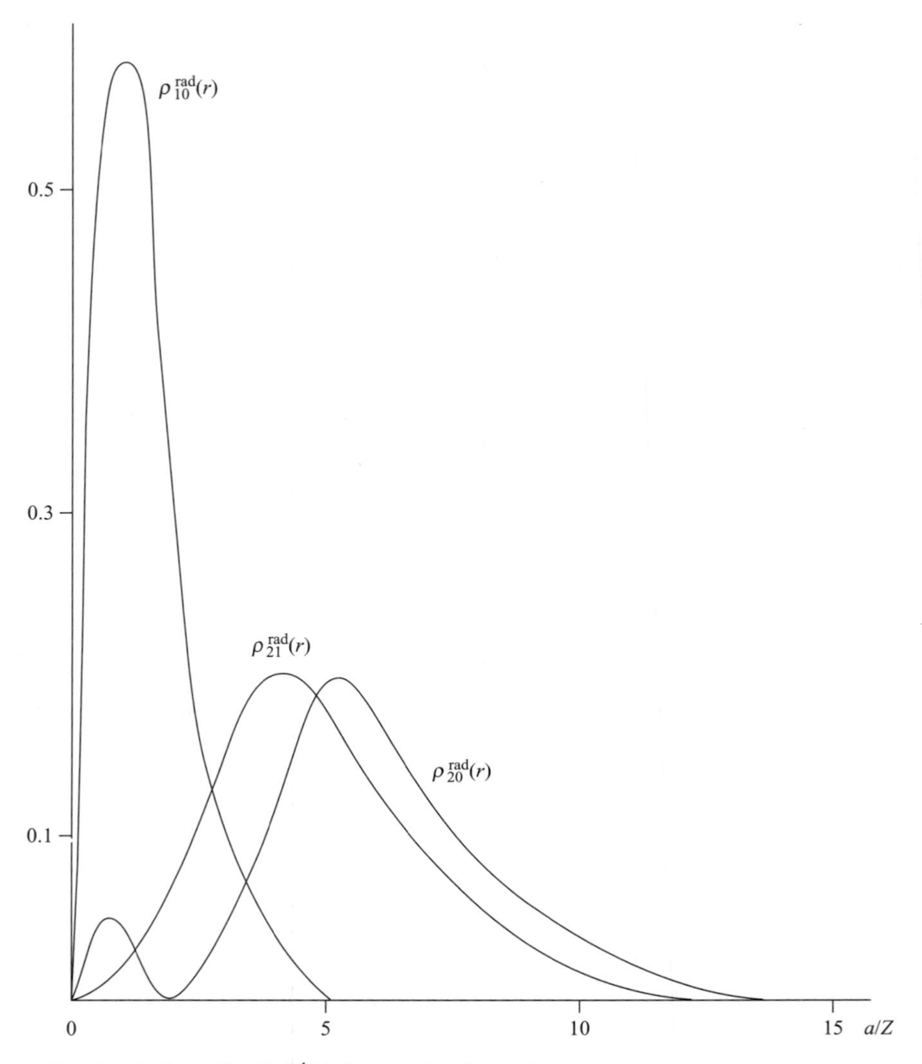

Fig. 11.11 Normalized $\rho_{n\ell}^{\mathrm{rad}}(r)$ for $n = 1$ and $n = 2$.

These last two quantities can be evaluated using Schrödinger's expressions (*op. cit.*). One finds

$$\bar{r}_{n\ell} = \tfrac{1}{2}[3n^2 - \ell(\ell + 1)]a/Z \tag{11.146a}$$

and

$$r_{n\ell}^{\mathrm{rms}} = \sqrt{\frac{n^2}{2}[5n^2 + 1 - 3\ell(\ell + 1)]}a/Z. \tag{11.146b}$$

The first of these is a measure of the average position, the second an estimate of the width of $\rho_{n\ell}^{\mathrm{rad}}(r)$; each grows as a power of n, thus bearing out the remarks above Eqs. (11.145).

More interesting, however, is the value at which the single peak occurs in $\rho^{\text{rad}}_{n,\ell=n-1}(r)$, since, for this value of ℓ, $L^{2\ell+1}_{n-\ell-1}$ has no nodes. In this case, use of Eq. (11.140) leads to[10]

$$rR_{n,\ell=n-1}(r) \propto r^n e^{-Zr/(na)},$$

from which one readily obtains for $r^{\max}_{n,\ell=n-1}$, the position of the maximum, the expression

$$r^{\max}_{n,\ell=n-1} = n^2 \frac{a}{Z} \; (!). \tag{11.147}$$

For $Z = 1$, this formula is identical to that defining the radii of the (circular) Bohr orbits, thus verifying a claim made in Chapter 5. The two equalities, viz., $r_n(Bohr) = r^{\max}_{n,\ell=n-1}$ and the finding of the same energy-level spectrum both for Bohr's *ad hoc* model and for the non-relativistic quantal theory of the H-atom, is uncanny, but is no more than fortuitous: Bohr's model, as stressed in Chapter 1, has no basis in a true microscopic description of the atom.

Schrödinger's original calculations on the H-atom were so momentous, and the wave functions and energies played such an important role in the early quantal attempts to understand atomic and molecular structure, that we present yet one more pictorial representation. In this one, the radial and the angular probability distribution are combined, leading most directly to the notion of the electron "cloud" characterizing the full probability density $r^2|\psi_{n\ell m_\ell}(r)|^2 = \rho^{\text{rad}}_{n\ell}((r)|Y^{m_\ell}_\ell(\Omega)|^2$. The result of combining Figs. 10.3 and 11.11 yields Fig. 11.12, in which the dependence on ℓ and m_ℓ is evident.

In Fig. 11.12(a), the two $\ell = 0$ densities are displayed, using the same radial scale. Since Y^0_0 is spherically symmetric, the densities are shown occupying full circles or circular shells. For $n = 1$, an arbitrary cut-off radius of $3a/Z$ has been chosen, even though there is a small non-zero probability density at larger radii. For $n = 2$, $\ell = 0$, the probability distribution is confined to two concentric shells – circles in Fig. 11.12(a) – wherein cut-off radii of $1.75a/Z$, $3.25a/Z$ and $8.25a/Z$ have been used to contain the vast majority of the probability density.

A contrast is provided by Fig. 11.12(b), in which the directionality of $|Y^{m_\ell}_1|^2$ results in two pairs of crescent-shaped probability distributions. In these cases, the y-axis is horizontal ($\phi = \pi/2$ and $3\pi/2$ as in Fig. 10.3) and the z-axis is vertical. As noted in Section 10.2, any directional character of the full probability density plays an important role in molecular structure, chemical bonds, and their physical interpretation (see Chapter 18 and also Exercise 11.24).

Degeneracy and the energy spectrum

Now that we have some understanding of the structure and behavior of $\psi_{n\ell m_\ell}$ and the related probability densities, we are ready to address the question of degeneracy, which is needed for plotting the energy spectrum. Spherically symmetric potentials produce energies that are independent of the magnetic quantum number m_ℓ: their bound-state energies are, *ab initio*, $2\ell + 1$ degenerate. The $1/r$ potential is much more highly degenerate than that, since its energies depend only on the principal quantum number n, Eq. (11.30). What is the total number of states $|n\ell m_\ell\rangle$ whose energy is E_n, i.e., what is

[10] It suffices to use $rR_{n\ell}(r) = [\rho^{\text{rad}}_{n\ell}(r)]^{1/2}$ to obtain the position of the peak in $\rho^{\text{rad}}_{n\ell}(r)$.

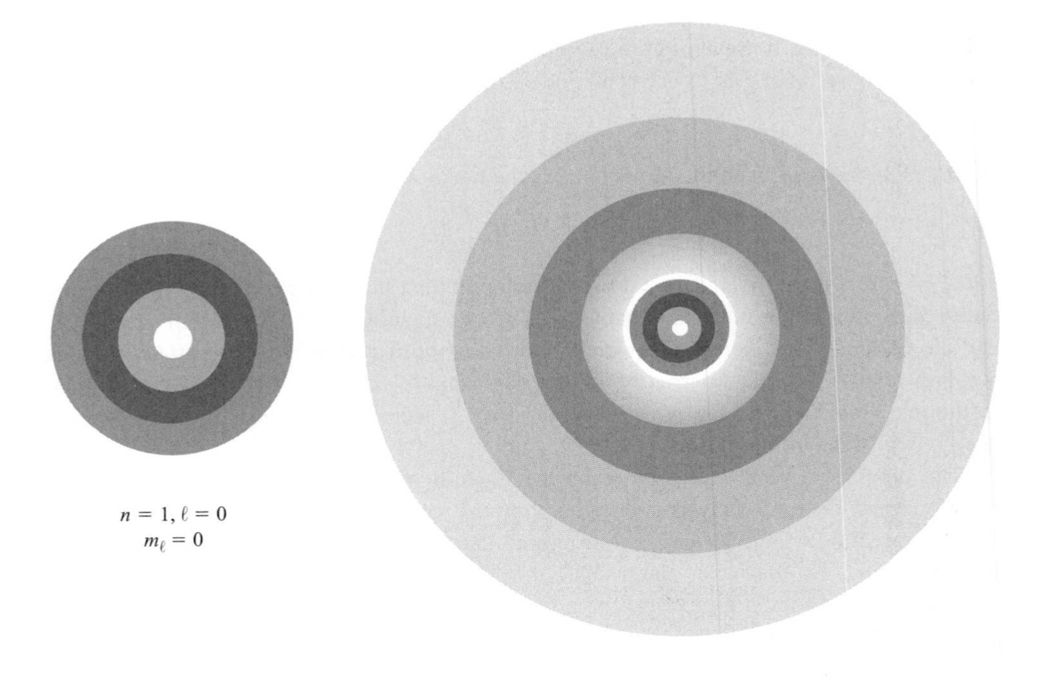

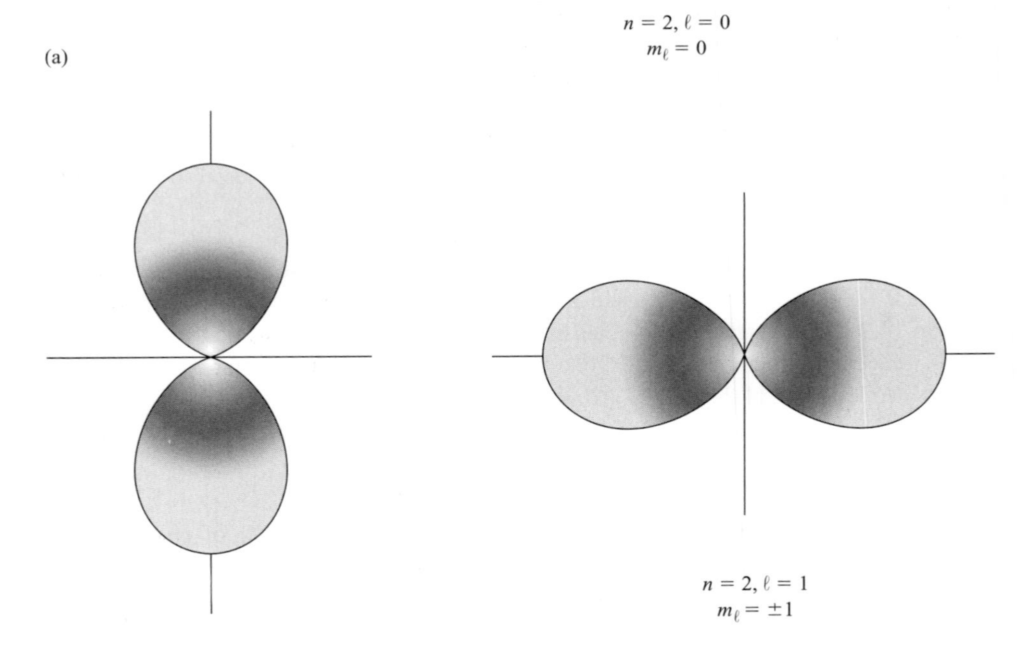

Fig. 11.12 (a) Full probability distributions $r^2|\psi_{n\ell m_\ell}(\mathbf{r})|^2 = \rho^{\text{rad}}_{n\ell}(r)|Y^{m_\ell}_\ell|^2$. Radii are measured to the same scale. (b) Full probability distributions $r^2|\psi_{n\ell m_\ell}(\mathbf{r})|^2 = \rho^{\text{rad}}_{n\ell}(r)|Y^{m_\ell}_\ell|^2$. Radii are measured to the same scale.

the degeneracy g_n? To determine g_n we sum all the m_ℓ values for each ℓ, and all the allowed ℓ values up to the maximum $n-1$:

$$g_n = \sum_{\ell=0}^{n-1} \sum_{m_\ell=-\ell}^{\ell} m_\ell = \sum_{\ell=0}^{n-1} (2\ell + 1).$$

The sum on ℓ is straightforwardly evaluated with the aid of the relation $\sum_{k=0}^{N} k = N(N+1)/2$; we get

$$g_n = n^2. \tag{11.148}$$

The n^2 dependence means a dramatic increase in g_n as n increases, and, for all but $|100\rangle$, the degeneracy mixes states of opposite parity, i.e., $|n\ell m_\ell\rangle$ and $|n, \ell - 1, m_\ell'\rangle$ have the same energy and opposite parities.

This parity degeneracy is made manifest by constructing a Grotrian diagram in which n increases vertically and ℓ horizontally. Writing

$$E_n = n^2 E_0(Z), \tag{11.149a}$$

with

$$E_0(Z) = \frac{Z^2}{2} \frac{\kappa_e e^2}{a}, \tag{11.149b}$$

the low-lying portion of the spectrum is shown in Fig. 11.13. Each horizontal line in the diagram is labeled using the spectroscopic state notation introduced in the preceding subsection.

The reader may wish to compare Fig. 11.13 with Figs. 11.5 and 11.7: spherical symmetry alone is not a guide to the structure of the spectrum. It is therefore reasonable to seek an understanding of the extraordinary degeneracy for the attractive (3-D) $1/r$ potential. It is an accidental degeneracy, caused by a hidden symmetry. There must be,

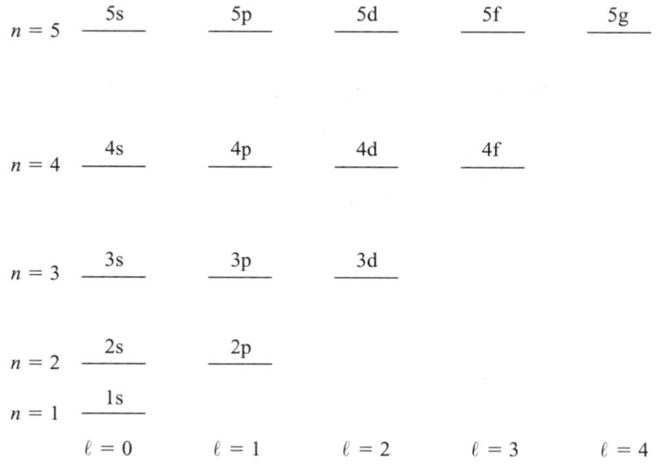

$n = 5$	5s	5p	5d	5f	5g
$n = 4$	4s	4p	4d	4f	
$n = 3$	3s	3p	3d		
$n = 2$	2s	2p			
$n = 1$	1s				
	$\ell = 0$	$\ell = 1$	$\ell = 2$	$\ell = 3$	$\ell = 4$

Fig. 11.13 The spectrum of the one-electron atom: $E_n = n^2 E_0(Z)$, $E_0(Z) = Z^2 \kappa_e e^2/(2a)$.

therefore, another quantal operator that commutes with $\hat{H}$ and is thus a constant of the motion. This new operator is the analog of a classical constant of the motion, but has not entered our solution of the eigenvalue problem. Classically, an attractive $1/r$ potential produces a trajectory that is an elliptical orbit, as in planetary motion under the gravitational attraction of a star. The orbit not only lies in a plane perpendicular to the (conserved) angular momentum, but, because it closes on itself, also has an orientation in space, for example the direction of its major axis: the ellipse is symmetric w.r.t. this axis. The direction of the major axis is given by the so-called Runge–Lenz vector, $\mathbf{R} = \mathbf{p} \times \mathbf{L}/\mu - \kappa_e e^2 \mathbf{r}/r$, where $\mathbf{p}$ and $\mathbf{L}$ are the classical momentum and angular momentum. $\mathbf{R}$ is a constant, i.e., it is a conserved quantity: $d\mathbf{R}/dt = 0$. Its quantal analog is the Hermitian operator $\hat{\mathbf{R}} = (\hat{\mathbf{P}} \times \hat{\mathbf{L}} - \hat{\mathbf{L}} \times \hat{\mathbf{P}})/(2\mu) - \kappa_e e^2 \mathbf{r}/r$ (where a coordinate representation is assumed). As has been noted, this new operator is a quantal constant of the motion: $[\hat{\mathbf{R}}, \hat{\mathrm{H}}] = 0$. Hence there is one more operator commuting with $\hat{H}$, in addition to $\hat{L}_z$ and $\hat{L}^2$, and this leads to the degeneracy, since, as remarked earlier, $\hat{\mathbf{R}}$ is not taken into account in solving the $1/r$-potential eigenvalue problem. One can, in fact, reformulate the problem, expressing $\hat{H}$ in terms of $\hat{L}^2$ and $\hat{R}^2$; the dependence of the energy solely on what turns out to be the principal quantum number n^2 is a result of this reformulation. For a relatively simple (mainly non-group-theoretical) discussion, the reader is referred to the text by Baym (1976), or, for a more detailed but not overly taxing account, to the review of Abarbanel in the collection of essays edited by Lieb, Simon, and Wightman (1976).

In 3-D, there are only two attractive, spherically symmetric potentials with an accidental degeneracy, i.e., with the degree of degeneracy being higher than that associated with the magnetic quantum number, viz., $2\ell + 1$. These are the $1/r$ and r^2 potentials, the latter being the 3-D symmetric oscillator. Some aspects of the latter potential and its eigenvalue/degeneracy problem are explored in the exercises. However, an extraordinary degeneracy also exists for the 2-D, circularly symmetric harmonic oscillator, and we close this chapter by examining it next. The wave functions, as we shall see, involve the associated Laguerre polynomials.

11.7. *The symmetric two-dimensional oscillator

For the final system considered in this chapter, we return to 2-D and examine the symmetric linear harmonic oscillator. As in Section 11.4 and Fig. 11.6, the 2-D coordinates are either x and y or r and ϕ, and in a coordinate representation the Hamiltonian takes either of the two forms

$$\hat{H}(x, y) = \hat{K}(x, y) + \tfrac{1}{2}\mu\omega^2(x^2 + y^2) \tag{11.150a}$$

and

$$\hat{H}(r, \phi) = \hat{K}(r, \phi) + \tfrac{1}{2}\mu\omega^2 r^2, \tag{11.150b}$$

with $\hat{K}(r, \phi)$ given by

$$\hat{K}(r, \phi) = -\frac{\hbar^2}{2\mu}\left(\frac{\partial^2}{\partial r^2} + \frac{1}{r}\frac{\partial}{\partial r} + \frac{1}{r^2}\frac{\partial^2}{\partial \phi^2}\right). \tag{11.151}$$

We shall solve the eigenvalue problem twice, using both sets of coordinates. An accidental degeneracy will be seen to occur in 2-D polar coordinates.

Rectangular coordinates

In Cartesian coordinates, the analysis is absolutely straightforward: there are no surprises. The Hamiltonian (11.150a) is the coordinate-space version of Eq. (11.19) with $\omega_x = \omega_y = \omega$. This relation does not change the fact that $\hat{H}(x, y) = \hat{H}_x + \hat{H}_y$, $[\hat{H}_x, \hat{H}_y] = 0$. Therefore, in a state-vector representation the eigenstates $|\Gamma\rangle$ are still products of 1-D harmonic oscillator states,

$$|\Gamma\rangle = |n_x\rangle|n_y\rangle, \tag{11.152}$$

where $\psi_{n_x}(x) = \langle x|n_x\rangle$ is given by Eq. (6.124) and likewise for $\psi_{n_y}(y) = \langle y|n_y\rangle$. The energies are again a sum of harmonic-oscillator energies: $E_\Gamma \to E_{n_x n_y}$, with

$$\begin{aligned} E_{n_x n_y} &= (n_x + \tfrac{1}{2})\hbar\omega + (n_y + \tfrac{1}{2})\hbar\omega \\ &= (n_x + n_y + 1)\hbar\omega, \qquad n_x,\, n_y = 0, 1, 2, \ldots, \end{aligned}$$

but now the individual energy units for the x and the y motion are the same.

The latter fact means that there is a degeneracy. To display it, we re-write the energy:

$$E_{n_x n_y} \to E_N = (N+1)\hbar\omega, \qquad N = n_x + n_y = 0, 1, 2, \ldots. \tag{11.153}$$

The degeneracy g_N is the number of ways that N can be written as a sum of two integers or zero. That number is $N + 1$, i.e.,

$$g_N = N + 1. \tag{11.154}$$

With the results (11.152)–(11.154) the 2-D symmetric oscillator eigenvalue problem is completely solved. The energy spectrum is so simple – just that of a harmonic oscillator with zero-point energy $\hbar\omega$ rather than $\hbar\omega/2$ – that we need not construct a Grotrian diagram. We stress that Eqs. (11.153) and (11.154) are valid independent of the representation used to express $\hat{H}$.

Polar coordinates

The solution of the eigenvalue problem expressed in rectangular coordinates is so simple that it is fundamentally uninteresting. With polar coordinates, however the fun begins.

The Schrödinger equation is

$$\hat{H}(r, \phi)\psi_\Gamma(r, \phi) = E_\Gamma \psi_\Gamma(r, \phi), \tag{11.155a}$$

with

$$\hat{H}(r, \phi) = -\frac{\hbar^2}{2\mu}\left(\frac{\partial^2}{\partial r^2} + \frac{1}{r}\frac{\partial}{\partial r} + \frac{1}{r^2}\frac{\partial^2}{\partial \phi^2}\right) + \frac{1}{2}\mu\omega^2 r^2. \tag{11.155b}$$

As with the confining-circle problem, this $\hat{H}$ commutes with $\hat{L}_z = -i\hbar\,\partial/\partial\phi$, and $\psi_\Gamma(r, \phi)$ becomes the same kind of product as that in Eq. (11.78a):

$$\psi_\Gamma(r,\phi) \rightarrow \psi_m(r,\phi) = R_m(r)\frac{e^{im\phi}}{\sqrt{2\pi}} \tag{11.156}$$

and

$$E_\Gamma \rightarrow E_m, \qquad m = 0, \pm 1, \pm 2, \ldots, \tag{11.157}$$

where the radial wave function $R_m(r)$ obeys

$$\left[-\frac{\hbar^2}{2\mu}\left(\frac{d^2}{dr^2} + \frac{1}{r}\frac{d}{dr} - \frac{m^2}{r^2}\right) + \frac{1}{2}\mu\omega^2 r^2 - E_m\right] R_m(r) = 0. \tag{11.158}$$

It is precisely here that we can see why in polar coordinates the symmetric oscillator provides some amusement: the structure of Eqs. (11.156)–(11.158) tells us that the (apparant) degeneracy of this system is two-fold, except for $m = 0$, for which there is seemingly no degeneracy at all! The contrast with $g_N = N + 1$ is evident. Clearly, the geometrically based invariance/degeneracy is only a fraction of the true story. As with the two 3-D potentials, there is a hidden symmetry that increases the degeneracy arising from rotational invariance.

We already know that $\hbar\omega$ is the natural energy unit, so on extracting this quantity from Eq. (11.158) and introducing both the natural length $\lambda = \sqrt{\hbar/(\mu\omega)}$ and the dimensionless ratio $\eta = r/\lambda$, we find that $R_m(\eta)$ obeys

$$R_m''(\eta) + \frac{1}{\eta}R_m'(\eta) - \left(\frac{m^2}{\eta^2} + \eta^2 - 2\epsilon_m\right) R_m(\eta) = 0, \tag{11.159}$$

where $' = d/d\eta$ and

$$E_m = \epsilon_m \hbar\omega. \tag{11.160}$$

An easy calculation shows that, for large η, $R_m(\eta) = \exp(-\eta^2/2)$, whereas for small η, $R_m(\eta) \rightarrow \eta^{|m|}$.[11] After extracting these factors, R_m is written as

$$R_m(\eta) = \eta^{|m|} e^{-\eta^2/2} G_m(\eta), \tag{11.161}$$

with $G_m(\eta)$ and the reduced eigenvalue ϵ_m to be determined. Inserting (11.161) into Eq. (11.159) and collecting terms, the new function $G_m(\eta)$ is found to satisfy

$$G_m''(\eta) + \left(\frac{2|m|+1}{\eta} - 2\eta\right) G_m' + 2[\epsilon_m - (|m|+1)]G_m(\eta) = 0. \tag{11.162}$$

Although (11.162) does not appear to be an equation leading to the associated Laguerre polynomials, it actually is such, only in disguise. The equation for these polynomials is masked in (11.162) by the form of the independent variable η. If instead of η one uses the new variable

$$z = \eta^2,$$

with

$$\frac{d}{d\eta} = 2\sqrt{z}\,\frac{d}{dz}, \qquad \frac{d^2}{d\eta^2} = 2\,\frac{d}{dz} + 4z\,\frac{d^2}{dz^2},$$

[11] $|m|$ is required in order that $R_m(\eta \rightarrow 0)$ be regular.

then $G_m(\eta) \to G_m(z)$ and (11.162) becomes

$$z\,\frac{d^2G_m(z)}{dz^2} + [(|m|+1) - z]\frac{dG_m(z)}{dz} + \left(\frac{\epsilon_m}{2} - \frac{|m|+1}{2}\right)G_m(z) = 0. \tag{11.163}$$

Comparison of (11.163) with Eqs. (11.124) and (11.134), plus the same kind of analysis as that used for (11.124), shows that $G_m(z)$ must be an associated Laguerre polynomial with $(\epsilon_m - |m| - 1)/2$ an integer or zero:

$$\frac{\epsilon_m - |m| - 1}{2} = n_r = 0,\, 1,\, 2,\, \ldots \tag{11.164a}$$

and

$$G_m(z) \to L_{n_r}^{|m|}(\eta^2) = L_{n_r}^{|m|}\left(\frac{r^2}{\lambda^2}\right). \tag{11.164b}$$

Equations (11.164) essentially solve the eigenvalue problem. In particular,

$$\epsilon_m = 2n_r + |m| + 1, \tag{11.165a}$$

so that the energy E_m becomes

$$\begin{aligned} E_m \to E_{n_r,m} &= (2n_r + |m| + 1)\hbar\omega \\ &\equiv (N+1)\hbar\omega, \end{aligned} \tag{11.165b}$$

$$N = 2n_r + |m| = 0,\, 1,\, 2,\, \ldots. \tag{11.166}$$

The same symbol N is used to characterize the energy in the polar representation as in the Cartesian one, though its dependence on the relevant quantum numbers is not the same in each case, and neither are the wave functions. Is the degeneracy the same in both cases? The answer, of course, is YES, as it must be. In the present case, g_N is the number of ways of choosing two non-negative integers n_r and $|m|$ such that $2n_r + |m| = N$. This total is $N+1$, as is easily seen by using the fact that $2n_r$ is always 0 or an even number. If N is even then there are $N/2+1$ ways of choosing $n_r \geqslant 0$ and $m \geqslant 0$. The remaining $N/2$ choices came from the negative values: $m = -|m|$. A similar argument holds for the case of N an odd integer.

A Grotrian diagram illustrating the spectrum is shown in Fig. 11.14; the normalized wave functions corresponding to the levels are

$$\psi_{n_r m}(r,\, \phi) = N_{n_r m} r^{|m|} e^{-r^2/2\lambda^2} L_{n_r}^{|m|}\left(\frac{r^2}{\lambda^2}\right) e^{im\phi}, \tag{11.167a}$$

with

$$N_{n_r m} = \frac{1}{\lambda^{|m|+1}}\sqrt{\frac{(n_r!)}{\pi(n_r + |m|)!}}, \tag{11.167b}$$

where the phase is chosen to make $N_{n_r m}$ real, as usual. Equation (11.167b) follows from the orthonormality relation (11.133) once the substitution $z = r^2/\lambda^2$ has been made in the radial part of the normalization integral for $\psi_{n_r m}(r,\, \phi)$.

We have thus proved that the apparent two-fold degeneracy (when $m \neq 0$) in polar coordinates is actually an $(N+1)$-fold degeneracy, the additional portion coming from

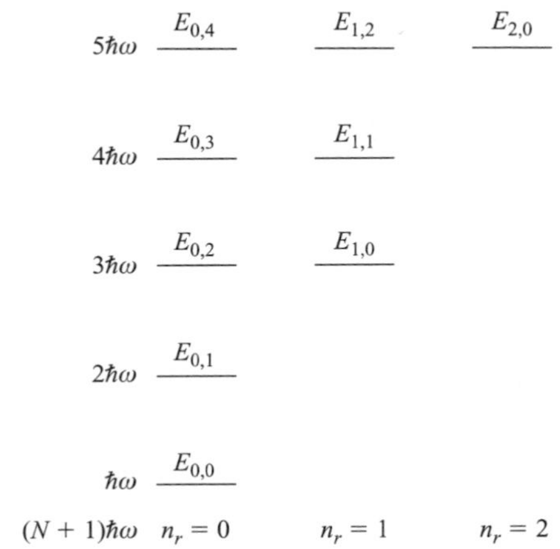

$(N+1)\hbar\omega$	$n_r = 0$	$n_r = 1$	$n_r = 2$
$5\hbar\omega$	$E_{0,4}$	$E_{1,2}$	$E_{2,0}$
$4\hbar\omega$	$E_{0,3}$	$E_{1,1}$	
$3\hbar\omega$	$E_{0,2}$	$E_{1,0}$	
$2\hbar\omega$	$E_{0,1}$		
$\hbar\omega$	$E_{0,0}$		

Fig. 11.14 The low-lying spectrum for the 2-D symmetric oscillator expressed in terms of polar-coordinate quantum numbers. Although only positive values for non-zero m are shown, every level with $m > 0$ is itself doubly degenerate.

the radial states. There is a hidden symmetry producing this accidental degeneracy, but it is more complex than the single (Runge–Lenz) vector that occurs in the $1/r$ potential case. The hidden symmetry is of quadrupole (rather than dipole) character, projected from 3-D onto the x–y plane. A brief discussion of it may be found in the text by Schiff (1968).

Exercises

11.1 (a) Determine whether the time-evolution operator for a system described by $\hat{H}_{\mathrm{IP}}(1, 2)$ of Eq. (11.5) factorizes.

(b) Given Eqs. (11.13)–(11.15) and Eq. (11.18), prove that there are $N!$ solutions of the form (11.17) corresponding to the energy E_Γ of (11.16), and that this energy is an eigenvalue of Eq. (11.13).

(c) Generalize the result of part (a) to the case of Eq. (11.13). If the time-evolution operator factorizes in this case, how many such factorizations (e.g., N, $N!$, etc.) are there?

11.2 Two non-interacting particles are in a 1-D box located between $-L$ and L. Their masses are $m_1 = (1+\delta)m$ and $m_2(1-\delta)m$, $\delta \ll 1$. Determine the wave functions and energy spectrum for this system, discussing in particular any degeneracies. (Hint: expand relevant results in powers of δ, keeping only appropriate low-order terms.)

11.3 A particle of mass m is in a 2-D rectangular box located in the region $x \in [0, L]$, $y \in [0, NL]$, N is an integer. Determine the wave functions and energy spectrum, and discuss the possibility of degeneracies.

11.4 A particle of mass m is acted on by the 2-D potential $V(x, y) = m[\omega_0^2(x^2 + y^2) + \omega_1^2 xy]/2$.

(a) Determine the condition(s) under which the eigenvalue problem for this system transforms into one corresponding to a sum of independent oscillator Hamiltonians.

(b) Assuming that the condition(s) of part (a) is (are) satisfied, find the wave functions and energy spectrum for the system. Under which circumstances, if any, can degeneracies occur? (Hint: carry out a rotation in the x–y plane to new coordinates x' and y'.)

11.5 Each of two interpenetrable rigid rotators has mass m_0 and each is at a distance a from the origin in the x–y plane, as in Section 9.7. Their angular positions are ϕ_1 and ϕ_2.

(a) Find the eigenfunctions and eigenvalues of this system and discuss possible degeneracies.

(b) At $t = 0$ the wave function is $\Psi(\phi_1, \phi_2, 0) = [2/(3\pi^2)]^{1/2} \sin^2 \phi_1 \exp(i\phi_2)$. What is the wave function for $t > 0$?

(c) The energy is measured at time $t > 0$. What are the possible outcomes of this measurement and what are the probabilities that each will occur?

11.6 A particle of mass m is acted on by the symmetric 2-D oscillator potential $V(\mathbf{r}) = m\omega^2(x^2 + y^2)/2$. The eigenfunctions are $\psi_{n_x n_y}(x, y) = \psi_{n_x}(x)\psi_{n_y}(y)$, with corresponding energies $E_{n_x n_y} = (n_x + n_y + 1)\hbar\omega = (N + 1)\hbar\omega$, where the ψ_n's are the usual (normalized) 1-D oscillator eigenfunctions and n_x and $n_y = 0, 1, 2, \ldots$. At time $t = 0$, the particle's wave function is $\Psi_\Gamma(x, y, 0) = [8/(\pi\lambda^4)]^{1/2} xy \exp[-(x^2 + y^2)/(2\lambda^2)]$ when both $x \geqslant 0$ and $y \geqslant 0$, and is zero otherwise.

(a) If $P_{++}(t)$ is the probability that the particle is in the quadrant $x \geqslant 0$, $y \geqslant 0$ at time t, determine dP_{++}/dt. (Hint: consider the equation of continuity.)

(b) Evaluate $\langle \Gamma, t|\hat{H}|\Gamma, t\rangle$.

(c) If $P_{n_x n_y}(t)$ is the probability for finding the state $\psi_{n_x n_y}(x, y)$ in $\Psi_\Gamma(x, y, t)$, determine $P_{00}(t)$, $P_{01}(t)$, $P_{10}(t)$, and $P_{11}(t)$.

(d) Explain why one should expect $P_{n_x n_y}(t) = 0$ if n_x or n_y are odd integers greater than unity.

(e) Determine the (2-D) expectation values $\langle \Gamma, t|\hat{\mathbf{Q}}|\Gamma, t\rangle$ and $\langle \Gamma, t|\hat{\mathbf{P}}|\Gamma, t\rangle$.

11.7 (a) Equations (11.29) are based on the definitions (11.22). Suppose that Eq. (11.22a) holds but that (11.22b) is replaced by $\mathbf{R} = a_1\mathbf{r}_1 + a_2\mathbf{r}_2$. Show that $a_j = m_j/M$, $j = 1, 2$, follows from requiring that (i) the analog of (11.29a) contain no cross terms in x and X and (ii) the Jacobian between $dx_1\, dx_2$ and $dx\, dX$ is unity.

(b) Prove the validity of Eq. (11.47c).

11.8 (a) Determine whether the time-evolution operator corresponding to Eq. (11.6) factorizes.

(b) In 1-D, the Hamiltonian for two mutually interacting charged particles placed in a constant electric field has the general form $\hat{H}(x_1, x_2) = \hat{K}_x(x_1) + \hat{K}_x(x_2) + \mathrm{V}(x) + A_1x_1 + A_2x_2$, with A_1 and A_2 each constant. Show that the time-evolution operator for this system factorizes (the resulting eigenvalue problem is non-trivial and is not considered here).

11.9 A 1-D example for which $\hat{H}(1, 2)$ for Eq. (11.1) leads to a soluble eigenvalue problem occurs when the three interactions are each of harmonic-oscillator form ($k > 0$): $\hat{H}(1, 2) = \sum_j [\hat{P}_x^2(x_j)/(2m_j) + \omega_0^2(m_1x_1^2 + m_2x_2^2)/2] + k(x_1 - x_2)^2/2$.

(a) Show that, by carrying out the transformation to relative and CM coordinates, $\hat{H}(1, 2)$ becomes a sum of harmonic-oscillator Hamiltonians, one involving the frequency ω_0 and the other a frequency ω you are to determine.

(b) Write out the general form of the eigenfunctions and eigenvalues that solve part (a), and discuss possible degeneracies.

(c) Construct a Grotrian diagram for the low-lying portions of the spectrum for the following two "extreme" cases: (i) $\omega \cong \omega_0$, for which you should find that the lowest-lying state is a singlet; the next two a doublet separated by $\hbar(\omega - \omega_0)$; then a triplet in which adjacent levels are also separated by $\hbar(\omega - \omega_0)$; then a quartet, etc. – all results being valid to an excellent approximation as long as $\omega - \omega_0 \ll \omega_0$; and (ii) $\omega \gg \omega_0$, for which you should find that, built on each normal oscillator level, there is an infinite series of levels separated from one another by a uniform spacing you are to determine.

11.10 (a) Fill in the details leading to the result $E > V_0$, Eq. (11.61).

(b) Evaluate the RHS of the Bargmann bound, Eq. (11.B) on p. 412, for the exponential potential $V_{\text{exp}}(\text{r}) = -V_0 \exp(-\alpha r)$, as well as for the Yukawa and Gaussian potentials, defined above Eq. (5.51).

(c) (i) Suppose that each of the potentials of part (b) supports just one $\ell = 0$ bound state. Are there circumstances in which each of them can support $\ell \neq 0$ excited bound states? Explain. (ii) Now suppose that each potential supports two $\ell = 0$ bound states. Are there circumstances in which any of them can support $\ell \neq 0$ excited bound states? Justify your conclusions.

11.11 The interior of a 3-D quantal box is located between $-L$ and L along each of $\mathbf{e}_1$, $\mathbf{e}_2$, and $\mathbf{e}_3$.

(a) Determine general expressions for the normalized eigenstates and the energies for a particle of mass m in the box.

(b) Specify the degeneracy of each level and the parity of each of its corresponding eigenstates.

(c) Construct a Grotrian diagram showing the first five levels, indicating the degeneracy of each of these levels. Comment on whether there is any parity mixing.

11.12 A 3-D quantal box is located between two concentric spheres of radii a and $b < a$.

(a) Derive the transcendental equation from which the eigenvalues are obtained.

(b) Determine the usual n dependence of the eigenvalues for the case $\ell = 0$.

(c) What are the normalized eigenfunctions when $\ell = 0$?

(d) Show that the form taken by the result of part (a) when $\ell = 1$ is $\tan[k(a - b)] = f(k, a, b)$, where you are to specify $f(k, a, b)$.

(e) Show that the result of part (a) goes over to Eq. (11.67b) when $b \to 0$.

11.13 The recursion relation of Exercise 10.24(b) also holds for the spherical Neumann functions $n_\ell(\rho)$.

(a) Use this recursion relation and Eqs. (11.97b) and (11.97c) to derive the functional form of $n_4(\rho)$. Compare your answer with the result obtained from Eq. (11.96).

(b) Obtain the $\rho \to 0$ and $\rho \to \infty$ limits of $n_3(\rho)$ and $n_4(\rho)$ and compare with Eq. (11.98).

11.14 Determine the form of the eigenfunctions and eigenvalues for a particle of mass m_0 confined to a cylinder of radius a located between $z = 0$ and $z = L$, i.e., confined by a potential that is infinite for $r \notin [0, L]$ and $r \geqslant a$. (You will need ∇^2 in cylindrical coordinates.)

11.15 Let $V(\mathrm{r})$ be $V_{\exp}(\mathrm{r})$ of Exercise 11.10(b), and set $\ell = 0$.
(a) As a procedure for determining the reduced radial wave function for this case, define $x = \exp(-\alpha r/2)$ and compare the resulting equation for u_0 with the radial equations of this chapter.
(b) State the relevant boundary conditions obeyed by $u_0(x)$ and show how these yield the corresponding eigenvalue equation.
(c) Using the Bargmann bound and the analysis of part (b), compare the minimum values of $V_0 > 0$ needed to obtain at least one bound state.

11.16 A simple model of the deuteron is an $\ell = 0$, spinless neutron–proton pair bound by a spherically symmetric square well with depth/range adjusted to yield the binding energy.
(a) Assume that the mass of each nucleon is the proton mass M_p and that the well depth is $V_0 = -52$ MeV. Determine the well-edge radius r_0 that yields the observed binding energy of 2.23 Mev.
(b) Obtain the normalized deuteron wave function in this model and sketch the resulting $\rho(r)$.
(c) Use the result of part (b) to calculate (i) the expectation value of the potential energy (and thus of the kinetic energy); and (ii) the average value of the neutron–proton separation, commenting on how to understand its size relative to r_0.
(Hints: use $\hbar^2/M_\mathrm{p} = 41.5$ MeV fm^2(1 fm $= 10^{-15}$ m) and the fact that $1.6\,\mathrm{fm} \leqslant r_0 \leqslant 1.7\,\mathrm{fm}$; keep at least three decimal places when evaluating r_0 since the numerical value of the product $\gamma_0 r_0$ – Eqs. (11.103a) and (11.111) – is expressed in radians.)

11.17 (a) Although they are not needed in the $1/r$ problem, values of $L_p^0(\rho)$ occur in the solutions for other potentials. Evaluate $L_p^0(\rho)$, $p = 1, 2, 3$ using Eq. (11.132a).
(b) Solve Eq. (11.127b) to obtain C_{m+1} in terms of C_0, substitute the result into the truncated version of (11.125) and show that, apart from normalization (the value of C_0), the resulting series is the same as that for $F_{n\ell}(\rho) = L_{n-\ell-1}^{2\ell+1}(\rho)$, Eq. (11.132b).

11.18 What is the probability that an electron in the ground state of hydrogen will be found in the classically forbidden region (i.e., in the region where $V(r) > E$)?

11.19 (a) Evaluate $\varphi_{100}(\mathbf{k}) = \langle \mathbf{k} | n = 1, \ell = 0, m_\ell = 0 \rangle$.
(b) From the result of part (a), find the average value of $\hat{\mathbf{P}}$ in the H-atom ground state. Set this value equal to $m_\mathrm{e} v$, and then use this result to estimate whether relativistic effects can be ignored.
(c) Compare the wave number of part (b) with the value producing the maximum in $\rho_{100}(\mathbf{k}) = |\phi_{100}(\mathbf{k})|^2$. (Hint: use Eq. (10.85) and integral tables.)

11.20 The electron in an H-atom whose proton mass is infinite is excited to the state $|21 m_\ell\rangle$. While in this state the Coulomb interaction is turned off, leaving the electron a free particle.
(a) Find the probability density $\rho_{21 m_\ell}(\mathbf{k})$ that the electron will have momentum $\hbar\mathbf{k}$ relative to an arbitrary set of coordinate axes. Express your answer in terms of the Bohr radius a_0, the wave number k, and the angles θ_k and ϕ_k of $\mathbf{k}$.
(b) If $m_\ell = 0$, what are the angles for which $\rho_{210}(\mathbf{k})$ is (i) maximal and (ii) minimal?

(c) What are the probability densities $\rho_{21m_\ell}(\mathbf{k})$ that the electron will have energy $E_k = \hbar^2 k^2/(2m_e)$?

(d) Using the result of part (a), evaluate the average momentum $\hbar\mathbf{k}_{av}$ for each value of m_ℓ.

11.21 Find the probability current density $\mathbf{J}_{21m_\ell}(\mathbf{r})$ for each of the three H-atom states $|21m_\ell\rangle$. Explain why each of these three results should or should not vanish.

11.22 At time $t = 0$, a hydrogen atom is described by the $(Z = 1)$ wave function $\Psi_\Gamma(\mathbf{r}, 0) = 2^{-1/2}\psi_{100}(\mathbf{r}) - 6^{-1/2}\psi_{210}(\mathbf{r}) + 3^{-1/2}\psi_{211}(\mathbf{r})$.

(a) What is $\Psi_\Gamma(\mathbf{r}, t)$, $t > 0$?

(b) Determine the probability that, at $t > 0$, measurements would find (i) $E = E_2$, (ii) $L^2 = 2\hbar^2$, and (iii) $L_z = 0$. In each case, what would $\Psi_\Gamma(\mathbf{r}, t)$ be after the measurement?

(d) Determine whether $\rho_\Gamma(\mathbf{r}, t)$ is periodic in time, and, if it is, find the period in seconds.

11.23 Let $R_{21}(r)$ be the hydrogenic 2p wave function of Eq. (11.142c), and define $\psi_{21j}(\mathbf{r}) = R_{21}(r)(j/r)[3/(4\pi)]^{1/2}$, with $j = x$, y, and z. Assume that, at $t = 0$, the wave function for an electron in a hydrogenic atom is $\Psi_\Gamma(\mathbf{r}, 0) = [\psi_{200}(\mathbf{r}) + \psi_{21x}(\mathbf{r}) + \psi_{21y}(\mathbf{r})]/3^{1/2}$.

(a) Is $\Psi_\Gamma(\mathbf{r}, 0)$ an eigenfunction of (i) $\hat{L}_z$, (ii) $\hat{L}^2$, (iii) $\hat{U}_P$, and (iv) $\hat{H}_Z(\mathbf{r}) = \hat{K}(\mathbf{r}) - \kappa_e Ze^2/r$? State the corresponding eigenvalue for each positive answer.

(b) Fine $\Psi_\Gamma(\mathbf{r}, t)$ and from it $\rho_\Gamma(\mathbf{r}, t)$. Determine whether $\rho_\Gamma(\mathbf{r}, t)$ is periodic in time, and, if so, state the period in seconds.

(c) Using $\Psi_\Gamma(\mathbf{r}, t)$ of part (b), calculate the expectation values of (i) $\hat{\mathbf{Q}}$, (ii) $\hat{\mathbf{P}}$, (iii) $\hat{\mathbf{L}}$, and (iv) the Coulomb force $F(r) = -\mathbf{e}_r\kappa_e Ze^2/r^2$. Express your answers, as needs be, in terms of the radial integrals $I_{n\ell n'\ell'}(\nu) = \int dr\, r^{2+\nu} R_{n\ell}(r)R_{n'\ell'}(r)$. (Hint: use the reality of $R_{n\ell}$ and Ehrenfest's theorem.)

11.24 One use of hydrogenic states is in the modeling of electron wave functions in molecules such as H_2O, NH_3 and CH_4. Among the states used are the four linearly independent combinations (in the notation of Exercise 11.23)

$$\psi_1(\mathbf{r}) = [\psi_{210}(\mathbf{r}) + \psi_{21x}(\mathbf{r}) + \psi_{21y}(\mathbf{r}) + \psi_{21z}(\mathbf{r})]/2,$$

$$\psi_2(\mathbf{r}) = [\psi_{210}(\mathbf{r}) + \psi_{21x}(\mathbf{r}) - \psi_{21y}(\mathbf{r}) - \psi_{21z}(\mathbf{r})]/2,$$

$$\psi_3(\mathbf{r}) = [\psi_{210}(\mathbf{r}) - \psi_{21x}(\mathbf{r}) + \psi_{21y}(\mathbf{r}) - \psi_{21z}(\mathbf{r})]/2,$$

$$\psi_4(\mathbf{r}) = [\psi_{210}(\mathbf{r}) - \psi_{21x}(\mathbf{r}) - \psi_{21y}(\mathbf{r}) + \psi_{21z}(\mathbf{r})]/2.$$

(a) Show that these wave functions are normalized and are mutually orthogonal.

(b) Determine the result of applying the rotation operators $\hat{U}_R(\mathbf{e}_j, \pi) = \exp(-i\pi\mathbf{e}_j \cdot \mathbf{L})$, $j = 1, 2, 3$, to each of the preceding four wave functions. (Hint: recall that $\hat{U}_R(\mathbf{e}, \pi)$ rotates by π about $\mathbf{e}$, so that for a function $f(x, y, z)$ one only needs to determine how x, y, or z changes.)

(c) Calculate the expectation values of $\hat{\mathbf{Q}}$ (in ångström units) using each of the four preceding ψ_j.

(d) Each of the four expectation values of part (c) defines a direction in space. Show that the angle between each of them and the remaining three is 109.5°. This is the bond angle in CH_4 (!), thereby implying that the four ψ_j provide a reasonable model for the electron states in that molecule, as well as a slightly poorer model for H_2O

and NH_3, whose bond angles are 105° and 107°, respectively. The four ψ_j are known as *hybridized orbitals*.

11.25 The hydrogenic atom in 2-D. In this case the relative-motion kinetic-energy operator is given by Eq. (11.151) and the electron–proton interaction has the same form as Eq. (11.114) except that, in 2-D, $r = (x^2 + y^2)^{1/2}$. In polar coordinates, the wave function takes the form $R_m(r)\exp(im\phi)/(2\pi)^{1/2}$, $m = 0, \pm1, \pm2, \ldots$.

(a) By writing $E = -\hbar^2\alpha^2/(2\mu)$, introducing $\rho = 2\alpha r$, and using the reduced-radial-function analysis of Section 11.3, prove that $R_m(\rho) = \rho^{|m|}\exp(-\rho/2)f_m(\rho)$, and then show that $f_m(\rho)$ satisfies a differential equation encountered in this chapter.

(b) From the last result of part (a) determine the discrete energy eigenvalues E and the $f_m(\rho)$. (Your analysis should yield the analog of the principal quantum number n of the 3-D case.)

(c) Sketch the r dependence of the ground-state "radial" probability density $r|R_m|^2$, and determine the position of its maximum value in terms of a/Z.

(d) State the degeneracy of each energy level.

(e) Set $Z = 1$ and $\mu = m_e$, and construct a Grotrian diagram for the low-lying states of the 2-D H-atom, labeling each energy level by its value in electron-volts, by m, and by its degeneracy.

11.26 The rectangular-coordinate states $\{|n_x, n_y\rangle\}$ and the polar-coordinate states $\{|n_r, m\rangle\}$ each form a complete set for the symmetric 2-D oscillator. The members of one set can therefore be expanded using the members of the other, e.g., $|n_x, n_y\rangle = \sum|n_r, m\rangle\{\langle n_r, m|n_x, n_y\rangle\}$. As simple examples illustrating this, consider $N = 1$ and then $N = 2$, Eq. (11.153).

(a) For each N, use coordinate transformations to express the corresponding wave functions $\psi_{n_x n_y}(x, y)$ in terms of the relevant $\psi_{n_r m}(r, \phi)$. Compare them with the complete-set expansions and identify which members of the complete set enter. In particular, comment for each N whether expansion states $|n_r m\rangle$ with differing values of N occur.

(b) Redo part (a) by reversing the roles of the two sets of wave functions.

(c) Discuss how you expect the preceding results to generalize to the case of arbitrary N.

11.27 The $t = 0$ wave function for a particle of mass m, acted on by a 2-D symmetric oscillator potential $m\omega^2(x^2 + y^2)/2$, is $\Psi_\Gamma(x, y, 0) = \exp(i\mathbf{k}\cdot\mathbf{r})\,\psi_0(x)\psi_0(y)$, where ψ_n is the nth (1-D) oscillator wave function and $\mathbf{k} = k(\mathbf{e}_1\cos\phi + \mathbf{e}_2\sin\phi)$. The $t = 0$ state evolves to $\Psi_\Gamma(x, y, t)$.

(a) Calculate the probability current density associated with $\Psi_\Gamma(x, y, 0)$.

(b) Calculate $\langle\Gamma, t|\hat{H}|\Gamma, t\rangle$, where $\hat{H}$ is the 2-D oscillator Hamiltonian.

(c) Let $P_{n_x n_y}(t) = |\langle n_x, n_y|\Gamma, t\rangle|^2$ denote the probability of finding the system at time t in the state $|n_x, n_y\rangle$. Calculate (i) $P_{00}(t)$, (ii) $P_{10}(t)$, (iii) $P_{01}(t)$, and (iv) $P_{11}(t)$.

11.28 Show that, as long as there are no degeneracies, $\langle\hat{\mathbf{P}}\rangle = 0 = \langle\hat{\mathbf{Q}}\rangle$ for a bound single particle acted on by a spherically symmetric potential. Explain clearly why the presence of an accidental degeneracy (e.g., for the H-atom or 2-D symmetric oscillator) can invalidate this result.

11.29 A particle of mass m is bound by the 3-D symmetric (isotropic) harmonic-oscillator potential $m\omega^2 r^2/2$, $r^2 = x^2 + y^2 + z^2$. In rectangular coordinates the eigenfunctions are products of 1-D oscillator wave functions, viz., $\psi_{n_x n_y n_z}(x, y, z) =$

$\psi_{n_x}(x)\psi_{n_y}(y)\psi_{n_z}(z)$; the energy is the usual sum: $E_{n_x n_y n_z} = (n_x + n_y + n_z + \frac{3}{2})\hbar\omega \equiv E_N$.

(a) Show that the degeneracy of the Nth level is $g_N = (N+1)(N+2)/2$.

(b) Prove that, in spherical polar coordinates, the eigenfunctions are $\psi_{n_r \ell m_\ell}(r, \theta, \phi) = N_{n_r \ell} r^\ell \exp[-r^2/(2\lambda^2)] L_{n_r}^{\ell+1/2}(r^2/\lambda^2) Y_\ell^{m_\ell}(\theta, \phi)$, where $n_r = 0, 1, 2, \ldots$, $N_{n_r \ell}$ is a normalization constant, $\lambda^2 = \hbar/(m\omega)$ and $L_{n_r}^{\ell+1/2}$ is an associated Laguerre polynomial.

(c) From the analysis establishing the result of part (b), determine the dependences of the energy $E_{n_r \ell} = (N + \frac{3}{2})\hbar\omega$ on n_r and ℓ. Prove that the degeneracy of the energy expressed in the $E_{n_r \ell}$ form is g_N of part (a).

(d) Construct a Grotrian diagram for the low-lying levels $E_{n_r \ell}$, analogous to Fig. 11.13, labeling each by its magnitude (in $\hbar\omega$), by the value of n_r and ℓ, and by the appropriate degeneracy.

11.30 (a) Determine the value of r at which the radial probability distribution for the ground state of the symmetric 3-D harmonic oscillator has its maximum and compare it with the ground-state expectation value of r, expressing your answers in units of $\lambda = [\hbar/(m\omega)]^{1/2}$.

(b) A particle of mass m is in the ground state of a symmetric 3-D oscillator. Calculate the probability density $|\varphi_{000}(\mathbf{k})|^2$ that in this state the particle will have momentum $\hbar\mathbf{k}$, and indicate the dependence on the polar angles of $\mathbf{k}$. Determine the probability that a measurement of the energy will yield the value $E_k = \hbar^2 k^2/(2m)$.

11.31 What are the probabilities that a particle of mass m bound by the (3-D) potential $V(r) = m\omega^2 r^2/2$ will be found in the classically forbidden region ($V(r) > E$) if it is in (i) the ground state and (ii) the first excited s state?

11.32 The initial wave function for a particle of mass m acted on by the 3-D oscillator potential $m\omega^2 r^2/2$ is $\Psi_\Gamma(\mathbf{r}, 0) = \exp(ikz)\,\psi_{100}(x, y, z)$, with $\psi_{n_x n_y n_z}(x, y, z)$ given in Exercise 11.29.

(a) What are the probabilities that, at time $t > 0$, a measurement of $\hat{L}_z$ will yield the values $m_\ell\hbar$, $m_\ell = 0, \pm 1, \pm 2, \ldots$?

(b) Calculate $\langle\Gamma, t|\hat{\mathbf{Q}}|\Gamma, t\rangle$, $t \geqslant 0$.

(c) Calculate $\langle\Gamma, t|\hat{\mathbf{L}}|\Gamma, t\rangle$, $t \geqslant 0$.

12

Electromagnetic Fields

In this chapter, we consider the behavior of a particle of charge q in electromagnetic fields. Electric ($\mathcal{E}$) and magnetic (**B**) fields are observables, and one might at first think a quantal description of electromagnetic phenomena should include operators for these fields. However, recalling that the classical Hamiltonian $H_{\text{cl}}^{(A)}$ for a charged particle acted on by $\mathcal{E}$ and/or **B** contains not these fields but the vector (**A**) and scalar (Φ) potentials, and also that $\hat{H}^{(A)}$ should be obtained by quantizing $H_{\text{cl}}^{(A)}$, one's revised thought might be to introduce operators for **A** and Φ in place of those for $\mathcal{E}$ and **B**. Following such a procedure leads to the non-relativistic version of quantum electrodynamics, an important subject but one beyond the scope of this book. Instead, we shall treat the fields as essentially classical entities. This means that we shall assume that the electromagnetic fields are generated externally to the quantum system and enter the description via unquantized (non-operator) vector and scalar potentials, as in $H_{\text{cl}}^{(A)}$. That is, only **p** will be quantized in going from $H_{\text{cl}}^{(A)}$ to $\hat{H}^{(A)}$.

The procedure just outlined is often referred to as a semi-classical treatment of electromagnetic phenomena.[1] It lacks two ingredients of the quantized electromagnetic-field analysis: complexity, and the concept of the photon. The latter is not a serious shortcoming, since the semi-classical treatment allows one to examine a very wide variety of phenomena, even including electromagnetic transitions in atoms, molecules, and nuclei, for which the photon concept can be imposed externally; see Chapter 16.

We begin the chapter with a brief review of the classical case, introducing the Lorentz force, Maxwell's equations for the fields, the scalar and vector potentials, and the Hamiltonian for a charged particle in external fields. As noted in the Preface, constants that allow one to express the foregoing quantities either in SI or in cgs units are included. One such, the constant κ_{e} appearing in Coulomb's law, has already made appearances in Chapters 1 and 11. The advantage of this slightly cumbersome notation is that it helps the reader to become adept at switching between the two sets of units, thereby preparing him/her to read various graduate-level texts that employ cgs units.

The transition to the quantal Hamiltonian $\hat{H}^{(A)}(\mathbf{r}, t)$ for a charged particle in external (classical) fields concludes Section 12.1. Before studying general properties of the quantum case, we first consider two important special situations, viz., a charged particle in a uniform magnetic field (Section 12.2) and in a uniform electric field (Section 12.3).

[1] The introduction of operators corresponding to the potentials **A** and Φ is known as "second quantization," first quantization refering to $\mathbf{p} \rightarrow \hat{\mathbf{P}}$, etc.

General properties – or quantum-theoretical aspects – of the charged-particle formalism are considered in the next two sections. Special emphasis is given to gauge invariance, the topic of Section 12.5, which is illustrated in Section 12.6 via a reconsideration of the uniform electric field case. Section 12.7 ends the chapter with a discussion of the remarkable prediction of Aharonov and Bohm (and, earlier, of Ehrenburg and Siday) that, unlike classical physics, the electromagnetic potentials can give rise quantally to observable effects.

12.1. Charged particle in external electromagnetic fields

Review of the classical case

The motion of a classical particle of charge q in external electric ($\mathcal{E}$) and magnetic (**B**) fields is governed by the Lorentz force

$$\mathbf{F}_{\mathrm{L}} = q(\mathcal{E} + \kappa_{\mathrm{c}}\mathbf{v} \times \mathbf{B}),$$

where $\mathbf{v} = d\mathbf{r}/dt$ is the particle's instantaneous velocity and the SI and cgs values of κ_{c} are given in Table 12.1. Solution of

$$\frac{d\mathbf{p}}{dt} = \mathbf{F}_{\mathrm{L}}$$

determines the trajectory, where $\mathbf{p} = m\mathbf{v}$.

Electromagnetic fields are generated from charge and current distributions. In the absence of material media, $\mathcal{E}$ and **B** are related to each other and to the charge (ρ_{ch}) and current (**j**) densities by Maxwell's equations:

$$\nabla \cdot \mathcal{E} = 4\pi\kappa_{\mathrm{e}}\rho_{\mathrm{ch}}, \tag{12.1a}$$

$$\nabla \cdot \mathbf{B} = 0, \tag{12.1b}$$

$$\nabla \times \mathcal{E} = -\kappa_{\mathrm{c}}\frac{\partial \mathbf{B}}{\partial t}, \tag{12.1c}$$

and

$$\nabla \times \mathrm{B} = 4\pi\kappa_{\mathrm{m}}\mathbf{j} + \frac{\kappa_{\mathrm{m}}}{\kappa_{\mathrm{e}}}\frac{\partial \mathcal{E}}{\partial t}. \tag{12.1d}$$

While no attempt will be made to solve these equations in our study of electromagnetic effects in quantum mechanics, one of their attributes is of the utmost importance for quantum theory. This is the fact that $\mathcal{E}$ and **B** can be derived from vector (**A**) and scalar (Φ) potentials:[2]

$$\mathbf{B} = \nabla \times \mathbf{A} \tag{12.2a}$$

and

$$\mathcal{E} = -\nabla\Phi - \kappa_{\mathrm{c}}\frac{\partial \mathbf{A}}{\partial t}. \tag{12.2b}$$

[2] The forms of Eqs. (12.2) follow from the structure of Eqs. (12.1b) and (12.1c), while the equations obeyed by **A** and Φ are obtained by substituting Eqs. (12.2) into the other two Maxwell equations.

Table 12.1 *Values of the constants appearing in Maxwell's equations (Eqs. (12.1))*

	SI units	cgs units
κ_{e}	$1/(4\pi\epsilon_0)$	1
κ_{c}	1	$1/c$
κ_{m}	$\mu_0/(4\pi)$	$1/c$
$\kappa_{\mathrm{m}}/\kappa_{\mathrm{e}}$	$\mu_0\epsilon_0 = 1/c^2$	$1/c$

Although $\mathcal{E}$ and $\mathbf{B}$ comprise six components and $\mathbf{A}$ and Φ only four, the potentials remain secondary quantities in classical physics: $\mathcal{E}$ and $\mathbf{B}$, not $\mathbf{A}$ and Φ, are the measurables and the direct determinants of the trajectory.

This last remark is especially pertinent in view of the structure of the classical Hamiltonian $H_{\mathrm{cl}}^{(A)}$ corresponding to $d\mathbf{p}/dt = \mathbf{F}_{\mathrm{L}}$:[3]

$$H_{\mathrm{cl}}^{(A)} = \frac{1}{2m}(\mathbf{p} - \kappa_{\mathrm{c}} q\mathbf{A})^2 + q\Phi \equiv \frac{1}{2m}\Pi^2 + q\Phi, \tag{12.3}$$

where $\mathbf{\Pi} = \mathbf{p} - \kappa_{\mathrm{c}} q\mathbf{A}$ is the kinetic momentum. Even though $\mathbf{A}$ and Φ rather than $\mathcal{E}$ and $\mathbf{B}$ enter $H_{\mathrm{cl}}^{(A)}$, the latter are the primary quantities. Not only are $\mathbf{A}$ and Φ not measurable, they are also not unique:[4] if $\mathbf{A}$ and Φ yield $\mathcal{E}$ and $\mathbf{B}$, then so will the pair

$$\mathbf{A}' = \mathbf{A} + \nabla\chi \tag{12.4a}$$

and

$$\Phi' = \Phi - \kappa_{\mathrm{c}}\frac{\partial\chi}{\partial t}, \tag{12.4b}$$

where χ is an arbitrary function of $\mathbf{r}$ and t. χ is known as a gauge function, and Eqs. (12.4) are referred to as a gauge transformation (of the first kind). The electromagnetic fields (and equations involving them) are *invariant* w.r.t. gauge transformations.

While $\mathbf{A}$ and Φ are not unique, they are not totally arbitrary. As discussed in texts on electromagnetic theory, $\nabla \cdot \mathbf{A}$ can be chosen in various ways, and the specification of $\nabla \cdot \mathbf{A}$ is one way of defining a "gauge." The two most widely used choices are the Lorentz and Coulomb gauges:

$$\nabla \cdot \mathbf{A} + \frac{1}{c^2\kappa_{\mathrm{e}}}\frac{\partial\Phi}{\partial t} = 0 \qquad \text{(the Lorentz gauge)} \tag{12.5a}$$

and

$$\nabla \cdot \mathbf{A} = 0 \qquad \text{(the Coulomb gauge).} \tag{12.5b}$$

The Coulomb gauge is often employed either when $\rho_{\mathrm{ch}} = 0$ or for static charge distribu-

[3] Derivations of this result are given in many texts on classical mechanics, e.g., Fetter and Walecka (1980), Marion (1965), and Goldstein (1980).

[4] This lack of uniqueness is another argument against $\mathbf{A}$ and Φ having no direct physical effects in the classical domain. As we shall see, this need not be true in the quantal domain.

tions, wherein $\rho_{\text{ch}} \neq \rho_{\text{ch}}(t)$, implying time-independent $\mathcal{E}$, Φ, and $\mathbf{A}$. The non-invariance of the Coulomb gauge w.r.t. Lorentz transformations is of no concern in the study of non-relativistic quantum mechanics.

Transition to the quantal Hamiltonian

The time-dependent Schrödinger equation for a 3-D particle of mass m and charge q in electromagnetic fields, and acted on as well by a non-electromagnetic potential $V(\mathbf{r})$, is

$$i\hbar \frac{\partial \Psi^{(A)}(\mathbf{r},\, t)}{\partial t} = \hat{H}^{(A)}(\mathbf{r},\, t)\Psi^{(A)}(\mathbf{r},\, t), \tag{12.6}$$

where

$$\hat{H}^{(A)}(\mathbf{r},\, t) = \frac{1}{2m}[\hat{\mathbf{P}}(\mathbf{r}) - \kappa_{\text{c}} q\mathbf{A}(\mathbf{r},\, t)]^2 + q\Phi(\mathbf{r},\, t) + V(\mathbf{r}). \tag{12.7}$$

Equation (12.7) is the quantal version of $H_{\text{cl}}^{(A)}$, generalized to include the non-electromagnetic (scalar) potential-energy term $V(\mathbf{r})$. $\hat{H}^{(A)}(\mathbf{r},\, t)$ contains the quantal version of the kinetic momentum $\mathbf{\Pi}$, viz.,

$$\mathbf{\Pi} \to \hat{\mathbf{\Pi}}(\mathbf{r},\, t) = \hat{\mathbf{P}}(\mathbf{r}) - \kappa_{\text{c}} q\mathbf{A}(\mathbf{r},\, t), \tag{12.8}$$

in terms of which $\hat{H}^{(A)}(\mathbf{r},\, t)$ becomes

$$\hat{H}^{(A)}(\mathbf{r},\, t) = \frac{1}{2m}\hat{\Pi}^2(\mathbf{r},\, t) + q\Phi + V(\mathbf{r}). \tag{12.9}$$

We shall show in Section 12.4 that $\hat{\mathbf{\Pi}}(\mathbf{r},\, t)$ is related to $\hat{\mathbf{P}}(\mathbf{r})$ by a gauge transformation of the second kind. Note well that only the classical momentum $\mathbf{p}$ is quantized when $\mathbf{\Pi} \to \hat{\mathbf{\Pi}}(\mathbf{r},\, t)$: $\hat{\mathbf{\Pi}}(\mathbf{r},\, t) \neq -i\hbar\nabla$.

The Hamiltonian $\hat{H}^{(A)}(\mathbf{r},\, t)$ can be written as the sum of two terms, one describing the system in the absence of the fields, the other incorporating their effects. This form is easily obtained by expanding the $\hat{\Pi}^2$ term in Eq. (12.9), leading to

$$\hat{H}^{(A)}(\mathbf{r},\, t) = \hat{H}^{(0)}(\mathbf{r}) - \frac{q\kappa_{\text{c}}}{m}\mathbf{A}(\mathbf{r})\cdot\hat{\mathbf{P}}(\mathbf{r}) + \frac{iq\hbar\kappa_{\text{c}}}{2m}[\nabla\cdot\mathbf{A}(\mathbf{r})] + \frac{q^2\kappa_{\text{c}}^2}{2m}A^2(\mathbf{r}) + q\Phi, \tag{12.10}$$

where

$$\hat{H}^{(0)}(\mathbf{r}) = \hat{K}(\mathbf{r}) + V(\mathbf{r}) \tag{12.11}$$

is the Hamiltonian in the absence of external fields, a reminder of which is the superscript "(0)." The term involving $\nabla\cdot\mathbf{A}(\mathbf{r})$ arises from the factor $\hat{\mathbf{P}}(\mathbf{r})\cdot\mathbf{A}(\mathbf{r}) = \mathbf{A}\cdot\hat{\mathbf{P}}(\mathbf{r}) - i\hbar[\nabla\cdot\mathbf{A}(\mathbf{r})]$. Although Eq. (12.10) may appear to be formidable, it is often straightforward to apply it, as we demonstrate in the next two sections. We present these applications prior to discussing general theoretical aspects of the quantal case, especially gauge invariance, so as to give the reader a feel for the physics involved in particular cases.

12.2. Quantal particle in a uniform magnetic field

In this section we study some effects of a uniform magnetic field (and zero electric field). Two cases – and two gauges – are considered: first, weak magnetic fields such that terms involving B^2 can be ignored, and second, stronger magnetic fields, which lead to the so-called Landau levels.

Since **A** fails to be unique because of gauge transformations, there is more than one **A** that will give rise to $\mathbf{B}$ = constant. Assuming that $\mathbf{B} = B\mathbf{e}_3$, three relatively simple vector potentials are

$$\mathbf{A}_1 = \tfrac{1}{2}\mathbf{B} \times \mathbf{r}, \tag{12.12a}$$

$$\mathbf{A}_2 = Bx\mathbf{e}_2, \tag{12.12b}$$

and

$$\mathbf{A}_3 = -By\mathbf{e}_1, \tag{12.12c}$$

all of which yield $\mathbf{B} = B\mathbf{e}_3$; for **B** constant but not along $\mathbf{e}_3$, we still have $\nabla \times \mathbf{A}_1 = \mathbf{B}$. Each member of (12.12) also obeys the Coulomb gauge condition: $\nabla \cdot \mathbf{A}_j = 0$, $j = 1, 2, 3$. The different $\mathbf{A}_j$ are related by gauge transformations.

$\mathbf{A}_1$ *and the weak-field approximation*

The first system to be studied involves $\Phi = 0$ and a spherically symmetric potential $V(r)$. We will be interested in those situations in which only the term linear in B is significant; this is the so-called weak-field limit, for which the choice $\mathbf{A} = \mathbf{A}_1$ is the most convenient. With this choice Eq. (12.10) becomes

$$\begin{aligned}\hat{H}^{(A_1)}(\mathbf{r}, t) &= \hat{H}^{(0)}(\mathbf{r}) - \frac{q\kappa_c}{2m}(\mathbf{B} \times \mathbf{r}) \cdot \hat{\mathbf{P}}(\mathbf{r}) + \frac{q^2\kappa_c^2}{8m}(\mathbf{r} \times \mathbf{B})^2 \\ &\equiv \hat{H}^{(A_1)}(\mathbf{r}).\end{aligned} \tag{12.13}$$

The second term in (12.13) can be re-written as

$$\begin{aligned}\frac{q\kappa_c}{2m}(\mathbf{B} \times \mathbf{r}) \cdot \hat{\mathbf{P}}(\mathbf{r}) &= \frac{q\kappa_c}{2m}\mathbf{B} \cdot (\mathbf{r} \times \hat{\mathbf{P}}(\mathbf{r})) \\ &= \frac{q\kappa_c}{2m}\mathbf{B} \cdot \hat{\mathbf{L}} = \frac{q\kappa_c}{2m}\hat{\mathbf{L}} \cdot \mathbf{B},\end{aligned} \tag{12.14}$$

while the third one is

$$\begin{aligned}\frac{q^2\kappa_c^2}{8m}(\mathbf{r} \times \mathbf{B})^2 &= \frac{q^2\kappa_c^2}{8m}[r^2B^2 - (\mathbf{r} \cdot \mathbf{B})^2] \\ &= \frac{q^2\kappa_c^2B^2}{8m}r_\perp^2,\end{aligned} \tag{12.15}$$

where $\mathbf{B} \cdot \hat{\mathbf{L}} = \hat{\mathbf{L}} \cdot \mathbf{B}$ due to the constancy of **B** and $\mathbf{r}_\perp$ is the portion of **r** perpendicular to **B**. In the present case, $\mathbf{B} \cdot \hat{\mathbf{L}} = B\hat{L}_z$ and $r_\perp^2 = x^2 + y^2$, but the more general forms of Eqs. (12.14) and (12.15) are used since our first goal is an interpretational one.

On substituting these results into Eq. (12.13), $\hat{H}^{(A_1)}(\mathbf{r})$ becomes

$$\hat{H}^{(A_1)}(\mathbf{r}) = \hat{H}^{(0)}(\mathbf{r}) - \frac{q\kappa_c}{2m}\hat{\mathbf{L}} \cdot \mathbf{B} + \frac{q^2r_\perp^2B^2\kappa_c^2}{8m}. \tag{12.16}$$

The second term in (12.16), which is proportional to the magnetic field, is the quantal analog of a classical term, viz., the energy $H_{\text{mag}} = -\boldsymbol{\mu} \cdot \mathbf{B}$ arising from the magnetic moment $\boldsymbol{\mu}$ of a charged particle in a magnetic field. For a current density $\mathbf{j}(\mathbf{r}')$ contained in a volume V, the magnetic moment $\boldsymbol{\mu}$ is defined as

$$\boldsymbol{\mu} = \frac{\kappa_c}{2} \int_V d^3 r' \, \mathbf{r}' \times \mathbf{j}(\mathbf{r}').$$

The current density for a particle of charge q and mass m at position $\mathbf{r}$ whose velocity is $\mathbf{v}$ is

$$\mathbf{j}(\mathbf{r}') = q\mathbf{v}\delta(\mathbf{r} - \mathbf{r}'),$$

for which $\boldsymbol{\mu}$ becomes

$$\boldsymbol{\mu} = \frac{q\kappa_c}{2} \mathbf{r} \times \mathbf{v} = \frac{q\kappa_c}{2m} \mathbf{L},$$

where $\mathbf{L}$ is the classical angular momentum. The units of $\boldsymbol{\mu}$ are energy/magnetic field strength, e.g., joules, ergs, or eV/tesla or gauss.

The classical energy H_{mag} due to the magnetic moment of a charged particle in an external magnetic field is therefore

$$H_{\text{mag}} = -\frac{q\kappa_c}{2m} \mathbf{L} \cdot \mathbf{B}.$$

Its quantal analog is

$$\hat{H}_{\text{mag}}^{(L)} = -\frac{q\kappa_c}{2m} \hat{\mathbf{L}} \cdot \mathbf{B}, \tag{12.17}$$

just the second term in Eq. (12.16).

It is customary to introduce a quantal analog for $\boldsymbol{\mu}$, viz., the (orbital) magnetic moment operator $\hat{\boldsymbol{\mu}}_{\text{L}}$:

$$\hat{\boldsymbol{\mu}}_L \equiv \frac{q\kappa_c}{2m} \hat{\mathbf{L}} = \frac{q\hbar\kappa_c}{2m} (\hat{\mathbf{L}}/\hbar). \tag{12.18}$$

Since $\hat{\mathbf{L}}/\hbar$ is dimensionless, the units of $\hat{\boldsymbol{\mu}}_{\text{L}}$ are carried by the quantity $|q|\hbar\kappa_c/(2m)$, denoted μ and known as a magneton:

$$\mu = \frac{|q|\hbar\kappa_c}{2m}.$$

In terms of μ, the orbital magnetic moment operator is

$$\hat{\boldsymbol{\mu}}_L = \frac{q}{|q|} \frac{\mu}{\hbar} \hat{\mathbf{L}} \tag{12.19}$$

and $H_{\text{mag}}^{(L)}$ becomes

$$\hat{H}_{\text{mag}}^{(L)} = -\frac{q}{|q|} \frac{\mu}{\hbar} \mathbf{B} \cdot \hat{\mathbf{L}}. \tag{12.20}$$

Values of μ depend on q and m. When the particle is an electron, $q \to -e$, $m \to m_e$, and μ is written μ_B, the Bohr magneton:

$$\mu_B = \frac{e\hbar\kappa_c}{2m_e} \cong 0.5788 \times 10^{-8} \text{ eV G}^{-1} = 0.5788 \times 10^{-4} \text{ eV T}^{-1}.$$

The nuclear magneton μ_N refers to the proton:

$$\mu_{\rm N} = \frac{e\hbar\kappa_{\rm c}}{2m_{\rm p}} = \frac{m_{\rm e}}{m_{\rm p}}\mu_{\rm B} \cong 3.152 \times 10^{-12}\ {\rm eV\ G^{-1}} = 3.152 \times 10^{-8}\ {\rm eV\ T^{-1}},$$

where $m_{\rm p}$ is the mass of the proton. The Bohr (nuclear) magneton is the unit in which the spin magnetic moment of the electron (proton and neutron) is measured.

The second term in (12.16) is a magnetic-moment contribution to $\hat{H}^{(A)}(\mathbf{r})$ due to the particle's "orbital" motion. This term often produces much greater effects than does the third term, in which case one is in the "weak-field" regime. Estimates of the relative importance of these two terms involve the size of typical matrix elements. Reverting to $\mathbf{B} = B\mathbf{e}_3$, the ratio of interest is

$$R = \frac{[q^2 B^2 \kappa_{\rm c}^2/(8m)]\langle x^2 + y^2\rangle}{[qB\kappa_{\rm c}/(2m)]\langle \hat{L}_z\rangle},$$

where the symbol $\langle\ \rangle$ refers to a diagonal matrix element. Let us assume that the particle is a single electron bound in an atom. Then, since atomic sizes are of the order of an ångström unit $\approx a_0$ ($a_0 = \hbar^2/(m_{\rm e}\kappa_{\rm e}e^2)$ is the Bohr radius) and $0 \leqslant \langle \hat{L}_z\rangle \leqslant 10\hbar$, we may estimate R by setting $\langle x^2 + y^2\rangle = a_0^2$ and $\langle \hat{L}_z\rangle = \hbar$:

$$\begin{aligned} R &\cong \frac{\kappa_{\rm c}^2\kappa_{\rm e}e^2 B^2 a_0^2/(4m_{\rm e})}{\mu_{\rm B} B} \\ &= B\mu_{\rm B}/[40(\kappa_{\rm e}e^2/a_0)] \cong 10^{-10}(B/a_0). \end{aligned}$$

However, the largest, non-superconducting-magnet magnetic fields produced in the laboratory are $\sim 1\ {\rm T} = 10^4\ {\rm G}$, so that $R_{\rm max} \cong 10^{-6}$ for states with non-excessively-large values of $\langle x^2 + y^2\rangle$ and a minimal value of $\langle \hat{L}_z\rangle = \hbar$. We take $R \precsim 10^{-4}$ as a reasonable estimate for defining the weak-field limit, for which case

$$\hat{H}^{(A_1)}(\mathbf{r}) \underset{\text{weak-field}}{\longrightarrow} \hat{H}^{(A_1)}_{\rm weak}(\mathbf{r}) = \hat{H}^{(0)}(\mathbf{r}) - \frac{q}{|q|}\frac{\mu B}{\hbar}\hat{L}_z, \tag{12.21}$$

where $\mathbf{B} = B\mathbf{e}_3$.

In the weak-field limit, the result of placing the charged, single-particle system in a constant magnetic field is therefore to add to $\hat{H}^{(0)}(\mathbf{r})$ a term proportional to $\hat{L}_z$. When the potential is spherically symmetric, the effect of this $\hat{L}_z$ term is to lift the $(2\ell + 1)$-fold degeneracy that we discussed near the end of Section 10.2 in connection with Eq. (10.66b). Let the energy in the absence of $\mathbf{B}$ be $E_{n\ell}$; the corresponding wave function is $\psi_{n\ell m_\ell}(\mathbf{r}) = R_{n\ell}(r)Y_\ell^{m_\ell}(\Omega)$, obeying

$$(\hat{H}^{(0)}(\mathbf{r}) - E_{n\ell})R_{n\ell}(r)Y_\ell^{m_\ell}\Omega = 0. \tag{10.66b}$$

Insertion of the system into $\mathbf{B}$ changes $\hat{H}^{(0)}(\mathbf{r})$ into $\hat{H}^{(A_1)}(\mathbf{r})$ but does not change the wave function: $R_{n\ell}(r)Y_\ell^{m_\ell}(\Omega)$ is also an eigenfunction of $\hat{H}^{(A_1)}_{\rm weak}(\mathbf{r})$, although the energy eigenvalue is changed:

$$(\hat{H}^{(A_1)}_{\rm weak}(\mathbf{r}) - E_{n\ell m_\ell})R_{n\ell}(r)Y_\ell^{m_\ell}(\Omega) = 0, \tag{12.22a}$$

with

$$E_{n\ell m_\ell} = E_{n\ell} - \frac{q}{|q|}\mu B m_\ell. \tag{12.22b}$$

Equation (12.22b) verifies our claim that the effect of the weak magnetic field is to lift the $(2\ell+1)$-fold, m_ℓ-degeneracy of $E_{n\ell}$. To illustrate this graphically, the particle is assumed to be an electron (whose spin is ignored). Then $q/|q| = -1$, $\mu \to \mu_{\rm B}$, and $E_{n\ell m_\ell} \to E_{n\ell} + \mu_{\rm B} B m_\ell$. A schematic representation is shown in the Grotrian diagram of Fig. 12.1, in which the degenerate level $E_{n\ell}$ appears on the left-hand side and its (weak) magnetic-field analog, split into $2\ell+1$ levels, each separated by $\mu_{\rm B}B$, is shown on the right-hand side. Note that, for a positive charge, the lowest-lying level would have $m_\ell = \ell$ rather than $m_\ell = -\ell$. Note also that the strength of the magnetic-field splitting in (12.22b) is proportional to m_ℓ, thus providing a justification for m_ℓ being referred to as the "magnetic quantum number" (see the comment above Eq. (10.63)).

This splitting of atomic energy levels in a magnetic field underlies the phenomenon known as the normal Zeeman effect. Consider a pair of atomic states with energies $E_{n\ell}$ and $E_{n,\ell+1} > E_{n\ell}$. In the absence of **B**, the higher-lying state will decay to the lower-lying one via emission of a single photon of energy $E_\gamma = E_{n,\ell+1} - E_{n\ell}$. This is in precise analogy to electromagnetic transitions in hydrogen. When the atom is put into a magnetic field and the two levels are split into $2\ell+1$ and $2\ell+3$ m_ℓ-levels, respectively, it turns out that only three transitions, rather than the much larger number one might expect, are observed. This is due to a selection rule governing electromagnetic radiation, which allows transitions only between magnetic substates whose m_ℓ values differ by $\Delta m = +1$, 0, or -1. We shall discuss this selection rule and the transitions to which it gives rise in Section 16.2.

A further consequence of the $\mathbf{B}\cdot\hat{\mathbf{L}}$ term, which is rather more profound than the splittings just noted, is that its structure becomes the paradigm by which the intrinsic spin (magnitude $\hbar/2$ for electrons, protons, and neutrons) is inferred from experiment, an

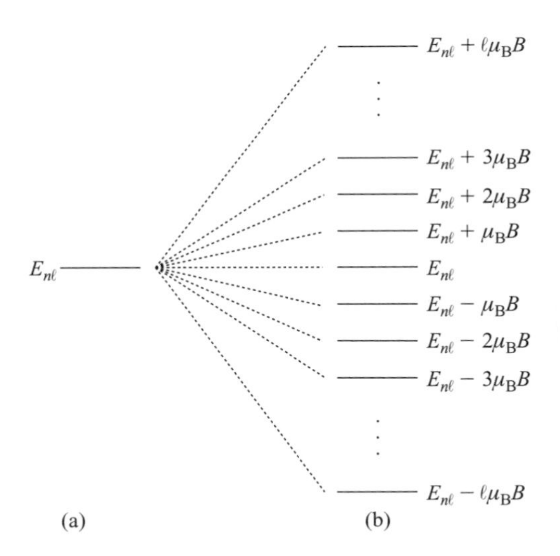

Fig. 12.1 (a) The $(2\ell+1)$-fold-degenerate level $E_{n\ell}$ of an electron in a spherically symmetric potential ($\mathbf{B}=0$). (b) The splitting of $E_{n\ell}$ into $(2\ell+1)$ distinct levels when the system of part (a) is immersed in a weak, constant magnetic field $\mathbf{B} = B\mathbf{e}_3$. Adjacent levels are separated by the splitting $\mu_{\rm B}B$. Since $q=-e$, the state with $m_\ell = -\ell$ lies lowest.

inference we examine in some detail in Chapter 13. This aspect, plus the others discussed here and in Chapter 9, should make clear the richness of the concept of angular momentum in quantum theory.

Landau levels

The vector potential $\mathbf{A}_1$ leads to the simplest treatment of the effect of a uniform but weak magnetic field on the bound states of a spherically symmetric potential. For the case of a free particle ($V = 0$) in a strong magnetic field, wherein the A^2 term cannot be ignored, L. D. Landau (1930) showed that either $\mathbf{A}_2$ or $\mathbf{A}_3$ is the more appropriate choice; they lead to essentially the same results. We follow Landau's original treatment in this subsection.

Choosing $\mathbf{A}_2$, Eq. (12.12b), and remembering that now $V = \Phi = 0$, Eq. (12.10) becomes $\hat{H}^{(A_2)}(\mathbf{r})$, given by[5]

$$\hat{H}^{(A_2)}(\mathbf{r}) = \frac{1}{2m}(\hat{P}_x^2 + \hat{P}_y^2 + \hat{P}_z^2) - \frac{q\kappa_c Bx}{m}\hat{P}_y + \frac{q^2\kappa_c^2 B^2 x^2}{2m}, \qquad V = \Phi = 0. \quad (12.23)$$

The eigenvalue problem to be solved is

$$[\hat{H}^{(A_2)}(\mathbf{r}) - E^{(A_2)}]\psi^{(A_2)}(\mathbf{r}) = 0. \quad (12.24)$$

Since x is the only coordinate in (12.23),

$$[\hat{P}_y, \hat{H}^{(A_2)}] = [\hat{P}_z, \hat{H}^{(A_2)}] = 0;$$

$\hat{P}_y$, $\hat{P}_z$, and $\hat{H}^{(A_2)}$ are the CSCO for this system, and $\psi^{(A_2)}(\mathbf{r})$ is therefore a product:

$$\psi^{(A_2)}(\mathbf{r}) = \frac{1}{2\pi} e^{i(k_y y + k_z z)} F(x). \quad (12.25)$$

Substitution of (12.25) into (12.24) leads to

$$-\frac{\hbar^2}{2m}F''(x) + \left(\frac{\hbar^2 k_y^2}{2m} - \frac{q\kappa_c B\hbar k_y}{m}x + \frac{q^2\kappa_c^2 B^2}{2m}x^2\right)F(x) + \frac{\hbar^2 k_z^2}{2m}F(x) = E^{(A_2)}F(x), \quad (12.26)$$

where $' = d/dx$.

Let us consider the term in large parentheses in (12.26). It is a perfect square, and, although its middle term is proportional to the magneton $\mu = q\kappa_c\hbar/(2m)$, μ actually plays no role in this eigenvalue problem. Instead, the operative parameter in the factor is $q^2\kappa_c^2 B^2/(2m)$. Suppose for the moment that k_y were zero, in which case the term in large parentheses would be $q^2\kappa_c^2 B^2 x^2/(2m)$. Combining this term with $-F''(x)/(2m)$ yields a 1-D harmonic oscillator Hamiltonian: $-(d^2F/dx^2)/(2m) + m\omega_c^2 x^2/2$, where the *cyclotron frequency* ω_c has been introduced:

$$\omega_c = q\kappa_c B/m. \quad (12.27)$$

ω_c is the angular frequency with which a classical particle of charge q and mass m

[5] Selection of $\mathbf{A}_3$ in place of $\mathbf{A}_2$ interchanges x and y but leaves the basic structure of (12.23) unchanged; hence, it suffices to consider only $\mathbf{A}_2$ for the free-particle case. Use of $\mathbf{A}_1$ in place of $\mathbf{A}_2$ complicates the analysis but leaves the energy spectrum essentially unchanged.

rotates in a constant magnetic field of strength B. Introducing it into (12.26), the term in large parentheses becomes

$$\tfrac{1}{2}m\omega_c^2(x - x_c)^2, \tag{12.28}$$

where

$$x_c = \frac{\hbar k_y}{m\omega_c} \tag{12.29}$$

is the displaced center of the effective 1-D oscillator. If the charge is negative, then x_c changes sign, while $\omega_c \to |q|\kappa_c B/m$.

With the result (12.28) and the definition $E = E^{(A_1)} - \hbar^2 k_z^2/(2m)$, Eq. (12.26) reads

$$\left(-\frac{\hbar^2}{2m}\frac{d^2}{dx^2} + \frac{1}{2}m\omega_c^2(x - x_c)^2\right)F(x) = EF(x). \tag{12.30}$$

This is just the eigenvalue equation for a 1-D linear harmonic oscillator whose center is at x_c. As in Chapter 6, the solution is

$$E \to E_n = (n + \tfrac{1}{2})\hbar\omega_c, \qquad n = 0, 1, 2, \ldots, \tag{12.31a}$$

$$F(x) \to F_n(x - x_c) = A_n H_n\left(\frac{x - x_c}{\lambda_c}\right)e^{-(x-x_c)^2/(2\lambda_c^2)}, \tag{12.31b}$$

where $H_n(\eta)$ is the nth-order Hermite polynomial – Eqs. (6.122) and (6.123) – and $\lambda_c = \sqrt{\hbar/(m\omega_c)}$.

The full solution to the eigenvalue problem (12.24) for an otherwise-free particle in the constant magnetic field $\mathbf{B} = B\mathbf{e}_3$ is thus

$$\psi^{(A_2)} \to \psi^{(A_2)}_{k_z k_y n}(\mathbf{r}) = \frac{A_n}{2\pi}e^{i(k_y y + k_z z)}H_n\left(\frac{x - x_c}{\lambda_c}\right)e^{-(x-x_c)^2/(2\lambda_c^2)} \tag{12.32a}$$

and

$$E^{(A_2)} \to E^{(A_2)}_{k_z n} = \frac{\hbar^2 k_z^2}{2m} + (n + \tfrac{1}{2})\hbar\omega_c, \tag{12.32b}$$

the latter being known as *Landau levels.*

There are several interesting features to this solution. First, there are no purely bound states, and all energies are positive definite. Second, the "motion" in the z- and y-directions is that of a free particle. The plane-wave nature of the z-motion is the analog of the classical case, wherein the force exerted by the magnetic field is normal to its direction. The plane-wave motion in the y-direction is an artifact of the gauge $\mathbf{A}_2$ (a property that persists when $\mathbf{A}_2$ is replaced by $\mathbf{A}_3$). It is an artifact because the energy does not contain a term depending on k_y^2: the k_y dependence enters only through the displaced center $x_c = \hbar k_y/(m\omega_c)$. This brings us to the third feature: the solution to the full eigenvalue problem is doubly degenerate. There is a denumerably infinite degeneracy in the total energy due to the $\hbar^2 k_z^2/(2m)$ term, and a continuously infinite degeneracy w.r.t. the oscillator levels, due to the $x - x_c$ dependence of the wave function. The latter degeneracy reflects the invariance of the system under translation along the y-axis: any point y may be chosen as the origin, and this turns into a degeneracy w.r.t. x_c. As has been remarked already, there is no basic change when $\mathbf{A}_3$ replaces $\mathbf{A}_2$, but this is not quite true when the gauge $\mathbf{A}_1$ is used, as in exercises 12.10 and 12.11.

Although the magneton μ does not play a direct role in formulating the eigenvalue problem for $F(x)$, it can be brought into the expression for $E_{k_z n}^{(A_2)}$ via

$$\hbar\omega_c = 2\frac{|q|\kappa_c\hbar}{2m}B = 2\mu B:$$

$$E_{k_z n}^{(A_2)} = \frac{\hbar^2 k_z^2}{2m} + \left(n + \frac{1}{2}\right)2\mu B. \tag{12.32c}$$

Which of Eqs. (12.32b) and (12.32c) is employed depends on whether one wants to emphasize cyclotron behavior or magnetic-moment interactions. The energy values do not change, of course. No matter which form is chosen, the eigenvalues $E_{k_z n}^{(A_2)}$ are known as Landau levels after their discoverer.

Not only are there two equivalent expressions for the Landau levels $E_{k_z n}^{(A_2)}$, there are also two different ways of constructing energy-level diagrams representing them. These are shown in Fig. 12.2. In Fig. 12.2(a), k_z is fixed and the usual Grotrian diagram for the 1-D oscillator is displayed. In Fig. 12.2(b), n is fixed at $n = 0$ and the parabolic dependence of $E_{k_z 0}^{(A_2)}$ is illustrated. Figure 12.2(c) shows several such parabolic curves, starting at $n = 0$.

Figures 12.1 and 12.2 help to characterize the different behaviors of bound and free particles in a uniform magnetic field. The situation is similar for the uniform-electric-field case, which is considered next.

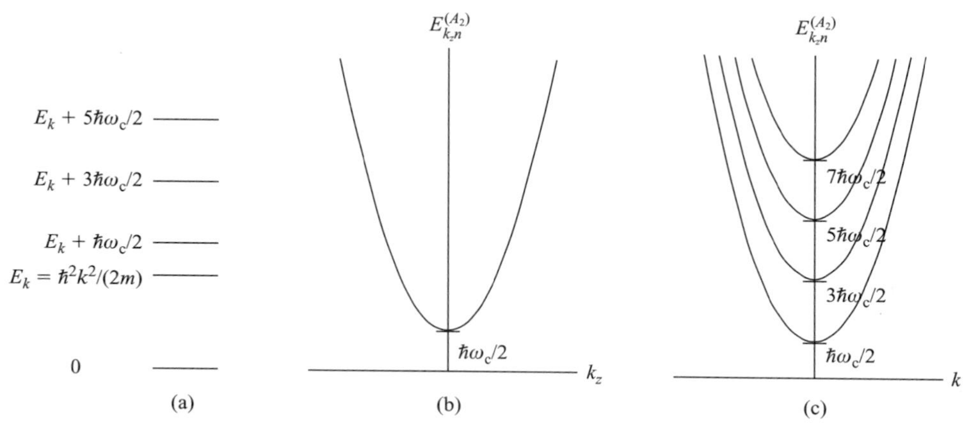

Fig 12.2 Representations of the Landau levels. (a) The Grotrian diagram for the 1-D oscillator levels built on the plane-wave value $\hbar^2 k_z^2/(2m)$. (b) The parabolic structure $E_{k_z n=0}^{(A_2)}$ built on the lowest-lying 1-D oscillator level $\hbar\omega_c/2$. (c) Four parabolic structures in k_z built on the first four 1-D oscillator levels $(n + \frac{1}{2})\hbar\omega_c$, $n = 0$, 1, 2, and 3.

12.3. Quantal particle in a uniform electric field

Electric fields in quantum mechanics are often considered as scalar-potential ($\mathbf{A} = 0$) problems. Suppose that $\mathcal{E}$ is a uniform electric field, constant in time. The simplest scalar potential producing it is

$$\Phi(\mathbf{r}) = -\mathcal{E} \cdot \mathbf{r}$$

and, if we choose $\mathcal{E} = \mathcal{E}\mathbf{e}_3$, then $\Phi(\mathbf{r})$ becomes

$$\Phi(\mathbf{r}) = -\mathcal{E}z = -\mathcal{E}r\cos\theta. \tag{12.33}$$

Unlike the magnetic-field case, there is no Φ^2 term, and the linear dependence on $\mathbf{r}$ or z leads in most instances to non-trivial eigenvalue problems. There are tractable examples, of course, and we examine two of them in this section. The first is a 1-D system: a charged particle bound by the linear harmonic-oscillator potential $m\omega_0^2 x^2/2$ is placed in the uniform electric field $\mathcal{E} = \mathcal{E}\mathbf{e}_1$. As we shall see, this system behaves like a displaced oscillator and so is analogous to the x-motion portion of the Landau-level problem. The second example is that of a charged particle initially in a 3-D plane-wave state that experiences a uniform electric field that is switched on at time $t = 0$. We will determine the wave function for all $t \geqslant 0$, and in Section 12.6 will not only re-analyze this system from the gauge-transformation aspect ($\Phi = 0$, $\mathbf{A} \neq 0$), but also allow $\mathcal{E}$ to become time-dependent.

1-D charged oscillator

For this system, $\mathbf{A} = 0$, while

$$V(\mathbf{r}) \to V(x) = m\omega_0^2 x^2/2,$$

$$\Phi(\mathbf{r}, t) \to \Phi = -q\mathcal{E}x,$$

and

$$\hat{K}(\mathbf{r}) \to \hat{K}(x) = \hat{P}_x^2(x)/(2m),$$

whereupon $\hat{H}^{(A)}(\mathbf{r}, t) \to \hat{H}^{(\mathcal{E})}(x)$, given by

$$\hat{H}^{(\mathcal{E})}(x) = -\frac{\hbar^2}{2m}\frac{d^2}{dx^2} + \frac{1}{2}m\omega_0^2 x^2 - q\mathcal{E}x. \tag{12.34}$$

This Hamiltonian can be turned into an effective 1-D oscillator Hamiltonian by combining the second and third terms to form a perfect square:

$$\frac{1}{2}m\omega_0^2 x^2 - q\mathcal{E}x = \frac{1}{2}m\omega_0^2 (x-a)^2 - \frac{q^2\mathcal{E}^2}{2m\omega_0^2}, \tag{12.35}$$

where

$$a = \frac{q\mathcal{E}}{m\omega_0^2}.$$

Using (12.35), $\hat{H}^{(\mathcal{E})}(\mathbf{r})$ becomes

$$\hat{H}^{(\mathcal{E})}(x) = -\frac{\hbar^2}{2m}\frac{d^2}{dx^2} + \frac{1}{2}m\omega_0^2(x-a)^2 - \frac{q^2\mathcal{E}^2}{2m\omega_0^2},$$

yielding

$$\left(-\frac{\hbar^2}{2m}\frac{d^2}{dx^2} + \frac{1}{2}m\omega_0^2(x-a)^2\right)\psi^{(\mathcal{E})}(x) = \left(E^{(\mathcal{E})} + \frac{q^2\mathcal{E}^2}{2m\omega_0^2}\right)\psi^{(\mathcal{E})}(x) \tag{12.36}$$

as the eigenvalue problem to be solved (note that $d/dx = d/d(x-a)$). As predicted, Eq. (12.36) corresponds to a 1-D linear harmonic oscillator whose center is displaced to the point a. The solution to (12.36) is

$$\psi^{(\mathcal{E})}(x) \to \psi_n^{(\mathcal{E})}(x-a) = A_n H_n\left(\frac{x-a}{\lambda_0}\right)e^{-(x-a)^2/(2\lambda_0^2)} \tag{12.37a}$$

and

$$E^{(\mathcal{E})} \to E_n^{(\mathcal{E})} = -\frac{q^2\mathcal{E}^2}{2m\omega_0^2} + \left(n + \frac{1}{2}\right)\hbar\omega_0, \qquad n = 0,\,1,\,2,\,\ldots, \tag{12.37b}$$

where $\lambda_0 = \sqrt{\hbar/(m\omega_0)}$ is the relevant oscillator-length parameter. The effect of the electric field is two-fold: (i) it displaces the center of the oscillator without altering the form of the eigenfunctions, and (ii) it shifts the energy spectrum, so that, instead of starting at the usual oscillator zero-point energy $\hbar\omega_0/2$, it begins at $-q^2\mathcal{E}^2/(2m\omega_0^2) + \hbar\omega_0/2$, an energy value that, depending on the magnitude of $q\mathcal{E}$, can be positive or negative. Apart from this negative possibility, the analogy with the x-motion in the Landau-level case is perfect.

Initially free particle in a uniform electric field

Up to time $t = 0$, let a particle of mass m, charge q, and energy $E_k = \hbar^2k^2/(2m)$ be in a potential-free region. Its Hamiltonian is $\hat{K}(\mathbf{r}) = \hat{\mathbf{P}}^2(\mathbf{r})/(2m)$, its wave function at time $t = 0$ is

$$\Psi_{\mathbf{k}}(\mathbf{r},\, t=0) = \left(\frac{1}{2\pi}\right)^{3/2} e^{i\mathbf{k}\cdot\mathbf{r}} = \langle\mathbf{r}|\mathbf{k}\rangle,$$

and, if a uniform electric field $\boldsymbol{\mathcal{E}}$ were not switched on at time $t = 0$, its wave function would evolve into

$$\Psi_{\mathbf{k}}(\mathbf{r},\, t) = \left(\frac{1}{2\pi}\right)^{3/2} e^{i(\mathbf{k}\cdot\mathbf{r} - E_k t/\hbar)}. \tag{12.38}$$

However, at time $t = 0$, a uniform electric field $\boldsymbol{\mathcal{E}}$ *is* turned on; we shall determine the wave function $\Psi^{(\mathcal{E})}(\mathbf{r},\, t)$, $t > 0$, to which $\langle\mathbf{r}|\mathbf{k}\rangle$ evolves. We follow the development presented by Fallieros and Friar (1982).

By assumption $\boldsymbol{\mathcal{E}}$ enters the Hamiltonian through the scalar potential $\Phi(\mathbf{r}) = -q\boldsymbol{\mathcal{E}}\cdot\mathbf{r}$:

$$\hat{H}^{(\mathcal{E})}(\mathbf{r}) = \frac{1}{2m}\hat{P}^2(\mathbf{r}) - q\boldsymbol{\mathcal{E}}\cdot\mathbf{r}. \tag{12.39}$$

Since $\hat{H}^{(\mathcal{E})}$ is time-independent, the time-evolution operator $\hat{U}^{(\mathcal{E})}(t)$ is

$$\hat{U}^{(\mathcal{E})}(t) = e^{-i\hat{H}^{(\mathcal{E})}(\mathbf{r})t/\hbar}, \tag{12.40}$$

where the $\mathbf{r}$ dependence in $\hat{U}^{(\mathcal{E})}(t)$ is suppressed. Application of (12.40) to $\langle \mathbf{r}|\mathbf{k}\rangle = \Psi_{\mathbf{k}}(\mathbf{r},\ t=0)$ gives the usual formal expression for the time-evolved state:

$$\Psi^{(\mathcal{E})}(\mathbf{r},\ t) = e^{-i\hat{H}^{(\mathcal{E})}(\mathbf{r})t/\hbar}\Psi_{\mathbf{k}}(\mathbf{r},\ t=0). \tag{12.41}$$

Although direct evaluation of the RHS of (12.41) appears formidable, an indirect method is at hand. It makes use of the fact that, like $\Psi_{\mathbf{k}}(\mathbf{r},\ t=0)$, $\Psi^{(\mathcal{E})}(\mathbf{r},\ t)$ is an eigenfunction of $\hat{\mathbf{P}}(\mathbf{r})$, with an eigenvalue we may conveniently denote as $\hbar\mathbf{k}(t)$:

$$\hat{\mathbf{P}}(\mathbf{r})\Psi^{(\mathcal{E})}(\mathbf{r},\ t) = \hbar\mathbf{k}(t)\Psi^{(\mathcal{E})}(\mathbf{r},\ t). \tag{12.42}$$

Given the fact that $[\hat{\mathbf{P}},\ \hat{H}^{(\mathcal{E})}] \neq 0$, Eq. (12.42) is an unexpected result; we shall discuss the reason underlying it after determining $\Psi^{(\mathcal{E})}(\mathbf{r},\ t)$.

To prove Eq. (12.42) and also determine $\mathbf{k}(t)$, we apply $\hat{\mathbf{P}}(\mathbf{r})$ to both sides of (12.41):

$$\begin{aligned}\hat{\mathbf{P}}(\mathbf{r})\Psi^{(\mathcal{E})}(\mathbf{r},\ t) &= \hat{\mathbf{P}}(\mathbf{r})e^{-i\hat{H}^{(\mathcal{E})}(\mathbf{r})t/\hbar}\Psi_{\mathbf{k}}(\mathbf{r},\ t=0)\\ &= e^{-i\hat{H}^{(\mathcal{E})}(\mathbf{r})t/\hbar}\hat{\mathbf{P}}^{(\mathrm{H})}\Psi_{\mathbf{k}}(\mathbf{r},\ t=0),\end{aligned} \tag{12.43}$$

where $\hat{\mathbf{P}}^{(\mathrm{H})}$ is the momentum operator in the Heisenberg picture:

$$\hat{\mathbf{P}}^{(\mathrm{H})} = \hat{U}^{(\mathcal{E})\dagger}(t)\hat{\mathbf{P}}(\mathbf{r})\hat{U}^{(\mathcal{E})}(t).$$

$\hat{\mathbf{P}}^{(\mathrm{H})}$ can be obtained by solving the Heisenberg equation of motion

$$\frac{d\hat{\mathbf{P}}^{(\mathrm{H})}}{dt} = \frac{1}{i\hbar}[\hat{\mathbf{P}}^{(\mathrm{H})},\ \hat{H}^{(\mathcal{E})}] = q\boldsymbol{\mathcal{E}};$$

it yields

$$\begin{aligned}\hat{\mathbf{P}}^{(\mathrm{H})} &= \hat{\mathbf{P}} + q\boldsymbol{\mathcal{E}}t\\ &= \hat{\mathbf{P}} + \mathbf{F}^{(\mathcal{E})}t,\end{aligned} \tag{12.44}$$

where

$$\mathbf{F}^{(\mathcal{E})} = q\boldsymbol{\mathcal{E}}$$

is the classical force on the particle. Insertion of (12.44) into (12.43) gives

$$\hat{\mathbf{P}}(\mathbf{r})\Psi^{(\mathcal{E})}(\mathbf{r},\ t) = (\hbar\mathbf{k} + \mathbf{F}^{(\mathcal{E})}t)\Psi^{(\mathcal{E})}(\mathbf{r},\ t),$$

thus establishing the validity of Eq. (12.42), with

$$\hbar\mathbf{k}(t) = \hbar\mathbf{k} + \mathbf{F}^{(\mathcal{E})}t. \tag{12.45}$$

While $\Psi^{(\mathcal{E})}(\mathbf{r},\ t)$ is an eigenfunction of $\hat{\mathbf{P}}(\mathbf{r})$, its eigenvalue changes linearly with time, the expression (12.45) for $\hbar\mathbf{k}(t)$ having exactly the form of the classical result. In analogy to $\Psi^{(\mathcal{E})}(\mathbf{r},\ t=0) = \Psi_{\mathbf{k}}(\mathbf{r},\ t=0) = \langle \mathbf{r}|\mathbf{k}\rangle$, the solution $\Psi^{(\mathcal{E})}(\mathbf{r},\ t)$, $t>0$, is proportional to the "momentum" wave function $\langle \mathbf{r}|\mathbf{k}(t)\rangle$, with $\mathbf{k}(t)$ given by (12.45):

$$\begin{aligned}\Psi^{(\mathcal{E})}(\mathbf{r},\ t) &= e^{i\beta(t)}\langle \mathbf{r}|\mathbf{k}(t)\rangle\\ &= e^{i\beta(t)}e^{i\mathbf{k}(t)\cdot\mathbf{r}}/(2\pi)^{3/2},\end{aligned} \tag{12.46}$$

where $\exp[i\beta(t)]$ is a (unitary) proportionality factor that can depend on t but not on $\mathbf{r}$.

The functional form of $\beta(t)$ can be obtained by substituting (12.46) into the time-dependent Schrödinger equation,

$$i\hbar \frac{\partial \Psi^{(\mathcal{E})}(\mathbf{r},\, t)}{\partial t} = \hat{H}^{(\mathcal{E})}(\mathbf{r})\Psi^{(\mathcal{E})}(\mathbf{r},\, t),$$

from which we find

$$\frac{d\beta(t)}{dt} = -\frac{\hbar}{2m}\left(k^2 + \frac{2\mathbf{k}\cdot\mathbf{F}^{(\mathcal{E})}}{\hbar}t + \frac{F^{(\mathcal{E})2}t^2}{\hbar^2}\right). \tag{12.47}$$

The solution to (12.47) that is consistent with $\beta(t \to 0) \to 0$, which is required in order that $\Psi^{(\mathcal{E})}(\mathbf{r},\, t \to 0) \to \langle \mathbf{r}|\mathbf{k}\rangle$, is

$$\beta(t) = -\frac{\hbar}{2m}\left(k^2 t + \mathbf{k}\cdot\mathbf{F}^{(\mathcal{E})}\frac{t^2}{\hbar} + \frac{F^{(\mathcal{E})2}t^3}{3\hbar^2}\right), \tag{12.48}$$

which, together with $\mathbf{k}(t) = \mathbf{k} + \mathbf{F}^{(\mathcal{E})}t/\hbar$, provides the complete solution for the time-dependent wave function of an initially free charged particle in a spatially uniform electric field turned on at time $t = 0$.

Let us return to the question raised earlier: how can $\Psi^{(\mathcal{E})}(\mathbf{r},\, t)$ be an eigenfunction of $\hat{\mathbf{P}}(\mathbf{r})$, albeit with a time-dependent eigenvalue, when $[\hat{\mathbf{P}},\, \hat{H}^{(\mathcal{E})}] \neq 0$? The answer relies on the relation $[\hat{H}^{(\mathcal{E})},\, [\hat{H}^{(\mathcal{E})},\, \hat{\mathbf{P}}]] = 0$. The double commutator enters the expansion

$$\begin{aligned} e^{i\hat{H}^{(\mathcal{E})}t/\hbar}\hat{\mathbf{P}}e^{-i\hat{H}^{(\mathcal{E})}t/\hbar} &= \hat{\mathbf{P}} + \frac{i}{\hbar}t[\hat{H}^{(\mathcal{E})},\, \hat{\mathbf{P}}] \\ &\quad + \left(\frac{it}{\hbar}\right)^2 [\hat{H}^{(\mathcal{E})},\, [\hat{H}^{(\mathcal{E})},\, \hat{\mathbf{P}}]]/2! \\ &\quad + \left(\frac{it}{\hbar}\right)^3 [\hat{H}^{(\mathcal{E})},\, [\hat{H}^{(\mathcal{E})},\, [\hat{H}^{(\mathcal{E})},\, \hat{\mathbf{P}}]]]/3! \\ &\quad + \ldots, \end{aligned}$$

a relation based on Exercise 4.21, where the power-series expansion of $\exp(\pm i\hat{H}^{(\mathcal{E})}t/\hbar)$ is assumed to be convergent. Since the double commutator vanishes, then

$$e^{i\hat{H}^{(\mathcal{E})}t/\hbar}\hat{\mathbf{P}}e^{i\hat{H}^{(\mathcal{E})}t/\hbar} = \hat{\mathbf{P}} + \mathbf{F}^{(\mathcal{E})}t,$$

which is Eq. (12.43). It is the vanishing of the double commutator $[\hat{H}^{(\mathcal{E})},\, [\hat{H}^{(\mathcal{E})},\, \hat{\mathbf{P}}]]$ that ensures that $\Psi^{(\mathcal{E})}(\mathbf{r},\, t)$ is an eigenfunction of $\hat{\mathbf{P}}$, while it is the relation $[\hat{\mathbf{P}},\, \hat{H}^{(\mathcal{E})}] = i\hbar q\boldsymbol{\mathcal{E}}$ that leads to the linear time dependence $\mathbf{F}^{(\mathcal{E})}t$ of the eigenvalue. A generalization of this result, pointed out by Fallieros and Friar (1982) is the subject of an end-of-chapter exercise.

12.4. Quantum-theoretic aspects of $\hat{H}^{(A)}(\mathbf{r},\, t)$

Apart from the relatively minor change $V(\mathbf{r}) \to V(\mathbf{r}) + q\Phi$, the essential difference between $\hat{H}(\mathbf{r},\, t)$ and $\hat{H}^{(A)}(\mathbf{r},\, t)$ is the replacement of $\hat{\mathbf{P}}$ by $\hat{\mathbf{\Pi}}$. In this section we examine the effect of this replacement on some commutation relations, on Ehrenfest's theorem and on the quantal equation of continuity. Even though $\mathbf{A}(\mathbf{r}, t)$ is not an operator, it will turn out to have a decided influence on almost all of these quantities. All evaluations will be made in the coordinate representation.

Some commutation relations

Three commutation relations are of particular interest. The first is trivial:

$$[x_j, \hat{\Pi}_k] = i\hbar\delta_{jk}, \tag{12.49a}$$

from which the second one readily follows:

$$[x_j, \hat{\Pi}_k^2] = 2i\hbar\hat{\Pi}_k\delta_{jk}. \tag{12.49b}$$

The third,

$$[\hat{\Pi}_j, \hat{\Pi}_k] = i\kappa_c\hbar q(\nabla \times \mathbf{A})_\ell = i\kappa_c\hbar qB_\ell, \qquad j,\ k,\ \ell \text{ in cyclic order}, \tag{12.49c}$$

where B_ℓ is the ℓth component of the magnetic field $\mathbf{B}$, is an essential ingredient in deriving Ehrenfest's theorem, which is considered next.

Ehrenfest's theorem

In the absence of electromagnetic fields, Ehrenfest's theorem is expressed by the two relations

$$\frac{d\langle\hat{\mathbf{Q}}\rangle_\Gamma(t)}{dt} = \frac{1}{m}\langle\hat{\mathbf{P}}\rangle_\Gamma(t) \tag{8.50a}$$

and

$$\frac{d\langle\hat{\mathbf{P}}\rangle_\Gamma(t)}{dt} = \langle\mathbf{F}\rangle_\Gamma(t), \tag{8.50b}$$

where

$$\mathbf{F}(\mathbf{r}) = -\nabla_\mathbf{r} V(\mathbf{r}) \tag{8.52}$$

is the force. If Ehrenfest's theorem is to hold in the electromagnetic-field case, then in place of (8.52) we should expect to find a quantal analog of the Lorentz force

$$\mathbf{F}_\text{L} = q(\mathcal{E} + \kappa_c\mathbf{v} \times \mathbf{B}). \tag{12.50}$$

In attempting to determine the quantal version of (12.50), two questions must be answered. First, is there a quantal operator associated with $\mathbf{v}$, and second, what is the operator corresponding to $\mathbf{v} \times \mathbf{B}$? These operators should, of course, be Hermitian. In the case in which $\mathcal{E} = \mathbf{B} = 0$, a likely candidate for the velocity operator, say $\hat{\mathbf{v}}$ (lower case to distinguish it from the potential-energy operator $\hat{V}$), is

$$\hat{\mathbf{v}} \stackrel{?}{=} \frac{1}{m}\hat{\mathbf{P}}, \qquad \mathcal{E} = \mathbf{B} = 0. \tag{12.51}$$

Comparison of (12.51) with (8.50) shows that $\hat{\mathbf{v}} = \hat{\mathbf{P}}/m$ is consistent with Ehrenfest's theorem. We shall use Eq. (8.50a) to define $\hat{\mathbf{v}}$:

$$\frac{d\langle\hat{\mathbf{Q}}\rangle_\Gamma(t)}{dt} \equiv \langle\hat{\mathbf{v}}\rangle_\Gamma(t). \tag{12.52}$$

Let us now apply this definition in the present situation of external fields. Using Eq. (8.4), with $\hat{H} \to \hat{H}^{(A)}$, it is readily found that

$$\frac{d\langle\hat{\mathbf{Q}}\rangle}{dt} = \frac{1}{m}\langle\hat{\mathbf{\Pi}}\rangle_\Gamma(t), \tag{12.53}$$

from which we conclude that, in the presence of external fields,

$$\hat{\mathbf{v}} \to \hat{\mathbf{v}}^{(A)} = \frac{1}{m}\hat{\mathbf{\Pi}}(\mathbf{r},\, t). \tag{12.54}$$

Like $\hat{\mathbf{v}}$, $\hat{\mathbf{v}}^{(A)}$ is Hermitian. It is also consistent with the corresponding classical quantity.

We next consider the quantal analog of the term $\mathbf{v} \times \mathbf{B}$. This term is of the general form $\mathbf{a} \times \mathbf{d}$, where $\mathbf{a}$ and $\mathbf{d}$ are classical quantities. If $\mathbf{a} \to \hat{\mathbf{A}}$ and $\mathbf{d} \to \hat{\mathbf{D}}$, with $\hat{\mathbf{A}}^\dagger = \hat{\mathbf{A}}$ and $\hat{\mathbf{D}}^\dagger = \hat{\mathbf{D}}$, then the quantal analog of $\mathbf{a} \times \mathbf{d}$ cannot be $\hat{\mathbf{A}} \times \hat{\mathbf{D}}$ because $\hat{\mathbf{A}} \times \hat{\mathbf{D}}$ is not Hermitian:

$$(\hat{\mathbf{A}} \times \hat{\mathbf{D}})^\dagger = -\hat{\mathbf{D}} \times \hat{\mathbf{A}} \neq \hat{\mathbf{A}} \times \hat{\mathbf{D}}.$$

The standard procedure for turning a non-Hermitian operator into a Hermitian one is by symmetrizing. That is, if $\hat{G}^\dagger \neq \hat{G}$, then

$$\hat{G} \to \hat{G}' = (\hat{G}')^\dagger,$$

with

$$\hat{G}' = \tfrac{1}{2}(\hat{G} + \hat{G}^\dagger).$$

Hence, the Hermitian version of $\mathbf{a} \times \mathbf{d}$ is

$$\mathbf{a} \times \mathbf{d} \to \tfrac{1}{2}(\hat{\mathbf{A}} \times \hat{\mathbf{D}} - \hat{\mathbf{D}} \times \hat{\mathbf{A}}),$$

from which it follows that

$$\mathbf{v} \times \mathbf{B} \to \tfrac{1}{2}(\hat{\mathbf{v}}^{(A)} \times \mathbf{B} - \mathbf{B} \times \hat{\mathbf{v}}^{(A)}). \tag{12.55}$$

Note that $\mathbf{B} = \nabla \times \mathbf{A}$ does not become an operator and that $\hat{\mathbf{v}}$ is necessarily replaced by $\hat{\mathbf{v}}^{(A)}$.

From the preceding analyses we are able to construct a quantal version of $\mathbf{F}_\mathrm{L}$:

$$\begin{aligned}\mathbf{F}_\mathrm{L} \to \hat{\mathbf{F}}_\mathrm{L} &= q[\mathcal{E} + \tfrac{1}{2}\kappa_\mathrm{c}(\hat{\mathbf{v}}^{(A)} \times \mathbf{B} - \mathbf{B} \times \hat{\mathbf{v}}^{(A)})] \\ &= q\left(\mathcal{E} + \frac{\kappa_\mathrm{c}}{2\mathrm{m}}(\hat{\mathbf{\Pi}} \times \mathbf{B} - \mathbf{B} \times \hat{\mathbf{\Pi}})\right).\end{aligned} \tag{12.56}$$

It remains to be seen whether this is a consistent definition of $\hat{\mathbf{F}}_\mathrm{L}$.

Consistency is checked by determining whether $d\langle\hat{\mathbf{\Pi}}\rangle_\Gamma(t)/dt$ equals $\langle\hat{\mathbf{F}}_\mathrm{L} + \mathbf{F}\rangle_\Gamma(t)$, with $\hat{\mathbf{F}}_\mathrm{L}$ given by (12.56) and $\mathbf{F} = -\nabla_\mathbf{r} V$. Employing (8.4), the time derivative becomes

$$\frac{d\langle\hat{\mathbf{\Pi}}\rangle_\Gamma(t)}{dt} = \left\langle \frac{1}{i\hbar}[\hat{\mathbf{\Pi}}, \hat{H}^{(A)}] + \frac{\partial \hat{\mathbf{\Pi}}}{\partial t}\right\rangle_\Gamma(t)$$

$$= \frac{1}{2i\hbar m}\langle[\hat{\mathbf{\Pi}}, \hat{\Pi}^2]\rangle_\Gamma(t) + \left\langle \frac{1}{i\hbar}[\hat{\mathbf{\Pi}}, q\Phi + V] - \kappa_c q\frac{\partial \mathbf{A}}{\partial t}\right\rangle_\Gamma(t). \tag{12.57}$$

Evaluation of the RHS of (12.57) makes use of the commutation relation (12.49c). One easily finds that

$$[\hat{\mathbf{\Pi}}, \hat{\Pi}^2] = [\hat{\mathbf{\Pi}}, \hat{\mathbf{\Pi}}\cdot\hat{\mathbf{\Pi}}]$$
$$= i\hbar q\kappa_c(\hat{\mathbf{\Pi}}\times\mathbf{B} - \mathbf{B}\times\hat{\mathbf{\Pi}}),$$

and

$$[\hat{\mathbf{\Pi}}, q\Phi + V] = -i\hbar(q\,\nabla\Phi + \nabla V).$$

In view of Eq. (12.2b), the preceding results transform Eq. (12.57) into

$$\frac{d\langle\hat{\mathbf{\Pi}}\rangle_\Gamma(t)}{dt} = \langle\hat{\mathbf{F}}_L + \mathbf{F}\rangle_\Gamma(t), \tag{12.58}$$

where $\mathbf{F} = -\nabla_{\mathbf{r}}V$ and $\hat{\mathbf{F}}_L$ is given by Eq. (12.56). The consistency we have sought is hereby demonstrated; it relies on the proper (consistent) definition of the velocity operator $\hat{\mathbf{v}}^{(A)}$ of Eq. (12.57).

The equation of continuity

Given the changes in the details of Ehrenfest's theorem when $\hat{H}\to\hat{H}^{(A)}$, the (relatively mild) change derived below in the quantal equation of continuity should be no surprise.

The derivation proceeds by multiplying both sides of (12.6) from the left by $\Psi^{(A)*}(\mathbf{r}, t)$, forming the Hermitian conjugate of the resulting equation, and then subtracting one of these two equations from the other – i.e., by following the same procedure as that which led to Eq. (7.58) we find the analogs of (7.65) and (7.66) to be

$$\frac{\partial\rho^{(A)}(\mathbf{r}, t)}{\partial t} + \nabla\cdot\mathbf{J}^{(A)}(\mathbf{r}, t) = 0, \tag{12.59}$$

where

$$\rho^{(A)}(\mathbf{r}, t) = |\Psi^{(A)}(\mathbf{r}, t)|^2 \tag{12.60}$$

and

$$\mathbf{J}^{(A)}(\mathbf{r}, t) = \frac{\hbar}{2mi}\Big(\Psi^{(A)*}(\mathbf{r}, t)\,\nabla\Psi^{(A)}(\mathbf{r}, t)$$
$$-\{\nabla\Psi^{(A)*}(\mathbf{r}, t)\}\Psi^{(A)}(\mathbf{r}, t) - \frac{2i}{\hbar}q\kappa_c\mathbf{A}\rho^{(A)}(\mathbf{r}, t)\Big), \tag{12.61}$$

which, since $\mathbf{A}$ is real, can be re-written as

$$\mathbf{J}^{(A)}(\mathbf{r}, t) = \mathrm{Im}\left\{\Psi^{(A)*}(\mathbf{r}, t)\left[\frac{1}{m}(\hat{\mathbf{P}}(\mathbf{r}) - iq\kappa_c\mathbf{A}(\mathbf{r}, t))\right]\Psi^{(A)}(\mathbf{r}, t)\right\}. \tag{12.62}$$

The direct effect of the external electromagnetic fields on the form of the continuity equation is through an additional term that is linear in the vector potential. The fields also (obviously) act indirectly, through the wave function $\Psi^{(A)}(\mathbf{r}, t)$. This direct effect of $\mathbf{A}$, given by $q\kappa_c\mathbf{A}\rho^{(A)}$, has the proper dimensions, as the reader can readily verify. Magnetic fields manifest themselves in a variety of ways in quantum mechanics and the term $q\kappa_c\mathbf{A}\rho^{(A)}$ may be thought of as the origin of them. Both $\rho^{(A)}$ and $\mathbf{J}^{(A)}(\mathbf{r}, t)$ will be shown to be invariant under gauge transformations, as is Ehrenfest's theorem.

12.5. Gauge invariance

Classical electrodynamics is invariant under the gauge transformation (12.4) because the relevant equations involve derivatives of the potentials $\mathbf{A}$ and Φ. An analogous invariance is desirable in quantum mechanics, but, because derivatives of the quantal Hamiltonian do not enter the time-dependent Schrödinger equation, gauge invariance in quantum theory will involve more than the replacements of Eqs. (12.4). We shall require that, under a gauge transformation of the first kind – Eqs. (12.4) – the time-dependent Schrödinger equation be unchanged. To ensure this requires transforming the wave function by multiplying it by a space- and time-dependent phase factor, a change known as a gauge transformation of the second kind. As we shall see, the effect of these two gauge transformations will be to leave observables invariant as long as two conditions are met: first, that "observables" involve gauge-invariant quantities – for example, $\hat{\mathbf{\Pi}}$ in place of $\hat{\mathbf{P}}$ – and second, that the spatial domains be singly connected. The Aharonov–Bohm effect, which is discussed in Section 12.7, occurs in a multiply connected domain. One must be careful.

Gauge invariance: general features

Under a gauge transformation of the first kind, $\mathbf{A}$ and Φ are replaced by

$$\mathbf{A}' = \mathbf{A} + \nabla\chi$$

and

$$\Phi' = \Phi - \kappa_c \frac{\partial\chi}{\partial t},$$

with $\chi = \chi(\mathbf{r}, t)$. The single-particle Hamiltonian $\hat{H}^{(A)}(\mathbf{r}, t)$ of (12.7) becomes $\hat{H}^{(A')}(\mathbf{r}, t)$, given by

$$\begin{aligned}\hat{H}^{(A')}(\mathbf{r}, t) &= \frac{1}{2m}[\hat{\mathbf{P}}(\mathbf{r}) - q\kappa_c\mathbf{A}(\mathbf{r}, t) - q\kappa_c\nabla\chi]^2 + q\Phi(\mathbf{r}, t) - q\kappa_c\frac{\partial\chi}{\partial t} + V(\mathbf{r})\\ &\equiv \frac{1}{2m}\hat{\Pi}'^2 + q\Phi(\mathbf{r}, t) - q\kappa_c\frac{\partial\chi}{\partial t} + V(\mathbf{r}).\end{aligned} \tag{12.63}$$

Replacing $\hat{H}^{(A)}$ by $\hat{H}^{(A')}$ of (12.63) obviously does not leave the Schrödinger equation (12.6) invariant. To achieve this invariance requires altering $\Psi^{(A)}(\mathbf{r}, t)$ in such a way as to cancel out the changes induced by the gauge transformation of the first kind. The desired result will occur if, when $\mathbf{A} \to \mathbf{A}'$ and $\Phi \to \Phi'$, $\Psi^{(A)} \to \Psi^{(A')}$ such that

$$\Psi^{(A')}(\mathbf{r},\, t) = \exp\left[i\left(\frac{q\kappa_c}{\hbar}\right)\chi(\mathbf{r},\, t)\right]\Psi^{(A)}(\mathbf{r},\, t). \tag{12.64}$$

Equation (12.64) is known as a gauge transformation of the second kind. It is a *local* transformation because χ depends on $\mathbf{r}$ and t, as opposed to a *global* transformation, for which χ is a constant for all $\mathbf{r}$ and t. Local gauge transformations play a fundamental role in elementary-particle physics, where so-called "gauge" theories are paramount. We shall return to the locality aspect shortly. For the present, we concentrate on the gauge-invariance feature, viz., that the Schrödinger equation

$$i\hbar\,\frac{\partial\Psi^{(A')}(\mathbf{r},\, t)}{\partial t} = \hat{H}^{(A')}(\mathbf{r},\, t)\Psi^{(A')}(\mathbf{r},\, t). \tag{12.65}$$

is actually identical to Eq. (12.6).

The proof of this claim is straightforward. On the one hand

$$i\hbar\,\frac{\partial}{\partial t}\Psi^{(A')} = -q\kappa_c\frac{\partial\chi}{\partial t}\Psi^{(A')} + i\hbar e^{i(q\kappa_c\hbar)\chi}\,\frac{\partial\Psi^{(A)}}{\partial t}, \tag{12.66}$$

while on the other hand

$$\begin{aligned}\hat{\mathbf{\Pi}}'\Psi^{(A')} &= [\hat{\mathbf{P}} - q\kappa_c\mathbf{A} - q\kappa_c(\nabla\chi)]\Psi^{(A')}\\ &= e^{i(q\kappa_c/\hbar)\chi}\hat{\mathbf{\Pi}}\Psi^{(A)},\end{aligned} \tag{12.67}$$

so that

$$\hat{\Pi}'^2\Psi^{(A')} = e^{i(q\kappa_c/\hbar)\chi}\hat{\Pi}^2\Psi^{(A)}, \tag{12.68}$$

where the $\mathbf{r}$ and t dependences are suppressed. On substituting Eqs. (12.63), (12.68), and (12.66) into (12.65), one easily finds

$$e^{i(q\kappa_c/\hbar)\chi}\left(i\hbar\,\frac{\partial}{\partial t}\Psi^{(A)}\right) = e^{i(q\kappa_c/\hbar)\chi}[\hat{H}^{(A)}\Psi^{(A)}];$$

cancelling out the multiplicative gauge factor $\exp[i(q\kappa_c/\hbar)\chi]$ in the last relation yields Eq. (12.6), thereby proving our claim.

Let us consider some other aspects of gauge invariance. For example, the probability density $\rho^{(A)}(\mathbf{r},\, t)$ remains invariant under a gauge transformation:

$$\rho^{(A)}(\mathbf{r},\, t) = |\Psi^{(A)}(\mathbf{r},\, t)|^2 \to \rho^{(A')}(\mathbf{r},\, t) = |\Psi^{(A')}(\mathbf{r},\, t)|^2 = |\Psi^{(A)}(\mathbf{r},\, t)|^2,$$

due to the multiplicative nature of the gauge factor. The same reasoning applies to Eq. (12.62) and to $\hat{\mathbf{\Pi}}$, for example

$$\langle\Psi^{(A')}|\hat{\mathbf{\Pi}}'|\Psi^{(A')}\rangle = \langle\Psi^{(A)}|\hat{\mathbf{\Pi}}|\Psi^{(A)}\rangle.$$

We see that $\hat{\mathbf{\Pi}}$, *not* $\hat{\mathbf{P}}$, has a gauge-invariant expectation value. Similarly, it is

$$\hat{\mathbf{L}}^{(A)} = \mathbf{r}\times\hat{\mathbf{\Pi}},$$

not $\hat{\mathbf{L}} = \mathbf{r}\times\hat{\mathbf{P}}$, whose matrix elements are gauge invariant.[6] Notice, however, that, if $\Phi \neq 0$, $\hat{H}^{(A)}$ and $\hat{H}^{(A')}$ lead to different eigenvalue problems because the term

[6] The classical analog is the angular momentum $\mathbf{L}^{(A)} = \mathbf{r}\times\mathbf{\Pi}$, rather than $\mathbf{L} = \mathbf{r}\times\mathbf{p}$.

$(-q\kappa_c)(\partial\chi/\partial t)\Psi^{(A')}$ does not cancel out. On the other hand, if $\chi \neq \chi(t)$, the latter term is zero, in which case $\hat{H}^{(A)}$ and $\hat{H}^{(A')}$ will yield the *same* energy spectrum but $\Psi^{(A)}$ and $\Psi^{(A')}$ corresponding to the same eigenvalue need *not* be the same (to within a phase factor) unless the state is non-degenerate. This point is explored in the exercises.

*Locality of gauge invariance and minimal coupling

As stressed in earlier portions of this book, the presence of an overall, constant phase factor multiplying a state vector or wave function cannot influence the results of measurements: only relative phases have measurable effects. This may be formalized as an invariance principle.

> All quantal systems are invariant w.r.t. a *global* phase or gauge transformation. For a spinless single-particle system this means that, if its wave function $\Psi(\mathbf{r}, t)$ obeys the time-dependent Schrödinger equation
>
> $$\left(i\hbar\frac{\partial}{\partial t} - \frac{1}{2m}\hat{P}^2(\mathbf{r}) - \hat{V}(\mathbf{r})\right)\Psi(\mathbf{r}, t) = 0, \tag{12.69}$$
>
> then so does $\exp(i\zeta)\Psi(\mathbf{r}, t)$, where ζ is a constant.

The transformation is global because ζ is a constant, and it is obvious that the invariance would be destroyed if ζ were to become a (local) function of $\mathbf{r}$ and/or t.

One of the beauties of the electromagnetic interaction is that, if the particle is charged, then by *requiring* invariance w.r.t. *local* gauge transformations, one can *derive* the general *form* of the electromagnetic interactions that occur in $\hat{H}^{(A)}(\mathbf{r}, t)$. To show this, we assume that the particle has charge q and also (incorrectly) that it obeys Eq. (12.69). Let $\Lambda(q, \mathbf{r}, t)$ be the gauge function and Ψ' be the gauge-transformed wave function:

$$\Psi(\mathbf{r}, t) \to \Psi'(\mathbf{r}, t) = e^{i\Lambda(q,\mathbf{r},t)}\Psi(\mathbf{r}, t). \tag{12.70}$$

Then, if (12.69) is the correct equation for $\Psi(\mathbf{r}, t)$, gauge invariance should mean that $\Psi'(\mathbf{r}, t)$ also obeys it: it is, after all, the only equation available at this point in our development. Substituting (12.70) into (12.69) and employing the results (suppressing the q, $\mathbf{r}$, and t dependences)

$$\frac{\partial\Psi'}{\partial t} = e^{i\Lambda}\left[\left(i\frac{\partial\Lambda}{\partial t}\right) + \frac{\partial}{\partial t}\right]\Psi \tag{12.71a}$$

and

$$\hat{P}^2\Psi' = e^{i\Lambda}[\hat{\mathbf{P}} + \hbar(\nabla\Lambda)]^2\Psi, \tag{12.71b}$$

we find

$$\left(i\hbar\frac{\partial}{\partial t} - \frac{1}{2m}[\hat{\mathbf{P}}(\mathbf{r}) + \hbar(\nabla\Lambda(q, \mathbf{r}, t))]^2 - V(\mathbf{r}) - \hbar\frac{\partial\Lambda(q, \mathbf{r}, t)}{\partial t}\right)\Psi(\mathbf{r}, t) = 0. \tag{12.71c}$$

This result is clearly not Eq. (12.69): the effect of the gauge transformation (12.70) has been to add $i(\partial\Lambda/\partial t)$ to the operator $\partial/\partial t$ and the term $-\hbar(\nabla\Lambda)$ to the operator $\hat{\mathbf{P}}(\mathbf{r})$ in the resulting equation for the original wave function $\Psi(\mathbf{r}, t)$. Gauge invariance, although desired, has failed to occur. Since we are demanding gauge invariance for a charged-particle system, the least complicated means for ensuring it is to modify the Hamiltonian $\hat{P}^2/(2m) + V$ of Eq. (12.69) by adding to it special terms. These must be terms that themselves will change under a gauge transformation in such a way as to cancel out the $i(\partial\Lambda/\partial t)$ and $-i\hbar(\nabla\Lambda)$ terms that the factor $\exp(i\Lambda)$ supplies.

We temporarily denote the new quantities to be added to the Hamiltonian portion of Eq. (12.69) by $\hat{\mathbf{W}}_1(q, \mathbf{r}, t)$ and $\hat{W}_2(q, \mathbf{r}, t)$, where, at least for the present, $\hat{\mathbf{W}}_1$ and $\hat{W}_2$ are arbitrary operator functions of q, $\mathbf{r}$, and t. To satisfy the conditions stated at the end of the preceding paragraph, we replace $\hat{\mathbf{P}}$ and $\hat{V}$ by

$$\hat{\mathbf{P}}(\mathbf{r}) \to \hat{\mathbf{P}}(\mathbf{r}) - \hat{\mathbf{W}}_1(q, \mathbf{r}, t) \tag{12.72a}$$

and

$$\hat{V}(\mathbf{r}) \to \hat{V}(\mathbf{r}) + \hat{W}_2(q, \mathbf{r}, t), \tag{12.72b}$$

in which case (12.69) becomes

$$\left(i\hbar\frac{\partial}{\partial t} - \frac{1}{2m}[\hat{\mathbf{P}}(\mathbf{r}) - \hat{\mathbf{W}}_1(q, \mathbf{r}, t)]^2 - [\hat{V}(\mathbf{r}) + \hat{W}_2(q, \mathbf{r}, t)]\right)\Psi(\mathbf{r}, t) = 0. \tag{12.73}$$

In order that the new form (12.73) of the time-dependent Schrödinger equation be invariant under the gauge transformation (12.70), we must additionally require that the new quantities $\hat{\mathbf{W}}_1$ and $\hat{W}_2$ transform as follows:

$$\hat{\mathbf{W}}_1(q, \mathbf{r}, t) \to \hat{\mathbf{W}}_1(q, \mathbf{r}, t) + \hbar\,\nabla\Lambda(q, \mathbf{r}, t) \tag{12.74a}$$

and

$$\hat{W}_2(q, \mathbf{r}, t) \to \hat{W}_2(q, \mathbf{r}, t) - \hbar\,\frac{\partial\Lambda(q, \mathbf{r}, t)}{\partial t}. \tag{12.74b}$$

Under the combined pair of gauge transformations (12.74) and (12.70), invariance of Eq. (12.73) is guaranteed.

Invariance of the theory under a local gauge transformation thus means more than altering $\Psi(\mathbf{r}, t)$ as in Eq. (12.70), it also requires that $\hat{\mathbf{P}}$ and $\hat{V}$ undergo the changes of Eq. (12.72) leading to Eq. (12.73), AND that the new quantities $\hat{\mathbf{W}}_1$ and $\hat{W}_2$ transform via Eqs. (12.74) when Ψ undergoes the transformation (12.70). Equation (12.73) has the same general structure as Eq. (12.6) with $\hat{H}^{(A)}$ given by Eq. (12.7). This structure follows from the requirement that, for a charged particle, the correct time-dependent Schrödinger equation be gauge-invariant. Since these results apply to the charged-particle case, then, to be consistent with the results of our semi-classical quantization of $H_{\mathrm{cl}}^{(A)}$, we drop the operator character and write

$$\hat{\mathbf{W}}_1 = q\kappa_{\mathrm{c}}\mathbf{A}(\mathbf{r}, t) \tag{12.75a}$$

and

$$\hbar\,\frac{\partial W_2}{\partial t} = q\Phi(\mathbf{r}, t); \tag{12.75b}$$

we also re-write $\Lambda(q, \mathbf{r}, t)$ as

$$\Lambda \to \frac{q\kappa_c}{\hbar}\chi(\mathbf{r},\, t). \tag{12.75c}$$

In all three cases the proportionality to q is manifest.

By comparison with classical physics, it is clear that the "new" quantities $\mathbf{A}$, Φ, and χ of Eqs. (12.75) are just the old ones, which we have encountered already. With them, the transformations corresponding to a charged particle and classical fields are

$$\hat{\mathbf{P}} \to \hat{\mathbf{P}} - q\kappa_c\mathbf{A}$$

and

$$V \to V + q\Phi$$

with

$$\mathbf{A} \to \mathbf{A} + \nabla\chi,$$

$$\Phi \to \Phi - q\kappa_c\frac{\partial\chi}{\partial t},$$

and

$$\Psi \to e^{i(q\kappa_c/\hbar)\chi}\Psi.$$

Thus, starting with the concept of invariance under a local gauge transformation, we have derived the quantal operators and their transformation properties governing the behavior of a spinless charged particle. This derivation embodies what is known as the principle of minimal electromagnetic coupling, i.e., the form $\hat{\mathbf{P}} - \hat{\mathbf{W}}_1(q,\, \mathbf{r},\, t)$ is the "minimal" one that local gauge invariance requires. The extension of the concept of invariance under a local gauge transformation is the foundation for much of modern elementary-particle theory, although a treatment of this particular topic is well beyond the scope of this book.

12.6. Gauge transformations: the spatially uniform electric field

In Section 12.3 we determined the behavior of an initially free charged particle in a constant electric field $\mathcal{E}$ that was turned on at time $t = 0$. The Hamiltonian was expressed in terms of the scalar potential Φ; $\mathbf{A}$ was zero, so that $\hat{\mathbf{\Pi}}$ became $\hat{\mathbf{P}}$. Use of the Heisenberg equations of motion obviated the need to determine the time-evolution operator directly. In the present section, we return to a version of this problem, treating it by means of a gauge transformation. We thereby illustrate the ideas of the preceding section. The problem we analyze is that of a spatially uniform, time-dependent electric field $\mathcal{E}(t)$; after deriving general expressions involving $\mathcal{E}(t)$, we shall work with a specific form for it. As in Section 12.3, our analysis follows that of Fallieros and Friar (1982).

There are two extreme ways of treating this system:

$$\text{Method (i): } \mathbf{A} = 0, \qquad \Phi(\mathbf{r},\, t) = -\mathbf{r}\cdot\mathcal{E}(t),$$

$$\hat{H}^{(\mathcal{E})}(\mathbf{r},\, t) = \frac{1}{2m}\hat{P}^2(\mathbf{r}) - q\mathbf{r}\cdot\mathcal{E}(t); \tag{12.76}$$

$$\text{Method (ii): } \Phi = 0, \qquad \mathcal{E}(\mathbf{t}) = -\kappa_c \frac{\partial \mathbf{A}(t)}{\partial t},$$

$$\hat{H}^{(A)}(\mathbf{r},\, t) = \frac{1}{2m}\hat{\Pi}^2(\mathbf{r},\, t) = \frac{1}{2m}[\hat{\mathbf{P}}(\mathbf{r}) - q\kappa_c \mathbf{A}(t)]^2, \tag{12.77}$$

where an $\mathbf{A}(t)$ suitable for our purposes is

$$\mathbf{A}(t) = -\frac{1}{\kappa_c}\int_0^t dt'\, \mathcal{E}(t'). \tag{12.78}$$

The electromagnetic potentials of Methods (i) and (ii) must be related by a gauge transformation. Given the form (12.78) of $\mathbf{A}(t)$, an appropriate gauge function $\chi(\mathbf{r},\, t)$ is

$$\begin{aligned} \chi(\mathbf{r},\, t) &= -\frac{1}{\kappa_c}\int_0^t dt'\, \mathbf{r}\cdot\mathcal{E}(t') \\ &= \frac{1}{\kappa_c}\int_0^t dt'\, \Phi(\mathbf{r},\, t'), \end{aligned} \tag{12.79}$$

so that

$$\mathbf{A}(t) = -\nabla\chi(\mathbf{r},\, t). \tag{12.80}$$

Equivalent descriptions of this system (a charged particle in $\mathcal{E}(t)$) are obtained from

$$i\hbar\, \frac{\partial \Psi^{(\mathcal{E})}(\mathbf{r},\, t)}{\partial t} = \hat{H}^{(\mathcal{E})}(\mathbf{r},\, t)\Psi^{(\mathcal{E})}(\mathbf{r},\, t) \tag{12.81}$$

and

$$i\hbar\, \frac{\partial \Psi^{(A)}(\mathbf{r},\, t)}{\partial t} = \hat{H}^{(A)}(\mathbf{r},\, t)\Psi^{(A)}(\mathbf{r},\, t), \tag{12.82}$$

where,[7] from the analysis of Section 12.5,

$$\Psi^{(A)}(\mathbf{r},\, t) = e^{i(q\kappa_c/\hbar)\chi(\mathbf{r},t)}\Psi^{(\mathcal{E})}(\mathbf{r},\, t). \tag{12.83}$$

As a final point before choosing a specific form for $\mathcal{E}(t)$, we remark that

$$[\hat{H}^{(j)}(\mathbf{r},\, t'),\, \hat{H}^{(j)}(\mathbf{r},\, t)] = 0, \qquad t' \neq t,\, j = \mathcal{E} \text{ or } A.$$

Hence, the time evolution operators are as stated in Eq. (5.90):

$$\hat{U}^{(j)}(t,\, t_0 = 0) = e^{-(i/\hbar)\int_0^t dt'\, \hat{H}^{(j)}(\mathbf{r},t')}, \qquad j = \mathcal{E} \text{ or } A, \tag{12.84}$$

from which we find

$$\Psi^{(j)}(\mathbf{r},\, t) = e^{-(i/\hbar)\int_0^t dt'\, \hat{H}^{(j)}(\mathbf{r},t')}\Psi^{(j)}(\mathbf{r},\, t = 0), \qquad j = \mathcal{E} \text{ or } A. \tag{12.85}$$

Since $\chi(\mathbf{r},\, t = 0) = 0$, the two initial states are equal, and we choose it to be a plane wave, just as in Section 12.3:

$$\Psi^{(\mathcal{E})}(\mathbf{r},\, t = 0) = \langle \mathbf{r}|\mathbf{k}\rangle = \Psi^{(A)}(\mathbf{r},\, t = 0). \tag{12.86}$$

[7] The reader is strongly urged to verify in detail that Eqs. (12.81) and (12.82) do yield, via (12.83), equivalent descriptions. In so doing, the validity of (12.79) as an appropriate gauge function will also be proved.

Although the initial states are the same, the wave functions for later times are not, owing to the phase factor in (12.83). We shall calculate $\Psi^{(A)}(\mathbf{r},\, t)$ using Eq. (12.85), and then use (12.83) to get $\Psi^{(\mathcal{E})}(\mathbf{r},\, t)$.

A harmonic time dependence

$$\mathcal{E}(t) = \mathcal{E}\cos(\omega t) \tag{12.87}$$

is chosen for $\mathcal{E}(t)$, leading to

$$\mathbf{A}(t) = -\frac{1}{\omega\kappa_c}\mathcal{E}\sin(\omega t), \tag{12.88a}$$

$$\chi(\mathbf{r},\, t) = -\frac{1}{\omega\kappa_c}\mathbf{r}\cdot\mathcal{E}\sin(\omega t) \tag{12.88b}$$

and

$$\int_0^t dt'\, \hat{H}^{(A)}(t') = \frac{1}{2m}\hat{P}^2(\mathbf{r})t + \frac{q}{m\omega^2}\mathcal{E}\cdot\hat{\mathbf{P}}[1-\cos(\omega t)] + \frac{q^2\mathcal{E}^2}{8m\omega^3}[2\omega t - \sin(2\omega t)]. \tag{12.88c}$$

The choice (12.86) permits a check on the calculation: the results obtained should go over to those of Section 12.3 when $\omega \to 0$, since in that limit $\mathcal{E}(t) \to \mathcal{E}$.

On setting $j = \mathcal{E}$ in Eq. (12.84) and making use of Eq. (12.88c), $\Psi^{(A)}(\mathbf{r},\, t)$ becomes

$$\Psi^{(A)}(\mathbf{r},\, t) = \left(\frac{1}{2\pi}\right)^{3/2} e^{i\beta_\omega(t)} e^{i\mathbf{k}\cdot\mathbf{r}}, \tag{12.89}$$

where

$$-\beta_\omega(t) = \frac{\hbar k^2 t}{2m} + \frac{q}{m\omega^2}\mathcal{E}\cdot\mathbf{k}[1-\cos(\omega t)] + \frac{q^2\mathcal{E}^2}{8m\hbar\omega^3}[2\omega t - \sin(2\omega t)]. \tag{12.90}$$

The only effect of the vector potential is to produce an $\mathbf{r}$-independent, time-dependent phase factor: the initial plane wave is unchanged. In contrast, the solution $\Psi^{(\mathcal{E})}$, which is obtained from $\Psi^{(A)}$ through multiplication by $\exp(-q\kappa_c\chi(\mathbf{r},\, t)/\hbar)$, gains an $\mathbf{r}$-dependent phase factor:

$$\Psi^{(\mathcal{E})}(\mathbf{r},\, t) = e^{[q/(\hbar\omega)]\mathbf{r}\cdot\mathcal{E}\sin(\omega t)}\Psi^{(A)}(\mathbf{r},\, t) = \left(\frac{1}{2\pi}\right)^{3/2} e^{i\beta_\omega(t)} e^{i\mathbf{k}_\omega(t)\cdot\mathbf{r}}, \tag{12.91}$$

where

$$\mathbf{k}_\omega(t) = \mathbf{k} + \mathcal{E}\frac{q\sin(\omega t)}{\hbar\omega}. \tag{12.92}$$

The quantities $\beta_\omega(t)$ and $\mathbf{k}_\omega$ are the ω-dependent analogs of $\beta(t)$ and $\mathbf{k}(t)$ introduced in Section 12.3. Since $\beta_\omega(t)$ appears in a phase factor, it might be thought that the gauge term $\exp[iq\mathbf{r}\cdot\mathcal{E}\sin(\omega t)/(\hbar\omega)]$ could just as well be lumped with $\exp[i\beta_\omega(t)]$. The reason we do *not* do this is that the r dependence makes the ω- and t-dependent wave vector $\mathbf{k}_\omega(t)$ the natural quantity rather than just $\mathbf{k}$ alone. Furthermore, $\mathbf{k}_\omega(t)$ is "natural" because $\Psi^{(\mathcal{E})}(\mathbf{r}, t)$ is an eigenfunction of the momentum operator $\hat{\mathbf{P}}(\mathbf{r})$ belonging to momentum $\hbar\mathbf{k}_\omega(t)$, not to $\hbar\mathbf{k}$. The analogy with Section 12.3 is evident.

On the other hand, it may seem that some physics would be lost were we to use $\Psi^{(A)}(\mathbf{r}, t)$ in place of $\Psi^{(\mathcal{E})}(\mathbf{r}, t)$: $\Psi^{(A)}$ is an eigenfunction of $\hat{\mathbf{P}}(\mathbf{r})$ whose eigenvalue is $\mathbf{k}$, not $\mathbf{k}_\omega(t)$. In fact, no physics would be lost, since, when $\mathbf{A} \neq 0$, the kinetic momentum operator $\hat{\mathbf{\Pi}}(\mathbf{r}, t) = \hat{\mathbf{P}}(\mathbf{r}) - q\kappa_c\mathbf{A}(t)$ rather than the canonical momentum $\hat{\mathbf{P}}(\mathbf{r})$ is appropriate. One readily finds that

$$\hat{\mathbf{\Pi}}(\mathbf{r}, t)\Psi^{(A)}(\mathbf{r}, t) = \hbar\mathbf{k}_\omega(t)\Psi^{(A)}(\mathbf{r}, t):$$

$\Psi^{(A)}$, like $\Psi^{(\mathcal{E})}$, *is* an eigenfunction of the momentum operator with eigenvalue $\hbar\mathbf{k}_\omega(t)$. This result makes it clear that caution is called for when one is dealing with non-zero vector potentials.

Several questions remain to be answered, e.g., does the $\omega \to 0$ limit of $\beta_\omega(t)$ or $\mathbf{k}_\omega(t)$ yield the results of Section 12.3, and does $\hat{U}^{(\mathcal{E})}\langle\mathbf{r}|\mathbf{k}\rangle$ produce the $\Psi^{(\mathcal{E})}(\mathbf{r}, t)$ of Eq. (12.91)? The providing of the answers to these questions is left to the exercises.

12.7. *The Aharonov–Bohm effect

In classical physics, gauge invariance is a means of expressing the fact that measurable results are independent of the choices of $\mathbf{A}$ and Φ, as long as they produce the same $\mathcal{E}$ and $\mathbf{B}$. To repeat an earlier comment, $\mathbf{A}$ and Φ are artifacts; only $\mathcal{E}$ and $\mathbf{B}$ have physical meaning. If there is no electric/magnetic field acting on the charged particle, then neither $\mathbf{A}$ nor Φ plays any role in the dynamics. That this characteristic applies to the gauge-invariant formulation of quantum mechanics seems obvious, especially in view of the discussion in the preceding two sections: as long as gauge-proper quantities are employed, the quantal description is independent of the gauge used. In particular, zero $\mathcal{E}/\mathbf{B}$ means constant $\Phi/\mathbf{A}$, whose effect is that of a global, i.e., of an unobservable, phase factor.

This viewpoint tended to be gospel until 1959, when Aharonov and Bohm (1959) published a paper pointing out that, if the physical domain was multiply connected, then the presence of a magnetic field in a region inaccessible to a charged particle could produce, via $\mathbf{A}$, an observable interference effect. Although the effect predicted by Aharonov and Bohm (A–B) had been noted ten years earlier by Ehrenburg and Siday (1949), it was the A–B paper that awakened the world of physics to the possibility that electromagnetic *potentials* could influence the outcome of experiments. It also engendered a controversy about the meaning and existence of the effect that was settled – in favor of the effect – only some 25 years later. We shall outline the underlying ideas of the A–B effect in this section; readers interested in more details and a summary of developments, including the controversy, are referred to the monograph of Peshkin and Tonomura (1989), and especially to the experimental work of the latter author and collaborators (cited in the monograph) that decisively established the validity of the A–B effect.

As pertinent background, let us recall the two-slit interference effect. A beam of electrons, generated coherently, falls on a screen S_1 that has two slits separated by a distance d; the electrons are detected on screen S_2, a large distance R from S_1. The experimental arrangement is depicted in Fig. 12.3. Since electrons can reach a point P on

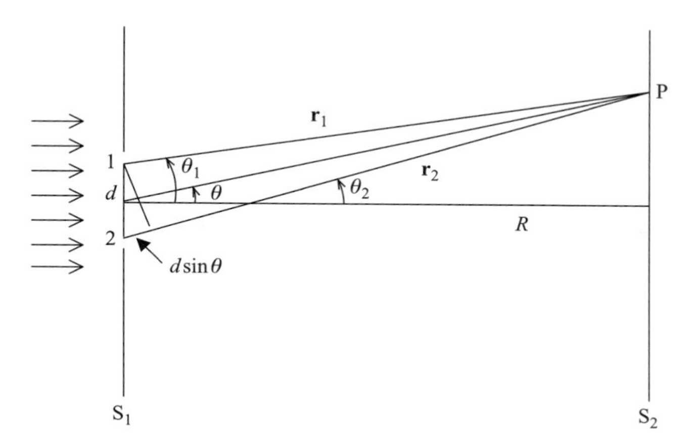

Fig 12.3 A sketch of the experimental arrangement for the two-slit interference experiment. Screen S_1 has the two slits; screen S_2 functions as the detector. Since the distances R, r_1, and r_2 are macroscopic (effectively asymptotic), the difference in path length $r_2 - r_1 \cong d \sin\theta$. Interference minima (and maxima) require non-zero values of θ.

screen S_2 by passing through either slit, the total wave function $\Psi(P,\, t)$ for the electron to be at P is a sum of the wave functions $\Psi_1(\mathbf{r}_1,\, t)$ and $\Psi_2(\mathbf{r}_2,\, t)$, corresponding to passage through slits 1 and 2, respectively:

$$\Psi(P,\, t) = \Psi_1(\mathbf{r}_1,\, t) + \Psi_2(\mathbf{r}_2,\, t), \tag{12.93}$$

where $\mathbf{r}_1$ and $\mathbf{r}_2$ are specified in Fig. 12.3.

The probability density $\rho(P,\, t)$ that an electron is at P is

$$\rho(P,\, t) = |\Psi|^2 = |\Psi_1 + \Psi_2|^2;$$

it contains the expected interference term, here $2\,\mathrm{Re}\,[\Psi_1^*\Psi_2]$. To evaluate the latter term, we assume the validity of a plane-wave-function description (as discussed in Section 7 for 1-D scattering), which yields[8] (see Chapter 15)

$$\Psi_j(\mathbf{r}_j,\, t) \cong f(\theta)\frac{e^{i(kr_j - E_k t/\hbar)}}{r}, \tag{12.94}$$

where $f(\theta)$ is the so-called scattering amplitude and $E_k = \hbar^2 k^2/(2m_\mathrm{e})$ is the incident energy. From Eq. (12.94) the interference term is

$$\begin{aligned} 2\,\mathrm{Re}[\Psi_1^*\Psi_2] &\cong 2\frac{|f(\theta)|^2}{r^2}\cos[k(r_2 - r_1)] \\ &= 2\frac{|f(\theta)|^2}{r^2}\cos\left(\frac{2\pi d\sin\theta}{\lambda}\right), \end{aligned} \tag{12.95}$$

where $k = 2\pi/\lambda$. The interference term is the same as that which appears in Section 1.1 for the two-slit (Young's) experiment with photons, Eq. (1.7). Notice that there is no

[8] Equation (12.94) is an asymptotic expression, employing $r_j^{-1} \cong r^{-1}$ and $f(\theta_j) \cong f(\theta)$, which are large-$r$ small-angle approximations.

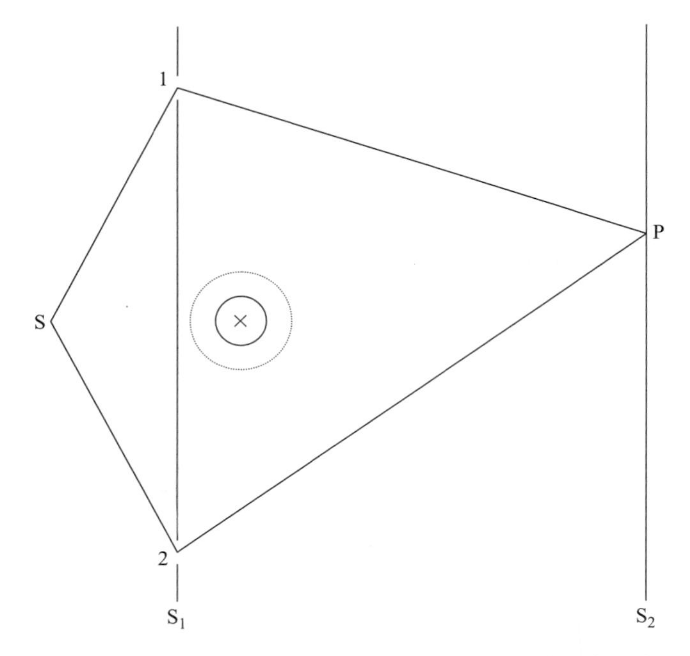

Fig 12.4 A sketch of an experimental arrangement illustrating the Aharonov–Bohm effect. An infinitely long solenoid, perpendicular to the plane of the paper, is placed between the slits on the far side of screen S_1 from the source of electrons S. A magnetic field **B** directed into the plane (symbolized by the "x") is generated by and is wholly contained within the solenoid. The inaccessibility of the solenoid to the electrons is indicated by the dotted circle. Because of this inaccessibility, the physical region is no longer singly connected; there is no magnetic field in the domain available to the electron. The path S1P is denoted ℓ_1, while the path S2P is denoted ℓ_2.

interference effect when $\theta = 0$, i.e., when $r_2 = r_1 = r$. On the other hand, in the A–B effect, examined next, there *is* a non-zero interference pattern at $\theta = 0$.

Conceptually, the difference between the standard two-slit experimental arrangement and that envisioned in the A–B case is simple: one adds, behind and between the slits, a very long solenoid that wholly contains a constant magnetic field **B** parallel to the solenoid's axis and is inaccessible to the electrons (which may go through either slit). A sketch of this set-up is shown in Fig. 12.4. (In reality, the experiments are much more complicated – see Peshkin and Tonomura (1989).) Classically, the absence of a magnetic field outside the solenoid – i.e., in the region accessible to the electrons – means that there is no influence on the electron's motion. This conclusion is independent of the fact that a vector potential extends into the exterior region,[9] since forces arise only from

[9] Continuity requries that an **A** producing the solenoidal field **B** be non-zero everywhere. One such – in the cylindrical coordinates ρ, ϕ, and z – is $A_\rho = A_z = 0$,

$$A_\phi = \begin{cases} B\rho/2 & \rho \leqslant a, \\ B\rho^2/(2a), & \rho > a, \end{cases}$$

where a is the radius of the solenoid. This **A** obeys $\nabla \cdot \mathbf{A} = 0$ and

$$\nabla \times \mathbf{A} = \begin{cases} B\mathbf{e}_z, & \rho \leqslant a, \\ 0, & \rho > a. \end{cases}$$

fields. Quantally, the conclusion is quite the opposite, as is shown next.

In the absence of the magnetic field, the Hamiltonian governing electrons in the two-slit experiment will be denoted $\hat{H}^{(0)}$:

$$\hat{H}^{(0)}(\mathbf{r}) = \frac{1}{2m_e}\hat{P}^2(\mathbf{r}) + V(\mathbf{r}), \tag{12.96}$$

as in Eq. (12.10). This is a Hamiltonian for which $\mathbf{A} = 0 = \Phi$. Corresponding to it is a wave function $\Psi^{(0)}(\mathbf{r}, t)$. When the field inside the solenoid is turned on (prior to emission of electrons from the source S), $\hat{H}^{(0)} \to \hat{H}^{(A')}$, given by

$$\hat{H}^{(A')}(\mathbf{r}) = \frac{1}{2m_e}[\mathbf{P}(\mathbf{r}) + e\kappa_c\mathbf{A}'(\mathbf{r})]^2 + V(\mathbf{r}), \tag{12.97}$$

where $\nabla \times \mathbf{A}' = \mathbf{B}$. Complementing $\hat{H}^{(A')}(\mathbf{r})$ is a wave function $\Psi^{(A')}(\mathbf{r}, t)$.

The potential is denoted $\mathbf{A}'$ as a reminder that it can be related to the previous potential $\mathbf{A}$ by a gauge transformation: $\mathbf{A}' = \mathbf{A} + \nabla\chi$. However, $\mathbf{A} = 0$, and $\mathbf{A}'$ is thus the gradient of a scalar gauge function:

$$\mathbf{A}'(\mathbf{r}) = \nabla\chi(\mathbf{r}). \tag{12.98}$$

The gauge function $\chi(\mathbf{r})$ is the link between $\Psi^{(0)}(\mathbf{r}, t)$ and $\Psi^{(A')}(\mathbf{r}, t)$, i.e.,

$$\Psi^{(A')}(\mathbf{r}, t) = e^{-i(e\kappa_c/\hbar)\chi(\mathbf{r})}\Psi^{(0)}(\mathbf{r}, t), \tag{12.99}$$

and our next task is to determine it. Since the gradient appears in (12.98), $\chi(\mathbf{r})$ is given as a line integral:

$$\chi(\mathbf{r}) = \int_{\ell(\mathbf{r})} d\boldsymbol{\ell} \cdot \mathbf{A}', \tag{12.100}$$

where $\ell(\mathbf{r})$ is any path ending at point $\mathbf{r}$, and an inconsequential constant has been set equal to zero. The simplest path $\ell(\mathbf{r})$ to choose is one that includes the electron's trajectory. For passage through slit j, $j = 1$ or 2, we therefore select $\ell(\mathbf{r})$ to be $\ell_j(\mathbf{r})$, where, from Fig. 12.4, $\ell 1(P)$ is the path S_1P and $\ell_2(P)$ is the path S2P.

Putting these results together, the wave function for the particle to be at P after passing through slit j is

$$\Psi_j^{(A')}(P, t) = e^{-i(e\kappa_c/\hbar)\mathrm{ds}\int_{\ell_j(P)} d\boldsymbol{\ell}\cdot\mathbf{A}'}\Psi_j^{(0)}(P, t), \tag{12.101}$$

where, in a plane-wave approximation, $\Psi_j^{(0)}(P, t)$ is essentially (12.94). The total wave function $\Psi^{(A)}(P, t)$ is the analog of Eq. (12.93),

$$\Psi^{(A')}(P, t) = \Psi_1^{(A')}(P, t) + \Psi_2^{(A')}(P, t).$$

The probability density $\rho^{(A')}(P, t) = |\Psi^{(A')}|^2$ contains a modified interference term given by

$$\begin{aligned} &2\mathrm{Re}\{\Psi_1^{(A')*}(P, t)\Psi_2^{(A')}(P, t)\} \\ &\quad = 2\,\mathrm{Re}(\Psi_1^{(0)*}(P, t)\Psi_2^{(0)}(P, t)e^{-i(e\kappa_c/\hbar)(\int_{\ell_1(P)} d\boldsymbol{\ell}\cdot\mathbf{A}' - \int_{\ell_2(P)} d\boldsymbol{\ell}\cdot\mathbf{A}')}). \end{aligned} \tag{12.102}$$

The modification due to the magnetic field resides in the difference of the two line integrals, which can be transformed as follows:

$$\begin{aligned}\int_{\ell_1(P)} d\boldsymbol{\ell}\cdot\mathbf{A}' - \int_{\ell_2(P)} d\boldsymbol{\ell}\cdot\mathbf{A}' &= \int_{\ell_1(P)} d\boldsymbol{\ell}\cdot\mathbf{A}' + \int_{-\ell_2(P)} d\boldsymbol{\ell}\cdot\mathbf{A}' \\ &= \oint d\boldsymbol{\ell}\cdot\mathbf{A}' = \int_{\text{Area}} d\mathbf{S}\cdot(\nabla\times A') \\ &= \int_{\text{Area}} d\mathbf{S}\cdot\mathbf{B} \equiv \Phi_{\text{mag}}. \end{aligned} \tag{12.103}$$

In the second equality in (12.103), $\oint$ means an integral around a closed loop encircling the solenoid and its magnetic field, while Stokes's theorem has been used to convert this integral into one over the surface area enclosed by the closed loop. Finally, Φ_{mag} is the magnetic flux due to the magnetic field *in the inaccessible region.*

The interference term is now seen to be

$$2\,\text{Re}[\Psi_1^{(A')*}\Psi_2^{(A')}] = 2\,\text{Re}(e^{-2\pi i(\Phi_{\text{mag}}/\Phi_0)}\Psi_1^{(0)*}\Psi_2^{(0)}),$$

where $\Phi_0 = 2\pi\hbar/(e\kappa_c) = 4.13\times10^{-7}\ \text{G cm}^2 = 4.13\times10^{-15}\ \text{T m}^2$ is a fundamental flux unit introduced by F. London (see, e.g., London (1961). As long as $\Phi_{\text{mag}}/\Phi_0 \neq$ integer, then the phase factor $\exp(2\pi i\Phi_{\text{mag}}/\Phi_0)$ gives rise to an interference pattern additional to that coming from $\text{Re}(\Psi_1^{(0)*}\Psi_2^{(0)})$. There will even be an interference effect at the angle $\theta = 0$ in Fig. 12.3. This new interference phenomenon comprises the A–B effect, and an observation of it was first reported by Chambers (1960). Because the effect is a manifestation through **A** of the magnetic field in the region forbidden to the electron, and also because of the difficulty of the experiments, the generation of a controversy might have been expected. As stated previously, the issue has been settled in favor of the A–B effect, largely through difficult experiments using magnetic fields confined to a toroidal region. The A–B interference phenomenon provides another beautiful and compelling illustration of the strange ways in which quantum physics can differ from classical physics.

Although the results of our analysis of the A–B effect are correct, the reader may have raised questions about the method, since quantum theory does not allow one to specify a particular trajectory – e.g., S1P – that a particle must follow. A summation over all paths, as in the Feynman path integral approach, is clearly more correct. The analysis can be carried out using this approach; not surprisingly, the conclusion is the same: one still obtains the additional A–B interference factor. Demonstration of this point is left to the exercises.

We have also stated that the A–B effect occurs because the physical domain is multiply connected. This is obvious, since an excluded region is needed in order to produce an interference phase factor involving Φ_{mag}. A less obvious consequence of this lack of a simply connected domain is that the wave function $\Psi^{(A')}(\mathbf{r},\,t)$ is no longer single-valued. Using Eq. (12.100) in (12.99), $\Psi^{(A')}(\mathbf{r},\,t)$ is

$$\Psi^{(A')}(\mathbf{r},\,t) = e^{-i(e\kappa_c/\hbar)\int_{\ell(r)} d\boldsymbol{\ell}\cdot\mathbf{A}'}\Psi^{(A)}(\mathbf{r},\,t). \tag{12.104}$$

As a final comment, we note that Aharonov and Casher (1984) have extended these ideas to the case of neutral particles having a non-zero magnetic moment. They suggested several experiments in which the effects of the Aharonov–Casher phase might be observed; the earliest successful measurements in this regard seem to be those of Sangster *et al.* (1993, 1994) and Görlitz *et al.* (1994). The emergence of a variety of measurable phase effects is one of the striking developments of quantum physics in the second half of the twentieth century – see Chapters 14 and 16 for additional discussion.

Exercises

12.1 An electron, whose $t = 0$ wave function is $\Psi_\Gamma(\mathbf{r}, 0) = N \exp[ikx - r^2/(2b^2)]$, where N, k, and b are real and positive, moves under the sole influence of a vector potential $\mathbf{A}(\mathbf{r}, t) = (x - vt)B\mathbf{e}_2$, with $v = \hbar k/m_e$ and B a positive constant.
 (a) Write out the equations of motion for the expectation values of $\hat{\mathbf{Q}}$ and $\hat{\mathbf{\Pi}}$ and identify the nature of the forces that these results imply are acting on the electron.
 (b) Determine the expectation values of $\hat{\mathbf{Q}}$ and $\hat{\mathbf{\Pi}}$ for all $t \geqslant 0$.

12.2 (a) Verify that $\psi_0(x, y) = N_0 \exp[-m\omega_c(x^2 + y^2)/(4\hbar)]$ is an eigenfunction of the Hamiltonian $\hat{H}^{(A)}(\mathbf{r}) = (\hat{\mathbf{P}} - q\kappa_c\mathbf{A})^2/(2m)$, where $\omega_c = q\kappa_c B/m$ (the cyclotron frequency, Eq. (12.27)), $\mathbf{A} = \mathbf{B} \times \mathbf{r}/2$ and $\mathbf{B} =$ constant, and also state the corresponding eigenvalue.
 (b) Ditto for $\psi_n(x, y) = N_n(x + iy)^n \exp[-m\omega_c(x^2 + y^2)/(4\hbar)]$.

12.3 (a) Assuming that a single electron is in the state $|211\rangle$ of a hydrogenic atom of nuclear charge Z, evaluate the Z-dependent proportionality constant relating R and B/a_0, where R is the ratio defined above Eq. (12.21). Discuss the validity of ignoring the B^2 term in Eq. (12.26) for this case.
 (b) Let the electron now be described by the 3-D oscillator wave function $\psi_{011}(r, \theta, \phi) = N_{01}\, r \exp[-r^2/(2\lambda^2)]\, Y_1^1(\theta, \phi)$ of Exercise 11.29 and redo the analysis of part (a) assuming that $\lambda = a_0$.

12.4 The wave function for an electron in a hydrogenic atom is initially $\Psi_\Gamma(\mathbf{r}, 0) = \psi_{21x}(\mathbf{r})$ (see Exercise 11.23 for an explanation of the notation). Assuming that a constant but weak magnetic field $\mathbf{B} = B\mathbf{e}_3$ is also acting, show that the time-evolved wave function $\Psi_\Gamma(\mathbf{r}, t)$ is an eigenfunction of the operator $\mathbf{n}(t) \cdot \hat{\mathbf{L}}$, provided that the unit vector $\mathbf{n}(t)$ has a specific time dependence that you are to determine.

12.5 A particle of charge q and mass m bound to a 3-D isotropic oscillator potential is acted on by a constant but weak magnetic field $\mathbf{B} = B\mathbf{e}_3$. At time $t = 0$ its wave function is $\Psi_\Gamma(\mathbf{r}, 0) = \psi_0(x - a)\psi_0(y)\psi_0(z)$, where a is a constant length and ψ_0 is the ground-state wave function for the 1-D oscillator.
 (a) Calculate $\Psi_\Gamma(\mathbf{r}, t)$, $t > 0$.
 (b) What is the probability that, for $t > 0$, the particle will be found in the oscillator product state whose wave function is $\psi_{n_x}(x)\psi_{n_y}(y)\psi_{n_z}(z)$?

12.6 The wave function for an H-atom electron is initially $\Psi_\Gamma(\mathbf{r}, 0) = R_{n\ell}(r)Y_\ell^\ell(\Omega)$. If a constant but weak magnetic field $\mathbf{B} = B\mathbf{e}_2$ also acts on the electron, what is the probability that, for all $t \geqslant 0$, the electron's wave function will be an eigenfunction of

$\hat{L}_x$ with eigenvalue $\hbar\ell$ yet will still have the same first two quantum numbers n and ℓ as those in $\Psi_\Gamma(\mathbf{r}, 0)$?

12.7 At time $t = 0$, an H-atom electron described by the wave function $\Psi_\Gamma(\mathbf{r}, 0)$ of Exercise 11.23 is subjected to a constant but weak magnetic field $\mathbf{B} = B\mathbf{e}_3$, for which the vector potential is $\mathbf{A} = \mathbf{B} \times \mathbf{r}/2$.
(a) Determine the wave function for all $t > 0$.
(b) Calculate the probability that, at times $t > 0$, the electron will be found in the initial state. At what time (in seconds) does this probability become zero?
(c) Calculate the expectation values of $\hat{\mathbf{Q}}$ and $\hat{\mathbf{\Pi}}$ for all $t > 0$.

12.8 At $t = 0$ an H-atom electron initially described by the wave function $\Psi_\Gamma(\mathbf{r}, 0) = [\psi_{100}(\mathbf{r}) + 2\psi_{21x}(\mathbf{r}) + i\psi_{21y}(\mathbf{r})]6^{-1/2}$ – in the notation of Exercise 11.23 – is subjected to a constant but weak magnetic field $\mathbf{B} = B\mathbf{e}_3$
(a) Determine $\Psi_\Gamma(\mathbf{r}, t)$, $t > 0$.
(b) Calculate $\langle\Gamma, t|\hat{\mathbf{Q}}|\Gamma, t\rangle$ and $\langle\Gamma, t|\hat{\mathbf{\Pi}}|\Gamma, t\rangle$ for all $t \geqslant 0$.
(c) If the field direction is changed to $\mathbf{e}_2$, what is the probability that a measurement of $\hat{L}_z$ at any $t \geqslant 0$ would yield the value zero?

12.9 The Hamiltonian for a particle of charge q and mass m is $\hat{H}(\mathbf{r}) = m\omega^2 z^2/2 + \hat{H}^{(A_2)}(\mathbf{r})$, the latter operator given by (12.23).
(a) Solve the eigenvalue problem for this Hamiltonian, obtaining general expressions for the eigenfunctions and eigenvalues. Compare your solution with the solutions and degeneracies of Eq. (12.24).
(b) Obtain and compare the current densities for the two cases of part (a).

12.10 Landau levels using $\mathbf{A}_1$: the coordinate-space solution. The $\hat{H}^{(A_1)}(\mathbf{r})$ eigenvalue problem is most conveniently solved in position space using the cylindrical coordinates ρ, ϕ, and z, where $x = \rho\cos\phi$ and $y = \rho\sin\phi$.
(a) Show that, when $V = 0$, the cylindrical-coordinate eigenvalue equation is

$$\{-[\hbar^2/(2m\rho)]\,\partial(\rho\partial/\partial\rho)/\partial\rho + m\omega^2\rho^2/2 + \hat{L}_z^2/(2m\rho^2)$$
$$+P_z^2/(2m) - \omega_L\hat{L}_z - E\}\psi(\rho, \phi, z) = 0.$$

(b) As a first orientation to the full solution, determine the eigenfunctions and eigenvalues by setting the two $\hat{L}_z$ terms to zero (thereby eliminating the ϕ dependence). Discuss the degeneracy of the solution and show that the splitting of the energy levels corresponding to motion in the x–y plane is $\hbar\omega_L$.
(c) Returning to the full eigenvalue problem of part (a), state the CSCOs for this system and use them to determine the functions $G(\phi)$ and $Z(z)$ (and the ranges of their appropriate quantum numbers) for the solution expressed in the form $\psi(\rho, \phi, z) = F(\rho)G(\phi)Z(z)$.
(d) Find an equation for $F(\eta)$, $\eta = \rho(m\omega_L/\hbar)^{1/2}$, setting the energy difference that occurs in it equal to $\beta\hbar\omega_L$, with β to be determined. The resulting eigenvalue equation, involving β and one of the quantum numbers from part (b), can be solved by comparison with a previously discussed eigenvalue problem. From this comparison, you should be able to deduce the allowed values of β and the functional form of $F(\eta)$ – which is a product. Each of these deduced quantities should contain an absolute value.

12.11 Landau levels using $\mathbf{A}_1$: algebraic solution via creation and annihilation operators in the presence of a symmetric, 3-D oscillator potential $\hat{V}(\hat{Q}) = m\omega_0^2\hat{Q}^2/2$. The goal is to find $|\Gamma\rangle$ and E_Γ in $[\hat{H}^{(A_1)} - E_\Gamma]|\Gamma\rangle = 0$.

(a) From the relation $\hat{H}^{(A_1)} = \sum_j[\hat{P}_j^2/(2m) + m\omega_j^2\hat{Q}_j^2/2)] + \omega_L\hat{L}_z$, where $\omega_3 = \omega_0$ and $\omega_1 = \omega_2 = \omega$, with ω to be determined, show that $\hat{H}^{(A_1)}$ can be expressed in the following form: $\hat{H}^{(A_1)} = \hat{H}_{xy} + \hat{H}_z$, where the subscripts and the requirement that $[\hat{H}_{xy}, \hat{H}_z] = 0$ uniquely determine $\hat{H}_{xy}$ and $\hat{H}_z$.

(b) Let $\hat{a}_x$, $\hat{a}_x^\dagger$, $\hat{a}_y$, and $\hat{a}_y^\dagger$ be the oscillator annihilation and creation operators for states corresponding to motion in the x- and y-directions, and define the left (L) and right (R) "circularly polarized" annihilation and creation operators $\hat{a}_L = (\hat{a}_x + i\hat{a}_y)/2^{1/2} = (\hat{a}_L^\dagger)^\dagger$ and $\hat{a}_R = (\hat{a}_x - i\hat{a}_y)/2^{1/2} = (\hat{a}_R^\dagger)^\dagger$. Solve the separate $\hat{H}_{xy}$ and $\hat{H}_z$ eigenvalue problems by re-expressing $\hat{H}_{xy}$ and $\hat{H}_z$ in terms of the latter operators and their corresponding number operators $\hat{N}_L = \hat{a}_L^\dagger\hat{a}_L$ and $\hat{N}_R = \hat{a}_R^\dagger\hat{a}_R$.

(c) From the results of part (b) determine E_Γ and $|\Gamma\rangle$.

(d) Set $\omega_0 = 0$, re-determine E_Γ and $|\Gamma\rangle$, and compare them with the corresponding quantities obtained in Exercise 12.10.

12.12 A particle of mass m and charge q is acted on both by a central potential $V(r)$ and by a constant electric field $\boldsymbol{\mathcal{E}} = \mathcal{E}\mathbf{e}_3$.

(a) If its wave function at time $t = 0$ is $\Psi(\mathbf{r}, 0) = \psi(r)$, where $\psi(r)$ is real and spherically symmetric, calculate $\langle\hat{\mathbf{L}}\rangle_t$ for all $t \geqslant 0$.

(b) Re-calculate $\langle\hat{\mathbf{L}}\rangle_t$ for all $t \geqslant 0$ if $V(r) = m\omega^2r^2/2$ and $\Psi(\mathbf{r}, 0) = \exp(ikx)\psi_0(x)\psi_0(y)\psi_0(z)$, where Ψ_0 is the 1-D ground-state wave function for an oscillator of frequency ω.

12.13 A particle of mass m and charge q moves under the sole influence of a constant electric field $\boldsymbol{\mathcal{E}} = \mathcal{E}\mathbf{e}_1$. At time $t = 0$ its wave function is $\Psi(\mathbf{r}, 0) = Nzf(r)$, where N is normalization constant and $f(r)$ is a real and spherically symmetric function such that, in the limit $r \to \infty$, $rf(r) \to 0$.

(a) Calculate $\langle\hat{\mathbf{L}}\rangle_0$, the expectation value of $\hat{\mathbf{L}}$ at $t = 0$.

(b) What is $\langle\hat{\mathbf{L}}\rangle_t$, the expectation value for $t > 0$?

(c) Determine whether $\Psi(\mathbf{r}, t)$ is an eigenfunction of either $\hat{L}^2$ or $\hat{L}_z$ at $t = 0$ and then at $t > 0$.

(d) What are the probabilities that measurements of $\hat{L}_x$ at $t = 0$ and at $t > 0$ will yield the values (i) $\hbar$, (ii) 0, and (iii) $-\hbar$?

12.14 A particle of charge q and mass m is acted on by constant electric and magnetic fields $\boldsymbol{\mathcal{E}}$ and $\mathbf{B}$, with $\mathbf{A} = \mathbf{B} \times \mathbf{r}/2$ and $\Phi = -\boldsymbol{\mathcal{E}} \cdot \mathbf{r}$. The Hamiltonian is $\hat{H} = \hat{H}_0 + q\Phi$, where $\hat{H}_0 = \hat{\Pi}^2/(2m)$.

(a) Evaluate $[\hat{H}_0, \hat{U}_P]$, and, if it is zero, determine whether the eigenstates of $\hat{H}_0$ will necessarily have a definite parity. Ditto for $\hat{H}$.

(b) Assume that $\mathbf{B} = B\mathbf{e}_3$ and $\boldsymbol{\mathcal{E}} = \mathcal{E}\mathbf{e}_1$, and that, at $t = 0$, the wave function of the system is an eigenfunction both of $\hat{H}_0$ and of $\hat{U}_P$. Calculate the expectation values of $\hat{Q}_x$, $\hat{Q}_y$, $\hat{\Pi}_x$, and $\hat{\Pi}_y$ at any $t > 0$.

(c) Identify the nature of the orbit described by the preceding values of $\langle\hat{Q}_x\rangle_t$ and $\langle\hat{Q}_y\rangle_t$ and compare with the orbit for the case $\mathcal{E} = 0$.

12.15 This exercise is inspired by the classical Hall effect. A particle of mass m and charge q is acted on by the constant electric and magnetic fields $\boldsymbol{\mathcal{E}} = \mathcal{E}\mathbf{e}_2$ and $\mathbf{B} = B\mathbf{e}_3$ (this system is an analog of that in the preceding exercise). You are to use the potentials $\mathbf{A}_3$ and $\Phi = -y\mathcal{E}$.

(a) Write out the Hamiltonian $\hat{H}^{(A_3)}$ for this system and determine which, if any, of the following operators are constants of the motion: $\hat{P}_x$, $\hat{P}_y$, $\hat{P}_z$, $\hat{L}^2$, and $\hat{L}_z$.

(b) Let $\psi_\Gamma(x, y, z)$ be the wave function for the system. Using the results of part (a), show that ψ_Γ is a product of familiar and possibly unfamiliar functions, specifying the familiar one(s).

(c) Introduce the frequency ω_c of Eq. (12.27) and the length $a = q\mathcal{E}/(m\omega^2)$ into the relevant Schrödinger equation as a means of identifying the possibly unfamiliar functions of part (b) as ones previously encountered. From this identification, determine the eigenvalues of $\hat{H}^{(A_3)}$.

(d) Calculate the probability current density $\mathbf{J}_\Gamma(\mathbf{r})$ corresponding to the solution $\psi_\Gamma(x, y, z)$, and then use it to evaluate the components $j_n(x, z) = q\int dy\, J_{\Gamma,n}(x, y, z)$, $n = 1, 2, 3$, of the electric current density. Compare these $j_n(x, z)$ with the classical expression for the case of a charge q and velocity components v_n, $n = 1, 2, 3$.

12.16 It is noted at the end of Section 12.3 that the result (12.43), with $\hbar\mathbf{k}(t)$ given by (12.45), follows from the relation $[\hat{H}^{(\mathcal{E})}, [\hat{H}^{(\mathcal{E})}, \hat{\mathbf{P}}] = 0$. This an example of a more general situation in which (i) $[\hat{H}, \hat{A}] \neq 0$ but $[\hat{H}, [\hat{H}, \hat{A}]] = 0$ and (ii) $\hat{A}\Psi_\Gamma(\mathbf{r}, t) = a\Psi_\Gamma(\mathbf{r}, t)$.

(a) Prove that the latter pair of relations implies that $\hat{A}\Psi_\Gamma(\mathbf{r}, t) = a(t)\Psi_\Gamma(\mathbf{r}, t)$.

(b) Assume that there is an operator $\hat{B}$ for which $[\hat{H}, \hat{B}] = \lambda\hat{B}$. Show that, if $(\hat{B} - b)\Psi_\Gamma(\mathbf{r}, t = 0) = 0$, then, for $t > 0$, the $b(t)$ which occurs in $[\hat{B} - b(t)]\Psi_\Gamma(\mathbf{r}, t) = 0$ is exponential in t.

12.17 An electron moves in the presence of a constant magnetic field $\mathbf{B} = B\mathbf{e}_3$. At time $t = 0$ a measurement of the x-component of the particle's velocity is made and the result v_x is obtained. Immediately afterward, the z-component of its velocity is measured and the value v_z is found.

(a) Is it true that, as soon as the second measurement is completed, the system is in an eignstate of $\hat{v}_x$? Justify your answer.

(b) Will the electron be in an eigenstate of $\hat{v}_z$ at any time $t > 0$ (i.e., always, never, or at some values of t to be determined)?

(c) Are there any values of $t > 0$ for which the electron is expected to be in an eigenstate of $\hat{v}_x$? If so, specify those values of t and the corresponding eigenvalues.

(d) Will the electron be in an eigenstate of $\hat{v}_y$ at any time $t > 0$? Explain your reasoning.

12.18 An H-atom electron is subjected to a weak, constant magnetic field $\mathbf{B} = B\mathbf{e}_1$, so that the H-atom Hamiltonian is supplemented by the term $\mu_B B\hat{L}_x = \omega_c\hat{L}_x/2$. In the notation of Exercise 11.23, the electron's wave function at $t = 0$ is $\Psi_\Gamma(\mathbf{r}, 0) = [\psi_{1s}(\mathbf{r}) + i\psi_{2s}(\mathbf{r}) + \psi_{21z}(\mathbf{r})]/3^{1/2}$.

(a) Write out the detailed form of $\Psi_\Gamma(\mathbf{r}, t)$, $t > 0$, and calculate the expectation value of the particle's momentum at times $t > 0$.

(b) Determine the expectation value of the particle's position at times $t > 0$.

(c) Investigate whether the preceding results satisfy the electromagnetic version of Ehrenfest's theorem. If they do not, explain why.

(d) Determine whether $\Psi_\Gamma(\mathbf{r}, t)$ satisfies the quantal equation of continuity. If it does not, explain why.

12.19 (a) Under a gauge transformation, the time-dependent wave function $\Psi^{(A)}(\mathbf{r}, t)$ is transformed into $\Psi^{(A')}(\mathbf{r}, t)$, Eq. (12.64). Calculate the expectation values of $\hat{\mathbf{L}}$ and of $\hat{\mathbf{L}}^{(A)} = \hat{\mathbf{Q}} \times \hat{\mathbf{\Pi}}$ using these two wave functions and analyze the results.

(b) Assume that the gauge function is time-independent: $\chi \neq \chi(t)$. Show that, if $\Psi^{(A)}(\mathbf{r}, t)$ and $\Psi^{(A')}(\mathbf{r}, t)$ correspond to the same eigenvalue of $\hat{H}^{(A)}$, they will not be equal (apart from a global phase factor) unless $\hat{H}^{(A)}$ is non-degenerate.

12.20 Although it is not stated, the analysis of the hydrogenic atom in Section 11.6 is based on a particular gauge, viz. $\mathbf{A} = 0$, $\Phi \neq 0$. One can also attempt to analyze this system using a gauge wherein $\Phi = 0$ and $\mathbf{A}'(\mathbf{r}, t) \propto t$.

(a) Determine a suitable $\mathbf{A}'(\mathbf{r}, t)$ and the corresponding gauge function $\chi(\mathbf{r}, t)$.

(b) Write out all the terms in $\hat{H}^{(A')}(\mathbf{r}, t)$ and show explicitly, using the $\chi(\mathbf{r}, t)$ of part (a), that the time-dependent Schrödinger equations involving $\hat{H}^{(A')}(\mathbf{r}, t)$ and the original hydrogenic Hamiltonian of Section 11.6 are equivalent.

(c) From this last result relate the time-dependent wave functions $\Psi_{n\ell m_\ell}(\mathbf{r}, t)$ and $\Psi^{(A')}_{n\ell m_\ell}(\mathbf{r}, t)$.

12.21 (a) Using the gauge $\mathbf{A}_3$ plus the analysis and results of Section 12.2, determine the detailed form of the solution $\Psi_{k_z k_x n}(\mathbf{r})$ to the Landau-level problem which corresponds to the energy analogous to $E_{k_z n}$ of Eq. (12.32b).

(b) Determine an appropriate gauge function $\chi(\mathbf{r})$ relating $\mathbf{A}_2(\mathbf{r})$ and $\mathbf{A}_3(\mathbf{r})$ and use it to relate the general solutions $\Psi^{(A_2)}(\mathbf{r}, t)$ and $\Psi^{(A_3)}(\mathbf{r}, t)$.

(c) Set $t = 0$ in the latter relation and choose for $\Psi^{(A_2)}(\mathbf{r}, 0)$ and $\Psi^{(A_3)}(\mathbf{r}, 0)$, respectively, the Landau-level solutions corresponding to the energy $E^{(A_2)}_{k_z n} = E^{(A_3)}_{k_z n}$. Despite the preceding "=" sign, the two $t = 0$ solutions are not equal. Explain why this lack of equality is not a violation of gauge invariance and show, therefore, how to convert your result into a true equality.

12.22 Show that the questions raised at the end of Section 12.6 are indeed answered in the affirmative, viz., that

(a) limits $\omega \to 0$ of $k_\omega(t)$ and $\beta_\omega(t)$ yield $k(t)$ and $\beta(t)$ of Section 12.3, and

(b) $\hat{U}^{(\mathcal{E})}\langle\mathbf{r}|\mathbf{k}\rangle = \Psi^{(\mathcal{E})}(\mathbf{r}, t)$ of Eq. (12.91).

12.23 In this exercise, the harmonic time dependence of Eq. (12.87) is replaced by an exponential one: $\cos(\omega t) \to \exp(-\omega t)$. Using this choice, carry through the remaining analysis of Eqs. (12.88)–(12.92). Discuss in particular the separate limits $\omega \to 0$ and $t \to \infty$.

12.24 A conceptual analog to the magnetic-field A–B effect is one involving spatially uniform electric potentials (see Peshkin and Tonomura (1989) for comments on the electric-field A–B effect). In this exercise a charged particle passes through one of two very long (macroscopic), hollow, conducting tubes. After emerging from a tube, the particle is directed to a screen, as in Fig. 12.5, where the tubes, which are electrically insulated from one another, are labeled 1 and 2. The electric potential in tube j is $\Phi_j(t)$, $j = 1, 2$. Each is turned on at time t_1 and off at time t_2, times such that classically the particle will be deep inside one of the tubes when $\Phi_j \neq 0$.

(a) Follow the analysis of Section 12.7 and determine the phase factor caused by interference effects due to the presence of non-vanishing Φ_j's.

(b) Suppose that the particles are 1-keV electrons described by the wave packet discussed in Section 7.1 (central wave number $k_0 \cong 16$ Å^{-1}). Let each Φ_j be constant, turned on for a time $t_2 - t_1 = 10^{-9}$ s. Derive an order-of-magnitude

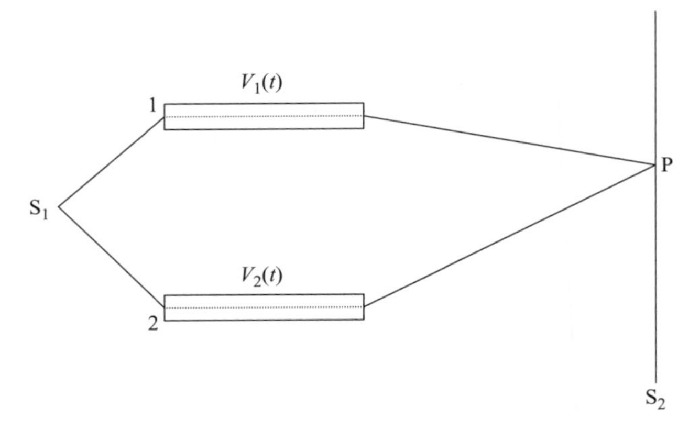

Fig 12.5 S_1 is the source of the charged particles; $V_1(t)$ and $V_2(t)$ are the spatially uniform electric potentials in tubes 1 and 2, respectively; and P is the interference point on screen S_2.

estimate for the difference $|\Phi_2 - \Phi_1|$ that is needed to produce (in principle) an observable shift in the interference pattern relative to that from the $\Phi_j = 0$ case. State any reasonable assumptions needed to obtain an answer in volts.

12.25 Use the Feynman path-integral method to validate the interference term "derived" below Eq. (12.103). (Hint: when the particle has charge q, the Lagrangian $L(t'')$ of Eq. (8.34) becomes 3-D and is changed by the addition of the term $q\kappa_c(d\mathbf{r}/dt) \cdot \mathbf{A}(\mathbf{r},\ t)$.]

13

Intrinsic Spin, Two-State Systems

Thus far we have concentrated on the postulates, on their ramifications, and on the means and methods by which they are applied. Connections with classical physics have been featured, and primary notions other than those of Chapter 5 have not been needed. We deviate from this path in the present chapter, introducing a new and extremely important concept; it is based on one of the most significant and fruitful experiments of the quantal era. This experiment was performed by O. Stern and W. Gerlach (1921) and subsequently interpreted by G. Uhlenbeck and S. Goudsmit (1925).

The experiment and its elucidation led to the new concept of "intrinsic spin" (in particular to spin "$\frac{1}{2}$"), a development that was one of the major turning points in modern physics. The consequences have been monumental, and include primary theoretical developments; the interpretation of data related to, and an understanding of, empirical rules concerning a wide variety of physical systems (e.g., the periodic table of the elements/atomic structure); and the generation of new experimental techniques and of innovative measurements. Our study of intrinsic spin and its correlative, two-state systems, introduces the reader to one of the most profound topics in quantum theory.

13.1. The Stern–Gerlach experiment and its interpretation

The experiment of Stern and Gerlach was designed to test the "spatial quantization" of neutral atoms, i.e., to determine whether the (orbital) angular momentum was quantized. The basic ideas are as follows. Associated with the atom's orbital angular momentum would be a magnetic moment, as in the Bohr model of hydrogen. Using classical physics, it was argued that, on subjecting a beam of neutral atoms – i.e., their magnetic moments – to a magnetic field, a change of motion would be observed. A constant field would cause a precession of the moment about the direction of the field but would not deflect the beam, whereas the force from an inhomogeneous field would deflect the beam. The classical-physics prediction (see below) for this behavior is a continuous spread of deflection angles, whereas, based on what was then (1921/22) believed about spatial quantization and what so far *we* know about quantization of orbital angular momentum (recall Eq. (12.22b) and the ensuing discussion), the magnetic field should split the beam into an *odd* number of deflected components. This number is $2\ell + 1$, depending on the orbital angular quantum momentum ℓ.

We first describe the experimental set-up, next briefly review the classical prediction, and then state the actual experimental result and the proposal of Uhlenbeck and

Goudsmit for interpreting it. Figure 13.1 is a schematic illustration of the experimental arrangement; the caption makes it self-explanatory. The coordinate system displayed is that used in the analysis. The inhomogeneous field's components satisfy

$$B_z > 0, \qquad \frac{\partial B_z}{\partial z} < 0; \qquad B_y = 0; \qquad B_x \neq 0, \text{ and slowly varying.} \tag{13.1}$$

Let μ be the neutral atom's magnetic moment.[1] In the confined region of the magnetic field, the classical potential energy is

$$V = -\mu \cdot \mathbf{B}, \tag{13.2}$$

giving rise to a classical force

$$\mathbf{F} = -\nabla V = \nabla(\mu \cdot \mathbf{B}).$$

Its component along z is

$$F_z = \frac{\partial}{\partial z}(\mu \cdot \mathbf{B}) \cong \mu_z \frac{\partial B_z}{\partial z},$$

where the slow variation of B_x leads to an ignorable derivative. For positive μ_z, F_z deflects the atom downward (negative z). The sign of μ_z depends on its co-latitude θ; in particular, if B_z is the dominant component, then

$$V \cong -\mu B_z \cos\theta,$$

and, for a classical Boltzmann distribution, the probability that μ is oriented at an angle θ to B_z is

$$P(\theta) = Z^{-1} e^{-V/(k_B T)} \cong e^{\mu B_z \cos\theta/(k_B T)}/Z, \tag{13.3}$$

where Z is the partition function. This verifies our claim that, classically, all possible deflections θ may occur. Quantally, quantization of space (i.e., orbital angular momentum) predicts an odd number of deflected beams, as noted above.

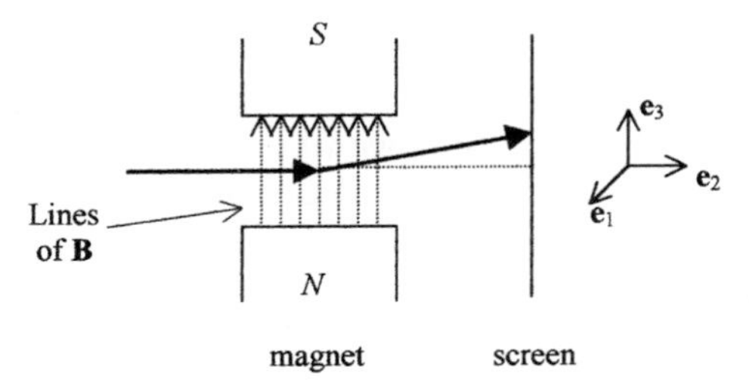

Fig. 13.1 A schematic illustration of the Stern–Gerlach experiment. A beam of previously collimated neutral atoms (represented by the dark arrows) is directed between the poles of the magnet. After deflection by the inhomogeneous magnetic field **B**, whose dominant direction is that of the dotted vertical arrows in the diagram, the atoms undergo impact on a detecting screen. The coordinate system used in the analysis is shown next to the screen.

[1] Neutral atoms are used because effects due to non-zero charges would overwhelm those on μ due to **B**.

Stern and Gerlach performed their experiment using neutral silver atoms. Neither the classical nor the space-quantization prediction was observed. Instead the beam split into two(!) components. The two theoretical predictions and the result of the measurement are shown schematically in Fig. 13.2. A result as unexpected and mysterious as that found by Stern and Gerlach calls for a new set of ideas, what in effect will be a paradigm shift (for analysis and comments on how previously accepted constructs are changed by new data or new ideas, see Kuhn (1970)). There are two linked aspects of the Stern–Gerlach (S–G) result that require explanation. Why is the number of deflected beams even rather than odd, and why is its value two?

Four years passed before G. Uhlenbeck and S. Goudsmit (1925), still graduate students(!!), answered both of the preceding questions by postulating the now universally accepted interpretation of the S–G measurement. Their proposal, couched in the language of Chapter 5, is that an additional pair of quantum numbers is needed in order to specify the state of an electron (and many other fundamental particles). These quantum numbers are related to the eigenvalues of a new Hermitian operator in the same way as ℓ and m_ℓ are related to $\hat{L}^2$ and $\hat{L}_z$. The new entity is a vector quantity named the "spin" angular momentum operator or just the spin operator, denoted $\hat{\mathbf{S}}$; it is an analog of orbital angular momentum, and its behavior is essentially the same as that of $\hat{\mathbf{L}}$. That is, $\hat{S}^2$ and its z-component $\hat{S}_z$ commute and share common eigenstates denoted $|S, m_S\rangle$, obeying

$$\hat{S}^2|S, m_S\rangle = S(S+1)\hbar^2|S, m_S\rangle \tag{13.4a}$$

and

$$\hat{S}_z|S, m_S\rangle = m_S\hbar|S, m_S\rangle, \tag{13.4b}$$

with the $2S+1$ values of m_S being

$$m_S = S,\ S-1,\ S-2,\ \ldots,\ -S. \tag{13.4c}$$

For electrons, Uhlenbeck and Goudsmit postulated that $S = \frac{1}{2}$, so that $m_S = \pm\frac{1}{2}$. The multiplicity, i.e., the number of values of m_S, is two. Furthermore, $\hat{\mathbf{S}}$ couples to magnetic fields in the same way that $\hat{\mathbf{L}}$ does, viz., by virtue of an added term in the Hamiltonian $\hat{H}^{(S)}_{\text{mag}}$ of the same form as (12.17):

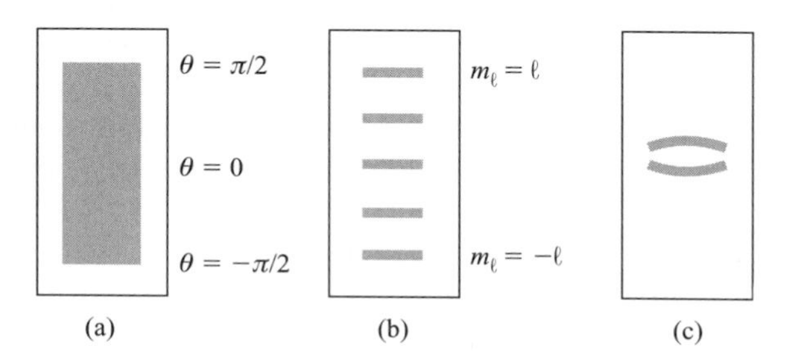

Fig. 13.2 A schematic representation of possible Stern–Gerlach experimental results: (a) a screen showing the classical prediction, (b) a screen showing the space-quantization prediction of $2\ell + 1$ components, and (c) a screen showing the results of Stern and Gerlach (1921).

$$\hat{H}^{(S)}_{\text{mag}} = -g\frac{q}{|q|}\frac{\mu}{\hbar}\mathbf{B}\cdot\hat{\mathbf{S}}, \tag{13.5}$$

where a factor "g" has been added: for electrons, g was long thought to be equal to 2, but differs slightly from that value due to relativistic quantum-electrodynamic effects, whereas for protons and even (uncharged) neutrons, $g \neq 2$. The magneton μ in (13.5) – not to be confused with the reduced-mass symbol – is defined in the usual way (Eq. (12.19)):

$$\mu = \frac{|q|\hbar\kappa_c}{2m},$$

where the particle is assumed to have charge q and mass m.

For a charged particle of spin $\frac{1}{2}$ in a constant or predominantly constant magnetic field along the z-axis, $\mathbf{B}\cdot\hat{\mathbf{S}} = B_z\hat{S}_z$, and therefore

$$\hat{H}^{(S)}_{\text{mag}}|\tfrac{1}{2}m_S\rangle = -g\frac{q\mu B_z}{|q|}m_S|\tfrac{1}{2}m_S\rangle. \tag{13.6}$$

The energy is split into the two values associated with $m_S = \pm\frac{1}{2}$, implying that, for a beam of single particles such as electrons or protons, the predominantly z-directed magnetic field of the S–G experiment would deflect the beam into two components, exactly as observed for the beam of Ag-atoms.

The only problem with the preceding analysis is that it applies to a beam of individual spin-$\frac{1}{2}$ particles, not to Ag-atoms, which contain 47 electrons. (Each Ag-atom also contains 47 protons, whose magnetic moment is about 10^{-3} times that of the electrons and thus whose contribution to the magnetic moment of Ag – and generally all atoms – can be neglected for S–G-type experiments.) Although modification of the preceding analysis is obviously needed, the basic idea still applies. The reason it applies is that Ag behaves *as if* it were a massive spin-$\frac{1}{2}$ particle. The following explanation of this behavior of Ag is correct, but is based on concepts yet to be introduced. As discussed in Chapter 18, 46 of the 47 electrons in Ag behave as if they were 23 pairs of spin-0 ($S = 0$) objects, for which $m_S = 0$ and whose contribution to $\hat{H}^{(S)}_{\text{mag}}$ vanishes: only the spin of the unpaired electron contributes. Furthermore, not only do the orbital angular momenta of 46 of the electrons also total to zero, but the value of ℓ for the 47th electron is zero as well. Hence, orbital angular-momentum effects in the B field are non-existent, leaving just a single electron's spin to account for the behavior of a neutral Ag-atom in a magnetic field. It should now be clear that the two components of the deflected Ag beam arise because the whole atom acts like a spin-$\frac{1}{2}$ particle, for which there are only two energy levels in a magnetic field. These two levels ($m_s = \pm\frac{1}{2}$) give rise to the two components seen by Stern and Gerlach.

In addition to electrons, all protons, neutrons, mu mesons, and a variety of other particles have spin $\frac{1}{2}$, as do many atoms and nuclei. Other spin values are found in nature: photons and deuterons have spin 1, pi mesons and α particles have spin 0, etc. Values of the g factor in $H^{(S)}_{\text{mag}}$ can be determined using "spin-resonance" techniques, a topic we explore in Section 13.3. Note: experimentally speaking, what one means by a spin-$\frac{1}{2}$ particle is that a S–G-type experiment was done and S was measured to be $\frac{1}{2}$. This is not what is meant theoretically, however (see Chapter 14).

13.2. Spin $\frac{1}{2}$: operators, states, properties

The possession of non-zero spin by quantal objects is a completely non-classical attribute. One should not visualize a particle or system of spin S as an object that is rotating (spinning) about some axis, much less at a rate related to S. There is no classical analog: spin is uniquely quantal.[2] Furthermore, there is no coordinate representation for $\hat{\mathbf{S}}$ or its eigenstates $|S, m_S\rangle$: $\langle\mathbf{r}|S, m_S\rangle$ is neither meaningful nor a scalar product – though it looks like one. In this section, we explore some features of spin $\frac{1}{2}$, some of which will turn out to be highly unusual, e.g., like the mythical bird the phoenix, extinguished spin components will be seen to rise from their own ashes – regeneration!

The eigenstate representation

In this first subsection, the $\hat{S}^2$ and $\hat{S}_z$ eigenstates are used to obtain the eigenstate representation for the entire set of spin-$\frac{1}{2}$ operators. Since the vector spin operator $\hat{\mathbf{S}}$ was introduced as an analog to $\hat{\mathbf{L}}$, its components obey identical commutation relations:[3]

$$[\hat{S}_j, \hat{S}_k] = i\hbar\epsilon_{jk\ell}\hat{S}_\ell. \tag{13.7}$$

From (13.7) it follows, as for the orbital angular momentum operator, that

$$[\hat{S}^2, \hat{S}_j] = 0, \qquad j = x, y, z. \tag{13.8}$$

Hence, $\hat{S}^2$ and one of its components can be simultaneously diagonalized. It is standard to choose $\hat{S}_z$ as that one component, leading to Eqs. (13.4). Henceforth, $|\frac{1}{2}m_s\rangle$ will mean a simultaneous (normalized) eigenstate of $\hat{S}^2$ and $\hat{S}_z$. When eigenstates of the other components are considered, we shall change notation so as to reflect that particular component, but the $\hat{S}_z$ eigenstates $|\frac{1}{2}\frac{1}{2}\rangle$ and $|\frac{1}{2}, -\frac{1}{2}\rangle$ will always be the basis states into which all other spin states can be expanded. As the basis consists of only two states, the spin-$\frac{1}{2}$ Hilbert space is 2-D.

Since $|\frac{1}{2}m_s\rangle$ is not an eigenstate of either $\hat{S}_x$ or $\hat{S}_y$, we need to determine the effect of these operators on $|\frac{1}{2}m_s\rangle$. To do so, we work with the spin raising and lowering operators $\hat{S}_+$ and $\hat{S}_-$,

$$\hat{S}_\pm = \hat{S}_x \pm i\hat{S}_y,$$

analogs to the $\hat{L}_\pm$ used in Section 10.4. By applying the analysis of Section 10.4 to spin $\frac{1}{2}$, we readily find that

$$[\hat{S}_z, \hat{S}_\pm] = \pm\hbar\hat{S}_\pm \tag{13.9}$$

[2] In addition to being non-classical, the electron's spin turns out to be a manifestation of relativity, although it is encountered here in a non-relativistic context. P. A. M. Dirac (1926) was the first physicist to successfully marry quantum theory and special relativity; spin $\frac{1}{2}$ "falls out of" his relativistic wave equation, one that bears his name. The reader interested in this topic has available a wealth of sources (listed, for example, in library catalogs), all of which assume a basic understanding of quantum theory (some of which this book supplies) and special relativity.

[3] Just as $\hat{\mathbf{L}}$ is an infinitesimal generator of rotations, so will $\hat{\mathbf{S}}$ also turn out to be, although this feature is not discussed in this chapter. A detailed study of rotations, spins, and generalized angular-momentum operators is included in Chapter 14.

and

$$\hat{S}_{\pm}|\tfrac{1}{2}m_s\rangle = \tilde{C}_{\pm}|\tfrac{1}{2},\ m_s \pm 1\rangle, \tag{13.10}$$

with

$$\tilde{C}_{\pm} = \sqrt{\tfrac{3}{4} - m_s(m_s \pm 1)} = \sqrt{(\tfrac{1}{2} \mp m_s)(\tfrac{1}{2} \pm m_s + 1)} \tag{13.11}$$

($\tilde{C}_{\pm}$ is chosen real and positive, as usual). Like the operators $\hat{L}_{\pm}$ and the states $|\ell,\ m_\ell\rangle$, $\hat{S}_{\pm}$ acting on $|\tfrac{1}{2}m_s\rangle$ cannot produce a state with $|m_s| > \tfrac{1}{2}$.

From the foregoing relations and

$$\hat{S}_x = \tfrac{1}{2}(\hat{S}_+ + \hat{S}_-)$$

and

$$\hat{S}_y = \frac{1}{2i}(\hat{S}_+ - \hat{S}_-),$$

we find the desired results:

$$\hat{S}_x|\tfrac{1}{2}m_s\rangle = \frac{\hbar}{2}[\sqrt{\tfrac{3}{4} - m_s(m_s + 1)}|\tfrac{1}{2},\ m_s + 1\rangle + \sqrt{\tfrac{3}{4} - m_s(m_s - 1)}|\tfrac{1}{2},\ m_s - 1\rangle] \tag{13.12a}$$

and

$$\hat{S}_y|\tfrac{1}{2}m_s\rangle = \frac{\hbar}{2i}[\sqrt{\tfrac{3}{4} - m_s(m_s + 1)}|\tfrac{1}{2},\ m_s + 1\rangle - \sqrt{\tfrac{3}{4} - m_s(m_s - 1)}|\tfrac{1}{2},\ m_s - 1\rangle]. \tag{13.12b}$$

For a typical spin-independent Hamiltonian, there are three representations for its Hilbert-space states: the coordinate representation, the momentum representation, and the eigenstate (matrix/vector) representation. The first two do not exist for spin states, but the third one does, and, now that the actions of all operators on the eigenstates are known, the eigenstate representation can be constructed.

The construction is straightforward, with all matrices being 2×2 and the states $|\tfrac{1}{2}m_s\rangle$ going over to two-rowed column vectors. Projecting both sides of Eqs. (13.4), (13.10), and (13.12) onto $\langle\tfrac{1}{2}m_s'|$, one gets the following matrices:

$$S^2_{m_s' m_s} = \langle\tfrac{1}{2}m_s'|\hat{S}^2|\tfrac{1}{2}m_s\rangle = \tfrac{3}{4}\hbar^2\delta_{m_s' m_s}, \tag{13.13a}$$

$$S_{z,m_s' m_s} = \langle\tfrac{1}{2}m_s'|\hat{S}_z|\tfrac{1}{2}m_s\rangle = m_s\hbar\delta_{m_s' m_s}, \tag{13.13b}$$

$$S_{\pm,m_s' m_s} = \langle\tfrac{1}{2}m_s'|\hat{S}_{\pm}|\tfrac{1}{2}m_s\rangle = \hbar\sqrt{\tfrac{3}{4} - m_s(m_s \pm 1)}\delta_{m_s',m_s\pm 1}, \tag{13.13c}$$

and

$$\left\{\begin{matrix} S_{x,m_s' m_s} \\ iS_{y,m_s' m_s} \end{matrix}\right\} = \langle\tfrac{1}{2}m_s'|\left\{\begin{matrix} \hat{S}_x \\ i\hat{S}_y \end{matrix}\right\}|\tfrac{1}{2}m_s\rangle = \frac{\hbar}{2}[\sqrt{\tfrac{3}{4} - m_s(m_s + 1)}$$
$$\times\, \delta_{m_s',m_s+1} \pm \sqrt{\tfrac{3}{4} - m_s(m_s - 1)}\delta_{m_s',m_{s-1}}]. \tag{13.13d}$$

As with previous eigenstate representations, the matrix version is indicated by the same capital letter but without an overlaid carat. Labeling the rows by m_s and the columns by m_s', with $m_s' = m_s = \frac{1}{2}$ being the upper left-hand element, the full matrices S^2, S_z, $S_\pm$, S_x, and S_y are

$$S^2 = \frac{3\hbar^2}{4}\begin{pmatrix} 1 & 0 \\ 0 & 1 \end{pmatrix}, \tag{13.14a}$$

$$S_z = \frac{\hbar}{2}\begin{pmatrix} 1 & 0 \\ 0 & -1 \end{pmatrix}, \tag{13.14b}$$

$$S_+ = \hbar\begin{pmatrix} 0 & 1 \\ 0 & 0 \end{pmatrix}, \tag{13.14c}$$

$$S_- = \hbar\begin{pmatrix} 0 & 0 \\ 1 & 0 \end{pmatrix}, \tag{13.14d}$$

$$S_x = \frac{\hbar}{2}\begin{pmatrix} 0 & 1 \\ 1 & 0 \end{pmatrix}, \tag{13.14e}$$

and

$$S_y = \frac{\hbar}{2}\begin{pmatrix} 0 & -i \\ i & 0 \end{pmatrix}, \tag{13.14f}$$

while the normalized real column vectors are $\binom{1}{0}$ and $\binom{0}{1}$:

$$|\tfrac{1}{2}\tfrac{1}{2}\rangle \rightarrow \begin{pmatrix} 1 \\ 0 \end{pmatrix}, \qquad |\tfrac{1}{2}, -\tfrac{1}{2}\rangle \rightarrow \begin{pmatrix} 0 \\ 1 \end{pmatrix}. \tag{13.14g}$$

These column vectors are sometimes denoted α and β, respectively; since they are eigenstates of S_z, we shall append the subscript z to them:

$$\alpha_z = \begin{pmatrix} 1 \\ 0 \end{pmatrix}, \qquad \beta_z = \begin{pmatrix} 0 \\ 1 \end{pmatrix}. \tag{13.14h}$$

EXAMPLE

A mixed state. Let the state $|\gamma\rangle$ of a spin-$\frac{1}{2}$ particle be a linear combination of the "spin up" ($|\frac{1}{2}\frac{1}{2}\rangle$ or α_z) and "spin down" ($|\frac{1}{2}, -\frac{1}{2}\rangle$ or β_z) states:

$$|\gamma\rangle = a|\tfrac{1}{2}\tfrac{1}{2}\rangle + b|\tfrac{1}{2}, -\tfrac{1}{2}\rangle, \tag{13.15a}$$

or, in the eigenstate representation,

$$\begin{pmatrix} \gamma_1 \\ \gamma_2 \end{pmatrix} = \begin{pmatrix} a \\ b \end{pmatrix}, \tag{13.15b}$$

where a and b are assumed known and $|a|^2 + |b|^2 = 1$. The expectation values of $\hat{S}^2$ and $\hat{S}_z$, $\langle\gamma|\hat{S}^2|\gamma\rangle$ and $\langle\gamma|\hat{S}_z|\gamma\rangle$, are readily found from Eqs. (13.4) to be

$$\langle\gamma|\hat{S}^2|\gamma\rangle = \tfrac{3}{4}\hbar^2$$

and

$$\langle\gamma|\hat{S}_z|\gamma\rangle = \frac{\hbar}{2}(|a|^2 - |b|^2). \tag{13.15c}$$

The eigenstate representation is an easy means of determining the effects of $\hat{S}_\pm$ on $|\gamma\rangle$:

$$S_+\begin{pmatrix}\gamma_1\\ \gamma_2\end{pmatrix} = \hbar\begin{pmatrix}\gamma_2\\ 0\end{pmatrix} = \gamma_2\hbar\alpha_z = b\hbar\alpha_z$$

and

$$S_-\begin{pmatrix}\gamma_1\\ \gamma_2\end{pmatrix} = \hbar\begin{pmatrix}0\\ \gamma_1\end{pmatrix} = \gamma_1\hbar\beta_z = a\hbar\beta_z.$$

Pauli spin operators/matrices

Since $\hbar$ is the unit in which $\hat{\mathbf{S}}$ is expressed, and $\pm\hbar/2$ are the eigenvalues of $\hat{S}_z$, Pauli (1927) introduced spin operators and matrices with the $\hbar/2$ factored out. These quantities are denoted by the symbols $\hat{\boldsymbol{\sigma}}$ and $\boldsymbol{\sigma}$:

$$\hat{\mathbf{S}} = \frac{\hbar}{2}\hat{\boldsymbol{\sigma}} \tag{13.16a}$$

and

$$\mathbf{S} = \frac{\hbar}{2}\boldsymbol{\sigma}, \tag{13.16b}$$

where, from Eqs. (13.14), the components of $\boldsymbol{\sigma}$ are

$$\sigma_x = \begin{pmatrix}0 & 1\\ 1 & 0\end{pmatrix}, \qquad \sigma_y = \begin{pmatrix}0 & -i\\ i & 0\end{pmatrix}, \qquad \sigma_z = \begin{pmatrix}1 & 0\\ 0 & -1\end{pmatrix}. \tag{13.17}$$

The Pauli spin matrices enjoy the property that each of their squares is the unit matrix:

$$\sigma_x^2 = \sigma_y^2 = \sigma_z^2 = I_2 = \begin{pmatrix}1 & 0\\ 0 & 1\end{pmatrix}.$$

A similar relation necessarily holds for the operators:

$$\hat{\sigma}_x^2 = \hat{\sigma}_y^2 = \hat{\sigma}_z^2 = \hat{I}.$$

The commutation relations for the $\hat{\sigma}_j$ follow from Eq. (13.7):

$$[\hat{\sigma}_j, \hat{\sigma}_k] = i\epsilon_{jk\ell}\hat{\sigma}_\ell; \tag{13.18}$$

these can be proved for the σ_j by matrix multiplication.

Equations (13.14) lead to simple results, for example,

$$\hat{\sigma}^2|\tfrac{1}{2}m_s\rangle = 3|\tfrac{1}{2}m_s\rangle, \tag{13.19a}$$

or

$$\sigma_z\begin{Bmatrix}\alpha_z\\ \beta_z\end{Bmatrix} = \begin{Bmatrix}\alpha_z\\ -\beta_z\end{Bmatrix}. \tag{13.19b}$$

Working with the σ_j often leads to some simplification in analysis or proofs, as will be seen in the next subsection. One example is the matrix form of $\hat{H}^{(S)}_{\text{mag}}$. Expressed in terms of $\hat{\boldsymbol{\sigma}}$, the magnetic moment operator $\hat{H}^{(S)}_{\text{mag}}$ – Eq. (13.5) – becomes

$$\hat{H}^{(S)}_{\text{mag}} = -\frac{g}{2}\frac{q}{|q|}\mu\mathbf{B}\cdot\hat{\boldsymbol{\sigma}}, \tag{13.20}$$

which in the eigenstate representation goes over to the matrix

$$H^{(S)}_{\text{mag}} = -\frac{g}{2}\frac{q\mu}{|q|}\begin{pmatrix} B_z & B_x - iB_y \\ B_x + iB_y & -B_z \end{pmatrix}, \tag{13.21}$$

a form we shall make use of later. An important relation, which we also apply subsequently, is

$$(\hat{\boldsymbol{\sigma}}\cdot\hat{\mathbf{A}})(\hat{\boldsymbol{\sigma}}\cdot\hat{\mathbf{B}}) = \hat{\mathbf{A}}\cdot\hat{\mathbf{B}}\hat{I} + i\hat{\boldsymbol{\sigma}}\cdot(\hat{\mathbf{A}}\times\hat{\mathbf{B}}), \tag{13.22}$$

where the unit operator is inserted into the first term on the RHS of (13.22) as a reminder that, in the eigenstate representation, (13.22) is a relation between matrices:

$$(\boldsymbol{\sigma}\cdot\hat{\mathbf{A}})(\boldsymbol{\sigma}\cdot\hat{\mathbf{B}}) = \hat{\mathbf{A}}\cdot\hat{\mathbf{B}}I_2 + \boldsymbol{\sigma}\cdot(\hat{\mathbf{A}}\times\hat{\mathbf{B}}). \tag{13.22$'$}$$

The latter form of the relation shows explicitly how one can mix spin matrices with non-spin operators. Of course, (13.22) holds also if the operators $\hat{\mathbf{A}}$ and $\hat{\mathbf{B}}$ are replaced by ordinary 3-D vectors **a** and **b**. The proof of these relations is left to the exercises.

Spin quantization along an arbitrary axis

While spin states have no coordinate representation, $\hat{\mathbf{S}}$ is an ordinary vector operator whose components can be referred to any set of coordinate axes. Often the axis of quantization will be defined as the direction of the dominant magnetic-field component in a S–G experiment. So far this has been B_z, and therefore the eigenstates have been those of $\hat{S}_z$. However, we shall later be interested in S–G experiments in which B_x, B_y or the component $B_n = \mathbf{n}\cdot\mathbf{B}$ predominates, where **n** is an arbitrary direction. The goal of this subsection is to obtain the eigenstates of $\hat{S}_n = \mathbf{n}\cdot\hat{\mathbf{S}}$, where **n** is the arbitrary unit vector just introduced, with co-latitude θ and azimuthal angle ϕ; see Fig. 13.3. Note that $\theta = 0 \Rightarrow \mathbf{n} = \mathbf{e}_3$; $\theta = \pi/2$, $\phi = 0 \Rightarrow \mathbf{n} = \mathbf{e}_1$; and $\theta = \phi = \pi/2 \Rightarrow \mathbf{n} = \mathbf{e}_2$: all possibilities are included in this general case.

By quantizing along **n**, the eigenvalue problem to be solved is

$$\hat{S}_n|\mathbf{n}, \lambda\rangle, = \frac{\lambda\hbar}{2}|\mathbf{n}, \lambda\rangle, \tag{13.23}$$

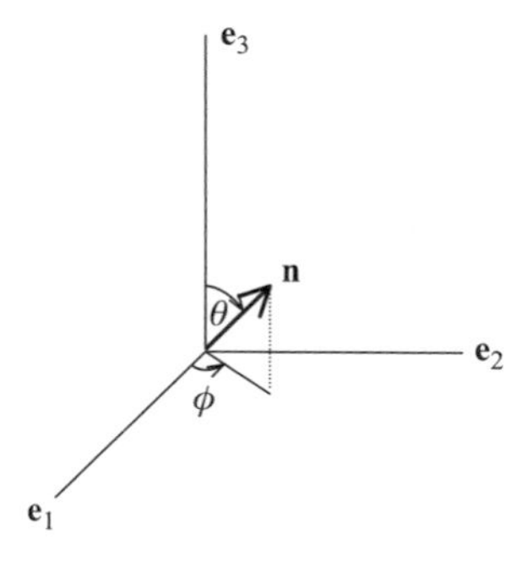

Fig. 13.3 The unit vector **n** and the polar angles θ and ϕ defining its orientation.

where

$$\hat{S}_n = \mathbf{n} \cdot \hat{\mathbf{S}} = \frac{\hbar}{2} \hat{\boldsymbol{\sigma}} \cdot \mathbf{n} = \frac{\hbar}{2} (n_1 \hat{\sigma}_1 + n_2 \hat{\sigma}_2 + n_3 \hat{\sigma}_3), \tag{13.24}$$

so that (13.23) becomes

$$\sum_{j=1}^{3} n_j \hat{\sigma}_j |\mathbf{n}, \lambda\rangle, = \lambda |\mathbf{n}, \lambda\rangle. \tag{13.25}$$

The usual way of proceeding is to expand $|\mathbf{n}, \lambda\rangle$ in terms of known basis states, here the $|\frac{1}{2} m_s\rangle$, and then determine the expansion coefficients and the eigenvalue λ. This is the procedure we shall follow, with one wrinkle: we shall use the eigenstate representation, in which $\hat{\sigma}_j \rightarrow \sigma_j$, and $|\mathbf{n}, \lambda\rangle$ becomes a two-rowed column vector:

$$|\mathbf{n}, \lambda\rangle \rightarrow \begin{pmatrix} a \\ b \end{pmatrix} \equiv a \begin{pmatrix} 1 \\ 0 \end{pmatrix} + b \begin{pmatrix} 0 \\ 1 \end{pmatrix}. \tag{13.26}$$

Recalling the definitions (13.17) and noting from Fig. 13.3 that

$$n_1 = \sin\theta \cos\phi, \qquad n_2 = \sin\theta \sin\phi, \qquad n_3 = \cos\theta,$$

the matrix form of the eigenvalue equation (13.25) is easily found to be

$$\begin{pmatrix} \cos\theta & \sin\theta\, e^{-i\phi} \\ \sin\theta\, e^{i\phi} & -\cos\theta \end{pmatrix} \begin{pmatrix} a \\ b \end{pmatrix} = \lambda \begin{pmatrix} a \\ b \end{pmatrix}. \tag{13.27}$$

(Equation (13.27) exemplifies one of the delights of dealing with spin $\frac{1}{2}$ and the Pauli matrices: the mathematics is simple – 2×2 matrix diagonalization, and special functions (Hermites, Laguerres, etc.) are absent.)

A non-trivial solution to (13.27) exists only if the secular determinant vanishes, viz.,

$$\begin{vmatrix} \cos\theta - \lambda & \sin\theta\, e^{-i\phi} \\ \sin\theta\, e^{i\phi} & -(\cos\theta + \lambda) \end{vmatrix} = 0,$$

from which we find

$$\lambda = \pm 1.$$

Hence, the eigenvalues of σ_n are the same as those of σ_z. Equivalently, the eigenvalues of $\hat{S}_n$ are $\pm\hbar/2$. Re-orientation (i.e., quantization along $\mathbf{n}$) has not changed the eigenvalues, which, of course, must be invariant w.r.t. rotations.

Corresponding to $\lambda = \pm 1$ will be coefficients $a^{(\pm)}$ and $b^{(\pm)}$. Choosing $\lambda = +1$ in (13.27) leads to

$$(\cos\theta - 1) a^{(+)} = -b^{(+)} \sin\theta\, e^{-i\phi}, \tag{13.28}$$

while normalization provides a second relation:

$$|a^{(+)}|^2 + |b^{(+)}|^2 = 1. \tag{13.29}$$

A nice simplification occurs if one works with $\theta/2$ in place of θ. Using $\cos\theta - 1 = -2\sin^2(\theta/2)$ and $\sin\theta = 2\sin(\theta/2)\cos(\theta/2)$, (13.28) becomes

$$b^{(+)} = \tan\left(\frac{\theta}{2}\right) e^{i\phi} a^{(+)},$$

which, when it is substituted into (13.29), leads to

$$a^{(+)} = e^{i\delta^{(+)}} \cos\left(\frac{\theta}{2}\right)$$

and

$$b^{(+)} = e^{i\delta^{(+)}} \sin\left(\frac{\theta}{2}\right) e^{i\phi},$$

where $\delta^{(+)}$ is an arbitrary phase. The standard convention is to choose $\delta^{(+)} = 0$, resulting in

$$a^{(+)} = \cos\left(\frac{\theta}{2}\right), \qquad b^{(+)} = \sin\left(\frac{\theta}{2}\right) e^{i\phi}. \tag{13.30}$$

The coefficients $a^{(-)}$ and $b^{(-)}$ can be obtained in the same way, but, since spin $\frac{1}{2}$ is a two-state system, it is simpler to infer the values from the requirement that the state $\binom{a^{(-)}}{b^{(-)}}$ be orthogonal to $\binom{a^{(+)}}{b^{(+)}}$. One finds, with the conventional choice of the phase factor, that

$$a^{(-)} = \sin\left(\frac{\theta}{2}\right), \qquad b^{(-)} = -\cos\left(\frac{\theta}{2}\right) e^{i\phi}. \tag{13.31}$$

Collecting these results, we have the following solutions to the σ_n eigenvalue problem:

$$\lambda = +1\colon \begin{pmatrix} \cos(\theta/2) \\ \sin(\theta/2)\, e^{i\phi} \end{pmatrix}; \qquad \lambda = -1\colon \begin{pmatrix} \sin(\theta/2) \\ -\cos(\theta/2)\, e^{i\phi} \end{pmatrix}. \tag{13.32}$$

A direct carry-over to the state-vector form yields ($|\mathbf{n}, \lambda = \pm 1\rangle \to |\mathbf{n}, \pm\rangle$)

$$|\mathbf{n}, +\rangle = \cos\left(\frac{\theta}{2}\right) |\tfrac{1}{2}\tfrac{1}{2}\rangle + \sin\left(\frac{\theta}{2}\right) e^{i\phi} |\tfrac{1}{2}, -\tfrac{1}{2}\rangle \tag{13.33a}$$

and

$$|\mathbf{n}, -\rangle = \sin\left(\frac{\theta}{2}\right) |\tfrac{1}{2}\tfrac{1}{2}\rangle - \cos\left(\frac{\theta}{2}\right) e^{i\phi} |\tfrac{1}{2}, -\tfrac{1}{2}\rangle. \tag{13.33b}$$

$|\mathbf{n}, +\rangle$ and $|\mathbf{n}, -\rangle$ are sometimes referred to as the "spin-up" and the "spin-down" eigenstates of $\hat{S}_n$, just as $|\frac{1}{2}\frac{1}{2}\rangle$ and $|\frac{1}{2}, -\frac{1}{2}\rangle$ are the spin-up and spin-down eigenstates of $\hat{S}_z$.

The $|\mathbf{n}, \pm\rangle$, are the normalized eigenstates of $\hat{S}_n$, expressed in terms of the standard spin-$\frac{1}{2}$ basis $\{|\frac{1}{2}m_s\rangle\}$. Equations (13.33) could be inverted to express $|\frac{1}{2}m_s\rangle$ in terms of the $|\mathbf{n}, \pm\rangle$, but there is no direct need because $\{|\frac{1}{2}m_s\rangle\}$ is the chosen basis. The states $|\mathbf{n}, \pm\rangle$ are those that would be produced by a S–G apparatus with **B** predominantly in the **n**, rather than in the z, direction. Thus, a S–G apparatus oriented along $\mathbf{e}_2$ would generate the states

$$|\mathbf{e}_2, +\rangle = |S_y, +\rangle = \sqrt{\tfrac{1}{2}}[|\tfrac{1}{2}\tfrac{1}{2}\rangle + i|\tfrac{1}{2}, -\tfrac{1}{2}\rangle] \tag{13.34a}$$

and

$$|\mathbf{e}_2, -\rangle = |S_y, -\rangle = \sqrt{\tfrac{1}{2}}[|\tfrac{1}{2}\tfrac{1}{2}\rangle - i|\tfrac{1}{2}, -\tfrac{1}{2}\rangle]. \tag{13.34b}$$

These states, which will be applied in the next subsection, should be compared with those of Eq. (4.110).

The spin-$\frac{1}{2}$ states encountered so far are eigenstates of $\hat{S}_z$, $\hat{S}_n$, etc. A general spin-$\frac{1}{2}$ state will be a linear combination of the base states $|\frac{1}{2}m_s\rangle$. We will denote the generic spin-$\frac{1}{2}$ state by the symbol $|\frac{1}{2}, \zeta\rangle$:

$$|\tfrac{1}{2}, \zeta\rangle = a_{1/2}|\tfrac{1}{2}, \tfrac{1}{2}\rangle + a_{-1/2}|\tfrac{1}{2}, -\tfrac{1}{2}\rangle, \tag{13.35}$$

where the $a_{\pm 1/2}$ obey $|a_{1/2}|^2 + |a_{-1/2}|^2 = 1$. The generic state $|\frac{1}{2}, \zeta\rangle$ is an eigenstate of $\hat{S}^2$ but need not be an eigenstate of any component:

$$\hat{S}^2|\tfrac{1}{2}, \zeta\rangle = \frac{3\hbar^2}{4}|\tfrac{1}{2}, \zeta\rangle.$$

We will use this generic state in Section 13.3.

Regeneration: now you see it, now you don't, now you do

Intrinsic spin is a remarkable attribute. In this subsection we establish one of its most remarkable features – regeneration. To begin, suppose that a beam of spin-$\frac{1}{2}$ particles is passed through an $\mathbf{e}_3$-oriented S–G apparatus. Two separated beams will be produced, one containing particles in the state $|\frac{1}{2}\frac{1}{2}\rangle$ $(=|\mathbf{e}_3, +\rangle)$, the other, particles in the state $|\frac{1}{2}, -\frac{1}{2}\rangle$. Next, the beam with particles in state $|\frac{1}{2}, -\frac{1}{2}\rangle$) is passed through a second S–G apparatus, this one oriented along $\mathbf{e}_2$. The particles in state $|\frac{1}{2}\frac{1}{2}\rangle$ are prevented from entering this second S–G apparatus, so the $|\frac{1}{2}\frac{1}{2}\rangle$ component of the $\hat{S}_z$ eignestates is extinguished w.r.t. the second apparatus.

The second S–G apparatus generates separated beams of particles in the states $|\mathbf{e}_2, \pm\rangle$, with probabilities $P_\pm$ given by (Postulate II (b))

$$P_\pm = |\langle\tfrac{1}{2}, -\tfrac{1}{2}|\mathbf{e}_2, \pm\rangle|^2 = \tfrac{1}{2}.$$

This is already a noteworthy result: particles in state $|\frac{1}{2}, -\frac{1}{2}\rangle$ have been transformed into particles in states $|\mathbf{e}_2, \pm\rangle$, even though there were none present in these states initially. Classically such a result cannot occur: red blocks do not turn into green blocks.

The final step is an analog of the preceding one: either – but only one – of the two beams ($|\mathbf{e}_2, +\rangle$ or $|\mathbf{e}_2, -\rangle$) is passed through a S–G apparatus with its predominant magnetic-field component being B_z. This apparatus also creates two beams of particles, but now the states are the same as those from the first S–G apparatus: $|\frac{1}{2}\frac{1}{2}\rangle$ and $|\frac{1}{2}, -\frac{1}{2}\rangle$! Quantum physics has allowed us to create, extinguish, and then recreate (regenerate) a particular state (in this case $|\frac{1}{2}\frac{1}{2}\rangle$): now you see it, now you don't, now you do. However, this is not a universal result: regeneration occurs because the intermediate states ($|\mathbf{e}_2, \pm\rangle$) belong to the operator $\hat{S}_y$ which does not commute with $\hat{S}_z$. This point is crucial. For instance, if a particle were to start in a linear combination of two eigenstates of an operator $\hat{A}$, and a measurement isolated one of them, then a subsequent attempt to measure the other would always give zero: the second state would remain extinguished until an operator not commuting with $\hat{A}$ were allowed to act on the system.

Regeneration is not limited to spin-$\frac{1}{2}$ systems: it also arises in the decay of K mesons, as discussed, e.g., by Baym (1976). Other remarkable features peculiar to spin $\frac{1}{2}$ will be discussed in Chapter 14.

13.3. Magnetic-field phenomena

The preceding section dealt, in effect, with the kinematics of spin $\frac{1}{2}$. We turn to dynamics in this section, in particular, to the behavior of spin-$\frac{1}{2}$ particles in magnetic fields. The main topics discussed are spin precession and magnetic resonance. Since this will be done in the framework of appropriate Hamiltonians, a general Hamiltonian for spin-$\frac{1}{2}$ particles in external electromagnetic fields is needed. The first subsection is concerned with this topic.

The Pauli Hamiltonian

It has been remarked that spin $\frac{1}{2}$ is an inherent feature of the relativistic Dirac equation. A non-relativistic reduction of this equation leads to the Schrödinger equation with a special Hamiltonian, known as the Pauli Hamiltonian, that involves $\hat{\boldsymbol{\sigma}}$ (see, e.g., Baym (1976)). This Hamiltonian will be denoted $\hat{H}_\sigma^{(A)}(\mathbf{r}, t)$, and is given by

$$\hat{H}_\sigma^{(A)}(\mathbf{r}, t) = \frac{1}{2m}(\hat{\boldsymbol{\sigma}} \cdot \hat{\mathbf{P}} - q\kappa_c \mathbf{A})^2 + q\Phi + V(\mathbf{r}), \tag{13.36}$$

where the particle has mass m and charge q.

Equation (13.36) can be simplified by applying (13.22) to the term in parentheses:

$$\begin{aligned}(\hat{\boldsymbol{\sigma}} \cdot \hat{\mathbf{P}} - q\kappa_c \mathbf{A})^2 &= [\hat{\boldsymbol{\sigma}} \cdot (\hat{\mathbf{P}} - q\kappa_c \mathbf{A})][\hat{\boldsymbol{\sigma}} \cdot (\hat{\mathbf{P}} - q\kappa_c \mathbf{A})] \\ &= (\hat{\mathbf{P}} - q\kappa_c \mathbf{A})^2 + i\hat{\boldsymbol{\sigma}} \cdot [(\hat{\mathbf{P}} - q\kappa_c \mathbf{A}) \times (\hat{\mathbf{P}} - q\kappa_c \mathbf{A})].\end{aligned}$$

The surviving terms in the cross product are

$$\begin{aligned}-q\kappa_c[(\hat{\mathbf{P}} \times \mathbf{A}) + \mathbf{A} \times \hat{\mathbf{P}}] &= iq\kappa_c\hbar[(\nabla \times \mathbf{A} + \mathbf{A} \times \nabla)] \\ &= iq\kappa_c\hbar[(\vec{\nabla} \times \mathbf{A}) + (\nabla) \times \mathbf{A} + \mathbf{A} \times \nabla] \\ &= iq\kappa_c\hbar(\nabla \times \mathbf{A}) = iq\kappa_c\hbar\mathbf{B},\end{aligned}$$

where $\mathbf{B}$ need not be constant. Hence, $\hat{H}_\sigma^{(A)}(\mathbf{r}, t)$ becomes

$$\hat{H}_\sigma^{(A)}(\mathbf{r}, t) = \frac{1}{2m}(\hat{\mathbf{P}} - q\kappa_c \mathbf{A})^2 - \frac{q\kappa_c\hbar}{2m}\hat{\boldsymbol{\sigma}} \cdot \mathbf{B} + q\Phi + V(\mathbf{r}). \tag{13.37}$$

With regard to the standard electromagnetic Hamiltonian, $\hat{H}_\sigma^{(A)}(\mathbf{r}, t)$ contains just the one new term

$$\begin{aligned}-\frac{q\kappa_c\hbar}{2m}\hat{\boldsymbol{\sigma}} \cdot \mathbf{B} &= -2\frac{q\kappa_c}{2m}\hat{\mathbf{S}} \cdot \mathbf{B} \\ &= -2\frac{q}{|q|}\frac{\mu}{\hbar}\hat{\mathbf{S}} \cdot \mathbf{B} \equiv \hat{H}_{\text{mag}}^{(\text{S})},\end{aligned} \tag{13.38}$$

where the predicted g factor is 2. This result was a triumph of the relativistic Dirac equation for the electron, and it was not until the experiments of Lamb and Retherford (1947) on the spectrum of the H-atom and the consequent development of quantum electrodynamics that it was realized that g for the electron, like that for the proton and the neutron, was not equal to 2. As with Eq. (13.5), we shall replace the factor 2 in (13.38) by g, thereby recovering Eq. (13.5) and turning (13.36) into

$$\hat{H}_\sigma^{(A)}(\mathbf{r},\, t) = \frac{1}{2m}(\hat{\mathbf{P}} - q\kappa_c\mathbf{A})^2 + \hat{\mathrm{H}}_{\mathrm{mag}}^{(\mathrm{S})} + q\Phi + V(\mathbf{r}). \tag{13.39}$$

Equation (13.39) defines a completely general Hamiltonian, which can enter either the time-dependent or the time-independent Schrödinger equation. Single-particle, time-dependent solutions obtained with $\hat{H}_\sigma^{(A)}(\mathbf{r},\, t)$ will be denoted $\Psi_\sigma^{(A)}(\mathbf{r},\, t)$:

$$i\hbar\frac{\partial}{\partial t}\Psi_\sigma^{(A)}(\mathbf{r},\, t) = \hat{H}_\sigma^{(A)}(\mathbf{r},\, t)\Psi_\sigma^{(A)}(\mathbf{r},\, t). \tag{13.40}$$

When $\mathbf{A}$ and Φ are time-independent, the eigenstates of $\hat{H}_\sigma^{(A)}(\mathbf{r},\, t)$ will be written $\psi_\sigma^{(A)}(\mathbf{r})$:

$$\hat{H}_\sigma^{(A)}(\mathbf{r})\psi_\sigma^{(A)}(\mathbf{r}) = E_\sigma^{(A)}\psi_\sigma^{(A)}(\mathbf{r}). \tag{13.41}$$

The spin dependence of $\psi_\sigma^{(A)}(\mathbf{r})$ can be obtained immediately, since $[\hat{S}^2,\, \hat{H}_\sigma^{(A)}(\mathbf{r})] = 0$:

$$\psi_\sigma^{(A)}(\mathbf{r}) = \psi_\zeta^{(A)}(\mathbf{r})|\tfrac{1}{2},\, \zeta\rangle. \tag{13.42}$$

This result clearly holds when $\mathbf{A} = \Phi = 0$: the space and the spin dependences always decouple into a product state, though the energy may depend on the (magnetic-type) spin quantum number ζ, so it has been appended to the space part $\psi_\zeta^{(A)}(\mathbf{r})$. As an example, consider an uncharged, free, spin-$\frac{1}{2}$ particle. Its spatial state is the plane wave $\exp(i\mathbf{k}\cdot\mathbf{r})/(2\pi)^{3/2}$ and its spin state is $|\frac{1}{2}m_s\rangle$, so that its overall "wave function" $\psi_{\mathbf{k},m_s}(\mathbf{r})$ is

$$\psi_{\mathbf{k},m_s}(\mathbf{r}) = \left(\frac{1}{2\pi}\right)^{3/2} e^{i\mathbf{k}\cdot\mathbf{r}}|\tfrac{1}{2}m_s\rangle. \tag{13.43}$$

The cases that will be of interest to us are those for which $\mathcal{E} = 0$ and $\mathbf{B}(\mathbf{r},\, t) = \mathbf{B}(t)$, i.e., spatially uniform magnetic fields. A convenient gauge for the latter situation is $\mathbf{A}_1 = \frac{1}{2}\mathbf{B}\times\mathbf{r}$, for which we use the results (12.16) and (12.17) to transform (13.39) into

$$\hat{H}_\sigma^{(A_1)}(\mathbf{r},\, t) = \hat{H}^{(0)}(\mathbf{r}) + \hat{H}_{\mathrm{mag}}^{(L)} + \frac{q^2\kappa_c^2 r_\perp^2 B^2}{8m} + \hat{H}_{\mathrm{mag}}^{(S)}(t) \tag{13.44a}$$

$$\equiv \hat{H}^{(A_1)}(\mathbf{r},\, t) + \hat{H}_{\mathrm{mag}}^{(S)}(t), \tag{13.44b}$$

where $\hat{H}^{(A_1)}$ is given by (12.13) with the proviso that $\mathbf{B}$ can be a function of t and a t dependence is specified in $\hat{H}_{\mathrm{mag}}^{(S)}$.

We have used $[\hat{\mathbf{S}},\, \hat{H}_\sigma^{(A)}] = 0$ in the preceding wave-function analysis, but we can now take that analysis much further: because $\mathbf{B} \neq \mathbf{B}(\mathbf{r})$, the two terms in (13.44b) commute:

$$[\hat{H}^{(A_1)}(\mathbf{r},\, t),\, \hat{H}_{\mathrm{mag}}^{(S)}(t)] = 0. \tag{13.45}$$

This is a most important relation, since, as will be demonstrated next, it allows us to concentrate exclusively on the behavior of the spin, without needing to know any details about the spatial behavior.

The demonstration is in two parts. We assume first that $\mathbf{B} \neq \mathbf{B}(t)$, and write $\hat{H}_\sigma^{(A_1)}(\mathbf{r},\, t) \to \hat{H}_\sigma^{(B)}(\mathbf{r})$. Then the time-evolution operator corresponding to $\hat{H}_\sigma^{(B)}(\mathbf{r})$ is

$$\begin{aligned}\hat{U}_\sigma^{(B)}(t) &= e^{-i\hat{H}_\sigma^{(B)}(\mathbf{r})t/\hbar} \\ &= e^{-i\hat{H}^{(A_1)}(\mathbf{r})t/\hbar}e^{-i\hat{H}_{\text{mag}}^{(S)}t/\hbar} \\ &\equiv \hat{U}^{(A_1)}(t)\hat{U}^{(S)}(t).\end{aligned} \tag{13.46}$$

In other words, the time evolutions of the space and the spin parts are unlinked: they occur independently and one has

$$\Psi_\sigma^{(A_1)}(\mathbf{r},\, t) = \Psi^{(A_1)}(\mathbf{r},\, t)|\zeta,\, t\rangle, \tag{13.47}$$

where

$$|\zeta,\, t\rangle = e^{-i\hat{H}_{\text{mag}}^{(S)}t/\hbar}|\tfrac{1}{2},\, \zeta\rangle, \tag{13.48}$$

$|\frac{1}{2},\, \zeta\rangle$ being the spin state at time $t = 0$. For this simple case of $\mathbf{B} \neq \mathbf{B}(t)$, we see that the time evolution of the spin state can indeed be analyzed independently of the space part.

An analogous result continues to hold for the case of interest, viz., $\mathbf{B} = \mathbf{B}(t)$. Noting that $[\hat{H}_\sigma^{(A_1)}(\mathbf{r},\, t_1),\, \hat{H}_\sigma^{(A_1)}(\mathbf{r},\, t_2)] = 0$, $t_1 \neq t_2$, Eq. (5.90) leads to ($t_0 = 0$)

$$\begin{aligned}\hat{U}_\sigma^{(A_1)}(t) &= e^{-(i/\hbar)\int_0^t dt'\, [\hat{H}^{(A_1)}(\mathbf{r},t') + \hat{H}_{\text{mag}}^{(S)}(t')]} \\ &= e^{-(i/\hbar)[\int_0^t dt'\, \hat{H}^{(A_1)}(\mathbf{r},t') + \int_0^t dt'\, \hat{H}_{\text{mag}}^{(S)}(t')]} \\ &= e^{-(i/\hbar)\int_0^t dt'\, \hat{H}^{(A_1)}(\mathbf{r},t')}e^{-(i/\hbar)\int_0^t dt'\, \hat{H}_{\text{mag}}^{(S)}(t')} \\ &= \hat{U}^{(A_1(t))}(t)\hat{U}^{(S(t))}(t).\end{aligned}$$

Although the time dependence is no longer a simple exponential, the time-evolution operator for the case of a time-dependent, spatially uniform magnetic field continues to be a product of time-evolution operators, one for the space part, the other for the spin part. As claimed, we can concentrate exclusively on the behavior of the spin state, which we do in the next three subsections. The underlying dynamics is based on the following Schrödinger equation:

$$i\hbar\frac{\partial|\zeta,\, t\rangle}{\partial t} = \hat{H}_{\text{mag}}^{(S)}|\zeta,\, t\rangle. \tag{13.49}$$

Magnetic-moment operators

The general form of $\hat{H}_{\text{mag}}^{(S)}$,

$$\hat{H}_{\text{mag}}^{(S)} = -g\frac{q}{|q|}\frac{\mu}{\hbar}\hat{\mathbf{S}}\cdot\mathbf{B},$$

is often written in one of two other ways, which the reader may well encounter in other studies of this subject. These are

$$\hat{H}_{\text{mag}}^{(S)} = -\frac{q}{|q|}\gamma\hat{\mathbf{S}}\cdot\mathbf{B}, \tag{13.50a}$$

where $\gamma = g\mu/\hbar$ is known as the gyromagnetic ratio, and

$$\hat{H}^{(S)}_{\rm mag} = -\hat{\boldsymbol{\mu}}_S \cdot \mathbf{B}, \tag{13.50b}$$

where, in analogy to the orbital magnetic-moment operator $\hat{\mu}_L$ – Eq. (12.19) – the spin-magnetic-moment operator $\hat{\mu}_S$ is defined by

$$\hat{\mu}_S \equiv g \frac{q}{|q|} \frac{\mu}{\hbar} \hat{\mathbf{S}}. \tag{13.51}$$

Any charged particle with non-zero spin has a measurable (spin) magnetic moment $\bar{\mu}$, defined by the following expectation value:

$$\bar{\mu} = \langle S, m_S = S|\hat{\mu}_{S,z}|S, m_S = S\rangle. \tag{13.52}$$

Three particles whose magnetic moments (g factors) have been measured carefully and analyzed in great detail theoretically are the electron (e), the proton (p) and the neutron (n). Even though $q_{\rm n} = 0$, $\mu_{\rm n}$ and $g_{\rm n}$ are non-zero! The Bohr magneton $\mu_{\rm B} = e\hbar\kappa_{\rm c}/(2m_{\rm e})$ is the unit for the electron, whereas the nuclear magneton $\mu_{\rm N} = e\hbar\kappa_{\rm c}/(2m_{\rm p})$ is the unit for the proton and neutron, where $m_{\rm p}$ is the proton mass. (The values of $\mu_{\rm B}$ and $\mu_{\rm N}$ are given below Eq. (12.20).) The magnetic-moment operators for these three spin-$\frac{1}{2}$ particles are

$$\hat{\boldsymbol{\mu}}_{\rm e} = -g_{\rm e} \frac{\mu_{\rm B}}{\hbar} \hat{\mathbf{S}}, \qquad g_{\rm e} \cong 2, \tag{13.53a}$$

$$\hat{\boldsymbol{\mu}}_{\rm p} = g_{\rm p} \frac{\mu_{\rm N}}{\hbar} \hat{\mathbf{S}}, \qquad g_{\rm p} \cong 5.58, \tag{13.53b}$$

and

$$\hat{\boldsymbol{\mu}}_{\rm n} = g_{\rm n} \frac{\mu_{\rm N}}{\hbar} \hat{\mathbf{S}}, \qquad g_{\rm n} \cong -3.82. \tag{13.53c}$$

We shall return to the problem of determining the g values shortly, but first we simply substitute the expressions (13.53) for the operators into (13.52) and obtain the values of the magnetic moments for the electron ($\bar{\mu}_{\rm e}$), the proton ($\bar{\mu}_{\rm p}$) and the neutron ($\bar{\mu}_{\rm n}$):

$$\bar{\mu}_{\rm e} \cong -\mu_{\rm B} \qquad (g_{\rm e} \cong 2), \tag{13.54a}$$

$$\bar{\mu}_{\rm p} \cong 2.79\mu_{\rm N}, \tag{13.54b}$$

and

$$\bar{\mu}_{\rm n} \cong -1.91\mu_{\rm N}. \tag{13.54c}$$

If $|\bar{\mu}_{\rm e}| \cong \mu_{\rm B}$ is taken as the norm, then one might naively have expected

$$\bar{\mu}_{\rm p} \cong \mu_{\rm N}, \qquad \bar{\mu}_{\rm n} = 0. \tag{13.55}$$

The difference between the values of $\bar{\mu}_{\rm p}$ and $\bar{\mu}_{\rm n}$ from (13.54) and from (13.55) is known as the *anomalous* magnetic moment, for which we use the symbol $\bar{\nu}$:

$$\bar{\mu}_{\rm p} = \mu_{\rm N} + \bar{\nu}_{\rm p},$$

and

$$\bar{\mu}_{\rm n} = \bar{\nu}_{\rm n},$$

with

$$\bar{\nu}_{\mathrm{p}} \cong 1.79\mu_{\mathrm{N}}, \qquad \bar{\nu}_{\mathrm{n}} \cong -1.91\mu_{\mathrm{N}}. \tag{13.56}$$

That $|\bar{\nu}_{\mathrm{p}}| \cong |\bar{\nu}_{\mathrm{n}}|$ intrigued theorists for many years. Various ideas were put forward in attempting to explain this near equality. The one that is most widely accepted is that protons, neutrons, and all other strongly interacting particles are composed of more fundamental objects known as quarks, which have fractional charges (of $\pm\frac{2}{3}$ and $\pm\frac{1}{3}$ – in units of the elementary charge e). Quarks are elements of a highly mathematical, relativistic framework that is a bit too complex to be discussed here, although interested readers can get a feeling for it through popular writings, such as those which have appeared in *Scientific American* and in popular books.[4] The only point concerning this framework that we mention here is that a non-relativistic version of it predicts the value of the ratio $|\bar{\mu}_{\mathrm{p}}/\bar{\mu}_{\mathrm{n}}|$ with astonishing accuracy. This certainly gives credence to the notion of an underlying substructure to particles that were themselves once thought to be fundamental/indivisible.

Let us now return to the postponed question: how does one determine the value of g for a given spin system? The answer is that one places the spin system in a magnetic field and attempts experimentally to extract g. In the next two subsections we explore aspects of this, starting with the constant (time-independent) case.

Constant magnetic field

For simplicity we assume that the spin-$\frac{1}{2}$ particle has positive charge $q = +|q|$ and that $\mathbf{B} = B\mathbf{e}_3$, with B constant. Then $\hat{H}^{(S)}_{\mathrm{mag}}$ is

$$\begin{aligned}\hat{H}^{(S)}_{\mathrm{mag}} &= -g\frac{\mu B}{\hbar}\hat{S}_z \\ &= -\frac{g\mu}{2}B\hat{\sigma}_z,\end{aligned} \tag{13.57a}$$

which, in the 2×2 eigenstate representation, becomes

$$H^{(S)}_{\mathrm{mag}} = -\frac{g}{2}\mu B\sigma_z. \tag{13.57b}$$

The state vector $|\zeta, t\rangle$ appearing in Eq. (13.49) thus goes over to a two-rowed column vector:

$$\begin{aligned}|\zeta, t\rangle &\to \zeta_+(t)\alpha_z + \zeta_-(t)\beta_z \\ &= \begin{pmatrix}\zeta_+(t)\\ \zeta_-(t)\end{pmatrix}.\end{aligned} \tag{13.58}$$

Hence, in the eigenstate representation, the time-dependent Schrödinger equation (13.49) reads

$$i\hbar\begin{pmatrix}\dot{\zeta}_+(t)\\ \dot{\zeta}_-(t)\end{pmatrix} = -\frac{g}{2}\mu B\begin{pmatrix}1 & 0\\ 0 & -1\end{pmatrix}\begin{pmatrix}\zeta_+(t)\\ \zeta_-(t)\end{pmatrix}, \tag{13.59}$$

[4] The theoretical framework of which quarks are a basic ingredient is known as the "standard model," and is currently the most successful theory of elementary particles. Qualitative and not-too-advanced quantitative material on quarks, etc., can be found, e.g., in Nambu (1981), Riordan (1987), Williams (1991) and Gottfried and Weisskopf (1984).

where the overlaid dot means d/dt. Carrying out the matrix multiplication on the RHS of (13.59) and recalling that $\mu = q\hbar\kappa_c/(2m)$, we get

$$\begin{pmatrix} \dot{\zeta}_+(t) \\ \dot{\zeta}_-(t) \end{pmatrix} = \frac{i\omega_L g}{2} \begin{pmatrix} \zeta_+(t) \\ -\zeta_-(t) \end{pmatrix}, \tag{13.60}$$

where

$$\omega_L = \frac{q\kappa_c B}{2m} = \frac{1}{2}\omega_c \tag{13.61}$$

is the Larmor frequency, equal to half of the cyclotron frequency $\omega_c = q\kappa_c B/m$, Eq. (12.27).

Equation (13.60) is actually a pair of uncoupled, first-order differential equations whose solutions are

$$\zeta_+(t) = \zeta_+(0)e^{ig\omega_L t/2} \tag{13.62a}$$

and

$$\zeta_-(t) = \zeta_-(0)e^{-ig\omega_L t/2}. \tag{13.62b}$$

If initially ($t = 0$) the state were an eigenstate of $\hat{S}_z$, say α_z, for which $\zeta_+(0) = 1$ and $\zeta_-(0) = 0$, then we learn nothing, since α_z remains the state for all t. Things are more interesting if an eigenstate of $\hat{S}_x$ or $\hat{S}_y$ is selected as the initial state. Let us choose α_y, the column-vector version of (13.34a), as the initial state:

$$\alpha_y = \sqrt{\frac{1}{2}} \begin{pmatrix} 1 \\ i \end{pmatrix}, \tag{13.63}$$

from which

$$\zeta_+(0) = \sqrt{\tfrac{1}{2}}, \qquad \zeta_-(0) = i\sqrt{\tfrac{1}{2}}. \tag{13.64}$$

Substituting into (13.62) leads to

$$\begin{pmatrix} \zeta_+(t) \\ \zeta_-(t) \end{pmatrix} = \sqrt{\frac{1}{2}} \begin{pmatrix} e^{ig\omega_L t/2} \\ ie^{-ig\omega_L t/2} \end{pmatrix}. \tag{13.65}$$

The exponential factors $\exp(\pm ig\omega_L t/2)$ have a period $T = 4\pi/(g\omega_L)$, and, as t ranges through the values $[0, T]$, the RHS of (13.65) ranges through the states α_y, α_x, β_y, and β_x (to within multiplicative factors), a claim whose verification is left as practice involving Eq. (13.32). This behavior leads to a *precession* of the spin values, in perfect analogy to the case of a classical magnetic moment in a constant magnetic field (an analogy dealt with in the exercises). Precession of spin is established by determining the expectation values of $\hat{S}_x$, $\hat{S}_y$, and $\hat{S}_z$ using (13.65). The generic relation is

$$\langle \hat{S}_j \rangle_\zeta(t) = \frac{\hbar}{4}(\zeta_+^*(t)\ \zeta_-^*(t))\sigma_j \begin{pmatrix} \zeta_+(t) \\ \zeta_-(t) \end{pmatrix}, \qquad j = x, y, z. \tag{13.66}$$

One readily finds

$$\langle \hat{S}_x \rangle_\zeta(t) = \frac{\hbar}{2} \sin(g\omega_\mathrm{L} t), \tag{13.67a}$$

$$\langle \hat{S}_y \rangle_\zeta(t) = \frac{\hbar}{2} \cos(g\omega_\mathrm{L} t), \tag{13.67b}$$

and

$$\langle \hat{S}_z \rangle_\zeta(t) = 0. \tag{13.67c}$$

By starting in α_y, there is never a component of $\langle \hat{\mathbf{S}} \rangle_\zeta(\mathrm{t})$ found along $\mathbf{e}_3$, while the expectation value precesses in time about the $\mathbf{e}_3$ direction in a negative sense ($y \to x$). The rate of precession for the spin-expectation values is $g\omega_\mathrm{L}$, twice that for the spin states, which, of course, are themselves not observable. On the other hand, spin *is* an observable, and, since ω_L is known, then, if the spin-precession rate could be measured, the value of g would be determined. The standard method for doing so is through use of spin magnetic resonance, a process described in the next subsection.

Magnetic resonance

The basic idea underlying spin magnetic resonance is classical in origin and simple in concept; the theoretical analysis is more complicated than that of pure precession, as will be seen. For positive charge and $\mathbf{B} = B\mathbf{e}_3$, spin precession occurs in a positive (counter-clockwise) direction at a rate $g\omega_\mathrm{L}$ in the x–y plane. Suppose, however, that the constant field $\mathbf{B} = B\mathbf{e}_3$ were applied in a reference frame rotating about $\mathbf{e}_3$ at the precession rate $g\omega_\mathrm{L}$. The precession of spin would then vanish, just as if the magnetic field $\mathbf{B}$ had become zero. This implies that the rotating frame acts on the spin like a magnetic field which cancels out the original one. If a real magnetic field pointing along $\mathbf{e}_1$ or $\mathbf{e}_2$ in the rotating frame were then turned on, and the initial direction of spin were up (α_z), this additional field would cause the spin to precess about $\mathbf{e}_1$ or $\mathbf{e}_2$, and eventually the direction of spin would be down (β_z), i.e., the spin would have flipped. This spin flipping could be observed and the value of g deduced. In place of a rotating frame, the ideal experiment would be performed with a rotating magnetic field. However, since it is hard to produce a rotating field, the experiment is done with a field fixed along either $\mathbf{e}_1$ or $\mathbf{e}_2$ and varying sinusoidally in time with, in principle, an adjustable angular frequency. There is not much difference between the effects produced by a linear and by a rotating field (see, e.g., comments by Baym (1976)), and, since the latter is a bit easier to analyze theoretically, it is the one we study here.

The magnetic field now has components in all three directions:

$$\begin{aligned} \mathbf{B} &= B_0\mathbf{e}_3 + \mathbf{B}_1(t) \\ &= B_0\mathbf{e}_3 + B_1[\cos(\omega t)\,\mathbf{e}_1 - \sin(\omega t)\,\mathbf{e}_2], \end{aligned} \tag{13.68}$$

rotating in a negative sense with frequency ω; it is sketched in Fig. 13.4. It gives rise to the following spin Hamiltonian (a positively charged particle is still assumed):

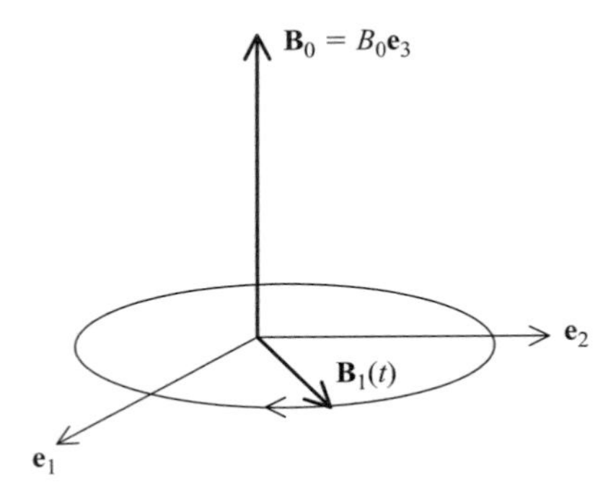

Fig. 13.4 Constant and rotating components of the magnetic field used in the analysis of the magnetic-resonance experiment.

$$H_{\text{mag}}^{(S)} = -\frac{g\mu}{2}[B_0\sigma_z + B_1\sigma_x\cos(\omega t) - B_1\sigma_y\sin(\omega t)]$$
$$= -\frac{g\mu}{2}\begin{pmatrix} B_0 & B_1e^{i\omega t} \\ B_1e^{-i\omega t} & -B_0 \end{pmatrix}. \tag{13.69}$$

Retaining (13.58) for the time-dependent state, the Schrödinger equation takes the form

$$\begin{pmatrix} \dot{\zeta}_+(t) \\ \dot{\zeta}_-(t) \end{pmatrix} = i\frac{g\mu}{2\hbar}\begin{pmatrix} B_0\zeta_+(t) & B_1e^{i\omega t}\zeta_-(t) \\ B_1e^{-i\omega t}\zeta_+(t) & -B_0\zeta_-(t) \end{pmatrix}. \tag{13.70}$$

In place of the uncoupled pair (13.60), Eq. (13.70) is a pair of coupled equations, involving two Larmor frequencies. They are defined as

$$\omega_0 = \frac{g\mu B_0}{2\hbar} \tag{13.71a}$$

and

$$\omega_1 = \frac{g\mu B_1}{2\hbar}, \tag{13.71b}$$

in terms of which (13.70) becomes

$$\dot{\zeta}_+(t) = i[\omega_0\zeta_+(t) + \omega_1e^{i\omega t}\zeta_-(t)] \tag{13.72a}$$

and

$$\dot{\zeta}_-(t) = i[\omega_1e^{-i\omega t}\zeta_+(t) - \omega_0\zeta_-(t)]. \tag{13.72b}$$

The frequency ω_1 provides the coupling between ζ_+ and ζ_- and causes the eventual (resonant) spin-flip.

To attempt to solve Eqs. (13.72) we try that ever-popular physics approach, the exponential (oscillatory) solution in the form

$$\zeta_+(t) = Ae^{i\omega_+ t} \tag{13.73a}$$

and

$$\zeta_-(t) = Be^{-i\omega_- t}, \tag{13.73b}$$

where $\omega_\pm$ are to be determined and A and B are constants, also to be determined. The signs in (13.73) are chosen to be consistent with the sign of the exponential and the ω_0 term in each equation.

Substitution into (13.72) plus a little rearranging yields

$$\omega_+ A = \omega_0 A + \omega_1 B e^{i(\omega-\omega_+-\omega_-)t} \tag{13.74a}$$

and

$$-\omega_- B = -\omega_0 B + \omega_1 A e^{-i(\omega-\omega_+-\omega_-)t}. \tag{13.74b}$$

This pair of equations violates the initial assumption that A and B are to be time-independent, a violation that can (and must) be prosecuted away by linking ω_+ and ω_- via $\omega - \omega_+ - \omega_- = 0$:

$$\omega_+ = \omega - \omega_- \qquad \text{or} \qquad \omega_- = \omega - \omega_+. \tag{13.75}$$

Using (13.75) to eliminate ω_- converts (13.74) into

$$(\omega_+ - \omega_0)A - \omega_1 B = 0 \tag{13.76a}$$

and

$$-\omega_1 A + (\omega_0 - \omega + \omega_+)B = 0. \tag{13.76b}$$

A non-trivial solution to the pair of coupled, homogeneous equations (13.76) exists only if its secular determinant vanishes; invoking this requirement leads to

$$\omega_+^2 - \omega\omega_+ + (\omega\omega_0 - \omega_0^2 - \omega_1^2) = 0,$$

whose solutions for the "eigenvalue" ω_+ are

$$\begin{aligned}\omega_+ &= \frac{\omega}{2} \pm \sqrt{\left(\omega_0 - \frac{\omega}{2}\right)^2 + \omega_1^2} \\ &\equiv \frac{\omega}{2} \pm \Omega,\end{aligned} \tag{13.77a}$$

and hence

$$\omega_- = \frac{\omega}{2} \mp \Omega. \tag{13.77b}$$

There are two sets of frequencies and therefore two sets of coefficients A and B. We shall label A and B by the sign on the RHS of (13.77a):

$$\omega_+ = \frac{\omega}{2} + \Omega \qquad \left(\omega_- = \frac{\omega}{2} - \Omega\right) \Rightarrow A^{(+)} \text{ and } B^{(+)}$$

and

$$\omega_+ = \frac{\omega}{2} - \Omega \qquad \left(\omega_- = \frac{\omega}{2} + \Omega\right) \Rightarrow A^{(-)} \text{ and } B^{(-)}.$$

Note that the superscripts in $A^{(\pm)}$ and $B^{(\pm)}$ designate the choice of the two values of the frequencies ω_+ (and thus of ω_-), whereas the subscripts in $\zeta_\pm(t)$ refer to the states α_z and β_z.

When Eq. (13.76) is used to obtain $B^{(\pm)}$ in terms of $A^{(\pm)}$, the result is

$$B^{(\pm)} = \frac{\omega/2 \pm \Omega - \omega_0}{\omega_1} A^{(\pm)}. \tag{13.78}$$

We are finally ready to construct the state-vector components $\zeta_\pm(t)$, Eqs. (13.73). Since there are two values of ω_+ (ω_-) and two values of A (B), the most general solutions are linear combinations of each of the preceding ones. For $\zeta_+(t)$ this is

$$\zeta_+(t) = A^{(+)} e^{i(\omega/2+\Omega)t} + A^{(-)} e^{i(\omega/2-\Omega)t}. \tag{13.79a}$$

A similar equation relates $\zeta_-(t)$ to $B^{(\pm)}$ and $\omega/2 \mp \Omega$, but Eq. (13.78) allows one to replace the coefficients $B^{(\pm)}$ by the *same* $A^{(\pm)}$ as those appearing in (13.79a), which gives

$$\zeta_-(t) = A^{(+)}\left(\frac{\omega/2 + \Omega - \omega_0}{\omega_1}\right) e^{-i(\omega/2-\Omega)t} + A^{(-)}\left(\frac{\omega/2 - \Omega - \omega_0}{\omega_1}\right) e^{-i(\omega/2+\Omega)t}. \tag{13.79b}$$

The components $\zeta_\pm(t)$ are quite clearly more complex than those of the preceding subsection for the case of no rotating field. Equations (13.79) are completely general, with the constants $A^{(\pm)}$ to be determined from the initial condition. The typical initial condition is that the spin is in an eigenstate of $\hat{S}_z$, which, as is more or less standard, we choose to be $\alpha_z = \binom{1}{0}$, i.e.,

$$\zeta_+(t=0) = 1 = A^{(+)} + A^{(-)} \tag{13.80a}$$

and

$$\zeta_-(t=0) = 0 = (A^{(+)} + A^{(-)})\left(\frac{\omega/2 - \omega_0}{\omega_1}\right) + (A^{(+)} - A^{(-)})\frac{\Omega}{\omega_1}. \tag{13.80b}$$

Substitution of (13.80a) into (13.80b) leads to

$$A^{(+)} = A^{(-)} - \frac{(\omega/2 - \omega_0)}{\Omega} = 1 - A^{(-)},$$

from which the solutions are

$$A^{(+)} = \frac{\Omega - \omega/2 + \omega_0}{2\Omega} \tag{13.81a}$$

and

$$A^{(-)} = \frac{\Omega + \omega/2 - \omega_0}{2\Omega}. \tag{13.81b}$$

Using these results, the full, time-dependent state vector is

$$\begin{pmatrix} \zeta_+(t) \\ \zeta_-(t) \end{pmatrix} = \begin{pmatrix} [\cos(\Omega t) - (i/\Omega)(\omega/2 - \omega_0)\sin(\Omega t)]e^{i\omega t/2} \\ i(\omega_1/\Omega)\sin(\Omega t)\, e^{-i\omega t/2} \end{pmatrix}. \tag{13.82}$$

Equation (13.82) is the time-evolved state that develops from the initial state $\binom{1}{0}$ in the presence of the magnetic field (13.68). A straightforward calculation will verify that the RHS of (13.82) fulfills the normalization condition $|\zeta_+(t)|^2 + |\zeta_-(t)|^2 = 1$.

Let us now recall that the purpose of the preceding analysis is to provide the theoretical underpinnings of the qualitative description, given at the beginning of this subsection, of a magnetic-spin-resonance (spin-flip) experiment designed to yield g values. We therefore calculate the spin-flip probability, that is, the probability $P_{\beta_z}(t)$ that, if α_z is the state at time $t = 0$, the system will be in state $\beta_z = \binom{0}{1}$ at time t. This probability is

$$\begin{aligned} P_{\beta_z}(t) &= \left|(0\ 1)\begin{pmatrix} \zeta_+(t) \\ \zeta_-(t) \end{pmatrix}\right|^2 \\ &= \frac{\omega_1^2}{\Omega^2}\sin^2(\Omega t). \end{aligned} \tag{13.83}$$

The oscillatory nature of P_{β_z} suggests working with the average of P_{β_z} over a period $2\pi/\Omega$:

$$\begin{aligned} \bar{P}_{\beta_z}(\omega) &= \frac{\Omega}{2\pi}\int_0^{2\pi/\Omega} P_{\beta_z}(t) = \frac{1}{2}\frac{\omega_1^2}{\Omega^2} \\ &= \frac{\omega_1^2/2}{(\omega/2 - \omega_0)^2 + \omega_1^2}, \end{aligned} \tag{13.84}$$

where Eq. (13.77a) has been used to transform the denominator Ω^2. That denominator is of standard, Lorentzian, *resonance* form, and is an expression analogous to ones occurring, e.g., in analyses of forced oscillations of a classical, damped, harmonic oscillator.

$\bar{P}_{\beta_z}(\omega)$ reaches its maximum of $\frac{1}{2}$ when $\omega = 2\omega_0$, independent of the value of ω_1, though ω_1 helps determine the width of the curve. A rough sketch of $\bar{P}_{\beta_z}(\omega)$ is displayed in Fig. 13.5.

In principle, one could carry out the experiment by first passing a beam of spin-$\frac{1}{2}$ particles through a S–G apparatus oriented along $\mathbf{e}_3$, and then placing the resulting pure α_z beam into the spin-resonance magnetic field. As $\omega/2$ varies, eventually getting closer and closer to the unknown value of $\omega_0 = g\mu B_0/(2\hbar)$ (unknown because g is not known),

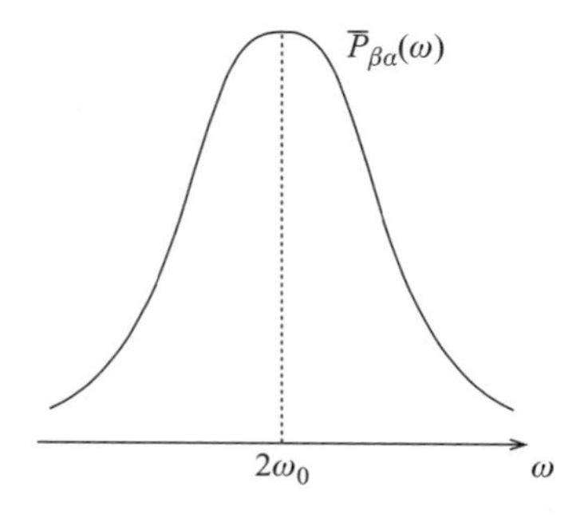

Fig. 13.5 A sketch of $\bar{P}_{\beta_z}(\omega) = (\omega_1^2/2)/[(\omega/2 - \omega_0)^2 + \omega_1^2]$ showing the peak and the symmetric resonance-type behavior.

the number of spin flips will increase, as will the energy absorbed from the rotating radio-frequency field $\mathbf{B}_1(t)$. The maximum energy will be supplied when $\omega/2 = \omega_0$, from which one can deduce g:

$$g = \frac{\hbar\omega}{\mu B_0}. \tag{13.85}$$

In practice it is often easier to fix ω and vary the strength B_0 of the field in the z-direction. Very sophisticated, careful experiments of this type established that g_e differed from the value 2 predicted by Dirac's relativistic equation but was in agreement with the predictions of quantum electrodynamics (the first theoretical calculation is that of Schwinger (1948), the original experiment establishing that $g_e \neq 2$ was performed by Kusch and Foley (1947)).

13.4. Other two-state systems

For spin $\frac{1}{2}$, the z-component eigenvalues are $m_s = \pm\hbar/2$: spin $\frac{1}{2}$ is the quintessential two-state system. There are others like it, two examples being polarization states of either classical electromagnetic radiation or quantal photons, and the states of the long- and short-lived K mesons $|K_L\rangle$ and $|K_S\rangle$.[5] These are very like spin $\frac{1}{2}$ because for each there is more than one basis in the 2-D space. A two-state system with simpler properties is that of a single operator with only two states ($\hat{\mathbf{S}}$ is a sum of three operators). Such operators are not readily found, but approximations to them are, and we consider an instance of one in this subsection.

Let $\hat{H}_0$ be a Hamiltonian with orthonormal, non-degenerate bound states $|n\rangle$ corresponding to energies E_n:

$$\hat{H}_0|n\rangle = E_n|n\rangle. \tag{13.86}$$

A resolution of the identity involves the complete set $\{|n\rangle\}$, but we here assume that E_1 and E_2 are relatively isolated from all the other energies $|E_n, n\rangle$:

$$E_n - E_j \gg E_1 \text{ or } E_2, \qquad j = 1, 2, \qquad n > 2.$$

Then, for energies in the vicinity of E_1 or E_2, the states $|1\rangle$ and $|2\rangle$ are taken to be an approximate basis:

$$\hat{I} \cong |1\rangle\langle 1| + |2\rangle\langle 2|. \tag{13.87}$$

We assume that the system of interest is governed by a Hamiltonian $\hat{H} \neq \hat{H}_0$, and we shall solve the $\hat{H}$ eigenvalue problem

$$\hat{H}|\gamma\rangle = E|\gamma\rangle \tag{13.88}$$

by expanding $|\gamma\rangle$ in the limited basis $\{|1\rangle, |2\rangle\}$. This will make $\hat{H}$ a two-state Hamiltonian. A necessary assumption, of course, is that the (two) states $|\gamma\rangle$ of $\hat{H}$ have energies in the vicinity of E_1 or E_2: the energies of all the other states are assumed to be very far away.

[5] Each of these is discussed in a manner accessible to readers of this text by Baym (1976); K meson states are also the subject of an exercise at the end of this chapter.

Equivalent to Eq. (13.87) is the expansion

$$|\gamma\rangle = C_1|1\rangle + C_2|2\rangle, \qquad C_j = \langle j|\gamma\rangle, \qquad j = 1, 2. \tag{13.89}$$

Substitution of (13.89) into (13.88) followed by projection of both sides of the result onto $\langle j|, j = 1, 2,$ yields the two-state eigenvalue problem

$$\sum_{k=1}^{2}(\langle j|\hat{H}|k\rangle - E\delta_{jk})C_k = 0. \tag{13.90}$$

Employing the notation $H_{jk} = \langle j|\hat{H}|k\rangle = H_{kj} =$ real, (13.90) has non-trivial solutions only when

$$\begin{vmatrix} H_{11} - E & H_{12} \\ H_{21} & H_{22} - E \end{vmatrix} = 0. \tag{13.91}$$

Solving Eq. (13.91) produces two eigenvalues labeled $E_\pm$:

$$E_\pm = \frac{H_{11} + H_{22}}{2} \pm \frac{1}{2}\sqrt{(H_{11} - H_{22})^2 + 4|H_{12}|^2}. \tag{13.92}$$

For the situation in which H_{11} and H_{22} are each positive, $E_+ > E_-$ unless $H_{11} = H_{22}$ AND $H_{12} = 0$ simultaneously occur. The vanishing of the off-diagonal matrix element H_{12} usually happens because of symmetry conditions, e.g., $\hat{H}$ is parity-conserving, but $|1\rangle$ and $|2\rangle$ have opposite parities, etc. Even if $H_{12} = 0$, it is unlikely that $H_{11} = H_{22}$, so that the normal situation is one in which $E_+ \neq E_-$.

The states that belong to these eigenvalues ($E_\pm$) will be labeled $|\pm\rangle$:

$$|\pm\rangle = C_1^{(\pm)}|1\rangle + C_2^{(\pm)}|2\rangle. \tag{13.93}$$

It is left to the exercises to show, for $C_1^{(\pm)}$ chosen real and positive, that

$$|+\rangle = \cos\theta\,|1\rangle + \sin\theta\,|2\rangle \tag{13.94}$$

and

$$|-\rangle = -\sin\theta\,|1\rangle + \cos\theta\,|2\rangle, \tag{13.95}$$

where $\tan\theta \propto H_{12}$, the proportionality constant to be determined. This result ensures the following: $\langle\pm|\mp\rangle = 0$, $\langle\pm|\pm\rangle = 1$, and $|+\rangle = |1\rangle$, $|-\rangle = 2$ when $H_{12} = 0$.

Equations (13.92), (13.94), and (13.95) provide a complete solution to the eigenvalue problem (13.88) as it is manifested in the two-state Hilbert space spanned by $\{|1\rangle, |2\rangle\}$. The analogy with the spin-$\frac{1}{2}$ case should be clear. Aspects and applications are left to the exercises.

Exercises

13.1 (a) Starting with Eq. (13.33), express $|S_x, \pm\rangle$ in terms of the S_z eigenkets.
(b) Using Eq. (5.107) and the results of part (a), express each of the operators $\hat{S}_x$, $\hat{S}_y$, and $\hat{S}_z$ in terms of the $\hat{S}_z$ eigenkets $|\frac{1}{2}m_s\rangle = |\pm\rangle$.
(c) Show that Eq. (13.7) is satisfied by the eigenket expressions of part (b).

13.2 A Stern–Gerlach apparatus (SGA), oriented in the direction $\mathbf{n}$ defined by the angles θ and ϕ, produces spin-$\frac{1}{2}$ particles in the states $|S_n, \pm\rangle$. If particles in each of these states were to be passed through either an $\mathbf{e}_1$- or an $\mathbf{e}_3$-oriented SGA, what percentages would be found, respectively, in the states $|S_x, \pm\rangle$ and $|S_z, \pm\rangle$?

13.3 Let $|\gamma\rangle$ be defined by Eqs. (13.15). Evaluate $\langle \hat{S}_j \rangle_\gamma$, $j = x, y, z$, in two different ways: first, using the eigenstate (matrix/column-vector) representation; second, by means of the $\hat{S}_z$-eigenket expressions of Exercise 13.1(b).

13.4 Prove the following relations involving the Pauli spin matrices:
(a) $\sigma_j \sigma_k = i\sigma_\ell$, j, k, ℓ in cyclic order,
(b) $\sigma_j \sigma_k + \sigma_k \sigma_j \equiv \{\sigma_j, \sigma_k\} = 0$, $j \neq k$; $j, k = 1, 2, 3$, and
(c) Eq. (13.22′).

13.5 Assume that the eigenstates of $\hat{S}^2$ and $\hat{S}_y$ are chosen as a basis.
(a) Determine the unitary transformation that links this basis with that of $\hat{S}^2$ and $\hat{S}_z$.
(b) Find the form taken by the Pauli spin matrices in the new basis.
(c) Express the $\hat{S}_x$ and $\hat{S}_z$ eigenkets in terms of those for $\hat{S}_y$.

13.6 A measurement of the spin-$\frac{1}{2}$ operator $\hat{S}_c = \hat{S}_x + \hat{S}_y$ is planned.
(a) What are the possible results of such a measurement?
(b) What are the eigenstates corresponding to the previous results? Express your answer in terms of the $\hat{S}_z$ states.
(c) After measuring $\hat{S}_c$, it is planned to measure $\hat{S}_x$. For each result of part (a), what is the probability that the second measurement will yield $-\hbar/2$?

13.7 Show that no state $|\gamma\rangle$ exists for a spin-$\frac{1}{2}$ particle such that $\langle \hat{S}_x \rangle_\gamma = \langle \hat{S}_y \rangle_\gamma = \langle \hat{S}_z \rangle_\gamma = 0$.

13.8 An $\mathbf{n}$-oriented Stern–Gerlach apparatus (SGA) produces equal numbers of spin-$\frac{1}{2}$ particles in the states $|S_n, +\rangle$ and $|S_n, -\rangle$. In this exercise, a beam of spin-$\frac{1}{2}$ particles is passed through three successive SGAs, as follows. SGA_1 is oriented along $\mathbf{e}_3$; its resulting $S_z = -\hbar/2$ particles are discarded. SGA_2 is oriented along $\mathbf{n}$, which lies in the x–z plane at an angle θ w.r.t. $\mathbf{e}_3$; its resulting $S_n = -\hbar/2$ particles are rejected. SGA_3 is oriented along $\mathbf{n}'$, also located in the x–z plane but at an angle θ' w.r.t. $\mathbf{e}_3$; its $S_n = -\hbar/2$ particles are discarded.
(a) If the number of $S_z = \hbar/2$ particles produced by SGA_1 is set equal to unity, how many final $S_n = -\hbar/2$ particles would be observed?
(b) How must SGA_2 be oriented so as to maximize the intensity of the beam of SGA_3 particles?

13.9 Each of the following four measurements is to be initiated by a beam of N spin-$\frac{1}{2}$ particles, all in the state $|S_x, +\rangle$, passing through a Stern–Gerlach apparatus SGA_1 that produces particles in states $|S_n, \pm\rangle$, where the polar angles of $\mathbf{n}$ are θ and ϕ as in Fig. 13.3.
(a) If a detector is placed in front of each of the two beams emerging from SGA_1, determine as a function of θ and ϕ the number of particles each detector would count.
(b) The two detectors used in the measurement of part (a) are removed, and the particles with $S_n = \hbar/2$ are to be passed through SGA_2 oriented along $\mathbf{e}_2$, while the $S_n = -\hbar/2$ particles would traverse SGA_3, similarly oriented. They would then be subjected to two further measurements: (i) In the first, those particles from SGA_2 and SGA_3 having $S_y = \hbar/2$ would be brought together on detector screen S_+, while those from SGA_2 and SGA_3 with $S_y = -\hbar/2$ would be brought together on detector

screen S_-. Find the number of particles that would strike screens $S_\pm$. (ii) In the second of the subsequent measurements, the SGA_2 particles having $S_y = \hbar/2$ and the SGA_3 particles having $S_y = -\hbar/2$ would be brought together on screen S_c. How many particles would be detected on S_c? (Note: all the detectors and screens in parts (i) and (ii) are sensitive only to the presence of the particles, not to their spin orientations.)

(d) Re-calculate the result of part (ii) of (c), now asuming that screen S_c is replaced by a detector that *is* sensitive to spin orientations.

13.10 Assuming that $\mathbf{A} = 0$ and $V(\mathbf{r}) = V(r)$, the simplest form taken by Eq. (13.42) is $|n\ell m_\ell m_s\rangle \equiv |n\ell m_\ell\rangle|\frac{1}{2}m_s\rangle$. Using this notation, the state of an electron in hydrogen at time $t = 0$ is $|\Gamma\rangle = N(2|100\frac{1}{2}\rangle - 3|211\frac{1}{2}\rangle + 4|200 - \frac{1}{2}\rangle - |320 - \frac{1}{2}\rangle)$.

(a) Determine $|N|$.

(b) Write out the detailed form of the time-evolved state $|\Gamma, t\rangle$, $t > 0$.

(c) Evaluate $\langle \hat{L}^2\rangle_\Gamma(t)$ and $\langle \hat{S}^2\rangle_\Gamma(t)$.

(d) What are the probabilities that a spin measurement will yield $S_z = \pm\hbar/2$?

(e) What is the average value of $\hat{S}_z$ at time $t = 0$?

13.11 As noted in Section 13.1, $_{47}$Ag behaves in a Stern–Gerlach experiment as if it were a one-electron atom. The behavior of $_{47}$Ag in this model is described by the one-electron Hamiltonian $\hat{H}(\mathbf{r}) = \hat{P}_2/(2\text{m}) - \mu\hat{\boldsymbol{\sigma}} \cdot \mathbf{B}(\text{r})$, where $\mathbf{B}(\mathbf{r}) = \beta x\mathbf{e}_1 + (B - \beta z)\mathbf{e}_3$, β is a small positive constant, and B is also constant. By solving the time-dependent Schrödinger equation in the 2×2 matrix form

$$i\hbar\frac{\partial}{\partial t}\begin{pmatrix}\Psi_+(\mathbf{r}, t)\\ \Psi_-(\mathbf{r}, t)\end{pmatrix} = \hat{H}\begin{pmatrix}\Psi_+(\mathbf{r}, t)\\ \Psi_-(\mathbf{r}, t)\end{pmatrix},$$

the model can account for the results obtained by Stern and Gerlach.

(a) The first calculation: let $\beta = 0$, assume that $\Psi_\pm(\mathbf{r}, t) = a_\pm^{(0)}\exp[i(ky - \omega_\pm t)]$, with $a_\pm^{(0)} =$ constant, and determine $\omega_\pm$ in terms of $\omega_k = \hbar k^2/(2m)$ and $\omega_c = \mu B\kappa_c/\hbar$, typical values for which are $\omega_k \sim 10^{14}$ Hz and $\omega_c \sim 10^{11}$ Hz.

(b) Next assume that $\beta \neq 0$ and write $\Psi_\pm(\mathbf{r}, t) = a_\pm\exp[i(ky - \omega_\pm t)]$, where the $\omega_\pm$ are the results from part (a) and $a_\pm = a_\pm(x, z, t)$. Here, small β means that the spatial derivatives of $a_\pm$ can be ignored. Using these assumptions, find the pair of coupled equations obeyed by the $\dot{a}_\pm = \partial a_\pm/\partial t$.

(c) Decouple the equations for $\dot{a}_\pm$ by assuming that ω_c is so large that the coupling terms either oscillate or average to zero. Decoupling implies that $a_\pm(x, z, t) = a_\pm^{(1)}\exp(i\Gamma_\pm)$, where $a_\pm^{(1)} =$ constant and the $\Gamma_\pm$ are to be determined.

(d) From part (c) it follows that $\Psi_\pm(\mathbf{r}, t) = a_\pm^{(1)}\exp[i(ky + \Gamma_\pm - \omega_\pm t)]$. Show that $ky + \Gamma_\pm = \mathbf{k}_\pm \cdot \mathbf{r}$, with the values of $\mathbf{k}_\pm$ implying a splitting of the beam into an upper and a lower portion. Find the time-dependent angle $\theta(t)$ between the upper and lower beams.

13.12 Let $\hat{H}(\mathbf{r})$ be the 2×2 Hamiltonian of Exercise 13.11, part (a). Using Ehrenfest's theorem, calculate the average force acting on the $_{47}$Ag atom if the wave function is the result of

(a) part (a) of Exercise 13.11, and

(b) part (d) of Exercise 13.11.

13.13 Let $V = 0$ in $\hat{H}_\sigma^{(A)}(\mathbf{r}, t)$ of Eqs. (13.36) and (13.37), and define the *helicity* operator $\hat{h} \equiv \boldsymbol{\sigma} \cdot \hat{\boldsymbol{\Pi}}/|\hat{\boldsymbol{\Pi}}|$, where $\hat{\boldsymbol{\Pi}} = (\hat{\mathbf{P}} - q\kappa_c\mathbf{A}) \equiv m\hat{\mathbf{v}}^{(A)}$, with $\hat{\mathbf{v}}^{(A)}$ the velocity operator of Eq.

(12.54). The observable $\hat{h}$ can be thought of as the component of spin along the direction of momentum or velocity; it has played an important role in certain weak-interaction phenomena.

(a) Show that $\hat{h}$ has eigenvalues ± 1.

(b) Prove that, as long as the g-factor of the particle described by the $V = 0$ version of $\hat{H}_{\sigma}^{(A)}(\mathbf{r}, t)$ is 2, then, if the particle is in an eigenstate of $\hat{h}$ at time t_0, it will be in the same eigenstate for $t > t_0$.

13.14 An alternate procedure for analyzing the case of a spatially uniform, time-dependent magnetic field is to use the Heisenberg picture, wherein $\hat{\mathbf{S}} \rightarrow \hat{\mathbf{S}}^{(\mathrm{H})}(t)$. Assuming a spin-$\frac{1}{2}$ particle with $q > 0$, the Hamiltonian of Eq. (13.50a) takes the Heisenberg-picture from $\hat{H}(t) = -\gamma \mathbf{B}(t) \cdot \hat{\mathbf{S}}^{(\mathrm{H})}(t)$.

(a) Using Eq. (8.8a) for $\hat{\mathbf{S}}^{(\mathrm{H})}(t)$, and recalling that $\hat{\mathbf{S}} \neq \hat{\mathbf{S}}(t)$, show that $d\hat{\mathbf{S}}^{(\mathrm{H})}(t)/dt = C\hat{\mathbf{S}}^{(\mathrm{H})}(t) \times \mathbf{B}(t)$, where C is a constant you are to determine.

(b) Now let $\mathbf{B} = B\mathbf{e}_3$, $B =$ constant, and write out the equations obeyed by each of the rectangular components of $\hat{\mathbf{S}}^{(\mathrm{H})}(t)$.

(c) Next, solve the equations of part (b) for the $\mathrm{S}_j^{(\mathrm{H})}(t)$, expressing them in terms of the initial values $\mathrm{S}_j = \mathrm{S}_j^{(\mathrm{H})}(t = 0)$, $j = 1, 2, 3$.

(d) The outcome of a large number of measurements of $\hat{S}_j$ is $\langle \Gamma | \hat{\mathrm{S}}^{(\mathrm{H})}(t) | \Gamma \rangle$, where $|\Gamma\rangle$ is the state at time $t = 0$. Set $|\Gamma\rangle = |S_y, +\rangle = \alpha_y$ and evaluate the three expectation values $\langle \hat{S}^{(\mathrm{H})}(t) \rangle_\Gamma$, comparing your answers with Eqs. (13.67).

(Hint: one way to solve part (c) is to form the linear combination $\hat{S}^{(\mathrm{H})}(t) = \hat{S}_x^{(\mathrm{H})}(t) + i\hat{S}_y^{(\mathrm{H})}(t)$, find an expression for it in terms of $\hat{S}^{(\mathrm{H})}(t = 0)$, and then extract $\hat{S}_x^{(\mathrm{H})}(t)$ and $\hat{S}_y^{(\mathrm{H})}(t)$. (If this doesn't seem straightforward, you should check your equations from part (c) for accuracy.))

13.15 The Hamiltonian for a structureless, $g = 2$, spin-$\frac{1}{2}$ particle in the magnetic field $\mathbf{B} = B_1\mathbf{e}_1 + B_3\mathbf{e}_3$ is $\hat{H}_{\mathrm{mag}}^{(\mathrm{H})} = -\mu\boldsymbol{\sigma} \cdot \mathbf{B}$, where $\boldsymbol{\sigma}$ is the vector of Pauli spin matrices and B_1 and B_3 are assumed to be constant.

(a) Determine the eigenvalues and normalized eigenvectors of this Hamiltonian.

(b) Using the results of part (a), find the time dependence of the Schrödinger eigenstate

$$|\Gamma, t\rangle = \begin{pmatrix} |\gamma_+, t\rangle \\ |\gamma_-, t\rangle \end{pmatrix},$$

assuming that, at $t = 0$, $|\Gamma, 0\rangle = \binom{0}{1}$, and then calculate the probability that the system will still be in the state $\binom{0}{1}$.

(c) If each of the components of $\hat{S}$ for this system were to be measured separately, what would be the time dependence of their expected values?

13.16 If the field in the magnetic resonance analysis of Section 13.3 were to rotate in the positive sense, then a unitary transformation plus the results of the preceding exercise would lead to an elegant solution, as demonstrated in this exercise. The Hamiltonian is still $\hat{H}_{\mathrm{mag}}^{(S)} = -\mu\boldsymbol{\sigma} \cdot \mathbf{B}$, but now $\mathbf{B} = B_0\mathbf{e}_3 + B_1[\mathbf{e}_1 \cos(\omega t) + \mathbf{e}_2 \sin(\omega t)]$, where $g = 2$ is assumed. The solution to the time-dependent Schrödinger equation is written in the following form:

$$|\Gamma, t\rangle = \begin{pmatrix} |\gamma_1, t\rangle \\ |\gamma_2, t\rangle \end{pmatrix}$$

(a) In place of $|\Gamma, t\rangle$, introduce the unitarily transformed state $|\Gamma, t\rangle' = \hat{U}(t)|\Gamma, t\rangle$, where

$$\hat{U}(t) = \begin{pmatrix} e^{i\omega t/2} & 0 \\ 0 & e^{-i\omega t/2} \end{pmatrix}$$

and show that the general solution to the $|\Gamma, t\rangle'$ problem is that of Exercise 13.15.

(b) Using the latter solution, find the time-averaged probability that the particle will be in state β_z if initially it was in state α_z. Compare/contrast your results with Eqs. (13.83) and (13.84).

13.17 Show that Eqs. (13.94) and (13.95) follow from the proper choice of the $C^{(\pm)}$ of Eq. (13.93) and obtain an exact expression for $\tan\theta_{12}$ in terms of H_{12}.

13.18 The ammonia molecule NH_3, which can be made to function as a MASER, has a tetrahedral structure and a magnetic moment that points from the N-atom at the apex of the tetrahedron towards the triangular base at whose corners are the three H-atoms. If no interaction coupled them, there would be in NH_3 a degenerate pair of ground states of energy E_0, say $|a\rangle$ and $|b\rangle$, distinguished, for example, by whether the N is above or below the H_3 triangular base. This degeneracy is lifted, however, by an interaction that can be modeled by constant off-diagonal terms in a 2×2 matrix description. In this model, the lowest-lying NH_3 states are governed (in the basis $\{|a\rangle, |b\rangle\}$) by the following 2×2 Hamiltonian matrix ($W > 0$):

$$H_0 = \begin{pmatrix} E_0 & -W \\ -W & E_0 \end{pmatrix}.$$

(a) Find the energy eigenvalues E_1 and $E_2 > E_1$ of H_0 (and thus of NH_3), and the corresponding normalized eigenstates $|E_1\rangle$ and $|E_2\rangle$, expressing them as linear combinations of $|a\rangle$ and $|b\rangle$ with real coefficients.

(b) Use the preceding H_0 to determine the solution $|\Gamma, t\rangle$ to the time-dependent Schrödinger equation, expressing it in terms of $|a\rangle$ and $|b\rangle$, assuming that $|\Gamma, t = 0\rangle = |a\rangle$.

(c) Evaluate the $t > 0$ probabilities $P_j(\mathrm{t}) = |\langle j|\Gamma, \mathrm{t}\rangle|^2$, $j = a, b$. Your results should display an oscillatory behavior, the period of which you are to determine.

(d) In addition to its magnetic moment, NH_3 has an electric dipole moment p, the operator for which in the $\{|a\rangle, |b\rangle\}$ basis is

$$D = \begin{pmatrix} p & 0 \\ 0 & -p \end{pmatrix}.$$

Let NH_3 be immersed in a constant electric field $\mathcal{E} > 0$, so that H_0 becomes $H = H_0 - \mathcal{E}D$. Find the eigenvalues E_{I} and $E_{\mathrm{II}} > E_{\mathrm{I}}$ of H, as well as approximations to them in the limit $p\mathcal{E} \ll W$. The latter results may be compared with those of a first-order pertubation calculation (Section 17.1).

13.19 The unstable, spinless, neutral K mesons K and $\overline{\mathrm{K}}$, created in collisions involving strongly interacting particles (protons and π mesons), are antiparticles of one another. Their formation is governed by the "strong" Hamiltonian $\hat{H}_s$, whose kets $|K\rangle$ and $|\bar{K}\rangle$ are also eigenstates of the so-called "strangeness" operator $\hat{S}$ ($\neq$ spin!): $\hat{S}|K\rangle = |\mathrm{K}\rangle$ and $\hat{S}|\overline{K}\rangle = -|\overline{\mathrm{K}}\rangle$. In contrast, the decaying states are not $|K\rangle$ and $|\overline{K}\rangle$, but instead are states denoted $|K_{\mathrm{L}}\rangle$ and $|K_{\mathrm{S}}\rangle$, which are here assumed to be eigenstates both of the "weak" Hamiltonian $\hat{H}_W$ and of an operator that we shall denote $\hat{T}$. (The subscripts "L" and "S" refer to the long-lived and short-lived decay modes.) Some relevant commutation relations are $[\hat{H}_S, \hat{S}] = 0 = [\hat{H}_W, \hat{T}]$, $[\hat{H}_S, \hat{T}] \neq 0 \neq [\hat{H}_W, \hat{S}]$, and $[\hat{T}, \hat{S}] \neq 0$. Although there are two sets of basis pairs in this model, the one chosen for use in this exercise is $\{|K\rangle, |\overline{K}\rangle\}$. The relations $\hat{T}|K\rangle = |\overline{K}\rangle$ and $\hat{T}|\overline{K}\rangle = |K\rangle$ are helpful for the aspects of decay considered here.

(a) In the eigenstate representation wherein $|\mathrm{K}\rangle$ ($|\overline{K}\rangle$) functions as an α_z (β_z) state, determine the 2×2 matrix representations S of $\hat{S}$ and T of $\hat{T}$.

(b) Determine the eigenvalues of $\hat{T}$ and the corresponding (normalized) eigenstates $|K_S\rangle$ and $|K_L\rangle$, where $|K_S\rangle$ corresponds to the larger of the two eigenvalues. Express the eigenstates in terms of the chosen basis.

(c) The average lifetimes for the decaying "particles" $\mathrm{K_S}$ and $\mathrm{K_L}$ are, respecively, $\tau_S \cong 10^{-10}$ s and $\tau_L \cong 10^{-8}$ s. To model the decays, $|K_S\rangle$ and $|K_L\rangle$ can be thought of as "eigenstates" of $\hat{H}_W$ corresponding to *complex* eigenvalues: $\hat{H}_W|K_j\rangle = \hbar(\omega_j - i\tau_j/2)|K_j\rangle$, $j = \mathrm{S, L}$. Suppose now that a K meson is created in the state $|K\rangle$ at $t = 0$. Solve the time dependent Schrödinger equation to determine the time dependence of the state vector describing the decaying meson, assuming that its behavior is governed by $\hat{H}_W$.

(d) From the result of part (c), calculate the probabilities $P_\mathrm{K}(t)$ and $P_{\overline{K}}(t)$ that, for $t > 0$, the decaying system will be observed as a K or a $\overline{\mathrm{K}}$ meson. Assuming $0 \leqslant t \leqslant 3_{\tau_S}$, use a reasonable approximation to express $P_\mathrm{K}(t)$ and $P_{\overline{K}}(t)$ in terms of τ_S and $\Delta\omega = \omega_L - \omega_S \cong c\,\Delta m/\hbar$, where the $\mathrm{K_L}$ and $\mathrm{K_S}$ mass difference is $\Delta m \cong 3.5 \times 10^{-8}\ \mathrm{eV}/c^2$. Make rough sketches of $P_\mathrm{K}(t)$ and $P_{\overline{K}}(t)$ for $t \in [0, 3_{\tau_S}]$; they sould display different time behaviors, which can be and has been used to determine Δm via *regeneration* of $\mathrm{K_S}$'s from a beam of pure $\mathrm{K_L}$'s.

14

Generalized Angular Momentum and the Coupling of Angular Momenta

The treatment of spin $\frac{1}{2}$ in the preceding chapter was based on hypothesis and analogy: spin was presented as a new type of angular momentum that behaves analogously to orbital angular momentum. This led to the positing of the eigenvalue equations (13.4) and the commutation relations (13.7). In this chapter, we explore the theoretical foundation of these ideas via the new concept of the generalized angular momentum operator $\hat{\mathbf{J}} = \hat{\mathbf{J}}^{\dagger}$. Just as $\hat{\mathbf{L}}$ is the infinitesimal generator of rotations for a scalar particle, so $\hat{\mathbf{J}}$ is defined to be the universal infinitesimal generator of rotations:

$$\hat{U}_{\mathrm{R}}(\mathbf{n}, \Theta) \equiv e^{-i\mathbf{n}\cdot\hat{\mathbf{J}}\Theta/\hbar}. \tag{14.1}$$

The commutation relations obeyed by the components of $\hat{\mathbf{J}}$ will be the same as those for $\hat{\mathbf{L}}$ and $\hat{\mathbf{S}}$, while the eigenvalue of $\hat{J}^2$ will be of now-standard form – $j(j+1)\hbar^2$, but with $j = n/2$: $j = 0, \frac{1}{2}, 1, \frac{3}{2}, 2, \ldots$. For $j = \frac{1}{2}$, $\hat{\mathbf{J}}$ is identical to the spin-$\frac{1}{2}$ operator $\hat{\mathbf{S}}$ of the preceding chapter. With $\hat{\mathbf{S}}$ identified as a generator of rotations (the theorist's meaning of spin), we shall be able to study the behavior of spin-$\frac{1}{2}$ eigenstates under rotations and, as with regeneration, will find a surprising result.

The second major topic treated in this chapter is the "coupling" of angular momenta. It addresses the question of how two particles with spin or one particle with both spin and orbital angular momentum will behave under rotations. This will lead, among other things, to the concept of the coupled representation, to singlet and triplet states of two spin-$\frac{1}{2}$ particles, to the claim of Einstein, Podolsky, and Rosen that quantum theory is incomplete, and to Bell's theorem.

14.1. Generalized angular momentum

We wish to determine the eigenvalues and eigenvectors of the generalized angular momentum operator $\hat{\mathbf{J}}$. Commutation relations (CRs) are a standard means for solving eigenvalue problems, as we saw with the harmonic oscillator and (in effect) with spin $\frac{1}{2}$. We did not need to work with CRs for orbital angular momentum because use of the coordination representation led to the eigenvalue problem being expressed in differential-equation form. However, since $\hat{\mathbf{J}}$ is intended to encompass both $\hat{\mathbf{L}}$ and $\hat{\mathbf{S}}$ as well as generalize them, and $\hat{\mathbf{S}}$ has no coordinate representation, then the CRs obeyed by the components $\hat{J}_j$ must provide the framework on which the eigenvalue analysis is based. However, we do not yet know what these CRs are. Our first task is to determine them.

Commutation relations

There are two ways to proceed. The first is postulational: by recognizing the parallel structure of the CRs for the components $\hat{L}_j$ and for the components $\hat{S}_j$ – Eqs. (9.172) or (10.2) and (13.7), respectively, one postulates that this CR structure holds for the $\hat{J}_j$ as well, i.e.,

$$[\hat{J}_j, \hat{J}_k] = i\hbar\epsilon_{jk\ell}\hat{J}_\ell. \tag{14.2}$$

Equation (14.2) is correct, and will be the basis for our analysis. Although it is correct, simply postulating these CRs seems inappropriate: we should be able to derive them from the structure of $\hat{U}_R(\mathbf{n}, \Theta)$ and a study of rotations. In fact, one can do so, as we demonstrate.

As with the analyses of Chapter 9, it is useful to work with infinitesimal changes, in the present case an infinitesimal angle of rotation; $\Theta \to \epsilon$, and $\hat{U}_R(\mathbf{n}, \epsilon)$ becomes

$$\hat{U}_R(\mathbf{n}, \epsilon) \cong \hat{I} - i\epsilon\mathbf{n}\cdot\hat{\mathbf{J}}/\hbar - \epsilon^2(\mathbf{n}\cdot\hat{\mathbf{J}})^2/\hbar^2. \tag{14.3}$$

Since the LHS of Eq. (14.2) is $\hat{J}_j\hat{J}_k - \hat{J}_k\hat{J}_j$ and $\hat{U}_R(\mathbf{n}, \epsilon)$ is given by (14.3), it is tempting to consider the action on the system of the difference of successive pairs of rotations, viz., $\hat{U}_R(\mathbf{e}_j, \epsilon)\hat{U}_R(\mathbf{e}_k, \epsilon) - \hat{U}_R(\mathbf{e}_k, \epsilon)\hat{U}_R(\mathbf{e}_j, \epsilon)$, since this difference is proportional to $\hat{J}_j\hat{J}_k - \hat{J}_k\hat{J}_j$. Unfortunately, we do not know how this difference is related to a rotation of the system about $\mathbf{e}_\ell$, the latter rotation presumably being needed in order to yield the factor $\hat{\mathbf{J}}_\ell$ occurring on the RHS of (14.2). Of course, if we already knew the eigenvalues and eigenvectors that the correct CR would yield, then we could determine the CR.

The way out of this dilemma is to change our perspective. Instead of considering active rotations of the system and the corresponding operator $\hat{U}_R(\mathbf{n}, \Theta)$, we study passive rotations of the coordinate axes, generated by $\hat{U}_R(\mathbf{n}, -\Theta) = \hat{U}_R^\dagger(\mathbf{n}, \Theta)$. The reason for doing this is that the action of $\hat{U}_R(\mathbf{n}, -\Theta)$ on the axes must produce the *same* effect as action by the classical rotation operator $\hat{R}(\mathbf{n}, -\Theta)$, whose effect on the axes we can easily evaluate. That is, we make use of two relations. First, the operator $\Delta\hat{U}_R(\mathbf{e}_j, \mathbf{e}_k, -\epsilon) = \hat{U}_R(\mathbf{e}_j, -\epsilon)\hat{U}_R(\mathbf{e}_k, -\epsilon) - \hat{U}_R(\mathbf{e}_k, -\epsilon)\hat{U}_R(\mathbf{e}_j, -\epsilon)$ is both proportional to $[\hat{J}_j, \hat{J}_k]$ by virtue of Eq. (14.3) and is expected to be related to $\hat{U}_R(\mathbf{e}_\ell, -\epsilon)$, although we do not yet know how. Second, the effect of $\Delta\hat{U}_R(\mathbf{e}_j, \mathbf{e}_k, -\epsilon)$ on the axes should be the same as that of $\Delta\hat{R}(\mathbf{e}_j, \mathbf{e}_k, -\epsilon) = \hat{R}(\mathbf{e}_j, -\epsilon)\hat{R}(\mathbf{e}_k, -\epsilon) - \hat{R}(\mathbf{e}_k, -\epsilon)\hat{R}(\mathbf{e}_j, -\epsilon)$. However, since the 3×3 rotation *matrices* are known, we can determine how the matrix $\Delta R(\mathbf{e}_j, \mathbf{e}_k, -\epsilon)$ is related to $R(\mathbf{e}_\ell, -\epsilon)$. This relation must be the same as that between $\Delta\hat{U}_R(\mathbf{e}_j, \mathbf{e}_k, -\epsilon)$ and $\hat{U}_R(\mathbf{e}_\ell, -\epsilon)$. Once it has been determined, the latter relation should then lead to Eq. (14.2).

It suffices to consider $j = 1$, $k = 2$, and $\ell = 3$. To start, we note that the rotation matrices $R(\mathbf{e}_1, -\epsilon)$ and $R(\mathbf{e}_2, -\epsilon)$ are obtained by cyclic permutation of the matrix elements of $R(\mathbf{e}_3, -\epsilon)$, Eq. (9.145b):

$$R(\mathbf{e}_1, -\epsilon) = \begin{pmatrix} 1 & 0 & 0 \\ 0 & \cos(-\epsilon) & \sin(-\epsilon) \\ 0 & -\sin(-\epsilon) & \cos(-\epsilon) \end{pmatrix} \cong \begin{pmatrix} 1 & 0 & 0 \\ 0 & 1-\epsilon^2/2 & -\epsilon \\ 0 & \epsilon & 1-\epsilon^2/2 \end{pmatrix}$$

and

$$R(\mathbf{e}_2, -\epsilon) = \begin{pmatrix} \cos(-\epsilon) & 0 & -\sin(-\epsilon) \\ 0 & 1 & 0 \\ \sin(-\epsilon) & 0 & \cos(-\epsilon) \end{pmatrix} \cong \begin{pmatrix} 1-\epsilon^2/2 & 0 & \epsilon \\ 0 & 1 & 0 \\ -\epsilon & 0 & 1-\epsilon^2/2 \end{pmatrix},$$

each to second order in ϵ (to first order, any pair of infinitesimal rotation operators commutes).

To second order, the difference $\Delta R(\mathbf{e}_1, \mathbf{e}_2, -\epsilon) = R(\mathbf{e}_1, -\epsilon)R(\mathbf{e}_2, -\epsilon) - R(\mathbf{e}_2, -\epsilon) \times R(\mathbf{e}_1, -\epsilon)$ is

$$\Delta R(\mathbf{e}_1, \mathbf{e}_2, -\epsilon) \cong \begin{pmatrix} 0 & \epsilon^2 & 0 \\ -\epsilon^2 & 0 & 0 \\ 0 & 0 & 0 \end{pmatrix}. \tag{14.4}$$

As noted above parenthetically, only ϵ^2 terms survive. The appearance of ϵ^2 suggests a rotation operator involving the angle ϵ^2, not ϵ. Neither $R(\mathbf{e}_1, \epsilon^2)$ nor $R(\mathbf{e}_2, \epsilon^2)$ has ϵ^2 appearing in the 1, 2 and 2, 1 positions, so the only remaining possibility would seem to be $R(\mathbf{e}_3, -\epsilon^2)$:

$$R(\mathbf{e}_3, -\epsilon^2) = \begin{pmatrix} \cos\epsilon^2 & \sin(-\epsilon^2) & 0 \\ -\sin(-\epsilon^2) & \cos\epsilon^2 & 0 \\ 0 & 0 & 1 \end{pmatrix} \cong \begin{pmatrix} 1 & -\epsilon^2 & 0 \\ \epsilon^2 & 1 & 0 \\ 0 & 0 & 1 \end{pmatrix}. \tag{14.5}$$

Comparison of Eqs. (14.4) and (14.5) shows that

$$\Delta R(\mathbf{e}_1, \mathbf{e}_2, -\epsilon) = I_3 - R(\mathbf{e}_3, -\epsilon^2), \tag{14.6}$$

where I_3 is the 3×3 unit matrix:

$$I_3 = \begin{pmatrix} 1 & 0 & 0 \\ 0 & 1 & 0 \\ 0 & 0 & 1 \end{pmatrix}.$$

The remaining steps are straightforward. First, since $\Delta R(\mathbf{e}_1, \mathbf{e}_2, -\epsilon)$ and $\Delta\hat{U}_{\mathrm{R}}(\mathbf{e}_1, \mathbf{e}_2, -\epsilon)$ must produce the same results, $I_3 - R(\mathbf{e}_3, -\epsilon^2)$ is the equivalent to $\hat{I} - \hat{U}_{\mathrm{R}}(\mathbf{e}_3, -\epsilon^2)$. These equivalences lead to

$$\Delta\hat{U}_{\mathrm{R}}(\mathbf{e}_1, \mathbf{e}_2, -\epsilon) = \hat{U}^{\dagger}_{\mathrm{R}}(\mathbf{e}_1, \epsilon)\hat{U}^{\dagger}_{\mathrm{R}}(\mathbf{e}_2, \epsilon) - \hat{U}^{\dagger}_{\mathrm{R}}(\mathbf{e}_2, \epsilon)\hat{U}^{\dagger}_{\mathrm{R}}(\mathbf{e}_1, \epsilon) = \hat{I} - \hat{U}^{\dagger}_{\mathrm{R}}(\mathbf{e}_3, \epsilon^2). \tag{14.7}$$

Use of Eq. (14.3) to expand the $\hat{U}^{\dagger}_{\mathrm{R}}$'s in (14.7) to second order in ϵ^2 gives

$$\left(\hat{I} + \frac{i\epsilon\hat{J}_x}{\hbar} - \frac{\epsilon^2}{2}\frac{\hat{J}_x^2}{\hbar^2}\right)\left(\hat{I} + \frac{i\epsilon\hat{J}_y}{\hbar} - \frac{\epsilon^2}{2}\frac{\hat{J}_y^2}{\hbar^2}\right)$$
$$- \left(\hat{I} + \frac{i\epsilon\hat{J}_y}{\hbar} - \frac{\epsilon^2 J_y^2}{2\hbar^2}\right)\left(\hat{I} + \frac{i\epsilon\hat{J}_x}{\hbar} - \frac{\epsilon^2 J_x^2}{2\hbar^2}\right) = -\frac{i\epsilon^2\hat{J}_z}{\hbar}. \tag{14.8}$$

Extracting the surviving ϵ^2 terms from the LHS of Eq. (14.8) finally yields the sought-after commutation relation:

$$[\hat{J}_x, \hat{J}_y] = i\hbar\hat{J}_z. \tag{14.9}$$

From the cyclic nature of this result we infer the correct complete set, Eq. (14.2):

$$[\hat{J}_j, \hat{J}_k] = i\hbar\epsilon_{jk\ell}\hat{J}_\ell.$$

Eigenvalues and eigenvectors for generalized angular momentum

As with $\hat{\mathbf{L}}$ and $\hat{\mathbf{S}}$, the relations (14.2) mean that only one component can be diagonalized along with $\hat{J}^2$. Standard convention selects $\hat{J}_z$ as this component. We write the eigenvalue equations as

$$\hat{J}^2|\lambda, \mu\rangle = \hbar^2\lambda|\lambda, \mu\rangle \tag{14.10a}$$

and

$$\hat{J}_z|\lambda, \mu\rangle = \hbar\mu|\lambda, \mu\rangle, \tag{14.10b}$$

with λ and μ to be determined. These eigenvalues obey the same inequality as that in Section 10.1:

$$0 \leqslant \mu^2 \leqslant \lambda, \tag{14.11}$$

derived from $\langle\hat{J}^2\rangle_{\lambda\mu} \geqslant \langle\hat{J}_z^2\rangle_{\lambda\mu}$, whose validity is established using the same methods as for Eq. (10.5) *ff*. Since $\hat{\mathbf{J}}$ encompasses $\hat{\mathbf{L}}$, one may properly expect that λ will be of the form $j(j+1)$, and that, for a given j, μ will take on the $j+1$ values $j, j-1, j-2, \ldots, -j$.

To establish these results and determine the allowed values of j, we use the relevant raising and lowering (ladder) operators, defined in analogy to $\hat{L}_\pm$ and $\hat{S}_\pm$:

$$\hat{J}_\pm = \hat{J}_x \pm i\hat{J}_y = \hat{J}^\dagger_\mp. \tag{14.12}$$

They obey the commutation relations

$$[\hat{J}_+, \hat{J}_-] = 2\hbar\hat{J}_z \tag{14.13a}$$

and

$$[\hat{J}_z, \hat{J}_\pm] = \pm\hbar\hat{J}_\pm, \tag{14.13b}$$

which are the analogs of Eqs. (10.89) and (10.103), and they also satisfy the relation

$$\begin{aligned}\hat{J}_\pm\hat{J}_\mp &= \hat{J}_x^2 + \hat{J}_y^2 \mp \hbar\hat{J}_z \\ &= \hat{J}^2 - \hat{J}_z^2 \mp \hbar\hat{J}_z.\end{aligned} \tag{14.13c}$$

Again in analogy with the orbital angular-momentum case, Eq. (14.13b) is the means for demonstrating that $\hat{J}_\pm|\lambda, \mu\rangle$ is an eigenstate of $\hat{J}_z$ with eigenvalue $(\mu \pm 1)\hbar$:

$$\hat{J}_z(\hat{J}_\pm|\lambda, \mu\rangle) = (\mu \pm 1)\hbar(\hat{J}_\pm|\lambda, \mu\rangle),$$

so that

$$\hat{J}_\pm|\lambda, \mu\rangle = C_\pm\hbar|\lambda, \mu \pm 1\rangle, \tag{14.14}$$

where this $C_\pm$ will have the same form as the $C_\pm$ of Eq. (10.96). We shall determine the new $C_\pm$ shortly.

For fixed λ, $|\mu| \leqslant \sqrt{\lambda}$ by Eq. (14.11), so that repeated application of $\hat{J}_\pm$ to $|\lambda, \mu\rangle$ will eventually produce states $|\lambda, \mu_{\text{max}}\rangle$ and $|\lambda, \mu_{\text{min}}\rangle$ such that

$$\hat{J}_+|\lambda, \mu_{\max}\rangle = 0 \tag{14.15a}$$

and

$$\hat{J}_-|\lambda, \mu_{\min}\rangle = 0, \tag{14.15b}$$

that is, $\mu_{\max} + 1 > \sqrt{\lambda}$ and $|\mu_{\min} - 1| > \sqrt{\lambda}$. Equations (14.15) suggest that for fixed λ there is a finite number of states $|\lambda, \mu\rangle$ and also that $\mu = \mu(\lambda)$.

The pair (14.15) is the means to determining the values of λ and μ. This is done by evaluating the action of $\hat{J}^2$ on $|\lambda, \mu_{\max}\rangle$ and $|\lambda, \mu_{\min}\rangle$ in two different ways and then equating the results. From Eq. (14.10a) we have

$$\hat{J}^2|\lambda, \mu_{\max}\rangle = \lambda\hbar^2|\lambda, \mu_{\max}\rangle \tag{14.16a}$$

and

$$\hat{J}^2|\lambda, \mu_{\min}\rangle = \lambda\hbar^2|\lambda, \mu_{\min}\rangle. \tag{14.16b}$$

However, Eq. (14.13c) tells us that

$$\hat{J}^2 = \hat{J}_z^2 + \hat{J}_\pm\hat{J}_\mp \mp \hbar\hat{J}_z, \tag{14.17}$$

and, acting with (14.17), the LHSs of Eqs. (14.16) are also given by

$$\hat{J}^2|\lambda, \mu_{\max}\rangle = (\mu_{max}^2 + \mu_{\max})\hbar^2|\lambda, \mu_{\max}\rangle \tag{14.18a}$$

and

$$\hat{J}^2|\lambda, \mu_{\min}\rangle = (\mu_{min}^2 - \mu_{\min})\hbar^2|\lambda, \mu_{\min}\rangle, \tag{14.18b}$$

where the terms $\hat{J}_-\hat{J}_+|\lambda, \mu_{\max}\rangle$ and $(\hat{J}_+\hat{J}_-)|\lambda, \mu_{\min}\rangle$ give no contribution.

The RHSs of the pair (14.18) must be equal:

$$\mu_{\max}(\mu_{\max} + 1) = \mu_{\min}(\mu_{\min} - 1).$$

This is a quadratic equation for $\mu_{\min}$, whose acceptable solution is

$$\mu_{\min} = -\mu_{\max} \tag{14.19}$$

(the other solution, $\mu_{\min} = \mu_{\max} + 1$, is a disallowed contradiction). It is standard to denote $\mu_{\max}$ by the symbol j:

$$\mu_{\max} = j = -\mu_{\min}, \tag{14.20a}$$

so that

$$\lambda = j(j + 1), \tag{14.20b}$$

as advertised. We now write $|\lambda, \mu\rangle$ as $|j, \mu\rangle$.

The next step is to determine the values that μ and j may assume. Since the state $|\lambda, \mu_{\max}\rangle = |j, j\rangle$ is the one with the largest possible value of μ, it follows that repeated application of $\hat{J}_-$ to $|j, j\rangle$ produces states $|j, \mu\rangle$ with $\mu = j - 1, j - 2, j - 3$, etc., until the state (to within a multiplicative constant) $|j, -j\rangle$ is reached. A further action of $\hat{J}_-$ (on $|j, -j\rangle$) gives zero, so $|j, -j\rangle$ is the last state this process can produce. A total of $2j$ states is obtained this way; adding to them $|j, j\rangle$ we get a total of $2j + 1$ states. The crucial question is whether there are any other states, i.e., ones not reached from $|j, j\rangle$ via repeated action of $\hat{J}_-$. For example, do states like $|j, j - s\rangle$, $0 < s < 1$ exist? The answer is NO, the previous $2j+1$ states form a complete set – a basis – for a given j. Proof of this statement follows on applying both sides of the equality $\hat{J}^2 = \hat{J}_z^2 + \hbar\hat{J}_z + \hat{J}_-\hat{J}_+$ to $|j, j-s\rangle$:

$$\hat{J}^2|j, j-s\rangle = j(j+1)\hbar^2|j, j-s\rangle, \tag{14.21a}$$

while

$$[J_z^2 + \hbar\hat{J}_z + \hat{J}_-\hat{J}_+]|j, j-s\rangle = [j(j+1) + s(s-1-2j)]\hbar^2|j, j-s\rangle. \tag{14.21b}$$

The only way in which the RHSs of (14.21a) and (14.21b) can be equal – and they must be – is for the non-integer s to vanish: $s = 0$, QED.

We have thus shown that, for a given j, the basis consists of the $2j+1$ states with μ varying in integer steps from j to $-j$. Therefore $2j+1$ is an integer:

$$2j+1 = N, \qquad N = 1, 2, 3, \ldots,$$

and thus

$$j = \frac{N-1}{2} = 0, \tfrac{1}{2}, 1, \tfrac{3}{2}, 2, \tfrac{5}{2}, \ldots$$

is the (infinite) set of allowed values for j.

The eigenvalue μ is referred to as a magnetic or projection quantum number, and, to be consistent with the notation $|\ell, m_\ell\rangle$ and $|S, m_S\rangle$, we relabel: $\mu \to m_j$, so that $|\lambda, \mu\rangle$ now reads $|j, m_j\rangle$, where

$$\hat{J}^2|j, m_j\rangle = j(j+1)\hbar^2|j, m_j\rangle, \qquad j = 0, \tfrac{1}{2}, 1, \tfrac{3}{2}, \ldots \tag{14.22a}$$

and

$$\hat{J}_z|j, m_j\rangle = m_j\hbar|j, m_j\rangle, \qquad m_j = j, j-1, j-2, \ldots, -j. \tag{14.22b}$$

With Eqs. (14.22) the generalized angular momentum eigenvalue problem is solved. Before evaluating $C_\pm$ of Eq. (14.14), a few points are worth mentioning. First, for $j = \frac{1}{2}, \frac{3}{2}, \frac{5}{2}$, etc., there is no coordinate representation for $\hat{\mathbf{J}}$ or $|j, m_j\rangle$; in this situation we are in the realm of half-odd-integer intrinsic spin states – see Section 14.2. Second, for integer values of j, $j = 0, 1, 2, \ldots$, there may but need not be a coordinate representation. When there is, then $\hat{\mathbf{J}} \equiv \hat{\mathbf{L}}$, the orbital angular momentum. When there is not, then $|j, m\rangle$, $j = 0, 1, 2, \ldots$, corresponds to integer intrinsic spin states. As noted earlier, pi mesons have zero intrinsic spin, photons have an intrinsic spin equal to unity, and electrons, protons, neutrons, etc., are our old friends with an intrinsic spin of $\frac{1}{2}$. In the case of intrinsic spin $\frac{1}{2}$, the states $|j = \frac{1}{2}, m_j\rangle \equiv |\frac{1}{2}, m_s\rangle$ $(= |\frac{1}{2}, m_j\rangle)$ are often referred to as "spinors," though this nomenclature could apply to any intrinsic spin state $|j, m_j\rangle$. We shall write $|S, m_S\rangle$ when intrinsic spin states are meant.

The remaining task is to evaluate $C_\pm$ of Eq. (14.14). Since all the states are assumed to be normalized, $\langle j', m_j'|j, m_j\rangle = \delta_{j'j}\delta_{m_j' m_j}$, then so is the state $|\lambda, \mu \pm 1\rangle \equiv |j, m_j \pm 1\rangle$ appearing on the RHS of (14.14). Therefore, on projecting both sides of (14.14) onto their Hermitian conjugate states, we have $(\lambda \to j, \mu \to m_j)$

$$\langle j, m_j|\hat{J}_\mp\hat{J}_\pm|j, m_j\rangle = |C_\pm|^2\hbar^2.$$

Equation (14.13c) permits easy evaluation of the preceding matrix element, and, choosing $C_\pm$ real and positive, we eventually find

$$\begin{aligned} C_\pm = C_\pm(j,\, m_j) &= \sqrt{j(j+1) - m_j(m_j \pm 1)} \\ &= \sqrt{(j \mp m_j)(j \pm m_j + 1)}. \end{aligned}$$

The eigenstate representation

The absence of a coordinate representation for intrinsic spin means that the generalized angular momentum operators are realized by means of the eigenstate representation. The relevant equations are (14.22) plus

$$\hat{J}_\pm|j,\, m_j\rangle = \hbar\sqrt{j(j+1) - m_j(m_j \pm 1)}|j,\, m_j \pm 1\rangle \tag{14.23a}$$

$$\begin{aligned} \hat{J}_x|j,\, m_j\rangle = \frac{\hbar}{2}\Big(&\sqrt{j(j+1) - m_j(m_j + 1)}|j,\, m_j + 1\rangle \\ &+\sqrt{j(j+1) - m_j(m_j - 1)}|j,\, m_j - 1\rangle\Big), \end{aligned} \tag{14.23b}$$

and

$$\begin{aligned} \hat{J}_y|j,\, m_j\rangle = \frac{\hbar}{2i}\Big(&\sqrt{j(j+1) - m_j(m_j + 1)}|j,\, m_j + 1\rangle \\ &-\sqrt{j(j+1) - m_j(m_j - 1)}|j,\, m_j - 1\rangle\Big). \end{aligned} \tag{14.23c}$$

In principle, the eigenstate representation is obtained by projecting these equations onto $\langle j'm_j'|$, but since only the $j' = j$ matrix elements are non-zero, it suffices to project with $\langle j,\, m_j'|$. The results are

$$\langle jm_j'|\hat{J}^2|jm_j\rangle = \hbar^2 j(j+1)\delta_{m_j' m_j}, \tag{14.24a}$$

$$\langle jm_j'|\hat{J}_z|jm_j\rangle = \hbar m_j \delta_{m_j' m_j} \tag{14.24b}$$

$$\langle jm_j'|\hat{J}_\pm|jm_j\rangle = \hbar\sqrt{j(j+1) - m_j(m_j \pm 1)}\delta_{m_j', m_j \pm 1} \tag{14.24c}$$

and

$$\begin{aligned} \langle jm_j'|\left\{\begin{matrix} \hat{J}_x \\ i\hat{J}_y \end{matrix}\right\}|jm_j\rangle = \frac{\hbar}{2}\Big(&\sqrt{j(j+1) - m_j(m_j + 1)}\delta_{m_j', m_j + 1} \\ &\pm \sqrt{j(j+1) - m_j(m_j - 1)}\delta_{m_j', m_j - 1}\Big). \end{aligned} \tag{14.24d}$$

For a given j, each of the matrices is $(2j+1) \times (2j+1)$, with the first two being diagonal and the last four having no diagonal elements. Since the Hilbert space, denoted $\mathcal{H}_j$, is infinite-dimensional, the collection of all these matrices forms an infinite matrix, made up of disjoint blocks which are each $(2j+1) \times (2j+1)$. A diagram of this will be shown after a few details for the $j = \frac{1}{2}$ and $j = 1$ cases have been discussed.

Each of Eqs. (14.24) provides the matrix elements of a $(2j+1) \times (2j+1)$ matrix, with m_j' referring to rows and m_j to columns. The largest values of m_j' and m_j, viz. j, label the upper-left-hand corner, i.e., the 1, 1 element. The smallest values of m'_j and m_j, viz. $-j$, label the lower-right-hand corner, i.e., the $2j+1, 2j+1$ element. For each successive row (column), the value of m'_j (m_j) is one unit less than that for the preceding row (column). In this representation the states $|j, m_j\rangle$ become real $(2j+1)$-rowed column vectors with unity at row m_j and zeros elsewhere, while $\langle jm'_j|$ becomes a real $(2j+1)$-columned row vector with unity at column m'_j and zeros elsewhere:

$$|j, m_j\rangle \rightarrow \begin{pmatrix} 0 \\ 0 \\ \vdots \\ 1 \\ 0 \\ \vdots \\ 0 \end{pmatrix} \leftarrow \text{unity at row } m_j \tag{14.25a}$$

and

$$\langle jm'_j| \rightarrow (00 \cdots 010 \cdots 0) \tag{14.25b}$$

$$\uparrow$$
$$\text{unity at column } m'_j.$$

These vectors correspond to the $(2j+1) \times (2j+1)$ matrices. Displayed below are the six matrices for the case $j = \frac{1}{2}$:

$$J^2 = \frac{3\hbar^2}{4}\begin{pmatrix} 1 & 0 \\ 0 & 1 \end{pmatrix} = S^2,$$

$$J_z = \frac{\hbar}{2}\begin{pmatrix} 1 & 0 \\ 0 & -1 \end{pmatrix} = S_z,$$

$$J_+ = \hbar\begin{pmatrix} 0 & 1 \\ 0 & 0 \end{pmatrix} = S_+,$$

$$J_- = \hbar\begin{pmatrix} 0 & 0 \\ 1 & 0 \end{pmatrix} = S_-,$$

$$J_x = \frac{\hbar}{2}\begin{pmatrix} 0 & 1 \\ 1 & 0 \end{pmatrix} = S_x,$$

and

$$J_y = \frac{\hbar}{2}\begin{pmatrix} 0 & -i \\ i & 0 \end{pmatrix} = S_y.$$

For the case $j = 1$, only J^2 and J_z are shown, the forms of the other four matrices being left to the exercises:

$$J^2 = 2\hbar^2\begin{pmatrix} 1 & 0 & 0 \\ 0 & 1 & 0 \\ 0 & 0 & 1 \end{pmatrix} \tag{14.26a}$$

and

$$J_z = \hbar \begin{pmatrix} 1 & 0 & 0 \\ 0 & 0 & 0 \\ 0 & 0 & -1 \end{pmatrix}. \tag{14.26b}$$

Far from their sources, classical radiation fields are transverse, i.e., $\mathcal{E}$ and $\mathbf{B}$ are perpendicular to the direction of propagation $\mathbf{k}$, see Section 16.1 The quanta of the (vector potential operator for the) radiation field are spin-1 photons, whose intrinsic states are the eigenstates of $\hat{J}_z$; that only two of the eigenvalues in (14.26b) are non-zero reflects the transverse nature of these photons.

The structure of the matrices for the higher values of j is a straightforward extrapolation of the results presented so far. Each of these generalized angular-momentum operators is represented by an infinite matrix whose detailed structure is a set of disjoint blocks, each $(2j+1) \times (2j+1)$, as remarked above. The generic structure of each matrix is shown below, starting with unity for $j = 0$ in the upper-left-hand corner, with the 2×2 spin-$\frac{1}{2}$ matrices next, then 3×3 spin-1 matrices next, etc:

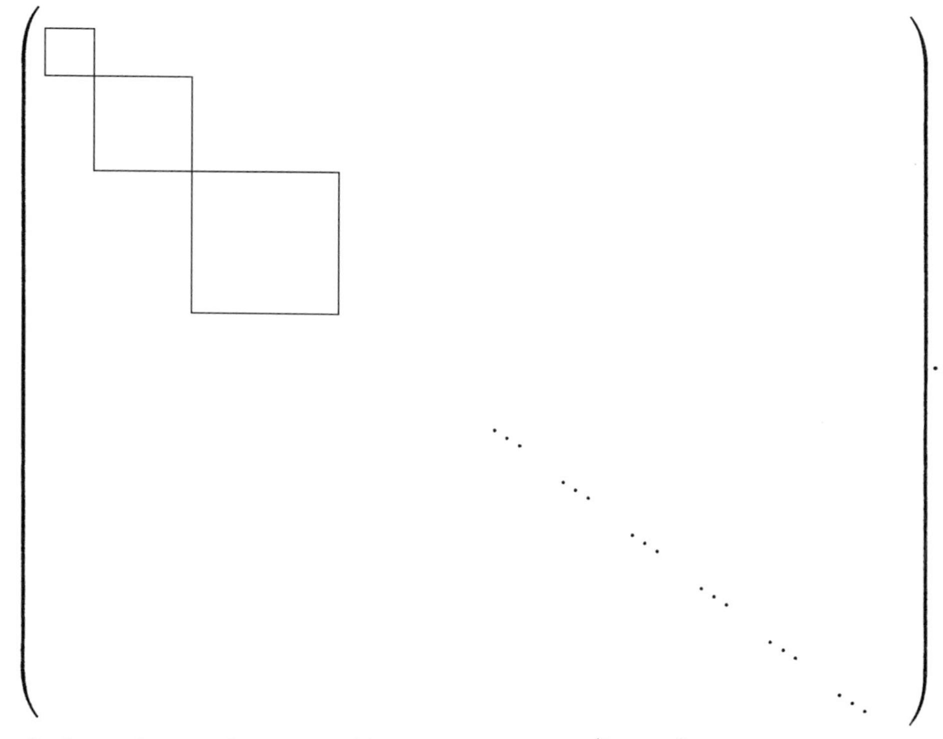

Such matrices are known as "block-diagonal." For $\hat{J}^2$ and $\hat{J}_z$ the block contains only diagonal elements, whereas for the four others the blocks have only off-diagonal elements, as should be evident from Eq. (14.24).

14.2. Rotations and spin $\frac{1}{2}$

Even though for $j = \frac{1}{2}$ the generalized angular-momentum operator $\hat{\mathbf{J}}$ is "just" the spin-$\frac{1}{2}$ operator $\hat{\mathbf{S}}$ of Chapter 13, we have actually learned an extremely important new aspect of this operator: it is the generator of rotations for spin-$\frac{1}{2}$ particles. Indeed, this is what a theorist means when referring to spin $\frac{1}{2}$: spin $\frac{1}{2}$ corresponds to that generator of rotations whose matrix representation is 2×2.

So, for spin-$\frac{1}{2}$ particles, the rotation operator is

$$\begin{aligned} \hat{U}_{\mathrm{R}}(\mathbf{n}, \Theta) &= e^{-i\Theta\mathbf{n}\cdot\hat{\mathbf{S}}/\hbar} \\ &= e^{-i\Theta\mathbf{n}\cdot\hat{\boldsymbol{\sigma}}/2}. \end{aligned} \tag{14.27}$$

It is convenient to work in the eigenstate representation, for which (14.27) reads

$$U_{\mathrm{R}}(\mathbf{n}, \Theta) = e^{-i\Theta\mathbf{n}\cdot\boldsymbol{\sigma}/2},$$

and in this form there is an especially simple expansion of the exponential. Invoking the Taylor series, which is valid because the eigenvalues of $\boldsymbol{\sigma}$ are finite, thus guaranteeing convergence, we have

$$U_{\mathrm{R}}(\mathbf{n}, \Theta) = \sum_{k=0}^{\infty} \left(\frac{(-i\Theta)}{2}\right)^k \frac{(\mathbf{n}\cdot\boldsymbol{\sigma})^k}{k!}. \tag{14.28}$$

When k is even, then $k = 2m$ and $(\mathbf{n}\cdot\boldsymbol{\sigma})^k = [(\mathbf{n}\cdot\boldsymbol{\sigma})^2]^m$, whereas for odd k, $k = 2m+1$ and $(\mathbf{n}\cdot\boldsymbol{\sigma})^k = (\mathbf{n}\cdot\boldsymbol{\sigma})[(\mathbf{n}\cdot\boldsymbol{\sigma})^2]^m$. In each case, we have separated out a factor $(\mathbf{n}\cdot\boldsymbol{\sigma})^2$. From Eq. (13.22′) this factor is the unit matrix I_2, in view of which (14.28) becomes

$$U_{\mathrm{R}}(\mathbf{n}, \Theta) = I_2 \sum_{m=0}^{\infty} \left(\frac{-i\Theta}{2}\right)^{2m} \frac{1}{(2m)!} + \mathbf{n}\cdot\boldsymbol{\sigma} \sum_{m=0}^{\infty} \left(\frac{-i\Theta}{2}\right)^{2m+1} \frac{1}{(2m+1)!}. \tag{14.29}$$

The summations in (14.29) are trigonometric functions: in particular, since

$$\cos\beta = \sum_{m=0}^{\infty} \frac{(-1)^m \beta^{2m}}{(2m)!}$$

and

$$\sin\beta = \sum_{m=0}^{\infty} \frac{(-1)^m \beta^{2m+1}}{(2m+1)!},$$

Eq. (14.29) reduces to

$$U_{\mathrm{R}}(\mathbf{n}, \Theta) = I_2 \cos\left(\frac{\Theta}{2}\right) - i\mathbf{n}\cdot\boldsymbol{\sigma} \sin\left(\frac{\Theta}{2}\right). \tag{14.30}$$

The appearance of the half angle is of great significance, as we shall see later.

Equation (14.30) is shorthand for a 2×2 matrix relation, one that is easy to write out. Let $\mathbf{n} = \sum_{j=1}^{3} n_j \mathbf{e}_j$, $\sum_{j=1}^{3} n_j^2 = 1$, in which case

$$\mathbf{n}\cdot\boldsymbol{\sigma} = \begin{pmatrix} n_3 & n_1 - in_2 \\ n_1 + in_2 & -n_3 \end{pmatrix}$$

and $U_{\mathrm{R}}(\mathbf{n}, \Theta)$ becomes

$$U_{\mathrm{R}}(\mathbf{n}, \Theta) = \begin{pmatrix} \cos\left(\frac{\Theta}{2}\right) - in_3 \sin\left(\frac{\Theta}{2}\right) & -i(n_1 - in_2)\sin\left(\frac{\Theta}{2}\right) \\ -i(n_1 + in_2)\sin\left(\frac{\Theta}{2}\right) & \cos\left(\frac{\Theta}{2}\right) + in_3 \sin\left(\frac{\Theta}{2}\right) \end{pmatrix}. \tag{14.31}$$

Whether one uses the abstract operator – $\hat{U}_{\mathrm{R}}(\mathbf{n}, \Theta)$ – or the matrix – $U_{\mathrm{R}}(\mathbf{n}, \Theta)$ – its effect on spinors is to rotate them. If $|\mathbf{n}', \pm\rangle$ is the initial state, then the rotated state is

$$|\mathbf{n}', \pm\rangle^{(\mathrm{R})} = \hat{U}_{\mathrm{R}}(\mathbf{n}, \Theta)|\mathbf{n}', \pm\rangle, \tag{14.32}$$

while, under rotations, the operators $\hat{S}_j$ become

$$\hat{S}_j^{(\mathrm{R})} = \hat{U}_{\mathrm{R}}(\mathbf{n}, \Theta)\hat{S}_j\hat{U}_{\mathrm{R}}^{\dagger}(\mathbf{n}, \Theta), \tag{14.33}$$

where $|\mathbf{n}', \pm\rangle$ is an eigenstate of $\mathbf{n}' \cdot \hat{\mathbf{S}} = \hat{\mathrm{S}}_{n'}$ and $\mathbf{n}$ is the rotation axis. We illustrate these relations with a few examples.

EXAMPLE 1

$|\mathbf{e}_1, +\rangle^{(\mathrm{R})}$ for $\mathbf{n} = \mathbf{e}_3$, $\Theta = \phi =$ the azimuthal angle. Writing $|\mathbf{e}_1, +\rangle$ in the eigenstate representation, we have

$$|\mathbf{e}_1, +\rangle \to \alpha_x = \sqrt{\frac{1}{2}}\left[\begin{pmatrix} 1 \\ 0 \end{pmatrix} + \begin{pmatrix} 0 \\ 1 \end{pmatrix}\right],$$

and its rotated partner is

$$\begin{aligned} \alpha_x^{(\mathrm{R})} &= \left[I_2 \cos\left(\frac{\phi}{2}\right) - i\sigma_3 \sin\left(\frac{\phi}{2}\right)\right]\alpha_x \\ &= \frac{e^{-i\phi/2}}{\sqrt{2}}\left[\begin{pmatrix} 1 \\ 0 \end{pmatrix} + e^{i\phi}\begin{pmatrix} 0 \\ 1 \end{pmatrix}\right]. \end{aligned} \tag{14.34}$$

Up to the phase factor $\exp(-i\phi/2)$, this is just the "up" state belonging to $\hat{\mathbf{S}} \cdot \mathbf{n}$ when the co-latitude is $\pi/2$. Furthermore, if $\phi = \pi/2$, corresponding to a rotation from $\mathbf{e}_1$ to $\mathbf{e}_2$, (14.34) becomes

$$\alpha_x^{(\mathrm{R})} = \frac{e^{-i\pi/4}}{\sqrt{2}}\left[\begin{pmatrix} 1 \\ 0 \end{pmatrix} + i\begin{pmatrix} 0 \\ 1 \end{pmatrix}\right] = e^{-i\pi/4}\alpha_y,$$

a result we could have expected.

EXAMPLE 2

$\alpha_y^{(\mathrm{R})}$ for $\mathbf{n} = \mathbf{e}_1$, $\Theta = \pi/2$. From the result of Example 1, we predict that, up to a phase factor, $\alpha_y^{(\mathrm{R})} = \alpha_z$. This is, indeed, the result:

$$\alpha_y^{(R)} = \left[I_2 \cos\left(\frac{\pi}{4}\right) - i\sigma_1 \sin\left(\frac{\pi}{4}\right)\right] \frac{1}{\sqrt{2}} \left[\begin{pmatrix} 1 \\ 0 \end{pmatrix} + i \begin{pmatrix} 0 \\ 1 \end{pmatrix} \right]$$
$$= \begin{pmatrix} 1 \\ 0 \end{pmatrix} = \alpha_z.$$

EXAMPLE 3

$S_x^{(R)}$ and $S_y^{(R)}$ for $\mathbf{n} = \mathbf{e}_3$ and $\Theta = \phi =$ the azimuthal angle. From $S_x = \hbar\sigma_x/2$ and Eq. (14.33), one has

$$S_x^{(R)} = \frac{\hbar}{2}\sigma_x^{(R)} = \frac{\hbar}{2}\left[I_2 \cos\left(\frac{\phi}{2}\right) - i\sigma_z \sin\left(\frac{\phi}{2}\right)\right]\sigma_x \left[I_2 \cos\left(\frac{\phi}{2}\right) + i\sigma_z \sin\left(\frac{\phi}{2}\right)\right]$$
$$= \frac{\hbar}{2}[\sigma_x \cos\phi + \sigma_y \sin\phi]$$
$$= S_x \cos\phi + S_y \sin\phi. \tag{14.35}$$

Similar reasoning, with $S_y = \hbar\sigma_y/2$, yields the result

$$S_y^{(R)} = -S_x \sin\phi + S_y \cos\phi. \tag{14.36}$$

We see that, under a rotation about $\mathbf{e}_3$, the originally separate x- and y-components of the spin operator become intertwined. Placing carats over the matrix symbols in Eqs. (14.35) and (14.36) turns them into relations involving the operators.

More than a pedagogical exercise is contained in this third example: it provides an alternate elegant way of approaching and generalizing the spin-precession system discussed in Section 13.3. In state-vector form, the time-dependent Schrödinger equation (13.59) reads

$$i\hbar \frac{\partial|\zeta, t\rangle}{\partial t} = -g\omega_L \hat{S}_z|\zeta, t\rangle, \tag{14.37}$$

whose formal solution is

$$|\zeta, t\rangle = e^{ig\omega_L t\hat{S}_z/\hbar}|\zeta, t = 0\rangle. \tag{14.38}$$

However, $\exp(ig\omega_L t\hat{S}_z/\hbar) = \hat{U}_R(\mathbf{e}_3, -g\omega_L t)$, so that the time-evolved state vector is just the initial state vector rotated through a time-dependent, negative angle. We can therefore apply rotation operator methods to find not only $|\zeta, t\rangle$ but also the rotated operators and eventually their matrix elements. This generalization of the spin-precession system is explored in the exercises. The magnetic-resonance analysis can be done in this way also, but the ensuing complications do not have sufficient pedagogical value in a book at this level for us to pursue this approach. The interested reader may consult more advanced texts, e.g., Baym (1976).

These examples and remarks do not, however, end our discussion of rotations for spin $\frac{1}{2}$: we have yet to consider the behavior of the spinor α_z (or β_z) under a rotation of 2π. To set the stage we first recall that, classically at least, rotating through 2π about any axis brings one back to the starting point, so no change should occur. This is certainly the case for the eigenstates $|\ell m_\ell\rangle$ of orbital angular momentum, for which the rotation operator is $\hat{U}_{\mathrm{R}}(\mathbf{n}, \Theta) = \exp(-i\Theta\mathbf{n}\cdot\hat{\mathbf{L}}/\hbar)$. Choosing $n = \mathbf{e}_3$ and $\Theta = 2\pi$, we have

$$|\ell m_\ell\rangle^{(\mathrm{R})} = e^{-2\pi i\hat{L}_z/\hbar}|\ell m_\ell\rangle. \tag{14.39}$$

However, $|\ell m_\ell\rangle$ is an eigenstate of $\hat{L}_z$, and therefore

$$|\ell m_\ell\rangle^{(\mathrm{R})} = e^{-2\pi m_\ell i}|\ell m_\ell\rangle = |\ell m_\ell\rangle,$$

since m_ℓ is a positive or negative integer or zero.

For the spinor states $|\frac{1}{2}m_s\rangle$, the rotation operator is $\exp(-i\Theta\mathbf{n}\cdot\hat{\mathbf{S}}/\hbar)$, and again choosing $\mathbf{n} = \mathbf{e}_3$ and $\Theta = 2\pi$, the rotated spinor is

$$\begin{aligned}|\tfrac{1}{2}m_s\rangle^{(\mathrm{R})} &= e^{-2\pi i\hat{S}_z/\hbar}|\tfrac{1}{2}m_s\rangle \\ &= e^{-2\pi m_s i}|\tfrac{1}{2}m_s\rangle.\end{aligned} \tag{14.40}$$

However, $m_s = \pm\frac{1}{2}$ so that the exponential in (14.40) is $\exp(\mp\pi i) = -1(!)$, yielding

$$|\tfrac{1}{2}m_s\rangle^{(\mathrm{R})} = -|\tfrac{1}{2}m_s\rangle. \tag{14.41}$$

Under a rotation of 2π the spinor changes sign: it takes a rotation of 4π to restore the initial state.

It was noted in Section 13.2 that spin $\frac{1}{2}$ is a remarkable attribute, in part because of the feature of regeneration. The non-invariance of spinors under a rotation of 2π about $\mathbf{e}_3$ is another reason for regarding spin $\frac{1}{2}$ as remarkable. Although this non-invariance had long been predicted, it remained a theoretical curiosity until 1975, when two groups (Werner *et al.* (1975), Rauch *et al.* (1975)) reported on experiments that demonstrated its existence, using a method known as neutron interferometry, since it involves the interference of two beams of neutrons. The underlying idea is simple: a beam of neutrons, all in state α_z, was split into two non-interacting portions. One of the resulting two beams was then passed through a constant magnetic field, which caused a precession through an angle of 2π, thereby changing the sign of the state of each particle in that beam. The two beams were then combined at a detector, where they produced the interference pattern predicted for a rotation through 2π. It is worth remarking that the article of Werner *et al.* was the second by that group on neutron interferometry, the first reporting on interference effects caused by the two neutron beams being in differing (Newtonian) gravitational fields (Collella, Overhauser, and Werner (1975)). The fascinated reader should consult the fine pedagogical review of Greenberger (1983), which describes both the theory and the (non-trivial) experiments. See also the exercises to this chapter.

14.3. The coupling of angular momenta

The generalized angular-momentum operator $\hat{\mathbf{J}}$ encompasses both orbital ($\hat{\mathbf{L}}$) and intrinsic spin ($\hat{\mathbf{S}}$) angular momentum. An isolated electron, being a spin-$\frac{1}{2}$ particle, is described by the spinor $|\frac{1}{2}m_s\rangle$. The state of the electron in a hydrogen atom will have both space and spin components, and, if its Hamiltonian is $\hat{H}(\mathbf{r}) = \hat{K}(\mathbf{r}) - \kappa_e e^2/r$, the non-relativistic, spin-independent one of Chapter 11, then its state vector is

$$|n\ell m_\ell m_s\rangle \equiv |n\ell m_\ell\rangle|\tfrac{1}{2}m_s\rangle. \tag{14.42}$$

For the preceding Hamiltonian, the spin aspect is entirely passive: it played no role in the dynamics of Chapter 11.

Equation (14.42) defines a state in the *uncoupled* or product representation, sometimes referred to as the "m-representation" because the state is labeled (in part) by the two magnetic quantum numbers m_ℓ and m_s. $|n\ell m_\ell m_s\rangle$ is an eigenstate both of $\hat{L}_z$ and of $\hat{S}_z$. Such states are easily generalized. Let $\hat{\mathbf{J}}_1$ and $\hat{\mathbf{J}}_2$ be two commuting angular-momentum operators with eigenstates $|j_1 m_1\rangle$ and $|j_2 m_2\rangle$, respectively:

$$\hat{J}_k^2|j_k m_k\rangle = j_k(j_k+1)\hbar^2|j_k m_k\rangle,\ k = 1 \text{ or } 2, \tag{14.43a}$$

and

$$\hat{J}_{kz}|j_k m_k\rangle = m_k\hbar|j_k m_k\rangle,\ k = 1 \text{ or } 2. \tag{14.43b}$$

Possibilities for $\hat{\mathbf{J}}_1$ and $\hat{\mathbf{J}}_2$ include $\hat{\mathbf{L}}$ and $\hat{\mathbf{S}}$ for a single particle; $\hat{\mathbf{L}}_1$ for particle 1 and $\hat{\mathbf{L}}_2$ for particle 2, and $\hat{\mathbf{S}}_1$ for particle 1 and $\hat{\mathbf{S}}_2$ for particle 2. In each of these examples, the two operators commute. The combined state of the system described by $\hat{\mathbf{J}}_1$ and $\hat{\mathbf{J}}_2$ is also an uncoupled state:

$$|j_1 m_1\rangle|j_2 m_2\rangle \equiv |j_1 j_2;\ m_1 m_2\rangle. \tag{14.44}$$

As with $|n\ell m_\ell m_s\rangle$, the two magnetic quantum numbers on the RHS of (14.44) appear next to each other but here they follow a semi-colon. This is a deliberate notational device, identifying the uncoupled or "m" representation.

In this section we are going to take states such as (14.44) and carry out a (unitary) transformation that "couples" them. "Coupling" of angular momentum is a technical term of great importance; it provides the only simple way of treating certain angular-momentum situations and also potentials involving interactions of the form $\hat{\mathbf{J}}_1 \cdot \hat{\mathbf{J}}_2$, examples of which are considered in the next section.

The coupled representation

We start with the physical system described by the uncoupled state (14.44) and then perform a rotation about $\mathbf{n}$ through an angle Θ. Rotations of the individual states $|j_k m_k\rangle$ are governed by the operator $\exp(-i\Theta\mathbf{n}\cdot\hat{\mathbf{J}}_k/\hbar)$, and, since the uncoupled state is a simple product, the rotated state is

$$\begin{aligned}|j_1 j_2;\ m_1 m_2\rangle^{(\mathrm{R})} &= e^{-i\Theta\mathbf{n}\cdot\hat{\mathbf{J}}_1/\hbar}|j_1 m_1\rangle e^{-i\Theta\mathbf{n}\cdot\hat{\mathbf{J}}_2/\hbar}|j_2 m_2\rangle \\ &= e^{-i\Theta\mathbf{n}\cdot(\hat{\mathbf{J}}_1+\hat{\mathbf{J}}_2)/\hbar}|j_1 m_1\rangle|j_2 m_2\rangle \\ &= e^{-i\Theta\mathbf{n}\cdot(\hat{\mathbf{J}}_1+\hat{\mathbf{J}}_2)/\hbar}|j_1 j_2;\ m_1 m_2\rangle. \end{aligned} \tag{14.45}$$

Crucial ingredients in reaching the final line of Eq. (14.45) are $[\hat{\mathbf{J}}_1, \hat{\mathbf{J}}_2] = 0$ and $[\hat{\mathbf{J}}_k, |j_\ell, m_\ell\rangle] = 0,\ k \neq \ell$, the latter result being based on the fact that the pairs $(\hat{\mathbf{J}}_1, |j_1 m_1\rangle)$ and $(\hat{\mathbf{J}}_2, |j_2 m_2\rangle)$ are in *different* Hilbert spaces: $|j_1 m_1\rangle \in \mathcal{H}_1$, $|j_2 m_2\rangle \in \mathcal{H}_2$.

The exponential in (14.45) contains a new operator, the *total angular momentum*, which we write as $\hat{\mathbf{J}}$:

$$\hat{\mathbf{J}} = \hat{\mathbf{J}}_1 + \hat{\mathbf{J}}_2. \tag{14.46}$$

The new operator is written as a sum, and we shall deduce the implications of adding $\hat{\mathbf{J}}_1$ and $\hat{\mathbf{J}}_2$. What we can infer at this point is that $\hat{\mathbf{J}}$, appearing where and how it does, must be a generalized angular-momentum operator since it generates rotations. To establish this inference it suffices to show that the components of $\hat{\mathbf{J}}$ obey angular-momentum-commutation rules. Clearly, (14.46) holds for components:

$$\hat{J}_k = \hat{J}_{1k} + \hat{J}_{2k}, \tag{14.47}$$

and, since $[\hat{J}_{1x}, \hat{J}_{1y}] = i\hbar\hat{J}_{1z}$ and $[\hat{J}_{2x}, \hat{J}_{2y}] = i\hbar\hat{J}_{2z}$ (plus cyclic permutations), a small bit of algebra leads to

$$[\hat{J}_x, \hat{J}_y] = i\hbar\hat{J}_z, \qquad \text{cyclically.} \tag{14.48a}$$

Furthermore, another easy calculation yields

$$[\hat{J}_z, \hat{J}^2] = 0. \tag{14.48b}$$

Equations (14.48) mean that there exists a set of states, temporarily labeled $|jm\rangle$, that satisfy

$$\hat{J}^2|jm\rangle = j(j+1)\hbar^2|jm\rangle \tag{14.49a}$$

and

$$\hat{J}_z|jm\rangle = m\hbar|jm\rangle, \tag{14.49b}$$

where

$$\hat{J}^2 = (\hat{\mathbf{J}}_1 + \hat{\mathbf{J}}_2)^2 = \hat{\mathbf{J}}_1^2 + \hat{\mathbf{J}}_2^2 + 2\hat{\mathbf{J}}_1 \cdot \hat{\mathbf{J}}_2. \tag{14.49c}$$

Unlike a single angular-momentum operator such as $\hat{\mathbf{J}}_k$, for which j_k in (14.43) can take on values in the range $0, \frac{1}{2}, 1, \frac{3}{2}$, etc., $\hat{\mathbf{J}}$ of (14.4) is a sum of two angular momenta, and therefore we do not yet know the values that j in $|jm\rangle$ of (14.49) can take. However, it is straightforward to determine the range of its values. We note first that $\hat{\mathbf{J}}$ as it appears in (14.45) acts in the combined Hilbert subspace of $\hat{\mathbf{J}}_1$ and $\hat{\mathbf{J}}_2$ spanned by the basis $|j_1 m_1\rangle|j_2 m_2\rangle = |j_1 j_2;\ m_1 m_2\rangle$. There are $(2j_1+1)(2j_2+1)$ states in this space – the totality of m values. We make the latter claim because the values of j_1 and j_2 do not change under the rotation (14.45) which defines $\hat{\mathbf{J}}$. Hence, j_1 and j_2 are quantum numbers that also characterize the states $|jm\rangle$ which ensue when we go from $\hat{\mathbf{J}}_k$ and the $|j_k m_k\rangle$ to $\hat{\mathbf{J}}$. To emphasize this aspect we relabel:

$$|jm\rangle \rightarrow |j_1 j_2;\ jm\rangle. \tag{14.50}$$

That is, the eigenstates of $\hat{J}^2$ and $\hat{J}_z$ are necessarily related to the eigenstates of $\hat{J}_1^2$ and $\hat{J}_2^2$.[1] Furthermore, since there are $(2j_1+1)(2j_2+1)$ states $|j_1 j_2;\ m_1 m_2\rangle$, there must be the same number of states $|j_1 j_2;\ jm\rangle$.

[1] The states of Eqs. (14.44) and (14.50) share a common and significant feature: each is characterized by four quantum numbers, implying that, for each, the CSCO contains four operators. In the uncoupled representation, this set is $((\hat{J}_1^2, \hat{J}_2^2, \hat{J}_{1z}, \hat{J}_{2z})$, whereas in the coupled representation $(|j_1 j_2, jm\rangle)$ the CSCO is $\{\hat{J}_1^2, \hat{J}_2^2, \hat{J}^2, \hat{J}_z\}$: if we start with four, then, after the transformation to the coupled representation, there must still be four. Note that two magnetic quantum numbers characterize $|j_1 j_2;\ m_1 m_2\rangle$, whereas only one is present in $|j_1 j_2;\ jm\rangle$: the quantum number j, corresponding to $\hat{J}^2$, has replaced a magnetic quantum number in the coupled representation. That is, only m is a good quantum number in $|j_1 j_2;\ jm\rangle$; m_1 and m_2 are not.

To determine which j (and m) values may occur in $|j_1j_2;\ jm\rangle$, we expand it in terms of the uncoupled states $|j_1j_2;\ m_1m_2\rangle$:

$$|j_1j_2;\ jm\rangle = \sum_{m_1,m_2} \langle j_1j_2;\ m_1m_2|j_1j_2;\ jm\rangle |j_1j_2;\ m_1m_2\rangle. \tag{14.51}$$

The $\{|j_1j_2;\ m_1m_2\rangle\}$ are thus the base states, in the same way that α_z and β_z are the base states for spin $\frac{1}{2}$. Note that m_1 and m_2 are no longer "good" quantum numbers in the coupled representation.

The transformation coefficients $\langle j_1j_2;\ m_1m_2|j_1j_2;\ jm\rangle$ are known as Clebsch–Gordan (C–G) or vector-coupling coefficients. They arise because the set $\{|j_1j_2;\ m_1m_2\rangle\}$ for fixed j_1 and j_2 is complete and thus allows a resolution of the identity in the joint Hilbert space:

$$\hat{I}_{(2j_1+1)\times(2j_2+1)} = \sum_{m_1\,m_2} |j_1j_2;\ m_1m_2\rangle\langle j_1j_2;\ m_1m_2|. \tag{14.52}$$

The scalar-product form appearing in Eq. (14.51) is the standard, generally recognized symbol for the C–G coefficients. Nonetheless, other notations have been introduced, partly as a means of eliminating the j_1j_2 redundancy of the scalar product. The notation of Rose (1957) is one of the more useful ones, and we shall employ it interchangeably with the scalar-product form:

$$\langle j_1j_2;\ m_1m_2|j_1j_2;\ jm\rangle \equiv C(j_1j_2j;\ m_1m_2m),$$

where, as we soon show, $m_1 = m - m_2$. The "C" notation is nicely suited for displaying the symmetry properties of the C–G coefficients. The C–G coefficients are probability amplitudes, with the following interpretation: $\langle j_1j_2;\ m_1m_2|j_1j_2;\ jm\rangle$ is the probability amplitude that the product state $|j_1m_1\rangle|j_2m_2\rangle$ will be found in $|j_1j_2;\ jm\rangle$, or equivalently, it is the probability amplitude that the coupled state $|j_1j_2;\ jm\rangle$ will be present in the uncoupled state $|j_1j_2;\ m_1m_2\rangle$. Either probability is given by $|\langle j_1j_2;\ m_1m_2|j_1j_2;\ jm\rangle|^2$.

From Eq. (14.51) one can show that $m = m_1 + m_2$. This is done by applying $\hat{J}_z$ to the state $|j_1j_2;\ jm\rangle$ on the LHS of (14.51) and $\hat{J}_{1z} + \hat{J}_{2z}(= \hat{J}_z)$ to the RHS (using $(|j_1j_2;\ m_1m_2\rangle = |j_1m_1\rangle|j_2m_2\rangle)$):

$$\begin{aligned}\hat{J}_z|j_1j_2;\ jm\rangle &= m\hbar|j_1j_2;\ jm\rangle \\ &= \sum_{m_1'\,m_2'} m\hbar\langle j_1j_2;\ m_1'm_2'|j_1j_2;\ jm\rangle|j_1m_1'\rangle|j_2m_2'\rangle \end{aligned} \tag{14.53a}$$

while

$$\begin{aligned}(\hat{J}_{1z} + \hat{J}_{2z}) &\sum_{m_1',m_2'} \langle j_1j_2;\ m_1'm_2'|j_1j_2;\ jm\rangle|j_1m_1'\rangle|j_2m_2'\rangle \\ &= \sum_{m_1',m_2'} \langle j_1j_2;\ m_1'm_2'|j_1j_2;\ jm\rangle(\hat{J}_{1z} + \hat{J}_{2z})|j_1m_1'\rangle|j_2m_2'\rangle \\ &= \sum_{m_1',m_2'} (m_1' + m_2')\hbar\langle j_1j_2;\ m_1'm_2'|j_1j_2;\ jm\rangle|j_1m_1'\rangle|j_2m_2'\rangle. \end{aligned} \tag{14.53b}$$

However, the RHSs of (14.53a) and (14.53b) are equal. Equating them and then projecting both sides of that equality onto $\langle j_1 m_1|\langle j_2 m_2|$ gives

$$(m - m_1 - m_2)\langle j_1 j_2;\ m_1 m_2|j_1 j_2;\ jm\rangle = 0. \tag{14.54}$$

Hence, if $m \neq m_1 + m_2$, the C–G coefficient vanishes:

$$\langle j_1 j_2;\ m_1 m_2|j_1 j_2;\ jm\rangle = 0, \qquad m \neq m_1 + m_2. \tag{14.55}$$

When m does equal $m_1 + m_2$, the C–G coefficient remains to be determined. We shall return to values and properties of the C–G coefficients after establishing the values that j can assume.

For a fixed j, m in $|j_1 j_2;\ jm\rangle$ takes on the $2j+1$ values

$$m = j,\ j-1,\ j-2,\ \ldots,\ -j, \tag{14.56}$$

since $|j_1 j_2;\ jm\rangle$ is an angular-momentum eigenstate. On the other hand, $m = m_1 + m_2$, $m_{1\,\max} = j_1$, and $m_{2\,\max} = j_2$, while the maximum value of j, viz. $j_{\max}$, cannot exceed $m_{\max} = m_{1\,\max} + m_{2\,\max}$. Hence we obtain an upper limit on j:

$$j \leqslant j_{\max} = j_1 + j_2. \tag{14.57}$$

Is the lower limit $j_{\min}$ equal to zero? The answer is: only if $j_1 = j_2$. To find $j_{\min}$, we evaluate the total number of states $|jm\rangle$ and set that number equal to $(2j_1+1)(2j_2+1)$:

$$\text{Number of states} = \sum_{j=j_{\min}}^{j_{\max}} \sum_{m=-j}^{j} = (2j_1+1)(2j_2+1).$$

However, $\sum_m = 2j+1$, so that the summation yielding the number of coupled states becomes

$$\sum_{j=j_{\min}}^{j_{\max}} (2j+1) = (2j_1+1)(2j_2+1). \tag{14.58}$$

Again employing the relation $\sum_{n=1}^{N} n = N(N+1)/2$, the LHS of (14.58) becomes

$$\sum_{j=j_{\min}}^{j_{\max}} (2j+1) = j_{\max}(j_{\max}+1) + j_{\max} - j_{\min}(j_{\min}-1) - (j_{\min}-1). \tag{14.59}$$

Finally, setting the RHSs of (14.58) and (14.59) equal to each other and recalling that $j_{\max} = j_1 + j_2$, we find

$$j_{min}^2 = (j_1 - j_2)^2$$

or

$$j_{\min} = |j_1 - j_2| \tag{14.60}$$

since $j \geqslant 0$.

Let us recapitulate. The largest value of j is $j_{\max} = j_1 + j_2$, with successively smaller values being obtained by subtracting first 1, then 2, etc., from $j_1 + j_2$ until the value $|j_1 - j_2|$ is reached:

$$j = j_1 + j_2,\ j_1 + j_2 - 1,\ j_1 + j_2 - 2,\ \ldots,\ |j_1 - j_2|, \tag{14.61a}$$

i.e.,

$$|j_1 - j_2| \leqslant j \leqslant j_1 + j_2. \tag{14.61b}$$

EXAMPLE 4

Ranges of j

(a) $j_1 = j_2 = \ell$: $j_{\max} = 2\ell$, $j_{\min} = 0$. Two particles with the same orbital angular-momentum quantum number ℓ can only be in states with total j of 2ℓ, $2\ell - 1$, $2\ell - 2, \ldots,$ down to 0.

(b) $j_1 = j_2 = \frac{1}{2}$: $j_{\max} = 1$, $j_{\min} = 0$. Two spin-$\frac{1}{2}$ particles can form only two coupled states, one with total spin 1, known as a triplet state (with three values of m), and the other with total spin 0, known as a singlet state (with only the value 0 for m).

(c) $j_1 = \ell$, $j_2 = \frac{1}{2}$: $j_{\max} = \ell + \frac{1}{2}$, $j_{\min} = \ell - \frac{1}{2}$ unless $\ell = 0$, in which case there is only the state with $j = \frac{1}{2}$.

We now have explored what are, in effect, the kinematic details concerning the coupled states $|j_1 j_2;\ jm\rangle$. The "dynamical" part involves their construction in terms of the uncoupled states $|j_1 j_2;\ m_1 m_2\rangle$. For this the C–G coefficients are necessary. They are the subject of the next subsection.

Clebsch–Gordan coefficients

One property of the C–G coefficients has already been established, viz., Eq. (14.55). A second is based on Eq. (14.61b): since the C–G coefficients produce $|j_1 j_2;\ jm\rangle$ from the $|j_1 j_2;\ m_1 m_2\rangle$, they must vanish if (14.61b) is not satisfied:

$$\langle j_1 j_2;\ m_1 m_2 | j_1 j_2;\ jm\rangle = 0, \qquad \text{when } j < |j_1 - j_2| \text{ or } j > j_1 + j_2. \tag{14.62}$$

The inequalities (14.61b) are sometimes written as $\triangle(j_1 j_2 j)$ since the same relation is also satisfied by the three sides of a triangle. Thus, an alternate form of Eq. (14.62) is

$$\langle j_1 j_2;\ m_1 m_2 | j_1 j_2;\ jm\rangle = 0 \qquad \text{if } \triangle(j_1 j_2 j) \text{ is not satisfied.} \tag{14.62$'$}$$

In addition to these limits, the C–G coefficients, being elements of a unitary transformation, obey two "sum rules" that follow from orthogonality. One is based on the relation

$$\langle j_1 j_2;\ jm | j_1 j_2;\ j'm'\rangle = \delta_{jj'}\delta_{mm'}. \tag{14.63}$$

The state $|j_1 j_2;\ j'm'\rangle$ will have an expansion just like Eq. (14.51), with $j'm'$ replacing jm. Substituting such an expansion for $\langle j_1 j_2;\ j'm'|$ as well as (14.51) for $|j_1 j_2;\ jm\rangle$ into Eq. (14.63) and using the orthogonality of the states $|j_1 j_2;\ m_1 m_2\rangle$ with differing values of m_1 and m_2 leads to

$$\sum_{m_1, m_2} \langle j_1 j_2;\ m_1 m_2 | j_1 j_2;\ jm\rangle \langle j_1 j_2;\ m_1 m_2 | j_1 j_2;\ j'm'\rangle = \delta_{jj'}\delta_{mm'}. \tag{14.64}$$

This "sum rule" is not only one of the two orthogonality conditions obeyed by a unitary transformation, but also plays an important role in certain angular-momentum analyses. We shall make use of it in Section 16.2.

The other sum rule is analogous to the foregoing, but starts with the relation inverse to (14.51), which is

$$|j_1 j_2;\ m_1 m_2\rangle = \sum_j \langle j_1 j_2;\ jm|j_1 j_2;\ m_1 m_2\rangle |j_1 j_2;\ jm\rangle. \tag{14.65}$$

Invoking the same set of steps for (14.65) as that which led to (14.64) yields

$$\sum_j \langle j_1 j_2;\ jm|j_1 j_2;\ m_1' m_2'\rangle \langle j_1 j_2;\ jm|j_1 j_2;\ m_1 m_2\rangle = \delta_{m_1' m_1} \delta_{m_2' m_2}. \tag{14.66}$$

It should be noted that the condition $m_1 + m_2 = m$ reduces the double sum in (14.64) and (14.66) to a single one. Each of these two sum rules also represents a normalization condition, one that we will see verified in the examples presented shortly.

Two other conditions are almost trivially written down. First, if $m = \pm(j_1 + j_2) = \pm(m_{1,max} + m_{2,max})$, then

$$\langle j_1 j_2;\ \pm j_1,\ \pm j_2|j_1 j_2;\ j,\ \pm(j_1 + j_2)\rangle = 1 \tag{14.67}$$

because we are dealing with the so-called "stretched state" and there is only way to achieve it:

$$|j_1 j_2;\ j,\ \pm(j_1 + j_2)\rangle = |j_1,\ \pm j_1\rangle |j_2,\ \pm j_2\rangle. \tag{14.68}$$

Hence, the correlations in Eq. (14.51) are absent: the stretched state is an uncorrelated product state. Second, if one of the two angular-momentum quantum numbers is zero, then the C–G coefficient must also be unity: either

$$\langle j_1 0;\ m_1 0|j_1 0;\ jm\rangle = \delta_{jj_1} \delta_{mm_1} \tag{14.69}$$

or

$$\langle 0 j_2;\ 0 m_2|0 j_2;\ jm\rangle = \delta_{jj_2} \delta_{mm_2}.$$

By proper choice of phases, the C–G coefficients can be chosen real, although not all of them are positive. They can be evaluated by means of recursion relations, which are discussed in the literature on angular-momentum theory (e.g., Rose (1957), Edmonds (1957), and Brink and Satchler (1971)). We display in Table 14.1 the values of the C–G coefficients for the cases $j_2 = \frac{1}{2}$ and $j_2 = 1$. Some of these are needed for the examples and also for a few of the exercises.

The order in which the angular momenta are coupled – i.e., which of the states refers to j_1 and which to j_2 – is that the state on the left corresponds to j_1. It may happen that j_1 rather than j_2 is $\frac{1}{2}$ or 1 (Table 14.1 refers to j_2 being $\frac{1}{2}$ or 1), so, for purposes of evaluation via Table 14.1, it may be necessary to reverse the positions of j_1 and j_2. This can be accomplished by means of the symmetry (interchange) relations obeyed by the $C(j_1 j_2 j;\ m_1 m_2 m)$. These relations are also useful in discussing the symmetry aspects of states describing two or more identical particles, as we shall see later. Table 14.2 lists the symmetry properties of the C–G coefficients.

The machinery allowing one to construct states in the coupled representation is now in place. This construction is illustrated in the following four examples. Notice that, in each

Table 14.1 *Selected values of Clebsch–Gordan coefficients*

$\langle j_1 \tfrac{1}{2};\, m-\sigma,\, \sigma | j_1 \tfrac{1}{2};\, jm \rangle$

j	$\sigma = \frac{1}{2}$	$\sigma = -\frac{1}{2}$
$j_1 + \frac{1}{2}$	$\left(\frac{j_1 + m + \frac{1}{2}}{2j_1 + 1}\right)^{1/2}$	$\left(\frac{j_1 - m + \frac{1}{2}}{2j_1 + 1}\right)^{1/2}$
$j_1 - \frac{1}{2}$	$-\left(\frac{j_1 - m + \frac{1}{2}}{2j_1 + 1}\right)^{1/2}$	$\left(\frac{j_1 + m + \frac{1}{2}}{2j_1 + 1}\right)^{1/2}$

$\langle j_1 1;\, m-\mu,\, \mu | j_1 1;\, jm \rangle$

j	$\mu = 1$	$\mu = 0$	$\mu = -1$
$j_1 + 1$	$\left(\frac{(j_1 + m)(j_1 + m + 1)}{(2j_1 + 1)(2j_1 + 2)}\right)^{1/2}$	$\left(\frac{(j_1 - m + 1)(j_1 + m + 1)}{(2j_1 + 1)(j_1 + 1)}\right)^{1/2}$	$\left(\frac{(j_1 - m)(j_1 - m + 1)}{(2j_1 + 1)(2j_1 + 2)}\right)^{1/2}$
j_1	$-\left(\frac{(j_1 + m)(j_1 - m + 1)}{2j_1(j_1 + 1)}\right)^{1/2}$	$\frac{m}{\sqrt{j_1(j_1 + 1)}}$	$\left(\frac{(j_1 - m)(j_1 + m + 1)}{2j_1(j_1 + 1)}\right)^{1/2}$
$j_1 - 1$	$\left(\frac{(j_1 - m)(j_1 - m + 1)}{2j_1(2j_1 + 1)}\right)^{1/2}$	$-\left(\frac{(j_1 - m)(j_1 + m)}{j_1(2j_1 + 1)}\right)^{1/2}$	$\left(\frac{(j_1 + m + 1)(j_1 + m)}{2j_1(2j_1 + 1)}\right)^{1/2}$

Table 14.2 *Some symmetry properties of C–G coefficients*

$$
\begin{aligned}
\langle j_1j_2;\ m_1m_2|j_1j_2;\ j_3m_3\rangle &= (-1)^{j_1+j_2-j_3}\langle j_1j_2;\ -m_1,\ -m_2|j_1j_2;\ j_3,\ -m_3\rangle \\
&= (-1)^{j_1+j_2-j_3}\langle j_2j_1;\ m_2m_1|j_2j_1;\ j_3m_3\rangle \\
&= (-1)^{j_1-m_1}\sqrt{\frac{2j_3+1}{2j_2+1}}\langle j_1j_3;\ m_1,\ -m_3|j_1j_3;\ j_2,\ -m_2\rangle \\
&= (-1)^{j_2+m_2}\sqrt{\frac{2j_3+1}{2j_1+1}}\langle j_3j_2;\ -m_3,\ m_2|j_3j_2;\ j_1,\ -m_1\rangle \\
&= (-1)^{j_1-m_1}\sqrt{\frac{2j_3+1}{2j_2+1}}\langle j_3j_1;\ m_3,\ -m_1|j_3j_1;\ j_2m_2\rangle \\
&= (-1)^{j_2+m_2}\sqrt{\frac{2j_3+1}{2j_1+1}}\langle j_2j_3;\ -m_2m_3|j_2j_3;\ j_1m_1\rangle
\end{aligned}
$$

case, the sum of the squares of the numerical (C–G) coefficients adds up to unity, as required by Eq. (14.64).

EXAMPLE 5

Two spinless particles in orbital angular momentum states with $\ell_1 = \ell_2 = 1$. The uncoupled representation is

$$|11;\ m_1m_2\rangle = |1m_1\rangle_1|1m_2\rangle_2,$$

where subscript k refers to particle k and the label k on m_ℓ has been suppressed. The coupled state is

$$|11;\ \ell m_\ell\rangle = \sum_{m_1,m_2} C(11\ell;\ m_1m_2m_\ell)|11;\ m_1m_2\rangle,$$

where $\ell = 0$, 1, or 2. We choose $\ell = 0$ (m_ℓ must be 0 also), for which the symmetry properties of the C–G coefficients are needed:

$$
\begin{aligned}
|11;\ 00\rangle &= \sum_{m_1,m_2} C(110;\ m_1m_20)|1m_1\rangle_1|1m_2\rangle_2 \\
&= \sum_{m_1,m_2} (-1)^{1-m_1}\sqrt{\tfrac{1}{3}}C(110;\ m_1,\ 0,\ -m_2)|1m_1\rangle_1|1m_2\rangle_2.
\end{aligned}
$$

However, $C(101;\ m_1,\ 0,\ -m_2) = \delta_{m_2,\ -m_1}$, so we get

$$
\begin{aligned}
|11;\ 00\rangle &= \sqrt{\tfrac{1}{3}}\sum_{m_1=-1}^{1}(-1)^{1-m_1}|1m_1\rangle_1|1,\ -m_1\rangle_2 \\
&= \sqrt{\tfrac{1}{3}}[|11\rangle_1|1,\ -1\rangle_2 - |10\rangle_1|10\rangle_2 + |1,\ -1\rangle_1|1,\ 1\rangle_2]. \qquad (14.70)
\end{aligned}
$$

This state is symmetric under interchange of the particle labels; and each of the three possible magnetic quantum-number substates contributes. In a coordinate representation the RHS of (14.70) is a sum of products of wave functions $Y_1^{m_1}(\Omega_1)Y_1^{-m_1}(\Omega_2)$ and therefore the LHS is also a function of Ω_1 and Ω_2.

EXAMPLE 6

An electron in the hydrogenic state $|n\ell m_\ell\rangle$, $\ell > 0$. Here the uncoupled representation is

$$|n\ell m_\ell\rangle|\tfrac{1}{2}m_s\rangle \equiv |n\ell\tfrac{1}{2};\, m_\ell m_s\rangle,$$

while the coupled representation is

$$|n\ell\tfrac{1}{2};\, jm_j\rangle = \sum_{m_s m_\ell} C(\ell\tfrac{1}{2}j;\, m_\ell m_s m_j)|n\ell m_\ell\rangle|\tfrac{1}{2}m_s\rangle.$$

The allowed values of j are $j = \ell \pm \frac{1}{2}$. For illustrative purposes we choose $j = \ell + \frac{1}{2}$ and $m_j = \ell - \frac{1}{2}$; using Table 14.1 it is readily shown that

$$|n\ell\tfrac{1}{2};\, \ell + \tfrac{1}{2},\, \ell - \tfrac{1}{2}\rangle = \sqrt{\frac{1}{\ell+1}}|n\ell\ell\rangle|\tfrac{1}{2},\, -\tfrac{1}{2}\rangle + \sqrt{\frac{2\ell}{2\ell+1}}|n\ell,\, \ell - 1\rangle|\tfrac{1}{2}\tfrac{1}{2}\rangle.$$

Both spin projections enter; the m_ℓ values are those which yield $m_j = \ell - \frac{1}{2}$.

EXAMPLE 7

Two spin-$\frac{1}{2}$ particles. The uncoupled representation is $|\frac{1}{2}\frac{1}{2};\, m_1 m_2\rangle = |\frac{1}{2}m_1\rangle_1|\frac{1}{2}m_2\rangle_2$ and the (total) spin values in the coupled state are $S = 0$ (singlet) and $S = 1$ (triplet):

$$|\tfrac{1}{2}\tfrac{1}{2};\, Sm_s\rangle = \sum_{m_1 m_2} C(\tfrac{1}{2}\tfrac{1}{2}S;\, m_1 m_2 m_s)|\tfrac{1}{2}m_1\rangle_1|\tfrac{1}{2}m_2\rangle_2.$$

Both the singlet and the triplet states will be determined, but we shall use C–G coefficients only to get the singlet state. The three members of the triplet state will instead be obtained using the lowering operator, starting with the stretched state.

We begin with the singlet state:

$$|\tfrac{1}{2}\tfrac{1}{2};\, 00\rangle = \sum_{m_1} C(\tfrac{1}{2}\tfrac{1}{2}0;\, m_1,\, -m_1,\, 0)|\tfrac{1}{2}m_1\rangle_1|\tfrac{1}{2},\, -m_1\rangle_2,$$

where the condition $m_s = 0$ imposes the requirement that $m_2 = -m_1$. The same symmetry condition as that used in Example 5 easily leads to

$$|\tfrac{1}{2}\tfrac{1}{2};\, 00\rangle = \sqrt{\tfrac{1}{2}}[|\tfrac{1}{2}\tfrac{1}{2}\rangle_1|\tfrac{1}{2},\, -\tfrac{1}{2}\rangle_2 - |\tfrac{1}{2},\, -\tfrac{1}{2}\rangle_1|\tfrac{1}{2}\tfrac{1}{2}\rangle_2]. \tag{14.71}$$

The minus sign in (14.71) means that the singlet state is *antisymmetric* (i.e., changes sign) under interchange of the particle labels.

Construction of the triplet state using the lowering operator is quite straightforward. The (unique) stretched state is

$$|\tfrac{1}{2}\tfrac{1}{2};\, 11\rangle = |\tfrac{1}{2}\tfrac{1}{2}\rangle_1|\tfrac{1}{2}\tfrac{1}{2}\rangle_2. \tag{14.72a}$$

Applying the lowering operator

$$\hat{J}_- = \hat{J}_{1-} + \hat{J}_{2-}$$

to both sides of (14.72a), using Eq. (14.23a), and canceling out the factor of $\hbar$ gives

$$|\tfrac{1}{2}\tfrac{1}{2};\ 10\rangle = \sqrt{\tfrac{1}{2}}[|\tfrac{1}{2}\tfrac{1}{2}\rangle_1|\tfrac{1}{2},\ -\tfrac{1}{2}\rangle_2 + |\tfrac{1}{2},\ -\tfrac{1}{2}\rangle_1|\tfrac{1}{2}\tfrac{1}{2}\rangle_2]. \tag{14.72b}$$

This state differs from the singlet state only by a change of the minus sign, but that change is of the greatest significance, since it distinguishes states of differing angular momentum and differing symmetry character: (14.72b) is *symmetric* under particle interchange. A second application of the lowering operator, this time to (14.72b), yields the third member of the triplet:

$$|\tfrac{1}{2}\tfrac{1}{2};\ 1,\ -1\rangle = |\tfrac{1}{2},\ -\tfrac{1}{2}\rangle_1|\tfrac{1}{2},\ -\tfrac{1}{2}\rangle_2. \tag{14.72c}$$

All three members of the triplet are obviously symmetric under particle-label interchange. We could have concluded this in advance of the construction by applying the label-interchange (two-particle-transposition) operator $\hat{P}_{12}$ to $|\tfrac{1}{2}\tfrac{1}{2};\ 1m_s\rangle$:

$$\begin{aligned}
\hat{P}_{12}|\tfrac{1}{2}\tfrac{1}{2};\ 1m_s\rangle &= \hat{P}_{12}\sum_{m_1,m_2} C(\tfrac{1}{2}\tfrac{1}{2}1;\ m_1m_2m_s)|\tfrac{1}{2}m_1\rangle_1|\tfrac{1}{2}m_2\rangle_2 \\
&= \sum_{m_1,m_2} C(\tfrac{1}{2}\tfrac{1}{2}1;\ m_1m_2m_s)\hat{P}_{12}[|\tfrac{1}{2}m_1\rangle_1|\tfrac{1}{2}m_2\rangle_2]. \\
&= \sum_{m_1,m_2} C(\tfrac{1}{2}\tfrac{1}{2}1;\ m_1m_2m_s)|\tfrac{1}{2}m_1\rangle_2|\tfrac{1}{2}m_2\rangle_1.
\end{aligned}$$

The simplest way to restore the original order of states is to relabel the (dummy) summation indices m_1 and m_2:

$$\begin{aligned}
m_1 &\longrightarrow m_2,\ m_2 \longrightarrow m_1, \\
C(\tfrac{1}{2}\tfrac{1}{2}1;\ m_1m_2m_s) &\rightarrow C(\tfrac{1}{2}\tfrac{1}{2}1;\ m_2m_1m_s) \\
&= (-1)^{\frac{1}{2}+\frac{1}{2}-1}C(\tfrac{1}{2}\tfrac{1}{2}1;\ m_1m_2m_s) \\
&= C(\tfrac{1}{2}\tfrac{1}{2}1;\ m_1m_2m_s),
\end{aligned}$$

by which we find

$$\begin{aligned}
\hat{P}_{12}|\tfrac{1}{2}\tfrac{1}{2};\ 1m_s\rangle &= \sum_{m_1\,m_2} C(\tfrac{1}{2}\tfrac{1}{2}1;\ m_1m_2m_s)|\tfrac{1}{2}m_1\rangle_1|\tfrac{1}{2}m_2\rangle_2 \\
&= |\tfrac{1}{2}\tfrac{1}{2};\ 1m_s\rangle.
\end{aligned}$$

QED. Thus, the triplet state is an eigenstate of $\hat{P}_{12}$ with eigenvalue 1; the singlet state is an eigenstate of $\hat{P}_{12}$ with eigenvalue -1, a fact we use later.

Equations (14.71) and (14.72b) are probably the simplest examples of correlated states. The singlet state will play a prominent role in the discussion of hidden variables and Bell's inequality, Section 14.5.

EXAMPLE 8

An inverse construction. We start with the uncoupled hydrogenic state $|\ell\frac{1}{2};\ \ell-1,\frac{1}{2}\rangle = |n\ell,\ \ell-1\rangle|\frac{1}{2}\frac{1}{2}\rangle$, $\ell \neq 0$, and express it in terms of the coupled states $|\ell\frac{1}{2};\ jm_j\rangle$, using Eq. (14.65):

$$|\ell\tfrac{1}{2};\ \ell-1,\tfrac{1}{2}\rangle = \sum_j C(\ell\tfrac{1}{2}j;\ \ell-1,\tfrac{1}{2},\ell-\tfrac{1}{2})|\ell\tfrac{1}{2};\ j,\ m_j=\ell-\tfrac{1}{2}\rangle.$$

The allowed values of j are $j=\ell\pm\frac{1}{2}$. Use of Table 14.1 leads to

$$\begin{aligned}|\ell\tfrac{1}{2};\ \ell-1,\tfrac{1}{2}\rangle =& \sqrt{\frac{2\ell}{2\ell+1}}|\ell\tfrac{1}{2};\ j=\ell+\tfrac{1}{2},\ m_j=\ell-\tfrac{1}{2}\rangle \\ & -\sqrt{\frac{1}{2\ell+1}}|\ell\tfrac{1}{2};\ j=\ell-\tfrac{1}{2},\ m_j=\ell-\tfrac{1}{2}\rangle.\end{aligned}$$

14.4. $\hat{\mathbf{J}}_1\cdot\hat{\mathbf{J}}_2$ interactions

Just as spin is a non-classical concept, so too is an interaction involving the scalar product of two angular-momentum operators, which we write generically as $\hat{\mathbf{J}}_1\cdot\hat{\mathbf{J}}_2$. Probably the best-known examples are the spin–orbit interaction, of the form

$$\hat{H}_{LS}(\mathbf{r}) = V_{LS}(r)\hat{\mathbf{L}}\cdot\hat{\mathbf{S}}/\hbar^2, \tag{14.73}$$

which acts on particles with non-zero spin (electrons in an atom, nucleons in a nucleus), and the spin–spin interaction

$$\hat{H}_{S_1S_2} = V_{S_1S_2}(\mathbf{r})\hat{\mathbf{S}}_1\cdot\hat{\mathbf{S}}_2/\hbar^2. \tag{14.74}$$

(The factor $\hbar^{-2}$ ensures that the V's have the dimension of energy.) In $\hat{H}_{S_1S_2}$, $\mathbf{r}$ is the relative distance between the two particles, labeled 1 and 2 (two nucleons, say, or the electron and the proton in the H-atom).

Interactions of these types occur in real systems. The other reason they are of interest, and why they are discussed in this chapter, is that any interaction of the form $\hat{\mathbf{J}}_1\cdot\hat{\mathbf{J}}_2$ is diagonal only in coupled-representation states: to understand its effects, the analysis of the preceding section must be applied. The reason why this statement is true also provides the method for dealing with such an interaction.

The method follows from Eq. (14.49c), which we re-write as

$$\hat{\mathbf{J}}_1\cdot\hat{\mathbf{J}}_2 = \tfrac{1}{2}[\hat{J}^2-\hat{J}_1^2-\hat{J}_2^2]. \tag{14.75}$$

Since (14.75) contains $\hat{J}^2$, whose eigenstates are the states $|j_1j_2;\ jm\rangle$, $\hat{\mathbf{J}}_1\cdot\hat{\mathbf{J}}_2$ can be easily evaluated (diagonalized) only by working in states of the coupled representation. In fact, $|j_1j_2;\ jm\rangle$ is an eigenstate of $\hat{\mathbf{J}}_1\cdot\hat{\mathbf{J}}_2$:

$$\begin{aligned}\hat{\mathbf{J}}_1\cdot\hat{\mathbf{J}}_2|j_1j_2;\ jm\rangle &= \tfrac{1}{2}[\hat{J}^2-\hat{J}_1^2-\hat{J}_2^2]|j_1j_2;\ jm\rangle \\ &= \frac{\hbar^2}{2}[j(j+1)-j_1(j_1+1)-j_2(j_2+1)]|j_1j_2;\ jm\rangle,\end{aligned} \tag{14.76}$$

where the eigenvalue depends on all three of the angular-momentum quantum numbers j, j_1, and j_2. Projection of both sides of Eq. (14.76) onto $\langle j_1' j_2';\, j'm'|$ produces only a diagonal matrix element:

$$\langle j_1' j_2';\, j'm'|\hat{\mathbf{J}}_1 \cdot \hat{\mathbf{J}}_2|j_1 j_2;\, jm\rangle = \delta_{j_1' j_1}\delta_{j_2' j_2}\delta_{j' j}\delta_{m' m}\frac{\hbar^2}{2}[j(j+1) - j_1(j_1+1) - j_2(j_2+1)]. \tag{14.77}$$

Equations (14.76) and (14.77) have proved to be very useful for systems described by rotationally invariant Hamiltonians, i.e., Hamiltonians whose eigenstates are labeled by coupled-state, angular-momentum quantum numbers. Since $\hat{\mathbf{J}}_1 \cdot \hat{\mathbf{J}}_2$ is a classical-type scalar product, it not only "looks" exactly like a scalar quantity, i.e., one that is invariant under rotations, but is one. This claim is verified by the fact that $\hat{\mathbf{J}}_1 \cdot \hat{\mathbf{J}}_2$ does not change the angular-momentum quantum numbers when it acts on $|j_1 j_2;\, jm\rangle$. Although there are non-scalar, non-rotationally-invariant operators, only scalar operators will concern us in this book. We examine spin–spin and spin–orbit interactions in separate subsections, using nuclear systems as the applications.

Spin–spin interaction

Early in the study of nuclear forces and nuclear structure, anomalous data were obtained from very low-energy neutron–proton (n–p) scattering experiments. The anomaly was caused by attempts to interpret the data by assuming that the interaction was solely central; the situation was soon clarified by E. P. Wigner (Bethe and Bacher (1936)), who showed that the data could be understood by adding a spin–spin interaction to the central one, thereby providing important evidence that nuclear forces are spin-dependent. We shall outline the argument in the following; no details of the scattering state – considered in Chapter 15 – are needed for the analysis.

Let $\mathbf{r}$ be the neutron–proton relative separation and the label 1 (2) refer to the neutron (proton). The Hamiltonian used in the successful analysis has the form

$$\hat{H} = K(\mathbf{r}) + V_c(r) + \hat{H}_{S_1 S_2}(r), \tag{14.78}$$

where $V_c(r)$ is the spin-independent central potential and the potential $V_{S_1 S_2}(r)$ appearing in $\hat{H}_{S_1 S_2}(r)$ is also assumed central. It follows from $[\hat{\mathbf{S}}_k, \hat{H}] = 0$, $k = 1$ or 2, that the simplest eigenstates of $\hat{H}$ are products of $|\frac{1}{2}\frac{1}{2};\, Sm_s\rangle$ times a scattering eigenstate of $\hat{L}^2$ and $\hat{L}_z$. For the low-energy situation considered here, the latter eigenstate has $\ell = 0$, but we shall be more general and write it as $|k\ell m_\ell;\, S\rangle$, with k the wave number; that the presence of $\hat{H}_{S_1 S_2}$ in $\hat{H}$ forces the scattering state to depend on S is the essence of Wigner's analysis.

The full eigenstate is $\langle \mathbf{r}|k\ell m_\ell;\, S\rangle|\frac{1}{2}\frac{1}{2};\, Sm_s\rangle$ and obeys

$$[\hat{K}(\mathbf{r}) + V_c(r) + \hat{H}_{S_1 S_2}(r)]\langle \mathbf{r}|k\ell m_\ell;\, S\rangle|\tfrac{1}{2}\tfrac{1}{2};\, Sm_s\rangle$$

$$= E_k\langle \mathbf{r}|k\ell m_\ell;\, S\rangle|\tfrac{1}{2}\tfrac{1}{2};\, Sm_s\rangle, \qquad E_k = \frac{\hbar^2 k^2}{2\mu}, \tag{14.79}$$

where E_k is the energy, of the order of a few kilovolts. Equation (14.79) is transformed into an S-dependent equation for $\langle \mathbf{r}|k\ell m_\ell;\, S\rangle$ by use of (14.76) (with $\hat{\mathbf{J}}_k \to \hat{\mathbf{S}}_k$, $j_k = \frac{1}{2}$), followed by projection onto $\langle \frac{1}{2}\frac{1}{2};\, Sm_s|$:

$$\{\hat{K}(\mathbf{r}) + V_c(r) + \tfrac{1}{2}[S(S+1) - \tfrac{3}{2}]V_{S_1S_2}(r)\}\langle \mathbf{r}|k\ell m_\ell;\, S\rangle = E_k\langle \mathbf{r}|k\ell m_\ell;\, S\rangle. \tag{14.80}$$

It is clear that the singlet ($S = 0$) and triplet ($S = 1$) wave-function components obey different (uncoupled) equations:

$$\{\hat{K}(\mathbf{r}) + [V_c(r) - \tfrac{3}{4}V_{S_1S_2}(r)] - E_k\}\langle \mathbf{r}|k\ell m_\ell;\, 0\rangle = 0 \tag{14.81a}$$

and

$$\{\hat{K}(\mathbf{r}) + [V_c(r) + \tfrac{1}{4}V_{S_1S_2}(r)] - E_k\}\langle \mathbf{r}|k\ell m_\ell;\, 1\rangle = 0. \tag{14.81b}$$

Note that the energy is the same in both members of (14.81). For both $V_c(r)$ and $V_{S_1S_2}(r)$ predominately attractive, we see that the overall potential acting on $\langle \mathbf{r}|k\ell m_\ell;\, 1\rangle$ is the more attractive of the two, and this shows up in the scattering data. It turns out that, for a random mixture of $S = 0$ and $S = 1$ states (a so-called "unpolarized" beam), the $S = 0$ and $S = 1$ states contribute to the measured scattering data incoherently – they add together without interference effects – with weights proportional to the spin multiplicity, i.e., the number of magnetic substates. Thus, a quarter of the contribution comes from the singlet state and three quarters from the triplet state.

Spin–spin interactions occur in bound-state systems as well. For example, one contribution to the ground-state energy of the real H-atom is a relativistic term in which the electron interacts with the magnetic moment of the proton, and this contains an $\hat{\mathbf{S}}_e \cdot \hat{\mathbf{S}}_p$ factor. This additional (very weak) interaction gives rise to what is known as the hyperfine splitting/structure of the H-atom energy levels, and we shall examine it in Chapter 17.

Spin–orbit interaction

Spin–orbit interactions occur both in bound-state and in scattering systems. The application in this subsection is to bound states, in particular to the splitting of energy levels which is produced by a purely central potential.

In abstract operator form, Eq. (14.73) is

$$\hat{H}_{LS}(\hat{Q}) = \hat{V}_{LS}(\hat{Q})\hat{\mathbf{L}} \cdot \hat{\mathbf{S}}/\hbar^2. \tag{14.82}$$

The relevant coupled states are the $|n\ell\frac{1}{2};\, jm\rangle$, and we shall assume the coordinate representation to be

$$\langle \mathbf{r}|n\ell\tfrac{1}{2};\, jm\rangle = R_{n\ell}(r)\langle \Omega|\ell\tfrac{1}{2};\, jm\rangle, \tag{14.83}$$

where Ω is the usual pair of spherical polar angles of $\mathbf{r}$. In contrast to the spin–spin case, the assumption underlying (14.83) is that the radial wave function $R_{n\ell}(r)$ does not depend on the total angular momentum j. This is often not the case, but it is a frequently made assumption, and it simplifies the analysis without trivializing it. We find using Eq. (14.77) that

$$\begin{aligned}\langle \hat{H}_{LS}\rangle_{n\ell j} &\equiv \langle n\ell\tfrac{1}{2};\, jm|\hat{V}_{LS}(\hat{Q})\hat{\mathbf{L}}\cdot\hat{\mathbf{S}}|n\ell\tfrac{1}{2};\, jm\rangle/\hbar^2 \\ &= \frac{1}{\hbar^2}(R_{n\ell}(r)|V_{LS}(r)|R_{n\ell}(r))\langle \ell\tfrac{1}{2};\, jm|\hat{\mathbf{L}}\cdot\hat{\mathbf{S}}|\ell\tfrac{1}{2};\, jm\rangle \\ &= [j(j+1) - \ell(\ell+1) - \tfrac{3}{4}]\langle V_{LS}\rangle_{n\ell}/2, \end{aligned} \tag{14.84}$$

where $(R_{n\ell}(r)|V_{LS}(r)|R_{n\ell}(r)) = \langle V_{LS}\rangle_{n\ell}$ means an integral over $r^2\, dr$.

In the present case, j takes on the values $\ell \pm \frac{1}{2}$, so that (14.84) becomes

$$\langle \hat{H}_{LS}\rangle_{n\ell j} = \begin{cases} (\ell/2)\langle V_{LS}\rangle_{n\ell}, & j = \ell + \frac{1}{2}, \\ -[(\ell+1)/2]\langle V_{LS}\rangle_{n\ell}, & j = \ell - \frac{1}{2}, \end{cases} \tag{14.85}$$

where the sign of $\langle \hat{H}_{LS}\rangle_{n\ell j}$ will depend on the sign of $\langle V_{LS}\rangle_{n\ell}$, typically on whether $V_{LS}(r)$ is predominantly attractive or repulsive. In hydrogen (Section 17.2) and the shell-model description of many atoms, $\langle V_{LS}\rangle_{n\ell}$ is repulsive (> 0), whereas in an independent-particle (shell-model) description for most nuclei, $\langle V_{LS}\rangle_{n\ell}$ is attractive (< 0).

As a specific example, we consider the nucleus of the oxygen isotope $^{17}_{8}\mathrm{O}$, which has eight protons and nine neutrons. The lowest-lying states of this nucleus are nicely approximated by the assumption that all nucleons but one neutron form "closed shells" that have total angular momentum zero (and contribute a constant energy), so that the angular momentum properties and energies of the low-lying states of $^{17}_{8}\mathrm{O}$ are determined by the one neutron (recall that, in the Stern–Gerlach experiment and analysis, the spin of just one electron determined the angular-momentum properties of Ag). The Schrödinger equation for this "valence" neutron is taken to be

$$\hat{H} = \hat{K} + V(\hat{Q}) + \hat{H}_{LS}(\hat{Q}), \tag{14.86}$$

and assuming still that $R_{n\ell}$ is independent of j, the eigenstates of (14.84) are $|n\ell\frac{1}{2};\, jm\rangle$ and the energies, obtained by diagonalizing (14.86), are

$$E_{n\ell j} = \mathcal{E}_{n\ell} + \langle \hat{H}_{LS}\rangle_{n\ell j}, \tag{14.87}$$

where

$$[\hat{K} + V(\hat{Q})]|n\ell\tfrac{1}{2};\, jm\rangle = \mathcal{E}_{n\ell}|n\ell\tfrac{1}{2};\, jm\rangle \tag{14.88}$$

defines a single-particle energy $\mathcal{E}_{n\ell}$ that is independent of j.

Substituting (14.84) into (14.87) we find

$$E_{n\ell j} = \mathcal{E}_{n\ell} + \begin{cases} (\ell/2)\langle V_{LS}\rangle_{n\ell}, & j = \ell + \frac{1}{2}, \\ -[(\ell+1)/2]\langle V_{LS}\rangle_{n\ell}, & j = \ell - \frac{1}{2}. \end{cases} \tag{14.89}$$

While a detailed model is needed to evaluate (14.89), it is not required to determine the difference in energy between the two $j = \ell \pm \frac{1}{2}$ states (fixed ℓ) and compare it with experiment, thereby obtaining an estimate for $\langle V_{LS}\rangle_{n\ell l}$. First we note that, since $\langle \hat{V}_{LS}\rangle_{n\ell}$ is negative (the nuclear case), the state with $j = \ell + \frac{1}{2}$ lies lower than the $j = \ell - \frac{1}{2}$ state. For $^{17}_{8}\mathrm{O}$, the ground state and the low-lying 5.08 MeV excited state each have $\ell = 2$; the ground state indeed has $j = \frac{5}{2}$, while the excited state has $j = \frac{3}{2}$: see Fig. 14.1 for a

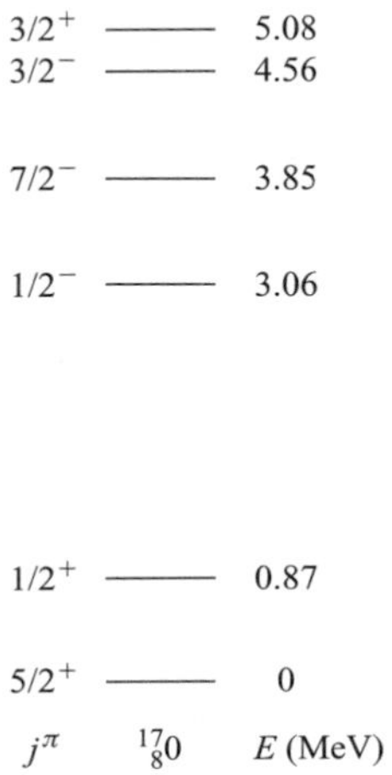

j^π	$^{17}_{8}$O	E (MeV)
$3/2^+$	———	5.08
$3/2^-$	———	4.56
$7/2^-$	———	3.85
$1/2^-$	———	3.06
$1/2^+$	———	0.87
$5/2^+$	———	0

Fig. 14.1 The six lowest-lying energy levels in $^{17}_{8}$O. The energies are to the right and the total angular momentum and parity (j^π) values are to the left. The $j = \frac{5}{3}, \frac{3}{2}$ spin–orbit pair have $\pi = (-1)^2 = +1$, since $\ell = 2$.

Grotrian diagram of the low-lying $^{17}_{8}$O spectrum. (From Ajzenberg-Selove and Lauritsen (1959). The $\pm$ superscripts on the values of j in the diagram denote the parities.)

The theoretical prediction for the difference in energies is

$$\Delta E^{(\text{th})}_{n2j} = E_{n2,3/2} - E_{n2,5/2} = -\tfrac{5}{2}\langle V_{LS}\rangle_{n2}.$$

From Fig. 14.1 the experimental difference in energy is

$$\Delta E^{(\text{ex})}_{n2j} = 5.08 \text{ MeV},$$

and a comparison of the two expressions yields

$$\langle V_{LS}\rangle_{n2} = -2.08 \text{ MeV},$$

a number that nuclear theorists have used to determine parameters of the nuclear shell model for $^{17}_{8}$O.

14.5. *Conceptual/interpretational controversies, Bell's inequality

As the theory of microscopic phenomena, quantum mechanics is phenomenally successful both in its predictive and in its data-interpretive aspects. Despite this success, doubts about the correctness and completeness of quantum theory have existed almost from its inception. The questions raised concern mainly the conceptual foundations of the theory and the adequacy of the time-dependent Schrödinger equation, in particular, Postulates II, III, and V. It is the author's belief that one should not learn quantum theory without being aware of these concerns – which are not shared, and especially not paid attention to, by every working physicist, essentially because the doubts address qualitative/philosophical/conceptual concerns: the quantitative aspects of the theory are not in question.

While awareness and eventual understanding of these concerns should be a part of every physicist's armory, a general discussion and analysis is well beyond the aims and scope of this book. We concentrate instead on Bell's theorem, whose underpinning is the analysis leading to what is now known as Bell's inequality. As background, we briefly discuss the idea of "hidden variables" as a means of restoring classical-type determinism to quantum mechanics, and the argument of Einstein, Podolsky, and Rosen concerning reality and their belief that quantum theory is incomplete.[2]

Background

Since *we* have seen how strange its results can be, it should not be surprising to learn that, even up to the present time, some physicists have been uneasy about quantum theory. Particularly disturbing has been the imprecision that is an integral part of the vision provided by quantum mechanics: statistical (average) quantities replace specific ones; probability and uncertainty abound; depending on what is measured, objects can be either wave-like or particle-like; the path followed by particles – say in an interference experiment – cannot be measured without destroying the original experimental results and replacing them by others; etc. In addition, the Copenhagen interpretation of Bohr, Heisenberg, and Born – which is partially embodied in Postulates II and III and statements on the probabilistic/uncertainty aspects – stresses the *absence* of a quantal-type of reality: not only does the theory provide no picture of the quantum world (since only the numerical results of measurements have meaning), but also Bohr claimed that there *is no* quantum world.

In the Copenhagen interpretation, probability/uncertainty is an incontestably inherent feature of the fundamental framework. There is, however, another viewpoint: in it, quantum mechanics is the "visible" component of a broader framework, whose "hidden" component provides a deterministic reality to the description of microscopic phenomena. Unfortunately, this hidden component is aptly named: it is inaccessible. From this viewpoint, the imprecision of quantum theory represents our ignorance concerning the details of the – unobservable – underlying reality. Characterizing this hypothetical underpinning for quantum theory would be one or more parameters, generically known as "hidden" (i.e., unobservable) variables. If this hidden aspect really existed, and if somehow the parameters could be specified (observed), then one would return to a fully deterministic, classical-physics type of theory, whose states would be "dispersion-free," i.e., there would no longer be any uncertainties. Hopes for the latter situation were effectively crushed by von Neumann's proof that hidden-variable theories were impossible, in the sense that no hidden-variable/deterministic theory could produce the same predictions/results as quantum theory (von Neumann (1932)).

Whether or not hidden-variable theories could be meaningful, von Neumann's 1932

[2] Well-written popular expositions of the foregoing topics (as well as others) are available, and we strongly recommend that the reader, at some point after his/her understanding has gone beyond the stage of "this course is a prerequisite for this course," enlarge his/her grasp of the subject by reading one or more of these qualitative and semi-quantitative treatments. Some that the author is familiar with and can recommend are the books by Polkinghorne (1984), von Baeyer (1992), Squires (1994) and Ellis and Amati (2000). These contain references the reader may wish to pursue. The collection of Bell's papers on some of these topics (Bell 1988), though it is rather more quantitative, is also recommended, as are the technical reviews of Clauser and Shimony (1978) and of Pipkin (1979). These books and reviews have myriad references both to the older and to more current literature; the shelves of good science libraries should contain enough material in this area for at least a winter's reading.

proof had eliminated them as a conceptual means for making quantum theory complete – if indeed it needs to be completed. That it needed completing was claimed a few years later by Einstein, Podolsky, and Rosen (1935), who analyzed the positions and momenta of a two-particle quantal system and, on the basis of their definition of "reality," concluded that quantum theory was incomplete. Since the Einstein, Podolsky, and Rosen (EPR) paper stimulated many further developments, including Bell's important work, we summarize the EPR ideas next, but do so using the system studied by David Bohm (1951), viz., two spin-$\frac{1}{2}$ particles in a spin-singlet ($S = 0$) state.

Consider then an unstable spin-0 particle that can decay into two spin-$\frac{1}{2}$ particles. They are produced in a singlet state due to conservation of angular momentum. Equation (14.71) is the simplest example of the singlet state:

$$|\tfrac{1}{2}\tfrac{1}{2};\, 00\rangle_{1,2} = \frac{1}{\sqrt{2}}[|\tfrac{1}{2}\tfrac{1}{2}\rangle_1|\tfrac{1}{2},\, -\tfrac{1}{2}\rangle_2 - |\tfrac{1}{2},\, -\tfrac{1}{2}\rangle_1|\tfrac{1}{2}\tfrac{1}{2}\rangle_2]. \tag{14.71}$$

This state necessarily preserves its $S = 0$ character under rotations, and therefore another correct way of writing it is

$$|\tfrac{1}{2}\tfrac{1}{2};\, 00\rangle_{1,2} = \frac{1}{\sqrt{2}}[|\mathbf{n},\, +\rangle_1|\mathbf{n},\, -\rangle_2 - |\mathbf{n},\, -\rangle_1|\mathbf{n},\, +\rangle_2], \tag{14.90}$$

where $\mathbf{n}$ is arbitrary.

The correlated state (14.90) tells us that, no matter which quantization direction is chosen, 1 and 2 always have opposite spin projections, so that, if 1 is measured (possibly by a Stern–Gerlach experiment) to have spin $\pm\hbar/2$, then 2 necessarily has spin $\mp\hbar/2$. For example, should measurement of $\hat{S}_{1x}$ yield $-\hbar/2$, then measurement of $\hat{S}_{2x}$ will yield $+\hbar/2$. On the other hand, if $S_{1x} = -\hbar/2$, no definite statement can be made for S_{2y} or S_{2z}: each will have probability $\frac{1}{2}$ that each of $\pm\hbar/2$ will be observed.

The foregoing is straightforward quantum mechanics. EPR used its analog for position and momentum to argue that "either (1) the quantum-mechanical description of reality given by the wave function[3] is not complete or (2) when the operators corresponding to two physical quantities do not commute the two quantities cannot have simultaneous reality." From their analysis they concluded that, if alternative (1) was false, i.e., if the quantal description *was* complete, then "two physical quantities, with non-commuting operators, can have simultaneous reality." This, of course, violates non-commutativity (in the present $S = 0$ example, measurement of both $\hat{S}_x$ and $\hat{S}_y$ simultaneously is not possible). Hence, EPR concluded that "the wave function does not provide a complete description of physical reality"; they left open the question of whether such a description exists, though they believed "that such a theory is possible."

Let us examine the EPR argument as it would apply in the spin-$\frac{1}{2}$, $S = 0$ case. First choose an $\mathbf{n}$ in (14.90), next let the separation between 1 and 2 become arbitrarily large, and then measure the component of $\hat{\mathbf{S}}_1$ along $\mathbf{n}$. By virtue of the structure of (14.90), that measurement instantaneously fixes $\hat{\mathbf{S}}_2 \cdot \mathbf{n}$ without any measurement of $\hat{\mathbf{S}}_2$ being needed. Indeed, one could argue that $\hat{\mathbf{S}}_2 \cdot \mathbf{n}$ must have had that value even if $\hat{\mathbf{S}}_1 \cdot \mathbf{n}$ had not been measured, though we, the observers, would not actually know what this value is. Furthermore, by choosing another value of $\mathbf{n}$, say $\mathbf{n}'$, and going through the same steps,

[3] In the present case of spin $\frac{1}{2}$, one should read "state vector" for "wave function."

particle 2 will then have a second, simultaneously known spin value $\hat{\mathbf{S}}_2 \cdot \mathbf{n}'$. For $\mathbf{n} = \mathbf{e}_1$ and $\mathbf{n}' = \mathbf{e}_2$, the non-commutativity of $\hat{S}_{2x}$ and $\hat{S}_{2y}$ renders this situation quantally impossible, and therefore the quantal description, violating "reality," must be incomplete.

Clearly, we must understand what EPR meant by reality. They proposed a mild sufficiency condition: "if, without in any way disturbing a system, we can predict with certainty (i.e., with probability equal to unity) the value of a physical quantity, then there exists an element of physical reality corresponding to this physical quantity." They also stated that these elements of physical reality "must be found by an appeal to results of experiments and measurements," and that "every element of the physical reality must have a counterpart in the physical theory," the latter statement constituting their notion of completeness. For discussion and comments on this, see, e.g., Jammer (1974), Polkinghorne (1984), or Squires (1994).

EPR's argument as applied to the singlet-spin case assigns reality to all quantization axes and concludes that quantum theory is incomplete. Quantum theory's response is, in essence, the statement that, in contradiction to the sufficiency condition, measurement of $\hat{\mathbf{S}}_1 \cdot \mathbf{n}$ *does* disturb the system: this measurement fixes the value of the spin of particle 2 along $\mathbf{n}$, no matter how far from 1 it is located. Measurement of $\hat{\mathbf{S}}_1 \cdot \mathbf{n}$ disturbs the system in a highly non-local fashion, since 2 can be an arbitrary distance from the locale of 1 (even farther than a light signal could reach). Thus, a correlated state implies non-locality.

This non-locality aspect was disturbing to Einstein, the founder of relativity and an absolute disbeliever in action at a distance. No method for completing quantum theory was presented in the EPR paper; the von Neumann argument against the hidden-variable extension having effectively eliminated it as a possibility.

The modern – Bohm/Bell – era

The basic elements of the foregoing discussion are Bohr's claim that there is no quantum world and no reality associated with state vectors/wave functions, EPR's conclusion that quantum theory needed completion to make it conform with their concept of reality, and von Neumann's proof that hidden-variable theories that will yield all quantal results cannot be constructed, thereby ruling out such theories as a means of "completing" quantum theory.

This is the background against which David Bohm (1952) produced a non-local hidden-variable theory whose predictions were in perfect agreement with the quantum ones(!!), thus conclusively establishing that von Neumann's "proof" needed re-examining. That Bohm's theory is non-local is of no consequence, since von Neumann's analysis said nothing about locality.[4]

Bohm's papers and the EPR argument were major factors motivating the beautiful work of John Bell (1964, 1966). He first showed (Bell (1966)) that von Neumann's analysis, while mathematically impeccable in the context of quantum theory, contained a then-to-fore unexamined assumption concerning its applicability to hidden-variable theories. That assumption was unnecessarily restrictive; dropping it was not compromising, as Bohm's hidden-variable theory clearly demonstrated.

[4] For discussions of Bohm's version of quantum theory, the interested reader might try consulting Cushing (1994) and Cushing, Fine, and Goldstein (1996). See also Exercise 14.21.

Bell's identification of the breakdown of von Neumann's proof was only one of two remarkable results that he derived in 1964, the second being far more important since it led to predictions, to additional (and crucial) theoretical developments, and, finally, to experiments establishing that the predictions of quantum theory rather than those of *local*-hidden-variable theories are in agreement with observation. This second work of Bell (1964), leading as it did to the experimental conclusion that quantum mechanics is complete, at least w.r.t. local-hidden-variable theories, is as monumental a theoretical achievement as that of Goudsmit and Uhlenbeck. The consequences of Bell's ideas and related developments are still being investigated.

In the remainder of this section we derive the version of the hidden-variable result first obtained by Bell (1964), using his notation and his definition of locality, and then close with comments about the experiments.

The system is still the pair of spin-$\frac{1}{2}$ particles in the singlet state; the measurements are of $\hat{\mathbf{S}}_1$ in the direction of the unit vector $\mathbf{a}$ and $\hat{\mathbf{S}}_2$ in the direction of the unit vector $\mathbf{b}$, viz., $\hat{\mathbf{S}}_1 \cdot \mathbf{a}$ and $\hat{\mathbf{S}}_2 \cdot \mathbf{b}$. To concentrate only on the signs, we replace $\hat{\mathbf{S}}$ by $\hat{\boldsymbol{\sigma}}$: $\langle \hat{\boldsymbol{\sigma}}_1 \cdot \mathbf{a} \rangle_1 = \pm 1$ and $\langle \hat{\boldsymbol{\sigma}}_2 \cdot \mathbf{a} \rangle_2 = \pm 1$, where the brackets $\langle \rangle_j$, $j = 1, 2$, refer to averages over single-particle spin states. Quantum theory states that, if $\mathbf{b} = \mathbf{a}$, then $\hat{\boldsymbol{\sigma}}_1 \cdot \mathbf{a}$ and $\hat{\boldsymbol{\sigma}}_2 \cdot \mathbf{a}$ have opposite signs in the singlet state. Hence measurements along the two different directions using the correlated $S = 0$ state should yield a result displaying correlations. It is straightforward to show (see the exercises) that the quantal expectation value – $P_{\mathrm{q}}(\mathbf{a}, \mathbf{b})$ – satisfies

$$P_{\mathrm{q}}(\mathbf{a}, \mathbf{b}) \equiv \langle (\hat{\boldsymbol{\sigma}}_1 \cdot \mathbf{a})(\hat{\boldsymbol{\sigma}}_2 \cdot \mathbf{b}) \rangle_{1,2} = -\mathbf{a} \cdot \mathbf{b} = -\cos \Theta_{\mathbf{ab}}, \tag{14.91}$$

where $\Theta_{\mathbf{ab}}$ is the angle between $\mathbf{a}$ and $\mathbf{b}$. Equation (14.91) displays an easily seen correlation. Let $\mathbf{a}$ and $\mathbf{b}$ be coplanar, in which case $\Theta_{\mathbf{ab}} = \Theta_{\mathbf{a}} - \Theta_{\mathbf{b}}$: the correlation resides in the fact that $\cos(\Theta_{\mathbf{a}} - \Theta_{\mathbf{b}}) \neq f(\Theta_{\mathbf{a}}) g(\Theta_{\mathbf{b}})$, where f and g are one-variable functions. Note that, for $\mathbf{b} = \mathbf{a}$, $-\mathbf{a} \cdot \mathbf{b} = -1$, the type of result that motivated the EPR paper.

The expression for $P_{\mathrm{h}}(\mathbf{a}, \mathbf{b})$, the analogous hidden-variable result, is constructed as follows. First, let the continuous variable $\lambda \in \Lambda$ denote the hidden variables (there is no change in the results if $\lambda \in \Lambda \rightarrow \lambda \in \{\lambda_1, \lambda_2, \ldots\}$). Second, let λ be weighted according to the normalized probability density $\rho(\lambda)$:

$$\int_{\Lambda} d\lambda \, \rho(\lambda) = 1.$$

Third, let $A = \pm 1$ be the outcomes of a measurement of $\hat{\boldsymbol{\sigma}}_1 \cdot \mathbf{a}$. In a hidden-variable theory, A must depend on λ and on $\mathbf{a}$, but since there are two spins in an $S = 0$ state and $\hat{\boldsymbol{\sigma}}_2 \cdot \mathbf{b}$ will also be measured, A could, at least in principle, also depend on $\mathbf{b}$:

$$A \stackrel{?}{=} A(\mathbf{a}, \mathbf{b}, \lambda) = \pm 1.$$

Similarly, the results $B = \pm 1$ of measuring $\hat{\boldsymbol{\sigma}}_2 \cdot \mathbf{b}$ could in principle be

$$B \stackrel{?}{=} B(\mathbf{a}, \mathbf{b}, \lambda) = \pm 1.$$

The dependence of each of A and B on $\mathbf{a}$, $\mathbf{b}$, and λ introduces the kind of non-local, classical action-at-a-distance property that Einstein found distasteful, and to eliminate it Bell introduced his *locality requirement*:

$$A = A(\mathbf{a}, \lambda) = \pm 1 \tag{14.92a}$$

and

$$B = B(\mathbf{b}, \lambda) = \pm 1. \tag{14.92b}$$

With Eqs. (14.92), the deterministic, local-hidden-variable expression for the expectation value $P_{\rm h}(\mathbf{a}, \mathbf{b})$ of the product $(\hat{\boldsymbol{\sigma}}_1 \cdot \mathbf{a})(\hat{\boldsymbol{\sigma}}_2 \cdot \mathbf{b})$ is

$$P_{\rm h}(\mathbf{a}, \mathbf{b}) = \int_\Lambda d\lambda\, \rho(\lambda) A(\mathbf{a}, \lambda) B(\mathbf{b}, \lambda). \tag{14.93}$$

Equation (14.93) is completely general. The question is whether it leads to results in agreement with those of quantum mechanics. Bell showed, using a third unit vector $\mathbf{c}$, that it did not. The argument is straightforward. First, Bell noted that min $P_{\rm h}(\mathbf{a}, \mathbf{b}) = -1$, a value reached only when $\mathbf{b} = \mathbf{a}$ and only if

$$B(\mathbf{a}, \lambda) = -A(\mathbf{a}, \lambda) \tag{14.94}$$

everywhere except perhaps at points of zero probability. The second step is to substitute (14.94) into (14.93):

$$P_{\rm h}(\mathbf{a}, \mathbf{b}) = -\int_\Lambda d\lambda\, \rho(\lambda) A(\mathbf{a}, \lambda) A(\mathbf{b}, \lambda). \tag{14.95}$$

The third, and crucial, step is to consider a third measurement direction specified by the unit vector $\mathbf{c}$ and then form the difference $P_{\rm h}(\mathbf{a}, \mathbf{b}) - P_{\rm h}(\mathbf{a}, \mathbf{c})$ using (14.95):

$$\begin{aligned} P_{\rm h}(\mathbf{a}, \mathbf{b}) - P_{\rm h}(\mathbf{a}, \mathbf{c}) &= -\int_\Lambda d\lambda\, \rho(\lambda)[A(\mathbf{a}, \lambda)A(\mathbf{b}, \lambda) - A(\mathbf{a}, \lambda)A(\mathbf{c}, \lambda)] \\ &= \int_\Lambda d\lambda\, \rho(\lambda) A(\mathbf{a}, \lambda) A(\mathbf{b}, \lambda)[A(\mathbf{b}, \lambda)A(\mathbf{c}, \lambda) - 1]. \end{aligned} \tag{14.96}$$

The final result is obtained by taking the absolute value of both sides of (14.96):

$$\begin{aligned} |P_{\rm h}(\mathbf{a}, \mathbf{b}) - P_{\rm h}(\mathbf{a}, \mathbf{c})| &\leqslant \int_\Lambda d\lambda\, \rho(\lambda) A(\mathbf{a}, \lambda) A(\mathbf{b}, \lambda) \int_\Lambda d\lambda\, \rho(\lambda)[1 - A(\mathbf{b}, \lambda)A(\mathbf{c}, \lambda)] \\ &\leqslant \int_\Lambda d\lambda\, \rho(\lambda)[1 - A(\mathbf{b}, \lambda)A(\mathbf{c}, \lambda)] = 1 + P_{\rm h}(\mathbf{b}, \mathbf{c}), \end{aligned}$$

i.e.,

$$1 + P_{\rm h}(\mathbf{b}, \mathbf{c}) \geqslant |P_{\rm h}(\mathbf{a}, \mathbf{b}) - P_{\rm h}(\mathbf{a}, \mathbf{c})|. \tag{14.97}$$

Equation (14.97) is known as Bell's inequality; it is one of several that have been derived. From it Bell concluded that $P_{\rm h}(\mathbf{a}, \mathbf{b})$ is not always equal to $P_{\rm q}(\mathbf{a}, \mathbf{b})$, since for small $|\mathbf{b} - \mathbf{c}|$ the RHS of (14.97) is of order of $|\mathbf{b} - \mathbf{c}|$, implying that $P_{\rm h}(\mathbf{b}, \mathbf{c})$ cannot be stationary at the minimum value -1, whereas $P_{\rm q}(\mathbf{b}, \mathbf{c})$ is always stationary there. That is, for small $|\mathbf{b} - \mathbf{c}|$, the LHS of (14.97) varies as $|\mathbf{b} - \mathbf{c}|^2$, whereas the RHS varies as $|\mathbf{b} - \mathbf{c}|$, recall Eq. (14.91).[5] Bell's inequality leads to a conclusion that is known as Bell's

[5] By choosing $\mathbf{a} = (\mathbf{b} - \mathbf{c})/|\mathbf{b} - \mathbf{c}|$, $\mathbf{b} \cdot \mathbf{c} = 0$, and assuming that $P_{\rm h}(\mathbf{b}, \mathbf{c}) = P_{\rm q}(\mathbf{b}, \mathbf{c}) = -\mathbf{b} \cdot \mathbf{c}$, Jammer (1974, p. 307) showed that (14.97) leads to $\sqrt{2} \leqslant 1(!)$, a clear contradiction, which establishes that $P_{\rm h}(\mathbf{b} \cdot \mathbf{c})$ cannot equal $P_{\rm q}(\mathbf{b}, \mathbf{c})$.

theorem, which we express as: *no local-hidden-variable theory can be constructed such that its results are always in agreement with those of quantum theory.* Or, in Bell's (1964) words,

> In a theory in which parameters are added to quantum mechanics to determine the results of individual measurements, without changing the statistical predictions, there must be a mechanism whereby the setting of one measuring device can influence the reading of another instrument, however remote. Moreover, the signal involved must propagate instantaneously, so that such a theory could not be Lorentz invariant.

Of course, the correlations built into $|\frac{1}{2}\frac{1}{2}; 00\rangle$ mean that quantum theory is non-local, but the non-locality is not one whereby signals travel instantaneously: because all possibilities are present in $|\frac{1}{2}\frac{1}{2}; 00\rangle$, a measurement of $\hat{\mathbf{S}}_1 \cdot \mathbf{n}$ which yields $\pm\hbar/2$ forces particle 2 into a state whereby $\hat{\mathbf{S}}_2 \cdot \mathbf{n}$ has a fixed value of $\mp\hbar/2$.

It is one thing to produce theoretical predictions, but quite another to perform experiments related to these predictions. Bell was aware of this, and commented in his 1964 paper that the crucial experiments were of the type proposed by Bohm and Aharonov (1957), viz., that photons be used in place of spin-$\frac{1}{2}$ particles and that the settings of the counters be changed while the photons are in flight. Bell's work dealt with ideal measurements; subsequent to the publication of his seminal papers, much theoretical analysis of realistic experiments was carried out, new inequalities were derived, and several experiments involving photons were performed. Summaries of work up to 1978–79 can be found in Clauser and Shimony (1978) and Pipkin (1979), while a review of hidden-variable theories is given by Belinfante (1973). The definitive experiments (of the type advocated by Bohm and Aharonov) are those of Aspect *et al.* (1981, 1982), whose observations were in agreement with the quantal predictions, but violated the relevant Bell inequality.

An extremely important conclusion has thus been reached: quantum mechanics cannot be completed by means of any local-hidden-variable theory. On the other hand, Bohm's 1952 hidden-variable theory (and others as well) is both non-local and in agreement with all the quantal predictions. One may rightly ask why "completion" should not be via non-local-hidden-variable theories, thus eliminating the uncertainty inherent in quantum mechanics. One claim is that, in contrast to the generality of quantum mechanics, non-local-hidden-variable theories always seem to be dependent on the specific measurement (system) for which they are constructed: changing the measurement means changing the theory. Discussions of this point, and of unresolved questions dealing with the meaning of reality at the microscopic level, with wave-packet reduction, etc., can be found both in technical journals and in the popularizations cited. In addition to the latter references, the author recommends the small book *Quantum Profiles* by J. Bernstein (1991), which provides both an excellent short history of this subject and fascinating brief biographical sketches, especially of Bell himself.

Exercises

14.1 (a) Construct the matrix representations of $\hat{J}_x$, $\hat{J}_y$, $\hat{J}_+$, and $\hat{J}_-$ for the case $j = 1$.

(b) Determine, from the preceding results, the eigenvalues and the normalized eigenvectors for J_+ and J_-. (The latter states correspond to the circular-polarization states of photons.)

14.2 (a) Construct the 4×4 matrices representing $\hat{J}_x$, $\hat{J}_y$, $\hat{J}_z$, $\hat{J}_+$ and $\hat{J}_-$ for the case $j = \frac{3}{2}$.

(b) Verify the correctness of your answers to part (a) by showing that they obey $[J_x, J_y] = i\hbar J_z$ as well as Eqs. (14.13a)–(14.13c).

14.3 Let

$$M = \begin{pmatrix} 1 & 1 \\ 1 & -3 \end{pmatrix}.$$

Express $\exp(i\pi M)$ as a 2×2 matrix. (Hint: write M as a sum of I_2 and the three Pauli spin matrices.)

14.4 Show that the time-evolution operator used to obtain the Heisenberg-picture results of Exercise 13.14 parts (b) and (c) corresponds to a rotation operator with $\mathbf{n} = \mathbf{e}_3$. Hence, show that the results of Exercise 13.14(c) can be obtained from Eqs. (14.35) and (14.36) by letting $\phi = \omega t$, with an appropriate choice of ω.

14.5 Let $\hat{H}^{(S)}_{\text{mag}} = -\gamma \mathbf{B} \cdot \hat{\mathbf{S}}$ in Eq. (13.49), with $\mathbf{B} = \sum_j B_j \mathbf{e}_j$ and $B_j = \text{constant}$, $j = 1, 2, 3$. If the spin-$\frac{1}{2}$ particle governed by this Hamiltonian is in state α_z at $t = 0$, compute the probabilities that it will be in the states α_z and β_z at $t > 0$. (Hint: express the time-evolution operator as a 2×2 matrix operator depending on the frequencies $\omega_j = \gamma B_j$.)

14.6 A coherent beam of neutrons, all in state α_z, is split into two portions that follow different paths of the same path length, and are eventually recombined at a detector. One path traverses a constant magnetic field $\mathbf{B} = B\mathbf{e}_3$ confined to a small volume, the other passes through a field-free region.

(a) If T is the length of time spent in the $B \neq 0$ region and Φ is the phase of the interference term that results when the two beams are recombined, determine the proportionality constant in the relation $\Phi \propto T$. Assume for simplicity that the only contribution to Φ is from the field.

(b) The interference pattern of part (a) can be made to vary by changing B. Show that the difference δB in the field strengths producing successive interference maxima can be expressed as $\delta B = 4\pi\beta\hbar/(g_n q \kappa_c \lambda d)$, where λ is the neutron wave number, d is the linear dimension (along the path) of the region where $B \neq 0$, and β is a constant you are to determine for the cases of a 2π and then a 4π rotation of the neutron's spin.

14.7 A neutron and a proton (spin-$\frac{1}{2}$) each have zero orbital angular momentum.

(a) If they have opposite values of m_s, what are the probabilities that they will be observed in a state of total angular momentum (i) 0, (ii) $\frac{1}{2}$, and (iii) 1?

(b) Ditto if their m_s values are the same.

14.8 An electron labeled 1 and a muon labeled 2, bound to the same nucleus, each have $\ell = 2$ (and spin $\frac{1}{2}$).

(a) Let $\hat{\mathbf{L}} = \hat{\mathbf{L}}_1 + \hat{\mathbf{L}}_2$ be the total orbital angular-momentum operator. What are the possible values of its quantum number ℓ?

(b) Let $\hat{\mathbf{S}}$ be the total spin operator. Determine the range of values of the total angular-momentum quantum number j corresponding to the total angular momentum defined by $\hat{\mathbf{J}} = \hat{\mathbf{L}} + \hat{\mathbf{S}}$.

(c) From the definition $\hat{\mathbf{J}}_k = \hat{\mathbf{L}}_k + \hat{\mathbf{S}}_k$ of the total angular-momentum operator for particle k, $k = 1, 2$, one can introduce the alternate definition $\hat{\mathbf{J}}_{\mathrm{T}} = \hat{\mathbf{J}}_1 + \hat{\mathbf{J}}_2$, with corresponding quantum number j_{T}. Calculate the possible values of j_{T} and state whether and why this set is the same as that obtained in part (b).

(d) Determine the total number of ways one can obtain (i) a state with $j = 0$ and (ii) a state with $j_{\mathrm{T}} = 0$. Explain why these two numbers should be equal or not.

14.9 Spinless particles 1 and 2 each have $\ell = 1$, and they may be in either of the coupled states $|11;\, 10\rangle$ and $|11;\, 20\rangle$. Determine in each case the probabilities that

(a) particle 1 will be found in the states (i) $|11\rangle$, (ii) $|10\rangle$, and (iii) $|1, -1\rangle$; and

(b) particle 1 will have polar angles (i) $\theta_1 = \phi_1 = 0$; (ii) $\theta_1 = \pi/2$, $\phi_1 = \pi/4$; and (iii) $\theta_1 = \pi$, $\phi_1 = \pi/2$.

14.10 A spin-$\frac{1}{2}$ particle is acted on by a spherically symmetric potential.

(a) It is first prepared in an eigenstate $|n, \ell, 1/2;\, j5/2\rangle$. (i) What are the possible results of a measurement of $\hat{J}^2 = (\hat{\mathbf{L}} + \hat{\mathbf{S}})^2$? (ii) What are the possible outcomes of a measurement of $\hat{J}_y$? (iii) If $\hat{J}_z$ is measured after the measurement of $\hat{J}^2$ of part (i), what are the possible results one would obtain?

(b) The particle is next prepared in the state $R(r)\sin\theta\,|\frac{1}{2}\frac{1}{2}\rangle$. (i) What results could occur if $\hat{J}^2$ were to be measured? (ii) Ditto for $\hat{J}_z$. (iii) Find the probabilities that $\hat{J}_z$ will be $\pm\hbar/2$.

14.11 Determine the probabilites that an electron in the state $|\Gamma\rangle$ of Exercise 13.10 will have

(a) j values equal to $\frac{1}{2}, \frac{3}{2}, \frac{5}{2}$, and $\frac{7}{2}$; and

(b) magnetic quantum numbers $m_j = \pm\frac{1}{2}, \pm\frac{3}{2}, \pm\frac{5}{2}$, and $\pm\frac{7}{2}$.

14.12 The Hamiltonian for a two-particle system is $\hat{H}(1, 2) = A(\hat{J}_1^2 + \hat{J}_2^2) + B\hat{\mathbf{J}}_1 \cdot \hat{\mathbf{J}}_2$, where A and B are constants while 1 and 2 label the particles.

(a) State whether either a single uncoupled state $|j_1 j_2;\, m_1 m_2\rangle$ or a single coupled state $|j_1 j_2;\, jm_j\rangle$ can be an eigenstate of $\hat{H}(1, 2)$, and justify your answer.

(b) Using the result of part (a), find expressions for the eigenstates of $\hat{H}(1, 2)$. State the conditions, if there are any, under which degeneracies can occur.

14.13 In contrast to Eq. (13.53a), the magnetic-moment operator $\hat{\boldsymbol{\mu}}_{\mathrm{e}}$ for an electron in a bound state is $\hat{\boldsymbol{\mu}}_{\mathrm{e}} = -\mu_{\mathrm{B}}(\hat{\mathbf{L}} + g_{\mathrm{e}}\hat{\mathbf{S}})/\hbar$. If the electron's state vector is $|n, \ell = 1, \frac{1}{2};\, jm_j\rangle$, determine the expectation value of $\hat{\boldsymbol{\mu}}_{\mathrm{e}}$ (i.e., the magnetic moment) when

(a) $j = \frac{3}{2}$, and $m_j = \pm\frac{3}{2}$ or $\pm\frac{1}{2}$; and

(b) $j = \frac{1}{2}$, $m_j = \pm\frac{1}{2}$.

(Hint: replace $\hat{\mathbf{L}}$ by $\hat{\mathbf{J}} - \hat{\mathbf{S}}$ and show that the expectation values of $\hat{\boldsymbol{\mu}}_{\mathrm{e}x}$ and $\hat{\boldsymbol{\mu}}_{\mathrm{e}y}$ vanish.)

14.14 Suppose that the spin-dependent interaction $\hat{H}_S = \hbar\omega_{\mathrm{e}}\hat{\sigma}_z^{(\mathrm{e})} + \hbar\omega_p\hat{\sigma}_z^{(\mathrm{p})}$ is added to the H-atom Hamiltonian of Section 11.6, and that the system is then placed in a constant magnetic field $\mathbf{B} = B\mathbf{e}_3$. At time $t = 0$, the system is in state $|\Gamma\rangle = (|S_z^{(\mathrm{e})}, +\rangle|S_z^{(\mathrm{p})}, -\rangle + |S_z^{(\mathrm{e})}, -\rangle|S_z^{(\mathrm{p})}, +\rangle)/2^{1/2}$, where e (p) refers to the electron (proton) and all spatial aspects of the system are ignored.

(a) What is the probability that at $t = 0$, the system will be in an eigenstate of S^2 with eigenvalue $2\hbar^2$? (Here, $\hat{\mathbf{S}} = \hat{\mathbf{S}}^{(e)} + \hat{\mathbf{S}}^{(p)}$.)

(b) If $|\Gamma\rangle$ evolves to $|\Gamma, t\rangle$, evaluate $\langle\Gamma, t|\hat{S}^2|\Gamma, t\rangle$ and $\langle\Gamma, t|\hat{S}_x^{(e)}|\Gamma, t\rangle$ for $t \geqslant 0$.

(c) Evaluate $\langle\Gamma, t|\hat{\boldsymbol{\mu}}_S^{(e)}|\Gamma, t\rangle$, $t \geqslant 0$, where $\hat{\boldsymbol{\mu}}_S^{(e)}$ is the spin part of the electron's magnetic-moment operator, Eq. (13.53a).

(d) If the directions of spin for the electron and the proton were measured at time $t \geqslant 0$, what is the probability that they would be aligned along $\mathbf{e}_1$ in the state $|S_x^{(e)}, +\rangle|S_x^{(p)}, +\rangle$?

14.15 The projection operators $\hat{P}_S$ for projection onto the singlet ($S = 0$) and triplet ($S = 1$) states of two spin-$\frac{1}{2}$ particles are defined by

$$\hat{P}_S = \sum_{M_S} |\tfrac{1}{2}\tfrac{1}{2};\ SM_S\rangle\langle\tfrac{1}{2}\tfrac{1}{2};\ SM_S|.$$

(a) Verify that $\hat{P}_S\hat{P}_{S'} = \hat{P}_S\delta_{SS'}$.

(b) Show that, for each value of S, $\hat{P}_S$ can be expressed as a linear combination of $\hat{I}$ and $\hat{\boldsymbol{\sigma}}_1 \cdot \hat{\boldsymbol{\sigma}}_2$, where 1 and 2 label the particles.

14.16 Let j and ℓ refer to the total and orbital angular-momentum quantum numbers for a spin-$\frac{1}{2}$ particle. Define the operators $\hat{P}_{\ell,\pm} = [\ell + \frac{1}{2} \pm (\hat{\mathbf{L}} \cdot \hat{\sigma}/\hbar + \frac{1}{2})]/(2\ell + 1)$, which act on the coupled states $|\ell\frac{1}{2};\ jm_j\rangle$.

(a) Prove that the $\hat{P}_{\ell,\pm}$ are projection operators.

(b) Determine the effects of $\hat{P}_{\ell,\pm}$ and of $\hat{P}_{\ell,\mp}$ on the states $|\ell\frac{1}{2};\ j = \ell \pm \frac{1}{2},\ m_j\rangle$.

(c) Verify the following relations:

$$\begin{aligned}\langle\ell\tfrac{1}{2};\ m_j - \tfrac{1}{2},\ \tfrac{1}{2}|\hat{P}_{\ell,\pm}|\ell\tfrac{1}{2};\ m_j - \tfrac{1}{2},\ \tfrac{1}{2}\rangle &= \langle\ell\tfrac{1}{2};\ m_j + \tfrac{1}{2},\ -\tfrac{1}{2}|\hat{P}_{\ell,\mp}|\ell\tfrac{1}{2};\ m_j + \tfrac{1}{2},\ -\tfrac{1}{2}\rangle \\ &= [2(\ell \pm m_j) + 1]/[2(2\ell + 1)],\end{aligned}$$

$$\langle\ell\tfrac{1}{2};\ m_j - \tfrac{1}{2},\ \tfrac{1}{2}|\hat{P}_{\ell,\pm}|\ell\tfrac{1}{2};\ m_j + \tfrac{1}{2},\ -\tfrac{1}{2}\rangle = \pm\tfrac{1}{2}\{[2(\ell - m_j) + 1]/(2\ell + 1)\}^{1/2}.$$

(d) Optional: show that

$$\hat{P}_{\ell,\pm} = \sum_{m_j} |\ell\tfrac{1}{2};\ j = \ell \pm \tfrac{1}{2},\ m_j\rangle\langle\ell\tfrac{1}{2};\ j = \ell \pm \tfrac{1}{2},\ m_j|.$$

14.17 Evaluate $\langle\ell\frac{1}{2};\ j'm_j|\hat{\boldsymbol{\sigma}}|\ell\frac{1}{2};\ jm_j\rangle$ for all j and j' consistent with a given ℓ.

14.18 A spin-$\frac{1}{2}$ particle of charge $q > 0$, acted on both by a central potential and by a weak, uniform magnetic field $\mathbf{B} = B\mathbf{e}_3$, is governed by the Hamiltonian $\hat{H} = \hat{H}_0 - \mu B(\hat{L}_z + g\hat{S}_z)/\hbar$, where $\mu(g)$ is the particle's magnetic moment (g-factor) and $\hat{H}_0 = \hat{K} + V(r)$. At time $t = 0$ the particle is in the $\hat{H}_0$ eigenstate $|\Gamma\rangle = |n\ell\frac{1}{2};\ jm_j\rangle$.

(a) Determine the state $|\Gamma, t\rangle$ to which $|\Gamma\rangle$ evolves for times $t > 0$.

(b) Calculate the probabilities that the particle will be in states $|n\ell'\frac{1}{2};\ j'm_j'\rangle$, with ℓ', j', m_j' equal to or different than, respectively, ℓ, j, m_j.

(c) What is the probability that, at $t > 0$, the particle will have a z-component of spin equal to $\hbar/2$?

14.19 The Hamiltonian for an electron acted on by a central potential and spin–orbit interaction is $\hat{H} = \hat{H}_0 + \hat{H}_{LS}$, with $H_0 = \hat{K}(\mathbf{r}) + V(r)$ and $\hat{H}_{LS} = V_{LS}(r)\hat{\mathbf{L}} \cdot \hat{\mathbf{S}}/\hbar^2$. At time $t = 0$, the electron is in the (uncoupled) eigenstate $|\Gamma\rangle = |n\ell\frac{1}{2};\ m_\ell m_S\rangle$ of $\hat{H}_0$.

(a) Determine the time-evolved state $|\Gamma, t\rangle$, $t > 0$.

(b) What are the probabilites that, for $t \geqslant 0$, the particle will have (i) $j = \ell \pm \frac{1}{2}$ and (ii) $m_S = -\frac{1}{2}$?

(c) Which, if any, of the probabilites of part (b) become equal to unity? Justify your answer.

14.20 (a) Prove the validity of Eq. (14.91).

(b) Fill in the steps leading to the inequality of Jammer stated in footnote 5 (p. 549) of this chapter.

14.21 In Bohm's version of quantum theory, the solution to the time-dependent Schrödinger equation for an uncharged spinless particle acted on by a potential $V(r)$ is written as the following product: $\Psi(\mathbf{r}, t) = A(\mathbf{r}, t)\exp[iS(\mathbf{r}, t)/\hbar]$, where A and S are real and $\rho(\mathbf{r}, t) = A^2(\mathbf{r}, t) = |\Psi|^2$ is also interpreted as the probability density. The particle's classical momentum $\mathbf{p} = m\mathbf{v}$ is *defined* by $\mathbf{p} = \nabla S$.

(a) Determine the equations obeyed by $\partial A/\partial t$ and $\partial S/\partial t$.

(b) From the equation for $\partial A/\partial t$, show that $\partial\rho/\partial t + \nabla\cdot(\rho\mathbf{v}) = 0$.

(c) Finally, using the relation $d\mathbf{p}/dt = \sum_j (v_j\,\partial\mathbf{p}/\partial x_j + \partial\mathbf{p}/\partial t)$, show that $d\mathbf{p}/dt = -\nabla(V + U) \equiv \mathbf{F}$, where U is a quantal potential you are to determine (U depends on A and $\nabla^2 A$). This equation establishes a quantal–classical connection, thereby obviating any need in Bohm's theory to find the classical limit.

15

Three-Dimensional Continuum States/Scattering

Collisions are a vital source of information for atomic, molecular, nuclear, elementary-particle and condensed-matter physics. In each of these cases, the fundamental process is the interaction of a projectile with a target, leading to a final state containing one or more ejectiles plus a residual system. All of the bodies involved – projectile, target, ejectiles, and residual systems – may themselves be structured, i.e., each one may be comprised of particles in a bound state. Because of this, the general formulation of collision theory is quite complicated. Since the intent of this chapter is to introduce the reader to basic concepts, we restrict ourselves to the simplest collision system: two unstructured *spinless* particles scattering off one another via short-ranged interactions. A realistic example is neutron–proton scattering ignoring spin, while a pedagogical example is the scattering of a spinless particle by a potential well. "Long-range" interactions, e.g., those behaving like $1/r$ or $1/r^4$, are excluded from formal considerations.

When only two bodies, structured or not, comprise the final state, the collision is known as a two-body collision. For such a collision, the simplest measurement consists of counting the number of ejectiles of a given energy that emerge at various angles relative to the direction of the incident beam of projectiles. This is shown schematically in Fig. 15.1. The ratio of the number of ejectiles scattered at angle θ to the number of projectiles incident on the target is a measure of the quantal probability for the process. As usual, that probability is the absolute square of a "scattering" amplitude that depends on the energies, the wave numbers, the spins (if any are non-zero), the masses, and also the details of the interaction. Because one usually measures the number of projectiles crossing a unit area but only the number (not the number per unit area) of ejectiles, the

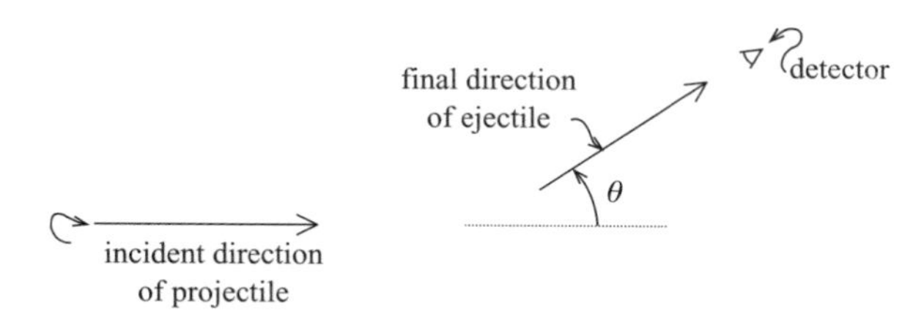

Fig. 15.1 A schematic illustration of a two-body collision. In the simplest experiment one measures the number of ejectiles that arrive at a counting device placed at an angle θ relative to the direction of the incident beam of projectiles.

quantal probability has the unusual dimension of area, i.e., $(\text{length})^2$. This collisional probability is known as a *differential cross section* or an *angular distribution*. The scattering amplitude has the dimension length, and its elucidation is one of the primary goals of this chapter. Since the amplitude occurs in the asymptotic form of the scattering wave function, analysis of the latter quantity is also a prime ingredient of this chapter.

Our approach to 3-D scattering is analogous to that of 1-D scattering in Chapter 7, viz., as a study of the time evolution of wave packets, with an ultimate transition to the case of improper (non-Hilbert-space) sharp-momentum states. Since the expression for the differential cross section depends on the time-evolved wave function, we begin with the wave-packet analysis.

15.1. Wave-packet analysis

The physical description of 3-D scattering is much like that of the 1-D case. An accelerator, located a macroscopic distance D from the target, generates a beam of particles (the projectiles) all of which have the same narrow distribution of kinetic energies. By means of collimating devices, the projectiles are directed towards the target particles: the projectiles thus have a relatively narrow distribution of momenta. The projectile–target interaction $V(\mathbf{r})$ is assumed short-ranged, so that, at early times, the initial wave packet is a linear combination of the solutions of the "free-particle" Hamiltonian, i.e., of the relative-motion kinetic energy operator $\hat{K}_{\text{rel}}(\mathbf{r}) \equiv \hat{K}(\mathbf{r})$. This free-particle wave packet evolves in time into a wave packet that is a linear combination of particular continuum states of the full Hamiltonian $\hat{H}(\mathbf{r})$, where

$$\hat{H}(\mathbf{r}) = \hat{K}(\mathbf{r}) + V(\mathbf{r}). \tag{15.1}$$

Just which type of states of $\hat{H}(\mathbf{r})$ enter will be determined by examining the early- and late-time behaviors of the packet. As in Chapter 7, we shall focus on wave packets that are narrowly peaked about a central momentum $\hbar\mathbf{k}_0$ and have an approximate energy $E_{k_0} = \hbar^2 k_0^2/2\mu$, where μ is the reduced mass.

The plane-wave packet

The improper plane-wave states $|\mathbf{k}\rangle$ are solutions of the relative-motion free-particle Schrödinger equation

$$\hat{K}|\mathbf{k}\rangle = E_k|\mathbf{k}\rangle = \frac{\hbar^2 k^2}{2\mu}|\mathbf{k}\rangle, \tag{15.2}$$

with wave functions

$$\psi_{\mathbf{k}}(\mathbf{r}) = \left(\frac{1}{2\pi}\right)^{3/2} e^{i\mathbf{k}\cdot\mathbf{r}}. \tag{15.3}$$

From them we form the free wave packet

$$\Psi^{(F)}(\mathbf{r},\ t) = \int d^3 k\, A(\mathbf{k})\psi_{\mathbf{k}}(\mathbf{r})e^{-iE_k t/\hbar}$$

$$= \left(\frac{1}{2\pi}\right)^{3/2} \int d^3 k\, A(\mathbf{k})e^{i(\mathbf{k}\cdot\mathbf{r}-\omega_k t)}, \tag{15.4}$$

where $\omega_k = E_k/\hbar$ and the profile function $A(\mathbf{k})$, which characterizes the energy/momentum resolution of the accelerator and collimators, is assumed to be sharply peaked around the central value $\mathbf{k}_0$:

$$A(\mathbf{k}) \cong 0 \begin{cases} \mathbf{k} < \mathbf{k}_0 - \Delta\mathbf{k} \\ \text{or} \\ \mathbf{k} > \mathbf{k}_0 + \Delta\mathbf{k}, \qquad \Delta k/k_0 \ll 1. \end{cases} \tag{15.5}$$

We shall also assume that the maximum of $|\Psi^{(F)}(\mathbf{r},\ 0)|$ is at $\mathbf{r} = 0$, so that $\Psi^{(F)}(\mathbf{r},\ t)$ is generated at a large negative time. $A(\mathbf{k})$ determines $\Psi^{(F)}(\mathbf{r},\ 0)$, and vice versa:

$$A(\mathbf{k}) = \left(\frac{1}{2\pi}\right)^{3/2} \int d^3 r\, e^{-i\mathbf{k}\cdot\mathbf{r}}\Psi^{(F)}(\mathbf{r},\ 0). \tag{15.6}$$

Furthermore, normalization of $\Psi^{(F)}(\mathbf{r},\ t)$ implies that

$$\int d^3 k\, |A(\mathbf{k})|^2 = 1, \tag{15.7}$$

just as in the 1-D case.

Another analog of the 1-D behavior, which is important for our later analysis, is the relation

$$\Psi^{(F)}(\mathbf{r},\ t) \cong e^{i\omega_{k_0} t}\Psi^{(F)}(\mathbf{r} - \mathbf{v}_0 t,\ 0), \qquad t \ll \frac{2\mu}{\hbar(\Delta k)^2}, \tag{15.8}$$

where the semi-classical velocity $\mathbf{v}_0$ is

$$\mathbf{v}_0 = \hbar\mathbf{k}_0/\mu. \tag{15.9}$$

That is, the 3-D wave packet $\Psi^{(F)}(\mathbf{r},\ t)$ propagates so that, at time t, its maximum is at the position $\mathbf{r} = \mathbf{v}_0 t$: its motion is classical for $t \ll 2\mu/[\hbar(\Delta k)^2]$. Hence, $\mathbf{k}_0$ is the wave vector characterizing the overall motion of the packet (see Fig. 15.2). Proof of Eq. (15.8), which uses Eq. (7.17), is left to the exercises. The comments below Eq. (7.17) concerning the description of the accelerator particles over the relevant macroscopic distances D

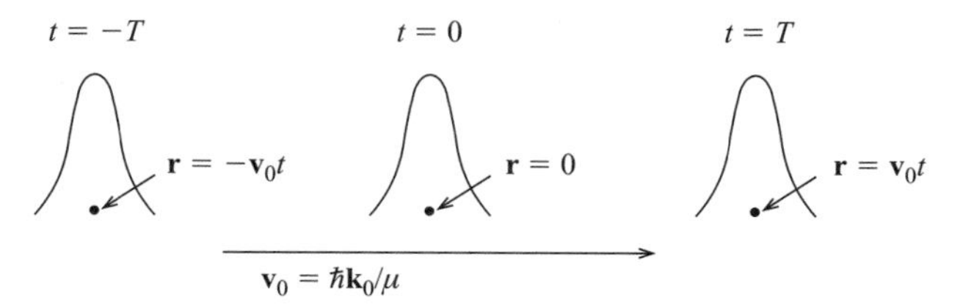

Fig. 15.2 An illustration of the behavior of $\Psi^{(F)}(\mathbf{r},\ t)$ for three different times, all being much less than the characteristic time $2\mu/[\hbar(\Delta k)^2]$.

apply in the present case, and the reader is urged to review them. In particular, the spread $\Delta\mathbf{r}$ implied by $\Delta\mathbf{k}$ is such that $|\Delta\mathbf{r}| \ll D$.

The full wave packet and outgoing-wave states

$\Psi^{(\mathrm{F})}(\mathbf{r}, t)$ describes the motion of a projectile after it has been generated by the accelerator and well before it feels the short-range interaction $V(\mathbf{r})$. Because $V(\mathbf{r})$ is zero during this part of the temporal evolution, the wave packet is constructed from eigenstates of $\hat{K}(\mathbf{r})$. Once the projectile has come closer to the target, so that they begin to interact, the wave packet should be constructed from continuum states of $\hat{H} = \hat{K} + V$. We shall deal with this behavior by seeking eigenstates $\psi_{\mathbf{k}}^{(\alpha)}(\mathbf{r})$ of $\hat{H}$ such that the full wave packet

$$\Psi^{(\alpha)}(\mathbf{r}, t) = \int d^3k\, A(\mathbf{k})\psi_{\mathbf{k}}^{(\alpha)}(\mathbf{r})e^{-i\omega_k t} \tag{15.10}$$

has the property that

$$\Psi^{(\alpha)}(\mathbf{r}, t) \underset{t\to-\infty}{\to} \Psi^{(\mathrm{F})}(\mathbf{r}, t), \tag{15.11}$$

where

$$[\hat{K}(\mathbf{r}) + V(\mathbf{r})]\psi_{\mathbf{k}}^{(\alpha)}(\mathbf{r}) = E_k\psi_{\mathbf{k}}^{(\alpha)}(\mathbf{r}). \tag{15.12}$$

The superscript α is an index that will eventually specify the appropriate asymptotic boundary condition obeyed by $\psi_k^{(\alpha)}(\mathbf{r})$; it also distinguishes the full scattering eigensolution of $\hat{H}(\mathbf{r})$ from the plane-wave eigensolution $\psi_{\mathbf{k}}(\mathbf{r})$ of $\hat{K}$. There are several aspects to (15.10) that we wish to emphasize. First, the same profile function $A(\mathbf{k})$ enters $\Psi^{(\alpha)}(\mathbf{r}, t)$ as is present in $\Psi^{(\mathrm{F})}(\mathbf{r}, t)$. $A(\mathbf{k})$ characterizes the momentum distribution of the beam as it was generated by the accelerator and collimators, and nothing subsequent can alter that distribution: the interaction changes $\psi_{\mathbf{k}}(\mathbf{r})$ to $\psi_{\mathbf{k}}^{(\alpha)}(\mathbf{r})$ but does not change $A(\mathbf{k})$. Correspondingly, the time-dependent factor $\exp(-i\omega_k t)$ is the same for $\Psi^{(\mathrm{F})}(\mathbf{r}, t)$ and $\Psi^{(\alpha)}(\mathbf{r}, t)$: the interaction produces no energy shifts – a feature characteristic of non-relativistic processes. This feature is reflected in Eq. (15.12): the energy of $\psi_{\mathbf{k}}^{(\alpha)}$ – Eq. (15.12) – is the same as that of $\psi_{\mathbf{k}}(\mathbf{r})$ – Eq. (15.3). Because of this, the yet-to-be-determined $\psi_k^{(\alpha)}(\mathbf{r})$ can loosely be thought of as the continuum eigenstates generated from the plane-wave state $\psi_{\mathbf{k}}(\mathbf{r})$, even though the concept is one strictly associated with the time evolution of wave packets.

The next aspect is connected with completeness. By Postulate VI, the eigenstates both of $\hat{H}$ and of $\hat{K}$ are complete. Let us assume that $V(\mathbf{r})$ supports bound states, which we denote by $|\gamma, E_\gamma\rangle$, $E_\gamma < 0$. With the continuum states denoted $|\mathbf{k}, \alpha\rangle$, the eigenvalue equation for $\hat{H}$ is

$$\hat{H}\left\{\begin{array}{c} |\mathbf{k}, \alpha\rangle \\ |\gamma, E_\gamma\rangle \end{array}\right\} = \left\{\begin{array}{c} E_k|\mathbf{k}, \alpha\rangle \\ E_\gamma|\gamma, E_\gamma\rangle \end{array}\right\}, \tag{15.13}$$

from which the completeness relation is

$$\hat{I} = \sum_\gamma |\gamma, E_\gamma\rangle\langle\gamma, E_\gamma| + \int d^3k\, |\mathbf{k}, \alpha\rangle\langle\mathbf{k}, \alpha|. \tag{15.14a}$$

Both the bound and the continuum states must be included to achieve (15.14a). On the other hand, because $\hat{K}$ has only continuum eigenstates, an alternate resolution of the identity is

$$\hat{I} = \int d^3k\, |\mathbf{k}\rangle\langle\mathbf{k}|. \tag{15.14b}$$

The states $\{|\mathbf{k}\rangle\}$ are an expansion basis, in particular for the bound eigenstates of $\hat{H}$:

$$|\gamma, E_\gamma\rangle = \int d^3k\, |\mathbf{k}\rangle\langle\mathbf{k}|\gamma, E_\gamma\rangle,$$

but *none* of the $|\gamma, E_\gamma\rangle$ are generated by *any* of the $|\mathbf{k}\rangle$, in the sense noted above that $|\mathbf{k}, \alpha\rangle$ is. This remark means that there is a one-to-one correspondence between $|\mathbf{k}, \alpha\rangle$ and $|\mathbf{k}\rangle$, even though $\{|\mathbf{k}, \alpha\rangle\}$ alone is not a complete set. Shown in Fig. 15.3 is a pictorial representation of this situation.

To say that $|\mathbf{k}\rangle$ generates $|\mathbf{k}, \alpha\rangle$ implies that $|\mathbf{k}\rangle$ is part of $|\mathbf{k}, \alpha\rangle$. We therefore write $|\mathbf{k}, \alpha\rangle$ as a sum of two terms:

$$|\mathbf{k}, \alpha\rangle = |\mathbf{k}\rangle + |\mathbf{k}, \mathrm{sc}\rangle, \tag{15.15}$$

where the "scattered wave" $|\mathbf{k}, \mathrm{sc}\rangle$ expresses the effect of the interaction V. Because of Eq. (15.11), the $|\mathbf{k}, \mathrm{sc}\rangle$ portion of (15.15) must be such that

$$\lim_{t\to-\infty} \Psi^{(\mathrm{sc})}(\mathbf{r}, t) \to 0, \tag{15.16}$$

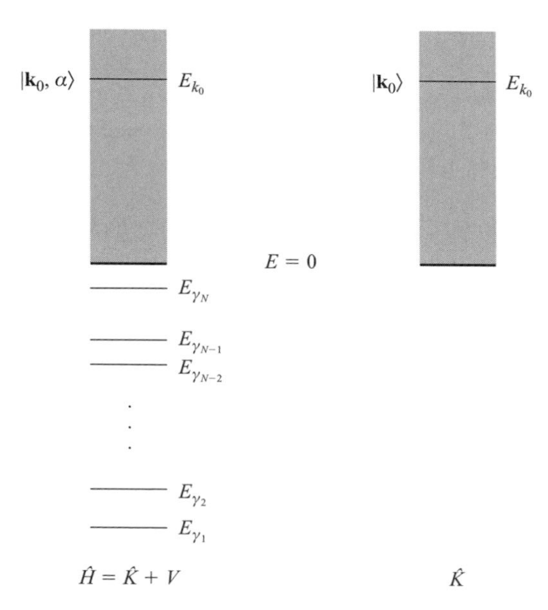

Fig. 15.3 Spectra of $\hat{H}$ and $\hat{K}$. The continuum states of $\hat{H}$ are in one-to-one correspondence to, and are generated by, the continuum states of $\hat{K}$, an example of which is indicated by the heavy line at E_{k_0}. Whereas the $\{|\mathbf{k}\rangle\}$ are complete, the $\{|\mathbf{k}, \alpha\rangle\}$ are not, and must be supplemented by the bound states $\{|\gamma, E_\gamma\rangle\}$, Eq. (15.14a).

where

$$\Psi^{(\text{sc})}(\mathbf{r}, t) = \int d^3k\, A(\mathbf{k})\psi_{\mathbf{k}}^{(\text{sc})}(\mathbf{r})e^{-i\omega_k t}, \tag{15.17}$$

with

$$\psi_{\mathbf{k}}^{(\text{sc})}(\mathbf{r}) = \langle \mathbf{r}|\mathbf{k}, \text{sc}\rangle. \tag{15.18}$$

To gain some understanding of $|\mathbf{k}, \text{sc}\rangle$ and thus of $|\mathbf{k}, \alpha\rangle$, we examine the spatially asymptotic part of $\Psi^{(\text{F})}(\mathbf{r}, t)$, obtained by substituting Eq. (10.83) into Eq. (15.4):

$$\lim_{r\to\infty}\Psi^{(\text{F})}(\mathbf{r}, t) \sim \left(\frac{1}{2\pi}\right)^{3/2}\frac{i}{k}\sum_{\ell}\sqrt{\pi(2\ell+1)}i^{\ell}Y_{\ell}^{0}(\Omega)$$
$$\times \int d^3k\, A(\mathbf{k})(e^{i\ell\pi/2}e^{-i(kr+\omega_k t)}/r - e^{-i\ell\pi/2}e^{i(kr-\omega_k t)}/r). \tag{15.19}$$

In its dependence on r and t, $\Psi^{(\text{F})}(\mathbf{r}, t)$ is asymptotically the difference of an *ingoing* wave packet involving $\exp[-i(kr+\omega_k t)]$ and an *outgoing* wave packet involving $\exp[i(kr-\omega_k t)]$.[1] Apart from the phase factors $\exp(\pm i\ell\pi/2)$, these two wave packets are present in equal amounts. To determine the asymptotic behavior of $\Psi^{(\text{sc})}(\mathbf{r}, t)$ we invoke the physics underlying Huyghen's principle and require that the effect of the target on the incident wave is to generate only *outgoing* waves asymptotically. That is, we demand that

$$\lim_{r\to\infty}\Psi^{(\text{sc})}(\mathbf{r}, t) \to \Psi^{(\text{out})}(\mathbf{r}, t) \equiv \left(\frac{1}{2\pi}\right)^{3/2}\int d^3k\, A(\mathbf{k})f_{\mathbf{k}}(\Omega)e^{i(kr-\omega_k t)}/r, \tag{15.20}$$

where the *scattering amplitude* $f_{\mathbf{k}}(\Omega)$ is an as-yet-unknown factor representing the effect of the potential V. This will turn out to be the same scattering amplitude as that mentioned on pages 555 and 556 of this chapter. Note that Ω is the pair of polar angles of $\mathbf{r}$.

Comparing Eqs. (15.20) and (15.17), we see that the asymptotic behavior of $\psi_{\mathbf{k}}^{(\text{sc})}(\mathbf{r})$ must be

$$\lim_{r\to\infty}\psi_{\mathbf{k}}^{(\text{sc})}(\mathbf{r}) \sim \left(\frac{1}{2\pi}\right)^{3/2} f_{\mathbf{k}}(\Omega)\frac{e^{ikr}}{r}. \tag{15.21}$$

We shall soon verify that the choice (15.21) leads to (15.16). Consequently, the correct solution $\psi_{\mathbf{k}}^{(\alpha)}(\mathbf{r})$ of (15.12) to use in forming the full wave packet $\Psi^{(\alpha)}(\mathbf{r}, t)$ is the one for which the scattered wave behaves asymptotically like (15.21). It is standard notational practice to replace the superscript α for this particular solution by a plus sign: "α"→"+", in which case

$$\Psi^{(\alpha)}(\mathbf{r}, t) \to \Psi^{(+)}(\mathbf{r}, t) = \int d^3k\, A(\mathbf{k})\psi_{\mathbf{k}}^{(+)}(\mathbf{r})e^{-i\omega_{\mathbf{k}} t}, \tag{15.22}$$

with

[1] The ingoing- and outgoing-spherical-wave aspects are discussed in the paragraph below Eq. (10.24b), to which the reader is directed for review.

$$\psi_{\mathbf{k}}^{(+)}(\mathbf{r}) \underset{r\to\infty}{\sim} \left(\frac{1}{2\pi}\right)^{3/2} \left(e^{i\mathbf{k}\cdot\mathbf{r}} + f_{\mathbf{k}}(\Omega)\frac{e^{ikr}}{r}\right). \tag{15.23}$$

This asymptotic form – a combination of a plane wave propagating along **k** and outgoing spherical waves forming concentric spheres centered on the target at $r = 0$ – is schematically illustrated in Fig. 15.4.

By also requiring that

$$\lim_{r\to 0} \psi_{\mathbf{k}}^{(+)}(\mathbf{r}) = \text{ finite}, \tag{15.24}$$

the combination of the two boundary conditions (15.23) and (15.24) uniquely determines the solution $\psi_{\mathbf{k}}^{(+)}(\mathbf{r})$ and thus the scattering amplitude $f_{\mathbf{k}}(\Omega)$. The actual evaluation of these quantities is, of course, another matter. We shall consider $\psi_{\mathbf{k}}^{(+)}(\mathbf{r})$ and $f_{\mathbf{k}}(\Omega)$ in Section 15.2.

Let us now verify that the boundary condition (15.21) leads to (15.11). The proof employs the property (15.5) assumed for $A(\mathbf{k})$. For early times, the wave packet must be at asymptotic distances, so, on using Eq. (15.20), the condition (15.11) reads

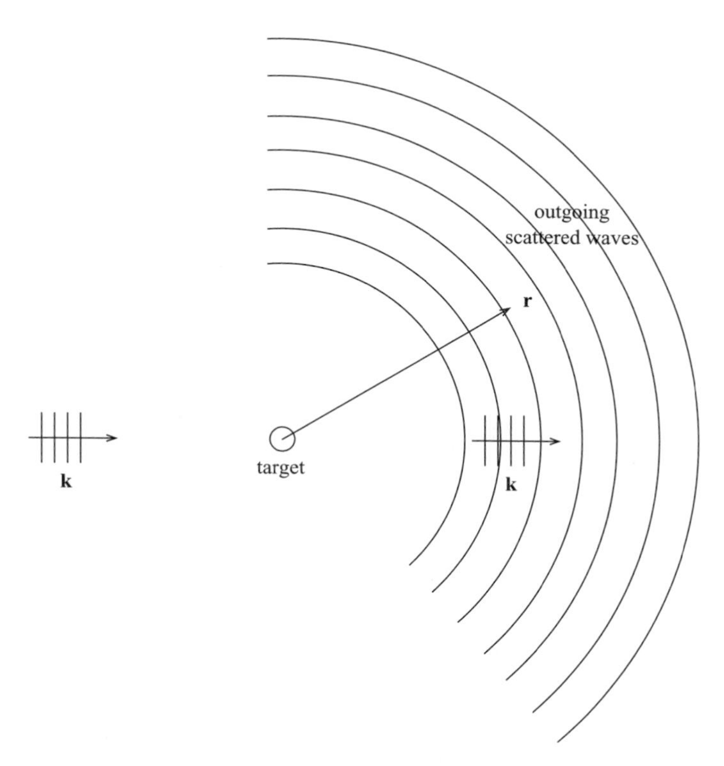

Fig. 15.4 A pictorial representation of the asymptotic behavior of $\psi_{\mathbf{k}}^{(+)}(\mathbf{r})$. Far to the left of the target particle one has the incident plane wave, which is transmitted far to the right as well. The target acts as a source of asymptotic outgoing waves, a portion of which is shown to the right of the target; these outgoing wave fronts lie on concentric spheres centered on the target. The vector **r** is in the direction of observation.

$$\lim_{t \to -\infty} \Psi^{(\text{out})}(\mathbf{r}, t) \to 0.$$

In the integral of (15.20), the magnitude k can be re-written as

$$\begin{aligned} k &= \sqrt{k^2} = |\mathbf{k}_0 + (\mathbf{k} - \mathbf{k}_0)| \\ &= \sqrt{k_0^2 + 2\mathbf{k}_0 \cdot (\mathbf{k} - \mathbf{k}_0) + (\mathbf{k} - \mathbf{k}_0)^2} \\ &\cong \mathbf{k}_0 + \mathbf{k}_0 \cdot (\mathbf{k} - \mathbf{k}_0)/k_0 \\ &= \mathbf{k}_0 \cdot \mathbf{k}/k_0 \\ &\equiv \mathbf{n}_0 \cdot \mathbf{k}, \end{aligned} \tag{15.25}$$

where $\mathbf{n}_0 = \mathbf{k}_0/k_0$ is a unit vector along the direction $\mathbf{k}_0$ and the "$\cong$" sign in the third line of (15.25) arises from the narrowness of the profile function $A(\mathbf{k})$, Eq. (15.5).

With (15.25), $\Psi^{(\text{out})}(\mathbf{r}, t)$ is given by

$$\Psi^{(\text{out})}(\mathbf{r}, t) \cong \left(\frac{1}{2\pi}\right)^{3/2} \int d^3k \, \mathrm{A}(\mathbf{k}) f_{\mathbf{k}}(\Omega) \frac{e^{i\mathbf{k}\cdot(\mathbf{n}_0 r)}}{r} e^{-i\omega_k t}.$$

To proceed further, we next assume that $f_{\mathbf{k}}(\Omega)$ is a slowly varying function of $\mathbf{k}$, so that the narrowness of $A(\mathbf{k})$ leads to

$$\begin{aligned} \Psi^{(\text{out})}(\mathbf{r}, t) &\cong \frac{f_{\mathbf{k}_0}(\Omega)}{r} \int d^3k \, A(\mathbf{k}) \psi_{\mathbf{k}}(\mathbf{n}_0 r) e^{-i\omega_k t} \\ &\equiv \frac{f_{\mathbf{k}_0}(\Omega)}{r} \Psi^{(\mathrm{F})}(\mathbf{n}_0 r, t). \end{aligned} \tag{15.26}$$

Equation (15.26) is the crucial result. It states that, for large r, $\Psi^{(\text{out})}(\mathbf{r}, t)$ is proportional to the pure plane-wave packet at position $\mathbf{r}_0 = \mathbf{n}_0 r$. However, r is always positive, and $\mathbf{n}_0$ points along $\mathbf{k}_0$, so that $\mathbf{n}_0 r$ is a point well to the right of the target, independent of the value of t. We now recall that $\Psi^{(\mathrm{F})}(\mathbf{r}, t)$ is constructed so that, for $T \ll 2\mu/[\hbar(\Delta k)^2]$ – the time taken for the projectile to propagate to the target, interact with it, be scattered, and finally be detected – $\Psi^{(\mathrm{F})}(\mathbf{r}, t) \cong \exp(-i\omega_{k_0} t)\Psi^{(\mathrm{F})}(\mathbf{r} - \mathbf{v}_0 t, 0)$. Hence, for $t \to -T$, $\mathbf{r} \to -\mathbf{v}_0 T$, a position well to the *left* of the target. Since $\mathbf{n}_0 r$ is always well to the right of the target, and $\Psi^{(\mathrm{F})}(\mathbf{r}, t)$ is reasonably well localized, we conclude that

$$\lim_{t \to -\infty} \Psi^{(\mathrm{F})}(\mathbf{n}_0 r, t) \to 0,$$

thus ensuring that Eq. (15.11) holds.[2]

We next ask what (15.26) tells us about the large-t behavior. From Eq. (15.8), used in the preceding analysis, we have

$$\begin{aligned} \Psi^{(\text{out})}(\mathbf{r}, t) &\cong e^{i\omega_{k_0} t} \Psi^{(\mathrm{F})}(\mathbf{n}_0 r - \mathbf{v}_0 t, 0) \\ &= e^{i\omega_{k_0} t} \Psi^{(\mathrm{F})}(\mathbf{n}_0 (r - v_0 t), 0), \end{aligned} \tag{15.27}$$

[2] The reader should keep in mind that the $t \to -\infty$ limit is a shorthand for $t \to -T$, $T \ll 2\mu/[\hbar(\Delta k)^2]$, which is always a quantity so large relative to typical transit times that $T \to \infty$ is a reasonable approximation.

which, for large t, states that the positions r where $\Psi^{(\text{out})}(\mathbf{r}, t)$ is non-zero lie on a sphere of radius $v_0 t$. Hence, for large $\mathbf{r}$ and t, $\Psi^{(\text{sc})}(\mathbf{r}, t) = \Psi^{(\text{out})}(\mathbf{r}, t)$ becomes an outgoing spherical wave packet, with a "strength" given by $f_{\mathbf{k}_0}(\Omega)$.

15.2. Angular distributions and the plane-wave limit

The projectile–target system is described theoretically by the wave packets of the preceding subsection. Experimentally one measures the number of ejectiles scattered into the infinitesimal angle $d\Omega$ at the angle Ω relative to the direction of the incident beam. From this number one obtains $d\sigma/d\Omega \equiv \sigma(\Omega)$, the differential cross section or angular distribution of the scattered particles. There are three equivalent definitions of $d\sigma/d\Omega$:

$$\frac{d\sigma}{d\Omega} = \frac{\text{Number of particles scattered into } d\Omega \text{ at } \Omega \text{ and } r \text{ per unit time}}{\text{Number of particles in the incident beam per area per unit time}} \tag{15.28a}$$

$$\frac{d\sigma}{d\Omega} = \frac{\text{Total probability that a particle is scattered into } d\Omega \text{ at } \Omega \text{ and } r}{\text{Total probability that an incident particle crosses a unit area at the target}}, \tag{15.28b}$$

and

$$\frac{d\sigma}{d\Omega} = \frac{(\mathbf{J}^{(\text{sc})} \cdot \mathbf{e}_r)\, dA}{(\mathbf{J}^{(\text{F})} \cdot \mathbf{J}^{(\text{F})}/J^{(\text{F})})\, d\Omega}, \tag{15.28c}$$

where $\mathbf{J}^{(\text{sc})}$ and $\mathbf{J}^{(\text{F})}$ are the probability current densities obtained, respectively, from $\Psi^{(\text{sc})}(\mathbf{r}, t)$ and $\Psi^{(\text{F})}(\mathbf{r}, t)$, and $dA = r^2\, d\Omega$ is an element of unit area at the detector.

Figure 15.5 shows schematically the various wave packets associated with the scattering event to which the three definitions of $d\sigma/d\Omega$ refer. These definitions, and related material, are discussed elsewhere in much greater detail than the reader will encounter here. For example, some introductory references are the books by Taylor (1972) – especially for Eqs. (15.28) – and by Rodberg and Thaler (1967); the chapter on potential scattering in the text by Baym (1976) is another possibility, though the reader should be aware that each of the foregoing references assumes a greater background of mathematics and quantum mechanics than has been covered so far in the present text.

Definition (15.28a) is essentially an experimental one, whereas the other two make the

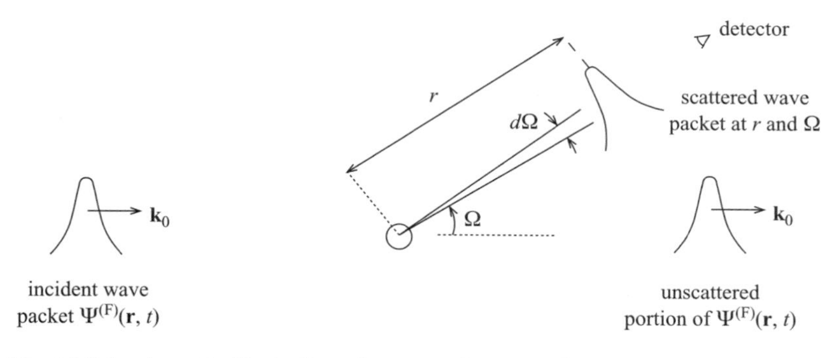

Fig. 15.5 A schematic illustration of wave-packet scattering.

connection with theory. We shall evaluate $d\sigma/d\Omega$ using both (15.28b) and (15.28c), starting with the latter definition and a sharp momentum state. Let $A(\mathbf{k})$ be infinitely sharp:

$$A(\mathbf{k}) = \delta(\mathbf{k} - \mathbf{k}_0), \tag{15.29}$$

in which case

$$\Psi^{(\mathrm{F})}(\mathbf{r},\, t) \rightarrow \Psi_{\mathbf{k}_0}(\mathbf{r},\, t) = \left(\frac{1}{2\pi}\right)^{3/2} e^{i(\mathbf{k}_0 \cdot \mathbf{r} - \omega_{k_0} t)} \tag{15.30a}$$

and

$$\Psi^{(\mathrm{sc})}(\mathbf{r},\, t) \rightarrow \Psi^{(\mathrm{sc})}_{\mathbf{k}_0}(\mathbf{r},\, t) = \psi^{(\mathrm{sc})}_{\mathbf{k}_0}(\mathbf{r}) e^{-i\omega_{k_0} t}. \tag{15.30b}$$

Although we do not know $\psi^{(\mathrm{sc})}_{\mathbf{k}_0}(\mathbf{r})$ anywhere but asymptotically, that knowledge suffices to evaluate the numerator in (15.28c). That is, because detection occurs at asymptotic distances r, we may use the asymptotic relation (15.21) to evaluate $\mathbf{J}^{(\mathrm{sc})}$.

From Eq. (15.21) for $\psi^{(\mathrm{sc})}_{\mathbf{k}_0}(\mathbf{r})$ and the definition (7.68b) – with $\Psi^{(\mathrm{sc})}_{\mathbf{k}_0}(\mathbf{r},\, t)$ and μ replacing $\Psi_\gamma(\mathbf{r},\, t)$ and m, respectively, we readily find that, to order r^{-2} in $\mathbf{J}^{(\mathrm{sc})}$,

$$(\mathbf{J}^{(\mathrm{sc})} \cdot \mathbf{e}_r)\, dA \cong v_0 |f_{\mathbf{k}_0}(\Omega)|^2\, d\Omega/(2\pi)^3. \tag{15.31a}$$

Similarly, we get for the denominator in (15.28c)

$$(\mathbf{J}^{(\mathrm{F})} \cdot \mathbf{J}^{(\mathrm{F})}/J^{(\mathrm{F})})\, d\Omega = v_0\, d\Omega/(2\pi)^3, \tag{15.31b}$$

and thus

$$\sigma(\Omega) = \frac{d\sigma}{d\Omega} = |f_{\mathbf{k}_0}(\Omega)|^2. \tag{15.32}$$

Equation (15.32) relates the differential cross section to the scattering amplitude for the case of an infinitely sharp momentum. To a very good approximation, the same result holds for a narrow wave packet with a profile function and a scattering amplitude that is slowly varying with $\mathbf{k}$, as we show next using (15.28b).

For asymptotic distances, the probability density associated with $\Psi^{(\mathrm{sc})}(\mathbf{r},\, t)$ can be expressed in terms of $\Psi^{(\mathrm{out})}(\mathbf{r},\, t)$, Eq. (15.20):

$$\rho^{\mathrm{sc}}(\mathbf{r},\, t) = |\Psi^{(\mathrm{out})}(\mathbf{r},\, t)|^2, \qquad \text{large } r, \tag{15.33a}$$

and the differential probability is

$$dP^{(\mathrm{sc})}(\mathbf{r},\, t) = \rho^{(\mathrm{sc})}(\mathbf{r},\, t)\, d^3 r. \tag{15.33b}$$

For $d^3 r$ we again use a volume element at a distance r along the direction Ω, viz.,

$$d^3 r = dr\, dA = v_0\, dt\, r^2\, d\Omega, \tag{15.33c}$$

where the distance $dr = v_0\, dt$ is the product of the average velocity with which the wave packet moves and the infinitesimal time interval dt.

The differential (scattered) probability per solid angle – Eq. (15.28c) – is $dP^{(\mathrm{sc})}/d\Omega$, which, using Eqs. (15.33), (15.20), and (15.26), becomes

$$\frac{dP^{(\mathrm{sc})}}{d\Omega} = v_0 |f_{\mathbf{k}_0}(\Omega)|^2 |\Psi^{(\mathrm{F})}(\mathbf{n}_0 r,\, t)|^2 dt. \tag{15.34}$$

The narrowness of the profile function and the slowly varying character of $f_{\mathbf{k}}(\Omega)$, which results in (15.26), here yield a differential probability/solid angle proportional to $|f_{\mathbf{k}_0}(\Omega)|^2$. The total probability that a particle is scattered into $d\Omega$ at angle Ω and distance r is the time integral of (15.34):

$$\frac{P^{(\mathrm{sc})}}{d\Omega} = \int dt \frac{dP^{(\mathrm{sc})}}{d\Omega} = v_0 |f_{\mathbf{k}_0}(\Omega)|^2 \int dt\, |\Psi^{(\mathrm{F})}(\mathbf{n}_0 r,\, t)|^2. \tag{15.35a}$$

An analogous analysis applies to the denominator of (15.28b). The differential probability for incidence is

$$dP^{(\mathrm{inc})} = v_0\, dt\, dA_{\mathrm{inc}}\, |\Psi^{(\mathrm{F})}(\mathbf{r},\, t)|^2,$$

where dA_{inc} is a differential of area perpendicular to the direction $\mathbf{n}_0 = \mathbf{k}_0/k_0$. The differential probability per unit area at the target is $dP^{(\mathrm{inc})}/dA_{\mathrm{inc}}$ and thus the total incidence probability per unit area at the target is

$$\frac{P^{(\mathrm{inc})}}{dA_{\mathrm{inc}}} = v_0 \int dt\, |\Psi^{(\mathrm{F})}(\mathbf{r},\, t)|^2. \tag{15.35b}$$

On substitution of Eqs. (15.35) into (15.28b) we find

$$\sigma(\Omega) = \frac{d\sigma}{d\Omega} = |f_{\mathbf{k}_0}(\Omega)|^2 \frac{\displaystyle\int dt\, |\Psi^{(\mathrm{F})}(\mathbf{n}_0 r,\, t)|^2}{\displaystyle\int dt\, |\Psi^{(\mathrm{F})}(\mathbf{r},\, t)|^2}. \tag{15.36}$$

Apart from the ratio of the two time integrals, the wave-packet analysis leads to the same result as Eq. (15.32), the sharp-momentum result. In general, the time-integral ratio of (15.36) is not a problem, since it is approximately unity. The reason for this is that the distance r in the integrals is an asymptotic one, so that only a small range of large t contributes. For these r, t values, the integrals over the two wave packets will not differ significantly as long as the spreading of the wave packets can be ignored. We shall assume that any spreading is small, an assumption based on the properties of the wave packets as discussed in Chapter 7. We therefore conclude that, as long as $A(\mathbf{k})$ is narrow and $f_{\mathbf{k}}(\Omega)$ is a slowly varying function of $\mathbf{k}$, we can by-pass the wave-packet treatment and work directly with the improper states $|\mathbf{k}_0\rangle$ and $|\mathbf{k}_0, +\rangle$, or their coordinate representations $\psi_{\mathbf{k}_0}(\mathbf{r})$ and $\psi^{(+)}_{\mathbf{k}_0}(\mathbf{r})$, so that the angular distribution is simply given by Eq. (15.32).

While the assumption that $A(\mathbf{k})$ is narrow usually holds, the assumption that $f_{\mathbf{k}}(\Omega)$ varies slowly with $\mathbf{k}$ will be violated when the energy E_{k_0} is in the vicinity of a so-called "narrow resonance" of the projectile-plus-target scattering system. The analysis of resonance behavior is beyond the scope of the present chapter, and the interested reader is referred to the texts cited above.

In addition to simply measuring the number of ejectiles emerging at angle Ω, one can attempt to measure the total number of particles scattered at all angles. This quantity is the total cross section σ_{T}, which in the present case is

$$\sigma_T \equiv \int \frac{d\sigma}{d\Omega}\, d\Omega = \int \sigma(\Omega)\, d\Omega.$$

$$= \int d\Omega\, |f_{\mathbf{k}_0}(\Omega)|^2. \tag{15.37}$$

There is an important and quite useful relation between σ_T and the imaginary part of the forward ($\Omega = 0$) part of the scattering amplitude. Its derivation involves a long but straightforward wave packet analysis, which we leave to the exercises. The result is

$$\sigma_T = \frac{4\pi}{k_0}\, \text{Im}[f_{\mathbf{k}_0}(\Omega = 0)], \tag{15.38}$$

often referred to as the "optical theorem." It was first obtained by Bohr, Peierls, and Placzek (1939) and is a flux-conservation result, the analog in 3-D of the 1-D relation $R + T = 1$ of Eq. (7.97), whose derivation also required a wave-packet, rather than a plane-wave, analysis. We shall employ (15.38) in the partial-wave analysis of $f_{\mathbf{k}_0}(\Omega)$.

15.3. The Lippmann–Schwinger equation and the scattering amplitude

From Eq. (15.23), the scattering amplitude $f_{\mathbf{k}}(\Omega)$ is obtained from the asymptotic form of the scattering state $\psi_{\mathbf{k}}^{(+)}(\mathbf{r})$, where we now use $\mathbf{k}$ rather than $\mathbf{k}_0$ as the incident wave number. As will be shown later, $f_{\mathbf{k}}(\Omega)$ can be written as a matrix element whose ket is $|\mathbf{k}, +\rangle$, so the connection between the scattering state and the scattering amplitude is even more intimate than that suggested by the asymptotic relation. Our aim in this section is to explore aspects of $\psi_{\mathbf{k}}^{(+)}(\mathbf{r})$ and $f_{\mathbf{k}}(\Omega)$ relevant to carrying out scattering calculations.

Because E_k is fixed in advance,

$$(\hat{H}(\mathbf{r}) - E_k)\psi_{\mathbf{k}}^{(+)}(\mathbf{r}) = 0 \tag{15.39}$$

is to be solved not as an eigenvalue equation but as a boundary-value problem for $\psi_{\mathbf{k}}^{(+)}(\mathbf{r})$, the boundary conditions being Eqs. (15.23) and (15.24). An alternate means of solving for $\psi_{\mathbf{k}}^{(+)}(\mathbf{r})$ is to turn the combination of (15.39) and the boundary conditions into an integral equation,[3] in the present instance the one known as the Lippmann–Schwinger (L–S) equation, after the physicists who first derived it (Lippmann and Schwinger (1950)). This equation is an indispensable element of the quantum theory of scattering and its derivation is our first task in this section.

We start with the abstract version of (15.39):

$$(E_k - \hat{K} - V)|\mathbf{k}, +\rangle = 0. \tag{15.40}$$

The ket $|\mathbf{k}, +\rangle$ is given by Eq. (15.15) with $\alpha \to$"+"; substituting that expression into (15.40) leads to

$$(E_k - \hat{K})[|\mathbf{k}\rangle + |\mathbf{k}, \text{sc}\rangle] = V|\mathbf{k}, +\rangle. \tag{15.41}$$

[3] An integral equation is one in which the unknown function – here $\psi_{\mathbf{k}}^{(+)}(\mathbf{r})$ – appears under an integral sign rather than being acted on by derivatives. Although we shall not solve any integral equations in this book, readers interested in types, properties, and solutions of integral equations will find readable introductions in, e.g., the texts by Mathews and Walker (1970) and by Wyld (1976), who also cite additional mathematical literature.

However, $|\mathbf{k}\rangle$ is an eigenket of $\hat{K}$ belonging to E_k, so that the $(E_k - \hat{K})|\mathbf{k}\rangle$ term in (15.41) vanishes, leaving

$$(E_k - \hat{K})|\mathbf{k}, \mathrm{sc}\rangle = V|\mathbf{k}, +\rangle. \tag{15.42}$$

A formal procedure for solving (15.42) is to act on both sides of it from the left with the inverse of $E_k - \hat{K}$, assuming that this inverse operator exists. Let us temporarily assume this to be the case and write

$$(E_k - \hat{K})^{-1} \equiv \hat{G}_k, \tag{15.43}$$

such that

$$(E_k - \hat{K})\hat{G}_k = \hat{G}_k(E_k - \hat{K}) = \hat{I}. \tag{15.44}$$

Thus, the presumed existence of $\hat{G}_k$ means that the formal relation

$$|\mathbf{k}, \mathrm{sc}\rangle = \hat{G}_k V|\mathbf{k}, +\rangle \tag{15.45}$$

holds, and hence that

$$|\mathbf{k}, +\rangle = |\mathbf{k}\rangle + \hat{G}_k V|\mathbf{k}, +\rangle. \tag{15.46}$$

The L–S equation (15.46) is an abstract version of an integral equation for $|\mathbf{k}, +\rangle$.

The procedure just followed is, unfortunately, not mathematically legitimate because $(E_k - \hat{K})^{-1}$ is not a well-defined operator. The latter statement can be verified by inserting Eq. (15.14b) to the right of $(E_k - \hat{K})^{-1}$ in (15.43):

$$\begin{aligned}\hat{G}_k &= (E_k - \hat{K})^{-1}\int d^3k'\, |\mathbf{k}'\rangle\langle\mathbf{k}'| \\ &= \int d^3k'\, \frac{|\mathbf{k}'\rangle\langle\mathbf{k}'|}{E_k - E_{k'}}.\end{aligned} \tag{15.47}$$

The integrand in (15.47) is singular when $k' = k$ and, though we shall not prove it, the singularity can be of delta-function character. It is possible, however, to turn $\hat{G}_k$ into a well-defined operator, although the general method for doing so – involving techniques from complex-variable theory – is beyond the scope of this book. Instead of describing this method, we shall go into a coordinate representation to express two of the proper inverses that result when these methods are used – Eq. (15.50) below.

In the coordinate representation, $E_k - \hat{K}$ is, of course, a local operator, whereas we see from (15.47) that $\hat{G}_k$ is non-local. Acting on both sides of (15.44) from the left with $\langle\mathbf{r}|$ and from the right with $|\mathbf{r}'\rangle$, we find

$$[E_k - \hat{K}(\mathbf{r})]G_k(\mathbf{r}, \mathbf{r}') = \delta(\mathbf{r} - \mathbf{r}'), \tag{15.48}$$

where

$$G_k(\mathbf{r}, \mathbf{r}') = \langle\mathbf{r}|\hat{G}_k|\mathbf{r}'\rangle. \tag{15.49}$$

Equation (15.49) is a *unit-source* inhomogeneous equation. Quantities that satisfy (15.48) are known as *Green's functions*, two of which are

$$G_k^{(\pm)}(\mathbf{r}, \mathbf{r}') = -\frac{2\mu}{\hbar^2}\frac{1}{4\pi|\mathbf{r} - \mathbf{r}'|}e^{\pm ik|\mathbf{r}-\mathbf{r}'|}. \tag{15.50}$$

These are the coordinate representations of the two proper inverses referred to above. In particular, $G_k^{(+)}(\mathbf{r}, \mathbf{r}')$ is an outgoing-wave Green function, while $G_k^{(-)}(\mathbf{r}, \mathbf{r}') = G_k^{(+)*}$ is its ingoing-wave counterpart. As we shall soon demonstrate, the outgoing-wave Green function $G_k^{(+)}(\mathbf{r}, \mathbf{r}')$ is the appropriate one to use with $\psi_{\mathbf{k}}^{(+)}(\mathbf{r})$: the matching of superscripts is not accidental.

That the $G_k^{(\pm)}(\mathbf{r}, \mathbf{r}') = G_k^{(\pm)}(|\mathbf{r} - \mathbf{r}'|)$ of (15.50) satisfy (15.48) is easily proved. Let $\rho = \mathbf{r} - \mathbf{r}'$, in which case $\nabla_{\mathbf{r}}^2 = \nabla_\rho^2$, $G_k^{(\pm)}(\mathbf{r}, \mathbf{r}') = G_k^{(\pm)}(\rho)$, and (15.48) becomes

$$(\nabla_\rho^2 + k^2)\left(-\frac{e^{\pm ik\rho}}{4\pi\rho}\right) \stackrel{?}{=} \delta(\rho). \tag{15.51}$$

The objective is to remove the question mark in (15.51). Assume first that $\rho \neq 0$. Then, in ∇_ρ^2, only the $\rho^{-2}(\partial/\partial\rho)(\rho^2\,\partial/\partial\rho)$ term is non-vanishing, and straightforward differentiation establishes that

$$(\nabla_\rho^2 + k^2)\left(-\frac{e^{\pm ik\rho}}{4\pi\rho}\right) = 0, \qquad \rho \neq 0,$$

as required.

To show that, in the limit $\rho = 0$, the RHS of (15.51) is indeed $\delta(\boldsymbol{\rho})$, we integrate both sides of (15.51) over a sphere of radius $\rho = R$, and then take the limit $R \rightarrow 0$. Integration yields

$$\int_{\substack{\text{Sphere,}\\ \text{radius } R}} d^3\rho\,(\nabla_\rho^2 + k^2)\left(-\frac{e^{\pm ik\rho}}{4\pi\rho}\right) \stackrel{?}{=} \int d^3\rho\,\delta(\boldsymbol{\rho}) = 1, \tag{15.52}$$

where the RHS value of unity is independent of R. In the limit $R \rightarrow 0$, the k^2 term on the LHS of (15.52) vanishes, while use of Green's theorem gives

$$\int_{\substack{\text{Sphere,}\\ \text{radius } R}} d^3\rho\,\nabla_\rho^2\left(-\frac{e^{\pm ik\rho}}{4\pi\rho}\right) = \int_{\substack{\text{Spherical}\\ \text{surface}}} d\mathbf{A}\cdot\nabla_\rho\left(-\frac{e^{\pm ik\rho}}{4\pi\rho}\right),$$

where $d\mathbf{A} = R^2\,d\Omega\,\mathbf{e}_\rho$. However,

$$\int_{\substack{\text{Spherical}\\ \text{surface}}} d\Omega = 4\pi$$

and $\mathbf{e}_\rho \cdot \nabla_\rho = d/d\rho$ so that

$$\int_{\substack{\text{Spherical}\\ \text{surface}}} d\mathbf{A}\cdot\nabla_\rho\left(-\frac{e^{\pm ik\rho}}{4\pi\rho}\right) = 1 \mp ikR\,e^{\pm ikR} \underset{R\rightarrow 0}{\longrightarrow} 1,$$

and we therefore conclude that the "?" can be dropped in (15.51), QED.

We have claimed that the appropriate Green function (inverse) to use in the coordinate-space version of the L–S equation (15.46) is $G_k^{(+)}(\mathbf{r}, \mathbf{r}')$, i.e., that

$$\psi_{\mathbf{k}}^{(+)}(\mathbf{r}) = \psi_{\mathbf{k}}(\mathbf{r}) - \frac{2\mu}{\hbar^2}\int d^3r'\,\frac{e^{ik|\mathbf{r}-\mathbf{r}'|}}{4\pi|\mathbf{r}-\mathbf{r}'|}V(\mathbf{r}')\psi_{\mathbf{k}}^{(+)}(\mathbf{r}'). \tag{15.53}$$

We establish this claim first, and then discuss a few properties of (15.53). Since the asymptotic form of $\psi_{\mathbf{k}}^{(+)}(\mathbf{r})$ is given by (15.23), we must examine the large r behavior of

$G_k^{(+)}(\mathbf{r}, \mathbf{r}')$; note that $\mathbf{r}'$, an integration variable in (15.53), can never be asymptotically large because of the assumption that $V(\mathbf{r}')$ is short-ranged. Thus, for $r \gg r'$, $|\mathbf{r} - \mathbf{r}'|^{-1} \cong r^{-1}$, while the large-$r$ form of $\exp(ik|\mathbf{r} - \mathbf{r}'|)$ is

$$e^{ik|\mathbf{r}-\mathbf{r}'|} \underset{r\to\infty}{\sim} e^{ikr} e^{-i\mathbf{k}'\cdot\mathbf{r}'},$$

which follows from

$$\begin{aligned} k|\mathbf{r} - \mathbf{r}'| &= k[r^2 + r'^2 - 2\mathbf{r}\cdot\mathbf{r}']^{1/2} \\ &\cong kr - (k\mathbf{r}/r)\cdot\mathbf{r}' \\ &\equiv kr - \mathbf{k}'\cdot\mathbf{r}', \end{aligned} \tag{15.54}$$

with $\mathbf{k}' \equiv k\mathbf{e}_r$ $(k' = k)$ being a wave vector that points along the direction of the observation point $\mathbf{r}$, whose polar-angle pair is Ω. Letting r become asymptotic and substituting (15.54) into (15.53), we find

$$\psi_{\mathbf{k}}^{(+)}(\mathbf{r}) \underset{r\to\infty}{\sim} \left(\frac{1}{2\pi}\right)^{3/2} \left[e^{i\mathbf{k}\cdot\mathbf{r}} + \frac{e^{ikr}}{r}\left(-\frac{\mu(2\pi)^3}{2\pi\hbar^2}\int d^3r' \, \frac{e^{-i\mathbf{k}'\cdot\mathbf{r}'}}{(2\pi)^{3/2}} V(\mathbf{r}')\psi_{\mathbf{k}}^{(+)}(\mathbf{r}')\right)\right], \tag{15.55}$$

thereby verifying the claim that use of $G_k^{(+)}(\mathbf{r}, \mathbf{r}')$ indeed leads to the outgoing-wave form for $\psi_{\mathbf{k}}^{(+)}(\mathbf{r})$.

Comparison of Eqs. (15.23) and (15.55) also yields an expression for $f_{\mathbf{k}}(\Omega)$:

$$\begin{aligned} f_{\mathbf{k}}(\Omega) &= -\frac{\mu(2\pi)^2}{\hbar^2}\int d^3r' \, \frac{e^{-i\mathbf{k}'\cdot\mathbf{r}'}}{(2\pi)^{3/2}} V(\mathbf{r}')\psi_{\mathbf{k}}^{(+)}(\mathbf{r}') \\ &\equiv -\frac{\mu(2\pi)^2}{\hbar^2}\langle\mathbf{k}'|V|\mathbf{k}, +\rangle, \end{aligned} \tag{15.56}$$

an expression that substantiates another claim, viz., that $f_{\mathbf{k}}(\Omega)$ can be written as a matrix element whose ket is $|\mathbf{k}, +\rangle$. While the matrix element contains the two wave vectors $\mathbf{k}$ and $\mathbf{k}'$, the fact that $k' = k$ means that only the angles Ω (i.e., of $\mathbf{k}' = k\mathbf{e}_r$), not the magnitude k', need to be displayed in the scattering-amplitude symbol $f_{\mathbf{k}}(\Omega)$. The direction Ω is that of $\mathbf{r}$ in Fig. 15.4.

The facts that $\psi_{\mathbf{k}}^{(+)}(\mathbf{r})$ asymptotically involves the unknown $f_{\mathbf{k}}(\Omega)$ and that $f_{\mathbf{k}}(\Omega)$ is a matrix element whose ket is the unknown $\psi_{\mathbf{k}}^{(+)}(\mathbf{r})$ may strike the reader as an insoluble paradox. However, the situation for scattering is not very different than that for bound states: one works with "normalized" plane waves and reads the scattering amplitude from the asymptotic form as the coefficient of $\exp(ikr)/r$. One could, at least in principle, *iterate* the integral equation (15.53) and obtain an infinite-series expansion for $\psi_{\mathbf{k}}^{(+)}(\mathbf{r})$. The procedure is straightforward, and is most easily worked out using the abstract form (15.46):

$$\begin{aligned} |\mathbf{k}, +\rangle &= |\mathbf{k}\rangle + \hat{G}_k^{(+)}V|\mathbf{k}\rangle + \hat{G}_k^{(+)}V\hat{G}_k^{(+)}V|\mathbf{k}\rangle + \cdots \\ &= \sum_{n=0}^{\infty}(\hat{G}_k^{(+)}V)^n|\mathbf{k}\rangle. \end{aligned} \tag{15.57}$$

The expression (15.57) is known as the Neumann series for $|\mathbf{k}, +\rangle$ and is meaningful only if it converges. Even if it does, the series expansion is usually not a practical procedure for determining $\psi_{\mathbf{k}}^{(+)}(\mathbf{r})$, since the integrals involved are normally very complicated, as is illustrated by the coordinate representation for the $n = 2$ term:

$$\langle \mathbf{r}|\hat{G}_k^{(+)} V \hat{G}_k^{(+)} V|\mathbf{k}\rangle = \left(\frac{\mu}{(2\pi\hbar)^2}\right)^2 \int d^3r'\, d^3r''\, \frac{e^{ik|\mathbf{r}-\mathbf{r}'|}}{|\mathbf{r}-\mathbf{r}'|} V(\mathbf{r}') \frac{e^{ik|\mathbf{r}'-\mathbf{r}''|}}{|\mathbf{r}'-\mathbf{r}''|} V(\mathbf{r}'') \frac{e^{i\mathbf{k}\cdot\mathbf{r}''}}{(2\pi)^{3/2}}. \tag{15.58}$$

Because of this complexity, the integral equation is far more often used as a means of formulating approximations, with exact (or numerically accurate) solutions being obtained from solving partial-wave equations, which is discussed in the next section.

The series (15.57) can also be used to generate an analogous one for the matrix-element portion of the scattering amplitude. Insertion into $\langle \mathbf{k}'|V|\mathbf{k}, +\rangle$ yields

$$\begin{aligned} \langle \mathbf{k}'|V|\mathbf{k}, +\rangle &= \langle \mathbf{k}'|V|\mathbf{k}\rangle + \langle \mathbf{k}'|V\hat{G}_k^{(+)} V|\mathbf{k}\rangle + \cdots \\ &= \sum_{n=0}^{\infty} \langle \mathbf{k}'|V(\hat{G}_k^{(+)} V)^n|\mathbf{k}\rangle \\ &= \sum_{n=0}^{\infty} \langle \mathbf{k}'|(V\hat{G}_k^{(+)})^n V|\mathbf{k}\rangle. \end{aligned} \tag{15.59}$$

The expansion on the RHS of (15.59) is known as the Born series, after Max Born, who won the Nobel prize for developing the probabilistic interpretation of the wave function. Born did this in the context of scattering, approximating the full matrix element of (15.59) by the first term of the series (Born 1926):

$$\langle \mathbf{k}'|V|\mathbf{k}, +\rangle \cong \langle \mathbf{k}'|V|\mathbf{k}\rangle \tag{15.60a}$$

and

$$f_{\mathbf{k}}(\Omega) \to f_{\mathbf{k}}^{(\mathrm{B})}(\Omega) = -\frac{\mu(2\pi)^2}{\hbar^2} \langle \mathbf{k}'|V|\mathbf{k}\rangle. \tag{15.60b}$$

This approximation is known as the (first) Born approximation; it is valid at high energies but also used at low energies in the hope of gaining insight into the behavior of the full amplitude. We shall consider the Born and other approximations in Section 15.6.

We close with a cautionary note: the L–S equation should NEVER be used for *exact* calculations on collision systems containing three or more particles! Only for scattering involving one or two particles does the L–S equation yield unique solutions: in the case of three or more particles, its solutions are non-unique, and one must employ other descriptions, collectively known under the general heading of many-particle or multi-channel scattering theories. Many-particle collision theory comprises its own research field and is well beyond the scope of this book, since it assumes as background material typically dealt with in graduate courses or in monographs on scattering theory (e.g., Sakurai (1994) and Taylor (1972)). For the reader whose curiosity is not damped by the foregoing comments, the logical starting point is three-particle collision theory, a "relatively easy-to-comprehend" source for which is the monograph by Glöckle (1983). A semi-qualitative review of three-particle collision theory and its applications is Chapter 1 of the collection of surveys edited by Ellis and Tang (1990).

15.4. Spherical symmetry, partial waves, phase shifts

The preceding section was concerned with elements of "formal" scattering theory, in particular the L–S equation and the infinite Neumann and Born series. Another approach to scattering is via angular-momentum/partial-wave expansions. Since $|\mathbf{k}\rangle$ generates $|\mathbf{k}, +\rangle$, and $|\mathbf{k}\rangle$ itself is represented by the infinite partial-wave series (10.81), both $|\mathbf{k}, +\rangle$ and $f_{\mathbf{k}}(\Omega) \propto \langle \mathbf{k}'|V|\mathbf{k}, +\rangle$ will also be represented by infinite partial-wave expansions. These different infinite-series expansions illustrate the inherent complexity of scattering in comparison with bound states, for which a single angular-momentum quantum number and its z-component can characterize eigenstates. Infinite-series representations suggest that exact calculations may be sufficiently difficult to carry out that one must rely either on numerical evaluations or on analytic approximations.

This pessimistic suggestion is, in fact, no more than a suggestion and is not a universal characteristic of scattering. As we shall see in this section, the infinite angular-momentum expansion can be sensibly truncated, often after only a few terms, because the scattering parameters associated with this expansion become zero for the higher partial waves. (The number of partial waves retained depends on the energy.) The aforementioned parameterization is in terms of quantities known as phase shifts, so the angular-momentum representation not only introduces these important quantities but also provides a means for carrying out essentially exact calculations.

For simplicity, we shall consider spherically symmetric potentials: $V(\mathbf{r}) \to V(r)$. This is not a major limitation and it does not preclude potentials of the type $V(r)\hat{\mathbf{J}}_1 \cdot \hat{\mathbf{J}}_2$. The restriction to $V(r)$ means that for spinless particles, the angular-momentum expansions involve only the $Y_\ell^{m_\ell}$'s; they do not involve coupled states such as $|\ell s;\, jm_j\rangle$. Once we have completed the present type of analysis, the machinery to consider more complex situations will be in place. In addition to expansions that involve only the spherical harmonics, spherically symmetric potentials – and the assumption of spin zero – provide two additional simplifications: first, by choosing $\mathbf{k} = k\mathbf{e}_3$, only $m_\ell = 0$ substates occur in the partial-wave expansions of $\psi_{\mathbf{k}}^{(+)}(\mathbf{r})$ and $f_{\mathbf{k}}(\Omega)$, and second, each partial-wave radial component of $\psi_{\mathbf{k}}^{(+)}(\mathbf{r})$ will be decoupled from all the others, will obey its own partial-wave radial equation, and will produce its own individual contribution to the scattering amplitude. We explore these points in the following subsections.

Partial-wave expansions

The arguments that led to $P_\ell(\cos\phi)$ or $Y_\ell^0(\Omega)$ yielding the full angular dependence of $\langle \mathbf{r}|\mathbf{k}\rangle$ for each partial wave also apply to $\langle \mathbf{r}|\mathbf{k}, +\rangle$ as long as $V(\mathbf{r}) = V(r)$ and the particles have zero spin. By choosing $\mathbf{k} = k\mathbf{e}_3$, no dependence on the azimuthal angle can occur when $V(\mathbf{r}) = V(r)$. Hence, the coordinate system can be oriented so that the final wave vector $\mathbf{k}' = k\mathbf{e}_r$ lies in the x–z plane, Fig. 15.6. As a result only $m_\ell = 0$ states will be present in the expansion of $\psi_{\mathbf{k}}^{(+)}(\mathbf{r})$, which we write, in analogy to (10.81), as[4]

$$\psi_{\mathbf{k}}^{(+)}(\mathbf{r}) = \sum_\ell \sqrt{4\pi(2\ell+1)}\, i^\ell F_\ell^{(+)}(kr) Y_\ell^0(\Omega), \tag{15.61}$$

[4] Elements of a mathematical proof that only $m_\ell = 0$ substates occur is left to the exercises.

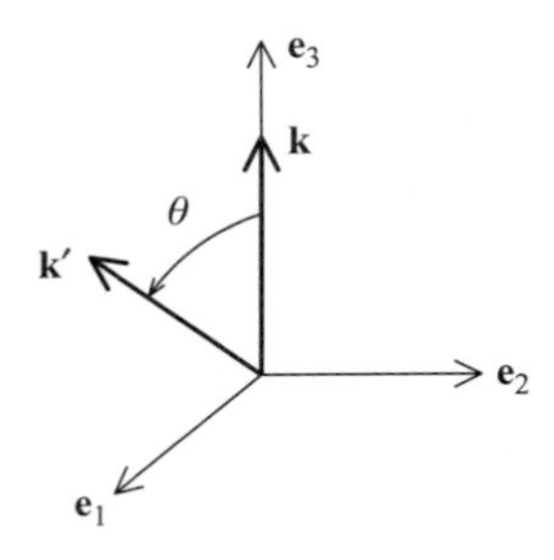

Fig. 15.6 A graphical illustration of the incident wave vector $\mathbf{k} = k\mathbf{e}_3$, the scattered wave vector $\mathbf{k}'$ lying in the x–z ($\mathbf{e}_1$–$\mathbf{e}_3$) plane and the scattering angle θ, also in the x–z plane.

where $\Omega = \theta, \phi$ are the angles of $\mathbf{r}$ w.r.t. $\mathbf{k}$. With the choice $\mathbf{k} = k\mathbf{e}_3$, $\psi_{\mathbf{k}}^{(+)}(\mathbf{r})$ could just as well be written $\psi_k^{(+)}(\mathbf{r})$, and we do so below for $f_{\mathbf{k}}(\Omega)$.

Subtituting (15.61) into (15.39) and projecting out the ℓth partial wave leads to the equation for the radial component $F_\ell^{(+)}(kr)$:

$$\left(\frac{1}{r^2}\frac{d}{dr}r^2\frac{d}{dr} - \frac{\ell(\ell+1)}{r^2} + k^2 - \frac{2\mu}{\hbar^2}V(r)\right)F_\ell^{(+)}(kr) = 0. \tag{15.62}$$

Division by k^2 shows that the derivatives are taken w.r.t. the product kr, thus justifying the kr dependence in $F_\ell^{(+)}(kr)$. At $r = 0$, $F_\ell^{(+)}(kr)$ must be finite, but, in order to solve (15.62), a large-r asymptotic boundary condition, analogous to (15.23), must be imposed. We shall determine the asymptotic behavior of $F_\ell^{(+)}(kr)$ after examining the partial-wave expansion of $f_{\mathbf{k}}(\Omega)$.

Spherical symmetry suggests that the partial-wave expansion of $f_{\mathbf{k}}(\Omega)$ also contains only the $Y_\ell^0(\Omega)$. This follows from Eqs. (15.61) and the asymptotic form (15.23) when the choice $\mathbf{k} = k\mathbf{e}_3$ is made. We shall also prove it below. The requisite expansion is written

$$f_{\mathbf{k}}(\Omega) = \sum_\ell \sqrt{4\pi(2\ell+1)}f_\ell(k)Y_\ell^0(\Omega) \equiv f_k(\theta), \tag{15.63}$$

where, in view of $\mathbf{k} = k\mathbf{e}_3$, the subscript $\mathbf{k}$ on f has been replaced by k. Factors like $(\pm i)^\ell$ have been absorbed into the ℓth partial-wave scattering amplitude $f_\ell(k)$. Our first goal is to show that the general form of $f_\ell(k)$ is

$$f_\ell(k) = \frac{e^{i\delta_\ell(k)}\sin[\delta_\ell(k)]}{k}, \tag{15.64}$$

where the real parameter $\delta_\ell(k)$ is known as the ℓth (partial-wave) *phase shift*. It expresses the full effect of the potential $V(r)$ on the ℓth partial waves both in $\psi_{\mathbf{k}}^{(+)}(\mathbf{r})$ and in $f_k(\theta)$. Furthermore, it can be determined by solving (15.62) subject to the asymptotic boundary condition we shall soon derive. We will calculate $\delta_\ell(k)$ for several potentials in Section 15.5.

Phase shifts and the asymptotic form of $F_\ell^{(+)}(kr)$

The phase shift form (15.64) for $f_\ell(k)$ is a direct consequence of the conservation of probability flux, as expressed via the optical theorem, Eq. (15.38):

$$\sigma_{\mathrm{T}} = \int d\Omega\, |f_{\mathbf{k}}(\Omega)|^2 = \frac{4\pi}{k}\, \mathrm{Im}[f_{\mathbf{k}}(\Omega = 0)]. \tag{15.65a}$$

Equation (15.63) is used in evaluating both sides of (15.65a). From $Y_\ell^0(\Omega = 0) = \sqrt{(2\ell+1)/(4\pi)}$ the term $f_{\mathbf{k}}(\Omega = 0)$ on the RHS of (15.65a) is

$$f_k(\theta = 0) = \sum_\ell (2\ell + 1) f_\ell(k),$$

while the integration defining σ_{T} yields

$$\sigma_{\mathrm{T}} = \sum_\ell 4\pi(2\ell + 1)|f_\ell(k)|^2. \tag{15.65b}$$

The optical theorem thus becomes

$$\sum_\ell 4\pi(2\ell + 1)\left(|f_\ell(k)|^2 - \frac{\mathrm{Im}[f_\ell(k)]}{k}\right) = 0. \tag{15.66}$$

A sufficient condition guaranteeing (15.66) is that the term in large parentheses vanishes for each ℓ separately, and we invoke this condition as the means of determining the form of $f_\ell(k)$:

$$|f_\ell(k)|^2 = \frac{1}{k}\, \mathrm{Im}[f_\ell(k)]. \tag{15.67}$$

Since $f_\ell(k)$ is a complex quantity, we introduce its phase $\delta_\ell(k)$ via

$$f_\ell(k) = |f_\ell(k)| e^{i\delta_\ell(k)}, \tag{15.68}$$

whereupon (15.67) leads to

$$|f_\ell(k)| = \frac{1}{k} \sin[\delta_\ell(k)]. \tag{15.69}$$

Combining Eqs. (15.69) and (15.67) produces the result (15.64), QED.

Apart from introducing one new concept, we postpone further discussion of the phase-shift description of scattering until after we have derived the asymptotic form of $F_\ell^{(+)}(kr)$. The new concept is the S-matrix, whose ℓth partial-wave (diagonal) element is

$$S_\ell = e^{2i\delta_\ell(k)}. \tag{15.70}$$

$S_\ell(k)$ is unitary, i.e., $|S_\ell(k)| = 1$, and is related to $f_\ell(k)$ by

$$f_\ell(k) = \frac{1}{2ik}(S_\ell(k) - 1) \tag{15.71a}$$

or

$$S_\ell(k) = 1 + 2ikf_\ell(k). \tag{15.71b}$$

As shown below, $S_\ell(k)$ is one measure among several of how $F_\ell^{(+)}(kr)$ differs from its plane-wave counterpart $j_\ell(kr)$.

It was remarked earlier that Eq. (15.62) can be solved once the asymptotic form of $F_\ell^{(+)}(kr)$ has been specified. We are now in a position to do so. The starting point is Eq. (15.23). On the LHS, $\psi_{\mathbf{k}}^{(+)}(\mathbf{r})$ is replaced by (15.61). On the RHS, Eq. (10.81) replaces $\exp(i\mathbf{k}\cdot\mathbf{r})$, while Eq. (15.63), with Eq. (15.64) for $f_\ell(k)$, is substituted for $f_{\mathbf{k}}(\Omega)$. On projecting out the ℓth partial wave and canceling out common factors we find

$$F_\ell^{(+)}(kr) \underset{r\to\infty}{\sim} \left(\frac{1}{2\pi}\right)^{3/2}\left(\frac{i}{2kr}(e^{-i(kr-\ell\pi/2)} - e^{i(kr-\ell\pi/2)}) + \frac{(-i)^\ell}{2ikr}e^{ikr}[e^{i\delta_\ell}(e^{i\delta_\ell} - e^{-i\delta_\ell})]\right),$$

where $\sin\delta_\ell = [\exp(i\delta_\ell) - \exp(-i\delta_\ell)]/(2i)$ has been used. Recalling that $(-i)^\ell = \exp(-i\ell\pi/2)$, two of the four terms in (15.74) are seen to cancel one another out, and a little rearranging yields

$$F_\ell^{(+)}(kr) \underset{r\to\infty}{\sim} \left(\frac{1}{2\pi}\right)^{3/2}\frac{1}{2ikr}[-e^{-i(kr-\ell\pi/2)} + S_\ell(k)e^{i(kr-\ell\pi/2)}] \tag{15.72a}$$

$$= \left(\frac{1}{2\pi}\right)^{3/2}\{e^{i\delta_\ell(k)}\sin[kr - \ell\pi/2 + \delta_\ell(k)]/kr\}, \tag{15.72b}$$

where $S_\ell(k) = \exp(2i\delta_\ell(k))$ has been used.

Because of the factor $\exp[i\delta_\ell(k)]$ in Eq.(15.72b), it is convenient to extract it from the solution $F_\ell^{(+)}(kr)$ by defining $R_\ell^{(+)}(kr)$ such that

$$F_\ell^{(+)}(kr) \equiv e^{i\delta_\ell(k)}R_\ell^{(+)}(kr), \tag{15.73a}$$

where $R_\ell^{(+)}(kr)$ also obeys (15.62), subject to

$$R_\ell^{(+)}(kr) \underset{r\to\infty}{\sim} \left(\frac{1}{2\pi}\right)^{3/2}\sin[kr - \ell\pi/2 + \delta_\ell(k)]/(kr). \tag{15.73b}$$

We shall henceforth work with $R_\ell^{(+)}(kr)$. Since $V(r)$ is short-ranged, then, for $r > b$, the range of $V(r)$, $R_\ell^{(+)}(kr)$ satisfies the free-particle equation. For $r \neq 0$, there are two pairs of linearly independent free-particle solutions: the spherical Bessel and Neumann functions $j_\ell(kr)$ and $n_\ell(kr)$, and the spherical Hankel functions $h_\ell^{(1)} = j_\ell(kr) + in_\ell(kr)$ and $h_\ell^{(1)^*}(kr)$, Eq. (11.100). In analogy to Eq. (15.72a), Eq. (15.73a) is an asymptotic form corresponding to

$$R_\ell^{(+)}(kr) = a_\ell h_\ell^{(1)*}(kr) + b_\ell h_\ell^{(1)}(kr), \qquad r > b, \tag{15.74a}$$

with

$$S_\ell(k) = e^{2i\delta_\ell(k)} = b_\ell/a_\ell \tag{15.74b}$$

and

$$h_\ell^{(1)}(kr) \underset{r\to\infty}{\sim} \frac{-ie^{i(kr-\ell\pi/2)}}{kr}. \tag{15.74c}$$

Furthermore, on writing

$$\sin(kr - \ell\pi/2 + \delta_\ell) = \cos\delta_\ell\sin(kr - \ell\pi/2) + \sin\delta_\ell\cos(kr - \ell\pi/2), \tag{15.75a}$$

we see that Eq. (15.73b) is an asymptotic form corresponding to

$$R_\ell^{(+)}(kr) \underset{r\to\infty}{\sim} c_\ell j_\ell(kr) - d_\ell n_\ell(kr), \qquad r > b, \tag{15.75b}$$

with

$$d_\ell / c_\ell = \tan[\delta_\ell(k)], \tag{15.75c}$$

where Eq. (11.98b) has been used. The relations (15.74) and (15.75) provide both analytic and numerical means for extracting the phase shift (mod π) from a solution to Eq. (15.62) – see Section 15.5.

It is evident from the preceding argument that, in the asymptotic region, the effect of the potential $V(r)$ is to change the free-particle argument $(kr - \ell\pi/2)$ to the scattering-state argument $(kr - \ell\pi/2 + \delta_\ell)$. Before calculating δ_ℓ's for specific potentials, it is helpful to determine some of their general properties.

Properties of the phase shifts and cross sections

The properties we shall discuss are essentially the answers to the questions "If a potential is predominantly attractive or repulsive, will δ_ℓ reflect this?", "How does $\delta_\ell(k)$ behave in the limiting cases of $k \to 0$ and $k \to \infty$?", and "Are there any special characteristics of $\delta_\ell(k)$ corresponding to cut-off potentials, i.e., potentials for which $V(r) = 0$ for $r > b$?"

The answer to the first question is based on a comparison between the zero-potential asymptotic form $\sin(kr - \ell\pi/2)$ and the $\sin[kr - \ell\pi/2 + \delta_\ell(k))$ asymptotic behavior of $R_\ell^{(+)}(kr)$. The analysis assumes, strictly for purposes of comparison, that these asymptotic forms are valid for all r, and contrasts the zero of the free wave $\sin(kr - \ell\pi/2)$ at $r_{\rm F} = \ell\pi/(2k)$ to the zero of the scattering-state wave $\sin[kr - \ell\pi/2 + \delta_\ell(k)]$ at $r_{\rm S} = [\ell\pi/2 - \delta_\ell(k)]/k$. First, suppose that $V(r)$ is predominantly repulsive. Its effect is to diminish $R_\ell^{(+)}(kr)$ in the region where $V(r) \neq 0$ relative to $j_\ell(kr)$. (It does this by pushing $R_\ell^{(+)}(kr)$ out to larger radii.) Smaller $R_\ell^{(+)}(kr)$ means that it should go to zero at a larger radius than does $j_\ell(kr)$. Representing j_ℓ and $R_\ell^{(+)}$ by their asymptotic forms, the latter argument implies that $r_{\rm S} > r_{\rm F}$, i.e., that $\delta_\ell(k)$ is negative. On the other hand, an attractive potential should pull $R_\ell^{(+)}$ in, causing it to be zero at a smaller radius than $r_{\rm F}$. Applying the preceding argument to this case, we conclude that $r_{\rm S} < r_{\rm F}$, i.e., that $\delta_\ell(k)$ is positive when $V(r)$ is predominantly attractive. Summarizing, we expect

$$\text{sign}[\delta_\ell(k)] = \begin{cases} \text{positive}, & V(r) \text{ predominantly attractive}, \\ \text{negative}, & V(r) \text{ predominantly repulsive}. \end{cases} \tag{15.76}$$

It is helpful in understanding these conclusions to relate $f_\ell(k)$, i.e., $\sin[\delta_\ell(k)]$, to an integral involving the potential $V(r)$ and the radial solution $R_\ell^{(+)}(kr)$. We start with Eq. (15.56). For $\psi_{\mathbf{k}}^{(+)}(\mathbf{r}')$ we use the expansion (15.61) and for $\exp(-i\mathbf{k}'\cdot\mathbf{r}')$ we employ the following version of (10.85):

$$e^{-i\mathbf{k}'\cdot\mathbf{r}'} = 4\pi \sum_{\ell, m_\ell} (-i)^\ell Y_\ell^{m_\ell^*}(\Omega_{r'}) Y_\ell^{m_\ell}(\Omega) j_\ell(kr'),$$

where the angles of $\mathbf{k}'$ are Ω ($\mathbf{k}' = k\mathbf{e}_r$). Substituting these expansions into (15.56) and noting that the integration on $\Omega_{r'}$ gives $\delta_{m_\ell 0}$, we find precisely the form (15.63) with $f_\ell(k)$ given by

$$f_\ell(k) = -\frac{2\mu}{\hbar^2}\int_0^\infty dr\, r^2 j_\ell(kr) V(r)[(2\pi)^{3/2} F_\ell^{(+)}(kr)]. \tag{15.77}$$

The analysis so far has yielded the proof that only $m_\ell = 0$ occurs in (15.63), exactly as claimed.

Comparison of (15.77) with (15.56) shows the similarity in form between the integrals for the full and the partial-wave scattering amplitudes. What is truly remarkable about (15.77) is that, apart from a factor of k, its absolute magnitude is never greater than unity! By equating (15.77) and (15.64) we obtain the desired relation:

$$\sin\delta_\ell = -\frac{2\mu k}{\hbar^2}\int_0^\infty dr\, r^2 j_\ell(kr) V(r)[(2\pi)^{3/2} R_\ell^{(+)}(kr)]. \tag{15.78}$$

The LHS of (15.78) is real and always $\leqslant \pm 1$. Hence, $R_\ell^{(+)}$ is real.

This last relation provides some insight into the conclusions stated in (15.76). Let us assume $V(r)$ to be a weak potential that causes little scattering, in which case $\delta_\ell(k)$ should be small. In this instance, we approximate $R_\ell^{(+)}(kr)$ in (15.78) by the free wave $j_\ell(kr)$, yielding

$$\sin[\delta_\ell(k)] \cong -\frac{2\mu k}{\hbar^2}\int dr\, |r j_\ell(kr)|^2 V(r); \tag{15.79}$$

the sign of $\delta_\ell(k)$ is clearly correlated to the assumed attractive or repulsive character of $V(r)$, exactly as predicted by (15.76). Equation (15.79) is the Born approximation for $\sin[\delta_\ell(k)]$.

The converse of (15.76) states that, if the sign of $\delta_\ell(k)$ is positive (negative), then the potential is predominantly attractive (repulsive). The nucleon–nucleon scattering system exemplifies this quite nicely. Although the nucleon–nucleon interaction contains both central and non-central portions and is also spin-dependent, analysis of the relevant experimental data allows one to extract a set of phase shifts that are analogous to the single $\delta_\ell(k)$ of this section. The energy dependence of these extracted phase shifts is determined by varying the bombarding energy (and thus k) and then re-analyzing the data. For laboratory energies of less than 300 MeV, a particular n–p phase shift is positive, implying that the potential experienced by the n–p system is attractive, but around 300 MeV this phase shift goes to zero and then for larger energies becomes negative. This behavior is interpreted as follows: since only for $E_k \geqslant 300$ MeV can n–p scattering probe the very-small-distance behavior of the potential, the negative value of the phase shift means that there is a repulsive core to the n–p interaction. Hence, for the relevant total angular momentum, the short-range nuclear potential is attractive until a very small (core) radius r_c is reached; for $r < r_c$ the potential changes from attractive to repulsive. Other data are consistent with this conclusion. For references to, and details on, the experiments and extraction of the phase shifts, the reader may consult, e.g., Preston and Bhaduri (1975).

Let us consider the last of the above three questions next: what effect does a true cut-off potential have on the phase shifts? The answer is a semi-quantitative one, based on the total-cross-section relation

$$\sigma_T = \frac{4\pi}{k^2}\sum_\ell (2\ell+1)\sin^2[\delta_\ell(k)] \equiv \sum_\ell \sigma_T(\ell), \tag{15.80}$$

where

$$|\sigma_T(\ell)| \leqslant \frac{4\pi}{k^2}(2\ell+1). \tag{15.81}$$

Equation (15.81) expresses the so-called unitarity limit on $\sigma_T(\ell)$. Equation (15.80) suggests that, for a given k, there should be a maximum value of ℓ, say $\ell_{max}(k)$, such that

$$\delta_\ell(k) = 0, \qquad \ell > \ell_{max}(k). \tag{15.82}$$

The argument underlying (15.82) is simple: if (15.82) did not hold, i.e., if all the δ_ℓ were non-zero, then in general the sum would diverge. We therefore conclude that

$$\sigma_T \cong \frac{4\pi}{k^2} \sum_{\ell=0}^{\ell_{max}(k)} (2\ell+1)\sin^2[\delta_\ell(k)]. \tag{15.83}$$

Correspondingly, we have

$$f_k(\theta) \cong \frac{1}{k} \sum_{\ell=0}^{\ell_{max}(k)} \sqrt{4\pi(2\ell+1)}e^{i\delta_\ell(k)}\sin[\delta_\ell(k)]\, Y_\ell^0(\Omega). \tag{15.84a}$$

Even if ℓ_{max} is as small as 3, the angular distribution obtained from (15.84a) is not simple, and, as ℓ_{max} increases, which it will do as k increases, $f_k(\theta)$ becomes more complicated, suggesting that expressions for $f_k(\theta)$ other than (15.84a) are worth seeking. Nonetheless, it is the existence of ℓ_{max} that makes calculation of $\delta_\ell(k)$, $\ell \leqslant \ell_{max}$, such an attractive method for obtaining exact scattering results, a point noted previously.

To estimate ℓ_{max}, we use a semi-classical argument, noting first that classically, only projectiles whose impact parameters are less than or equal to the cut-off radius b of the potential will be deflected. Hence, for a given momentum p, the maximum angular momentum a deflected classical projectile may have is $l_{max} = pb$. Quantally, the relation between p and k is $p = \hbar k$; furthermore we replace ℓ by $\sqrt{\ell(\ell+1)}\hbar$, the square root of the eigenvalue of $\hat{L}^2$. Hence, semi-classically we infer that $\ell_{max}(k)$ is related to k and b by

$$\sqrt{\ell_{max}(k)[\ell_{max}(k)+1]} \cong bk. \tag{15.84b}$$

For non-cut-off, short-range potentials such as $V_0\exp(-r/b)$ and $V_0\exp(-r^2/b^2)$, the relation (15.84b) will still give a reasonable estimate of ℓ_{max}. Thus, for $k \ll b^{-1}$, we would expect that $\ell_{max} \cong 0$, i.e., that only S-wave scattering is significant. This is the case for neutron–proton scattering up to laboratory energies of about 5 MeV.

The low-energy regime is always of interest, so we ask next for the k dependence of $\delta_\ell(k)$ when k is very small (the second question above), in particular $k \ll b^{-1}$, where b characterizes the range of the potential. Referring to Eq. (15.78), the short-range nature of $V(r)$ means that only for values of $r \lesssim b$ will the integrand be non-zero, in which case the product kr is always very small. We therefore approximate $j_\ell(kr)$ and $R_\ell^{(+)}(kr)$ by their small-argument values. The analysis leading to Eqs. (11.59) applies to Eq. (15.62), – which $R_\ell^{(+)}(kr)$ obeys – and, since $R_\ell^{(+)}(kr)$ is well-behaved at the origin, we conclude that

$$R_\ell^{(+)}(kr) \propto (kr)^\ell, \qquad kr \ll 1.$$

The small-argument behavior of $j_\ell(kr)$ is given by (10.80):

$$j_\ell(kr) \propto (kr)^\ell, \qquad kr \ll 1,$$

the same as $R_\ell^{(+)}(kr)$. Hence for $k \ll b^{-1}$, these results plus (15.78) lead to

$$\sin[\delta_\ell(k)] \propto k^{2\ell+1}, \qquad k \ll b^{-1}. \tag{15.85}$$

Because $\sin\theta \cong 0$ suggests that $\theta \cong 0$, it is tempting to conclude from (15.85) that

$$\lim_{k\to 0} \delta_\ell(k) \to 0,$$

However the limit $k \to 0$ in (15.85) is satisfied also if $\delta_\ell(k=0) = n_\ell \pi$, $n_\ell = 0, 1, 2, \ldots$, and the latter zero-energy behavior does occur, albeit in a somewhat modified form. We consider only the $\ell = 0$ case (see, e.g., Newton (1982) for $\ell > 0$). The following result was first proved by Levinson (1949); it is known as Levinson's theorem:

$$\delta_0(k=0) = n_0 \pi \ (+\tfrac{1}{2}\pi), \tag{15.86}$$

where n_0 is the number of S-wave bound states supported by the potential $V(r)$ and the parenthetical factor of $\pi/2$ is to be added ONLY when there is a bound state with exactly zero energy, which is a rare situation. Three conditions must be satisfied for (15.86) to hold: $\int_0^\infty dr\, rV(r) < \infty$, $\int_0^\infty dr\, r^2 V(r) < \infty$, and $\delta_0(k \to \infty) \to 0$. We shall discuss the third condtion shortly.

Suppose that there is only one bound state, with $E < 0$. Then $\delta_0(k)$ is π at $k = 0$ and generally will fall monotonically to zero as $k \to \infty$. On the other hand, if there are no bound states and $V(r)$ is attractive, then, since δ_0 is zero both at $k = 0$ and at $k = \infty$, it must rise to some maximum value at $k > 0$ before falling towards zero at $k = \infty$.

Again suppose that the term $+\pi/2$ is absent from (15.86). The small-k behavior of (15.85) then implies that

$$\delta_0(k) - n_0\pi \propto k \tag{15.87}$$

and therefore that

$$\begin{aligned} f_0(k) &= \frac{1}{k}\sin[\delta_0(k)]\, e^{i\delta_0(k)} \\ &\underset{k\to 0}{\longrightarrow} \frac{1}{k}\sin[\delta_0(k)] \\ &\underset{k\to 0}{\longrightarrow} -a, \end{aligned} \tag{15.88}$$

where the quantity a, known as the scattering length,[5] is a real number characteristic of the particular potential. As a consequence of (15.88), $d\sigma/d\Omega$ is isotropic (angle-independent) at very low energies and both $d\sigma/d\Omega$ and σ_T are proportional to a^2, so

[5] An alternate form for $f_0(k)$ is $f_0(k) = 1/\{k\{\cot[\delta_0(k)] - i\}\}$, and the scattering length is often defined by

$$\lim_{k\to 0} k\cot[\delta_0(k)] = -1/a.$$

The appearance of the minus sign reflects the nuclear physics origin of the scattering length (see, e.g., Bethe and Morrison (1956)): when $V(r)$ produces a bound state, $-1/a < 0 \Rightarrow a > 0$.

that, for spinless particles, the magnitude of a can be determined in principle by one measurement:

$$\frac{d\sigma}{d\Omega} = a^2 \tag{15.89a}$$

and

$$\sigma_{\mathrm{T}} = 4\pi a^2. \tag{15.89b}$$

Analogous but more complex results obtain in the case of two spin-$\frac{1}{2}$ particles, see, e.g., Taylor (1972), Rodberg and Thaler (1967) or Newton (1982).

The final question we take up in this subsection concerns the behavior of $\delta_\ell(k)$ in the limit $k \to \infty$. This is an unachievable limit both practically and theoretically, the significance of the latter comment being that, as the energy becomes increasingly large, the non-relativistic theory treated in this text becomes increasingly inapplicable, and an alternate relativistic framework must be invoked. These remarks notwithstanding, it is reasonable mathematically to investigate the limit $k \to \infty$, one of whose consequences is Levinson's theorem.

The result

$$\delta_\ell(k \to \infty) \to 0 \tag{15.90}$$

is to be expected on physical grounds: as $k \to \infty$ so does E_k, and an extremely large kinetic energy implies negligible time spent in the vicinity of the short-range potential with a consequent lack of deflection, implying that $f_k(\infty) \to 0$, $\mathrm{R}_\ell^{(+)}(kr) \to j_\ell(kr)$.

A heuristic mathematical analysis suggesting Eq. (15.90) is based on the Neumann series and Eq. (15.78). Each term in the Neumann series (15.57) involves powers of $\hat{G}_k V$. However,

$$\hat{G}_k V = (E_k - \hat{K})^{-1} V,$$

and we expect that, in a coordinate representation, the matrix elements of V arising from the extremely-high-momentum components in $\hat{K}$ should average out to zero because of the rapid variation in $\exp(i\mathbf{q}' \cdot \mathbf{r})$ for q' large. Hence, only the non-infinite values of q' in $(E_k - \hat{K})^{-1} = \int d^3q' \, |q'\rangle\langle q'|[E_k - \hbar^2 q'^2/(2\mu)]$ will be significant, so that

$$\lim_{k\to\infty} \hat{G}_k V \to E_k^{-1} V \to 0,$$

for every factor $\hat{G}_k V$ in every power of $\hat{G}_k V$. Hence, all terms in the Neumann series but the leading one should go to zero, implying that $|\mathbf{k}, +\rangle \to |\mathbf{k}\rangle$ as $k \to \infty$. Thus, in Eq. (15.78), $R_\ell^{(+)}(kr) \to j_\ell(kr) \to \sin(kr - \ell\pi/2)/(kr)$ and we get

$$\sin[\delta_\ell(k)] \underset{k\to\infty}{\longrightarrow} \frac{1}{k}\left(-\frac{2\mu}{\hbar^2}\int_0^\infty dr \sin^2(kr - \ell\pi/2)\, V(r)\right) \to 0 \tag{15.90a}$$

as k^{-1}. The inference now is that, if $\sin[\delta_\ell(k)] \to 0$, then $\delta_\ell(k) \to 0$, but this leaves open the same possibility as that in Levinson's theorem, viz., that $\delta_\ell(k) \to n_\ell\pi + \beta k^{-1}$. However, any factors of $n_\ell\pi$ can be arranged to occur either at $k = 0$ OR at $k = \infty$, and the standard procedure is to *define* $\delta_\ell(k \to \infty) \to 0$, which then leads to the factors of π appearing only at $k = 0$. For details (and an involved mathematical analysis) the interested reader may consult, e.g., Newton (1982).

15.5. Some phase-shift calculations

Only a few potentials $V(r)$ allow exact scattering calculations; we consider two of the simplest ones in this section. They are the hard sphere or infinite repulsive core and its finite converse, the attractive square well, each of which is an example of a pure cut-off potential. We begin with a general description of scattering by an arbitrary cut-off potential.

The general cut-off potential

The only assumption now imposed on the otherwise well-behaved potential $V(r)$ of Eq. (15.62) is that it vanishes for $r > b$:

$$V(r) = 0, \qquad r > b. \tag{15.91}$$

It gives rise to an interior solution $R_\ell^{(+)}(kr)$, $r \leqslant b$. For $r > b$, $R_\ell^{(+)}(kr)$ obeys the free-particle equation, whose solution, from Eqs. (15.75), can be written as

$$R_\ell^{(+)}(kr) = \cos\delta_\ell\, j_\ell(kr) - \sin\delta_\ell\, n_\ell(kr). \tag{15.92}$$

The phase shift $\delta_\ell = \delta_\ell(k)$ is determined by matching the logarithmic derivatives of the inner and outer solutions, exactly as in the 3-D bound-state case:

$$\lim_{\epsilon\to 0}\left(\frac{1}{R_\ell^{(+)}(kr)}\frac{dR_\ell^{(+)}(kr)}{dr}\right)_{b-\epsilon} = \lim_{\epsilon\to 0}\left(\frac{1}{R_\ell^{(+)}(kr)}\frac{dR_\ell^{(+)}(kr)}{dr}\right)_{b+\epsilon}. \tag{15.93}$$

Since $V(r)$ is unspecified, the LHS of Eq. (15.93) is an unknown parameter depending on k, ℓ, and b. We write it as $\beta_\ell(k, b)$:

$$\beta_\ell(k, b) = \lim_{\epsilon\to 0}\left(\frac{1}{R_\ell^{(+)}(kr)}\frac{d}{dr}R_\ell^{(+)}(kr)\right)_{b-\epsilon}. \tag{15.94}$$

The RHS of Eq. (15.93) is evaluated using the exterior solution (15.92), yielding

$$\beta_\ell(k, b) = \frac{k\left(\cos\delta_\ell\,\dfrac{dj_\ell(\rho_b)}{d\rho_b} - \sin\delta_\ell\,\dfrac{dn_\ell(\rho_b)}{d\rho_b}\right)}{\cos\delta_\ell\, j_\ell(\rho_b) - \sin\delta_\ell\, n_\ell(\rho_b)}, \tag{15.95}$$

where $\rho_b = kb$. From (15.95), $\delta_\ell(k)$ can be determined, but only[6] mod π since $\tan(\theta \pm \pi) = \tan\theta$. Levinson's theorem can be used to fix $\delta_\ell(k)$, as long as $V(r)$ is sufficiently well-behaved.

Hard-sphere scattering

The hard sphere or infinitely repulsive potential is

$$V(r) = \begin{cases} \infty, & r \leqslant b, \\ 0, & r > b, \end{cases} \tag{15.96}$$

[6] $\tan\theta$ has an infinite number of branches, so in principle $\delta_\ell(k)$ is defined only to within $\pm n\pi$, $n = 1, 2, \ldots$. It is standard convention to include only the $n = 1$ branch.

for which $R_\ell^{(+)}(r) = 0$, $r \leqslant b$, and hence $\beta_\ell(k, b) = \infty$. Substituting the latter value into (15.95) gives

$$\tan[\delta_\ell^{(\mathrm{HS})}(k)] = \frac{j_\ell(kb)}{n_\ell(kb)}, \qquad \text{hard sphere.} \tag{15.97}$$

The S-wave case is especially easy:

$$\tan[\delta_0^{(\mathrm{HS})}(k)] = -\tan(kb),$$

and the absence of bound states suggests that

$$\delta_0^{(\mathrm{HS})}(k) = -kb, \tag{15.98}$$

although Levinson's theorem does not apply, since $\int dr\, r^n V(r) = \infty$, $n = 1, 2$. Equation (15.98) is the conventionally defined result. It yields a scattering length $a^{\mathrm{HS}} = -b$.

The S-wave scattering state is also readily obtained:

$$\begin{aligned} R_0^{(+)}(kr) &= \left(\frac{1}{2\pi}\right)^{3/2} e^{i\delta_0^{(\mathrm{HS})}}[\cos\delta_0^{(\mathrm{HS})}\, j_0(kr) - \sin\delta_0^{(\mathrm{HS})}\, n_0(kr)], \\ &= \left(\frac{1}{2\pi}\right)^{3/2} e^{-ikb}\frac{\sin(kr - kb)}{kr}, \qquad r > b, \end{aligned} \tag{15.99a}$$

whose $k \to 0$ limit is

$$R_0^{(+)}(kr) \underset{k\to 0}{\longrightarrow} \left(\frac{1}{2\pi}\right)^{3/2}(1 - b/r), \qquad r > b. \tag{15.99b}$$

For arbitrary ℓ the low-energy behavior of $\delta_\ell^{(\mathrm{HS})}(k)$ is determined by the small-argument behavior of $j_\ell(kb)$ and $n_\ell(kb)$:

$$\begin{aligned} \lim_{k\to 0}\tan[\delta_\ell^{(\mathrm{HS})}(k)] &\to -\frac{2^\ell(\ell!)}{(2\ell+1)!}(kb)^\ell \bigg/ \left(\frac{(2\ell)!}{2^\ell \ell!}\frac{1}{(kb)^{\ell+1}}\right) \\ &= -(kb)^{2\ell+1}\frac{2^{2\ell}(\ell!)^2}{(2\ell+1)[(2\ell)!]^2}, \end{aligned} \tag{15.100}$$

in agreement with Eq. (15.85). If Levinson's theorem held, then the RHS of (15.100) would be equal to the $k \to 0$ limit $\delta_\ell^{(\mathrm{HS})}(k)$. A direct demonstration that Levinson's theorem fails follows from examining the $k \to \infty$ limit of (15.97):

$$\lim_{k\to\infty}\tan[\delta_\ell^{(\mathrm{HS})}(k)] = -\frac{\sin(kb - \ell\pi/2)}{\cos(kb - \ell\pi/2)},$$

implying that

$$\delta_\ell^{(\mathrm{HS})}(k) \underset{k\to\infty}{\longrightarrow} -kb + \frac{\ell\pi}{2} + n\pi.$$

The failure is manifest, since the high-energy limit of $\delta_\ell^{(\mathrm{HS})}$ is ∞, not zero.

The attractive square well

With the well radius changed from r_0 to b, the square-well potential is that of Eq. (11.91) and Fig. 11.8:

$$V(r) = \begin{cases} -V_0, & r \leqslant b \\ 0, & r \geqslant b. \end{cases} \tag{15.101}$$

The interior-region, partial-wave equation (15.62) thus becomes

$$\left(\frac{1}{r^2}\frac{d}{dr}r^2\frac{d}{dr} - \frac{\ell(\ell+1)}{r^2} + \kappa^2\right)R_\ell^{(+)}(kr) = 0, \qquad r \leqslant b, \tag{15.102}$$

where

$$\kappa^2 = k^2 + \frac{2\mu}{\hbar^2}V_0. \tag{15.103}$$

Equation (15.102) has as solutions $j_\ell(\kappa r)$ and $n_\ell(\kappa r)$, but as with the bound-state problem, only $j_\ell(\kappa r)$ is allowed:

$$R_\ell^{(+)}(kr) = A_\ell j_\ell(\kappa r), \qquad r \leqslant b, \tag{15.104}$$

and we get

$$\beta_\ell(k, b) = \frac{\kappa[dj_\ell(\eta_b)/d\eta_b]}{j_\ell(\eta_b)}, \tag{15.105}$$

where

$$\eta_b = \kappa b.$$

Finally, letting a prime denote differentation with respect to the arguments $\rho_b = kb$ and $\eta_b = \kappa b$, we have

$$\tan[\delta_\ell(k)] = \frac{kj_\ell'(\rho_b)j_\ell(\eta_b) - \kappa j_\ell'(\eta_b)j_\ell(\rho_b)}{kj_\ell(\eta_b)n_\ell'(\rho_b) - \kappa j_\ell'(\eta_b)n_\ell(\rho_b)}. \tag{15.106}$$

A detailed evaluation of the RHS of (15.106) requires specification of V_0, b, and k, as in the exercises. However, we can explore several limiting situations. First, let $k \to \infty$, in which case $\kappa \to k$ (i.e., $E_k \gg V_0$), and $\eta_b \to \rho_b$. It is straightforward to show (see the exercises) that, in this limit, the RHS of (15.106) goes to zero, i.e.,

$$\lim_{k\to\infty} \tan[\delta_\ell(k)] \to 0,$$

thus leading to Eq. (15.90) by virtue of Levinson's theorem.

Second, let $\ell = 0$ and $k \to 0$, in which case $\rho_b \to 0$, $\kappa \to k_0$, and $\eta_b \to k_0 b$, with

$$k_0 = \sqrt{\frac{2\mu}{\hbar^2}V_0}, \tag{15.107}$$

while

$$\beta_0(k, b) \to k_0\left(\cot[k_0 b] - \frac{1}{k_0 b}\right). \tag{15.108}$$

Retaining only terms of order k, the RHS becomes $-k\beta_0 b^2/(1 + \beta_0 b)$ from which we get

$$\lim_{k\to 0} k\cot[\delta_0(k)] = -\frac{\beta_0 b^2}{1+\beta_0 b}. \tag{15.109}$$

Comparing with the standard definition of the scattering length[5] (p. 578), for the square well one has

$$a = \frac{\beta_0 b^2}{1+\beta_0 b} = \frac{k_0 b\cot(k_0 b) - 1}{k_0\cot(k_0 b)}, \tag{15.110a}$$

or equivalently

$$k_0\cot(k_0 b) = \frac{1}{b-a}. \tag{15.110b}$$

As noted in Section 15.4, the scattering length depends only on the properties of the potential. Now, from Eq. (11.112) and the analysis of the 1-D square well, Section 6.4, if $V_0 < \hbar^2\pi^2/(8\mu b^2)$ there are no bound states in the square well, whereas if $\hbar^2\pi^2/(8\mu b^2) < V_0 < 9\hbar^2\pi^2/(8\mu b^2)$ there is one bound state. In terms of $k_0 b$, we have ($\epsilon > 0$)

$$k_0 b = \begin{cases} \pi/2+\epsilon \Rightarrow 1 \text{ bound state}, \\ \pi/2-\epsilon \Rightarrow \text{ no bound states}, \end{cases}$$

and substituting into (15.110a) yields

$$a = \begin{cases} \text{positive, one bound state}, \\ \text{negative, no bound state}. \end{cases} \tag{15.111}$$

These results hold in general, not just for the square well, and are a motivation for the definition $k\cot[\delta_0(k)] \to -1/a$.

A third limit is that of $k_0 b = \pi/2$, i.e., $V_0 = \hbar^2\pi^2/(8\mu b^2)$. This is the limiting strength, approached downward from $V_0 > \hbar^2\pi^2/(8\mu b^2)$, for the existence of a zero-energy bound state. However, at $k_0 b = \pi/2$, $\cot(k_0 b) = 0$ and the scattering length becomes infinite:

$$a = \infty, \quad \text{when } V_0 \text{ is just strong enough to support a zero-energy bound state.} \tag{15.112}$$

Again, this result holds in general, not just for a square well.

Finally, we ask for the value of $k\cot[\delta_0(k)]$ when powers up to k^2 are retained in evaluating the RHS of (15.106). Interestingly enough, $k\cot[\delta_0(k)]$ has no term linear in k and we find

$$k\cot[\delta_0(k)] \cong -\frac{1}{a} + \frac{1}{2}r_0 k^2, \tag{15.113}$$

where the length r_0 is known as the *effective range*. As with Eqs. (15.111) and (15.112), (15.113) is valid for any well-behaved short-range potential, not just the square well (see, e.g., Galindo and Pascual (1990)), although the value of r_0 is not as easily calculated analytically for general $V(r)$ as it is for the square-well case (see the exercises). The effective range is not a mere theoretical construct: for neutron–proton scattering, there are spin-singlet and spin-triplet values of a and r_0, and all four can be (and have been) determined experimentally. Their values are used in helping to specify parameters of the neutron–proton interaction.

15.6. A weak-potential/high-energy approximation

Because it is difficult to carry out scattering calculations analytically, almost all analyses of collisions are based on approximations; the actual evaluations are typically numerical. We consider one of several popular approximation schemes in this chapter, viz., the Born approximation. Although it is formulated for weak potentials or high energies, this and other high-energy approximations have been applied outside their assumed regimes of validity – often because there is nothing else available. The reader should keep this caveat in mind, particularly when encountering analyses of collisions involving transfer of mass, especially when they are successful in fitting data!

The Born approximation is probably the most famous approximation in scattering theory and is valid when either the interaction is so weak or the energy (i.e., k) is so large that all terms in the Born series involving $(\hat{G}_k V)^n$, $n \geqslant 1$, can be safely ignored. The scattering amplitude then goes over to $f_{\mathbf{k}}^{(\mathrm{B})}(\theta)$,

$$f_k(\theta) \to f_k^{(\mathrm{B})}(\theta) = -\frac{(2\pi)^2\mu}{\hbar^2}\langle \mathbf{k}'|V|\mathbf{k}\rangle, \tag{15.60b}$$

while $\sin[\delta_\ell(k)]$ is approximated by $\sin[\delta_\ell^{(\mathrm{B})}(k)]$:

$$\sin[\delta_\ell(k)] \to \sin[\delta_\ell^{(\mathrm{B})}(k)] = -\frac{2\mu k}{\hbar^2}\int dr\,|rj_\ell(kr)|^2 V(r). \tag{15.79}$$

The partial-wave expansion of (15.60b) leads, of course, to (15.79). We shall concentrate on $f_k^{(\mathrm{B})}$.

It was noted in Section 5.4 (Eq. (5.69)) that the matrix element $\langle \mathbf{k}'|V|\mathbf{k}\rangle$ is related to the Fourier transform $\tilde{V}(\mathbf{k}-\mathbf{k}')$. In the present context, the difference $\mathbf{k}-\mathbf{k}'$ is the momentum transferred during the collision, a quantity we denote $\mathbf{q}$:

$$\mathbf{q} \equiv \mathbf{k} - \mathbf{k}', \tag{15.114a}$$

with $(k' = k)$

$$\begin{aligned} q &= \sqrt{2k^2(1-\cos\theta)} \\ &= 2k\sin\left(\frac{\theta}{2}\right), \end{aligned} \tag{15.114b}$$

where $\theta = \cos^{-1}(\mathbf{k}\cdot\mathbf{k}'/k^2)$ is the scattering angle; see Fig. 15.7.

For central potentials, the dependence on θ of $f_k^{(\mathrm{B})}$ occurs only through q:

$$f_k^{(\mathrm{B})}(\theta) = f^{(\mathrm{B})}(q) = -\frac{\mu}{2\pi\hbar^2}\int d^3r\, e^{i\mathbf{q}\cdot\mathbf{r}}\, V(r) = -\frac{2\mu}{\hbar^2 q}\int_0^\infty dr\, rV(r)\sin(qr). \tag{15.115}$$

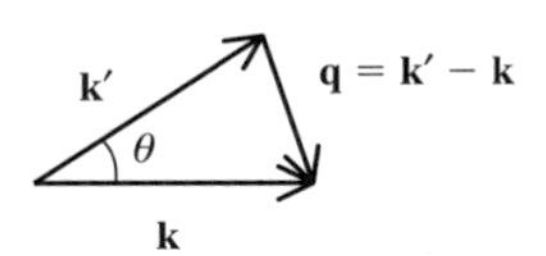

Fig. 15.7 An illustration of the momentum transfer $\mathbf{q} = \mathbf{k} - \mathbf{k}'$, where $k' = k$ and θ is the scattering angle.

There are three noteworthy features to this result. The first, mentioned just above (15.115), is that the entire angular dependence enters through q: an expansion in partial waves is unnecessary. Second, $f^{(\mathrm{B})}(q)$ is real, and, although σ_{T} can be calculated, the optical theorem is not satisfied. Third, evaluation of $f_k^{(\mathrm{B})}(\theta)$ requires only that a Fourier sine transform (of $rV(r)$) be carried out, i.e., evaluation of $f_k^{(\mathrm{B})}(\theta)$ has been reduced to a quadrature. In contrast to $\delta_\ell(k)$, which can be determined analytically for only a few potentials, $f_k^{(\mathrm{B})}(\theta)$ can be evaluated for a wide range of potentials, a few examples of which are given in the following.

EXAMPLE 1

The $1/r$ potential. Probably the most famous instance of this potential is the Coulomb interaction between charges $\pm Z_1 e$ and $\mp Z_2 e$:

$$V^{(\mathrm{Coul})}(r) = -\kappa_{\mathrm{e}} \frac{Z_1 Z_2 e^2}{r}. \tag{15.116a}$$

Substitution of Eq. (15.116a) into Eq. (15.115) leads to the ill-defined integral $\int dr \sin(qr)$. The standard method for dealing with this problem is to alter $V^{(\mathrm{Coul})}(r)$ by means of a convergence factor, typically a decaying exponential, thereby turning $V^{(\mathrm{Coul})}(r)$ into a screened Coulomb or Yukawa potential for the purpose of carrying out the integral:

$$V^{(\mathrm{Coul})}(r) \to V_{\mathrm{s.c.}}^{(R)}(r) = e^{-r/R} V^{(\mathrm{Coul})}(r). \tag{15.116b}$$

For any finite R, $V_{\mathrm{s.c.}}^{(R)}(r)$ is short-ranged, the integrand is well-behaved, and one finds

$$\begin{aligned} f_{\mathrm{s.c.}}^{(\mathrm{B})}(q) &= -\frac{2\mu}{\hbar^2 q} \int_0^\infty dr\, r V_{\mathrm{s.c.}}^{(R)}(r) \sin(qr) \\ &= \left(\frac{\kappa_{\mathrm{e}} Z_1 Z_2 e^2}{\hbar^2/(2\mu)} \right) \frac{1}{q^2 + 1/R^2} \\ &= \left(\frac{\kappa_{\mathrm{e}} Z_1 Z_2 e^2}{4E_k} \right) \frac{1}{\sin^2(\theta/2) + 1/(4k^2R^2)}, \end{aligned} \tag{15.117}$$

where $E_k = \hbar^2 k^2/(2\mu)$ is the kinetic energy.

As long as R is finite, the amplitude (15.117) and its differential cross section are each a finite, monotonically decreasing function of θ. The most physically interesting case, however, is that of pure Coulomb scattering, which is obtained in the zero-screening limit, $R \to 0$:

$$\frac{d\sigma^{(\mathrm{Coul})}}{d\Omega} = \left(\frac{\kappa Z_1 Z_2 e^2}{4E_k} \right)^2 \frac{1}{\sin^4(\theta/2)}, \tag{15.118}$$

which is identical to the classical Coulomb-cross-section result first obtained by Rutherford (1911)! Even more remarkable is the fact that (15.118) is also the *exact* quantal result!! The "only" effect of adding all the higher-order Born-approximation contributions to (15.117) is to multiply the (real) Born amplitude by a phase factor, whose absolute square is unity, but whose argument becomes infinite in the zero-

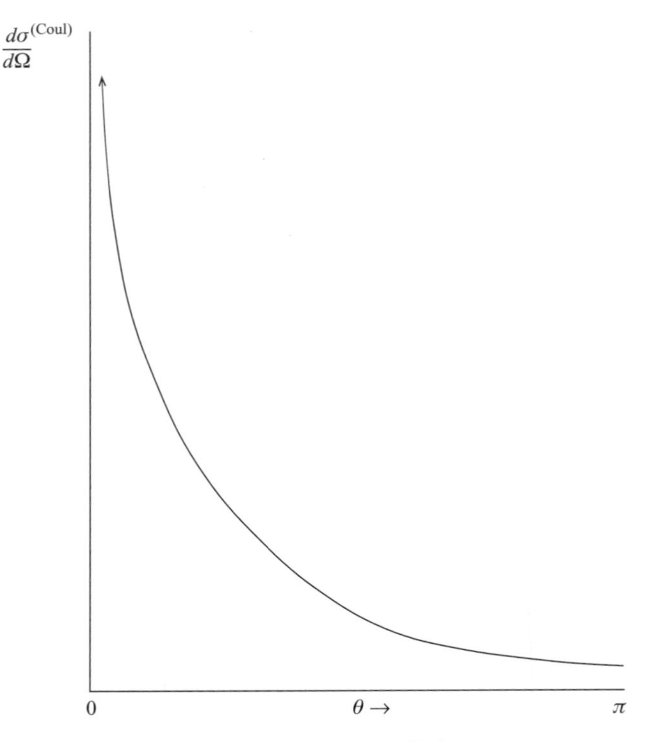

Fig. 15.8 A sketch of the θ dependence of $d\sigma^{(\mathrm{Coul})}/d\Omega$. The infinity at $\theta = 0$ is suggested by the arrow on the upper-left-hand portion of the curve.

screening limit. This feature was first discussed by Dalitz (1951). The interested reader is referred to this paper or to Newton (1982), where the infinity in the second Born approximation as well as the exact amplitude and cross section for Coulomb scattering are examined.

In contrast to the $R \neq 0$ case, the Coulomb cross section (15.118) is infinite in the forward direction ($\theta = 0$); the total cross section $\sigma_{\mathrm{T}}^{(\mathrm{Coul})}$ is also infinite. (See Fig. 15.8 for a sketch of $d\sigma^{(\mathrm{Coul})}/d\Omega$.) These infinities arise from the long-range nature of the $1/r$ potential: there is never a distance beyond which the effect of the interaction is negligible and therefore scattering occurs at all r. Further consequences are that, asymptotically, the incident wave is not a plane wave, and the scattering state does not obey a L–S equation for Coulomb scattering.

EXAMPLE 2

The square well. As our second example we return to the square-well potential, Eq. (15.101). Like the delta-shell potential, $V(r) = -V_0\delta(r - b)$, the square well is a true cut-off potential, in contrast to smooth potentials such as the Yukawa, exponential, and Gaussian ones. As with the Yukawa case, the Born amplitudes for the latter two potentials are monotonically decreasing functions of θ (see the exercises), whereas the delta-shell and square-well potentials can, in principle, display a diffraction-like behavior due to their sharp-edged cut-off nature, as we now show for the square well.

Substituting (15.101) into (15.115) and using $\int dy\, y \sin y = \sin y - y \cos y$, the Born approximation for attractive square-well scattering is

$$f_k^{(\mathrm{B})}(\theta) = \frac{V_0}{\hbar^2/(2\mu b^2)}\, b\, \frac{j_1(qb)}{qb}. \tag{15.119}$$

The occurrence of the dimensionless factor $2\mu V_0 b^2/\hbar^2$ is typical, as is the combination $V_0 b^2$, which is a standard measure of the strength of the interaction w.r.t. the scattering amplitude. The angular dependence is contained in $qb = 2bk \sin(\theta/2)$, which is a monotonically increasing function of θ. Although $qb = 0$ at $\theta = 0$, $j_1(qb)/(qb)$ is finite there:

$$\frac{j_1(qb)}{qb} \underset{qb\to 0}{\longrightarrow} \frac{1}{3}.$$

Furthermore, from Table 11.1, $j_1(qb)$ vanishes for $qb \cong 4.5$, 8.0, 10.9, and 14.1, as well as at larger values. Thus, depending on the magnitude of kb, the square-well differential cross section $d\sigma^{\mathrm{sq.w}}/d\Omega$ in the Born approximation may go to zero one or more times.

The foregoing analysis clearly suggests diffraction-like behavior (or, more generally, an interference phenomenon), as noted above. However, $j_1(qb)/(qb)$ decreases sufficiently rapidly that, even if kb allows the second peak at $qb \cong 5.94$ to occur, almost all of the cross section will be contained at angles less than $\theta_1 \cong 2\sin^{-1}[4.5/(kb)]$. The third peak occurs at $qb \cong 9.2$, and the values of $|f_k^{(\mathrm{B})}(\Omega)|$ for the first, second, and third peaks are in the ratios $1 : 0.028 : 0.012$, so that the angular distribution decreases by factors of approximately 140 and 770 for the second and third peaks, respectively.

A sketch of $j_1(qb)/(qb)$ versus qb illustrating these remarks is shown in Fig. 15.9. Even though the cross section decreases by about three orders of magnitude at the third peak, such decreases are normally well within reach of experimental measurement. Indeed, measurements of cross sections over a range of eight or more orders of magnitude have routinely been carried out.

To get a feeling for the angles involved, we first choose $k = 4b^{-1}$ and then $k = 6b^{-1}$, for which $\theta_1 \cong 68.5^\circ$ and $\theta_1 \cong 44^\circ$, respectively. For a square-well model of neutron–proton scattering with $b = 1$ fm $= 10^{-15}$ m, the $k = 4b^{-1}$ and $k = 6b^{-1}$ values yield (CM) kinetic energies of approximately 670 and 1500 MeV(!!), values for which the Born approximation should be valid, although non-relativistic quantum mechanics is not. On the other hand, if the electron–atom interaction is modeled by a square well – which is not a good choice by the way – of size $b = 1$ Å, then, for $k = 6b^{-1}$, the incident kinetic energy is roughly 135 eV, for which the Born approximation is less likely to be accurate.

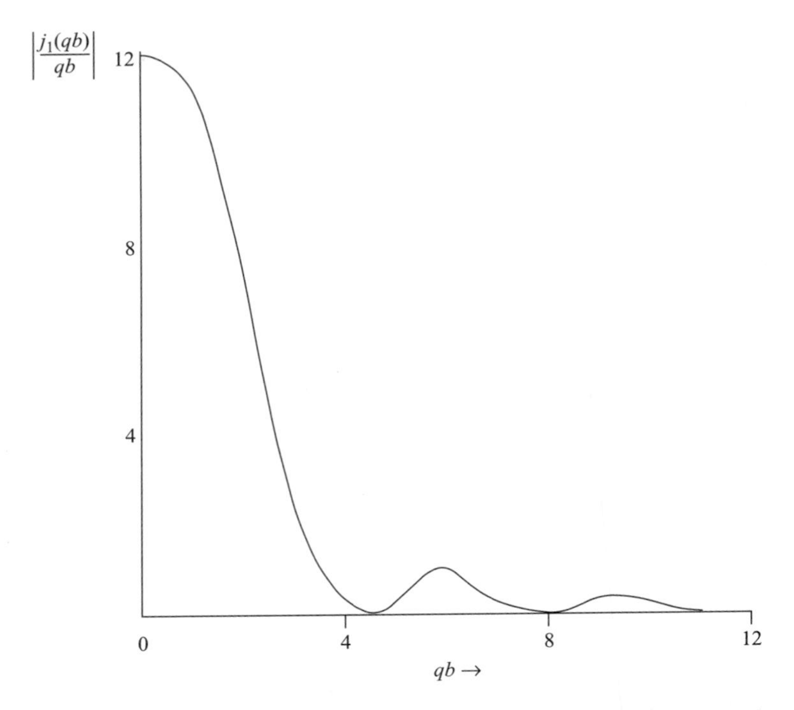

Fig. 15.9 A sketch of the qb behavior of $|j_1(qb)/qb|$, the angle-dependent portion of $f_k^{(B)}$ for scattering by a 3-D square-well of range b; $q = 2k\sin(\theta/2)$ is the momentum transfer. The diffraction-like character is evident.

15.7. *Spin-$\frac{1}{2}$ projectiles

In most collision experiments the projectiles and targets have non-zero spin. As an introduction to this aspect of the theory we consider the simplest case: the scattering of a spin-$\frac{1}{2}$ projectile by a spin-0 target. For the general analysis of the first subsection we impose the following restrictions: Hamiltonians that are invariant under rotations and reflections; random (equal) distribution of the states α_z and β_z in the incident beam; and no measurements of spin-projection (i.e., of m_s in $|\frac{1}{2}m_s\rangle$). The last restriction means that we shall not be concerned with the polarization vector (a quantity related to the expectation value of $\hat{\boldsymbol{\sigma}}$), proper consideration of which involves the density operator, a quantity beyond the scope of this book.[7]

General considerations

Suppose first that the Hamiltonian is spin-independent. In this instance the initial state $|\mathbf{k}\rangle|\frac{1}{2}m_s\rangle$ evolves into $|\mathbf{k}, +\rangle|\frac{1}{2}m_s\rangle$: $\hat{U}(t) = \exp(-i\hat{H}t/\hbar)$, being spin-independent, cannot

[7] Polarization analyses are dealt with in more advanced quantum texts and scattering-theory monographs, e.g., Sakurai (1994), Galindo and Pascual (1990), Newton (1982), Merzbacher (1970), Rodberg and Thaler (1967), Gottfried (1966), etc.

change the initial spin state; spin thus plays no dynamical role, and the preceding results apply with no change.

Hence, for spin to influence the scattering, $\hat{H}$ must include $\hat{\mathbf{S}}$. Rotational invariance requires that $\hat{\mathbf{S}}$ occur in a scalar product, and, since $|\frac{1}{2}m_s\rangle$ is an eigenstate of $\hat{S}^2 = \hat{\mathbf{S}} \cdot \hat{\mathbf{S}}$, the only other operator possibilities are $\hat{\mathbf{L}} \cdot \hat{\mathbf{S}}$, $\hat{\mathbf{Q}} \cdot \hat{\mathbf{S}}$, and $\hat{\mathbf{P}} \cdot \hat{\mathbf{S}}$. Since $\hat{\mathbf{L}} \cdot \hat{\mathbf{S}}$ as well as the nuclear and electromagnetic interactions of interest in this book are invariant under reflection but $\hat{\mathbf{P}} \cdot \hat{\mathbf{S}}$ and $\hat{\mathbf{Q}} \cdot \hat{\mathbf{S}}$ are not, rotational and reflection invariance restricts the spin-dependent interaction to the spin–orbit form $\hat{H}_{LS} = V_{LS}(r)\hat{\mathbf{L}} \cdot \hat{\mathbf{S}}/\hbar^2$. Hence the Hamiltonian we shall work with is

$$\hat{H}(\mathbf{r}) = \hat{K}(\mathbf{r}) + V(r) + V_{LS}(r)\hat{\mathbf{L}} \cdot \hat{\mathbf{S}}/\hbar^2; \tag{15.120}$$

compare this with Eq. (14.80) *et seq.*

Both $V(r)$ and $V_{LS}(r)$ are assumed to be short-ranged, and thus all of the spin-independent analysis leading to $\psi_{\mathbf{k}}^{(+)}(r)$ and $f_{\mathbf{k}}(\theta)$ can be carried over to the present case. We shall not carry out this analysis in detail, but instead shall use physical and further invariance arguments in deriving the results.

Let us assume the projectile to be in the initial spin state $|\frac{1}{2}m_s\rangle$. We begin by noting that, in contrast to the incident wave $|\mathbf{k}\rangle|\frac{1}{2}m_s\rangle$, the full scattering state will contain both $|\frac{1}{2}m_s\rangle$ and $|\frac{1}{2}, -m_s\rangle$. This follows from the fact that $\hat{\mathbf{L}} \cdot \hat{\mathbf{S}}$ requires working with the coupled states, and, for $\ell \neq 0$, each of $|\frac{1}{2}, \pm m_s\rangle$ can be used with an appropriate $|\ell, m_\ell\rangle$ to form $|\ell\frac{1}{2}; jm_j\rangle$. As a consequence, in the asymptotic region where the scattered part of the wave function contains only outgoing waves, there will be not one but two scattering amplitudes, e.g., $f_k(\theta)$ corresponding to a final spin state equal to the initial spin state $|\frac{1}{2}m_s\rangle$, and $g_k(\theta)$ corresponding to a final spin state $|\frac{1}{2}, -m_s\rangle$. This will be true irrespective of whether $m_s = +\frac{1}{2}$ or $m_s = -\frac{1}{2}$. $f_k(\theta)$ is known as the "no-flip" amplitude, while $g_k(\theta)$ is the "spin-flip" amplitude.

Still assuming $|\frac{1}{2}m_s\rangle$ as the initial spin state, we write the full scattering state as $|\mathbf{k}, m_s, +\rangle$, where the label m_s is a book-keeping device, not a final-state quantum number. Then the mathematical statement of the foregoing asymptotic argument is

$$(2\pi)^{3/2}\{\langle\tfrac{1}{2}, m_s|\langle\mathbf{r}|\}|\mathbf{k}, m_s, +\rangle \underset{r\to\infty}{\sim} e^{i\mathbf{k}\cdot\mathbf{r}} + f_k(\theta)e^{ikr}/r \tag{15.121a}$$

and

$$(2\pi)^{3/2}\{\langle\tfrac{1}{2}, -m_s|\langle\mathbf{r}|\}|\mathbf{k}, m_s, +\rangle \underset{r\to\infty}{\sim} g_k(\theta)e^{ikr}/r. \tag{15.121b}$$

There can be no plane-wave term in (15.121b), since $\langle\mathbf{r}|\mathbf{k}\rangle$ occurs only in connection with $|\frac{1}{2}m_s\rangle$. Given that $f_k(\theta)$ and $g_k(\theta)$ correspond, respectively, to no spin flip and spin flip, then $\sigma^{\text{nf}}(\theta) == |f_k(\theta)|^2$ and $\sigma^{\text{f}}(\theta) = |g_k(\theta)|^2$ are the corresponding angular distributions. They refer to final spin states $|\frac{1}{2}m_s\rangle$ and $|\frac{1}{2}, -m_s\rangle$. However, in this subsection we are assuming that the detectors are not sensitive to the magnetic quantum number. Hence, the detector will count both the m_s and the $-m_s$ particles. Therefore the measured cross section is the (incoherent) sum of the no-flip and the spin-flip cross sections:

$$\sigma(\theta) = \sigma^{\text{nf}}(\theta) + \sigma^{\text{f}}(\theta) = |f_k(\theta)|^2 + |g_k(\theta)|^2. \tag{15.121c}$$

Next, let the initial spin state be the linear combination

$$|\gamma\rangle = a|\tfrac{1}{2}\tfrac{1}{2}\rangle + b|\tfrac{1}{2}, -\tfrac{1}{2}\rangle. \qquad (13.15a)$$

(If $|a|^2 = |b|^2 = \frac{1}{2}$, then $\langle \hat{S}_z \rangle = 0$, Eq. (13.15c).) The initial state is $|\mathbf{k}\rangle|\gamma\rangle$ and the scattering state is $|\mathbf{k}, \gamma, +\rangle$. Let us use the eigenstate representation, wherein

$$|\gamma\rangle \rightarrow \begin{pmatrix} \gamma_1 \\ \gamma_2 \end{pmatrix} = \begin{pmatrix} a \\ b \end{pmatrix}. \qquad (13.15b)$$

The asymptotic forms (15.121) will now be replaced by a spin-operator expression:

$$(2\pi)^{3/2}\langle \mathbf{r}|\mathbf{k}, \gamma, +\rangle \underset{r\rightarrow\infty}{\sim} e^{i\mathbf{k}\cdot\mathbf{r}} \begin{pmatrix} a \\ b \end{pmatrix} + \frac{e^{ikr}}{r} M_k(\theta) \begin{pmatrix} a \\ b \end{pmatrix}, \qquad (15.122)$$

where $M_k(\theta)$ is a 2×2, spin-dependent, scattering-amplitude matrix.

We shall determine the general form of $M_k(\theta)$ from the requirement that it, like $\hat{H}$, be invariant under rotations and reflections. M_k can depend on the initial and final wave vectors $\mathbf{k}$ and $\mathbf{k}'$ and on the spin matrix $\mathbf{S}$, which latter dependence we write as $\boldsymbol{\sigma}$ $(= 2\mathbf{S}/\hbar)$:

$$M_k(\theta) = M_k(\theta, \mathbf{k}, \mathbf{k}', \boldsymbol{\sigma}).$$

Rotational invariance means that the dependences on the three vectors $\mathbf{k}$, $\mathbf{k}'$, and $\boldsymbol{\sigma}$ may occur only via scalar products, and, since the k and θ dependences have already been indicated, $\mathbf{k}\cdot\mathbf{k}'$ can be eliminated, as can $\mathbf{k}\cdot\boldsymbol{\sigma}$ and $\mathbf{k}'\cdot\boldsymbol{\sigma}$ since they each change sign under reflections. Given that $\boldsymbol{\sigma}$ is an axial vector, as is $\mathbf{n} = \mathbf{k}' \times \mathbf{k}/k^2$, then only the scalar product $\boldsymbol{\sigma}\cdot\mathbf{n}$ is allowed. We therefore write $M_k(\theta)$ in the form

$$M_k(\theta) = I_2 f_k(\theta) + g_k(\theta)\mathbf{n}\cdot\boldsymbol{\sigma}, \qquad (15.123)$$

where I_2 is the 2×2 unit matrix.

The choice of symbols $f_k(\theta)$ and $g_k(\theta)$ for the k- and θ-dependent amplitudes is deliberate, since they play the same role, in effect, as the amplitudes occurring in (15.121). To prove this, we choose $\mathbf{k}$ and $\mathbf{k}'$ as before, viz., $\mathbf{k} = k\mathbf{e}_3$ and $\mathbf{k}'$ in the $\mathbf{e}_1$–$\mathbf{e}_3$ plane at an angle θ w.r.t. $\mathbf{e}_3$. Then $\mathbf{n} = \mathbf{e}_2$, $\mathbf{n}\cdot\boldsymbol{\sigma} = \boldsymbol{\sigma}_2$ and we have

$$M_k(\theta) = \begin{pmatrix} f_k(\theta) & -ig_k(\theta) \\ ig_k(\theta) & f_k(\theta) \end{pmatrix}. \qquad (15.124)$$

The probability amplitude $A_\alpha(k, \theta)$ that, after scattering the particle is in spin state α_z is

$$A_\alpha(k, \theta) = (1\ 0)M_k(\theta)\begin{pmatrix} a \\ b \end{pmatrix}$$

$$= f_k(\theta)a - ig_k(\theta)b. \qquad (15.125a)$$

Analogously, the amplitude $A_\beta(k, \theta)$ that the final spin state is measured to be β_z is

$$A_\beta(k, \theta) = (0\ 1)M_k(\theta)\begin{pmatrix} a \\ b \end{pmatrix}$$

$$= ig_k(\theta)a + f_k(\theta)b. \qquad (15.125b)$$

A_α and A_β are the spin-up and the spin-down amplitudes; the corresponding cross sections are

$$\sigma^{\rm u}(\theta) = |A_\alpha(k,\,\theta)|^2 = |a|^2|f_k(\theta)|^2 + |b|^2|g_k(\theta)|^2 \tag{15.125c}$$

and

$$\sigma^{\rm d}(\theta) = |A_\beta(k,\,\theta)|^2 = |b|^2|f_k(\theta)|^2 + |a|^2|g_k(\theta)|^2. \tag{15.125d}$$

These reduce to the previous no-flip and spin-flip cross sections when the initial state is either α_z ($a \to 1,\ b \to 0$) or β_z. The cross section when spin magnetic quantum numbers are not measured is the sum of $\sigma^{\rm u}$ and $\sigma^{\rm d}$:

$$\sigma(\theta) = \sigma^{\rm u}(\theta) + \sigma^{\rm d}(\theta) = |f_k(\theta)|^2 + |g_k(\theta)|^2, \tag{15.125e}$$

exactly as before, where the normalization condition $|a|^2 + |b|^2 = 1$ has been used.

The L–S equation and the Born approximations

A L–S equation for the case of spin-$\frac{1}{2}$ projectiles is readily obtained, using essentially the same procedures as in Section 15.3. Let us assume a definite spin state for the projectile: $|\frac{1}{2}m_s\rangle$. We write $|\mathbf{k},\,m_s,\,+\rangle$ as the sum of a plane wave plus a scattered wave,

$$|\mathbf{k},\,m_s,\,+\rangle = |\mathbf{k}\rangle|\tfrac{1}{2}m_s\rangle + |\mathbf{k},\,m_s,\,\mathrm{sc}\rangle; \tag{15.126}$$

substitute (15.126) into $(E_k - \hat{H})|\mathbf{k},\,\gamma,\,+\rangle = 0$, where $\hat{H}$ is given by (15.120); and then extract the following equation for $|\mathbf{k},\,\gamma,\,\mathrm{sc}\rangle$:

$$(E_k - \hat{K})|\mathbf{k},\,m_s,\,\mathrm{sc}\rangle = (V + V_{LS}\hat{\mathbf{L}}\cdot\hat{\mathbf{S}}/\hbar^2)|\mathbf{k},\,m_s,\,+\rangle. \tag{15.127a}$$

Since $\langle\mathbf{r}|\mathbf{k},\,m_s,\,+\rangle$ must be outgoing for r asymptotic, the relevant inverse for (15.127a) is the operator $\hat{G}_k^{(+)}$ whose coordinate-space matrix element is given by (15.50). That is, the presence of spin does not affect the choice of Green's function:

$$|\mathbf{k},\,m_s,\,\mathrm{sc}\rangle = \hat{G}_k^{(+)}(V + V_{LS}\hat{\mathbf{L}}\cdot\hat{\mathbf{S}}/\hbar^2)|\mathbf{k},\,m_s,\,+\rangle. \tag{15.127b}$$

With the solution (15.127b) replacing $|\mathbf{k},\,m_s,\,\mathrm{sc}\rangle$ in Eq. (15.126) we get the abstract form of the L–S equation:

$$|\mathbf{k},\,m_s,\,+\rangle = |\mathbf{k}\rangle|\tfrac{1}{2}m_s\rangle + \hat{G}_k^{(+)}(V + V_{LS}\hat{\mathbf{L}}\cdot\hat{\mathbf{S}}/\hbar^2)|\mathbf{k},\,m_s,\,+\rangle. \tag{15.128}$$

Its coordinate-space version is

$$\langle\mathbf{r}|\mathbf{k},\,m_s,\,+\rangle = \langle\mathbf{r}|\mathbf{k}\rangle|\tfrac{1}{2}m_s\rangle - \frac{\mu}{2\pi\hbar^2}\int d^3r'\,\frac{e^{ik|\mathbf{r}-\mathbf{r}'|}}{|\mathbf{r}-\mathbf{r}'|}[V(r') + V_{LS}(r')\hat{\mathbf{L}}\cdot\hat{\mathbf{S}}/\hbar^2]\langle\mathbf{r}'|\mathbf{k},\,m_s,\,+\rangle \tag{15.129a}$$

$$\underset{r\to\infty}{\sim}\left(\frac{1}{2\pi}\right)^{3/2}\left[e^{i\mathbf{k}\cdot\mathbf{r}}|\tfrac{1}{2}m_s\rangle + \frac{e^{ikr}}{r}\left(-\frac{(2\pi)^2\mu}{\hbar^2}\int d^3r'\,\frac{e^{-i\mathbf{k}'\cdot\mathbf{r}'}}{(2\pi)^{3/2}}\right.\right.$$

$$\left.\left.\times\,[V(r') + V_{LS}(r')\hat{\mathbf{L}}\cdot\hat{\mathbf{S}}/\hbar^2]\langle\mathbf{r}'|\mathbf{k},\,m_s,\,+\rangle\right)\right]. \tag{15.129b}$$

Comparing Eqs. (15.121) and (15.129), integral expressions for $f_k(\theta)$ and $g_k(\theta)$ are easily obtained. We write them in matrix-element form:

$$f_k(\theta) = -\frac{(2\pi)^2\mu}{\hbar^2}\langle\mathbf{k}',\,\tfrac{1}{2}m_s|(V + V_{LS}\mathbf{L}\cdot\mathbf{S}/\hbar^2)|\mathbf{k},\,m_s,\,+\rangle \tag{15.130a}$$

and

$$g_k(\theta) = -\frac{(2\pi)^2\mu}{\hbar^2}\langle \mathbf{k}', \tfrac{1}{2}, -m_s|(V + V_{LS}\hat{\mathbf{L}}\cdot\hat{\mathbf{S}}/\hbar^2)|\mathbf{k}, m_s, +\rangle. \tag{15.130b}$$

These expressions are such straightforward analogs to (15.56) that they could almost have been written down without recourse to the analysis which led to them. In fact, the whole analysis is paradigmatic, readily generalized to the scattering of two particles with spins S_1 and S_2, a topic considered in the exercises.

For sufficiently high energies or weak potentials (V and V_{LS}), the Born approximations, obtained via $|\mathbf{k}, m_s, +\rangle \to |\mathbf{k}\rangle|\frac{1}{2}m_s\rangle$, are

$$f_k(\theta) \to f_k^{(\mathrm{B})}(\theta) = -\frac{(2\pi)^2\mu}{\hbar}\langle \mathbf{k}', \tfrac{1}{2}m_s|(V + V_{LS}\hat{\mathbf{L}}\cdot\hat{\mathbf{S}}/\hbar^2)|\mathbf{k}, \tfrac{1}{2}m_s\rangle \tag{15.131a}$$

and

$$g_k(\theta) \to g_k^{(\mathrm{B})}(\theta) = -\frac{(2\pi)^2\mu}{\hbar}\langle \mathbf{k}', \tfrac{1}{2}, -m_s|(V_{LS}\hat{\mathbf{L}}\cdot\hat{\mathbf{S}}/\hbar^2)|\mathbf{k}, \tfrac{1}{2}m_s\rangle. \tag{15.131b}$$

Just as in the spinless-projectile case, one can introduce partial waves and phase shifts for spin-$\frac{1}{2}$ projectiles, and we refer the reader to Exercise 15.22 and the references already cited for details.

Exercises

15.1 Determine $\Psi^{\mathrm{F}}(\mathbf{r}, t = 0)$ for the following 3-D Gaussian profile functions:
(a) $A(\mathbf{k}) = N\exp[-(\mathbf{k} - \mathbf{k}_0)^2 d^2/2]$
(b) $A(\mathbf{k}) = N\exp[-(k - k_0)^2 d^2/2]$,
where in each case N is a real normalization to be evaluated. Compare and contrast your results with each other and with the 1-D case of Section 7.1. (Hints: use the analysis leading to Eq. (7.35) and differentiate under the integral sign if you cannot find the relevant integrals in an integral table.)

15.2 Derive Eq. (15.8) and from it obtain Eq. (15.27).

15.3 Calculate and compare with one another the probability current densities corresponding to the two $\Psi^{\mathrm{F}}(\mathbf{r}, t = 0)$ of Exercise 15.1.

15.4 In addition to the outgoing-wave solution $\psi_{\mathbf{k}}^{(+)}(\mathbf{r})$ obeying Eq. (15.23), there is a second solution denoted $\psi_{\mathbf{k}}^{(-)}(\mathbf{r})$ whose asymptotic form is

$$\psi_{\mathbf{k}}^{(-)}(\mathbf{r}) \underset{r\to\infty}{\sim} (2\pi)^{-3/2}[\exp(i\mathbf{k}\cdot\mathbf{r}) + g_{\mathbf{k}}(\Omega)\exp(-ikr)/r].$$

It gives rise to the wave packet

$$\Psi^{(-)}(\mathbf{r}, t) = \int d^3k\, A(\mathbf{k})\psi_{\mathbf{k}}^{(-)}(\mathbf{r})\exp(-i\omega_k t).$$

Show that $\Psi^{(-)}(\mathbf{r}, t)$ does not obey Eq. (15.11) and is therefore unacceptable as a possible wave packet describing the collision.

15.5 It can be shown that the ingoing-wave solution $\psi_{\mathbf{k}}^{(-)}(\mathbf{r})$ of the preceding exercise obeys $\psi_{\mathbf{k}}^{(-)*}(\mathbf{r}) = \psi_{-\mathbf{k}}^{(+)}(\mathbf{r})$.

(a) Use the latter relation to determine the partial-wave expansion of the amplitude $g_{\mathbf{k}}(\Omega)$ via the phase shifts $\delta_\ell(k)$. (See Exercise 15.4.)

(b) Let $F_\ell^{(-)}(kr)$ be the partial-wave analog in $\psi_{\mathbf{k}}^{(-)}(\mathbf{r})$ of the quantity $F_\ell^{(+)}(kr)$ appearing in Eqs. (15.61). Determine the asymptotic form of $R_\ell^{(-)}(kr)$, analogous to Eqs. (15.73).

15.6 Proof of the optical theorem, Eq. (15.38). The key to the proof is that the total probability $\int d^3r\,|\Psi^{(+)}(\mathbf{r},\,t)|^2$ is independent of time as long as $\hat{H} \neq \hat{H}(t)$.

(a) From the facts that $\Psi^{(+)}(\mathbf{r},\,t) \to \Psi^{(\mathrm{F})}(\mathbf{r},\,t)$ in the limit $t \to -\infty$ and $\Psi^{(+)}(\mathbf{r},\,t) \to \Psi^{(\mathrm{F})}(\mathbf{r},\,t) + \Psi^{(\mathrm{out})}(\mathbf{r},\,t)$ in the limit $t \to \infty$, show that, as t becomes very large,

$$\int d^3r\,|\Psi^{(\mathrm{out})}(\mathbf{r},\,t)|^2 = -\int d^3r\,[\Psi^{(\mathrm{F})*}(\mathbf{r},\,t)\Psi^{(\mathrm{out})}(\mathbf{r},\,t) + \Psi^{(\mathrm{F})}(\mathbf{r},\,t)\Psi^{(\mathrm{out})*}(\mathbf{r},\,t)].$$

(b) Making use of the facts that $\Psi^{(\mathrm{out})}(\mathbf{r},\,t)$ is a wave packet spreading radially out from $r = 0$ while $\Psi^{(\mathrm{F})}(\mathbf{r},\,t)$ travels along the direction $\mathbf{k}_0$, so that only the $\theta = 0$ portion of $f_{k_0}(\Omega)$ survives asymptotically when $t \to \infty$, show that the LHS of the above equation equals $\sigma_{\mathrm{T}}\int dr\,|\Psi^{(\mathrm{F})}(\mathbf{n}_0 r,\,t)|^2$ while the RHS becomes $(4\pi/k_0)\,\mathrm{Re}[(f_{k_0}(\Omega = 0)/i)]\int dr\,|\Psi^{(\mathrm{F})}(\mathbf{n}_0 r,\,t)|^2$, which immediately yields (15.38).

15.7 The absence of a ϕ dependence and thus the presence of only $m_\ell = 0$ terms in Eq. (15.61) follow from the choice of $\mathbf{k}$ as the z-axis and the fact that $V(\mathbf{r}) = V(r)$.

(a) Show that $\hat{U}_{\mathrm{R}}(\mathbf{n},\,\Theta)\psi_{\mathbf{k}}(\mathbf{r})$ is also a solution of (15.39) and therefore that Eq. (15.61) holds for any orientation of the coordinate system.

(b) Use the partial-wave expansion $G_k^{(+)}(\mathbf{r},\,\mathbf{r}') = \sum_\ell [4\pi(2\ell + 1)]^{1/2} G_\ell^{(+)}(k;\,r,\,r') Y_\ell^0(\mathbf{e}_r \cdot \mathbf{e}_{r'})$ and the assumed convergence of the Born series (15.57) as applied to Eq. (15.53) to show that only $m_\ell = 0$ terms can appear in the partial wave expansion of $\psi_{\mathbf{k}}^{(+)}(\mathbf{r})$. (Hint: analyze the behavior of Eq. (15.58) in this regard and then extrapolate the result to the general term of the series.)

15.8 An experimentalist analyzing her scattering data, taken at wave number $k = 10^{13}\ \mathrm{cm}^{-1}$, deduces that $\delta_0 = 0.887\,79$, $\delta_1 = -0.566\,40$, $\delta_2 = 0.234\,47$ and all other $\delta_\ell = 0$. Using Eqs. (15.84a) and (15.32)

(a) plot $\mathrm{d}\sigma/\mathrm{d}\Omega$ as a function of $\cos\theta$,

(b) plot $\mathrm{d}\sigma/\mathrm{d}\Omega$ as a function θ, and

(c) compute σ_{T} – Eq. (15.83) – and compare it with the result of the optical theorem.

15.9 A second experimentalist repeats the measurement alluded to in Exercise 15.8, from which she deduces that $\delta_0 = 0.109\,30$, $\delta_1 = -0.769\,93$, $\delta_2 = 0.234\,47$, and all other $\delta_\ell = 0$. Carry out the same calculations as those in Exercise 15.8, and compare and comment on the corresponding results (Kok and Visser (1987)).

15.10 An important quantity in scattering theory is the transition operator $\hat{T}$, defined through the relation $\hat{T}|\mathbf{k}\rangle \equiv V|\mathbf{k},\,+\rangle$.

(a) Use the fact that $|\mathbf{k}\rangle$ is an arbitrary ket to derive (i) the L–S equation obeyed by $\hat{T}$, and (ii) the Born series for $\hat{T}$.

(b) Show that, for $k' = k$, the following partial-wave expansion is valid: $\langle\mathbf{k}'|\hat{T}|\mathbf{k}\rangle = \sum_\ell [4\pi(2\ell + 1)]^{1/2} T_\ell(k) Y_\ell^0(\theta)$, where $\theta = \cos^{-1}(\mathbf{k}\cdot\mathbf{k}'/k^2)$. State the assumptions regarding $\hat{T}$ that lead to this result, in particular relate $T_\ell(k)$ to the matrix element $\langle k',\,\ell' m_\ell'|\hat{T}|k,\,\ell m_\ell\rangle$.

(c) Express $T_\ell(k)$ first in terms of the phase shifts $\delta_\ell(k)$ and then in terms of the S-

matrix element $S_\ell(k)$. State the implication for $T_\ell(k)$ of the equation $|S_\ell(k)| = 1$ and relate $\text{Im}[T_\ell(k)]$ to $|T_\ell(k)|^2$.

15.11 Certain properties of scattering amplitudes have been deduced by analyzing their behavior when E or k is allowed to be a complex number. Aspects of this are explored in the following for negative values of k, $k \propto -E^{1/2}$.

(a) Use the fact that $R_\ell^{(+)}(-kr)$ is a regular solution to Eq. (15.62) to evaluate the factor β_ℓ in the relation $R_\ell^{(+)}(-kr) = (-1)^{\beta_\ell} R_\ell^{(+)}(kr)$.

(b) From the result of part (a) and the structure of Eq. (15.78) relate $\sin[\delta_\ell(-k)]$ to $\sin[\delta_\ell(k)]$.

(c) Use the result of part (b) and the asymptotic form of (15.73) to relate $\delta_\ell(-k)$ to $\delta_\ell(k)$.

(d) Relate $f_\ell(-k)$ to $f_\ell(k)$ and relate $T_\ell(-k)$ of the preceding exercise to $T_\ell(k)$.

15.12 This exercise and the next illustrate some of the many uses of Green's theorem in scattering theory.

(a) Define $\rho = kr$, $u_\ell(\rho) = \rho^{-1}R^{(+)}(\rho)$ and $f_\ell(\rho) = \rho^{-1}j_\ell(\rho)$. Write out the equations obeyed by $u_\ell(\rho)$ and $f_\ell(\rho)$, multiply the former by $f_\ell(\rho)$ and the latter by $u_\ell(\rho)$ and then subtract one of the resulting equations from the other. You should obtain an expression in which the derivatives can be transformed to read $d(f'_\ell u_\ell - f_\ell u'_\ell)/d\rho$, where $' \equiv d/d\rho$. Finally, derive Eq. (15.78) by integrating on ρ, from 0 to ∞, all members of the final equation you have obtained.

(b) Assume that Eq. (15.91) holds, set $\ell = 0$, and define $u_j(k_jr) = (k_jr)^{-1}R^{(+)}(k_jr)$, where $E_j = \hbar^2 k_j^2/(2\mu)$, $j = 1, 2$. Multiply the equation obeyed by each u_j by u_i, $i \neq j$, subtract one set of the resulting equations from the other, and then integrate the equation obtained by subtraction on r from 0 to b. Next, take the limit $k_2 \to k_1 = k(\Rightarrow u_2 = u_1 = u)$, normalize so that, for $r \geqslant b$, $u(kr) = (2\pi)^{-3/2}\sin[kr + \delta_0(k)]$, and then derive the relation $\partial\delta_0(k)/\partial k \geqslant -b + \sin[2(kb + \delta_0)]/(2k)$, an inequality first obtained by Wigner. (Hints: use the reality of u, that an intermediate step in the analysis involves the expression $(\partial u/\partial r)(\partial u/\partial k) - u(\partial^2 u/\partial r\,\partial k)$, and the asymptotic form of $u(kr)$.)

15.13 Derivation of the effective-range expansion. Set $\ell = 0$, $E_j = \hbar^2 k_j^2/(2\mu)$, $j = 1, 2$, and assume that $V(r)$ is short-ranged. Define $u(k_jr) = (k_jr)^{-1}R^{(+)}(k_jr)$ and the following solution to the $\ell = 0$ free-particle, partial-wave equation: $v(k_jr) = \{(2\pi)^{3/2}\sin[\delta_0(k_j)]\}^{-1}\sin[k_jr + \delta_0(k_j)]$, $j = 1, 2$.

(a) Using the multiplication/subtraction procedure of the preceding exercise, derive two equations. One will involve $u(k_1r)$, $u(k_2r)$, and their derivatives, the other will involve $v(k_1r)$, $v(k_2r)$, and *their* derivatives. The results of each pair of manipulations should be integrated on r from 0 to a large radius R much greater than the range of $V(r)$.

(b) Subtract from one another the final equations obtained in part (a), make use of the behavior of V and of u at $r = 0$ and $r = R$, and then let $R \to \infty$.

(c) Set $k_1 = 0$ and $k_2 = k$ in the result of part (b) and show that this leads to $k\cot[\delta_0(k)] = -1/a + k^2\rho(0, k)/2$, where $\rho(0, k)$ is an integral you are to determine. (Note: the effective range r_0 of Eq. (15.113) is the $k \to 0$ limit of $\rho(0, k)$.)

15.14 (a) Starting with Eq. (15.106) and using small-k expansions ($kb \ll 1$), find expressions for a and r_0 in the case of a square-well potential.

(b) Refer to the results of Exercise 11.16 and use them to evaluate numerically the n–p scattering length and effective range for that model of the deuteron (this will require additional calculation if you have not solved Exercise 11.16). Compare your answers with the spin-triplet experimental values $a = 5.424 \times 10^{-15}$ m and $r_0 = 1.759 \times 10^{-15}$ m. Since the latter comparison is used because the deuteron is predominantly in a spin-triplet state, comment on the quality of a square-well-model description of the low-energy n–p system.

15.15 S-wave scattering by the square well of Exercise 11.16 for k corresponding to 100 keV. Compare/contrast the values of the phase shift computed in the following ways.
(a) Calculate δ_0 directly from Eq. (15.106).
(b) Use the values of a and r_0 from Exercise 15.14 and the approximation (15.113) to find δ_0.
(c) Determine δ_0 from the Born approximation, Eq. (15.79).
(d) Optional. Solve the time-independent Schrödinger equation numerically for $(kr)^{-1}R_0^{(+)}(kr) \equiv u_0(kr)$ and extract δ_0 from its asymptotic form.

15.16 Calculate $k\cot[\delta_\ell(k)]$ for the delta-shell potential $V_{\rm sh} = (-[g\hbar^2/(2\mu b^2)]\delta(r-b)$, $g > 0$, and compare your answer with that obtained from the Born approximation, Eq. (15.79).

15.17 Sketch and compare the Born-approximation amplitudes for the potentials $-V_0\exp(-r^2/b^2)$ and $-V_0\exp(-r/b)$, and contrast them to the Born amplitude for the attractive 3-D square well. Explain why the amplitudes from the first two potentials do or do not exhibit diffraction peaks. Note that $V_0 > 0$ is the same for all three potentials.

15.18 The first two terms in the Born series (15.59) are $V^{(1)}(\mathbf{k}', \mathbf{k}) = \langle \mathbf{k}'|V|\mathbf{k}\rangle$ and $V^{(2)}(\mathbf{k}', \mathbf{k}) = \langle \mathbf{k}'|VG_k^{(+)}V|\mathbf{k}\rangle$. Let $V(r) = -(g/r)\exp(-r/b)$, the Yukawa potential.
(a) Choose $\mathbf{k}' = \mathbf{k}$ (forward scattering) and evaluate $V^{(2)}(\mathbf{k}, \mathbf{k})$.
(b) Calculate the ratio $R = V^{(2)}(\mathbf{k}, \mathbf{k})/V^{(1)}(\mathbf{k}, \mathbf{k})$ and then determine the values of g, b, $E \propto k^2$, and the particle's mass m for which $R \ll 1$, i.e., for which the first Born approximation is valid.

15.19 (a) Verify Eqs. (15.130).
(b) Instead of two particles with spins 0 and $\frac{1}{2}$, let the particles in the scattering event have spins S_1 and S_2; however, the Hamiltonian governing their behavior is still $\hat{H}(\mathbf{r})$ of (15.120). If the collision is initiated with the particles in the spin states $|S_1M_1\rangle$ and $|S_2M_2\rangle$, write out the analogs for this situation of the amplitudes (15.130) and the cross-section expression analogous to (15.121c).
(c) Finally, let $S_2 = 0$ and $S_1 = S > \frac{1}{2}$. Assume that the initial spin state is the relevant analog to (13.15a) and state the appropriate normalization condition. Then generalize Eqs. (15.122), (15.123), etc, to the present case, writing out in particular the analog to Eq. (15.125e).

15.20 Evaluate $f_k^{(\rm B)}(\theta)$ and $g_k^{(\rm B)}(\theta)$ – Eqs. (15.131) – for the case $m_s = \frac{1}{2}$ and $V(r) = V_{LS}(r) = -[g\hbar^2/(2\mu b^2)]\delta(r-b)$, $g > 0$.

15.21 A spin-$\frac{1}{2}$ particle of mass μ, initially in state $|\mathbf{k}\rangle|\frac{1}{2}\frac{1}{2}\rangle$ is scattered to the final state $|\mathbf{k}'\rangle|\frac{1}{2}m_s\rangle$ by the parity-non-conserving interaction $V(r) = V_0\delta(r-b)\mathbf{r}\cdot\hat{\boldsymbol{\sigma}}$, where $\hat{\boldsymbol{\sigma}}$ is the vector of Pauli spin operators.
(a) Calculate the Born-approximation amplitude $f_{k,m}(\theta) = \langle \mathbf{k}', \frac{1}{2}m|V|\mathbf{k}, \frac{1}{2}\frac{1}{2}\rangle$.
(b) Evaluate the differential cross-section $d\sigma/d\Omega = \frac{1}{2}\sum_m |f_{k,m}(\theta)|^2$, and obtain its value in the limit of zero momentum transfer.

15.22 For each $\ell > 0$, there are two partial waves when a spin-$\frac{1}{2}$ particle is scattered from a spin-0 target by the potential in Eq. (15.120). These correspond to $j = \ell \pm \frac{1}{2}$, for which the partial-wave solution $F_\ell^{(+)}$ of (15.61) generalizes to $F_{\ell,\pm}$. The solution $\langle \mathbf{r}|\mathbf{k},\, m_s,\, +\rangle$ has the partial-wave expansion

$$(2\pi)^{3/2}\langle \mathbf{r}|\mathbf{k},\, m_s,\, +\rangle = \sum_\ell [4\pi(2\ell+1)]^{1/2} i^\ell [F_{\ell,+}(kr)\hat{P}_{\ell,+} + F_{\ell,-}(kr)\hat{P}_{\ell,-}] Y_\ell(\Omega)|\tfrac{1}{2}m_s\rangle,$$

where

$$F_{\ell,\pm}(kr) \underset{r\to\infty}{\sim} [1/(kr)] \exp[i\delta_{\ell,\pm}(k)] \sin[kr - \ell\pi/2 + \delta_{\ell,\pm}(k)],$$

$\pm$ refers to $j = \ell \pm \frac{1}{2}$, and the $\hat{P}_{\ell,\pm}$ are the projection operators of Exercise 14.16.

(a) Using the asymptotic experession (15.122) with $\binom{a}{b} \to |\frac{1}{2}m_s\rangle$, and an expansion of $\langle \mathbf{r}|\mathbf{k}\rangle|\frac{1}{2}m_s\rangle$ involving the $\hat{P}_{\ell,\pm}$, show that $(M_k(\theta) \to \hat{M}_k(\theta))$

$$\hat{M}_k(\theta) = \sum_\ell [4\pi(2\ell+1)]^{1/2} [a_{\ell,+}(k)\hat{P}_{\ell,+} + a_{\ell,-}(k)\hat{P}_{\ell,-}] Y_\ell^0(\Omega),$$

where the $a_{\ell,\pm}(k)$ are to be expressed in terms of the $\delta_{\ell,\pm}(k)$.

(b) Use the fact that $\mathbf{k}'$ lies in the x–z plane to show that $\hat{\boldsymbol{\sigma}} \cdot \hat{\mathbf{L}} Y_\ell^0(\theta) = i\hat{\boldsymbol{\sigma}} \cdot \mathbf{n} \sin\theta \, dY_\ell^0(\mu)/d\mu$, where $\mu = \cos\theta$ and $\mathbf{n} = \mathbf{k} \times \mathbf{k}'/k^2$.

(c) From the result of part (b), derive an expression for $\hat{M}_k(\theta)$ involving $\hat{\boldsymbol{\sigma}} \cdot \mathbf{n}$, the $a_{\ell,\pm}$, Y_ℓ^0, and $dY_\ell^0(\mu)/d\mu$, $\mu = \cos\theta$.

(d) From a comparison of the result of part (c) with Eq. (15.123), express $f_k(\theta)$ and $g_k(\theta)$ in terms of the $a_{\ell,\pm}(k)$.

Part IV

COMPLEX SYSTEMS

16

Time-Dependent Approximation Methods

This is the first of three chapters concerned with complex systems, typified by those containing many particles. These many-degree-of-freedom systems may often be thought of as a complex of subsystems, e.g., atoms forming molecules or closed shells forming structureless cores in nuclei and atoms. However, "complex" also carries the connotation of "difficult," "not amenable to easy solution or interpretation," and from this perspective a complex system can be a single particle acted on by an intractable Hamiltonian. Indeed, the latter feature characterizes all complex systems: the Schrödinger equations cannot be solved exactly, thereby necessitating the use of approximations, some analogous to the Born approximation to the scattering amplitude.

In this chapter we consider time-dependent Hamiltonians and some of the various methods for treating them. The primary emphasis is on time-dependent perturbation theory (Section 16.1) and its application in Section 16.2 to electromagnetic transitions between bound states. In Section 16.3 we examine two additional procedures: the sudden approximation and the adiabatic approximation. The latter approximation leads naturally to a study of what are sometimes denoted "geometric" phases, which were first discovered by Berry (1984) and then extended by others. Like the Aharonov–Bohm effect and the sign change of spinors under 2π rotations, these geometric phases can be and have been observed – another remarkable, purely quantal phenomenon. The chapter ends with a brief discussion of the energy–time uncertainty relation, whose important role in some transition rates is noted in Section 16.2.

16.1. Time-dependent perturbation theory

The physics we wish to describe is conceptually simple: a quantal system initially in an eigenstate of a Hamiltonian $\hat{H}_0$ is acted on – or perturbed – at time t_0 by a weak time-dependent Hamiltonian $\hat{H}_1(t)$, so that the full Hamiltonian becomes

$$\hat{H}(t) = \hat{H}_0 + \hat{H}_1(t), \qquad t > t_0. \tag{16.1}$$

The perturbation $\hat{H}_1(t)$ induces transitions among eigenstates of $\hat{H}_0$, and the aim of the analysis is to calculate the probabilities for such transitions. Because $\hat{H}(t)$ cannot be solved exactly, approximations are required; the one considered here involves an expansion in powers of the weak perturbation $\hat{H}_1(t)$, which is analogous to the Neumann series for the scattering state $|\mathbf{k}, +\rangle$, Eq. (15.57).

Small-parameter expansion

That $\hat{H}_1(t)$ is weak means that its matrix elements taken w.r.t. the eigenstates of $\hat{H}_0$ are small relative to the eigenenergies of $\hat{H}_0$. This suggests that, for each system described by (16.1), there is a small parameter (of "weakness"), e.g., $\alpha = \kappa_e e^2/(\hbar c) \cong 1/137$ in the case of electromagnetic perturbations. To make the assumed weakness of $\hat{H}_1(t)$ manifest, we multiply $\hat{H}_1(t)$ by the parameter λ, $0 \leqslant \lambda \leqslant 1$:

$$\begin{aligned} \hat{H}_1(t) &\to \lambda \hat{H}_1(t) \\ \hat{H}(t) &\to \hat{H}(t;\,\lambda) = \hat{H}_0 + \lambda \hat{H}_1(t). \end{aligned} \tag{16.2}$$

Obvious limits are

$$\hat{H}(t;\,\lambda = 0) = \hat{H}_0 \tag{16.3a}$$

and

$$\hat{H}(t;\,\lambda = 1) = \hat{H}(t). \tag{16.3b}$$

In addition to emphasizing the weakness of $\hat{H}_1(t)$, λ serves as the expansion parameter in the infinite-series expansion that underlies the theory developed next.

The aim is to solve

$$\begin{aligned} i\hbar \frac{\partial |\Gamma,\, t;\, \lambda\rangle}{\partial t} &= \hat{H}(t;\,\lambda)|\Gamma,\, t;\, \lambda\rangle \\ &= [\hat{H}_0 + \lambda \hat{H}_1(t)]|\Gamma,\, t;\, \lambda\rangle, \end{aligned} \tag{16.4}$$

with λ eventually set equal to unity; the dependence of the eigenvector on λ has been made manifest. We assume that the unperturbed eigenvalue problem

$$\hat{H}_0|\gamma\rangle = E_\gamma|\gamma\rangle \tag{16.5}$$

has been solved; the set $\{|\gamma\rangle\}$ will be the Hilbert-space basis. The vectors $|\gamma\rangle$ can include continuum states, although we do not explicitly refer to them, writing simply

$$\sum_\gamma |\gamma\rangle\langle\gamma| = \hat{I}. \tag{16.6}$$

Corresponding to $|\gamma\rangle$ is its time-dependent counterpart $|\gamma,\, t\rangle$:

$$|\gamma,\, t\rangle = |\gamma\rangle e^{-iE_\gamma t/\hbar}. \tag{16.7}$$

The $\{|\gamma,\, t\rangle\}$ will be used as the actual expansion basis. This just means that $\exp(-iE_\gamma t/\hbar)$ is extracted from the relevant expansion coefficient $C_\gamma(t;\,\lambda)$:

$$|\Gamma,\, t;\, \lambda\rangle = \sum_\gamma C_\gamma(t;\,\lambda)|\gamma,\, E_\gamma,\, t\rangle, \tag{16.8}$$

where

$$C_\gamma(t;\,\lambda) = \langle\gamma,\, t|\Gamma,\, t;\, \lambda\rangle \tag{16.9}$$

is the amplitude for observing the system in state $|\gamma\rangle$ at time t.

An equation for $\dot{C}_\beta(t;\,\lambda) \equiv dC_\beta(t;\,\lambda)/dt$ is obtained by substituting (16.8) into (16.4) and projecting both sides of the resulting expression onto $\langle\beta,\, t|$:

$$i\hbar\dot{C}_\beta(t;\,\lambda) = \lambda\sum_\gamma C_\gamma(t;\,\lambda)\langle\beta,\, t|\hat{H}_1(t)|\gamma,\, t\rangle$$

$$\equiv \lambda\sum_\lambda H_{1,\beta\gamma}(t)C_\gamma(t;\,\lambda), \tag{16.10}$$

where $(i\hbar\,\partial/\partial t - \hat{H}_0)|\gamma,\, t\rangle = 0$ has been used and the second equality defines the matrix element $H_{1,\beta\gamma}(t)$.

To formulate a means for solving the infinite set of coupled equations (16.10), we first assume that $C_\gamma(t;\,\lambda)$ is a well-behaved function of λ with a power-series expansion that converges for all $\lambda \in [0,\, 1]$:

$$C_\gamma(t;\,\lambda) = C_\gamma^{(0)}(t) + \lambda C_\gamma^{(1)}(t) + \lambda^2 C_\gamma^{(2)}(t) + \cdots$$

$$= \sum_{n=0}^{\infty}\lambda^n C_\gamma^{(n)}(t). \tag{16.11}$$

Insertion of (16.11) into (16.10) gives

$$i\hbar\sum_{n=0}^{\infty}\lambda^n \dot{C}_\beta^{(n)}(t) = \sum_{n=0}^{\infty}\lambda^{n+1}\sum_\gamma H_{1,\beta\gamma}(t)C_\gamma^{(n)}(t); \tag{16.12}$$

by equating terms with like powers of λ on the LHS and RHS of (16.12) we get

$$i\hbar\dot{C}_\beta^{(0)} = 0, \tag{16.13a}$$

$$i\hbar\dot{C}_\beta^{(1)}(t) = \sum_\gamma H_{1,\beta\gamma}(t)C_\gamma^{(0)}(t), \tag{16.13b}$$

$$i\hbar\dot{C}_\beta^{(2)}(t) = \sum_\gamma H_{1,\beta\gamma}(t)C_\gamma^{(1)}(t), \tag{16.13c}$$

$$\vdots$$

which is an infinite set of coupled first-order differential equations.

Unless additional assumptions are made, only Eq. (16.13a) has a simple solution:

$$C_\beta^{(0)}(t) = C_\beta^{(0)} = \text{constant}. \tag{16.14}$$

Therefore, to continue, we make the standard assumption that, at the time t_0 when $\hat{H}_1(t)$ is "turned on," the system is in an eigenstate of $\hat{H}_0$, say $|\alpha\rangle$, which means that (16.14) becomes

$$C_\beta^{(0)} = \delta_{\beta\alpha}. \tag{16.15a}$$

This assumption allows an immediate solution to (16.13b):

$$C_\beta^{(1)}(t) \to C_{\beta\alpha}^{(1)}(t) = \frac{1}{i\hbar}\int_{t_0}^{t} dt'\,\langle\beta,\, t'|\hat{H}_1(t')|\alpha,\, t'\rangle, \tag{16.15b}$$

from which we find $\left(C_\beta^{(2)}(t) \to C_{\beta\alpha}^{(2)}(t)\right)$

$$C_{\beta\alpha}^{(2)}(t) = \left(\frac{1}{i\hbar}\right)^2 \sum_\gamma \int_{t_0}^{t} dt' \langle \beta, t' | \hat{H}_1(t') | \gamma, t' \rangle \int_{t_0}^{t'} dt'' \langle \gamma, t'' | \hat{H}_1(t'') | \alpha, t'' \rangle, \qquad (16.15c)$$

etc.

The earlier claim that our solution is expressed as an expansion in powers of the perturbation $\hat{H}_1(t)$, like the series (15.57) for $|\mathbf{k}, +\rangle$, is verified. As with the Born (weak-potential/high-energy) approximation to the scattering amplitude, we shall retain only the first-order approximation $C_{\beta\alpha}^{(1)}(t)$ to $C_\beta(t; \lambda)$, the amplitude that if the system starts in $|\alpha\rangle$ it makes a transition via $\hat{H}_1(t)$ to the state $|\beta\rangle$. The corresponding probability is approximated by

$$P_{\beta\alpha}^{(1)}(t) = |C_{\beta\alpha}^{(1)}(t)|^2, \qquad \beta \neq \alpha. \qquad (16.16)$$

(When β refers to a continuum state, $|C_{\beta\alpha}^{(1)}(t)|^2$ is a probability density.)

As the final step in our general analysis, we now assume that the time dependence of $\hat{H}_1(t)$ is multiplicative:

$$\hat{H}_1(t) = \hat{H}_1 f(t). \qquad (16.17)$$

The expression (16.17) is neither universal nor overly restrictive; it includes perturbations that are turned on and off, over either a finite or an infinite interval, as well as ones with a harmonic time dependence as in the case of electromagnetic radiation, Section 16.2.

With (16.17), $C_{\beta\alpha}^{(1)}(t)$ becomes

$$C_{\beta\alpha}^{(1)}(t) = \frac{1}{i\hbar} H_{1,\beta\alpha} \int_{t_0}^{t} dt'\, e^{i\omega_{\beta\alpha} t'} f(t'), \qquad (16.18a)$$

where the time-independent matrix element is

$$H_{1,\beta\alpha} = \langle \beta | \hat{H}_1 | \alpha \rangle, \qquad (16.18b)$$

and

$$\omega_{\beta\alpha} = (E_\beta - E_\alpha)/\hbar. \qquad (16.18c)$$

Evaluation of the first-order probability $P_{\beta\alpha}^{(1)}(t)$ is carried out in the next two subsections: first for a constant time dependence and then for a harmonic time dependence.

Constant time dependence

We assume here that the perturbation is turned on at time t_0 and off at time t:

$$f(t') = \begin{cases} 0, & t' < t_0, \\ 1, & t' \in [t_0, t], \\ 0, & t' > t. \end{cases} \qquad (16.19)$$

Then, for $t' \leqslant t$, the time integral in (16.18a) can be put into the form

$$\int_{t_0}^{t} dt'\, e^{i\omega_{\beta\alpha} t'} f(t') = e^{i\omega_{\beta\alpha}(t_0 + t)/2} \frac{\sin(\omega_{\beta\alpha} \Delta t/2)}{\omega_{\beta\alpha}/2}, \qquad (16.20a)$$

where

$$\Delta t = t - t_0, \tag{16.20b}$$

and the probability $P^{(1)}_{\beta\alpha}(t)$ becomes

$$P^{(1)}_{\beta\alpha} = \left| \frac{H_{1,\beta\alpha}}{\hbar} \right|^2 \frac{\sin^2(\omega_{\beta\alpha}\,\Delta t/2)}{(\omega_{\beta\alpha}/2)^2}. \tag{16.21}$$

In evaluating the temporal behavior, we shall consider two extreme cases: Δt much smaller and Δt much larger than the intrinsic time interval $|\omega^{-1}_{\beta\alpha}|$.

CASE (i).

$|\omega_{\beta\alpha}\,\Delta t| \ll 1$. Let $\theta = |\omega_{\beta\alpha}\,\Delta t/2| \ll 1$, so that

$$\begin{aligned} P^{(1)}_{\beta\alpha}(t) &= (\Delta t)^2 \left| \frac{H_{1,\beta\alpha}}{\hbar} \right|^2 \frac{\sin^2\theta}{\theta^2} \\ &\cong \Delta t^2 \left| \frac{H_{1,\beta\alpha}}{\hbar} \right|^2. \end{aligned}$$

Next, we introduce the (first-order) probability per unit time interval or *transition rate* $w^{(1)}_{\beta\alpha}$:

$$w^{(1)}_{\beta\alpha} \equiv \frac{P^{(1)}_{\beta\alpha}(t)}{\Delta t} = \Delta t \left| \frac{H_{1,\beta\alpha}}{\hbar} \right|^2. \tag{16.22}$$

Hence, for very short time intervals (w.r.t. $|\omega_{\beta\alpha}|^{-1}$), the transition rate grows linearly with Δt. The strength of $w^{(1)}_{\beta\alpha}$ is determined by the square of the perturbation-matrix element $H_{1,\beta\alpha}$: it is assumed to be non-zero but small, so that higher-order corrections involving $C^{(n)}_{\beta}(t)$, $n \geqslant 2$, can be neglected. We shall not investigate the $H_{1,\beta\alpha} = 0$ case, although we issue a caveat concerning it: should the spectrum of $\hat{H}_0$ contain degeneracies, then special procedures must be invoked when evaluating $C^{(n)}_{\beta}(t)$, $n \geqslant 2$. We consider these in Chapter 17, in the context of time-independent, degenerate perturbation theory.

CASE (ii)

$|\omega_{\beta\alpha}\,\Delta t| \gg 1$. In this situation, the behavior of $P^{(1)}_{\beta\alpha}(t)$, Eq. (16.21), depends crucially on $\omega_{\beta\alpha}$. Shown in Fig. 16.1 is a sketch of $P^{(1)}_{\beta\alpha}(t)$ as a function of $\omega_{\beta\alpha}$. The vast majority of the probability is contained under the first peak, so the most likely transitions are to those states $|\beta\rangle$ with $|E_\beta - E_\alpha| < 2\pi\hbar/\Delta t$. The less interesting cases occur when E_β differs significantly from E_α, for which the probabilities are expected to be small. In addition, because of the perodicity of $P^{(1)}_{\beta\alpha}(t)$ due to the $\sin\theta$ factor ($\theta = |\omega_{\beta\alpha}\,\Delta t|/2$), the accumulated probability will eventually become greater than unity and higher-order contributions must be taken into account. Before that time, of course, the state will undoubtedly become depleted, and the theory becomes inapplicable.

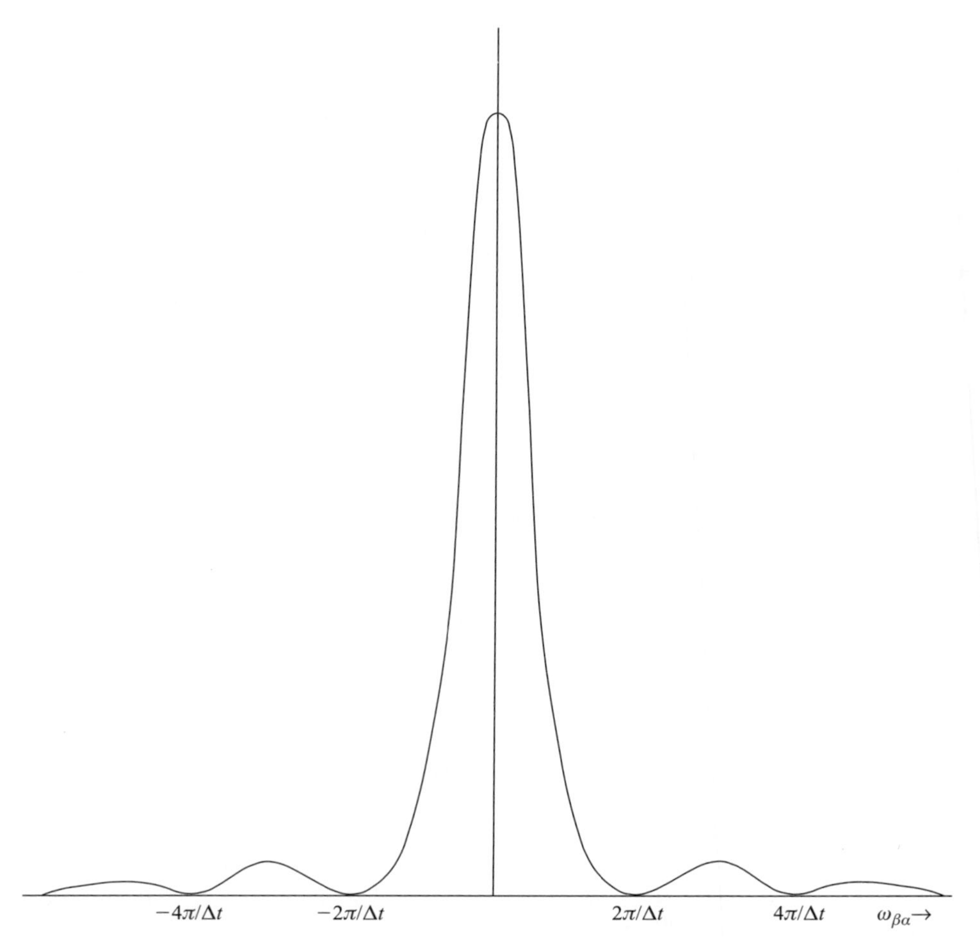

Fig. 16.1 A sketch of $P_{\beta\alpha}(t)$ as a function of $\omega_{\beta\alpha}$.

The most interesting situation occurs when $\Delta t \to \infty$, $\omega_{\beta\alpha} \to 0$ in such a way that $|\omega_{\beta\alpha}\,\Delta t| \gg 1$ holds, with $|\beta\rangle$ being one state among a group G_f of final states $|f\rangle$, all having energies E_f so close to E_β as to be indistinguishable from it. In this situation we must replace $P^{(1)}_{\beta\alpha}(t)$ by a sum over all the states in the group, a sum we denote $\overline{P}^{(1)}_{\beta\alpha}(t)$:

$$P^{(1)}_{\beta\alpha}(t) \to \overline{P}^{(1)}_{\beta\alpha}(t) = \sum_{f\in G_f} P^{(1)}_{f\alpha}(t), \tag{16.23a}$$

where

$$\begin{aligned} P^{(1)}_{f\alpha}(t) &= |C^{(1)}_{f\alpha}(t)|^2 \\ &= \left|\frac{H_{1,f\alpha}}{\hbar}\right|^2 \frac{\sin^2(\omega_{f\alpha}\,\Delta t/2)}{(\omega_{f\alpha}/2)^2}, \end{aligned} \tag{16.23b}$$

with $\omega_{f\alpha} = (E_f - E_\alpha)/\hbar$. Examples of this situation are transitions leading to a single-particle continuum state, say $|\mathbf{k}\rangle$, examined below, or to a final state containing a photon, considered in Section 16.2.

To determine $\overline{P}^{(1)}_{\beta\alpha}(t)$, we first convert $\sum_{f\in G_f}$ to an integral over energies E_f:

$$\overline{P}^{(1)}_{\beta\alpha}(t) = \int dE_f\, \rho_f(E_f) \left|\frac{H_{1,f\alpha}}{\hbar}\right|^2 \frac{\sin^2(\omega_{f\alpha}\,\Delta t/2)}{(\omega_{f\alpha}/2)^2}, \tag{16.24}$$

where $\rho_f(E_f)$, the *density of (final) states*,[1] is the number of states $|f,\, E_f\rangle$ per energy interval dE_f. Next, we introduce a standard and necessary assumption: both $\rho_f(E_f)$ and $H_{1,f\alpha}$ are assumed to be so slowly varying with f that each may be brought outside the integral and evaluated at $E_f = E_\beta$ and $|f,\, E_f\rangle = |\beta,\, E_\beta\rangle$:

$$\overline{P}^{(1)}_{\beta\alpha}(t) \cong \rho_f(E_\beta) \left|\frac{H_{1,\beta\alpha}}{\hbar}\right|^2 \int dE_f\, \frac{\sin^2(\omega_{f\alpha}\,\Delta t/2)}{(\omega_{f\alpha}/2)^2}.$$

The third step in the evaluation utilizes the fact that, when $\Delta t \gg |\omega_{f\alpha}|$, the main peak is so strongly centered around $\omega_{f\alpha} = 0$ (see Fig. 16.1) that the limits on the integral can be extended to $\pm\infty$ while incurring negligible error:

$$\begin{aligned}
\overline{P}^{(1)}_{\beta\alpha}(t) &\cong \rho_f(E_\beta) \left|\frac{H_{1,\beta\alpha}}{\hbar}\right|^2 \int_{-\infty}^{\infty} dE_f\, \frac{\sin^2(\omega_{f\alpha}\,\Delta t/2)}{(\omega_{f\alpha}/2)^2} \\
&= \rho_f(E_\beta) \left|\frac{H_{1,\beta\alpha}}{\hbar}\right|^2 \hbar \int_{-\infty}^{\infty} d\omega_{f\alpha}\, \frac{\sin^2(\omega_{f\alpha}\,\Delta t/2)}{(\omega_{f\alpha}/2)^2} \\
&= \frac{2\pi\,\Delta t}{\hbar}\rho_f(E_\beta)|H_{1,\beta\alpha}|^2, \qquad E_\beta = E_\alpha,
\end{aligned} \tag{16.25}$$

where we have used $dE_f = \hbar\, d\omega_{f\alpha}$ and $\int_{-\infty}^{\infty} d\theta\,(\sin^2\theta)/\theta^2 = \pi$. Finally, since $\overline{P}^{(1)}_{\beta\alpha}(t)$ grows linearly with Δt, we again introduce the transition rate $w^{(1)}_{\beta\alpha}$, equal to the probability per unit time that the constant perturbation $\hat{H}_1(t)$ induces the transition $|\alpha,\, E_\alpha\rangle \to |\beta,\, E_\beta\rangle$:

$$w^{(1)}_{\beta\alpha} = \frac{2\pi}{\hbar}\rho_f(E_\alpha)|\langle\beta|H_1|\alpha\rangle|^2. \tag{16.26}$$

Note that, in (16.26), conservation of energy, $E_\beta = E_\alpha$, has been made explicit in ρ_f. This expression for the transition rate – which is constant in time – is known as Fermi's golden rule number 2, named by the physicist Enrico Fermi (1950).

Along with the facts that $w^{(1)}_{\beta\alpha}$ is first order in $\hat{H}_1$ and that the initial state can become depleted, there are two conditions on Δt to bear in mind when using Eq. (16.26).

(a) Let ΔE_f be the range of energies around E_β. Then we must have $\Delta E_f > 2\pi\hbar/\Delta t$ in order that the central peak falls within this range, that is,

$$\Delta t > 2\pi\hbar/\Delta E_f. \tag{16.27a}$$

(b) Let δE_f be the spacing between adjacent values of E_f. Then, for $\sum_f \to \int dE_f$ to be valid, there must be many states within the main peak, i.e., $\delta E_f \ll 2\pi\hbar/\Delta t$, or

$$\Delta t \ll 2\pi\hbar/\delta E_f. \tag{16.27b}$$

[1] The symbol $\rho_f(E_f)$ should not be confused with the probability density $\rho_{\beta\alpha}$.

The rate $w_{\beta\alpha}^{(1)}$ is the number of transitions per unit time; its dimension is t^{-1}. Its inverse, which has the dimension time, is conventionally regarded as the approximate lifetime $\tau_{\beta\alpha}^{(1)}$ of the state $|\alpha\rangle$ against decay to the state $|\beta\rangle$, where $E_\beta = E_\alpha$:

$$\tau_{\beta\alpha}^{(1)} \equiv \frac{1}{w_{\beta\alpha}^{(1)}}. \tag{16.27c}$$

This and related definitions of lifetime as well as the time–energy uncertainty relation are examined in Section 16.4 and in the exercises.

EXAMPLE

Transitions in the continuum. We consider spinless-particle scattering from the present viewpoint: $\hat{H}_0(\mathbf{r}) = \hat{K}(\mathbf{r})$; $\hat{H}_1(\mathbf{r}) = V(r)$, a short-ranged central potential; $|\alpha\rangle = |\mathbf{k}\rangle$; $|\beta\rangle = |\mathbf{k}'\rangle$, with $k' = k$; and $E_\alpha = E_\beta = E_k = \hbar^2 k^2/(2m)$. Since the final state is in the continuum, $\sum_f$ is actually an integral on final momenta: $\sum_f \rightarrow \int d^3k'$. This must be converted into an integral on $dE_{k'}$. Writing $d^3k' = d\Omega_{k'} k'^2 dk'$, we have

$$\sum_f \rightarrow \int d^3k' = \int d\Omega_{k'}\, k'^2\, dk'. \tag{16.28a}$$

However, since $E_{k'} = \hbar^2 k'^2/(2m)$, $dE_{k'} = \hbar^2 k'\, dk'/m$, and the integral in (16.28a) becomes ($k' = k$)

$$\int d^3k' = \int dE_{k'} \left(\frac{d\Omega_{k'}\, mk}{\hbar^2}\right). \tag{16.28b}$$

Comparing with Eq. (16.24), we see that the quantity in large parentheses in (16.28b) is the density-of-states factor, here denoted $\rho_f(E_k)$:

$$\rho_f(E_k) = \frac{d\Omega_{k'}\, mk}{\hbar^2}. \tag{16.29}$$

The transition rate, being proportional to the infinitesimal angle $d\Omega_{k'}$, is itself a differential, which we write as $dw_{\mathbf{k}'\mathbf{k}}^{(1)}$:

$$dw_{\mathbf{k}'\mathbf{k}}^{(1)} = d\Omega_{k'} \frac{2\pi mk}{\hbar^3} |\langle \mathbf{k}'|V|\mathbf{k}\rangle|^2. \tag{16.30}$$

$dw_{\mathbf{k}'\mathbf{k}}^{(1)}$ is the number of transitions or scatterings into the infinitesimal solid angle $d\Omega_{k'}$ (in the direction $\mathbf{k}' = k\mathbf{e}_r$). Dividing $dw_{\mathbf{k}'\mathbf{k}}^{(1)}$ by the incident probability current (i.e., the incident flux) $\hbar k/[(2\pi)^3 m]$ gives the differential of cross section $d\sigma$ – see Eq. (15.28b):

$$\frac{dw_{\mathbf{k}'\mathbf{k}}^{(1)}}{\text{flux}} = d\sigma = d\Omega_{k'} \frac{(2\pi)^4 m^2}{\hbar^4} |\langle \mathbf{k}'|V|\mathbf{k}\rangle|^2;$$

dropping the subscript k' on $\Omega_{k'}$, we find the differential cross section to be

$$\frac{d\sigma}{d\Omega} = \frac{(2\pi)^4 m^2}{\hbar^4} |\langle \mathbf{k}'|V|\mathbf{k}\rangle|^2. \tag{16.31}$$

Equation (16.31) is just the Born approximation to the cross section, a result that should not be surprising since the perturbation – here the scattering potential $V(r)$ – is treated only to first order.

Several points about this analysis are worth noting. First, one could work with unnormalized plane waves $\exp(i\mathbf{k}\cdot\mathbf{r})$ in place of $(2\pi)^{-3/2}\exp(i\mathbf{k}\cdot\mathbf{r})$. This changes both $\rho_f(E_k)$ and the formal expression for $d\sigma/d\Omega$. Second, one could quantize in a finite volume V rather than in all of space; this also changes the states and $\rho_f(E_k)$. Third, (16.29) and its relatives to which we alluded in the preceding comments are only one example of a variety of density-of-states factors. These three points will be explored via the exercises; see also Section 16.2.

Harmonic time dependence

In this subsection we consider the harmonic time dependence[2]

$$f(t') = \begin{cases} 2\cos(\omega t'), & t_0 \leqslant t' \leqslant t, \\ 0, & t' \notin [t_0, t]. \end{cases} \tag{16.32}$$

The results of our analysis will provide the framework for the treatment of electromagnetic transitions in the next section.

With (16.32) the perturbation is

$$\begin{aligned} \hat{H}_1(t') &= 2\hat{H}_1\cos(\omega t') \\ &= \hat{H}_1(e^{i\omega t'} + e^{-i\omega t'}), \qquad t' \in [t_0, t]. \end{aligned} \tag{16.33}$$

The $\cos(\omega t')$ dependence is chosen since it reduces to the static case implied by Eq. (16.19) when $\omega \to 0$. Because the exponentials are complex conjugates of one another, Eq. (16.33) is sometimes generalized to $\hat{H}_1(t') = [\hat{B}^\dagger\exp(i\omega t') + \hat{B}\exp(-i\omega t')]$, where $\hat{B}$ need not be Hermitian; this form occurs when the electromagnetic field is quantized: see, e.g., Baym (1976) or Sakurai (1994). It is the form used in Section 16.2. The structure of Eq. (16.33) suffices for our present purposes.

We shall again consider only the first order and the limit of large $\Delta t = t - t_0$. Use of (16.33) in (16.15b) yields

$$\begin{aligned} C_{\beta\alpha}^{(1)}(t) &= \frac{H_{1,\beta\alpha}}{i\hbar}\int_{t_0}^{t} dt'\,(e^{i(\omega_{\beta\alpha}+\omega)t'} + e^{i(\omega_{\beta\alpha}-\omega)t'}) \\ &= \frac{H_{1,\beta\alpha}}{i\hbar}\left(\frac{e^{i(\omega_{\beta\alpha}+\omega)t_0}(e^{i(\omega_{\beta\alpha}+\omega)\Delta t} - 1)}{\omega_{\beta\alpha}+\omega}\right. \\ &\qquad\left. + \frac{e^{i(\omega_{\beta\alpha}-\omega)t_0}(e^{i(\omega_{\beta\alpha}-\omega)\Delta t} - 1)}{\omega_{\beta\alpha}-\omega}\right). \end{aligned} \tag{16.34}$$

There are significant contributions to $C_{\beta\alpha}^{(1)}(t)$ in the large-Δt limit only when $\omega \cong \omega_{\beta\alpha}$ or $\omega \cong -\omega_{\beta\alpha}$. For $\omega \cong \omega_{\beta\alpha}$, the second term is larger than the first by $\Delta t \propto$

[2] We write $2\cos(\omega t)$ in order that the form of the final result is essentially the same as Eq. (16.26).

$(\omega - \omega_{\beta\alpha})^{-1}$, whereas for $\omega \cong -\omega_{\beta\alpha}$, the first term is larger than the second by $\Delta t \propto (\omega - \omega_{\beta\alpha})^{-1}$. That is,

$$C^{(1)}_{\beta\alpha}(t) \underset{\omega\to\omega_{\beta\alpha}}{\longrightarrow} \frac{H_{1,\beta\alpha}}{i\hbar}\left(\frac{e^{2i\omega_{\beta\alpha}t_0}(e^{2i\omega_{\beta\alpha}\Delta t}-1)}{2\omega_{\beta\alpha}} + i\,\Delta t\right),$$

and similarly for $\omega \to -\omega_{\beta\alpha}$.

As an aid in determining $P^{(1)}_{\beta\alpha}(t) = |C^{(1)}_{\beta\alpha}(t)|^2$, we re-write Eq. (16.34) as

$$C^{(1)}_{\beta\alpha}(t) = \frac{H_{1,\beta\alpha}}{i\hbar}\left(\exp\left[i\Omega_+\left(\frac{t_0+t}{2}\right)\right]\frac{\sin(\Omega_+\,\Delta t/2)}{\Omega_+/2} + \exp\left[\mathrm{i}\Omega_-\left(\frac{t_0+t}{2}\right)\right]\frac{\sin(\Omega_-\,\Delta t/2)}{\Omega_-/2}\right),$$

and, on dropping the interference term in $|C^{(1)}_{\beta\alpha}(t)|^2$, $P^{(1)}_{\beta\alpha}(t)$ becomes

$$P^{(1)}_{\beta\alpha}(t) = \left|\frac{H_{1,\beta\alpha}}{\hbar}\right|^2\left(\frac{\sin^2(\Omega_+\Delta t/2)}{(\Omega_+/2)^2} + \frac{\sin^2(\Omega_-\Delta t/2)}{(\Omega_-/2)^2}\right), \tag{16.35}$$

where

$$\Omega_\pm = \omega_{\beta\alpha} \pm \omega. \tag{16.36}$$

For purposes of display, both $\sin^2$ terms are retained in (16.35), but only one or the other makes an actual contribution in the large-Δt limit (which is why the interference term was dropped).

For large Δt, we will find either $\Omega_+ = 0$ and the second term in (16.36) to be negligible, or $\Omega_- = 0$ and the first term to be negligible. Let us examine the physical meanings of these conditions: $\Omega_\pm = 0 \Rightarrow$

$$\omega_{\beta\alpha} \pm \omega = 0,$$

or

$$E_\beta = E_\alpha \mp \hbar\omega, \tag{16.37}$$

since $\omega_{\beta\alpha} = (E_\beta - E_\alpha)/\hbar$. In other words, when the frequency ω is such that $\hbar\omega = |E_\beta - E_\alpha|$, the harmonic perturbation $\hat{H}_1\cos(\omega t)$ can cause transitions from $|\alpha\rangle$ either to a state with the higher energy $E_\alpha + \hbar\omega$ or to a state with the lower energy $E_\alpha - \hbar\omega$.[3] The former process is known as *absorption* (Ω_-), the latter as *emission* (Ω_+). One may interpret $\hat{H}_1\cos(\omega t)$ as carrying a quantum of energy $\hbar\omega$; when the perturbation is an electromagnetic one, that quantum is interpreted as a photon; see Section 16.2.

To put (16.35) in the desired form, we use the following delta-function expression:

$$\lim_{\Delta t\to\infty}\frac{\sin^2(\Omega\,\Delta t/2)}{\Omega^2} = \frac{\pi}{2}\,\Delta t\,\delta(\Omega),$$

whereupon $P^{(1)}_{\beta\alpha}(t)$ becomes

[3] For such transitions to occur, of course, the matrix element $\langle\beta|\hat{H}_1|\alpha\rangle$, must be non-zero.

$$P^{(1)}_{\beta\alpha}(t) = \frac{2\pi}{\hbar^2}|H_{1,\beta\alpha}|^2[\delta(\Omega_+) + \delta(\Omega_-)]\,\Delta t$$
$$= \frac{2\pi}{\hbar}|H_{1,\beta\alpha}|^2\big[\delta\big(E_\beta - E_\alpha + \hbar\omega\big) + \delta\big(E_\beta - E_\alpha - \hbar\omega\big)\big]\,\Delta t$$
$$\equiv \Gamma_{\beta\alpha}(\omega)\,\Delta t. \tag{16.38a}$$

The quantity $\Gamma_{\beta\alpha}(\omega)$ is the "unsummed" transition rate, given by

$$\Gamma_{\beta\alpha}(\omega) = \frac{2\pi}{\hbar}|H_{1,\beta\alpha}|^2\big[\delta\big(E_\beta - E_\alpha + \hbar\omega\big) + \delta\big(E_\beta - E_\alpha - \hbar\omega\big)\big]; \tag{16.38b}$$

it is constant in time. In deriving (16.38), we have taken the limit $\Delta t \to \infty$ *before* summing over the states in the group G_f, thus reversing the order followed in the preceding subsection. It will become evident that the result is independent of the order in which we perform the $\sum_f$, $\Delta t \to \infty$ limits.

For $\omega \neq 0$, only one of the two delta functions in (16.38b) can contribute to the rate $w^{(1)}_{\beta\alpha}$ for any particular transition. The factor $\delta\big(E_\beta - E_\alpha + \hbar\omega\big)$ corresponds to absorption, $E_\beta > E_\alpha$, while $\delta\big(E_\beta - E_\alpha - \hbar\omega\big)$ corresponds to emission, $E_\beta < E_\alpha$. The transition rate $w^{(1)}_{\beta\alpha}(\omega)$ is to be obtained from (16.38b) by again replacing $|\beta\rangle$ by an integral over the energies in the set $\{|f\rangle\} \in G_f$, so that $\Gamma_{\beta\alpha}(\omega) \to \sum_f \Gamma_{f\alpha}(\omega)$, as in Eq. (16.23a). When $E_f > E_\beta$, $\forall\ E_f$, the term $\delta\big(E_\beta - E_\alpha + \hbar\omega\big)$ makes no contribution, whereas the term $\delta\big(E_\beta - E_\alpha - \hbar\omega\big)$ makes no contribution when $E_f < E_\alpha$, $\forall\ E_f$. In either case, the resulting $w^{(1)}_{\beta\alpha}(\omega)$ will have the same structure as the $\omega = 0$ transition rate, Eq. (16.26).

The preceding comments are almost obvious. What is not obvious and must be discussed before writing out the expression for $w^{(1)}_{\beta\alpha}(\omega)$ is the meaning of the sum $\sum_f$ when $|\beta\rangle$ is an energy-isolated bound state, since, in such a case, there are no adjacent states to include in the $\sum_f$ sum. There are two ways to deal with this situation, each of them leading to an integral like (16.24), although it need not directly involve energies of $\hat{H}_0$.

In the first, it is assumed that the argument of the cosine is $\mathbf{k}\cdot\mathbf{r} - \omega t$, rather than ωt; $\cos(\mathbf{k}\cdot\mathbf{r} - \omega t)$ is a plane wave propagating in the direction $\mathbf{k}$. This is the electromagnetic case. Instead of $\exp(i\omega t) + \exp(-i\omega t)$, one now has $\exp(i\mathbf{k}\cdot\mathbf{r})\exp(-i\omega t) + \exp(-i\mathbf{k}\cdot\mathbf{r})\exp(i\omega t)$ and, as discussed in the next section, the $\exp(\pm i\mathbf{k}\cdot\mathbf{r})$ behave as photon wave functions. Since they are continuum states, whether initially or finally, there is a phase-space density-of-final-states factor associated with them, and $\sum_f$ corresponds to $\int dE_k = \int d(\hbar\omega) = \hbar\int d(ck)$. The range of final states is that associated with the photon plane wave, not with eigenstates of $\hat{H}_0$, at least in our semi-classical description of electromagnetic transitions between bound states.

In the second way of dealing with energy-isolated states and a strict $\cos(\omega t)$ time dependence, we invoke the fact that either the initial state or the final state (or both) is an *unstable* state that can later *decay* by emission of particles or of electromagnetic radiation. Let $|\beta\rangle$ be such a state. A characteristic of each decaying state (heretofore unstated) is that its energy is not sharp, i.e., is not infinitely narrow (is not a single point along the real energy axis). Instead, it is spread out slightly; we denote this spread or uncertainty in $|\beta\rangle$ by ΔE_β. The energy spread ΔE_β exists because the state is unstable; ΔE_β is related to the lifetime τ_β of the decaying state $|\beta\rangle$ by a type of uncertainty relation:

$$\Delta E_\beta \tau_\beta \sim \hbar. \tag{16.39}$$

This is one form of the so-called "time–energy uncertainty relation" discussed in Section 16.4. It is not a true uncertainty relation, since it is not based on Eq. (9.17): time, being a parameter rather than a dynamical variable, seemingly has no operator image in Hilbert space (but see remarks in Section 16.4).

The existence of energy spreads or widths fundamentally alters our notions of eigenstates and eigenvalues for unstable states. A schematic representation of this new viewpoint is shown in Fig. 16.2, whose caption is intended to help the reader to grasp the new concept. In practice, unstable states occur only as intermediaries (possibly long-lived ones) in processes that begin and end with stable ground states. The formation of the unstable state requires an external mechanism, e.g., a collision or a previous decay. Hence, in describing an unstable state one should take account of its mechanism of formation; this typically means that $\hat{H}_0$ must be supplemented by additional terms; these govern both the state of the external mechanism and the interaction which produces the decaying state. One consequence of this enlargement is that E_β is replaced by $E_\beta + \Delta E_\beta$. For atoms, for which excited-state energies are normally in the range of electron-volts, ΔE_β can be as small 10^{-6} or 10^{-7} eV, whereas for nuclei, whose excited state energies are the order of keVs to MeVs, ΔE_β can be as small as a few eVs.

No matter how small ΔE_β, its existence means that, for a pure $\cos(\omega t)$ time dependence, one must integrate over the ΔE_β range when calculating amplitudes, unless possible uncertainties in ω are larger than $\Delta E_\beta/\hbar$, when we can use $d(\hbar\omega)$ as the integration variable. As an important aside, we note that, for most considerations, the existence of non-zero ΔE_β introduces negligible effects, so that the excited state can be treated as if its energy were infinitely sharp. However, if ΔE_β must be taken into account, then the sum on final states in the present case leads to

$$\Gamma_{\beta\alpha}(\omega) \to \int_{E_\beta - \Delta E_\beta/2}^{E_\beta + \Delta E_\beta/2} dE_\beta\, \rho_f(E_\beta)\Gamma_{\beta\alpha}(\omega), \tag{16.40}$$

With Eq. (16.40) as needed plus the analysis of the preceding subsection, we are readily able to extend the expression for the transition rate from that for a constant time dependence to the case of a harmonic time dependence including both continuum and energy-isolated discrete final states:

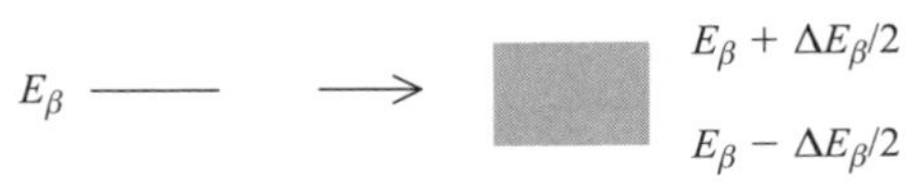

E_β ——— ⟶ $E_\beta + \Delta E_\beta/2$ / $E_\beta - \Delta E_\beta/2$

(a) (b)

Fig. 16.2 Energies of an excited state $|\beta, E_\beta\rangle$. (a) The old way of thinking: E_β is a sharp energy, possessing no width or spread. (b) The new understanding: because $|\beta, E_\beta\rangle$ can decay, then associated with its lifetime τ_β is an energy spread or width or uncertainty $\Delta E_\beta \sim \hbar/\tau_\beta$. Any probe that excites the system to state $|\beta, E_\beta\rangle$ can do so even if its energy is uncertain, as long as the uncertainty falls in the range ΔE_β.

$$w^{(1)}_{\beta\alpha} = \frac{2\pi}{\hbar}\rho_f(E_\beta)|H_{1,\beta\alpha}|^2, \qquad \begin{cases} E_\beta = E_\alpha + \hbar\omega \text{ (``absorption'')} \\ E_\beta = E_\alpha - \hbar\omega \text{ (``emission'')}. \end{cases} \tag{16.41}$$

For continuum final states, Eq. (16.41) is to be interpreted in the same way as Eq. (16.26), the only difference being that $E_\beta = E_\alpha \pm \hbar\omega$, as in the following example.

EXAMPLE

Electric-field ionization of an H-atom with $m_p = \infty$. For $t > t_0$, an H-atom composed of an electron and an infinitely heavy proton is acted on by a spatially uniform electric field whose variation with time is $\cos(\omega t)$:

$$\mathcal{E}(t) = \mathcal{E}\mathbf{e}_3\cos(\omega t), \qquad t > t_0.$$

We wish to estimate the rate $\omega^{(1)}_{\mathbf{k},100}$ at which $\mathcal{E}(t)$ induces transitions from the atom's ground state to a continuum state of energy $E_k = \hbar^2k^2/(2m_e) = -\alpha^2 m_e c^2/2 + \hbar\omega > 0$. The calculation will be an estimate rather than an exact determination of $\omega^{(1)}_{\mathbf{k},100}$ because we shall replace the exact H-atom continuum state $|\mathbf{k}, +\rangle$ by the plane-wave state $|\mathbf{k}\rangle$: Coulomb-potential continuum states are beyond our capacity to treat in this book. Spin is also ignored.

Thus, for our purposes the initial and final states are $|\alpha\rangle = |100\rangle$ and $|\beta\rangle = |\mathbf{k}\rangle$, while the time-independent portion of the perturbation is

$$\hat{H}_1 = e\mathcal{E}z/2 = e\mathcal{E}r\cos(\theta)/2.$$

The density of final states is again given by Eq. (16.29):

$$\rho_f(E_k) = \frac{m_e k}{\hbar^2}\,d\Omega_k = \frac{m_e k}{\hbar^2}\sin\theta_k\,d\theta_k\,d\phi_k,$$

in view of which $w^{(1)} \to dw^{(1)}$.

The transition matrix element $H_{1,\beta\alpha}$ takes the form

$$\begin{aligned} H_{1,\beta\alpha} &= \langle\mathbf{k}|H_1|100\rangle \\ &= \sqrt{\frac{\pi}{3}}e\mathcal{E}\langle\mathbf{k}|\hat{Q}_r Y_1^0|100\rangle \\ &= \sqrt{\frac{1}{24\pi^3 a_0^3}}e\mathcal{E}\int d^3r\, e^{-i\mathbf{k}\cdot\mathbf{r}}rY_1^0(\Omega)e^{-r/a_0}, \end{aligned} \tag{16.42}$$

where $\cos\theta = \sqrt{4\pi/3}Y_1^0(\Omega)$ has been used. Since $\rho_f(E_k) \propto d\Omega_k$, we employ Eq. (10.85) as the spherical-harmonics expansion of $\exp(-i\mathbf{k}\cdot\mathbf{r})$. Because of the factor $Y_1^0(\Omega)$, Eq. (16.42) reduces to

$$H_{1,\beta\alpha} = -2\sqrt{\frac{1}{6\pi a_0^3}}e\mathcal{E}iY_1^0(\Omega_k)\int_0^\infty dr\,r^3 j_1(kr)e^{-r/a_0}.$$

Using (10.79b) for $j_1(kr)$, the radial integral becomes

$$I_1 \equiv \int_0^\infty dr\,r^3 j_1(kr)e^{-r/a_0} = \frac{1}{k^2}\int_0^\infty dr\,r\sin(kr)\,e^{-r/a_0} - \frac{1}{k}\int_0^\infty dr\,r^2\cos(kr)\,e^{-r/a_0}.$$

The sine and cosine integrals can be evaluated using integral tables, e.g., Gradshteyn and Ryzhik (1965):

$$\frac{1}{k^2}\int_0^\infty dr\, r\sin(kr)\, e^{-r/a_0} = \frac{2}{k}\frac{a_0^3}{\left(1+k^2a_0^2\right)^2}$$

and

$$\frac{1}{k}\int_0^\infty dr\, r^2\cos(kr)\, e^{-r/a_0} = \frac{2}{k}\frac{a_0^3\left(1-3k^2a_0^2\right)}{\left(1+k^2a_0^2\right)^3},$$

where the product ka_0 is dimensionless. With these results the radial integral becomes

$$I_1 = 8a_0^4\frac{ka_0}{(1+k^2a_0^2)^3},$$

leading to

$$H_{1,\beta\alpha} = -\frac{16i(a_0e\mathcal{E})a_0^{3/2}}{\sqrt{6\pi}}\frac{ka_0}{(1+k^2a_0^2)^3}\cos\theta_k. \tag{16.43}$$

The matrix element $H_{1,\beta\alpha}$ has the dimension energy $\times$ (length)$^{3/2}$ and varies with the emission angles Ω_k only through the factor $\cos\theta_k$. The first-order differential transition rate will therefore depend on $\cos^2\theta_k$; we find

$$\begin{aligned} dw^{(1)}_{\mathbf{k},100} &= \frac{128}{3\pi}\frac{m_e a_0^2(a_0e\mathcal{E})^2}{\hbar^3}\left(\frac{(ka_0)^{1/2}}{1+k^2a_0^2}\right)^6\cos^2\theta_k\, d\Omega_k \\ &= \frac{128}{3\pi}\frac{(a_0e\mathcal{E})^2}{\hbar\kappa_e e^2/a_0}\left(\frac{(ka_0)^{1/2}}{1+k^2a_0^2}\right)^6\cos^2\theta_k\, d\Omega_k, \end{aligned} \tag{16.44}$$

where the dimension of the ratio $(a_0e\mathcal{E})^2/[\hbar(\kappa_e e^2/a_0)]$ is (time)$^{-1}$. The transition rate per unit solid angle, $dw^{(1)}_{\mathbf{k},100}/d\Omega_k$, is symmetric about $\theta_k = \pi/2$, the angle at which it vanishes. The total transition rate $w^{(1)}_{k,100}$ is the integral of $dw^{(1)}_{\mathbf{k},100}$ over all angles:

$$\begin{aligned} w^{(1)}_{k,100} &= \frac{128}{3\pi}\frac{(a_0e\mathcal{E})^2}{\hbar\kappa_e e^2/a_0}\left(\frac{(ka_0)^{1/2}}{1+k^2a_0^2}\right)^6\int d\Omega_k\cos^2\theta_k \\ &= \frac{512}{9}\frac{(a_0e\mathcal{E})^2}{\hbar\kappa_e e^2/a_0}\left(\frac{(ka_0)^{1/2}}{1+k^2a_0^2}\right)^6. \end{aligned} \tag{16.45}$$

When $\omega \to \kappa_e e^2/(2a_0\hbar)$, i.e., when $k\to 0$, the total transition rate $w^{(1)}_{k,100}$ goes to zero as k^3, as does the differential rate $dw^{(1)}_{\mathbf{k},100}$.

16.2. Electromagnetic transitions between bound states

The emission and absorption of electromagnetic radiation – Fig. 16.3 – is one of the primary quantal processes. In this section we present an introduction to the topic of

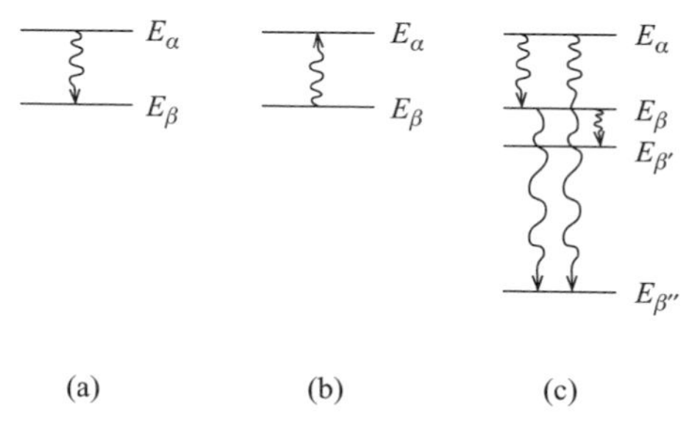

Fig. 16.3 Grotrian-type diagrams illustrating emission and absorption of electromagnetic radiation. (a) Emission of a photon between a single pair of levels: the wavy line with a down-facing arrowhead represents the emitted photon. (b) Absorption of a single photon, with the wavy line and its up-facing arrowhead indicating the absorbed photon. (c) Multiple-emission possibilities. In each case, the photon energy $\hbar\omega$ equals the absolute value of the difference in energies of the two levels that the photon wavy line connects.

electromagnetic transitions between bound states. To help gain a basic understanding of this phenomenon, we employ the semi-classical approach of Chapter 12 plus first-order time-dependent perturbation theory. Spin is ignored and the quantal system undergoing these transitions is assumed to consist of a single particle. (Even with those restrictions, the analysis is far from simple, as we shall see.) Some extensions to multi-particle systems are considered in the exercises. The photon nature of the radiation is imposed by means of the normalization of the vector potential and the Planck distribution law embodied in Eq. (1.12). An essential ingredient in our development is the semi-classical analysis of emission and absorption rates due to Einstein. We begin by summarizing this work.

Photons and spontaneous emission

In general, an arbitrary vector potential $\mathbf{A}(\mathbf{r}, t)$ can be expanded in a Fourier series or integral, a representation in which all wave vectors and thus all frequencies are present. Rather than follow this general procedure, we shall work with only one component, the monochromatic, real, plane wave $\mathbf{A}_0 \cos(\mathbf{k}\cdot\mathbf{r} - \omega t)$, $\omega = kc$, which will form part of the perturbing Hamiltonian. As such, its $\exp(i\omega t)$ and $\exp(-i\omega t)$ portions will be associated with emission and absorption, respectively, as described in Section 16.1. The presence of an external field *induces* or *stimulates* emission as well as absorption. However, suppose that the system is in an excited state and there is no external field: we still expect the system to decay, only in this case it does so *spontaneously*, via creation of a photon and its associated electromagnetic field. Spontaneous emission is a true quantum-electrodynamic event, essentially caused by the fields arising from the phenomenon denoted vacuum polarization, which is the momentary creation and annihilation of particle–antiparticle pairs, e.g., e^+e^- pairs. Our previous semi-classical formalism is not able to deal with this phenomenon (e.g., it does not include photon-creation and -annihilation operators), but we can and do inject both photons and spontaneous emission into the semi-classical description using the mechanism of the Einstein A and B

coefficients. The following analysis will yield a relation between the spontaneous and induced emission rates.

As with Einstein (1917), we consider an enclosure of volume V containing N identical quantum systems in equilibrium with electromagnetic radiation at temperature T. For simplicity, it is assumed that each quantum system can be in only one of two states, $|\alpha\rangle$ and $|\beta\rangle$, $E_\alpha > E_\beta$, with degeneracies g_α and g_β, respectively. Of the N systems, N_α are in state $|\alpha\rangle$ and $N_\beta = N - N_\alpha$ are in state $|\beta\rangle$. We also assume that the Boltzmann distribution governs the number of systems in state $|\gamma\rangle$, $\gamma = \alpha$ or β:

$$N_\gamma = N g_\gamma e^{-E_\gamma/(k_B T)}/Z_{\alpha\beta}, \qquad \gamma = \alpha \text{ or } \beta, \tag{16.46}$$

where $Z_{\alpha\beta} = g_\alpha \exp[-E_\alpha/(k_B T)] + g_\beta \exp[-E_\beta/(k_B T)]$ and k_B is Boltzmann's constant.

Following Planck and Einstein, the radiation is assumed to consist of photons of energy $\hbar\omega = E_\alpha - E_\beta$. A state of n photons with total energy $E_n = n\hbar\omega$ occurs with probability

$$\begin{aligned} P_n &= e^{-E_n/(k_B T)}/Z = e^{-n\hbar\omega/(k_B T)}/Z \\ &= e^{-n\hbar\omega/(k_B T)}(1 - e^{-\hbar\omega/(k_B T)}), \end{aligned} \tag{16.47a}$$

where the photon partition function is

$$Z = \sum_{n=0}^{\infty} e^{-n\hbar\omega/(k_B T)} = (1 - e^{-\hbar\omega/(k_B T)})^{-1}. \tag{16.47b}$$

Since the total number of photons is not fixed, we shall work with their average number $\bar{n}$:

$$\bar{n} = \sum_{n=0}^{\infty} n P_n = \frac{1}{e^{\hbar\omega/(k_B T)} - 1}. \tag{16.48}$$

Proofs of the second equality in each of (16.47b) and (16.48) are left to the exercises.

The distributions (16.47) and (16.48) are those of Planck (1900), originally formulated for blackbody radiation and here (as with Einstein) extended to all electromagnetic radiation. We now ask how $\bar{n}$ changes in time. There are contributions to $d\bar{n}/dt$ from increases in $\bar{n}$ due both to induced and to spontaneous emission as well as from decreases in $\bar{n}$ due to (induced) absorption. In the case of emission, N_α decreases, whereas for absorption, N_β (like $\bar{n}$) decreases; the relations between these rates of change are

$$\left.\frac{d\bar{n}}{dt}\right|_{\text{ind. em.}} = -\left.\frac{dN_\alpha}{dt}\right|_{\text{ind. em.}}, \tag{16.49a}$$

$$\left.\frac{d\bar{n}}{dt}\right|_{\text{sp. em.}} = -\left.\frac{dN_\alpha}{dt}\right|_{\text{sp. em.}}, \tag{16.49b}$$

and

$$\left.\frac{d\bar{n}}{dt}\right|_{\text{abs.}} = \left.\frac{dN_\beta}{dt}\right|_{\text{abs.}}, \tag{16.49c}$$

where "ind." means induced, "sp." stands for spontaneous, "em." means emission, and "abs." stands for absorption. We shall evaluate the $d\bar{n}/dt$'s by determining the dN/dt's.

Expressions for the three rates of change dN/dt appearing in (16.49) are obtained as follows. First, we make the standard assumption that the rate is proportional to the number of systems decaying, i.e., $dN/dt \propto -N$. For a population containing only one decaying state, this leads to the usual radioactive decay formula $N(t) = N_0 \exp(-t/\tau)$, where τ in the proportionality constant $1/\tau$ is the lifetime. Second, we use the fact that induced emission and absorption occur because the quantum systems are acted on by external electromagnetic fields. We shall characterize these fields by the vector potential $2\mathbf{A}_0 \cos(\mathbf{k} \cdot \mathbf{r} - \omega t)$, which will be treated perturbatively. From Section 16.1, the resulting transition rate will be proportional to A_0^2. But, as will be shown later, A_0^2 is proportional to the energy density per frequency $u(\nu)$, with $\nu = \omega/2\pi$, while $u(\nu)$ is proportional to $\bar{n}$, Eq. (1.12). Hence, the transition rates for induced emission and absorption are proportional to $\bar{n}$, the average number of photons. The overall proportionality constants in the expressions for the dN/dt are the individual transition rates.

Putting the foregoing comments into quantitative form we have

$$\left.\frac{dN_\alpha}{dt}\right|_{\text{ind.em.}} = -B_{\beta\alpha} N_\alpha \bar{n}, \tag{16.50a}$$

$$\left.\frac{dN_\alpha}{dt}\right|_{\text{sp.em.}} = -A_{\beta\alpha} N_\alpha, \tag{16.50b}$$

and

$$\left.\frac{dN_\beta}{dt}\right|_{\text{abs.}} = -B_{\alpha\beta} N_\beta \bar{n}, \tag{16.50c}$$

where $B_{\beta\alpha}$ and $B_{\alpha\beta}$ are the individual (i.e., the one-photon) transition rates for induced emission ($\alpha \to \beta$) and absorption ($\beta \to \alpha$),[4] while $A_{\beta\alpha}$ is the rate of spontaneous emission. On writing $dN/dt = -wN$, where w is the overall (emission or absorption) transition rate, we deduce from Eqs. (16.50) that

$$w_{\beta\alpha}(\text{em.}) = \bar{n} B_{\beta\alpha} + A_{\beta\alpha} \tag{16.51a}$$

and

$$w_{\alpha\beta}(\text{abs.}) = \bar{n} B_{\alpha\beta}. \tag{16.51b}$$

To obtain relations between the individual transition rates, we first substitute Eqs. (16.50) into Eqs. (16.49). Next we note that the total rate $d\bar{n}/dt$ is the sum of the three contributions of Eqs. (16.49). However, at equilibrium $\bar{n}$ is constant, so that $d\bar{n}/dt = 0$, which leads to

$$B_{\beta\alpha} N_\alpha \bar{n} + A_{\beta\alpha} N_\alpha = B_{\alpha\beta} N_\beta \bar{n},$$

or

$$\begin{aligned} \bar{n} &= \frac{A_{\beta\alpha} N_\alpha}{B_{\alpha\beta} N_\beta - B_{\beta\alpha} N_\alpha} \\ &= \frac{A_{\beta\alpha}}{(g_\beta/g_\alpha) e^{\hbar\omega/(k_{\mathrm{B}} T)} B_{\alpha\beta} - B_{\beta\alpha}}, \end{aligned} \tag{16.52}$$

where (16.46) has been used. Comparison of Eqs. (16.52) and (16.48) finally yields

[4] Our $B_{\beta\alpha}$ and $B_{\alpha\beta}$ are related but not identical to the corresponding symbols introduced by Einstein (1917).

$$g_\alpha B_{\beta\alpha} = g_\beta B_{\alpha\beta} \tag{16.53a}$$

and

$$A_{\beta\alpha} = B_{\beta\alpha}. \tag{16.53b}$$

Equation (16.53b) is the sought relation between the individual transition rates for spontaneous and stimulated emission. The rate $B_{\beta\alpha}$ (and thus $A_{\beta\alpha}$) will be obtained from a calculation of the quantal induced-emission rate $w^{(1)}_{\beta\alpha}$. Equation (16.53a) is a bonus of our analysis: it is a statement of *detailed balance*, relating the induced transition rates for the inverse processes $\alpha \to \beta$ and $\beta \to \alpha$.

Although quantum theory is needed to determine $B_{\beta\alpha}$, Eqs. (16.51a) and (16.48) suffice for estimating which of spontaneous emission and induced emission will be dominant. Since the rate of induced emission $w_{\alpha\beta}(\text{ind. em.}) = \bar{n}B_{\alpha\beta}$,

$$\frac{w_{\beta\alpha}(\text{ind. em.})}{w_{\beta\alpha}(\text{sp. em.})} = \frac{\bar{n}B_{\beta\alpha}}{A_{\beta\alpha}} = \bar{n} = \frac{1}{e^{\hbar\omega/(k_B T)} - 1}.$$

The two rates are equal when $\bar{n} = 1$. As a specific example, let T correspond to room temperature, $T = 300$ K, in which case $k_B T \cong 0.025$ eV. Then, at this value of T, $\bar{n} \cong 1$ when $\ln[\hbar\omega_0/(k_B T)] \cong 2$, or $\omega_0 \cong 2.6 \times 10^{13}$ rad s^{-1}. Hence, at 300 K $w_{\beta\alpha}(\text{ind. em.}) \gg w_{\beta\alpha}(\text{sp. em.})$ when $\omega \ll \omega_0$ (microwaves, radiofrequencies), while $w_{\beta\alpha}(\text{sp. em.}) \gg w_{\beta\alpha}(\text{ind. em.})$ when $\omega \gg \omega_0$ (visible light, ultraviolet, x-rays, γ-rays).

Perturbation expression for the induced-emission transition rate

In the absence of spin, the Hamiltonian for a particle of mass m and charge q in external electromagnetic fields is that of Eqs. (12.10) and (12.11), here written as

$$\hat{H}^{(A)}(\mathbf{r}, t) = \hat{H}_0(\mathbf{r}) - \frac{q\kappa_c}{m}\mathbf{A}(\mathbf{r}, t)\cdot\hat{\mathbf{P}}(\mathbf{r}) + \frac{iq\hbar\kappa_c}{2m}[\nabla\cdot\mathbf{A}(\mathbf{r}, t)] + \frac{q^2\kappa_c^2}{2m}A^2(\mathbf{r}, t) + q\Phi(\mathbf{r}, t),$$

where

$$\hat{H}_0(\mathbf{r}) = \hat{K}(\mathbf{r}) + V(\mathbf{r}),$$

and we have allowed the electromagnetic potentials to be time-dependent. We shall assume that the charges and currents generating the fields are a very large distance from the quantum system described by $\hat{H}_0$, in which case we can set $\Phi = 0$ and choose the gauge such that $\nabla\cdot\mathbf{A}(\mathbf{r}) = 0$.[5] $\hat{H}_0$ will be the unperturbed Hamiltonian, while

$$-q\kappa_c\mathbf{A}(\mathbf{r}, t)\cdot\hat{\mathbf{P}}(\mathbf{r})/m + q^2\kappa_c^2A^2(\mathbf{r}, t)/(2m)$$

is the perturbation. However, since the perturbation will be treated only to first order, i.e., since $\mathbf{A}(\mathbf{r}, t)$ is assumed to be weak, we can drop the A^2 term relative to the $\mathbf{A}\cdot\hat{\mathbf{P}}$ term. In this approximation $\hat{H}^{(A)}(\mathbf{r}, t)$ becomes

$$\hat{H}^{(A)}(\mathbf{r}, t) = \hat{H}_0(\mathbf{r}) + \hat{H}_1(\mathbf{r}, t), \tag{16.54}$$

with

[5] For comments concerning the validity of these choices, see, e.g., Jackson (1975) or Marion (1965). The choice $\nabla\cdot\mathbf{A} = 0$ is variously known as the Coulomb, radiation, or transverse gauge.

$$\hat{H}_1(\mathbf{r},\ t) = -\frac{q\kappa_c}{m}\mathbf{A}(\mathbf{r},\ t)\cdot\hat{\mathbf{P}}(\mathbf{r}). \tag{16.55}$$

As remarked in the preceding subsection, the vector potential is

$$\begin{aligned}\mathbf{A}(\mathbf{r},\ t) &= 2\mathbf{A}_0\cos(\mathbf{k}\cdot\mathbf{r}-\omega t)\\ &= \mathbf{A}_0(e^{i(\mathbf{k}\cdot\mathbf{r}-\omega t)} + e^{-i(\mathbf{k}\cdot\mathbf{r}-\omega t)}),\end{aligned} \tag{16.56}$$

where $k = \omega/c$, $\omega = |E_\alpha - E_\beta|/\hbar$, and the vector $\mathbf{k}$ defines the propagation direction. The gauge condition $\nabla\cdot\mathbf{A}(\mathbf{r},\ t) = 0$ leads to

$$\mathbf{k}\cdot\mathbf{A}_0 = 0, \tag{16.57a}$$

i.e., the wave is transverse. We let $\mathbf{n}$ be the polarization vector, in which case

$$\mathbf{A}_0 = A_0\mathbf{n}, \qquad \mathbf{n}\cdot\mathbf{k} = 0, \qquad n^2 = 1. \tag{16.57b}$$

The constant A_0 is chosen to make the electromagnetic energy density u equal to the average photon energy $\bar{n}\hbar\omega$ divided by the volume V of the enclosure: $u = \bar{n}\hbar\omega/V$. In terms of $\mathcal{E}$ and B, u is given by

$$u = \frac{1}{8\pi}\left(\overline{\frac{1}{\kappa_e}\mathcal{E}^2 + \frac{\kappa_c}{\kappa_m}B^2}\right),$$

where the bar signifies a time average over the oscillating fields. Using Eq. (16.56) in $\mathcal{E} = -\kappa_c\,\partial\mathbf{A}/\partial t$ and $\mathbf{B} = \nabla\cdot\mathbf{A}$, plus $\overline{\sin^2(\mathbf{k}\cdot\mathbf{r}-\omega t)} = \frac{1}{2}$, we get

$$\overline{\mathcal{E}^2} = c^2\kappa_c^2\overline{B^2} = 4A_0^2\omega^2\kappa_c^2\,\overline{\sin^2(\mathbf{k}\cdot\mathbf{r}-\omega t)} = 2A_0^2\omega^2\kappa_c^2.$$

Hence, the electromagnetic expression for u becomes

$$u = \frac{1}{2\pi}\frac{\kappa_c^2}{\kappa_e}\omega^2A_0^2,$$

and, equating this to $\bar{n}\hbar\omega/V$, we find

$$A_0 = \frac{1}{\omega\kappa_c}\sqrt{2\pi\kappa_e\bar{n}\hbar\omega/V}. \tag{16.58}$$

With Eq. (16.58), the perturbing Hamiltonian becomes

$$\begin{aligned}\hat{H}_1(\mathbf{r},\ t) &= -\frac{q}{m\omega}\sqrt{\frac{2\pi\kappa_e\bar{n}\hbar\omega}{V}}(e^{i(\mathbf{k}\cdot\mathbf{r}-\omega t)} + e^{-i(\mathbf{k}\cdot\mathbf{r}-\omega t)})\mathbf{n}\cdot\hat{\mathbf{P}}(\mathbf{r})\\ &\to -\frac{q}{m\omega}\sqrt{\frac{2\pi\kappa_e\bar{n}\hbar\omega}{V}}(e^{i(\mathbf{k}\cdot\hat{\mathbf{Q}}-\omega t)} + e^{-i(\mathbf{k}\cdot\hat{\mathbf{Q}}-\omega t)})\mathbf{n}\cdot\hat{\mathbf{P}}.\end{aligned} \tag{16.59}$$

The second term in the parentheses in (16.59) is responsible for induced emission, for which $\hbar\omega = E_\alpha - E_\beta$, and we therefore introduce $\hat{H}_1^{\text{ind.em.}}(t)$:

$$\hat{H}_1^{\text{ind.em.}}(t) = -\frac{q}{m\omega}\sqrt{\frac{2\pi\kappa_e\bar{n}\hbar\omega}{V}}e^{-i\hat{\mathbf{Q}}\cdot\mathbf{k}}\mathbf{n}\cdot\hat{\mathbf{P}}e^{i\omega t}. \tag{16.60a}$$

The time-independent operator for induced emission is therefore

$$\hat{H}_1^{\text{ind. em.}} = -\frac{q}{m\omega}\sqrt{\frac{2\pi\kappa_e \bar{n}\hbar\omega}{V}} e^{-i\mathbf{k}\cdot\hat{\mathbf{Q}}} \mathbf{n}\cdot\hat{\mathbf{P}}(\mathbf{r}), \tag{16.60b}$$

and the transition rate $w_{\beta\alpha}^{(1)}(\text{ind. em.})$, given by (16.41), becomes

$$w_{\beta\alpha}^{(1)}(\text{ind. em.}) = \frac{4\pi^2 q^2 \bar{n}\kappa_e}{m^2\omega V}\rho_f(E_\alpha)|\langle\beta|e^{-i\mathbf{k}\cdot\hat{\mathbf{Q}}}\mathbf{n}\cdot\hat{\mathbf{P}}|\alpha\rangle|^2, \qquad E_\beta + \hbar\omega = E_\alpha. \tag{16.61}$$

Obtaining the density of final states involves a re-interpretation of the matrix element in (16.61), one associated with the injecting of photons into the semi-classical approach. Injection of photons into the description means that the final state should be a product of $|\beta\rangle$ times a one-photon state. Is there a quantity present in (16.61) that can function as a one-photon state? The answer is YES: $\mathbf{n}\exp(-i\mathbf{k}\cdot\mathbf{r})$ is the photon state.[6] It is essentially a renormalized version of $\mathbf{n}\langle\mathbf{k}|\mathbf{r}\rangle$, and use of it plus periodic boundary conditions (box normalization) will be the key to determining $\rho_f(E_\alpha)$, as described in the next two paragraphs.

Since the field is contained in the volume V, the photon wave function must also be restricted to V, taken to be a very large box of sides L_x, L_y, and L_z: $V = L_xL_yL_z$. Rather than have the wave function go to zero at the surfaces, we shall use periodic boundary conditions: the components k_x, k_y, and k_z will be chosen so that $\exp[ik_x(x+L_x)] = \exp(ik_xx)$ and similarly for the y- and z-components. This requirement means that $k_x \to 2\pi n_x/L_x$, $k_y \to 2\pi n_y/L_y$, and $k_z \to 2\pi n_z/L_z$, where n_x, n_y, and n_z each equals $1, 2, 3, \ldots$. With these choices, we readily find that $\int_0^{L_x} dx \exp[i(k_x - k_x')x] = L_x\delta_{n_xn_x'}$ and similarly for the y and z normalization integrals.

The density of final states is the number of final states per final state energy: $\rho_f(E_\alpha) = dN/dE_f$, where $dE_f = d(E_\beta + \hbar\omega) = \hbar\, d\omega$. The number of photons is $N = n_xn_yn_z$, with $dN \cong \Delta n_x\Delta n_y\Delta n_z$, where, because the n's are integers, we write Δn in place of dn. However, $\Delta n_j = L_j\Delta k_j/(2\pi)$, $j = x, y, z$, so that $dN \cong L_xL_yL_z\Delta k_x\Delta k_y\Delta k_z/(2\pi)^3 \cong V\,d^3k/(2\pi)^3$, where L_j is assumed to be so large that Δk_j can be approximated by dk_j. A final comment is that, for each d^3k, there are two polarization directions in the plane perpendicular to $\mathbf{k}$, and, since we are counting the number of states rather than taking directions into account, we must include a factor of 2 in dN. Putting these remarks and results together, we get

$$\rho_f(E_\alpha) = \frac{2V\,d^3k}{(2\pi)^3\hbar\,d\omega} = \frac{2Vk^2\,dk\,d\Omega_k}{(2\pi)^3\hbar\,d\omega}$$
$$= \frac{2V\omega^2\,d\Omega_k}{(2\pi)^3c^3\hbar}, \tag{16.62}$$

where $k = \omega/c$ has been used.

With (16.62), $w_{\beta\alpha}^{(1)}$ (ind. em.) becomes a differential rate:

$$dw_{\beta\alpha}^{(1)}(\text{ind. em.}) = \bar{n}\frac{q^2\kappa_e}{\hbar c}\frac{\omega\,d\Omega_k}{\pi m^2c^2}|\langle\beta|e^{-i\mathbf{k}\cdot\hat{\mathbf{Q}}}\mathbf{n}\cdot\hat{\mathbf{P}}|\alpha\rangle|^2, \tag{16.63}$$

whose dimension is easily shown to be $(\text{time})^{-1}$. We also see that the "parameter of

[6] Since photons have spin 1 and the polarization vector $\mathbf{n}$ behaves like a rank-1 spinor under rotations, $\mathbf{n}$ must be included as part of the photon state.

weakness" is $q^2\kappa_e/(\hbar c)$, approximately equal to $1/137$ when $q = \pm e$, and that $dw^{(1)}_{\beta\alpha}$(ind. em.) $\propto \bar{n}$. Equation (16.63) is a general, first-order expression for the rate of stimulated emission, which is valid beyond the confines of our semi-classical approach. The individual emission rates $B_{\beta\alpha}$ and $A_{\beta\alpha}$ of the preceding subsection are obtained from (16.63) by dividing $dw^{(1)}_{\beta\alpha}$(ind. em.) by $\bar{n}$ and then integrating over the angles Ω_k:[7]

$$B_{\beta\alpha} = A_{\beta\alpha} = \frac{q^2\kappa_e}{\hbar c}\frac{\omega}{\pi m^2 c^2}\int d\Omega_k\,|\langle\beta|e^{-i\mathbf{k}\cdot\hat{\mathbf{Q}}}\mathbf{n}\cdot\hat{\mathbf{P}}|\alpha\rangle|^2. \tag{16.64}$$

The matrix element in (16.64) is evaluated in the next subsection.

The electric-dipole approximation

Since we are considering only single-particle systems and $V(\mathbf{r}) = V(r)$, the initial and final states can be written as

$$|\alpha\rangle = |n_\alpha \ell_\alpha m_\alpha\rangle, \qquad E_\alpha = E_{n_\alpha \ell_\alpha} \tag{16.65a}$$

and

$$|\beta\rangle = |n_\beta \ell_\beta m_\beta\rangle, \qquad E_\beta = E_{n_\beta \ell_\beta}, \tag{16.65b}$$

where any spin degrees of freedom are suppressed. The matrix element in (16.64) will then depend on the six quantum numbers of (16.65), plus $\mathbf{k}$ and $\mathbf{n}$. We write this matrix element as

$$M_{\beta\alpha}(\mathbf{n}, \mathbf{k}) = \mathbf{n}\cdot\langle n_\beta \ell_\beta m_\beta|e^{-i\mathbf{k}\cdot\hat{\mathbf{Q}}}\hat{\mathbf{P}}|n_\alpha \ell_\alpha m_\alpha\rangle. \tag{16.66}$$

In a coordinate representation the exponential in (16.66) is $\exp(-i\mathbf{k}\cdot\mathbf{r})$, and one possible approach to evaluating $M_{\beta\alpha}(\mathbf{n}, \mathbf{k})$ would be to employ Eq. (10.85). This is a standard procedure in the nuclear case, for which the electric-dipole approximation, which is typically used for atomic transitions, is often not applicable. However, we shall restrict our considerations to the atomic case, and we will show that, for many atomic transitions the maximum value of $|\mathbf{k}\cdot\mathbf{r}|$ is very small, thereby justifying the approximation $\exp(-i\mathbf{k}\cdot\mathbf{r}) \cong 1$.

To estimate k, we choose $\hbar\omega = 40$ eV, corresponding to the $n = 2 \to n = 1$ transition in He^+. This yields $k \cong 10^{-3}$ Å^{-1}. For the maximum r, we choose twice the value appearing in the exponential of the hydrogen wave function, viz., $2na_0$. Hence, $|\mathbf{k}\cdot\mathbf{r}|$ is very often bounded by

$$|\mathbf{k}\cdot\mathbf{r}| \lesssim 10^{-3}n, \tag{16.67a}$$

so that, in the expansion

$$e^{i\mathbf{k}\cdot\mathbf{r}} = 1 - i\mathbf{k}\cdot\mathbf{r} - (\mathbf{k}\cdot\mathbf{r})^2 + \cdots, \tag{16.67b}$$

it is reasonable to retain only the first term on the RHS of (16.67b):[8]

[7] $B_{\beta\alpha}$ and $A_{\beta\alpha}$ are the total transition rates; hence the integration over angles.

[8] For $Z > 1$ the RHS of (16.67a) should be replaced by $10^{-3}Zn$, which shows that, as Z or n increases, it becomes less justified to use (16.67c). See the exercises.

$$e^{i\mathbf{k}\cdot\mathbf{r}} \cong 1. \tag{16.67c}$$

Equation (16.67c) leads to what is known as the *electric-dipole* approximation to $M_{\beta\alpha}$; it is basically a long-wavelength approximation: $r_{\max} \ll \lambda = 2\pi/k$.

With Eq. (16.67c), $M_{\beta\alpha}(\mathbf{n}, \mathbf{k})$ becomes independent of $\mathbf{k}$:

$$M_{\beta\alpha}(\mathbf{n}, \mathbf{k}) \rightarrow M_{\beta\alpha}(\mathbf{n}) = \mathbf{n} \cdot \langle n_\beta \ell_\beta m_\beta | \hat{\mathbf{P}} | n_\alpha \ell_\alpha m_\alpha \rangle. \tag{16.68}$$

Ehrenfest's theorem, Eq. (8.50a), can now be applied to the RHS of (16.68), where we revert to a $|\beta\rangle$, $|\alpha\rangle$ notation:

$$\begin{aligned}\langle\beta|\hat{\mathbf{P}}|\alpha\rangle &= \frac{m}{i\hbar}\langle\beta|[\hat{\mathbf{Q}}, \hat{H}_0]|\alpha\rangle \\ &= \frac{m(E_\alpha - E_\beta)}{i\hbar}\langle\beta|\hat{\mathbf{Q}}|\alpha\rangle \\ &\equiv \frac{m\omega}{i}\langle\hat{\mathbf{Q}}\rangle_{\beta\alpha}.\end{aligned} \tag{16.69}$$

Here, $\langle\hat{\mathbf{Q}}\rangle_{\beta\alpha}$ is the *electric-dipole* matrix element

$$\langle\hat{\mathbf{Q}}\rangle_{\beta\alpha} = \langle n_\beta \ell_\beta m_\beta | \hat{\mathbf{Q}} | n_\alpha \ell_\alpha m_\alpha \rangle,$$

so named because $\langle q\hat{\mathbf{Q}}\rangle_{\beta\alpha}$ defines the generalized expectation value of the particle's electric-dipole moment: $\hat{\mathbf{d}} = q\hat{\mathbf{Q}}$ is the electric-dipole operator.

Substituting back into (16.68), we find

$$M_{\beta\alpha}(\mathbf{n}) = \frac{m\omega}{i}\mathbf{n} \cdot \langle\mathbf{Q}\rangle_{\beta\alpha},$$

and the differential rate becomes

$$dw^{(1)}_{\beta\alpha}(\text{ind. em.}) \cong \bar{n}\frac{q^2\kappa_e}{\hbar c}\frac{\omega^3 d\Omega_k}{\pi c^2}|\mathbf{n} \cdot \langle\mathbf{Q}\rangle_{\beta\alpha}|^2. \tag{16.70}$$

Since $M_{\beta\alpha}$ no longer depends on $\mathbf{k}$, we can determine the total first-order transition rate by integrating over the angles Ω_k:

$$\begin{aligned}w^{(1)}_{\beta\alpha}(\text{ind. em}) &= \int_{\Omega_k} d\omega^{(1)}_{\beta\alpha}(\text{ind. em.}) \\ &\cong \bar{n}\left(\frac{q^2\kappa_e}{\hbar c}\right)\frac{4\omega^3}{c^2}|\mathbf{n} \cdot \langle\hat{\mathbf{Q}}\rangle_{\beta\alpha}|^2.\end{aligned} \tag{16.71}$$

Hence, to first order and in the electric-dipole approximation, the rate of spontaneous emission $w^{(1)}_{\beta\alpha}(\text{sp. em.}) \equiv A_{\beta\alpha} = w^{(1)}_{\beta\alpha}(\text{ind. em.})/\bar{n}$ is

$$w^{(1)}_{\beta\alpha}(\text{sp. em.}) \cong \left(\frac{q^2\kappa_e}{\hbar c}\right)\frac{4\omega^3}{c^2}|\mathbf{n} \cdot \langle\hat{\mathbf{Q}}\rangle_{\beta\alpha}|^2. \tag{16.72}$$

This rate will be related to the lifetime $\tau_{\beta\alpha}$ of the state $|\alpha\rangle$ against decay to the state $|\beta\rangle$.

Equation (16.72) is a first-order result. The parameter of weakness is $q^2\kappa_e/(\hbar c) \cong 1/137$ when $q = e$. It will have a Z dependence when $Z > 1$, a feature explored in the exercises.

Selection rules and evaluation of the transition rate

We next examine $\mathbf{n}\cdot\langle\hat{\mathbf{Q}}\rangle_{\beta\alpha}$. The matrix element of $\hat{\mathbf{Q}}$ will vanish unless certain conditions are met; these conditions, known as *selection rules,* are a manifestation of the symmetries presumed to be inherent in the system. The simplest is the parity selection rule, which follows from the experimentally based conclusions that the electromagnetic and nuclear interactions responsible for atomic, molecular, and nuclear structure are invariant under reflections, in which case Eq. (9.85b) applies. Let π_α and π_β be the parities of $|\alpha\rangle$ and $|\beta\rangle$. The parity of $\hat{\mathbf{Q}}$ is $\pi_{\hat{\mathbf{Q}}}=-1$. Hence, applied to $\langle\hat{\mathbf{Q}}\rangle_{\beta\alpha}$, Eq. (9.85b) yields the *parity selection rule*:

$$\langle\hat{\mathbf{Q}}\rangle_{\alpha\beta}=0 \qquad \text{if } \pi_\alpha\pi_\beta=+1. \tag{16.73}$$

That is, electric-dipole transitions can connect only states of opposite parity. For single-particle states of the form $|n_\gamma\ell_\gamma m_\gamma\rangle$, $\pi_\gamma=(-1)^{\ell_\gamma}$, and (16.73) becomes $\langle\hat{\mathbf{Q}}\rangle_\beta=0$ if $\ell_\alpha+\ell_\beta=\text{even}$.

The other selection rules are based on angular-momentum considerations, and their form is simplest in the present case of a single-particle system in which spin is being ignored. Going to the coordinate representation, $\mathbf{n}\cdot\langle\hat{\mathbf{Q}}\rangle_{\alpha\beta}$ becomes

$$\mathbf{n}\cdot\langle\hat{\mathbf{Q}}\rangle_{\alpha\beta}=\int d^3r\, R_{n_\beta\ell_\beta}(r)Y_{\ell_\beta}^{m_\beta *}(\Omega)\mathbf{n}\cdot\mathbf{r}R_{n_\alpha\ell_\alpha}(r)Y_{\ell_\alpha}^{m_\alpha}(\Omega).$$

The scalar product $\mathbf{n}\cdot\mathbf{r}$ can be expressed in terms of r and spherical harmonics using Eqs. (10.68):

$$\mathbf{n}\cdot\mathbf{r}=r\sqrt{\frac{4\pi}{3}}\left(\frac{n_x}{\sqrt{2}}\left(-Y_1^1+Y_1^{-1}\right)+\frac{in_y}{\sqrt{2}}\left(Y_1^1+Y_1^{-1}\right)+n_zY_1^0\right), \tag{16.74}$$

where

$$n_x=\mathbf{e}_1\cdot\mathbf{n}, \qquad n_y=\mathbf{e}_2\cdot\mathbf{n}, \qquad n_z=\mathbf{e}_3\cdot\mathbf{n}$$

are the components of the polarization vector.

Use of (16.74) transforms $\mathbf{n}\cdot\langle\hat{\mathbf{Q}}\rangle_{\beta\alpha}$ into a sum of products of radial and angular integrals:

$$\begin{aligned}\mathbf{n}\cdot\langle\hat{\mathbf{Q}}\rangle_{\beta\alpha}= \sqrt{\frac{4\pi}{3}}\int dr\, R_{n_\beta\ell_\beta}(r)r^3R_{n_\alpha\ell_\alpha}(r)\int d\Omega\,\Bigg[Y_{\ell_\beta}^{m_\beta *}(\Omega)\Bigg(\frac{(-n_x+in_y)}{\sqrt{2}}\,Y_1^1(\Omega)\\ +n_zY_1^0(\Omega)+\frac{(n_x+in_y)}{\sqrt{2}}Y_1^{-1}(\Omega)\Bigg)Y_{\ell_\alpha}^{m_\alpha}(\Omega)\Bigg].\end{aligned} \tag{16.75}$$

The three angular integrals are each of the general form

$$I_{\ell_\beta\ell_\alpha}^{m_\beta\mu m_\alpha}=\int d\Omega\, Y_{\ell_\beta}^{m_\beta *}(\Omega)Y_1^\mu(\Omega)Y_{\ell_\alpha}^{m_\alpha}(\Omega), \tag{16.76a}$$

with $\mu=1$, 0, and -1. Integrals of this type can be evaluated using the methods of angular-momentum theory; the result is[9]

[9] See, e.g., Rose (1957), Edmonds (1957), Brink and Satchler (1971), Baym (1976), or Galindo and Pascual (1990).

$$I_{\ell_\beta \ell_\alpha}^{m_\beta \mu m_\alpha} = \sqrt{\frac{3(2\ell_\alpha + 1)}{4\pi(2\ell_\beta + 1)}} C(\ell_\alpha 1 \ell_\beta;\ m_\alpha \mu m_\beta) C(\ell_\alpha 1 \ell_\beta;\ 000). \tag{16.76b}$$

Since $\mathbf{n} \cdot \langle \hat{\mathbf{Q}} \rangle_{\beta\alpha} \propto I_{\ell_\beta \ell_\alpha}^{m_\beta \mu m_\alpha}$, the properties of the C–G coefficients in (16.76b) yield the *angular-momentum selection rules*:

$$\langle \hat{\mathbf{Q}} \rangle_{\beta\alpha} = 0 \qquad \text{if} \begin{cases} \ell_\beta < |\ell_\alpha - 1| & \text{or} \qquad \ell_\beta > \ell_\alpha + 1, \\ m_\beta \neq \mu + m_\alpha & (\mu = 1,\ 0,\ -1), \\ \text{both } \ell_\alpha = 0 \text{ and } \ell_\beta = 0. \end{cases} \tag{16.77}$$

The third criterion in (16.77) states that there are no $\ell_\alpha = 0 \to \ell_\beta = 0$ transitions: they violate conservation of angular momentum. Furthermore, the parity selection rule $\langle \hat{\mathbf{Q}} \rangle_{\beta\alpha} = 0$ if $\ell_\alpha + \ell_\beta =$ even, which holds for single-particle states of a central potential, also follows from a symmetry property of the C–G coefficient $C(\ell_\alpha 1 \ell_\beta;\ 000)$, proof of which is left as an exercise for the reader.

In general, if a selection rule does not forbid it, a transition can and will occur. In the present case of the dipole approximation, should $\langle \hat{\mathbf{Q}} \rangle_{\beta\alpha}$ vanish, then at least the first correction to (16.67c), $-i\mathbf{k} \cdot \mathbf{r}$, needs to be taken into account, investigation of which is beyond the scope of this book: we are considering only the dipole case.

On using Eqs. (16.76) and defining the complex numbers n^μ by

$$n^{\pm 1} = \frac{\pm n_x + i n_y}{\sqrt{2}}, \qquad n^0 = n_z,$$

Eq. (16.75) can be expressed in a more compact form:

$$\mathbf{n} \cdot \langle \hat{\mathbf{Q}} \rangle_{\beta\alpha} = \sqrt{\frac{2\ell_\alpha + 1}{2\ell_\beta + 1}} \langle r \rangle_{\beta\alpha} C(\ell_\alpha 1 \ell_\beta;\ 000) \sum_\mu n^\mu C(\ell_\alpha 1 \ell_\beta;\ m_\alpha \mu m_\beta), \tag{16.78}$$

where

$$\langle r \rangle_{\beta\alpha} \equiv \int dr\, r^2 R_{n_\beta \ell_\beta}(r) r R_{n_\alpha \ell_\alpha}(r).$$

The summation on μ in (16.78) is redundant: μ is limited to the single value $\mu = m_\beta - m_\alpha$, and m_α and m_β are fixed. However, the μ sum is retained because we shall eventually sum on m_α and m_β. The latter sums will take account of the degeneracies, but, before considering them, let us see how Eq. (16.78) explains why there are only three transitions observed when the degenerate m-levels are split by a weak magnetic field – recall the comments in the second paragraph below Eq. (12.22b) concerning the Zeeman effect. When the m-levels are split, the only transitions that occur will be between those states $|n_\alpha \ell_\alpha m_\alpha\rangle$ and $|n_\beta \ell_\beta m_\beta\rangle$ satisfying one of the relations $m_\beta = m_\alpha$, $m_\beta = m_\alpha - 1$, and $m_\beta = m_\alpha + 1$, corresponding to $\mu = 0, -1$, and 1. For each value of μ there is a unique difference in energy for all the values of m_α and m_β which satisfy $\mu = m_\beta - m_\alpha$, and therefore only the three spectral lines noted in Chapter 12 will be seen – assuming that the transition is an electric-dipole one.

In view of the dependence of $\mathbf{n} \cdot \langle \hat{\mathbf{Q}} \rangle_{\beta\alpha}$ on the magnetic quantum numbers, we display them (while still suppressing the n, ℓ dependence) in the expression for $w_{\beta\alpha}^{(1)}(\text{sp. em.})$ obtained when (16.78) is substituted into (16.72):

$$w_{\beta\alpha}^{(1)}(\text{sp. em.}) \to w_{m_\beta m_\alpha}^{(1)}(\text{sp. em.}) \cong \left(\frac{q^2\kappa_e}{\hbar c}\right)\frac{4\omega^3}{c^2}\langle r\rangle_{\beta\alpha}^2 \frac{2\ell_\alpha+1}{2\ell_\beta+1}C^2(\ell_\alpha 1\ell_\beta;\ 000)$$
$$\times\left|\sum_\mu n^\mu C(\ell_\alpha 1\ell_\beta;\ m_\alpha\mu m_\beta)\right|^2. \qquad (16.79)$$

Equation (16.79) is the first-order transition rate for the decay $n_\alpha\ell_\alpha m_\alpha \to n_\beta\ell_\beta m_\beta$. In a typical experiment involving N_α decaying systems all of the magnetic substates will be populated. If the decaying systems are neither aligned nor polarized, as is often the case, then the magnetic substates are populated uniformly with probability $(2\ell_\alpha+1)^{-1}$, and we must average $w_{m_\beta m_\alpha}^{(1)}$ (sp. em.) over the initial set of magnetic substates; in so doing, we also sum over the final set of magnetic quantum numbers. Hence, in this typical situation, the actual transition rate is obtained by averaging over the m_α's of the initial set of degenerate states and summing over the final set of m_β's, leading to

$$w_{n_\beta\ell_\beta,n_\alpha\ell_\alpha}^{(1)}(\text{sp. em.}) = \frac{1}{2\ell_\alpha+1}\sum_{m_\alpha,m_\beta} w_{m_\beta m_\alpha}^{(1)}(\text{sp. em.})$$
$$\cong \frac{1}{2\ell_\beta+1}\left(\frac{q^2\kappa_e}{\hbar c}\right)\frac{4\omega^3\langle r\rangle_{n_\beta\ell_\beta,n_\alpha\ell_\alpha}^2}{c^2}C^2(\ell_\alpha 1\ell_\beta;\ 000)$$
$$\times\sum_{m_\alpha m_\beta}\left|\sum_\mu n^\mu C(\ell_\alpha 1\ell_\beta;\ m_\alpha\mu m_\beta)\right|^2, \qquad (16.80)$$

where the n, ℓ dependence has been restored.

The sum on magnetic quantum numbers in (16.80) can be evaluated with the aid of the orthogonality relation (14.64). The sum of interest is

$$\sum \equiv \sum_{m_\alpha m_\beta}\left|\sum_\mu n^\mu C(\ell_\alpha 1\ell_\beta;\ m_\alpha\mu m_\beta)\right|^2$$
$$= \sum_{m_\alpha m_\beta}\sum_{\mu,\mu'} n^\mu n^{\mu'*} C(\ell_\alpha 1\ell_\beta;\ m_\alpha\mu m_\beta)C(\ell_\alpha 1\ell_\beta;\ m_\alpha\mu' m_\beta). \qquad (16.81)$$

To make use of (16.64) we need to interchange the second and third quantum numbers in the C–G coefficients, which is accomplished via the symmetry relation (Table 14.2)

$$C(j_1 j_2 j_3;\ m_1 m_2 m_3) = (-1)^{j_1-m}\sqrt{\frac{2j_3+1}{2j_2+1}}C(j_1 j_3 j_2;\ m_1 - m_3 - m_2). \qquad (16.82)$$

Applying (16.82) to (16.81), carrying out the sums on m_α and m_β first, and employing (14.64) yields

$$\sum = \sum_{\mu,\mu'} n^\mu n^{\mu'*}\delta_{\mu\mu'}\left(\frac{2\ell_\beta+1}{3}\right) = \frac{2\ell_\beta+1}{3}, \qquad (16.83)$$

since $\sum_\mu |n^\mu|^2 = 1$. Thus, the final expression for $w_{n_\beta\ell_\beta n_\alpha\ell_\alpha}^{(1)}(\text{sp. em.})$ in the dipole approximation is

$$w^{(1)}_{n_\beta \ell_\beta n_\alpha \ell_\alpha}(\text{sp. em.}) \cong \left(\frac{q^2 \kappa_e}{\hbar c}\right) \frac{4\omega^3 \langle r\rangle^2_{n_\beta \ell_\beta n_\alpha \ell_\alpha}}{3c^2} C^2(\ell_\alpha 1 \ell_\beta;\, 000), \tag{16.84}$$

which is valid for a single-particle system in the approximation that effects of spin can be ignored. The "$\cong$" sign in (16.84) is a reminder that the electric-dipole approximation has been made.

Equation (16.84) is the first-order, electric-dipole transition rate for a transition between any pair of single-particle states with opposite parities and orbital angular momenta obeying the restrictions contained in the triangle relation $\Delta(\ell_\alpha 1 \ell_\beta)$ associated with $C(\ell_\alpha 1 \ell_\beta;\, 000)$. For example, transitions between s and p states and between p and d states are allowed, whereas transitions between any pair of states with $\ell_\alpha = \ell_\beta$ (p $\to$ p or d $\to$ d) or $\pi_\alpha = \pi_\beta$ in general are forbidden in the electric-dipole approximation. The lifetime of the state $|\alpha\rangle$ against decay to the state $|\beta\rangle$, denoted $\tau_{\beta\alpha}$, is the inverse of the transition rate:

$$\tau_{\beta\alpha} = [w^{(1)}_{n_\beta \ell_\beta, n_\alpha \ell_\alpha}(\text{sp. em.})]^{-1}. \tag{16.85}$$

EXAMPLE

The 2p $\to$ 1s transition in hydrogen. In the non-relativistic approximation, the difference in energy $E_2 - E_1$ between the 2p and 1s states is 10.2 eV, for which $\omega = 1.55 \times 10^{16}\ \text{s}^{-1}$. The C–G coefficient $C(110;\, 000)$ is $-\sqrt{1/3}$, while

$$\begin{aligned}\langle r\rangle_{1s,2p} &= \int dr\, R_{10}(r) r^3 R_{21}(r) \\ &= \frac{1}{\sqrt{6}a^4}\int_0^\infty dr\, r^4 e^{-3r/(2a)} \\ &= \frac{256}{81\sqrt{6}} a,\end{aligned}$$

with $a \cong a_0 = 0.53$ Å. Substituting these values into (16.84) with $q = e$, we find for the theoretical lifetime the value

$$\tau^{(1)}_{1s,2p}(H) \cong 1.59 \times 10^{-9}\ \text{s},$$

in excellent agreement with the measured value[10]

$$\tau^{(\text{exp})}_{1s,2p}(H) \cong 1.6 \times 10^{-9}\ \text{s}.$$

The agreement is no accident, of course: corrections to the H-atom energy levels, spin effects, etc., play no essential role in the calculation, and neither do higher-order corrections to the pure dipole-matrix element. The agreement between the theoretical and experimental transition rates once again affirms the validity of the quantal description of microscopic phenomena.

[10] Experimental information on atomic transitions are available from various compilations. See, e.g., Wiese, Smith, and Glennon (1966).

16.3. *Sudden and adiabatic approximations, geometric phases

The sudden and the adiabatic approximations are at opposite ends of the time spectrum, the former describing what in essence is an instantaneous change, while the latter is applicable to those events for which $d\hat{H}(t)/dt \cong 0$. That it took more than 50 years before Berry (1984) discovered the existence of a new quantal phase associated with the adiabatic approximation is another of those sublime breakthroughs that have punctuated the development of quantum theory.

The sudden approximation

The sudden approximation is an example of approximations that are connected with discontinuous changes in time-dependent Hamiltonians. We confine our considerations to the following situation:

$$\hat{H}(t) = \begin{cases} \hat{H}^{(1)}, & t < t_0, \\ \hat{H}^{(2)}, & t > t_0, \end{cases} \tag{16.86}$$

where $\hat{H}^{(1)}$ and $\hat{H}^{(2)}$ are time-independent. For simplicity we shall set $t_0 = 0$. The change from $\hat{H}^{(1)}$ to $\hat{H}^{(2)}$ is assumed to occur instantaneously at time $t = 0$. Suppose that, at time $t_1 = 0 - \delta t$, the state of the system is $|\Gamma,\, t_1\rangle$:

$$|\Gamma,\, t_1\rangle = e^{-i\hat{H}^{(1)}t_1/\hbar}|\Gamma\rangle. \tag{16.87}$$

At time $t_2 = 0 + \delta t$, the state is $|\Gamma',\, t_2\rangle$, given by

$$|\Gamma',\, t_2\rangle = e^{-i\hat{H}^{(2)}t_2/\hbar}e^{-i\hat{H}^{(1)}t_1/\hbar}|\Gamma\rangle. \tag{16.88}$$

The assumption of instantaneous change is taken to mean that

$$|\Gamma',\, t_2\rangle \cong |\Gamma,\, t_2\rangle = e^{-i\hat{H}^{(1)}t_2/\hbar}|\Gamma\rangle, \tag{16.89}$$

i.e., the change $\hat{H}^{(1)} \to \hat{H}^{(2)}$ is so rapid and t_2 so small that the state has no time to adjust. We are thus assuming that

$$|\Gamma'\rangle = |\Gamma\rangle. \tag{16.90}$$

Let the eigenstate/eigenvalue relations be

$$\hat{H}^{(j)}|\gamma^{(j)},\, E_\gamma^{(j)}\rangle = E_\gamma^{(j)}|\gamma^{(j)},\, E_\gamma^{(j)}\rangle, \qquad j = 1, 2. \tag{16.91}$$

Then under the assumption (16.90),

$$P_{\beta^{(2)}}(\Gamma) = |\langle\beta^{(2)},\, E_\beta^{(2)}|\Gamma\rangle|^2 \tag{16.92}$$

is the probability (or probability density when $|\Gamma\rangle$ or $|\beta^{(2)},\, E_\beta^{(2)}\rangle$ is in the continuum) that the system will be found in the state $|\beta^{(2)},\, E_\beta^{(2)}\rangle$.

An example neatly illustrating (16.92) concerns the β-decay of the tritium nucleus ^{3}H, wherein one of the two neutrons turns into a proton and simultaneously emits an electron and an electron antineutrino $\bar{\nu}_e$, with the result that

$$^3\mathrm{H} \to {}^3\mathrm{He} + \mathrm{e}^- + \bar{\nu}_\mathrm{e}.$$

In this process, the neutral, one-electron, $Z = 1$ tritium atom becomes a one-electron,

$Z = 2$, singly ionized atom of He, viz., He^+. If the tritium atom is in its ground state, here denoted $|100;\ Z = 1\rangle$, the probability that the He ion will be in its ground state, denoted $|100;\ Z = 2\rangle$, after the β-decay is

$$P_{100}(100) = |\langle 100;\ Z = 2|100;\ Z = 1\rangle|^2 \cong 0.70.$$

There is thus a 30% probability that the He^+ ion will be in an excited state. As is clear from this example, the procedure is straightforward; other applications of the method are explored in the exercises.

The adiabatic approximation and Berry's phase

We consider a system governed by a time-dependent Hamiltonian $\hat{H}(t)$, with $\hat{H}(t)$ changing slowly with time: $\partial\hat{H}(t)/\partial t$ is assumed small in comparison with E/τ, where E and τ are an energy and a time typically characteristic of the system. At any instant t the time-dependent eigenvalue problem is

$$\hat{H}(t)|\gamma_n(t)\rangle = E_{\gamma_n}(t)|\gamma_n(t)\rangle, \tag{16.93a}$$

$$\langle\gamma_n(t)|\gamma_m(t)\rangle = \delta_{mn}, \tag{16.93b}$$

and since $\hat{H}(t)$ is a quantal operator, then by Postulate VI its eigenstates constitute a complete set:[11]

$$\sum_{\gamma_n} |\gamma_n(t)\rangle\langle\gamma_n(t)| = \hat{I}. \tag{16.94}$$

As t changes so may the quantum numbers also change, and we have therefore expressed the time-dependent ket as $|\gamma_n(t)\rangle$ rather than $|\gamma_n,\ t\rangle$ to indicate that it is a solution of (16.93) rather than of the time-dependent Schrödinger equation.

The premise underlying the adiabatic approximation is that, for $\Delta t = t - t_0$ smaller than some (long) time or time interval that is characteristic of the system, the initial state $|\Gamma,\ t_0\rangle \equiv |\gamma_n(t_0)\rangle$ transforms smoothly, or *adiabatically*, into $|\gamma_n(t)\rangle$, *without* admixtures from other states:

$$|\gamma_n(t_0)\rangle \to e^{i\phi_{\gamma_n}(t,t_0)}|\gamma_n(t)\rangle\,, \tag{16.95}$$

where $\phi_{\gamma_n}(t,\ t_0)$ is a real phase factor. More formally, if $\hat{U}(t,\ t_0)$ is the time-evolution operator, then

$$\begin{aligned}|\Gamma,\ t\rangle &= \hat{U}(t,\ t_0)|\gamma_n(t_0)\rangle \\ &= \sum_k |\gamma_k(t)\rangle\langle\gamma_k(t)|\hat{U}(t,\ t_0)|\gamma_n(t_0)\rangle,\end{aligned} \tag{16.96}$$

and the adiabatic approximation is expressed as

$$\langle\gamma_k(t)|\hat{U}(t,\ t_0)|\gamma_n(t_0)\rangle \cong e^{i\phi_{\gamma_n}(t,t_0)}\delta_{kn}, \tag{16.97}$$

whereby $|\Gamma,\ t\rangle$ becomes

[11] For simplicity, we suppress the integral on the LHS of (16.94) corresponding to the contribution from any continuum states.

$$|\Gamma,\ t\rangle \cong e^{i\phi_{\gamma_n}(t,t_0)}|\gamma_n(t)\rangle. \tag{16.98}$$

The absence in (16.98) of the "off-diagonal" terms in (16.96) means that, for all $n \neq k$, $|\langle\gamma_k(t)|\hat{U}(t,\ t_0)|\gamma_n(t_0)\rangle|$ is negligible. We now derive a criterion for the adiabatic approximation to hold; in doing so we will also obtain an expression for the phase $\phi_{\gamma_n}(t,\ t_0)$.

The solution $|\Gamma,\ t\rangle$ to the time-dependent Schrödinger equation

$$i\hbar\,\frac{\partial|\Gamma,\ t\rangle}{\partial t} = \hat{H}(t)|\Gamma,\ t\rangle \tag{16.99}$$

can be expanded using the instantaneous eigenstates $\{|\gamma_k(t)\rangle\}$:

$$|\Gamma,\ t\rangle = \sum_{\gamma_k} C_{\gamma_k}(t)|\gamma_k(t)\rangle e^{-(i/\hbar)\int_0^t E_{\gamma_k}(t')\,dt'}, \tag{16.100}$$

where we chose $t_0 = 0$ and have extracted the "dynamical phase" $\hbar^{-1}\int_0^t E_{\gamma_k}(t')\,dt'$ from the time dependence of the coefficients $C_{\gamma_k}(t)$. Substituting (16.100) into (16.99) and projecting both sides of the resulting equation onto $\langle\gamma_m(t)|$ yields

$$\dot{C}_{\gamma_m}(t) = -\sum_{\gamma_k}\left\langle \gamma_m(t)\left|\frac{\partial}{\partial t}\right|\gamma_k(t)\right\rangle C_{\gamma_k}(t) e^{-(i/\hbar)\int_0^t [E_{\gamma_k}(t') - E_{\gamma_m}(t')]\,dt'} \tag{16.101}$$

where $\dot{C}_{\gamma_m}(t) \equiv \partial C_{\gamma_m}(t)/\partial t$.

Since our interest is in investigating conditions under which the adiabatic approximation $|\gamma_n(t=0)\rangle \to |\gamma_n(t)\rangle$ holds, we assume that, at $t = 0$,

$$C_{\gamma_k}(t=0) = \delta_{kn}. \tag{16.102a}$$

Furthermore, we also assume that C_{γ_n} will remain the dominant coefficient in $|\Gamma,\ t\rangle$ for $t > 0$:

$$C_{\gamma_k}(t) \cong \delta_{kn}, \tag{16.102b}$$

an approximation we shall justify *a posteriori*. Of course, rather than $C_{\gamma_k}(t) = 0$, $k \neq n$, we expect that, if adiabaticity occurs, then $C_{\gamma_k}(t)$ will be small relative to $C_{\gamma_n}(t)$. Use of Eq. (16.102b) in (16.101) will help us estimate $C_{\gamma_k}(t)$, $k \neq n$. Employing (16.102b) we find

$$\dot{C}_{\gamma_m}(t) \cong -\left\langle \gamma_m(t)\left|\frac{\partial}{\partial t}\right|\gamma_n(t)\right\rangle e^{-(i/\hbar)\int_0^t [E_{\gamma_n}(t') - E_{\gamma_m}(t')]\,dt'}, \tag{16.103}$$

where we specifically assume that $m \neq n$.

An expression for $\langle\gamma_m(t)|\partial/\partial t|\gamma_n(t)\rangle$ can be obtained by differentiating both sides of Eq. (16.93a) w.r.t. t and then projecting both sides of the resulting equation onto $\langle\gamma_m(t)|$. This leads to

$$\begin{aligned}\left\langle \gamma_m(t)\left|\frac{\partial}{\partial t}\right|\gamma_n(t)\right\rangle &= \frac{\langle\gamma_m(t)|\partial\hat{H}(t)/\partial t|\gamma_n(t)\rangle}{E_{\gamma n}(t) - E_{\gamma m}(t)} \\ &= -\frac{\langle\gamma_m(t)|\partial\hat{H}/\partial t|\gamma_n(t)\rangle}{\hbar\omega_{mn}(t)}, \qquad m \neq n,\end{aligned} \tag{16.104}$$

where

$$\omega_{mn}(t) = [E_{\gamma_m}(t) - E_{\gamma_n}(t)]/\hbar. \tag{16.105}$$

Using Eq. (16.104), $\dot{C}_{\gamma_m}(t)$ becomes

$$\dot{C}_{\gamma_m}(t) \cong \frac{\langle \gamma_m(t)|\partial \hat{H}(t)/\partial t|\gamma_n(t)\rangle}{\hbar\omega_{mn}(t)} e^{i\int_0^t \omega_{mn}(t')\,dt'}. \tag{16.106}$$

The next step is the crucial one. We assume that adiabaticity means that t is small enough and the variation with time slow enough that

$$\omega_{mn}(t) \cong \omega_{mn}, \tag{16.107}$$

independent of time, and also that the t dependence of $\langle \gamma_m(t)|\partial \hat{H}/\partial t|\gamma_n(t)\rangle$ can be ignored. Equation (16.107) leads to

$$\int_0^t \omega_{mn}(t')\,dt' \cong \omega_{mn} t,$$

while the approximate constancy of the matrix element in (16.106) means that $\dot{C}_{\gamma_m}(t)$ can be integrated:

$$C_{\gamma_m}(t) \cong \frac{\langle \gamma_m(t)|\partial \hat{H}(t)/\partial t|\gamma_n(t)\rangle}{i\hbar\omega_{mn}^2}(e^{i\omega_{mn}t} - 1), \qquad m \neq n. \tag{16.108}$$

Equation (16.108) leads to the criterion for $C_{\gamma_m}(t)$ being sufficiently small, and correspondingly, to the circumstances justifying Eq. (16.102b), viz.,

$$\left|\left\langle \gamma_m(t)\left|\frac{\partial \hat{H}(t)}{\partial t}\right|\gamma_n(t)\right\rangle\right| \ll \hbar\omega_{mn}^2. \tag{16.109}$$

When the inequality (16.109) is satisfied, the adiabatic approximation is justified. This criterion means that the $|\gamma_n(t)\rangle \rightarrow |\gamma_m(t)\rangle$ matrix element of the operator $\partial\hat{H}/\partial t$ is much smaller than the difference in energy $|E_{\gamma_m}(t) - E_{\gamma_n}(t)|$ times the rate $|\omega_{mn}|$, for all states $|\gamma_m(t)\rangle$. Inherent in this analysis is the assumption of non-degeneracy: $E_{\gamma_m}(t) \neq E_{\gamma_n}(t)$, $m \neq n$. That $\partial\hat{H}(t)/\partial t$ is to be small relative to some characteristic ratio E/τ means that $E \sim |E_{\gamma_m} - E_{\gamma_n}|$ and $\tau^{-1} \sim |\omega_{nm}|$.

As long as (16.109) is satisfied, then $C_{\gamma_m}(t)| \ll 1$, and $|\Gamma, t = 0\rangle = |\gamma_n(t = 0)\rangle$ is transformed smoothly into $|\Gamma, t\rangle \propto |\gamma_n(t)\rangle$. Let us now determine the proportionality factor, $\exp(i\phi_{\gamma_n}(t, t_0 = 0))$ of Eq. (16.98). Setting $m = n$ in (16.101) yields

$$\dot{C}_{\gamma_n}(t) = -\left\langle \gamma_n(t)\left|\frac{\partial}{\partial t}\right|\gamma_n(t)\right\rangle C_{\gamma_n}(t) - \sum_{\gamma_k \neq \gamma_n}\left(\left\langle \gamma_n(t)\left|\frac{\partial}{\partial t}\right|\gamma_k(t)\right\rangle e^{i\int_0^t \omega_{kn}(t')\,dt'}\right) C_{\gamma_k}(t). \tag{16.110}$$

The assumption of an adiabatic process means that $|C_{\gamma_k}(t)| \ll 1$, $k \neq n$ and therefore that the summation on the RHS of (16.110) can be neglected, leading to

$$\dot{C}_{\gamma_n}(t) \cong -\left\langle \gamma_n(t)\left|\frac{\partial}{\partial t}\right|\gamma_n(t)\right\rangle C_{\gamma_n}(t),$$

or

$$C_{\gamma_n}(t) \cong \exp\left(-\int_0^t \left\langle \gamma_n(t') \middle| \frac{\partial}{\partial t'} \middle| \gamma_n(t') \right\rangle dt'\right). \tag{16.111}$$

Comparing Eqs. (16.98), (16.100), and (16.111), we find that, in the relation

$$|\Gamma,\, t\rangle \cong e^{i\phi_{\gamma_n}(t)}|\gamma_n(t)\rangle, \tag{16.112a}$$

the phase $\phi_{\gamma_n}(t) \equiv \phi_{\gamma_n}(t,\, t_0 = 0)$ is given by

$$\phi_{\gamma_n}(t) = -\frac{1}{\hbar}\int_0^t E_{\gamma_n}(t')\, dt' + i\int_0^t \left\langle \gamma_n(t') \middle| \frac{\partial}{\partial t'} \middle| \gamma_n(t') \right\rangle dt'. \tag{16.112b}$$

The first term in (16.112b) is the dynamical phase; the second is one form of Berry's phase (Berry 1984). Note that the matrix element in Berry's phase is pure imaginary, so that $i\langle \gamma_n(t')|\partial/\partial t'|\gamma_n(t')\rangle$ is real. This last statement follows from normalization:

$$\frac{\partial}{\partial t}\langle \gamma_n(t)|\gamma_n(t)\rangle = 0,$$

or

$$\left\langle \frac{\partial}{\partial t}\gamma_n(t)|\gamma_n(t)\right\rangle + \left\langle \gamma_n(t) \middle| \frac{\partial}{\partial t} \middle| \gamma_n(t) \right\rangle = 0,$$

or

$$\left\langle \gamma_n(t) \middle| \frac{\partial}{\partial t} \middle| \gamma_n(t) \right\rangle + \left\langle \gamma_n(t) \middle| \frac{\partial}{\partial t} \middle| \gamma_n(t) \right\rangle^* = 0,$$

a relation that can be satisfied only if $\mathrm{Re}[\langle \gamma_n(t)|\partial/\partial t|\gamma_n(t)\rangle] = 0$, QED.

To recapitulate, the following result has been derived: in an adiabatic situation, a state $|\gamma_n(t = 0)\rangle$ evolves under $\hat{H}(t)$ to the state $|\Gamma,\, t\rangle = \exp[i\phi_{\gamma_n}(t)]\,|\gamma_n(t)\rangle$. The appearance of the simple phase factor is analogous to the result that occurs when $\hat{H}$ is time-independent. Quite apart from having determined the condition under which this result is valid – Eq. (16.109) – the reader may reasonably inquire into its significance. Is it not true that a phase factor is a phase factor is a phase factor: it is present but unobservable, no? The answer is NO(!): the dynamical part of $\phi_{\gamma_n}(t)$ is unobservable, but under the circumstance of a cyclic evolution, the Berry portion not only is observable but *has been observed.* Berry (1984) suggested two possible experments, one involving a magnetic field whose direction varies slowly in time, the other involving an optical fiber whose cross-sectional area is slowly altered. The first experiments establishing that the Berry phase was indeed observable were those of Chiao and Wu (1986) using optical fibers; other experiments were reported soon after.

The result is doubly remarkable. First, for $\hat{H} \neq \hat{H}(t)$, the phase – just the relevant dynamical one, $E_{\gamma_n}t/\hbar$ – is not observable, but for an $\hat{H}(t)$ changing adiabatically, the non-dynamical part of the phase – Berry's phase – can lead to observable effects. The second remarkable aspect is that, beginning with the very earliest formulation of the adiabatic approximation,[12] it had been assumed that Berry's phase could be eliminated

[12] See, e.g., Born and Fock (1928), Kemble (1958), and Schiff (1968.)

from $\phi_{\gamma_n}(t)$ by a "renormalization" procedure. This consisted of multiplying $|\gamma_n(t)\rangle$ by the complex conjugate of the Berry phase factor, in which case only the dynamical part of the phase remains. Such a procedure seems to be the analog of eliminating the arbitrary phase multiplying the eigenstates of an $\hat{H} \neq \hat{H}(t)$, as in the cases of the bound states studied in Chapters 3, 6, and 11. However, $\phi_{\gamma_n}(t)$ is not a global phase factor and, as Berry (1984) has commented, his phase cannot be eliminated by a gauge transformation.

One can associate a "geometric" character with Berry's phase. To understand this, we follow Berry's original analysis and suppose that the time dependence enters $\hat{H}(t)$ through some parameters, say $\{X_j(t)\}_{j=1}^N$. For example, if $N = 3$, X_1, X_2, and X_3 could be the Cartesian components of a vector $\mathbf{R} = \mathbf{R}(t)$. $\hat{H}(t)$ then depends on the $\{X_j\}$:

$$\hat{H}(t) = \hat{H}(\{X_j(t)\})$$

and

$$|\gamma_n(t)\rangle = |\gamma_n(\{X_j(t)\})\rangle,$$

whereupon the integral defining Berry's phase can be transformed via

$$\int_0^t \left\langle \gamma_n(t') \middle| \frac{\partial}{\partial t'} \middle| \gamma_n(t') \right\rangle dt' \rightarrow \int_C \sum_j \left\langle \gamma_n(\{X_k\}) \middle| \frac{\partial}{\partial X_j} \middle| \gamma_n(\{X_k\}) \right\rangle dX_j, \tag{16.113}$$

with C being the path along which the parameter-space integration occurs.

The second term on the RHS of (16.112b) is Berry's phase. Using (16.113), we write it as

$$\begin{aligned} \beta(t) - \beta(0) &= i\int_0^t \left\langle \gamma_n(t') \middle| \frac{\partial}{\partial t'} \middle| \gamma_n(t') \right\rangle dt' \\ &= i\int_C \sum_j \left\langle \gamma_n(\{X_k\}) \middle| \frac{\partial}{\partial X_j} \middle| \gamma_n(\{X_k\}) \right\rangle dX_j \\ &\equiv \beta(C). \end{aligned} \tag{16.114}$$

The notation $\beta(C)$ underscores the fact that Berry's phase depends only on the path, i.e., the relevant "geometry" in parameter space. Berry has emphasized that, for a cyclic adiabatic process, with t set equal to the appropriate period T, $\beta(C)$ will not vanish.

The Aharonov–Anadan phase

A few years after Berry's analysis, Aharonov and Anandan (1987) extended the concept of a phase associated with adiabatic evolution to a geometric phase defined for any cyclic evolution of a quantum system governed by a time-dependent Hamiltonian. The Berry and Aharonov–Anandan (A–A) phases arise in the context of $\hat{H} = \hat{H}(t)$. More recently, Zeng and Lei (1995) have shown that a geometric phase can occur for non-stationary states that are linear combinations of the eigenstates of a time-independent Hamiltonian. We restrict our discussion to that of the A–A phase.

Aharonov and Anandan's extension of Berry's phase is simple and elegant. Let $\hat{H}(t)$

govern a system in which cyclic temporal evolution occurs. If τ is the cycle time (i.e., the relevant period), then, after an interval τ, the system must return to its initial configuration. Hence, the state vector $|\Gamma,\, t_0+\tau\rangle$ can differ from $|\Gamma,\, t_0\rangle$ by at most a phase factor:

$$|\Gamma,\, \tau\rangle = e^{i\phi^{\mathrm{AA}}}|\Gamma,\, t=0\rangle, \tag{16.115}$$

where t_0 has been set to zero, the A–A phase ϕ^{AA} is real, and $|\Gamma,\, t\rangle$ is normalized, $\langle\Gamma,\, t|\Gamma,\, t\rangle = 1$.

To show that ϕ^{AA} is made up of a dynamical part and a portion analogous to the Berry phase, $|\Gamma,\, t\rangle$ is transformed in a particular way:

$$|\Gamma,\, t\rangle \to |\Gamma,\, t\rangle^{(f)} = e^{-if(t)}|\Gamma,\, t\rangle, \tag{16.116a}$$

with the real phase $f(t)$ chosen so that

$$f(\tau) - f(0) = \phi^{\mathrm{AA}}, \tag{16.116b}$$

from which it follows that

$$|\Gamma,\, \tau\rangle^{(f)} = |\Gamma,\, 0\rangle^{(f)}. \tag{16.116c}$$

An expression for $\dot{f}(t)$ can be obtained by using (16.116a) in (16.99):

$$\dot{f} = \frac{df}{dt} = -\frac{1}{\hbar}\langle\Gamma,\, t|\hat{H}(t)|\Gamma,\, t\rangle + {}^{(f)}\!\left\langle\Gamma,\, t\left|i\frac{\partial}{\partial t}\right|\Gamma,\, t\right\rangle^{(f)},$$

which can be integrated to yield

$$f(\tau) - f(0) = \phi^{\mathrm{AA}} = \alpha^{\mathrm{AA}} + \beta^{\mathrm{AA}}, \tag{16.117a}$$

where

$$\alpha^{\mathrm{AA}} = -\frac{1}{\hbar}\int_0^\tau dt\langle\Gamma,\, t|\hat{H}(t)|\Gamma,\, t\rangle \tag{16.117b}$$

is the dynamical phase and

$$\beta^{\mathrm{AA}} = \int_0^\tau dt\; {}^{(f)}\!\left\langle\Gamma,\, t\left|i\frac{\partial}{\partial t}\right|\Gamma,\, t\right\rangle^{(f)} \tag{16.117c}$$

is the analog of Berry's phase. However, the results (16.117) are exact, not approximate.

Aharonov and Anandan (1987) used a projective-geometry analysis to show that ϕ^{AA} is geometric in nature, depending only on a closed curve in a projective space. This feature can be understood using the language of the preceding subsection: if the t dependence is expressed through a set of parameters $\{X_j(t)\}$, then ϕ^{AA} will depend only on a closed curve traced out in parameter space during the cycle.

Aharonov and Anandan discussed several situations in which β^{AA} emerges naturally and is observable in principle. Their phase was first observed by Suter, Mueller, and Pines (1988) using nuclear-magnetic-resonance interferometry. For a compilation of references and lectures on the Berry and A–A phases, the well-prepared reader may consult the volumes edited by Shapere and Wilczek (1989) and Anandan (1990).

16.4. *Exponential decay and the time–energy uncertainty relation

In this section we establish a connection between two properties often associated with an unstable state. These are the decaying exponential character of its time evolution and a time–energy uncertainty-type expression relating its lifetime and its energy spread (or uncertainty). Neither property is exact; both are very useful.

Background: exponential decay

In Section 16.2 we employed a standard relation between the time rate of change of the population $N_\alpha(t)$ of a decaying state $|\alpha\rangle$ and the population at that time, viz.,

$$\frac{dN_\alpha(t)}{dt} = -w_\alpha N_\alpha(t). \tag{16.118}$$

This is a widely used relation. In the quantal context, in which $N_\alpha(t) = P_\alpha(t)N_0$, with $N_0 = N_\alpha(t=0)$ and $P_\alpha(t)$ being the decay probability, it leads to

$$P_\alpha(t) = e^{-w_\alpha t}, \tag{16.119}$$

where $w_\alpha^{-1} = \tau_\alpha$ is the lifetime of the decaying state.

Equations (16.118) and (16.119) apply only to an isolated population of the unstable states $|\alpha\rangle$. In reality, such isolation is not found, e.g., repopulation can occur or $|\alpha\rangle$ can decay to more than one final state, each process having its own decay probability. Hence, Eq. (16.119) is at best a potent approximation, valuable even though exponential decay can never occur.

That pure exponential decay is actually forbidden on theoretical grounds has been shown by Kabir and Pilaftsis (1996) to follow from the fact that the time-evolution operator is unitary. The proof is simple. Let $C_{\beta\alpha}(t)$ be the amplitude that an initial state $|\alpha\rangle$ will be observed to be in state $|\beta\rangle$ at time t and let the Hamiltonian $\hat{H}$ governing the system be time-independent. Then

$$C_{\beta\alpha}(t) = \langle\beta|e^{-i\hat{H}t/\hbar}|\alpha\rangle, \tag{16.120}$$

with

$$P_{\beta\alpha}(t) = |C_{\beta\alpha}(t)|^2. \tag{16.121}$$

Since $\hat{H}^\dagger = \hat{H}$, it follows from (16.120) that

$$C_{\beta\alpha}(t) = \langle\alpha|e^{i\hat{H}t/\hbar}|\beta\rangle^* = C^*_{\alpha\beta}(-t). \tag{16.122}$$

Now set $\beta = \alpha$, with $P_{\alpha\alpha}(t)$ being the probability that the state does not decay, while $1 - P_{\alpha\alpha}(t)$ is the probability that it does. Equation (16.122) leads to

$$P_{\alpha\alpha}(t) = P_{\alpha\alpha}(-t),$$

i.e., $P_{\alpha\alpha}(t)$ is an even function of t. It then follows that $\dot{P}_{\alpha\alpha}(t) = dP_{\alpha\alpha}(t)/dt$ is an *odd* function of time and therefore that

$$\dot{P}_{\alpha\alpha}(t=0) = 0. \tag{16.123}$$

Equation (16.123) is incompatible with an exponential probability of decay. That is, Eq. (16.119) cannot be true for all t.

While (16.119) cannot hold in general and in particular not for $t \cong 0$ (and not for $t \to \infty$, either), it is an important and useful approximation, one that characterizes the long-time evolution of a certain type of unstable state, as discussed later in this section.

Background: time–energy uncertainty relations

Since time is a parameter rather than a dynamical variable in quantum mechanics, it was long believed that neither a "time" operator $\hat{T}$ obeying $[\hat{T}, \hat{H}] = i\hbar$ nor a formal time–energy uncertainty relation analogous to Eq. (9.17) could exist. Pauli (1958) argued against the existence of such a $\hat{T}$ on the ground that time is unbounded ($t \to \pm\infty$ is a standard pair of limits in quantum theory) while $\hat{H}$ is always bounded from below (zero energy for a free particle, minimum negative energy for bound states).

An operator $\hat{T}$ being absent (but see below), two approaches to formulating time–energy uncertainty relations have been followed. (A reasonably comprehensive set of references in this regard can be found in Kobe and Aguilera-Novarro (1994).) In one approach, various formal definitions that led to quantities ΔT (or Δt) to be used with ΔH in setting up a $\Delta T \, \Delta H \geqslant \hbar/2$ relation have been proposed. Some of these definitions can lead to infinities, as noted, for example by Bauer and Mello (1978), whose own definitions were designed to avoid infinities. In the other approach, quantities denoted Δt and ΔE that satisfy a $\Delta E \, \Delta t \geqslant \hbar$ relation are obtained. Although the latter are somewhat *ad hoc*, they have led to some very useful expressions, in particular Eq. (16.39), which states that the energy spread or "width" ΔE_β of an unstable state is related to its lifetime τ_β via $\Delta E_\beta \tau_\beta \sim \hbar$. In Section 16.1 we mentioned that this relation is a possible means for justifying the sum over final states used in deriving Eq. (16.41). A "derivation" of Eq. (16.39) is considered later in this section.

Various heuristic definitions of Δt and time–energy uncertainty-type relations have been proposed. If a quantal time operator $\hat{T}$ existed, it might be the means to establish the primacy of one of the heuristic definitions. The first possibility in this direction became available with the introduction by Kobe and Aguilero-Navarro (1994) of a quantal operator $\hat{T}$ that obeys

$$[\hat{T}, \hat{H}] = i\hbar, \tag{16.124a}$$

leading to

$$\Delta T \, \Delta H \geqslant \tfrac{1}{2} |\langle [\hat{T}, \hat{H}] \rangle| = \hbar/2, \tag{16.124b}$$

although these relations are not valid on all of $\mathcal{H}$.

The Kobe–Aguilero-Navarro (K–AN) operator $\hat{T}$ is denoted *tempus,* and is the quantal version of a classical quantity T defined for conservative systems. Classically, *tempus* is a generalized coordinate, which in the single-particle, 1-D case is obtained from the coordinate q and momentum p by a canonical transformation. The new pair of generalized coordinates p' and q' is chosen so that p' is the energy E ($= H$ for a conservative system); its canonical conjugate partner q' is the (classical) quantity *tempus* and has the dimension time: $p' = E$, $q' = T$. In the case of a conservative system it is shown by K–AN that $\dot{T} = 1$ and $\dot{E} = 0$, so that $T = t + C_1$ and $E =$ constant, where C_1 is a constant.

Quantization makes use of the fact that the classical Poisson bracket $[\ ,\ ]_{\text{cl}}$ cannot change value under a canonical transformation, so that $[q, p]_{\text{cl}} = 1$ means that $[T, E]_{\text{cl}} = 1$. On writing the quantal counterparts as $\hat{T}$ and $E = \hat{H}$, the Poisson-bracket relation for T and E becomes Eq. (16.124a), thus leading to Eq. (16.124b). However, for Eq. (16.124b) to be meaningful, only those Hilbert-space states for which $\langle \hat{T} \rangle$ and $\langle \hat{T} \rangle^2$ are defined are allowed. In addition to this latter restriction, it is essential to realize that, in contrast to $\hat{\mathbf{P}}$, $\hat{\mathbf{Q}}$, and $\hat{\mathbf{L}}$, $\hat{T}$ is *not* a universal operator: for each conservative H and its quantal equivalent $\hat{H}$ there will be a corresponding pair T and $\hat{T}$. That is, due to T and H being a pair of canonically related generalized "coordinates," $\hat{T} = \hat{T}(\hat{H})$. Since we are demanding a $\hat{T}$ such that $[\hat{T}, \hat{H}] = i\hbar$, this might be no surprise, but it undoubtedly means that the determination of $\hat{T}$ in particular cases will be formidable.

K–AN evaluate $\hat{T}$ for two examples, one of which we recount here, namely a 1-D free particle of mass m. The classical energy is $E = p^2/(2m)$, for which T was found to be

$$T = mq/p, \tag{16.125a}$$

where q is the position. Quantization of (16.125a) yields

$$\hat{T} = \frac{m}{2}(\hat{Q}\hat{P}^{-1} + \hat{P}^{-1}\hat{Q}) = \hat{T}^\dagger, \tag{16.125b}$$

and we can predict that the presence of $\hat{P}^{-1}$ might be troublesome.

It is convenient to work in momentum space, wherein $\hat{H}(p) = p^2/(2m)$. One of the simplest normalizable wave functions that one might consider using to evaluate ΔT and ΔH is a pure Gaussian, but this choice leads to problems due to the $\hat{P}^{-1}$ factor in $\hat{T}$. Because of this K–AN chose a modified Gaussian packet:

$$\varphi_\alpha(p) = N_n p^n e^{-p^2 a_0}, \tag{16.126a}$$

where it will be shown that the integer n must be greater than unity. For later use, we express the factor a_0 of Eq. (16.126a) in terms of the "width" δp of the modified packet:

$$a_0 \equiv \frac{1}{2(\delta p)^2}. \tag{16.126b}$$

The time-evolved wave function is

$$\Phi_\alpha(p, t) = N_n p^n e^{-p^2 a}, \tag{16.127a}$$

where

$$a = a(t) = a_0 + i\frac{t}{2m\hbar}. \tag{16.127b}$$

Recalling that $\hat{Q}(p) = i\hbar\,\partial/\partial p$, the expectation value of $\hat{T}(p)$ at time t is, using Eqs. (16.125b) and (16.127a),

$$\langle \hat{T} \rangle_\alpha(t) = t, \tag{16.128}$$

where t is the current value of time for the system (i.e., the elapsed time of development from $t_0 = 0$). Note that (16.128) is independent of n. That $\hat{T}$ is indeed a quantal time operator is borne out by this last result.

Rather more interesting than (16.128) is ΔT, the uncertainty at time t:

$$\Delta T = \frac{2ma_0\hbar}{\sqrt{n-\frac{3}{2}}} = \frac{\hbar}{2}\frac{1}{(\delta p)^2/(2m)}\frac{1}{\sqrt{n-\frac{3}{2}}}, \tag{16.129}$$

where we have used Eq. (16.126b). Several features of this result are worthy of comment. First, it is independent of t: the uncertainty in $\hat{T}$ is constant for all time. Second, its dependence on n means that it slowly diminishes as n increases, going to zero as $n^{-1/2}$. Third, it is inversely proportional to the energy δE associated with the momentum spread δp:

$$\delta E = \frac{(\delta p)^2}{2m}; \tag{16.130a}$$

that is,

$$\Delta T = \frac{\hbar}{2}\frac{1}{\delta E}\frac{1}{\sqrt{n-\frac{3}{2}}}. \tag{16.130b}$$

We shall remark shortly on the meaning of both ΔT and δE. Fourth, the appearance of the factor $(n-\frac{3}{2})^{-1/2}$ means that $\hat{T}$ is not defined over all of $\mathcal{H}$: states $\varphi_\alpha(p)$ with $n=0$ and $n=1$ are excluded, thus verifying our earlier remark. In other words, because it is a singular operator, $\hat{T}$ has meaning only on a subspace of $\mathcal{H}$.

The next step is to evaluate ΔH. Both $\langle\hat{H}\rangle_\alpha(t)$ and ΔH have the dimension energy, and, because the only quantity with this dimension that can be constructed from the parameter δp and the mass m is $(\delta p)^2/m \propto \delta E$ of (16.130a), we may expect that each of $\langle\hat{H}\rangle_\alpha(t)$ and ΔH will be proportional to δE. As it turns out, not only is this the case, but also ΔH and $\langle\hat{H}\rangle_\alpha(t)$ are equal:

$$\Delta H = \langle\hat{H}\rangle_\alpha(t) = \frac{(\delta p)^2}{2m}(n+\tfrac{1}{2}) \equiv (n+\tfrac{1}{2})\,\delta E. \tag{16.131}$$

It follows from Eqs. (16.130b) and (16.131) that

$$\Delta H\,\Delta T = \frac{\hbar}{2}\sqrt{\frac{(n+\frac{1}{2})^2}{n-\frac{3}{2}}} > \frac{\hbar}{2}, \qquad n>1. \tag{16.132}$$

We see that, in the present case, the equality in (16.124b) is never achieved. This is a consequence of the singular nature of $\hat{T}$ and the use of modified Gaussian wave packets for free particles.

It is amusing to note that, while ΔT approaches zero as $n^{-1/2}$, both $\langle\hat{H}\rangle_\alpha$ and ΔH go to infinity like n, so that $\Delta H\,\Delta T \to \infty$ as $n^{1/2}$. Apart from the factors of n, we also see that (16.132) can be characterized as

$$\delta E\,\Delta T \sim \hbar, \tag{16.133a}$$

or, equivalently

$$\Delta T \sim \frac{\hbar}{\delta E}. \tag{16.133b}$$

Equations (16.133) exemplify a more general qualitative relation that we shall explore in detail below and apply in the next subsection, but we first seek a more direct interpretation of ΔT. That is, does ΔT have a meaning beyond the qualitative one expressed by Eq. (16.133b)? The answer is YES, but the meaning is extrinsic to the K–AN calculation that produced the result (16.129). Because $\Phi_\alpha(p, t)$ is a modified Gaussian packet, then, apart from the factor $(n - \frac{3}{2})^{-1/2}$, ΔT is the analog of the "spreading time" t_d appearing in Eqs. (7.35) and (7.36). In other words, $d^2/2$ of (7.23) is to be identified with a_0 of (16.126a), whereupon $t_d = md^2/\hbar$ becomes

$$t_d = \frac{\hbar}{2} \frac{1}{\delta E} \tag{16.134}$$

($a_0^{-1} = 2(\delta p)^2$). Furthermore, the normalization factor of $\Psi_\alpha(x, t)$, the coordinate-space counterpart to $\Phi_\alpha(p, t)$ of (16.127a), contains the quantity

$$r_n = \sqrt{\frac{1}{a_0(1 + it/t_d)^{m+1}}}, \qquad n = 2m \text{ or } n = 2m + 1, \qquad m = 0, 1, 2, \ldots, \tag{16.135}$$

with t_d given by (16.134). Hence, exactly as in the simple Gaussian case of Chapter 7,

$$\Delta T = t_d \Big/ \sqrt{n - \tfrac{3}{2}}$$

is a measure of the spreading of the coordinate-space wave packet $\Psi_\alpha(x, t)$: the larger ΔT the longer the time interval over which the packet's shape remains unchanged.

This result required knowledge of $\Psi_\alpha(x, t)$: it is not intrinsic to the definition of $\hat{T}$. In most cases, however, neither the analog of $\Psi_\alpha(x, t)$ nor $\hat{T}$ will be easy or even possible to obtain. In the general absence of the latter quantities, we see that the theoretical existence of $\hat{T}$ seems not to lead to the primacy of any of the heuristic time–energy uncertainty relations. Nevertheless, one expects that the qualitative relation (16.133) will generally be valid, and we therefore still need a means for obtaining it. As noted, for example, by Cohen-Tannoudji, Diu, and Laloë (1977), one means for doing so makes use of a general property of Fourier transforms. A sketch of their analysis follows. We work with (1-D) energy eigenstates, labeled $|E\rangle$:

$$(\hat{H} - E)|E\rangle = 0, \qquad \hat{H} \neq \hat{H}(t).$$

These can be used to express an arbitrary state $|\Gamma, t_0\rangle$:

$$|\Gamma, t_0\rangle = \int dE\, A_\Gamma(E) e^{-iEt_0/\hbar} |E\rangle,$$

where $A_\Gamma(E)$ is centered on E_0 and has a width δE.

At a later time t we have

$$|\Gamma, t\rangle = \int dE\, A_\Gamma(E) e^{-iE(t-t_0)/\hbar} |E\rangle,$$

and the amplitude for finding an arbitrary state $|\alpha\rangle$ in $|\Gamma, t\rangle$ is

$$C_\alpha(t) = \langle\alpha|\Gamma, t\rangle = \int dE\, A_\Gamma(E) e^{-iE\,\delta t/\hbar} \langle\alpha|E\rangle, \qquad \delta t = t - t_0.$$

In order to obtain the desired result we make the crucial assumption that $\langle\alpha|E\rangle$ is slowly varying for E in a range δE around E_0, by which $C_\alpha(t)$ becomes

$$C_\alpha(t) \cong \langle\alpha|\mathrm{E}_0\rangle \int dE\, A_\Gamma(E) e^{-iE\,\delta t/\hbar}.$$

However, a general property of Fourier transforms states that the range δt over which $C_\alpha(t)$ is significant is related to the range δE by

$$\delta E\,\delta t \sim \hbar, \tag{16.136}$$

which is a generalization of (16.133). Equation (16.136) is a qualitative statement, based on wave packets, of the formal relation (16.124b). In the next subsection we describe some features of a model of a decaying state that leads to an exponential decay law and a qualitative time–energy uncertainty relation like (16.136).

Resonance model of a decaying state

Rather than consider an excited, particle-stable *bound* state that decays by emitting electromagnetic radiation, we examine aspects of a particle-unstable *continuum* state formed in a collision and assumed to exist for a relatively long time in a spatially confined region, thus mimicking, over its lifetime, a true bound state. The occurrence of this particular type of spatially confined, long-lived continuum state has been inferred from experiments in many different areas of physics (e.g., atomic, nuclear, and elementary-particle physics); it is known as a *resonance.* Analyzing it quantitatively even for a single-particle system is too complex for this text, and our discussion will therefore be qualitative and descriptive. It is also somewhat simplified compared with a quantitative analysis, but the main features noted below do arise in the more accurate description.

Resonances are characterized experimentally as relatively narrow bumps or peaks occurring when collision cross sections are plotted as a function of the bombarding energy. In particular, in the vicinity of the resonance energy, the (elastic) scattering phase shift rises rapidly through $\pi/2$ as the energy increases. Figure 16.4 is a schematic rendering of resonance behavior in a collision cross section $\sigma(E)$, while Figs. 16.5

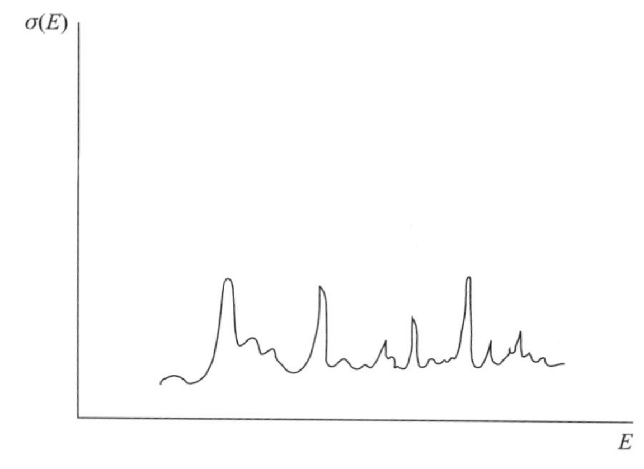

Fig. 16.4 Resonance peaks in a cross section $\sigma(E)$.

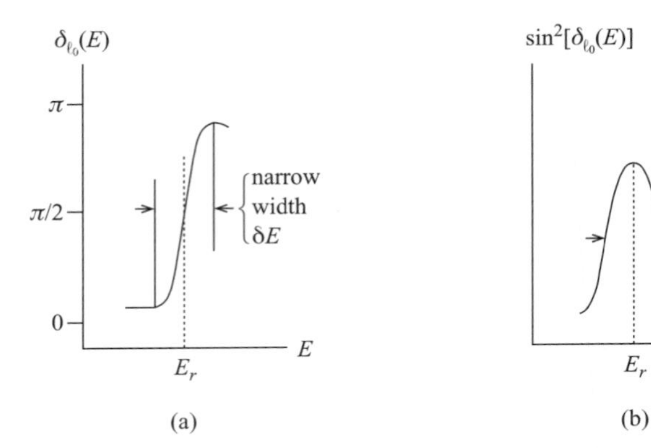

Fig. 16.5 Behaviors of $\delta_{\ell_0}(E)$ and $\sin^2[\delta_{\ell_0}(E)]$ in the vicinity of a narrow resonance of width δE, which occurs at $E \cong E_r$. (a) $\delta_{l_0}(E)$ rising rapidly up through $\pi/2$ at $E = E_r$. (b) $\sin^2[\delta_{\ell_0}(E)]$ rapidly reaching its maximum of unity at $E = E_r$.

illustrate the behaviors of $\delta_{\ell_0}(E)$ and $\sin^2[\delta_{\ell_0}(E)]$ when a resonance is assumed to occur at an energy E_r in the ℓ_0th partial wave of a spinless system. We assume rotational invariance, so that there is a unique orbital angular momentum ℓ_0 associated with the resonance state.

Since $\delta_{\ell_0}(E)$ rises rapidly through $\pi/2$ at $E = E_r$, it can be parameterized for E near to E_r by

$$\delta_{\ell_0}(E) = \frac{\Gamma_{\ell_0}/2}{E_r - E}, \tag{16.137}$$

where $\Gamma_{\ell_0} > 0$ is the width of resonance (δE in Fig. 16.5), which is assumed to be narrow: $\Gamma_{\ell_0} \ll E_r$. It then follows that the ℓ_0th S-matrix takes the form (E near to E_r)

$$S_{\ell_0}(E) = \frac{E_r - E + i\Gamma_{\ell_0}/2}{E_r - E - i\Gamma_{\ell_0}/2}, \tag{16.138}$$

while the partial-wave scattering amplitude becomes (E near E_r)

$$f_{\ell_0}(k) = f_{\ell_0}(E) = \frac{1}{2k}\frac{\Gamma_{\ell_0}/2}{E_r - E - i\Gamma_{\ell_0}/2}, \tag{16.139}$$

an expression known as the Breit–Wigner resonance amplitude.

To derive an exponential decay law and a relation like (16.136), Eq. (16.139) is substituted into Eq. (15.20) for $\Psi^{(\text{out})}(\mathbf{r},\, t)$, a projection onto $Y^0_{\ell_0}$ is taken, and the relevant integral is evaluated by means of a technique from the mathematical theory of complex variables. The time dependence of the resulting probability density $P_{\ell_0}(r,\, t)$ is found to be a decaying exponential:

$$P_{\ell_0}(r,\, t) \propto e^{-\Gamma_{\ell_0} t/\hbar}. \tag{16.140}$$

Equation (16.140) is the desired result: the particle-unstable resonance state decays exponentially with a lifetime τ_{ℓ_0} given by

$$\tau_{\ell_0} = \hbar/\Gamma_{\ell_0},$$

or

$$\Gamma_{\ell_0} \tau_{\ell_0} = \hbar, \tag{16.141}$$

We see that, in this case, the product of the energy width or spread of the decaying resonance state and its lifetime is equal to $\hbar$, thereby justifying Eq. (16.136). While it may be possible to obtain (16.141) using the tempus operator T, it is not clear how to do so, since detailed analytic expressions for the resonance states of even a one-particle Hamiltonian are not readily available. On the other hand, the scattering-theory analysis, and its generalization to many-particle systems, even though parameters are introduced, provides a straightforward procedure, as in the foregoing analysis.

It should also be clear why the exponential decay form for $P_{\ell_0}(r, t)$ is approximate: the form (15.20) for $\Psi^{(\text{out})}(\mathbf{r}, t)$ holds only for r (and t) asymptotic. For r not asymptotic, we need $\Psi^{(\text{sc})}(\mathbf{r}, t)$, whose behavior cannot be analyzed in the relatively straightforward fashion used with $\Psi^{(\text{out})}(\mathbf{r}, t)$. Nonetheless, as we have stated several times, exponential decay is not only a very useful approximation, it is also a very good one empirically.

Readers who wish to investigate resonances and resonance phenomena further may consult Baym (1976), Bethe and Morrison (1956), Cottingham and Greenwood (1986), or Elton (1966). The literature on decays of unstable quantum systems is vast. The well-prepared reader who wishes to delve into this topic could start with Vol. II of Galindo and Pascual (1990), Sudbery (1984), Peres (1980), and the review of Fonda, Ghirardi, and Rimini (1978).

Exercises

16.1 In (16.1) let $t_0 = 0$ and $\hat{H}_1(t) = \hat{H}_1$, with $\hat{H}_1$ time-independent. By retaining all terms in the perturbation expansion through $|\hat{H}_{1,\beta\alpha}|^2$, show that

$$\sum_{\gamma \neq \alpha} |C_\gamma^{(1)}(t)|^2 + \left| \sum_{n=0}^{2} C_\gamma^{(n)}(t) \right|^2 = 1,$$

where $C_\beta^{(0)} = \delta_{\beta\alpha}, \forall\, \beta$.

16.2 A spin-$\frac{1}{2}$ particle of mass m and charge $q > 0$, acted on by a weak, uniform magnetic field $\mathbf{B}_1 = B\mathbf{e}_2$, is in state α_y. A second weak, uniform magnetic field $\mathbf{B}_2 = B\mathbf{e}_3$ is turned on at time $t_0 = 0$ and off at time t. Ignoring all spatial behaviour,

(a) find the first-order rate at which, for very short times t, the particle will make a transition to state β_y;

(b) show that, for very short times t, the second-order contribution to $P_{\beta_y \alpha_y}(t) = 0$;

(c) use the magnetic-field analysis of Section 13.3 to calculate the exact $\alpha_y \to \beta_y$ probability and from it the short-time transition rate, comparing it with the answer to part (a);

(d) explain clearly into which parts of each of the three preceding calculations $\mathbf{B}_1$ and $\mathbf{B}_2$ enter;

(e) finally, compare the exact and the first-order probabilities that the system will be in state β_y at time t and comment on the similarities/differences between the two results.

16.3 (a) A particle of mass m and charge q is in the ground state of the 1-D oscillator potential $V(x) = m\omega^2 x^2/2$. If at time $t_0 = 0$ the spatially uniform electric field $\boldsymbol{\mathcal{E}} = \mathcal{E}\mathbf{e}_1 \cos(\omega t)$ is switched on, calculate the first-order probability that the particle will make a transition to the first excited state.

(b) The particle of part (a) is now in the ground state of the 3-D symmetric oscillator potential $V(r) = m\omega^2 r^2/2$, and the same electric field as that of part (a) is switched on at time $t_0 = 0$. Calculate the first-order probabilities that the particle will end up in each of the four lowest-lying degenerate excited states (in the notation of Exercise 11.29), these correspond to $n_r = 1, \ell = 0$ and $n_r = 0, \ell = 1, m_\ell = 0, \pm 1$). Compare these results with that of part (a) and comment on similarities/differences.

16.4 Redo the H-atom ionization analysis of the example that ends Section 16.1 (p. 611), dropping the restriction $m_{\mathrm{p}} = \infty$, but retaining all other assumptions. Define all relevant quantities and specify not only to which coordinate system your calculation pertains, but also why you chose it.

16.5 The quantity $\rho_{\mathrm{f}}(E_\beta)$ is the density of available final states per unit energy E_β. One can introduce a more general density of states $\rho(E)$, which is the number of states per unit energy for an arbitrary system. Find expressions for $\rho(E)$ when a particle of mass m is confined to the interior of quantal boxes in 1-D, 2-D, and 3-D, with the sides of all the boxes of length L.

16.6 The β-decay process $\mathrm{n} \rightarrow \mathrm{p} + \mathrm{e} + \nu_{\mathrm{e}}$, where n is a neutron, p is a proton, and ν_{e} is the electron anti-neutrino, involves a three-body final state, so that the total energy E_β is shared among three final particles. Assuming $m_\nu = 0$, plane-wave final states, and negligible Coulomb effects, determine $\rho_{\mathrm{f}}(E_\beta)$ as a function of E_{e}, $E^{\max}$, and p_{e}, where e refers to the electron, $E^{\max}$ is the maximum energy that can be imparted to the electron, and p_{e} is the magnitude of its momentum. Express your final result in the form $\rho_{\mathrm{f}}(E_\beta) \propto [p_{\mathrm{e}}(E_{\max} - E_{\mathrm{e}})]^n \, dp_{\mathrm{e}} \, d\Omega_{\mathrm{e}} \, d\Omega_\nu$, where n is to be determined and Ω_{e} (Ω_ν) are the solid angles of $\mathbf{p}_{\mathrm{e}}$ ($\mathbf{p}_\nu$). (Hints: use relativistic kinematics, consistent approximations, and constraints due to conservation of energy and momentum.)

16.7 The ground-state H-atom ($m_{\mathrm{p}} = \infty$) is placed in a spatially uniform electric field $\boldsymbol{\mathcal{E}}(t) = \mathcal{E}(t)\mathbf{e}_3$, where $\mathcal{E}(t) = 0$ for $t < 0$ and $\mathcal{E}(t) = \mathcal{E}_0 \exp(-\varepsilon t)$, $\varepsilon > 0$, for $t \geqslant 0$.

(a) What are the first order probabilities that, as $t \rightarrow \infty$, the atom makes transitions to (i) the 2s state, (ii) the $21x$ state, (iii) the $21y$ state, and (iv) the $21z$ state? (See Exercise 11.23 for an explanation of the notation.)

(b) Repeat the calculation of part (a) with the non-zero portion of $\mathcal{E}(t)$ changed to $\mathcal{E}(t) = \mathcal{E}_0 \exp(-\eta^2 t^2)$, $\eta > 0$, and discuss your result in the limits when the variation of the field with time is (i) very slow and (ii) very fast.

16.8 Replace the H-atom and the four $n = 2$ states of Exercise 16.7 by the charged particle of part (b) of Exercise 16.3 and the four states listed therein, and then carry out the same calculations as in the preceding exercise.

16.9 Re-calculate $d\sigma/d\Omega$ of Eq. (16.31) assuming that the particle is enclosed in a very large cube of side L. (For this you will need the 3-D value of $\rho(E)$ from Exercise 16.5.)

16.10 Prove the validity of the delta-function relation above Eq. (16.38a), viz., that $\pi x \delta(y)/2 = \lim_{x \rightarrow \infty}(y^{-2} \sin^2(xy/2))$.

16.11 Prove the validity of the second equality in Eqs. (16.47b) and (16.48).

16.12 Instead of being bound in an H-atom, assume the electron is acted on by a 3-D, symmetric oscillator potential whose frequency ω is adjusted so that the degenerate quartet of first excited states lies 3 Ryd/4 above the ground state. Use the electric-dipole approximation to calculate the lifetime for decay from the $\ell = 1$ first excited states to the ground state via spontaneous emission, and comment on its comparison with $\tau_{1s,2p}(H)$.

16.13 Consider decay by spontaneous emission from the 2p to the 1s state of hydrogenic atoms, whose nuclear charge is Ze and whose reduced mass is $\mu \neq m_e$.
 (a) Verify the comment in footnote 8 of this chapter.
 (b) Determine the Z and μ dependences of $\tau_{1s,2p}$ in the electric-dipole approximation. Specify in your result the dependence on the ionization energy.
 (c) Using the result of part (b), evaluate numerically the 2p $\to$ 1s lifetimes for (i) positronium (e^+e^-) in the singlet-spin state, (ii) singly ionized He, (iii) doubly ionized Li, and (iv) muonic hydrogen ($\mu^- p$).

16.14 The 3p level in hydrogen can decay via electric-dipole transitions to either the 2s or the 1s states. Using Eq. (16.84), calculate
 (a) the 3p $\to$ 2s transition rate,
 (b) the 3p $\to$ 1s transition rate,
 (c) the (3p $\to$ 2s)/(3p $\to$ 1s) branching ratio, and
 (d) the inverse of the sum of the rates from parts (a) and (b) – which is an approximation to the theoretical lifetime $\tau_{3p}(H)$ of the 3p state – and compare the result with the experimental value $\tau_{3p}^{(\mathrm{exp})}(H) \cong 5.3 \times 10^{-9}$ s.

16.15 Assume that the bound system on which the electromagnetic field acts consists of N spinless, charged particles with masses m_j and charges q_j, $j = 1, \ldots, N$. The Hamiltonian governing the bound system is $\hat{H}_0(\{\mathbf{r}_j\}) = \sum_j \hat{K}_j(\{\mathbf{r}_j\}) + V(\{\mathbf{r}_j\})$; the vector potential is $\mathbf{A}(\{\mathbf{r}_j\}) = \sum_j \mathbf{A}(\mathbf{r}_j, t)$, where each $\mathbf{A}(\mathbf{r}_j, t)$ is given by Eq. (16.56) with $\mathbf{r} \to \mathbf{r}_j$; and $\Phi = \nabla_j \cdot \mathbf{A}(\mathbf{r}_j, t) = 0$. Using these assumptions, derive the (spontaneous-emission) analog of Eq. (16.72) for the N-particle case.

16.16 The matrix element occurring in the TRK sum rule of Exercise 9.10 also occurs in the dipole approximation to the transition rates, Eqs. (16.71) and (16.72). Choosing $\mathbf{e} = \mathbf{n} = \mathbf{e}_3$, determine how much of the sum rule is satisfied by the individual 2p $\to$ 1s and 3p $\to$ 2s (dipole) transitions, as well as by their sum, all in the spinless-electron model of the H-atom.

16.17 Two particles of equal mass m and opposite charges $q_1 = -q_2 = q$ rotate freely at a fixed separation $|\mathbf{r}| = a$. The Hamiltonian of this rigid-rotator system is $\hat{H}(\theta, \phi) = \hat{L}^2(\theta, \phi)/(ma^2)$; its eigenstates are the $|\ell m_\ell\rangle$ and its eigenvalues are denoted E_ℓ. Let $\hat{\mathbf{d}} = qa\mathbf{e}_r$ denote the dipole operator of the system, with z-component $\hat{d}_z = qa\cos\theta$.
 (a) Calculate the energy-weighted sums

$$S_n = \sum_{\ell, m_\ell} E_\ell^n |\langle \ell m_\ell | \hat{d}_z | 0 \rangle|^2, \qquad n = 0, \pm 1,$$

 where $|0\rangle$ is the ground state.
 (b) Calculate the generalized energy-weighted sums

$$W_n = \sum_{\ell, m_\ell} (E_\ell - E_\lambda)^n |\langle \ell m_\ell | \hat{d}_z | \lambda \mu \rangle|^2, \qquad n = 0, \pm 1,$$

where $|\lambda\mu\rangle$ is an arbitrary eigenstate of $\hat{H}(\theta, \phi)$. (Note that S_0 and W_0 are the sums appearing in the *TRK* sum rule.)

16.18 (a) Let $\hat{A} = \exp(i\mathbf{k}\cdot\hat{\mathbf{Q}})$ in Exercise 9.10(c) and retain the Hamiltonian specified therein, and then evaluate the RHS of the resulting analog to the TRK sum rule.
(b) Now assume a bound system of N spinless particles as in Exercise 16.15, and derive the N-particle generalizations of the TRK sum rule and its analog of part (a).

16.19 Muonic hydrogen is a bound system consisting of a proton and a μ^- meson, the latter decaying into an electron and two neutrinos. Assuming that the $p\mu^-$ pair is in the ground state, determine the probabilities that the pe^- pair resulting from the μ^- decay will be found in
(a) the 1s state
(b) each of the 2p states.
List and justify all of the assumptions used in your calculations.

16.20 A particle of charge q and mass m in 1-D is acted on by the harmonic-oscillator potential $m\omega^2x^2/2$ and a constant electric field $\boldsymbol{\mathcal{E}} = \mathcal{E}\mathbf{e}_1$. For $t<0$ the system is in its ground state, and at time $t = 0$ the electric field is switched off.
(a) What is the probability that, for positive $t \cong 0$, it will be found in the $n = 0$ oscillator state?
(b) Generalize this result to an arbitrary state $|n\rangle$ of the potential $m\omega^2x^2/2$.

16.21 The Hamiltonian for a particle of mass m acted on by a 1-D oscillator potential with a time-dependent, displaced center $a(t)$ is $\hat{H}(x, t) = \hat{K}_x(x) + (m\omega^2x^2/2)[x - a(t)]^2$.
(a) Determine the eigenstates $|n(t)\rangle$ and eigenvalues $E_n(t)$ of $\hat{H}(x, t)$.
(b) Suppose that the system is in its ground state at $t = 0$. Establish a criterion for the validity of the adiabatic approximation, expressing your result in terms of $\dot{a}(t) = da(t)/dt$.
(c) If the preceding result is to hold for all t, and $a(t) = a[1 - \exp(-t/\tau)]$, with a and τ being positive constants, express the general criterion of part (b) in terms of an inequality involving τ.
(d) Evaluate Berry's phase for this Hamiltonian and the $a(t)$ of part (c), and discuss the result.
(e) With the choice $a(t) = a\theta(t)$, where $\theta(t) = 0$ for $t<0$ and 1 for $t \geqslant 0$ $(d\theta/dt = \delta(t))$, the sudden approximation will apply to this Hamiltonian. Calculate the probability that, if the system is in the appropriate ground state for $t<0$, it will be found in the corresponding ground state for $t>0$. Relate your answer to the result obtained in part (a) of the preceding exercise.

16.22 A spin-$\frac{1}{2}$ particle with a g-factor of 2 and magnetic moment μ is acted on by a constant magnetic field $\mathbf{B} = B\mathbf{e}_3$. At time $t = 0$ it is in the $|\mathbf{n}, +\rangle$ eigenstate (13.33a) with $\phi = 0$.
(a) Determine the period associated with its cyclic behavior.
(b) Calculate Berry's phase. (Note: the correct result is an exact expression for the phase of Aharonov and Anandan.)

16.23 Prove that (16.127a) is the time-evolved version of (16.126a) with $a(t)$ given by (16.127b), and then show that Eqs. (16.128, 16.129), and (16.131) follow from it.

16.24 Determine the partial-wave total cross-section $\sigma_T(\ell_0)$ arising from (16.139), and plot the result as a function of E for E in the neighborhood of E_r, assuming that $\Gamma_{\ell_0}/2 \ll E_r$.

16.25 A more realistic resonance parameterization for $\delta_{\ell_0}(E)$ than (16.137) is $\delta_{\ell_0}(E) = \eta_{\ell_0}(E) + \phi_{\ell_0}$, where $\eta_{\ell_0}(E) = (\Gamma_{\ell_0}/2)/(E_\mathrm{r} - E)$, $E_r \gg \Gamma_{\ell_0}$, and ϕ_{ℓ_0} is positive and varies very slowly with E.

(a) Show that $f_{\ell_0}(k)$ can be expressed as $\exp(2i\phi_{\ell_0})/(2ik)$ times a sum of two factors, one of which is the Breit–Wigner expression $(i\Gamma_{\ell_0}/2)/(E_\mathrm{r} - E - i\Gamma_{\ell_0}/2)$.

(b) Determine the partial-wave total cross section $\sigma_\mathrm{T}(\ell_0)$ implied by part (a) and plot it as a function of E. Be certain to show whether the interference-effect dip in $\sigma_\mathrm{T}(\ell_0)$ occurs for $E < E_r$ or for $E > E_r$. Such interference effects are seen in very-low-energy neutron–nucleus total cross sections.

17

Time-Independent Approximation Methods

As with the time-dependent case, the stationary-state Schrödinger equation for complex systems is generally too difficult to solve exactly, and one must again rely on approximation methods. We study three approximation procedures in this chapter: Rayleigh–Schrödinger perturbation theory; the variational method; and the WKB or semi-classical approximation, aspects of which were touched on in Section 8.3. While each method is formulated analytically, we emphasize that numerical procedures are often needed for the final evaluations. In addition, there are purely numerical approximation methods which are of great utility for systems with relatively few degrees of freedom. Although numerical techniques will not be discussed in this book, the practicing physicist should at least know where to find discussions of them.[1]

17.1. Rayleigh–Schrödinger perturbation theory: the non-degenerate case

As in Sections 16.1 and 16.2, the (time-independent) Hamiltonian is here assumed to be composed of a soluble part $\hat{H}_0$ and a weak perturbation $\hat{H}_1$:

$$\hat{H} = \hat{H}_0 + \hat{H}_1. \tag{17.1}$$

The $\hat{H}_0$ eigenvalue problem is

$$\hat{H}_0|\gamma\rangle = E_\gamma|\gamma\rangle, \tag{17.2}$$

and its solutions are presumed known; $\{|\gamma\rangle\}$ provides the relevant basis in $\mathcal{H}$:

$$\sum_\gamma |\gamma\rangle\langle\gamma| = \hat{I}. \tag{17.3}$$

An integration on any continuum states is suppressed in (17.3).

Lambda-parameterization

The assumed weakness of $\hat{H}_1$ means that the magnitude of matrix elements such as $\langle\alpha|\hat{H}_1|\beta\rangle$ is small relative to $|E_\alpha - E_\beta|$. Furthermore, the weakness of $\hat{H}_1$ suggests that

[1] In addition to books such as those by Acton (1970), Rice (1983), and Press *et al.* (1986), which concern standard methods, we recommend that the reader eventually become familiar with "local methods," e.g., the finite-element method for solving differential equations. See, e.g., Zienkiewicz and Taylor (1989) and Ram-Mohan *et al.* (1990) – the latter for application to eigenvalue problems.

both the eigenvectors and the eigenvalues of $\hat{H}$ can be expressed as series involving powers of $\hat{H}_1$. One means of keeping track of such powers is to multiply $\hat{H}_1$ by a parameter of weakness λ, $0 \leqslant \lambda \leqslant 1$. As in Section 16.1, λ will be the actual expansion parameter, whose value will eventually be set equal to unity, and we again write

$$\hat{H}_1 \to \lambda \hat{H}_1,$$
$$\hat{H} \to \hat{H}(\lambda) = \hat{H}_0 + \lambda \hat{H}_1; \tag{17.4}$$

the limits of $\hat{H}(\lambda)$ are

$$\hat{H}(\lambda = 0) = \hat{H}_0 \tag{17.4a}$$

and

$$\hat{H}(\lambda = 1) = \hat{H}. \tag{17.4b}$$

Even for a very weak perturbation, its non-zero value means that each eigenvector and eigenvalue of $\hat{H}_0$ will be altered. The goal of time-independent perturbation theory is to calculate these changes. In doing so, it is assumed that each state $|\alpha\rangle$ will be transformed into a unique[2] solution of $\hat{H}(\lambda)$ denoted by $|\Gamma, \lambda; \alpha\rangle$:

$$\hat{H}(\lambda)|\Gamma, \lambda; \alpha\rangle = E_\Gamma(\lambda; \alpha)|\Gamma, \lambda; \alpha\rangle, \tag{17.5}$$

with the expectation that

$$\lim_{\lambda \to 0} \begin{cases} |\Gamma, \lambda; \alpha\rangle \to |\alpha\rangle & (17.6\text{a}) \\ E_\Gamma(\lambda; \alpha) \to E_\alpha. & (17.6\text{b}) \end{cases}$$

The full solutions are the $\lambda \to 1$ limits:

$$|\Gamma, \lambda = 1; \alpha\rangle \equiv |\Gamma; \alpha\rangle \tag{17.7a}$$

and

$$E_\Gamma(\lambda = 1; \alpha) \equiv E_\Gamma(\alpha). \tag{17.7b}$$

State-vector normalizations and perturbation expansions

The perturbation series are obtained by assuming that $|\Gamma, \lambda; \alpha\rangle$ and $E_\Gamma(\lambda; \alpha)$ can each be expanded in a *convergent* power series[3] in λ for $0 \leqslant \lambda \leqslant 1$:

$$|\Gamma, \lambda; \alpha\rangle = \sum_{n=0}^{\infty} \lambda^n |\Gamma; \alpha\rangle^{(n)} \tag{17.8a}$$

and

[2] The case of degenerate states of $\hat{H}_0$ is discussed in Section 17.2.
[3] Note well: although convergence is assumed when $\lambda = 1$, it need not hold for either of Eqs. (17.8), and care must be exercised in using the results derived below. For example, the bound-state character of a $|\Gamma; \alpha\rangle$ is not a guarantee of convergence, an example being $|\alpha\rangle = |\mathbf{k}\rangle$, $\hat{H}_0 = \hat{K}$, $H_1 = V =$ a short-ranged potential: the Born series diverges for a bound state. Another divergent example concerns the "Cooper pairs" of superconductivity theory: see, e.g., Baym (1976). On the other hand, even if the series diverges, its first few terms may provide a very good approximation, as discussed later.

$$E_\Gamma(\lambda;\,\alpha)=\sum_{n=0}^{\infty}\lambda^n E_\Gamma^{(n)}(\alpha), \tag{17.8b}$$

where

$$|\Gamma;\,\alpha\rangle^{(0)}=|\alpha\rangle \tag{17.6a}$$

and

$$E_\Gamma^{(0)}(\alpha)=E_\alpha. \tag{17.6b}$$

In deriving expressions for $|\Gamma;\,\alpha\rangle^{(n)}$ and $E_\Gamma^{(n)}(\alpha)$, a normalization convention is needed. The unperturbed states form an orthonormal set:

$$\langle\alpha|\beta\rangle=\delta_{\alpha\beta}; \tag{17.9a}$$

following standard practice we shall require that

$$\langle\alpha|\Gamma,\,\lambda;\,\alpha\rangle\rangle=1. \tag{17.9b}$$

The latter condition means that

$$\langle\Gamma,\,\lambda;\,\alpha|\Gamma,\,\lambda;\,\alpha\rangle\equiv Z\neq 1,$$

while substitution of (17.8a) into (17.9b) plus use of (17.9a) leads to the important consequence that

$$\langle\alpha|\Gamma;\,\alpha\rangle^{(n)}=0,\qquad n\neq 0, \tag{17.9c}$$

a relation we shall employ later.

As a result of Eqs. (17.9), projection of both sides of

$$E_\Gamma(\lambda;\,\alpha)|\Gamma,\,\lambda;\,\alpha\rangle=(\hat{H}_0+\lambda\hat{H}_1)|\Gamma,\,\lambda;\,\alpha\rangle \tag{17.10}$$

onto $\langle\alpha|$ yields ($\lambda=1$)

$$E_\Gamma(\alpha)=E_\alpha+\langle\alpha|\hat{H}_1|\Gamma;\,\alpha\rangle. \tag{17.11}$$

The exact energy differs from the unperturbed energy by a matrix element of $\hat{H}_1$ taken between the unperturbed state and the (as-yet-unknown) exact eigenstate.

Expressions for $|\Gamma;\,\alpha\rangle^{(n)}$ and $E_\Gamma^{(n)}(\alpha)$ are obtained by substituting Eqs. (17.8) into (17.10). On doing so and rearranging, we find

$$(\hat{H}_0+\lambda\hat{H}_1)\left(|\alpha\rangle+\sum_{n=1}^{\infty}\lambda^n|\Gamma;\,\alpha\rangle^{(n)}\right)$$
$$=\left(E_\alpha+\sum_{m=1}^{\infty}\lambda^m E_\Gamma^{(m)}(\alpha)\right)\left(|\alpha\rangle+\sum_{p=1}^{\infty}\lambda^p|\Gamma;\,\alpha\rangle^{(p)}\right). \tag{17.12}$$

Equation (17.12) is the fundamental relation. Equating like powers of λ on both sides of (17.12) produces an infinite set of coupled equations that can be solved iteratively:

$$\hat{H}_0|\alpha\rangle=E_\alpha|\alpha\rangle; \tag{17.2}$$

$$\hat{H}_0|\Gamma;\,\alpha\rangle^{(1)} + \hat{H}_1|\alpha\rangle = E_\alpha|\Gamma;\,\alpha\rangle^{(1)} + E_\Gamma^{(1)}(\alpha)|\alpha\rangle, \tag{17.13a}$$

$$\vdots \qquad\qquad \vdots$$

$$\hat{H}_0|\Gamma;\,\alpha\rangle^{(n)} + \hat{H}_1|\Gamma;\,\alpha\rangle^{(n-1)} = E_\alpha|\Gamma;\,\alpha\rangle^{(n)} + \sum_{\substack{m+p=n\\ m\geqslant 1\\ p\geqslant 1}} E_\Gamma^{(m)}(\alpha)|\Gamma;\,\alpha\rangle^{(p)} + E_\Gamma^{(n)}(\alpha)|\alpha\rangle, \tag{17.13b}$$

etc.

All the $E_\Gamma^{(j)}(\alpha)$ and $|\Gamma;\,\alpha\rangle^{(j)}$, $j = 1, \ldots, n$, are needed in order to determine $E_\Gamma^{(n+1)}(\alpha)$ and $|\Gamma;\,\alpha\rangle^{(n+1)}$. We start with the case $n = 1$. Projecting both sides of (17.13a) onto $\langle\alpha|$ leads, after canceling out of common terms, to

$$E_\Gamma^{(1)}(\alpha) = \langle\alpha|\hat{H}_1|\alpha\rangle. \tag{17.14}$$

That is, the first-order correction to the energy is the diagonal matrix element of the perturbation taken between the unperturbed, or zeroth-order, states.

To obtain the first-order correction $|\Gamma;\,\alpha\rangle^{(1)}$, we make use of the resolution of the identity (17.3):

$$\begin{aligned}|\Gamma;\,\alpha\rangle^{(1)} &= \sum_\gamma |\gamma\rangle\langle\gamma|\Gamma;\,\alpha\rangle^{(1)}\\ &= \sum_{\gamma\neq\alpha} |\gamma\rangle\langle\gamma|\Gamma;\,\alpha\rangle^{(1)}\\ &\equiv \sum_\gamma{}' |\gamma\rangle\langle\gamma|\Gamma;\,\alpha\rangle^{(1)},\end{aligned} \tag{17.15}$$

where the second equality in (17.15) follows from (17.9c), and $\sum'_\gamma = \sum_{\gamma\neq\alpha}$. An expression for the scalar product in (17.15) is found by projecting both sides of (17.13a) onto $\langle\gamma|$, $\gamma \neq \alpha$. A short analysis yields

$$(E_\alpha - E_\gamma)\langle\gamma|\Gamma;\,\alpha\rangle^{(1)} = \langle\gamma|\hat{H}_1|\alpha\rangle, \qquad \gamma \neq \alpha,$$

or, assuming that $E_\gamma \neq E_\alpha$, $\gamma \neq \alpha$, so that division by $E_\alpha - E_\gamma$ is allowed,

$$\langle\gamma|\Gamma;\,\alpha\rangle^{(1)} = \frac{\langle\gamma|\hat{H}_1|\alpha\rangle}{E_\alpha - E_\gamma},$$

and therefore

$$|\Gamma;\,\alpha\rangle^{(1)} = \sum_\gamma{}' \frac{|\gamma\rangle\langle\gamma|\hat{H}_1|\alpha\rangle}{E_\alpha - E_\gamma}. \tag{17.16}$$

The presence of $E_\alpha - E_\gamma$ in the denominators of these equations is a reminder that we are working with the non-degenerate case.

These procedures also apply to the general expression (17.13b). The only survivors from projecting both sides of it onto $\langle\alpha|$ are the $\hat{H}_1$ term from the LHS and $E_\Gamma^{(n)}(\alpha)$ from the RHS, from which we get

$$E_\Gamma^{(n)}(\alpha) = \langle\alpha|\hat{H}_1|\Gamma;\,\alpha\rangle^{(n-1)}, \tag{17.17}$$

a result that also follows from substituting Eq. (17.8a) into Eq. (17.11). Setting $n = 1$ in (17.17) reproduces Eq. (17.14).

To determine $|\Gamma;\,\alpha\rangle^{(n)}$, we shall calculate $\langle\gamma|\Gamma;\,\alpha\rangle^{(n)}$ and then substitute the resulting expression into the following analog of (17.15):

$$\begin{aligned}|\Gamma;\,\alpha\rangle^{(n)} &= \sum_\gamma |\gamma\rangle\langle\gamma|\Gamma;\,\alpha\rangle^{(n)} \\ &\equiv \sum_\gamma{}' |\gamma\rangle\langle\gamma|\Gamma;\,\alpha\rangle^{(n)}.\end{aligned} \tag{17.18a}$$

The desired expression for $\langle\gamma|\Gamma;\,\alpha\rangle^{(n)}$ is obtained by projecting both sides of (17.13b) onto $\langle\gamma|$, $\gamma \neq \alpha$, which ultimately leads to

$$(E_\alpha - E_\gamma)\langle\gamma|\Gamma;\,\alpha\rangle^{(n)} = \langle\gamma|\hat{H}_1|\Gamma;\,\alpha\rangle^{(n-1)} - \sum_{\substack{m+p=n \\ m\geqslant 1 \\ p\geqslant 1}} E_\Gamma^{(m)}\langle\gamma|\Gamma;\,\alpha\rangle^{(p)}, \tag{17.18b}$$

or

$$\langle\gamma|\Gamma;\,\alpha\rangle^{(n)} = \frac{1}{E_\alpha - E_\gamma}\left(\langle\gamma|\hat{H}_1|\Gamma;\,\alpha\rangle^{(n-1)} - \sum_{\substack{m+p=n \\ m,p\geqslant 1}} E_\Gamma^{(m)}(\alpha)\langle\gamma|\Gamma;\,\alpha\rangle^{(p)}\right). \tag{17.18c}$$

The iterative character of this result is apparent. Furthermore, that there must be no degeneracies is again emphasized in the transition from Eq. (17.18b) to Eq. (17.18c), since only if $E_\gamma \neq E_\alpha$, $\forall\ \gamma$, is the division by $E_\alpha - E_\gamma$ allowed.

Substitution of (17.18c) into (17.18a) yields the nth-order approximation to $|\Gamma;\,\alpha\rangle$:

$$|\Gamma;\,\alpha\rangle^{(n)} = \sum_\gamma{}' \frac{|\gamma\rangle\langle\gamma|}{E_\alpha - E_\gamma}\left(\hat{H}_1|\Gamma;\,\alpha\rangle^{(n-1)} - \sum_{\substack{m+p=n \\ m,p\geqslant 1}} E_\Gamma^{(m)}(\alpha)|\Gamma;\,\alpha\rangle^{(p)}\right). \tag{17.19}$$

Equations (17.17) and (17.19) are the general results, functioning together iteratively. For example, setting $n = 2$ in Eq. (17.17) and then substituting into it the expression for $|\Gamma;\,\alpha\rangle^{(1)}$ obtained from (17.19) gives the second-order correction to the energy:

$$E_\Gamma^{(2)}(\alpha) = \sum_\gamma{}' \frac{|\langle\alpha|\hat{H}_1|\gamma\rangle|^2}{E_\alpha - E_\gamma}. \tag{17.20}$$

The first-order correction to $E_\Gamma(\alpha)$ is the diagonal matrix element of $\hat{H}_1$ – Eq. (17.14) – whereas the second-order correction involves the absolute square of all the off-diagonal matrix elements of $\hat{H}_1$. $E_\Gamma^{(3)}(\alpha)$ contains a sum over various products of three matrix elements of $\hat{H}_1$, etc. Note that, if $|\alpha\rangle$ is the ground state (g.s.), then $E_\Gamma^{(2)}(\alpha = \text{g.s.})$ is negative, so that the second-order correction to the ground state always lowers the previous value $E_\alpha + E_\Gamma^{(1)}(\alpha)$.

From Eq. (17.17) and the structure of (17.19) it should be clear that the expression for $E_\Gamma^{(n)}(\alpha)$ will always be simpler in form than that for $|\Gamma;\,\alpha\rangle^{(n)}$. Equations (17.14) and (17.16) illustrate this, as does a comparison of (17.20) with the result for $|\Gamma;\,\alpha\rangle^{(2)}$:

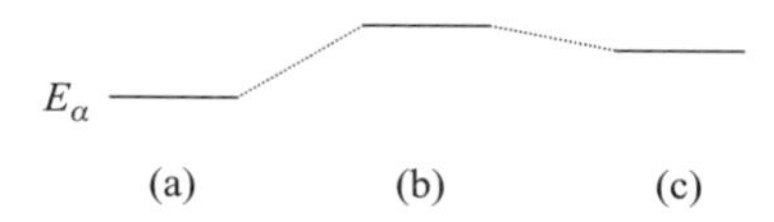

Fig. 17.1 A Grotrian-type diagram illustrating successive perturbation approximations to a ground-state energy $E_\Gamma(\alpha)$. (a) $E_\Gamma(\alpha) \cong E_\alpha$; (b) $E_\Gamma(\alpha) \cong E_\alpha + E_\Gamma^{(1)}(\alpha)$, with $E_\Gamma^{(1)}(\alpha)$ taken to be positive, thereby increasing the energy relative to the zeroth-order result; and (c) $E_\Gamma(\alpha) \cong E_\Gamma^{(0)}(\alpha) + E_\Gamma^{(1)}(\alpha) + E_\Gamma^{(2)}(\alpha)$, where $E_\Gamma^{(2)}(\alpha)$, being negative, has reduced the value obtained in (b). $|E_\Gamma^{(2)}(\alpha)|$ is smaller than $E_\Gamma^{(1)}(\alpha)$, as must be the case if the series is to converge.

$$|\Gamma;\,\alpha\rangle^{(2)} = \sum_{\gamma,\beta}{}' \frac{|\gamma\rangle\langle\gamma|\hat{H}_1|\beta\rangle\langle\beta|\hat{H}_1|\alpha\rangle}{(E_\alpha - E_\gamma)(E_\alpha - E_\beta)} - \sum_{\gamma}{}' \frac{|\gamma\rangle\langle\gamma|\hat{H}_1|\alpha\rangle\langle\alpha|\hat{H}_1|\alpha\rangle}{(E_\alpha - E_\gamma)^2}, \tag{17.21}$$

where the prime in the first sum means that neither γ nor β equals α.

Through second order, the preceding results yield

$$E_\Gamma(\alpha) \cong E_\alpha + \langle\alpha|\hat{H}_1|\alpha\rangle + \sum_{\gamma}{}' \frac{|\langle\alpha|\hat{H}_1|\gamma\rangle|^2}{E_\alpha - E_\gamma}. \tag{17.22}$$

The basic idea that $\hat{H}_1$ is weak such that its higher-order matrix elements make progressively smaller contributions to $E_\Gamma(\alpha)$ is consistent with Eq. (17.22). However, the nature of the sums that enter the second (and higher)-order corrections suggests a modification of this basic idea: successive orders will make progressively smaller contributions if the ratio $|\langle\gamma|\hat{H}_1|\alpha\rangle|/|E_\alpha - E_\gamma|$, $\forall\, \gamma \neq \alpha$, is itself small. Unfortunately, there is no guarantee that the perturbation series will converge, much less converge rapidly, even if the latter criterion is satisfied. Typically, one hopes that the first few terms in the perturbation series for $E_\Gamma(\alpha)$ will provide an accurate result. When this occurs, for instance when Eq. (17.22) suffices, we get a result schematically represented in Fig. 17.1.

EXAMPLES

We illustrate the foregoing analysis with four examples. The two with which we conclude this section involve perturbation of the 1-D linear harmonic oscillator Hamiltonian and are chosen for differing pedagogical purposes, as discussed below. Two "realistic" ones constitute the next section, where each of the two perturbations is a component of the leading corrections to the H-atom Hamiltonian of Chapter 11.

EXAMPLE 1

The 1-D charged oscillator in a uniform electric field $\mathcal{E}$. The exact eigenstates and energies of this system were determined in Section 12.3. Here we show that second-order perturbation theory yields the exact energy.

The Hamiltonian $\hat{H}_0$ is that of the unperturbed oscillator:

$$\hat{H}_0 = \frac{1}{2m}\hat{P}_x^2 + \frac{1}{2}m\omega_0^2\hat{Q}_x^2, \tag{17.23a}$$

while the perturbation is (see Eq. (12.34))

$$\hat{H}_1 = -q\mathcal{E}\hat{Q}_x. \tag{17.23b}$$

The eigenstates and energies of $\hat{H}_0$ are $|\alpha\rangle = |n\rangle$ and $E_\alpha = E_n = (n + \frac{1}{2})\hbar\omega_0$; we choose $|n\rangle$ as the unperturbed state. Since the perturbation is linear in $\hat{Q}_x$, Eq. (6.95) leads to the conclusion that the first-order correction to the energy, viz., $-e\mathcal{E}\langle n|\hat{Q}_x|n\rangle$, vanishes.

The first non-zero correction to the energy is

$$E_\Gamma^{(2)}(n) = q^2\mathcal{E}^2 \sum_{m \neq n} \frac{|\langle n|\hat{Q}_x|m\rangle|^2}{E_n - E_m}.$$

Referring again to Eq. (6.95), we see that only the terms with $m = n \pm 1$ contribute to $E_\Gamma^{(2)}(n)$, and an easy calculation leads to

$$E_\Gamma^{(2)}(n) = -\frac{q^2\mathcal{E}^2}{2m\omega_0^2}; \tag{17.24}$$

comparison with (12.37b) shows that this is the exact result!

That (17.24) is the exact result suggests that all higher-order corrections $E_\Gamma^{(k)}(n)$, $k > 2$, are zero. While this is true, a moment's reflection concerning $\langle x|\Gamma;\, n\rangle$ should be sufficient to realize that all higher-order corrections to it cannot vanish. Rather than attempting to calculate each one separately, we shall use Eq. (17.11) and the exact result (12.37a) for $\langle x|\Gamma;\, n\rangle$ to provide a relatively simple way to understand the role played by the higher-order terms. Equation (12.37a) states that $\langle x|\Gamma;\, n\rangle$ is a standard oscillator wave function but with a center displaced from $x = 0$ to $x = a = q\mathcal{E}/(m\omega_0^2)$. Displaced states can be obtained by means of the translation operator $\hat{U}_\mathrm{T}(a) = \exp(-ia\hat{P}_x/\hbar)$. Hence,

$$|\Gamma;\, n\rangle = e^{-ia\hat{P}_x/\hbar}|n\rangle,$$

and, on substituting this expression into (17.11), we get

$$E_\Gamma(n) = (n + \tfrac{1}{2})\hbar\omega_0 - q\mathcal{E}\langle n|\hat{Q}_x e^{-ia\hat{P}_x/\hbar}|n\rangle. \tag{17.25}$$

For this perturbed oscillator example, Eq. (17.25) provides a direct means for obtaining the corrections $E_\Gamma^{(k)}(n)$. Every term $E_\Gamma^{(k)}(n)$ can be obtained from (17.25) by first expanding the operator $\exp(-ia\hat{P}_x/\hbar)$ in a power series and then evaluating the results term by term. To "work backwards" and obtain Eq. (17.25) from the perturbation expression that results on substituting (17.19) into (17.17) is very non-trivial. To do so, one must employ the 1-D version of Ehrenfest's theorem as it occurs, for example, in Eq. (16.69), which in the present case is

$$\begin{aligned}\langle k|\hat{P}_x|n\rangle &= \frac{am\omega_0(n-k)}{i}\langle k|\hat{Q}_x|n\rangle \\ &= \frac{q\mathcal{E}(n-k)}{i\omega_0}\langle k|\hat{Q}_x|n\rangle,\end{aligned} \tag{17.26}$$

but even if one thought of employing (17.26), without foreknowledge of the exact result it is unlikely that one could go easily from the perturbation expressions for $E_{\Gamma}^{(k)}(n)$ to the "summed" result (17.25).

EXAMPLE 2

The quartically perturbed 1-D linear oscillator. In view of the fact that the 1-D linear oscillator is both a soluble quantal system and a model for real systems, its use in supplying an $\hat{H}_0$ for illustrating perturbation theory might seem almost foreordained. Precisely because of this we cannot caution the reader too strongly that, even though this next example is widely used, insofar as the present text is concerned a quartic perturbation to the linear oscillator Hamiltonian serves only as an exercise in calculating individual correction terms: despite one's ability in this case to calculate $E_{\Gamma}^{(n)}(\alpha)$ for n much larger than 2, the perturbation series is not only divergent but almost certainly has a *zero* radius of convergence(!!), as discussed, e.g., in the review of Simon (1982). A similar divergence and likely zero radius of convergence also holds for the Stark-effect problem (Section 17.3) and for the H-atom in a constant magnetic field. Despite these pathologies, a meaning can be given to the perturbation series (via asymptotic series and special summation techniques), and numerically accurate eigenvalues can be obtained for these three physical systems. The means for doing so is beyond the scope of this book, which is why we treat the quartically (and also the cubically) perturbed oscillator only as a pedagogical illustration of the perturbation method.[4]

For this system, $\hat{H}_0$ is given by (17.23a), while

$$\hat{H}_1 = \beta\left(\frac{\hat{Q}_x}{a}\right)^4 \hbar\omega_0, \tag{17.23c}$$

where the length a and the energy $\hbar\omega_0$ are introduced so as to make the strength β dimensionless.

The unperturbed state $|\alpha\rangle$ is again chosen to be the nth oscillator state $|n\rangle$, for which $E_\alpha = E_n = (n+\frac{1}{2})\hbar\omega_0$. The first-order correction $E_{\Gamma}^{(1)}(n)$ is

$$E_{\Gamma}^{(1)}(n) = \frac{\beta\hbar\omega_0}{a^4}\langle n|\hat{Q}_x^4|n\rangle,$$

and, to evaluate the matrix element of $\hat{Q}_x^4 = \hat{Q}_x\hat{Q}_x\hat{Q}_x\hat{Q}_x$, three complete sets of oscillator states are employed, i.e.,

$$\langle n|\hat{Q}_x^4|n\rangle = \sum_{m,p,k}\langle n|\hat{Q}_x|m\rangle\langle m|\hat{Q}_x|p\rangle\langle p|\hat{Q}_x|k\rangle\langle k|\hat{Q}_x|n\rangle.$$

Use of Eq. (6.95) plus some straightforward and not too lengthy algebra leads to

$$E_{\Gamma}^{(1)}(n) = \frac{3\beta}{4}\left(\frac{\lambda_0}{a}\right)^4 (2n^2+2n+1)\hbar\omega_0, \tag{17.27}$$

[4] The mathematically-sophisticated reader who wishes to delve into the nature of these perturbation and eigenvalue analyses can start by consulting Simon (1982) and Galindo and Pascual (1990) and references cited therein.

where $\lambda_0 = \sqrt{\hbar/(m\omega_0)}$ is the oscillator's natural length. The rapid growth of $E_\Gamma^{(1)}(n)$ with n is consistent with $\psi_n(x)$ extending further from the origin with increasing n and the fact that $\hat{Q}_x^4$ samples the large-x portions of $\psi_n(x)$.

Although we have calculated only $E_\Gamma^{(1)}$, it is easy to see that $E_\Gamma^{(k)}(n) \neq 0, \forall k$; indeed, $E_\Gamma^{(k)}(n) \propto \beta^k$. Evaluation of $E_\Gamma^{(2)}(n)$ and $|\Gamma;\, n\rangle^{(1)}$ for the present case of the quartic perturbation is left as an exercise, as is the determination of the lowest non-vanishing correction to $E_n = (n + \frac{1}{2})\hbar\omega_0$ when $\hat{H}_1 \propto \hat{Q}_x^3$.

17.2. Fine structure and hyperfine structure in the spectrum of hydrogen

Non-relativistic quantum theory, represented by the Hamiltonian of Section 11.6, provides a model for the hydrogen atom that is adequate for many purposes. However, the spectrum it yields is not in perfect agreement with highly accurate measurements on H-atom energy levels. A better description of H is obtained by using the relativistic Dirac equation (Dirac (1928)); the best one is based on relativistic quantum-field theory. While these approaches are beyond the scope of this book, it is possible to introduce a non-relativistic reduction of these theories, particularly of the Dirac-equation description; this leads to a Hamiltonian that involves corrections to that of Chapter 11. The leading corrections of this type can be divided into two portions, each of which can be treated perturbatively. We do so in this section, via first-order perturbation theory.

The first of the two correction terms which we consider is the fine-structure interaction. It constitutes the lowest-order (v/c) relativistic contribution to the non-relativistic Hamiltonian. As will be shown, it yields a modification roughly $\alpha^2 \sim 10^{-4}$ smaller than the non-relativistic energy level ($E_n = -1$ Ryd$/n^2$) being perturbed. The second part of the correction term, the hyperfine-structure interaction, leads to a modification approximately three orders of magnitude smaller than the preceding one. The hyperfine structure (HFS) correction will be evaluated for the case of the ground state, which it splits into two levels, in contrast to the fine structure (FS) correction, which only lowers the ground-state energy. As we shall see, the HFS splitting is important astrophysically.

Both the FS effect and the HFS effect are large enough to have measurable consequences, so, by including them in the hydrogenic Hamiltonian, we will have taken a major step along the road leading to the "real" H-atom. Because these additional interactions are structurally complicated and their evaluation is more complex than that of the quartically perturbed 1-D oscillator, we assume in this section that the H-atom is in a region free of external electromagnetic fields: only the fields internal to the atom are included. The effects of bathing the atom in a uniform magnetic field are considered in the exercises (Zeeman and Paschen–Back effects).

The unperturbed H-atom Hamiltonian is that of Section 11.6:

$$\hat{H}_0(\mathbf{r}) = \frac{1}{2\mu}\hat{P}^2(\mathbf{r}) - e\Phi(r), \tag{17.28}$$

where

$$\Phi(r) = \kappa_e e/r \tag{17.29}$$

is the usual scalar potential and μ is the electron–proton reduced mass. Including the electron and proton spins, the eigenstates and energies of (17.28) are

$$|\gamma\rangle = |n\ell m_\ell\rangle|\tfrac{1}{2}m_s^{(\mathrm{e})}\rangle|\tfrac{1}{2}m_s^{(\mathrm{p})}\rangle, \tag{17.30a}$$

where $m_s^{(\mathrm{e})}\,(m_s^{(\mathrm{p})})$ is the electron (proton) spin-projection quantum number, and

$$E_\gamma \equiv E_n = -\alpha^2\mu c^2/(2n^2), \tag{17.30b}$$

the latter measured w.r.t. the rest-mass energy μc^2, while $\alpha = \kappa_e e^2/(\hbar c)$ is the fine-structure(!) constant – see Eq. (11.130′).

The perturbation $\hat{H}_1$ consists of a fine-structure and a hyperfine-structure portion:

$$\hat{H}_1 = \hat{H}^{(\mathrm{fs})} + \hat{H}^{(\mathrm{hfs})}, \tag{17.31}$$

where

$$\hat{H}^{(\mathrm{fs})} = -\frac{1}{8m_\mathrm{e}^3c^2}\hat{P}^4 - \frac{1}{2m_\mathrm{e}^2c^2}\frac{e}{r}\frac{d\Phi}{dr}\hat{\mathbf{L}}\cdot\hat{\mathbf{S}}_\mathrm{e} - \frac{\hbar^2 e}{8m_\mathrm{e}^2c^2}\nabla^2\Phi \tag{17.32}$$

and

$$\hat{H}^{(\mathrm{hfs})} = -\kappa_m\left(\frac{e}{m_\mathrm{e}r^3}\hat{\mathbf{L}}\cdot\hat{\boldsymbol{\mu}}_\mathrm{p} + \frac{1}{r^3}[3(\hat{\boldsymbol{\mu}}_\mathrm{e}\cdot\mathbf{e}_r)(\hat{\boldsymbol{\mu}}_\mathrm{p}\cdot\mathbf{e}_r) - (\hat{\boldsymbol{\mu}}_\mathrm{e}\cdot\hat{\boldsymbol{\mu}}_\mathrm{p})] + \frac{8\pi}{3}\hat{\boldsymbol{\mu}}_\mathrm{e}\cdot\hat{\boldsymbol{\mu}}_\mathrm{p}\delta(\mathbf{r})\right). \tag{17.33}$$

In these expressions, $\hat{\mathbf{S}}_\mathrm{e}$ is the electron spin operator, $\hat{\boldsymbol{\mu}}_\mathrm{e}(\hat{\boldsymbol{\mu}}_\mathrm{p})$ is the electron (proton) magnetic moment operator, and κ_m is given in Table 12.1.

Fine-structure effects

Equation (17.32) is the result of making a non-relativistic reduction of Dirac's (relativistic) equation for the electron, some accounts of which can be found, e.g., in Baym (1976), Das (1973), Feshbach and Villars (1958), and Holstein (1992). The first two terms in $\hat{H}^{(\mathrm{fs})}$ have straightforward classical interpretations: $\hat{P}^4$ is the quantal version of the leading correction to the familiar approximation $\sqrt{p_\mathrm{e}^2c^2 + m_\mathrm{e}^2c^4} \cong m_\mathrm{e}c^2 + p_\mathrm{e}^2/(2m_\mathrm{e})$, while the spin–orbit coupling term, involving $d\Phi/dr$, arises from the interaction of the electron's magnetic moment with the magnetic field into which $\boldsymbol{\mathcal{E}} = -(d\Phi/dr)\mathbf{e}_r$ is transformed when a Lorentz transformation is applied to the rest frame of the electron. For discussion of these points, and also the portion containing $\nabla^2\Phi$, the so-called Darwin term, whose interpretation is non-classical, see e.g., Baym (1976), Cohen-Tannoudji, Diu, and Laloë (1977), or Galindo and Pascual (1990).

In evaluating the first-order FS corrections perturbatively, two points are immediately evident from (17.32): first, the absence of terms involving $\hat{\mathbf{S}}_\mathrm{p}$ ($\propto \hat{\boldsymbol{\mu}}_\mathrm{p}$) means that the proton's spin state $|\tfrac{1}{2}m_s^{(\mathrm{p})}\rangle$ can be suppressed, and second, the presence of the factor $\hat{\mathbf{L}}\cdot\hat{\mathbf{S}}_\mathrm{e}$ requires use of the coupled representation for the electron states, rather than the uncoupled representation of Eq. (17.30a). That is, in place of (17.30a), we write

$$|\gamma\rangle \to |n\ell\tfrac{1}{2};\, jm_j^{(\mathrm{e})}\rangle, \qquad j = \ell \pm \tfrac{1}{2}, \tag{17.34}$$

where the superscript – (e) – serves as a reminder that (17.34) refers only to the electron's state. With (17.34) the first-order FS correction is

$$E_\Gamma^{(1)}(n\ell j;\,\mathrm{fs}) = \langle n\ell\tfrac{1}{2};\, jm_j^{(\mathrm{e})}|\hat{H}^{(\mathrm{fs})}|n\ell\tfrac{1}{2};\, jm_j^{(\mathrm{e})}\rangle. \tag{17.35}$$

Each of the three terms in (17.32) makes a non-vanishing contribution to (17.35). We consider the $\hat{P}^4$ term first. Because it acts on unperturbed states in (17.35), one can make the following replacement in the matrix element: $\hat{P}^4/(4m_e^2) = [\hat{P}^2/(2m_e)]^2 \to (\hat{H}_0 + e\Phi)^2$, where, due to the α^2 nature of the correction, the substitution $\mu \to m_e$ in $\hat{H}_0$ is insignificant. Using an abbreviated notation, the matrix element of the first term on the RHS of (17.32) becomes

$$-\frac{1}{2m_e c^2}\left\langle\left(\frac{\hat{P}^2}{2m_e}\right)^2\right\rangle_{n\ell;jm_j^{(e)}} = -\frac{1}{2m_e c^2}\langle n\ell\tfrac{1}{2};\, jm_j^{(e)}|(\hat{H}_0 + e\Phi)(\hat{H}_0 + e\Phi)|n\ell\tfrac{1}{2};\, jm_j^{(e)}\rangle$$

$$= -\frac{1}{2m_e c^2}\left(E_n^2 + 2E_n\kappa_e e^2\left\langle\frac{1}{\hat{Q}}\right\rangle_{n\ell} + \kappa_e^2 e^4\left\langle\frac{1}{\hat{Q}^2}\right\rangle_{n\ell}\right), \quad (17.36a)$$

where $\hat{Q} = |\hat{\mathbf{Q}}|$ and matrix elements of $\hat{Q}^{\pm n}$ are independent of j and $m_j^{(e)}$.

Next, using $d\Phi/dr = -\kappa_e e^2/r^2$ and Eq. (14.84) for the $\hat{\mathbf{L}}\cdot\hat{\mathbf{S}}_e$ portion, and reverting to a $(\,)_{n\ell;jm_j^{(e)}}$ form so as to display the r dependence of the matrix element, we find for the contribution of the second term in $\hat{H}^{(fs)}$

$$-\frac{e}{2m_e c^2}\left(\frac{1}{r}\frac{d\Phi}{dr}\,\hat{\mathbf{L}}\cdot\hat{\mathbf{S}}_e\right)_{n\ell;jm_j^{(e)}} = \frac{\kappa_e e^2\hbar^2}{2m_e^2 c^2}\left\langle\frac{1}{\hat{Q}^3}\right\rangle_{n\ell} \times \begin{cases} \ell/2, & j = \ell + \frac{1}{2}, \\ -(\ell+1)/2, & j = \ell - \frac{1}{2}, \end{cases} \quad (17.36b)$$

which holds for $\ell > 0$. When $\ell = 0$, the contribution of this second term is *zero*.

In the third term of (17.32), the factor $\nabla^2\Phi$ leads, via $\nabla^2(1/r) = -4\pi\delta(\mathbf{r})$, to the final form $[4\pi\hbar^2\kappa_e e^2/(8m_e^2c^2)]\delta(\mathbf{r})$, and thus the last contribution to the matrix element in (17.35) becomes

$$-\frac{\hbar^2 e}{8m_e^2 c^2}(\nabla^2\Phi)_{n\ell;jm_j^{(e)}} = \frac{\pi\hbar^2\kappa_e e^2}{2m_e^2 c^2}|\psi_{n\ell m_\ell}(\mathbf{r}=0)|^2,$$

which is zero unless $\ell = 0$, since only for $\ell = 0$ is $R_{n\ell}(r = 0)$ non-vanishing. Using Eqs. (11.139) and (11.140), we find $|\psi_{n\ell m_\ell}(\mathbf{r}=0)|^2 = [1/(\pi n^3 a^3)]\delta_{\ell 0}$ and therefore

$$-\frac{\hbar^2 e}{8m_e^2 c^2}(\nabla^2\Phi)_{n\ell;jm_j^{(e)}} = \frac{\hbar^2\kappa_e e^2}{2m_e^2 c^2 n^3 a^3}\delta_{\ell 0}$$

$$= -\left(\frac{\mu}{m_e}\right)\frac{\alpha^2}{n}E_n\delta_{\ell 0}$$

$$= -\frac{\alpha^2}{n}E_n\delta_{\ell 0}, \quad (17.36c)$$

where the α^2 factor allows us to set $(\mu/m_e)^2$ equal to unity.

The final step is the determination of the matrix elements $\langle\hat{Q}^{-p}\rangle_{n\ell}$, $p = 1, 2, 3$. As in Section 11.6, these quantities can be evaluated using the integral expressions derived by Schrödinger (1928), but, since their values are listed by Galindo and Pascual (1990), we refer to that compilation:

$$\left\langle \frac{1}{\hat{Q}} \right\rangle_{n\ell} = \frac{1}{n^2 a}, \tag{17.37a}$$

$$\left\langle \frac{1}{\hat{Q}^2} \right\rangle_{n\ell} = \frac{1}{n^3 a^2} \frac{1}{\ell + \frac{1}{2}}, \tag{17.37b}$$

and

$$\left\langle \frac{1}{\hat{Q}^3} \right\rangle_{n\ell} = \frac{1}{n^3 a^3} \frac{1}{\ell(\ell + \frac{1}{2})(\ell + 1)}, \qquad \ell > 0. \tag{17.37c}$$

Using the first two of these expressions in (17.36a) it is readily shown that

$$-\frac{1}{2m_e c^2} \left\langle \left(\frac{\hat{P}}{2m_e} \right) \right\rangle_{n\ell j, m_j^{(e)}} = \frac{\alpha^2}{n} E_n \left(\frac{1}{\ell + \frac{1}{2}} - \frac{3}{4n} \right), \tag{17.38a}$$

where the factor μ/m_e has again been set equal to unity. Also, substitution of (17.37c) into (17.36b) leads to

$$-\frac{e}{2m_e^2 c^2} \left(\frac{1}{r} \frac{d\Phi}{dr} \hat{\mathbf{L}} \cdot \hat{\mathbf{S}}_e \right)_{n\ell j, m_j^{(e)}} = -\frac{\alpha^2}{n} E_n \frac{1 - \delta_{\ell 0}}{\ell(\ell+1)(2\ell+1)} \times \begin{cases} \ell, & j = \ell + \frac{1}{2}, \\ -(\ell+1), & j = \ell - \frac{1}{2}, \end{cases} \tag{17.38b}$$

where $(\mu/m_e)^2 = 1$ has been used and the vanishing of the spin–orbit term when $\ell = 0$ has been taken into account by the factor $1 - \delta_{\ell 0}$: $(1 - \delta_{\ell 0})/\ell$ is set equal to zero when $\ell = 0$.

Putting these various results into (17.35), we get

$$E_\Gamma^{(1)}(n\ell j;\ \mathrm{fs}) = \frac{\alpha^2}{n} E_n \left(\frac{1}{\ell + \frac{1}{2}} - \frac{3}{4n} - \delta_{\ell 0} - \frac{1 - \delta_{\ell 0}}{2\ell + 1} \times \begin{cases} 1/(\ell+1), & j = \ell + \frac{1}{2} \\ -1/\ell, & j = \ell - \frac{1}{2} \end{cases} \right). \tag{17.39}$$

However, irrespective of whether $\ell = 0$ or $\ell > 0$, the ℓ-dependent terms in (17.39) always combine so as to form $(j + \frac{1}{2})^{-1}$, where $j = \frac{1}{2}$ for $\ell = 0$ and $j = \ell \pm \frac{1}{2}$ otherwise. Hence, the final expression for the first-order FS correction to the unperturbed energy E_n is

$$E_\Gamma^{(1)}(n\ell j;\ \mathrm{fs}) = \frac{\alpha^2}{n} E_n \left(\frac{1}{j + \frac{1}{2}} - \frac{3}{4n} \right). \tag{17.40}$$

Since $E_\Gamma^{(1)}$ is proportional to $\alpha^2 E_n$, the claim made below Eq. (17.35) is substantiated:

$$\frac{E_\Gamma^{(1)}(n\ell j;\ \mathrm{fs})}{E_n} \sim \alpha^2 \sim 10^{-4}.$$

With first-order FS corrections taken into account, the energy spectrum of hydrogen becomes

$$\begin{aligned} E_\Gamma(n\ell j) &= E_n \left[1 + \frac{\alpha^2}{n} \left(\frac{1}{j + \frac{1}{2}} - \frac{3}{4n} \right) \right] \\ &\equiv E_n[1 + \Delta(n, j)], \end{aligned} \tag{17.41}$$

the second equality defining $\Delta(n, j)$.

Without the FS term but including the electron spin, the degeneracy of the nth level is $g_n = 2n^2$. With the FS correction, the degeneracy is apparently reduced to $2j + 1$, the multiplicity of $m_j^{(e)}$. However, there is an additional, accidental degeneracy for each $j < n - \frac{1}{2}$, since in this case there are two ℓ values, $\ell_<$ and $\ell_>$, such that $\ell_< + \frac{1}{2} = j = \ell_> - \frac{1}{2}$. Therefore the actual degeneracy is

$$g_{n\ell j} = 2(2j + 1).$$

The origin of this accidental degeneracy was discovered by Johnson and Lippmann (1950); see Baym (1976) for analysis and comments.

For $n = 1$, the FS correction produces only a lowering of the non-relativistic energy $E_1 = -1$ Ryd by the fractional amount $\alpha^2/4 \cong 1.33 \times 10^{-5}$. For $n = 2$, the levels defined by the $n\ell j$ quantum numbers $20\frac{1}{2}$ and $21\frac{1}{2}$ coincide; their energy value is reduced from $E_2 = -1$ Ryd/4 by the fractional amount $5\alpha^2/16 \cong 1.66 \times 10^{-5}$. There are two pairs of coincident energy levels for $n = 3$, three for $n = 4$, etc.

Before displaying some of these shifts in a Grotrian diagram, it is convenient to introduce standard spectroscopic notation. The unperturbed states $|n\ell m_\ell\rangle$ of Section 11.6 were designated by the number–letter pairs 1s, 2s, 2p, 3s, 3p, 3d, etc., as in Table 11.3. In the present situation, there are three analogous quantum numbers, viz., $n\ell j$. The new spectroscopic notation is that of Section 11.6 with the j value added as a subscript to the ℓ-valued letter. That is, one writes $n\ell_j$, specific instances of which are $1s_{1/2}$, $2s_{1/2}$, $2p_{1/2}$, $2p_{3/2}$, etc. Using this notation, Eq. (17.41) predicts the $2s_{1/2}$ and $2p_{1/2}$ energy levels to be the same, and similarly for the $3p_{3/2}$ and $3d_{3/2}$ levels, the $3s_{1/2}$ and $3p_{1/2}$ pairs, etc. The change in energy-level structure on going from $E_n = -1$ Ryd$/n^2$ to $E_\Gamma(n\ell j;\text{ hf})$ of Eq. (17.41) is illustrated in Fig. 17.2 for the cases $n = 1$ and $n = 2$. It is, of course, known as the *fine structure* in the hydrogen spectrum. Note that, in reality, the exaggerated shifts are only of the order of 10^{-4} eV.

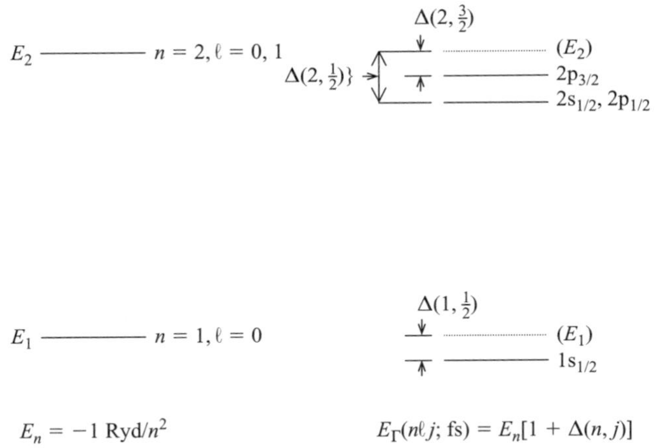

$\Delta(2, \frac{3}{2})$
E_2 $n = 2, \ell = 0, 1$
$\Delta(2, \frac{1}{2})\}$
(E_2)
$2p_{3/2}$
$2s_{1/2}, 2p_{1/2}$
$\Delta(1, \frac{1}{2})$
E_1 $n = 1, \ell = 0$
(E_1)
$1s_{1/2}$
$E_n = -1$ Ryd$/n^2$
$E_\Gamma(n\ell j;\text{ fs}) = E_n[1 + \Delta(n, j)]$

Fig. 17.2 A Grotrian diagram illustrating the unperturbed (non-relativistic) and the hyperfine-perturbed levels for the H-atom. The energy shifts in the latter case are given by $E_n\Delta(n, j)$, where $\Delta(n, j) = (\alpha^2/n)[(j + \frac{1}{2})^{-1} - 3/(4n)]$, and are indicated – in a highly exaggerated manner – by the positions of the solid lines below the dotted ones representing the unperturbed E_n values. Each perturbed level is designated using the $n\ell_j$ spectroscopic notation.

The equality of the two energy levels for which $n\ell_<$ and $n\ell_>$ produce the same j is also a result of solving Dirac's relativistic equation for the case of hydrogen, as discussed, e.g., by Baym (1976). However, nature is NOT correctly described by this elegant relativistic wave equation for the H-atom, as was discovered in the Nobel-prize-winning experiment of Lamb and Retherford (1947), who found that the $2p_{1/2}$ level is shifted a tiny amount ($\cong 4.4 \times 10^{-6}$ eV or 1057 MHz in frequency units) below the $2s_{1/2}$ level. This, another of those exceedingly influential and important twentieth-century measurements, led first to the theory of quantum electrodynamics (theoretically, the "Lamb shift" can be understood as a consequence of electromagnetic-field quantization), for whose elucidation and calculations R. P. Feynman, J. Schwinger, and S. Tomonaga received the Nobel prize (See, e.g., Schwinger (1958) for references); eventually it led to the gauge field theories of the last quarter of the twentieth century. Well-prepared readers who are interested in this topic could try consulting Gottfried and Weisskopf (1984), Griffiths (1987), or Holstein (1992).

Hyperfine-structure effects

The four terms comprising the hyperfine interaction of Eq. (17.33) are each quantal versions of "internal" classical magnetic effects. The factor $\kappa_m e\hat{\mathbf{L}}/(m_e r^3)$ in the first term is the quantal magnetic field generated by the orbital motion of the electron; it interacts in the usual way with the magnetic moment of the proton. The second pair of terms is the interaction of the electron's magnetic moment with the dipole field generated by the non-vanishing magnetic moment of the proton. The final term is a magnetic field energy arising from the fact that the "current density," i.e., the magnetic moment, is contained within a sphere of arbitrarily small radius (see Jackson (1975) for comments on the classical origin of this and the other three terms in $\hat{H}^{(\text{hfs})}$ and Cohen-Tannoudji, Diu, and Laloë (1977) for further discussion of them).

Because the hyperfine interaction contains the three angular-momentum operators $\hat{\mathbf{L}}$, $\hat{\mathbf{S}}_e$, and $\hat{\mathbf{S}}_p$, evaluation of its first-order shift generally requires knowledge of angular momentum algebra beyond that already covered in this book. Because of this, we shall determine HFS effects only for the $\ell = 0$ ground state of the H-atom. As shown below, the sole non-vanishing contribution to an $\ell = 0$, diagonal matrix element of $\hat{H}^{(\text{hfs})}$ comes from the term containing $\delta(\mathbf{r})$. Since that term also involves $\boldsymbol{\mu}_e \cdot \boldsymbol{\mu}_p \propto \hat{\mathbf{S}}_e \cdot \hat{\mathbf{S}}_p$, the relevant states to work with are the electron–proton coupled-spin states. They are the usual singlet and triplet eigenstates of the total spin operator, which, following traditional notation, is denoted $\hat{\mathbf{F}}$:

$$\hat{\mathbf{F}} = \hat{\mathbf{S}}_e + \hat{\mathbf{S}}_p. \tag{17.42}$$

The relevant spin eigenstates are labeled $|S_e, S_p; FM_F\rangle$:

$$|S_e, S_p; FM_F\rangle = \sum_{m_s^{(e)}, m_s^{(p)}} C(\tfrac{1}{2}\tfrac{1}{2}F;\, m_s^{(e)} m_s^{(p)} M_F)|\tfrac{1}{2}m_s^{(e)}\rangle|\tfrac{1}{2}m_s^{(p)}\rangle, \tag{17.43}$$

where $S_e = \frac{1}{2} = S_p$ and $F = 0$ or $F = 1$. In place of Eqs. (17.30a) or (17.34), we have

$$|\alpha\rangle = |100\rangle|S_e, S_p; FM_F\rangle \equiv |10; FM_F\rangle, \tag{17.44}$$

where only the ground-state quantum numbers $n = 1$, $\ell = 0$ have been indicated in the shorthand notation $|10; FM_F\rangle$.

In view of the fact that we are working with $\ell = 0$, the diagonal matrix element of the spin–orbit term in $\hat{H}^{(\text{fs})}$ vanishes – Eq. (17.38b) – and we can calculate the corrections from $\hat{H}^{(\text{fs})}$ and $\hat{H}^{(\text{hfs})}$ together. That is, we shall evaluate

$$E_\Gamma^{(1)}(10;\ F) = \langle 10;\ FM_F|\hat{H}^{(\text{fs})} + \hat{H}^{\text{hfs}}|10;\ FM_F\rangle. \tag{17.45}$$

We can do this because the values of the non-zero contributions from $\hat{H}^{(\text{fs})}$, Eqs. (17.36c) and (17.38a), are independent of the spin states. Hence we may import the result (17.39) into (17.45), leading to

$$E_\Gamma^{(1)}(10;\ F) = \frac{\alpha^2}{4}E_1 + \langle 10;\ FM_F|\hat{H}^{(\text{hfs})}|10;\ FM_F\rangle. \tag{17.46}$$

Let us now consider the contributions to (17.46) from the four terms of (17.33). As with the $\hat{\mathbf{L}} \cdot \hat{\mathbf{S}}_e$ term of $\hat{H}^{(\text{fs})}$, the matrix element of $\hat{\mathbf{L}} \cdot \hat{\boldsymbol{\mu}}_p$ vanishes when $\ell = 0$. The pair of terms in the square brackets of (17.33) can be evaluated using a result from Section 3-2 of the text by Mathews and Walker (1970). To apply that result, we invoke the coordinate representation and the ( ) notation:

$$\begin{aligned}
&(10;\ FM_F|[3(\hat{\boldsymbol{\mu}}_e \cdot \mathbf{e}_r)(\hat{\boldsymbol{\mu}}_p \cdot \mathbf{e}_r) - (\hat{\boldsymbol{\mu}}_e \cdot \hat{\boldsymbol{\mu}}_p)]|10;\ FM_F) \\
&\quad = \int d^3r\, |R_{10}(r)|^2 |Y_0^0|^2 \langle FM_F|[3(\hat{\boldsymbol{\mu}}_e \cdot \mathbf{e}_r)(\hat{\boldsymbol{\mu}}_p \cdot \mathbf{e}_r) - (\hat{\boldsymbol{\mu}}_e \cdot \hat{\boldsymbol{\mu}}_p)]|FM_F\rangle \\
&\quad = \frac{1}{4\pi}\langle FM_F|\int d\Omega_r\, [3(\hat{\boldsymbol{\mu}}_e \cdot \mathbf{e}_r)(\hat{\boldsymbol{\mu}}_p \cdot \mathbf{e}_r) - (\hat{\boldsymbol{\mu}}_e \cdot \hat{\boldsymbol{\mu}}_p)]|FM_F\rangle,
\end{aligned} \tag{17.47}$$

where the angles of $\mathbf{e}_r$ are Ω_r and the matrix element on the RHS of (17.47) involves only spin variables. Then, as shown by Mathews and Walker, the angular integral of $3(\hat{\boldsymbol{\mu}}_e \cdot \mathbf{e}_r)(\hat{\boldsymbol{\mu}}_p \cdot \mathbf{e}_r)/(4\pi)$ is just $\hat{\boldsymbol{\mu}}_e \cdot \hat{\boldsymbol{\mu}}_p$, which cancels out the contribution coming from the integral of $(\hat{\boldsymbol{\mu}}_e \cdot \hat{\boldsymbol{\mu}}_p)/(4\pi)$. Hence, the RHS of (17.47) is zero.

The first-order correction therefore reduces to

$$E_\Gamma^{(1)}(10;\ F) = \frac{\alpha^2}{4}E_1 - \frac{8\pi\kappa_m}{3}\langle 10;\ FM_F|\hat{\boldsymbol{\mu}}_e \cdot \hat{\boldsymbol{\mu}}_p \delta(\hat{\mathbf{Q}})|10;\ FM_F\rangle,$$

which, on use of Eqs. (13.53), becomes

$$E_\Gamma^{(1)}(10;\ F) = \frac{\alpha^2}{4}E_1 + \frac{8\pi\kappa_m g_e\mu_e g_p\mu_p}{3\hbar^2}\langle 100|\delta(\hat{\mathbf{Q}})|100\rangle\langle FM_F|\hat{\mathbf{S}}_e \cdot \hat{\mathbf{S}}_p|FM_F\rangle. \tag{17.48}$$

The $\delta(\hat{\mathbf{Q}})$ matrix element is equal to $|\psi_{100}(\mathbf{r} = 0)|^2 = |R_{10}(r = 0)|^2/(4\pi) = 1/(\pi a^3)$, while, from Eq. (14.77) with $\hat{\mathbf{J}}_1 = \hat{\mathbf{S}}_e$ and $\hat{\mathbf{J}}_2 = \hat{\mathbf{S}}_p$,

$$\langle FM_F|\hat{\mathbf{S}}_e \cdot \hat{\mathbf{S}}_p|FM_F\rangle = \frac{\hbar^2}{2}\left(F(F+1) - \frac{3}{2}\right).$$

With these results the first-order shift (17.48) becomes

$$E_\Gamma^{(1)}(10;\ F) = \frac{\alpha^2}{4}E_1 + \frac{4\kappa_m g_e\mu_e g_p\mu_p}{3a^3}\left(F(F+1) - \frac{3}{2}\right). \tag{17.49}$$

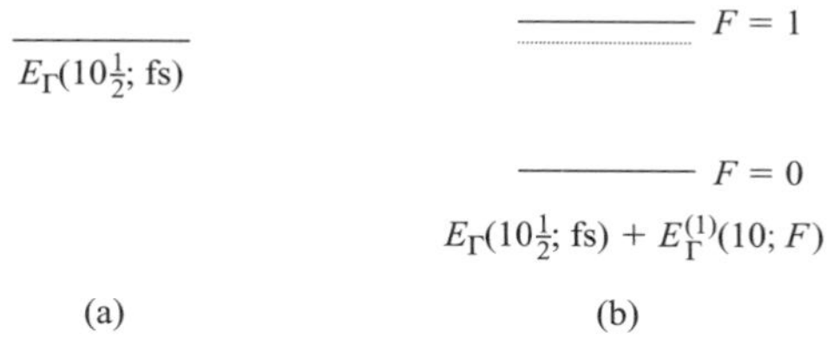

$E_\Gamma(10\frac{1}{2};\text{ fs})$ — $F = 1$, $F = 0$ — $E_\Gamma(10\frac{1}{2};\text{ fs}) + E_\Gamma^{(1)}(10; F)$

(a) (b)

Fig. 17.3 A Grotrian diagram showing the splitting of the original fine-structure ground-state energy (a) into the two hyperfine-structure levels (b). The dotted line in (b) is the original $E_\Gamma(10\frac{1}{2};\text{ fs})$ level.

Since the term in large parentheses in (17.49) is positive for the triplet state ($F = 1$) and negative for the singlet state, it follows that, due to the hyperfine interaction, the ground state of hydrogen is actually a spin singlet. This splitting into $F = 0$ and $F = 1$ states, and similar ones for higher n, is known collectively as the hyperfine structure.

Extracting E_1 as an overall factor, Eq. (17.49) becomes

$$E_\Gamma^{(1)}(10;\, F) = \frac{\alpha^2}{4} E_1 \left[1 - \frac{8 g_\mathrm{p} g_\mathrm{e}}{3} \frac{m_\mathrm{e}}{m_\mathrm{p}} \left(F(F+1) - \frac{3}{2}\right)\right], \tag{17.50}$$

so that the difference in energy between the triplet and singlet $n = 1$ states is

$$E_\Gamma(10;\, F = 1) - E_\Gamma(10;\, F = 0) \equiv \Delta_{10} = \frac{4}{3} \frac{m_\mathrm{e}}{m_\mathrm{p}} \alpha^2 g_\mathrm{p} g_\mathrm{e} |E_1| \cong 5.87 \times 10^{-6} \text{ eV},$$

where we have used $g_\mathrm{e} = 2$ and $|E_1| = 13.6$ eV. The wavelength of the radiation emitted in the transition from the $F = 1$ to the $F = 0$ state is

$$\lambda \cong 21.1 \text{ cm},$$

while the frequency is

$$\nu \cong 1420 \text{ MHz}.$$

These numbers are in very good agreement with the experimental ones. Observation of radiation with this wavelength is a unique means of determining the presence of hydrogen in intergalactic space, and such determinations have been used to map the positions of hydrogen gas in the universe. Excitation from the normal $F = 0$ ground state to the excited $F = 1$ state occurs by inelastic collisions. A Grotrian diagram showing the hyperfine splitting of the fine-structure ground state of Fig. 17.2 is displayed in Fig. 17.3.

17.3. Rayleigh–Schrödinger perturbation theory: the degenerate case

In this section the theory is reformulated to take account of degeneracies. The existence of degeneracies requires at least two degrees of freedom; that is, the collective quantum number must consist of at least two labels, with the energy being independent of one (or more) of the labels. To emphasize the degeneracy, we relabel as follows:

$$\alpha \to d_i, \qquad i = 0, 1, \ldots, M, \tag{17.51}$$

where d_i refers to the degeneracy, of which there are $M + 1$ states. As a simple illustration of (17.51), consider a particle bound by a spherically symmetric square well subjected to a perturbation. The quantum numbers for which d_i stands are $n\ell m_\ell$, with $E_d = E_{n\ell}$, so that the degeneracy is $(2\ell+1)$-fold: $E_{n\ell} \neq E_{n\ell m_\ell}$. Hence, in this case, i refers to the m_ℓ values. Note that the unperturbed energy is E_d, not E_{d_i}.

Equation (17.51) means that the unperturbed state becomes a *set* of zeroth-order states:

$$|\alpha\rangle \to \{|d_i,\, E_d\rangle\}_{i=0}^{M}, \tag{17.52}$$

where the $|d_i,\, E_d\rangle$ notation makes explicit the fact that the $M+1$ unperturbed states all have the same energy E_d. These states span an $(M+1)$-dimensional subspace denoted $\mathcal{D}_M$, with

$$\sum_{i=0}^{M} |d_i,\, E_d\rangle\langle d_i,\, E_d| = \hat{I}_M(d), \tag{17.53}$$

where

$$\langle d_j,\, E_d|d_i,\, E_d\rangle = \delta_{ij}. \tag{17.54}$$

The presence of degeneracies not only negates the transition from Eq. (17.18b) to (17.18c) but also means that, while $E_\Gamma(\lambda \to 0;\, d) \to E_d$, we do not know to which of the $M+1$ states $|d_i,\, E_d\rangle$ the perturbed state $|\Gamma,\, \lambda;\, d\rangle$ goes in the limit $\lambda \to 0$. This situation will be clarified by a diagonalization, leading to linear combinations of the $|d_i,\, E_d\rangle$. Rather than confronting the need for a diagonalization as it would arise in a straightforward development of the perturbation equations, we introduce the diagonalization first, with the understanding that it is motivated by the preceding considerations.[5]

The diagonalization produces $M+1$ states denoted $|\alpha_n,\, E_d\rangle$:

$$|\alpha_n,\, E_d\rangle = \sum_{i=0}^{M} C_{ni}|d_i,\, E_d\rangle, \qquad n = 0,\, 1,\, 2,\, \ldots,\, M, \tag{17.55}$$

with the C_{ni} chosen so that

$$\langle \alpha_m,\, E_d|\hat{H}_1|\alpha_n,\, E_d\rangle = E_{\alpha_n}^{(1)}\delta_{mn}, \tag{17.56a}$$

and

$$\langle \alpha_m,\, E_d|\alpha_n,\, E_d\rangle = \delta_{mn}. \tag{17.56b}$$

That is, we seek solutions of

$$\hat{H}_1|\alpha,\, E_d\rangle = E^{(1)}|\alpha,\, E_d\rangle \tag{17.57a}$$

that are valid in $\mathcal{D}_M$, and we obtain them by means of the expansion

$$|\alpha,\, E_d\rangle = \sum_i C_i|d_i,\, E_d\rangle. \tag{17.57b}$$

[5] The analysis rests on the hope that the perturbation $\hat{H}_1$ will completely lift the degeneracy among the states $|d_i,\, E_d\rangle$. We shall assume that this hope is fulfilled; if it is not, the remaining degeneracies must be dealt with by additional diagonalizations using higher-order perturbation corrections. We shall not discuss the latter situation, referring the reader instead to the treatment of Schiff (1968).

Substitution of (17.57b) into (17.57a) followed by projection of both sides of the resulting equation onto $\langle d_j,\, E_d|$ leads to

$$\sum_{i=0}^{M}[H_{1,ji} - E^{(1)}\delta_{ji}]C_i = 0, \tag{17.57c}$$

where $H_{1,ji} = \langle d_j,\, E_d|\hat{H}_1|d_i,\, E_d\rangle$. Since $\hat{H}_1$ is assumed to lift the degeneracy, the secular equation will yield $M+1$ (distinct) energies labeled $E^{(1)}_{\alpha_n}$, $M+1$ sets of coefficients $\{C_{ni}\}_{i=0}^{M}$, and the $M+1$ orthonormal states $|\alpha_n,\, E_d\rangle$ of Eqs. (17.55) and (17.56).[6] The $E^{(1)}_{\alpha_n}$ will turn out to be the $M+1$ first-order perturbation corrections.

Equation (17.57a) is a "miniature" time-independent Schrödinger equation, and as with its full-fledged counterpart, there is a possibility of degeneracies. Our assumption is that none occur. However, should some of the $|\alpha_n,\, E_d\rangle$ actually be degenerate, i.e., belong to the same $E^{(1)}_{\alpha_n}$, our (first-order) analysis will apply only to the non-degenerate ones.

Let us now see how the foregoing diagonalization fits into the degenerate perturbation structure. We first replace the states $|d_i,\, E_d\rangle$ by the $|\alpha_n,\, E_d\rangle$ in the resolution of the identity:

$$\sum_{n=0}^{M}|\alpha_n,\, E_d\rangle\langle\alpha_n,\, E_d| + \sum_{\gamma\neq\{\alpha_n\}}|\gamma\rangle\langle\gamma| = \hat{I}. \tag{17.58}$$

Next, one of the diagonalized states, say $|\alpha_0,\, E_d\rangle$, is chosen as the unperturbed state. This induces slight notational changes from that of Section 17.1:

$$|\Gamma,\, \alpha\rangle^{(m)} \rightarrow |\Gamma,\, \alpha_0\rangle^{(m)}$$

and

$$E^{(m)}_{\Gamma}(\alpha) \rightarrow E^{(m)}_{\Gamma}(\alpha_0),$$

with

$$|\Gamma;\, \alpha_0\rangle^{(0)} = |\alpha_0,\, E_d\rangle, \tag{17.59a}$$

$$E^{(0)}_{\Gamma}(\alpha_0) = E_d, \tag{17.59b}$$

and

$$\langle\alpha_0,\, E_d|\Gamma;\, \alpha_0\rangle^{(m)} = 0, \qquad m>0. \tag{17.59c}$$

As a result, the perturbation equations of Section 17.1 for $m = 0$, 1, and 2 become

$$[E^{(0)}_{\Gamma}(\alpha_0) - \hat{H}_0]|\Gamma;\, \alpha_0\rangle^{(0)} = 0, \tag{17.60a}$$

$$[E^{(0)}_{\Gamma}(\alpha_0) - \hat{H}_0]|\Gamma;\, \alpha_0\rangle^{(1)} = [\hat{H}_1 - E^{(1)}_{\Gamma}(\alpha_0)]|\Gamma;\, \alpha_0\rangle^{(0)}, \tag{17.60b}$$

[6] We emphasize that each state $|\alpha_n,\, E_d\rangle$, being a linear combination of the states $|d_i,\, E_d\rangle$, is an eigenstate of $\hat{H}_0$ whose eigenvalue is the unperturbed energy E_d. Hence that energy appears as a quantum number in the ket $|\alpha_n,\, E_d\rangle$. At the same time, however, each of the states $|\alpha_n,\, E_d\rangle$ is also an eigenstate of $\hat{H}_1$ – in the subspace $\mathcal{D}_M$ – with eigenvalue $E^{(1)}_{\alpha_n}$. (If the degeneracy is fully lifted (as we assume), then no more than one of the $E^{(1)}_{\alpha_n}$ could be equal to E_d.) Thus, $|\alpha_n,\, E_d\rangle$ should really have two energy labels, viz., E_d and $E^{(1)}_{\alpha_n}$. The latter is suppressed for notational simplicity, but the reader is cautioned that a second energy quantum number is associated with $|\alpha_n,\, E_d\rangle$.

and

$$[E_\Gamma^{(0)}(\alpha_0) - \hat{H}_0]|\Gamma;\, \alpha_0\rangle^{(2)} = [\hat{H}_1 - E_\Gamma^{(1)}(\alpha_0)]|\Gamma;\, \alpha_0\rangle^{(1)} - E_\Gamma^{(2)}(\alpha_0)|\Gamma;\, \alpha_0\rangle^{(0)}. \quad (17.60c)$$

The first-order correction can be obtained by projecting both sides of (17.60b) onto $\langle\alpha_0,\, E_d|$:

$$\begin{aligned}\langle\alpha_0,\, E_d|E_d - \hat{H}_0|\Gamma;\, \alpha_0\rangle^{(1)} &= \langle\alpha_0,\, E_d|\hat{H}_1|\alpha_0,\, E_d\rangle - E_\Gamma^{(1)}(\alpha_0)\\ &= E_{\alpha_0}^{(1)} - E_\Gamma^{(1)}(\alpha_0),\end{aligned}$$

the second equality following from (17.56a), i.e., from our *ab initio* diagonalization. This, of course, is the *a posteriori* reason for having introduced the diagonalization procedure. However, the LHS of this last result vanishes, and we get as the first-order corrrection

$$E_\Gamma^{(1)}(\alpha_0) = E_{\alpha_0}^{(1)}. \quad (17.61)$$

Similarly, $E_\Gamma^{(1)}(\alpha_n) = E_{\alpha_n}^{(1)}$ and, as announced above, the set $\{E_{\alpha_n}^{(1)}\}$ provides the first-order perturbation corrections.

A further projection of both sides of (17.60b) onto $\langle\alpha_n,\, E_d|$, $n \neq 0$, leads to

$$\begin{aligned}(E_d - E_d)\langle\alpha_n,\, E_d|\Gamma;\, \alpha_0\rangle^{(1)} &= \langle\alpha_n,\, E_d|\hat{H}_1|\alpha_0,\, E_d\rangle\\ &= 0, \qquad n > 0.\end{aligned}$$

Since the factor $E_d - E_d$ on the LHS is zero, this last equation is satisfied for any choice of the as-yet-unspecified value of the scalar products $\langle\alpha_n,\, E_d|\Gamma;\, \alpha_0\rangle^{(1)}$, $n > 0$. We choose them all to be zero:

$$\langle\alpha_n,\, E_d|\Gamma;\, \alpha_0\rangle^{(1)} = 0, \qquad n \neq 0. \quad (17.62)$$

Equation (17.62) is our second revised result. Combining it with the unchanged relation (17.59c), we have

$$|\Gamma;\, \alpha_0\rangle^{(1)} = \sum_\gamma{}'' |\gamma\rangle\langle\gamma|\Gamma;\, \alpha_0\rangle^{(1)}, \quad (17.63)$$

where the double prime in (17.63) means $\gamma \neq \alpha_n$, $n = 0, 1, 2, \ldots, M$. In view of the facts (i) that only the degeneracy associated with E_d is being addressed, and (ii) that (17.62) holds, we can simply take over the other result of the non-degenerate case, i.e.,

$$|\Gamma;\, \alpha_0\rangle^{(1)} = \sum_\gamma{}'' \frac{|\gamma\rangle\langle\gamma|\hat{H}_1|\alpha_0,\, E_d\rangle}{E_d - E_\gamma} \quad (17.64a)$$

and

$$E_\gamma^{(2)}(\alpha_0) = \sum_\gamma{}'' \frac{|\langle\gamma|\hat{H}_1|\alpha_0,\, E_d\rangle|^2}{E_d - E_\gamma}. \quad (17.64b)$$

By assumption, none of the E_γ are equal to E_d, so all denominators in Eqs. (17.64) are non-zero.

Collecting the various orders, the energy through $E_\Gamma^{(2)}(\alpha_0)$ is given by

$$E_\Gamma(\alpha_0) \cong E_d + E^{(1)}_{\alpha_0} + \sum_\gamma{}'' \frac{|\langle\gamma|\hat{H}_1|\alpha_0, E_d\rangle|^2}{E_d - E_\gamma}, \tag{17.65}$$

where $E^{(1)}_{\alpha_0}$ is the energy chosen from the set obtained by diagonalizing $\hat{H}_1$ in the original degenerate basis $\{|d_i, E_d\rangle\}$. We shall work through two examples below that yield $E_\Gamma(\alpha_0)$ through first order, but before doing so, we call the reader's attention to the fact that the diagonalization which leads to the $E^{(1)}_{\alpha_n}$ has already been discussed for the case $M = 2$: it is the analysis of Section 13.4. Those results and accompanying comments apply to the present situation (and also to the "almost" degenerate case, the latter being one that is non-trivial to analyze otherwise (see Baym (1976) for remarks about this case)).

EXAMPLE 1

The symmetric 2-D oscillator perturbed by an $x-y$ potential. This is an example treated in many texts. Our analysis differs from the usual ones, since it is based on the polar-coordinate representation of Section 11.7. This representation, which introduces an angular coordinate, may be regarded as a precursor to the 3-D Stark-effect calculation of Example 2.

The unperturbed Hamiltonian is given by Eq. (11.155b); its eigenfunctions and eigenenergies are – Eqs. (11.167) and (11.165) –

$$\psi_{n_r m}(r, \phi) = N_{n_r m} r^{|m|} e^{-r^2/(2\lambda^2)} L^{|m|}_{n_r}\left(\frac{r^2}{\lambda^2}\right) e^{im\phi}$$

and

$$\begin{aligned} E_{n_r m} &= (2n_r + |m| + 1)\hbar\omega \\ &= (N + 1)\hbar\omega, \end{aligned}$$

where

$$n_r = 0, 1, 2, \ldots,$$

$$m = 0, \pm 1, \pm 2, \ldots,$$

and

$$\lambda = \sqrt{\frac{\hbar}{\mu\omega}}.$$

The perturbation is

$$\begin{aligned} \hat{H}_1 &= V_0 xy/b^2 \\ &= V_0\left(\frac{r^2}{b^2}\right) \sin\phi \cos\phi, \end{aligned} \tag{17.66}$$

with the length b introduced to give V_0 the dimension of energy.

We examine the simplest situation in this example, namely $N = 1$, in which case

$n_r = 0$ and $m = \pm 1$. The label d_i thus corresponds to the pair (n_r, m); the two degenerate wave functions $\langle r\phi|0m\rangle = \psi_{0m}(r, \phi)$ are $(m = \pm 1)$:

$$\psi_{0,\pm 1}(r, \phi) = \frac{\sqrt{2}}{\lambda^2} r e^{-r^2/(2\lambda^2)} \frac{e^{\pm i\phi}}{\sqrt{2\pi}}. \tag{17.67}$$

The 2×2 secular determinant – recall Eq. (17.57c) – contains the matrix elements $H_{1,mn} = \langle 0m|\hat{H}_1|0n\rangle$, $m = \pm 1$, $n = \pm 1$. Since there is no degeneracy in the radial part of the wave function, its contribution to the matrix element need be evaluated only once. That is,

$$\langle 0m|\hat{H}_1|0n\rangle = \frac{I_R}{2\pi}\int_0^{2\pi} d\phi\, e^{-im\phi} \sin\phi \cos\phi\, e^{in\phi}, \qquad m = \pm 1,\ n = \pm 1, \tag{17.68a}$$

where

$$\begin{aligned} I_R &= \frac{2V_0}{\lambda^4 b^2}\int_0^\infty dr\, r^5 e^{-r^2/\lambda^2} \\ &= \frac{2V_0\lambda^2}{b^2}\int_0^\infty dy\, y^5 e^{-y^2} = \frac{2V_0\lambda^2}{b^2} \end{aligned} \tag{17.68b}$$

is the contribution from the radial part.

The angular part of the matrix element will be denoted I_{mn}:

$$I_{mn} = \frac{1}{2\pi}\int_0^{2\pi} d\phi\, e^{-im\phi} \sin\phi \cos\phi\, e^{in\phi}. \tag{17.68c}$$

Since m labels rows and n specifies columns, we let $m = n = 1$ correspond to the upper left-hand entry in the secular determinant and $m = n = -1$ correspond to the lower right-hand entry. It is easily shown that $I_{mm} = 0$, while

$$I_{1,-1} = -I_{-1,1} = -i/4. \tag{17.68d}$$

With these results the matrix elements are

$$\langle 0, \pm 1|\hat{H}_1|0, \mp 1\rangle = \mp\frac{iV_0\lambda^2}{2b^2} \equiv \mp iE_0$$

and

$$\langle 0, \pm 1|\hat{H}_1|0, \pm 1\rangle = 0,$$

leading to

$$\begin{vmatrix} -E^{(1)} & -iE_0 \\ iE_0 & -E^{(1)} \end{vmatrix} = 0,$$

whose solutions are

$$E^{(1)} \to E^{(1)}_\pm = \pm E_0 = \pm\frac{V_0}{2}\left(\frac{\lambda}{b}\right)^2.$$

That these energies are equal and opposite follows from Eq. (13.92) and the fact that $H_{11} = H_{22} = 0$ in that equation. The index α_n of Eq. (17.56a) appears here as the subscript "±."

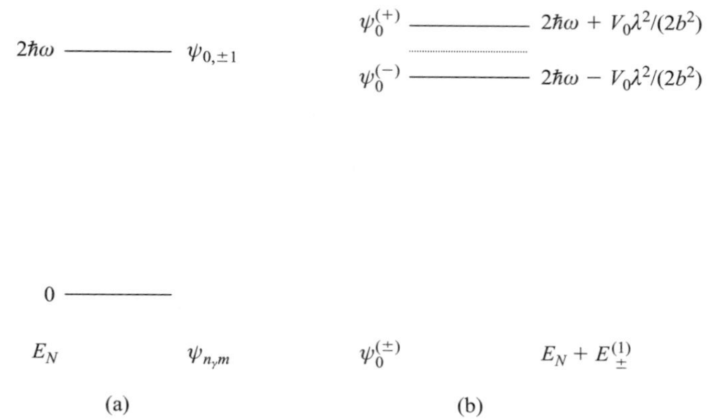

Fig. 17.4 A Grotrian diagram showing (a) the two-fold degenerate level $2\hbar\omega$ of the unperturbed 2-D oscillator, and (b) the effect on the energy – to first order – of the perturbation $\hat{H}_1 = V_0 xy/b^2$, which fully lifts the degeneracy. Shown also are the relevant unperturbed ($\psi_{0,\,\pm1}$) and perturbed ($\psi_0^{(\pm)}$) states.

Corresponding to the energies $E_\pm^{(1)}$ are real, orthonormal eigensolutions that we denote by a superscript to distinguish them from the $\psi_{0,\pm1}$:

$$\psi^{(\pm)}_{n_r=0}(r,\,\phi) = \frac{\sqrt{2}}{\lambda^2}\,re^{-r^2/(2\lambda^2)}\,\frac{\cos\phi \pm \sin\phi}{\sqrt{2\pi}}. \tag{17.69}$$

Equation (17.69) defines, for this example, the polar-coordinate representation of the states $|\alpha_n,\,E_d^{(0)}\rangle$ of Eq. (17.55). They are obtained by forming the linear combinations $C_1\exp(i\phi) + C_2\exp(-i\phi)$ and solving for the pair of C_1 and C_2 from the analog of Eq. (17.57c). Equation (17.69) takes on a simple form in rectangular coordinates, just a sum or a difference of first-excited-state wave functions, one in x, the other in y.

From $E_+^{(1)} = -E_-^{(1)}$, it is seen that the x–y perturbation fully lifts the degeneracy of the two unperturbed states $\psi_{0,\pm1}(r,\,\phi)$. Through first order, the perturbed energies are

$$E_\Gamma(0,\,\pm) = 2\hbar\omega \pm \frac{V_0}{2}\left(\frac{\lambda}{b}\right)^2,$$

whose values are shown schematically in Fig. 17.4.

EXAMPLE 2.

The Stark effect in hydrogen. The perturbation in this example is a constant electric field, whose action on the electron alters the spectrum. Choosing the field $\mathcal{E}$ to be in the z-direction, the perturbation is

$$\hat{H}_1 = e\mathcal{E}z = e\mathcal{E}r\cos\theta, \tag{17.70}$$

where r is the usual electron–proton relative separation and θ is the co-latitude of $\mathbf{r}$.

Before undertaking the analysis we remind the reader of the comments preceding the example of the quartically perturbed, 1-D oscillator in Section 17.1: the perturbation series for the Stark-effect problem diverges (Simon 1982). An examination of the total potential of the Stark problem for H indicates why. Let us set $x = y = 0$, in which case the total potential is

$$V(z) \equiv V(x = y = 0,\, z) = -\frac{\kappa_e e^2}{|z|} + e\mathcal{E}z, \qquad (17.71)$$

whose z dependence is shown in Fig. 17.5. The fact that $V(z \to -\infty) \to -\infty$ means that an electron in any initially bound state will eventually either tunnel into the region of large, negative z or simply "flow" there when its energy is greater than $-2e\sqrt{\kappa_e e\mathcal{E}}$. As a result, the electron–proton separation becomes infinite, leading to an ultimate absence of bound states. Since the perturbation series starts with bound states, it can never directly yield the metastable states produced by the Stark effect, "states" whose energies are actually complex (Simon 1982, Galindo and Pascual 1990). BUT, there is a flip side to this ultimate situation: the low-order perturbation results turn out to be in good agreement with the experimental ones for weak-field perturbations of the low-lying states! The reason is that the tunneling time is so gigantically large that the perturbed low-lying states remain effectively bound for all experimental times. So, in this weak-field, low-lying-state case one may say, after all, that perturbation theory is alive and well.

Let us now return to the analysis, choosing as $\hat{H}_0$ just the spin-independent Hamiltonian $\hat{P}^2/(2\mu) - \kappa_e e^2/r$, since the FS and HFS corrections to it of Section 17.2 are very small. Recall that this Hamiltonian is even under reflection, so that its eigenvectors $|n\ell m_\ell\rangle$ are also eigenstates of the parity operator $\hat{U}_p$ with parity $\pi_\ell = (-1)^\ell$. Since spin plays no role in either $\hat{H}_0$ or $\hat{H}_1$, we suppress this degree of

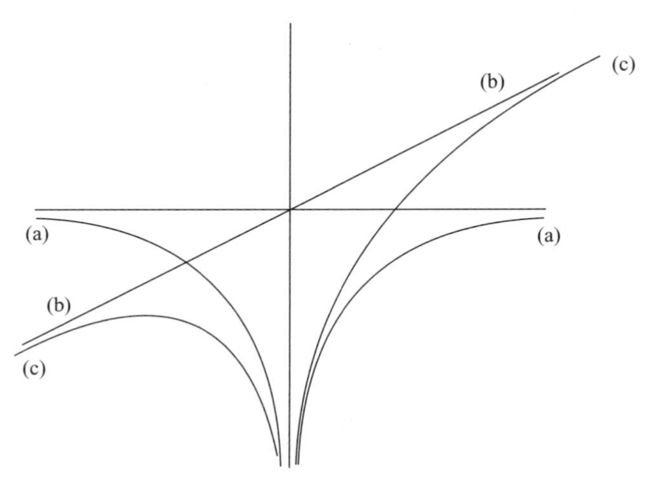

Fig. 17.5 A sketch of the $x = y = 0$ potential $V(z) = -\kappa_e e^2/|z| + e\mathcal{E}z$ of the Stark effect in hydrogen. Curve (a) is the attractive Coulomb interaction, curve (b) is the constant electric field term, and curve (c) is their sum.

freedom, employing the parity eigenstates $|n\ell m_\ell\rangle$ as the (initial) unperturbed states. Because $\hat{H}_1$ is odd under parity, we can use the results of Section 9.5 to conclude that all diagonal matrix elements of $\hat{H}_1$ vanish:

$$\langle n\ell m_\ell|\hat{H}_1|n\ell m_\ell\rangle = 0. \tag{17.72}$$

Equation (17.72) tells us that there is no linear (first-order) Stark-effect shift for the (non-degenerate) ground state. Hence, the lowest-order Stark shift for $|100\rangle$ is $E_\Gamma^{(2)}(100)$, which is second order in $\mathcal{E}$. All the matrix elements in $E_\Gamma^{(2)}(100)$ can be evaluated, leaving $E_\Gamma^{(2)}(100)$ expressed as an infinite sum. A procedure for attempting to compute this infinite sum, not only for $E_\Gamma^{(2)}(100)$ but also for any $E_\Gamma^{(2)}(\alpha)$ as long as $|\alpha\rangle$ is non-degenerate, is the method pioneered by Dalgarno and Lewis (1956). We refer the interested reader to the original paper plus those of Schwartz (1959) and the discussion in Merzbacher (1970).

The lowest-lying states for which there is a *linear* Stark shift are the set of degenerate $n = 2$ states $|200\rangle$ and $|21m_\ell\rangle$: these are the $|d_i, E_d\rangle$ of Eq. (17.52). As we shall see, the perturbation lifts the degeneracy among only two rather than all four of these states.

Since $\langle \mathbf{r}|n\ell m_\ell\rangle$ involves spherical harmonics, we re-express (17.70) in terms of Y_1^0:

$$\hat{H}_1 = \sqrt{\frac{4\pi}{3}}e\mathcal{E}rY_1^0(\Omega). \tag{17.73}$$

The diagonalizing states $|\alpha_n\rangle$ and the energies $E_{\alpha_n}^{(1)}$ are obtained by solving the 4×4 secular determinant equation:

$$\det|\langle 2\ell' m_\ell'|\hat{H}_1|2\ell m_\ell\rangle - E^{(1)}\delta_{\ell\ell'}\delta_{m_\ell m_\ell'}| = 0. \tag{17.74}$$

Because of (17.72) and the Hermiticity of the matrix elements in (17.74) we need only evaluate

$$\langle 200|\hat{H}_1|21m_\ell\rangle = \sqrt{\frac{4\pi}{3}}e\mathcal{E}\int_0^\infty dr\, R_{20}(r)r^3R_{21}(r)\int d\Omega\, Y_0^0(\Omega)Y_1^0(\Omega)Y_1^{m_\ell}(\Omega). \tag{17.75}$$

In view of $Y_0^0 = \sqrt{1/(4\pi)}$ and $Y_1^0(\Omega) = Y_1^{0^*}(\Omega)$, the angular integral in (17.75) is

$$\int d\Omega\, Y_0^0 Y_1^0 Y_1^{m_\ell} = \frac{1}{\sqrt{4\pi}}\delta_{m_\ell 0},$$

from which we conclude that the perturbation affects only the two states with $m_\ell = 0$.

Making use of Eqs. (11.142b) and (11.142c) the radial integral is readily evaluated:

$$\int_0^\infty dr\, R_{20}(r)r^3R_{21}(r) = -3\sqrt{3}a,$$

where a is the reduced-mass, $Z = 1$, Bohr radius.

These results lead to

$$\langle 200|\hat{H}_1|21m_\ell\rangle = -3e\mathcal{E}a\delta_{m_\ell 0} \equiv -W\delta_{m_\ell 0}. \tag{17.76}$$

Since only two matrix elements are non-vanishing ($\langle 200|\hat{H}_1|210\rangle$ and its Hermitian conjugate), the 4×4 secular-determinant equation reduces to a 2×2 equation. This becomes evident if we arrange rows and columns such that the $\ell = 0$ state is the first entry and the $\ell = 1$, $m_\ell = 0$ state is the second, whereupon (17.74) becomes

$$\begin{vmatrix} -E^{(1)} & -W & 0 & 0 \\ -W & -E^{(1)} & 0 & 0 \\ 0 & 0 & -E^{(1)} & 0 \\ 0 & 0 & 0 & -E^{(1)} \end{vmatrix} = 0,$$

the solutions of which are

$$E^{(1)} \to E^{(1)}_{\pm} = \pm W = \pm 3e\mathcal{E}a. \tag{17.77}$$

Although the unperturbed states are specified by three quantum numbers, only two are needed to characterize the states corresponding to $E^{(1)}_{\pm}$. They are written $|2, \pm\rangle$ – i.e., $\{\alpha_n\} = \{2, \pm\}$ – and it is straightforward to show that

$$|2, \pm\rangle = \sqrt{\frac{1}{2}}[|200\rangle \mp |210\rangle].$$

The effect of the interaction (17.73), which is odd under inversion, is to mix states of opposite parity. (Similar parity mixtures occur in some molecular states: see the references in Chapter 18.) Notice also that, under interchange of labels, the higher-lying perturbed state is antisymmetric (odd) and the lower-lying one is symmetric (even). Of course, neither of the $m_\ell = \pm 1$ eigenstates $|21, \pm 1\rangle$ is altered. The effect (to first order) of the perturbation on the four states $|2\ell m_\ell\rangle$ is illustrated in Fig. 17.6.

As a final remark, we note that, if the atom were excited to its long-lived $|200\rangle$

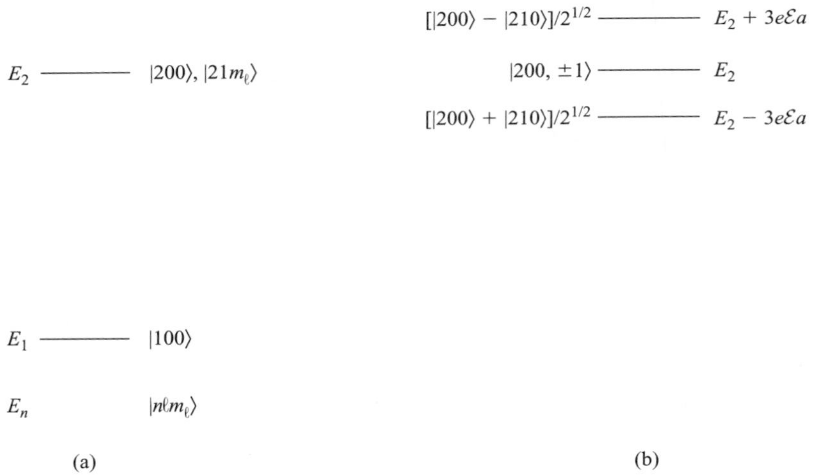

(a)

E_2 ——— $|200\rangle, |21m_\ell\rangle$

E_1 ——— $|100\rangle$

E_n $|n\ell m_\ell\rangle$

(b)

$[|200\rangle - |210\rangle]/2^{1/2}$ ——— $E_2 + 3e\mathcal{E}a$

$|200, \pm 1\rangle$ ——— E_2

$[|200\rangle + |210\rangle]/2^{1/2}$ ——— $E_2 - 3e\mathcal{E}a$

Fig. 17.6 Unperturbed (a) and perturbed (b) energies for the four $|2\ell m_\ell\rangle$ states comprising the degenerate quartet of $n = 2$ hydrogenic states. The relevant states are shown next to the lines denoting the energy values.

state and then immersed in a weak, uniform electric field, it would subsequently be found in either $|2, +\rangle$ or $|2, -\rangle$ with equal probability.

17.4. The variational method for bound states

Perturbation theory is based on a separation of $\hat{H}$ into a "solved" portion $\hat{H}_0$ and a "weak" remainder $\hat{H}_1$. When such a separation is not feasible, e.g., no suitable $\hat{H}_0$ might exist, or $\hat{H}_1$ might be too strong, one needs other approximate means for attempting to solve the time-independent Schrödinger equation. Probably the most powerful, general procedure is the Rayleigh–Ritz variational method discussed in this section (see, e.g., Ritz (1909)).

The variational method can be thought of as the converse of perturbation theory, which fixes a set of (unperturbed) states and uses them plus the perturbation to construct a highly formalized, tightly structured set of approximations. In contrast, the variational method is more loosely constructed: instead of always working with a fixed set of states, one can introduce a guess for the state or states – a "trial" state – and then determine the corresponding energy (or energies). The variational method can be likened to an inventor's paradise: *any* well-behaved trial state that one introduces will yield an approximate energy, although some trial states will be better than others, but, whatever the guess, one can always make systematic improvements.

The idea of being able to construct, often by dint of physical insight and ingenuity, a reasonably accurate wave function has appealed to physicists and chemists since early in the development of quantum theory. Variational methods underlie the analysis of many atomic, molecular, and nuclear bound states, ranging from calculations on two-electron atoms to the central-field/single-particle-state shell model of atomic physics, topics to be discussed in the next chapter.[7]

The variational method is based on the fact that the time-independent Schrödinger equation is the Euler–Lagrange equation that minimizes the value of a particular functional, and we begin by deriving this result and its profound consequence that use of any (well-behaved) state or wave function in the variational expression provides an upper bound to the ground-state energy of the system described by $\hat{H}$. We conclude by extending the analysis to an expansion of the unknown state in terms of arbitrary eigenstates; the resulting set of energies obeys the Hylleraas–Undheim (matrix-interleaving) theorem, one aspect of which is that, in the limit in which the set of arbitrary eigenstates becomes complete, the resulting energies *monotonically* approach the exact eigenenergies of $\hat{H}$.

The variational principle for the Schrödinger equation

For the purpose of orientation, we start with the cooordinate representation and a single particle in 1-D subject to a potential $V(x)$:

[7] Variational (stationary) methods also apply to scattering states, scattering amplitudes, and phase shifts, although this topic is beyond the scope of this book. The interested reader may consult, e.g., the monographs of Mikhlin (1964), Moiseiwitsch (1966), and Newton (1982), or the review of Gerjuoy, Rau, and Spruch (1983). It is an amusing historical footnote that, prior to the advent of electronic computers – and there was such a time – almost all atomic and many nuclear scattering analyses were carried out variationally.

$$\hat{H}(x) = -\frac{\hbar^2}{2m}\frac{d^2}{dx^2} + V(x). \tag{17.78}$$

We seek those functions $\psi_\alpha(x)$ which vanish at the boundaries and for which the functional

$$\begin{aligned} I[\psi_\alpha] &= \int dx\, \psi_\alpha^*(x)\hat{H}(x)\psi_\alpha(x) \\ &\to \int dx \left(\frac{\hbar^2}{2m}\frac{d\psi_\alpha^*}{dx}\frac{d\psi_\alpha}{dx} + \psi_\alpha^*(x)V(x)\psi_\alpha(x)\right) \\ &\equiv \int dx\, F(x,\, \psi_\alpha(x),\, \psi_\alpha^*(x),\, d\psi_\alpha/dx,\, d\psi_\alpha^*/dx) \end{aligned} \tag{17.79}$$

is stationary (minimal) w.r.t. to variations of $\psi_\alpha(x)$ and $\psi_\alpha^*(x)$ subject to the constraint

$$\int dx\, |\psi_\alpha(x)|^2 = 1. \tag{17.80}$$

Although the functional is written $I[\psi_\alpha]$, it depends on both of $\psi_\alpha(x)$ and $\psi_\alpha^*(x)$, each of which can be varied separately. Using standard procedures from the calculus of variations[8] in which (17.80) is incorporated into (17.79) by means of a Lagrange multiplier denoted $E_\alpha(!)$, so that $F \to F - E_\alpha|\psi_\alpha|^2$, one obtains both

$$-\frac{\hbar^2}{2m}\frac{d^2}{dx^2}\psi_\alpha(x) + V(x)\psi_\alpha(x) = E_\alpha\psi_\alpha(x)$$

and its complex conjugate. That is, $(\hat{H} - E)\psi(x) = 0$ is the relevant Euler–Lagrange equation.

The functional (17.79) can be generalized to include 3-D, many particles, spin, etc., by working with state vectors, viz.,

$$I[|\alpha\rangle] = \langle\alpha|\hat{H}|\alpha\rangle.$$

Requiring first-order variations of $I[|\alpha\rangle]$ ($|\alpha\rangle \to |\alpha\rangle + \delta|\alpha\rangle$) to vanish for those states $|\alpha\rangle$ for which

$$\langle\alpha|\alpha\rangle = 1$$

leads to

$$\hat{H}|\alpha\rangle = E_\alpha|\alpha\rangle.$$

These results establish the variational nature of the time-independent Schrödinger equation. Furthermore, it is relatively straightforward to show (e.g., Mathews and Walker (1970) or Dettman (1988)) that the miminum value of the functional

$$E[|\alpha\rangle] \equiv E_\alpha = \frac{\langle\alpha|\hat{H}|\alpha\rangle}{\langle\alpha|\alpha\rangle} \tag{17.81}$$

is just the ground-state energy of the system, with $|\alpha\rangle$ then becoming the ground-state eigenvector. The latter conclusion underlies the variational method of approximation.

[8] See e.g., Mathews and Walker (1970), Dettman (1988), and Fowles and Cassidy (1990).

The variational upper bound on the ground-state energy

Let $|\Gamma_t\rangle$ be an arbitrary but normalized ket (in the Hilbert space of $\hat{H}$) that satisfies suitable conditions in coordinate space. It is a "trial" state. Furthermore, let $\{|\gamma\rangle\}$ be the complete set of orthonormal eigenstates of $\hat{H}$ – they exist as a formal set even if we cannot calculate them – and let γ_0 and E_{γ_0} refer to the ground state. Finally, define the functional

$$E[|\Gamma_t\rangle] = \langle\Gamma_t|\hat{H}|\Gamma_t\rangle. \tag{17.82}$$

Then, as long as $E[|\Gamma_t\rangle]$ exists, the following result is valid:

$$E_{\gamma_0} \leqslant E[|\Gamma_t\rangle], \tag{17.83}$$

with the equality holding only when $|\Gamma_t\rangle = |\gamma_0\rangle$. In other words, the expectation value of $\hat{H}$ taken w.r.t. *any* arbitrary (well-behaved) state yields an upper bound to the ground-state energy. Clearly some $|\Gamma_t\rangle$'s will yield bounds closer to E_{γ_0}; the game is to find them. With (17.83) we have entered the inventor's paradise noted above.

We prove (17.83) next and later will show how it can be extended. The proof uses the expansion of $|\Gamma_t\rangle$ via the $|\gamma\rangle$:

$$|\Gamma_t\rangle = \sum_\gamma C_\gamma|\gamma\rangle, \tag{17.84a}$$

with

$$\sum_\gamma |C_\gamma|^2 = 1. \tag{17.84b}$$

Substitution of (17.84a) into (17.82) leads to

$$E[|\Gamma_t\rangle] = \sum_\gamma E_\gamma|C_\gamma|^2, \tag{17.85}$$

where $\langle\beta|\hat{H}|\gamma\rangle = E_\gamma\delta_{\alpha\beta}$ has been used. However, since the ground-state energy E_{γ_0} is the smallest of the E_γ's, replacement of every E_γ in (17.85) by E_{γ_0} either does not change the magnitude of each term in the sum or else decreases it, i.e.,

$$\sum_\gamma E_\gamma|C_\gamma|^2 \geqslant \sum_\gamma E_{\gamma_0}|C_\gamma|^2 = E_{\gamma_0}, \tag{17.86}$$

the equality following from Eq. (17.84b). Comparison of (17.86) with (17.85) yields the inequality (17.83), QED.

Not only does (17.82) always yield an upper bound to E_{γ_0} but also, in cases in which $|\gamma_0\rangle$ is known, use of unsophisticated trial states can often give a value of $E[|\Gamma_t\rangle]$ much closer to E_{γ_0} than might naively have been expected. This happens because the lowest-order contribution to $E[|\Gamma_t\rangle] - E_{\gamma_0}$ is of second order (not first order) in the difference $|\Gamma_t\rangle - |\gamma_0\rangle$ (proof of this claim is left to the exercises). This helps us to understand why, for real systems for which $|\gamma_0\rangle$ cannot be determined, a seemingly crude trial state can yield a result close to the measured energy.

As a simple example illustrating the procedure we apply it to the 1-D quantal box problem solved in Chapter 3. The ground-state wave function is

$$\psi_1(x) = \begin{cases} \sqrt{2/L}\sin(\pi x/L), & x \in [0, L], \\ 0, & \text{otherwise}, \end{cases}$$

while the ground-state energy is $E_1 = \pi^2\hbar^2/(2mL^2)$. If we use a normalizable trial state $\psi_{\Gamma_t}(x)$ that does not vanish for $x \notin [0, L]$, it will yield the useless bound $E[|\Gamma_t\rangle] \equiv E[\psi_{\Gamma_t}] = \infty$, so a better choice requires $\psi_{\Gamma_t}(x)$ to vanish outside the box. We choose as the trial function the lowest-order real polynomial that vanishes at $x = 0$ and $x = L$, viz.,

$$\psi_{\Gamma_t}(x) = \begin{cases} \sqrt{30/L^5}\,x(L-x), & x \in [0, L], \\ 0, & \text{otherwise}. \end{cases} \tag{17.87}$$

This function, like $\psi_1(x)$, is symmetric about the point $L/2$.

With (17.87) for $\psi_{\Gamma_t}(x)$, the trial energy $E[\psi_{\Gamma_t}]$ is

$$E[\psi_{\Gamma_t}] = -\frac{30\hbar^2}{2mL^2}\int_0^L dx\, x(L-x)\,\frac{d^2}{dx^2}[x(L-x)]$$

$$= \frac{10\hbar^2}{2mL^2}, \tag{17.88a}$$

yielding the relative error

$$\frac{E[\psi_{\Gamma_t}] - E_1}{E_1} = 1.3\%. \tag{17.88b}$$

How are we to regard this result? The answer is considered in the next subsection.

Assessing the results

The production of a bound, which also functions as an approximate energy, is not the end of the analysis. As with any and all approximate calculations, one must also assess the result. That is, one must try to determine whether the relative error is small enough and, by implication, whether the chosen trial function is sufficiently accurate that no further calculations are needed. For perturbation theory, a criterion for undertaking a second-order calculation is that the ratio $E_\Gamma^{(1)}(\alpha)/E_\alpha$ is greater than some prescribed amount. The very small value of this ratio for the H-atom computations of Section 17.2 indicated that the first-order correction should suffice. For variational analyses of real systems, for which the energy is measured (to some accuracy) but the state/wave function is not, the initial assessment concerns the size of the relative error of the approximate energy. How small should the relative error be for the calculation to be considered satisfactory (and thus "final")? It should be no surprise that there is no universally accepted value for the relative error: the true answer to the question is "It depends on the purpose of the calculation."

The reader should not be put off by this answer: it reflects the real world as opposed to typical pedagogical analyses, even those concerning what might be referred to as realistic but "toy" problems, although these are capable of providing useful insights. Sometimes real-world calculations may be undertaken for order-of-magnitude purposes, e.g., to show that the Hamiltonian – often a model one for a complex system – contains the proper types of terms. In such cases a relative accuracy of 10%–50% might suffice, an example being the initial sets of variational calculations in the late 1920s applying quantum mechanics to molecules. Those not-very-accurate results established that quantum theory could account for molecular structure (see Chapter 18 for quantitative remarks). Many calculations, especially for complex systems for which one must resort to models (nuclear-structure computations being an instance), are deemed satisfactory if the accuracy of a variational or a variationally related analysis yields an energy accurate to as many significant figures (three or four, say) as can be determined by measurement. On the other hand, if there were reason (theoretical and/or experimental) to think that an improved measurement should or could be undertaken, then a decision to do so might follow from a variational calculation yielding an $E[|\Gamma_t\rangle]$ more accurate than the existing energy.

The ultimate in variational calculations is an energy that has "converged," i.e., is accurate to so many significant figures that all the physics is presumably accounted for. This is not a common occurrence. Possibly the most famous example of such calculations is the two-electron atom, in particular He, which we study in Chapter 18 after discussing identical-particle-symmetry effects. General procedures for attempting to achieve convergence will be discussed in the next two subsections, along with a method for approximating energies of excited state.

In addition to assessing the approximate energy $E[|\Gamma_t\rangle]$ one would also like to evaluate the trial state $|\Gamma_t\rangle$. A reasonably accurate $E[|\Gamma_t\rangle]$, being of second order in the difference $|\Gamma_t\rangle - |\gamma_0\rangle$, need not imply that $|\Gamma_t\rangle$ is of the same accuracy. Since for real systems the exact state is unavailable for comparison, some other criterion is needed, for example one that globally assesses the coordinate representation. . Only one such exists: use of $|\Gamma_t\rangle$ to calculate matrix elements of other observables, followed by comparison with the measured values. Typical here are evaluations of electromagnetic effects, three of which are transition rates, permanent electric-dipole or quadrupole moments, and electron-scattering cross sections; one may also attempt calculations of collisions involving projectiles heavier than electrons. Although all of these are topics beyond the scope of this book, they are listed in order that the reader will be aware of specific processes by which a $|\Gamma_t\rangle$ might be assessed.

Let us now return to the illustrative example of the preceding subsection, Eqs. (17.87) and (17.88), for which $\psi_1(x)$ is known. The number of significant figures is of less interest than the relative error of 1.3%, which is small but not very small. How does $\psi_{\Gamma_t}(x)$ relate to this? Lacking other observables whose matrix elements might test $\psi_{\Gamma_t}(x)$ globally (i.e., over its full range of x), we compare in Fig. 17.7 the two functions $\sqrt{L/2}\psi_1(\theta)$ and $\sqrt{L/2}\psi_{\Gamma_t}(\theta)$ over the half range $0 \leqslant \theta \leqslant 90°$, where $\theta = \pi x/L$ is expressed in degrees. Although the two curves cross, they are not very dissimilar, and the absolute value of their difference is no larger than about 0.04. Interestingly, the square of the maximum of the relative error

$$\frac{\psi_{\Gamma_t}(\theta) - \psi_1(\theta)}{\psi_1(\theta)} \tag{17.88c}$$

is consistent with the relative error in the energy of about 1%.[9]

[9] Recall that the correction to $E[|\Gamma_t\rangle]$ is of second order in the difference $|\delta\Gamma\rangle \equiv |\Gamma_t\rangle - |\gamma_0, E_{\gamma_0}\rangle$, i.e., it involves $\langle\delta\Gamma|\hat{H} - \mathrm{E}_{\gamma_0}|\delta\Gamma\rangle$.

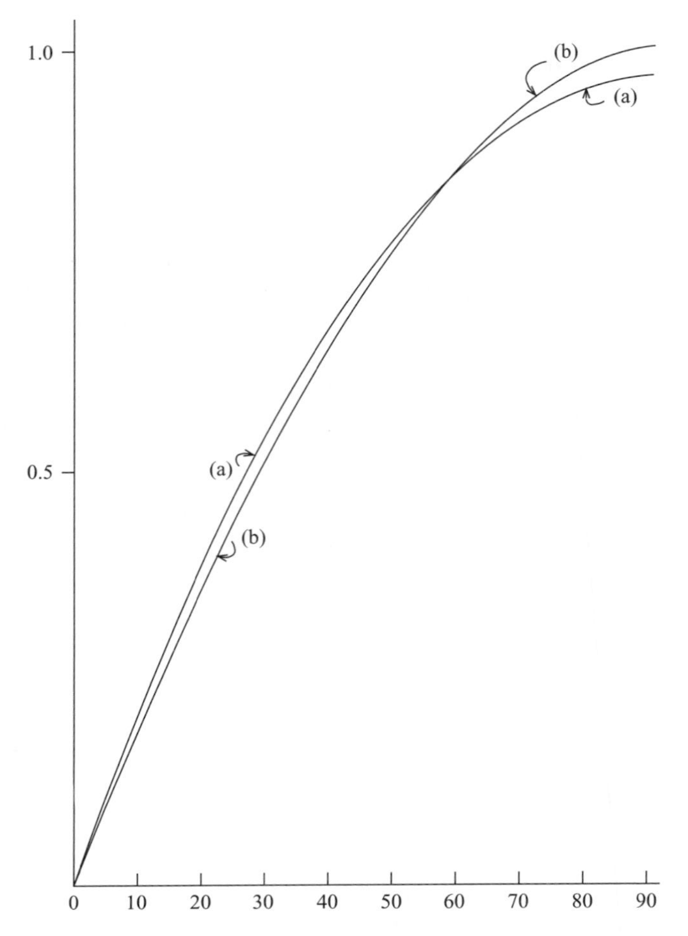

Fig. 17.7 Sketches of $\sqrt{L/2}\psi_1(\theta)$, curve (a), and $\sqrt{L/2}\psi_{\Gamma_t}(\theta)$, curve (b). Since these functions are symmetric about 90°, only values for $\theta \in [0, 90°]$ are shown.

The visual comparison of Fig. 17.7 accompanied by Eq. (17.88c) is one of the best measures for assessing this particular ψ_{Γ_t}. Notice that, if there were analogs for this system of a permanent dipole moment ($\propto \langle \hat{Q}_x \rangle$) or quadrupole moment ($\propto \langle \hat{Q}_x^2 \rangle$), then their evaluations using ψ_1 and ψ_{Γ_t} could even lead to an erroneous conclusion, namely that ψ_{Γ_t} is an excellent rather than a reasonable approximation to ψ_1, since the relative errors for these quantities are zero ($\langle \hat{Q}_x \rangle$) and 1% ($\langle \hat{Q}_x^2 \rangle$).

In view of our inability to make a truly quantitative statement about the quality of the approximation (17.87), the next step would be to improve (17.87) and then re-evaluate $E[\psi_{\Gamma_t}]$. To set the stage for this development, we may consider different ways of trying to improve ψ_{Γ_t}. Within the context of simple polynomials, one possibility is to add to ψ_{Γ_t} a second term containing either a symmetric or an asymmetric polynomial, e.g., $x^2(L-x)^2$ or $x(L-x)^2$. This leads to a calculation involving an unknown parameter, viz., the relative strength of the added term. Or one might try another functional form, e.g., $\sin(\beta x)(!)$ – which contains the undetermined parameter β. Terms involving more parameters could be part of the improvement. We show next how to deal with the presence of "variational" parameters in $|\Gamma_t\rangle$.

Improvements: single states, variational parameters

In general, a trial state $|\Gamma_t\rangle$ should encompass the symmetries, boundary conditions, angular-momentum properties, etc. appropriate to the system. The choice (17.87) for the 1-D box does this and, as we have seen, to improve the estimate (17.88a) requires the introduction of a variational parameter. We first assume that $|\Gamma_t\rangle$ depends on only one parameter, denoted β. Then $|\Gamma_t\rangle \to |\Gamma_t(\beta)\rangle$ and $E[|\Gamma_t\rangle]$ becomes a function of β:

$$E[|\Gamma_t\rangle] \to E_t(\beta) = \langle\Gamma_t(\beta)|\hat{H}|\Gamma_t(\beta)\rangle \geqslant E_{\gamma_0}, \tag{17.89}$$

where it is assumed that $\langle\Gamma_t(\beta)|\Gamma_t(\beta)\rangle = 1$. Since the inequality in (17.89) holds for all β, the value to be used is the one which minimizes $E_t(\beta)$. That is, the appropriate β is the solution β_0 of

$$\frac{dE_t(\beta_0)}{d\beta_0} = 0. \tag{17.90}$$

In the case of distinct multiple roots, the correct choice β_0 yields the lowest real minimum. As simple but uninteresting illustrations, we leave it to the reader to prove that the values of β which minimize the ground-state energies for the 1-D box, the 1-D oscillator, and the spin-independent H-atom (with $m_\mathrm{p} = \infty$) using the respective trial functions $\psi_{\Gamma_t}(x, \beta) = N_b \sin(\beta x)$, $\psi_{\Gamma_t}(x, \beta) = N_0 \exp[-x^2/(2\beta^2)]$, and $\psi_{\Gamma_t}(r, \beta) = N_a \exp(-r/\beta)\, Y_0^0$ are the exact values $\beta_0 = \pi/L$ for the box, $\beta_0 = \lambda = \sqrt{\hbar/(m\omega)}$ for the oscillator, and $\beta_0 = a_0 = \hbar^2/(m\kappa_\mathrm{e} e^2)$ for hydrogen.

In general, one's preference is to work with those trial functions which permit relatively easy evaluation of matrix elements in coordinate space. In the early days of atomic-structure calculations, especially prior to the advent of electronic calculators, single-particle trial functions of the form of parameterized hydrogenic wave functions – known as Slater orbitals – were used to help simplify certain variational calculations, and we consider some in the exercises to Chapter 18. Another functional form, still used in large-scale molecular-structure calculations, is the Gaussian function. In the following examples, we employ a parameterized 1s hydrogenic state to estimate the energy of a 3-D symmetric oscillator, and a Gaussian – the parameterized ground-state wave function for the symmetric 3-D oscillator – to estimate the energy of the non-relativistic hydrogen atom with $m_\mathrm{p} = \infty$.

EXAMPLE 1

The 3-D symmetric oscillator. Here,

$$\hat{H} = \frac{\hat{P}^2}{2m} + \frac{1}{2}m\omega^2 r^2$$

and the trial function is the parameterized 1s hydrogenic orbital

$$\psi_t(r, \beta) = \frac{2}{\beta^{3/2}} e^{-r/\beta}, \tag{17.91}$$

with the exact (3-D zero-point) energy being

$$E_{\text{g.s.}} = \tfrac{3}{2}\hbar\omega.$$

Using Eq. (17.91), the expectation values of the kinetic- and potential-energy operators are

$$\langle \hat{K}_x \rangle_t(\beta) = \frac{\hbar^2}{2m\beta^2} \qquad \text{and} \qquad \langle \tfrac{1}{2} m\omega^2 \hat{Q}_x^2 \rangle_t(\beta) = \tfrac{3}{2} m\omega^2 \beta^2,$$

so that

$$E_t(\beta) \equiv \frac{\hbar^2}{2m\beta^2} + \frac{3}{2} m\omega^2 \beta^2. \tag{17.92}$$

From (17.92) we find

$$\beta_0 = \frac{1}{3^{1/4}} \lambda, \qquad \lambda = \sqrt{\frac{\hbar}{m\omega}},$$

for which the minimum energy is

$$E_t(\beta_0) = \frac{2}{\sqrt{3}} \cdot \frac{3}{2} \hbar\omega \cong 1.15 E_{\text{g.s.}}$$

The crude estimate (17.91) using a 1s hydrogenic wave function yields a relative error of 15%.

EXAMPLE 2

The non-relativistic hydrogen atom. In this case

$$\hat{H} = \frac{\hat{P}^2}{2m_e} - \frac{\kappa_e e^2}{r},$$

and the trial function is the parameterized Gaussian

$$\psi_t(r, \beta) = \sqrt[4]{\frac{2}{\pi}} \left(\frac{2}{\beta}\right)^{3/2} e^{-r^2/\beta^2} Y_0^0, \tag{17.93}$$

while the exact energy is

$$E_{\text{g.s.}} = -\frac{1}{2} \frac{\kappa_e e^2}{a_0} \equiv -1 \text{ Ryd}, \qquad a_0 = \frac{\hbar^2}{m_e \kappa_e e^2}.$$

Following the same steps as in Example 1, one finds

$$\langle \hat{K} \rangle_t(\beta) = \frac{3}{2} \frac{\hbar^2}{m_e \beta^2} \qquad \text{and} \qquad \left\langle -\frac{\kappa_e e^2}{|\hat{\mathbf{Q}}|} \right\rangle_t (\beta) = -\sqrt{\frac{2}{\pi}} \frac{\kappa_e e^2}{\beta},$$

with

$$E_t(\beta) = -\sqrt{\frac{2}{\pi}}\frac{\kappa_e e^2}{\beta} + \frac{3}{2}\frac{\hbar^2}{m_e \beta^2}. \tag{17.94}$$

The minimum of (17.94) occurs at

$$\beta_0 = 3\sqrt{\frac{\pi}{2}}a_0,$$

for which we get the bound

$$E_t(\beta_0) = -\frac{2}{3\pi}\ \text{Ryd} \cong -0.212\ \text{Ryd}.$$

This poor result, lying slightly higher than the exact $n = 2$ level, is a consequence of the short-range nature of the Gaussian trial function (17.93), which is unable to sample enough of the Coulomb attraction to produce a sufficiently negative estimate of the potential energy. Relative ease of evaluation of matrix elements is thus no guarantee of an accurate result.

Before proceding to the case of multiple variational parameters, it is worth noting that use of a variational parameter β that functions as a length scale, as in the preceding two examples, leads to a dimensional necessity: no combination of m and $\hbar$ alone will yield a quantity having the dimension of energy, whereas $\hbar^2/(m\beta^2)$ is dimensionally correct. The final expression for $E_t(\beta)$ must therefore be of the form $N\hbar^2/(m\beta^2)$, and the role of the variational analysis is to evaluate the proportionality constant N.

The simplicity of the preceding calculations will not necessarily occur when multi-term, multi-parameter trial states are employed. Suppose that the trial state depends on a set of parameters $\{\beta_1, \beta_2, \ldots, \beta_n\}$. Then the trial energy is a function of these parameters,

$$E_t(\{\beta_j\}_{j=1}^n) = \langle \Gamma_t(\beta_1 \cdots \beta_n)|\hat{H}|\Gamma_t(\beta_1 \cdots \beta_n)\rangle,$$

and it must be minimized w.r.t. each of the n β_j's:

$$\frac{\partial E_t(\{\beta_j\}_{j=1}^n)}{\partial \beta_k} = 0, \qquad k = 1, \ldots, n. \tag{17.95}$$

Equation (17.95) is a set of n coupled equations for the n variational parameters. Their determination can be a formidable task: one must find the global minimum of an n-dimensional function that may have at least several local minima. The evaluation can become numerically tedious if several of the parameters occur individually in exponentials. Although we shall not carry out any evaluations related to (17.95) here (they are discussed in Chapter 18, see also the exercises to this chapter for calculational opportunities), there is an extremely important point to be made in this connection: substitution of the n minimizing values $\{\beta_j^{(0)}\}$ into $|\Gamma_t\{\beta_j^{(0)}\}\rangle$ produces only *one* state, not n of them. This single state is an approximation to the ground-state, and the single energy $E_t(\{\beta_j^{(0)}\})$ is an upper bound to the true ground-state energy of the Hamiltonian $\hat{H}$.

The latter result is in contrast to that of the next subsection, where the introduction of variational parameters as expansion coefficients leads to excited-state energies as well. Nonetheless, we may ask whether the procedure of this subsection can be used to derive a bound for, say, the energy of the first excited-state. One's first thought might be to constuct a second trial state orthogonal to the first. Unfortunately, this does not guarantee the desired excited-state-energy bound, essentially because the first trial state is not the true ground state. Thus, the second trial state, while orthogonal to the first one, need not be orthogonal to the true ground state, so that, if it is not, the second trial state actually produces another estimate for the ground-state energy. Despite this caveat, calculations leading to approximate excited-state energies and wave functions have long been carried out: see Chapter 18 for examples.

As a final point, we remind the reader that the analysis so far has dealt only with *upper* bounds. Numerous approaches to obtaining lower bounds also exist. We refer the interested reader to Volume II of Galindo and Pascual (1990) and to the review of Reid (1976) for discussion and citations of literature on this topic.

Improvements: finite basis-set expansions, excited states

Probably the simplest general procedure for obtaining approximations to excited-state energies is to expand $|\Gamma_t\rangle$ using states taken from a complete set. We consider this topic in the present subsection. For the unknown states of $\hat{H}$ we continue to use the set $|\gamma, E_\gamma\rangle$, but now with a sequential index appended to γ:

$$\{|\gamma, E_\gamma\rangle\} \rightarrow \{|\gamma_j, E_{\gamma_j}\rangle\},$$

where[10]

$$E_{\gamma_j} \leqslant E_{\gamma_{j+1}} \tag{17.96}$$

and $|\gamma_0, E_{\gamma_0}\rangle$ is still the ground state.

The members of the complete expansion set are denoted $|\tau_n, E_{\tau_n}\rangle$, arranged so that

$$E_{\tau_n} \leqslant E_{\tau_{n+1}},$$

with $|\tau_0, E_{\tau_0}\rangle$ the state of lowest energy. These states need not be orthogonal, an example being atomic states used in calculations for diatomic (or polyatomic) molecules (see Chapter 18 for examples). To encompass the possibility of non-orthogonality we introduce the overlap integral S_{nm} in place of the usual Kronecker delta:

$$S_{nm} = \langle \tau_n, E_{\tau_n} | \tau_m, E_{\tau_m} \rangle. \tag{17.97}$$

The finite set of coupled equations we will derive consists of variational approximations to the infinite set of coupled equations that represent the diagonalization of $\hat{H}$ in the complete basis $\{|\tau_n, E_{\tau_n}\rangle\}$. The latter set is obtained from the variational principle

$$\delta(\langle \alpha | \hat{H} | \alpha \rangle - E\langle \alpha | \alpha \rangle) = 0 \tag{17.98a}$$

by first expanding $|\alpha\rangle$,

$$|\alpha\rangle = \sum_n C_n |\tau_n, E_{\tau_n}\rangle, \tag{17.98b}$$

then substituting (17.98b) into (17.98a) to produce

[10] The equality sign in (17.96) is meant to take account of possible degeneracies.

$$\delta\left(\sum_{n,m} C_n^* C_m(H_{nm} - ES_{nm})\right) = 0, \tag{17.99}$$

where

$$H_{nm} = \langle\tau_n, E_{\tau_n}|\hat{H}|\tau_m, E_{\tau_m}\rangle, \tag{17.100}$$

and finally regarding the coefficients C_n^* and C_m as variational parameters.

Varying w.r.t. C_n^* and C_m separately leads to

$$\frac{\partial}{\partial C_n^*}\sum_{j,m} C_j^* C_m(H_{jm} - ES_{jm}) = 0$$

plus its complex conjugate; the preceding equation becomes

$$\sum_m C_m(H_{nm} - ES_{nm}) = 0. \tag{17.101}$$

The solutions of the infinite secular determinant

$$\det|H_{nm} - ES_{nm}| = 0 \tag{17.102}$$

are the exact energies E_{γ_j}, and the linear combinations $\sum_m C_{\gamma_j m}|\tau_m, E_{\tau_m}\rangle$ are just the exact eigenstates $|\gamma_j, E_{\gamma_j}\rangle$. One must actually be careful here: see, e.g., Galindo and Pascual (1990), Vol. II, for comments about convergence, spaces, etc.

The preceding, as has already been remarked, is just the diagonalization of $\hat{H}$ in the basis $\{|\tau_n\rangle, E_{\tau_n}\}$, a diagonalization whose origin we now have seen is variational. In the variational approximation to these results, the complete expansion (17.98) is replaced by the finite one

$$|\alpha\rangle = \sum_{n=0}^{N} C_n|\tau_n, E_{\tau_n}\rangle, \tag{17.103}$$

where the sum is over the states corresponding to the first $N + 1$ energies. The variational analysis now yields

$$\sum_{n=0}^{N} C_m^{(N)}(H_{nm} - E(N)S_{nm}) = 0, \tag{17.104}$$

where the fact that states $|\tau_n, E_{\tau_n}\rangle$ with $n > N$ have been excluded is indicated by the N dependence of $C_m^{(N)}$ and $E(N)$.

Solution of the secular equation (17.104) yields a set of $N + 1$ real energies $\{E_n(N)\}$ and $N + 1$ eigenstates labeled $|\gamma_n(N), E_n(N)\rangle$, where the variational energies are arranged sequentially:

$$E_n(N) \leqslant E_{n+1}(N). \tag{17.105}$$

Suppose that we repeat the analysis by adding to the expansion (17.103) the state $|\tau_{N+1}, E_{\tau_{N+1}}\rangle$ corresponding to the next energy $E_{\tau_{N+1}}$ in the sum (17.98b). This leads to an enlarged version of (17.104):

$$\sum_{m=0}^{N+1} C_m^{(N+1)}(H_{nm} - E(N+1)S_{nm}) = 0. \tag{17.106}$$

The result of solving the enlarged secular determinant (larger by one row and one column) is a new set of $N+2$ real energies $\{E_n(N+1)\}$, as well as $N+2$ orthonormal eigenstates $|\gamma_n(N+1),\ E_n(N+1)\rangle$, where the ordering is again sequential:

$$E_n(N+1) \leqslant E_{n+1}(N+1). \tag{17.107}$$

From the results embodied in Eqs. (17.98)–(17.102), and ignoring the fact that $\hat{H}$ is generally an unbounded operator, we conclude that

$$\lim_{N\to\infty} E_n(N) \to E_{\gamma_n}.$$

Because $\hat{H}$ is Hermitian, the convergence to E_{γ_n} is not only achieved monotonically but also occurs in an especially significant way, viz.,

$$E_n(N+1) \leqslant E_n(N) \leqslant E_{n+1}(N+1). \tag{17.108}$$

In physics this set of inequalities is known as the Hylleraas–Undheim theorem, named after the physicists who first derived it (Hylleraas and Undheim (1933)). In mathematics it is referred to as the matrix-interleaving or separation theorem. For proof and discussion, see Newton (1982) and Spruch (1962), whose proof uses graphical means.

Although any well-behaved set of states may be used to carry out the diagonalization $\hat{H}$, one set is particularly noteworthy. It is based on a perturbation-theory partition of $\hat{H}$ into a soluble portion $\hat{H}_0$ and a remainder $\hat{H}_1$ that need not be weak. As the expansion set we choose the eigenstates of $\hat{H}_0$, now denoted $|\gamma_n,\ E_{\gamma_n}\rangle$; they obey

$$[\hat{H}_0 - E_{\gamma_n}]|\gamma_n,\ E_{\gamma_n}\rangle = 0.$$

If only the ground state $|\gamma_0,\ E_{\gamma_0}\rangle$ is included (no diagonalization), then the variational bound to the exact ground-state energy E_{γ_0} is just the first-order perturbation result(!)

$$E_{\gamma_0} \leqslant E_0(1) = E_{\gamma_0} + \langle\gamma_0,\ E_{\gamma_0}|\hat{H}_1|\gamma_0,\ E_{\gamma_0}\rangle. \tag{17.109}$$

This relation is often referred to as a perturbation-variation estimate. It is left as an exercise to show that, if the expansion includes both the ground state and (an assumed non-degenerate) first excited state $|\gamma_1,\ E_{\gamma_1}\rangle$, then the inequality (17.108) is satisfied.

In view of the variational-bound nature of the first-order perturbation result for the ground state, the reader may be curious as to how the results of perturbation theory and the variational method compare in a real-world case. Such comparisons exist and we draw the reader's attention to analyses on the so-called "Rydberg states" of He. Reviews of each of the two approximation methods as applied to this system have been written by the persons mainly responsible for the calculations: Drake (1993) on the variational approach and Drachman (1993) for the perturbation method, though its application is not quite as straightforward as outlined here. Interested readers may wish to master the material of Sections 18.1 and 18.2 before turning to these references, however.

17.5. *The WKB approximation

The semi-classical or WKB approximation was introduced into physics independently by G. Wentzel (1926) and L. Brillouin (1926) and was soon improved upon by H. A. Kramers (1926); the mathematical method underlying it was pioneered by H. Jeffreys (1923). In addition to its role in the limit $h \to 0$ (see Eqs. (8.63) and (8.64)), it has been used in a variety of capacities, e.g., for the calculation of bound-state energies and transmission coefficients (tunneling phenomena such as field emission of electrons and α-particle decays); for the derivation of semi-classical quantization rules; and in analyzing convergence and summation properties of perturbation series. However, since the approximation is mathematically much more complicated than it initially appears, we shall only outline the method, referring the reader to other discussions for detailed analyses, e.g., Kemble (1958), Mathews and Walker (1970), Galindo and Pascual (1990), and Sakurai (1994), plus references cited therein.

In addition to presenting only an outline of the WKB method, we assume both a stationary-state situation and motion in a single dimension,[11] for which the Hamiltonian is

$$\hat{H}(x) = -\frac{\hbar^2}{2m}\frac{d^2}{dx^2} + V(x)$$

and Eq. (8.61) becomes

$$\Psi(x, t) = Ne^{iS(x,t)/\hbar} = Ne^{i[S(x)-Et]/\hbar}. \tag{17.110}$$

In place of Eq. (8.62) for $S(\mathbf{r}, t)$ we now have

$$\begin{aligned}\left(\frac{dS(x)}{dx}\right)^2 - i\hbar\frac{d^2S(x)}{dx^2} &= 2m[E - V(x)] \\ &\equiv p^2(x).\end{aligned} \tag{17.111}$$

The second equality defines $p(x)$, referred to as the quasi-momentum; we examine it next. Suppose first that $V(x) = 0$. Then $p(x) = p = \sqrt{2mE}$ is the classical free-particle momentum. Now let $V(x)$ be a constant such that $V < E$, in which case $p = \sqrt{2m(E - V)}$ is again the classical momentum. In each of these cases, the corresponding de Broglie wavelength is $\lambda = h/p$. Generalizing to $V(x) \neq$ constant, we can regard the quasi-momentum $p(x)$ as a position-dependent momentum and

$$\lambda(x) = \frac{\hbar}{p(x)} = \frac{\hbar}{\sqrt{2m[E - V(x)]}} \tag{17.112}$$

as a position-dependent wavelength (it is assumed that $E > V(x)$). The more slowly varying $V(x)$, the closer to an actual momentum or wavelength will these quantities be.

The presumption of a slow variation of $V(x)$ with x lies at the heart of the WKB method. To understand this, we note that, if the factor $d^2S(x)/dx^2$ in (17.111) is small, then its smallness as well as that of $\hbar$ should render

$$\left(\frac{dS(x)}{dx}\right)^2 \cong p^2(x) \tag{17.113}$$

and its solution

[11] The stationary-state assumption is standard, as is the restriction to 1-D. For 3-D, in which $V(\mathbf{r}) = V(r)$, the method developed here can be applied to the radial wave function: see, e.g., the treatment in Galindo and Pascual (1990).

$$S(x) \cong \pm \int^x dx'\, p(x') = \pm \int^x dx' \sqrt{2m[E - V(x)]} \tag{17.114}$$

good approximations. For this to be the case the value of $|\hbar\, d^2S/dx^2|$ obtained from (17.114) should be small. This latter quantity can be written

$$\left| \frac{\hbar\, d^2S(x)}{dx^2} \right| \cong \frac{m}{2\pi} \left| \lambda(x) \frac{dV(x)}{dx} \right|, \tag{17.115}$$

and the criterion is satisfied as long as $V(x)$ is sufficiently slowly varying that $dV(x)/dx$ can be neglected. Alternately, if $dV(x)/dx$ is approximated by $\Delta V(x)/\lambda(x)$, then the criterion is that $|\Delta V(x)|$, the change in $V(x)$ over a wavelength, be small.

Although (17.113) is the form Eq. (8.64) takes when $S_{\rm cl}(\mathbf{r}, t) = S_{\rm cl}(\mathbf{r}) - Et$, that form and the (1-D) solution (17.114) do not provide an adequate representation of $S(x)$, despite the preceding comments. An improved approximation is obtained by expanding $S(x)$ in powers of h and determining the first $h \neq 0$ correction. That is, writing

$$S(x) = \sum_{n=0}^{\infty} h^n S^{(n)}(x), \tag{17.116}$$

substituting into (17.111), and then equating coefficients of like powers of h, we find for $n = 0$ and $n = 1$ the relations

$$\left(\frac{dS^{(0)}(x)}{dx} \right)^2 = p^2(x) \tag{17.117a}$$

and

$$\frac{dS^{(0)}(x)}{dx} \frac{dS^{(1)}(x)}{dx} = \frac{i}{2} \frac{d^2S^{(0)}(x)}{dx^2}. \tag{17.117b}$$

The solutions to (17.117) are

$$S^{(0)}(x) = \pm \int^x dx'\, p(x') \tag{17.118a}$$

and

$$S^{(1)}(x) = i \ln[p^{1/2}(x)], \tag{17.118b}$$

where the lower limit in (17.118a) can be any fixed point x_0 and an inconsequential constant has been set equal to zero in $S^{(1)}(x)$. Now approximating $S(x) \cong S^{(0)}(x) + S^{(1)}(x)$, we find from

$$\Psi(x, t) = \psi(x) e^{-iEt/\hbar} = e^{i[S(x) - Et]/\hbar}$$

that $\psi(x) \rightarrow \psi_{\pm}^{(\rm WKB)}(x)$ with

$$\psi_{\pm}^{(\rm WKB)}(x) \cong \frac{N_{\pm}}{\sqrt[4]{2m[E - V(x)]}} e^{\pm(i/\hbar)\int^x dx' \sqrt{2m[E - V(x')]}}. \tag{17.119}$$

In the WKB approximation the general solution for $\psi(x)$ is a linear combination of the two functions (17.119). They generally provide reasonable approximations when $V(x)$ is slowly varying, i.e., everywhere except at those positions x_i where $E = V(x_i)$. This is illustrated schematically in Fig. 17.8 for simple attractive and repulsive potentials, for

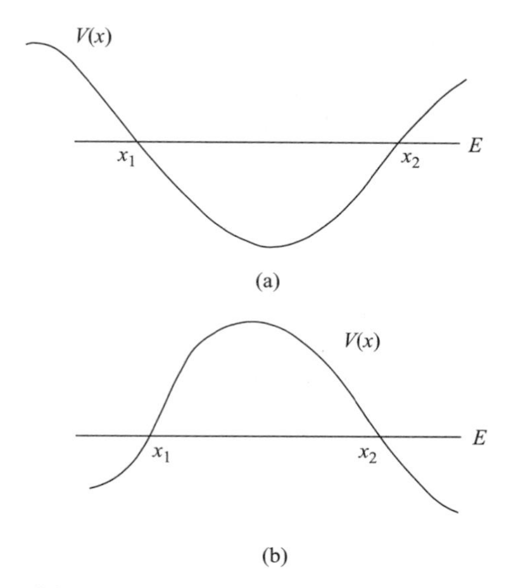

Fig. 17.8 Values of the energy E (horizontal lines) for an attractive (a) and a repulsive (b) pair of potentials $V(x)$. In each case the points x_1 and x_2 are the classical turning points, such that a classical particle in (a) is confined to the interior region $x \in [x_1, x_2]$, whereas in (b) it is excluded from the interior region: $x \notin [x_1, x_2]$.

which there are only two such points, labeled x_1 and x_2. These are the classical turning points: at x_1 or x_2 the particle's velocity becomes zero and its classical motion reverses. Quantally there are no turning points for a smoothly varying potential (recall the 1-D harmonic oscillator) and their appearance in (17.119) is an artifact of the WKB approximation. Artifact or not, if the results (17.119) are to be used, a means for treating the singularity in $[E - V(x)]^{-1/4}$ is required.

Futhermore, the foregoing requirement also means that, if

$$\psi^{(>)}(x) = \frac{1}{\sqrt{p(x)}}(N_+^{(>)}e^{i\int^x dx'\, p(x')/\hbar} + N_-^{(>)}e^{-i\int^x dx'\, p(x')/\hbar}) \tag{17.120a}$$

is the WKB solution for those x such that $E > V(x)$ and

$$\psi^{(<)}(x) = \frac{1}{\sqrt{p(x)}}(N_+^{(<)}e^{+\int^x dx'\, |p(x')|/\hbar} + N_-^{(<)}e^{-\int^x dx'\, |p(x')|/\hbar}) \tag{17.120b}$$

is the solution for those x wherein $V(x) > E$, then one needs a procedure to connect the two WKB solutions across the turning points in order to ensure continuity. The results of such procedures are known as connection formulae, and derivations of them have been the subject of much mathematical investigation. The mathematically knowledgeable reader may wish to consult Kemble (1958), Mathews and Walker (1970), and Galindo and Pascual (1990) for discussions, for application to specific potentials, and for citations to the literature on this topic. A nice (essentially WKB) treatment of the physics associated with barrier penetration/tunneling/transmission coefficients for problems such as α-particle decay of radioactive nuclei can be found in French and Taylor (1978), whereas WKB analyses of some simple bound-state problems can be found in Park (1964) and Merzbacher (1970).

In addition to applications such as those just mentioned, one can derive from the connection formulae a quantization condition known as the WKB (bound-state) quantization formula:

$$\int_{x_1}^{x_2} dx\, \sqrt{2m[E - V(x)]} = (n + \tfrac{1}{2})\pi\hbar, \qquad n = 0, 1, \ldots, \tag{17.121}$$

where x_1 and x_2 are the turning points for the attractive potential $V(x)$ – as in Fig. 17.8(a). The quantized energy levels yielded by (17.121) – which is very similar to quantization formulae predating quantum theory (see Galindo and Pascual (1990) for references) – can be reasonably accurate (they are exact for the oscillator, as examined in the exercises), although the WKB wave functions are not. Details are given in the various references cited.

Exercises

17.1 Assume that the Hamiltonian for a system depends on a parameter λ: $\hat{H} = \hat{H}(\lambda)$, so that the the eigenvalue problem is $[\hat{H}(\lambda) - E_\Gamma(\lambda)]|\Gamma, \lambda\rangle = 0$, with $\langle\Gamma, \lambda|\Gamma, \lambda\rangle = 1, \forall\, \lambda$.
   (a) Prove the Hellmann–Feynman theorem, viz., that $\partial E_\Gamma(\lambda)/\partial\lambda = \langle\Gamma, \lambda|\partial\hat{H}(\lambda)/\partial\lambda|\Gamma, \lambda\rangle$.
   (b) Let the λ dependence be that of Eq. (17.4), with the eigenvalue problem defined by Eq. (17.5). Prove that

$$E_\Gamma(\alpha) - E_\alpha = \int d\lambda\, \langle\Gamma, \lambda;\, \alpha|\hat{H}_1|\Gamma, \lambda;\, \alpha\rangle.$$

17.2 Show that, if $Z_\Gamma(\alpha) = \langle\Gamma, \lambda;\, \alpha|\Gamma, \lambda;\, \alpha\rangle$ and there are no degeneracies, then, to second order,

$$[Z_\Gamma(\alpha)]^{-1} = 1 - \lambda^2 \sum_\beta{}' |\langle\beta|\hat{H}_1|\alpha\rangle|^2/(E_\beta - E_\alpha) = \partial E_\Gamma(\alpha)/\partial E_\alpha.$$

17.3 Again assuming that there are no degeneracies, prove that the energy, through second order, is the expectation value of the Hamiltonian taken between the first-order normalized states. You may use the result of Exercise 17.2 without proving it.

17.4 Let the 1-D oscillator of Eq. (17.23a) be perturbed by the quadratic interaction $\hat{H}_1 = \lambda m\omega_0^2 \hat{Q}_x^2/2$, where λ is a small parameter.
   (a) Use perturbation theory to determine (i) the perturbed ground state through terms in λ, and (ii) the perturbed ground-state energy through terms in λ^2.
   (b) Solve the problem exactly for the ground state and its energy, and then show that a power-series expansion of each expression contains the results of part (a).

17.5 Extend the result (17.27) by calculating $|\Gamma;\, n\rangle^{(1)}$ and $E_\Gamma^{(2)}(n)$ for the linear oscillator perturbed by the interaction (17.23c).

17.6 Determine the lowest non-vanishing correction to the nth energy level of a 1-D oscillator when it is perturbed by an $\hat{H}_1 \propto \hat{Q}_x^3$.

17.7 A particle of mass is in a 2-D quantal box whose infinite sides are at $x = y = 0$, $x = a$ and $y = b$, with $b > a$. Also acting on the particle is the perturbation $\lambda\hat{Q}_x\hat{Q}_y$. Through order λ, determine the perturbed ground and first-excited states and their corresponding energies.

17.8 The K-electron energy shift in a heavy nucleus. The radius of a proton is roughly $r_\text{p} = 10^{-15}$ m; that of the $n = 1$ state in hydrogen is approximately $a_0 \sim 10^5 r_\text{p}$. Hence, for the H-atom it is an excellent approximation to assume a point proton. In a heavy nucleus ($Z \sim 80$), $a_0 \to a = a_0/Z \sim 10^3 r_\text{p}$, while the radius is $R \sim 10 r_\text{p}$, so that the assumption of a point-like nucleus is no longer valid. There is therefore a shift in the $n = 1$ energy level due to the finite nuclear size, which can be estimated by first-order perturbation theory. For the purposes of this calculation, assume that a single electron bound to a nuclear charge $Q = Ze$ is distributed throughout a structureless sphere of radius R.

(a) Use classical electrostatics to determine the perturbation $\hat{H}_1 = \hat{H}_1(Z, R, r)$, where r is the electron's position coordinate.

(b) The integral yielding the first-order correction $E_\Gamma^{(1)}(n = 1)$ to the $n = 1$ energy is trivial to calculate if an approximation, based on the facts that $R \sim 7r_\text{p}$ and $Z \sim 80$, is made. Find this approximation, show why it is valid, and then use it to evaluate $E_\Gamma^{(1)}(n = 1)$.

(c) Now set $R = 7r_\text{p}$ and $Z = 81$ and calculate the numerical values of $E_1(Z)$ and $E_\Gamma^{(1)}(n = 1)$. Discuss whether first-order perturbation theory is reliable for this situation.

17.9 Classically, when a static electric field $\mathcal{E}$ is applied to a system of charged particles, the energy becomes a function of $\mathcal{E}$, say $E(\mathcal{E})$, and, in the power-series expansion $E(\mathcal{E}) = E(0) + \mathcal{E}(\partial E/\partial\mathcal{E})_{\mathcal{E}=0} + (\mathcal{E}^2/2)(\partial^2 E/\partial\mathcal{E}^2)_{\mathcal{E}=0} + \cdots$, the electric polarizability α_el is defined as the negative of the second-order derivative: $\alpha_\text{el} = -(\partial^2 E/\partial\mathcal{E}^2)_{\mathcal{E}=0}$. An identical expansion can be used in the quantal case, to which perturbation theory is often applied, as in this exercise. Assuming constant $\boldsymbol{\mathcal{E}} = \mathcal{E}\mathbf{e}_3$, calculate α_el and the first-order corrected eigenstate for a particle of charge q and mass m

(a) in the ground-state of a symmetric 3-D oscillator of frequency ω, and

(b) freely rotating in 3-D at a fixed distance a from the origin.

17.10 Using the definition of α_el given in Exercise 17.9 and assuming a constant electric field oriented along $\mathbf{e}_3$, show, using cgs units, that, for an electron in the ground state of an H-atom with infinitely massive proton

$$2e^2 d_{21}/\Delta E < \alpha_\text{el} < 2e^2\langle\hat{Q}^2\rangle_{1\text{s}}/(3\,\Delta E),$$

where $\Delta E = E_2 - E_1$, $\langle\hat{Q}^2\rangle_{1\text{s}} = \langle 100|\hat{Q}^2|100\rangle$, and $d_{21} = \langle 21z|\hat{Q}_z|100\rangle$, with $|21z\rangle$ defined in Exercise 11.23. What is the form of the inequality in MKS units?

17.11 Derive the results stated in Eqs. (17.36).

17.12 In the weak-field approximation, the energy levels of a spinless particle bound to a central potential and acted on by a constant magnetic field $\mathbf{B} = B\mathbf{e}_3$ are given by Eq. (12.22b). Assuming that the spinless particle is an electron bound to an infinitely massive charge center Ze, use first-order perturbation theory to calculate the shift in energy due to the neglected term $[q^2B^2\kappa_\text{c}^2/(8m)]\langle x^2 + y^2\rangle$ when the electron is

(a) in the 2s level, and

(b) in the 2p, $\text{m}_\ell = \pm 1$ levels.

Choose $B = 1$ T and, if your shifts display a Z dependence differing from that of $E_n(Z)$, select representative values of Z in evaluating the magnitude of the shift. Compare your values of $\langle x^2 + y^2 \rangle$ with a_0, the value used in Chapter 12.

17.13 Consider an H-atom in the presence of a uniform magnetic field $\mathbf{B} = B\mathbf{e}_3$, assumed weak enough that the B^2 term can be ignored. In addition to the term linear in B, however, the hyperfine interaction $\hat{H}_1(\mathbf{r}) = -8\pi\kappa_{\mathrm{m}}\hat{\boldsymbol{\mu}}_{\mathrm{e}} \cdot \hat{\boldsymbol{\mu}}_{\mathrm{p}}\delta(\mathbf{r})/3$ of Section 17.2 is present. The Hamiltonian for this system ($\mu = m_{\mathrm{e}}$, $g_{\mathrm{e}} = 2$) is $\hat{H}(\mathbf{r}) = \hat{K}_{\mathrm{e}}(\mathbf{r}) - \kappa_{\mathrm{e}}e^2/r + \omega_{\mathrm{L}}(\hat{L}_z + 2\hat{S}_{ez}) + \hat{H}_1(\mathbf{r})$, where $\omega_{\mathrm{L}} = e\kappa_{\mathrm{m}}B/(2m_{\mathrm{e}})$ is the Larmor frequency.

(a) Determine an expression for the lowest-lying unperturbed energies $E_\alpha = E_{1\mathrm{s},M}$, where M is the electron's spin projection, and evaluate the difference $E_{1\mathrm{s},1/2} - E_{1\mathrm{s},-1/2}$ assuming that $B = 1$ T.

(b) Calculate the first-order shift in the energy $E_{1\mathrm{s},M}$ due to $\hat{H}_1(\mathbf{r})$.

(c) Now ignoring $\hat{H}_1(\mathbf{r})$ for the time evolution, assume that, at time $t = 0$, the system is the state $|\Gamma\rangle = |10;\, F = 0,\, M_F = 0\rangle$ of Eq. (17.44), and calculate the probabilities that, for $t > 0$, the atom will be found in state with $n = 1$, $\ell = 0$, $F = 1$, and $M_F = \pm 1, 0$.

17.14 Redo parts (a) and (b) of Exercise 17.7, now assuming that $a = b$.

17.15 Consider the symmetric 2-D oscillator perturbed by the x–y interaction discussed in Example 1 of Section 17.3 (p. 663), *only now you are to work in rectangular coordinates*.

(a) Set $N = 2$ (three-fold degeneracy) and determine the first-order corrections to the energy due to $\hat{H}_1$ of Eq. (17.66), and also the first-order corrected eigenstates.

(b) Compare the answer to part (a) with the exact solution and its expansion in powers of V_0/b^2. Ditto for the result $E_\Gamma(0, \pm)$ obtained in Example 1.

17.16 The symmetric oscillator of Example 1 in Section 17.3 is here perturbed by the interaction $\hat{H}_1 = V_0\hat{Q}_x/b$. Use first-order perturbation theory to calculate the energy shift and the new eigenstates that correspond to the degenerate $N = 2$ states. Carry out your analysis both in rectangular and in polar coordinates, and compare the results with those of an exact calculation.

17.17 In matrix form, a Hamiltonian is the sum $H = H_0 + \lambda H_1$, where

$$H_0 = \begin{pmatrix} 1 & 0 & 0 & 0 \\ 0 & 1 & 0 & 0 \\ 0 & 0 & 1 & 0 \\ 0 & 0 & 0 & 2 \end{pmatrix}, \qquad H_1 = \begin{pmatrix} 0 & 1 & 0 & 0 \\ 1 & 0 & 1 & 0 \\ 0 & 1 & 0 & 1 \\ 0 & 0 & 1 & 2 \end{pmatrix}.$$

(a) Use first-order degenerate perturbation theory to calculate the shifted energies and eigenstates corresponding to the first three levels.

(b) Calculate the energy shift of the fourth level through second order in λ (Kok and Visser 1987).

17.18 A one-electron atom in a constant magnetic field. In the absence of spin–orbit coupling and magnetic fields, the Hamiltonian for a one-electron atom is $\hat{H}_0(\mathbf{r}) = \hat{K}(\mathbf{r}) + V(r)$, where $(\hat{H}_0 - E_{n\ell})|n\ell\tfrac{1}{2};\, jm_j\rangle = 0$. Reinstating the spin–orbit interaction $\hat{H}_{LS} = V_{LS}(\mathbf{r})\hat{\mathbf{L}} \cdot \hat{\mathbf{S}}/\hbar^2$ and placing the atom in a constant magnetic field $\mathbf{B} = B\mathbf{e}_3$, which gives rise to the additional interaction $\hat{H}_B = \omega_{\mathrm{L}}(g_{\mathrm{L}}\hat{L}_z + g_S\hat{S}_z)$, the complete Hamiltonian is $\hat{H} = \hat{H}_0 + \hat{H}_1$, with $\hat{H}_1 = \hat{H}_{LS} + \hat{H}_B$, and $\omega_{\mathrm{L}} = eB\kappa_{\mathrm{m}}/(2m_{\mathrm{e}})$.

(a) Using the degenerate coupled states as a basis ($j = \ell \pm \frac{1}{2}$), calculate the first-order energy shift due to $\hat{H}_1$. (You may use without derivation the results $\langle n\ell\frac{1}{2};\ \ell \pm \frac{1}{2},\ m_j|\hat{S}_z|n\ell\frac{1}{2};\ \ell \pm \frac{1}{2},\ m_j\rangle = [\pm m_j\hbar/(2\ell+1)]$ and $\langle n\ell\frac{1}{2};\ \ell - \frac{1}{2},\ m_j|\hat{S}_z|n\ell\frac{1}{2};\ \ell + \frac{1}{2},\ m_j\rangle = [\hbar/(2\ell+1)][(\ell+\frac{1}{2})^2 - m_j]^{1/2}$.)

(b) The Zeeman effect occurs when $[(\ell+\frac{1}{2})\langle V_{LS}\rangle] \gg \hbar\omega_{\mathrm{L}}$. Prove that, in this situation, the first-order correction to $E_{n\ell}$ is

$$E^{(1)}_{n\ell m_j} = \begin{cases} \ell\langle V_{LS}\rangle/2 + 2m_j\hbar\omega_{\mathrm{L}}(\ell+1)/(2\ell+1), & j = \ell+\frac{1}{2}, \\ -(\ell+1)\langle V_{LS}\rangle/2 + 2m_j\hbar\omega_{\mathrm{L}}\ell/(2\ell+1), & j = \ell - \frac{1}{2}, \end{cases}$$

and specify in a Grotrian diagram the m_j dependence of the Zeeman splittings. (Note: $\langle V_{LS}\rangle = \langle n\ell|V_{LS}|n\ell\rangle$ is assumed to be independent of j.)

(c) The Paschen–Back effect occurs when $\hbar\omega_{\mathrm{L}} \gg [(2\ell+1)\langle V_{LS}\rangle/2]$. Show that, for this case, $E^{(1)}_{n\ell m_j} \cong [(m_j \pm \frac{1}{2})(\hbar\omega_{\mathrm{L}} \pm \langle V_{LS}\rangle/2) - \langle V_{LS}\rangle/2]$, $j = \ell \pm \frac{1}{2}$, and sketch, for fixed m_j, the B dependence of $E^{(1)}_{n\ell m_j}$.

17.19 The premises of this exercise are those stated in the preceding one.

(a) Evaluate the first-order energy shift obtained for part (a) of Exercise 17.18 for the case $\ell = 1$.

(b) There are six degenerate $n = 1$ levels for $\ell = 1$. Use first-order perturbation theory to obtain the energies, expressing them in terms of ω_{L} and $\langle V_{LS}\rangle$, and show explicitly that $\hat{H}_1$ fully lifts the degeneracy.

17.20 Prove that the lowest-order correction to $E[|\Gamma_t\rangle] - E_\gamma$ is of second order on the difference $|\Gamma_t\rangle - |\gamma_0\rangle$.

17.21 Let $V(x)$ be an arbitrary 1-D potential that is nowhere positive in the range $x \in [\infty, -\infty]$. Use the variational method to prove that such a $V(x)$ will always support at least one bound state (exclude the trivial case $V(x) = 0$).

17.22 An alternate to the simple trial function (17.87) for the 1-D box is

$$\psi_{\Gamma_t}(x) = \begin{cases} x/(\beta L), & 0 \leqslant x \leqslant \beta L, \\ (L-x)/[(1-\beta)L], & \beta L \leqslant x \leqslant L. \end{cases}$$

Calculate the optimal value of the variational parameter β and the corresponding energy, comparing your results both with (17.88a) and with the exact result.

17.23 One way to try to improve $\psi_{\Gamma_t}(x)$ of (17.87) is by adding to it the term $x^2(L-x)^2$. Determine the optimal value of the variational parameter thus introduced and the energy it yields, comparing your results with (17.88a) and the exact result.

17.24 A more complicated improvement to $\psi_{\Gamma_t}(x)$ of (17.87) is the trial function $\psi_{\Gamma_t}(x) = N[x(L-x) + \beta_1 x^2(L-x) + \beta_2 x(L-x)^2]$. Calculate N, the optimal values of β_1 and β_2, and the new variational estimate of the energy, comparing your results with (17.88a) and the exact answer, and also with the result of Exercise 17.22 if you have obtained it.

17.25 A particle of mass m is acted on by the 3-D potential $V(r) = -V_0\exp(-r/a)$, where $\hbar^2 V_0 a^2/m = \frac{3}{4}$. Use the trial function $\exp(-r/\beta)$ to obtain a bound on the energy.

17.26 One can extend the analysis based on Eq. (17.93) to an excited-state calculation by means of the trial function $r[\exp(-r^2/\beta^2)]\cos\theta$. Determine the optimal value of β and the resulting variational energy. Specify which excited-state energy is thus approximated, and justify your claim.

17.27 The ground-state wave function and energy for a particle of mass m in the symmetric 1-D box of Eq. (9.86) are the $n = 0$ values of (9.90a) and (9.90b). Determine the variational upper bound to the latter energy resulting from the trial function $\psi_{\Gamma_t}(x) \propto [1 - (x/L)^2]$, for $|x| \leqslant L$, and $\psi_{\Gamma_t}(x) = 0$ for all other values of x.

17.28 To the Hamiltonian of the 1-D box of Chapter 3 is added the term

$$\hat{H}_1(x) = \begin{cases} V_0 x/L, & x \in [0/L], \\ 0, & \text{otherwise}. \end{cases}$$

(a) Determine upper bounds to the ground- and first-excited-state energies by using a linear combination of the two lowest box eigenfunctions ψ_0 and ψ_1 as a variational trial function.

(b) Evaluate the improvement in the preceding two energies as well as the bound on the second excited-state energy obtained by adding in the next box eigenfunction ψ_2.

17.29 Show that the WKB quantization formula (17.121) yields the exact energies when $V(x)$ is the 1-D oscillator potential $m\omega^2 x^2/2$.

18

Many Degrees of Freedom: Atoms and Molecules

Quantum mechanics was accepted early on as the correct framework for describing all microscopic physical phenomena because it was able to account for the properties of few-particle atoms (H and He), molecules (H_2^+ and H_2), and later the neutron–proton scattering system and the deuteron (^2H). Of course, it has subsequently been applied successfully to the structures of more complex atoms, nuclei, and molecules, and to larger systems such as solids, liquids, and gases, as well as to collisions involving all of them. In this concluding chapter, we introduce the reader to aspects of the many-degree-of-freedom problem by considering several multi-particle systems. The first of these is the two-electron atom, exemplified by He. Although the two-electron eigenvalue problem can only be solved approximately, the system is simple enough that energies of astonishing accuracy have been determined. We examine both primitive and highly sophisticated approximations in Section 18.2. The next system studied is the multi-electron atom. While atoms in general are too complex for the super-accurate, He-type calculations to be undertaken, the collection of atoms considered together display regularities – e.g., those of the periodic table – whose features quantum theory can explain. It does so through the medium of the independent-particle, shell-model of atomic structure, which we shall explore in some detail. The final system considered is the diatomic molecule, whose treatment in Section 18.5 is based on the Born–Oppenheimer approximation.

Shell models (described later in this chapter) arise in part from a particular kind of approximation to a many-particle Hamiltonian and its associated eigenvalue problem. Such approximations are essential in order to gain even a qualitative understanding of the relevant phenomena, since the systems involved contain N particles, where, typically, $3 \leqslant N \leqslant 100$, values that are generally too large for the 3-D Schrödinger equations to be solved exactly. Additionally, it is of paramount importance for the shell model of atoms (and of nuclei) that these systems are composed of identical particles: electrons in atoms or neutrons/protons in nuclei. The major consequence of this indistinguishability of particles is a new symmetry property, one associated with the behavior of $\hat{H}$ and its eigenstates under interchange of particle labels.

We begin by considering the new symmetry property, the fundamental result for which is the so-called spin-statistics theorem. It is stated and illustrated by various two-particle examples in Section 18.1, whereas in Section 18.2 it is applied to the He-atom using variational approximations. In Section 18.3 we introduce the N-particle Hartree–Fock approximation, which is the paradigm for all identical-particle shell models. The remaining sections are concerned with the structures of atoms and of molecules.

18.1. Identical-particle symmetries

The spin-statistics theorem

About a year prior to Uhlenbeck and Goudsmit's interpretation of the Stern–Gerlach experiment, Wolfgang Pauli (1924) inferred the existence of a previously unsuspected, two-valued attribute of electrons. He then used this attribute – later identified as spin $\frac{1}{2}$ – to formulate a universal principle governing the behavior of electrons in atoms. Now known as the Pauli exclusion principle, it is the key ingredient in the quantal explanation of the periodic table of the elements. Following Pauli, we state the principle for an independent-particle model of an atom, the Hamiltonian for which is $\hat{H}_{\text{IP}}(1, 2, \ldots, N)$ of Section 11.1. In such a model, each of the N electrons in an atom is in a central-potential, single-particle state, referred to as an atomic orbital. Each atomic orbital is specified by the values of the four quantum numbers n, ℓ, m_ℓ, and m_S $(= \pm\frac{1}{2})$. Under this assumption concerning atomic structure, the exclusion principle asserts that no electron can be in an orbital each of whose four quantum numbers is identical to those of an orbital occupied by another electron. That is, if $n\ell m_\ell m_S$ and $n'\ell' m'_\ell m'_S$ specify states occupied by two electrons, then in at least one of the four pairs (nn'), $(\ell\ell')$, $(m_\ell m'_\ell)$, and $(m_S m'_S)$, the two numbers must be different. Should this requirement not be satisfied, i.e., if all four equalities $n = n'$, $\ell = \ell'$, $m_\ell = m'_\ell$, and $m_S = m'_S$ hold, then the N-electron state vector vanishes.

As shown below, the form taken by a two-particle state conforming to the preceding behavior is a 2×2 determinant; it generalizes for N particles to an $N \times N$ determinant. One may either use the particle labels $(1, 2, \ldots, N)$ to specify the rows of the determinant and the N sets of four quantum numbers $(n\ell m_\ell m_S)$ to specify its columns, or vice versa. Either way, by virtue of the properties of determinants,[1] the determinantal state changes sign when any two rows or any two columns are interchanged, that is, when any pair of particle labels are interchanged.[2] Now, a state that changes sign on interchange of particle labels is *antisymmetric* – recall the behavior of the two-particle spin-singlet state, Eq. (14.71) *et seq.* Hence we conclude from the Pauli principle that an N-electron atomic state composed of single-particle orbitals is antisymmetric in the particle labels.

The preceding statement is a remarkable result. Even more remarkable is the fact that *every* N-electron state must similarly be antisymmetric irrespective of whether or not it is composed of single-particle orbitals. Most remarkable of all is the result that all states of N identical, half-odd-integer-spin particles (not just electrons) must be antisymmetric under interchange of any pair of the identical-particle labels $(1, 2, \ldots, N)$. Furthermore, states of N identical, integer- or zero-spin particles must be *symmetric* under interchange of any pair of the identical-particle labels.[3] Two-particle examples of such symmetric

[1] Properties of determinants are discussed, e.g., by Byron and Fuller (1970) and Hassani (1991).

[2] Specific examples of this behavior are given below.

[3] In addition, the Hamiltonian for a system of N identical particles must be *symmetric* in the particle labels, otherwise it cannot yield fully symmetric or antisymmetric states, a point we explore in the exercises. Furthermore, although nature employs only symmetric (antisymmetric) states for identical-particle systems of zero or integer (half-odd-integer) spins, one can carry out calculations using states of either type of symmetry for either type of spin without warnings being set off: it is up to the scientist to impose the proper symmetry requirement *ab initio*.

states are the three spin-1 components of Eqs. (14.72). An amusing consequence of the latter symmetry condition is that, if the state of N identical, zero-spin or integer-spin particles consists only of orbitals $(n\ell m_\ell m_S)$, then not only can any two of the particles be in the same orbital but also all N particles can be in that one orbital!

The foregoing statements comprise the spin-statistics theorem, although a more appropriate name in the present context might be the state-vector-symmetry theorem. The theorem, which follows from certain results of quantum-field theory first established by Pauli (1940), may be regarded as the seventh postulate of non-relativistic quantum theory:

THE SPIN-STATISTICS THEOREM

The state vectors/wavefunctions for any system of N identical particles having half-odd-integer spin must be antisymmetric under interchange of the labels of any pair of the N particles. No others are allowed. Correspondingly, the state vectors/wave functions of any system of N identical, zero-spin or integer-spin particles must be symmetric under interchange of the labels of any pair of the N particles.

In addition to systems composed of identical single particles, the spin-statistics theorem has been applied to the states of an ensemble of identical, *multi-particle*, bound systems, themselves each composed of identical particles. Let there be N systems in the ensemble, and let J be the total angular momentum of each system. Then, when $J = 0, 1, 2, \ldots$, the state vector for the N systems must be symmetric under interchange of the label of any pair of *systems,* while it must be similarly antisymmetric when $J = \frac{1}{2}, \frac{3}{2}, \frac{5}{2}, \ldots$. The preceding application will be valid when the probability that any of the particles comprising one system is in the vicinity of any of the other $N - 1$ systems is exceedingly small, since only in this situation can one refer to a well-localized, identifiable system. Otherwise, it may be impossible to claim that there are N viable systems forming the ensemble, in which case one would need to deal with an overall symmetric or antisymmetric state vector for all the identical particles considered together. It should be evident that dilute gases composed of identical multi-particle bodies could be systems to which the spin-statistics theorem would apply, although it is often assumed to be true for convenience, if not for absolute correctness, in other circumstances. Thus, the unstable nucleus ^{8}Be has been considered as composed of two α-particles (in symmetric two-body states), and similarly the stable nucleus ^{12}C has been described as three α-particles, even though each nucleus is made up of separate neutrons and protons.[4]

As a final comment, we note that the "statistics" part of the theorem refers to the (statistical-mechanical) distribution functions describing systems composed of identical bodies – usually identical particles. Let us consider the case of individual particles (or equivalently, multi-particle objects depicted as structureless bodies) of spin S. For $S = \frac{1}{2}, \frac{3}{2}, \frac{5}{2}, \ldots$, the distribution function to be used is the one derived by Fermi (1926)

[4] On an even finer scale, with each nucleon composed of three quarks, these two nuclei could be thought of as stable collections of 24 and 36 quarks, whose behavior is described by quantum chromodynamics (see, e.g., Griffiths (1987) or Gottfried and Weisskopf (1984)). These are intractable problems: there is no reliable solution of the deuteron-as-six-quarks problem.

and Dirac (1926), denoted (of course!) the Fermi–Dirac distribution function. In view of this, single particles and multi-particle bound systems with half-odd-integer spin are known as "fermions." Thus, all electrons, protons, neutrons, quarks, neutrinos, ^{3}He nuclei, etc., are fermions. The distribution function to be employed with particles and multi-particle bound systems having $S = 0, 1, 2, \ldots$, is that of Bose (1924) and Einstein (1924, 1925), and the latter entities are referred to collectively as "bosons." Examples of bosons are photons, α-particles, π mesons, ^{12}C nuclei, etc. Discussions of the connection between the symmetry properties of the identical-particle state vectors and the appropriate statistical distribution functions can be found in many places, a few of which are Schrödinger (1952), Fermi (1966), and Galindo and Pascual (1990). Another treatment, with nicely worked-out applications, can be found in Park (1964).

The systems examined in this chapter are single atoms and molecules containing N identical electrons. Because electrons are fermions, we focus our attention on antisymmetric state vectors and wave functions, leaving developments concerning bosons to the exercises. Scattering systems *per se* will not be considered since the application of these particular symmetry concepts to collision phenomena is a formidable undertaking in general, well beyond the scope of this book, although the exercises will deal with aspects of the scattering of a pair of identical particles.

As a means of introducing the reader to the consequences and intricacies connected with identical-particle symmetries, we focus in the remainder of this section on systems containing two fermions: such systems are rich enough in detail to provide the background needed for the analyses of Sections 18.2–18.5.

A pair of identical fermion spinors

In this first subsection we examine the pedagogical example of two spin-$\frac{1}{2}$ fermions that have no spatial attributes, particles we shall refer to as pure "spinors." Each particle is described solely by the spin-$\frac{1}{2}$ eigenstates $|\frac{1}{2}m_s\rangle$. As in Chapters 11 and 14, we add the numerical labels 1 and 2 as subscripts to the particles, to their eigenstates, and to their Hilbert spaces. This is done for bookkeeping purposes, i.e., to facilitate enumeration and calculation. The assignment of the labels 1 and 2 is arbitrary, as discussed below.

Labeling the individual Hilbert spaces as $\mathcal{H}_j$, $j = 1$ and 2, the overall Hilbert space $\mathcal{H}$ is an outer product:

$$\mathcal{H} = \mathcal{H}_1 \otimes \mathcal{H}_2. \tag{18.1}$$

Correspondingly, there are two sets of states $|\frac{1}{2}m_S\rangle_j$, $j = 1, 2$, so that a basis spanning $\mathcal{H}$ is

$$\{|\tfrac{1}{2}m_S\rangle_1|\tfrac{1}{2}m_S'\rangle_2\}, \tag{18.2}$$

for which the resolution of the identity $\hat{I}_{1,2}$ is

$$\hat{I}_{1,2} = \sum_{m_S, m_S'} [|\tfrac{1}{2}m_S\rangle_1|\tfrac{1}{2}m_S'\rangle_2][{}_1\langle\tfrac{1}{2}m_S|{}_2\langle\tfrac{1}{2}m_S'|]. \tag{18.3}$$

Since m_S and m_S' each take on the values $\pm\frac{1}{2}$, the ordering of the labels 1 and 2 in the set (18.2) and the expansion (18.3) is inconsequential.

In this idealized situation of no spatial attributes, a proper state vector for the two particles will be composed of spinors selected from the set (18.2). All other things being equal, one could consider as a candidate the uncoupled (product) state vector

$$|m_1, m_2\rangle_{1,2} \equiv |\tfrac{1}{2}m_1\rangle_1 |\tfrac{1}{2}m_2\rangle_2. \tag{18.4}$$

However, all things are NOT equal, for the particles are identical fermions, and according to the spin-statistics theorem, their state vector must be antisymmetric under the exchange of the labels 1 and 2. Equation (18.4), not being antisymmetric, must therefore be rejected.

Although (18.4) is not allowed, it can be used as the basis for obtaining a correctly antisymmetrized state. One way of accomplishing this is by applying to (18.4) the appropriate *antisymmetrizing* operator, i.e., the two-particle *antisymmetrizer* $\hat{\mathcal{A}}_{12}$, given by

$$\hat{\mathcal{A}}_{12} = \hat{I} - \hat{P}_{12}, \tag{18.5}$$

where $\hat{P}_{12}$ is the two-particle interchange operator. Applying $\hat{\mathcal{A}}_{12}$ to (18.4), we get

$$\hat{\mathcal{A}}_{12}|m_1, m_2\rangle_{1,2} = [|\tfrac{1}{2}m_1\rangle_1 |\tfrac{1}{2}m_2\rangle_2 - |\tfrac{1}{2}m_1\rangle_2 |\tfrac{1}{2}m_2\rangle_1], \tag{18.6}$$

where

$$\hat{P}_{12}|m_1, m_2\rangle_{1,2} = |m_1, m_2\rangle_{2,1} = |\tfrac{1}{2}m_1\rangle_2 |\tfrac{1}{2}m_2\rangle_1 \tag{18.7}$$

has been used (particle labels, rather than magnetic quantum numbers, have been interchanged). The minus sign in $\hat{\mathcal{A}}_{12}$ ensures that $\hat{\mathcal{A}}_{12}|m_1, m_2\rangle_{1,2}$ is properly antisymmetrized:

$$\hat{P}_{12}(\hat{\mathcal{A}}_{12}|m_1, m_2\rangle_{1,2}) = -\hat{\mathcal{A}}_{12}|m_1, m_2\rangle_{1,2}; \tag{18.8a}$$

in other words, $\hat{P}_{12}$ and $\hat{\mathcal{A}}_{12}$ anticommute:

$$\hat{P}_{12}\hat{\mathcal{A}}_{12} + \hat{\mathcal{A}}_{12}\hat{P}_{12} = 0. \tag{18.8b}$$

The reader should be aware that the result (18.6) achieved using $\hat{\mathcal{A}}_{12}$ is identical to that which is obtained by forming the appropriate linear combination of basis states.

One consequence of the minus sign in $\mathcal{A}_{12}$ is that it eliminates all states for which $m_2 = m_1$, since for these choices (18.6) is zero. Since m_1 and m_2 are limited to the values $\pm\frac{1}{2}$, $m_2 = -m_1$ and (18.6) ends up being proportional to the coupled, singlet-spin state vector $|\frac{1}{2}\frac{1}{2}; 00\rangle$ of Eq. (14.71). Invoking the normalization of that state, we have therefore found the following evolution:

$$\begin{aligned}
\hat{\mathcal{A}}_{12}|m_1, m_2\rangle_{1,2} &\to \hat{\mathcal{A}}_{12}|m_1, m_2\rangle_{1,2}\delta_{m_2,-m_1} \\
&\to \sqrt{\tfrac{1}{2}}\hat{\mathcal{A}}_{12}|m_1 = \tfrac{1}{2}, m_2 = -\tfrac{1}{2}\rangle_{1,2} \\
&= \sqrt{\tfrac{1}{2}}[|\tfrac{1}{2}\tfrac{1}{2}\rangle_1 |\tfrac{1}{2}, -\tfrac{1}{2}\rangle_2 - |\tfrac{1}{2}, -\tfrac{1}{2}\rangle_1 |\tfrac{1}{2} \cdot \tfrac{1}{2}\rangle_2] \equiv |\tfrac{1}{2}\tfrac{1}{2}; 00\rangle_{1,2}.
\end{aligned} \tag{18.9}$$

This is a striking progression: of the four two-particle spinor states generated by the Clebsch–Gordon angular-momentum-coupling procedure, four states that are themselves a basis in $\mathcal{H}$, only the singlet state is appropriate for describing two identical spin-$\frac{1}{2}$ fermions with no spatial attributes.

As remarked earlier, an antisymmetric state of two (or N) identical particles, when each particle is itself in a single-particle state, has a determinantal structure and Eq. (18.9) is, in fact, equivalent to a determinant:

$$|\tfrac{1}{2}\tfrac{1}{2};\,00\rangle_{1,2} = \sqrt{\frac{1}{2}}\begin{vmatrix} |\tfrac{1}{2}\tfrac{1}{2}\rangle_1 & |\tfrac{1}{2},-\tfrac{1}{2}\rangle_1 \\ |\tfrac{1}{2}\tfrac{1}{2}\rangle_2 & |\tfrac{1}{2},-\tfrac{1}{2}\rangle_2 \end{vmatrix} \tag{18.10a}$$

$$= \sqrt{\frac{1}{2}}\begin{vmatrix} |\tfrac{1}{2}\tfrac{1}{2}\rangle_1 & |\tfrac{1}{2}\tfrac{1}{2}\rangle_2 \\ |\tfrac{1}{2},-\tfrac{1}{2}\rangle_1 & |\tfrac{1}{2},-\tfrac{1}{2}\rangle_2 \end{vmatrix}. \tag{18.10b}$$

These two determinants are related by a reflection about the main diagonal, a non-sign-changing action. In Eq. (18.10a), the rows are specified by the particle labels and the columns by the state labels (here $m_S = \pm\frac{1}{2}$), whereas in (18.10b) the opposite is true. In each, there is an overall change of sign on interchange of the two rows or the two columns – a standard characteristic of determinants. We remind the reader of a point made earlier in a different context (Chapter 14): the antisymmetric state – (18.9) or (18.10) – is a correlated one, not a simple product of single-particle states. The Pauli principle forces states of identical fermions to be correlated.

That identical-particle systems must be described by symmetric or antisymmetric states is a conclusion based on deep-seated theoretical tenets. This conclusion is expressed through the spin-statistics theorem, which tells us the type of symmetry property appropriate for particular kinds of identical particles. In this regard, it is essential to understand that the state vector and its quantum numbers cannot be used to distinguish identical particles, since quantum numbers, like particle labels, are extrinsic to the particles. If this point is not clear, the following analogy may help. Consider truly identical, but mute sisters, Betty and Bettye, one of whom is wearing blue clothing while the other is dressed in green. Since it is not previously known to an observer which twin is which, the different clothing cannot be used by the observer to make an identification. Neither could the names: the twins could be called Cindy and Cindi, but they would remain indistinguishable. The comparison with the pair of identical particles is straightforward: the names are the analogs of the labels 1 and 2, whereas the different colored clothing is the analog of the different magnetic quantum numbers. To paraphrase Shakespeare: clothes do not identify the identical twin.

A pair of isolated spin-$\frac{1}{2}$ fermions

In contrast to the preceding subsection, where a Hamiltonian was not needed, we work with one from here on. The system of interest is a pair of isolated particles, so that there are no external interactions. For simplicity the interparticle interaction is taken to be spin-independent, for which the relevant Hamiltonian is $\hat{H}_{12}$ of Section 11.6, whose coordinate representation is assumed to be

$$\hat{H}_{12}(\mathbf{r}_1, \mathbf{r}_1) = \hat{K}_1(\mathbf{r}_1) + \hat{K}_2(\mathbf{r}_2) + V(|\mathbf{r}_1 - \mathbf{r}_2|). \tag{18.11}$$

That is, the interparticle potential is a scalar, depending only on $|\mathbf{r}_1 - \mathbf{r}_2|$. Because the fermions are identical, the same mass enters each kinetic-energy operator: $\hat{K}_j = [1/(2m)]\hat{P}_j^2$.

The Hamiltonian (18.11) is invariant, i.e., symmetric, under particle interchange:

$$\hat{P}_{12}\hat{H}_{12}\hat{P}_{12} = \hat{H}_{21} = \hat{H}_{12},$$

or

$$[\hat{P}_{12},\ \hat{H}_{12}] = 0. \tag{18.12a}$$

$\hat{H}_{12}$ also commutes with the total spin operator $\hat{\mathbf{S}} = \hat{\mathbf{S}}_1 + \hat{\mathbf{S}}_2$:

$$[\hat{\mathbf{S}},\ \hat{H}_{12}] = 0, \tag{18.12b}$$

and, since $\hat{\mathbf{S}}$ also commutes with $\hat{P}_{12}$, the eigenstates of $\hat{H}_{12}$ are simultaneous eigenstates of $\hat{\mathbf{S}}$ and $\hat{P}_{12}$.

As in Section 11.2, it is advantageous to work with CM and relative coordinates $\mathbf{R}$ and $\mathbf{r} = \mathbf{r}_1 - \mathbf{r}_2$, in which case $\hat{H}_{12}(\mathbf{r}_1,\ \mathbf{r}_2)$ becomes

$$\hat{H}_{12}(\mathbf{r},\ \mathbf{R}) = \hat{H}_{\text{rel}}(\mathbf{r}) + \hat{K}_{\text{CM}}(\mathbf{R}), \tag{18.13a}$$

with

$$\hat{H}_{\text{rel}} = \hat{K}_{\text{rel}}(\mathbf{r}) + V(r) = \frac{1}{2\mu}\hat{P}^2_{\text{rel}}(\mathbf{r}) + V(r) \tag{18.13b}$$

and

$$\hat{K}_{\text{CM}}(\mathbf{R}) = \frac{1}{2M}\hat{P}^2_{\text{CM}}(\mathbf{R}).$$

From the definitions (11.22), one finds $\hat{P}_{12}\mathbf{R} = \mathbf{R}$ and $\hat{P}_{12}\mathbf{r} = -\mathbf{r}$, while the structure of $\hat{H}_{\text{rel}}$ implies that it and $\hat{K}_{\text{CM}}$ are each invariant under interchange of labels:

$$[\hat{P}_{12},\ \hat{K}_{\text{CM}}] = 0 = [\hat{P}_{12},\ \hat{H}_{\text{rel}}]. \tag{18.14}$$

In this case of two particles, Eqs. (18.14) mean that the eigenstates of $\hat{K}_{\text{CM}}$ and $\hat{H}_{\text{rel}}$ can be either symmetric or antisymmetric under action of $\hat{P}_{12}$, as proved in the exercises. However, the eigenstates of $\hat{K}_{\text{CM}}(\mathbf{R})$, given by $\exp(i\mathbf{k}_{\text{CM}}\cdot\mathbf{R}) = \exp[i\mathbf{k}_{\text{CM}}\cdot(\mathbf{r}_1 + \mathbf{r}_2)]$, are already symmetric under interchange of the labels 1 and 2, and thus the symmetry character of the overall eigenstate of $\hat{H}_{12}$ is determined by the eigenstates of $\hat{H}_{\text{rel}}$.

The eigenstates of $\hat{H}_{\text{rel}}$ are products of spatial and spin states. Since $[\hat{\mathbf{S}},\ \hat{H}_{\text{rel}}] = 0$, we can work immediately with the two-particle singlet and triplet spin eigenstates of Eqs. (14.71) and (14.72), here denoted $|\frac{1}{2}\frac{1}{2};\ SM_S\rangle_{1,2}$. The singlet ($S = 0$) and triplet ($S = 1$) spin states are antisymmetric (odd, -1) and symmetric (even, $+1$) under interchange of labels, respectively:

$$\hat{P}_{12}|\tfrac{1}{2}\tfrac{1}{2};\ SM_S\rangle_{1,2} = (-1)^{1+S}|\tfrac{1}{2}\tfrac{1}{2};\ SM_S\rangle_{1,2}, \qquad S = 0,\ 1. \tag{18.15}$$

The spatial eigenstates of $\hat{H}_{\text{rel}}$ are the usual $|n\ell m_\ell\rangle_{\text{rel}}$, where the subscript "rel" reminds us that the coordinate dependence is $\mathbf{r} = \mathbf{r}_1 - \mathbf{r}_2 = r\theta_{\text{rel}}\phi_{\text{rel}}$. To determine the behavior of $|n\ell m_\ell\rangle_{\text{rel}}$ under $\hat{P}_{12}$, we note that $\hat{P}_{12}$ and the parity operator $\hat{U}_{\text{P}}$ produce the same results when acting on a position ket $|\mathbf{r}_{12}\rangle = |\mathbf{r}\rangle$, from which we conclude that

$$\hat{P}_{12}|n\ell m_\ell\rangle_{\text{rel}} = \hat{U}_{\text{P}}|n\ell m_\ell\rangle_{\text{rel}} = (-1)^\ell|n\ell m_\ell\rangle_{\text{rel}}. \tag{18.16}$$

Equations (18.15) and (18.16) can now be used to determine the behavior under $\hat{P}_{12}$ of the eigensolutions $|\Gamma\rangle^{\text{A}}_{1,2}$ of

$$(\hat{H}_{\rm rel} - E_\Gamma^{\rm A})|\Gamma\rangle_{1,2}^{\rm A} = 0 \tag{18.17}$$

where the superscript "A" is a reminder of the required antisymmetry. The $|\Gamma\rangle_{1,2}^{\rm A}$ are products of $|n\ell m_\ell\rangle_{\rm rel}$ and $|\frac{1}{2}\frac{1}{2};\, SM_S\rangle_{1,2}$. On replacing the collective symbol Γ by the set $\{n\ell m_\ell,\, SM_S\}$, we have

$$|\Gamma\rangle_{1,2}^{\rm A} \to |n\ell m_\ell,\, SM_S\rangle_{1,2}^{\rm A} \equiv |n\ell m_\ell\rangle_{\rm rel}|\tfrac{1}{2}\tfrac{1}{2};\, SM_S\rangle_{1,2}. \tag{18.18}$$

Notice that, apart from the $S = 0$ spin state, the structure of (18.18) is non-determinantal. The antisymmetry requirement,

$$\hat{P}_{12}|\Gamma\rangle_{1,2}^{\rm A} = -|\Gamma\rangle_{1,2}^{\rm A}, \tag{18.19a}$$

becomes

$$\begin{aligned}\hat{P}_{12}|n\ell m_\ell\rangle_{\rm rel}|\tfrac{1}{2}\tfrac{1}{2};\, SM_S\rangle_{1,2} &= (-1)^{\ell+S+1}|n\ell m_\ell\rangle_{\rm rel}|\tfrac{1}{2}\tfrac{1}{2};\, SM_S\rangle_{1,2} \\ &= -|n\ell m_\ell\rangle_{\rm rel}|\tfrac{1}{2}\tfrac{1}{2};\, SM_S\rangle_{1,2},\end{aligned} \tag{18.19b}$$

where Eqs. (18.18), (18.15), and (18.16) have been employed. From (18.19b) we conclude that, in the present case, the antisymmetry requirement leads to

$$\ell + S = \text{even}, \tag{18.20a}$$

i.e.,

$$\begin{aligned}\ell &= \text{even}, \qquad S = 0, \\ \ell &= \text{odd}, \qquad S = 1.\end{aligned} \tag{18.20b}$$

That is, for an isolated pair of spin-$\frac{1}{2}$ fermions whose mutual interaction is a central potential, singlet-spin states require symmetric spatial states, whereas triplet-spin states require antisymmetric spatial states. This is a profound and important conclusion, which will be seen in Section 18.2 to apply to the He-atom. An analogous result can be derived for a pair of isolated bosons (see the exercises).

A pair of non-interacting spin-$\frac{1}{2}$ fermions

In this final example of constructing antisymmetric state vectors for a pair of spin-$\frac{1}{2}$ fermions, we turn to the independent-particle system of Section 11.1, wherein the two particles each experience an external (central) potential but do not interact with one another. The Hamiltonian is $\hat{H}_{\rm IP}(1, 2)$, given by

$$\hat{H}_{\rm IP}(1, 2) = \hat{H}_1 + \hat{H}_2, \tag{18.21a}$$

where

$$\hat{H}_k(\mathbf{r}_k) = \frac{\hat{P}_k^2(\mathbf{r}_k)}{2m} + V(r_k), \qquad k = 1, 2, \tag{18.21b}$$

and

$$[\hat{H}_1,\, \hat{H}_2] = 0, \tag{18.21c}$$

with $\hat{P}_{12}\hat{H}_1\hat{P}_{12} = \hat{H}_2$. In view of $V(\mathbf{r}_k) = V(r_k)$, the eigenvalue equation

$$(\hat{H}_k - E_\gamma)|\gamma, E_\gamma\rangle_k = 0 \tag{18.22a}$$

has solutions characterized by the four quantum numbers n, ℓ, m_ℓ, and m_S:

$$|\gamma, E_\gamma\rangle_k = |n\ell m_\ell\rangle_k|\tfrac{1}{2}m_S\rangle_k \equiv |n\ell m_\ell m_S\rangle_k, \tag{18.22b}$$

where the subscript k attached to the ket – $|\,\rangle_k$ – is again the particle label. As in Section 11.1, the simplest solutions to

$$(\hat{H}_{\text{IP}}(1, 2) - E_\Gamma)|\Gamma\rangle = 0$$

are either the product

$$|\Gamma\rangle_{1,2} = |\gamma E_\gamma\rangle_1|\gamma' E_{\gamma'}\rangle_2 \tag{18.23a}$$

or its label-transformed sibling

$$|\Gamma\rangle_{2,1} = \hat{P}_{12}|\Gamma\rangle_{1,2} = |\gamma E_\gamma\rangle_2|\gamma' E_{\gamma'}\rangle_1; \tag{18.23b}$$

each has the same energy E_Γ:

$$E_\Gamma = E_\gamma + E_{\gamma'}. \tag{18.23c}$$

The structure of $\hat{H}_{\text{IP}}(1, 2)$ – a sum of commuting Hamiltonians – as well as the product structure of $|\Gamma\rangle_{1,2}$ and $|\Gamma\rangle_{2,1}$ inevitably yield the sum in (18.23c). However, neither $|\Gamma\rangle_{1,2}$ nor $|\Gamma\rangle_{2,1}$ is an allowed state vector for this system of identical fermions: an antisymmetric state vector is required. We obtain (a normalized) one by applying the antisymmetrizer $\sqrt{\frac{1}{2}}\hat{\mathcal{A}}_{12}$ to $|\Gamma\rangle_{1,2}$, producing $|\Gamma\rangle_{1,2}^{\text{A}}$:

$$\begin{aligned}
|\Gamma\rangle_{1,2}^{\text{A}} &= \sqrt{\tfrac{1}{2}}\hat{\mathcal{A}}_{12}|\Gamma\rangle_{1,2} \\
&= \sqrt{\tfrac{1}{2}}[|n\ell m_\ell m_S\rangle_1|n'\ell' m_\ell' m_S'\rangle_2 - |n'\ell' m_\ell' m_S'\rangle_1|n\ell m_\ell m_S\rangle_2] &\text{(18.24a)} \\
&= \sqrt{\frac{1}{2}}\begin{vmatrix} |n\ell m_\ell m_S\rangle_1 & |n'\ell' m_\ell' m_S'\rangle_1 \\ |n\ell m_\ell m_S\rangle_2 & |n'\ell' m_\ell' m_S'\rangle_2 \end{vmatrix}. &\text{(18.24b)}
\end{aligned}$$

The superscript "A" plays the same role here as in the preceding subsection. Note that, if $\sqrt{\frac{1}{2}}\hat{\mathcal{A}}_{12}$ had been applied to $|\Gamma\rangle_{2,1}$, it would have produced $-|\Gamma\rangle_{1,2}^{\text{A}}$, a result that follows from the relation $\hat{P}_{12}|\Gamma\rangle_{1,2} = |\Gamma\rangle_{2,1}$. Given our earlier discussion, the determinantal form of (18.24b) could have been written immediately, rather than as a consequence of using $\hat{\mathcal{A}}_{12}$, and this immediate use of the determinantal form will be applied in Section 18.3 where we deal with the case of arbitrary N, since, as N becomes progressively larger than 2, the structure of the antisymmetrizer becomes progressively more complex.

Equation (18.24b) is a 2×2 version of what is known as a Slater determinant (Slater 1929). Even in the simple 2×2 form, the need to have differing values for at least one number among the two sets of four pairs of quantum numbers is evident. Equation (18.24b) is an example of an "unrestricted" determinant, one for which total spin is not a good quantum number, in contrast to Eq. (18.18). In fact, Eq. (18.24b) contains both singlet and triplet total-spin components. They can be extracted individually using the expansion

$$|\Gamma\rangle_{1,2}^{\mathrm{A}} = \sum_{S,M_S} |\tfrac{1}{2}\tfrac{1}{2};\ SM_S\rangle_{1,2}\langle\tfrac{1}{2}\tfrac{1}{2};\ SM_S|\Gamma\rangle_{1,2}^{\mathrm{A}}. \tag{18.25}$$

However, although Eq. (18.25) is correct, the complete symmetric or antisymmetric portions are obtained from it only if there is a sum on magnetic quantum numbers in $|\Gamma\rangle_{1,2}^{\mathrm{A}}$, a point developed in the exercises. Rather than inserting such a sum into Eq. (18.24b), it turns out to be simpler to start over and replace the determinant of (18.24b) with overall-antisymmetric states in which S is a good quantum number. The analysis of the preceding subsection permits the relevant result to be obtained immediately. Corresponding to each $|\tfrac{1}{2}\tfrac{1}{2};\ SM_S\rangle_{1,2}^{\mathrm{A}}$ will be an overall solution $|\Gamma,\ SM_S\rangle_{1,2}^{\mathrm{A}}$:

$$|\Gamma,\ SM_S\rangle_{1,2}^{\mathrm{A}} = |\gamma\rangle_{1,2}^{(S)}|\tfrac{1}{2}\tfrac{1}{2};\ SM_S\rangle_{1,2}^{\mathrm{A}}, \tag{18.26}$$

where the superscript (S) refers to the total spin. Recalling the label-transforming property (18.15) and the comments of the preceding subsection, it follows that

$$\hat{P}_{12}|\gamma\rangle_{1,2}^{(S)} = (-1)^S|\gamma\rangle_{1,2}^{(S)}. \tag{18.27}$$

Since the spatial basis states are the $|n\ell m_\ell\rangle$, we can use them to construct $|\gamma\rangle_{1,2}^{(S)}$, analogously to Eq. (18.24a).

The simplest such construction involves only one pair of states, and one readily finds

$$|\gamma\rangle_{1,2}^{(S=0)} = \sqrt{\tfrac{1}{2}}[|n\ell m_\ell\rangle_1|n'\ell' m_\ell'\rangle_2 + |n'\ell' m_\ell'\rangle_1|n\ell m_\ell\rangle_2] \tag{18.28a}$$

and

$$|\gamma\rangle_{1,2}^{(S=1)} = \sqrt{\tfrac{1}{2}}[|n\ell m_\ell\rangle_1|n'\ell' m_\ell'\rangle_2 - |n'\ell' m_\ell'\rangle_1|n\ell m_\ell\rangle_2]. \tag{18.28b}$$

That is, the singlet-spin (triplet-spin) state is multiplied by a symmetric (antisymmetric) space state. Notice that $|\Gamma,\ 00\rangle_{1,2}^{\mathrm{A}}$ and $|\Gamma,\ 1M_S\rangle_{1,2}^{\mathrm{A}}$ have the same energy. This degeneracy can be lifted only by adding to $\hat{H}_{\mathrm{IP}}(1,\ 2)$ an interparticle potential whose diagonal matrix elements taken w.r.t. $|\gamma\rangle_{1,2}^{(S=0)}$ and $|\gamma\rangle_{1,2}^{(S=1)}$ will differ from one another.

Let us now assume that $\hat{H}_{\mathrm{IP}}(1,\ 2)$ provides a model for more realistic situations. In this model each of the (degenerate) states has a unique pictorial representation, which we shall specify shortly. For these states one particle is in orbital $|n\ell m_\ell\rangle$ and the other is in orbital $|n'\ell' m_\ell'\rangle$. We shall assume that the energy of an orbital depends on $n\ell$: $E_\gamma \rightarrow E_{n\ell}$. Then the energy of the states $|\gamma\rangle_{1,2}^{(S=0)}$ and $|\gamma\rangle_{1,2}^{(S=1)}$ is

$$E_\Gamma \rightarrow E_{n\ell,n'\ell'} = E_{n\ell} + E_{n'\ell'}. \tag{18.29a}$$

For the values of n and ℓ we use the hydrogenic and atomic specifications: $n = 1,\ 2,\ \ldots,$ with $\ell = 1,\ 2,\ \ldots,\ n - 1$. While we know (Section 11.3) that $E_{1\ell=0}$ lies lowest, the relative ordering of the other single-particle energies generally depends on the choice of $V(r)$. For purposes of illustration, we postulate an ordering like that used in the atomic-shell model. This ordering is indicated in Fig. 18.1(a), which shows a few of the lowest-lying levels. The lowest energy is $E_{1\mathrm{s},1\mathrm{s}} = E_{1\mathrm{s}} + E_{1\mathrm{s}}$; the next lowest, corresponding to the first excited state, is $E_{1\mathrm{s},2\mathrm{s}}$; the third level, corresponding to the second excited state, is $E_{1\mathrm{s},2\mathrm{p}}$; etc. This mixed notation employs the letter designations of Table 11.3. For each energy level, two specific orbitals are occupied, as exemplified below:

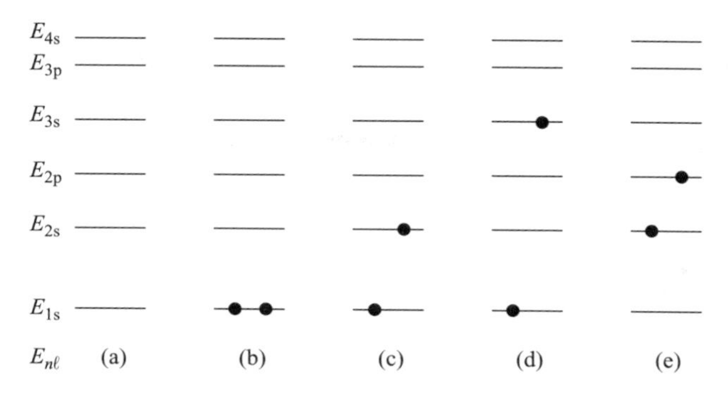

Fig. 18.1 (a) A Grotrian diagram illustrating the single-particle set of energy levels $E_{n\ell}$. (b)–(d) Orbital/energy-level occupation diagrams for the two-particle system. The filled circle represents a particle in the orbital $|n\ell m_\ell\rangle$ corresponding to the single-particle energy $E_{n\ell}$. (b) The two-particle ground state. (c) The lowest-lying excited state. (d) A higher-lying excited state resulting from a single-particle excitation. (e) A two-particle excited state formed by promoting both particles out of the E_{1s} level.

$$E_{1s,1s}\colon |100\rangle|100\rangle, \tag{18.29b}$$

$$E_{1s,2s}\colon |100\rangle|200\rangle, \tag{18.29c}$$

and

$$E_{1s,2p}\colon |100\rangle|21m_\ell\rangle. \tag{18.29d}$$

Apart from $|100\rangle|100\rangle$, the product of single-particle eigenstates in (18.29b)–(18.29d) is not any of the allowed two-particle eigenstates of Eqs. (18.28). Instead, the single-particle-eigenstate product simply specifies which single-particle states are used in constructing the relevant two-particle eigenstate. Once we know the associations expressed in Eqs. (18.29b)–(18.29d) and how the total (two-particle) energy is obtained – Eq. (18.29a) – we can replace them by the pictorial representation illustrated by Figs. 18.1(b)–(e). A small black circle is used to represent each particle (two in the present case), and one such circle is placed on the line representing the energy level $E_{n\ell}$ when a particle occupies the single-particle state $|n\ell m_\ell\rangle$. Since the single-particle energy-level lines of Fig. 18.1 are arranged in sequential order, Fig. 18.1(b) corresponds to the ground state, (18.29b); Fig. 18.1(c) represents the first excited state (18.29c), generated by promoting one of the two particles from the 1s to the 2s level; a higher-lying, single-particle excited state is depicted in Fig. 18.1(d); and a two-particle excited state is shown in Fig. 18.1(e). The total energy in each case is the sum of the single-particle energies for each of the two occupied single-particle eigenstates.

A single-particle energy scheme like that of Fig. 18.1(a) can accommodate more than two particles in some of the energy levels, and we shall consider how many are allowed in Section 18.4. Notice that, for $\ell > 0$, neither the m_ℓ values nor the value of S is specified, although, for $E_{1s,1s}$, the two-particle ground state necessarily has $S = 0$.

18.2. Two-electron atoms

He typifies the two-electron atom: all of its siblings, starting with H^-, have particle-stable ground states, and all but H^- have particle-stable excited states. One of the goals of quantum theory is to determine the eigenstates and eigenenergies of these two-electron systems. Because the generic Hamiltonian is of the intractable form (11.1), it is impossible to obtain analytically exact results, and one must rely on approximations, a wide-ranging introduction to which is the subject of this section.

Shown in Fig. 18.2 are the coordinates for the generic two-electron atom, whose nucleus has positive charge Ze. Although this is a three-particle system for which the nucleus in principle can move, we will treat it as if the nucleus were stationary and infinitely massive. We therefore do not include in the Hamiltonian a kinetic-energy operator for the nucleus. Hence, in this spin-independent, non-relativistic approximation, the three-body Hamiltonian is

$$\hat{H}_Z(\mathbf{r}_1, \mathbf{r}_2) = \hat{K}_1(\mathbf{r}_1) + \hat{K}_2(\mathbf{r}_2) - Ze^2\kappa_e\left(\frac{1}{r_1} + \frac{1}{r_2}\right) + \frac{e^2\kappa_e}{r_{12}}, \tag{18.30}$$

where $\hat{K}_j = \hat{P}_j^2/(2m_e)$, $j = 1, 2$, and $r_{12} = |\mathbf{r}_1 - \mathbf{r}_2|$. Each of the two electrons feels a Coulomb attraction to the positive charge Ze and a Coulomb repulsion due to the other electron.

In terms of the hydrogenic Hamiltonians

$$\hat{H}_Z(\mathbf{r}_j) = \hat{K}_j(\mathbf{r}_j) - \frac{Ze^2\kappa_e}{r_j}, \qquad j = 1, 2, \tag{18.31}$$

Eq. (18.30) becomes

$$\hat{H}_Z(\mathbf{r}_1, \mathbf{r}_2) = \hat{H}_Z(\mathbf{r}_1) + \hat{H}_Z(\mathbf{r}_2) + \frac{e^2\kappa_e}{r_{12}}, \tag{18.32}$$

and in this form the system described by $\hat{H}_Z(\mathbf{r}_1, \mathbf{r}_2)$ is a pair of hydrogenic electrons perturbed by a Coulomb repulsion. Classically, the two electrons will tend to be on opposite sides of the positive charge center, thus minimizing the role of the repulsion, and, as a first orientation to the actual ground-state energy, we assume that matrix elements of the r_{12}^{-1} term are small relative to the hydrogenic ground-state energy $-Z^2$ Ryd, in which case the actual ground state energy $E_{\text{g.s.}}(Z)$ is approximated by

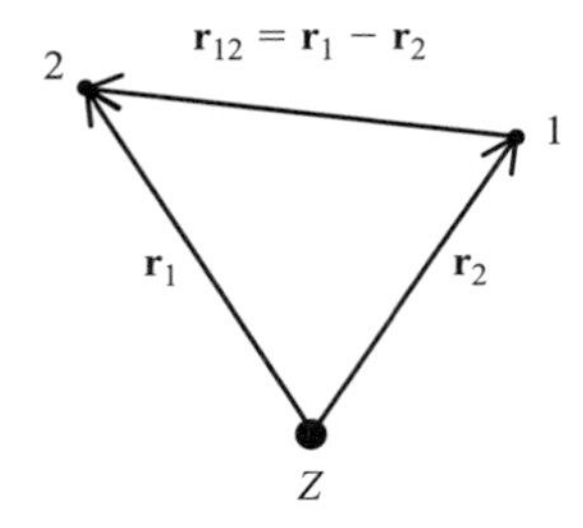

Fig. 18.2 Coordinates for the generic two-electron atom. The electrons are labeled 1 and 2; the nucleus is indicated by its charge Z.

$$E_{\text{g.s.}}(Z) \cong -2Z^2 \text{ Ryd, no Coulomb repulsion.} \tag{18.33}$$

For He, with $Z = 2$, this gives

$$E_{\text{g.s.}}(\text{He}) \cong -4 \text{ a.u.,} \tag{18.34a}$$

which is to be compared with the experimental energy

$$E^{(\text{exp})}_{\text{g.s.}}(\text{He}) = -2.9037 \text{ a.u.,} \tag{18.34b}$$

where one atomic unit (1 a.u.) = 2 Ryd = $\kappa_e e^2/a_0 \cong -27.2$ eV. The comparison clearly shows that one cannot ignore the contribution of the Coulomb repulsion to the ground-state energy. On the other hand, the contribution of the repulsion to the energy for any specific calculation depends on the details, as shown below.

The neglect of (very small) relativistic/spin-dependent contributions to $\hat{H}_Z(\mathbf{r}_1, \mathbf{r}_2)$ leads, through $[\hat{\mathbf{S}}, \hat{H}_Z] = 0$, to the conclusion that total spin S is a good quantum number. Therefore, the antisymmetric solution to the time-independent Schrödinger equation for two-electron atoms can be written as

$$\langle \mathbf{r}_1, \mathbf{r}_2|\Gamma\rangle^{\text{A}}_{1,2} = \psi^{(S)}_\Gamma(\mathbf{r}_1, \mathbf{r}_2)|\tfrac{1}{2}\tfrac{1}{2}; SM_S\rangle_{1,2}, \qquad S = 0, 1,$$

where the wave function $\psi^{(S)}_\Gamma(\mathbf{r}_1, \mathbf{r}_2)$ satisfies

$$(\hat{H}_Z(\mathbf{r}_1, \mathbf{r}_2) - E^{(S)}_\Gamma)\psi^{(S)}_\Gamma(\mathbf{r}_1, \mathbf{r}_2) = 0$$

such that

$$\hat{P}_{12}\psi^{(S)}_\Gamma(\mathbf{r}_1, \mathbf{r}_2) = (-1)^S\psi^{(S)}_\Gamma(\mathbf{r}_1, \mathbf{r}_2).$$

In this approximation it is clear that the spectrum of two-electron atoms consists of two disparate portions. These are the singlet-spin portion, for which $\psi^{(S=0)}_\Gamma(\mathbf{r}_1, \mathbf{r}_2)$ is symmetric, and the triplet-spin portion, for which $\psi^{(S=1)}_\Gamma(\mathbf{r}_1, \mathbf{r}_2)$ is antisymmetric. In the case of He ($Z = 2$), the atom in singlet spin states is often referred to as *parahelium*, while the designation of the triplet spin state is *orthohelium*. This unique separation into singlet and triplet spin states obviously carries over to any approximate versions of $\psi^{(S)}_\Gamma(\mathbf{r}_1, \mathbf{r}_2)$, and is also an appropriate designation for portions of the spectra of those atoms whose shell structure consists of closed shells plus two valence electrons (see Section 18.4). The reader should also be aware that there are measurable corrections that alter this picture, e.g., those due to fine-structure effects, although they are not treated in this text.

The perturbation/variation approximation

Our initial attempt to understand the spectra of two-electron atoms uses the perturbation/variation approach mentioned at the end of Section 17.4. In this procedure, the Hamiltonian is written as the sum $\hat{H}_0 + \hat{H}_1$, with eigenstates of $\hat{H}_0$ used as the trial functions. For the present two-electron case, we choose

$$\hat{H}_0(\mathbf{r}_1, \mathbf{r}_2) = \hat{H}_Z(\mathbf{r}_1) + \hat{H}_Z(\mathbf{r}_2), \tag{18.35}$$

and

$$\hat{H}_1(\mathbf{r}_1, \mathbf{r}_2) = e^2\kappa_e/r_{12}. \tag{18.36}$$

Since $\hat{H}_0$ is a sum of commuting Hamiltonians, its description of H^-, He, Li^+, etc. is that of an independent-particle model. The bound-state part of the basis consists of products of hydrogenic states $\psi_{n\ell m_\ell}(\mathbf{r}_j)$, $j = 1, 2$, with $n = 1, 2, \ldots$ etc.[5] (the Z dependence is suppressed). There are two types of properly symmetrized, independent-particle trial functions. The first, which is valid only for singlet-spin states, is the symmetric product

$$\psi^{(S=0)}_{n\ell m_\ell, n\ell m_\ell}(\mathbf{r}_1, \mathbf{r}_2) = \psi_{n\ell m_\ell}(\mathbf{r}_1)\psi_{n\ell m_\ell}(\mathbf{r}_2), \tag{18.37a}$$

i.e., $n'\ell' m'_\ell = n\ell m_\ell$. The second type consists of the symmetric or antisymmetric linear combinations

$$\psi^{(\pm)}_{n\ell m_\ell, n'\ell' m'_\ell}(\mathbf{r}_1, \mathbf{r}_2) = \sqrt{\tfrac{1}{2}}[\psi_{n\ell m_\ell}(\mathbf{r}_1)\psi_{n'\ell' m_\ell}(\mathbf{r}_2) \pm \psi_{n\ell m_\ell}(\mathbf{r}_2)\psi_{n'\ell' m_\ell}(\mathbf{r}_1)], \tag{18.37b}$$

with the superscript $+$ $(-)$ now standing for the singlet (triplet) spin state:

$$(+) \Leftrightarrow (S = 0),$$
$$(-) \Leftrightarrow (S = 1).$$

The division of the theoretical spectrum into two disjoint portions ($S = 0$ and $S = 1$) is mirrored in the experimental one, as will be seen shortly.

The unperturbed energies are

$$E^{(0)}_{n,n'} = \frac{-Z^2}{2}\left(\frac{1}{n^2} + \frac{1}{n'^2}\right) \text{ a.u.} \tag{18.38}$$

These energies are independent of S and embody the n^2 degeneracy of the one-particle hydrogenic states as well as the label degeneracy $E^{(0)}_{n,n'}(Z) = E^{(0)}_{n',n}(Z)$.

Only first-order corrections to (18.38) will be considered. Their general form is

$$E^{(1)}_{n\ell m_\ell, n'\ell' m'_\ell} = \left\langle \frac{e^2\kappa_e}{\hat{Q}_{12}} \right\rangle_{n\ell m_\ell, n'\ell' m'_\ell}, \tag{18.39}$$

where the six subscripts on the diagonal matrix element allow for states from either of Eqs. (18.37) and both the S and the Z dependences are suppressed in (18.39). From these last two equations, the perturbation/variation approximation to the energy is

$$E^{(t)}_{n\ell m_\ell, n'\ell' m'_\ell}(Z, S) = -\frac{Z^2}{2}\left(\frac{1}{n^2} + \frac{1}{n'^2}\right) \text{ a.u.} + \left\langle \frac{e^2\kappa_e}{\hat{Q}_{12}} \right\rangle_{n\ell m_\ell, n'\ell' m'_\ell}, \tag{18.40}$$

some qualitative features of which we consider next.

For the simple approximations (18.37) that produce the energies (18.40), the Z^2 dependence of $E^{(0)}_{n,n'}(Z)$ suggests, at least for the low-lying states, that the ordering of the two-electron energy levels will primarily be that from $E^{(0)}_{n,n'}(Z)$ itself, i.e., it will be that of the independent-particle model. This model predicts that the ground state corresponds

[5] Continuum states will not be included among the unperturbed basis functions, even though they can make important contributions.

to $n = n' = 1$, with all other states lying higher. The lowest-lying excited states will correspond to $n = 1$, $n' = 2$, with the $n' = 2$, $\ell = 0, 1$ degeneracy being split by the $\langle e^2\kappa_e/\hat{Q}_{12}\rangle$ matrix elements. The spatial eigenstates will be those of Eqs. (18.37), to be multiplied by the relevant spin states. Because the approximate ground-state wave function $\psi_{100}(\mathbf{r}_1)\psi_{100}(\mathbf{r}_2)$ is symmetric under action of $\hat{P}_{12}$, its spin part must be a singlet. Experimentally, the ground state of He, for example, *is* a spin singlet. On the other hand, the higher-lying states whose approximate spatial portion is given by (18.37b) can be either singlet or triplet, depending on the $(\pm)$ sign in (18.37b). As the following qualitative argument shows, for a given set of $n\ell m_\ell$ and $n'\ell' m'_\ell$ values, the triplet form of (18.37b) yields a lower-lying state than does its spatially symmetric sibling.

The determining factor is the behavior of the non-negative matrix element $\langle e^2\kappa_e/\hat{Q}_{12}\rangle_{100,n'\ell' m'_\ell}$. The maximum contribution to this matrix element occurs in the vicinity of $r_{12} = 0$, which requires $\mathbf{r}_1 \cong \mathbf{r}_2$. However, for $\mathbf{r}_1 \cong \mathbf{r}_2$, $\psi^{(-)}_{100,n'\ell' m'_\ell} \cong 0$ whereas $\psi^{(+)}_{100,n'\ell' m'_\ell} \cong 2\psi_{100}(\mathbf{r}_1)\psi_{n'\ell' m'_\ell}(\mathbf{r}_1)$. We therefore expect $\langle e^2\kappa_e/\hat{Q}_{12}\rangle_{100,n'\ell' m'_\ell}$ to be larger for singlet than for triplet spin states, implying that $E^{(t)}_{100,n'\ell' m'_\ell}(Z, S = 1) < E^{(t)}_{100,n'\ell' m'_\ell}(Z, S = 0)$. Indeed, it is only because the ground state has no triplet counterpart that it is a singlet. Notice, by the way, that the foregoing argument relies only on the spatial symmetry, not on the form of the approximations (18.37). That triplet states should lie lower than singlets is an example of one of Hund's (1927) empirical rules, which will be discussed later.

A similar qualitative argument can be used to show that $E^{(t)}_{1s,2s}(Z, S) < E^{(t)}_{1s,2p}(Z, S)$. In this instance, the radial part of the 2s single-particle state has a node (at $r = 2a_0/Z$) whereas the analogous 2p state has none between $r = 0$ and $r = \infty$. The presence of the node means that more canceling out occurs in $\langle e^2\kappa_e/\hat{Q}_{12}\rangle_{1s,2s}$ than in $\langle e^2\kappa_e/\hat{Q}_{12}\rangle_{1s,2p}$ and hence the latter matrix element is larger than the former, leading to the conclusion that $E^{(t)}_{1s,2s}(Z, S) < E^{(t)}_{1s,2p}(Z, S)$.

As a final point before turning to the evaluations, we introduce some standard spectroscopic nomenclature and notation. The first is the concept of "configuration." This refers to the set of $n\ell$ orbitals that are used to construct the independent-particle wave functions, which for the two-electron case are the *single-configuration* states (18.37). Their configurations are written as $n\ell n'\ell'$, for example, $1s1s \equiv (1s)^2$, 1s2s, 1s3p, $2s2s \equiv (2s)^2$, etc. A multi-configuration approximation means a sum of states of the forms (18.37), with coefficients to be determined; the multi-configuration notation is $n_1\ell_1 n'_1\ell'_1 + n_2\ell_2 n'_2\ell'_2 + \cdots$

The other concept is that of the spectroscopic term value. Let S, L, and J be the total spin, the total orbital angular momentum, and the total angular momentum of the state. Then the spectroscopic term value is written as ${}^{2S+1}L_J$, with numerical values of L being represented by capital letters, as exemplified in Table 18.1, the analog of Table 11.3. The left-hand superscript $2S + 1$ is the spin multiplicity, 1 in the case of singlets ($S = 0$), 2 in the case of doublets ($S = \frac{1}{2}$, which does not occur for two-fermion configurations), 3 in the case of triplets ($S = 1$), etc. Allowed J values are the usual ones, viz., $|L - S| \leqslant J \leqslant L + S$. The possible spectroscopic term values are inferred from the configuration, e.g., $(1s)^2$ implies ${}^1S_0$; 1sns yields both ${}^1S_0$ and ${}^3S_1$, which are sometimes written n^1S_0 and n^3S_1, with n indicating the particular excited state; 1snp leads to ${}^1P_1$ (or n^1P_1) and ${}^3P_J$ (or n^3P_J), with $J = 0, 1$, and 2; etc. Configurations such as 2p3d give

Table 18.1 *Spectroscopic notation for total orbital angular momentum L values, L = 0, ..., 6*

	L						
	0	1	2	3	4	5	6
Letter designations	S	P	D	F	G	H	I

rise to a variety of L and J values; each corresponds to a different state. Clearly, the number of states can rapidly become large, especially for N-particle configurations.

We now turn to the quantitative analysis, starting with evaluation of $\langle e^2\kappa_e/\hat{Q}_{12}\rangle_{1s,1s}$ for the approximate $(1s)^2\,{}^1S_0$ ground state. In a coordinate representation this matrix element is

$$\left\langle \frac{e^2\kappa_e}{\hat{Q}_{12}} \right\rangle_{1s,1s} \equiv {}_{1,2}\left\langle 100,\, 100 \left| \frac{e^2\kappa_e}{\hat{Q}_{12}} \right| 100,\, 100 \right\rangle_{1,2}$$
$$= 16\frac{Z^6 e^2\kappa_e}{a_0^6}\int d^3r_1\, d^3r_2\, e^{-Zr_1/a_0} e^{-Zr_2/a_0}\, Y_0^0(\Omega_1)Y_0^0(\Omega_2)$$
$$\times \frac{1}{r_{12}} e^{-Zr_1/a_0} e^{-Zr_2/a_0}\, Y_0^0(\Omega_1)Y_0^0(\Omega_2), \tag{18.41}$$

where Eq. (11.142a) has been used.

The denominator can be expressed in terms of r_1, r_2, and the angles Ω_1 and Ω_2 by first invoking Eq. (10.41). Applying it to $r_{12}^{-1} = (r_1^2 + r_2^2 - 2r_1r_2\cos\theta_{12})^{1/2}$, where θ_{12} is the angle between $\mathbf{r}_1$ and $\mathbf{r}_2$, we find

$$\frac{1}{r_{12}} = \begin{cases} \displaystyle\sum_{\ell=0}^{\infty} (r_1^\ell/r_2^{\ell+1}) P_\ell(\cos\theta_{12}), & r_2 > r_1, \\ \displaystyle\sum_{\ell=0}^{\infty} (r_2^\ell/r_1^{\ell+1}) P_\ell(\cos\theta_{12}), & r_1 > r_2. \end{cases}$$

Combining these expressions and using Eq. (10.72a), r_{12}^{-1} becomes

$$\frac{1}{r_{12}} = \sum_{\ell,m_\ell} \frac{4\pi}{2\ell+1}\frac{r_<^\ell}{r_>^{\ell+1}} Y_\ell^{m_\ell}(\Omega_1) Y_\ell^{m_\ell^*}(\Omega_2), \tag{18.42}$$

where $r_<$ ($r_>$) is the lesser (greater) of r_1 and r_2. As we shall see, the singularity that arises when $\mathbf{r}_1 = \mathbf{r}_2$ is not a problem in the integral in (18.41).

On substituting (18.42) into (18.41), it is readily shown that the angular integrals yield $\delta_{\ell 0}\delta_{m_\ell 0}/(4\pi)$, leading to

$$\left\langle \frac{e^2\kappa_e}{\hat{Q}_{12}} \right\rangle_{1s,1s} = \frac{16 Z^6 e^2\kappa_e}{a_0^6}\int_0^\infty dr_1\, dr_2\, r_1^2 r_2^2 \frac{1}{r_>} e^{-2Z(r_1+r_2)/a_0}. \tag{18.43}$$

The presence of $r_>$ means that the integral in (18.43) is split into two parts, one with $r_1 > r_2$ and one with $r_2 > r_1$:

$$\int_0^\infty dr_1\, dr_2\, r_1^2 r_2^2 \frac{1}{r_>} e^{-2Z(r_1+r_2)/a_0} = \int_0^\infty dr_1\, r_1 e^{-2Zr_1/a_0} \int_0^{r_1} dr_2\, r_2^2 e^{-2Zr_2/a_0}$$
$$+ \int_0^\infty dr_2\, r_2 e^{-2Zr_2/a_0} \int_0^{r_2} dr_1\, r_1^2 e^{-2Zr_1/a_0}. \tag{18.44}$$

On introducing the dimensionless variables

$$x = \frac{Zr_1}{a_0}, \qquad y = \frac{Zr_2}{a_0}$$

and then using

$$\int dx\, x^2 e^{-2x} = \frac{x^2 e^{-2x}}{-2} + \int dx\, x e^{-2x}$$

and

$$\int dx\, x e^{-2x} = -\frac{e^{-2x}}{4}(2x+1)$$

in the evaluation of (18.44), one eventually finds

$$\left\langle \frac{e^2 \kappa_e}{\hat{Q}_{12}} \right\rangle_{1s,1s} = \frac{5Z}{8} \text{ a.u.} \tag{18.45}$$

Inserting this result into Eq. (18.40), the perturbation/variation bound on the ground-state energy of a two-electron atom with nuclear charge Ze is

$$E^{(t)}_{1s,1s}(Z) = \left(-Z^2 + \frac{5}{8}Z\right) \text{ a.u.}, \tag{18.46}$$

an expression first obtained by Unsöld (1927). The bound (18.46) is compared with the experimental energies $E^{\text{exp}}_{\text{g.s.}}(Z)$ for the cases $1 \leqslant Z \leqslant 5$ in Table 18.2, wherein the value 1 a.u. $= 27.211$ eV has been used. Results for $Z > 5$ can be found in various references, e.g., Galindo and Pascual (1990).

In contrast to Bohr's H-atom model, which could not be extended to the He-atom, and apart from H^-, the results of Table 18.2 establish that even this very simple approximation to the Schrödinger-equation eigenvalue problem accounts quite well for the trend of the two-electron ground-state energies. Numerically, all the $E^{(t)}_{\text{g.s.}}(Z)$ values are in error by about 4.2 eV, an amount associated with a percentage deviation that decreases significantly with increasing Z. Nonetheless, 4.2 eV is a much larger error than the experimental uncertainties in each $E^{\text{exp}}_{\text{g.s.}}(Z)$, and better approximations are called for. We shall consider them in the next subsection.

The present approach is a total failure for the H^- ion, for which Hill (1977) has proved that no particle-stable excited states exist. Of all the two-electron atoms/ions, the nuclear attraction is weakest in H^-, leading to the very weak binding energy – or *affinity* – for the second electron of only 0.754 eV (measured by Lykke, Murray, and Lineberger (1991)). With such a small value, it is not surprising that particle-stable excited states of

Table 18.2 *A comparison of the experimental and perturbation/variation ground-state energies for various two-electron atoms*

Z	Atom	$-E^{(\text{exp})}_{\text{g.s.}}(Z)^a$ (a.u.)	$-E^{(t)}_{1\text{s},1\text{s}}(Z)^{a,d}$ (a.u.)	$E^{(t)}_{1\text{s},1\text{s}}(Z) - E^{(\text{exp})}_{\text{g.s.}}(Z)$ (eV)
1	H^-	0.5277^b	0.375	4.16
2	He	2.9037^c	2.750	4.18
3	Li^+	7.2794^c	7.125	4.20
4	Be^{2+}	13.654^c	13.50	4.19
5	B^{3+}	22.031^c	21.875	4.27

[a] 1 a.u. $\simeq$ 27.21 eV.
[b] Determined from the electron affinity (second-electron binding energy) of 0.754 195 eV measured by Lykke, Murray, and Lineberger (1991).
[c] Obtained from the data of Kuhn (1962). See also Radzig and Smirnov (1985).
[d] $E^{(t)}_{1\text{s},1\text{s}}(Z)$ values calculated from Eq. (18.46).

H^- are non-existent; what is surprising is that only highly correlated spatial wave functions yield a positive binding energy for the second electron: neither the product $\psi_{100}(\mathbf{r}_1)\psi_{100}(r_2)$ nor the simplest, screened variational wave function discussed in the next subsection, nor the Hartree–Fock-determinantal wave function of Section 18.3 lead to binding of the second electron in H^-. For example, the $E^{(t)}_{1\text{s},1\text{s}}(Z = 1)$ calculation of this subsection (see Table 18.2) predicts that the second electron is unbound by 0.125 a.u. $\equiv$ 3.4 eV! While improved approximations are called for, the reader should not assume that H^- is of interest only because it is a challenging system for quantum mechanics to describe: far from it! For instance, H^- is an important ingredient in determining the opacity of the outer portion of stars, an aspect briefly summarized by Bethe and Salpeter (1957). Furthermore, an extremely important and exciting tool in the study of the $e^- + H$ system has been relativistic beams ($\sim$ 800 MeV) of H^- ions, experimental and theoretical aspects of which are reviewed by Bryant and Halka (1996). H^- ions also play an essential role at the Tevatron accelerator of the Fermilab facility near Chicago, where their acceleration is the first stage in the eventual production of protons of tevatron energies (10^3 GeV).

Before considering improved ground-state approximations, we use the approximation (18.37b) to estimate some excited-state energies. Numerical results will be presented only for those states of He described in the present approximation by the two-electron configurations 1s2s and 1s2p. These correspond to promotion of a single electron out of the $(1\text{s})^2$ ground-state configuration to an $n = 2$ hydrogenic excited state. Both singlet (*para*) and triplet (*ortho*) states occur. From Eq. (18.37b) the spatial wave functions for the configurations $1\text{s}2\ell$ are

$$\psi^{(\pm)}_{1\text{s},2\ell m_\ell}(\mathbf{r}_1, \mathbf{r}_2) = \sqrt{\tfrac{1}{2}}[\psi_{100}(\mathbf{r}_1)\psi_{2\ell m_\ell}(\mathbf{r}_2) \pm \psi_{2\ell m_\ell}(\mathbf{r}_1)\psi_{100}(\mathbf{r}_2)], \tag{18.47}$$

and the correction to $E^{(0)}_{1,2}(Z)$ is the matrix element

$$
\begin{aligned}
E^{(1)\pm}_{100,2\ell m_\ell} &= \left\langle \frac{e^2\kappa_e}{\hat{Q}_{12}} \right\rangle_{1s,2\ell} \\
&= \tfrac{1}{2}[{}_{1,2}\langle 100,\, 2\ell m_\ell| \pm {}_{1,2}\langle 2\ell m_\ell,\, 100|] \frac{e^2\kappa_e}{\hat{Q}_{12}} [|100,\, 2\ell m_\ell\rangle_{1,2} \pm |2\ell m_\ell,\, 100\rangle_{1,2}] \\
&= {}_{1,2}\left\langle 100,\, 2\ell m_\ell \left| \frac{e^2\kappa_e}{\hat{Q}_{12}} \right| 100,\, 2\ell m_\ell \right\rangle_{1,2} \pm {}_{1,2}\left\langle 100,\, 2\ell m_\ell \left| \frac{e^2\kappa_e}{\hat{Q}_{12}} \right| 2\ell m_\ell,\, 100 \right\rangle_{1,2} .
\end{aligned}
\tag{18.48}
$$

The third equality in (18.48) is a consequence of the label-transforming properties of each of the four matrix elements on the RHS of the second equality. Recalling that the first set of quantum numbers in $|n\ell m_\ell,\, n'\ell' m'_\ell\rangle_{1,2}$ belongs to the first label, here particle 1, one can show that the third equality in (18.48) follows from the second in one of two ways. One procedure uses the coordinate representation and recognizes that the integration variables $\mathbf{r}_1$ and $\mathbf{r}_2$ are dummy variables whose interchange ($\mathbf{r}_1 \to \mathbf{r}_2$, $\mathbf{r}_2 \to \mathbf{r}_1$) leaves the integral unchanged. The other uses the two-particle transposition-operator relations $\hat{P}_{12}\hat{P}_{12} = \hat{I}$, $\hat{P}_{12}\hat{Q}_{12}\hat{P}_{12} = \hat{Q}_{21}$, and $\hat{P}_{12}|n\ell m_\ell,\, n'\ell' m'_\ell\rangle_{1,2} = |n\ell m,\, n'\ell' m'_\ell\rangle_{2,1}$ in the matrix elements.

The final two matrix elements in (18.48) are specific instances of more general quantities typically referred to as "Coulomb" and "exchange" integrals (see Section 18.3). Various notations have been used for them, among which are $J_{n\ell,n'\ell'}$ for the Coulomb term and $K_{n\ell,n'\ell'}$ for the exchange term:

$$
J_{n\ell,n'\ell'} = {}_{1,2}\left\langle n\ell m_\ell,\, n'\ell' m'_\ell \left| \frac{e^2\kappa_e}{\hat{Q}_{12}} \right| n\ell m_\ell,\, n'\ell' m'_\ell \right\rangle_{1,2} \tag{18.49a}
$$

and

$$
K_{n\ell,n'\ell'} = {}_{1,2}\left\langle n\ell m_\ell,\, n'\ell' m'_\ell \left| \frac{e^2\kappa_e}{\hat{Q}_{12}} \right| n'\ell' m'_\ell,\, n\ell m_\ell \right\rangle_{1,2} . \tag{18.49b}
$$

The subscript notation on J and K takes account of the fact that the matrix elements are independent of the magnetic quantum numbers. This independence is a consequence of $\hat{Q}_{12} = |\hat{\mathbf{Q}}_1 - \hat{\mathbf{Q}}_2|$ being a scalar, that is, $\hat{Q}_{12}$ is invariant under rotations. For discussion, and proof and applications of the relevant result – the so-called Wigner–Eckart theorem – see Rose (1957) or Edmonds (1957).

With the notation of Eqs. (18.49), the excited-state energies in the independent-particle approximation are

$$
E^{(t)\pm}_{1s,2\ell}(Z) = -\tfrac{5}{8}Z^2 \text{ a.u.} + J_{1s,2\ell} \pm K_{1s,2\ell}. \tag{18.50}
$$

Since it can be shown (e.g., Pilar (1968)) that $0 \leqslant K_{n\ell,n'\ell'} \leqslant J_{n\ell,n'\ell'}$, it follows from (18.50) that

$$
E^{(t)+}_{1s,2\ell}(Z) > E^{(t)-}_{1s,2\ell}(Z).
$$

In other words, the previous qualitative prediction that the triplet state lies lower than its singlet counterpart is substantiated in the present independent-particle case.

The procedures used to evaluate $\langle e^2\kappa_e/\hat{Q}_{12}\rangle_{1s,1s}$ also apply to Eqs. (18.49). As remarked

Table 18.3 *Selected values of* $J_{1s,2\ell}$ *and* $K_{1s,2\ell}$ *in a.u. for He*

	ℓ	
	0	1
$J_{1s,2\ell}$	0.419	0.486
$K_{1s,2\ell}$	0.044	0.034

Table 18.4 *A Comparison of the four lowest-lying perturbation/variation and experimental excited-state energies in He*

		Triplets	
ℓ	$-E^{(t)-}_{1s,2\ell}(Z=2)$ (a.u.)[a]	$-E^{(\exp)}(Z=2)$ (a.u.)[a,b]	$E^{(t)-}_{1s,2\ell}(Z=2) - E^{(\exp)}(Z=2)$ (eV)
0	2.125	2.176	1.36
1	2.048	2.133	2.31
		Singlets	
ℓ	$-E^{(t)+}_{1s,2\ell}(Z=2)$ (a.u.)[a]	$-E^{(\exp)}(Z=2)$ (a.u.)[a,b]	$E^{(t)+}_{1s,2\ell}(Z=2) - E^{(\exp)}(Z=2)$ (eV)
0	2.037	2.147	2.99
1	1.980	2.124	3.92

[a]The value 1 a.u. $\cong$ 27.211 eV is used.
[b]See, e.g., Kuhn (1962) or Radzig and Smirnov (1985)

earlier, we are considering only the He-atom ($Z = 2$), the values of $J_{1s,2\ell}$ and $K_{1s,2\ell}$ for which are given in Table 18.3, while the values of $E^{(t)\pm}_{1s,2\ell}(Z=2)$ are stated and compared with experiment in Table 18.4. The comparison is with the two lowest-lying singlet and triplet experimental energies, the (standard) assumption being that the $1s2\ell$ states (18.37) are the major components in the exact theoretical energies. The agreement between theory and experiment is slightly better than the agreement of Table 18.2, all differences in Table 18.4 being less than 4 eV, a number still too large for the perturbation/variation approximation to be considered more than a good starting point. (Even errors as "small" as 1 eV are too large for the approximation to be considered sufficiently accurate.)

It is helpful to see where on an energy-level plot these various energies lie, and a Grotrian diagram is presented in Fig. 18.3. The comparable experimental and theoretical energy levels from Tables 18.2 and 18.4 are connected by arrows in this diagram; other measured energy levels are also shown. All these levels, both theoretical and experimental, can be characterized by term values, and these are indicated at the top of each column in Fig. 18.3. It is clear from Fig. 18.3 that further improvements in the theory are needed. Rather than consider multi-configuration approximations, we will turn instead to other variational approximations. We note in passing, however, that, if multi-configuration approximations were examined, the low-lying states would still be characterized as "$1sn\ell$" states because the $1sn\ell$ configuration is the main component, as indicated by the spectrum of Fig. 18.3. The "regularities" of Fig. 18.3 do not carry over to those higher-lying states of He which correspond to multi-configuration excited levels, and from

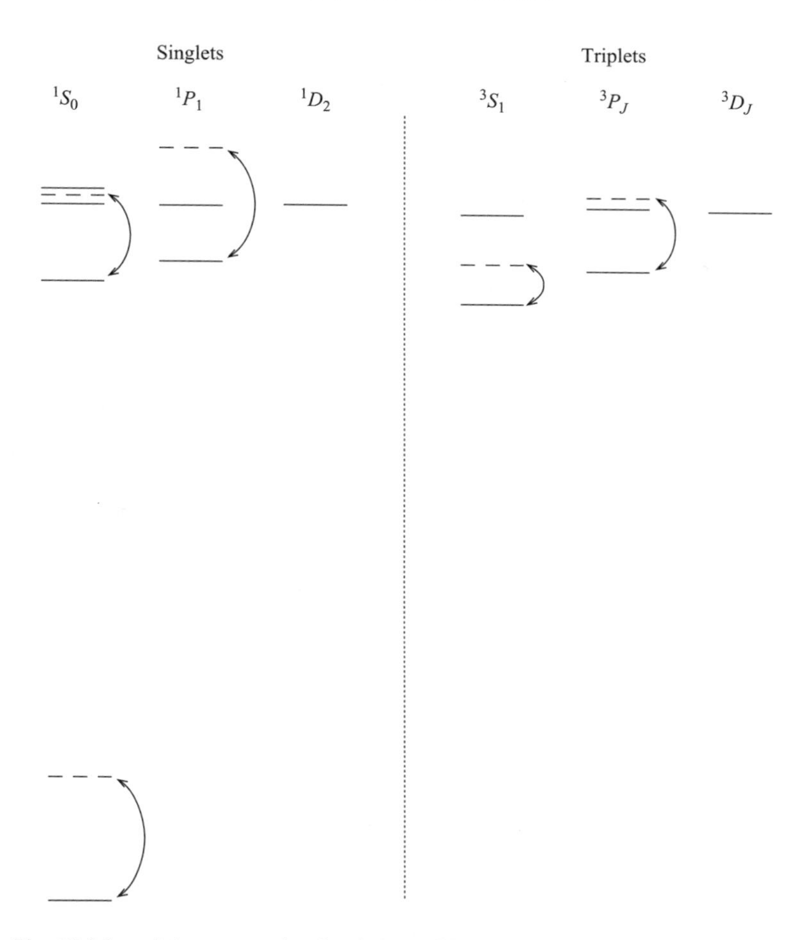

Fig. 18.3 Low-lying energy levels of He. Solid lines are measured values; dashed lines are the values from the perturbation/variation calculations of the text. Arrows indicate comparable levels.

which it is very difficult to extract patterns. Nonetheless there are patterns and regularities to certain of these two-particle excited states, and these have been beautifully accounted for by using hyperspherical coordinates, "collective"-motion analysis, and group theory. The initial work in this direction is that of Wulfman and Sukeyuki (1973) and Sinanoglu and Herrick (1973), with the connection to hyperspherical coordinates being made by Lin (1984). Some reviews, at a fairly sophisticated level, are those of Herrick (1983) and Briggs (1996). It should be evident that some aspects of two-electron atoms were understood only many years after the early days of quantum theory; some remain to be explained (e.g., Rydberg states and $e^- + H$ ionization).

Approximations involving variational parameters

For systems as relatively uncomplicated as two-electron atoms, for which the (non-relativistic) Hamiltonians are known, the use of wave functions containing variational parameters has yielded the greatest accuracy. We shall concentrate here on the ground states of H^- and He, first outlining a few of the simpler calculations and then summarizing some of the computations for He based on trial functions containing hundreds of variational parameters. Although high-speed, large-memory computers can carry out calculations of enormous complexity, it should be stressed at the outset that the use of physical insight is still called for in formulating trial functions, if for no other reason than the need to carefully probe and treat correlation effects. As with all numerical efforts, one should engage the mind before crunching numbers.

Case (i). Uncorrelated, symmetric-product trial functions. A symmetric trial function suggests a classical-physics argument: classically, the electron placement that would produce the lowest energy is a linear or "stretched" configuration, for which the position coordinates of electrons obey $\mathbf{r}_2 = -\mathbf{r}_1$. Quantally, of course, the electron positions are not fixed (but see the exercises): the positions can take on almost any value, with the distance of one sometimes larger and sometimes smaller than that of the other. Physically, whenever one electron is closer to the nucleus, it partially screens the nuclear charge Ze, resulting in an effective or reduced charge being seen by the other electron. Although the Hamiltonian cannot be changed to take this effect into account, the trial wave function *can* incorporate it. If we consider the hydrogenic ground state as an example, the nuclear charge Z enters the wave function via $\exp(-Zr(a_0))$. To take account of screening, one can replace Z by an effective charge ξ, thereby producing a single-particle, parameterized, 1s-type radial function $\exp(-\xi r/a_0)$.

The simplest (singlet, normalized) trial function one can construct using $\exp(-\xi r/a_0)$ is the symmetric product

$$\psi^{(t)}(\mathbf{r}_1, \mathbf{r}_2; \xi) = 4\left(\frac{\xi}{a_0}\right)^3 e^{-\xi(r_1+r_2)/a} Y_0^0(\Omega_1) Y_0^0(\Omega_2). \tag{18.51}$$

This function is simplest because ξ is the same for both electrons. It was introduced by Kellner (1927). The variational ground-state energy produced by (18.51) and (18.30) is

$$\begin{aligned} E^{(t)}_{\text{g.s.}}(Z, \xi) &= (\psi^{(t)}(\mathbf{r}_1, \mathbf{r}_2; \xi)|\hat{H}_Z(\mathbf{r}_1, \mathbf{r}_2)|\psi^{(t)}(\mathbf{r}_1, \mathbf{r}_2; \xi)) \\ &= \left(\xi^2 - 2Z\xi + \frac{5}{8}\xi\right) \text{ a.u.}, \end{aligned} \tag{18.52}$$

whose minimum value $\xi_0(Z)$ is

$$\xi_0(Z) = Z - 5/16. \tag{18.53}$$

The fact that $\xi_0(Z) < Z$ is consistent with the physical picture of screening.

For H^- and He ($Z = 1$ and $Z = 2$, respectively) the appropriate screened charges are

$$\xi_0(1) = 11/16$$

and

$$\xi_0(2) = 27/16,$$

which yield the ground-state energy bounds

$$E^{(t)}_{\text{g.s.}}(\text{H}^-, \xi_0(1)) = -0.473 \text{ a.u.}$$

and

$$E_{\text{g.s.}}^{(t)}(\text{He}, \xi_0(2)) = -2.85 \text{ a.u.}$$

Comparison with the results of Table 18.3 shows that use of the simplest screened trial function improves the theoretical ground-state energies by 0.1 a.u. $\cong$ 2.72 eV. For He, the discrepancy between theory and experiment is reduced to 1.46 eV, an error of only 1.8%. While the improvement is significant, the energy is still not very accurate. On the other hand, the situation for H^- remains a disaster: the screened variational calculation does not yield a bound two-electron state.

So far, we have considered two uncorrelated approximate ground-state wave functions, the products (18.37a) and (18.51). Before examining trial functions that are not symmetric products, e.g., ones that depend also on the correlation variable r_{12}, we ask whether there is an optimum symmetric-product trial function, say

$$\psi_F^{(t)}(\mathbf{r}_1, \mathbf{r}_2) = F(r_1)F(r_2)Y_0^0(\Omega_1)Y_0^0(\Omega_2). \tag{18.54}$$

Should we be able to determine this optimum function, it would be that member of the set of symmetric-product functions which yields the smallest expectation value of $\hat{H}_Z(\mathbf{r}_1, \mathbf{r}_2)$. The desired function $F(r)$ *does* exist: it is the solution of the so-called Hartree–Fock (H–F) equation for the two-electron atom. The H–F approach is the subject of the next section, and, other than stating the two H–F energies here, we defer all details concerning it to Section 18.3. Denoting the H–F energies by $E_{\text{g.s.}}^{(\text{HF})}(Z)$, the results of the H–F calculations (Fischer 1977) are

$$E_{\text{g.s.}}^{(\text{HF})}(\text{H}^-) = -0.488 \text{ a.u.}$$

and

$$E_{\text{g.s.}}^{(\text{HF})}(\text{He}) = -2.862 \text{ a.u.}$$

This is only an improvement of 0.01 a.u. = 0.27 eV for He, or a discrepancy of 1.45 eV, whereas H^- still remains unbound. If we define the correlation energy $E_{\text{corr}}(Z)$ by

$$E_{\text{corr}}(Z) = E_{\text{g.s.}}^{(\text{HF})}(Z) - E_{\text{g.s.}}^{\text{exact}}(Z), \tag{18.55}$$

where $E_{\text{g.s.}}^{\text{exact}}(Z)$ is the exact ground-state energy for $\hat{H}_Z(\mathbf{r}_1, \mathbf{r}_2)$, then it is clear that correlation effects are consequential and need to be taken into account.

Case (ii). Correlated trial functions. The simplest correlated function is

$$\psi^{(t)}(\mathbf{r}_1, \mathbf{r}_2; \xi_1, \xi_2) = N(e^{-\xi_1 r_1/a_0}e^{-\xi_2 r_2/a_0} + e^{-\xi_2 r_1/a_0}e^{-\xi_1 r_2/a_0})Y_0^0(\Omega_1)Y_0^0(\Omega_2). \tag{18.56}$$

It was first used for He by Eckart (1930) and for H^- by Chandrasekhar (1944). The correlation – due to the fact that $\xi_1 \neq \xi_2$ – is a radial one. Evaluation of the variational energy and its subsequent minimization is straightforward if a bit tedious. The minimizing values $\xi_j^{(0)}$, $j = 1, 2$, and the corresponding variational energies $E_{\text{g.s.}}^{(t)}(Z, \xi_1^{(0)}, \xi_2^{(0)})$ are listed in Table 18.5. All aspects of Table 18.5 are noteworthy. First, for both values of Z, the larger of the two effective charges is approximately equal to the nuclear charge Z, whereas the other one is approximately equal to $Z - 1$. The correlation in radius is not only evident but also the crucial factor that finally leads to particle-stable bound-state status for H^- even though the affinity is given by the relatively poor value 0.0135 a.u. $\cong$ 0.37 eV. The binding of 2.8757 a.u. for He is an improvement over the H–F value by 0.024 a.u. $\cong$ 0.64 eV. The latter quantity is an approximation to E_{corr}; the

Table 18.5 *Minimization values of the parameters ξ_1 and ξ_2 and the corresponding variational ground-state energies for H^- ($Z = 1$) and He ($Z = 2$) obtained using Eq. (18.56)*

Z	$\xi_1^{(0)}(Z)$	$\xi_2^{(0)}(Z)$	$-E_{\text{g.s.}}^{(t)}(Z, \xi_1^{(0)}, \xi_2^{(0)})$ (a.u.)
1	1.04	0.28	0.5135
2	2.18	1.19	2.8757

energy itself differs from the "exact" value of 2.903 724 a.u. by 0.028 a.u. $\cong$ 0.76 eV, which is a relative error of 0.96%.

The preceding two-parameter, radially correlated, 1s-type trial function has led to errors of 2.6% and 1% in the ground-state energies of H^- and He. To do significantly better, one must use functions that also contain angular correlations, e.g., through the factor r_{12}. It has been found that functions expressed in terms of the so-called Hylleraas coordinates

$$s = r_1 + r_2, \qquad t = r_1 - r_2, \qquad u = r_{12}, \tag{18.57}$$

e.g.,

$$\psi_P^{(t)}(s, t, u; \xi) = e^{-\xi s} \sum_{i=1}^{N} [1 + P_i(s, t, u)], \tag{18.58}$$

where $P_i(s, t, u)$ is a sum of polynomials in each of the variables and contains coefficients that are variational parameters, can yield significant improvements. Functions of the form (18.58) were employed very early on for H^- by Bethe (1929) and Hylleraas (1930) and for He by Hylleraas in a multi-year series of papers starting in 1930: see Hylleraas (1963, 1964) for a personal review of his own contributions and a summary of work on the two-electron system. Both ground states and low-lying, single-particle-type excited states have been approximated in this way.

The two-electron system, especially He, has been a theoretical laboratory in the quest for accurate computation of measurable energies, and Hylleraas coordinates and functions like (18.58) have been important tools in this quest. For example, three-parameter functions of this type yield ground-state values of −2.9024 a.u. for He and −0.5253 a.u. for H^-, errors of about 0.5% and 0.05%. By going to a 39-parameter, Hylleraas-type, trial function, the ground-state energy of He was found to be −2.903 722 5 a.u., to be compared with −2.903 724 a.u.! Such a calculation is formidable. The result of a truly astonishing calculation on He, employing so-called perimetric coordinates and 1078 variational parameters(!!!), carried out by Pekeris (1958), is −2.903 724 375 a.u.!! (relativistic effects excluded). For references to these and other calculations (e.g., Kinoshita (1959)), including low-lying excited states, the interested reader may consult the references cited, especially Bethe and Salpeter (1977), Hylleraas (1963, 1964), Moiseiwitsch (1966), and (for a simpler approach with pedagogical analyses), Pilar (1968). The reader should also be aware that non-variational calculations, also of very high accuracy, have been carried out. One such is the hyperspherical-method analysis of Haftel and Mandelzweig (1988), whose ground and $2\,^1S$ excited-state energies are accurate to a few parts in 10^9.

Extensions of the variational method to slightly heavier atoms have been carried out but such calculations are tedious. It is simply difficult to obtain high accuracy, even for low-lying excited states of He. As a specific instance we consider Rydberg states of He, which were mentioned earlier. (These are states in which one electron is excited out of the dominant $(1s)^2$ ground-state configuration to a large-n excited state, say $n \geqslant 5$.) Drachman (1993) has noted that the Rydberg-state formula $E_{n\ell} = -[2 + 1/(2n^2)]$ a.u., corresponding to one electron in the $Z = 2$, $|100\rangle$ ground state and the other in a $Z = 1$, n, ℓ excited state, differs by about three parts in 10^7(!) from the accurate value for $n = 5$ (the difference between $S = 1$ and $S = 0$ for this level is tiny, about 2.5×10^{-11} a.u.). The product of hydrogenic wave functions $(|100\rangle|n\ell m_\ell\rangle)$ is a very poor state-vector approximation even though the "crude" estimate $E_{n\ell}$ itself is highly accurate. To achieve the $E_{n\ell}$ accuracy with Hylleraas-type functions requires a major effort for each Rydberg level (even with the computing facilities available during the 1990s). A breakthrough in carrying out variational calculations for Rydberg states was made by Drake (1987). This led to the exquisite calculations which he later reported (Drake 1993).

The goal of obtaining highly accurate energy levels and wave functions is a fundamental one in determining properties such as transition rates and electromagnetic moments as well as the effects of small corrections, etc. These are topics to which most attention is paid by specialists in the various fields of atomic, nuclear, molecular, and elementary-particle physics. There is, however, a less detailed but equally important task, particularly from a semi-quantitative point of view, and that is accounting for general features, for example, those of the periodic table of the elements. We consider in the next two sections the standard theoretical frameworks which provide the explanation for the existence of these organizing principles. In so doing, we will gain an understanding of some features of atomic structure. That understanding, of course, is another of the triumphs of quantum theory.

18.3. The Hartree–Fock approximation for atoms

The explanation of the regularities embodied in the periodic table of the elements is based on a shell-model description of the relevant ground states. Shell models are the organizing principle noted at the end of the preceding section, and they in turn are based on an independent-particle approximation. For atoms, the fundamental independent-particle approximation is that of the self-consistent field, first introduced by Hartree (1928) and later modified to include effects of antisymmetry by Fock (1930) and Slater (1930). For a system of N identical spin-$\frac{1}{2}$ fermions, the H–F states are single determinants formed from single-particle orbitals of the form $|n\ell m_\ell m_s\rangle$ or $|n\ell\frac{1}{2};\, jm_j\rangle$. Furthermore, because it is the H–F approximation, the single-particle orbitals are those for which the corresponding determinant yields the lowest possible determinantally based, N-particle energy. These H–F energies are sufficiently accurate to justify the picture of independent-particle orbitals and thereby explain the regularities in the periodic table mentioned above.

In the simplest non-relativistic approximation the Hamiltonian for an atom consisting of N electrons and a nucleus of charge Z is

$$\hat{H}_Z(1, 2, \ldots, N) = \sum_{j=1}^{N} \hat{K}_j - \sum_{j=1}^{N} \frac{\kappa_e e^2 Z}{\hat{Q}_j} + \sum_{i>j=1}^{N} \frac{\kappa_e e^2}{\hat{Q}_{ij}}, \tag{18.59}$$

where the nuclear mass is assumed to be stationary, $\hat{K}_j = \hat{P}_j^2/(2m_e)$ is the kinetic-energy operator for electron j, $\hat{Q}_j = |\hat{\mathbf{Q}}_j|$, and $\hat{Q}_{ij} = |\hat{\mathbf{Q}}_i - \hat{\mathbf{Q}}_j|$, with $\hat{\mathbf{Q}}_j$ being the position operator for electron j. The "simplicity" of this Hamiltonian resides in the absence of spin operators as much as it does in the assumption of an infinitely massive nucleus. We shall see later that there are instances in which the former condition must be relaxed; however, we shall retain the latter assumption.

An independent-particle approximation requires the use of a Hamiltonian that is a sum of commuting one-particle Hamiltonians. Although the first two terms on the RHS *are* such a sum, the two-electron studies of Section 18.2 should make it unsurprising that the energies to which $\sum_j(\hat{K}_j - \kappa_e e^2/\hat{Q}_j)$ leads are insufficiently accurate. An alternate independent-particle Hamiltonian is required.

Use of a sufficiently accurate independent-particle Hamiltonian is one requirement of a shell model, another is imposing the antisymmetry condition obeyed by fermion state vectors and wave functions. Imposition of antisymmetry has a remarkable effect on the eigenstates of an independent-particle Hamiltonian $\hat{H}_{\mathrm{IP}}(1, 2, \ldots, N)$ describing a system of identical fermions: it eliminates the $(N!)$-fold degeneracy of states such as $|\Gamma\rangle$ of (11.17). (That there are a total of $N!$ such states is a straightforward combinatorial result: there are $N!$ ways to order the integers 1, 2, ..., N.) All the other degenerate states are of the form ($|\gamma\rangle \to |\Gamma\rangle$)

$$\hat{P}_{ij}|\Gamma\rangle = |\Gamma_1\rangle_1|\Gamma_2\rangle_2 \cdots |\Gamma_{i-1}\rangle_{i-1}|\Gamma_i\rangle_j|\Gamma_{i+1}\rangle_{i+1} \cdots |\Gamma_{j-1}\rangle_{j-1}|\Gamma_j\rangle_i|\Gamma_{j+1}\rangle_{j+1} \cdots |\Gamma_N\rangle_N; \tag{18.60}$$

the energy of each is $E_\Gamma^{(N)}$ of Eq. (11.16). However, among all possible linear combinations of these $N!$ degenerate states there is only one that is antisymmetric. It is that combination which is the full expansion of an $N \times N$ determinant made up of the single-particle states. This determinant is

$$|\Gamma\rangle^{\mathrm{A}} = \sqrt{\frac{1}{N!}} \begin{vmatrix} |\Gamma_1\rangle_1 & |\Gamma_1\rangle_2 & |\Gamma_1\rangle_3 & \cdots & |\Gamma_1\rangle_N \\ |\Gamma_2\rangle_1 & |\Gamma_2\rangle_2 & |\Gamma_2\rangle_3 & \cdots & |\Gamma_2\rangle_N \\ \vdots & & & & \\ |\Gamma_N\rangle_1 & |\Gamma_N\rangle_2 & |\Gamma_N\rangle_3 & \cdots & |\Gamma_N\rangle_N \end{vmatrix}$$

$$\equiv \sqrt{\frac{1}{N!}} \det||\Gamma_1\rangle_1 \cdots |\Gamma_N\rangle_N|. \tag{18.61}$$

No other linear combinations of the N states $\{|\Gamma_i\rangle_j\}_1^N$ are antisymmetric in all N particle labels, and therefore no others are allowed as states describing a system of N identical fermions.

Other than obeying

$$_k\langle\Gamma_i|\Gamma_j\rangle_k = \delta_{\Gamma_i\Gamma_j}, \tag{18.62}$$

where $\delta_{\Gamma_i\Gamma_j}$ stands for a product of Kronecker deltas, one for each of the quantum numbers comprising the collective index Γ_i (e.g., $\Gamma_i = n_i\ell_i m_{\ell_i} m_{s_i}$), the $|\Gamma_i\rangle$ appearing in (18.61) are arbitrary. Our interest, however, is in a special class of single-particle states, viz., the *atomic H–F* states $|\Gamma_i\rangle^{(\mathrm{HF})}$ which minimize the value of

$$E_t^{(N)}(\{|\Gamma_i\rangle\}, \{\langle\Gamma_j|\}) = {}^{\mathrm{A}}\langle\Gamma|\hat{H}_Z(1, 2, \ldots, N)|\Gamma\rangle^{\mathrm{A}}, \tag{18.63}$$

subject to the constraint

$${}^{\mathrm{A}}\langle\Gamma|\Gamma\rangle^{\mathrm{A}} = 1. \tag{18.64}$$

Equation (18.64) is a consequence of (18.62) and the normalizing factor $(N!)^{-1/2}$ of Eq. (18.61); its proof is left to the exercises.

The minimization (including a diagonalization) leads to a non-linear, state-dependent, non-local eigenvalue equation for each $|\Gamma_i\rangle^{(\mathrm{HF})}$. Deriving the H–F equation is an exercise in variational analysis and the manipulation of determinants, the details of which can be found in a variety of references, e.g., Pilar (1968), Baym (1976), Steiner (1976), Volume II of Galindo and Pascual (1990), and Atkins and Friedman (1997); our concern here is only with the final result.

The version of the H–F equation given below assumes that $\langle\mathbf{r}_k|\Gamma_i\rangle_k^{(\mathrm{HF})}$ is a product ($|\Gamma_i\rangle \to |\gamma_i\sigma_i\rangle$):

$$\langle\mathbf{r}_k|\Gamma_i\rangle_k^{(\mathrm{HF})} = \psi_{\gamma_i\sigma_i}^{(\mathrm{HF})}(\mathbf{r}_k)|\tfrac{1}{2}\sigma_i\rangle_k, \tag{18.65}$$

where γ_i denotes the spatial quantum numbers, $|\frac{1}{2}\sigma_i\rangle_k$ is a general spinor of the form $a_i^{(+)}|\frac{1}{2}\frac{1}{2}\rangle_k + a_i^{(-)}|\frac{1}{2}, -\frac{1}{2}\rangle_k$, and a possible dependence of the coordinate-space wave function on σ_i is specified. With (18.65) one finds

$$\hat{H}_k^{(\mathrm{HF})}\psi_{\gamma_i\sigma_i}^{(\mathrm{HF})}(\mathbf{r}_k) = \epsilon_{\gamma_i\sigma_i}^{(\mathrm{HF})}\psi_{\gamma_i\sigma_i}^{(\mathrm{HF})}(\mathbf{r}_k), \qquad i, k = 1, 2, \ldots, N, \tag{18.66}$$

where

$$\hat{H}_k^{(\mathrm{HF})} = \hat{H}_Z(\mathbf{r}_k) + \hat{J}_k - \hat{K}_k^{(\mathrm{ex})}, \tag{18.67}$$

$$\hat{J}_k\psi_{\gamma_i\sigma_i}^{(\mathrm{HF})}(\mathbf{r}_k) = \left(\sum_{\substack{j=1\\ j\neq k}}^{N}\int d^3r_j\,\psi_{\gamma_j\sigma_j}^{(\mathrm{HF})*}(\mathbf{r}_j)\frac{e^2\kappa_{\mathrm{e}}}{|\mathbf{r}_j - \mathbf{r}_k|}\psi_{\gamma_j\sigma_j}^{(\mathrm{HF})}(\mathbf{r}_j)\right)\psi_{\gamma_i\sigma_i}^{(\mathrm{HF})}(\mathbf{r}_k). \tag{18.68a}$$

and

$$[\hat{K}_k^{(\mathrm{ex})}\psi_{\gamma_i\sigma_i}^{(\mathrm{HF})}](\mathbf{r}_k) = \sum_{\substack{j=1\\ j\neq k}}^{N}\int d^3r_j\,\psi_{\gamma_j\sigma_i}^{(\mathrm{HF})*}(\mathbf{r}_j)\frac{e^2\kappa_{\mathrm{e}}}{|\mathbf{r}_j - \mathbf{r}_k|}\psi_{\gamma_i\sigma_i}^{(\mathrm{HF})}(\mathbf{r}_j)\psi_{\gamma_j\sigma_i}^{(\mathrm{HF})}(\mathbf{r}_k). \tag{18.68b}$$

Of the three terms comprising $\hat{H}_k^{(\mathrm{HF})}$, the first is just our old friend, the one-electron, hydrogenic Hamiltonian with nuclear charge Z, exemplified by Eq. (18.31). The second term, the direct or Coulomb operator $\hat{J}_k$, is the average Coulomb potential felt by electron k, where the average is over all the orbitals. These first two terms form the *Hartree* single-particle Hamiltonian $\mathrm{H}_k^{(\mathrm{H})}$:

$$\hat{H}_k^{(H)} = \hat{H}_Z(\mathbf{r}_k) + \hat{J}_k. \tag{18.69}$$

When it is summed on k it yields an N-particle Hamiltonian from which Pauli-principle effects are excluded, although from it one may gain an orientation to the structure of the more complex $\hat{H}_k^{(\mathrm{HF})}$ and the determinantal state $|\Gamma\rangle^{(\mathrm{HF})}$. For example, calculations based on $\hat{H}_k^{(\mathrm{H})}$ yield atomic ground-state energies that are typically 10%–20% larger than those obtained from use of $\hat{H}_k^{(\mathrm{HF})}$.

The exchange operator[6] $\hat{K}_k^{(\mathrm{ex})}$ contains the effects of the Pauli principle. It is an *integral* operator, since the coordinates of the function $\psi_{\gamma_i\sigma_i}^{(\mathrm{HF})}$ on which it acts appear under the integral sign (i.e., they are integration variables), whereas $\mathbf{r}_k$, the coordinates of interest, occur in $\psi_{\gamma_j\sigma_i}^{(\mathrm{HF})}$, $j \neq i$. $\hat{K}_k^{(\mathrm{ex})}$ is non-local in the same way as $\hat{G}_k = (E_k - \hat{K})^{-1}$ of Eqs. (15.43): both are integral operators. In addition to its being an integral operator, $\hat{K}_k^{(\mathrm{ex})}$ is distinguished by the fact that it is diagonal in the spin quantum number, i.e., it is non-vanishing only when $\sigma_j = \sigma_i$. This means that, insofar as the i–j pair is concerned, their total spin state is a *triplet*. This feature has important consequences for the H–F energy, as discussed below.

The non-linear aspect of the eigenvalue relation (18.66) arises from $\hat{J}_k$ and $\hat{K}_k^{(\mathrm{ex})}$ being state-dependent: they are single-particle operators defined in terms of the unknown functions that we are seeking: to calculate $\psi_{\gamma_i\sigma_i}^{(\mathrm{HF})}$ requires our knowing it beforehand! This is a classic "Catch 22" or "this course is a prerequisite for this course" dilemma. It is resolved by the iterative method known as self-consistency, carried out in a sequence of steps. The initial or zeroth-order step can be taken in one of two ways: either one introduces N orthonormal guesses $\psi_{\gamma_i\sigma_i}^{(\mathrm{HF})_0}$ and uses them to construct zeroth-order operators $\hat{J}_k^{(0)}$ and $\hat{K}_k^{(\mathrm{ex})_0}$, or else one chooses $\hat{J}_k^{(0)}$ and $\hat{K}_k^{(\mathrm{ex})_0}$ (perhaps via the Thomas–Fermi average-field method of Thomas (1926) and Fermi (1927), described, e.g., in Galindo and Pascual (1990)). These zeroth-order operators are substituted into (18.66), the solutions of which are a once-iterated (orthonormalized) set $\{\psi_{\gamma_i\sigma_i}^{(\mathrm{HF})_1}\}$ and (real) energies $\{\epsilon_{\gamma_i\sigma_i}^{(\mathrm{HF})_1}\}$, which in turn lead to $\hat{J}_k^{(1)}$ and $\hat{K}_k^{(\mathrm{ex})_1}$, etc. When the differences between the $(n+1)$st and the nth iterated H–F functions and energies are less than some prescribed value, the process is halted, $\{\psi_{\gamma_i\sigma_i}^{(\mathrm{HF})_n}\}_{i=1}^N$ and $\{\epsilon_{\gamma_i\sigma_i}^{(\mathrm{HF})_n}\}_{i=1}^N$ are accepted as the H–F states and energies, and $\hat{J}_k^{(n)}$ and $\hat{K}_k^{(\mathrm{ex})_n}$ are denoted the "self-consistent" fields. Although we shall review the results of some self-consistent H–F calculations below, we remind the reader of the H–F results for He, e.g., the remarks and equations leading up to Eq. (18.55).

Let us now suppose that the solutions $\{\psi_{\gamma_i\sigma_i}^{(\mathrm{HF})}\}$ and $\{\epsilon_{\gamma_i\sigma_i}^{(\mathrm{HF})}\}$ have been determined. Then forming the scalar product of both sides of (18.66) with $\psi_{\gamma_i\sigma_i}^{(\mathrm{HF})}(\mathbf{r}_k)$ yields an expression for the eigenvalue $\epsilon_{\gamma_i\sigma_i}^{(\mathrm{HF})}$:

$$\epsilon_{\gamma_i\sigma_i}^{(\mathrm{HF})} = {}_k\langle\gamma_i\sigma_i|\hat{H}_Z(k)|\gamma_i\sigma_i\rangle_k + \sum_{\substack{j=1\\ j\neq i}}^{N}(J_{\gamma_i\sigma_i,\gamma_j\sigma_j} - K_{\gamma_i\sigma_i,\gamma_j\sigma_i}), \tag{18.70}$$

where

$$J_{\gamma_i\sigma_i,\gamma_j\sigma_j} = \int d^3r_k\, d^3r_j\, |\psi_{\gamma_i\sigma_i}^{(\mathrm{HF})}(\mathbf{r}_k)|^2 \frac{e^2\kappa_\mathrm{e}}{r_{kj}} |\psi_{\gamma_j\sigma_j}^{(\mathrm{HF})}(\mathbf{r}_j)|^2 \tag{18.71a}$$

and

[6] The superscript "(ex)" has been appended to the symbol for the exchange operator so as to distinguish it from the kinetic-energy operator.

$$K_{\gamma_i\sigma_i,\gamma_j\sigma_i} = \int d^3r_k\, d^3r_j\, \psi^{(\mathrm{HF})*}_{\gamma_i\sigma_i}(\mathbf{r}_k)\psi^{(\mathrm{HF})}_{\gamma_j\sigma_i}(\mathbf{r}_k)\frac{e^2\kappa_e}{r_{kj}}\psi^{(\mathrm{HF})*}_{\gamma_j\sigma_i}(\mathbf{r}_j)\psi^{(\mathrm{HF})}_{\gamma_i\sigma_i}(\mathbf{r}_j). \tag{18.71b}$$

The latter two matrix elements are known, respectively, as Coulomb and exchange integrals, specific instances of which were encountered in Section 18.2 – Eqs. (18.49) – and, like those examples, can be written as abstract matrix elements. Although the sums in Eq. (18.70) do not include $j = i$, the equality $J_{\gamma_i\sigma_i,\gamma_i\sigma_i} = K_{\gamma_i\sigma_i,\gamma_i\sigma_i}$ means that there is no change in the value of $\epsilon^{(\mathrm{HF})}_{\gamma_i\sigma_i}$ if these terms are included.

In view of $\hat{H}^{(\mathrm{HF})}_k$ being a single-particle operator, one may at first think that the energy of the H–F state

$$|\Gamma\rangle^{(\mathrm{HF})} = \sqrt{\frac{1}{N!}}\det|\,|\gamma_1\sigma_1\rangle^{(\mathrm{HF})}_1 \cdots |\gamma_N\sigma_N\rangle^{(\mathrm{HF})}_N| \tag{18.72}$$

is $\sum_i \epsilon^{(\mathrm{HF})}_{\gamma_i\sigma_i}$. However, this conclusion is not correct: since $\{|\gamma_i\sigma_i\rangle^{(\mathrm{HF})}\}$ and $\{\epsilon^{(\mathrm{HF})}_{\gamma_i\sigma_i}\}$ are the solutions to the Euler–Lagrange equation (18.66) derived from minimization of (18.63), the actual H–F energy $E^{(N)}_{\mathrm{HF}}$ is the value of $E^{(N)}_t(\{|\gamma_i\sigma_i\rangle^{(\mathrm{HF})}\}, \{{}^{(\mathrm{HF})}\langle\gamma_j\sigma_j|\})$ obtained from Eq. (18.63):

$$E^{(N)}_{\mathrm{HF}} = {}^{(\mathrm{HF})}\langle\Gamma|\hat{H}_Z(1, 2, \ldots, N)|\Gamma\rangle^{(\mathrm{HF})}. \tag{18.73}$$

Substitution of (18.72) into (18.73) eventually leads (Pilar 1968, or Galindo and Pascual 1990) to

$$\begin{aligned} E^{(N)}_{\mathrm{HF}} &= \sum_{i=1}^{N}\left({}_k\langle\gamma_i\sigma_i|\hat{H}_Z(k)|\gamma_i\sigma_i\rangle_k + \frac{1}{2}\sum_{j=1}^{N}[J_{\gamma_i\sigma_i,\gamma_j\sigma_j} - K_{\gamma_i\sigma_i,\gamma_j\sigma_i}]\right) \\ &= \frac{1}{2}\sum_i[{}_k\langle\gamma_i\sigma_i|\hat{H}_Z(k)|\gamma_i\sigma_i\rangle_k + \epsilon^{(\mathrm{HF})}_{\gamma_i\sigma_i}], \end{aligned} \tag{18.74}$$

which differs from $\sum_i \epsilon^{(\mathrm{HF})}_{\gamma_i\sigma_i}$, as claimed.

Although $\sum_i \epsilon^{(\mathrm{HF})}_{\gamma_i\sigma_i}$ is not the H–F energy, a rough physical meaning can be given to each $\epsilon^{(\mathrm{HF})}_{\gamma_j\sigma_j}$, as follows. Suppose that the atom were ionized by removal both of an electron and of the state $|\gamma_j\sigma_j\rangle^{(\mathrm{HF})}$. Furthermore, suppose also that the state of the resulting $(N-1)$-electron ion were $|\Gamma\rangle^{(\mathrm{HF})}$ with one electron and $|\gamma_j\sigma_j\rangle^{(\mathrm{HF})}$ absent, an assumption that might be justified instantaneously via the sudden approximation but not otherwise. Then the ion's energy $E^{(N-1)}_{\mathrm{ion}}$ would be

$$E^{(N-1)}_{\mathrm{ion}} = E^{(N)}_{\mathrm{HF}} - \epsilon^{(\mathrm{HF})}_{\gamma_j\sigma_j},$$

from which we conclude that $-\epsilon^{(\mathrm{HF})}_{\gamma_j\sigma_j}$ is approximately the energy needed to ionize the atom by removal of an electron from the jth orbital $|\gamma_j\sigma_j\rangle^{(\mathrm{HF})}$. This conclusion is part of Koopmans' theorem (Koopmans (1933)), whose limited validity is discussed, e.g., by Steiner (1976); see also Pilar (1968), Fischer (1977), Galindo and Pascual (1990), and Atkins and Friedman (1997).

It was noted in Section 18.2 that the two-electron exchange integral $K_{n\ell,n'\ell'}$ is non-negative. This is also true for the H–F exchange integral $K_{\gamma_i\sigma_i,\gamma_j\sigma_i}$ occurring in $\epsilon^{(\mathrm{HF})}_{\gamma_i\sigma_i}$ and $E^{(N)}_{\mathrm{HF}}$ (see, e.g., Pilar (1968) for the analysis). Hence, the change from the Hartree to the H–F approximation results in a lowering of the calculated energies. Furthermore, the fact that $\sigma_j = \sigma_i$ in the exchange integral means that, all other things being equal, states

Table 18.6 *Hartree–Fock ground-state energies for selected atoms*

Z	Atom	$-E^{(Z)}_{\mathrm{HF}}$ (a.u.)
2	He	2.862
3	Li	7.433
6	C	37.689
8	O	74.809
13	Al	241.877
18	Ar	526.818
30	Zn	1777.848
50	Sn	6022.932
79	Au	17 865.212
82	Pb	19 524.008

based on triplet pairs will lie lower in energy than those based on singlet pairs. This argument is the foundation for one of Hund's rules, mentioned earlier and discussed in the next section.

Among the ingredients one associates with the periodic table and atomic structure are the concepts of "closed" shells, "open" shells and "valence electrons," and "holes" in closed shells. These are not concepts that arise naturally from (numerical) solution of the H–F equations: they must be inserted as part of the solution. Without any of these ingredients, one refers to the unrestricted H–F model; forcing the equations to produce, e.g., solutions of a closed-shell nature (when appropriate) means working within the "restricted H–F" (RH–F) model. A closed-shell structure can arise only if the same $n\ell$ orbitals are used for both the "up" and the "down" spin states (i.e., for both $|\frac{1}{2} \pm \frac{1}{2}\rangle$); a closed shell occurs when all $2\ell + 1\, m_\ell$-substates for a given n and ℓ are occupied. There are then $2(2\ell + 1)$ electrons in the (closed) shell. Even this description relies on the generally false assumption that $\psi^{(\mathrm{HF})}_{\gamma_i\sigma_i}(\mathbf{r})$ has the simplest possible orbital structure $R_{n_i\ell_i}(r)Y^{m_\ell}_{\ell_i}(\Omega_r)$, i.e., that only one value of ℓ enters. In other words, $\psi^{(\mathrm{HF})}_{\gamma_i\sigma_i}(\mathbf{r})$ is conjectured to behave as if it were generated by a central field, whereas neither $\hat{J}_i$ nor $\hat{K}^{(\mathrm{ex})}_i$ in general is central.[7]

With or without the single-ℓ-value assumption, the non-linear H–F calculations are very complicated (see Fischer (1977) for details). Fortunately, this complexity does not concern us, since our interest is in the results of such calculations and especially how a *local* central-field Hamiltonian can be used to describe atomic structure and the periodic table. A few results of the H–F calculations carried out by Fischer (1977) are displayed in Table 18.6; these use configurations containing, as appropriate, only one open shell, with the shell fillings underlying these calculations enumerated in Section 18.4. The results presented in Table 18.6 are in agreement with those of other investigators, e.g., Veillard and Clementi (1968) and Mann and Johnson (1971). The latter authors also state the values both of the correlation energy E_{c} and of the relativistic corrections E_{R}. The former is the difference between $E^{(N)}_{\mathrm{HF}}$ and the (numerically) exact ground-state energy

[7] For discussion of this in the context of closed shells, open shells, and excited states, see Pilar (1968), Steiner (1976), and Fischer (1977). See also Section 18.4.

$E^{(N)}_{\text{g.s.}}$ computed for the Hamiltonian (18.59), whereas the latter consists of a variety of effects including those due to spin–orbit and spin–spin coupling and to finite nuclear size. In general, $-E_c$ is about 1 eV per electron (e.g., it is 16.2 eV for S ($Z = 16$) and 19.9 for Ar ($Z = 18$)). For small Z, $E_R \leqslant E_c$, but, for $Z \geqslant 14$, $E_R > E_c$, so that relativistic effects become increasingly important. They turn out to be greatest for the inner shells, and, since the chemical properties of elements are determined by the outermost (largest-ℓ) electrons, for which E_R is very small (see Steiner (1976) for a compilation of the original results of Clementi (1965) on this topic), it follows that the non-relativistic approximation suffices for many applications to chemistry.

The H–F approximation gives a good account of the trends of atomic ground-state structure, but it has drawbacks that lead one to seek other descriptions. First, as Z changes, so do the H–F states: in principle, a different set of states must be calculated for each Z. Second, a new H–F calculation should be carried out for each excited state of a given atom. Thus, because every calculation is numerical, and each state requires a new calculation, the trends and features of atomic ground-state spectra cannot be easily understood within the theoretical framework, but only as individual numerical consequences. Third, while the relative size of the corrections from E_c decreases with increasing Z, the errors in the difference between ground- and excited-state energies can be so large as to render calculations of transition rates, etc., wholly inadequate. The latter feature is especially frustrating since the generating of H–F states and energies for each excited state is rather tedious. A non-frustrating procedure, one that readily leads to understanding, is to use a local central field approximation and then treat the difference between it and $\hat{H}_Z(1, 2, \ldots, N)$ as either a perturbation or a variational quantity. With the local central-field approximation, the generation of excited-state wave functions and energies is much simpler than it is in the H–F approximation.

18.4. The atomic central-field approximation and the periodic table of the elements

The local central-field approximation sidesteps some of the H–F features such as state-dependent interactions and the need for generating new orbitals for excited states and for different Z. With a central interaction one has not only an orbital picture but also total energies that *are* a sum of the single-particle energies. Furthermore, it provides a basis for treating ground and excited states together. Corrections are straightforward to formulate, although they are not necessarily easy to evaluate.

An added impetus for considering a central-field approximation is that, for a "closed-shell" atom, the H–F equations are those of a (non-local) central field.[8] The configurations for closed-shell atoms are $(n_1\ell_1)^{2(2\ell_1+1)}(n_2\ell_2)^{2(2\ell_2+1)} \cdots (n_f\ell_f)^{2(2\ell_f+1)}$, where the total number of electrons is $N = 2\sum_1^f(2\ell_j + 1)$, and the factors of two take account of the fact that there are two m_S values for each m_ℓ. Furthermore, for a system of spin-$\frac{1}{2}$ particles, angular-momentum considerations arise, and closed shells play a significant role here also: their total angular momentum J is zero, as will be proved later. This feature will be very useful in helping us to understand the possible angular momenta for systems with only one or a few particles in open shells.

[8] A proof that $\hat{J}_k$ and $\hat{K}_k^{(\text{ex})}$ for atomic closed shells are each central can be found in Vol. II of Galindo and Pascual (1990).

The generic central-field Hamiltonian for particle k is

$$\hat{H}_k(\mathbf{r}_k) = \hat{K}_k(\mathbf{r}_k) + V_c(r_k) + V_{LS}(r_k)\hat{\mathbf{L}} \cdot \hat{\mathbf{S}}/\hbar^2. \tag{18.75}$$

Although our considerations are limited to atoms, it is interesting to compare the way Eq. (18.75) is used in the atomic and nuclear shell models. In the case of open-shell atoms $V_c(r)$ represents both a weighted averaging over the open shells and a local approximation to the effects of the exchange operator. It can be thought of as a modified Coulomb interaction that leads to energies $\mathcal{E}_{n\ell}$, a Grotrian diagram for which is presented later (Fig. 18.4). In the nuclear case, $V_c(r)$, whose shape tends to follow the nuclear radial density distribution, is a potential well the ordering of whose lower-lying energy levels $\mathcal{E}_{n\ell}$ is intermediate between those of a 3-D symmetric oscillator and an infinite square well. The spin–orbit interaction plays very different roles in atoms and in nuclei. In atoms, it is a reasonable zeroth-order approximation throughout the periodic table to treat it as weak relative to the electrostatic repulsion, although spin–orbit effects are more significant for heavier atoms and are also important in many cases for determining angular-momentum values. For the nuclear case, however, it is necessary to include the spin–orbit term as part of the central-field potential for essentially all nuclei. In view of these comments, single-particle states for atoms have the uncoupled form $|n\ell m_\ell m_s\rangle$, whereas for nuclei they are the coupled states $|n\ell\frac{1}{2};\, jm_j\rangle$.

Single-particle energy levels and the **Aufbau** *principle*

The N-electron central-field Hamiltonian is

$$\hat{H}_{\rm IP}^{(\rm atom)}(1, 2, \ldots, N) = \sum_{j=1}^{N} [\hat{K}_j(\mathbf{r}_j) + V_{\rm c}^{(\rm atom)}(r_j)]. \tag{18.76}$$

It is treated as the unperturbed portion of a more accurate but still approximate Hamiltonian $\hat{H}_Z^{(\rm atom)}(1, 2, \ldots, N)$, where

$$\hat{H}_Z^{(\rm atom)}(1, 2, \ldots, N) = \hat{H}_{\rm IP}^{(\rm atom)}(1, 2, \ldots, N) + \hat{H}_1^{(\rm atom)}(1, 2, \ldots, N), \tag{18.77a}$$

with

$$\hat{H}_1^{(\rm atom)}(1, 2, \ldots, N) = \sum_{k=1}^{N} \left(-\frac{Ze^2\kappa_{\rm e}}{r_k} - V_{\rm c}^{(\rm atom)}(r_k) + \frac{V_{LS}(r_k)\hat{\mathbf{L}}_k \cdot \hat{\mathbf{S}}_k}{\hbar^2} + \sum_{j \neq k} \frac{e^2\kappa_{\rm e}}{r_{jk}} \right). \tag{18.77b}$$

Since $V_{\rm c}^{(\rm atom)}$ is supposed to take average effects of the electrostatic interactions into account, and spin–orbit effects are generally presumed to be small, the corrections arising from $\hat{H}_1^{(\rm atom)}(1, 2 \ldots, N)$ are assumed not to change significantly the conclusions based on $\hat{H}_{\rm IP}^{(\rm atom)}$.

The potential $V_{\rm c}^{(\rm atom)}(r)$ produces single-particle energies $\mathcal{E}_{n\ell}$, where $n = 1, 2, 3, \ldots$ and $\ell \leqslant n - 1$, as in the hydrogenic case. The standard number/letter designation is used for these levels: 1s, 2s, 2p, 3s, etc. The ordering of the single-particle energies $\mathcal{E}_{n\ell}$ through the 7p level is shown in the Grotrian diagram of Fig. 18.4. While the separations and especially the magnitudes of individual levels are Z-dependent, the relatively large

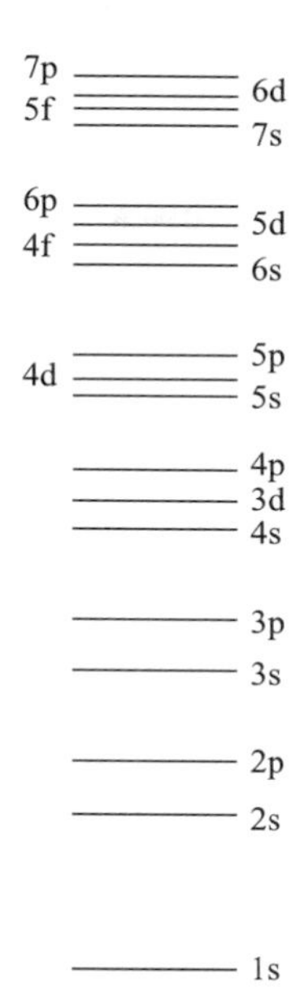

Fig. 18.4 The first nineteen (1s → 7p) single-particle atomic energy levels. An added electron will occupy the lowest-lying energy level consistent with the Pauli principle, except for the anomalous shell fillings discussed in the text. Note well: the actual level separations and magnitudes are Z-dependent.

spacing between an ns level and the one just preceding it, compared with the spacings among the set of energies grouped with each ns level, is typical. For example,

$$\mathcal{E}_{4s} - \mathcal{E}_{3p} > \max\{\mathcal{E}_{4p} - \mathcal{E}_{3d}, \mathcal{E}_{4p} - \mathcal{E}_{4s}, \mathcal{E}_{3d} - \mathcal{E}_{4s}\},$$

etc.

As noted in the preceding section, each level $\mathcal{E}_{n\ell}$ can contain at most $2(2\ell + 1)$ electrons. An energy level $\mathcal{E}_{n\ell}$ is referred to as an $n\ell$ shell; it is filled, or closed, when it holds $2(2\ell + 1)$ electrons. The relatively large separations in Fig. 18.4 lead, as discussed below, to especially stable configurations when the shell preceding an ns level is filled. This is indicated by the sequencing

$$1s|2s,\ 2p|3s,\ 3p|4s,\ 3d,\ 4p|5s,\ 4d,\ 5p|6s,\ 4f,\ 5d,\ 6p|7s,\ 5f,\ 6d,\ 7p|\cdots, \tag{18.78}$$

where a vertical line in (18.78) signifies that the level to its left is particularly stable when it is filled. As we shall see, certain of the atoms whose ground states consist solely of closed shells are so stable that they are inert: they do not combine with other atoms, except perhaps at exceedingly low temperatures, at which they may form dimers, trimers, etc.

For those states which consist of a single determinant, as in the H–F approximation for all atomic ground states, the general configuration is of the form $(n_1\ell_1)^{N_1}(n_2\ell_2)^{N_2} \cdots (n_f\ell_f)^{N_f}$, where $N = \sum_j N_j$, and $N_j \leqslant 2(2\ell_j + 1)$, the equality holding when the shell is full. Such a configuration leads to a total energy given by

$$E = \sum_j \mathcal{E}_{n_j \ell_j}, \tag{18.79}$$

a true central-field, independent-particle result.

Equation (18.79) is the foundation of all shell models, in particular the single-configuration, shell-model description of atomic ground states. Ground states have the lowest energies, and, to achieve this via (18.79), electrons are placed first in the lowest level ($\mathcal{E}_{1s}$) until it is filled, then into the next lowest level ($\mathcal{E}_{2s}$) until it also is filled, etc., going up through the level sequencing of Fig. 18.4 or the ordering (18.78), specific examples of which will be given shortly. This approach is denoted the "building-up" or *Aufbau* principle, by which the ground state of each successive atom is built on the ground state of the preceding atom by adding to it an electron in the lowest-lying single-particle state consistent with the Pauli principle. This building-up principle is the key to understanding the ground-state structures of atoms (and of some nuclei) and thus the regularities of the periodic table.

In general, the orderings of (18.78) and Fig. 18.4 mean that orbits/energy levels are filled via increasing values of the sum $n + \ell$, and for equal $n + \ell$ values in order of increasing n, as illustrated by the orderings $\mathcal{E}_{2s} < \mathcal{E}_{2p}$ and $\mathcal{E}_{4p} < \mathcal{E}_{5s}$. There are some exceptions to this order of shell fillings, which we consider below, but, since they begin at $Z = 24$ (Cr) it is the normal ordering which determines ground-state atomic structure up to $Z = 24$. It is also the operative factor for understanding both shell closings with their consequent stability and the small ionization energies of atoms with all closed shells plus one extra electron.

Ground-state configurations

The *Aufbau* principle is, in effect, an instruction to place electrons in the energy levels described in the foregoing, consistent, of course, with the Pauli principle. Alternately, it is a procedure for assembling atoms from electrons, single-particle energy levels, and the exclusion principle. As we describe the results of doing so, it is essential for the reader to keep two points in mind. First, as the number of electrons increases so does the nuclear charge Z: except for ions, which do not concern us here, all (neutral) atoms have equal numbers of protons and electrons. Second, the values and spacings of the energy levels of Fig. 18.4 are both Z- and state-dependent. These caveats in no way detract from Fig. 18.4 being a truly marvelous instrument that not only allows one to interpret the periodic table but also provides the means for creating it *ab initio*. It also provides a means for understanding some excited states of atoms.

Listed in Table 18.7 are the values of Z, the chemical symbols, the ground-state configurations, and the single-electron ionization energies for the first 90 elements.[9] It is one of several forms, each of which one may call the "periodic table of the elements." The elements from La through Lu ($Z = 57$ through $Z = 71$) are known as the lanthanide series (or the rare earths), and all those starting from Ac form the actinide series. The transuranic elements are those with $Z > 92$. With some exceptions, the configurations of this table are those arising from the *Aufbau* principle combined with the sequencing of

[9] Configurations and ionization energies are tabulated in various places, e.g., Martin and Wiese (1996) or Sobelman (1979).

Table 18.7 *Some ground-state properties of atoms*

Z	Element	Ground-state configuration	Ionization energy (eV)	Z	Element	Ground-state configuration	Ionization energy (eV)
1	H	1s	13.598	46	Pd	Kr + $4d^{10}$	8.337
2	He	$1s^2$	24.587	47	Ag	Kr + $4d^{10}5s$	7.576
3	Li	He + 2s	5.392	48	Cd	Kr + $4d^{10}5s^2$	8.994
4	Be	He + $2s^2$	9.323	49	In	Cd + 5p	5.786
5	B	Be + 2p	8.298	50	Sn	Cd + $5p^2$	7.344
6	C	Be + $2p^2$	11.260	51	Sb	Cd + $5p^3$	8.608
7	N	Be + $2p^3$	14.534	52	Te	Cd + $5p^4$	9.010
8	O	Be + $2p^4$	13.618	53	I	Cd + $5p^5$	10.451
9	F	Be + $2p^5$	17.423	54	Xe	Cd + $5p^6$	12.130
10	Ne	Be + $2p^6$	21.565	55	Cs	Xe + 6s	3.894
11	Na	Ne + 3s	5.139	56	Ba	Xe + $6s^2$	5.212
12	Mg	Ne + $3s^2$	7.646	57	La	Ba + 5d	5.577
13	Al	Mg + 3p	5.986	58	Ce	Ba + 4f5d	5.539
14	Si	Mg + $3p^2$	8.152	59	Pr	Ba + $4f^3$	5.464
15	P	Mg + $3p^3$	10.487	60	Nd	Ba + $4f^4$	5.525
16	S	Mg + $3p^4$	10.360	61	Pm	Ba + $4f^5$	5.580
17	Cl	Mg + $3p^5$	12.968	62	Sm	Ba + $4f^6$	5.644
18	Ar	Mg + $3p^6$	15.760	63	Eu	Ba + $4f^7$	5.670
19	K	Ar + 4s	4.341	64	Gd	Ba + $4f^75d$	6.150
20	Ca	Ar + $4s^2$	6.113	65	Tb	Ba + $4f^9$	5.864
21	Sc	Ar + $3d4s^2$	6.562	66	Dy	Ba + $4f^{10}$	5.939
22	Ti	Ar + $3d^24s^2$	6.828	67	Ho	Ba + $4f^{11}$	6.022
23	V	Ar + $3d^34s^2$	6.746	68	Er	Ba + $4f^{12}$	6.108
24	Cr	Ar + $3d^54s$	6.767	69	Tm	Ba + $4f^{13}$	6.184
25	Mn	Ar + $3d^54s^2$	7.434	70	Yb	Ba + $4f^{14}$	6.254
26	Fe	Ar + $3d^64s^2$	7.902	71	Lu	Yb + 5d	5.426
27	Co	Ar + $3d^74s^2$	7.881	72	Hf	Yb + $5d^2$	6.825
28	Ni	Ar + $3d^84s^2$	7.640	73	Ta	Yb + $5d^3$	7.550
29	Cu	Ar + $3d^{10}4s$	7.726	74	W	Yb + $5d^4$	7.864
30	Zn	Ar + $3d^{10}4s^2$	9.394	75	Re	Yb + $5d^5$	7.834
31	Ga	Zn + 4p	5.999	76	Os	Yb + $5d^6$	8.438
32	Ge	Zn + $4p^2$	7.899	77	Ir	Yb + $5d^7$	8.967
33	As	Zn + $4p^3$	9.789	78	Pt	Xe + $4f^{14}5d^96s$	8.959
34	Se	Zn + $4p^4$	9.752	79	Au	Xe + $4f^{14}5d^{10}6s$	9.226
35	Br	Zn + $4p^5$	11.814	80	Hg	Xe + $4f^{14}5d^{10}6s^2$	10.438
36	Kr	Zn + $4p^6$	14.000	81	Tl	Hg + 6p	6.108
37	Rb	Kr + 5s	4.177	82	Pb	Hg + $6p^2$	7.417
38	Sr	Kr + $5s^2$	5.695	83	Bi	Hg + $6p^3$	7.286
39	Y	Kr + $4d5s^2$	6.217	84	Po	Hg + $6p^4$	8.417
40	Zr	Kr + $4d^25s^2$	6.634	85	At	Hg + $6p^5$	9.5
41	Nb	Kr + $4d^45s$	6.759	86	Rn	Hg + $6p^6$	10.749
42	Mo	Kr + $4d^55s$	7.092	87	Fr	Rn + 7s	4.073
43	Tc	Kr + $4d^55s^2$	7.280	88	Ra	Rn + $7s^2$	5.278
44	Ru	Kr + $4d^75s$	7.361	89	Ac	Ra + 6d	5.170
45	Rh	Kr + $4d^85s$	7.459	90	Th	Ra + $6d^2$	6.307

(18.78) and Fig. 18.4, and reference to it is implied in connection with the following remarks.

As we have noted, ground states are constructed theoretically by placing electrons into the lowest level consistent with the exclusion principle. The $Z = 1$, single-electron case is hydrogen, for which the central field is that of Section 11.6 and H–F is not relevant. Adding a second electron (and increasing Z by one) yields the ground state of the inert gas He as the closed-shell configuration $(1s)^2$. Since only two electrons can occupy the 1s level, the $Z = 3$ neutral atom Li has the third electron in the 2s level; its configuration is $(1s)^2 2s$, making it the first of the alkali metals, all of which have a single ns electron (and the term value ${}^2S_{1/2}$, discussed later).

For Be, with $Z = 4$ and the configuration $(1s)^2(2s)^2$, the 2s shell is closed. Given that He is an inert gas that does not combine chemically with other elements, one's first thought might be that Be behaves in a similar fashion. However, Be is not inert, as the existence of the triatomic molecule BeH_2 demonstrates. This is emphasized by reference to the ordering/stability sequence (18.78), wherein the greatest stability occurs for the 2p closed shell (Ne) rather than the 2s closed shell. Comparison of the ionization energies for He and Be indicates why this is so: it requires 24.59 eV to break up the $(2s)^2$ pair in He, but only 9.32 eV to sever the $(2s)^2$ bond in Be. There is clearly more to understanding chemical properties than mere knowledge of the electronic configuration.

For the six atoms with $5 \leqslant Z \leqslant 10$, the *Aufbau* principle and the sequencing of (18.78) directs us to place the added electrons in the 2p shell without disturbing the underlying $(1s)^2(2s)^2$ closed-shell structure of Be. It is customary to indicate the configurations following those of a closed shell by abstracting all the closed-shell configurations into the chemical symbol for the closed-shell element. The configurations of the atoms built from the Be core plus 2p electrons may therefore be written as $\mathrm{Be} + (2p)^k$, $k = 1, 2, \ldots, 6$. B is thus Be plus a single 2p electron; C has two 2p electrons; and so forth up through F with a 2p *hole* and finally Ne, whose sixth 2p electron closes the 2p shell. B and Li are mild analogs in that each is the first atom following one with a closed-shell configuration and it is of interest to compare their ionization energies: 5.39 eV for Li and 8.3 eV for B. The former is quite small compared with that of He, but the latter is almost equal to the value of 9.32 eV for Be, thus indicating that the 1s pair shields the nucleus reasonably well in Li whereas the 2s pair in B, being much farther from the nucleus on average, provides less shielding and therefore more positive charge to be seen by the 2p electron. From Table 18.7 we see that, apart from a slight decrease at O, the ionization energies in the 2p shell increase monotonically, reaching the large value of 21.57 eV for the all-closed-shell inert-gas atom Ne, whose configuration is $\mathrm{Be} + (2p)^6$.

This sequential filling of the $n = 1$ and $n = 2$ energy levels can be represented by the pictorial analog of Fig. 18.1, which is shown in Fig. 18.5 for four of the first ten members of the periodic table. This figure makes it easy to recognize that the inert He and Ne atoms, which correspond to the first two vertical lines in the sequence (18.78), contain two and ten electrons, respectively. The atoms which follow each of them, viz. Li and Na, have much lower ionization energies and are the beginnings of new shells. The numbers 2 and 10 are the first two atomic "magic numbers," i.e., the numbers of electrons corresponding to atoms with filled shells, large ionization energies, and zero angular momentum. We shall enumerate the others as we continue our journey through the

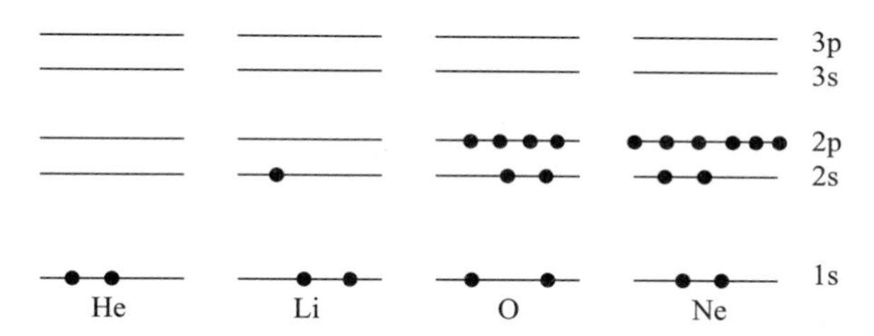

Fig. 18.5 A pictorial representation of shell occupations for the atoms He, $(1s)^2$; Li, $(1s)^2 2s$; O, Be + $(2p)^4$; and Ne, Be + $(2p)^6$. O may be thought of as having two holes in the 2p shell; it is the particle–hole analog of C, which has two particles in the 2p shell. The heavy circles, of course, represent electrons.

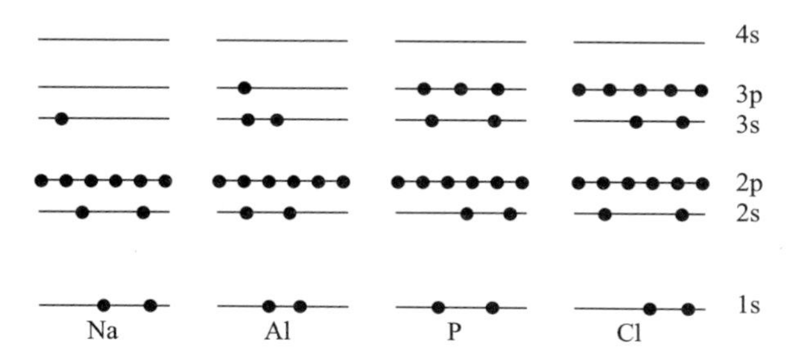

Fig. 18.6 A pictorial representation of shell occupations for the atoms Na = Ne + 3s; Al = Ne + $(3s)^2 3p$; P = Ne + $(3s)^2(3p)^3$; and Cl = Ne + $(3s)^2(3p)^5$. Cl, which has one hole in the 3p shell, is the particle–hole partner of Al.

periodic table (the existence of their nuclear analogs was an important factor in the genesis of the nuclear-shell model – see, e.g., Goeppert-Mayer and Jensen (1955)).

To reach the next atom of greatest stability, the inert gas Ar, we must add electrons first in the 3s and then in the 3p shell, increasing Z each time another electron is added. Mg, which is a $(3s)^2$ atom, has a slightly greater ionization energy than that of the alkali atom Na (7.65 versus 5.14 eV), whereas the first 3p electron (in Al) has the slightly lower ionization energy of 6 eV. As more electrons (and protons) are included the ionization energies monotonically increase to the relatively large value of 15.76 eV for Ar, whose configuration is Ne + $(3s)^2(3p)^6$. Other than filling the $n = 3$ levels, the construction here is exactly the same as in the $n = 2$, B $\rightarrow$ Ne case, and, apart from the assigning of J values for B and F and L, S, and J values for the ground states of C, N, and O, considered in the next subsection, there is little more to learn at the qualitative level that we are employing in our analysis. This should be clear from Fig. 18.6, in which the configurations for four of the atoms corresponding to $n = 3$ are shown pictorially. The atomic magic number for Ar is 18, the total number of electrons forming this all-closed-shell atom.

So far the shell-filling procedure has been somewhat mechanical: we have followed the rules in which the lowest $n + \ell$ levels are filled first, and, for equal $n + \ell$ values, one begins with the lowest-n level. However, for some of the atoms heavier than Ar one finds deviations from the preceding sequencing rules. They occur first among the 4s – 3d – 4p group of levels and are a consequence of the very small difference between $\mathcal{E}_{4s}$ and $\mathcal{E}_{3d}$ for

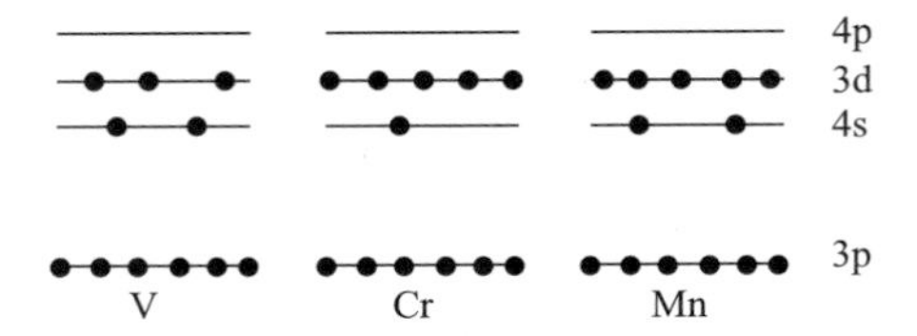

Fig. 18.7 Anomalous shell filling in the 4s–3d shells. The deviation occurs in Cr, for which a 4s electron has been promoted to the 3d level. The atoms V and Mn, which precede and follow Cr, display the normal shell-filling sequencing. The filled 1s, 2s, 2p, and 3s levels are not shown.

this group. Since the more accurate Hamiltonian is that of Eq. (18.77a), one needs to take account of the perturbation term (18.77b) in deciding among competing configurations and occasionally an out-of-sequencing-order configuration will be the dominant one.

The first five atoms of the 4s–3d–4p group, K through V, have the expected configurations. K, the third alkali atom, has the ground-state structure $\mathrm{Ar} + 4\mathrm{s}$; its ionization energy is relatively small: 4.34 eV. The 4s shell closes at Ca, and one then adds 3d electrons to form the ground states of the next three atoms: $\mathrm{Sc} = \mathrm{Ar} + (4\mathrm{s})^2 3\mathrm{d}$, $\mathrm{Ti} = \mathrm{Ar} + (4\mathrm{s})^2(3\mathrm{d})^2$ and $\mathrm{V} = \mathrm{Ar} + (4\mathrm{s})^2(3\mathrm{d})^3$. As noted earlier, with Cr the first shell-filling deviation occurs: its configuration is $\mathrm{Ar} + 4\mathrm{s}(3\mathrm{d})^5$ rather than $\mathrm{Ar} + (4\mathrm{s})^2(3\mathrm{d})^4$. There is scarcely a change in ionization energy between V and Cr: the value for each is roughly 6.75 eV. For Mn, the next atom after Cr, one returns to the normal sequencing: $\mathrm{Mn} = \mathrm{Ar} + (4\mathrm{s})^2(3\mathrm{d})^5$. Figure 18.7 illustrates the deviation by presenting occupations of the relevant levels in a Grotrian diagram for V, Cr, and Mn.

The normal sequencing is maintained up to Ni, whose configuration is $\mathrm{Ar} + (4\mathrm{s})^2(3\mathrm{d})^8$. The only other deviation in the 4s–3d–4p grouping – the fourth row of atoms (for row designations see the next subsection and Fig. 18.8 later) – then occurs at Cu, in which one of the two 4s electrons is promoted to the 3d shell, yielding the configuration $\mathrm{Cu} = \mathrm{Ar} + (4\mathrm{s})(3\mathrm{d})^{10}$. A return to the normal ordering occurs at Zn, for which both the 4s shell and the 3d shell are closed, and they remain so in the ground states of all the remaining atoms in the group. The filling of the 4p shell starts with Ga, proceeds without anomalies, and ends at Kr ($Z = 36$), which, as the fourth of the inert gases, has the largish ionization energy of 14 eV, and the atomic magic number 36.

The theoretical construction of the remaining atoms is carried out in the same manner as in the preceding discussion. About half a dozen shell-filling anomalies occur for each of the 5s, 6s and 7s groupings; these, like the ones already discussed, are identified in Table 18.7. The atomic magic numbers associated with the other two inert gases, viz., Xe and Rn, are 54 and 86, respectively, while their single-electron ionization energies are at least twice as large as those of each of the next following atoms: 12.13 eV for Xe and 10.75 eV for Rn. Other features may be gleaned by examining Table 18.7.

Angular momenta, term values

As the last topic in our qualitative study of atomic ground states, we consider angular momenta and term values. All the interactions in (18.77) are scalars, i.e., are invariant under rotations, so that, if $\hat{\mathbf{J}}$ is the total angular-momentum operator – to be defined shortly – then this invariance means that

$$[\hat{\mathbf{J}},\ \hat{H}_Z^{(\text{atom})}] = 0. \tag{18.80}$$

Hence, all eigenstates of $\hat{H}_Z^{(\text{atom})}$ and $\hat{H}_{\text{IP}}^{(\text{atom})}$ are also eigenstates of $\hat{J}^2$ and $\hat{J}_z$, with corresponding quantum numbers J and M_J, respectively.

By generalizing the analysis of Section 14.3, the total angular-momentum operator $\hat{\mathbf{J}}$ for an N-electron atom is the sum of all the individual orbital and spin-angular momenta:

$$\hat{\mathbf{J}} = \sum_k \hat{\mathbf{J}}_k, \tag{18.81a}$$

with $\hat{\mathbf{J}}_k = \hat{\mathbf{L}}_k + \hat{\mathbf{S}}_k$, so that (18.81a) becomes

$$\hat{\mathbf{J}} = \sum_k (\hat{\mathbf{L}}_k + \hat{\mathbf{S}}_k). \tag{18.81b}$$

Corresponding to this operator is an enormous number of ways to construct a state having the quantum numbers J and M_J, each associated with different couplings of the individual angular momenta. When the description is in terms of $\hat{H}_{\text{IP}}^{(\text{atom})}$, there is a standard procedure for accomplishing this. It is based on the absence from $\hat{H}_{\text{IP}}^{(\text{atom})}$ of spin–orbit interactions. This procedure, known as LS or Russell–Saunders coupling (Russell and Saunders (1925)), couples all the orbital angular momenta to form a total orbital angular-momentum operator $\hat{\mathbf{L}}$, couples all the spins to form a total spin angular-momentum operator $\hat{\mathbf{S}}$, and then couples $\hat{\mathbf{L}}$ and $\hat{\mathbf{S}}$ to form $\hat{\mathbf{J}}$:

$$\hat{\mathbf{L}} = \sum_{k=1}^{N} \hat{\mathbf{L}}_k, \tag{18.82a}$$

$$\hat{\mathbf{S}} = \sum_{k=1}^{N} \hat{\mathbf{S}}_k, \tag{18.82b}$$

and

$$\hat{\mathbf{J}} = \hat{\mathbf{L}} + \hat{\mathbf{S}}. \tag{18.82c}$$

The paired eigenvalues corresponding to these operators are $(L,\ M_L)$, $(S,\ M_S)$, and $(J,\ M_J)$.

To illustrate this method, suppose that the individual orbital angular momenta are $\ell_1, \ell_2 \ldots, \ell_N$. Then, by first coupling $\hat{\mathbf{L}}_1$ and $\hat{\mathbf{L}}_2$ to form $\hat{\mathbf{L}}_{12}$, with quantum number ℓ_{12}, one finds $|\ell_1 - \ell_2| \leqslant \ell_{12} \leqslant \ell_1 + \ell_2$. Next, one could couple $\hat{\mathbf{L}}_{12}$ to $\hat{\mathbf{L}}_3$ to form $\hat{\mathbf{L}}_{123}$, such that its quantum number ℓ_{123} takes on the values in the range $|\ell_{12} - \ell_3| \leqslant \ell_{123} \leqslant \ell_{12} + \ell_3$. This process is continued until a final quantum number L is obtained, where $\hat{L}^2 \Rightarrow L(L+1)\hbar^2$. An analogous procedure holds for spin: for $n = 2M$, we find $0 \leqslant S \leqslant M$, whereas for $N = 2M + 1$, we get $\frac{1}{2} \leqslant S \leqslant M + \frac{1}{2}$. As a specific example, for $N = 3$ and $\ell_1 = \ell_2 = \ell_3 = \ell$, one finds $0 \leqslant L \leqslant 3\ell$, whereas the allowed values of S are $S = \frac{1}{2}$ and $S = \frac{3}{2}$. As noted, there is usually a large variety of ways of achieving these results. For instance, when $N = 3$, the state with $S = \frac{1}{2}$ can be obtained by coupling $\frac{1}{2}$ to either $S_{12} = 0$ or $S_{12} = 1$. We explore some of this variety and structure in the exercises.

For economy of notation one often labels an N-particle state in terms of its L, S, and J values plus a collective index, say α: $|\alpha,\ LSJ\rangle$. It can be referred to using the generalized spectroscopic-term-value notation ${}^{2S+1}L_J$, where capital letters are employed for the numerical values of L, as in Table 18.1. States with the same L and S but differing J are said to be members of the same LS *multiplet*.

Determining J values for a particular state can be non-trivial, and it is in this context that all-closed-shell states play such an essential role. For N identical electrons governed by $\hat{H}_{\text{IP}}^{(\text{atom})}$, an all-closed-shell determinant $|\Gamma_{\text{cs}}\rangle^{\text{A}}$ has the structure

$$|\Gamma_{\text{cs}}\rangle^{\text{A}} = (N!)^{-1/2} \det\left| \prod_{j=1}^{f} [|n_j\ell_j\ell_j\tfrac{1}{2}\rangle|n_j\ell_j\ell_j, -\tfrac{1}{2}\rangle|n_j\ell_j, \ell_j - 1, \tfrac{1}{2}\rangle \right.$$
$$\left. \times\ |n_j\ell_j, \ell_j - 1, -\tfrac{1}{2}\rangle \cdots |n_j\ell_j, -\ell_j, \tfrac{1}{2}\rangle|n_j\ell_j, -\ell_j, -\tfrac{1}{2}\rangle] \right|, \quad (18.83)$$

where the usual particle-label subscript has been suppressed in each single-particle state $|n_j\ell_j m_{\ell_j} m_{s_j}\rangle$. Because all allowed m_{ℓ_j} values occur for each ℓ_j, applying $\hat{L}_z$ (the z-component from Eq. (18.82a)) to (18.83) yields zero:

$$\hat{L}_z|\Gamma_{\text{cs}}\rangle^{\text{A}} = 0. \quad (18.84a)$$

Similarly,

$$\hat{S}_z|\Gamma_{\text{cs}}\rangle^{\text{A}} = 0, \quad (18.84b)$$

from which we conclude that

$$\hat{J}_z|\Gamma_{\text{cs}}\rangle^{\text{A}} = (\hat{L}_z + \hat{S}_z)|\Gamma_{\text{cs}}\rangle^{\text{A}} = 0. \quad (18.84c)$$

That is, for a pure closed-shell state, $M_{\text{cs}} = 0$, where M_{cs} is the eigenvalue of $\hat{J}_z$.

The foregoing is a nice result. A wonderful result is that the total angular momentum J_{cs} of $|\Gamma_{\text{cs}}\rangle^{\text{A}}$ is also zero! To prove this, we assume that $J_{\text{cs}} \neq 0$, and write $|\Gamma_{\text{cs}}\rangle^{\text{A}}$ as $|\Gamma_{\text{cs}}, J_{\text{cs}}, M_{\text{cs}} = 0\rangle^{\text{A}}$. We then rotate the N-electron atom about the direction $\mathbf{n}$ through an angle θ. As a consequence, $|\Gamma_{\text{cs}}, J_{\text{cs}}, M_{\text{cs}} = 0\rangle^{\text{A}}$ goes over to $|\Gamma_{\text{cs}}, J_{\text{cs}}, M_{\text{cs}} = 0\rangle_{\text{R}}^{\text{A}}$, given by

$$|\Gamma_{\text{cs}}, J_{\text{cs}}, M_{\text{cs}} = 0\rangle_{\text{R}}^{\text{A}} = e^{-i\theta\mathbf{n}\cdot\hat{\mathbf{J}}/\hbar}|\Gamma_{\text{cs}}, J_{\text{cs}}, M_{\text{cs}} = 0\rangle^{\text{A}}. \quad (18.85)$$

Next, we introduce the complete set of angular-momentum states $\{|JM_J\rangle\}$, by which the resolution of the identity is

$$\hat{I} = \sum_{J,M_J} |JM_J\rangle\langle JM_J|.$$

Inserting this expression to the left of the operator $\exp(-i\theta\mathbf{n}\cdot\hat{\mathbf{J}}/\hbar)$ in (18.85), and recalling that $\hat{\mathbf{J}}$ is diagonal in the quantum number J, the rotated state becomes

$$|\Gamma_{\text{cs}}, J_{\text{cs}}, M_{\text{cs}} = 0\rangle_{\text{R}}^{\text{A}} = \sum_{M_J} |J_{\text{cs}}M_J\rangle\langle J_{\text{cs}}M_J|e^{-i\theta\mathbf{n}\cdot\hat{\mathbf{J}}/\hbar}|\Gamma_{\text{cs}}, J_{\text{cs}}, M_{\text{cs}} = 0\rangle^{\text{A}}. \quad (18.86)$$

In other words, the rotated state is a weighted sum over all the M_J values associated with J_{cs}, which is assumed to be non-zero. However, inspection of Eq. (18.83) shows that $|\Gamma_{\text{cs}}\rangle^{\text{A}}$ is *unique* and therefore it cannot be changed under any rotation. Hence, the sum on M_J in (18.86) must collapse to one term, which is possible only if $J_{\text{cs}} = 0$, since only then can there be a unique, single-magnetic-quantum-number state. As a corollary, it is clear that we must also have $L = 0$ and $S = 0$ for $|\Gamma_{cs}\rangle^A$. Writing $|\Gamma_{cs}\rangle^A$ to display the *LSJ* quantum numbers, we have

$$|\Gamma_{cs}\rangle^A \rightarrow |\Gamma_{cs};\, L_{cs} = 0,\, S_{cs} = 0,\, J_{cs} = 0\rangle^A.$$

As noted, the foregoing result plays a key role in shell-model analyses. By application of the *Aufbau* principle, the ground state of an $(N+1)$-electron atom is obtained by *retaining* the shell structure of the N-electron atom and adding to it an electron in whichever shell leads to the lowest energy (consistent with the exclusion principle). For $N = 2\sum_j (2\ell_j + 1)$, the atom is an all-closed-shell atom and therefore the succeeding $(N+1)$-electron atom is formed in its ground state by adding an electron in a single-particle state, say $|n\ell m_\ell m_S\rangle$, to the preceding closed-shell configuration. The angular-momentum properties of such a closed-shell-plus-one-electron state are just those of the added single-particle state, viz., $L = \ell$, $S = \frac{1}{2}$, and $J = \ell \pm \frac{1}{2}$. (As an example, recall the discussion below Eq. (13.6) concerning the fact that, in its ground state, Ag behaves like a single spin-$\frac{1}{2}$ particle.) The closed shells in this instance behave like a structureless core w.r.t. angular-momentum properties. Similarly, by going from an all-closed-shell atom with N electrons to the ground state of an atom with $N+2$ electrons, each in single-particle states with orbital angular momentum ℓ, the angular-momentum properties are determined solely by the two additional single-particle states, or, as one loosely says, by the two "outermost" or "valence" electrons. In this particular case of $N+2$ electrons, the allowed range of L values is $0 \leqslant L \leqslant 2\ell$, while $S = 0$ or 1, and $J = 0$, 1, 2, ..., $2\ell + 1$.

The preceding type of analysis is used until a shell is half filled, whereupon it is advantageous to think in terms of the unfilled portion of the shells, referred to as "holes" in a filled shell. The angular-momentum properties of a one-hole system are the same as if the hole, represented by an absence of a particle in a state $|n\ell m_\ell m_S\rangle$, were replaced by a closed shell plus a particle in the state $|n\ell m_\ell m_S\rangle$. The same is true for two holes, etc. Thus, an atom with all closed shells plus the one-hole configuration $(n_j\ell_j)^{4\ell_j+1}$ has $L = \ell_j$, $S = \frac{1}{2}$, and $J = \ell_j \pm \frac{1}{2}$, and so forth for two or more holes.

Although $J_{cs} = 0$ leads to great simplifications, determination of L, S, and J for a non-closed-shell atom is far from a simple problem. Only for the configuration $(n\mathrm{s})^1$ is the corresponding term value $^2S_{1/2}$ obvious. Even the case of a single non-s electron outside a closed shell is ambiguous: $(n\ell)^1$ allows both of $J = \ell \pm \frac{1}{2}$, and without further information there is no means for deciding which is the ground-state J value. The situation becomes increasingly more complicated as the number of open-shell particles increases, since there are then many values of L, S, and J that can be constructed. However, there is a further, probably unexpected, complication: not all values of S, L, and J permitted by angular-momentum coupling are allowed by the Pauli principle! One must not simply crank out angular momenta. A famous example is the $(n\mathrm{p})^3$ configuration, which characterizes the ground states of N, P, As, Sb, and Bi. Clebsch–Gordanry leads to the possibilities $S = \frac{1}{2}$ and $\frac{3}{2}$; $L = 0$, 1, 2, and 3; and $\frac{1}{2} \leqslant J \leqslant \frac{9}{2}$. However, the set of term values $\{^{2S+1}F_J\}$ cannot occur. The reason is that, when $L = 3$, the three p electrons are in a spatial state that is symmetric under interchange of each of the three pairs of particle labels. To satisfy the Pauli principle, the spin state of the three p electrons must be antisymmetric. However, one *cannot* construct a fully antisymmetric spin state for three spin-$\frac{1}{2}$ particles and hence states with $^{2S+1}F_J$ are thus forbidden.

Since the $(n\mathrm{p})^3$ example is typical, we consider it further. It turns out that additional restrictions, analogous to that of the $L = 3$ case, occur and one ends up with the following as the only permitted term values: ${}^4S_{3/2}$, ${}^2P_J$, and ${}^2D_J$. To determine which one of these is the ground-state term value requires a further analysis, for instance one involving $\hat{H}_1^{(\mathrm{atom})}$ of (18.77b). The $\hat{\mathbf{L}} \cdot \hat{\mathbf{S}}$ term will split ${}^2P_J$ into ${}^2P_{3/2}$ and ${}^2P_{1/2}$ levels while ${}^2D_J$ goes over to ${}^2D_{5/2}$ and ${}^2D_{3/2}$ levels, but this is still not sufficient to solve the ground-state term-value problem: one needs either detailed calculations or, in the absence of such, empirical rules.

Empirical rules for assigning ground-state quantum numbers in the case of L–S coupling do exist. They were formulated by Hund (1927), and are the following.

1a. The L–S multiplet with the largest value of S lies lowest.

1b. If the largest S is associated with several L values, the state with the largest L lies lowest.

2. In atoms with a single unfilled shell, the ground-state value of J arising from the spin–orbit splitting of an L–S multiplet is determined as follows.
   (a) If the shell is no more than half filled, then $J = |L - S|$, the minimum value.
   (b) If the shell is more than half filled, then $J = L + S$, the maximum value.

Hund's rules work quite well for atoms (and also for molecular ground states); they are, however, not successful for most excited states. The following physical arguments underlie them. First, recall that, in the exchange-matrix element (18.71b), the spins are aligned, leading to $S = 1$, so there is a tendency to have spin-aligned pairs. If the aligned spins add together to form the maximum allowed S, then the resulting spin state is the most symmetric, and therefore the spatial state is the most antisymmetric. However, the greater the degree of antisymmetry the less likely is it that any two electrons will be close together, in which case the positive contribution of the exchange term (involving $1/r_{jk}$) is minimized. For Rule 1(b), a similar argument, based on the maximum value of all the possible M_L's plus the proximity of electrons to $\hat{\mathbf{e}}_z$ and thus each other, leads to minimization of the exchange term for the maximum value of L.

The second rule can be understood as arising from the following three properties. First, for a single electron outside a closed shell, the matrix elements $\langle V_{LS} \rangle_{n\ell}$ are positive, so that the coupled state with $j = \ell - \frac{1}{2}$ lies lower than the one with $j = \ell + \frac{1}{2}$ (recall Eq. (14.85) and the comments below it). Second, holes in filled shells behave like particles with positive charge, so that $\langle V_{LS} \rangle_{n\ell}$ for a single hole is negative, leading to hole states with $j = \ell + \frac{1}{2}$ lying lower than those with $j = \ell - \frac{1}{2}$. Third, shells that are no more than half filled are particle-like, whereas those that are more than half filled are hole-like.

The ground-state term values for most of the atoms are correctly obtained from Hund's rules, as long as the L, S, and J values considered are consistent with the Pauli principle. Thus, the ground states of all atoms which are closed-shell atoms plus or minus a single $n\ell$ electron have the values of J predicted by the rules. The rules also predict correctly the term values for the $(n\mathrm{p})^3$ and $(n\mathrm{p})^4$ configurations: $(n\mathrm{p})^3 \Rightarrow {}^4S_{3/2}$, while $(n\mathrm{p})^4 \Rightarrow {}^3P_2$. When a configuration contains two or more open shells, these empirical rules no longer suffice and detailed analysis is called for. Such is beyond the scope of this book; the interested reader is referred to monographs such as those by Condon and Shortley (1963), Mizushima (1970), and Sobelman (1979) among others. Some qualitative and

	I	II											III	IV	V	VI	VII	
1	H^{1} $^{2}S_{1/2}$																	He^{2} $^{1}S_{0}$
2	Li^{3} $^{2}S_{1/2}$	Be^{4} $^{1}S_{0}$											B^{5} $^{2}P_{1/2}$	C^{6} $^{3}P_{0}$	N^{7} $^{4}S_{3/2}$	O^{8} $^{3}P_{2}$	F^{9} $^{2}P_{3/2}$	Ne^{10} $^{1}S_{0}$
3	Na^{11} $^{2}S_{1/2}$	Mg^{12} $^{1}S_{0}$											Al^{13} $^{2}P_{1/2}$	Si^{14} $^{3}P_{0}$	P^{15} $^{4}S_{3/2}$	S^{16} $^{3}P_{2}$	Cl^{17} $^{2}P_{3/2}$	A^{18} $^{1}S_{0}$
4	K^{19} $^{2}S_{1/2}$	Ca^{20} $^{1}S_{0}$	Sc^{21} $^{2}D_{3/2}$	Ti^{22} $^{3}F_{2}$	V^{23} $^{4}F_{3/2}$	Cr^{24} $^{7}S_{3}$	Mn^{25} $^{6}S_{5/2}$	Fe^{26} $^{5}D_{4}$	Co^{27} $^{4}F_{9/2}$	Ni^{28} $^{3}F_{4}$	Cu^{29} $^{2}S_{1/2}$	Zn^{30} $^{1}S_{0}$	Ga^{31} $^{2}P_{1/2}$	Ge^{32} $^{3}P_{0}$	As^{33} $^{4}S_{3/2}$	Se^{34} $^{3}P_{2}$	Br^{35} $^{2}P_{3/2}$	K^{36} $^{1}S_{0}$
5	Rb^{37} $^{2}S_{1/2}$	Sr^{38} $^{1}S_{0}$	Y^{39} $^{2}D_{3/2}$	Zr^{40} $^{3}F_{2}$	Nb^{41} $^{6}D_{1/2}$	Mo^{42} $^{7}S_{3}$	Tc^{43} $^{6}S_{5/2}$	Ru^{44} $^{5}F_{5}$	Rh^{45} $^{4}F_{9/2}$	Pd^{46} $^{1}S_{0}$	Ag^{47} $^{2}S_{1/2}$	Cd^{48} $^{1}S_{0}$	In^{49} $^{2}P_{1/2}$	Sn^{50} $^{3}P_{0}$	Sb^{51} $^{4}S_{3/2}$	Te^{52} $^{3}P_{2}$	I^{53} $^{2}P_{3/2}$	Xe^{54} $^{1}S_{0}$
6	Cs^{55} $^{2}S_{1/2}$	Ba^{56} $^{1}S_{0}$	La^{57} $^{2}D_{3/2}$	Hf^{72} $^{3}F_{2}$	Ta^{73} $^{4}F_{3/2}$	W^{74} $^{5}D_{0}$	Re^{75} $^{6}S_{5/2}$	Os^{76} $^{5}D_{4}$	Ir^{77} $^{4}F_{9/2}$	Pt^{78} $^{3}D_{3}$	Au^{79} $^{2}S_{1/2}$	Hg^{80} $^{1}S_{0}$	Ti^{81} $^{2}P_{1/2}$	Pb^{82} $^{3}P_{0}$	Bi^{83} $^{4}S_{3/2}$	Po^{84} $^{3}P_{2}$	At^{85} $^{2}P_{3/2}$	Rn^{86} $^{1}S_{0}$
7	Fr^{87} $^{2}S_{1/2}$	Ra^{88} $^{1}S_{0}$	Ac^{89} $^{2}D_{3/2}$															
Lanthanides (Rare Earths)			La^{57} $^{2}D_{3/2}$	Ce^{58} $^{1}G_{4}$	Pr^{59} $^{4}I_{9/2}$	Nd^{60} $^{5}I_{4}$	Pm^{61} $^{6}H_{5/2}$	Sm^{62} $^{7}F_{0}$	Eu^{63} $^{8}S_{7/2}$	Gd^{64} $^{9}D_{2}$	Tb^{65} $^{6}H_{15/2}$	Dy^{66} $^{5}I_{8}$	Ho^{67} $^{4}I_{15/2}$	Er^{68} $^{3}H_{6}$	Tm^{69} $^{2}F_{7/2}$	Yb^{70} $^{1}S_{0}$	Lu^{71} $^{2}D_{3/2}$	
Actinides			Ac^{89} $^{2}D_{3/2}$	Th^{90} $^{3}F_{2}$	Pa^{91} $J=11/2$	U^{92} $J=6$	Np^{93} $J=11/2$	Pu^{94} $^{7}F_{0}$	Am^{95} $^{8}S_{7/2}$	Cm^{96} $^{9}D_{2}$	Bk^{97} $^{6}H_{15/2}$	Cf^{98} $^{5}I_{8}$	E^{99} $^{4}I_{15/2}$	Fm^{100} $^{3}H_{6}$	Md^{101} $^{2}F_{7/2}$	No^{102} $^{1}S_{0}$	Lr^{103} $^{2}P_{1/2(?)}$	Rf^{104} $^{3}F_{2(?)}$

Fig. 18.8 The periodic table of the elements. Each box contains the chemical symbol, the atomic number as a superscript, and the ground-state spectroscopic term value where known (or the spin).

semi-quantitative discussions can be found, for instance, in the books by Sobelman (1979), Baym (1976), and Galindo and Pascual (1990).

The collections of term values and configurations for the atoms H through Rf are presented in Fig. 18.8, which is the traditional form of the periodic table of the elements and specifies the traditional row and group designations. The row designations $n = 1, 2, \ldots, 7$ correspond to the ns beginning of each of the seven groupings in the sequence (18.78). The order of shell filling is specified by the sequencing and configurations in each row, with the inert gases being the last elements of rows. Atoms whose outermost electrons belong to s, to s^2, or to p^N ($N \leqslant 6$) are the column elements of each group, with He being included with the inert p^6 atoms of group 0. The sets of transition elements, which include the lanthanide and actinide series, and which tend to have similar chemical properties, occur in rows 6 and 7 and lie between groups II and III.

18.5. Elements of molecular structure

Molecules consist of two or more atoms bound together in a particle-stable configuration. Probably the most important qualitative feature of any molecule is the non-zero separation between its positive charge centers: the nuclei of the atoms which form the molecule do not coalesce. Molecules are thus multi-centered systems of electrons, in contrast to atoms, which are single-centered. The existence of more than one positive charge center means, as we shall see later, that, in addition to the electronic component, molecular spectra also display effects arising from nuclear rotations and vibrations.

This spectral variety means that molecules are rather more challenging to understand quantally than are their individual atomic consituents. For instance, the non-zero separation between the positive charge centers underlies the chemical-symbol notation for molecules (CO, H_2O, etc.). However, the mere fact of one's writing such a symbol is no guarantee that an accurate quantal description of a molecular ground state will involve just the ground states of the atoms from which it is formed, even though the chemical-symbol notation suggests it. Indeed, one should expect that the molecular environment will alter the wave functions and thus the electronic charge distributions of the isolated atomic constituents, perhaps unrecognizably. In fact, in the transition era of the 1960s, when computer-based computations were replacing hand calculations, it was sometimes claimed that there were no atoms in molecules, only electrons and nuclei. The credibility of this claim depends in part on the degree of accuracy one is willing to accept as satisfactory: the more accurate the results of approximate calculations the less the electronic parts of the ground-state molecular wave functions resemble combinations of isolated atomic ground-state wave functions. Nonetheless, it is amusing, to say the least, that the measured electronic charge distributions of molecules are often close to a superposition of the isolated-atom charge distributions, i.e., when the latter are simply placed over the former, the agreement can be very good. In this regard, see, e.g., Steiner (1976).

Molecular physics, like atomic or nuclear physics, is a field unto itself, and in this section we shall provide only a bare introduction to the complex subject of molecular structure. Our aim is to show, through the medium of diatomic molecules, how the challenge of understanding molecular structure is met by quantum theory. Our treatment will be variational, and the framework in which it will be applied is that of the Born–

Oppenheimer (B–O) approximation, wherein the motion of the much heavier nuclei is assumed to be very slow relative to the electronic motion. As a consequence of this assumption, the electronic and nuclear motions become decoupled, leading to a separate Schrödinger equation for the electrons, one in which the nuclear positions enter only as parameters. The derivation of the B–O equation is the subject of the first subsection. We then examine electronic-structure concepts via the one-electron ion H_2^+, which is the molecular equivalent of the H-atom. In the concluding subsection we consider vibrations and rotations of the two nuclei in a diatomic molecule.

The Born–Oppenheimer approximation

The generic diatomic molecule consists of two nuclei and a total of N electrons; it is sketched in Fig. 18.9, in which the labels are a and b for the nuclei, and 1, 2, ..., N for the electrons, with corresponding position vectors $\mathbf{r}_a$, $\mathbf{r}_b$, $\mathbf{r}_1$, $\mathbf{r}_2$, ..., $\mathbf{r}_N$. The charges and masses of the nuclei are denoted M_a, M_b, $Z_a e$, and $Z_b e$ in a self-evident notation. In terms of these quantities, the spin-independent, non-relativistic Hamiltonian governing the system is

$$\begin{aligned}\hat{H}(\mathbf{r}_1, \mathbf{r}_2, \ldots, \mathbf{r}_N, \mathbf{R}_a, \mathbf{R}_b) = &-\frac{\hbar^2}{2M_a}\nabla_a^2 - \frac{\hbar^2}{2M_b}\nabla_b^2 - \frac{\hbar^2}{2m_e}\sum_{j=1}^{N}\nabla_j^2 \\ &+ \frac{\kappa_e Z_a Z_b e^2}{R} + \kappa_e e^2 \sum_{j\neq k}\frac{1}{r_{jk}} \\ &- \kappa_e e^2 \sum_{j=1}^{N}\frac{1}{|\mathbf{R}_a - \mathbf{r}_j|} - \kappa_e e^2 \sum_{j=1}^{N}\frac{1}{|\mathbf{R}_b - \mathbf{r}_j|},\end{aligned} \tag{18.87}$$

where $R = |\mathbf{R}_a - \mathbf{R}_b|$ and $r_{jk} = |\mathbf{r}_j - \mathbf{r}_k|$.

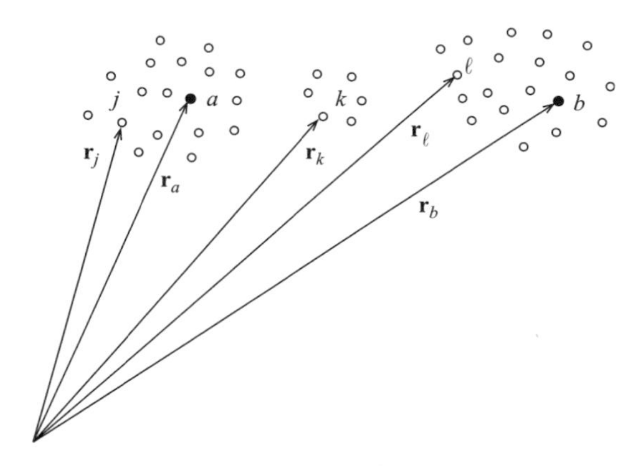

Fig. 18.9 A representation of the generic diatomic molecule. The black circles indicate the two nuclei, labeled a and b; their position vectors are $\mathbf{r}_a$ and $\mathbf{r}_b$, respectively. The small open circles signify the electrons, not all of which need be close to one nucleus or the other. Electrons are labeled 1, 2, ... etc.; three of them, viz., j, k, and l, have their position vectors $\mathbf{r}_j$, $\mathbf{r}_k$, and $\mathbf{r}_\ell$ specified.

To gain some insight into this Hamiltonian and prepare the ground for the B–O approximation (Born and Oppenheimer 1927), we shall argue qualitatively, using empirical information. For example, because the molecule ab exists, the average value of the nuclear separation R must be non-zero (otherwise the system would be an atom with nuclear charge $(Z_a + Z_b)e$). Furthermore, the fact that the binding energies of diatomic molecules are of the same order of magnitude as the binding energies of electrons in atoms, namely, electron-volts, suggests not only that the electrons somehow act as the "glue" which binds the two positively charged nuclei together but also that, in general, the motion of the much heavier nuclei should not drastically influence this binding. Finally, we note that the "sizes" of atoms and molecules are of the same order of magnitude: a few ångström units $\sim a_0$ (the Bohr radius).

Since the binding energy and size of molecules are atomic in character, the appropriate order of magnitude for energy is $\hbar^2/(m_e a_0^2)$ rather than $\hbar^2/(M_a a_0^2)$, consistent with our expectation that nuclear motion is not a crucial factor in determining molecular binding energies. However, as the following qualitative analysis indicates, there should be excitations of the molecule that involve the nuclei. For very small R, the nuclei should be repelled by the Coulomb interaction $\kappa_e Z_a Z_b e^2/R$. For neutral atoms, this interaction should go to zero for asymptotic values of R. That a and b are bound and maintain a non-zero average separation suggests that the electronic "glue" plus the Coulomb repulsion combine to yield a potential-energy function, say $V(R)$, governing the behavior of the nuclei. At small R it should be repulsive, at large R it should become a constant or zero, and at intermediate R it must be attractive in order to provide both binding and an average nuclear separation.

The simplest potential having the general features just described is sketched in Fig. 18.10; the attractive portion has a minimum value $-V_0$ at the average separation R_{eq}, where V_0 will be of the order of the a–b binding energy, viz., $\hbar^2/(m_e a_0^2)$. One way of trying to determine whether this potential can support excitations is by asking for the frequencies at which the two nuclei can oscillate about R_{eq}. To estimate this, we expand $V(R_{eq})$ about the equilibrium or average position, keeping terms quadratic in the variable $x = R - R_{eq}$, i.e., we make the small-oscillation approximation:

$$V(R) \simeq -V_0 + \frac{1}{2}x^2 \left.\frac{\partial^2 V(R)}{\partial R^2}\right|_{R_{eq}}, \tag{18.88}$$

where $[\partial V(R)/\partial R]_{R_{eq}} = 0$.

The x dependence, of course, is that of a harmonic oscillator with angular frequency

$$\omega = \left(\frac{1}{M}\left.\frac{\partial^2 V(R)}{\partial R^2}\right|_{R_{eq}}\right)^{1/2}, \tag{18.89}$$

where M is a nuclear mass (M_a, M_b, or their reduced mass). A crude estimate of the second derivative is

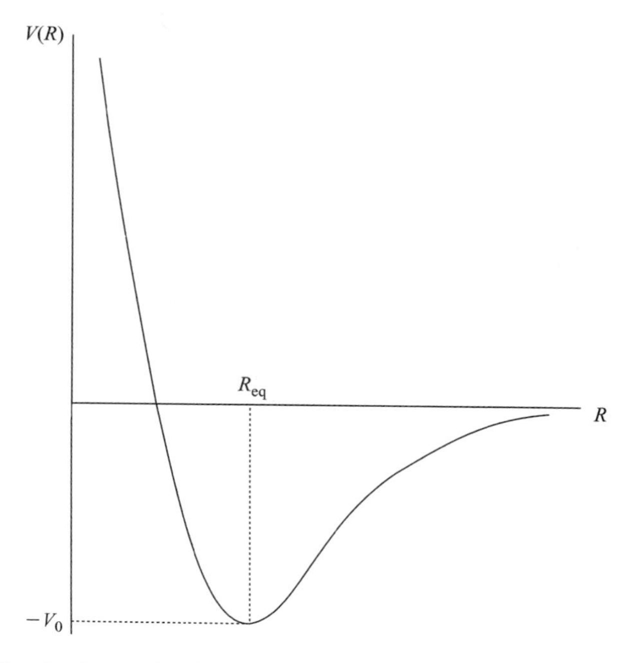

Fig. 18.10 The simplest molecular potential-energy function: its features are a short-range repulsion ($\propto 1/R$), an intermediate-range attraction, and a constant or zero value at asymptotic distances. The average or equilibrium separation $R_{\rm eq}$ occurs at the position of the (single) minimum in the intermediate-range, attractive portion.

$$\left.\frac{\partial^2 V(R)}{\partial R^2}\right|_{R_{\rm eq}} \simeq \frac{V_0}{R_{\rm eq}^2} \simeq \frac{\hbar^2}{m_{\rm e} a_0^4}, \tag{18.90a}$$

which leads to

$$\omega \simeq \left(\frac{m_{\rm e}}{M}\right)^{1/2} \frac{\hbar}{m_{\rm e} a_0^2}, \tag{18.90b}$$

and thus to

$$\hbar\omega \simeq \left(\frac{m_{\rm e}}{M}\right)^{1/2} \frac{\hbar^2}{m_{\rm e} a_0^2}. \tag{18.91}$$

Thus, the nuclear vibrational energy in a diatomic molecule is estimated to be of the order of 10^{-2} times the electronic energies, or about 0.01–0.1 eV. We may draw two conclusions from this estimate. First, small oscillations of the nuclei are allowed, since they do not lead to dissociation. Second, the vibrational energy is small enough to justify qualitatively the decoupling of the electronic and nuclear (vibrational) motions.

If the nuclei can oscillate about their average positions, semi-classical thinking suggests that they could also rotate about their CM. The Hamiltonian for rotational motion with a fixed separation R_{eq} can be approximated as $[-1/(2MR_{eq}^2)]\hat{L}^2$, and the energy unit corresponding to this is

$$E^{(Rot)} \sim \frac{\hbar^2}{2MR_{eq}^2} \simeq \frac{\hbar^2}{Ma_0^2} = \left(\frac{m_e}{M}\right)\frac{\hbar^2}{m_e a_0^2}, \tag{18.92}$$

whose values are in the range 0.01 to 10^{-4} eV. As with the (low-lying) vibrational excitations, we find that the relatively small magnitudes of the energies of rotational excitations (for ℓ not excessively large) continue to justify the decoupling of the electronic and nuclear motions.

Let us now consider the quantitative expression of these ideas. We choose an overall CM coordinate system, whose origin, because $m_e \ll M_a$ and M_b, is taken to be at the CM of a and b. Then the analysis of Section 11.2 transforms Eq. (18.87) into

$$\hat{H}(\{\mathbf{r}_j\}, \mathbf{R}_a, \mathbf{R}_b) \to \hat{H}(\{\mathbf{r}_j\}, \mathbf{R}) = -\frac{\hbar^2}{2\mu_{ab}}\nabla_R^2 - \frac{\hbar^2}{2m_e}\sum_{j=1}^{N}\nabla_j^2 + \frac{\kappa_e Z_a Z_b e^2}{R} + \kappa_e e^2 \sum_{j\neq k}\frac{1}{r_{jk}}$$

$$- \kappa_e e^2 \sum_{j=1}^{N}\left(\frac{1}{|\mathbf{R}_a - \mathbf{r}_j|} + \frac{1}{|\mathbf{R}_b - \mathbf{r}_j|}\right), \tag{18.93}$$

where $\mu_{ab} = M_a M_b/(M_a + M_b)$.

The time-independent Schrödinger equation governing the diatomic molecule is

$$\hat{H}(\{\mathbf{r}_j\}, \mathbf{R})\psi(\{\mathbf{r}_j\}, \mathbf{R}) = E\psi(\{\mathbf{r}_j\}, \mathbf{R}), \tag{18.94}$$

and the decoupling of the electronic and nuclear motions that underlies the B–O approximation resides in the form assumed for $\psi(\{\mathbf{r}_j\}, \mathbf{R})$. It is first written as a product,

$$\psi(\{\mathbf{r}_j\}, \mathbf{R}) = \psi^{(e)}(\{\mathbf{r}_j\}, \mathbf{R})\chi(\mathbf{R}), \tag{18.95}$$

with $\psi^{(e)}$ being a solution of

$$\hat{H}^{(e)}(\{\mathbf{r}_j\}, \mathbf{R})\psi^{(e)}(\{\mathbf{r}_j\}, \mathbf{R}) = E^{(e)}(\mathbf{R})\psi^{(e)}(\{\mathbf{r}_j\}, \mathbf{R}), \tag{18.96a}$$

where

$$\hat{H}^{(e)}(\{\mathbf{r}_j\}, \mathbf{R}) = \hat{H}(\{\mathbf{r}_j\}, \mathbf{R}) - \hat{K}_{ab}(\hat{\mathbf{R}}) \tag{18.96b}$$

and

$$\hat{K}_{ab}(\mathbf{R}) = -\frac{\hbar^2}{2\mu_{ab}}\nabla_{\mathbf{R}}^2 \tag{18.96c}$$

is the nuclear (relative-motion) kinetic-energy operator.

Removal of $\hat{K}_{ab}(\mathbf{R})$ from $\hat{H}$ means that $\mathbf{R}$ enters the "electronic" Hamiltonian $\hat{H}^{(e)}$ as a parameter rather than as a dynamical variable. Hence it is a parameter in $\psi^{(e)}$ and in the "electronic" energy $E^{(e)}$: $\mathbf{R}$ is a dynamical variable only in the equation for the nuclear motion. To find the latter equation we observe that the presence of the parameter $\mathbf{R}$ in $\hat{H}^{(e)}$ in no way alters its Hermitian character, and therefore (18.96a) is a standard quantal-type eigenvalue equation. It has a complete, orthonormal set of eigensolutions $\psi_\Gamma^{(e)}$ and corresponding eigenenergies $E^{(e)}$:

$$\sum_{\Gamma'} \psi^{(e)}_{\Gamma'}(\{\mathbf{r}_j\}, \mathbf{R}) \psi^{(e)\dagger}_{\Gamma'}(\{\mathbf{r}'_j\}, \mathbf{R}) = \prod_{j=1}^{N} \delta(\mathbf{r}_j - \mathbf{r}'_j) \tag{18.97a}$$

and

$$(\psi^{(e)}_{\Gamma}(\{\mathbf{r}_j\}, \mathbf{R}) | \psi^{(e)}_{\Gamma'}(\{\mathbf{r}_j\}, \mathbf{R})) = \delta_{\Gamma\Gamma'}. \tag{18.97b}$$

Notice that the same value of $\mathbf{R}$ (necessarily) appears in Eqs. (18.97), that there is no integration on $\mathbf{R}$ on the LHS of (18.97b), and that an integration on any continuum quantum numbers in $\sum_{\Gamma'}$ is suppressed, as usual.

With (18.97a), the single-product form of (18.95) goes over to a sum of products:

$$\psi(\{\mathbf{r}_j\}, \mathbf{R}) = \sum_{\Gamma} \psi^{(e)}_{\Gamma}(\{\mathbf{r}_j\}, \mathbf{R}) \chi_{\Gamma}(\mathbf{R}), \tag{18.98}$$

where $\chi_\Gamma(\mathbf{R})$ is the probability amplitude that a and b are at relative position $\mathbf{R}$ when the electronic state is $\psi^{(e)}_{\Gamma}$. Substituting Eq. (18.98) into the eigenvalue relation (18.94) and making use of Eqs. (18.96a) and (18.97b) when taking the scalar product with $\psi^{\dagger}_{\Gamma}(\{\mathbf{r}_j\}, \mathbf{R})$ yields the following set of coupled equations for the $\chi_\Gamma(\mathbf{R})$:

$$\sum_{\Gamma'} (\psi^{(e)}_{\Gamma}(\{\mathbf{r}_j\}, \mathbf{R}) | \hat{K}_{ab}(\mathbf{R}) | \psi^{(e)}_{\Gamma'}(\{\mathbf{r}_j\}, \mathbf{R}) \chi_{\Gamma'}(\mathbf{R})) + E^{(e)}_{\Gamma}(R) \chi_{\Gamma}(\mathbf{R}) = E_{\Gamma} \chi_{\Gamma}(\mathbf{R}), \tag{18.99}$$

where again there is no integration over $\mathbf{R}$ in the matrix element of $\hat{K}_{ab}(\mathbf{R})$.

The $\hat{K}_{ab}$ matrix element is readily seen to take the form

$$(\psi^{e}_{\Gamma} | \hat{K}_{ab} | \psi^{(e)}_{\Gamma'} \chi_{\Gamma}) = \delta_{\Gamma\Gamma'} \hat{K}_{ab}(\mathbf{R}) \chi_{\Gamma}(R) + M^{(e)}_{\Gamma\Gamma'}(R), \tag{18.100}$$

where the coupling term is

$$M^{(e)}_{\Gamma\Gamma'}(\mathbf{R}) = -\frac{\hbar^2}{\mu_{ab}} (\psi^{(e)}_{\Gamma}(\{\mathbf{r}_j\}, \mathbf{R}) | [\nabla_{\mathbf{R}} \psi^{(e)}_{\Gamma'}(\{\mathbf{r}_j\}, \mathbf{R})] \cdot [\nabla_{\mathbf{R}} \chi_{\Gamma'}(\mathbf{R})]$$
$$+ \; [(\psi^{(e)}_{\Gamma} | \hat{K}_{ab} | \psi^{(e)}_{\Gamma'})] | \chi_{\Gamma'}(\mathbf{R})), \tag{18.101}$$

and in the second term $\hat{K}_{ab}$ acts only on $\psi^{(e)}_{\Gamma'}$, not on $\chi_{\Gamma'}(\mathbf{R})$.

The B–O approximation consists of neglecting the coupling term $M^{(e)}_{\Gamma\Gamma'}(\mathbf{R})$, thereby reducing (18.99) to the simple Schrödinger form

$$[\hat{K}_{ab}(\mathbf{R}) + E^{(e)}_{\Gamma}(R)] \chi_{\Gamma}(\mathbf{R}) = E \chi_{\Gamma}(\mathbf{R}). \tag{18.102}$$

The B–O approximation leads to a one-way decoupling: the electronic motion is treated as independent of the nuclear motion, whereas the behavior of the nuclei depends on the particular electronic state through $E^{(e)}_{\Gamma}(R)$, which functions as the generic potential function $V(R)$ of Fig. 18.10.

Born and Oppenheimer (1927) argued that the corrections due to $M^{(e)}_{\Gamma\Gamma'}(\mathbf{R})$ could be treated perturbatively, with the small parameter being essentially the ratio $(m_e/\mu_{ab})^{1/4}$, whose size is typically 0.1 or smaller – see Merzbacher (1970) for a qualitative discussion. Others have argued that the small parameter should be the order of $(m_e/\mu_{ab})^{1/2} \sim 0.01$ or smaller, e.g., Baym (1976). Although theoretical studies on this topic as well as on ways to perform calculations without using the B–O approximation have intrigued chemists for years, this is a topic well beyond the scope of this book. The interested reader could consult, e.g., the end-of-the-chapter references in Atkins and Friedman (1997). The one theoretical discussion on the B–O approximation to which we do draw the reader's attention is a three-body-model study (Fonseca and Shanley (1979)) on the accuracy of B–O energy levels as a function of the ratio m/M, where m is the mass of the light particle and M is the mass of either of the two heavy particles. Although the model employs non-local, separable potentials rather than Coulomb interactions, and must therefore be treated with some caution (the interactions are S-wave and short-range), the results are non-trivial: they show that, for this model, the B–O approximation is valid for much larger mass ratios m/M than had previously been suspected.

The B–O approximation allows one to treat electronic structure separately from the nuclear dynamics. Our illustrative example of electronic structure is H_2^+.

The H_2^+ ion

The molecular equivalent of the H-atom is the H_2^+ ion, a three-body system for which numerically exact solutions can be obtained. It is thus an excellent testing ground for the study of a variety of molecular-physics concepts and approximations. We shall examine its electronic structure using simple approximations within the B–O picture.

Shown in Fig. 18.11 is a schematic representation of the ion, with a and b labeling the protons, $\mathbf{r}_a$ and $\mathbf{r}_b$ specifying the position vectors of the single electron from the protons, and $\mathbf{R}$ being the vector from a to b.

For this system, Eqs. (18.87) and (18.96b) reduce to

$$\hat{H}(\mathbf{r}_a, \mathbf{r}_b, \mathbf{R}) = \hat{K}_e + \hat{K}_{ab} - \frac{\kappa_e e^2}{r_a} - \frac{\kappa_e e^2}{r_b} + \frac{\kappa_e e^2}{R} \tag{18.103}$$

and

$$\hat{H}^{(e)}(\mathbf{r}_a, \mathbf{r}_b, \mathbf{R}) = \hat{K}_e - \kappa_e e^2 \left(\frac{1}{r_a} + \frac{1}{r_b} - \frac{1}{R} \right), \tag{18.104}$$

where $\hat{K}_e$ is the electron's kinetic-energy operator, which can be expressed using either $\nabla^2_{\mathbf{r}_a}$ or $\nabla^2_{\mathbf{r}_b}$. Since H_2^+ is a two-center system, there are two obvious electron coordinates – they are related by $\mathbf{r}_b = \mathbf{R} - \mathbf{r}_a$ – even though there is only one electron. Furthermore, our concentration on the electronic motion should not minimize the fact that, because the protons are identical spin-$\frac{1}{2}$ fermions and $\hat{H}$ contains no spin operators, the analysis of Section 18.1 also applies to H_2^+. We can therefore draw immediate conclusions concerning the proton portion of the H_2^+ eigenstates. Let $\hat{\mathbf{S}}_a$ and $\hat{\mathbf{S}}_b$ be the individual protons' spin operators, $\hat{\mathbf{S}}_{ab} = \hat{\mathbf{S}}_a + \hat{\mathbf{S}}_b$ be the total proton spin operator, and $\hat{P}_{ab}$ be the proton-

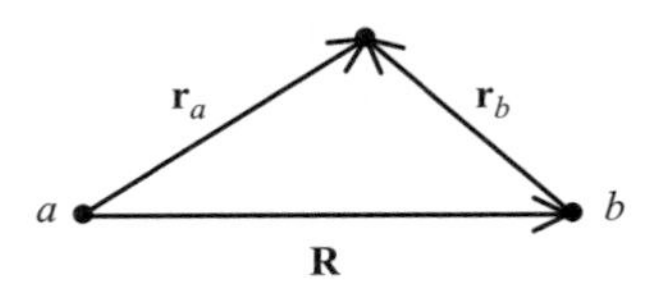

Fig. 18.11 Coordinates for the H_2^+ ion. The protons are labeled a and b while the single electron, represented by the heavy black dot, needs no label.

interchange operator. According to the spin-statistics theorem, the overall eigenstates of $\hat{H}$ must be antisymmetric in the labels a and b, and, because $[\hat{\mathbf{S}}_{\text{ab}}, \hat{H}] = 0 = [\hat{P}_{\text{ab}}, \hat{H}]$, these eigenstates can be written as products of spin and space states, just as in Section 18.2. Corresponding to the singlet (triplet) proton spin state will be a symmetric (antisymmetric) proton spatial state.

In view of these remarks, the H_2^+ eigenstate $|\Gamma_{H_2^+}\rangle$ has the following coordinate-space structure:

$$\langle \mathbf{r}_{\text{a}}, \mathbf{r}_{\text{b}}, \mathbf{R}|\Gamma_{H_2^+}\rangle = \psi_\Gamma^{(S_{\text{ab}})}(\mathbf{r}_{\text{a}}, \mathbf{r}_{\text{b}}, \mathbf{R})|\tfrac{1}{2}\tfrac{1}{2};\ S_{\text{ab}}M_{\text{ab}}\rangle_{\text{a,b}}|\tfrac{1}{2},\ m_S\rangle_{\text{e}}, \tag{18.105}$$

where $S_{\text{ab}} = 0$ or 1, $m_S = \pm\frac{1}{2}$, and

$$\hat{P}_{\text{ab}}\psi_\Gamma^{(S_{\text{ab}})}(\mathbf{r}_{\text{a}}, \mathbf{r}_{\text{b}}, \mathbf{R}) = (-1)^{S_{\text{ab}}}\psi_\Gamma^{(S_{\text{ab}})}(\mathbf{r}_{\text{a}}, \mathbf{r}_{\text{b}}, \mathbf{R}). \tag{18.106}$$

The eigenvalues $(-1)^{S_{\text{ab}}}$ are ± 1, and it is of some importance, especially for purposes of spectroscopic classification, that the action of $\hat{P}_{\text{ab}}$, i.e., of interchanging a and b, produces the same result as does a reflection of H_2^+ across a plane at the midpoint of and perpendicular to $\mathbf{R}$: see, e.g., Pilar (1968) or Atkins and Friedman (1997). Let us denote the operator for this type of reflection as $\hat{U}'_{\text{P}}$. From the structure of $\hat{H}$, it follows that $[\hat{U}'_{\text{P}}, \hat{H}] = 0$, so that eigenstates of $\hat{H}$ are simultaneous eigenstates of $\hat{U}'_{\text{P}}$. The eigenvalues of $\hat{U}'_{\text{P}}$ are $\pi' = \pm 1$: $\hat{U}'_{\text{P}}|\Gamma_{H_2^+}\rangle = \pm|\Gamma_{H_2^+}\rangle$, and comparison with Eq. (18.106) tells us that $\pi'_{S_{\text{ab}}} = (-1)^{S_{\text{ab}}}$. It is standard in chemical/molecular physics to denote states for which $\pi'_{S_{\text{ab}}} = +1$ as "*gerade*" states, whereas those for which $\pi'_{S_{\text{ab}}} = -1$ are referred to as "*ungerade*" states (the language is German, wherein *gerade* (*ungerade*) means "even" ("odd")). Hence, spin singlets (triplets) are *gerade* (*ungerade*); as we shall see using the simplest variational wave function in the B–O approximation, the molecular ground state is *gerade*.

For simplicity of notation we replace the superscript S_{ab} by the superscript "e." Numerically exact solutions of

$$[\hat{H}^{(\text{e})}(\mathbf{r}_{\text{a}}, \mathbf{r}_{\text{b}}, \mathbf{R}) - E_\Gamma^{(\text{e})}]\psi_\Gamma^{(\text{e})}(\mathbf{r}_{\text{a}}, \mathbf{r}_{\text{b}}, \mathbf{R}) = 0 \tag{18.107}$$

are most easily obtained using confocal elliptical coordinates μ, ν, and ϕ, where

$$\mu = \frac{r_{\text{a}} + r_{\text{b}}}{R}, \qquad 1 \leqslant \mu \leqslant \infty,$$

$$\nu = \frac{r_{\text{a}} - r_{\text{b}}}{R}, \qquad -1 \leqslant \nu \leqslant 1,$$

and ϕ is the rotational angle about the internuclear axis $\mathbf{R}$. Burrau (1927) showed that (18.107) is separable in this coordinate system, its use leading to $\psi^{(\text{e})}(\mu, \nu, \phi) = f(\mu)g(\nu)\varphi(\phi)$, where each of f, g, and φ obeys equations independent of the other two functions. The reader is referred, e.g., to Pilar (1968) for discussion of the resulting equations. For the *gerade* ground state, $\psi^{(\text{e})}$ is independent of ϕ; numerically exact ground-state results obtained using electronic computers were first reported by Wind (1965), and we shall refer to them after carrying out some approximate ground-state calculations.

Approximate solutions to (18.107) serve two purposes. The first is obvious: H_2^+, since it can be described exactly, functions as a laboratory for more complex molecules, which can only be treated approximately. The second concerns those approximations which yield analytic results, since these allow one to gain insight into the behavior of the system. Sometimes the behavior of $\hat{H}^{(e)}$ in certain limits will suggest an approximation, as in the following.

Let us assume that H_2^+ is in its ground state and that no energy is pumped into the system. We then ask for the behavior of H_2^+ when $R \to \infty$. From Fig. 18.11, the ion should dissociate into a ground state H-atom and a free proton. The atom could be either a + e or b + e, and, since the two protons are identical, each of these two H-atoms should occur with equal probability. The simplest variational approximation based on this separated-atom concept is a linear combination of the two ground-state H-atom wave functions. Denoting it by $\psi_0^{(\pm)}(\mathbf{r}_a, \mathbf{r}_b, \mathbf{R})$, we have

$$\begin{aligned}\psi_\Gamma^{(e)}(\mathbf{r}_a, \mathbf{r}_b, \mathbf{R}) &\to \psi_0^{(\pm)}(\mathbf{r}_a, \mathbf{r}_b, \mathbf{R}) \\ &= N_0(\mathbf{R})[\psi_{100}(\mathbf{r}_a) \pm \psi_{100}(\mathbf{r}_b)],\end{aligned} \tag{18.108}$$

where $\psi_{100}(\mathbf{r})$ is the $Z = 1$ hydrogenic ground state, Eqs. (11.141) and (11.142a). The RHS of (18.108) gives the only linear combination of hydrogenic ground states that are eigenstates of $\hat{P}_{ab}$ with the appropriate eigenvalues; for large R, this combination yields equal probabilities for dissociation into either an a + e or a b + e H-atom.

Equation (18.108) is an expansion in a non-orthogonal basis, a possibility foreseen in Section 17.4. It is an expansion typically referred to as a linear combination of atomic orbitals (LCAO), wherein the atomic orbital in the present case is the 1s state of hydrogen. Because the two 1s states are centered on different protons, their overlap is a (real) function of $\mathbf{R}$:

$$(\psi_{100}(\mathbf{r}_a)|\psi_{100}(\mathbf{r}_b)) \equiv S_0(\mathbf{R}), \tag{18.109}$$

using the generic notation introduced in Eq. (17.97). It will be shown below that $S_0(\mathbf{R}) = S_0(R)$ and that $S_0(R \to \infty) \to 0$. Choosing $N_0(\mathbf{R})$ to be real and positive, and requiring $\psi_0^{(\pm)}$ to be normalized, yields

$$N_0(\mathbf{R}) \to N_0^{(\pm)}(R) = \frac{1}{\sqrt{2[1 \pm S_0(R)]}}. \tag{18.110}$$

The variational approximations to the energy are obtained by solving the 2×2 version of (17.102) appropriate to the present case. In place of the H_{nm} of Eq. (17.100), we have the following four matrix elements:

$$H_{aa}(R) = H_{bb}(R) = (\psi_{100}(\mathbf{r}_a)|\hat{H}^{(e)}|\psi_{100}(\mathbf{r}_a)) \tag{18.110a}$$

and

$$H_{ab}(R) = H_{ba}(R) = (\psi_{100}(\mathbf{r}_a)|\hat{H}^{(e)}|\psi_{100}(\mathbf{r}_b)), \tag{18.110b}$$

while

$$S_{aa} = S_{bb} = (\psi_{100}(\mathbf{r}_a)|\psi_{100}(\mathbf{r}_a)) = 1$$

and

$$S_{ab}(R) = S_{ba}(R) = S_0(R).$$

With these definitions/replacements, the generic secular equation (17.102) becomes

$$\begin{vmatrix} H_{aa} - E_{\Gamma}^{(e)} & H_{ab} - E_{\Gamma}^{(e)} S_0 \\ H_{ba} - E_{\Gamma}^{(e)} S_0 & H_{bb} - E_{\Gamma}^{(e)} \end{vmatrix} = 0. \tag{18.111}$$

Its solutions are

$$E_{\Gamma}^{(e)} \to E_0^{(\pm)}(R) = \frac{H_{aa}(R) \pm H_{ab}(R)}{1 \pm S_0(R)}, \tag{18.112}$$

where the $\pm$ superscripts on $E_0^{(\pm)}$ and $\psi_0^{(\pm)}$ go together. These energies are the same as those obtained from $E_0^{(\pm)}(R) = (\psi_0^{(\pm)}|H^{(e)}|\psi_0^{(\pm)})$. Furthermore, whenever one makes any two-state approximation to $\psi^{(e)}$ whose structure is the same as that of Eq. (18.108), the resulting energy has the form of Eq. (18.112).

Substituting Eq. (18.104) into Eqs. (18.110) and writing $-\frac{1}{2}$ a.u. $= E_{1s}$ for the hydrogenic ground-state energy, we readily find

$$H_{aa}(R) = \left(-\frac{1}{2} + \frac{1}{R/a_0} - (\psi_{100}(\mathbf{r}_a) \left| \frac{1}{r_b/a_0} \right| \psi_{100}(\mathbf{r}_a)) \right) \text{ a.u.} \tag{18.113a}$$

and

$$H_{ab}(R) = \left[\left(-\frac{1}{2} + \frac{1}{R/a_0} \right) S_0(R) - (\psi_{100}(\mathbf{r}_a) \left| \frac{1}{r_a/a_0} \right| \psi_{100}(\mathbf{r}_b)) \right] \text{ a.u.}, \tag{18.113b}$$

where we use a_0 in place of $a_0/(1 + m_e/M_p)$.

The as-yet-undetermined quantities needed to evaluate $E_0^{(\pm)}(R)$ are the two matrix elements in Eqs. (18.113) and the overlap integral $S_0(R)$. The integrations are straightforward using confocal elliptical coordinates and the relation

$$\int d^3 r_a = \int d^3 r_b = \frac{R^3}{8} \int_1^\infty d\mu \int_{-1}^1 d\nu \int_0^{2\pi} d\phi (\mu^2 - \nu^2). \tag{18.114}$$

One finds

$$(\psi_{100}(\mathbf{r}_a) \left| \frac{1}{r_b/a_0} \right| \psi_{100}(\mathbf{r}_a)) = \frac{1}{R/a_0} \left[1 - e^{-2R/a_0} \left(1 + \frac{R}{a_0} \right) \right], \tag{18.115a}$$

$$(\psi_{100}(\mathbf{r}_a) \left| \frac{1}{r_a/a_0} \right| \psi_{100}(\mathbf{r}_b)) = e^{-R/a_0} \left(1 + \frac{R}{a_0} \right), \tag{18.115b}$$

and

$$S_0(\mathbf{R}) = S_0(R) = e^{-R/a_0} \left(1 + \frac{R}{a_0} + \frac{R^2}{3a_0^3} \right), \tag{18.115c}$$

verification of which is left to the exercises.

Combining these various results, the desired expression for $E_0^{(\pm)}(R)$ is

$$E_0^{(\pm)}(R) = \left(-\frac{1}{2} + \frac{1}{R/a_0} \right) \text{ a.u.} \\ - \frac{[1 - e^{-2R/a_0}(1 + R/a_0)]/(R/a_0) \pm e^{-R/a_0}(1 + R/a_0)}{1 \pm e^{-R/a_0}[1 + R/a_0 + R^2/(3a_0^3)]} \text{ a.u.} \tag{18.116}$$

The behavior of $E_0^{(\pm)}(R)$ for very small and very large R can be read off (18.116):

$$\lim_{R\to 0} E_0^{(\pm)}(R) \to +\infty \tag{18.117a}$$

and

$$\lim_{R\to\infty} E_0^{(\pm)} \to -\tfrac{1}{2}\ \text{a.u.} \tag{18.117b}$$

Equation (18.117a) is, of course, a reflection of the $1/R$ nuclear repulsion, whereas Eq. (18.117b) expresses the physical boundary condition that, as the protons are separated asymptotically, the system consists of a free proton plus an H-atom in its ground state: the asymptotic value of $E_0^{(\pm)}(R)$ is not zero.

Although $E_0^{(+)}(R)$ and $E_0^{(-)}(R)$ behave the same way for very small and very large R, for R between these limits the *gerade* and *ungerade* energy curves are completely different. Portions of them and of the exact results of Wind (1965) are shown in Fig. 18.12.

The *gerade* curve $E_0^{(+)}(R)$ and its exact counterpart $E_{\text{exact}}^{(+)}(R)$ follow the general structure outlined in Fig. 18.10. Note that, in the *gerade* cases, the displayed curves are the sum $E^{(+)}(R) + \kappa_e e^2/(2a_0)$; the magnitude $D_e \equiv D_e(R_{eq}) = |E^{(+)}(R_{eq}) + \kappa_e e^2/(2a_0)|$, known as the "spectroscopic dissociation energy," occurs at the equilibrium separation R_{eq}. The value D_e is a measure of the binding energy of the system[10] – even the crudest

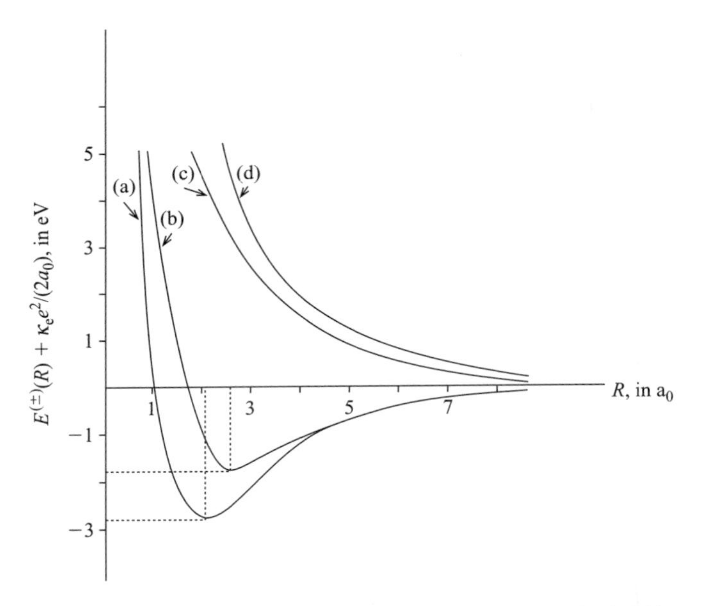

Fig. 18.12 Portions of the exact (Wind 1965) and the simplest LCAO-MO potential-energy curves for the H_2^+ ion. Curves (a) and (b) are the exact and approximate results for the *gerade* case, respectively, while curves (c) and (d) are the exact and approximate *ungerade* results, respectively.

[10] The theoretical binding energy D_0 is obtained by subtracting from D_e the lowest-lying vibrational energy E_{vib} of the molecule, since the nuclei do not "sit" at D_e; the energy E_{vib} need not be in harmonic-oscillator form. See, e.g., Pilar (1968) or Atkins and Friedman (1997) and discussion later in this section.

Table 18.8 *A comparison of spectroscopic dissociation energies and equilibrium separations of H_2^+ for the gerade ground state*

	D_e (eV)	R_{eq}
Exact results[a]	2.79	$2.02a_0$
LCAO approximation, Eq. (18.116)	1.78	$2.50a_0$

[a]Wind (1965).

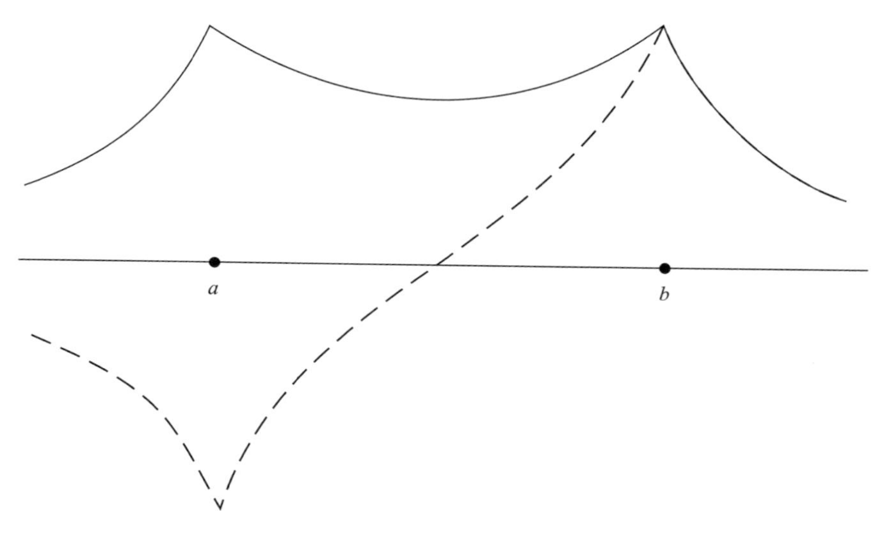

Fig. 18.13 A sketch of the LCAO-MO wave functions $\psi_0^{(\pm)}$ of Eq. (18.108): (a) and (b) refer to the two protons in H_2^+, while the solid curve is the bonding orbital $\psi_0^{(+)}$ and the dashed curve is the anti-bonding orbital $\psi_0^{(-)}$.

(LCAO) approximation $\psi_0^{(\pm)}$ yields a bound-state description of H_2^+. Table 18.8 gives a comparison of the values of D_e and R_{eq} for the exact and approximate *gerade* energy curves. Although the correct physics is generated from this LCAO calculation, the approximate results are not very accurate, the relative errors in D_e and R_{eq} being 36% and 25%, respectively. If one is to rely on an approximate (analytic) description, then improvements are clearly called for and we shall discuss a few shortly.

The *ungerade* energy curves of Fig. 18.12 are, except at $R = \infty$, everywhere greater than the asymptotic value of $-\frac{1}{2}$ a.u.; they do not correspond to a bound state. The bound and unbound characters of the $E_0^{(\pm)}(R)$ are mirrored in the behaviors of the wave functions $\psi_0^{(\pm)}$. Wave functions in the LCAO form are often referred to as "molecular orbitals" (MOs), and the $\psi_0^{(\pm)}$ are a pair of LCAO MOs. $\psi_0^{(+)}$ is a "bonding" orbital and $\psi_0^{(-)}$ is an "anti-bonding" orbital. A 2-D sketch of them is given in Fig. 18.13 for $R = R_{eq}$; it was obtained by passing a plane through the vector **R**. Rotation of Fig. 18.13 about **R** yields the probability amplitudes. The difference between $\psi_0^{(+)}$ and $\psi_0^{(-)}$ is perhaps even more dramatic than that between $E_0^{(+)}$ and $E_0^{(-)}$. If the electrons act as the "glue" binding the ion together (basically shielding the two protons from one another, thereby mitigating the effect of their mutual repulsion), then Fig. 18.13 indicates a plenitude of such glue between the protons for the *gerade* case and its relative absence (none at $R/2$) in the *ungerade* case. These comments concerning electronic "glue," while physically appealing, are a little too superficial to provide a quantitative explanation: the reasons why $\psi_0^{(+)}$ leads to bonding are connected with delicate energy balances, and the interested reader may consult Atkins and Friedman (1997) for a qualitative analysis of this point.

The LCAO MO $\psi_0^{(+)}$ of Eq. (18.108) is the molecular analog of the $n = 1$, $n' = 1$ $(1s)^2$ approximation (18.37a) for the ground state of the He-atom. These simplest of approximations yield similar results; viz., bound states whose energies are accurate enough to assure one that quantum mechanics is the correct theoretical framework, yet these energies are sufficiently inaccurate to demand better approximations. We shall consider only one. As with He and H^-, our procedure is to replace the exact hydrogen ground state by the function $(\xi/a_0)^{3/2}[\exp(-\xi r/a_0)]/\sqrt{\pi}$, where ξ is a scale factor. Its use produces the "scaled" LCAO MOs

$$\psi_1^{(\pm)}(\mathbf{r}_a, \mathbf{r}_b, \mathbf{R}; \xi) = N(\mathbf{R}; \xi)(e^{-\xi r_a/a_0} \pm e^{-\xi r_b/a_0}) \tag{18.118}$$

first introduced by Finkelstein and Horowitz (1928).

Just as Eq. (18.51) led to the analytic result (18.52), so Eq. (18.118) leads to an energy curve $E_1^{(\pm)}(R; \xi)$ that is the analytic analog of Eq. (18.116). Allowing the internuclear distance R to be scaled also, so that $R \to \xi R$, the minimizing value of ξ is found to be $\xi_e = 1.238$. For this value of ξ, the equilibrium separation R_{eq} becomes $(\xi_e R_{eq} = 2.5a_0)$ $R_{eq} = 2.02a_0$ (!), and the spectroscopic dissociation energy D_e is improved to 2.37 eV, which cuts the relative error in the energy from 37% to 15%.

The scaling approximation is the simplest of the improved variational ones for which the H_2^+ ion has been a theoretical laboratory. Discussion of and references to some of the others can be found in a variety of texts and monographs, e.g., those of Pilar (1968) and Hurley (1976), where not only two-center but also single-center analyses are considered.

The goal of exploring quantal aspects of molecular electronic structure has been achieved with the preceding analysis of H_2^+: the study of other diatomics, even one as "simple" as H_2, is a step towards molecular physics that is beyond the scope of this book. The step we do wish to take is towards some understanding of the vibrational and rotational motion of the nuclei forming a diatomic molecule, a topic we consider in the remaining subsection.

Nuclear motion: rotations and vibrations

The motion of the two nuclei forming a diatomic molecule is governed by Eq. (18.102), in particular the "potential" $E_\Gamma^{(e)}(R)$, whose general shape is that of Fig. 18.10. This potential will produce at most a finite number of states whose energies lie between $-D_e$ and 0; energies greater than 0 correspond to dissociation. Because of this, the oscillator approximation of Eq. (18.88) can be expected to be valid, if at all, only for the few lowest-lying energy levels in the potential $E_\Gamma^{(e)}(R)$; recall the vibrational spectrum of HCl in Fig. 6.6. That is, we can expect neither an infinite set of levels nor uniform spacing. In order to understand the structures of realistic spectra that could arise from equations for which (18.102) is the generic form, qualitative semi-classical arguments and also model potentials have been examined. The quantitative analysis that we summarize below is based on one of the most famous of these model potentials, that due to P. M. Morse (1929).

If we use the spherical polar coordinates R, θ, and ϕ and write

$$\chi_\Gamma(\mathbf{R}) = \frac{u_{\gamma\ell}(R)}{R} Y_\ell^{m_\ell}(\Omega), \tag{18.119}$$

then substitution of (18.119) into (18.102) leads to

$$\left[-\frac{\hbar^2}{2\mu_{\mathrm{ab}}} \left(\frac{d^2}{dR^2} - \frac{\ell(\ell+1)}{R^2} \right) + E_\Gamma^{(\mathrm{e})}(R) \right] u_{\gamma\ell}(R) = E_{\gamma\ell} u_{\gamma\ell}(R). \tag{18.120}$$

Let us first assume that a quadratic approximation to $E_\Gamma^{(\mathrm{e})}(R)$ is adequate for describing the very-low-lying states. Following the ideas embodied in Eqs. (18.88) and (18.89), we assume that

$$E_\Gamma^{(\mathrm{e})}(R) = -D_{\mathrm{e}}(\Gamma) + \tfrac{1}{2}\mu_{\mathrm{ab}}\omega_\Gamma^2 x^2, \tag{18.121}$$

where

$$D_{\mathrm{e}} = -E_\Gamma^{(\mathrm{e})}(R_{\mathrm{eq}}), \tag{18.122a}$$

$$\omega_\Gamma^2 = \frac{1}{\mu_{\mathrm{ab}}} \left. \frac{\partial^2 E_\Gamma^{(\mathrm{e})}(R)}{\partial R^2} \right|_{R_{\mathrm{eq}}}, \tag{18.122b}$$

and

$$x = R - R_{\mathrm{eq}}. \tag{18.122c}$$

To simplify a bit more, we also assume that $E_\Gamma^{(\mathrm{e})}(R \to \infty) \to 0$, in which case $E_{\gamma\ell} = -|E_{\gamma\ell}|$.

With the preceding definitions, Eq. (18.120) becomes

$$\left[-\frac{\hbar^2}{2\mu_{\mathrm{ab}}} \left(\frac{d^2}{dx^2} - \frac{\ell(\ell+1)}{(R_{\mathrm{eq}}+x)^2} \right) + \frac{1}{2}\mu_{\mathrm{ab}}\omega_\Gamma^2 x^2 \right] u_{\gamma\ell}(x) = [D_{\mathrm{e}}(\Gamma) - |E_{\gamma\ell}|] u_{\gamma\ell}(x). \tag{18.123}$$

Although a quadratic approximation has been used, the factor $(R_{\mathrm{e}} + x)^{-2}$ means that Eq. (18.123) is not in the standard symmetric-oscillator form. To deal with this situation, we use the fact that $x \ll R_{\mathrm{eq}}$ to introduce a truncated Taylor expansion for $(R_{\mathrm{eq}} + x)^{-2}$:

$$(R_{\mathrm{eq}} + x)^{-2} \simeq R_{\mathrm{eq}}^{-2} \left(1 - \frac{2x}{R_{\mathrm{eq}}} + \frac{3x^2}{R_{\mathrm{eq}}^2} \right). \tag{18.124}$$

On making use of the approximation (18.124), Eq. (18.123) can be put in the form of a 1-D oscillator equation with a displaced center:

$$\left(\frac{-\hbar^2}{2\mu_{\mathrm{ab}}} \frac{d^2}{dx^2} + \tfrac{1}{2}\mu_{\mathrm{ab}}\omega_{\mathrm{eff}}^2 (x - x_0)^2 \right) u_{\gamma\ell}(x - x_0) = \mathcal{E}_{\gamma\ell} u_{\gamma\ell}(x - x_0), \tag{18.125}$$

where

$$x_0 = \frac{\hbar^2 \ell(\ell+1) R_{eq}}{I_{ab}^2 \omega_\Gamma^2 + 3\hbar^2 \ell(\ell+1)}, \tag{18.126a}$$

$$\omega_{eff}^2 = \omega_\Gamma^2 + \frac{3\hbar^2 \ell(\ell+1)}{I_{ab}^2}, \tag{18.126b}$$

$$\mathcal{E}_{\gamma\ell} = D_e(\Gamma) - \frac{\hbar^2 \ell(\ell+1)}{2I_{ab}} + \frac{1}{2}\mu_{ab}\omega_{eff}^2 x_0^2 - |E_{\gamma\ell}|, \tag{18.126c}$$

and $I_{ab} = \mu_{ab} R_{eq}^2$ is the classical moment of inertia for the molecule considered as a rigid rotator of fixed separation R_{eq}. With this interpretation of I_{ab}, the term $\hbar^2\ell(\ell+1)/(2I_{ab})$ that enters $\mathcal{E}_{\gamma\ell}$ is just the rotational energy of the molecule considered as a rigid rotator with moment of inertia I_{ab} and kinetic energy $\hat{L}^2/(2I_{ab})$.

Although x has been assumed to be small, we shall drop this assumption and temporarily allow x to lie in the range $[-\infty, \infty]$. Then the solutions to (18.125) are the oscillator wave functions and energies; in particular,

$$\mathcal{E}_{\gamma\ell} \to \mathcal{E}_{\upsilon} = (\upsilon + \tfrac{1}{2})\hbar\omega_{eff}, \qquad \upsilon = 0, 1, 2, \ldots, \tag{18.127a}$$

or

$$E_{\gamma\ell} \to E_{\upsilon\ell} = -D_e(\Gamma) + \left(\upsilon + \frac{1}{2}\right)\hbar\omega_{eff} + \frac{\hbar^2\ell(\ell+1)}{2I_{ab}} - \frac{1}{2}\mu_{ab}\omega_{eff}^2 x_0^2, \tag{18.127b}$$

where $\upsilon = 0, 1, 2, \ldots$.

Since the exponential part of the oscillator wave function is $\exp[-x^2/(2\lambda^2)]$, where

$$\lambda = \sqrt{\frac{\hbar}{\mu_{ab}\omega_{eff}}}, \tag{18.128}$$

then, if λ is small, the exponential factor will force x also to be small, thereby justifying the use of the full oscillator solution. If not, the eigenvalue expression (18.128) provides an orientation to the quantum-number dependence of $E_{\gamma\ell} = E_{\upsilon\ell}$.

In order to evaluate λ, we first need a value, or at least an estimate, of ω_{eff}. This in turn requires estimating ω_Γ and $I_{ab} = \mu_{ab} R_{eq}^2$. For μ_{ab} we use the proton mass m_p, while we set R_{eq} equal to $1.5a_0$. Combining the analysis leading to (18.90a) with the definition (18.122), we estimate ω_Γ^2 via

$$\omega_\Gamma^2 \simeq \frac{1}{m_p}\frac{D_e(\Gamma)}{R_{eq}^2},$$

with a typical value for $D_e(\Gamma)$ taken to be 3 eV. These estimates lead to

$$\frac{\hbar^2}{I_{ab}^2} \simeq 10^{26} \text{ rad s}^{-1} \tag{18.129a}$$

and

$$\omega_\Gamma^2 \simeq 10^{29} \text{ rad}^2 \text{ s}^{-2}. \tag{18.129b}$$

These lead to $\lambda \simeq 0.27a_0$ and, for $\ell = 1$ or 2, to

$$\omega_{eff}^2 \simeq \omega_\Gamma^2, \tag{18.129c}$$

as well as to

$$\mu_{ab}\omega_{eff}^2 x_0^2 \simeq \frac{\hbar^2}{I_{ab}}\left(\frac{\hbar^2/I_{ab}}{\hbar\omega_\Gamma}\right)^2 \ell^2(\ell+1)^2. \qquad (18.129d)$$

In comparison with $R_{eq} \simeq 1.5a_0$, the value of λ tends to keep $|x| \leqslant 0.5a_0$, which is not very small, but perhaps small enough that we can regard the result (18.127b) as at least containing the right kinds of ingredients even if the numerical coefficients of the various terms are no more than order-of-magnitude estimates. Using (18.129c) and (18.129d), the eigenvalue expression $E_{v\ell}$ becomes

$$E_{v\ell} \simeq -D_e(\Gamma) + \left(v+\frac{1}{2}\right)\hbar\omega_\Gamma + \frac{\hbar^2}{2I_{ab}}\ell(\ell+1) - \frac{\hbar^2}{2I_{ab}}\left(\frac{\hbar^2/I_{ab}}{\hbar\omega_\Gamma}\right)^2 \ell^2(\ell+1)^2. \quad (18.130)$$

The second and third terms in (18.130) are the expected vibrational and rotational energy levels. The fourth term, involving $\ell^2(\ell+1)^2$, can be interpreted as an effect arising from the fact that the molecule is not a rigid rotator but can stretch: for small deviations from rigidity, semi-classical arguments lead to a correction proportional to $\ell^2(\ell+1)^2$ (see, e.g., Pilar (1968) or Atkins and Friedman (1997)).

Even for low-lying vibrational excitations, the shape of the parabola yielded by the quadratic approximation can deviate significantly from the shape of the true potential-energy curve $E_\Gamma^{(e)}(R)$, an exaggerated example of which is sketched in Fig. 18.14. In order to gain some understanding of the spectrum that a typical (non-oscillator) $E_\Gamma^{(e)}(R)$ would yield, we examine the Morse-type potential[11]

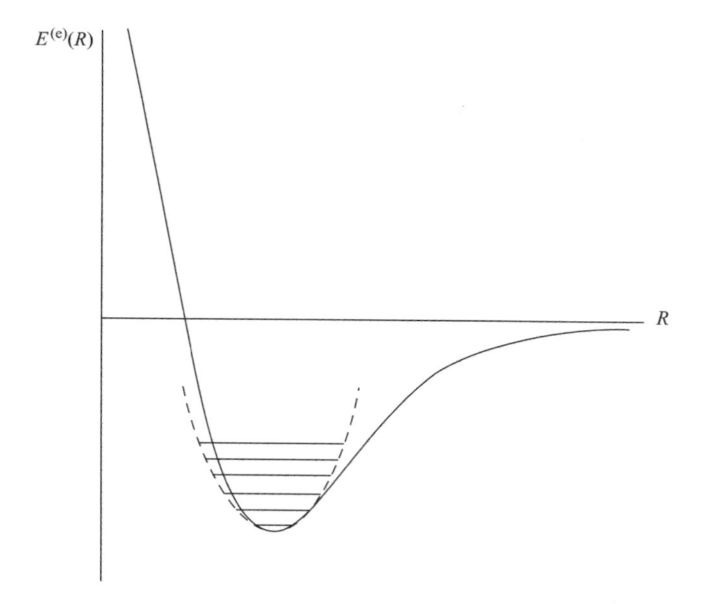

Fig. 18.14 A sketch of the harmonic-oscillator approximation to a typical potential-energy curve $E_\Gamma^{(e)}(R)$. A few of the low-lying oscillator levels are indicated.

[11] The potential introduced by Morse (1929) is $D_e[1 - \exp(-\beta x/R_{eq})]^2$, which differs from (18.131) by the term D_e. Hence, at $x = 0$, the original Morse potential is zero, whereas (18.131) equals $-D_e$ at $x = 0$.

$$E_{\Gamma}^{(e)}(R) \to V_{\rm M}(R) = D_{\rm e}(e^{-2\beta x/R_{\rm eq}} - 2e^{-\beta x/R_{\rm eq}}), \tag{18.131}$$

where $x = R - R_{\rm eq}$ as before and β is a parameter. Except at $R = 0$, where it equals the potentially large, but finite, value $D_{\rm e}[\exp(2\beta) - 2\exp\beta]$, $V_{\rm M}(R)$ has the desired shape, as shown in Fig. 18.15.

An easy calculation leads to $V_{\rm M}(R) = 0$ at $R_0 = R_{\rm eq}(1 - 0.639/\beta)$. Since $V_{\rm M}(R)$ has been used to characterize diatomic ground states, we show in Table 18.9 values of $D_{\rm e}$, $R_{\rm eq}$, β, R_0, and $V_{\rm M}(R = 0)$ for H_2 and HCl. The values of $D_{\rm e}$ and $R_{\rm eq}$ have been calculated from relevant quantities tabulated by Flügge (1974); the values of β are the same as his tabulated values α.

The quadratic approximation to the potential $V_{\rm M}(R)$ has a structure identical to that of (18.121):

$$V_{\rm M}(R) \simeq -D_{\rm e} + \tfrac{1}{2}\mu_{\rm ab}\omega_{\rm M}^2 x^2, \tag{18.132a}$$

with

$$\omega_{\rm M}^2 = \frac{2D_{\rm e}\beta^2}{\mu_{\rm ab} R_{\rm eq}^2}, \tag{18.132b}$$

so we can check the estimate of Eq. (18.129b). On substituting the numbers from Table 18.9 for H_2 and HCl into (18.132b), we find $\omega_{\rm M}^2(H_2) = 6.41 \times 10^{29}$ rad^2 s^{-2} and $\omega_{\rm M}^2({\rm HCl}) = 6.22 \times 10^{29}$ rad^2 s^{-2}, numbers that nicely justify the order-of-magnitude estimate of Eq. (18.129b).

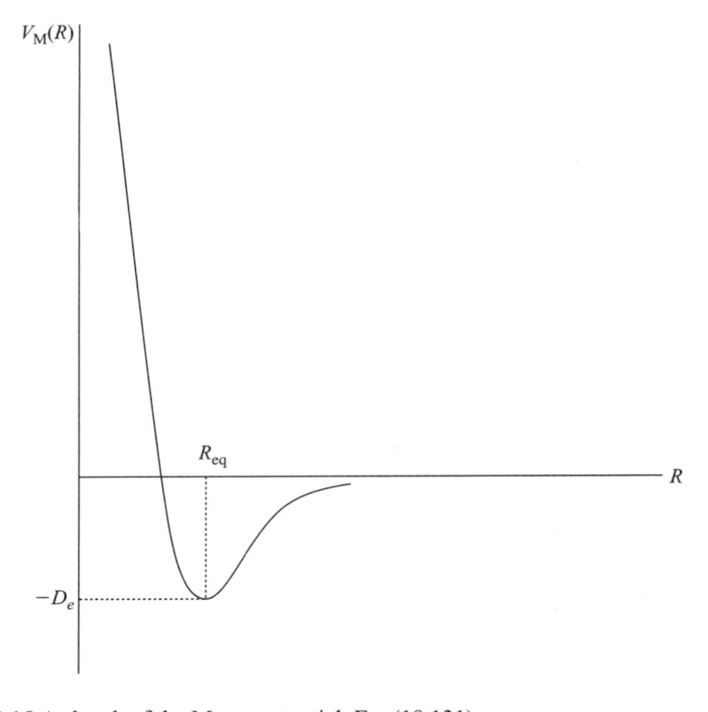

Fig. 18.15 A sketch of the Morse potential, Eq. (18.131).

Table 18.9 *Characteristic values of the Morse-type potential $V_M(R)$ for H_2 and HCl*

	D_e (eV)	R_{eq}	β	R_0	$V_M(R=0)$ (eV)
H_2	4.747	$1.40a_0$	1.44	$0.78a_0$	44.48
HCl	4.618	$2.42a_0$	2.38	$1.78a_0$	439.36

Next, we assess the approximate eigenvalue formula (18.130). To do so requires solving Eq. (18.120) with $E_\Gamma^{(e)}(R)$ replaced by $V_M(R)$. Only for $\ell = 0$ is this a relatively straightforward procedure, albeit one involving the confluent hypergeometric functions discussed, e.g., by Mathews and Walker (1970), Wyld (1976), and Hassani (1991). The eigenvalues for the $\ell = 0$ case are the solutions of a transcendental equation involving this function. We shall only quote the result, which relies on an accurate approximation; the interested reader may consult Flügge (1974). Using the notation already introduced, the $\ell = 0$ eigenvalues $E_{v,\ell=0}^{(M)}$ of $V_M(R)$ are

$$E_{v,\ell=0}^{(M)} = -D_e + \left(v + \frac{1}{2}\right)\hbar\omega_M - \frac{\beta}{2}\sqrt{\frac{\hbar^2/(2I_{ab})}{D_e}}\left(v + \frac{1}{2}\right)^2 \hbar\omega_M, \tag{18.133a}$$

where

$$v = 0,\ 1,\ 2,\ \ldots,\ v_0, \tag{18.133b}$$

with

$$v_0 = \left[\frac{1}{\beta}\sqrt{\frac{D_e}{\hbar^2/(2I_{ab})}} - \frac{1}{2}\right]_{\text{int}}, \tag{18.133c}$$

the $[\ ]_{\text{int}}$ notation signifying the largest integer contained in the expression in the square brackets.

The first two terms in $E_{v,\ell=0}^{(M)}$ agree with the result (18.130) when $\ell = 0$ and $\omega_\Gamma \to \omega_M$, while the third term is obviously an anharmonicity correction. The relation (18.133c) not only yields a finite spectrum but also ensures that the magnitude of the anharmonicity term cannot exceed that of the linear term $(v + \frac{1}{2})\hbar\omega_M$. Furthermore, the spacing between adjacent levels decreases as v increases, so that the $\ell = 0$ spectrum is not uniform.

For non-zero ℓ, the Schrödinger equation with the potential $V_M(R)$ is no longer tractable. The standard procedure is to assume that ℓ is not too large and x/R_{eq} is small, and then expand the term $\ell(\ell+1)/[2\mu_{ab}(x + R_{eq})^2]$ exactly as we did in Eq. (18.124) *et seq*. By keeping only the first two terms, one can replace $V_M(R)$ by a potential $V'_M(R)$ whose R dependence involves just the same factors $\exp(-2\beta x/R_{eq})$ and $\exp(-\beta x/R_{eq})$ as those occurring in $V_M(R)$, although there are new constants in $V'_M(R)$. As Flügge (1974) shows, the Schrödinger equation with $V'_M(R)$ can also be solved via use of confluent hypergeometric functions. The resulting eigenvalue expression is an ℓ-dependent extension of Eq. (18.133a):

$$E_{v\ell}^{(\mathrm{M})} = -D_e + \left(v + \frac{1}{2}\right)\hbar\omega_{\mathrm{M}} - \frac{\beta}{2}\sqrt{\frac{\hbar^2/(2I_{\mathrm{ab}})}{D_e}}\left(v + \frac{1}{2}\right)^2 \hbar\omega_{\mathrm{M}}$$
$$+ \frac{\hbar^2}{2I_{\mathrm{ab}}}\ell(\ell+1) - \frac{9(\beta-1)^2}{4\beta^4}D_e\ell^2(\ell+1)^2$$
$$- \frac{3(\beta-1)}{\beta}\sqrt{\frac{\hbar^2/(2I_{\mathrm{ab}})}{D_e}}\frac{\hbar^2}{2I_{\mathrm{ab}}}\left(v+\frac{1}{2}\right)\ell(\ell+1). \quad (18.134)$$

Not surprisingly, the first three terms in this equation comprise the $\ell = 0$ result (18.133a). The fourth term is the rigid-rotator energy, and the fifth term is the contribution to the rotational energy arising from the "stretchiness" of the diatomic molecule, as in Eq. (18.130). The last term represents a coupling between the rotational and vibrational degrees of freedom. It can be thought of as a consequence of vibrational motion leading to a larger internuclear separation and therefore to an increase in I_{ab}, with a corresponding decrease in the rotational energy, which is reflected in the minus signs (of both the fifth term and the sixth term). Although the coefficients in $E_{v\ell}^{(\mathrm{M})}$ are based on the Morse-type potential $V_{\mathrm{M}}(R)$, the v, ℓ structure of (18.134) is typical of empirical expressions used to analyze experimental data and intermolecular attributes. Discussions and references to the literature on this topic can be found in Pilar (1968), Flygare (1978), and Atkins and Friedman (1997), which interested readers are recommended to consult.

Exercises

18.1 Derive the analogs of Eqs. (18.20) for a system of two identical bosons of spin S described by the Hamiltonian (18.11).

18.2 Two particles, labeled 1 and 2, each of mass m, are fixed at the ends of a rigid rod of length a, which rotates about its midpoint in the x–y plane. In the CM system, the Hamiltonian for this system is $\hat{H}_{12}(\phi) = \hat{L}_z^2(\phi)/(ma^2)$, where ϕ is the azimuthal angle of the rod. Determine the eigenfunctions, eigenvalues, and degeneracies if the particles are
(a) distinguishable,
(b) identical spin-0 bosons, and
(c) identical spin-$\frac{1}{2}$ fermions, each of which is in spin state $|S_z, +\rangle$.

18.3 Two identical, non-interacting particles of mass m are in a symmetric 1-D quantal box located between $x = -a$ and $x = a$ (Section 9.5). Determine the lowest five energy levels, the corresponding wave functions, and the degrees of degeneracy if the particles are
(a) spin-0 bosons, and
(b) spin-$\frac{1}{2}$ fermions, each in spin state $|S_z, +\rangle$.
Construct Grotrian diagrams for each case.

18.4 Consider two-particle scattering in the CM system for the case in which the particles are either identical bosons or spin-$\frac{1}{2}$ fermions in the singlet spin state. Let $\psi_{\mathbf{k}}^{(+)}(\mathbf{r})$ denote the relative-motion scattering state.

(a) Prove that $\psi_{\mathbf{k}}^{(+)}(\mathbf{r})$ has even parity.

(b) Show that, up to a normalization factor,

$$\psi_{\mathbf{k}}^{(+)}(\mathbf{r}) \underset{r\to\infty}{\sim} e^{i\mathbf{k}\cdot\mathbf{r}} + e^{-i\mathbf{k}\cdot\mathbf{r}} + [f(\theta) + f(\pi - \theta)][\exp(ikr)]/r,$$

where $\exp(i\mathbf{k}\cdot\mathbf{r}) + f(\theta)[\exp(ikr)]/r$ is the asymptotic form in the absence of symmetry effects.

(c) Show that, for $\theta = \pi/2$, the probability of observing either particle is twice that for the case in which the particles are distinguishable.

(d) Show that only even ℓ occurs in the partial-wave expansion of the scattering amplitude.

18.5 Let the scattering system of Exercise 18.4 comprise two identical spin-$\frac{1}{2}$ particles in the spin triplet state.

(a) Prove that $\psi_{\mathbf{k}}^{(+)}(\mathbf{r})$ has odd parity.

(b) Show that, to within a normalization factor,

$$\psi_{\mathbf{k}}^{(+)}(\mathbf{r}) \underset{r\to\infty}{\sim} e^{i\mathbf{k}\cdot\mathbf{r}} - e^{-i\mathbf{k}\cdot\mathbf{r}} + [f(\theta) - f(\pi - \theta)][\exp(ikr)]/r,$$

where $f(\theta)$ is the scattering amplitude in the absence of symmetry effects.

(c) Determine the effect of these results on the $\theta = \pi/2$ cross section and on the partial-wave expansion of the symmetrized scattering amplitude.

18.6 In the CM system, two spin-$\frac{1}{2}$ particles are in a state whose total angular momentum is $J = 1$ and whose parity is positive.

(a) Determine the total spin.

(b) Does the state exist if the particles are identical? Justify your answer.

(c) What are the possible values of the orbital angular momentum of the system?

18.7 Each of two non-interacting electrons, labeled 1 and 2, has $\ell = 1$ and $m_\ell = 0$.

(a) Determine the possible values of the total angular momentum for the system and calculate the probability of measuring each of them.

(b) Now suppose that each electron is described by the same wave function. Determine the total spin and the total angular momentum of the system.

18.8 Two non-interacting electrons are bound in hydrogenic states $|n\ell\frac{1}{2};\, jm_j\rangle_i$, where $i = 1, 2$ labels the electrons. Assume that $n = 2$ and $j = \frac{1}{2}$ for each electron. Make predictions (probabilistic or otherwise) concerning measurements of the total angular momentum of the system when

(a) each particle has negative parity,

(b) each particle has positive parity, and

(c) the particles have opposite parity.

18.9 Prove that the completely symmetric and the completely antisymmetric portions of $|\Gamma\rangle_{1,2}^{\mathrm{A}}$ can be obtained from the RHS of (18.25) only if there is a sum on magnetic quantum numbers in $|\Gamma\rangle_{1,2}^{\mathrm{A}}$.

18.10 Three equal-mass particles labeled 1, 2, and 3 are described in the CM system by the two Jacobi coordinates $\mathbf{r} = \mathbf{r}_1 - \mathbf{r}_2$ and $\rho = \mathbf{r}_3 - (\mathbf{r}_1 + \mathbf{r}_2)/2$, where $\mathbf{r}_j$ is the coordinate vector for particle j, $j = 1, 2, 3$.

(a) Assume that the particles are spinless and distinguishable, with wave function

$$\psi_1(\mathbf{r}, \rho) = \chi(\mathbf{r})\xi(\rho) = [R_{n\ell}(r)Y_\ell^{m_\ell}(\Omega_r)][R_{n'\ell'}(\rho)Y_{\ell'}^{m'_\ell}(\Omega_\rho)],$$

where χ describes particles 1 and 2, and ξ describes particle 3. (i) State the set of possible values of the total orbital angular momentum L of the system. (ii) If $L = \ell$, what is the parity π_1 of the system?

(b) Now let 1 and 2 be identical spin-$\frac{1}{2}$ fermions, with 3 unchanged and $\psi_1(\mathbf{r}, \rho)$ replaced by $\psi_1(\mathbf{r}, \rho)|\frac{1}{2}\frac{1}{2}; SM_S\rangle$, with χ and ξ of ψ_1 as in part (a). Determine the dependence on ℓ' of the parity π_2 of this system for the cases $S = 0$ and $S = 1$, i.e., find the functional form of $\pi_2 = \pi_2(\ell', S)$.

(c) Next, let particle 3 also have spin-$\frac{1}{2}$, but be distinguishable from 1 and 2 (e.g., is a proton, whereas 1 and 2 are neutrons). If the total angular momentum is $J = \frac{1}{2}$, and the parity of the system is positive, what are the allowed values of the sets $\{S, \ell, \ell'\}$ if ℓ and ℓ' are each $\leqslant 2$?

(d) Finally, let 1, 2, and 3 be identical spin-$\frac{1}{2}$ fermions, and define $|\Gamma\rangle_{1,2;3} \equiv |\gamma\rangle^{\mathrm{A}}_{1,2}|\alpha\rangle_3$, where each ket contains the relevant space and spin quantum numbers for the particles and $\hat{P}_{12}|\gamma\rangle^{\mathrm{A}}_{1,2} = -|\gamma\rangle^{\mathrm{A}}_{1,2}$. Construct a fully antisymmetrized state vector for this system using a linear combination of $|\Gamma\rangle_{1,2;3}$, $|\Gamma\rangle_{2,3;1}$, and $|\Gamma\rangle_{1,3;2}$, bearing in mind that $|\gamma\rangle_{i,j}$ is not a 2×2 determinant.

18.11 A system is composed of two, identical, spin-S fermions labeled 1 and 2. In the absence of a mutual interaction, each is governed by the Hamiltonian (18.21), so that the unperturbed Hamiltonian for the system is $\hat{H}_0(1, 2) = \hat{H}_1(\mathbf{r}_1) + \hat{H}_2(\mathbf{r}_2)$. The eigenstates and energies of $\hat{H}_k(\mathbf{r}_k)$ are $|n\ell S;\, jm_j\rangle_k$ and $E_{n\ell j}$.

(a) If each particle has the same values of n, ℓ, S, and j, what are the allowed values of the total angular momentum J in the coupled state $|1, 2;\, JM\rangle = |[n\ell S;\, j]_1, [n\ell S;\, j]_2;\, JM\rangle$?

(b) Let $\ell = 0$ and $S = \frac{3}{2}$, add to $H_0(1, 2)$ the mutual interaction $\hat{H}_{12}(\mathbf{r}_1, \mathbf{r}_2) = g\delta(\mathbf{r}_1 - \mathbf{r}_2)\hat{\mathbf{S}}_1 \cdot \hat{\mathbf{S}}_2$, $g > 0$, and use first-order perturbation theory to calculate the energy shifts among the degenerate states $|1, 2;\, JM\rangle$ due to $\hat{H}_{12}$. Express your answer in terms of J and $\beta = g \int d^3r\, |\psi_{n00}(\mathbf{r})|^4$ and specify the ordering of levels as a function of J.

18.12 Two non-interacting electrons are bound by the Coulomb interaction to an infinitely massive center of charge Ze, with one in a 1s state and the other in a 2s state. Calculate the first-order energy shift when the perturbation $\kappa_e e^2/r_{12} + \hat{H}_{\mathrm{mag}}$ is turned on, where $\hat{H}_{\mathrm{mag}}$ is the interaction due to a constant magnetic field $\mathbf{B} = B\mathbf{e}_3$. Determine whether the initial four-fold degeneracy is completely lifted in first order, and specify the structure of all states for which the perturbation does lift the degeneracy.

18.13 The following model was communicated to the author by H. Krüger. It is based on the classical-physics observation that the effect of the repulsion $\kappa_e e^2/r_{12}$ between electrons 1 and 2 bound to a massive charge center Ze will be minimized when they are at equal distances from, and are on opposite sides of a line passing through, the charge center Ze, i.e., when $\mathbf{r}_2 = -\mathbf{r}_1$.

(a) Set $\mathbf{r}_2 = -\mathbf{r}_1$ in (18.30), show that the result is a one-body Schrödinger equation with an effective charge, and obtain an expression for the energy eigenvalues.

(b) Set $Z = 1$ (H^-) in the energy-level formula of part (a) and evaluate numerically the energies of the ground and the degenerate $n = 2$ levels. From these results determine the affinity of the second electron, comparing it with the experimental value of 0.754 eV. Also indicate whether the $n = 2$ levels are bound.

(c) Let $Z = 2$ in the formula of part (a) and compare the values of its two lowest-lying energies with the relevant ($S = 0$ or $S = 1$) levels in He.

(d) Construct arguments to explain why this truncated Schrödinger-equation model works better for the more accurate of the ground-state energies calculated in parts (b) and (c).

18.14 Show by direct calculation that the values of $J_{1s,2\ell}$ and $K_{1s,2\ell}$ in Table 18.3 (p. 708) are correct.

18.15 Prove the second equality in (18.52).

18.16 As discussed in Section 17.4, one can obtain variational estimates of excited states by using as trial functions linear combinations of states from an orthogonal basis. Calculate the ground and the first-excited singlet states in He by using a linear combination of the H-atom ($Z = 1$) 1s and 2s states. That is, form linear combinations of the $Z = 1$, $1s^2$ and 1s2s configurations, taking symmetry effects into account. Compare your results with the experimental energies as well as with the energies from the various uncorrelated approximations used in Section 18.2.

18.17 Show that Eq. (18.64) requires the normalization factor $(N!)^{-1/2}$ in Eq. (18.61).

18.18 One way of truncating infinite systems to finite size without introducing edge effects is to impose periodic boundary conditions, a method that is applied in this exercise to the infinite array of delta functions treated in Section 9.6. Assume that the delta-function potentials (of strength $\hbar^2/(mx_0)$ are spaced equally at intervals a along a line stretching from $x = 0$ to $x = L = Na$, where N is even. The periodic boundary condition is imposed via $\psi(x = 0) = \psi(x = L)$.

(a) Show that the exponent k appearing in the Bloch function (9.119a) now takes on a finite number of values k_n (which you are to enumerate).

(b) Prove that the energies of the system still fall into allowed bands, and find the edge values.

(c) Let N_e non-interacting (spin-$\frac{1}{2}$) electrons be inserted into the energy levels of the lowest-lying band. Determine (i) the maximum value of N_e; (ii) the maximum energy an electron can have; and (iii) the explicit structure of the wave function when N_e is maximal.

(d) Indicate how, if at all, the preceding results will change if N is odd.

(e) Finally, comment on how the nature of this finite model containing real electrons may alter the analysis of conductivity, etc., at the end of Section 9.6.

18.19 The ground-state configuration of the carbon atom is $1s^2 2s^2 2p^2$. Determine for this configuration the set of term values $^{2S+1}L_J$ that are consistent with the Pauli principle.

18.20 The spectroscopic term values for some of the low-lying excited states of Li are $^2P_{1/2}$, $^2S_{1/2}$, and $^2D_{3/2}$.

(a) Assuming that only an outermost electron has undergone excitation from the ground state, and only to the lowest possible level in each case, what are the corresponding configurations?

(b) Now assume that, in another universe, there exists an alternate periodic table, constructed in the same way as in our universe, *except that the 1s shell is absent.* State the electron configurations and spectroscopic term values of the ground and first four excited states for a Li atom in the alternate universe, such that the latter states are formed by promoting an electron from the outermost shell.

(c) In the alternate universe of part (b) it is found that the lowest-lying, two-electron excited state occurs when the electrons are promoted into the 3s level. Write out the configuration and term value for this state, justifying your conclusions.

18.21 Prove that the term values $^{2S+1}F_J$ cannot occur for the np^3 configuration.

18.22 The extension of the electric-dipole approximation to the case of an atom containing N spin-$\frac{1}{2}$ electrons is obtained by neglecting spin-dependent terms, e.g., quantities such as $(g_L\hat{\mathbf{L}} + g_S\hat{\mathbf{S}}) \cdot \mathbf{B}$. The result obtained in Exercise 16.14 can then be taken over directly (readers who have not obtained this result will need to work it out for this exercise). The assumption to be used in deriving the analog of Eq. (16.72) is that $|\alpha\rangle$ and $|\beta\rangle$ are exact eigenstates of $\hat{H}_0$. However, the application will be to states approximated by single-shell-model determinants whose single-particle states are $|n\ell m_\ell m_S\rangle$. In the determinantal approximation to $|\alpha\rangle$ and $|\beta\rangle$, $N-1$ of the single-particle states are assumed to be the same, while the state $|n_\alpha \ell_\alpha m_\alpha m_S\rangle$ in $|\alpha\rangle$ is replaced by $|n_\beta \ell_\beta m_\beta m_S\rangle$ in $|\beta\rangle$.

(a) Using the preceding information, derive the analog of (16.84) for the case of electric-dipole transitions in atoms whose states are single-shell-model determinants differing by one orbital.

(b) As an application, calculate the 3p $\rightarrow$ 2s transition in Li, for which the energy of the $1s^2 3p$ excited state is approximately 3.83 eV above the ground state. The measured lifetime is $\approx 8.5 \times 10^{-7}$ s. You should find the calculated lifetime to be within a factor of two of the measured one by using $Z = 1$ H-atom wave functions, i.e., by assuming that the $1s^2$ pair of electrons shields two of the protons from the outermost electron.

18.23 The zeroth-order approximation to the ground and some low-lying excited states of many nuclei is a shell model whose single-particle structure is based on Eq. (18.75), i.e., the basis is formed from the coupled states $|n\ell\frac{1}{2};\, jm_j\rangle$, with energies $E_{n\ell j}$ such that $E_{n\ell\ell+1/2}$ lies lower than $E_{n\ell\ell-1/2}$, Section 14.4. Unlike the atomic-physics nomenclature, in the nuclear case n denotes the sequential occurrence of each ℓ value, while the $n\ell$ atomic-physics symbology is replaced by $n\ell_j$, e.g., $1s_{1/2}$, $1p_{1/2}$, $1p_{3/2}$, $1d_{3/2}$, ..., $2s_{1/2}$, $2p_{1/2}$, etc. In analogy to the closed-shell numbers 2, 10, 18, etc., the corresponding neutron or proton "magic" (closed-shell) numbers are 2, 8, 20, 28, 50, 82, etc. At these numbers, particular groups of orbitals are fully occupied, although each group typically contains nucleon orbitals with differing values of n. By using the magic numbers; the facts that the $2s_{1/2}$ level lies between the split $1d_j$ levels, the lower-lying $2p_j$ level lies between the split $1f_j$ levels, the two $1g_j$ levels lie between the upper $2p_j$ and lower $2d_j$ levels; and the ordering $E_{22j} < E_{22j'} < E_{301/2} < E_{15j''}$ (with j, j', and j'' to be deduced, and all $E_{n\ell j} > 0$), determine

(a) the ordering of the single-neutron or -proton energy levels $E_{n\ell j}$ through $1h_{j''}$ and construct the Grotrian diagram analogous to Fig. 18.4

(b) the ground-state configurations of ^{3}He, ^{4}He, ^{10}B, ^{11}B, ^{12}C, ^{16}O, ^{20}Ne, ^{37}Ar, ^{40}Ca, and ^{57}Fe.

18.24 In ^{6}Li, which contains three neutrons and three protons, spin–orbit effects are small and the $n\ell_j$ orbitals of the preceding exercise can be replaced by the L–S coupling orbitals of the atomic-shell model. In this approach, the ground-state configuration of ^{6}Li is $(1s_n)^2(1s_p)^2 2p_n 2p_p$, where the subscript n (p) denotes neutron (proton), and the angular-momentum properties are determined by the two p-shell nucleons. Determine all the allowed multiplets $^{2S+1}L_J$ for ^{6}Li.

18.25 ^{42}Ca contains 20 neutrons and 20 protons forming closed shells plus two additional neutrons in the $\ell = 3$ orbit. Since spin–orbit forces are important, the non-closed-shell part of the configuration is $(1f_{7/2})^2$ (see Exercise 18.23 for an explanation of the nomenclature). Here you are to assume that the 40 nucleons in closed shells form a *structureless* $L = S = J = 0$ core, so that ^{42}Ca is to be treated as an effective two-neutron nucleus.

(a) If each of the two neutrons (labeled 1 and 2) is in the state $|13\,\frac{1}{2};\,\frac{7}{2}m_j\rangle$, determine the allowed values of J in the coupled state $|\frac{7}{2},\frac{7}{2};\,JM\rangle$.

(b) The proper ordering of the various $|\frac{7}{2},\frac{7}{2};\,JM\rangle$ states of part (a) can be obtained from a perturbation/variation analysis of the neutron–neutron interaction $A\hat{\mathbf{J}}_1\cdot\hat{\mathbf{J}}_2$, $A>0$. Using $|\frac{7}{2},\frac{7}{2};\,JM\rangle$ as the trial state, determine the energies E_J and their ordering, specifying in particular the value of J for the ground state.

18.26 Verify the RHSs of Eqs. (18.110) and (18.115).

18.27 Using Eq. (18.118) in place of Eq. (18.108), and following the same analysis as that which led to (18.116), verify that the minimizing value of ξ is $\xi_e = 1.238$, that the new equilibrium separation is $R_{eq} = 2.02a_0$, and that D_e becomes 2.37 eV.

Appendix A. Elements of Probability Theory

Basic concepts

Probabilities and probability densities are an intrinsic feature of quantum theory. One cannot fully understand the theory without some knowledge of these concepts. In this appendix we provide a brief introduction to these topics, sufficient for grasping the ideas as they are encountered in this text. For the reader who wishes to pursue this subject further, a short list of references is cited at the end of this appendix.

Probability theory deals with the likelihood that an event of some kind will occur. An event, or action, typically has a variety of outcomes, i.e., it can be realized in a variety of ways. Examples abound. A few are achieving a head when a coin is tossed (and is neither lost nor lands on its edge), obtaining a total of 7 when a pair of dice is thrown, drawing a diamond from a standard deck of 52 cards, and obtaining a velocity $v = \sqrt{k_B T/m}$ from a Boltzmann gas of molecules of mass m at temperature T. The first three of these are examples in which the probabilities are discrete, whereas for the last one, the probability is a continuous function, i.e., it is a continuous probability *distribution* (or probability density).

Suppose that A is a possible outcome of an event. The probability that A will occur (out of all the possibilities) is denoted $P(A)$. If A is certain to occur, then it is assigned a probability of 1: $P(A \text{ is certain}) = 1$. If A cannot occur, then it is assigned a probability of 0: $P(A \text{ cannot occur}) = 0$. All other probabilities lie between these two limits, viz.,

$$0 \leqslant P(A) \leqslant 1. \tag{A1}$$

Normalization of quantal states to unity refers to the upper limit of 1 in Eq. (A1).

Consider now an event in which outcomes A and B can occur. We restrict ourselves to those cases for which the outcomes are statistically independent of one another. That is, the possibility of conditional probabilities is excluded. A consequence of this restriction is that the probability $P(AB)$ of *both* A and B occurring becomes the product of the two probabilities:

$$P(AB) = P(A)P(B) \qquad \text{(statistically independent outcomes).} \tag{A2}$$

There are also situations in which outcomes A and B are mutually exclusive. In this case, $P(AB) = 0$, from which one can show that the probability of *either* A or B (or both) occurring, denoted $P(A + B)$, is

$$P(A + B) = P(A) + P(B) \qquad \text{(mutually exclusive outcomes).} \tag{A3}$$

Relations (A2) and (A3) are easily generalized. Let $\{A_i\}_{i=1}^n$ be the set of the n discrete outcomes of a particular event. If A_i and A_j are statistically independent outcomes, $\forall$ i and j, then the probability that all of the outcomes (A_1 and A_2 and A_3 and ... A_n) will occur is

$$P(A_1 A_2 \cdots A_n) = P\left(\prod_{i=1}^{n} A_i\right) = \prod_{i=1}^{n} P(A_i) \qquad \text{(independence).} \tag{A4}$$

Furthermore, when the A_i are mutually exclusive, the probability of any one of the A_i (i.e., of either A_1 or A_2 or ... or A_n, plus any pair plus any triplet,..., etc.) occurring is

$$P\left(\sum_{i=1}^{n} A_i\right) = \sum_i P(A_i) \qquad \text{(mutual exclusivity).} \tag{A5}$$

Mutual exclusivity is often encountered in quantal situations. Let $\mathbf{r}$ be the position coordinate of an electron in an atom and $\rho(\mathbf{r}) = \rho(r, \theta, \phi)$ be the *probability density* that it will be found between $\mathbf{r}$ and $\mathbf{r} + d\mathbf{r}$, with $P(\mathbf{r}) = \rho(\mathbf{r})\, d^3 r$ the corresponding probability (as noted in the next subsection). Since the electron cannot be simultaneously at $\mathbf{r}$ and $\mathbf{r}' \neq \mathbf{r}$, mutual exclusivity in the present case means that $\rho(\mathbf{r}\mathbf{r}') = 0$. Hence, to find the probability that the electron can be anywhere in a sphere of radius a, we apply Eq. (A5), replacing the sum by an integral since $\mathbf{r}$ is a continuous variable:

$$P(r \leqslant a) = \int_0^a r^2\, dr \int_{\text{angles}} d\Omega\, \rho(r, \theta, \phi), \tag{A6}$$

where $\Omega = (\theta, \phi)$ are the polar angles. For an unbound electron (or other quantal particle), the notion of mutual exclusivity need not apply, as suggested by the double slit experiment for electrons discussed in Section 1.3: in this case, the distribution of electrons on the detector screen indicates that the electron has a non-zero probability of passing *through both* slits. (See the last subsection in this appendix.)

Definitions – the discrete case

We turn next to the definitions of probability, examining the discrete case first. Let us assume that the various outcomes of an event are all equally likely to occur, e.g., for an unweighted or unbiased coin, the occurrence of a head or a tail is equally likely. Or, with a deck of 52 distinct, well-shuffled cards, it is equally likely that a 2 or a 6 will be the nth card from the top of the deck. In this equal-likelihood case, if the total number of outcomes of the given event is N and the number of ways outcome A_i can occur is n_i, then

$$P(A_i) \equiv n_i / N. \tag{A7}$$

Equation (A7) is a statement of the *a priori* definition of probability. Since $\sum_i n_i = N$ (by definition of the total number of outcomes), it follows that

$$P \equiv \sum_i P(A_i) = 1, \tag{A8}$$

i.e., this definition yields a total probability P equal to unity. If the outcomes A_i are mutually exclusive, then (A8) also states that $P = P(\sum_i A_i) = 1$.

Equation (A7) is easily exemplified. Suppose that an unbiased coin is tossed, then $P(\text{head}) = P(\text{tail}) = \frac{1}{2}$, since $N = 2$, while $n_{\text{head}} = n_{\text{tail}} = 1$. Correspondingly, the probability of drawing any specified card from a standard deck of 52 is P(any specific card) $= 1/52$, while the probability of achieving any one of the six face values $(1, 2, 3 \ldots, 6)$ when a die is rolled is P(specified value) $= \frac{1}{6}$. If two dice are rolled the probability that the sum of the face values will be 7 is $P(\text{sum} = 7) = 6/36 = \frac{1}{6}$. If human births are uniformly distributed over the calendar year (of 365 days), then the probability P(birthday) for having a birthday on any particular day is 1/365. The probability $P_N^{(2)}$(same day) that among N randomly chosen persons in a room two will have the same date of birth evidently depends on N. Since $P(\text{birthday}) = 1/365$, it might be thought that $P_N^{(2)}(\text{same day}) \approx 0.5$ would require $N \approx 180$. The surprising result is that $P_N^{(2)}$(same day) $\cong 0.5$ when $N = 23$, while it becomes $0.99 \cong 1$ for $N = 57$!

In all but the simplest cases determination of these discrete probabilities requires counting techniques, especially combinatorial methods, as encountered, for instance, in statistical physics. These are not normally needed for the applications of probability theory to quantum mechanics and we shall not study them here.

The *a priori* definition of probability just described, is one of two; we now briefly summarize the other, known as the *a posteriori* definition. The two definitions yield the same results in principle; in practice, however, the latter allows for the possibility of fluctuations in assigning values, and is thus worth knowing about. In the *a posteriori* procedure, N occurrences of the event are initiated. Suppose the outcome A_i happens n_i times. The probability $P(A_i)$ is defined via a limiting process:

$$P(A_i) = \lim_{N\to\infty} \frac{n_i}{N}. \tag{A9}$$

Thus, if a coin is tossed N times and n_{h} is the number of heads that results, then $P(\text{head}) = \lim_{N\to\infty}(n_{\text{h}}/N)$. This should lead to $P(\text{head}) = \frac{1}{2}$ if the coin is uniform.

Equation (A9) is, of course, an idealization. In practice, $\lim N \to \infty$ is never attained. In place of the infinite limit, one tries to ensure that N is large enough that fluctuations are small. Since the size of a fluctuation tends to be proportional[1] to $\sqrt{N}$, the ratio of the size of a fluctuation to the number N is $\sim N^{-1/2}$. The smaller the value of $N^{-1/2}$ the more accurate the value of n_i/N.[2] For N large but finite, Eq. (A9) is supposed to yield essentially the same result as (A7); the latter relation is the definition used herein.

The collection $\{P(A_i)\}$ for the outcomes A_i of some event is known as a (discrete) probability distribution. The *average* value of the outcome, often denoted $\bar{A}$, is defined as

$$\bar{A} \equiv \sum_i A_i P(A_i). \tag{A10a}$$

Average values are sometimes termed *expected* values (in quantum mechanics, the "expectation" value; see Chapter 5), for which the symbol $\langle A \rangle$ is used:

[1] This result is derived in books on probability and statistics, e.g., Feller (1961, 1966).

[2] As we discuss in connection with Postulate III, the connection between this *a posteriori* determination of $P(A_i)$ and its *a priori* definition via (A7) is the analog in quantum theory of the connection between the *average* or *expected* value of an observable calculated theoretically and that measured in an experiment.

$$\langle A \rangle \equiv \sum_i A_i P(A_i) = \bar{A}. \tag{A10b}$$

$\langle A \rangle$ is the notation typically employed in quantum theory.

The average value is one measure of the 'center' of the distribution. Another measure is the *root mean square*, $A_{\text{rms}} \equiv \langle A^2 \rangle^{1/2}$, where

$$\langle A^2 \rangle \equiv \sum_i A_i^2 P(A_i) \tag{A11}$$

is the *second moment* of the distribution. (Higher moments $\langle A^n \rangle$ are defined analogously.) From (A10) and (A11) one defines the *disperson* σ^2 and the *standard deviation* σ:

$$\sigma^2 = \langle (A - \bar{A})^2 \rangle = \langle A^2 \rangle - \langle A \rangle^2. \tag{A12}$$

Analogs of these quantities enter the Heisenberg uncertainty relations.

As an example, suppose that the $\{A_i\}$ are the scores obtained by the N students who have taken an exam. If n_i students have score A_i, then

$$P(A_i) = n_i/N,$$

$$\bar{A} = \sum_i A_i P(A_i), \tag{A13}$$

and $\{P(A_i)\}$ is the discrete probability distribution. Histograms of the scores are sometimes made, as a means of graphically displaying the results. The histogram itself is not the probability distribution, though it is based on and related to it. Similar comments hold if the A_i are the results of measuring the same quantity N times.

Continuous probability distributions

Many of the distributions that occur in physics are continuous, such as the distribution of energies in a beam of projectiles to be used in a scattering experiment and the positions or velocities of particles in a gas in equilibrium at temperature T. The distributions of heights and ages in a large population may also be considered continuous. In such cases, the discrete set of outcomes $\{A_i\}$ is replaced by a continuous set $\{A | A \in [a, b]\}$, while the collection of discrete probabilities $\{P(A_i)\}$ is replaced by the continuous distribution of probability densities $\rho(A)$. The probability $P(A)$ that the outcome will lie in the range A to $A + dA$ is

$$P(A) = \rho(A)\, dA. \tag{A14}$$

We shall explicitly assume for the continuous case that the outcomes A are mutually exclusive. Hence, the probability that the outcomes lie between $c_1 > a$ and $c_2 < b$ is given by the sum of $P(A)$ over this range. For a continuous distribution, the sum becomes an integral:

$$P(A \in [c_1, c_2]) = \int_{c_1}^{c_2} \rho(A)\, dA. \tag{A15}$$

Since A lies in the interval $[a, b]$, it follows from (A15) that

$$P(A \in [a, b]) = \int_a^b \rho(A)\, dA = 1. \tag{A16}$$

In order to satisfy (A16) a probability density $\rho(A)$ will contain a multiplicative normalization constant.

The outcome A can be a scalar quantity such as energy or position in 1-D, a 3-D quantity like momentum, or a multi-dimensional object referring to the simultaneous positions of N particles. Because the fundamental ideas do not change with the dimensionality, our discussion of the generalization to the continuous case is in terms of a 1-D distribution.

Let x be the outcome, with $x \in [a, b]$ distributed according to the probability density $\rho(x)$. Then, the probability $P(x)$ that the outcome lies between x and $x + dx$ is

$$P(x) = \rho(x)\, dx, \tag{A17}$$

such that

$$\int_a^b \rho(x)\, dx = 1. \tag{A18}$$

The number $N(x)$ of objects per unit interval in x that have characteristic x out of the total number N is

$$N(x) = N\rho(x), \tag{A19}$$

which, from (A18), leads to

$$N = \int_a^b N(x)\, dx. \tag{A20}$$

The expected or average value of x, again denoted either $\langle x \rangle$ or $\bar{x}$, is

$$\langle x \rangle = \int_a^b x\rho(x)\, dx; \tag{A21}$$

$\langle x \rangle$ is the first moment of the distribution. The second moment or mean-square value of x is

$$\langle x^2 \rangle = \int_a^b x^2 \rho(x)\, dx, \tag{A22}$$

with

$$x_{\mathrm{rms}} = \langle x^2 \rangle^{1/2}. \tag{A23}$$

The dispersion σ^2 (standard deviation $\sigma = \sqrt{\sigma^2}$) is defined in strict analogy to the discrete case:

$$\sigma^2 = \langle (x - \langle x \rangle)^2 \rangle \tag{A24a}$$

$$= \langle x^2 \rangle - \langle x \rangle^2. \tag{A24b}$$

Higher moments $\langle x^n \rangle$ are defined analogously:

$$\langle x^n \rangle = \int_a^b x^n \rho(x)\, dx. \tag{A25}$$

If $a \geqslant 0$ and $b = \infty$, the existence of these moments requires that $\rho(x \to \infty) \to 0$ with sufficient rapidity.

As a specific example of a continuous distribution, consider the Gaussian case:

$$\rho(x) = \frac{1}{d\sqrt{2\pi}} e^{-(x-x_0)^2/(2d^2)}, \qquad x \in [-\infty, \infty]; \tag{A26}$$

the factor $(d\sqrt{2\pi})^{-1}$ is the normalization constant. It is straightforward to show (via tabulated integrals, see Chapter 6), that $\rho(x)$ is indeed normalized to unity and that

$$\langle x \rangle = x_0 \tag{A27}$$

and

$$\sigma^2 = \langle x^2 \rangle - \langle x \rangle^2 = d^2, \tag{A28}$$

so that the standard deviation σ is equal to d.

For a Boltzmann gas at temperature T (formed of N structureless, non-interacting particles of mass m), the jth component v_j of velocity ($j = x$ or y or z) is governed by a normalized Gaussian probability distribution, $f(v_j)$:

$$f(v_j) = \sqrt{\frac{m}{2\pi k_{\rm B} T}} e^{-mv_j^2/(2k_{\rm B}T)}\, dv_x, \tag{A29}$$

where $k_{\rm B}$ is Boltzmann's constant. From (A29) one obtains[3] the distribution $F(v^2)$ for the square of the velocity, $v^2 = v_x^2 + v_y^2 + v_z^2$, given by ($v \in [0, \infty]$)

$$F(v^2) = 4\pi \left(\frac{m}{2\pi k_{\rm B} T} \right)^{3/2} v^2 e^{-mv^2/(2k_{\rm B}T)}\, dv, \tag{A30}$$

which is no longer Gaussian. This distribution leads to the standard result that the average kinetic energy of the gas, $\langle K \rangle$, is equal to $\frac{3}{2}k_{\rm B}T$, a result easily derived.

The two-slit experiment

When two outcomes A and B are mutually exclusive, the probability that either or both will occur is given by Eq. (A3). Consider the two-slit experiment discussed in Chapter 1 (Fig. 1.1). A particle is directed from the left towards the slitted screen. Outcome A (B) is that it reaches the point P after passing through slit 1 (2). If the particle is a classical object, it can pass through only one of the two slits, thus guaranteeing mutual exclusivity; hence,

$$P(A + B) = P(A) + P(B), \qquad \text{classical particle.}$$

There is no interference, just as in the case discussed in Section 1.1, in which the intensities were added. One of the many interesting features of quantum theory is that the probability is obtained from the absolute square of a quantal amplitude. If there is more than one available path to follow, the relevant amplitude becomes a sum of amplitudes,

[3] See, for example, Ohanian (1995).

one for each "open" path, and squaring such an amplitude *always* gives rise to interference effects, as in the two-slit experiment for electrons. This aspect of quantum theory is formulated as Postulate II in Chapter 5.

Probability References: Feller (1961, 1966); Gasiorowicz (1979); Hausner (1971); Margenau and Murphy (1943); Mathews and Walker (1970); Weaver (1963).

Appendix B. Fourier Series and Integrals

One result of Section 3.1 is that a continuous function $y(x)$, defined for $x \in [0, L]$ and obeying $y(0) = y(L) = 0$, can be expressed as a Fourier sine series

$$y(x) = \sum_{n=1}^{\infty} b_n \sin\left(\frac{n\pi x}{L}\right), \tag{B1}$$

with

$$b_n = \frac{2}{L}\int_0^L dx \sin\left(\frac{n\pi x}{L}\right) y(x). \tag{B2}$$

Equation (B2) follows from two conditions: the assumption that (B1) is a convergent series, thus allowing interchange of summation and integration, and the orthogonality relation (3.43).

The set $\{\sqrt{2/L}\sin(n\pi x/L)\}_{n=1}^{\infty}$ is both orthonormal (Eq. (3.43)) and *complete*, which means that it provides a unique expansion set for any sufficiently well-behaved function $y(x)$ satisfying appropriate boundary and symmetry conditions, not just $y(0) = y(L) = 0$.

In the next subsection, we shall generalize the sine series to the more general Fourier series involving both sines and cosines. The notions of orthonormality and completeness will occur here as well; they both provide additional examples of these important concepts for quantum mechanics.

Fourier series

Let $f(x)$ be a piecewise continuous function having a piecewise continuous derivative in the interval $x \in [a, a + 2L]$ such that $f(a) = f(a + 2L)$. Then, as shown in many books (e.g., Dettman (1988) or Mathews and Walker (1970)), $f(x)$ has the expansion

$$f(x) = \frac{a_0}{2} + \sum_{n=1}^{\infty}\left[a_n \cos\left(\frac{n\pi x}{L}\right) + b_n \sin\left(\frac{n\pi x}{L}\right)\right], \tag{B3}$$

where

$$a_n = \frac{1}{L}\int_a^{a+2L} dx \cos\left(\frac{n\pi x}{L}\right) f(x)$$

and

$$b_n = \frac{1}{L}\int_a^{a+2L} dx \sin\left(\frac{n\pi x}{L}\right) f(x).$$

Equation (B3) is the Fourier series for $f(x)$ and the set of functions

$$\left\{\sqrt{\frac{1}{L}} + \left\{\sqrt{\frac{1}{L}}\cos\left(\frac{n\pi x}{L}\right), \sqrt{\frac{1}{L}}\sin\left(\frac{n\pi x}{L}\right)\right\}_{n=1}^{\infty}\right\}$$

is complete and orthonormal in the intervals $[0, 2L]$ and $[a, a+2L]$. Orthonormality here means that, if $\varphi_n(x)$ and $\varphi_m(x)$ are any two functions from this set (two sines, a sine and a cosine, etc.), then

$$\int_a^{a+2L} dx\, \varphi_n(x)\varphi_m(x) = \delta_{nm}, \tag{B4}$$

where δ_{nm} is the Kronecker delta, Eq. (3.44). As in the case of Eq. (3.43), (B4) can be proved using various trigonometric identities for the products of two sines, of two cosines, or of a sine and cosine.

An enormous variety of functions can be expanded in Fourier series. We shall consider two special classes: functions that are even and those which are odd, categories of particular interest in quantum mechanics. To simplify the discussion, the arbitrary point a will be set equal to $-L$, so that

$$x \in [-L, L],$$

$$f(-L) = f(L),$$

$$a_n = \frac{1}{L}\int_{-L}^{L} dx \cos\left(\frac{n\pi x}{L}\right) f(x), \tag{B5a}$$

$$b_n = \frac{1}{L}\int_{-L}^{L} dx \sin\left(\frac{n\pi x}{L}\right) f(x), \tag{B5b}$$

and

$$\int_{-L}^{L} dx\, \varphi_n(x)\varphi_m(x) = \delta_{nm}. \tag{B6}$$

Setting $a = -L$ can be thought of as a relocation or translation of the origin of coordinates to the point $a + L$. There is no change in the expansion (B3).

Even functions: $f(-x) = f(x)$. In this case, (B5a) can be transformed as follows:

$$\begin{aligned} a_n &= \frac{1}{L}\int_0^L dx \cos\left(\frac{n\pi x}{L}\right) f(x) - \frac{1}{L}\int_L^0 dx \cos\left(\frac{n\pi x}{L}\right) f(-y) \\ &= \frac{2}{L}\int_0^L dx \cos\left(\frac{n\pi x}{L}\right) f(x), \end{aligned} \tag{B7a}$$

where the changes of variables $x \to -y$ and then $y \to x$, and the relation $f(-y) = f(y)$ have been employed. An analogous procedure carried out on Eq. (B5b) yields

$$b_n = \frac{1}{L}\int_0^L dx \sin\left(\frac{n\pi x}{L}\right) f(x) + \frac{1}{L}\int_L^0 dx \sin\left(\frac{n\pi x}{L}\right) f(x) = 0, \tag{B7b}$$

the absence of the minus sign in the second integral resulting from $\sin(-\theta) = -\sin\theta$. Hence, for an even function $f(x)$, Eq. (B3) reduces to a cosine series:

$$f(x) = \frac{a_0}{2} + \sum_{n=1}^{\infty} a_n \cos\left(\frac{n\pi x}{L}\right), \tag{B8}$$

with a_n given by Eq. (B7a). Examples of even functions are $f(x) = \text{constant}$; $f(x) = x^{2m}$; $f(x) = \cos(kx)$, $f(x) = \exp(|x|)$, etc. As an illustration of (B8), we choose $f(x) = x^2$. Then

$$\begin{aligned} a_n &= \frac{1}{L}\int_0^L x^2 \cos\left(\frac{n\pi x}{L}\right) dx \\ &= \begin{cases} 2L^2/3, & n = 0, \\ (-1)^n 4L^2/(n^2\pi^2), & n \geqslant 1, \end{cases} \end{aligned}$$

where the relations $\int y^2 \cos y \, dy = 2y\cos y + (y^2 - 2)\sin y$ and $\cos(n\pi) = (-1)^n$ have been employed. The series for x^2 is therefore

$$x^2 = \frac{L^2}{3} + \frac{4L^2}{\pi^2}\sum_{n=1}^{\infty}\frac{(-1)^n}{n^2}\cos\left(\frac{n\pi x}{L}\right), \qquad x \in [-L, L]. \tag{B9}$$

Odd functions: $f(-x) = -f(x)$. The same analyses as those performed in the case in which $f(x)$ is even now yield

$$a_n = 0 \tag{B10a}$$

and

$$b_n = \frac{2}{L}\int_0^L dx \sin\left(\frac{n\pi x}{L}\right) f(x). \tag{B10b}$$

Equation (B10b) is identical in form to Eq. (3.46) for the coefficients in the Fourier-sine-series representation of the $t = 0$ shape of the stretched string. This implies that, if the string were to be extended to include the interval $[-L, 0]$, then its shape there would be opposite (in sign) to that in $[0, L]$. It is this implied "odd" behavior that underlies the pure sine-series expansion of the initial form of the string.

Fourier integrals

Fourier series are examples of expansions in terms of a discretely indexed, orthonormal, complete set. Analogous, but *continuously indexed*, expansion sets also exist; they are vital to quantum mechanics. An example is the Fourier-integral or Fourier-transform representation, which is considered next. We begin by noting that, if $k_n = n\pi/L$, $n = 0$, $\pm 1, \pm 2, \ldots$, then $\exp(ik_n x)$ solves the differential equation (3.84) with $k \to k_n$. Furthermore, if the orthogonality integral (or "scalar" product) is changed by including the complex conjugate of one of the two functions – a requirement because $\exp(ik_n x)$ is complex – then $\{\psi_{k_n}(x) = \sqrt{1/(2L)}\exp(ik_n x)\}$ is an orthonormal set on $[-L, L]$:

$$\int_{-L}^{L} dx\,\psi_{k_n}^*(x)\psi_{k_m}(x) = \frac{1}{2L}\int_{-L}^{L} dx\,e^{i\pi(m-n)x/L} = \delta_{nm}. \tag{B11}$$

Finally, $\{\psi_{k_n}(x)\}_{-\infty}^{\infty}$ is complete and periodic w.r.t. $2L$,

$$\psi_{k_n}(x) = \psi_{k_n}(x+2L). \tag{B12}$$

Therefore $\{\psi_{k_n}(x)\}$ can be used to form a complex Fourier-series expansion for any sufficiently well-behaved function $f(x)$ obeying $f(-L) = f(L)$:

$$f(x) = \sum_{n=-\infty}^{\infty} c_n \frac{e^{in\pi x/L}}{\sqrt{2L}} \qquad \left(= \sum c_n \psi_{k_n}(x)\right), \tag{B13}$$

with

$$c_n = \sqrt{\frac{1}{2L}}\int_{-L}^{L} dx\, f(x) e^{-in\pi x/L}. \tag{B14}$$

Note that the integral in (B14) involves $\exp(-in\pi x/L)$, the complex conjugate of $\psi_{k_n}(x)$.

Expansions such as (B13) are often employed when one works in a finite rather than an infinite volume (or on a line) and requires that the behavior of the system at one edge be the same as that at the other edge; i.e., that $f(-L) = f(L)$. Our interest in (B13), however, is as an intermediary for deriving the Fourier-integral representation, which applies to the case of the infinite interval $[-\infty, \infty]$, where a physicist's convention $[-\infty, \infty]$ rather than $(-\infty, \infty)$ is used for the infinite interval.

To obtain the desired form, several limits must be taken, e.g., $L \to \infty$ and $\sum_{-\infty}^{\infty} \to \int_{-\infty}^{\infty} dk$. Our argument is heuristic; rigor can be found in the mathematical literature, e.g., Carslaw (1930) and Titchmarsh (1937).

Before taking the limit $L \to \infty$, we must absorb the factors $\sqrt{1/L}$ and also the $1/L$ in $\exp(in\pi x/L)$. This is done by means of new variables:

$$\xi = \sqrt{\frac{\pi}{L}}x,$$

$$q_n = \sqrt{\frac{\pi}{L}}n = \sqrt{\frac{L}{\pi}}k_n,$$

and

$$\Delta q_n = q_{n+1} - q_n = \sqrt{\frac{\pi}{L}};$$

we also introduce g for c_n:

$$c_n \to g(q_n).$$

With these changes, Eqs. (B13) and (B14) become

$$f(\xi) = \sqrt{\frac{1}{2\pi}} \sum_{q_n=-\infty}^{\infty} \Delta q_n\, g(q_n) e^{iq_n\xi} \tag{B15}$$

and

$$g(q_n) = \sqrt{\frac{1}{2\pi}} \int_{-\sqrt{L\pi}}^{\sqrt{L\pi}} d\xi\, e^{-iq_n\xi} f(\xi), \tag{B16}$$

where, in the new sum on q_n in (B15), the steps are in units of $\Delta q_n = \sqrt{\pi/L}$.

The limit $L \to \infty$ can now be taken: $q_n \to q$, $\sum_{q_n} \Delta q_n \to \int_{-\infty}^{\infty} dq$, and $\sqrt{L\pi} \to \infty$, so that

$$f(\xi) = \sqrt{\frac{1}{2\pi}} \int_{-\infty}^{\infty} dq\, e^{iq\xi} g(q)$$

and

$$g(q) = \sqrt{\frac{L}{2\pi}} \int_{-\infty}^{\infty} d\xi\, e^{-iq\xi} f(\xi).$$

Re-naming via $\xi \to x$ and $q \to k$, we finally get the standard Fourier-integral relations

$$f(x) = \sqrt{\frac{1}{2\pi}} \int_{-\infty}^{\infty} dk\, e^{ikx} g(k) \tag{B17}$$

and

$$g(k) = \sqrt{\frac{1}{2\pi}} \int_{-\infty}^{\infty} dx\, e^{-ikx} f(x). \tag{B18}$$

The function $g(k)$ is known as the Fourier transform of $f(x)$, $f(x)$ is referred to as the inverse (Fourier) transform, and the two functions together are denoted Fourier-transform pairs. It should be evident that $f(x)$ (and $g(k)$) must be sufficiently well behaved that the Fourier-integral representation is meaningful: the relevant integral must exist. We shall assume this to be the case, and refer the reader to the mathematical literature for details.

As a preliminary to considering the complete-set-expansion aspect of (B17) and its application to a wave packet, we mention a few instances of Fourier-transform pairs. Let $f(x)$ be an even function, $f(-x) = f(x)$. Then, precisely as in the Fourier series case (Eq. (B8)), only cosine terms enter the Fourier integrals. For example, if we choose $f(x)$ to be a Gaussian function centered at $x = 0$,

$$f(x) = e^{-x^2/(2d^2)}, \qquad d > 0, \tag{B19a}$$

then

$$g(k) = \frac{d}{2} e^{-k^2 d^2/2}. \tag{B19b}$$

The Fourier transform of a Gaussian is itself a Gaussian, a fact we shall make use of in Chapter 6.

As a second example involving a physical system, let $k \to t = \text{time}$ and $g(k) \to g(t)$ such that

$$g(t) = \begin{cases} 0, & t < 0, \\ e^{-t/\tau}, & t \geqslant 0. \end{cases} \tag{B20a}$$

Equation (B20a) describes the decay of a radio-active population of lifetime τ that has been created at time $t = 0$ by some process. In place of $f(x)$ one determines $f(\omega)$, where ω is the frequency variable conjugate to time:

$$f(\omega) = \sqrt{\frac{1}{2\pi}} \int_{-\infty}^{\infty} dt\, e^{i\omega t} g(t)$$
$$= \frac{\tau}{\sqrt{2\pi}} \frac{1 + i\omega\tau}{1 + \omega^2\tau^2}$$
$$= \frac{\tau/\sqrt{2\pi}}{1 - i\omega\tau}. \qquad \text{(B20b)}$$

Here, $f(\omega)$ is necessarily complex, in order that its inverse Fourier transform produce a function $g(t)$ which is zero for $t < 0$. This vanishing of $g(t)$, $t < 0$, is a statement of *causality*, viz., that no system can begin decaying before it comes into existence. The general manifestation in frequency space of this requirement is that $f(\omega)$, considered as a function of complex ω, can have poles only for $\operatorname{Im}\omega < 0$ (the lower half plane). In this present example the pole occurs, i.e., $f(\omega) = \infty$, when $\omega = -i/\tau$, as required.[1]

Fourier transforms are applicable in many areas of physics and mathematics, e.g., they are often a useful means for solving differential equations. One way of regarding the Fourier transform is as an expansion of a function $f(x)$ in terms of the complete set of continuously indexed functions

$$\psi_k(x) = \sqrt{\frac{1}{2\pi}} e^{ikx}. \qquad \text{(B21)}$$

Comparison of Eqs. (3.20) and (B21) shows that $\psi_k(x)$ is the $t = 0$ or spatial portion of the complex plane wave

$$\Psi_k(x, t) = \sqrt{\frac{1}{2\pi}} e^{i(kx - \omega t)}. \qquad \text{(B22)}$$

$\psi_k(x)$ is also referred to as a plane wave. It is a function of some significance in quantum mechanics; e.g., it is the coordinate-space eigenfunction of the 1-D momentum operator that corresponds to momentum eigenvalue $p = \hbar k$.

We now use these plane waves to represent a quantum wave packet. The wave form we choose to examine has been studied, for example, by French (1971). It is

$$\Psi(x, t) = \frac{N}{b^2 + (x - vt)^2}, \qquad b > 0, \qquad \text{(B23)}$$

where N is a normalization constant such that $\int_{-\infty}^{\infty} |\Psi(x, t)|^2\, dx = 1$. (Determination of N is left as an exercise.)

In contrast to pure sine or exponential functions of $x - vt$, $\Psi(x, t)$ of (B23) is *localized in space* for any fixed t. Choosing $t = 0$, (B23) is

$$\Psi(x, 0) = \frac{N}{b^2 + x^2}, \qquad \text{(B24)}$$

[1] The subject of causality, in the sense of "no output prior to input," has been the subject of much study. It is a topic for an advanced course in quantum mechanics, and is thus only an incidental example here. Some discussion and references can be found, for instance, in Mathews and Walker (1970).

whose functional form is shown in Fig. B.1(a). Figure B.1(b) depicts $\Psi(x, \Delta t)$, thus establishing how the peak in Ψ has moved from $x = 0$ to $x = v\,\Delta t$.

$\Psi(x, t)$ can be expressed as a Fourier integral. Setting $\xi = x - vt$, we write

$$\Psi(x, t) \equiv \frac{N}{b^2 + \xi^2} = \sqrt{\frac{1}{2\pi}} \int_{-\infty}^{\infty} dk\, e^{ik\xi} A(k); \tag{B25}$$

$A(k)$ is known as the "profile function," and is determined from the inverse transform:

$$A(k) = \sqrt{\frac{1}{2\pi}} \int_{-\infty}^{\infty} dx\, e^{-ikx} \frac{N}{b^2 + x^2}. \tag{B26}$$

Since $b^2 + x^2$ is an even function of x, only the $\cos(kx)$ portion of $\exp(-ikx)$ contributes to the integral, whose value (from integral tables) is

$$A(k) = \frac{N}{b} \sqrt{\frac{\pi}{2}} e^{-b|k|}. \tag{B27}$$

Equation (B27) represents the wave packet in momentum (or wave-number) space. Like $\Psi(x, t)$, it too is localized, with a cusp at $k = 0$; it is plotted in Fig. B.2.

There is an interesting relation between the "spread" Δx of a localized function in coordinate space and the spread Δk of the Fourier transform. It is

$$\Delta x\, \Delta k \sim 1, \tag{B28}$$

reminiscent of the uncertainty relation $\delta x\, \delta p_x \sim \hbar$ deduced for the γ-ray microscope, or the more general inequality

$$\Delta x\, \Delta P_x \geqslant \hbar/2$$

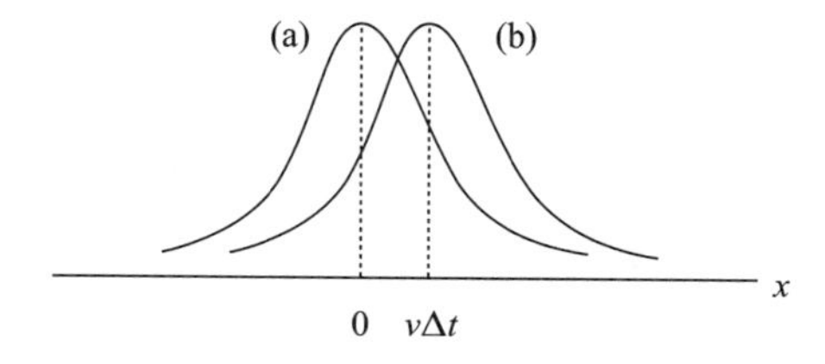

Fig. B1 Short-time propagation of the wave packet (B23): (a) $\Psi(x, 0)$ and (b) $\Psi(x, \Delta t)$.

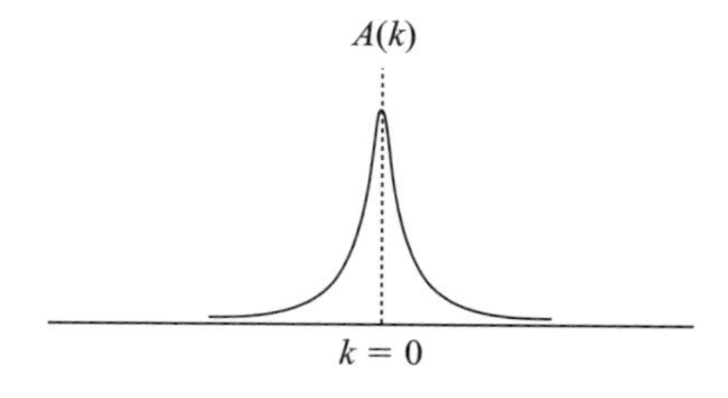

Fig. B2 A sketch of the profile function $A(k)$ of Eq. (B27) corresponding to the wave packet (B23).

which we will define and derive in Chapter 9. Equation (B28) is an order-of-magnitude estimate holding true for (localized) Fourier-transform pairs (e.g., Mathews and Walker (1970)); it is not, however, equivalent to the uncertainty relation but only like it. In the present case, we choose as the spreads Δx and Δk the full widths at half maximum of $\Psi(x, 0)$ and $A(k)$, respectively. They are $\Delta x = 2b$ and $\Delta k = (\ln 2)/b$, from which we get

$$\Delta x \, \Delta k \cong 3.8;$$

3.8 is certainly $\sim$1, and the prediction (B28) is thus verified in the present case.

Appendix C. Solution of Legendre's Equation

The solution of Legendre's equation, viz.,

$$(1-x^2)P_\lambda^{m''}(x) - 2xP_\lambda^{m'}(x) + \left(\lambda - \frac{m^2}{1-x^2}\right)P_\lambda^m(x) = 0, \qquad (10.38)$$

with $' = d/dx$, is carried out in two stages, first for $m = 0$ using a power-series expansion and then for $m \neq 0$ via differentiation, as noted below Eq. (10.40). We start with $m = 0$.

Legendre polynomials

When $m = 0$, Eq. (10.38) becomes ($P_\lambda^0 \to P_\lambda$)

$$(1-x^2)P_\lambda''(x) - 2xP_\lambda'(x) + \lambda P_\lambda(x) = 0. \qquad (C1)$$

We seek a solution in the form

$$P_\lambda(x) = \sum_{n=0}^{\infty} a_n x^n \qquad (C2)$$

(rather than $\sum a_n x^{n+s}$, since the index s is easily shown to be zero). Substituting (C2) into (10.38) eventually leads to

$$\sum_{n=0}^{\infty}[(n+1)(n+2)a_{n+2} + \lambda a_n]x^n - \sum_{n=2}^{\infty} n(n-1)a_n x^n - \sum_{n=1}^{\infty} 2na_n x^n = 0,$$

and equating to zero the coefficient multiplying each x^n, $n = 0, 1, 2, \ldots$, yields a general two-term recursion relation:

$$a_{n+2} = \frac{n(n+1)-\lambda}{(n+1)(n+2)} a_n. \qquad (C3)$$

From Eq. (C3) it follows that the a_{2n} are proportional to a_0, whereas the a_{2n+1} are proportional to a_1. The generic even and odd solutions, $P_\lambda^{(e)}(x)$ and $P_\lambda^{(o)}(x)$, can therefore be written as

$$P_\lambda^{(e)}(x) = a_0 \sum_{n=0}^{\infty}\left(\frac{a_{2n}}{a_0}\right)x^{2n} \qquad (C4a)$$

and

$$P_\lambda^{(o)}(x) = a_1 x \sum_{n=0}^{\infty} \left(\frac{a_{2n+1}}{a_1} \right) x^{2n}. \tag{C4b}$$

Equations (C4) are a valid representation as long as each infinite series converges for all $x \in [-1, 1]$. Using the ratio test we find

$$\lim_{n\to\infty} \frac{a_{n+2}x^{n+2}}{a_n x^n} \longrightarrow x^2, \tag{C5}$$

which guarantees convergence as long as $|x| < 1$. However, when $|x| = 1$, each series behaves asymptotically like[1] $\ln(1 - x^2)$, which diverges at $|x| = +1$. The only way to avoid this singular behavior and ensure a finite probability density for all $x = \cos\theta$ is for the infinite series (C4) to become a polynomial, i.e., for all coefficients a_{n+2} to vanish when n exceeds a selected value. The freedom to accomplish this resides in the as-yet-unspecified value of λ, the eigenvalue of $\hat{L}^2$. Choosing $\lambda = \ell(\ell + 1)$, where ℓ is an integer or zero, then $a_n \longrightarrow a_n^{(\ell)}$ and (C3) becomes

$$a_{n+2}^{(\ell)} = \frac{n(n+1) - \ell(\ell+1)}{(n+1)(n+2)} a_n^{(\ell)}, \tag{C6a}$$

and it is obvious that

$$a_{n+2}^{(\ell)} = 0, \qquad n \geqslant \ell,\ \ell = 0, 1, 2, \ldots. \tag{C6b}$$

Once again imposition of a boundary condition has determined eigenvalues and the form of the eigensolutions. The relations (C6) imply that the solution is even or odd as ℓ is even or odd, with ℓ being the highest power of x that occurs in the polynomial. Subject to a normalization condition, the $P_\ell(x)$ are the *Legendre* polynomials:

$$P_\ell(x) = \sum_{n=0}^{N} c_n^{(\ell)} x^{\ell-2n}, \qquad N = \begin{cases} \ell/2, & \ell = \text{even}, \\ (\ell-1)/2, & \ell = \text{odd}. \end{cases} \tag{C7}$$

Comparing Eqs. (C6) and (C7), the connection between the $c_n^{(\ell)}$ and the $a_n^{(\ell)}$ is found to be

$$\frac{c_n^{(\ell)}}{c_{n-1}^{(\ell)}} = \frac{a_{\ell-2n}^{(\ell)}}{a_{\ell-2n+2}^{(\ell)}} = \frac{(\ell-2n+1)(\ell-2n+2)}{(\ell-2n)(\ell-2n+1) - \ell(\ell+1)}$$
$$= \frac{(\ell-2n+1)(\ell-2n+2)}{2n(2\ell-2n+1)}.$$

The constant $c_0^{(\ell)}$ combines the roles of a_0 and a_1 in Eqs. (C4); it specifies the normalization of $P_\ell(x)$. Although $P_\ell(x)$ is an orthogonal set on the interval $[-1, 1]$, $c_0^{(\ell)}$ is chosen so that $P_\ell(x = 1) = 1$, rather than making $\int_{-1}^{1} dx\, P_\ell^2(x) = 1$, as discussed, e.g., in Copson (1935) and Wyld (1976):

[1] The Taylor series for $\ln(1 - x^2)$ is

$$\ln(1 - x^2) = x^2 + \frac{x^4}{2} + \frac{x^6}{3} + \cdots + \frac{x^{2n}}{n} + \cdots;$$

the asymptotic ratio of successive terms is x^2 as claimed above.

$$c_0^{(\ell)} = \frac{(2\ell)!}{2^\ell(\ell!)^2} \Rightarrow P_\ell(1) = 1. \tag{C8}$$

With the (mathematician's) choice (C8) for $c_0^{(\ell)}$, $P_\ell(x)$ can be put in the following form:

$$P_\ell(x) = \frac{1}{2^\ell \ell!}\frac{d^\ell}{dx^\ell}(x^2-1)^\ell, \tag{C9}$$

an expression known as Rodrigues' formula.

Associated Legendre functions

Rather than solve (10.38) directly when $m \neq 0$, we first claim that, if a power series approach were taken, then solutions finite throughout $[-1, 1]$ would be obtained only if $\lambda = \ell(\ell+1)$, $\ell = 0, 1, 2, \ldots$. We shall derive this result in Chapter 14 using "algebraic" methods, but its plausibility is related to the fact that the $m \neq 0$ solutions have $\phi \neq 0$ whereas the $m = 0$ ones (i.e., the $P_\ell(\cos\theta)$) correspond to $\phi = 0$. Going from $\phi = 0$ to $\phi \neq 0$ can be accomplished by a rotation about $\mathbf{e}_3$. Such a rotation cannot change the value of λ, essentially because λ is the $\hat{L}^2$ eigenvalue and $[\hat{L}_z, \hat{L}^2] = 0$. Therefore, if $\lambda = \ell(\ell+1)$ when $\phi = 0$, it must also be $\ell(\ell+1)$ when $\phi \neq 0$.

Setting $\lambda = \ell(\ell+1)$, Eq. (10.38) becomes

$$(1-x^2)P_\ell^{m\prime\prime}(x) - 2xP_\ell^{m\prime}(x) + \left(\ell(\ell+1) - \frac{m^2}{1-x^2}\right)P_\ell^m(x) = 0. \tag{C10}$$

As noted in Chapter 10, the $P_\ell^m(x)$ are the associated Legendre functions, which we now determine. Since $m \neq 0$, there is an additional singular term in (C10) as compared with (C1), viz., $m^2/[(1-x)(1+x)]$, where $(1 \pm x)^{-1}$ is singular at $x = \mp 1$. If a power-series solution were to be used in the vicinity of either one of these points, say $P_\ell^m(x) = \sum_{n=0}^\infty b_n(x \pm 1)^{n+\alpha}$, then an easy calculation would give $\alpha = |m|/2$ as the indicial equation. The form of solution we consider takes account of the $(x \pm 1)^{|m|/2}$ factors by explicitly extracting them both. That is, we write

$$\begin{aligned} P_\ell^m(x) &= (1-x)^{|m|/2}(1+x)^{|m|/2}u(x) \\ &= (1-x^2)^{|m|/2}u(x). \end{aligned} \tag{C11}$$

Substituting (C11) into (C10), we find

$$(1-x^2)u''(x) - 2(|m|+1)xu'(x) + [(\ell+1) - |m|(|m|+1)]u(x) = 0. \tag{C12}$$

Instead of solving this equation directly, say by power series expansion of $u(x)$, we compare it with the equation that results when the operator $d^{|m|}/dx^{|m|}$ is applied to each term of the $\lambda = \ell(\ell+1)$ version of Eq. (C1). The latter equation is

$$\begin{aligned} (1-x^2)\left(\frac{d^{|m|}P_\ell(x)}{dx^{|m|}}\right)'' - 2(|m|+1)x\left(\frac{d^{|m|}P_\ell(x)}{dx^{|m|}}\right)' \\ +[\ell(\ell+1) - |m|(|m|+1)]\frac{d^{|m|}P_\ell(x)}{dx^{|m|}} = 0, \end{aligned} \tag{C13}$$

and comparison of it with Eq. (C12) immediately provides the function $u(x)$:

$$u(x) = \frac{d^{|m|} P_\ell(x)}{dx^{|m|}}. \tag{C14}$$

The trick of differentiating (C1) and then comparing with (C12) is standard in the mathematical literature. The proof that (C13) is the result of such differentiation is left as an exercise for readers who wish to carry out the derivation rather than look it up.

Combining Eqs. (C11) and (C13), we have

$$P_\ell^m(x) = (1 - x^2)^{|m|/2} \frac{d^{|m|} P_\ell(x)}{dx^{|m|}},$$

which appears in the main text as Eq. (10.48).

Further information can be gained by inserting Rodrigues' formula (C9) into (C14), which yields

$$u(x) = \frac{1}{2^\ell \ell!} \frac{d^{\ell+|m|}}{dx^{\ell+|m|}} (x^2 - 1)^\ell, \tag{C15}$$

thereby showing that $u(x)$ is a polynomial of degree $\ell - |m|$. Hence, non-trivial solutions $u(x)$ exist only when $|m| \leqslant \ell$. This conclusion may also be reached from the inequality (10.31) with $\lambda = \ell(\ell + 1)$ and $\mu = m$, viz.,

$$\ell(\ell + 1) \geqslant m^2. \tag{C16}$$

Since $|m|$ must be an integer or zero, (C16) can be satisfied only if

$$|m| \leqslant \ell, \tag{C17}$$

i.e., $m = -\ell, -\ell + 1, \ldots, -1, 0, 1, 2, \ldots, \ell$, a result referred to above Eq. (10.52).

Appendix D. Fundamental and Derived Quantities: Conversion Factors

Quantity (Symbol)	SI Unit (symbol)	Factor	cgs Unit (symbol)
Length	Meter (m)	10^2	Centimeter (cm)
Time (t)	Second (s)		Second (s)
Mass (M, m)	Kilogram (kg)	10^3	Gram (g)
Force (**F**)	Newton (N)	10^5	Dyne (d)
Energy (E)	Joule (J)	10^7	Erg
	$1 \text{ eV} = 1.6 \times 10^{-19}$ J	$1 \text{ MeV} = 10^6$ eV	$1 \text{ eV} = 1.6 \times 10^{-12}$ erg
Power	Watt (W)	10^7	erg s^{-1}
Charge (q)	Coulomb (C)	3×10^9	esu
Electric potential (Φ)	Volt (V)	$1/300$	Stat volt (SV)
Electric field ($\mathcal{E}$)	V m^{-1}	$10^{-4}/3$	SV cm^{-1}
Magnetic induction (B)	Tesla (T)	10^4	Gauss (G)
Speed of light (c)	2.9979×10^8 m s^{-1}		2.9979×10^{10} cm s^{-1}
Unit of charge (e)	1.6×10^{-19} C		4.8×10^{-10} esu
Femtometer (fm)/fermi (F)	10^{-15} m		10^{-13} cm
Ångström unit (Å)	10^{-10} m		10^{-8} cm
Fine-structure constant (α)	$e^2/(4\pi\varepsilon_0\hbar c)$	$= 1/137.036$	$= e^2/(\hbar c)$
Planck's constant (h)	6.626×10^{-34} J s		6.626×10^{-27} erg s
Reduced Planck constant ($\hbar = h/(2\pi)$)	1.0546×10^{-34} J s		6.582×10^{-22} MeV s
$\hbar c$	3.1615×10^{-26} J m	$= 197.327$ MeV fm	
Electron mass (m_e)	9.109×10^{-31} kg	$= 0.511 \text{ MeV}/c^2$	
Proton mass (M_p or m_p)	1.673×10^{-27} kg	$= 938.272 \text{ MeV}/c^2$	
Neutron mass (M_n or m_n)	1.675×10^{-27} kg	$= 939.566 \text{ MeV}/c^2$	

References

Abramowitz, M. and Stegun, I. (1965): *Handbook of Mathematical Functions* (Dover, New York)
Acton, F. S. (1970): *Numerical Methods That Work* (Harper and Row, New York)
Aharonov, Y. and Anandan, J. (1987): *Phys. Rev. Lett.* **58**, 1593
Aharonov, Y. and Bohm, D. (1959): *Phys. Rev.* **115**, 485
Aharonov, Y. and Casher, A. (1984): *Phys. Rev. Lett.* **53**, 319
Ajzenberg-Selove, F. and Lauritsen, T. (1959): *Nucl. Phys.* **11**, 225
Anandan, J. S. (1990): *Quantum Coherence* (World Scientific, Singapore)
Aspect, A., Dalibard, J., Grangier, P., and Roger, G. (1981): *Phys. Rev. Lett.* **47**, 460
Aspect, A., Dalibard, J., and Roger, G. (1982): *Phys. Rev. Lett.* **49**, 91, 1804
Atkins, P. W. and Friedman, R. S. (1997): *Molecular Quantum Mechanics* (Oxford University Press, New York)
Bargmann, V. (1952): *Proc. Nat. Acad. Sci.* **38**, 961
Barrow, G. H. (1962): *Introduction to Molecular Spectroscopy* (McGraw-Hill, New York)
Bauer, M. and Mello, P. A. (1978): *Ann. Phys. (NY)* **111**, 38
Baym, G. (1976): *Lectures on Quantum Mechanics* (Benjamin, Reading)
Belinfante, F. J. (1973): *A Survey of Hidden-Variable Theories* (Pergamon, Oxford)
Bell, J. S. (1964): *Physics* **1**, 195
Bell, J. S. (1966): *Rev. Mod. Phys.* **38**, 447
Bell, J. S. (1988): *Speakable and Unspeakable in Quantum Mechanics* (Cambridge University Press, Cambridge)
Bernstein, J. (1991): *Quantum Profiles* (Princeton University Press, Princeton)
Berry, M. V. (1984): *Proc. Roy. Soc. (London)* A **392**, 45
Bethe, H. A. (1929): *Z. Phys.* **57**, 815
Bethe, H. A. and Bacher, R. F. (1936): *Rev. Mod. Phys.* **8**, 193
Bethe, H. A. and Morrison, P. (1956): *Elementary Nuclear Theory*, 2nd Ed. (Wiley, New York)
Bethe, H. A. and Salpeter, E. E. (1957): *Quantum Mechanics of One and Two Electron Atoms* (Plenum, New York)
Bloch, F. (1928): *Z. Phys.* **52**, 555
Bohm, D. (1951): *Quantum Theory* (Prentice Hall, Englewood Cliffs)
Bohm, D. (1952): *Phys. Rev.* **85**, 166, 180
Bohm, D. and Aharonov, Y. (1957): *Phys. Rev.* **108**, 1070
Bohr, N., Peierls, R. E., and Placzek, G. (1939): *Nature* **144**, 200
Born, M. (1926): *Z. Phys.* **37**, 863; **38**, 803
Born, M. (1957): *Atomic Physics*, 5th Ed. (Hafner, New York)
Born, M. and Fock, V. (1928): *Z. Phys.* **51**, 165
Born, M., Heisenberg, W., and Jordan, P. (1925): *Z. Phys.* **35**, 557
Born, M. and Oppenheimer, J. R. (1927): *Ann. Phys. (Leipzig)* **84**, 457
Born, M. and Wolf, E. (1980): *Principles of Optics* (Pergamon, Oxford)
Bose, S. N. (1924): *Z. Phys.* **26**, 178
Briggs, J. S. (1996): Coulomb Forces in Three-Particle Atomic and Molecular Systems, in *Coulomb Interactions in Nuclear and Atomic Few-Body Collisions*, ed. by Levin, F. S. and Micha, D. A. (Plenum, New York)
Brillouin, L. (1926): *Comptes Rendus* **183**, 24
Brink, D. and Satchler, G. R. (1971): *Angular Momentum*, 2nd Ed., (Oxford University Press, Oxford)

Bryant, H. C. and Halka, M. (1996): H^- Spectroscopy, in *Coulomb Interactions in Nuclear and Atomic Few-Body Collisions*, ed. by Levin, F. S. and Micha, D. A. (Plenum, New York)

Burrau, Ø. (1927): *Det. Kgl. Danske Videnskab. Selskab.* **7**, 1

Byron, F. W., Jr and Fuller, R. W. (1970): *Mathematics of Classical and Quantum Physics* (Addison-Wesley, Reading)

Carslaw, H. S. (1930): *Introduction to the Theory of Fourier's Series and Integrals* (Macmillan, London)

Chambers, R. G. (1960): *Phys. Rev. Lett.* **5**, 3

Chandrasekhar, S. (1944): *Astrophys. J.* **100**, 176

Chiao, R. and Wu, Y. (1986): *Phys. Rev. Lett.* **57**, 933

Cini, M. and Levy-Leblond, J.-M. (1990): *Quantum Theory without Reduction* (Hilger, New York)

Clauser, J. F. and Shimony, A. (1978): *Rep. Prog. Phys.* **41**, 1881

Clementi, E. (1965): *IBM J. Res. Develop.* **9**, 2

Cohen-Tannoudji, C., Diu, B., and Laloë, F. (1977): *Quantum Mechanics*, Vol. I (Wiley, New York)

Colella, R., Overhauser, A., and Werner, S. A. (1975): *Phys. Rev. Lett.* **34**, 1472

Condon, E. U. and Shortley, E. H. (1963): *The Theory of Atomic Spectra* (Cambridge University Press, Cambridge)

Copson, E. T. (1935): *An Introduction to the Theory of Functions of a Complex Variable* (Oxford University Press, Oxford)

Cottingham, W. N. and Greenwood, D. A. (1986): *An Introduction to Nuclear Physics* (Cambridge University Press, Cambridge)

Cushing, J. T. (1994): *Quantum Mechanics* (University of Chicago Press, Chicago)

Cushing, J. T., Fine, A., and Goldstein, S. (1996) eds.: *Bohmian Mechanics and Quantum Theory: An Appraisal* (Kluwer, Dordrecht)

Dalgarno, A. and Lewis, J. T. (1956): *Proc. Roy. Soc. (London)* A **233**, 70

Dalitz, R. (1951): *Proc. Roy. Soc. (London)* A **206**, 509

Das, T. P. (1973): *Relativistic Quantum Mechanics of Electrons* (Harper and Row, New York)

Dettman, J. W. (1988): *Mathematical Methods in Physics and Engineering* (Dover, New York)

Dirac, P. A. M. (1926): *Proc. Roy. Soc. (London)* A **112**, 661

Dirac, P. A. M. (1928): *Proc. Roy. Soc. (London)* A **117**, 610

Dirac, P. A. M. (1947): *Quantum Mechanics*, 3rd Ed. (Oxford University Press, London)

Drachman, R. J. (1993): High Rydberg States of Two-Electron Atoms in Perturbation Theory, in *Long Range Casimir Forces*, ed. by Levin, F. S. and Micha, D. A. (Plenum, New York)

Drake, G. W. F. (1987): *Phys. Rev. Lett.* **59**, 1549

Drake, G. W. F. (1993): High Precision Calculations for the Rydberg States of Helium, in *Long Range Casimir Forces*, ed. by Levin, F. S. and Micha, D. A. (Plenum, New York)

Dyson, F. J. (1949): *Phys. Rev.* **75**, 486

Eckart, C. (1926): *Phys. Rev.* **28**, 711

Eckart, C. (1930): *Phys. Rev.* **36**, 878

Edmonds, A. R. (1957): *Angular Momentum in Quantum Mechanics* (Princeton University Press, Princeton)

Ehrenburg, W. and Siday, R. W. (1949): *Proc. Roy. Soc. (London)* B **62**, 8

Einstein, A. (1917): *Phys. Z.* **18**, 121

Einstein, A. (1924): *Berliner Ber.*, p. 21

Einstein, A. (1925): *Berliner Ber.*, p. 18

Einstein, A., Podolsky, B., and Rosen, N. (1935): *Phys. Rev.* **47**, 777

Ellis, P. J. and Amati, E. (2000), eds.: *Quantum Reflections* (Cambridge University Press, Cambridge)

Ellis, P. J. and Tang, Y. C. (1990), eds.: *Trends in Theoretical Physics*, Vol. I (Addison-Wesley, Redwood City)

Elton, L. R. B. (1966): *Introductory Nuclear Theory* (Saunders, Philadelphia)

Fallieros, S. and Friar, J. L. (1982): *Am. J. Phys.* **50**, 1001

Feller, W. (1961, 1966): *An Introduction to Probability Theory and its Applications*, 2 Vols. (Wiley, New York)

Fermi, E. (1926): *Lincei Rend.* (6) **3**, 145; *Z. Phys.* **36**, 902

Fermi, E. (1927): *Rend. Accad. Naz. Lincei* **6**, 602

Fermi, E. (1950): *Nuclear Physics*, Revised Ed. (University of Chicago Press, Chicago)

Fermi, E. (1966): *Notes on Thermodynamics and Statistics* (University of Chicago Press, Chicago)

Feshbach, H. and Villars, F. (1958): *Rev. Mod. Phys.* **30**, 24

Fetter, A. L. and Walecka, J. D. (1980): *Theoretical Mechanics of Particles and Continua* (McGraw-Hill, New York)

Feynman, R. P. (1948): *Rev. Mod. Phys.* **20**, 267

Feynman, R. P. (1962): *Quantum Electrodynamics* (Benjamin, Reading)

Feynman, R. P. and Hibbs, A. R. (1965): *Quantum Mechanics and Path Integrals* (McGraw-Hill, New York)

Feynman, R. P., Leighton, R. B., and Sands, M. (1965): *The Feynman Lectures on Physics*, Vol. III (Addison-Wesley, Reading)

Finkelstein, B. N. and Horowitz, G. E. (1928): *Z. Phys.* **48**, 118, 448

Fischer, C. F. (1977): *The Hartree–Fock Method for Atoms* (Wiley, New York)

Flügge, S. (1974): *Practical Quantum Mechanics* (Springer, New York)

Flygare, W. H. (1978): *Molecular Structure and Dynamics* (Prentice Hall, Englewood Cliffs)

Fock, V. (1930): *Z. Phys.* **61**, 126; **63**, 795

Fonda, L., Ghirardi, G. C., and Rimini, A. (1978): *Rep. Prog. Phys.* **47**, 587

Fonseca, A. C. and Shanley, P. E. (1979): *Ann. Phys. (NY)* **117**, 268

Fowles, G. and Cassidy, G. L. (1990): *Analytical Mechanics*, 5th Ed., (Saunders, Fort Worth)

Frauenfelder, H. and Henley, E. M. (1991): *Subatomic Physics*, 2nd Ed. (Prentice Hall, Englewood Cliffs)

French, A. P. (1968): *Special Relativity* (Norton, New York)

French, A. P. (1971): *Vibrations and Waves* (Norton, New York)

French, A. P. and Taylor, E. F. (1978): *An Introduction to Quantum Physics* (Norton, New York)

Galindo, A. and Pascual, P. (1990): *Quantum Mechanics*, Vols. I and II (Springer, Berlin)

Gasiorowicz, S. (1974): *Quantum Physics* (Wiley, New York)

Gasiorowicz, S. (1979): *The Structure of Matter* (Addison-Wesley, Reading)

Gerjuoy, E., Rau, A. R. P., and Spruch, L. (1983): *Rev. Mod. Phys.* **55**, 725

Glauber, R. J. (1963): *Phys. Rev.* **130**, 2529; **131**, 2766

Glöckle, W. (1983): *The Quantum Mechanical Few-Body Problem* (Springer, Berlin)

Goeppert-Mayer, M. and Jensen, J. H. D. (1955): *Elementary Theory of Nuclear Shell Structure* (Wiley, New York)

Goldstein, H. (1980): *Classical Mechanics* (Addison-Wesley, Reading)

Görlitz, A., Schuh, B., and Weiss, A. (1994): *Phys. Rev.* A **51**, R4305

Gottfried, K. (1966): *Quantum Mechanics* (Benjamin, New York)

Gottfried, K. and Weisskopf, V. F. (1984): *Concepts of Particle Physics*, Vol. I (Clarendon Press, Oxford)

Gradshteyn, I. S. and Ryzhik, I. M. (1965): *Table of Integrals, Series and Products* (Academic, New York)

Greenberger, D. (1983): *Rev. Mod. Phys.* **55**, 85

Griffiths, D. (1987): *Introduction to Elementary Particles* (Harper and Row, New York)

Haftel, M. I. and Mandelzweig, V. B. (1988): *Phys. Rev.* A **38**, 5995

Hartree, D. R. (1928): *Proc. Camb. Phil. Soc.* **24**, 89, 111

Hassani, M. (1991): *Foundations of Mathematical Physics* (Allyn and Bacon, Needham Heights)

Hausner, M. (1971): *Elementary Probability Theory* (Plenum, New York)

Heisenberg, W. (1925): *Z. Phys.* **33**, 879

Heisenberg, W. (1927): *Z. Phys.* **43**, 172

Heisenberg, W. (1949): *The Physical Principles of the Quantum Theory* (Dover, New York)

Herrick, D. (1983): *Adv. Chem. Phys.* **52**, 1

Hill, R. N. (1977): *Phys. Rev. Lett.* **38**, 643

Hodgson, P. E. (1973): *Nuclear Reactions and Nuclear Structure* (Clarendon Press, Oxford)

Holstein, B. R. (1992): *Topics in Advanced Quantum Mechanics* (Addison-Wesley, Reading)

Hund, F. (1927): *Linienspektren und periodisches System der Elemente* (Springer, Berlin)

Hurley, A. C. (1976): *Introduction to the Electron Theory of Small Molecules* (Academic, London)

Hylleraas, E. A. (1930): *Z. Phys.* **63**, 291

Hylleraas, E. A. (1963): *Rev. Mod. Phys.* **35**, 421

Hylleraas, E. A. (1964): *Adv. Quant. Chem.* **1**, 1

Hylleraas, E. A. and Undheim, B. (1933): *Z. Phys.* **65**, 759

Ikegami, H., Ewbank, W. B., and Way, K. (1966): *Nucl. Data Sheets* B1-6-103

Jackson, J. D. (1975): *Classical Electrodynamics* (Wiley, New York)

Jammer, M. (1974): *The Philosophy of Quantum Mechanics* (Wiley, New York)

Jeffreys, H. (1923): *Proc. London Math. Soc.* (2) **23**, 428

Johnson, M. H. and Lippmann, B. A. (1950): *Phys. Rev.* **78**, 329 (A)

Jones, L. M. (1979): *An Introduction to*

Mathematical Methods of Physics (Benjamin-Cummings, Menlo Park)

Jönsson, C. (1961): *Z. Phys.* **161**, 454

Kabir, P. K. and Pilaftsis, A. (1996): *Phys. Rev.* A **56**, 66

Kellner, G. W. (1927): *Z. Phys.* **44**, 91, 110

Kemble, E. C. (1958): *The Fundamental Principles of Quantum Mechanics* (Dover, New York)

Kinoshita, T. (1959): *Phys. Rev.* **115**, 366

Kobe, D. H. and Aguilera-Novarro, V. C. (1994): *Phys. Rev.* A **50**, 933

Kok, L. P. and Visser, J. (1987): *Quantum Mechanics Problems and the Solutions* (Coulomb Press, Leiden)

Koopmans, T. A. (1933): *Physica* **1**, 104

Kramers, H. A. (1926): *Z. Phys.* **39**, 828

Kuhn, H. G. (1962): *Atomic Spectra* (Longmans, London)

Kuhn, T. (1970): *The Structure of Scientific Revolutions*, 2nd Ed. (University of Chicago Press, Chicago)

Kuruoglu, Z. C. and Levin, F. S. (1990): *Phys. Rev. Lett.* **64**, 1701

Kuruoglu, Z. C. and Levin, F. S. (1992): *Phys. Rev.* A **46**, 2304

Kusch, P. and Foley, H. (1947): *Phys. Rev.* **72**, 1256

Lamb, W. E., Jr and Retherford, R. C. (1947): *Phys. Rev.* **72**, 241

Landau, L. D. (1930): *Z. Phys.* **64**, 629

Landau, L. D. and Lifshitz, E. M. (1958): *Quantum Mechanics* (Addison-Wesley, Reading)

Lebedev, M. N. (1972): *Special Functions and their Applications*, translated by R. R. Silverman (Dover, New York)

Lee, T. D. and Yang, C. N. (1956): *Phys. Rev.* **104**, 254

Levinson, N. (1949): *Det. Kgl. Danske Videnskab. Selskab., Mat-fys. Medd.* **25** (9)

Lieb, E. H., Simon, B., and Wightman, A. (1976), eds.: *Studies in Mathematical Physics* (Princeton University Press, Princeton)

Lighthill, M. J. (1958): *An Introduction to Fourier Analysis and Generalized Functions* (Cambridge University Press, Cambridge)

Lin, C. D. (1984): *Phys. Rev.* A **21**, 1019

Lippmann, B. A. and Schwinger, J. (1950): *Phys. Rev.* **79**, 469

London, F. (1961): *Superfluids* (Dover, New York)

Lykke, K. R., Murray, K. K., and Lineberger, W. C. (1991): *Phys. Rev.* A **43**, 6104

Mann, J. B. and Johnson, W. R. (1971): *Phys. Rev.* A **4**, 41

Margenau, M. and Murphy, G. (1943): *The Mathematics of Physics and Chemistry* (Van Nostrand, New York)

Marion, J. B. (1965): *Classical Electromagnetic Radiation* (Academic, New York)

Marion, J. B. (1970): *Classical Dynamics of Particles and Systems* (Addison-Wesley, Reading)

Martin, W. C. and Wiese, W. L. (1996): Atomic Spectroscopy, in *Atomic, Molecular and Optical Physics Handbook*, ed. by Drake, G. W. F. (American Institute of Physics, Woodbury)

Massey, H. S. W. and Burhop, E. H. (1952): *Electronic and Ionic Impact Phenomena* (Clarendon Press, Oxford)

Mathews, J. and Walker, R. L. (1970): *Mathematical Methods of Physics* (Addison-Wesley, Reading)

Merzbacher, E. (1970): *Quantum Mechanics*, 2nd Ed. (Wiley, New York)

Mikhlin, S. G. (1964): *Variational Methods in Mathematical Physics* (Pergamon, Oxford)

Mizushima, M. (1970): *Quantum Mechanics of Atomic Spectra and Atomic Structure* (Benjamin, New York)

Moiseiwitsch, B. L. (1966): *Variational Principles* (Interscience, New York)

Morse, P. M. (1929): *Phys. Rev.* **34**, 57

Morse, P. M. and Feshbach, H. (1953): *Methods of Theoretical Physics*, Vols. I and II (McGraw-Hill, New York)

Nambu, Y. (1981): *Quarks: Frontiers in Elementary Particle Physics* (World Scientific, Singapore)

Newton, R. G. (1982): *Scattering Theory of Waves and Particles*, 2nd Ed. (Springer, New York)

Ohanian, H. C. (1995): *Modern Physics* (Prentice-Hall, Englewood Cliffs)

Orihara, M., Zafiratos, C. D., Nishihara, S., Furukawa, K., Kabasawa, M., Nakagawa, T., Maeda, K., Miura, K., Kiang, G. C., and Ohnuma, H. (1983): *Proc. 1983 RCNP International Symposium on Light Ion Reaction Mechanism*, ed. by Ogata, H., Kammuri, T., and Katayama, I. (Research Center for Nuclear Physics, Osaka University, Osaka)

Panofsky, W. K. H. and Phillips, M. (1955): *Classical Electricity and Magnetism* (Addison-Wesley, Cambridge)

Park, D. (1964): *Introduction to the Quantum Theory* (McGraw-Hill, New York)

Park, D. (1997): *The Fire Within the Eye* (Princeton University Press, Princeton)

Pauli, W. (1924): *Z. Phys.* **31**, 373, 765

Pauli, W. (1927): *Z. Phys.* **43**, 601

Pauli, W. (1940): *Phys. Rev.* **58**, 713

Pauli, W. (1958): *Die allgemeinen Prinzipien der Wellenmechanik*, Vol. V of the *Encyclopedia of Physics*, ed. by Flügge, S. (Springer, Berlin)

Peierls, R. E. (1979): *Surprises in Theoretical Physics* (Princeton University Press, Princeton)

Peierls, R. E. (1991): *More Surprises in Theoretical Physics* (Princeton University Press, Princeton)

Pekeris, C. L. (1958): *Phys. Rev.* **112**, 1649

Peres, A. (1980): *Ann. Phys. (NY)* **129**, 33

Peshkin, M. and Tonomura, A. (1989): *The Aharonov–Bohm Effect* (Springer, Berlin)

Pilar, F. L. (1968): *Elementary Quantum Chemistry* (McGraw-Hill, New York)

Pipkin, F. (1979): *Adv. Atom. Mol. Phys.* **14**, 281

Planck, M. (1900): *Verh. Deutsche Phys. Ges.* **2**, 202, 237

Polkinghorne, J. C. (1984): *The Quantum World* (Longmans, London)

Press, W. H., Flannery, B. P., Teukolsky, S. A., and Vetterling, W. T. (1986): *Numerical Recipes* (Cambridge University Press, Cambridge)

Preston, M. A. and Bhaduri, R. K. (1975): *Structure of the Nucleus* (Addison-Wesley, Reading)

Radzig, A. A. and Smirnov, B. M. (1985): *Reference Data on Atoms, Molecules and Ions* (Springer, Berlin)

Ram-Mohan, L. R., Saigal, S., Dassa, D., and Shertzer, J. (1990): *Comput. Phys.* **50**, 1

Rauch, H., Zeilinger, A., Badurek, G., Wilfing, A., Bauspiess, W., and Bonse, U. (1975): *Phys. Lett.* A **45**, 425

Reed, M. and Simon, B. (1972): *Methods of Modern Mathematical Physics Vol. I, Functional Analysis* (Academic, New York)

Reid, C. (1976): in *Quantum Science*, ed. by Calais, J. L. *et al.* (Plenum, New York)

Rice, J. R. (1983): *Numerical Methods, Software and Analysis* (McGraw-Hill, New York)

Richtmyer, F. D. (1978): *Principles of Advanced Mathematical Physics* (Springer, New York)

Riordan, M. (1987): *The Hunting of the Quark* (Simon and Schuster, New York)

Ritz, W. (1909): *J. Reine Angew. Math.* **135**, 1

Rodberg, L. S. and Thaler, R. M. (1967): *Introduction to the Quantum Theory of Scattering* (Academic, New York)

Rose, M. E. (1957): *Elementary Theory of Angular Momentum* (Wiley, New York)

Russell, H. N. and Saunders, F. A. (1925): *Astrophys. J.* **61**, 38

Rutherford, E. (1911): *Phil. Mag.* **21**, 669

Sakurai, J. J. (1994): *Modern Quantum Mechanics* (Addison-Wesley, Reading)

Sangster, K., Hinds, E. A., Barnett, S. M., and Riis, E. (1993): *Phys. Rev. Lett.* **71**, 3641

Sangster, K., Hinds, E. A., Barnett, S. M., Riis, E., and Sinclair, A. G. (1994): *Phys. Rev.* A **51**, 1776

Schiff, L. I. (1968): *Quantum Mechanics*, 3rd Ed. (McGraw-Hill, New York)

Schrödinger, E. (1928): *Collected Papers on Wave Mechanics* (Blackie, London)

Schrödinger, E. (1952): *Statistical Thermodynamics*, 2nd Ed. (Cambridge University Press, Cambridge)

Schwartz, C. (1959): *Ann. Phys. (NY)* **6**, 156, 170, 178

Schwartz, L. (1950–51): *Théorie des distributions*, Vols. I and II (Hermann, Paris)

Schwinger, J. (1948): *Phys. Rev.* **73**, 416

Schwinger, J. (1958): *Quantum Electrodynamics* (Dover, New York)

Shapere, A. and Wilczek, F. (1989): *Geometric Phases in Physics* (World Scientific, Singapore)

Simon, B. (1982): *J. Quant. Chem.* **21**, 3

Sinanoglu, O. and Herrick, D. (1973): *J. Chem. Phys. Lett.* **62**, 886

Slater, J. C. (1929): *Phys. Rev.* **34**, 1293

Slater, J. C. (1930): *Phys. Rev.* **35**, 210

Sobelman, I. I. (1979): *Atomic Spectra and Radiative Transitions* (Springer, Berlin)

Spruch, L. (1962): *Lectures in Theoretical Physics*, Vol. IV, ed. by Brittin, W. (Interscience, New York)

Squires, E. J. (1994): *The Mystery of the Quantum World* (Institute of Physics, Bristol)

Steiner, E. (1976): *The Determination and Interpretation of Molecular Wave Functions* (Cambridge University Press, Cambridge)

Stern, O. and Gerlach, W. (1921): *Z. Phys.* **8**, 110

Sudbery, A. (1984): *Ann. Phys. (NY)* **157**, 512

Suter, D., Mueller, K. T., and Pines, A. (1988): *Phys. Rev. Lett.* **60**, 1218

Taylor, J. R. (1972): *Scattering Theory* (Wiley, New York)

Thomas, L. H. (1926): *Proc. Camb. Phil. Soc.* **23**, 542

Titchmarsh, E. C. (1937): *Introduction to the Theory of Fourier Integrals* (Clarendon Press, Oxford)

Uhlenbeck, G. E. and Goudsmit, S. (1925): *Naturwissenschaft* **13**, 953

Unsöld, A. (1927): *Ann. Phys. (Leipzig)* **82**, 355

Veillard, A. and Clementi, E. (1968): *J. Chem. Phys.* **49**, 2415

von Baeyer, H. C. (1992): *Taming the Atom* (Random House, New York)

von Neumann, J. (1932): *Mathematische Grundlagen der Quanten-mechanik* (Springer, Berlin), English translation by R. T. Beyer (1955): *Mathematical Foundations of Quantum Mechanics* (Princeton University Press, Princeton)

Weaver, W. (1963): *Lady Luck – The Theory of Probability* (Heinemann, London)

Wentzel, G. (1926): *Z. Phys.* **38**, 518

Werner, S. A., Colella, R., Overhauser, A. W., and Eagen, C. F. (1975): *Phys. Rev. Lett.* **35**, 1053

Wheeler, J. A. and Zurek, W. H. (1983), eds.: *Quantum Theory and Measurement* (Princeton University Press, Princeton)

Whittaker, E. T. and Watson, G. N. (1965): *A Course of Modern Analysis* (Cambridge University Press, Cambridge)

Wiese, W. L., Smith, M. W., and Glennon, B. M. (1966): *Atomic Transition Probabilities, Vol. I, Hydrogen through Neon* (National Bureau of Standards of the USA Department of Commerce, Washington)

Williams, W. S. C. (1991): *Nuclear and Particle Physics* (Clarendon Press, Oxford)

Wind, H. (1965): *J. Chem. Phys.* **42**, 2371

Wu, C. S., Ambler, E., Hayward, R. W., Hoppes, D. D. and Hudson, R. P. (1957): *Phys. Rev.* **105**, 1413

Wulfman, C. and Sukeyuki, K. (1973): *Chem. Phys. Lett.* **23**, 367

Wyld, H. W. (1976): *Mathematical Methods for Physics* (Benjamin, Reading)

Zeng, J. Y. and Lei, Y. A. (1995): *Phys. Rev.* A **51**, 4415

Zienkiewicz, O. C. and Taylor, R. L. (1989): *The Finite Element Method*, 4th Ed. (McGraw-Hill, London)

Index